New Perspectives on the Old Red Sandstone

Special Publication reviewing procedure

The Society makes every effort to ensure that the scientific and production quality of its books matches that of its journals. Since 1997, all book proposals have been refereed by specialist reviewers as well as by the Society's Publications Committee. If the referees identify weaknesses in the proposal, these must be addressed before the proposal is accepted.

Once the book is accepted, the Society has a team of series editors (listed above) who ensure that the volume editors follow strict guidelines on refereeing and quality control. We insist that individual papers can only be accepted after satisfactory review by two independent referees. The questions on the review forms are similar to those for *Journal of the Geological Society*. The referees' forms and comments must be available to the Society's series editors on request.

Although many of the books result from meetings, the editors are expected to commission papers that were not presented at the meeting to ensure that the book provides a balanced coverage of the subject. Being accepted for presentation at the meeting does not guarantee inclusion in the book.

Geological Society Special Publications are included in the ISI Science Citation Index, but they do not have an impact factor, the latter being applicable only to journals.

More information about submitting a proposal and producing a Special Publication can be found on the Society's web site: www.geolsoc.org.uk

It is recommended that reference to all or part of this book should be made in one of the following ways.

FRIEND, P. F. & WILLIAMS, B. P. J. (eds) 2000. *New Perspectives on the Old Red Sandstone*. Geological Society, London, Special Publications, **180**.

HIGGS, K. T., MACCARTHY, I. A. J. & O'BRIEN, M. M. 2000. A mid-Frasnian marine incursion into the southern part of the Munster Basin: evidence from the Foilcoagh Bay Beds, Sherkin Formation, SW County Cork, Ireland. *In*: FRIEND, P. F. & WILLIAMS, B. P. J. (eds) *New Perspectives on the Old Red Sandstone*. Geological Society, London, Special Publications, **180**, 319–332.

GEOLOGICAL SOCIETY SPECIAL PUBLICATION NO. 180

New Perspectives on the Old Red Sandstone

EDITED BY

P. F. FRIEND
University of Cambridge, UK

and

B. P. J. WILLIAMS
University of Aberdeen, UK

2000
Published by
The Geological Society
London

THE GEOLOGICAL SOCIETY

The Geological Society of London was founded in 1807 and is the oldest geological society in the world. It received its Royal Charter in 1825 for the purpose of 'investigating the mineral structure of the Earth' and is now Britain's national society for geology.

Both a learned society and a professional body, the Geological Society is recognized by the Department of Trade and Industry (DTI) as the chartering authority for geoscience, able to award Chartered Geologist status upon appropriately qualified Fellows. The Society has a membership of 9099, of whom about 1500 live outside the UK.

Fellowship of the Society is open to persons holding a recognized honours degree in geology or a cognate subject, or not less than six years' relevant experience in geology or a cognate subject. A Fellow with a minimum of five years' relevant postgraduate experience in the practice of geology may apply for chartered status. Successful applicants are entitled to use the designatory postnominal CGeol (Chartered Geologist). Fellows of the Society may use the letters FGS. Other grades of membership are available to members not yet qualifying for Fellowship.

The Society has its own Publishing House based in Bath, UK. It produces the Society's international journals, books and maps, and is the European distributor for publications of the American Association of Petroleum Geologists (AAPG), the Society for Sedimentary Geology (SEPM) and the Geological Society of America (GSA). Members of the Society can buy books at considerable discounts. The Publishing House has an online bookshop (http://bookshop.geolsoc.org.uk).

Further information on Society membership may be obtained from the Membership Services Manager, The Geological Society, Burlington House, Piccadilly, London W1V 0JU (E-mail: enquiries@geolsoc.org.uk; tel: +44 (0) 207 434 9944).

The Society's Web Site can be found at http://www.geolsoc.org.uk/. The Society is a Registered Charity, number 210161.

Published by The Geological Society from:
The Geological Society Publishing House
Unit 7, Brassmill Enterprise Centre
Brassmill Lane
Bath BA1 3JN, UK
Orders: Tel. +44 (0)1225 445046
Fax +44 (0)1225 442836
Online bookshop: http://bookshop.geolsoc.org.uk

British Library Cataloguing in Publication Data

A catalogue record for this book is available from the British Library.

ISBN 1-86239-071-1

Typeset by Alden Multimedia, Westonzoyland, UK

Printed by Hobbs the Printers, Southampton, UK

Distributors

USA
AAPG Bookstore
PO Box 979
Tulsa
OK 74101-0979
USA
Orders: Tel. +1 918 584-2555
Fax +1 918 560-2652
Email bookstore@aapg.org

Australia
Australian Mineral Foundation Bookshop
63 Conyngham Street
Glenside
South Australia 5065
Australia
Orders: Tel. +61 88 379-0444
Fax +61 88 379-4634
Email bookshop@amf.com.au

India
Affiliated East-West Press PVT Ltd
G-1/16 Ansari Road, Daryaganj,
New Delhi 110 002
India
Orders: Tel. +91 11 327-9113
Fax +91 11 326-0538
Email affiliat@nda.vsnl.net.in

Japan
Kanda Book Trading Co.
Cityhouse Tama 204
Tsurumaki 1-3-10
Tama-shi
Tokyo 206-0034
Japan
Orders: Tel. +81 (0)423 57-7650
Fax +81 (0)423 57-7651

Contents

Foreword vii

List of Referees ix

List of Sponsors ix

Review

WILLIAMS, E. A., FRIEND, P. F. & WILLIAMS, B. P. J. A review of Devonian time scales: databases, construction and new data 1

HOUSE, M. R. Chronostratigraphic framework for the Devonian and Old Red Sandstone 23

FRIEND, P. F., WILLIAMS, B. P. J., FORD, M. & WILLIAMS, E. A. Kinematics and dynamics of Old Red Sandstone basins 29

Eastern North America

GRIFFING, D. H., BRIDGE, J. S. & HOTTON, C. L. Coastal-fluvial palaeoenvironments and plant palaeoecology of the Lower Devonian (Emsian), Gaspé Bay, Québec, Canada 61

BRIDGE, J. S. The geometry, flow patterns and sedimentary processes of Devonian rivers and coasts, New York and Pennsylvania, USA 85

Ireland-Dingle and North

McSHERRY, M., PARNELL, J., LESLIE, A. G. & HAGGAN, T. Depositional and structural setting of the (?) Lower Old Red Sandstone sediments of Ballymastocker, Co. Donegal 109

BOYD, J. D. & SLOAN, R. J. Initiation and early development of the Dingle Basin, SW Ireland, in the context of the closure of the Iapetus Ocean 123

RICHMOND, L. K. & WILLIAMS, B. P. J. A new terrane in the Old Red Sandstone of the Dingle Peninsula, SW Ireland 147

TODD, S. P. Taking the roof off a suture zone: basin setting and provenance of conglomerates in the ORS Dingle Basin of SW Ireland 185

Ireland-Munster

VERMEULEN, N. J., SHANNON, P. M., MASSON, F. & LANDES, M. Wide-angle seismic control on the development of the Munster Basin, SW Ireland 223

WILLIAMS, E. A. Flexural cantilever models of extensional subsidence in the Munster Basin (SW Ireland) and Old Red Sandstone fluvial dispersal systems 239

WILLIAMS, E. A., SERGEEV, S. A., STÖSSEL, I., FORD, M. & HIGGS, K. T. U–Pb zircon geochronology of silicic tuffs and chronostratigraphy of the earliest Old Red Sandstone in the Munster Basin, SW Ireland 269

PRACHT, M. Controls on magmatism in the Munster Basin, SW Ireland 303

HIGGS, K. T., MACCARTHY, I. A. J. & O'BRIEN, M. M. A mid-Frasnian marine incursion into the southern part of the Munster Basin: evidence from the Foilcoagh Bay Beds, Sherkin Formation, SW County Cork, Ireland 319

JARVIS, D. E. Palaeoenvironment of the plant bearing horizons of the Devonian–Carboniferous Kiltorcan Formation, Kiltorcan Hill, Co. Kilkenny, Ireland 333

Wales

HILLIER, R. D. Silurian marginal marine sedimentation and the anatomy of the marine — Old Red Sandstone transition in Pembrokeshire, SW Wales 343

EDWARDS, D. & RICHARDSON, J. B. Progress in reconstructing vegetation on the Old Red Sandstone Continent: two *Emphanisporites* producers from the Lochkovian sequence of the Welsh Borderland 355

LOVE, S. E. & WILLIAMS, B. P. J. Sedimentology, cyclicity and floodplain architecture in the Lower Old Red Sandstone of SW Wales 371

OWEN, G. & HAWLEY, D. Depositional setting of the Lower Old Red Sandstone at Pantymaes Quarry, central South Wales: new perspectives on the significance and occurrence of 'Senni Beds' facies 389

MARSHALL, J. D. Fault-bounded basin fill: fluvial response to tectonic controls in the Skrinkle Sandstones of SW Pembrokeshire, Wales

Scotland

401

BLUCK, B. J. Old Red Sandstone basins and alluvial systems of Midland Scotland 417

POWELL, C. L., TREWIN, N. H. & EDWARDS, D. Palaeoecology and plant succession in a borehole through the Rhynie cherts, Lower Old Red Sandstone, Scotland 439

ARMSTRONG, H. A. & OWEN, A. W. Age and provenance of limestone clasts in Lower Old Red Sandstone conglomerates: implications for the geological history of the Midland Valley Terrane 459

MARSHALL, J. E. A. Devonian (Givetian) miospores from the Walls Group, Shetland 473

BALIN, D. F. Calcrete morphology and karst development in the Upper Old Red Sandstone at Milton Ness, Scotland 485

Norway and the Arctic

OSMUNDSEN, P. T., BAKKE, B., SVENDBY, A. K. & ANDERSEN, T. B. Architecture of the Middle Devonian Kvamshesten Group, western Norway: sedimentary response to deformation above a ramp-flat extensional fault 503

HARTZ, E. Early syndepositional tectonics of East Greenland's Old Red Sandstone basin 537

CLACK, J. A. & NEININGER, S. L. Fossils from the Celsius Bjerg Group, Late Devonian sequence, East Greenland; significance and sedimentological distribution 557

MCCANN, A. J. Deformation of the Old Red Sandstone of NW Spitsbergen; links to the Ellesmerian and Caledonian orogenies 567

PIEPJOHN, K. The Svalbardian–Ellesmerian deformation of the Old Red Sandstone and the pre-Devonian basement in NW Spitsbergen (Svalbard) 585

PIEPJOHN, K., BRINKMANN, L., GREWING, A. & KERP, H. New data on the age of the uppermost ORS and the lowermost post-ORS strata in Dickson Land (Spitsbergen) and implications for the age of the Svalbardian deformation 603

Index 611

Foreword

The Old Red Sandstone (mainly Siluro-Devonian) successions of the North Atlantic Region have probably stimulated more research, and generated more ideas on facies analysis in continental basin fills, than any other comparable stratigraphy. The Old Red Sandstone played an important early role in the history of stratigraphic research in Western Europe and Eastern North America, where it was realized quite quickly that it had special significance in linking stratigraphy and mountain building. It also attracted attention because it clearly provided basic information on the early evolution of animals and plants in non-marine (terrestrial) environments.

However understanding of the sedimentary processes that formed the Old Red Sandstone was very limited until the 1960s. Modern analysis on Old Red Sandstone facies architecture was mainly initiated by the pioneering research of J. R. L. Allen (eg 1979), in the early-sixties to mid-seventies, on the Welsh Basin. This innovative work set new standards of detailed analysis that have been applied widely to all the main ORS basins in the years that have followed.

The first international symposium on the Devonian System, held in Calgary in 1967 (Oswald 1968), started the process of setting the tectonics and facies patterns of the Old Red Sandstone in their global context, and encouraged further investigation of the biostratigraphical attributes of its floras, vertebrates and invertebrates.

In the UK and Ireland, an informal and unusual grouping of academic researchers, with a common interest in the ORS, was formed in the early 1970s and named the 'Friends of the Old Red Sandstone'. The 'Friends' convened annual field meetings in diverse areas of active ORS research, initially in Wales (Borders and South Pembrokeshire–1972 and 1973; then in Scotland (Borders–1975; Strathmore Basin–1978; Midland Valley–1982) and southern Ireland (Munster Basin and Dingle Basin–1976, 1980, 1983). The average attendance on these excellent weekend trips was about 20 which peaked with 40 on the spectacular Ardennes trip of 1979 when our European mainland colleagues from France, Belgium and Germany joined the 'Friends'. Key University departments supporting the 'Friends', and generating substantial ORS research output, included Reading, Cambridge, Bristol, Glasgow, Newcastle, Cardiff, London, Cork and Trinity College, Dublin.

Research on the Devonian system was further focused in the late 1970s by the Palaeontological Association International Symposium on the Devonian system (1979). PADS was held in Bristol in 1978 and a splendid volume was published (House *et al.* 1979). Four field excursions, two of them to ORS basins in Wales and

Fig. 1. Professor J. R. L. Allen.

Fig. 2. Professor D. L. Dineley.

From: Friend, P. F. & Williams, B. P. J. (eds). *New Perspectives on the Old Red Sandstone.* Geological Society, London, Special Publications, **180**, vii–ix. 0305-8719/00/$15.00

Scotland (Friend & Williams 1978), accompanied this Symposium and the published guides reflect much of the ongoing research of that time on those basins.

Old Red Sandstone research gathered much impetus in the 1980s with research published on Scandinavia, Greenland, Scotland, the NE American Appalachian Belt and Ireland. An admirably compact, well-integrated survey of the global Devonian, with some special focus on the Old Red Sandstone, was provided by D. L. Dineley (1984). As Head of the Geology Department at the University of Bristol, he was a major factor in the success of Old Red studies, on both sides of the Atlantic, and in both vertebrate palaeontology and the field study of the sediments. A further step forward in integration was achieved by the second international Devonian System symposium held in Calgary 1987 (McMillan *et al.* 1988) which was published in a remarkable three-volume work entitled 'Devonian of the World'. In parallel with University ORS research, the Irish and British Geological Surveys were re-mapping several areas that contained Siluro-Devonian red bed sequences, some of which were projected into offshore areas where the ORS locally acts as either potential hydrocarbon reservoirs or source rocks (e.g. Trewin 1989).

In recent years, key palaeontological discoveries have been made in the ORS which throw much new light on the early evolution of land plants and their sporangia, and the evolutionary pathway of fish to tetrapods (Westenberg 1999). Refinement of the chronostratigraphy of the Siluro–Devonian ORS into the basal Carboniferous has been made possible by new advances in palynology and U-Pb zircon geochronology of ashfall tuffs in several ORS basins using new methodologies.

With the above advances in knowledge of ORS basins it occurred to the Editors at a recent meeting of the British Sedimentological Research Group, that a special meeting on 'New Perspectives on the Old Red Sandstone' was timely and thus a 2-day conference was convened at Burlington House, London, where we were hosted by the Geological Society of London. Some forty talks and posters were presented over the two days, sessions being chaired by many of the contributors to this volume, and also by J. R. L. Allen and D. L. Dineley. The presentations embraced a wide spectrum of Old Red Sandstone themes from the dynamic evolution and seismic interpretation of ORS basins to evolution of early land floras, and from new advances in chronostratigraphy to displaced terranes and early amphibians. Twenty-six of those talks are published herein together with five additional papers added after the conference.

Some time ago a science fiction novel, entitled 'Cryptozoic' was published in which a time traveller is able to move 'like a ghost through the dim vistas of geological time' (Aldiss 1969). This time traveller's favourite time-slots for investigation are the Devonian and the Jurassic. During his journeys he is haunted by 'phantoms and unsolved questions'. We are confident that our new volume shows important progress in answering some of those questions. How appropriate then that Chapter 1 in the 'Cryptozoic' (Aldiss 1969) is 'A Bed in the Old Red Sandstone'!

The editors dedicate this volume to Professors J. R. L. Allen and D. L. Dineley (see figs 1 & 2).

References

Aldiss, B. W. 1969. *Cryptozoic*. Sphere Books.

Allen, J. R. L. 1979. Old Red Sandstone facies in external basins with particular reference to southern Britain. *Special Papers in Palaeontology*, **23**, 65–80.

Dineley, D. L. 1984. *Aspects of a Stratigraphic System: The Devonian*. Macmillan.

Friend, P. F. & Williams, B. P. J. (eds) 1978. *Devonian of Scotland, the Welsh Borderland and south Wales*. Palaeontogical Association.

House, M. R., Scrutton, C. T. & Bassett, M. G. (eds) 1979. The Devonian system. *Special Papers in Palaeontology*, **23**.

McMillan, N. J., Embry, A. F. & Glass, D. J. (eds) 1988. *The Devonian of the World.* Canadian Society of Petroleum Geologists, Memoir **14**, vol. 1, vol 2 & vol 3.

Oswald, D. H. (ed.) 1968. *International symposium on the Devonian system*. Alberta Society of Petroleum Geologists.

Trewin, N. H. 1989. The petroleum potential of the Old Red Sandstone of northern Scotland. *Scottish Journal of Geology*, **25**, 201–225.

Westenberg, K. 1999. From Fins to Feet. *National Geographic*. **195**, 114–127.

Our warmest thanks go to Helen Knapp, Angharad Hills and other staff of the Geological Society Publishing House for doing so much to lighten our editorial work. We would also like to thank all the many colleagues who attended the meeting in December 1998, and/or contributed to the volume as authors and referees. We are also most grateful to the organisations listed below who supported this project.

List of referees

Almond, J.
Ashcroft, W. A.
Balin, D. F.
Bluck, B. J.
Boyd, J. D.
Dineley, D. L.
Edwards, D.
Evans, J. A.
Friend, P. F.
Gee, D. G.
Graham, J. R.
Haughton, P. D. W.
Higgs, K. T.
Hillier, R. D.
Hole, M. J.
Kelly, S. B.
Lawrence, D. A.
Love, S. E.
Marshall, J. D.
Marshall, J. E. A.
McCann, A. J.
Nilsen, T. H.
Osmundsen, P. T.
Piepjohn, K.
Preston, R. J.
Shannon, P. M.
Sloan, R. J.
Smith, A. G.
Todd, S. P.
Trewin, N. H.
Wellman, C. H.
Williams, E. A.
Williams, B. P. J.
Woodcock, N. H.
Woodrow, D. L.
Wright, V. P.

List of sponsors

(a) The Conference, held under the auspices of the British Sedimentological Research Group, with the support of the Geological Society of London, was also supported by the following companies:

Amerada Hess
ARCO British
Badley Ashton and Associates
B.P. Exploration
PMGeos Ltd
Shell (U.K.) Exploration and Production
Total Oil Marine

(b) The following companies also generously funded the use of colour illustrations in this volume:

BP Amoco
Conoco (U.K.) Limited
Exxon Mobil

Peter. F. Friend & Brian P. J. Williams

A review of Devonian time scales: databases, construction and new data

E. A. WILLIAMS[1], P. F. FRIEND[2] & B. P. J. WILLIAMS[3]

[1]*CRPG–CNRS, B.P. 20, 54501 Vandoeuvre-lès-Nancy cedex, France*

[2]*Department of Earth Sciences, University of Cambridge, Downing Street, Cambridge CB2 3EQ, UK*

[3]*Department of Geology and Petroleum Geology, University of Aberdeen, Meston Building, King's College, Aberdeen AB24 3UE, UK*

Abstract: Aspects of the isotopic age and stratigraphical databases underpinning Devonian geological time scales are reviewed to assess differences in recent U–Pb zircon-based schemes and older schemes based on Rb–Sr, K–Ar, ^{40}Ar–^{39}Ar dating of minerals and whole-rock samples. The various methods of time-scale construction are described and, with their databases, 14 calibrations of Devonian time are discussed. Finally, the most recent data are collated and compared against current U–Pb-based time scales.

Developments in the Palaeozoic chronometric time scale during the last decade have been profound, and have been driven largely by the provision of new U–Pb high-precision dates from magmatic zircon crystals. These chronometric data have almost exclusively been derived from altered, often thin, airfall volcanic beds (K-bentonites) within marine successions that are well constrained biostratigraphically, allowing numerical dates to be assigned at the level of biozones. This trend has encompassed the Devonian System, and has more recently included volcanic rocks in continental settings from parts of the Old Red Sandstone of the North Atlantic region.

The generation of new U–Pb dates has been driven in turn by technical advances in instrumental analysis (e.g. Compston 1999) and other analytical techniques (see Tucker *et al.* (1990) and references therein). Two methods have dominated recent provision of zircon dates: (1) isotope dilution mass spectrometry (ID-MS) of single grains and small multigrain populations; (2) ablation of very small, specifically targeted, areas of single grains by ion beam, measured by the sensitive high-mass resolution ion microprobe (SHRIMP), pioneered at the Australian National University.

The publication of suites of U–Pb dates generated by these methods has given rise to two trends. First, new, largely zircon-based, geological time scales for various parts of the Palaeozoic eon (in particular the Cambrian–Silurian and Devonian periods) have been recently produced. Second, isolated, high-precision dates (especially SHRIMP-derived) have been incorporated into pre-existing time scales to constrain (and modify) particular boundaries, which were often calibrated originally on a different basis. The explicitly zircon-based time scales do not assess or account for the isotopic database on which the immediately preceding generation of time scales depended. This database predominantly comprises numerical ages from Rb–Sr, ^{40}Ar/^{39}Ar, K–Ar decay schemes for whole-rock samples and mineral separates, and utilizes a wider context of geological constraints, such as intrusion age relationships, conglomerate clast ages, glauconite mineral ages from sediments, etc. As the latest zircon-based time scales for the Devonian period (Tucker *et al.* 1998; Compston 2000*b*) differ significantly between themselves, and from earlier scales (e.g. Harland *et al.* 1990), it is an aim of this paper to compare the current isotopic database underpinning coherent major (non-zircon) time scales with the latest zircon-based schemes. Thus it may be possible to improve the calibration of stage boundaries within the Devonian period by using a combination of (U–Pb) zircon-based information and well-constrained dates from different decay schemes.

The evident differences in recent time scales are a combined function of the isotopic database, new statistical techniques to evaluate U–Pb zircon ages (Compston 2000*a*, *b*) and also the method employed in scale construction. A further aim is therefore to document methods

From: Friend, P. F. & Williams, B. P. J. (eds). *New Perspectives on the Old Red Sandstone.* Geological Society, London, Special Publications, **180**, 1–21. 0305-8719/00/$15.00

of construction of recent time scales, and to highlight the assumptions used, such as the proportioning of biozones and stages by palaeontological methods (e.g. Boucot 1975), and more recently applied techniques such as graphical correlation (Shaw 1964; Fordham 1992, 1998; Mann & Lane 1995) and orbitally modulated cyclostratigraphy (House 1991, 1995). Finally, we aim to collate stratigraphically and isotopically well-constrained published dates for the Devonian period, vital for evaluation of issues affecting the Old Red Sandstone (ORS).

Devonian chronostratigraphy

Since the 1989 review of the Devonian chronostratigraphic scale given by Harland *et al.* (1990, pp. 40–42), all of the remaining stage boundaries have been assigned stratotype sections and points, as detailed by House (this volume), calibrated largely by ammonoid, conodont, miospore and graptolite biozones. Except for the Pragian–Emsian boundary (Yolkin *et al.* 1997), modifications to stage bases in stratotype sections have been relatively minor (see House 1988; this volume, fig. 1; Harland *et al.* 1990, fig. 3.5). The changes have been sufficiently small as to be probably unresolvable given the analytical uncertainties of the isotopic techniques employed (see below). Young (1996) pointed out that the number and inter-calibration of Devonian biozones provides a finer-scale division of time than does modern isotopic dating, where the greatest precision is *c.* 1%, which is equivalent to several conodont or ammonoid zones (Young 1996). For major inter-regional spore zones, particularly relevant for Old Red Sandstone (ORS) calibration, Richardson *et al.* (1984) estimated durations of 3–4 Ma. It should be stressed that the conodont zonation employed in the time scale of Harland *et al.* (1990), which was used to subdivide Devonian time into (30) chrons, was superseded by a revised scale before the time of publication (see Ziegler & Sandberg (1994), McGhee (1996) and Sandberg & Ziegler (1996), for discussion), and should not be used as a basis for new time-scale calibration.

The correlation and recognition of system and particularly stage boundaries in continental ORS successions remains a major problem (House this volume). For the ORS, perhaps the most comprehensive biostratigraphic scheme is the microfloral (miospore) zonation, of which there are two established versions, those of Richardson & McGregor (1986) and Streel *et al.* (1987). Major works tackling the correlation of the standard conodont zonation to particularly Middle and Upper Devonian miospore zones are by Loboziak *et al.* (1990) and Streel & Loboziak (1996). The comprehensive correlation chart of Young (1996) for Devonian marine and continental fossil groups in Australia indicates that the correlation between microfloral zonations applicable to Australia and Europe with the standard conodont zonation remains provisional. The correlation of microfloral and conodont biozones in terms of chronostratigraphy is given in Fig. 1.

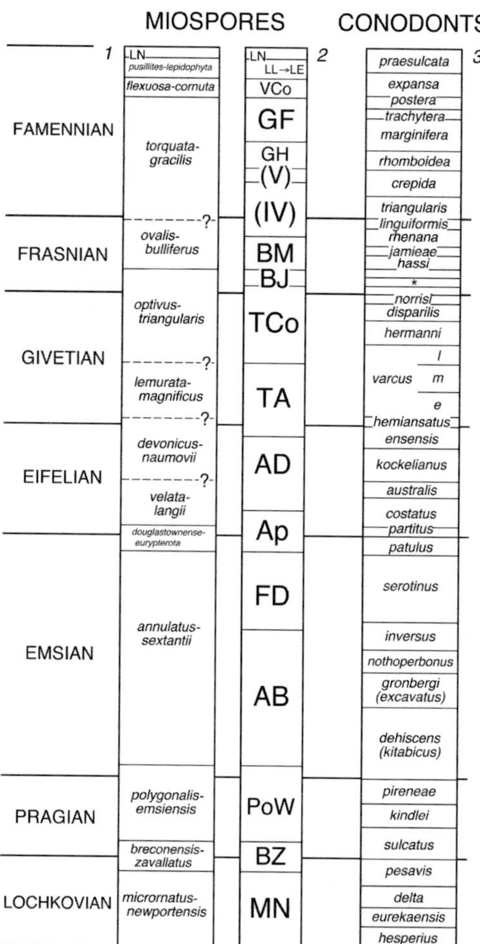

Fig. 1. Correlation of miospore zonation schemes (1, Richardson & McGregor 1986; 2, Streel *et al.* 1987) with the (generalized) standard conodont zonation (3, Sandberg & Ziegler 1996), and Devonian chronostratigraphy (see House *this volume*). *conodont zones not labelled *falsiovalis* (*rotundiloba*), *transitans* and *punctata* (in ascending order). No relative durations of stratigraphical units are implied.

Database of Devonian geological time scales

Examination of the literature on 'non-zircon-based' (pre-1995) time scales shows that most rely on individual isotopic ages from a suite of major compilations (Harland & Francis 1971; Armstrong 1978; Odin 1982; Snelling 1985) dating back to and including that of Harland *et al.* (1964, the Holmes Symposium on the Phanerozoic time scale). A process of selective elimination has occurred, in which authors of subsequent time scales have rejected certain published dates for inclusion (usually on the grounds of large analytical uncertainty, or evidence of the likelihood of isotopic system disturbance) while retaining others. Despite this, in the widely quoted time scale of Harland *et al.* (1990) for example, 26% of the Devonian data points originated in the compilation of Harland *et al.* (1964). However, a feature of these pre-1995 databases is that isotopic data have been (1) recalculated using the recommended decay constants of Steiger & Jäger (1977) and (2) frequently reassessed using an improved generation of statistical data reduction techniques (regression procedures and error propagation in particular, e.g. Gale *et al.* 1980; Odin 1982; Harland *et al.* 1990). This has produced uniformity of reliability in terms of analytical precision (generally quoted as two standard deviations or 2σ) and mean ages from original data. However, these techniques cannot offset the less precise analytical and instrumental standards inherent in some of the earlier age determinations undertaken (Gale *et al.* 1980; Forster & Warrington 1985).

Despite the fact that many of the analytical points have been superseded, and have been rejected by some workers, Figs 2 and 3 include Devonian-age data points from previous time scales, to facilitate comparison with new U–Pb ages on zircons (Tucker *et al.* 1998) and to identify any age convergence of different data. Most of these data are taken from the widely quoted compilations of Odin & Gale (1982, fig. 2, p. 492) and Harland *et al.* (1990, table 4.2, pp. 96–97), which contain 27 and 38 Devonian data points respectively. However, it should be noted that there are only 28 individual data in the compilation by Harland *et al.* (1990) for the Devonian period, on account of repeat dates on the same rock unit using different isotopic decay schemes and/or different material measured (whole rock or minerals). Some 42% (i.e. $n = 16$) of Odin & Gale's (1982) data points were used by Harland *et al.* (1990). Other publications that present selected isotopic ages to construct a time scale contain very few Devonian data (Fig. 4), ranging from $n = 4$ to $n = 10$. These scales (see below) rely on extrapolation from usually the Lower Palaeozoic era, for which considerably more data exist.

Most of the time scales discussed in this review are based on a mixture of whole-rock Rb–Sr and Rb–Sr, Ar/Ar, K/Ar and Sm–Nd mineral ages, which have uncertainties, expressed as 2σ relative standard deviations, of $<3\%$ according to McKerrow *et al.* (1985) to approaching 4% according to Tucker *et al.* (1990). These mixed database time scales can be considered as the traditional form that have developed to utilize all of the available analytical data as they became available. However, several issues arise in the new situation of increasingly available U–Pb (zircon) dates, which have better relative standard deviations of *c.* 1% for Palaeozoic time (Young 1996). For example, should a variety of isotopic systems be used for Palaeozoic time-scale database construction (e.g. Tucker & McKerrow 1995), or is it preferable to use one only (e.g. U–Pb; Tucker *et al.* 1998), and should a particular U–Pb analytical technique (e.g. ID-MS, Tucker *et al.* 1998; SHRIMP) be used for consistency in time-scale construction, or a combination (e.g. Compston 2000*b*)? We explore these points below.

Geological context of data points

One clear trend emerging with the publication of U–Pb dates on zircons is that the rock types studied are strata-bound (stratigraphically concordant, i.e. thin airfall K-bentonites, other tuffs, lavas), rather than from a mixture of geologically diverse situations (e.g. intrusions). For example, in the databases underpinning the Devonian time scales of Odin & Gale (1982) and Harland *et al.* (1990) intrusion ages comprise 44% and 63%, respectively, in addition to minor clast, glauconite and metamorphic ages. The remainder are composed of age determinations on dominantly lavas, and bentonites (Fig. 4). Similarly, the Devonian sections of the time scales of Gale (1985) and McKerrow *et al.* (1985), although containing low numbers of data, rely on a majority of intrusion ages (Fig. 4).

Plutons and other intrusions provide complex 'bracketed' ages, as a result of cross-cutting, structural and unconformable relationships (Fig. 3). In general, bracketed ages ought to be less precise than stratigraphically concordant dates. However, because strata-bound isotopic dates of earlier time scales were generally known only to stage or series level (Fig. 2), the bracketed ages provided by intrusions were of a similar level of precision. This is markedly different from recent dating of bentonites from marine successions, which can normally be correlated with biozones. There is no reason why felsic plutons, commonly used in previous time scales, could not be considered using, for example, U–Pb dating to provide higher precisions. However, it is clearly desirable to combine high analytical precision with narrow biostratigraphical control, to provide the most useful points for time-scale calibration.

The environmental setting of airfall tuffs and lavas is a significant factor in the stratigraphical precision with which they can be characterized. Tuffs and lavas in continental ORS successions generally have much poorer calibration, as a result of the low incidence of biostratigraphy (normally microfloral and/or micro- and macrovertebrates). Errors are also potentially propagated when intercalibrating with the marine record, where current correlations are uncertain. This general problem was identified by Odin & Gale (1982, p. 494), who suggested it was one reason for the relatively inaccurate series ages they proposed.

Glauconite ages are rare in the Devonian period; indeed, only two data points have been

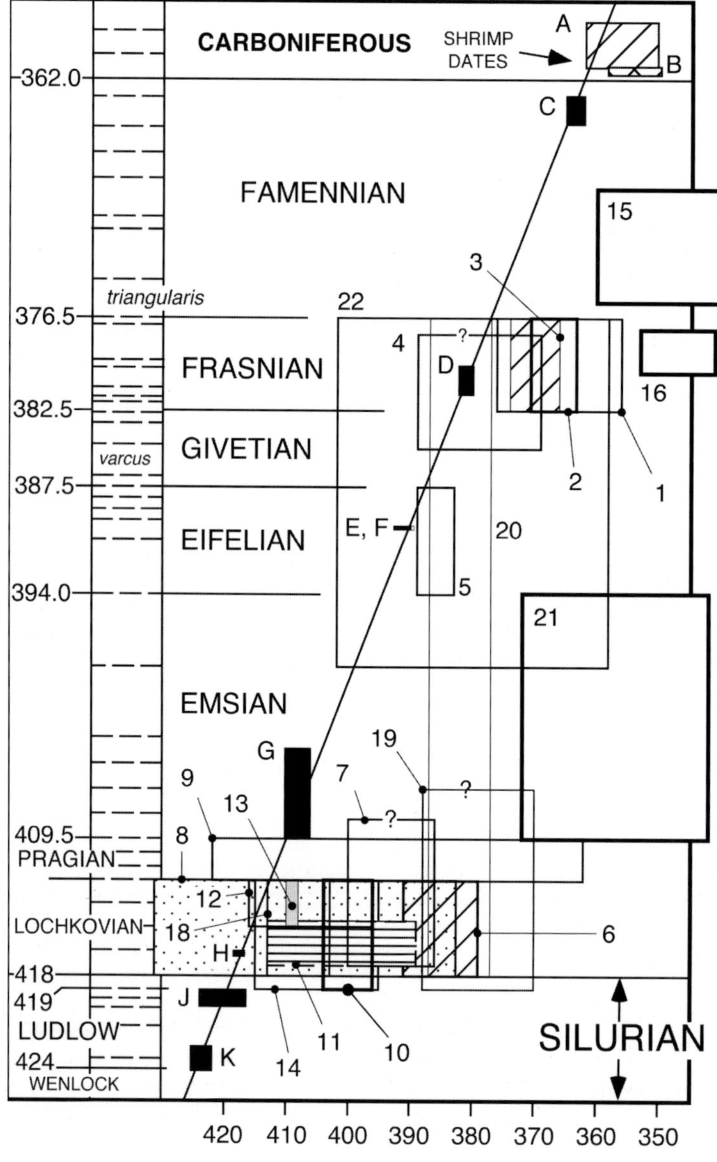

considered robust enough for inclusion by Odin & Gale (1982) and Harland *et al.* (1990; Fig. 2). In a study of glauconite ages, Craig *et al.* (1989) showed that these give systematically younger dates than real for beds older than 115 Ma. This may be due to non-closed isotopic systems, ease of weathering or the high probability of reworking. Harland *et al.* (1990) treated data from this category as minimum ages.

Non U–Pb isotopic systems

There are well-known isotopic and petrographic criteria on which to suspect ages intended for geochronology given by K–Ar, $^{40}Ar/^{39}Ar$ and Rb–Sr methods (see review by Harland *et al.* (1990)). These include Ar loss and low K in minerals for the K–Ar method, chemical mobility of Rb and Sr, variable initial Sr isotope ratios for the Rb–Sr method, and excess Ar in the $^{40}Ar/^{39}Ar$ technique. Both Rb–Sr and K–Ar systems are prone to resetting, following weathering, fluid migration, thermal perturbation and transformations as a result of volcanic mineral and glass instability. Fine-grained and glassy rocks, as well as coarser rocks, are prone to resetting, and K–Ar whole-rock ages are unsuitable for time-scale calibration. Harland *et al.* (1990) interpreted well-constrained Rb–Sr isochrons as minimum values for rock ages, and K–Ar dates on intrusions as minimum dates for the biostratigraphic age of the country rock.

Even when the above problems are considered negligible, opinion varies as to the inclusion or exclusion of certain data. For example, there has been a considerable debate over the validity of apparently well-defined whole-rock Rb–Sr isochron dates on acid volcanic rocks (Gale *et al.* 1979, 1980; McKerrow *et al.* 1980, 1985; Compston *et al.* 1982). The balance of opinion seems to suggest that they are frequently reset as a result of alteration, giving apparent ages up to 30 Ma younger than extrusion ages (Thirlwall 1988, p. 956). This affects several Přídolí and Early Devonian data points common to several time scales. Furthermore, opinions vary as to whether it is viable to use single K–Ar determinations for time-scale data points (Gale *et al.* 1979; see Odin 1982). Important points on the Devonian database (e.g. the Scottish Hoy lavas) are also affected by this.

Ambiguities in individual dates from one method have led to several studies that have attempted to establish concordant dating of the same material by different radiometric methods (e.g. Williams *et al.* 1982; Wyborn *et al.* 1982; Claoué-Long *et al.* 1995). All of these studies contribute to the Silurian–Carboniferous time-scale database, but some are not without inconsistencies when their biostratigraphic ages are considered against other data (e.g. the Cerberean Volcanics, Williams *et al.* 1982; Compston 2000*b*; see below).

Finally, fission-track ages have been used extensively in some time scales (e.g. McKerrow

Fig. 2. Plot of stratigraphically concordant isotopic data points (K-bentonites, other tuff beds, lavas, glauconite-rich beds) from recent non-zircon-based time scales against that of Tucker *et al.* (1998). The x-axis is the numerical age (Ma) and y-axis is the stratigraphical age calibrated by the scheme of Tucker *et al.* (1998). Data of Tucker *et al.* (1998, and references therein) shown as filled and hachured boxes labelled A–K are derived from bentonites and volcanics from the following units: (A) Kingsfield Formation; (B) Hasselbachtal bentonite; (C) Piskahegan Group; (D) Chattanooga Shale, Little War Gap; (E) Tioga Ash zircon; (F) Tioga Ash monazite (Roden *et al.* 1990); (G) Esopus Formation; (H) Kalkberg Formation; (J) Whitcliffe Formation; (K) Middle Elton Formation. Analytical uncertainties all plotted at 2σ; stratigraphic constraints of non-zircon data from Jones *et al.* (1981), Odin & Gale (1982), McKerrow *et al.* (1985) and Harland *et al.* (1990), where the narrowest range indicated has been used. Data plotted are as follows where the codes used in Harland *et al.* (1990) for cross referencing with major compilations are given (A, Armstrong 1978; NDS, Odin 1982; Odin & Gale 1982; J, Jones *et al.* 1981; PTS, Harland *et al.* 1964 or Harland & Francis 1971; MLC, McKerrow *et al.* 1985). (1) Kurskiy Glauconite (A425); (2) Cerberean Volcanics (NDS 234) (K–Ar); (3) Cerberean Volcanics (NDS 234) (Rb–Sr)Bi/WR; (4) Hoy Lavas (NDS 244); (5) Stribling Formation (NDS 151) (Glauconite age); (6) Vinalhaven Rhyolite (J29b), 385 ± 6 Ma (2σ) (Rb–Sr/Biotite); (7) Pembroke Formation Volcanics (NDS 237); (8) Vinalhaven Rhyolite (J29a) (K–Ar/Biotite). Table 2 of Jones *et al.* (1981) shows the Vinalhaven rhyolite with a range restricted to Gedinnian time; in contrast, Harland *et al.* (1990, table 4.2, p. 96) showed it as one of their Emsian items. (9) Shiphead Bentonites (PTS3); (10) Lorne Lavas (MLC22); (11) Eastport Volcanics (NDS222); (12) Hedgehog Volcanics (NDS 223); (13) Costigan Mountain Rhyolite Volcanics (FC1) (Rb–Sr WR isochron), referred to as Gulquac Lake area by Harland *et al.* (1990). This determination has a very large uncertainty (not shown) of ± 35 Ma (Fyffe & Cormier 1979); (14) Mareuil-Sur-Lay Volcanics (Harland *et al.* 1990; PPC9); (15) Chattanooga Shale bentonite (PTS2; Odin & Gale 1982 p. 488); (16) Chattanooga Shale bentonite (PTS94; Odin & Gale 1982); (17) Famennian sub-alkali basalt (A440, Armstrong 1978; Odin & Gale 1982); (18) Wormit Bay Lava (MCL21) (Rb–Sr); (19) Cranberry Island Series/Maine (J25) (Rb–Sr); (20) Castine Volcanics (J31) (Rb–Sr); (21) Kineo Volcanic Sequence (PTS96/J43) (Rb–Sr); (22) Snobs Creek Volcanics (PTS95/J49) (Rb–Sr).

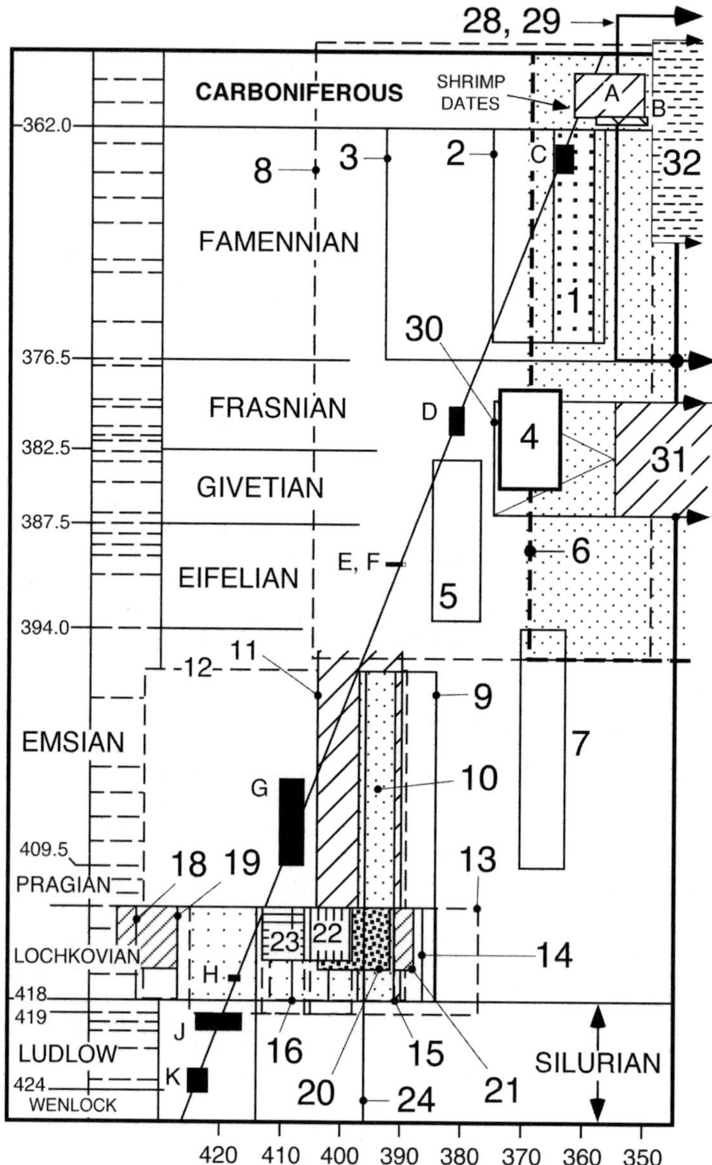

Fig. 3. Plot of stratigraphically discordant intrusion ages (frequently felsic plutons) from recent non zircon based time scales (Harland *et al.* 1990, table 4.2; Odin & Gale 1982) on the scale of Tucker *et al.* (1998, fig. 2). Plot x and y axes as detailed in Fig. 2. (1) Post-Cerberean intrusion (NDS235; K–Ar); (2) Post-Cerberean intrusion (NDS235; Rb–Sr); (3) Kennack Gneiss (Harland *et al.* 1990); (4) Mount Morgan tonalite (MLC26); (5) Mirimbah/Mnt. Stirling granodiorite (NDS233); (6) Bear River granite (PTS98a); (7) Nova Scotia granitoid (Harland *et al.* 1990); (8) Bear River (PTS98b; Rb–Sr); (9) Shap granite (NDS241c/PTS6; U–Pb); (10) Shap granite (NDS241b; Rb–Sr); (11) Shap granite (NDS241a; K–Ar); (12) Shap granite (PTS6; Rb–Sr); (13) Snowy River granite (PTS97/J22); (14) Three Mile Pond pluton (Harland *et al.* 1990); (15) St. George pluton (NDS236/A472; K–Ar); (16) NE Maine plutons (Harland *et al.* 1990; Ar–Ar); (17) St. George pluton (NDS236/A471; Rb–Sr); (18) Gulquac Lake granitoid (Fyffe & Cormier 1979); (19) Calais granite, Maine (PTS5); (20) Red Beach pluton (NDS236b/A488; Rb–Sr); (21) Red Beach pluton (NDS236a; K–Ar); (22) Gocup granite (NDS210b; Rb–Sr); (23) Gocup granite (NDS210a); (24) Fongen-Hillingen complex (Harland *et al.* 1990); (25) Heckla Hoek mica-schists (PTS4; not shown); (26) Creetown granite (PTS93; not shown); (27) South Mountain batholith (Odin & Gale 1982; not shown); (28–29) Central Kazakhstan granites (A424/A442); (30) Dahut River amphibole (A441); (31) Ugum Range pebbles (A443); (32) Huelgoat granite (NDS229).

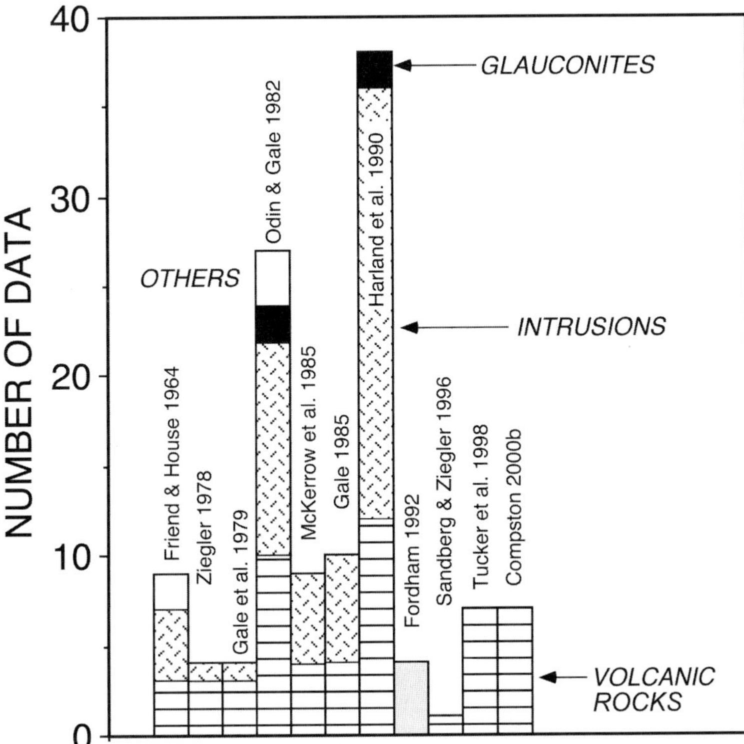

Fig. 4. Bar chart of the number of Devonian data points in selected recent time scales, and break-down of data type (as a proportion) in the same scales.

et al. 1980; Harland *et al.* 1990), although not currently for data points within the Devonian period. However, there has been a general consensus that the large uncertainties about the central age limit their use in accurate time-scale production (Gale *et al.* 1980; McKerrow *et al.* 1985). The Ordovician and Silurian K-bentonites originally dated by apatite and zircon fission-track methods (Ross *et al.* 1976, 1982) have been in many cases redated by U–Pb ID-MS on igneous zircon phenocrysts (Tucker *et al.* 1990; Tucker & McKerrow 1995; see also Compston 2000*a*, *b*).

U–Pb zircon dating

Recently published Devonian time scales that rely heavily on U–Pb zircon ages are those of Tucker & McKerrow (1995), Tucker *et al.* (1998) and Compston (2000*b*). U–Pb dating is highly advantageous because of the well-known property of its two independent radioactive decay schemes ($^{238}U \rightarrow {}^{206}Pb$, $^{235}U \rightarrow {}^{207}Pb$), allowing three dates to be calculated for concordant analytical points ($^{206}Pb/^{238}U$–$^{207}Pb/^{235}U$–$^{207}Pb/^{206}Pb$). For the more common situation of discordant analytical points that form a linear array on a conventional concordia diagram, upper concordia intercept ages and weighted mean $^{207}Pb/^{206}Pb$ ages can be calculated to give the mineral crystallization age. See Faure (1986) for a review of U–Pb geochronology.

The technique is predominantly carried out on zircon because of its abundance in silicic igneous rocks. The technique is also fully viable for monazite and, although this mineral is comparatively rare in volcanic rocks, requiring specific magma chemistry conditions, notable (Devonian) determinations have proved to be essentially concordant and very precise (Roden *et al.* 1990). Zircon is also an advantageous mineral as it is exceptionally stable, has a very high closure temperature (*c.* 800 °C), and has refractory properties. These factors render it relatively safe from common geochemical events that affect other systems (particularly Rb–Sr, K–Ar) and thermal resetting. Complicating factors involving zircons, however, include radiogenic lead loss, inherited (xenocrystic) zircon (often occurring as grain cores), complex overgrowth zonation and captured mineral inclusions that result in high 'common' lead contents reducing analytical precision. The complex structure of some zircon grains can be addressed only by the use of the ion probe (SHRIMP), which can make *in situ* measurements of genetically different zones. This can avoid old cores and sample zircon overgrowths caused by magmatic crystallization, and thus obtain information pertinent to time-scale construction. Limitations of SHRIMP instruments, however, include their inability to precisely measure ^{235}U–^{207}Pb ages (Sambridge & Compston 1994). Because of this, multiple determinations on single grains are necessary to assess the validity of the derived age; for this reason, related to the amount of material consumed, ion probe results are currently less precise than ID-MS (Compston 1999). A systematic (younger age) technical bias in the SHRIMP instrument has also been claimed (e.g. Tucker & McKerrow 1995), although this has been refuted in detail by Compston (1999, 2000*a*, *b*). Finally, SHRIMP instruments require calibration by a reference material zircon. Compston (1999, 2000*a*) has recently identified previously unrecognized heterogeneity in the previously used standard zircon, which will have affected the accuracy of some published SHRIMP ages. New techniques to correct for this problem have been developed (see Compston (2000*a*, *b*) for full discussion). Other zircon reference materials that can be used (see Jagodzinski & Black (1999)) give differing results (see below).

U–Pb isotope dilution-mass spectrometry (Rollinson 1993) is widely regarded as the most accurate and sensitive method available to modern geochronology. Single grain or small (typically <25; Tucker *et al.* 1990) multi-grain zircon samples are currently analysed after careful selection and treatment (see e.g. Tucker *et al.* 1990, 1998). For discordant suites of analytical points, two principal interpretative problems arise: (1) the choice of correction for initial common lead content, where the model of Stacey & Kramers (1975) is commonly used; (2) the identification of analytical points containing a component of inherited zircon material. The first can lead to the overestimation of $^{207}Pb/^{206}Pb$ mean weighted ages, because of sensitivity to the common Pb correction. Compston (2000*a*, *b*) has argued that different choices for the correction can render slightly discordant analytical points concordant and, in conjunction with the presence of inherited zircon samples, lead to alternative interpretations of U–Pb ages (see also Young *et al.* (1999)). Accurate lead corrections can be applied if Pb data can be directly measured from co-magmatic minerals (e.g. sanidine) in the target rock (W. Compston, pers. comm.). Sambridge & Compston (1994) have developed new statistical techniques for the detection and estimation of ages and uncertainties of multiple components in a population of U–Pb analyses. Under the above conditions, one or more components may be interpreted as containing inherited (i.e. older) zircon grains. This is the basis of the recent Devonian time scale of Compston (2000*b*), which reassesses the ID-MS data of Tucker *et al.* (1998; see below).

Some problems with apparently suitable target rock types for time-scale purposes have recently been highlighted as a consequence of the above detailed research into the interpretation of SHRIMP and ID-MS zircon ages. Sambridge & Compston (1994) pointed out that, although K-bentonites and tuffs in marine sequences are ideal targets because of their potentially excellent biostratigraphical control, they may suffer from having their magmatic zircon population contaminated by exotic zircon from background sedimentation and those derived from wall rocks during eruption. Glass-rich tuffs also undergo transformation to clay minerals, during which zircon may be affected by Pb loss or U gain. Compston (2000*b*) has pointed out that S-type volcanic rocks, derived from melting of sedimentary or metasedimentary source rocks, will be restite rich and will thus very probably contain several populations of inherited zircons (e.g. the Laidlaw Volcanics, Wyborn *et al.* (1982)), making interpretation of the magmatic age of the rock exceptionally difficult.

Methods of time-scale construction

Early methods of time-scale construction included using palaeontologically calibrated sediment thicknesses and maximum known thicknesses at stage level to extrapolate from isotopically dated points (see Hudson 1964; Friend & House 1964). Although these methods are generally not now used (House 1991), the principle behind them has been locally employed (Churkin *et al.* 1977; Murphy 1987). Some scales employed the hypothesis of equal stage lengths (Harland *et al.* 1982), where local calibration by isotopic data was considered inadequate or absent. There are seven principal methods of construction employed exclusively or in combination in the time scales discussed below. The validity of these methods is not universally agreed.

(1) The most straightforward method is that of simple inspection of isotopically derived data spanning stage (or other) boundaries. Ideally the age data should be sequential (or ‘internally consistent’), that is, not inconsistent with their chronostratigraphic position. Discrepant data

points in this respect are discarded. This method has been used for the Devonian period by Armstrong (1978), Odin & Gale (1982), and for the Devonian–Carboniferous boundary and later systems by Forster & Warrington (1985).

(2) A mathematical development of the simple assessment of mean ages described above is the 'scanning age procedure', which utilizes isotopic age determinations (and their analytical uncertainties) from the stratigraphical units above and below a target boundary (normally a stage boundary, defined by a stratotype). This method was developed and used by Harland *et al.* (1982, 1990), from an original application in palaeomagnetic reversal dating (Cox & Dalrymple 1967; Mankinen & Dalrymple 1979). The method estimates the age of a boundary by finding the minimum error function (E in equation (1); equation (5.1) of Harland *et al.* (1990)), derived from a series of isotopic dates from units adjacent to the boundary, for a range of 'scanning ages' (Fig. 5a) spanning the boundary. The range of the scanning age is estimated from the values of the isotopic dates.

$$E = \sum(\mathrm{A}i - t)^2/s_{\mathrm{A}i}^2 + \sum(\mathrm{B}i - t)^2/s_{\mathrm{B}i}^2 \quad (1)$$

where Ai are individual isotopic dates in unit A (Fig. 5a), Bi are dates in unit B, $s_{\mathrm{A}i}$ and $s_{\mathrm{B}i}$ are the respective 1σ analytical uncertainties of the dates, and t is the scanning age.

Harland *et al.* (1990) pointed out that this is a reproducible method, unlike others in use, which are subjective. Figure 5a shows three Devonian error function curves from Harland *et al.* (1990), which illustrate cases where there are (i) gaps in the isotopic age database, in this example unevenly distributed between stratigraphic units (non-overlapping, asymmetrically distributed data), (ii) isotopic ages inconsistent with their stratigraphical position (overlapping data) and (iii) well-distributed, internally consistent isotopic ages (perfect age distribution). Ideally, with a large number of analytically high-precision data, which are well constrained biostratigraphically, all stage boundaries could be calibrated by this method.

(3) Perhaps the most commonly used time-scale method is the 'best-fit' time line procedure (Gale *et al.* 1980; McKerrow *et al.* 1980, 1985; Gale 1985; Tucker & McKerrow 1995; Tucker *et al.* 1998). This involves fitting a straight line through error boxes defined by the 2σ analytical uncertainty of the isotopic age and the chronostratigraphical range of the dated rock unit (Fig. 5b). This is an iterative method, in which the error boxes are freely adjusted on the plot

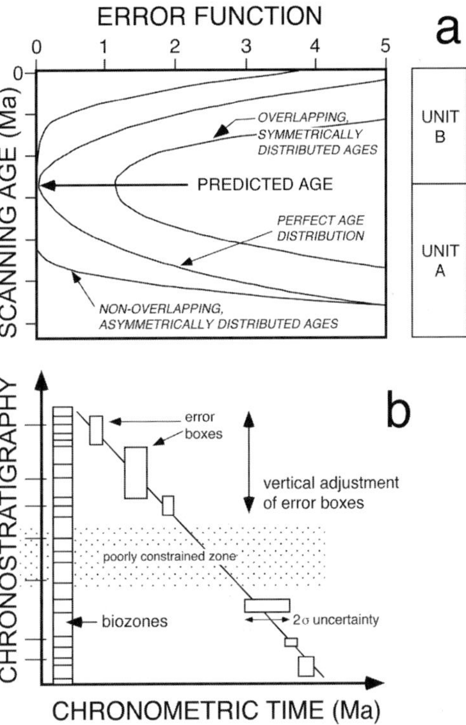

Fig. 5. (**a**) Diagram to illustrate the derivation of a boundary age from error function curves (from Harland *et al.* 1990). (**b**) Diagram of the iterative best-fit time line method of time-scale construction.

y-axis until they allow intersection with a line. Stage boundary positions are therefore shifted according to constraints imposed by the biostratigraphic ranges of the data. Boundary ages are estimated by intersection with the best-fit line.

This method has been (justifiably) criticized for being subjective (Forster & Warrington 1985; Harland *et al.* 1990). For example, it is possible to construct alternative scales to include selected data points (see, e.g. Gale 1985, figs 1 and 2; compare McKerrow *et al.* 1985). Tucker *et al.* (1998) produced two scales from their U–Pb zircon data, one assuming equal biozone lengths, the other with unequal biozone lengths that were scaled to palaeontological estimates (see below, (4)). This highlights the necessity to make such assumptions when the density of isotopic age data is low.

(4) Several notable studies have estimated the relative durations of Devonian chronostratigraphic stages using evolution estimates of different fossil groups (Boucot 1975; Ziegler 1978; Bayer & McGhee 1986, 1989; Ziegler &

Table 1. *Data from Boucot (1975) and Ziegler (1978) on palaeontologically estimated stage durations for the Devonian period (values are percentages with original indices of relative duration given in parentheses)*

Stage	Boucot (1975)	Ziegler (1978)
Famennian	(1.4) 18.92	(1.5) 20.27
Frasnian	(1.2) 16.22	(1.3) 17.57
Givetian	(1.2) 16.22	(1.1) 14.86
Eifelian	(1.0) 13.51	(0.9) 12.16
Emsian	(1.0) 13.51	(1.3) 17.57
Pragian	(0.8) 10.81	(0.6) 8.11
Lochkovian	(0.8) 10.81	(0.7) 9.46

Sandberg 1990; Sandberg & Ziegler 1996; Table 1). Employing local calibration by isolated isotopic dates, it is then possible to interpolate the boundary ages of successive stages. However, there is neither agreement that rates of faunal evolution can be used successfully to interpolate between data points (e.g. Tucker & McKerrow 1995) nor agreement on whether biozones have equal durations (Ziegler & Sandberg 1990), a thesis considered unlikely by House (1991). Harland *et al.* (1990) assigned stratigraphic *chrons* on the basis of the number of biozones in a period. For age interpolation in parts of the time scale that were poorly constrained radiometrically, it was further assumed that chrons were of equal duration.

(5) A variant of method (3) (above) employing regression is possible using assumptions regarding the stratigraphic position of the isotopic data. The relative durations of stages, estimated by faunal evolution (Boucot 1975; Ziegler 1978), can be used to assign a numerical value to stage boundaries and thus quantify the stratigraphic position of isotopic data points (Gale *et al.* 1979; Jones *et al.* 1981). Line fitting by regression is then used to calibrate stage boundaries.

(6) The graphic correlation technique (Shaw 1964), which classically correlates palaeontologically calibrated successions on a cross-plot, one of which is regarded as a standard reference section, has been exploited to proportion or estimate biozone durations (Fordham 1992, 1998). In conjunction with isotopically dated horizons that can be linked to correlated sections (e.g. Bergström 1989), it is possible to calibrate stratigraphical boundaries. Work on some Devonian conodont biostratigraphy and graphic correlation has been detailed by Klapper *et al.* (1995).

(7) Well-developed microrhythmicity throughout a near-complete Givetian section in marine pelagic facies was analysed by House (1991,1995) in terms of control by orbital precession and obliquity cycles. Taking into account estimates of Mid-Devonian orbital signatures, and utilizing graphical correlation procedures, House (1995) used the cyclostratigraphy to calculate a duration of 6.5 Ma for the Givetian stage, and also specified durations for conodont zones and sub-zones. The results compare well with palaeobiological estimates of Givetian time (Boucot 1975; Ziegler 1978) and also strongly indicate that the assumption of equal biozone lengths used in some time scales is false (House 1995). These techniques are independent of isotopic control, and potentially provide a finer-scale calibration of geological time than can be currently achieved by the latter method. Although general orbital forcing time scales are not currently available, a convergence between their development and improved radiometric scales is clearly desirable for geological time-scale refinement. Other cyclostratigraphic studies on Devonian (Old Red Sandstone) successions include those by Kelly (1992), Olsen (1994), Marshall (1996) and Marshall *et al.* (1996).

Discussion of the evolution of the Devonian time scale

Which time scales should be considered? It is worth drawing attention to the issue of what Harland *et al.* (1990) referred to as 'eclectic' time scales; that is, those scales, often for the whole Phanerozoic eon, that combine different, individually constructed scales (Haq & Van Eysinga 1987; Gradstein & Ogg 1996), repeat large sections of other works, use isolated single dates to constrain major boundaries (Palmer 1983; the 1999 Geological Society of America chart) or serve as compromise scales between conflicting schemes (e.g. Snelling 1985, 1987). For example, the widely quoted International Union of Geological Sciences time scale (Cowie & Bassett 1989) is based on Snelling (1987), which (for the Devonian period) is in turn based on individual contributions (Forster & Warrington 1985; Gale 1985; McKerrow *et al.* 1985) to the Subcommision on Geochronology (Snelling 1985). More recently, the time scale of Gradstein & Ogg (1996) is based on Harland *et al.* (1990), Tucker & McKerrow (1995) for the Silurian–Lower Devonian section, and the SHRIMP ages of Roberts *et al.* (1995) for the uppermost Devonian–lower Carboniferous section. The Devonian, upper Silurian and Carboniferous periods of the 1999 Geological Society of America time scale are based verbatim on Gradstein & Ogg (1996).

Comprehensive comparisons of various time scales have been given by Odin (1982) and Harland *et al.* (1990, figs 1.5 and 1.6), including some of the scales excluded here, as well as the important pioneering scales of Holmes (1937, 1947, 1960) and Kulp (1961). We briefly discuss below 14 selected 'primary' works of time-scale research (compared in Fig. 6) that (in general) rely on a coherent method of construction and use a specified and selected isotopic database.

The early Devonian scale of Friend & House (1964; Fig. 6) stemmed from selected data (Fig. 4), from the compilation of Harland *et al.* (1964), located at the top and bottom of the system. Boundary estimates of series subdivisions were made using arguments regarding sediment thicknesses. The relatively young top and base of this scale is consistent with contemporary and earlier scales (e.g. Kulp 1961). The time scale of Armstrong (1978) is of interest because it uses an expanded isotopic age database, largely including data from Harland *et al.* (1964), and was the first scale to follow the revision of decay constants (Steiger & Jäger 1977). The scale gives the oldest estimate of the top and second oldest of the base of the Devonian section of any comparable scale (Fig. 6). In the same year, Ziegler (1978) published a Devonian time scale that combined his palaeontologically based estimates of stage duration and the radiometric age estimates of Boucot (1975). These were based on only four Devonian data points (Eastport Volcanics, Shiphead bentonites, Acadian granites and the Cerberean Volcanics; see Fig. 2). Although not shown in Fig. 6, a time scale estimated from Boucot's (1975) faunal analysis would be very similar to that of Ziegler (1978); both are marked by a young age for the top of the Devonian sequence (350 Ma).

Three subsequent scales (Gale *et al.* 1979, 1985; McKerrow *et al.* 1985) and an earlier scale (not shown in Fig. 6) by McKerrow *et al.* (1980; see Gale *et al.* 1980) rely heavily on extrapolation of time lines from the Lower Palaeozoic section, although Gale *et al.* (1979) used averaged fossil group durations combined with a regression method. These scales highlight the different ages generated, from few, essentially similar, data using the best-fit line technique. The scales agree closely on the Devonian–Carboniferous boundary (354–356 Ma), although not on the base of the Devonian period, nor internally (Fig. 6). The time scale of Odin & Gale (1982), based on a much larger Devonian database (Odin 1982) and a different method of construction, indicated large uncertainties for series boundaries ($\pm$ 5–10 Ma).

The time scale of Harland *et al.* (1982) employed the database of Armstrong (1978) and introduced an original construction method (chronograms). However, for the Devonian period no tie points were defined, and the scale utilized an artificial equal stage duration hypothesis to interpolate between Silurian and Carboniferous data points to derive stage ages. However, the age of the top of the Devonian system was similar to that given by Odin & Gale (1982) and that suggested by de Souza (1982, 360–365 $\pm$ 5 Ma). The revised version of Harland *et al.* (1982), published in 1990, used a revised database (including data of Odin (1982)), and was able to define five Devonian 'tie-points' (stage and intra-stage boundaries). However, linear interpolation was performed between these where data were lacking, assuming equal duration of chrons (biozones) between tie-points. Harland *et al.* (1990) considered this assumption to hold well on average, but not in detail (but see House (1991, 1995)). Sandberg & Ziegler (1996) pointed out that the number of conodont biozones employed as chrons (30) by Harland *et al.* (1990) is incorrect. Despite this, their estimate of the age of the top of the Devonian system (362.5 Ma) is again in line with other studies (e.g. Forster & Warrington 1985, 365 $\pm$ 5 Ma) that are based on different construction methods, though common data.

Different approaches were taken in the work of Fordham (1992, Fig. 6) and Sandberg & Ziegler (1996). The former used conodont biozone-based graphic correlation, calibrated by the tie- and pseudo-tie-points of Harland *et al.* (1990) and the *c.* 354 Ma SHRIMP age for the Devonian–Carboniferous boundary (Claoué-Long *et al.* 1992, 1993), to estimate the remaining unconstrained stage boundaries (Fig. 6). Sandberg & Ziegler (1996) relied entirely on the SHRIMP age (Claoué-Long *et al.* 1992, 1993), and palaeontological estimates of stage durations to calculate stage boundary ages. The scale of Young (1996), which accompanies comprehensive biostratigraphic correlations of the Australian Devonian system, is calibrated against the *c.* 354 Ma (SHRIMP-derived) age for the Devonian–Carboniferous (D–C) boundary, and an age of 410 Ma for the base of the Devonian system, similar to that of Cowie & Bassett (1989). It is not clear how the Devonian stage boundaries were derived.

The relatively recent trend to largely or exclusively based U–Pb zircon time scales, with implications for the Devonian period, dates from the paper of Tucker & McKerrow (1995). Devonian information in this paper (not shown in Fig. 6) is, however, limited, because of a large

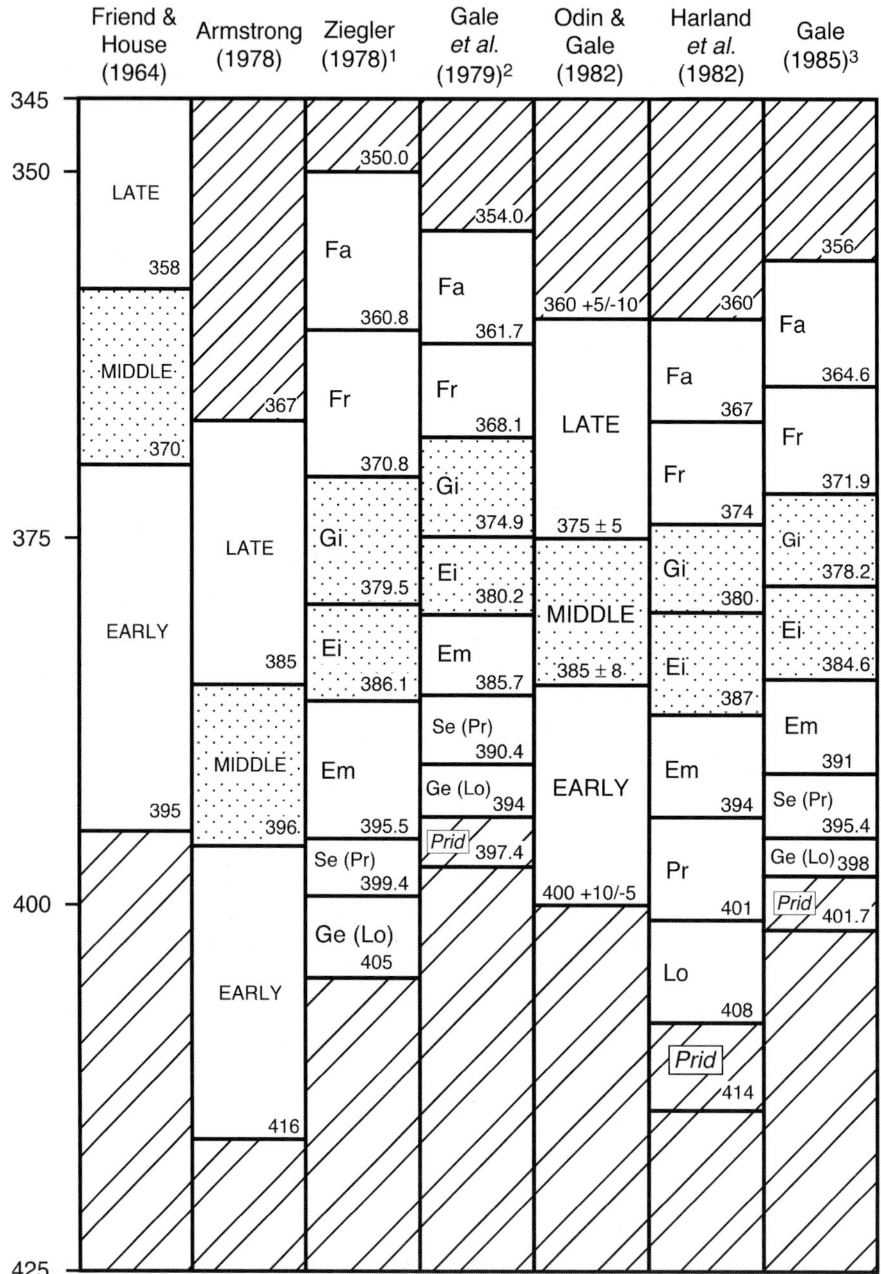

Fig. 6(a).

time-line extrapolation from Lower Palaeozoic time, extending as far as Eifelian time, where it is constrained by the precise data point for the *costatus* biozone Tioga ash bed (Roden *et al.* 1990). The two Devonian data points in conjunction with those from the Lower Palaeozoic section led to an (imprecise) estimate of 417 Ma for the base of the Devonian system. The largely zircon-based scale of Tucker *et al.* (1998; Figs 2, 3 and 6) is based on five Devonian data points, plus the Lower Carboniferous Kingsfield SHRIMP age (Claoué-Long *et al.* 1995), and two Silurian data from Tucker & McKerrow (1995). The Hasselbachtal SHRIMP date

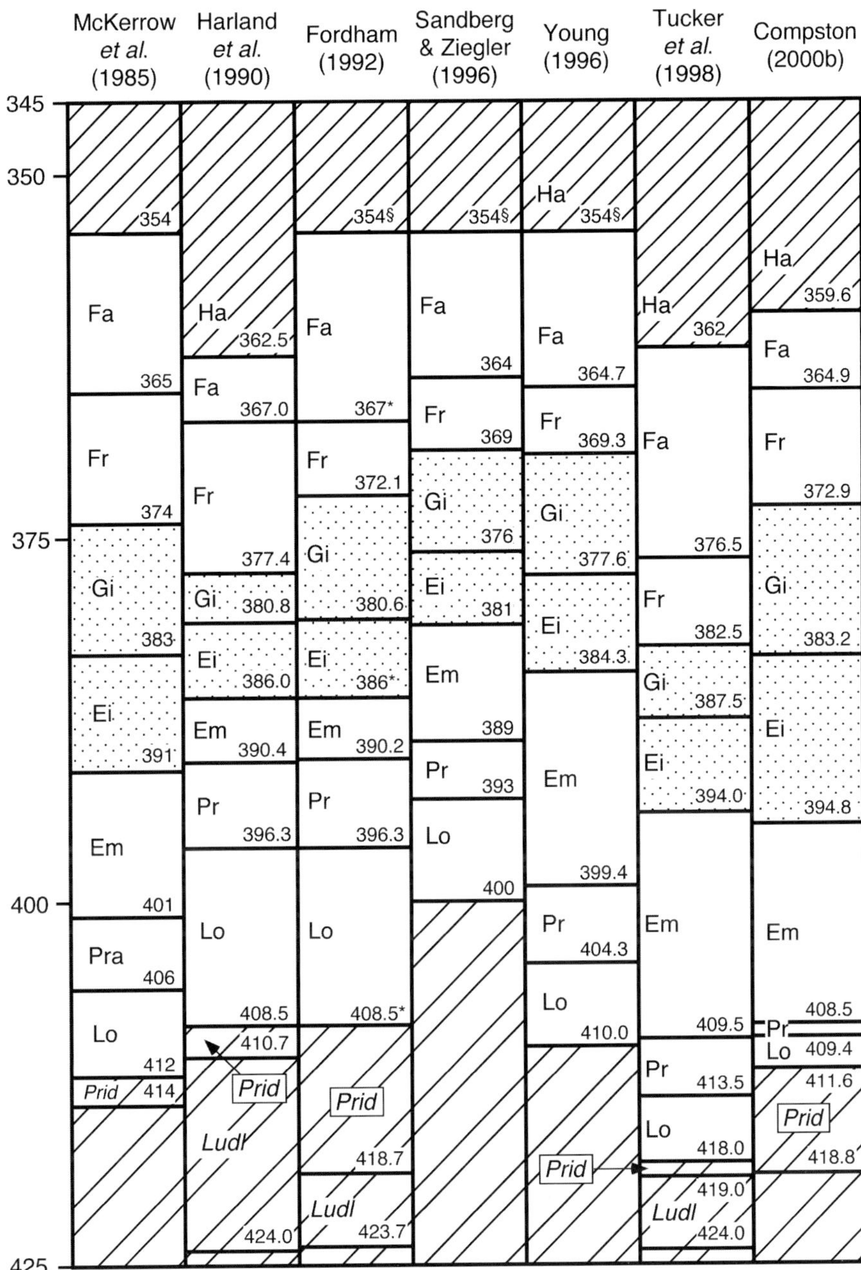

Fig. 6(b).

Fig. 6. Chart comparing 14 original Devonian time scales. Superscripts 1–3 indicate that stage boundary dates of the time scales of Ziegler (1978, fig. 1), Gale *et al.* (1979, fig. 3) and Gale (1985, fig. 1) are digitized from the specified figures in these publications. *in the column for Fordham (1992) are the tie- and pseudo tie-points of Harland *et al.* (1990) used to calibrate this scheme. §, indicates boundaries constrained or influenced by the SHRIMP age for the Devonian–Carboniferous boundary auxiliary stratotype at Hasselbachtal, Germany (Claoué-Long *et al.* 1992, 1993). Ha, Hastarian; Fa, Famennian; Fr, Frasnian; Gi, Givetian; Ei, Eifelian; Em, Emsian; Pr, Pragian; Se, Siegenian; Lo, Lochkovian; Ge, Gedinnian; Prid, Přídolí; Ludl, Ludlow. Silurian is diagonally shaded at the base of the diagram. Middle Devonian series and stages are stippled for clarity.

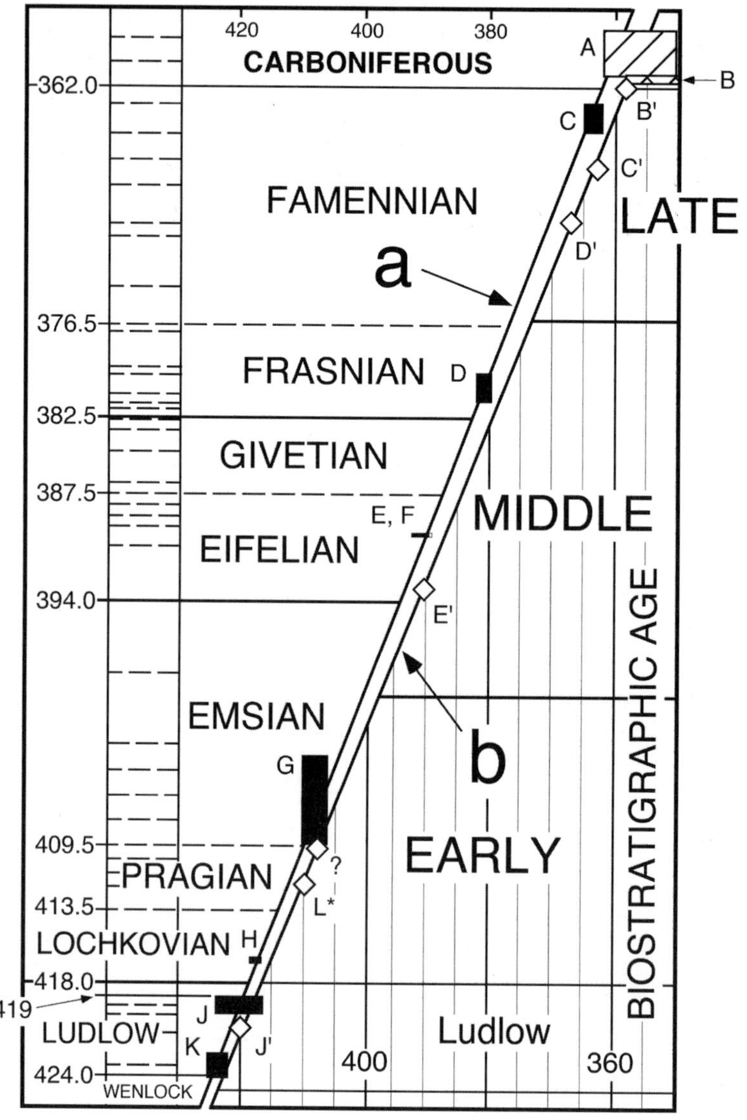

Fig. 7. Comparison of the schemes of (**a**) Tucker *et al.* (1998) and (**b**) Compston (2000*b*). x axes are numerical ages (Ma) and y (vertical) axes are the stratigraphical ages calibrated by each scheme. Analytical points of Tucker *et al.* (1998) shown as filled error boxes, coded as in Fig. 2; data reinterpreted by Compston (2000*b*) are shown as open diamonds, with the corresponding code letter from (**a**); point L* is based on Jagodzinski & Black (1999; see Compston 2000*b*, table 4 for full discussion); the reinterpreted age of the Kalkberg bentonite (409.3 ± 1.1 Ma 1σ), not plotted by Compston (2000*b*, fig. 15), has a similar numerical age to L* but a different biostratigraphic age. SHRIMP data of Claoué-Long *et al.* (1992, 1995) are shown by hachured and cross-hatched boxes, coded A and B.

(353 ± 4.3 Ma, Claoué-Long *et al.* 1992, 1993) is plotted, but does not lie on the best-fit line. The final line represents a solution using proportioned biozones (based on Fordham's (1992) study, and that of Johnson *et al.* (1991)), although only small discrepancies were observed with a version based on equal biozone lengths (Tucker *et al.* 1998, p. 183).

The time scale of Compston (2000*b*) is based on a time line fitted to (zircon-based ID-MS and SHRIMP) data from the Lower Palaeozoic section, reinterpreted data points from Tucker *et al.* (1998) and one Lower Devonian point from Jagodzinski & Black (1999). Two of the data from Tucker *et al.* (1998) are significantly different (Fig. 7), resulting in a very much younger base to the Devonian sequence (411.6 Ma) and several different stage boundaries from those given by Tucker *et al.* (1998; Fig. 6). Nearly all the reinterpreted data are younger based on Compston's (2000*b*) discrimination of inherited components in previous dates (by mixture modelling; Sambridge & Compston 1994), and departure from the accepted model of Stacey & Kramers (1975) for Pb evolution. The value of the Hasselbachtal SHRIMP age is adjusted to be *c.* 1% older, but it is suggested that this unit needs to be reprocessed to resolve a more accurate date (see also Young *et al.* (1999)).

Meaningful trends that emerge from these time scales are difficult to identify and, possibly, any trend may be spurious because of the redundancy of many of the (older) analytical data. An increasing age for the top of the Devonian system is crudely visible from the earliest scales (Friend & House 1964), and the larger more precise studies suggest convergence to *c.* 360–362 Ma. The reliance of several studies on the *c.* 354 Ma SHRIMP age (Fordham 1992; Sandberg & Ziegler 1996; Young 1996) may have been premature (Tucker *et al.* 1998; Young *et al.* 1999; Compston 2000*b*). Little correspondence exists between the scales of Harland *et al.* (1990) and Tucker *et al.* (1998), except for the D–C boundary; the age of the Middle Devonian series is highly variable in all recent scales. There is considerable variation in the age of the base of the Devonian system, even in supposedly more reliable recent time scales (Figs 6 and 7), and some exceptionally short estimates of stage durations (e.g. 1 Ma for the Přídolí stage by Tucker *et al.* (1998); 0.9 Ma for the Pragian stage by Compston (2000*b*)).

Recent data compilation

Here we consider the latest available U–Pb zircon ages with other critically selected isotopic data points for the Devonian, Siluro-Devonian and Early Carboniferous periods, to move towards a new time scale for the Devonian period. New data, previously published information that compares well with the time line of Tucker *et al.* (1998), as well as stratigraphically and isotopically corrected dates are plotted in Fig. 8. The base of the Pragian stage is modified in Fig. 8 (compare Tucker *et al.* 1998, fig. 2), where it is placed at the base of the *sulcata* conodont zone (Chlupác & Oliver 1989). Using the time line intersection of Tucker *et al.* (1998), this gives an age of *c.* 412.6 Ma for the Lochkovian–Pragian boundary (Fig. 8).

Improved isotopic determinations as well as stratigraphical constraints apply to the Lorne Lavas (Marshall 1991), Glencoe Volcanics, Wormit Bay Lavas and Arbuthnot Group rocks (Richardson *et al.* 1984; Thirlwall 1988; Marshall 1991; Tucker & McKerrow 1995) from the ORS of Scotland. The Lorne Lavas are considered to be approximately at the Silurian–Devonian boundary based on latest Silurian to earliest Devonian palynomorphs located <50–60 m stratigraphically beneath (Marshall 1991). Error boxes for the Lorne Lavas and Glencoe Volcanics intersect the time line of Tucker *et al.* (1998; Fig. 8); this however, requires a late Přídolí to early Lochkovian age for the Glencoe Volcanics. A single determination for the Wormit Bay Lavas (Arbuthnott Group ORS of the Midland Valley of Scotland) by Thirlwall (1988) from the lower and middle sub-zone of the *microcornatus–newportensis* biozone (Richardson *et al.* 1984; Fig. 1), and a mean age of intrusion and lava ages (by Tucker & McKerrow (1995), from data of Thirlwall (1988)) from the same group, both plot younger than the time line (Fig. 8). These points are, however, congruent with the recalculated Kalkberg bentonite age (Compston 2000*b*; Figs 7 and 8, point C2).

U–Pb ID-MS ages from Lower, Middle and Upper Devonian tuffs (Williams *et al.* 1997, this volume, in prep.) provide stratigraphically moderately well constrained points for early Emsian (?AB miospore biozone, see Higgs 1999), Givetian and mid-Frasnian times, respectively. The Cooscrawn Tuff bed (Fig. 8, provisional) from the Dingle Basin (SW Ireland, Williams *et al.* in prep.) would suggest an older age (⩾411 Ma) for the base of the Emsian stage. SHRIMP determinations from Australian Lochkovian (*delta* conodont zone) and Pragian (approximately late *delta* to *dehiscens* zones) marine sequences (Jagodzinski & Black 1999) are well constrained by biostratigraphy. However, age assignment is complicated in this instance by the fact that two sets of ages are quoted, consequent upon the use of different standard reference zircons. Figure 8 shows the preferred dates of Jagodzinski & Black (1999), based on standard zircon QGNG. Compston (2000*b*) preferred the alternative dates, because of lack of published data on zircon QGNG and a known disturbance in its isotopic system. The QGNG-calibrated data compare well with the time line of Tucker *et al.* (1998), and led Jagodzinski & Black (1999) to prefer this time scale over the current Australian Devonian scheme (Young 1996, Fig. 6; see Young *et al.*

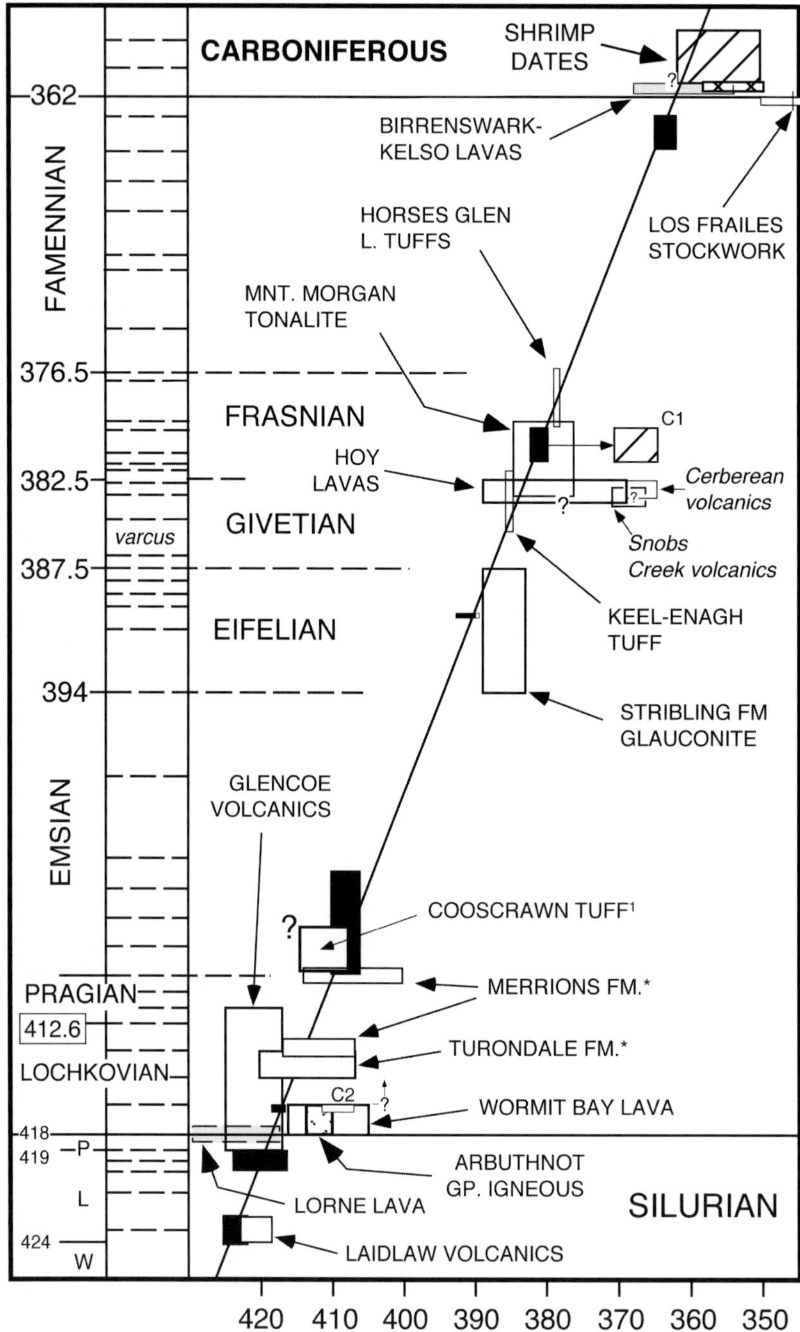

Fig. 8. Plot showing new data along with selected, well-calibrated data from previous time scales, superimposed on the time scale of Tucker *et al.* (1998). The x axis is the numerical age (Ma) and the y axis is the stratigraphical age from Tucker *et al.* (1998, fig. 2). Note the corrected base of the Pragian (cf. Tucker *et al.* 1998). Points C1 and C2 are, respectively, recalculated ages by Compston (2000*b*) of the Little War Gap/Centre Hill and Kalkberg bentonites of Tucker *et al.* (1998). *The lowermost and uppermost ages determinations from the Merrions Formation are shown (based on the preferred QGNG-based calibrated ages of Jagodzinski & Black 1999). Los Failes stockwork zircons age (Strunian) from Nesbitt *et al.* (1999). (1) Cooscrawn Tuff age is currently provisional. Laidlaw volcanics, based on Wyborn *et al.* (1982).

(1999) and Compston (2000*b*) for a full discussion). Compston (2000*b*) reinterpreted the age of the Kalkberg bentonite (Tucker *et al.* 1998), as shown by point C2 in Fig. 8 (see also Fig. 7). This reduces the age of the base of the Devonian system and does not require a very short Přídolí series (see Fordham (1998) for discussion). Compston's (2000*b*) time scale did, however, compress the duration of the Pragian stage significantly (Fig. 6), which is nevertheless regarded as a short stage (Boucot 1975; Ziegler 1978; McGhee 1996).

A new SHRIMP determination of the previously dated (McKerrow *et al.* 1985) Mount Morgan trondhjemite–tonalite, New South Wales (Golding *et al.* 1994), is the only intrusion age included, because of its relatively good constraints. This plots similarly to the Little War Gap–Center Hill K-bentonite of Tucker *et al.* (1998). The stratigraphic age of the Hoy Lavas from the ORS of the Orcadian Basin (Halliday *et al.* in Odin 1982; Harland *et al.* 1990) is considered here to be late Givetian (see Rogers *et al.* 1989, fig. 2), the 2σ error box of which intersects the time line (Fig. 8). New isotopic determinations are available for the Cerberean (cauldron) and the underlying Snobs Creek Volcanics (Williams *et al.* 1982), as well as a reappraised biostratigraphical assignment to Givetian, rather than Frasnian age (Young 1999; pers. comm.). Although Compston's (2000*b*) reinterpretation of the Little War Gap K-bentonite of Tucker *et al.* (1998; point C1 in Fig. 8) allowed a Frasnian-aged designation of the Cerberean Volcanics to fit his revised time scale (Fig. 7), the Givetian age now considered likely makes fitting this date on both scales more difficult.

The Kelso–Birrenswark lavas at the top of the Upper ORS of the Scottish Border Basin, were interpreted by Gawthorpe *et al.* (1989) as being of Famennian–Hastarian age, whereas they were given a stratigraphical assignment of Ivorian age by Harland *et al.* (1990). When replotted (Fig. 8) their central age (de Souza 1982) lies close to the time line of Tucker *et al.* (1998). Recent SHRIMP dating of stockwork zircons related to regional mineralization of sediments in the Iberian Pyrite Belt are calibrated by palynology as being of Strunian (latest Famennian) age (Nesbitt *et al.* 1999). The error box of this point (Fig. 8) is significantly different from the time line, and just overlaps with that of the SHRIMP age for the Hasselbachtal tuff (Claoué-Long *et al.* 1992, 1993). The (non-volcanic) paragenesis of the zircon minerals in this case contrasts significantly with all of the other quoted data, and may be responsible for the younger age implied for the D–C boundary.

Summary

Single zircons and small populations apparently provide the most precise data points by ID-MS. However, these apparently precise age determinations are not without problems: (1) mixtures involving xenocrystic-cored zircons may yield anomalously old ages; (2) different choices for the common Pb content correction for zircons having suffered post-magmatic Pb loss can yield younger ages (Compston 2000*a*, *b*), and with (1) may show that zircon ages are multi-generational; (3) recently quoted precisions may be overestimates because of uncertainties in U series decay constants (Ludwig 1998).

Published U–Pb zircon ages by SHRIMP are probably affected by defects in the zircon standard material used to calibrate the determinations, although new statistical techniques can be used in future experiments to combat this (Compston 1999, e.g. 2000*a*, *b*). The widely quoted SHRIMP age for the Devonian–Carboniferous boundary auxiliary stratotype at Hasselbachtal (Claoué-Long *et al.* 1992, 1993) is contradicted by the study of Tucker *et al.* (1998) and to a lesser extent that of Compston (2000*b*). This perhaps shows the danger of placing too much reliance on one determination: the studies of Fordham (1992), Gradstein & Ogg (1996), Sandberg & Ziegler (1996) and Young (1996) were all influenced by this determination. Different choices of zircon calibration standard reference material for the SHRIMP instrument can produce significantly different geological ages (e.g. Jagodzinski & Black 1999).

Data points from pre-U–Pb-based time scales, based on Rb–Sr whole-rock or mineral and K–Ar mineral ages, tend to plot significantly younger than the best-fit time line of Tucker *et al.* (1998), or that of Compston (2000*b*). These older data points tend to have stratigraphical constraints no better than the stage level, and except near the Silurian–Devonian boundary, are not congruent with U–Pb zircon data. This is also true of intrusive rocks, which are no longer significant targets for time-scale research, because of their relatively poor stratigraphical control. There are no new data (U–Pb or other types), nor reliable older data for most of the Famennian, late Emsian nor parts of the Eifelian–Givetian stages. With the possible exceptions of the Silurian–Devonian boundary and some Early Devonian stage boundaries, data density is too low for the application of mathematical assessment of ages.

New advances in time-scale research will be instrumental, with further enhancement of the ion-microprobe (Compston 1999) leading to higher precision, and the development of laser ablation techniques (Jackson *et al.* 1997; Nesbitt *et al.* 1997; Horn *et al.* 2000), which has already yielded age determinations of (unzoned) zircon (Fernández-Suárez *et al.* 1999). In parallel, advances must also be stratigraphical, with refinement of the continental–marine biostratigraphic correlation, particularly in the field of palynology.

Sincere thanks are due to W. Compston (for preprints and discussion), B. G. Fordham, J. C. Claoué-Long, G. C. Young and P. J. Jones for discussions on Devonian geochronology, as well as to J. E. A. Marshall, M. R. House and K. T. Higgs for valuable information on biostratigraphical and chronostratigraphical aspects of the ORS and Devonian system. Particular thanks are due (from E.A.W.) to S. A. Sergeev for his expert introduction to the world of zircon geochronology. None of these colleagues, however, are responsible for the views expressed in this paper.

References

ARMSTRONG, R. L. 1978. Pre-Cenozoic Phanerozoic time scale—computer file of critical dates and consequences of new and in-progress decay constant revisions. *In*: COHEE, G. V., GLAESSNER, M. F. & HEDBERG, H. D. (eds) *Contributions to the Geological Time Scale*. American Association of Petroleum Geologists Studies in Geology, **6**, 73–93.

BAYER, U. & MCGHEE, G. R. JR 1986. Cyclic patterns in the Paleozoic and Mesozoic: implications for time scale calibrations. *Paleoceanography*, **1**, 383–402.

—— & —— 1989. Periodicity of Devonian sedimentary and biological perturbations: implications for the Devonian time scale. *Neues Jahrbuch für Geologie und Paläontologie*, **1**, 1–16.

BERGSTRÖM, S. M. 1989. Use of graphic correlation for assessing event-stratigraphic significance and trans-Atlantic relationships of Ordovician K-bentonites. *Proceedings of the Estonian Academy of Sciences*, **38**, 55–59.

BOUCOT, A. J. 1975. *Evolution and Extinction Rate Controls*. Elsevier, Amsterdam.

CHLUPÁC, I. & OLIVER, W. A. JR 1989. Decision on the Lochkovian–Pragian boundary stratotype (Lower Devonian). *Episodes*, **12**, 109–113.

CHURKIN, M. C. JR, CARTER, C. & JOHNSON, B. 1977. Subdivision of Ordovician and Silurian time using accumulation rates of graptolitic shale. *Geology*, **5**, 452–456.

CLAOUÉ-LONG, J. C., COMPSTON, W., ROBERTS, J. & FANNING, C. M. 1995. Two Carboniferous ages: a comparison of SHRIMP zircon dating with conventional zircon ages and $^{40}Ar/^{39}Ar$ analysis. *In*: BERGGREN, W. A., KENT, D. V., AUBRY, M.-P. & HARDENBOL, J. (eds) *Geochronology, Time Scales and Global Stratigraphic Correlation*. SEPM (Society for Sedimentary Geology) Special Publications, **54**, 3–21.

——, JONES, P. J. & ROBERTS, J. 1993. The age of the Devonian–Carboniferous boundary. *Annales de la Société Géologique de Belgique*, **T115**, 531–549.

——, ——, —— & MAXWELL, S. 1992. The numerical age of the Devonian–Carboniferous boundary. *Geological Magazine*, **129**, 281–291.

COMPSTON, W. 1999. Geological age by instrumental analysis: the 29th Hallimond Lecture. *Mineralogical Magazine*, **63**, 297–311.

—— 2000*a*. Interpretations of SHRIMP and isotope dilution zircon ages for the geological time-scale: Part 1, the early Ordovician and late Cambrian. *Mineralogical Magazine*, **64**, 43–57.

—— 2000*b*. Interpretation of SHRIMP and isotope dilution zircon ages for the Palaeozoic time-scale: Part 2, Silurian to Devonian. *Mineralogical Magazine*, in press.

——, MCDOUGALL, I. & WYBORN, D. 1982. Possible two stage ^{87}Sr evolution in the Stockdale Rhyolite. *Earth and Planetary Science Letters*, **61**, 297–302.

COWIE, J. W. & BASSETT, M. G. 1989. International Union of Geological Sciences 1989 Global Stratigraphic Chart with geochronometric and magnetostratigraphic calibration. *Episodes*, **12**(s).

COX, A. V. & DALRYMPLE, G. B. 1967. Statistical analysis of geomagnetic reversal data and the precision of potassium–argon dating. *Journal of Geophysical Research*, **72**, 2603–2614.

CRAIG, L. E., SMITH, A. G. & ARMSTRONG, R. L. 1989. Calibration of the geologic time scale: Cenozoic and Late Cretaceous glauconite and non-glauconite dates compared. *Geology*, **17**, 830–832.

DE SOUZA, H. A. F. 1982. Age data from Scotland and the Carboniferous time scale. *In*: ODIN, G. S. (ed.) *Numerical Dating in Stratigraphy*. Wiley, Chichester, 455–465.

FAURE, G. 1986. *Principles of Isotope Geology*, 2nd edn. Wiley, New York.

FERNÁNDEZ-SUÁREZ, J., GUTIÉRREZ-ALONSO, G., JENNER, G. A. & TUBRETT, M. N. 1999. Crustal sources in Lower Palaeozoic rocks from NW Iberia: insights from laser ablation U–Pb ages of detrital zircons. *Journal of the Geological Society, London*, **156**, 1065–1068.

FORDHAM, B. G. 1992. Chronometric calibration of mid-Ordovician to Tournaisian conodont zones: a compilation from recent graphic correlation and isotope studies. *Geological Magazine*, **129**, 709–721.

—— 1998. Silurian time: how much of it was Přídolí? *In*: GUTIÉRREZ-MARCO, J. C. & RÁBANO, I. (eds) *Proceedings of the 6th International Graptolite Conference and Field Meeting, IUGS Subcommission on Silurian Stratigraphy*. Temas Geológico-Mineros ITGE, **23**, 80–84.

Forster, S. C. & Warrington, G. 1985. Geochronology of the Carboniferous, Permian and Triassic. *In*: Snelling, N. J. (ed.) *The Chronology of the Geological Record*. Geological Society of London, Memoirs, **10**, 99–113.

Friend, P. F. & House, M. R. 1964. The Devonian period. *In*: Harland, W. B., Smith, A. G. & Wilcock, B. (eds) *The Phanerozoic Time-Scale. Quarterly Journal of the Geological Society, London*, **120s**, 233–236.

Fyffe, L. R. & Cormier, R. F. 1979. The significance of radiometric ages from the Gulquac Lake area of New Brunswick. *Canadian Journal of Earth Sciences*, **16**, 2046–2052.

Gale, N. H. 1985. Numerical calibration of the Palaeozoic time-scale; Ordovician, Silurian and Devonian periods. *In*: Snelling, N. J. (ed.) *The Chronology of the Geological Record*. Geological Society, London, Memoirs, **10**, 81–88.

——, Beckinsale, R. D. & Wadge, A. J. 1979. A Rb–Sr whole rock isochron for the Stockdale Rhyolite of the English Lake District and a revised mid-Palaeozoic time scale. *Journal of the Geological Society, London*, **136**, 235–242.

——, —— & —— 1980. Discussion of a paper by McKerrow, Lambert and Chamberlain on the Ordovician, Silurian and Devonian time scales. *Earth and Planetary Science Letters*, **51**, 9–17.

Gawthorpe, R. L., Gutteridge, P. & Leeder, M. R. 1989. Late Devonian and Dinantian basin evolution in northern England and North Wales. *In*: Arthurton, R. S., Gutteridge, P. & Nolan, S. C. (eds) *The Role of Tectonics in Devonian and Carboniferous Sedimentation in the British Isles*. Occasional Publications of the Yorkshire Geological Society, **6**, 1–23.

Golding, S. D., Messenger, P. R., Dean, J. A., Perkins, C., Huston, D. A. & White, A. H. 1994. Mount Morgan gold–copper deposit: geochemical constraints on the source of volatiles and lead and the age of mineralization. *In*: Henderson, R. A. & Davis, B. K. (eds) *New Developments in Geology and Metallogeny: Northern Tasman Orogenic Zone*. Economic Geology Research Unit, James Cook University, Townsville, 89–95.

Gradstein, F. M. & Ogg, J. G. 1996. A Phanerozoic time scale. *Episodes*, **19**, 3–5.

Harland, W. B. & Francis, H. (eds) 1971. *The Phanerozoic Time-Scale—a Supplement*. Geological Society, London, Special Publications, **5**.

——, Armstrong, R. L., Cox, A. V., Craig, L. E., Smith, A. G. & Smith, D. G. 1990. *A Geologic Time Scale 1989*. Cambridge University Press, Cambridge.

——, Cox, A. V., Llewellyn, P. G., Pickton, C. A. G., Smith, A. G. & Walters, R. 1982. *A Geologic Time Scale*. Cambridge University Press, Cambridge.

——, Smith, A. G. & Wilcock, B. (eds) 1964. *The Phanerozoic Time-Scale*. Quarterly Journal of the Geological Society, London, **120s**.

Haq, B. U. & Van Eysinga, F. W. B. 1987. Geological Time Table, 4th edn (wall chart). Elsevier, Amsterdam.

Higgs, K. T. 1999. Early Devonian spore assemblages from the Dingle Group, County Kerry, Ireland. *Bolletino della Societa Paleontologica Italiania*, **38**, 187–197.

Holmes, A. 1937. The Age of the Earth. Nelson (London), new edition.

—— 1947. The construction of a geological time-scale. *Transactions of the Geological Society, Glasgow*, **32**, 117–152.

—— 1960. A revised geological time-scale. *Transactions of the Edinburgh Geological Society*, **17**, 183–216.

Horn, I., Rudnick, R. L. & McDonough, W. F. 2000. Precise elemental and isotope ratio determination by simultaneous solution rebulization and laser ablation-ICP-MS: application to U-Pb geochronology. *Chemical Geology*, **164**, 281–301.

House, M. R. 1988. International definition of Devonian System boundaries. *Proceedings of the Ussher Society*, **7**, 41–46.

—— 1991. Devonian sedimentary microrhythms and a Givetian time scale. *Proceedings of the Ussher Society*, **7**, 392–395.

—— 1995. Devonian precessional and other signatures for establishing a Givetian timescale. *In*: House, M. R. & Gale, A. S. (eds) *Orbital Forcing Timescales and Cyclostratigraphy*. Geological Society, London, Special Publications, **85**, 37–49.

—— 2000. Chronostratigraphic framework for the Devonian and Old Red Sandstone. *This volume*.

Hudson, J. D. 1964. Sedimentation rates in relation to the Phanerozoic time scale. *In*: Harland, W. B., Smith, A. G. & Wilcock, B. (eds) *The Phanerozoic Time-Scale*. Quarterly Journal of the Geological Society, London, **120s**, 37–42.

Jackson, S. E., Longerich, H. P., Horn, I. & Dunning, G. R. 1997. The application of laser ablation microprobe (LAM)–ICP-MS to in situ U–Pb zircon geochronology. *In*: Proceedings of the Goldschmidt Conference, Abstract volume, 283–284.

Jagodzinski, E. A. & Black, L. P. 1999. U–Pb dating of silicic lavas, sills and syn-eruptive resedimented volcaniclastic deposits of the Lower Devonian Crudine Group, Hill End Trough, New South Wales. *Australian Journal of Earth Sciences*, **46**, 749–764.

Jones, B. G., Carr, P. F. & Wright, A. J. 1981. Silurian and Early Devonian geochronology—a reappraisal, with new evidence from the Bungonia Limestone. *Alcheringa*, **5**, 197–207.

Johnson, J. G., Sandberg, C. A. & Poole, F. G. 1991. Devonian lithofacies of western United States. *In*: Cooper, J. D. & Stevens, C. H. (eds) Paleozoic Paleogeography of the Western United States, II, Vol. 1. SEPM Pacific Section, **67**, 83–105.

Kelly, S. B. 1992. Milankovitch cyclicity recorded from Devonian non-marine sediments. *Terra Nova*, **4**, 578–584.

Klapper, G., Kirchgasser, W. T. & Baesemann, J. F. 1995. Graphic correlation of a Frasnian (Upper Devonian) composite standard. *In*: Mann, K. O. & Lane, H. R. (eds) *Graphic Correlation*. Society for Sedimentary Geology, Special Publications, **53**, 177–184.

KULP, J. L. 1961. Geologic time-scale. *Science*, **133**, 1105–1114.

LOBOZIAK, S., STREEL, M. & WEDDIGE, K. 1990. Miospores, the *lemurata* and *triangulatus* levels and their faunal indices near the Eifelian–Givetian boundary in the Eifel (F.R.G.). *Annales de la Société Géologique de Beligique*, **T.113**, 299–313.

LUDWIG, K. 1998. On the treatment of concordant uranium–lead ages. *Geochimica et Cosmochimica Acta*, **62**, 665–676.

MANKINEN, E. A & DALRYMPLE, G. B. 1979. Revised geomagnetic polarity time scale for the interval 0–5 m.y. B.P. *Journal of Geophysical Research*, **84**(B2), 615–626.

MANN, K. O. & LANE, H. R. (eds) 1995. *Graphic Correlation*. Society for Sedimentary Geology, Special Publications, **53**.

MARSHALL, J. E. A. 1991. Palynology of the Stonehaven Group, Scotland: evidence for a Mid Silurian age and its geological implications. *Geological Magazine*, **128**, 283–286.

—— 1996. *Rhabdosporites langii*, *Geminospora lemurata* and *Contagisporites optivus*: an origin for heterospory within the Progymnosperms. *Review of Palaeobotany and Palynology*, **93**, 159–189.

——, ROGERS, D. A. & WHITELEY, M. J. 1996. Devonian marine incursions into the Orcadian Basin, Scotland. *Journal of the Geological Society, London*, **153**, 451–466.

MCGHEE, G. R. JR 1996. *The Late Devonian Mass Extinction, The Frasnian/Famennian Crisis*. Columbia University Press, New York.

MCKERROW, W. S., LAMBERT, R. S. J. & CHAMBERLAIN, V. E. 1980. The Ordovician, Silurian and Devonian time scales. *Earth and Planetary Science Letters*, **51**, 1–8.

——, —— & COCKS, L. R. M. 1985. The Ordovician, Silurian and Devonian Periods. *In*: SNELLING, N. J. (ed.) *The Chronology of the Geological Record*. Geological Society, London, Memoirs, **10**, 73–80.

MURPHY, M. A. 1987. The possibility of a Lower Devonian equal-increment time scale based on lineages in Lower Devonian conodonts. *In*: AUSTIN, R. L. (ed.) *Conodonts—Investigative Techniques and Applications. British Micropalaeontological Society Series*. Ellis Horwood, Chichester, 284–293.

NESBITT, R. W., HIRATA, T., BUTLER, I. B. & MILTON, J. A. 1997. UV laser ablation *ICP*-MS: some applications in the Earth sciences. *Geostandards Newsletter: The Journal of Geostandards and Geoanalysis*, **20**, 231–243

——, PASCUAL, E., FANNING, C. M., TOSCANO, M., SÁEZ, R. & ALMODÓVAR, G. R. 1999. U–Pb dating of stockwork zircons from the eastern Iberian Pyrite Belt. *Journal of the Geological Society, London*, **156**, 7–10.

ODIN, G. S. (ed.) 1982. *Numerical Dating in Stratigraphy*. Wiley–Interscience, New York.

—— & GALE, N. H. 1982. Numerical dating of Hercynian times (Devonian to Permian). *In*: ODIN, G. S. (ed.) *Numerical Dating in Stratigraphy*. Wiley, Chichester, 487–500.

OLSEN, H. 1994. Orbital forcing on continental depositional systems—lacustrine and fluvial cyclicity in the Devonian of East Greenland. *In*: DE BOER, P. L. & SMITH, D. G. (eds) *Orbital Forcing and Cyclic Sequences*. International Association of Sedimentologists, Special Publications, **19**, 429–438.

PALMER, A. R. 1983. Decade of North American Geology (DNAG), Geologic time scale. *Geology*, **11**, 503–504.

RICHARDSON, J. B. & MCGREGOR, D. C. 1986. Silurian and Devonian spore zones of the Old Red Sandstone continent and adjacent regions. *Bulletin of the Geological Survey of Canada*, **364**, 1–79.

——, FORD, J. H. & PARKER, F. 1984. Miospores, correlation and age of some Scottish Lower Old Red Sandstone sediments from the Strathmore region (Fife and Angus). *Journal of Micropalaeontology*, **3**, 109–124.

ROBERTS, J., CLAOUÉ-LONG, J. C. & JONES, P. J. 1995. Australian Early Carboniferous time. *In*: BERGGREN, W. A., KENT, D. V., AUBRY, M.-P. & HARDENBOL, J. (eds) *Geochronology, Time Scales and Global Stratigraphic Correlation*. SEPM (Society for Sedimentary Geology), Special Publications, **54**, 23–40.

RODEN, M. K., PARRISH, R. R. & MILLER, D. S. 1990. The absolute age of the Eifelian Tioga ash bed, Pennsylvania. *Journal of Geology*, **98**, 282–285.

ROGERS, D. A., MARSHALL, J. E. A. & ASTIN, T. R. 1989. Devonian and later movements on the Great Glen fault system, Scotland. *Journal of the Geological Society, London*, **146**, 369–372.

ROLLINSON, H. R. 1993. *Using Geochemical Data: Evaluation, Presentation, Interpretation*. Longman, Harlow, UK.

ROSS, R. J. JR, NAESER, C. W. & IZETT, G. A. 1976. Apatitie fission-track dating of a sample from the type Caradoc (Ordovician) series in England. *Geology*, **4**, 505–506.

——, —— & —— ET AL.1982. Fission-track dating of British Ordovician and Silurian stratotypes. *Geological Magazine*, **119**, 135–153.

SAMBRIDGE, M. S. & COMPSTON, W. 1994. Mixture modeling of multi-component data sets with application to ion-probe zircon ages. *Earth and Planetary Science Letters*, **128**, 373–390.

SANDBERG, C. A. & ZIEGLER, W. 1996. Devonian conodont biochronology in geologic time calibration. *Senckenbergiana Lethaea*, **76**, 259–265.

SHAW, A. B. 1964. *Time in Stratigraphy*. McGraw–Hill, New York.

SNELLING, N. J. 1985. An interim time-scale. *In*: SNELLING, N. J. (ed.) *The Chronology of the Geological Record*. Geological Society, London, Memoirs, **10**, 261–265.

—— 1987. Measurement of geological time and the geological time scale. *Modern Geology*, **11**, 365–374.

STACEY, J. S. & KRAMERS, J. D. 1975. Approximation of terrestrial lead isotope evolution by a two-stage model. *Earth and Planetary Sciences Letters*, **26**, 207–221.

STEIGER, R. H. & JÄGER, E. 1977. Subcommission on geochronology: convention on the use of decay constants in geo- and cosmochronology. *Earth and Planetary Sciences Letters*, **36**, 359–362.

STREEL, M. & LOBOZIAK, S. 1996. Middle and Upper Devonian miospores. *In*: JANSONIUS, J. & MCGREGOR, D. C. (eds) *Palynology: Principles and Applications*, American Association of Stratigraphic Palynologists Foundation, **2**, 575–587.

——, HIGGS, K., LOBOZIAK, S., RIEGEL, W. & STEEMANS, P. 1987. Spore stratigraphy and correlation with faunas and floras in the type marine Devonian of the Ardenne–Rhenish regions. *Review of Palaeobotany and Palynology*, **50**, 211–229.

THIRLWALL, M. F. 1988. Geochronology of Late Caledonian magmatism in northern Britain. *Journal of the Geological Society, London*, **145**, 951–967.

TUCKER, R. D. & MCKERROW, W. S. 1995. Early Paleozoic chronology: a review in light of new U–Pb zircon ages from Newfoundland and Britain. *Canadian Journal of Earth Sciences*, **32**, 368–379.

——, BRADLEY, D. C., VER STRAETEN, C. A., HARRIS, A. G., EBERT, J. R. & MCCUTCHEON, S. R. 1998. New U–Pb zircon ages and the duration and division of Devonian time. *Earth and Planetary Science Letters*, **158**, 175–186.

——, KROGH, T. E., ROSS, R. J. JR & WILLIAMS, S. H. 1990. Time-scale calibration by high-precision U–Pb zircon dating of interstratified volcanic ashes in the Ordovician and Lower Silurian stratotypes of Britain. *Earth and Planetary Science Letters*, **100**, 51–58.

WILLIAMS, E. A., SERGEEV, S. A., STÖSSEL, I. & FORD, M. 1997. An Eifelian U–Pb zircon date for the Enagh Tuff Bed from the Old Red Sandstone of the Munster Basin in NW Iveragh, SW Ireland. *Journal of the Geological Society, London*, **154**, 189–193.

——, ——, ——, —— & HIGGS, K. T. 2000. U–Pb zircon geochronology of silicic tuffs and chronostratigraphy of the earliest Old Red Sandstone in the Munster Basin, SW Ireland. *This volume*.

WILLIAMS, I. S., TETLEY, N. W., COMPSTON, W. & MCDOUGALL, I. 1982. A comparison of K–Ar and Rb–Sr ages of rapidly cooled igneous rocks: two points in the Palaeozoic time scale re-evaluated. *Journal of the Geological Society, London*, **139**, 557–568.

WYBORN, D., OWEN, M., COMPSTON, W. & MCDOUGALL, I. 1982. The Laidlaw Volcanics: a Late Silurian point on the geological time scale. *Earth and Planetary Science Letters*, **59**, 90–100.

YOLKIN, E. A., KIM, A. I., WEDDIGE, K., TALENT, J. A. & HOUSE, M. R. 1997. Definition of the Pragian/Emsian stage boundary. *Episodes*, **20**, 235–240.

YOUNG, G. C. 1996. Devonian (Chart 4). *In*: YOUNG, G. C. & LAURIE, J. R. (eds) *An Australian Phanerozoic Timescale*. Oxford University Press, Oxford, 96–109.

—— 1999. Preliminary report on the biostratigraphy of new placoderm discoveries in the Hervey Group (Upper Devonian) of central New South Wales. *Records of the Western Australian Museum, Supplement 57*, 139–150.

——, JONES, P., CLAOUÉ-LONG, J. C., COMPSTON, W. & ROBERTS, J. 1999. Letter: Were Queensland and Victoria in different Devonian time zones? *Australian Geologist*, **113**, 10–12.

ZIEGLER, W. 1978. Devonian. *In*: COHEE, G. V., GLAESSNER, M. F. & HEDBERG, H. D. (eds) *Contributions to the Geological Time Scale*. American Association of Petroleum Geologists, Studies in Geology, **6**, 337–339.

—— & SANDBERG, C. A. 1994. Conodont phylogenetic-zone concept. *Newsletters in Stratigraphy*, **30**, 105–123.

Chronostratigraphic framework for the Devonian and Old Red Sandstone

MICHAEL R. HOUSE

School of Ocean and Earth Science, Southampton Oceanography Centre, European Way, Southampton SO14 3ZH, UK

Abstract: Boundaries for all the Devonian series and stages have now been proposed by the Subcommission on Devonian Stratigraphy (SDS) of the International Commission on Stratigraphy (ICS) and have been ratified by the International Union of Geological Sciences (IUGS). A summary of these decisions is given. All the defined Global Stratigraphic Section and Points (GSSPs) are in the marine realm, and in pelagic facies. Much work is still needed to correlate the new boundaries into neritic and terrestrial facies.

The history of an internationally agreed stratigraphic framework for the Palaeozoic eon has been one of fluctuating opinion, recommendation and practice. The first Congress of Geologists (now IGC, International Geological Congress) was held in Paris in 1879 with a major aim of clarifying the definition of stratigraphical terms (Clarke 1923; Hancock 1977). At Bologna in 1880 it was decided that all systems should be divided into Lower, Middle and Upper series. The Devonian System is one of few that is still divided in this way, although the definitions both of the upper and lower limits of the system, and of the series themselves has varied considerably over the last century. The introduction of local stages has similarly been haphazard. At the IGC in 1910 it was agreed that each country should produce definitions of terms originating in their countries; those for Belgium (Waterlot 1957) and Germany (Kutscher & Schmidt 1958) were those most relevant for Devonian subdivisions. Although southwest England is the *locus typicus* for the Devonian System (Sedgwick & Murchison 1839, 1840) and it was early recognized that the marine deposits of Devon were the time equivalents of the Old Red Sandstone terrestrial facies of Wales and Scotland (Rudwick 1985), the area has not provided so precise a biostratigraphy as the Ardennes, Eifel and Schiefergebirge, where many zonations have been established.

By 1960 it was recognized that an enormous international anomaly was associated with levels drawn for the base of the Devonian System and base of the Middle Devonian Series in major areas of the world. Because graptolites disappear from the Welsh borders stratigraphic record near the Ludlow Bone Bed, the extinction of graptolites had been taken elsewhere as the definition of the base of the system. The discovery of graptolites associated with brachiopod zonal fossils from high in the Lower Devonian succession led to the IGC at Copenhagen in 1960 establishing a committee to investigate. This was followed in 1972 with the establishment of a programme that led to the erection of subcommissions for each stratigraphic system under the International Commission on Stratigraphy (ICS) empowered to make recommendations to the IUGS on boundary definition. Later working parties were established to deal with system boundaries. Thus a chronostratigraphy of defined GSSPs for the stratigraphical column was proposed. For the Devonian System all series and stage boundaries have now been ratified by the IUGS. A brief review of the decisions is given here. Earlier summaries were given by House (1988) and Oliver & Chlupáč (1991) but these were before all decisions had been made. A substantial summary of GSSPs, zonal schemes in relation to them, and surveys of how the new boundaries fall in many countries is currently in course of publication in the *Courier Forschungsinstitut Senckenberg*. Therefore this account is reduced to essentials.

Devonian boundary definitions

Devonian Standard Series and Stages

The international chronostratigraphic standard divisions for the subdivision of the Devonian System are given in Fig. 1. This leaves many

From: FRIEND, P. F. & WILLIAMS, B. P. J. (eds). *New Perspectives on the Old Red Sandstone.*
Geological Society, London, Special Publications, **180**, 23–27. 0305-8719/00/$15.00

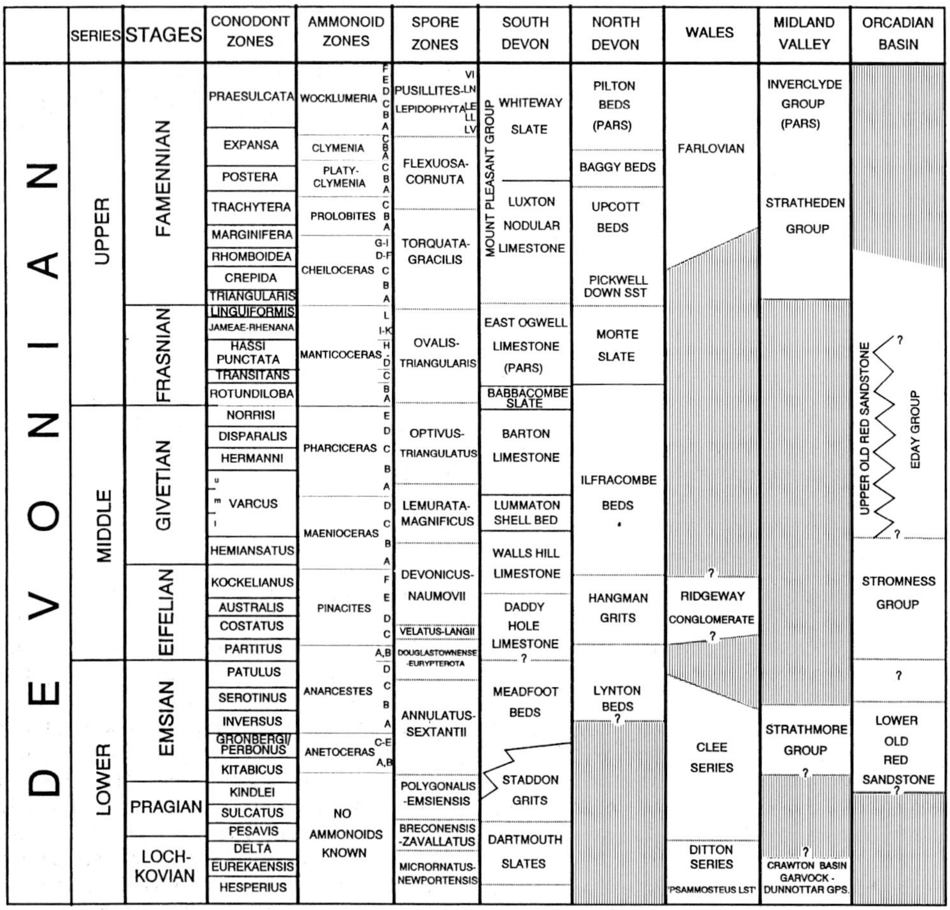

	SERIES	STAGES	CONODONT ZONES	AMMONOID ZONES	SPORE ZONES	SOUTH DEVON	NORTH DEVON	WALES	MIDLAND VALLEY	ORCADIAN BASIN
DEVONIAN	UPPER	FAMENNIAN	PRAESULCATA	WOCKLUMERIA F E D C B A	VI PUSILLITES-LN LEPIDOPHYTA LE LL LV	MOUNT PLEASANT GROUP: WHITEWAY SLATE	PILTON BEDS (PARS)	FARLOVIAN	INVERCLYDE GROUP (PARS)	
			EXPANSA	CLYMENIA C B A	FLEXUOSA-CORNUTA		BAGGY BEDS			
			POSTERA	PLATY-CLYMENIA C B A						
			TRACHYTERA	PROLOBITES C B A		LUXTON NODULAR LIMESTONE	UPCOTT BEDS		STRATHEDEN GROUP	
			MARGINIFERA	G-I	TORQUATA-GRACILIS					
			RHOMBOIDEA	D-F						
			CREPIDA	CHEILOCERAS C B A			PICKWELL DOWN SST			
			TRIANGULARIS							
		FRASNIAN	LINGUIFORMIS	L		EAST OGWELL LIMESTONE (PARS)	MORTE SLATE			
			JAMEAE-RHENANA	I-K	OVALIS-TRIANGULARIS					
			HASSI PUNCTATA	MANTICOCERAS H D-G						UPPER OLD RED SANDSTONE ? / EDAY GROUP
			TRANSITANS	C						
			ROTUNDILOBA	B A		BABBACOMBE SLATE				
	MIDDLE	GIVETIAN	NORRISI	E						
			DISPARALIS	PHARCICERAS D C	OPTIVUS-TRIANGULATUS	BARTON LIMESTONE	ILFRACOMBE BEDS			
			HERMANNI	B						
			VARCUS (u, m, l)	A						
				MAENIOCERAS D C	LEMURATA-MAGNIFICUS	LUMMATON SHELL BED				
			HEMIANSATUS	B A		WALLS HILL LIMESTONE				?
		EIFELIAN	KOCKELIANUS	PINACITES F E	DEVONICUS-NAUMOVII			?		STROMNESS GROUP
			AUSTRALIS	D		DADDY HOLE LIMESTONE	HANGMAN GRITS	RIDGEWAY CONGLOMERATE		
			COSTATUS	C	VELATUS-LANGII			?		
			PARTITUS	A,B	DOUGLASTOWNENSE-EURYPTEROTA	?				
	LOWER	EMSIAN	PATULUS	ANARCESTES D C						?
			SEROTINUS	B	ANNULATUS-SEXTANTII	MEADFOOT BEDS	LYNTON BEDS			
			INVERSUS	A			?		STRATHMORE GROUP	LOWER OLD RED SANDSTONE
			GRONBERGI/PERBONUS	ANETOCERAS C-E				CLEE SERIES		
			KITABICUS	A,B					?	
		PRAGIAN	KINDLEI		POLYGONALIS-EMSIENSIS	STADDON GRITS				?
			SULCATUS	NO AMMONOIDS KNOWN						
			PESAVIS		BRECONENSIS-ZAVALLATUS	DARTMOUTH SLATES				
		LOCHKOVIAN	DELTA					DITTON SERIES	?	
			EUREKAENSIS		MICRORNATUS-NEWPORTENSIS				CRAWTON BASIN GARVOCK - DUNNOTTAR GPS.	
			HESPERIUS					'PSAMMOSTEUS LST'		

Fig. 1. Table showing the internationally agreed Series and Stage boundaries for the Devonian System as ratified by the IUGS and a correlation chart showing the broad stratigraphical terminology used in five areas of the UK. (From Marshall & House (2000).)

widely used terms as local and regional stages, such as, for the Lower Devonian Series, the Gedinnian and Siegenian Stages. For the Emsian Stage, a subdivision into two is currently under consideration by the SDS: the Czech terms Zlichovian and Dalejan are at present local terms. It was the miscorrelation of the Dalejan deepening event with the Eifelian deepening event that led to many complications in the definition of the Lower–Middle Devonian boundary. The term Couvinian is now a regional term only. The Givetian Stage also may come to be formally subdivided, as the base of the Upper Devonian Series, and Frasnian Stage, is drawn at a level above what was formerly regarded by many as early Frasnian Stage. The German Upper Devonian Stufen, Adorf, Nehden, Hemberg, Dasberg and Wocklum Stages are now regional terms, but a formal subdivision of the Famennian Stage is likely. The Strunian Stage falls within the late Famennian Stages of the revised terminology.

Base of the Devonian Series and Lochkovian Stage

The GSSP for the Silurian–Devonian boundary, basal Lower Devonian Series, and basal Lochkovian Stage is at Klonk in the Czech Republic as documented in Martinsson (1977), where McLaren recounted the scientific and political problems in reaching a decision. The Ludlow Bone Bed had been one of several levels used in this country and by raising the boundary to near the *Psammosteus* Limestone (Holland & Richardson 1977; Richardson *et al.* 1981), at least some accommodation was made to the very high level previously used in central and eastern Europe and Asia.

Base of the Pragian Stage

The GSSP for this boundary is at Velka Chuchle, near Prague, Czech Republic (Chlupáč & Oliver 1989) operationally where the conodont *Eognathodus sulcatus* enters. As the spore zonation is currently defined (Steemans 1989) the base of the Pragian Stage falls within a spore zone such that it cannot be accurately defined using palynology.

Base of the Emsian Stage

The GSSP for the basal Emsian Stage is in the Zinzalban Gorge of the Kitab National Park in Uzbekistan (Yolkin *et al.* 1998). The entry of the conodont *Polygnathus dehiscens*, probably a synonym for *Poly. kitabicus*, was the key form in determining the level. Coiled ammonoids enter just above the boundary and uniserial graptolites become extinct within early Emsian time.

Base of the Middle Devonian Series and Eifelian Stage

The GSSP for the base of the Middle Devonian Series and for the Eifelian Stage is at the Wetteldorf Richtschnitt in the Eifel District of Germany (Ziegler & Werner 1982, pp. 13–84), and was drawn to correspond to the base of the *partitus* conodont Zone.

Base of the Givetian Stage

The GSSP for this boundary is at Mech Irdane, Tafilalt, Morocco (Walliser *et al.* 1996), where the boundary is taken at the base of the conodont *hemiansatus* Zone, which falls within the *ensensis* Zone of an earlier terminology. The ammonoid *Maenioceras* zones commence a little below the boundary and it is thought that the level approximates to the entry of the spore *Geminospora lemurata*.

Base of the Upper Devonian Series and Frasnian Stage

The GSSP for base of the Upper Devonian Series and of the Frasnian Stage is at Puech de la Suque, near Cessenon in the Montagne Noire, France (Klapper *et al.* 1987). The boundary was chosen at the entry of a key form, *Ancyrodella rotundiloba* early morph Klapper. But there has been dispute over taxonomy. In more modern terms, the base is drawn at the base of Montagne Noire zone MN 1 (Klapper 1989), or of the *soluta* Zone (Iudina 1995), or falls within the *falsiovalis* Zone (Ziegler & Sandberg 1990) according to the terminology followed. The entry of the goniatite genus *Neopharciceras* occurs immediately above.

Base of the Famennian Stage

The basal Famennian GSSP is in the Upper Coumiac Quarry, Near Cessenon, Montagne Noire, France (Klapper *et al.* 1993). The level follows the major extinctions of the Upper Kellwasser Event. The boundary falls at the junction of the *linguiformis* and Lower *triangularis* conodont Zones. The goniatites *Manticoceras* and *Beloceras* become extinct at the boundary; *Cheiloceras* enters significantly higher.

Top of the Devonian Series

The GSSP for the Devonian–Carboniferous boundary is at La Serre, near Clermont l'Herault, Montagne Noire, France (Paproth *et al.* 1991). The guide to the base is the conodont *Siphonodella sulcata* but there has been some dispute on the definition of this form. Palynologically the Devonian–Carboniferous boundary is well defined by the last appearance of *Retispora lepidophyta*, which disappears (Higgs & Streel 1984) just beneath the first occurrence of *Siphonodella sulcata* in the Hasselbachtal auxiliary section in Germany. Major extinctions of the ammonoid goniatites and clymenids occur below the boundary, which corresponds to the Hangenberg Event.

Recognition of these boundaries in the British Isles

A review of the levels in the marine Devonian Series and Old Red Sandstone where these boundaries are thought to lie (Marshall & House 2000) is summarised in Fig. 1: it is not thought appropriate to repeat that review. As decisions have been made, statements of their relevance in the UK have been published (Holland & Richardson 1977; House & Sevastopulo 1984; House & Dineley 1985; House 1988). These mostly post-date a review of Devonian and Old Red correlations for Britain (House *et al.* 1977). Because most of the new GSSP levels were chosen on conodont criteria, the summary of the conodont records from the British Isles by Austin *et al.* (1985) is particularly valuable.

Problems

The major problems lie in the lack of precise biostratigraphic correlation between the GSSP levels established in marine and pelagic regimes. This is currently one of the main concerns of the SDS and it is expected that palynological, vertebrate and microvertebrate evidence will become increasingly important. The many extinction events associated with sedimentary perturbations in the marine realm can be expected to be reflected in terrestrial facies. Magnetosusceptibility, cyclostratigraphy and orbital forcing timescale studies are commencing and are likely to aid correlations in the future. Finally, in recent years refined age-dating using zircons has given a precision to parts of the Devonian radiometric time scale in several parts of the world. Occurrences in the Old Red Sandstone of Britain are urgently in need of study.

References

AUSTIN, R. L., ORCHARD, M. J. & STEWART, I. J. 1985. Conodonts of the Devonian System from Great Britain. *In*: HIGGINS, A. C. & AUSTIN, R. L. (eds) *A Stratigraphical Index of Conodonts*. Ellis Horwood, Chichester, 93–166.

CHLUPÁČ, I. & OLIVER, W. A. JR 1989. Decision on the Lochkovian–Pragian boundary stratotype (Lower Devonian). *Episodes*, **12**, 109–113.

CLARKE, J. M. 1923. *James Hall of Albany, Geologist and Palaeontologist, 1811–1898*. Albany, NY.

HANCOCK, J. M. 1977. The historic development of concepts of biostratigraphical correlation. *In*: KAUFFMAN, E. G. & HAZEL, J. E. (eds) *Concepts and Methods of Biostratigraphy*. Dowden, Hutchinson and Ross, Stroudsburg, PA, 3–22.

HIGGS, K. & STREEL, M. 1984. Spore stratigraphy at the Devonian–Carboniferous boundary in the northern 'Rheinisches Schiefergebirge', Germany. *Courier Forschungsinstitut Senckenberg*, **67**, 157–179.

——, CLAYTON, G. & KEEGAN, J. B. 1988. *Stratigraphic and Systematic Palynology of the Tournaisian Rocks of Ireland*. Geological Survey of Ireland Special Paper, **7**, 1–93.

HOLLAND, C. H. & RICHARDSON, J. B. 1977. The British Isles. *In*: MARTINSSON, A. (ed.) *The Siluro-Devonian boundary*. International Union of Geological Sciences, Ser. A, **5**, Schweizerbart'sche, Stuttgart, 35–44.

HOUSE, M. R. 1988. International definition of Devonian System boundaries. *Proceedings of the Ussher Society*, **7**, 41–46.

—— & DINELEY, D. L. 1985. Devonian Series boundaries in Britain. *Courier Forschungsinstitut Senckenberg*, **75**, 301–310.

—— & SEVASTOPOLO, G. D. 1984. The Devonian–Carboniferous boundary in the British Isles. *Courier Forschungsinstitut Senckenberg*, **67**, 51–52.

——, FEIST, R. & KORN, D. 2000. The Middle/Upper Devonian boundary at Puech de la Suque, Southern France. *Courier Forschungsinstitut Senckenberg*, in press.

——, RICHARDSON, J. B., CHALONER, W. G., ALLEN, J. R. L., HOLLAND, C. H. & WESTOLL, T. S. 1977. *A Correlation of the Devonian Rocks of the British Isles*. Geological Society, London, Special Report, **7**.

IUDINA, A. B. 1995. Genus *Ancyrodella* succession in the earliest Frasnian (?) of the northern Chernyshev Swell. *Geolines*, **3**, 17–20.

KLAPPER, G. 1989. The Montagne Noire Frasnian (Upper Devonian) conodont succession. *In*: MCMILLAN, N. J., EMBRY, A. F. & GLASS, D. J. (eds) *Devonian of the World*. Canadian Society of Petroleum Geology, Memoirs, **14**(3), 449–468.

——, FEIST, R. & HOUSE, M. R. 1987. Decision on the boundary stratotype for the Middle/Upper Devonian Series boundary. *Episodes*, **10**, 97–101.

——, ——, BECKER, R. T. & HOUSE, M. R. 1993. Definition of the Frasnian/Famennian Stage boundary. *Episodes*, **16**, 433–441.

KUTSCHER, F. & SCHMIDT, H. 1958. *Lexique Stratigraphique Internationale, 1, Europe Fascicule 5 Allemagne, 5b Dévonien*. Louis-Jean, Paris.

MARSHALL, J. E. A. & HOUSE, M. R. 2000. Devonian Stage boundaries in England, Wales and Scotland. *Courier Forschungsinstitut Senckenberg*, in press.

MARTINSSON, A. (ed.) 1977. *The Siluro-Devonian boundary*. International Union of Geological Sciences, Series A, 5. Schweizerbart'sche, Stuttgart.

OLIVER, W. A. JR & CHLUPÁČ, I. 1991. Defining the Devonian: 1979–89. *Lethaia*, **24**, 119–122.

PAPROTH, E., FEIST, R. & FLAJS, G. 1991. Decision on the Devonian–Carboniferous boundary stratotype. *Episodes*, **14**, 331–336.

RICHARDSON, J. B., RASUL, S. M. & AL-AMERI, T. 1981. Acritarchs, miospores and correlation of the Ludlovian-Downtonian and Silurian–Devonian boundaries. *Reviews of Palaeobotany and Palynology*, **34**, 209–224.

RUDWICK, M. J. S. 1985. *The Great Devonian Controversy: the Shaping of Scientific Knowledge among Gentlemanly Specialists*. University of Chicago Press, Chicago, IL.

SEDGWICK, A. & MURCHISON, R. I. 1839. Stratification of the older stratified deposits of Devonshire and Cornwall. *Philosophical Magazine, Journal of Science, Series 3*, **14**, 241–260.

—— & —— 1840. On the physical structure of Devonshire, and on the subdivisions and geological relations of its older stratified deposits. *Transactions of the Geological Society, London, Series 2*, **5**, 633–704.

STEEMANS, P. 1989. *Etude palynostratigraphique du Devonien Inférieur dans l'ouest de l'Europe*. Mémoire Explication des Cartes Géologiques et Minéralogiques de Belgique, **27**, 1–453.

WALLISER, O. H., BULTYNCK, P., WEDDIGE, K., BECKER, R. T. & HOUSE, M. R. 1996. Definition of the Eifelian–Givetian Stage boundary. *Episodes*, **18**, 107–115.

WATERLOT, G. (ed.) 1957. *Lexique Stratigraphique Internationale, 1, Europe Fascicule. 4a France, Belgique, Pays Bas, Luxembourg, 4a1 Anté-cambrien Paléozoique Inférieur*. Louis-Jean, Paris.

YOLKIN, E. A., KIM, A. I., WEDDIGE, K., TALENT, J. A. & HOUSE, M. R. 1998. Definition of the Pragian/Emsian Stage Boundary. *Episodes*, **20**, 235–240.

ZIEGLER, W. & WERNER, R. (eds) 1982. On Devonian Stratigraphy of the Ardenno–Rhenisch Maintains and related Devonian matters. Courier Forshung-sinstitut Senchenberg, **75**.

—— & SANDBERG, C. A. 1990. The Late Devonian Standard Zonation. *Courier Forschungsinstitut Senckenberg*, **121**.

Kinematics and dynamics of Old Red Sandstone basins

P. F. FRIEND[1], B. P. J. WILLIAMS[2], M. FORD[3] & E. A. WILLIAMS[4]

[1]*Department of Earth Sciences, University of Cambridge, Downing Street, Cambridge CB2 3EQ, UK*

[2]*Department of Geology and Petroleum Geology, University of Aberdeen, Meston Building, King's College, Aberdeen AB24 3UE, UK*

[3]*École Nationale Supérieure de Géologie, CRPG–CNRS, Rue du Doyen Marcel Roubault, B.P. 40, 54501 Vandoeuvre-lès-Nancy cedex, France*

[4]*CRPG–CNRS, B.P. 20, 54501 Vandoeuvre-lès-Nancy cedex, France*

Abstract: The Old Red Sandstone basins of the North Atlantic borderlands provide a record of diverse dynamics in very different settings, related to the Variscan, Caledonian and Ellesmerian orogenies. This paper is a first attempt to review much new information on the basins, including information presented, for the first time, in this book. Five basin groupings are distinguished: (1) Scandinavian basins of, syn- to post-Scandian (Caledonian) age, formed on greatly thickened crust by extension or transtension (Western Norway, East Greenland, Spitsbergen); (2) NE Scotland, Orcadian Basin, mid Caledonian setting, formed by extension; (3) Scotland (Midland Valley) and related Irish basins, north of the Caledonian Iapetus Suture Zone, formed by extension; (4) southern Britain and Ireland, basins south of the Iapetus Suture Zone, related to collision of Eastern Avalonia with Laurentia, and Maritime Canada and the Catskills related to collision of Western Avalonia; these are load-induced flexural basins; (5) Southern margin of Eastern Avalonia, (Munster, South Wales, SW England), of Late Devonian age, extensional basins of various (Early to Late) Devonian ages.

The Old Red Sandstone (ORS) of the lands bordering the North Atlantic Ocean ranges in age from mid- Silurian to Carboniferous time. It provides the fill of many basins, which range in location from the Appalachians at 40° N, to Spitsbergen, at 80° N, a distance of some 4500 km.

The Old Red Sandstone has long been seen as a stratigraphic response to major Palaeozoic (Caledonian) mountain-building, and particularly as a late- or post-orogenic (molasse) magnafacies. However, the basin-forming mechanisms have not been reviewed recently enough to reflect new research into geodynamics. There is now much greater knowledge of the basins involved, and much new work has been performed on the orogens. One object of this paper is to provide a modern overview of the basins in general tectono-stratigraphic terms. The second object is to discuss the extent to which current ideas on basin geodynamics can be applied to the ORS of the North Atlantic region. In this, we attempt to specify (1) ORS basin kinematics, based on a review of published information on mainly fault and related fold structures that have been traditionally used to characterize kinematics, and (2) ORS basin dynamics, from available geohistory (subsidence) analyses and tectono-stratigraphic context of individual basins. Finally, in the discussion section, we attempt a synthesis of ORS basin geodynamics in relation to plate-scale and body forces related to the temporally and spatially overlapping Caledonian, Variscan and Ellesmerian Orogenies. First, we briefly review the principal basin-forming mechanisms.

Basin-forming mechanisms

Basin research developed in an important direction with the numerical modelling of lithosphere dynamics (e.g. Turcotte & Schubert 1982; summarized by Allen & Allen (1990), and references therein). These models typically assume an initial state of isostatic equilibrium of the lithosphere and a crustal surface at sea level. They then consider vertical perturbation of this initial state by, and the isostatic response to, various mechanisms that are classified in the following sections.

From: FRIEND, P. F. & WILLIAMS, B. P. J. (eds). *New Perspectives on the Old Red Sandstone.* Geological Society, London, Special Publications, **180**, 29–60. 0305-8719/00/$15.00

Lithospheric stretching

The kinematic process of extension of continental lithosphere, usually across a linear rift zone, causes syn-rift subsidence (Fig. 1f) accommodated in the rheologically brittle upper-crustal layer by active normal fault systems (Fig. 1a), as seen today in the East African rift. This process of stretching may continue until the continental lithosphere is thinned to such an extent that active sea-floor spreading ensues and a new divergent plate boundary is created. If extension ends before 'break-up' is achieved, the basin is sometimes described as a failed rift (e.g. the North Sea basin). Mechanical stretching of the continental lithosphere is accompanied by a shallowing of the Moho so that a thermal perturbation develops (i.e. heat flow increases) as the mantle lithosphere rises. When stretching terminates, whether before or after 'break-up' conditions are reached, the stretched continental lithosphere will begin to thermally re-equilibrate. As the hot, thinned lithosphere cools it subsides further and this phase of 'thermal subsidence' is known as the post-rift phase of basin development (Fig. 1f). The process is the same as that observed in ageing oceanic lithosphere that cools, thickens and subsides as it moves away from the mid-ocean ridge (Fig. 1b; Parsons & Sclater 1977). Thermally re-equilibrated rift basins typically show a 'steer's head' geometry in profile (White & McKenzie 1988), comprising the syn-rift deposits overlain by the broader post-rift sequence (Fig. 1c).

Following a qualitative conceptualization of rift formation by Salveson (1978), quantitative analyses of the process were pioneered by McKenzie (1978) and Jarvis & McKenzie (1980) for instantaneous and finite uniform stretching of the lithosphere, respectively. Important developments in extensional basin modelling have included depth-dependent stretching (Hellinger & Sclater 1983; Kusznir *et al.* 1987; Kusznir & Ziegler 1992; Friedmann & Burbank 1995), asymmetrical stretching (Wernicke 1985; Buck 1988; Lister & Davis 1989), and the recognition of the differences between narrow and wide rifts (Bassi *et al.* 1993).

Flexural mechanisms

It has been shown that when a load is emplaced upon a lithospheric plate, it will respond by flexing downward (Watts & Cochran 1974; Karner & Watts 1983). This behaviour can be equated in two dimensions to the bending of an elastic beam with a finite elastic strength (e.g. Beaumont 1981; Turcotte & Schubert 1982;

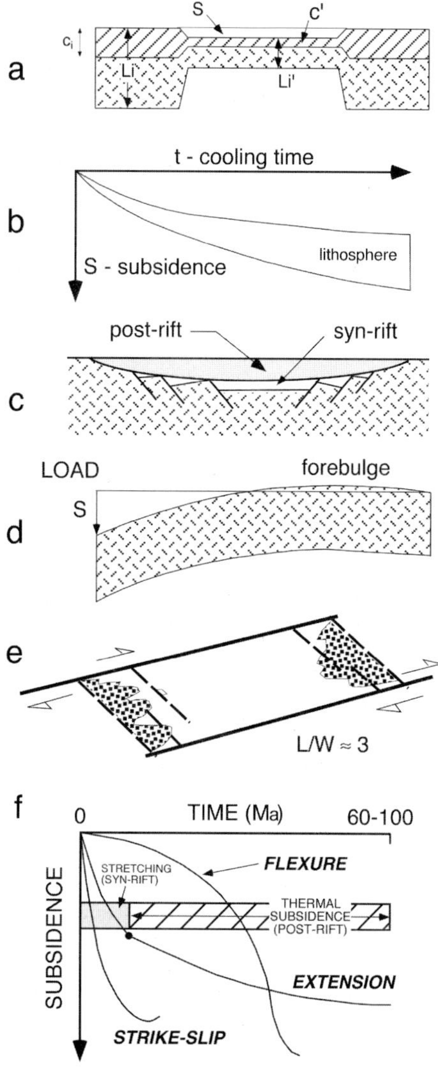

Fig. 1. Principal mechanisms of subsidence. (**a**) Thinning of continental lithosphere as a result of stretching, accommodated in the upper crust by normal faulting and in the lower crust and mantle lithosphere by ductile thinning. (**b**) Thermal cooling and thickening of the oceanic lithosphere as it moves away from the mid-ocean ridge causes subsidence so that depth of ocean floor is a function of age. (**c**) Geometry of a thermally equilibrated rift basin showing the rift basin below a more broadly distributed sequence deposited during thermal re-equilibration. (**d**) Flexural response of lithosphere to loading generates an asymmetrical trough and a forebulge. (**e**) Plan view of a releasing overstep on a strike-slip fault system, generating local extensional conditions causing the development of a pull-apart basin. This is typically a smaller-scale structure than in (**c**) or (**d**). (**f**) Typical subsidence curves for stretching, flexural and strike-slip pull-apart basin mechanisms.

Jordan 1995; Fig. 1d). Examples of geological loads that cause flexural responses are volcanic islands on oceanic lithosphere (Watts & Cochran 1974), and orogenic belts that can load both the lower and overriding plates in a collision zone (Price 1973; Karner & Watts 1983; Stockmal & Beaumont 1987). Down-flexing of these plates creates asymmetrical troughs around the periphery of the mountain belt that are typically filled with detritus derived from the uplifting orogen (Price 1973; Dickinson 1974; Flemings & Jordan 1989; Jordan 1995). These foreland basins are generally represented as wedge shaped in cross-section, delimited by the thrust front and thinning toward the craton onto a forebulge. As the orogen evolves, the thrust front, the depocentre and forebulge migrate towards the craton. The typical (accelerating) subsidence signature of flexural loading is shown in Fig. 1f. Depending on sediment supply to the basin, tectonic subsidence may be significantly enhanced by sediment loading. Dickinson (1974) defined two principal types of foreland basins based on their position with respect to the subducting plate. Retro-arc foreland basins are situated behind a magmatic arc and linked to subduction of oceanic lithosphere (e.g. Canadian Rockies foreland basin and the Andean foreland basin). Peripheral foreland basins develop on the outer arc of collisional orogens during continent–continent collision (e.g. Alpine Molasse Basin, Indus–Ganges Basin).

Wrench-zone mechanisms

Pull-apart basins may develop within strike-slip zones (Fig. 1e; Wilcox *et al.* 1973). Lithospheric-scale strike-slip zones can represent plate boundaries if they are transform faults or can be associated with oblique subduction, oblique collision or indentor type collision (Woodcock 1986). Localized zones of small-scale compression or extension can develop along strike-slip zones as a result of curves in fault orientation, braiding of faults or side-stepping of faults. Woodcock & Fischer (1986) defined releasing and restraining bends or jogs and the structures that can develop in each. At a restraining bend contractional duplexes and uplift are dominant (e.g. the Transverse Ranges, USA). These compressional belts may have associated flexural basins because of loading. At a releasing bend, extensional structures and subsidence are dominant and pull-apart basins typically develop (e.g. the Dead Sea). Pull-apart basins record rapid subsidence (Fig. 1f) and sediment accumulation. They have complex stratigraphies (Christie-Blick & Biddle 1985) with rapid lateral facies changes (Hempton & Dunne 1983). Their width-to-length ratio is on average around 1:3 (Aydin & Nur 1982), making them deep and narrow, often with syndepositional relief. Their thermal evolution depends on whether the mantle is involved or not; that is, whether the controlling strike-slip faults cut through the whole lithosphere. If the mantle is involved then the basin will have a thermal cooling phase; if not, then it will simply be a deep, fault-controlled basin.

Strike-slip systems rarely accommodate pure strike-parallel movement. Usually displacements are oblique with components of either compression (transpression; Harland 1971) or extension (transtension; Woodcock 1986). Transpression will enhance the development of folds and thrusts whereas transtension will enhance the development of normal faults. These regimes can grade into purely extensional or compressional regimes and can be difficult to distinguish. In these zones flower structures can develop, in which all faults converge downwards into a single vertical fault zone. The amount of 'flowering' depends on the degree of obliquity of displacement (Wilcox *et al.* 1973).

Developments in the understanding of the Old Red Sandstone

Up to and including the 1960s, most work on the Old Red Sandstone tended to be devoted to parts of individual basins, with a necessary focus on fossil faunas and floras, stratigraphy, structure and sedimentology. Old Red Sandstone successions were recognized and separated from others because they were clastic, of non-marine character, generally younger than marine Lower Palaeozoic successions and generally older than marine Carboniferous sediments. Outcrop areas were divided between those that contained marine and non-marine sediments, and those that contained only non-marine sediments (Friend 1969). A distinction was also drawn between successions that rested unconformably on a basement of metamorphic rocks, and those that rested on a conformable transition from marine sediments. The ORS was widely considered as evidence for uplift of a source area. Whether the onset of ORS accumulation signals an episode involving a new system of sediment transfer and retention, or simply a change from accumulation below sea level to accumulation above sea level, both situations provided evidence for crustal uplift kinematics.

Most of the earlier attempts to review the Old Red Sandstone of the Atlantic borderlands (Friend *et al.* 1967; Friend 1969, 1981, 1985) involved comparison of the histories of basin sediment accumulation, and resulted, for example, in a distinction, using basin patterns of palaeocurrents, between extramontane and intramontane basins (Friend 1969). It was also pointed out that many Old Red Sandstone basinal episodes involved very high average sediment accumulation rates of between 200 and 600 m/Ma^{-1}. There was some consideration of the likely kinematics of structures, particularly faults, but these considerations were dynamic only in the sense that the movements were regarded qualitatively as responses to largely horizontal stresses, generated by the tectonic plate-scale convergence and continental collision that drove the orogenesis.

It has become clear that greater understanding of these kinematic episodes will require the determination of the extent and scale of various effects, and it is one of the modern challenges to establish the necessary spatial correlations and time constraints. For example, most earlier workers have regarded collisional convergence, and the resulting compressional dynamics, to be the paramount control of Old Red Sandstone orogenic evolution. However, many workers also recognized evidence for strike-slip faulting, and, although some interpreted this as a basin-scale kinematic phenomenon, others regarded it as a much more widespread feature, extending for thousands of kilometres, with the same sense of lateral movement, along much of the Old Red Sandstone basin belt. One of the early proponents of this 'sinistral megashear hypothesis' (the term used by Norton *et al.* (1987)) was Harland (1967; see also 1997), but there have been many others (e.g. Ziegler 1982). Distinguishing whether there was an orogen-wide megashear, or whether the strike-slip faulting varied in sense of movement and timing in different parts of the orogen, perhaps during indentor convergence (e.g. Coward 1990, 1993), is a difficult task, which must be examined from the standpoint of the contemporary orogenies, in the following section.

Orogenic framework and timing

Following the pre-drift reconstruction of Bullard *et al.* (1965) it became clear that the present Atlantic borderlands had been part of a single Old Red land (the ORS Landmass), which had close contiguity with the Caledonian orogen. This template was widely used for much of the primary work synthesizing ORS palaeogeography (e.g. Friend 1969), and differs surprisingly little from more recent reconstructions based on new palaeomagnetic pole determinations and also work on ocean closure between the major continents (Le Pichon *et al.* 1977; Channell *et al.* 1992; Torsvik *et al.* 1996; Smith, pers. comm.). Three large continents on lithospheric plates were involved: (1) Laurentia, consisting of the cratonic block of North America and Greenland; (2) Baltica (the Baltic shield); (3) Gondwana to the south, consisting principally of the African craton. A number of microplates also played key roles. Particularly important for ORS basin evolution are the plates of Eastern and Western Avalonia and Armorica.

In all these reconstructions a broad belt of Palaeozoic orogenic activity is defined between the three cratons. Furthermore, the present northern margin of Laurentia consists of the Palaeozoic Ellesmerian orogenic belt (Fig. 2), the junction of which with the northern Scandinavian Caledonides occurs in the Spitsbergen region. Individual orogenic events or phases, limited in space and time, are given local names but are divided broadly into two groups, the early Palaeozoic 'Caledonian' events and the late Palaeozoic 'Variscan' events. Old Red Sandstone basin formation straddles these two periods of orogenesis in Europe and eastern North America, which themselves overlap significantly in space (Fig. 2). Consequently, although ORS sedimentation is classically related to the Caledonian orogen, the role of Variscan events to the south must also be considered, as discussed by Dewey (1982), Leeder (1982, 1988*a*) and Rey *et al.* (1997).

Caledonides

In mid-Ordovician and Silurian times the Caledonian Orogeny was most active in the northern Old Red Sandstone areas. In the northern British Isles and Ireland in mid-Ordovician time the convergent Grampian Orogeny, caused by arc accretion (Ryan & Dewey 1991), was short lived (around 470 Ma; Soper *et al.* 1999) and was accommodated by an estimated 700 km of shortening across the orthotectonic Caledonides. The Grampian Orogeny was associated with the longer-lived Taconic orogen in the Appalachians to the west, also generated by arc accretion (Dewey & Shackleton 1984).

The main Caledonian event recorded in the British Isles took place in Silurian time and involved the oblique closure of the Iapetus Ocean along the Iapetus Suture Zone (Fig. 2) as a result of the oblique docking of the Eastern Avalonia terrane against the Laurentian margin (Soper & Woodcock 1990). The size of this ocean is ambiguous, as is the direction and timing of subduction, as evidence exists for subduction toward the NW below the Southern Uplands and toward the SE below the Lake District. The

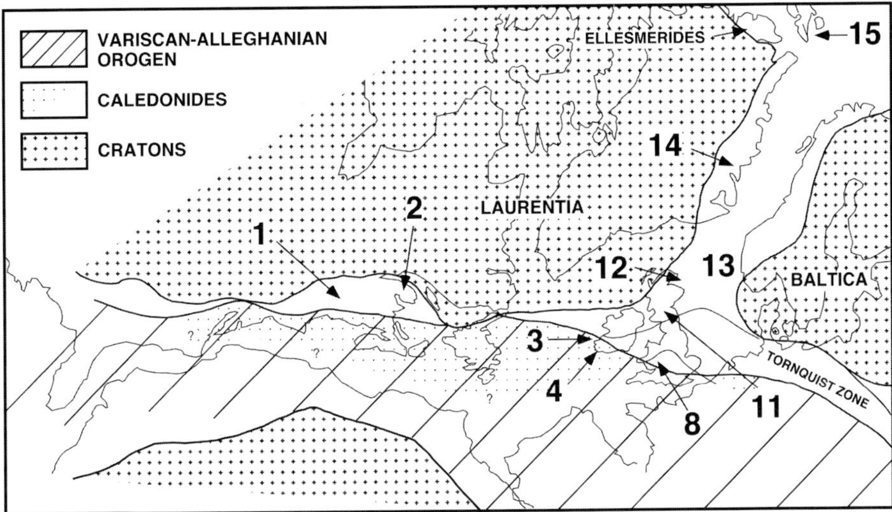

Fig. 2. The Caledonian–Appalachian, Variscan and Ellesmerian orogens and cratonic blocks. Major basins discussed in the text: 1, Catskills; 2, Maritime Canada; 3, Dingle Basin; 4, Munster Basin; 8, Isle of Man–Lake District; 11, Midland Valley, Scotland; 12, Orcadian Basin; 13, West Norway; 14, East Greenland; 15 Spitsbergen.

importance of substantial sinistral displacements on NE–SW-trending faults between crustal blocks is controversial (Fig. 2; Soper & Hutton 1984; Thirlwall 1989). A model of oblique collision and the accretion of many small-scale terranes onto the southern margin of Laurentia has been proposed (Hutton 1987, 1989) and the displaced terrane hypothesis may help to explain the absence in present outcrops of certain crustal blocks whose existence is recorded only by eroded clasts (Bluck 1983; Haughton & Bluck 1988). Deformed Proterozoic granite clasts found in the Ordovician sediments of the Southern Uplands have been correlated with source rocks in Newfoundland (Elders 1987).

South of the Iapetus Suture the principal deformation in England, Wales, Ireland and the Maritime Provinces of Canada occurred during the Acadian Orogeny, which is now dated to late Early to Mid-Devonian time (385–400 Ma; McKerrow 1988). This transpressive deformation is associated with widespread intrusion of granitoid bodies in Ireland, Britain and the Canadian Maritime Provinces (Keppie 1989), all of which have been dated between 410 and 385 Ma. The event is related to the docking of the Eastern Avalonian terrane (Soper *et al.* 1987) in the Irish–British sector of the orogen. In the Northern Appalachian sector, the colliding crustal block is referred to as Western Avalonia, and is considered as a separate microplate in several reconstructions (Torsvik *et al.* 1996). Hutton (1987, 1989) proposed that the sinistral shear in the British Isles–Maritime Provinces could have been caused either by oblique collision of the whole of Eastern Avalonia against Laurentia or by the collision of Eastern Avalonia as a rigid indentor into the Laurentia–Baltica continent causing sinistral shear on its western flank (British Isles–Appalachians) and dextral shear on its eastern flank, along the Tornquist Line (e.g. the Holy Cross Mountains).

Further to the north, the Iapetus Ocean was closed in Mid–Late Silurian time by relatively orthogonal collision between Laurentia and Baltica (Roberts & Gee 1985). This caused the Scandian Orogeny in Scandinavia, involving (hundreds of kilometres of) crustal shortening and thickening by SE-directed nappe stacking (Roberts & Gee 1985; Dewey *et al.* 1993; Milnes *et al.* 1997). Collision commenced by obduction in mid-Wenlock time and crustal thicknesses reached over 150 km as evidenced by the eclogites of Western Norway, dated as 425–400 Ma (Andersen *et al.* 1991; Dewey *et al.* 1993; Milnes *et al.* 1997). Exhumation of these eclogites took place throughout Early Devonian time along the Nordfjord–Sogn detachment, a west-dipping, low-angle crustal shear zone.

In East Greenland, Caledonian deformation (Haller 1971; Henriksen 1985) is constrained by a syn- to post-tectonic intrusion (445 ± 5 Ma) that cuts the sole thrust, to late Ordovician time. Upper Proterozoic–middle Ordovician miogeocline sediments were folded with N–S axes, and thrust westwards with a minimum translation of 25 km; peak metamorphism and uplift occurred at *c*. 420 Ma (summarized by Larsen & Bengaard (1991)). A significant component of sinistral transpression is indicated by the work of Strachan *et al.* (1992). Crustal extension, by orogenic collapse, is suggested to have occurred at *c*. 425 Ma (Hartz this volume). Clear differences in timing are therefore evident between East Greenland and Western Norway (see Dewey *et al.* 1993, fig. 4).

In Spitsbergen, there was a major Caledonian orogenic episode with Silurian-aged metamorphic and plutonic events. These were broadly coeval with contractional structures indicating shortening perpendicular to the general north–south direction that is sub-parallel to the main structural trends in Western Norway and East Greenland.

Variscides

The Variscan belt of Europe resulted from the Early Devonian to Mid-Carboniferous collision between Laurentia–Baltica and Gondwana, and involved a series of microplates caught between these continents (principally Eastern Avalonia and Armorica; Matte 1986; Franke 1989; Rey *et al.* 1997). These more southerly events were therefore partly contemporaneous with, and partly a continuation of, the Caledonian evolution outlined above. Plate tectonic models for the Variscan Orogeny are diverse, differing in the number of microplates and oceans implicated, and the importance of strike-slip. As a simple summary, Rey *et al.* (1997) proposed that two oceans of Cambrian to Ordovician age, each of contentious width, were closed: the Rheic ocean between Eastern Avalonia and Armorica and the Theic ocean between Armorica and Gondwana (including the Iberian promontory; McKerrow & Ziegler 1972; Matte 1991; Rey *et al.* 1997). The collision of the Iberian microplate with Avalonia–Laurentia–Baltica caused the Ibero-Armorican orocline (Burg *et al.* 1987). In Early Devonian time most of the oceanic lithospheres were subducted (Matte 1986; Franke 1989). Variscan high-pressure metamorphism is of Silurian to Early Devonian age and overall is older than 380 Ma (Rey *et al.* 1997). Emplacement of high-grade metamorphic nappes onto the forelands to the south and north occurred from Early Devonian to the Early–Mid Carboniferous time. Metamorphism, magmatism and flysch sedimentation migrated outward onto the two forelands with time (Matte *et al.* 1990). Compressional Variscan deformation culminated in the British Isles in Late Carboniferous and early Permian time. Major dextral strike-slip along crustal boundary faults occurred in these later stages of orogenesis (Arthaud & Matte 1977). Whereas compression was occurring in the external zones, the thickened crust of the internal Variscides was undergoing extensional collapse to produce coeval migmatitic domes with granitic intrusions and intracontinental coal-bearing basins in a metamorphic core complex style (Van der Driessche & Brun 1991–1992).

Ellesmerides

This orogenic belt (also referred to as the Innuitian and North Greenland fold belt) extends across northernmost parts of Greenland and Ellesmere Island (Fig. 2) and then continues along the northern margin of the Canadian Arctic archipelago, forming the present northern margin of Laurentia. In Palaeozoic times, the present Arctic Ocean did not exist, and its position was occupied by continental crust now forming the North Slope of Alaska and the Chukotka area of northeastern-most Russia.

Pearya is a distinct tectono-stratigraphic area of northernmost Ellesmere Island which provides evidence of Ordovician and Silurian orogeny with similar events to those of Caledonian Taconic areas of the Appalachians, far to the south. However, the main Ellesmerian episodes involved folding and thrusting in Late Devonian to early Carboniferous times (Ziegler 1988). These compressional structures were linked to uplift and the growth of a large clastic wedge complex, up to 4 km thick, that prograded southwestwards across much of the area of the Arctic Islands. Proximal deposits of the wedge complex were of Old Red Sandstone facies, but they pass distally into marine deposits and are thought to have accumulated in a foreland basin setting (Embry 1988).

Kinematics and dynamics of North Atlantic ORS basins

Published information on all the outcrop areas is of an increasingly high standard, and this is providing more information about the kinematics and dynamics of the basins, although it also serves to confirm the incomplete nature of the record. We now provide a review of the main ORS basins of the Atlantic borderlands, with the object of assessing information that they can provide on the dynamics of basin formation, and of providing some comparative review in relation to the evolution of the whole orogen. This review is arranged in terms of the two major time intervals into which ORS basin development in

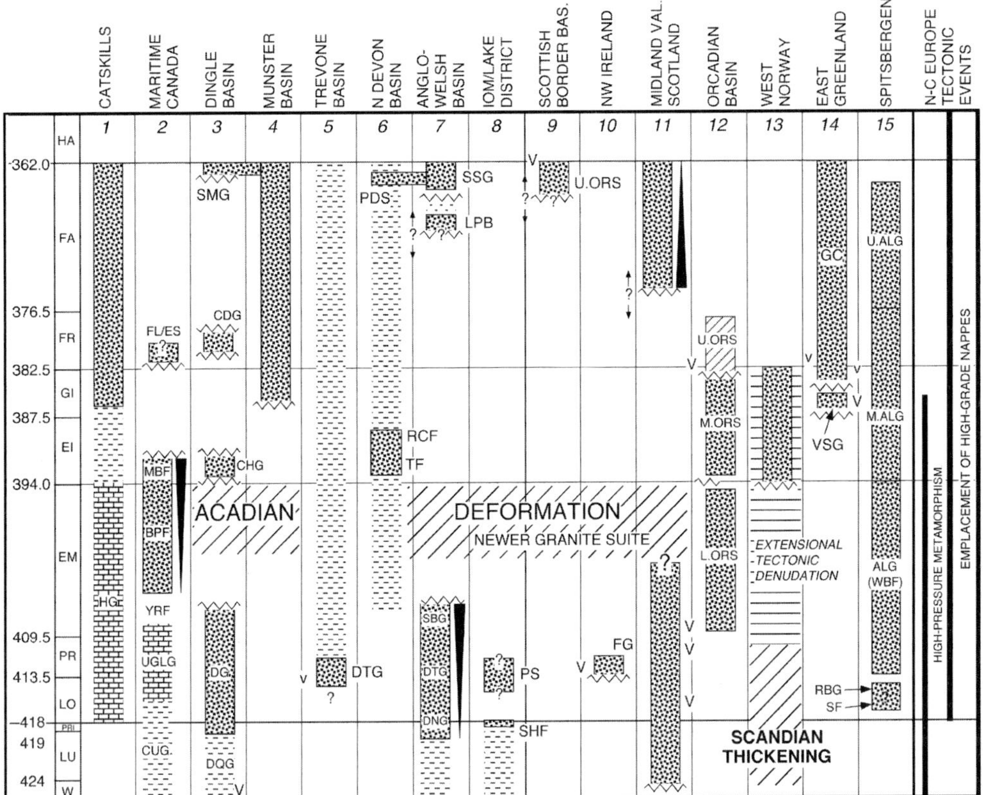

Fig. 3. Chart showing the timing of the principal Old Red Sandstone basins along strike of the Caledonian orogen. Timing of major events are also shown. Old Red Sandstone magnafacies are shown by coarse stipple. HG, Helderberg Group; CUG, Chaleurs Group; UGLG, Upper Gaspé Limestone Group; YRF, York River Formation; BPF, Battery Point Formation; MBF, Malbaie Formation; FL/ES, Fleurant and Escuminac Group; DQG, Dunquin Group; DG, Dingle Group; CHG, Caherbla Group; CDG, Carrigduff Group; SMG, Slieve Mish Group; FG, Fintona Group; DTG, Dartmouth Group; TF, Trentishoe Formation; RCF, Rawns Conglomerate Formation; PDS, Pickwell Down Sandstone; DNG, Downton Group; DTG, Ditton Group; SBG, Senni Beds–Brownstones Group; LPB, Lower Plateau Beds; SSG, Skrinkle Sandstone Group; SHF, Scout Hill Flags; PS, Peel Sandstone; U.ORS, Upper Old Red Sandstone; VSG, Vildaal Supergoup; GC, Gauss Complex; RBG, Red Bay Group; SF, Siktefjellet Group; ALG (WBF), Wood Bay Formation (Andrée Land Group); M.ALG, Middle Andrée Land Group; U.ALG, Upper Andrée Land Group. Sources not quoted in the text are Leeder (1973, 1976) for the Scottish Border Basin, and Allen & Crowley (1983) for the Isle of Man and Lake District ORS. Major episodes of volcanism (V) are shown approximately. Timing of generalized tectono-metamorphic Variscan events is shown for comparison.

most regions can be conveniently divided (late Silurian–Early Devonian and Mid–Late Devonian time). Stratigraphically constrained tectono-sedimentary events have been dated using the time scale of Tucker *et al.* (1998). In each interval, the basins are discussed from south to north along the Appalachian–Caledonian orogen (Fig. 2, numbering the areas or basins from one to 15; Fig. 3). Areas with basins that existed in both time intervals are treated in both sections, except where the time overlap is small.

Late Silurian–Early Devonian ORS basin development

Maritime Canada. The Devonian succession in the Northern (Canadian) Appalachians forms part of an upper Ordovician–Carboniferous sequence that unconformably oversteps Early Palaeozoic terranes (Keppie 1989, fig. 1; see Hibbard *et al.* 1996). The Devonian clastic wedge extends for >200 km along strike in the Gaspé synclinorium, which lies within one of a series of

NNW-(foreland-) verging thrust sheets. The successions were deposited on the NW side of the Iapetus Ocean (McKerrow 1988). The stratigraphical record in eastern Gaspé is continuous from Mid-Silurian to early Mid-Devonian time (Rust 1981; see McKerrow 1988). After deposition of Early Devonian (Pragian) marine platform carbonates (U. Gaspé Limestone Group), a deepening event in early Emsian time preceded a clastic wedge of 5 km thickness, which culminated in the Maritime Canada ORS magnafacies, and the late Mid- and Late Devonian Acadian unconformity. This clastic wedge is a coarsening-upwards succession (Fig. 3) comprising the shallow-marine clastic York River (Late Pragian–Early Emsian), as well as the fluviatile Battery Point (Emsian) and Malbaie (early–mid Eifelian) Formations (Rust 1981).

After establishment during York River Formation time, an approximately E–W-trending marine embayment, located west of the present Gaspé peninsula, persisted throughout the early Battery Point Formation (Lawrence & Rust 1988; Rust *et al.* 1989). Earliest alluvial sedimentation (Petit Gaspé Member) was characterized by a west-flowing axial sand-bed braided river that was fed by north-flowing tributaries from an evolving source area to the south. River channel style was transformed in the succeeding Cap-aux-Os Member, with mixed load, probably single-thread, rivers passing downstream to a vegetated coastal plain environment to the west (Lawrence & Rust 1988; Griffing *et al.* this volume). This river system remained axial, and supplied by south-flowing tributary rivers. In the late Battery Point Formation (Fort Prével Member) alluvial dispersal was dramatically altered with a reversal of the drainage direction, which remained, however, axial with respect to the basin. Alluvial style also changed significantly to ephemeral braided river channels, flanked by desiccated mud flats (Lawrence & Williams 1987). The Fort Prével to Malbaie Formation transition, in latest Emsian time, is recorded by increasingly gravelly alluvial systems with northward dispersal and the establishment of the axial, east-flowing sand-bed braided river system. This dispersal was maintained throughout accumulation of the 1.45 km thick Malbaie Formation, with gravel supply from the south becoming increasingly important with time (Rust 1984). The persistent interbedding of the sand-rich axial and the exotic gravel-rich transversely draining components comprising the Malbaie Formation (Rust 1984), suggests repetitive relocation of the axial river system with respect to the proximal gravel 'braidplain' envisaged by Rust (1984) and Rust *et al.* (1989).

Contemporaneous ORS magnafacies alluvium in southern Gaspé (Lagarde and Pirate Cove Formations) was gravel rich, and interpreted to be confined to a small intermontane basin (Dineley & Williams 1968*a*). Post-Acadian (Frasnian) ORS sediments in the same region commenced with limited gravelly proximal alluvium (Fleurant Formation), succeeded by lacustrine or tidally-influenced estuarine sediments of the Escuminac Formation (Dineley & Williams 1968*b*; Hesse & Saah 1992; Cloutier *et al.* 1998).

Although no quantitative subsidence analysis of the succession is currently available, the stratigraphy and sedimentology of the basin allows interpretation in terms of one-sided sediment supply into a foreland basin that was subsiding under the load of the Acadian orogen. Early Emsian marine deepening, dating the initiation of probable flexural subsidence, followed by shallow-marine clastic deposits and increasingly proximal fluvial sediments, indicates that the foreland basin evolved from an underfilled to an overfilled state with time. In Maritime Canada, sediment transfer was grossly from south to north, feeding an axial system, located to the north of the principal source area, which was determined by the local orientation of the Acadian orogen.

South Devon–Trevone Basin. The Old Red Sandstone magnafacies in SW England, confined to the Trevone (South Devon) Basin, is represented by the ?Lochkovian to lower Pragian Dartmouth Group (House *et al.* 1977; Bluck *et al.* 1992), which is *c.* 3–4 km thick. The structural setting and stratigraphy have been most recently described by Seago & Chapman (1988). Overall the Dartmouth Group is silt and mud rich (80–90%), and in the east it contains distinctive (intraformational) gravel-bearing mudstones (interpreted as mass-flows), with soft-sediment deformation features (Smith & Humphreys 1991) as well as minor tuffs. Fine-grained sandstone (distal sheetflood) facies interbedded with red siltstones occur west of Plymouth, and comprise the bulk of the Dartmouth Group (Smith & Humphreys 1991). Although generally described as 'distal alluvium' and attributed to a coastal mud flat setting, Smith & Humphreys (1989, 1991) argued for local lacustrine conditions periodically interrupted by muddy–sandy terminal-fan (fluvial) inputs. There is no direct evidence of connection to South Wales (see Bluck *et al.* 1992), and there is an age difference between the marine-influenced mud flat sediments of the Downton Group and equivalents in South Wales. The local association with marine Devonian sequences and presence of volcanic rocks suggests a common setting and a probable extensional origin. This is part of a very thick sequence (> 12 km) of marine (mostly clastic) sediments which infilled discrete extensional basins in SW England throughout Devonian time (Holder & Leveridge 1986).

Anglo-Welsh Basin. The exposed part of the Lower ORS Anglo-Welsh Basin overlies the western part of the Midland microcraton (Fig. 4) and continues in the subsurface across SE England, where marine intercalations become more important (Chaloner & Richardson 1977). The main outcrop is confined to the SE of the Welsh Borderland Fault System (Woodcock & Gibbons 1988), in particular the component Church Stretton–Carreg Cennen–Llandefaelog Fault zone (Fig. 4). In the SW of the basin (SW Dyfed) the ESE–WNW-trending extensional Benton and Ritec Faults affected Lower Devonian ORS (Powell 1987, 1989; Woodcock & Gibbons 1988). The footwall to the (northernmost) Benton Fault, the Johnston Block, was a long-lived stable region (from Ordovician to Devonian time), indicating that this fault formed an important local (northern) margin.

The basin is characterized by a regionally developed, coarsening-upward Přídolí to Emsian basin-fill sequence (Allen 1974, 1983, 1985), which culminated in the late Emsian–Mid-Devonian (late Caledonian) Acadian unconformity (McKerrow 1988; Woodcock & Gibbons 1988; Fig. 3). The basin preserves relatively uniform thicknesses up to *c.* 2.4 km (mean 1.73 km) in central–eastern regions, whereas in the SW 3.4–4.4 km of ORS accumulated in the hanging wall of the Benton Fault. Slip on the latter allowed a decompacted accumulation rate of *c.* 0.2 mm a^{-1} for Cosheston Group alluvium (Powell 1989). In the hanging wall of the synthetic Ritec Fault the ORS is thin (0.4–1.4 km) and of contrasting character, with more complex dispersal (Allen & Williams 1978; Allen *et al.* 1982; Williams *et al.* 1982). The basin lacks volcanic rocks, other than far-travelled Plinian air-fall tuffs (Allen & Williams 1981).

The basin fill was punctuated by widespread intervals of hiatus, represented, initially, by an erosive-based phosphate-bearing bone bed (Ludlow Bone Bed) traditionally the base of the ORS and, later, by very thick pedogenic limestones (calcretes) at two levels in the ORS magnafacies (the *Psammosteus* and Abdon–Ffynnon–Ruperra Limestones). The Ludlow Bone Bed is mainly restricted to the SE of the Church Stretton–Carreg Cennen–Llandefaelog Fault zone (Fig. 4; Allen 1974, fig. 8). To the NW of this structure, but within the Welsh Borderland Fault zone, the basal ORS is conformable from marine Silurian strata (of the Welsh Basin), indicating a fundamental spatial control on the subsidence rate of the early basin. Regional alluvial palaeodispersal was SE- (and S-) directed for the bulk of the basin history (Allen 1974). However, north-dispersing exotic conglomerates, the Pragian Llanishen and ? latest Emsian to Middle Devonian Ridgeway Conglomerates, in SE and SW Wales (Allen 1975; Williams *et al.* 1982) indicate proximal source areas in the Bristol Channel, inferred to be (strike-slip) fault controlled (see Tunbridge (1986) for a review), although without direct evidence of fault kinematics. The unconformity-bounded Ridgeway Conglomerate thickens northward into the hanging wall of the Ritec Fault (Powell 1987).

Initial shallow-marine sandstones (in central–eastern areas) are replaced by basin-wide marine-influenced (Přídolí–Lochkovian) calcretized mud flats of the regressive Downton Group (and equivalents), with externally and internally sourced minor rivers. In contrast, in SW Wales there was a disconformable transition from the marine Wenlock (Bassett 1982) Grey Sandstone Group to red beds of the earliest Milford Haven Group (Hillier this volume), which show fault-controlled local axial and transverse dispersal (Allen & Williams 1978, 1982; Allen *et al.* 1982). The Downton Group (and equivalents) was supplied by a metamorphic rock-dominated source area that was fundamentally reorganized during sediment cut-off in the basin (*Psammosteus* Limestone time), as the subsequent Ditton and Brownstones Groups (and equivalents) had a Lower Palaeozoic sedimentary and reworked Lower ORS source (Allen 1983, 1985) from NW of the Welsh Borderland Fault System. The Ditton and Brownstones Groups, separated in the NE by the lower Abdon–Ffynnon calcretes (Earp & Hains 1971), comprise a crudely coarsening-upward megasequence characterized respectively by mixed low- and high-sinuosity rivers and moderate- to large-scale sand-bed braided rivers (e.g. Allen 1983). Although the coarsening-upward basin fill is interpreted as progradational, the major lithostratigraphic units are punctuated by horizons of hiatus (Love & Williams this volume) suggesting an episodic accumulation history, with relatively rapid progradation likely for the earliest Senni Beds–Brownstones braided alluvium in SE Wales.

On the basis of backstripped subsidence analyses of a composite section for South Wales, James (1987) inferred load-generated flexural subsidence for the Lower Devonian

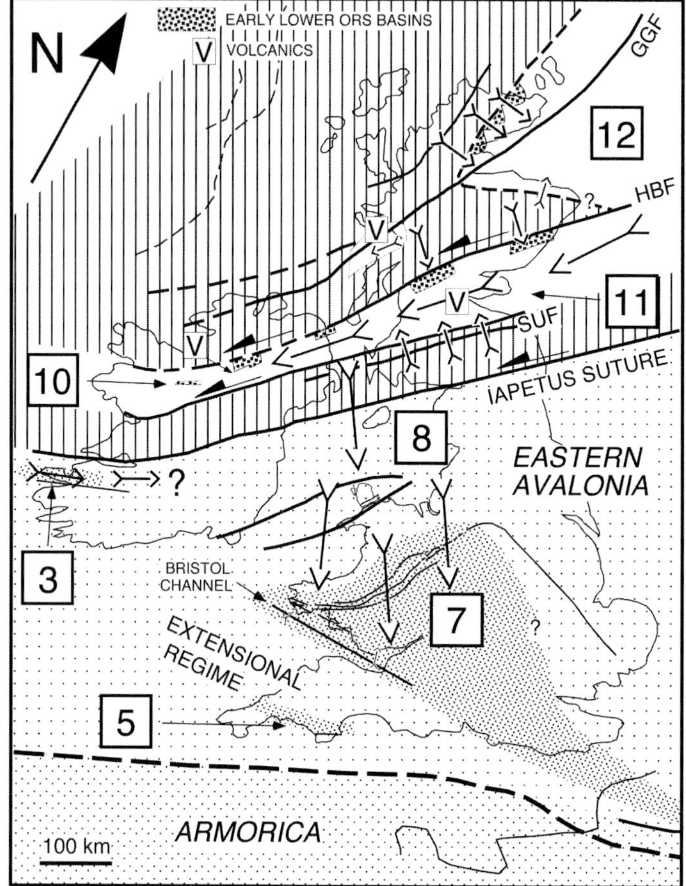

Fig. 4(a). Late Silurian–Early Devonian basin kinematics.

(ORS) section. In a major study of Early Palaeozoic subsidence, King (1994) also suggested probable flexural subsidence for the bulk of South Wales Lower ORS. This is consistent with the foreland location and timing of the Anglo-Welsh Basin with respect to Silurian flexural subsidence in NW England (King 1994), the increasingly proximal ORS alluvium with time and the evolving source area. An exception to this may be the Lower ORS of SW Dyfed, where dip-slip synsedimentary faults suggest, at least local, extension.

Dingle Basin. The elongate Dingle Basin is located immediately south of the trace of the Iapetus Suture in SW Ireland on Leinster Terrane crust (Fig. 4; Murphy *et al.* 1991), and is bounded to the SSE and NNW by the Dingle Bay and North Kerry Lineaments, respectively (Todd 1989; Richmond & Williams this volume). The earliest basin-fill material exposed in the basin (Dunquin Group), discussed by Boyd & Sloan (this volume), is of late Wenlock to mid-Ludlow age. This consists of marine and non-marine siliciclastic sediments and sub-alkaline basaltic to rhyolitic volcanic rocks, interpreted as forming in a volcanic island setting. The volcanism had a destructive plate margin signature, and is considered to represent the last phases of SE subduction before collision along the Iapetus Suture Zone (Boyd & Sloan this volume).

The transition to the ORS magnafacies (Dingle Group) is conformable in the central regions of the basin, but may be unconformable or faulted towards the margins (Holland 1987; Todd 1991). The continental fill of the Dingle Basin (approximate maximum thickness of 4.15 km) can be divided into two phases: an earlier phase characterized by lacustrine sedimentation (Bulls Head Formation, Horne 1974; Boyd & Sloan this volume) coeval with transversely dispersing sand-rich ORS fluvial sediments from the basin margins; and a later phase, dominating the remaining fill of the basin,

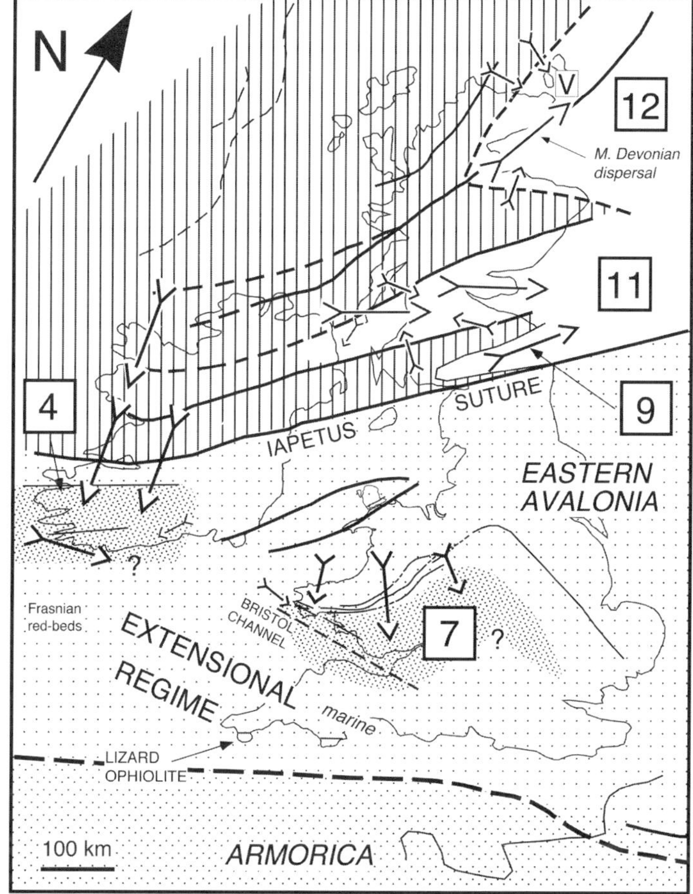

Fig. 4(b). Middle–Late Devonian basinal kinematics.

Fig. 4. Map of Old Red Sandstone basins in Ireland and Britain, showing terrane-bounding faults. Sediment dispersal directions are shown by generalized vector arrows. (See Fig. 3 for identification of basins.)

characterized by axial rivers flowing to the NE. The destination of these major rivers is unconstrained. Early Pragian and late Pragian–early Emsian microfloras (Higgs 1999), within the ?middle Lower ORS of the basin, date the switch to axial dispersal in the basin (Coumeenoule Formation) to approximately mid-Pragian time. Lower ORS sedimentation probably terminated in late Emsian time (Fig. 3). The later phase of dominantly through-basin axial drainage was augmented by a transverse gravelly system from the southern basin margin (Trabeg Conglomerate). The axial Slea Head river system is considered to have entered the basin from the SW, whereas the Glashabeg system entered from the NW, from a volcanic-rich source area (Todd this volume). These gravel-rich inputs reflect spasmodic unroofing in different domains within the Iapetus Suture Zone. Todd (1989, this volume) attributed Dingle Basin subsidence to Acadian sinistral strike-slip on ENE–WSW-trending bounding faults.

Following the main (Lower ORS) basin fill, the basin history was punctuated by a series of fluvio-aeolian sediment accumulations defining thin unconformity-bounded groups developed at the basin margins. These are currently undated, and conflicting models of subsidence history and tectonics have them ranging into Late Devonian time, and caused by extensional or transpressive subsidence regimes (Todd *et al.* 1988; Todd 1989, this volume; Richmond & Williams this volume). Recent work interprets the unconformably based Smerwick Group to have accumulated in a pull-apart basin (the Smerwick Basin) in the NW of the Dingle Basin, before Acadian deformation. Undated post-Acadian fluvio-aeolian sediments of the Pointagare and Caherbla Groups (Fig. 3) are considered contemporaneous and to have been deposited in hydrologically connected sub-basins in NW and SE Dingle, respectively (Richmond & Williams this volume). The Caherbla Group contains alluvial fan breccio-conglomerates eroded from high-grade metamorphic basement south of the Dingle Bay Lineament (Todd this volume, and references therein), the accumulation of which is considered to be pre-Munster Basin initiation (pre-385 Ma; Williams *et al.* 1997, this volume *a*).

Midland Valley of Scotland. Most of the Old Red Sandstone outcrops are within the Midland Valley, as defined by the Highland Boundary Fault to the north, and the Southern Uplands Fault to the south (Fig. 4). Significant outcrops also occur to the north (Lorne Basin) and south of these terrane-bounding faults, and record the history of the relationships between Caledonian terranes (Bluck *et al.* 1988, 1992). This bundle of narrow terranes, deformed in Caledonian times, occupies the present crust immediately north of the Iapetus (Caledonian) Suture, which marks the NW margin of the Avalonian crust that underlies England to the SE.

The first deposits of ORS facies in the Midland Valley are of Early Llandovery to Wenlock age (Marshall 1991). The earliest ORS accumulated in small sinistral strike-slip pull-apart (sub)basin(s) in the NE of the Midland Valley, for example the Crawton Basin (Haughton & Bluck 1988). These early basins were filled by gravel-dominant alluvium, interpreted to be derived from large antecedent drainage basins, tapping previously eroded Caledonian basement (Haughton 1989). The coarse alluvium is not all thought to be a result of local high relief (and conventional alluvial fan deposits). Source areas for the basin fills were complex (Haughton 1989, 1993), and dispersal patterns in the pull-aparts were complex, although the gross supply appears to have been transverse with respect to the Midland Valley basin and the smaller sub-basins. Source areas that are not now represented in present-day outcrop are considered to have been lost to view after displacement by strike-slip faulting (cryptic terranes of Haughton (1988) and Haughton & Bluck (1988)).

Northern (Strathmore) and southern (Lanark) sub-basins are distinguished in the Lower ORS (Bluck this volume), separated by a major volume of arc-related basalts and andesites (Thirlwall 1981, 1988), emplaced from a series of volcanic centres that formed down the central axis of the Midland Valley. Gravels supplied to the Lanark sub-basin from the SSE, that is, from the Southern Uplands region, cannot be simply correlated with the Ordovician–Silurian deep marine clastic rocks at present exposed (Bluck 1983, this volume). Arc-type volcanic and plutonic lithologies derived from the south led Bluck (1983) to characterize the Midland Valley basin as a whole as an inter-arc basin, related to NW subduction. Ordovician limestones discovered in the Midland Valley basin gravelly alluvium (Armstrong & Owen this volume) further complicate the source terrane evolution of the basin.

Lower Old Red Sandstone deposition continued into Emsian times (Fig. 3), and is characterized by finer-grained sand-rich fluvial deposits, with prominent axial dispersal to the SW. The late Lower ORS sedimentation was extensive, occupying the width of the Midland Valley basin, and draping earlier sub-basins. The aggregate thickness of the thicker sub-basins fills and the later Lower ORS is *c.* 9 km in the Strathmore region. The later (Emsian) Lower ORS was characterized by major sand-bed rivers (>10 m deep), sourced from the NE, from a drainage basin interpreted to be developed on thickened and elevated orogenic crust in the region of the Scandinavian Caledonides (Bluck this volume).

Midland Valley continuation (northern Ireland). Northern areas of Ireland contain a series of areally small Lower ORS basins (see review by Graham & Clayton (1988)), in each case adjacent to major Caledonian faults or splays (Fig. 4). The Cushendall, Fintona and Curlew Basins are located immediately south of the continuation of the Highland Boundary Fault (HBF), developed on crust of the Midland Valley terrane (Hutton 1987, 1989; Fig. 4). These basins are thus in equivalent positions to the northern sub-basin (Strathmore Basin) of the Midland Valley of Scotland. The exception to this is the small Ballymastocker Basin, located to the NW of the Leannan Fault (related to the Great Glen Fault). This fault-bounded basin contains an undated conglomeratic Lower ORS fill, thought to succeed a basal lacustrine sequence (McSherry *et al.* this volume). The succession is *c.* 0.25 km thick, and dominantly derived from local Dalradian metamorphic basement.

The Curlew Mountains Basin, bounded to the north by an ENE–WSW splay (the Curlews Fault) from the HBF, contains a fill of *c.* 1.5 km of bedded conglomeratic alluvium (Moygara Formation) and 0.28 km of interbedded sandstones and mudstones with localized andesitic lavas and volcanogenic conglomerates (Keadew Formation; Simon 1984). Miospores from the Keadew Formation are consistent with an Early Devonian age (Graham & Clayton 1988). The Moygara gravels are interpreted as alluvial fan deposits transversely dispersing from a source area north of the boundary fault and interacting with basin-floor playas, whereas the later finer-grained Keadew Formation preserves a basin axial dispersal of sand-rich sheetfloods (Simon 1984).

The complex Fintona Basin is bounded to the north by the Castle Archdale–Omagh Fault (equivalent to the HBF), although the sediments of the basin are cut by sub-parallel major Caledonide and cross faults. The basin stratigraphy has been redefined by Mitchell & Owens (1990), who identified Carboniferous red beds (using miospores) previously considered to be (Devonian) Lower ORS. The redefined Lower ORS Fintona Group (Fig. 3) occurs in two fault-bounded blocks. Probably the older, dated by Pragian miospores (Mitchell & Owens 1990), is located south of the Castle Archdale–Omagh Fault and represents fine-grained floodplain–playa conditions (Mitchell & Owens 1990) with dispersal transverse to the trend of the basin (Simon 1984, Fig. 2b). The second block, in the SE, comprises >3.5 km of volcanogenic boulder conglomerates, sandstones and andesite lavas. Conglomerate clast fining directions and palaeocurrent data suggest that the gravels (interpreted as alluvial fan sediments) and the sandstone lithosomes involved transport axially westward (Simon 1984). Although lavas within the conglomerates have been dated (K–Ar whole rock) to 376 ± 12 Ma (Rundle unpublished, in Mitchell & Owens (1990, p. 410)), the central age of which is Late Devonian on the time scale of Tucker *et al.* (1998), the stratigraphical character of this block is in common with the other Lower ORS basins in northern Ireland. Thirlwall (1988) reports and Rb/Sr augire, whole rock age of 437 ± 6 Ma for a dacitic volcanic rock, considered to be anomalously old due to possible matrix alteration.

The Cushendall Basin (Fig. 4) preserves a 1.27 km thick, predominantly conglomeratic ?Lower ORS alluvial succession (Simon & Bluck 1982; summarized by Graham & Clayton 1988). The basin probably overlies Dalradian (Grampian terrane) basement, and is located to the SE of the projected trace of the HBF. The early basin fill fines up (from alluvial fan sediments) and shows SSE transverse dispersal of polycyclic quartzite gravel, which is not locally sourced from the Dalradian basement. Overlying cross-stratified sandstones, interpreted as >3 m deep river channel deposits, have a contrasting (sub-)axial NNE dispersal. This succession is capped by a volcaniclastic conglomeratic unit with a source area to the south (Simon & Bluck 1982).

Other, areally limited, ORS successions are associated with the extension of the HBF (Hutton 1987, 1989) in Clew Bay (Fig. 4). Silurian-aged (Wenlock) fine-grained red beds in Clew Bay–Clare Island are separate from the upper Pragian–lower Emsian Lower ORS Islandeady Group (Graham 1981; Graham *et al.* 1982). The *c.* 1 km thick Islandeady Group is crudely coarsening upwards, being dominantly conglomeratic above a sand–mud-rich basal division.

Orcadian Basin. The Orcadian Basin was centred in Caithness and the Orkneys, but extends to the north to include the Shetland Islands, and southeastwards to the southern shores of the Moray Firth. The basin fill overlies Northern Highland (Moine) and Grampian (Dalradian) crustal terranes juxtaposed by the Great Glen Fault (GGF; Fig. 4).

The Lower Old Red Sandstone in this basin is Emsian in age (Rogers *et al.* 1989) and is mainly restricted to the western Moray Firth region. Outcrops provide evidence for a number of small, fault-defined basins, often characterized by sediment-fills consisting of alluvial fan marginal deposits and basin-centre, fine-grained, locally lacustrine deposits. Dispersal in the main outcrops was east directed, toward the GGF (Bluck *et al.* 1992). Field relationships to onshore faults indicate an extensional origin for the Lower ORS subsidence.

Offshore seismic data, west of the Orkney Islands and in the Moray Firth, image arrays of small half-graben basins, suggested to be reactivated Caledonian thrusts (Enfield & Coward 1987; Norton *et al.* 1987; Coward *et al.* 1989). These basins have previously been interpreted as containing thick ORS sediments (summarized by Coward 1993); however, much of this fill is now known to be Permo-Triassic in age (Roberts & Holdsworth 1999, and references therein), thus previous interpretations involving major extensional collapse must be reconsidered.

Norway. An Upper Silurian and probably Lower Devonian succession of sediments has been preserved in the Permian or later Oslo graben of southern Norway, on the foreland basement to the southern Norwegian Caledonides (Fig. 2). The lower part of the succession comprises Llandovery to lower Wenlock marine carbonates, and the upper part comprises non-marine fluvial red beds (the Ringerike Group; Bjorlykke 1983). The Ringerike Group was deposited by a prograding system involving rivers flowing SE (Turner & Whittaker 1976), away from the orogenic hinterland. The Wenlock to Emsian Ringerike Group is considered to be a foreland basin succession (Bjorlykke 1983). As pointed out by Dewey *et al.* (1993), the Wenlock–Přídolí section of the Ringerike Group was coeval with Scandian crustal thickening, whereas the Lower Devonian section was synchronous with extensional tectonic denudation of the orogen.

In the coastal Trondheimsfjord region of central Norway, the Hitra Formation (Siedlecka 1975) has been considered to be of Late Silurian age, and consists of conglomerates, sandstones and shales, probably of fluvial and lacustrine origin. No evidence on transport directions is available. The basinal tectonics of the outer Trondheim ORS outcrop areas have been reviewed by Steel *et al.* (1985). Most of the ORS sediment was deposited in response to the activity of steep faults (Møre–Hitra Fault), striking parallel to the main orogen trend, that formed as dip-slip structures (Norton *et al.* 1987) in two different episodes (of latest Silurian–earliest Devonian and Mid-Devonian age).

Spitsbergen. The pre-ORS basement of Spitsbergen has been divided into tectonostratigraphic terranes (Harland 1997). Although there are late-stage granitic intrusions, the basement is otherwise metamorphic, and is characterized by a thermal event indicated by Silurian isotopic mineral ages (Harland 1997). Where the base of the ORS is seen, it is marked by a well-developed unconformity. The two older Groups of the ORS are the Siktefjellet and Red Bay Groups, and they will be considered in this Late Silurian–Early Devonian section, whereas the younger Andree Land Group will be discussed in the section on Mid–Late Devonian basin development, because much of its upper part is of that age.

The Siktefjellet Group is of limited areal extent, and is up to 1.4 km thick. It consists of conglomerates and fine-grained sandstones with coarse siltstones. The Group has yielded no fossil evidence of its age, but its stratigraphic context shows it to be of Late Silurian or Lochkovian age. Its variation in thickness and facies has been suggested to indicate deposition in a pull-apart basin under sinistral transtensional wrench fault zone conditions, and similar kinematics have been interpreted to have produced the subsequent folding (Haakonian phase; McCann this volume).

The Red Bay Group is exposed over a considerably larger area, although it is exposed only on the western edge of the main graben of ORS exposures in Spitsbergen. Rich vertebrate faunas provide evidence of a Lochkovian age. A total maximum thickness for the Group of 3.25 km has been estimated (McCann this volume). Most of the Group is considered to have been fluvial, although some elements of the fauna appear to hint at a marine influence. The palaeocurrent pattern involves northward flow in general, with a westwards component in the north of the outcrop area (McCann, pers. comm.). Spectacular conglomerates at the base of the Group are interpreted to have been sourced from nearby fault scarps, which may have been generated as extensional normal faults, and southern outcrop areas of the Group have revealed a pattern of numerous tilted fault blocks considered to be features of north–south, sinistral, strike-slip kinematics during the Monacobreen wrench phase (McCann this volume).

Mid–Late Devonian ORS basin development

Catskill foreland basin (Northern Appalachians, USA). The extensive coverage of largely undeformed Devonian sediments, particularly in New York State (Fig. 2, area 1), provides an excellent example of prograding facies belts (Fig. 5a). The Old Red Sandstone of this area has long been interpreted as the proximal sedimentary component of the so-called Catskill 'Delta' (Woodrow & Sevon 1985; Bridge this volume). The asymmetry of the thickness of the total Devonian succession (Faill 1985) is clear (Fig. 5b). Palaeocurrent data and stratigraphic geometries confirm the westward (and north-westward) build-out of sediment (and westward progradation of facies belts) from source areas to the east of the present outcrops, which must have been actively uplifted in Mid- and Late Devonian times. In dynamic terms, it is generally accepted that the basin was a flexural foreland type, caused by loading by the Mid- and Late Devonian Acadian fold and thrust belt in the east. Eustatic fluctuations (Johnson *et al.* 1985) and shore zone transgression and regressions (Bridge & Willis 1994) are recorded throughout the succession, which affect the dispersal style and organization of the distal reaches of river channels and systems (Willis & Bridge 1988; Bridge this volume). Quantitative analysis of the Frasnian Catskill single-channel, low-sinuosity rivers, however, suggests that dependent changes in Acadian uplift, sediment flux and flexural subsidence exerted the principal controls on the observed transgressive–regressive sequences (Gordon & Bridge 1987).

Munster Basin. The Munster Basin was initiated across a large area of southern Ireland at the latest in late Givetian times, before *c.* 385 Ma (Williams *et al.* this volume *b*; Fig. 3), <15 Ma after Acadian deformation

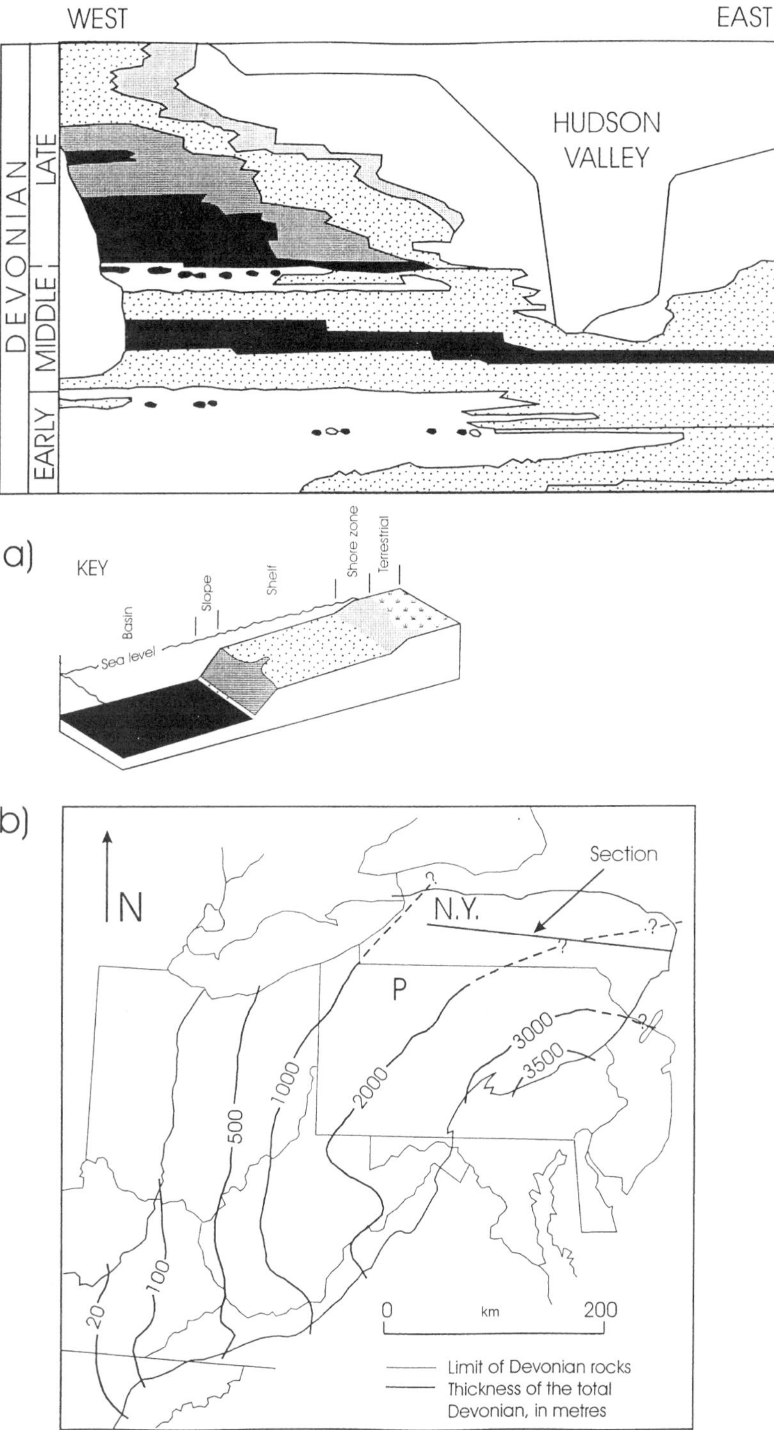

Fig. 5. The Old Red Sandstone of the Catskill 'delta' of the Appalachian USA consists of a Mid- and Late Devonian westward prograding sedimentary wedge. (**a**) Generalized cross-section of the facies belts along a line shown on the key map, redrawn from Bridge (this volume). (**b**) An isopach map for the Devonian succession in Appalachian USA (redrawn from Faill (1985)).

and plutonism in Ireland. Subsidence continued into Carboniferous time, with a latest Devonian–early Carboniferous rifting event forming the smaller marine South Munster Basin. The basin boundary fault (the Dingle Bay–Galtee Fault Zone) controlled syn-rift subsidence to the south until late Famennian times, when ORS footwall sequences (Slieve Mish Group) accumulated. A thick basin-wide ORS magnafacies ($\gg$1 km) filled the basin, with up to *c.* 6 km preserved in the depocentre (Williams this volume). The basin is considered to overlie deformed Lower Palaeozoic crust of the paratectonic Caledonides (exposed in the Leinster Massif of SE Ireland), although the base of the ORS is not exposed south and west of the exposed basin boundaries.

The ORS principally represents a transverse fill from a degrading Caledonian source area in north–central Ireland. Two large-scale low-gradient (sand-dominant) fluvial dispersal systems, with lengths of *c.* 120 and 70 km entered the basin in western (Iveragh) and eastern (Galtee Mountains) regions. The longevity of these fixed inputs allowed stratigraphical differentiation into temporally distinct fluvial systems. A third major system (the Sherkin System) entered in the SW of the basin, demarcated by antithetic extension faults south of the depocentre. This supplied sediment of a different provenance to the ESE. Although basin sedimentation has been characterized as internal (Graham 1983; Leeder 1988*b*), distal reaches of the western and SW dispersal systems cannot be examined. The early basin fill contains silicic lavas and volcaniclastic tuffs, and rarer basaltic lavas and shallow intrusions, located towards the northern basin margin fault zone, which have geochemical signatures of continental stretching (Graham *et al.* 1995; Williams *et al.* this volume, *b*). Overall the volume of contemporaneous volcanicity is very low in relation to basin size. One-dimensional decompaction and (Airy isostatic) backstripping analyses of the ORS and the Late Palaeozoic records have exploited recent isotopic dating of tuff beds in the basin fill (Williams *et al.* this volume *b*; Williams this volume). These confirm ORS rift-related subsidence, and an extensional mode of origin for the basin.

Although the basin has been widely considered as a simple extensional half-graben, the pre-Variscan width of the basin ($>$170 km) implies a more complex situation. Flexural modelling of this scale of crustal subsidence suggests that multiple planar, upper-crustal extension faults are necessary (Williams this volume). The models across the basin suggest stretching factors $\beta = 1.3$–1.48 for an initially thickened post-Acadian crust 40 km thick (for realistic effective elastic thickness values of 7–8 km). Models involving low-angle, listric boundary faults and detachments are not thought to be viable.

North Devon–South Wales. The ORS of the North Devon Basin is represented by the *c.* 1.25 km thick Trentishoe Formation of the Eifelian Hangman Sandstone Group (Tunbridge 1980, 1984, 1986). The Hangman Sandstone Group is marine regressive, and upward coarsening to a sandy ephemeral stream–clay playa complex, forming the southward-dispersing Trentishoe Formation (Tunbridge 1984). These fluvial sediments were sourced in central–south Wales from eroding Lower ORS of the Anglo-Welsh Basin, uplifted by Acadian deformation. Thus in early Mid-Devonian time the fall line–coastline was demarcated by the Bristol Channel region, inferred to be fault related or controlled. A decline in sandy sediment flux in late Trentishoe Formation times resulted in the fine-grained calcretized Yes Tor Member (Tunbridge 1980). Abruptly following this interval of sediment cut-off, the Rawns Conglomerate Formation, a thin exotic breccio-conglomerate sequence, accumulated in the North Devon Basin, from a proximal source in the Bristol Channel, with no contribution from the degrading Anglo-Welsh Basin (Tunbridge 1986). This reorganization of drainage basins feeding the North Devon Basin, involving Bristol Channel uplift, may represent the southernmost effect of the SE-verging Acadian deformation that affected South Wales.

Very thin ($<$0.3 km) Upper ORS (Famennian) successions accumulated in SW and south–central Wales following the Acadian unconformity. In SW Wales the Skrinkle Sandstone Group was restricted to the south of the Ritec Fault, which formed the late Famennian basin margin. An axial braided river system flowed west into a non-saline lake that was also restricted by the fault, and possibly structures in the western Bristol Channel (Marshall 2000*a*). Later, immature fluvial clastic deposits were supplied SW across the Ritec Fault, before marine transgression in Early Carboniferous time (Marshall this volume). Regional palaeoflow for the Upper ORS of South Wales was from N–NW sources (Allen 1965). The alluvial plain in south–central Wales was marine transgressed, as represented by the thin Upper Plateau Beds, followed by an intra-Famennian unconformity (Allen 1965). Regression subsequent to this unconformity in late Famennian time is recorded by the fluvial Pickwell Down Sandstone of North Devon (Allen 1965, 1974) and the Grey Grits in South Wales, demonstrating the clear linkage between SW England–North Devon and South Wales during Late Devonian time (Fig. 3).

Midland Valley of Scotland. Upper Old Red Sandstone outcrops of Late Devonian age occur widely in the Midland Valley and provide evidence of syndepositional Fault activity, some with evidence of strike-slip movement (Bluck 1980, 1983, 2000; Bluck *et al.* 1988, 1992).

Orcadian Basin. The Middle ORS succeeds the Lower with conformable to locally minor angular unconformable relationships (Mykura 1976, 1983), attributed to extensional fault-block tilting (Rogers *et al.* 1989) or an increase in extension rate–accelerated subsidence (Norton *et al.* 1987). The Middle ORS is Eifelian to late Givetian in age (summarized by Rogers *et al.* (1989)), and comprises a thick (4–9 km) and widespread largely permanent lacustrine fill, found from the Shetlands to the SW basin margins around the Moray Firth. Coward (1993) suggested that the Middle ORS represented a widespread post-rift phase of subsidence. Important evidence for marine incursions into the Orcadian Basin during mid–late Givetian time (and latest Givetian–early Frasnian time) from the SE (Marshall *et al.* 1996), suggests that the basin floor was not elevated in an intermontane setting, on thickened crust. Minor fluvial deposits were associated with sediment transport NE–NNE, axially with respect to the GGF; local proximal alluvium was transported from the NW (Bluck *et al.* 1992).

The Middle ORS was deformed before the deposition of the Upper ORS, which nevertheless did not result in a major time gap in the sedimentary record (Rogers *et al.* 1989, fig. 2). The predominantly fluvial Upper Old Red Sandstone accumulated from late Givetian to late Frasnian time (Rogers *et al.* 1989), in areally restricted regions of the basin, notably at Hoy in the Orkneys where thick calc-alkaline lavas were erupted, and in the SW Moray Firth. Marginal aeolian sediments and marine influence in the basin centre also developed.

Norway. Two ORS outcrop areas in Norway will be discussed, which reveal distinct basin styles (Steel *et al.* 1985). These are in the Nordfjord–Sogn area (including the Solund, Kvamshesten, Håsteinen and Hornelen Basins) and the Røragen area.

Mapping of the relationships between the thick ORS proximal conglomerates, and the more distal sandstones in the outcrop areas of the Nordfjord–Sogn area, has shown that the outcrop areas are relicts of a number of different, relatively small (5–20 km by 20–40 km), highly tectonically active basins (Osmundsen *et al.* this volume). These ORS basins are fault-defined structures on the top of a stack of allochthonous sheets of high-grade Caledonian and basement crustal rocks. They formed as the last expression of east–west extension above a west-dipping low-angle, regional crustal detachment, the Nordfjord–Sogn shear zone. This shear zone was active during the exhumation of high-pressure (Western Gneiss Region) eclogites throughout Early Devonian time in the waning stages of the Scandian Orogeny (Fig. 3; Andersen *et al.* 1991; Dewey *et al.* 1993; Milnes *et al.* 1997). A system of syn-extensional east–west upright folds affects the basal detachment and the ORS sedimentary basins (Serrane & Seguret 1987; Chauvet & Seranne 1994). The commencement of ORS sedimentation is not precisely constrained but is generally placed at the boundary between the Early and Mid-Devonian time (Andersen 1998). Sedimentation continued throughout Mid-Devonian time as recorded by scattered fossil evidence. Several basins developed above local bounding extensional detachments lying above and paralleling the principal detachment. The ORS basins lack clasts from the deep allochthons that crop out close to some of the basin perimeters (see Cuthbert 1991) and must have moved by further extension into their present proximity after the last of the preserved Mid-Devonian sedimentation had taken place. Measurement of a cumulative thickness of 25 km of basin-fill sediment in the Hornelen Basin, and its well-exposed, largely complete preservation, has made it a target for research interest both structurally and sedimentologically (Steel *et al.* 1977). It is now generally accepted that the 25 km of stratigraphic thickness is an aggregate thickness resulting from the stacking (or shingling) along the basin axis of many different basin-fill increments, and that the maximum thickness accumulated at any one point was no more than a few kilometres. The stacking was the result of the movement of the basin westwards relative to sediment sources that were north, east and south of boundary faults that were primarily strike-slip on the northern and southern boundaries.

The Old Red Sandstone outcrops at Røragen extend for only some 6 km horizontally at their greatest, but seem to be a relict of a much larger basin, some 300 km along strike. They now represent a thickness of about 1.5 km of the early stages of alluvial fan sediment accumulation (Steel *et al.* 1985), but the basin-fill may originally have been 8 km in thickness (Norton *et al.* 1987). The basin is interpreted to have been a half-graben formed during NW–SE extension on the Røragen detachment (Norton *et al.* 1987).

Baltic States. Plink-Bjorklund & Bjorklund (1999) have presented new data on the sedimentology of the Devonian sediments of Estonia and Latvia. They have proposed that the area formed part of a foreland bulge in Early Devonian times, which then subsided in Emsian and Mid-Devonian times when 0.5 km of siliciclastic sediments were transported into the area from the NW and deposited in fluvial, tidal and deltaic sub-environments. Sandstones in the upper part of this siliciclastic succession are increasingly mature, and this is interpreted as a result of the cannibalization of a Caledonian foreland basin, which was being eroded upstream. In Late Devonian times, sedimentation was dominated by evaporites.

East Greenland. The ORS of East Greenland rests unconformably on a succession of Late Proterozoic to Ordovician sediments that is up to 15 km thick, and has been described as 'pre-Caledonian' and 'supracrustal'. The contact between these supracrustal rocks and the underlying, infracrustal, crystalline basement of the Caledonides has been recently interpreted (Hartz & Andresen 1995) as a low-angle extensional detachment traceable for at least 200 km along the length of the orogen. The new dynamic interpretation of this fault is that it acted as the basal movement surface or zone when the supracrustal rocks moved eastwards over the infracrustal rocks, during the late-stage, gravitational collapse of the thickening Caledonian orogen (Hartz & Andresen 1995). It is proposed that this collapse began in Silurian time and continued in Devonian time, and that both the subsidence and the accumulation of the ORS basinal sequences were responses, at the Earth's surface, to this major tectonic episode.

Two distinct ORS basinal sequences, the Vilddal Supergroup and the Gauss Complex (Friend *et al.* 1983; Fig. 3), were formed with very different configurations of basinal subsidence. Net thicknesses for the two sequences are estimated as 3 km and 6.5 km, which provide estimates of sediment accumulation rates of hundreds of metres per million years in both cases. The Vilddal Supergroup consists of conglomerates, sandstones and siltstones formed in fluvial and lacustrine environments, and the main direction of fluvial transport was towards the east. These early Givetian sediments were preceded by and intercalated with volcanic rocks, both lavas and pyroclastic, and generally of rhyolitic composition.

The later Gauss Complex sediments consist of conglomerates and sandstones, which were overwhelmingly fluvial, although aeolian intervals occurred in the late Givetian lower parts of the Complex and a distinct lacustrine interval is a feature of the Famennian later levels (Clack & Neininger this volume). Palaeocurrents measured in the Gauss Complex show that the basin-floor topography was lowest in the centre of the present ORS outcrop area; marginal transport transverse to the north–south overall folding was eastwards in the west, and westwards in the east. Axial flow developed in the basin centre and was strongly southwards early in the evolution of the Complex, and less strongly northwards in its later history.

There is evidence of various episodes of faulting and local folding, as well as volcanism and granite emplacement in different areas during the accumulation of these sequences. Recent structural work (Hartz this volume) recognizes the importance of east–west syndepositional faulting, in response to north–south extension during orogen-normal shortening, followed by west–east transpression. Orogen-wide late-stage dynamics are claimed to have been responsible for convergent structures, deforming the later Gauss Complex of the western margin of the ORS succession at some time during the Carboniferous or early Permian period.

Spitsbergen. The earlier two of the three stratigraphic groups recognized in the ORS of Spitsbergen have been discussed in the Late Silurian–Early Devonian section of this paper. The third of these, the Andree Land Group (Fig. 3), crops out across the central and eastern parts of the main ORS graben of northern Spitsbergen. Fluvial deposition was well established, with strongly developed northwards flow in Pragian times (Friend & Moody-Stuart 1972), axial to the basin as defined by its graben-bounding faults and the gross folding in the basement. Deposition then appears to have continued, with distinct evidence of marine influence in Mid-Devonian time, before coarsening upwards culminated in conglomerates that have recently been dated to Late Famennian time (Piepjohn *et al.* this volume). These conglomerates have been interpreted as evidence of the onset of the local Svalbardian phase of compressive deformation, which has been interpreted as a transpressive, wrench-related episode (Harland 1997) that appears to have been over by Late Tournaisian or Viséan time (Piepjohn this volume; Piepjohn *et al.* this volume). The subsidence that accompanied the deposition of the great thickness of the Andree Land Group, estimated as a maximum of 4.6 km, is most likely to have been a result of extensional dynamics, but the time resolution is too poor to allow subsidence modelling. There is no evidence of a pattern of flexural loading that could explain this Andree Land Group subsidence.

Discussion: plate boundary and orogenic context

Although there is a paucity of decompacted-backstripped subsidence analyses of Palaeozoic basins, the information compiled above for the North Atlantic ORS basins, in terms of stratigraphic geometry, sedimentology, duration and patterns of subsidence, allows confident assessments to be made of geodynamic mechanisms. ORS depocentres were generated by each of the mechanisms summarized at the beginning of this paper. Ambiguity and disagreement remain, however, for several whole basins or for specific stratigraphic subdivisions of particular basins. On the larger scale, there is disagreement over the application of unifying hypotheses for groups of basins, such as the 'indentor–escape tectonics' model (Coward 1993) or the 'extensional collapse' model (McClay *et al.* 1986) for the Scandinavian–Scottish Caledonides. Details of Caledonian and Variscan orogenic development are also contentious, in particular the role of (sinistral) strike-slip (and the displaced terrane concept), the timing of Iapetus closure, the cause of the Acadian Orogeny and effects of far-field (Variscan–Gondwanan) collision events.

Apart from being infilled with ORS magnafacies, the generation of all of these basins can be linked to late-stage tectonic activity within the Caledonian and Variscan orogens. As outlined, Variscan orogenesis partly overlapped with late Caledonian activity although it mostly occurred to the south (Fig. 2). The diversity of ORS basin positions, ages and types directly reflects the complexity of crustal accretion between the three main plates of Gondwana, Laurentia and Baltica, and associated microplates. The British Isles lies on the triple junction between these three major plates. In this section we discuss the possible origin of the forces responsible for basin location and subsidence regime. Orogenic forces can be divided into body forces, for example, gravitational instability of overthickened crust, and plate boundary forces. It is therefore important to identify as far as possible the positions of, and displacements along plate boundaries during the formation of the ORS basins, that is, from Mid-Silurian to Early Carboniferous time. Palaeomagnetic reconstructions of the North Atlantic region for this period are somewhat problematical because, although the polar wander paths of Laurentia and Baltica are fairly well constrained, the position of Gondwana in Palaeozoic time remains poorly known. The timing of collision or docking of microplates, such as Eastern Avalonia, along the southern margin of Laurentia–Baltica is also poorly constrained by palaeomagnetic data. In detail, reconstructions must therefore depend more on interpretation of geological observations than on palaeomagnetic constraints.

We divide the ORS basins into five groups based on their orogenic context: (1) the syn- to post-Scandian basins of the northernmost North Atlantic region (Greenland, Scandinavia and Spitsbergen); (2) the extensional Orcadian Basin; (3) the basins of the Midland Valley and Ireland north of the Iapetus Suture Zone; (4) the flexural basins of Eastern Avalonia and Western Avalonia; (5) Late Devonian extensional basins on the southern rim of Eastern Avalonia.

Syn- to post-Scandian basins of the Scandinavian North Atlantic

The closure of the northern arm of Iapetus is usually presented in terms of an approximately orthogonal collision between Baltica and Laurentia (Roberts & Gee 1985; Fig. 6) although oblique convergence has also been proposed (e.g. Strachan *et al.* 1992). Scandian orogenesis in mid-Late Silurian time caused the greatest crustal thickening seen in the Caledonides. Sinistral strike-slip between the two plates is also frequently indicated (e.g. Ziegler 1986). ORS depocentres of Scandian age were generated by strike-slip in Spitsbergen (McCann this volume) and possibly in the Hitra Basin (Coward 1993). The distinctive basins of SW Norway record the late stages of crustal extension of the internal Scandian orogen above low-angle extensional detachments along which high-pressure eclogites were exhumed. Late- to post-Scandian extension is also seen in East Greenland and possibly in Spitsbergen. Internal extension in Norway was arguably and locally contemporaneous with final compressive deformation in the external zones to the SE (Gee 1978; Dewey *et al.* 1993; Milnes *et al.* 1997). In the foreland to the SE in the Oslo region, ORS accumulation (Ringerike Group) in a limited flexural basin was coeval with (Ludlow–Pragian) Scandian crustal thickening and later (in Emsian time) with tectonic extensional denudation (Dewey *et al.* 1993).

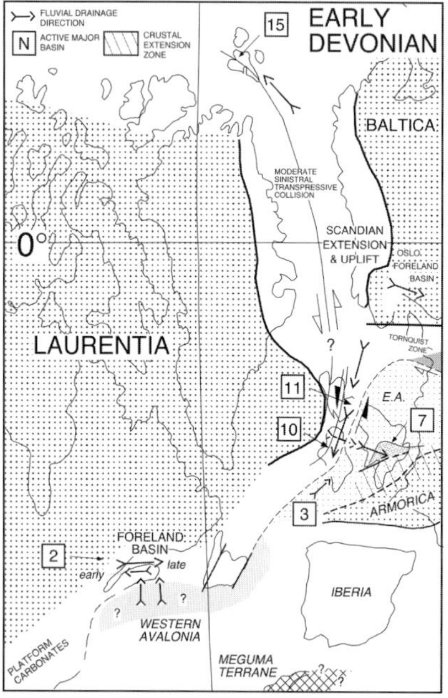

Fig. 6. Palaeomagnetic continental reconstruction for Early Devonian time (400 Ma), showing tectonic–kinematic settings of active Old Red Sandstone basins. Global fits of the major continental areas, indicated by present-day coastlines (North America, Greenland, Baltica, UK), using ocean opening and palaeomagnetic global database to provide the main fit and latitudinal control (data assembled and compiled on a global basis by A. G. Smith, University of Cambridge). Smaller areas, influenced by subsequent tectonics are, in some cases, positioned arbitrarily and other data may require them to be repositioned. Numbers mark the positions of the main ORS basins discussed in the text (see Fig. 3 for identification). Arrows mark some major vectors of sediment transfer. E.A., Eastern Avalonia.

Despite being intensively studied, a consensus has not been reached on the origin of extensional tectonics in SW Norway. The ORS basins here have been widely quoted as classic examples of late orogenic lithospheric collapse basins caused by gravitational instability of overthickened crust (e.g. Hossack 1984; McClay *et al.* 1986; Norton *et al.* 1987; Coward *et al.* 1989; Andersen & Jamtveit 1990; Andersen 1998). However, other workers have argued that the extensional collapse model is problematical because of the lack of Scandian magmatism (high heat flow), the small scale of the observed fold and thrust belt compared with that predicted by a gravitational collapse model (e.g. Fossen 1992; Wilks & Cuthbert 1994) and the relative timing of extension and compression on some transects (Milnes *et al.* 1997). Alternatively, it has been proposed that lithospheric extension was driven by changes in far-field plate boundary forces (Séranne *et al.* 1991; Fossen 1992; Wilks & Cuthbert 1994). Plate reconstructions indicate sinistral strike-slip within the Caledonian orogen during Devonian time (e.g. Ziegler 1986) and extensional conditions may have occurred at a stepover between the Great Glen Fault–Møre–Hitra and Highland Boundary Faults (Séranne *et al.* 1991; Coward 1993). Rey *et al.* (1997) argued that these wrench movements were preceded by a phase of pure extension at 395 Ma driven by north–south Variscan collision. Most recently, Krabbendam & Dewey (1998) have suggested an active sinistral transtensional setting for the Nordfjord–Sogn crustal shear zone, presenting a strain model accounting for the folded detachment, and deformed fill of the SW Norway basins.

The tectonics of the East Greenland basin remains less well known than that of SW Norway and a number of conflicting models have been published to explain its formation. Subsidence has been attributed to north–south (orogen-parallel) sinistral wrenching (e.g. Friend *et al.* 1983), extensional tectonics of differing polarity (McClay *et al.* 1986; Norton *et al.* 1987), and to transtensional kinematics (Larsen & Bengaard 1991). Recent work (Hartz & Andresen 1995) proposed that basin development was synchronous with post-collision collapse on a low-angle, east-dipping detachment. However, Hartz (this volume) presents evidence for syn-depositional deformation of the ORS basin fill by east–west compression. The East Greenland basin initiated during Mid-Devonian time, somewhat later than the Scandinavian Devonian basins on the opposite side of the orogen (Fig. 3). However it continued to develop throughout Late Devonian time. The internal stratigraphic geometries of this basin differ fundamentally from the supradetachment basins in the hanging wall of the Nordfjord–Sogn crustal shear zone in Norway. In addition, the East Greenland extension was not associated with exhumation of eclogites.

The position of East Greenland with respect to Norway during the Devonian period is controversial, because of the poor control on the amount and timing of orogen-parallel strike-slip displacements during and after the Caledonian Orogeny. The palaeomagnetic reconstruction of Kent & Keppie (1988), for example, shows Scandinavia opposite North America in Devonian time. The more conservative models of Ziegler (1986) and Scotese & McKerrow (1990) indicate over 1000 km displacement between Greenland and Scandinavia, whereas other models indicate considerably less displacement between the two continents (Fig. 6).

Sinistral strike-slip generated the Early Devonian subsidence in Spitsbergen. Basins here developed on accreted basement terranes. The more important Devonian subsidence appears to have been extensional in origin.

The extensional Orcadian Basin, NE Scotland

The origin of the long-lived Orcadian subsidence is problematical. Reconstructions of the Orcadian Basin have involved major late Caledonian (and syn-ORS) sinistral displacements on the GGF (Bluck *et al.* 1992), and large Carboniferous dextral strike-slip on the same fault (Coward 1993). Rogers *et al.* (1989), however, maintained that the stratigraphy and sedimentology of the Lower to Upper ORS reflects contiguity of the basin, and allow a maximum of 20 km post-ORS dextral slip on the GGF. Rogers *et al.* (1989) indicated an extensional origin for the subsidence of the basin, having a radically different disposition to the 600 km by 200 km pull-apart or releasing overstep basin suggested by Séranne (1992) and Coward (1993).

The Orcadian Basin has also been included in the orogenic extensional collapse models of McClay *et al.* (1986) encompassing the Scandinavian–East Greenland Caledonides. However, this rift basin developed on orthotectonic Caledonian crust that had undergone the Grampian Orogeny in Ordovician time and was already deeply eroded by Early Devonian time. Basin generation was clearly not associated with exhumation of eclogites in overthickened crust and thus cannot be equated with the supradetachment basins in SW Norway. Norton *et al.* (1987), like Rogers *et al* (1989), argued for regional NW–SE extension of the Orcadian region throughout Devonian time. Far-field plate boundary forces must have driven this extension.

Basins of the Scottish Midland Valley and Ireland, north of the Iapetus Suture Zone

Along with the Fintona–Cushundall Basins in Ireland, the Mid–Late Silurian–Early Devonian Crawton Basins (and others) within the northern Midland Valley of Scotland (Haughton & Bluck 1988; Haughton 1989) were generated as pull-apart basins within a belt of sinistral strike-slip along the HBF. Evidence summarized earlier also allows strike-slip on the SUF. A similar, sinistral (transpressive) origin is proposed (Todd this volume) for the Dingle Basin, which is of the same age and lies within the Iapetus Suture Zone further south. This episode of strike-slip tectonics was active during the anticlockwise docking of Eastern Avalonia (dated at mid-Silurian to early Devonian time; Soper & Woodcock 1990), and is also synchronous with Scandian deformation in Western Norway (Fig. 3).

The more extensive late Lower ORS (Emsian) of the Midland Valley (the Strathmore Basin of Haughton & Bluck (1988)), was filled axially by finer-grained (sandy) alluvium containing major rivers. This late phase overlaps in time with Lower ORS (mid–late Emsian) extension in the Orcadian Basin, and may be due to this mechanism, rather than thermal subsidence following pull-apart basin development as suggested by Haughton & Bluck (1988). Haughton & Bluck (1988) also suggested that the later Lower ORS may have been due to flexural subsidence related to uplift of the Grampian source block before Acadian-age deformation. Bluck (1983) proposed that the entire Lower ORS Midland Valley Basin has an inter-arc character, with contemporaneous volcanic rocks of destructive plate margin type (Thirlwall 1981, 1988; Soper 1986) and unroofing arc rock and metamorphic detritus from north and south of the basin. Up to Emsian time, oceanic crust attached to the northern border of Eastern Avalonia continued to be subducted very slowly below the Southern Uplands (McKerrow & Soper 1989) even though the Iapetus Ocean had closed by Early Devonian time (Soper & Woodcock 1990).

Late Caledonian flexural basins

British Isles. By reviewing faunal evidence and new palaeomagnetic results, Soper & Woodcock (1990) demonstrated that Eastern Avalonia docked obliquely during Silurian and Early Devonian time with a component of anticlockwise rotation. Sediment dispersal patterns and facies distributions indicate that along the southern margin of the Southern Uplands sedimentation occurred in a trench environment related to northward subduction. This trench gradually converted to a foreland basin during Wenlock time. Soper & Woodcock (1990) suggested that in Ludlow time, the marine foreland basin depocentres of the Lake District (Kneller 1991) and the north Welsh Basin became overfilled with sediment derived from the rising Scandian Caledonides to the NE leading to a transition to non-marine (ORS) conditions in the Přídolí–Lochkovian time (Scout Hill Flags, Lake District). As outlined above, both James (1987) and King (1994) demonstrate a flexural style subsidence for the Lower ORS of South Wales, implying that the late Caledonian flexural basin migrated southward across Eastern Avalonia as far as the Bristol Channel.

In late Emsian time, Acadian deformation (Fig. 3), the last phase of Caledonian orogenesis, affected central Britain and Ireland, mainly between the Midland Valley of Scotland and the Midland Platform (Soper *et al.* 1987; McKerrow 1988; Keppie 1989; Soper & Woodcock 1990). This deformation is characterized by SSE-directed shortening and sinistral transpression associated with low-grade regional metamorphism (Murphy 1985; Soper *et al.* 1987; Woodcock *et al.* 1988; Soper & Woodcock 1990). In the Scottish Midland Valley, Mid-Devonian shortening and uplift is recognized as the most northerly signature of the Acadian event. Acadian deformation is closely associated with a thermal peak recorded by extensive intrusion of the Newer Granites (Soper *et al.* 1987; Leake 1990). The continued southward migration of Acadian uplift led to the late Early to Mid-Devonian unconformity in South Wales, and possibly the Eifelian ORS clastic wedge in North Devon.

With the possible exception of the Eifelian ORS in the North Devon Basin, Acadian deformation post-dates, and therefore was not responsible for, the formation of the late Caledonian foreland basins. As suggested by Soper & Woodcock (1990), lithospheric flexure was probably generated as the northern margin of Eastern Avalonia was underthrust below the Southern Uplands in a 'soft' collision. At the same time, to the south of the Midland Platform the southern rim of Eastern Avalonia was undergoing significant extension as recorded by the dominantly marine basins of Devon and Cornwall (Holder & Leveridge 1986). The cause of the Acadian sinistral transpressive event, clearly limited in time and space, and the related intrusion of the Newer Granites is as yet unclear. Invoking a far-field plate boundary source somewhere to the south is problematical because the major extensional regime along the southern margin of Eastern Avalonia may have been active throughout Devonian time. It may be more probable that this event was generated by the much more substantial Acadian deformation of the Northern Appalachians and Canadian Maritime Provinces that was occurring along strike, and thus closer to the separate Western Avalonian continental fragment.

Northern Appalachians: Canada–Catskills. In the Northern Appalachians and the Maritime Provinces of Canada, the Acadian Orogeny had a very different character, being a major metamorphic and deformational event that affected almost the entire Appalachian chain (Williams & Hatcher 1982). Peak metamorphic conditions and maximum plutonic activity coincide with this orogenic phase. It was caused by the collision during mid-Devonian time of the Avalon microplate (Western Avalonia) and possibly the Meguma terrane (Williams & Hatcher 1982; Keppie 1989) with the North American cratonic margin. High-temperature and relatively high-pressure metamorphism and plutonic magmatism were largely restricted to central and southern New England. The Gaspé and Catskill ORS successions form part of the Appalachian foreland basin. The Middle Ordovician to Lower Pennsylvanian stratigraphy of this enormous flexural basin records diachronous episodes of Taconic, Acadian and Alleghenian orogenesis (Tankard 1986). The thick prograding clastic wedges of the Gaspé and Catskill units record overfilled conditions that can be linked directly to the Acadian event. However, their timing is rather different, implying that overfilled conditions and thus collision-related uplift of the hinterland were achieved at different times along strike. Lawrence & Rust (1988) and Rust *et al.* (1989) argued for an early collision with the St Lawrence Promontory and subsequent dextral strike-slip along its SW edge, to account for the reversal of the axial river systems preserved in the basin. Keppie (1989), however, on the basis of structural arguments, suggested an oblique sinistral transpressional docking motion for Western Avalonia.

Syn-Variscan extensional basins

In SW Ireland, southernmost Wales (SW Dyfed) and SW England Mid- to Late ORS basins either post-date Acadian deformation or formed on crust unaffected by this phase of deformation (SW England). In SW Ireland the Munster Basin underwent major syn-rift subsidence from at least mid-Givetian to late Famennian time, with probable post-rift subsidence commencing before the end of the Devonian period, and continuing (at least) into late Viséan time. The generation of the Munster Basin cannot be realistically linked to late Caledonian processes, as there is a significant time gap ($\leqslant$ 15 Ma) between its initiation (pre-385 Ma) and peak Acadian deformation and syn-tectonic plutonism (c. 400 Ma). Extension was oriented NNW–SSE, from major boundary and intra-basinal extension faults, and no evidence has been adduced for either local or regional strike-slip associated with the basin (see Coward 1993). The basin formed on moderately deformed Caledonian crust, composed of low-grade volcano-sedimentary accreted terranes, which is interpreted to be modestly thickened. An estimated initial crustal thickness of 40 km successfully reproduces, in flexural cantilever models, major ORS subsidence in Munster (Williams this volume). Mid-Frasnian marine facies restricted to the exposed base of the ORS succession in the SW of the onshore Munster Basin (Higgs *et al.* this volume) indicate that extension had lowered the crust to sea level by this time. The regional extent of the subsiding–extending crustal tract undergoing Late Devonian extension is shown by probably Frasnian ORS magnafacies 155 km south of SW Ireland. The nature of the basement, which was undergoing at least local erosion before basin initiation (represented by the NNW supply of sediment into the Dingle Basin during ?early Mid-Devonian time), and these factors strongly imply that the cause of extension cannot have been gravitationally unstable (overthickened) lithosphere.

In Early Devonian time, the ORS magnafacies accumulating in the Trevone–South Devon Basin (Dartmouth Group) is most likely to have been controlled by extensional subsidence. This continental episode was succeeded by thick marine Devonian sediments and voluminous volcanism related to continental stretching. Extension was continuous and advanced in SW England, culminating with the creation of the Lizard ophiolite at 375 ± 34 Ma (Davies 1984) in Mid-Devonian time, possibly in a pull-apart basin along a suggested WNW–ESE transform fault (Barnes & Andrews 1986).

Further east, in mainland Europe, thick Devonian shelf and hemi-pelagic deposits associated with Emsian to early Carboniferous bimodal volcanism accumulated in the Rheno-Hercynian basin and clearly record considerable crustal extension (Franke 1989). Thus the Devonian extensional basins in southern Britain and Ireland were part of a larger extensional regime along the southern rim of the Eastern Avalonian plate that was active from Silurian to mid-Viséan time (Franke 1989). Mid-ocean ridge basalt type mafic rocks preserved in the Giessen, Harz and Lizard nappes suggest the formation of ocean floor in early–mid Devonian time. Towards the west the amount of extension appears to decrease and the timing of its initiation becomes younger during the Devonian period. During Early Carboniferous time extensional conditions spread northward across Britain and Ireland. The origin of this Rheno-Hercynian basin is not clear. It was opening while all other parts of the Variscan orogen were undergoing convergence. In addition, it occupies the boundary between the Eastern Avalonia and Armorica microplates. This was the site of the major Rheic ocean, which was opened in Ordovician time and closed by Early Devonian time (Matte 1986; Franke 1989), leading Franke & Oncken (1995) to argue that, like the North Atlantic, the Rheic ocean closed and then reopened. Several workers including Leeder (1982, 1988*a*) and Ziegler (1986) have proposed that the Rheno-Hercynian basin represents back-arc extension related to northward subduction along the southern margin of the Armorica plate. However, as reviewed by Rey *et al.* (1997) and Franke (1989), most geological evidence from the internal Variscides indicates that the Variscan oceans were closed by Early Devonian time. Although a back-arc stretching mechanism cannot be completely ruled out, alternative explanations for the Rheno-Hercynian basin extension are being sought (e.g. Franke 1989). This basin was closed in late Carboniferous time by northward-migrating Variscan compression.

To explain the mid-Devonian extensional regime responsible for eclogite exhumation and ORS basin subsidence in Scandinavia, Rey *et al.* (1997) proposed a tectonic model in which northward compression was transmitted from Central Europe through Eastern Avalonia to the site of the collision between Baltica and Laurentia. However, the presence of a major zone of north–south extension along the southern margin of Eastern Avalonia implies that Gondwanan collision cannot have been transmitted as far north as the Scandinavian Caledonides.

Orogen-scale sediment dispersal

Bluck (1983, this volume) has suggested that sediment was transferred from Norway to the Midland Valley of Scotland, hundreds of kilometres, by rivers running along the trend of the orogen, sourced from a large drainage basin on the uplifted Scandian orogen (the Norway–Greenland mountains). The dominant northward sediment dispersal in the approximately synchronous Wood Bay Formation of Spitsbergen represents river flow in the opposite axial direction, and our survey of many basins shows that flow patterns axial to the general orogenic trend are common, and that some of them may involve distances of transport much greater than the length of the sedimentary basins. We suggest that this situation is analogous in scale and sense to the Himalayan patterns of the westward-flowing upper Indus River in India and Pakistan, and eastward-flowing Tsangpo–Brahmaputra from Tibet through India (Friend 1998).

One controversial detail, however, also demonstrates the importance of provenance studies. The detailed question is the destination of the major SW-flowing, through-going, rivers in the late Midland Valley Basin, which have been interpreted as terminating internally by Allen & Crowley (1983) and Leeder (1988*b*), or turning SSE, across orogenic strike, to ultimately supply the Anglo-Welsh Basin (Simon & Bluck 1982). Haughton & Farrow (1989), using detrital garnet compositions to assess provenance, have indicated that garnets of the late Midland Valley axial fluvial sediments resemble those fed to the basin from the NW across the HBF, and are unlike garnets expected from the high-pressure eclogitic basement in SW Norway, although it is possible they were derived from the cover to these rocks. As pointed out by Haughton & Farrow (1989), the hydrological scale of the axial systems in the Midland Valley makes it unlikely that they terminated in an intermontane setting. Haughton & Farrow (1989) further concluded that differences in garnet type preclude a coeval (post early Pragian) link between the Midland Valley and the Anglo-Welsh Basin (see Simon & Bluck 1982). An earlier (Lochkovian) link is also considered unlikely, but the evidence is equivocal. We suggest that the axial system of the Midland Valley Basin continued SW into northern Ireland, but that its sediment has not been preserved. The (undated) NE fluvial dispersal in the upper Cushendall Basin (Fig. 4) is not evidence that the Midland Valley system was diverted SSE as shown by Simon & Bluck (1982), as these sediments may be part of the early small pull apart basins along the HBF. Earlier, Přídolí–Lochkovian, drainage from garnetiferous Caledonian basement in this region may have been permitted to supply the earliest fill (Downton Group) of the Anglo-Welsh Basin.

Summary

In Figs 6 and 7 we use continental configurations kindly supplied from his general global database by A. G. Smith (pers comm.). These provide a very generalized summary of the dispositions of the major continental areas, based simply on ocean-opening data and available palaeomagnetic data. The positions of the principal ORS basins are shown with respect to plate boundaries in Early and Late Devonian time. Gondwana is not shown, nor is the activity along the Tornquist Line considered.

The late Silurian–earliest Devonian strike-slip regimes seen in Spitsbergen, the Midland Valley of Scotland and in Ireland were short-lived and probably related to the oblique 'soft' docking of Eastern Avalonia with Laurentia. The docking of Eastern Avalonia coincides with the Scandian

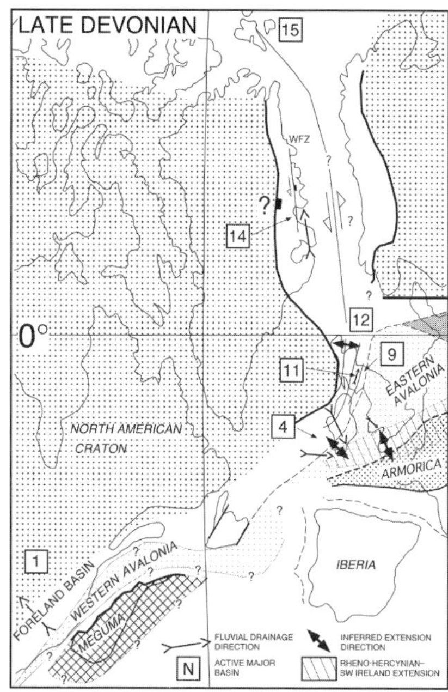

Fig. 7. Reconstruction for Late Devonian time (360 Ma). Numbers mark the locations of the main Old Red Sandstone basins discussed in the text (see Fig. 3 for identification), and detailed in this volume. Arrows mark some major vectors of sediment transfer.

Orogeny further north. By Early Devonian time, Scandian orogenesis changed from a compressional regime to one of strike-normal extension. Extensional regimes took over everywhere north of the Iapetus Suture (Midland Valley of Scotland, Orcadian Basin, East Greenland, Norway, Spitsbergen) and were active (sometimes sporadically), possibly in some places in association with sinistral strike-slip (e.g. East Greenland), throughout the rest of the Devonian period.

To the south of the Iapetus Suture in Britain and Ireland the ORS facies record the last phase of development of a dominantly marine Caledonian foreland basin that migrated south across Eastern Avalonia to the Midland microcraton. At the same time, to the south of the Midland microcraton, the extensional Rheno-Hercynian basin developed throughout Devonian time with ORS facies deposited particularly at its western and northern extremities (Munster Basin and SW Wales, where the amount of extension was relatively low), and during early rifting in SW England (Trevone Basin).

To the west, in North America, ORS facies represent overfilled phases of the Appalachian foreland basin related to the Acadian Orogeny, a major oblique collisional event.

Problems

Although we conclude that ORS basins are intimately related to processes involved in the Caledonian and Variscan Orogenies, and are not simply a function of erosion of the Caledonides, the traditional view, melding different synchronous geodynamic processes together remains difficult. We highlight several areas which need renewed research to achieve breakthroughs. (1) Geohistory work (decompaction-backstripping, with careful evaluation of uncertainties) although difficult in Palaeozoic basins, is lacking for the majority of ORS basins, and would clarify subsidence mechanisms in many of the basins discussed. (2) For this, better local geochronology calibrated to local and global continental biostratigraphy is needed. (3) The palaeomagnetic definition of microplates, so important in the Acadian and Variscan orogens needs to be improved, and particularly for Western and Eastern Avalonia, and Gondwana. (4) Traditional provenance studies for fine-grained ORS basin fills, as well as more sophisticated techniques using single mineral typing (e.g. Haughton & Farrow 1989), and U–Pb and Sm–Nd dating, are badly needed in many areas. (5) Outstanding problematic geodynamic processes include that highlighted by Early Devonian extensional subsidence in SW England (?SW Wales) and the Rheno-Hercynian basin v. simultaneous flexural subsidence for the bulk of Anglo-Welsh Basin. (6) Quantification, and even identification, of major strike-slip movements along the Greenland–Baltica collision, the northern British Caledonides and in the region of the Western Avalonian collision remains exceptionally difficult and controversial. Palaeomagnetic data for critical areas is still sparse and resolutions coarse, and these are needed to provide a truly independent test of tectonic models, and an overall framework for the Old Red Sandstone.

Our thanks go to A. G. Smith (Cambridge), N. H. Woodcock (Cambridge), T. H. Torsvik (Norges Geologiscke Undersøkelse), T. B. Andersen (Oslo), N. J. Soper (Sheffield) and D. H. Tarling (Plymouth), for discussion and information recently and over years past.

References

ALLEN, J. R. L. 1965. Upper Old Red Sandstone (Farlovian) paleogeography in South Wales and the Welsh Borderland. *Journal of Sedimentary Petrology*, **35**, 167–195.

—— 1974. The Devonian rocks of Wales and the Welsh Borderland. *In*: OWEN, T. R. (ed.) *The Upper Palaeozoic and Post-Palaeozoic Rocks of Wales*. University of Wales Press, Cardiff, 47–84.

—— 1975. Source rocks of the Lower Old Red Sandstone: the Llanishen Conglomerate of the Cardiff area, South Wales. *Proceedings of the Geologists' Association*, **86**, 63–76.

—— 1983. Studies in fluviatile sedimentation: bars, bar-complexes and sandstone sheets (low sinuosity braided streams) in the Brownstones (L. Devonian), Welsh Borders. *Sedimentary Geology*, **33**, 237–293.

—— 1985. Marine to freshwater: the sedimentology of the interrupted environmental transition (Ludlow–Siegenian) in the Anglo-Welsh area. *Philosophical Transactions of the Royal Society of London, Series B*, **309**, 85–104.

—— & CROWLEY, S. F. 1983. Lower Old Red Sandstone fluvial dispersal systems in the British Isles. *Transactions of the Royal Society of Edinburgh: Earth Sciences*, **74**, 61–68.

—— & WILLIAMS, B. P. J. 1978. The sequence of the earlier Lower Old Red Sandstone (Siluro-Devonian) north of Milford Haven (south-west Dyfed), Wales. *Geological Journal*, **13**, 113–136.

—— & —— 1981. Sedimentology and stratigraphy of the Townsend Tuff Bed (Lower Old Red Sandstone) in South Wales and the Welsh Borders. *Journal of the Geological Society, London*, **138**, 15–29.

——, THOMAS, R. G. & WILLIAMS, B. P. J. 1982. The Old Red Sandstone north of Milford Haven. *In*: BASSETT, M. G. (ed.) *Geological Excursions in Dyfed, South-west Wales*. National Museum of Wales, Cardiff, 123–149.

ALLEN, P. A. & ALLEN, J. R. 1990. *Basin Analysis, Principles and Applications*. Blackwell Scientific, Oxford.

ANDERSEN, T. B. 1998. Extensional tectonics in the Caledonides of Southern Norway, an overview. *Tectonophysics*, **285**, 333–351.

—— & JAMTVEIT, B. 1990. Uplift of deep crust during orogenic collapse: a model based on field studies in the Sogn–Sunnfjord region of west Norway. *Tectonics*, **9**, 1097–1112.

——, ——, DEWEY, J. F. & SWENSSON, E. 1991. Subduction and eduction of continental crust: major mechanisms during continent–continent collision and orogenic extensional collapse, a model based on the south Norwegian Caledonides. *Terra Nova*, **3**, 303–310.

ARMSTRONG, H. A. & OWEN, A. W. 2000. Age and provenance of limestone clasts in Lower Old Red Sandstone conglomerates: implications for the geological history of the Midland Valley Terrane. *This volume*.

ARTHAUD, F. & MATTE, P. 1977. Late Paleozoic strike-slip faulting in southern Europe and northern Africa: result of a right-lateral shear zone beween the Appalachians and the Urals. *Geological Society of America Bulletin*, **88**, 1305–1320.

AYDIN, A. & NUR, A. 1982. Evolution of pull-apart basins and their scale independence. *Tectonics*, **1**, 91–105.

BARNES, R. P. & ANDREWS, J. R. 1986. Upper Palaeozoic ophiolite generation and obduction in south Cornwall. *Journal of the Geological Society, London*, **143**, 117–124.

BASSETT, M. G. 1982. Silurian rocks of the Marloes and Pembroke peninsulas. *In*: BASSETT, M. G. (ed.) *Geological Excursions in Dyfed, South-west Wales*. National Museum of Wales, Cardiff, 103–122.

——, BLUCK, B. J., CAVE, R., HOLLAND, C. H. & LAWSON, J. D. 1992. Silurian. *In*: COPE, J. C. W., INGHAM, J. K. & RAWSON, P. F. (eds) *Atlas of Palaeogeography and Lithofacies*. Geological Society, London, Memoirs, **13**, 37–56.

BASSI, G., KEEN, C. E. & POTTER, P. 1993. Contrasting styles of rifting: models and examples from the eastern Canadian margin. *Tectonics*, **12**, 639–655.

BEAUMONT, C. 1981. Foreland basins. *Geophysical Journal of the Royal Astronomical Society*, **65**, 291–329.

BJØRLYKKE, K. 1983. Subsidence and tectonics in late Precambrian and Palaeozoic sedimentary basins of Southern Norway. *Norges Geologiske Undersøkelse*, **380**, 159–172.

BLUCK, B. J. 1980. Evolution of a strike-slip fault-controlled basin, Upper Old Red Sandstone, Scotland. *In*: BALLANCE, P. F. & READING, H. G. (eds) *Sedimentation in Oblique-slip Mobile Zones*. International Association of Sedimentologists, Special Publications, **4**, 63–78.

—— 1983. Role of the Midland Valley of Scotland in the Caledonian orogeny. *Transactions of the Royal Society of Edinburgh: Earth Sciences*, **74**, 119–136.

—— 2000. Old Red Sandstone basins and alluvial systems of Midland Scotland. *This volume*.

——, COPE, J. C. W. & SCRUTTON, C. T. 1992. Devonian. *In*: COPE, J. C. W., INGHAM, J. K. & RAWSON, P. F. (eds) *Atlas of Palaeogeography and Lithofacies*. Geological Society, London, Memoirs, **13**, 57–66.

——, HAUGHTON, P. D. W., HOUSE, M. R., SELWOOD, E. B. & TUNBRIDGE, I. P. 1988. Devonian of England, Wales and Scotland. *In*: MCMILLAN, N. J., EMBRY, A. F. & GLASS, D. J. (eds) *The Devonian of the World, Volume 1*. Canadian Society of Petroleum Geologists, Memoirs, **14**, 305–324.

BOYD, J. D. & SLOAN, R. J. 2000. Initiation and early development of the Dingle Basin, southwest Ireland, in the context of the closure of the Iapetus Ocean. *This volume*.

BRIDGE, J. S. 2000. The geometry, flow patterns and sedimentary processes of Devonian rivers and coasts, New York and Pennsylvania, USA. *This volume*.

—— & WILLIS, B. J. 1994. Marine transgressions and regressions recorded in Middle Devonian shore-zone deposits of the Catskill clastic wedge. *Geological Society of America Bulletin*, **106**, 1440–1458.

BUCK, W. R. 1988. Flexural rotation of normal faults. *Tectonics*, **7**, 959–973.

BULLARD, E. C., EVERETT, J. E. & SMITH, A. G. 1965. The fit of the continents around the Atlantic. *Philosophical Transactions of the Royal Society, London, Series A*, **258**, 41–51.

BURG, J.-P., BALÉ, P., BRUN, J.-P. & GIRARDEAU, J. 1987. Stretching lineations and transport direction in the Ibero-Armorican arc during the Siluro-Devonian collision. *Geodinamica Acta*, **1**, 71–87.

CHALONER, W. G. & RICHARDSON, J. B. 1977. South-east England. *In*: HOUSE, M. R., RICHARDSON, J. B., CHALONER, W. G., ALLEN, J. R. L., HOLLAND, C. H. & WESTOLL, T. S. (eds) *A Correlation of Devonian Rocks of the British Isles*. Geological Society, London, Special Report, **8**, 26–40.

CHANNELL, J. E. T., MCCABE, C. & WOODCOCK, N. H. 1992. Early Devonian (pre-Acadian) magnetization directions in Lower Old Red Sandstone of south Wales (UK). *Geophysical Journal International*, **108**, 883–894.

CHAUVET, A. & SERANNE, M. 1994. Microtectonic evidence of Devonian extensional westward shearing in Southwest Norway. *In*: GAYER, R. (ed.) *The Caledonides Geology of Scandinavia*. Graham & Trotman, London, 245–254.

CHRISTIE-BLICK, N. & BIDDLE, K. T. 1985. Deformation and basin formation along strike-slip faults. *In*: BIDDLE, K. T. & CHRISTIE-BLICK, N. (eds) *Strike-slip Deformation, Basin Formation and Sedimentation*. Society of Economic Paleontologists and Mineralogists, Special Publications, **37**, 1–34.

CLACK, J. A. & NEININGER, S. L. 2000. Fossils from the Celsius Bjerg Group, Late Devonian Sequence, East Greenland: significance and sedimentological distribution. *This volume*.

CLOUTIER, R., EL ALBANI, A., CANDILIER, A.-M. & TESSIER, B. 1998. Paleoenvironmental analysis of the Upper Devonian Escuminac Formation, Quebec (Canada), based on clay mineralogy, organic matter and fish taphonomy. Abstract, BSRG Research Meeting, Geological Society, London, 7.

COPE, J. C. W., GUION, P. D., SEVASTOPULO, G. D. & SWAN, A. R. H. 1992. Carboniferous. *In*: COPE, J. C. W., INGHAM, J. K. & RAWSON, P. F. (eds) *Atlas of Palaeogeography and Lithofacies*. Geological Society, London, Memoirs, **13**, 67–86.

COWARD, M. P. 1990. The Precambrian, Caledonian and Variscan framework to NW Europe. *In*: HARDMAN, R. F. P. & BROOKS, J. (eds) *Tectonic Events Responsible for Britain's Oil and Gas Reserves*. Geological Society, London, Special Publications, **55**, 1–34.

—— 1993. The effect of Late Caledonian and Variscan continental escape tectonics on basement structure, Paleozoic basin kinematics and subsequent Mesozoic basin development in NW Europe. *In*: PARKER, J. R. (ed.) *Petroleum Geology of Northwest Europe: Proceedings of the 4th Conference*. The Geological Society, London, 1095–1108.

——, ENFIELD, M. A. & FISCHER, M. W. 1989. Devonian basins of Northern Scotland: extension and inversion related to Late Caledonian–Variscan tectonics. *In*: COOPER, M. A. & WILLIAMS, G. D. (eds) *Inversion Tectonics*.

Geological Society, London, Special Publications, **44**, 275–308.

CUTHBERT, S. J. 1991. Evolution of the Devonian Hornelen Basin, west Norway: new constraints from petrological studies of metamorphic clasts. *In*: MORTON, A. C., TODD, S. P. & HAUGHTON, P. D. W. (eds) *Developments in Sedimentary Provenance Studies*. Geological Society, London, Special Publications, **57**, 343–360.

DAVIES, G. R. 1984. Isotopic evolution of the Lizard Complex. *Journal of the Geological Society, London*, **141**, 3–14.

DEWEY, J. F. 1982. Plate tectonics and the evolution of the British Isles. *Journal of the Geological Society, London*, **139**, 371–412.

—— & SHACKLETON, R. M. 1984. A model for the evolution of the Grampian tract in the early Caledonides and Appalachians. *Nature*, **312**, 115–121.

——, RYAN, P. D. & ANDERSEN, T. B. 1993. Orogenic uplift and collapse, crustal thickness, fabrics and metamorphic phase changes: the role of eclogites. *In*: PRITCHARD, H. M., ALABASTER, T., HARRIS, N. B. W. & NEARY, C. R. (eds) *Magmatic Processes and Plate Tectonics*. Geological Society, London, Special Publications, 76, 325–343.

DINELEY, D. L. & WILLIAMS, B. P. J. 1968*a*. The Devonian continental rocks of the lower Restigouche River Quebec. Canadian Journal of Earth Sciences, **5**, 945–953.

—— & —— 1968*b* Sedimentation and Paleoecology of the Devonian Escuminac Formation and related strata, Escuminac Bay, Quebec. *Geological Society of America Special Paper*, **106**, 241–264.

DICKINSON, W. R. 1974. Plate tectonics and sedimentation. *In*: DICKINSON, W. R. (ed.) *Tectonics and Sedimentation*. SEPM Special Publications, **22**, 1–27.

EARP, J. R. & HAINS, B. A. 1971. *The Welsh Borderland*, British Regional Geology handbook series, 3rd edn. Institute of Geological Sciences, London.

ELDERS, C. F.1987. The provenance of granite boulders in conglomerates of the Northern and Central Belts of the Southern Uplands of Scotland. *Journal of the Geological Society, London*, **144**, 853–863

EMBRY, A. F. 1988, Middle–Upper Devonian sedimentation in the Canadian Arctic Islands and the Ellesmerian orogeny. *In*: MCMILLAN, N. J., EMBRY, A. F. & GLASS, D. J. (eds) *The Devonian of the World, Volume II*. Canadian Society of Petroleum Geologists, Memoir **14**(2), 15–28

ENFIELD, M. A. & COWARD, M. P. 1987. The structure of the West Orkney Basin, northern Scotland. *Journal of the Geological Society, London*, **144**, 871–884.

FAILL, R. T. 1985. The Acadian orogeny and the Catskill Delta. *In*: WOODROW, D. L. & SEVON, W. D. (eds) *The Catskill Delta*. Geological Society of America, Special Papers, **201**, 15–37.

FLEMINGS, P. B. & JORDAN, T. E. 1989. A synthetic model of foreland basin development. *Journal of Geophysical Research*, **94**, 3851–3866.

FOSSEN, H. 1992. The role of extensional tectonics in the Caledonides of southern Norway. *Journal of Structural Geology*, **14**, 1003–1046.

FRANKE, W. 1989. Tectonostratigraphic units in the Variscan belt of Central Europe. *In*: DALLMEYER, R. D. (ed.) *Terranes in the Circum-Atlantic Palaeozoic Orogens*. Geological Society of America, Special Papers, **230**, 221–228.

—— & ONCKEN, O. 1995. Zur prädevonischen Geschichte des Rheno hercynischen Beckens. *Nova Acta Leopoldina, NF71*, **291**, 53–72.

FRIEDMANN, S. J. & BURBANK, D. W. 1995. Rift basins and supradetachment basins: intracontinental extensional end-members. *Basin Research*, **7**, 109–127.

FRIEND, P. F. 1969. Tectonic features of Old Red sedimentation in North Atlantic borders. *In*: KAY, M. (ed.) *North Atlantic Geology and Continental Drift*. American Association of Petroleum Geologists, Memoirs, **12**, 703–710.

—— 1981. Devonian sedimentary basins and deep faults of the northernmost Atlantic borderlands. *In*: KERR, J. W. & FERGUSSON, A. J. (eds) *Geology of the North Atlantic Borderlands*. Canadian Society of Petroleum Geologists, Memoirs, **7**, 149–165.

—— 1985. Molasse basins of Europe: a tectonic assessment. *Transactions of the Royal Society of Edinburgh*, **76**, 451–462.

—— 1998. General form and age of the denudation pattern of the Himalaya. *GGF*, **120**, 231–236

—— & MOODY-STUART, M. 1972. Sedimentation of the Wood Bay Formation (Devonian) of Spitsbergen: regional analysis of a late orogenic basin. *Norsk Polarinstitutt Skrifter*, **157**, 1–77.

——, ALEXANDER-MARRACK, P. D., ALLEN, K. C., NICHOLSON, J. & YEATS, A. K. 1983 Devonian sediments of East Greenland, VI, Review of results. *Meddelelser om Gronland*, **206**(6), 1–96.

——, ALLEN, J. R. L. & DINELEY, D. L. 1967. The Old Red Sandstone of North America and N.W. Europe. *International Symposium on Devonian System*, Calgary, **1**, 68–98.

GEE, D. G. 1978. Nappe displacement in the Scandinavian Caledonides. *Tectonophysics*, **47**, 393–419.

GORDON, E. A. & BRIDGE, J. S. 1987. Evolution of Catskill (Upper Devonian) rivers systems: intra- and extrabasinal controls. *Journal of Sedimentary Petrology*, **57**, 234–248.

GRAHAM, J. R. 1981. The 'Old Red Sandstone' of County Mayo, northwest Ireland. *Geological Journal*, **16**, 157–173.

—— 1983. Analysis of the Upper Devonian Munster Basin, an example of a fluvial distributary system. *In*: COLLINSON, J. D. & LEWIN, J. (eds) *Modern and Ancient Fluvial Systems*. International Association of Sedimentologists, Special Publications, **6**, 473–483.

—— & CLAYTON, G. 1988. Devonian rocks in Ireland and their relation to adjacent regions. *In*: MCMILLAN, N. J., EMBRY, A. F. & GLASS, D. J. (eds) *The Devonian of the World, Volume 1*. Canadian Society of Petroleum Geologists, Memoirs, **14**, 325–340.

——, Richardson, J. B. & Clayton, G. 1982. Age and significance of the Old Red Sandstone around Clew Bay, NW Ireland. *Transactions of the Royal Society of Edinburgh: Earth Sciences*, **73**, 245–249.

——, Russell, K. J. & Stillman, C. J. 1995. Late Devonian magmatism in west Kerry and its relationship to the development of the Munster Basin. *Irish Journal of Earth Sciences*, **14**, 7–23.

Griffing, D. H., Bridge, J. S. & Hotton, C. L. 2000. Coastal–fluvial palaeoenvironments and plant palaeoecology of the Early Devonian (Emsian). Gaspé Bay, Quebec, Canada. *This volume.*

Haller, J. 1971. *Geology of the East Greenland Caledonides.* Wiley, New York.

Harland, W. B. 1967. The tectonic evolution of the Arctic–North Atlantic region. *Philosophical Transactions of the Royal Society, London, Series A*, **258**, 59–75.

—— 1971. Tectonic transpression in Caledonian Spitsbergen. *Geological Magazine*, **108**, 27–42.

—— (ed.) 1997. *The Geology of Svalbard.* Geological Society, London, Memoirs, **17**, 1–521.

Hartz, E. H. 2000. Early syndepositional tectonics of East Greenland's Old Red Sandstone basin. *This volume.*

—— & Andresen, A. 1995. Caledonian sole thrust of central East Greenland: a crustal-scale Devonian extensional detachment? *Geology*, **23**, 637–640.

Haughton, P. D. W. 1988. A cryptic Caledonian flysch terrane in Scotland. *Journal of the Geological Society, London*, **145**, 685–703.

—— 1989. Structure of some Lower Old Red Sandstone conglomerates, Kincardineshire, Scotland: deposition from late orogenic antecedent streams? *Journal of the Geological Society, London*, **146**, 509–525.

—— 1993. Simultaneous dispersal of volcaniclastic and non-volcanic sediment in fluvial basins: examples from the Lower Old Red Sandstone, east–central Scotland. *In*: Marzo, M. & Puigdefabregas, C. (eds) *Alluvial Sedimentation.* International Association of Sedimentologists, Special Publications, **17**, 451–471.

—— & Bluck, B. J. 1988. Diverse alluvial sequences from the Lower Old Red Sandstone of the Strathmore region, Scotland—implications for the relationship between late Caledonian tectonics and sedimentation. *In*: McMillan, N. J., Embry, A. F. & Glass, D. J. (eds) *The Devonian of the World, Volume 2.* Canadian Society of Petroleum Geologists, Memoirs, **14**, 269–293.

—— & Farrow, C. M. 1989. Compositional variation in Lower Old Red Sandstone detrital garnets from the Midland Valley of Scotland and the Anglo-Welsh Basin. *Geological Magazine*, **126**, 373–396.

Hellinger, S. J. & Sclater, J. G. 1983. Some comments on two-layer extensional models for the evolution of sedimentary basins. *Journal of Geophysical Research*, **88**, 8251–8269.

Hempton, M. R. & Dunne, L. A. 1983. Sedimentation in pull-apart basins: active examples in eastern Turkey. *Journal of Geology*, **92**, 513–530.

Henriksen, N. 1985. The Caledonides of central East Greenland, 70–76°N. *In*: Gee, D. G. & Sturt, B. A. (eds) *The Caledonide Orogen, Scandinavia and Related Areas.* Wiley, Chichester, 1095–1113.

Hesse, R. & Sawh, H. 1992. Geology and sedimentology of the Upper Devonian Escuminac Formations, Quebec, and evaluation of its paleoenvironment: lacustrine versus estuarine Turbidite sequence. *Atlantic Geology*, **28**, 257–275.

Hibbard, J. P., Van Staal, C. R. & Cawood, P. A. (eds) 1996. *Current Perspectives in the Appalachian–Caledonian Orogen.* Geological Association of Canada, Special Papers, **41**.

Higgs, K. T. 1999. Early Devonian spore assemblages from the Dingle Group, County Kerry, Ireland. *Bolletino della Societa Paleontologica Italiana*, **38**, 187–197.

——, MacCarthy, I. A. J. & O'Brien, M. M. 2000. A mid-Frasnian marine incursion into the southern part of the Munster basin; evidence from the Foilcoagh Bay Beds, Sherkin Formation, southernmost County Cork, Ireland. *This volume.*

Hillier, R. D. 2000. Silurian marginal marine sedimentation and the anatomy of the marine–Old Red Sandstone transition in Pembrokeshire, Southwest Wales. *This volume.*

Holder, M. J. & Leveridge, B. E. 1986. A model for the tectonic evolution of south Cornwall. *Journal of the Geological Society, London*, **143**, 125–134.

Holland, C. H. 1987. Stratigraphical and structural relationships of the Dingle Group (Silurian), County Kerry, Ireland. *Geological Magazine*, **124**, 33–42.

Horne, R. R. 1974. The lithostratigraphy of the Late Silurian to Early Carboniferous of the Dingle peninsula, Co. Kerry. *Geological Survey of Ireland Bulletin*, **1**, 395–428.

Hossack, J. R. 1984. The geometry of listric growth faults in the Devonian Basins of Sunnfjord, W. Norway. *Journal of Geological Society, London*, **141**, 629–637.

House, M. R., Richardson, J. B., Chaloner, W. G., Allen, J. R. L., Holland, C. H. & Westoll, T. S. 1977. *A correlation of Devonian rocks of the British Isles.* Geological Society, London Special Report **8**.

Hutton, D. H. W. 1987. Strike-slip terranes and a model for the evolution of the British Caledonides. *Geological Magazine*, **124**, 405–425.

—— 1989. Pre-Alleghenian terrane tectonics in the British and Irish Caledonides. *In*: Dallmeyer, R. D. (ed.) *Terranes in the Circum-Atlantic Paleozoic Orogens.* Geological Society of America, Special Papers, **230**, 47–57.

James, D. M. D. 1987. Tectonics and sedimentation in the Lower Palaeozoic back-arc basin of South Wales, UK: some quantitative aspects of basin development. *Norsk Geologisk Tidsskrift*, **67**, 419–426.

Jarvis, G. T. & McKenzie, D. P. 1980. The development of sedimentary basins with finite extension rates. *Earth and Planetary Science Letters*, **48**, 42–52.

JOHNSON, J. G., KLAPPER, G. & SANDBERG, C. A. 1985. Devonian eustatic fluctuations in Euramerica. *Geological Society of America Bulletin*, **96**, 567–587.

JORDAN, T. E. 1995. Retroarc foreland and related basins. *In*: BUSBY, C. J. & INGERSOLL, R. V. (eds) *Tectonics of Sedimentary Basins*. Blackwell Science, Oxford, 331–361.

KARNER, G. D. & WATTS, A. B. 1983. Gravity anomalies and flexure of the lithosphere at mountain ranges. *Journal of Geophysical Research*, **88**(B12), 10449–10477.

KENT, D. V. & KEPPIE, J. D. 1988. Silurian–Permian paleocontinental reconstructions and circum-Atlantic tectonics. *In*: HARRIS, A. L. & FETTES, D. J. (eds) *The Caledonian–Appalachian Orogen*. Geological Society, Special Publications **38**, 469–480.

KEPPIE, J. D. 1989. Northern Appalachian terranes and their accretionary history. *In*: DALLMEYER, R. D. (ed.) *Terranes in the Circum-Atlantic Paleozoic Orogens*. Geological Society of America, Special Papers, **230**, 159–192.

KING, L. M. 1994. Subsidence analysis of Eastern Avalonian sequences: implications for Iapetus closure. *Journal of the Geological Society, London*, **151**, 647–657.

KNELLER, B. C. 1991. A foreland basin on the southern margin of Iapetus. *Journal of the Geological Society, London*, **148**, 207–210.

KRABBENDAM, M. & DEWEY, J. F. 1998. Exhumation of UHP rocks by transtension in the Western Gneiss Region, Scandinavian Caledonides. *In*: HOLDSWORTH, R. E., STRACHAN, R. A. & DEWEY, J. F. (eds) *Continental Transpressional and Transtensional Tectonics*. Geological Society, London, Special Publications, **135**, 159–181.

KUSZNIR, N. J. & ZIEGLER, P. A. 1992. The mechanics of continental extension and sedimentary basin formation: a simple-shear/pure-shear flexural cantilever model. *Tectonophysics*, **215**, 117–131.

——, KARNER, G. D. & EGAN, S. S. 1987. Geometric, thermal and isostatic consequences of detachments in continental lithosphere extension and basin formation. *In*: BEAUMOUNT, C. & TANKARD, A. J. (eds) *Sedimentary basins and basin-forming mechanisms*. Canadian Society of Petroleum Geologists, Memoirs, **12**, 185–203.

LARSEN, P.-H. & BENGAARD, H.-J. 1991. Devonian basin initiation in East Greenland: a result of sinistral wrench faulting and Caledonian extensional collapse. *Journal of the Geological Society, London*, **148**, 355–368.

LAWRENCE, D. A. & RUST, B. R. 1988. The Devonian clastic wedge of eastern Gaspé and the Acadian orogeny. *In*: MCMILLAN, N. J., EMBRY, A. F. & GLASS, D. J. (eds) *The Devonian of the World, Volume*. Canadian Society of Petroleum Geologists, Memoirs, **14**, 53–64.

—— & WILLIAMS, B. P. J. 1987. Evolution of drainage systems in response to Acadian deformation: the Devonian Battery Point Formation, eastern Canada. *In*: ETHRIDGE, F. G., FLORES, R. M. & HARVEY, M. D. (eds) *Recent Developments in Fluvial Sedimentology*. Society of Economic Paleontologists and Mineralogists, Special Publications, **39**, 269–277.

LEAKE, B. E. 1990. Granite magmas: their sources, initiation and consequences of emplacement. *Journal of the Geological Society, London*, **147**, 579–589.

LEEDER, M. R. 1973. Sedimentology and palaeogeography of the Upper Old Red Sandstone in the Scottish Border Basin. *Scottish Journal of Geology*, **9**, 117–144.

—— 1976. Palaeogeographic significance of pedogenic carbonates in the topmost Upper Old Red Sandstone of the Scottish border basin. *Geological Journal*, **11**, 21–28.

—— 1982. Upper Palaeozoic basins of the British Isles—Caledonide inheritance versus Hercynian plate margin processes. *Journal of the Geological Society, London*, **139**, 479–491.

—— 1988*a*. Recent developments in Carboniferous geology: a critical review with implications for the British Isles and N.W. Europe. *Proceedings of the Geologists' Association*, **99**, 73–100.

—— 1988*b*. Devono-Carboniferous river systems and sediment dispersal from the orogenic belts and cratons of NW Europe. *In*: HARRIS, A. L. & FETTES, D. J. (eds) *The Caledonian–Appalachian Orogen*. Geological Society, London, Special Publications, **38**, 549–558.

LE PICHON, X., SIBUET, J. C. & FRANCHETEAU, J. 1977. The fit of the continents around the North Atlantic Ocean. *Tectonophysics*, **38**, 169–209.

LISTER, G. S. & DAVIS, G. A. 1989. The origin of metamorphic core complexes and detachment faults formed during Tertiary continental extension in the northern Colorado River region, USA. *Journal of Structural Geology*, **11**, 65–94.

LOVE, S. E. & WILLIAMS, B. P. J. 2000. Sedimentology, cyclicity and floodplain architecture in the Lower Old Red Sandstone of SW Wales. *This volume*.

MARSHALL, J. D. 2000*a*. Sedimentology of a Devonian fault-bounded braidplain and lacustrine fill in the lower part of the Skrinkle Sandstones, Dyfed, Wales. *Sedimentology*, **47**, 325–342.

—— 2000*b*. Fault-bounded basin-fill: fluvial response to tectonic controls in the Skrinkle Sandstone of SW Pembrokeshire, Wales. *This volume*.

MARSHALL, J. E. A. 1991. Palynology of the Stonehaven Group, Scotland: evidence for a Mid Silurian age and its geological implications. *Geological Magazine*, **128**, 283–286.

—— 2000. Devonian (Givetian) miospores from the Walls Group, Shetland. *This volume*.

——, ROGERS, D. A. & WHITELEY, M. J. 1996. Devonian marine incursions into the Orcadian Basin, Scotland. *Journal of the Geological Society, London*, **153**, 451–466.

MATTE, P. 1986. Tectonics and plate tectonic model for the Variscan belt of Europe. *Tectonophysics*, **126**, 329–374.

—— 1991. Accretionary history and crustal evolution of the Variscan belt in Western Europe. *Tectonophysics*, **196**, 309–337.

——, Maluski, H., Rajlich, P. & Franke, W. 1990. Terrane boundaries in the Bohemian Massif: result of large-scale Variscan shearing. *Tectonophysics*, **177**, 151–170.

McClay, K. R., Norton, M. G., Coney, P. & Davis, G. H. 1986. Collapse of the Caledonian orogen and the Old Red Sandstone. *Nature*, **323**, 147–149.

McKenzie, D. (1978). Some remarks on the development of sedimentary basins. *Earth and Planetary Science Letters*, **40**, 25–32.

McKerrow, W. S. 1988. Wenlock to Givetian deformation in the British Isles and the Canadian Appalachians. *In*: Harris, A. L. & Fettes, D. J. (eds) *The Caledonian–Appalachian Orogen*. Geological Society, London, Special Publications, **38**, 437–448.

—— & Soper, N. J. 1989. The Iapetus Suture in the British Isles. *Geological Magazine*, **126**, 1–8.

—— & Ziegler, A. M. 1972. Palaeozoic oceans. *Nature (Physical Sciences)*, **240**, 92–94.

McCann, A. J. 2000. Deformation of the Old Red Sandstone of NW Spitsbergen; links to the Ellesmerian and Caledonian orogenies. *This volume.*

McClay, K. R., Norton, M. G., Coney, P. & Davis, G. H. 1986. Collapse of the Caledonide orogen and the Old Red Sandstone. *Nature*, **343**, 147–149.

McSherry, M., Parnell, J., Leslie, A. G. & Haggan, T. 2000. Depositional and structural setting of the (?)Lower Old Red Sandstone sediments of Ballymastocker, Co. Donegal. *This volume.*

Milnes, A. G., Wennberg, O. P., Skar, O. & Koestler, A. G. 1997. Contraction, extension and timing in the South Norwegian Caledonides: the Sognefjord transect. *In*: Burg, J.-P. & Ford, M. (eds) *Orogeny through Time*. Geological Society, London, Special Publications, **121**, 123–148.

Mitchell, W. I. & Owens, B. 1990. The geology of the western part of the Fintona Block, Northern Ireland: evolution of Carboniferous basins. *Geological Magazine*, **127**, 407–426.

Murphy, F. C. 1985. Non-axial planar cleavage and Caledonian sinistral transpression in eastern Ireland. *Geological Journal*, **20**, 257–279.

——, Anderson, T. B., Daly, J. S. *et al.* 1991. An appraisal of Caledonian suspect terranes in Ireland. *Irish Journal of Earth Sciences*, **11**, 11–41.

Mykura, W. 1976. *Orkney and Shetland.* British Regional Geology. Institute of Geological Sciences, London.

Norton, M. G., McClay, K. R. & Way, N. A. 1987. Tectonic evolution of Devonian basins in northern Scotland and Southern Norway. *Norsk Geologisk Tidsskrift*, **67**, 323–338.

Osmundsen, P. T., Andersen, T. B., Markussen, S. & Svendby, A. K. 1998. Tectonics and sedimentation in the hanging wall of a major extensional detachment: the Devonian Kvamshesten Basin, western Norway. *Basin Research*, **10**, 213–234.

——, Bakke, B., Svendby, A. K. & Andersen. T. B. 2000. Architecture of the Middle Devonian Kvamshesten Group, Western Norway: sedimentary response to deformation above a ramp-flat extensional fault. *This volume.*

Parsons, B. & Sclater, J. G. 1977. An analysis of the variation of ocean floor bathymetry with heat flow and age. *Journal of Geophysical Research*, **82**, 803–827.

Piepjohn, K. 2000. The Svalbardian–Ellesmerian deformation of the Old Red Sandstone and the pre-Devonian basement in NW Spitsbergen (Svalbard). *This volume.*

——, Brinkman, L., Grewing, A. & Karp, H. 2000. New data on the age of the uppermost ORS and the lowermost post-ORS strata in Dickson Land and implications for the age of the Svalbardian deformation. *This volume.*

Plink-Bjorklund, P. & Bjorklund, L. 1999. Sedimentary record of post-collisional events of the Scandinavian Caledonides in the Baltic Devonian Basin in Estonia. *Journal of Conference Abstracts* **4**(1), 286

Powell, C. M. 1987. Inversion tectonics in S.W. Dyfed. *Proceedings of the Geologists' Association*, **98**, 193–203.

—— 1989. Structural controls on Palaeozoic basin evolution and inversion in southwest Wales. *Journal of the Geological Society, London*, **146**, 439–446.

Price, R. A. 1973. Large-scale gravitational flow of supra-crustal rocks, southern Canadian Rockies. *In*: De Jong, K. A. & Scholten, R. (eds) *Gravity and Tectonics*. Wiley, New York.

Rey, P., Burg, J.-P. & Casey, M. 1997. The Scandinavian Caledonides and their relationship to the Variscan belt. *In*: Burg, J.-P. & Ford, M. (eds) *Orogeny through Time*. Geological Society, London, Special Publications, **121**, 179–200.

Richmond, L. K. & Williams, B. P. J. 2000. A new terrane in the Old Red Sandstone of the Dingle Peninsula, SW Ireland. *This volume.*

Roberts, A. M. & Holdsworth, R. E. 1999. Linking onshore and offshore structures: Mesozoic extension in the Scottish Highlands. *Journal of the Geological Society, London*, **156**, 1061–1064.

Roberts, D. & Gee, D. G. 1985. An introduction to the structure of the Scandinavian Caledonides. *In*: Gee, D. G. & Sturt, B. A. (eds) *The Caledonide Orogen: Scandinavia and Related Areas*. Wiley, Chichester, 55–68.

Rogers, D. A., Marshall, J. E. A. & Astin, T. R. 1989. Devonian and later movements on the Great Glen Fault System. *Journal of the Geological Society, London*, **146**, 369–372.

Rust, B. R. 1981. Alluvial deposits and tectonic style: Devonian and Carboniferous successions in eastern Gaspé. *In*: Miall, A. D. (ed.) *Sedimentation and Tectonics in Alluvial Basins*. Geological Association of Canada, Special Papers, **23**, 49–76.

—— 1984. Proximal braidplain deposits in the Middle Devonian Malbaie Formation of Eastern Gaspé, Quebec, Canada. *Sedimentology*, **31**, 675–695.

——, Lawrence, D. A. & Zaitlin, B. A. 1989. The sedimentology and tectonic significance of

Devonian and Carboniferous terrestrial successions in Gaspé, Quebec. *Atlantic Geology*, **25**, 1–13.

RYAN, P. D. & DEWEY, J. F. 1991. A geological and tectonic cross section of the Caledonides of western Ireland. *Journal of the Geological Society, London*, **151**, 579–582.

SALVESON, J. O. 1978. Variations in the geology of rift basins—a tectonic model. Paper presented at Rio Grande Rift Symposium, Santa Fe, NM.

SCOTESE, C. R. & MCKERROW, W. S. 1990. Revised world maps and introduction. *In*: MCKERROW, W. S. & SCOTESE, C. R. (eds) *Palaeozoic Palaeogeography and Biogeography*. Geological Society, London, Memoirs, **12**, 1–21.

SEAGO, R. D. & CHAPMAN, T. J. 1988. The confrontation of structural styles and the evolution of the foreland basin in central SW England. *Journal of the Geological Society, London*, **145**, 789–800

SERRANE, M. 1992. Devonian extensional tectonics versus Carboniferous inversion in the northern Orcadian basin. *Journal of the Geological Society, London*, **149**, 27–37.

—— & SEGURET, M. 1987. The Devonian basins of western Norway: tectonics and kinematics of an extending crust. *In*: COWARD, M. P., DEWEY, J. F. & HANCOCK, P. L. (eds) *Continental Extensional Tectonics*. Geological Society, London, Special Publications, **28**, 537–548.

——, CHAUVET, A. & FAURE, J.-L. 1991. Cinématique de l'extension tardi-orogénique (Dévonien) dans les Caledonides Scandinaves et Britanniques. *Comptes Rendus de l'Academie de Sciences, Série II*, **313**, 1305–1312.

SIEDLECKA, A. 1975. Old Red Sandstone lithostratigraphy and sedimentation of the Outer Fosen area, Trondheim Region. *Norges Geologiske Undersokelse*, **321**, 1–35.

SIMON, J. B. 1984. Sedimentation and tectonic setting of the Lower Old Red Sandstone of the Fintona and Curlew Mountain districts. *Irish Journal of Earth Sciences*, **6**, 213–228.

—— & BLUCK, B. J. 1982. Palaeodrainage of the southern margin of the Caledonian mountain chain in the northern British Isles. *Transactions of the Royal Society of Edinburgh: Earth Sciences*, **73**, 11–15.

SMITH, S. A. & HUMPHREYS, B. 1989. Lakes and alluvial sandflat–playas in the Dartmouth Group, southwest England. *Proceedings of the Ussher Society*, **7**, 112–117.

—— & 1991. Sedimentology and depositional setting of the Dartmouth Group, Bigbury Bay, south Devon. *Journal of the Geological Society, London*, **148**, 235–244.

SOPER, N. J. 1986. The Newer Granite problem: a geotectonic view. *Geological Magazine*, **123**, 227–236.

—— & HUTTON, D. H. W. 1984. Late Caledonian sinistral displacements in Britain: implications for a three plate collision model. *Tectonics*, **3**, 781–794.

—— & WOODCOCK, N. H. 1990. Silurian collision and sediment dispersal patterns in southern Britain. *Geological Magazine*, **127**, 527–542.

——, RYAN, P. D. & DEWEY, J. F. 1999. Age of the Grampian orogeny in Scotland and Ireland. *Journal of the Geological Society, London*, **156**, 1231–1236.

——, WEBB, B. C. & WOODCOCK, N. H. 1987. Late Caledonian (Acadian) transpression in north-west England: timing, geometry and geotectonic significance. *Proceedings of the Yorkshire Geological Society*, **46**, 175–192.

STEEL, R., SIEDLECKA, A. & ROBERTS, D. 1985. The Old Red Sandstone basins of Norway and their deformation: a review, Part 1. *In*: GEE, D. G. & STURT, B. A. (eds) *The Caledonide Orogen, Scandinavia and Related Areas*. Wiley, Chichester, 293–315.

STEEL, R. J., MŒHLE, S., NILSEN, H., RØE, S. L. & SPINNANGR, Å. 1977. Coarsening-upward cycles in the alluvium of Hornelen Basin (Devonian) Norway: sedimentary response to tectonic events. *Geological Society of America Bulletin*, **88**, 1124–1134.

STOCKMAL, G. S. & BEAUMONT, C. 1987. Geodynamic models of convergent margin tectonics: the southern Canadian Cordillera and the Swiss Alps. Canadian Society of Petroleum Geologists, Memoirs, **12**, 393–412.

STRACHAN, R. A., HOLDSWORTH, R. E., FRIDERICHSEN, J. D. & JEPSEN, H. F. 1992. Regional Cadedonian structure within an oblique convergence zone, Dronning Louise Land, NE Greenland. *Journal of the Geological Society, London*, **149**, 359–371.

TANKARD, A. J. 1986. On the depositional response to thrusting and lithospheric flexure: examples from the Appalachian and Rocky Mountain basins. *In*: ALLEN, P. A. & HOMEWOOD, P. (eds) *Foreland Basins*. International Association of Sedimentologists, Special Publications, **8**, 369–392.

THIRLWALL, M. F. 1981. Implications for Caledonian plate tectonic models of chemical data from volcanic rocks of the British Old Red Sandstone. *Journal of the Geological Society, London*, **145**, 951–967.

—— 1988. Geochronology of Late Caledonian magmatism in northern Britain. *Journal of the Geological Society, London*, **145**, 951–967.

—— 1989. Movement on proposed terrane boundaries in northern Britain: constraints from Ordovician–Devonian igneous rocks. *Journal of the Geological Society, London*, **146**, 373–376.

TODD, S. P. 1989. Role of the Dingle Bay Lineament in the evolution of the Old Red Sandstone of southwest Ireland. *In*: ARTHURTON, R. S., GUTTERIDGE, P. & NOLAN, S. C. (eds) *The Role of Tectonics in Devonian and Carboniferous Sedimentation in the British Isles*. Occasional Publication of the Yorkshire Geological Society, **6**, 35–54.

—— 1991. The Silurian rocks of Inishnabro, Blasket Islands, County Kerry and their regional significance. *Irish Journal of Earth Sciences*, **11**, 91–98.

—— 2000. Taking the roof off a suture zone: basin setting and provenance of conglomerates in the ORS Dingle Basin of SW Ireland. *This volume.*

——, BOYD, J. D. & DODD, C. D. 1988. Old Red Sandstone sedimentation and basin development in the Dingle Peninsula, southwest Ireland. *In*: MCMILLAN, N. J., EMBRY, A. F. & GLASS, D. J. (eds) *The Devonian of the World, Volume II*. Canadian Society of Petroleum Geologists, Memoirs, **14**, 251–268.

TORSVIK, T. H., SMETHURST, M. A., MEERT, J. G. *et al.* 1996. Continental break-up and collision in the Neoproterozoic and Palaeozoic—a tale of Baltica and Laurentia. *Earth Science Reviews*, **40**, 229–258.

TUCKER, R. D., BRADLEY, D. C., VER STRAETEN, C. A., HARRIS, A. G., EBERT, J. R. & MCCUTCHEON, S. R. 1998. New U–Pb zircon ages and the duration and division of Devonian time. *Earth and Planetary Science Letters*, **158**, 175–186.

TUNBRIDGE, I. P. 1980. The Yes Tor Member of the Hangman Sandstone Group (North Devon). *Proceedings of the Ussher Society*, **5**, 7–12.

—— 1984. Facies model for a sandy ephemeral stream and clay playa complex; the Middle Devonian Trentishoe Formation of North Devon, UK *Sedimentology*, **31**, 697–715.

—— 1986. Mid-Devonian tectonics and sedimentation in the Bristol Channel area. *Journal of the Geological Society, London*, **143**, 107–115.

TURCOTTE, D. L. & SCHUBERT, G. 1982. *Geodynamics: Applications of Continuum Mechanics to Geological Problems.* Wiley, New York.

TURNER, P. & WHITTAKER, J. H. McD. 1976. Petrology and provenance of late Silurian fluviatile sandstones from the Ringerike Group of Norway. *Sedimentary Geology*, **16**, 45–68.

VAN DEN DRIESSCHE & BRUN, J.-P. 1991–1992. Tectonic evolution of the Montagne Noire (French Massif Central): a model of extensional gneiss dome. *Geodinamica Acta*, **5**, 85–99.

WATTS, A. B. & COCHRAN, J. R. 1974. Gravity anomalies and flexure of the lithosphere along the Hawaiian–Emperor seamount chain. *Geophysical Journal of the Royal Astronomical Society*, **38**, 119–141.

WERNICKE, B. 1985. Uniform sense simple shear of the continental lithosphere. *Canadian Journal of Earth Sciences*, **22**, 108–125.

WHITE, N. J. & MCKENZIE, D. P. 1988. Formation of the 'steer's head' geometry of sedimentary basins by differential stretching of the crust and mantle. *Geology* **16**, 250–253

WILCOX, R. E., HARDING, P. & SEELY, D. R. 1973. Basic wrench tectonics. *AAPG Bulletin*, **57**, 74–96.

WILKS, W. J. & CUTHBERT, S. J. 1994. The evolution of the Hornelen basin detachment system, western Norway: implications for the style of late orogenic extension in the southern Scandinavian Caledonides. *Journal of Structural Geology*, **16**, 1023–1027.

WILLIAMS, B. P. J., ALLEN, J. R. L. & MARSHALL, J. D. 1982. Old Red Sandstone facies of the Pembroke peninsula, south of the Ritec Fault. *In*: BASSETT, M. G. (ed.) *Geological Excursions in Dyfed, South-west Wales*. National Museum of Wales, Cardiff, 151–174.

WILLIAMS, E. A. 2000. Flexural cantilever models of extensional subsidence in the Munster Basin (SW Ireland) and Old Red Sandstone dispersal systems. *This volume.*

——, FRIEND, P. F. & WILLIAMS, B. P. J. 2000*a*. A review of Devonian time scales: databases, construction and new data. *This volume.*

——, SERGEEV, S. A., STÖSSEL, I. & FORD, M. 1997. An Eifelian U–Pb zircon date for the Enagh Tuff Bed from the Old Red Sandstone of the Munster Basin in NW Iveragh, SW Ireland. *Journal of the Geological Society, London*, **154**, 189–193.

——, ——, ——, ——, & HIGGS, K. T. 2000*b*. U–Pb zircon geochronology of silicic tuffs and chronostratigraphy of the earliest Old Red Sandstone in the Munster Basin, SW Ireland. *This volume.*

WILLIS, B. J. & BRIDGE, J. S. 1988. Evolution of Catskill River Systems, New York State. *In*: MCMILLAN, N. J., EMBRY, A. F. & GLASS, D. J. (eds) *Devonian of the World, Volume II.* Canadian Society of Petroleum Geologists, Memoirs, **14**, 85–106.

WOODCOCK, N. H. 1986. The role of strike-slip fault systems at plate boundaries. *Philosphical Transactions of the Royal Society, London, Series A*, **317**, 13–29.

—— & FISCHER, M. W. 1986. Strike-slip duplexes. *Journal of Structural Geology* **8**, 725–735

—— & GIBBONS, W. 1988. Is the Welsh Borderland fault system a terrane boundary? *Journal of the Geological Society, London*, **145**, 915–923.

——, AWAN, M. A., JOHNSON, T. E., MACKIE, A. H. & SMITH, R. D. A. 1988. Acadian tectonics of Wales during Avalonia/Laurentia convergence. *Tectonics*, **7**, 483–495.

WOODROW, D. L. & SEVON, W. D. (eds) 1985. *The Catskill Delta.* Geological Society of America, Special Papers, **201**.

ZIEGLER, P. A. 1982. *Geological Atlas of Western and Central Europe*. Shell, The Hague.

—— 1986. Geodynamic model for the Palaeozoic crustal consolidation of western and central Europe. *Tectonophysics*, **126**, 303–328.

—— 1988. Laurussia—the Old Red Continent. *In*: MCMILLAN, N. J., EMBRY, A. F. & GLASS, D. J. (eds) *The Devonian of the World, Volume 1.* Canadian Society of Petroleum Geologists, Memoirs, **14**, 15–48.

Coastal–fluvial palaeoenvironments and plant palaeoecology of the Lower Devonian (Emsian), Gaspé Bay, Québec, Canada

DAVID H. GRIFFING[1], JOHN S. BRIDGE[1] & CAROL L. HOTTON[2]

[1]*Department of Geological Sciences, Binghamton University, Binghamton, NY 13902-6000, USA (e-mail: jbridge@binghamton.edu)*

[2]*National Center for Biotechnology Information, NCBI/NLM/NIH, Bethesda, MD 20894; and National Museum of Natural History, Smithsonian Institution, Washington, DC 20560, USA (e-mail: hotton@ncbi.nlm.nih.gov)*

Abstract: The Cap-aux-Os Member (of Emsian age) of the Battery Point Formation, Gaspé Bay, Québec, comprises coastal and fluvial deposits containing abundant, well-preserved remains of early land plants (embryophytes). Metres-thick sandstone bodies represent the deposits of the main river channels, including some that were tidally influenced. Thinner sandstone bodies within mudstone successions represent deposits of crevasse splays–levees, lacustrine deltas, tidal channels, flood-tidal deltas and washovers. Mudstone-dominant strata represent backswamps and marshes, lakes, coastal bays and tidal flats. Certain plants (trimerophytes, *Huvenia* and *Sciadophyton*) are particularly common on the moist, upper parts of near-coastal channel bars. They apparently germinated and grew rapidly on freshly exposed muds deposited during floods or exceptionally high tides. *In situ* zosterophylls are most common in backswamp and marshy areas, where they formed extensive, relatively long-lived stands. It is uncertain whether these plants were tolerant of brackish water or killed by it. However, absence of acritarchs in these *in situ* plant horizons suggests that they occupied primarily freshwater habitats. Transported *Prototaxites* and *Spongiophyton*, both interpreted as possible fungi, are the main components of fully non-marine channel deposits. It is no coincidence that the best-preserved plants are those that lived in moist, oxygen-poor settings. However, the occurrence of root traces in palaeosols and sparse fossil evidence suggests that embryophytes also occupied drier settings.

The evolution of land plants in the Silurian and Devonian periods had profound consequences for the history of life. In addition to their numerous ameliorating effects on terrestrial habitats (increased substrate stability and nutrient availability, damping of temperature and humidity extremes), land plants drew down CO_2 from the atmosphere and probably played some part in the climatic changes that took place in the latter part of Palaeozoic time (Berner 1993, 1994; Mora *et al.* 1996; Retallack 1997). Although the evolutionary history of early land plants has been under study for decades, their ecological setting remains poorly understood. Detailed studies of early land-plant palaeoecology and taphonomy are rare (Andrews *et al.* 1977; Schweitzer 1983; Edwards & Fanning 1985), and are hampered by absence of sedimentological detail. For this reason, we undertook a detailed study of the sedimentary context of well-preserved Emsian land plant assemblages from Gaspé Bay, Québec, Canada (Hotton *et al.* 2001; see also Elick *et al.* 1998). The fossil floras of the Gaspé region, studied since the time of Dawson (1859, 1871), are extensive, well preserved, and (relatively) well understood.

Our understanding of early land plant ecology is as primitive and undeveloped as the plants themselves, and is based primarily on inferences drawn from their morphology rather than on analysis of the sedimentological context of their fossil occurrence. Early land plants are believed to have been ecological generalists (exhibiting little niche specialization), to have reproduced primarily by clonal, vegetative growth, and to have been unable to occupy xeric, hydric and other types of high-stress environments because of their free-sporing habit (DiMichele *et al.* 1992). Our study of the Gaspé flora (Hotton *et al.* 2001) showed that this view

From: FRIEND, P. F. & WILLIAMS, B. P. J. (eds). *New Perspectives on the Old Red Sandstone.* Geological Society, London, Special Publications, **180**, 61–84. 0305-8719/00/$15.00

is over-simplified. By Emsian times, at least two clades of plants had apparently evolved distinct life-history strategies with distinctive suites of morphological characters adapted toward different parts of the available landscape. A complete account of the palaeoecological results has been given by Hotton *et al.* (2001), with only a brief synopsis of the sedimentological work. The present paper provides a more complete account of the sedimentological evidence. Previous descriptions of these plant-bearing strata and interpretations of their depositional setting (e.g. Lawrence 1986; Lawrence & Williams 1987; Lawrence & Rust 1988) formed the basis for our more detailed sedimentological work.

Study area and methods

The study area is on the northern shore of Gaspé Bay, Québec, Canada (Fig. 1). Strata are exposed in coastal cliffs of 20–30 m height, and dip to the SSW at between 30 and 45°. A detailed, continuous sedimentological log, encompassing 186 m of strata, was measured. Lateral exposure of the sedimentary units is limited by the stratal dips and cliff height: however, coastline physiography made it possible to measure some stratigraphic intervals in several places, thus giving more information on lateral variation of strata. Many samples were taken for laboratory

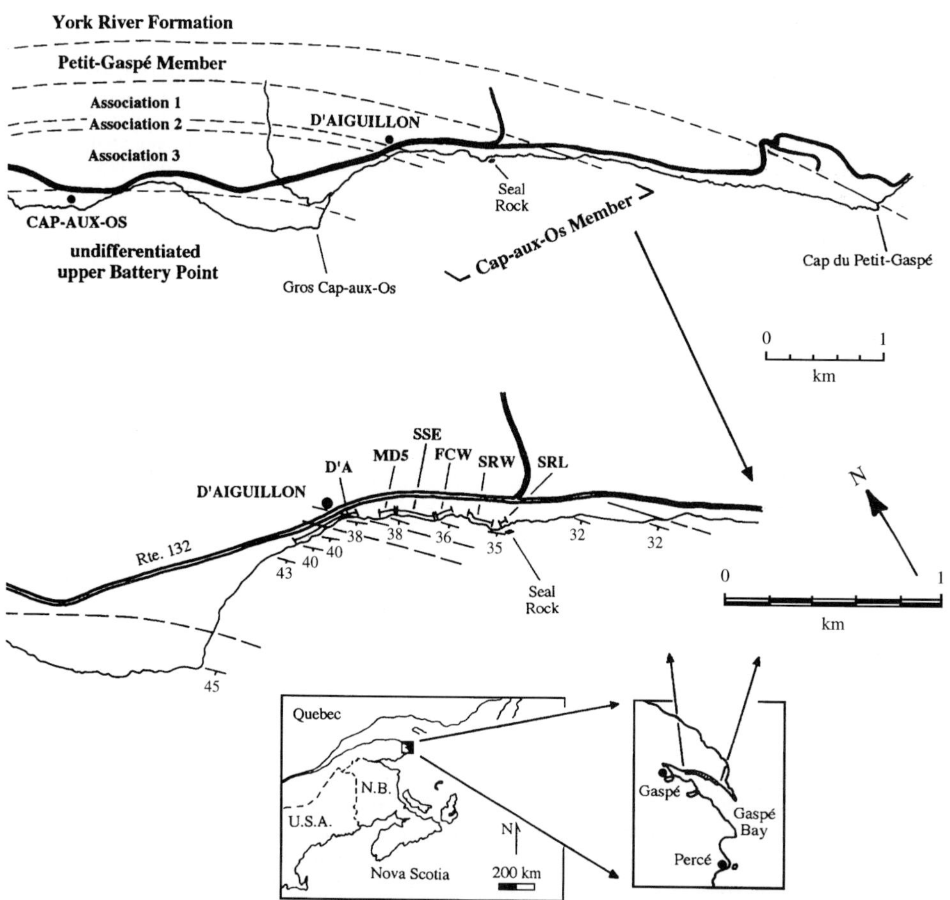

Fig. 1. Location map of the Cap-aux-Os Member exposures along the north shore of Gaspé Bay. Individual mudstone intervals and the strike and dip of strata are diplayed in the lower enlargement of the study area. SRL, easternmost exposure of Lawrence's Mudstone 4 at Seal Rock Landing; SRW, Lawrence's Mudstone 4 at west side of Seal Rock Point; FCW, westernmost exposures of Lawrence's Mudstone 4; SSE, Lawrence's Sandstone E (multistorey sandstone body); MD5, Lawrence's Mudstone 5; D'A, Lawrence's Association 2 and 3 along beach at D'Aiguillon. Modified from Lawrence (1986). (from *Plants Invade the Land*, eds. P. G. Gensel and D. Edwards. ©2001 Columbia University Press. Reprinted by permission of the publisher).

analysis (petrography, ichnology, taphonomy, palaeoecology and taxonomy).

Stratigraphic setting

The strata are part of the Catskill clastic wedge, deposited in a foreland basin to the west of the Acadian orogen (Rust 1981, 1984; Lawrence & Williams 1987; Lawrence & Rust 1988), and represent deposition in response to the early stages of the Acadian Orogeny. Although the overall basin fill is regressive, there is evidence of frequent marine transgressions of various scales (Pageau & Prichonnet 1976; Lawrence 1986). The sequence studied is close to the palaeoshoreline, and therefore contains evidence for these marine transgressions and regressions. The palaeolatitude was approximately 10–20° S in Emsian times (Ziegler 1988; Scotese & McKerrow 1990), so climate was presumably tropical (Witzke & Heckel 1988; Witzke 1990), possibly monsoonal.

Lithostratigraphic subdivision, by McGerrigle (1950) and Brisebois (1981), is shown in Fig. 2. The Battery Point Formation is divided into three members. The strata studied are from the middle, Cap-aux-Os Member. The Cap-aux-Os Member was interpreted by Lawrence & Rust (1988) as the deposits of meander belts of moderate- to high-sinuosity rivers that flowed to the north and NNW across a vegetated, muddy coastal plain adjacent to a marine embayment. Lawrence (1986) subdivided these strata into three facies associations. The lowermost, Association 1 is dominated by relatively thick, multistorey sandstones with subordinate, thinner red and grey mudstones. Association 2 comprises mainly grey and red mudstones with numerous, relatively thin sheet sandstones and less common single-storey, channel-form sandstone bodies. Association 3 is similar to Association 1, but the sandstones are coarser grained, contain greater quantities of extraformational pebbles, and nearly all mudstones are red. Our log encompasses the top part of Association 1, Association 2 and the lower part of Association 3.

The Battery Point Formation is of early Emsian to early Eifelian age based on brachiopods (Boucot *et al.* 1967) and spores (McGregor 1973, 1977; Richardson & McGregor 1986). The measured section is in the *annulatus–sextantii* spore assemblage zone of Richardson & McGregor (1986), or from early to late Emsian time.

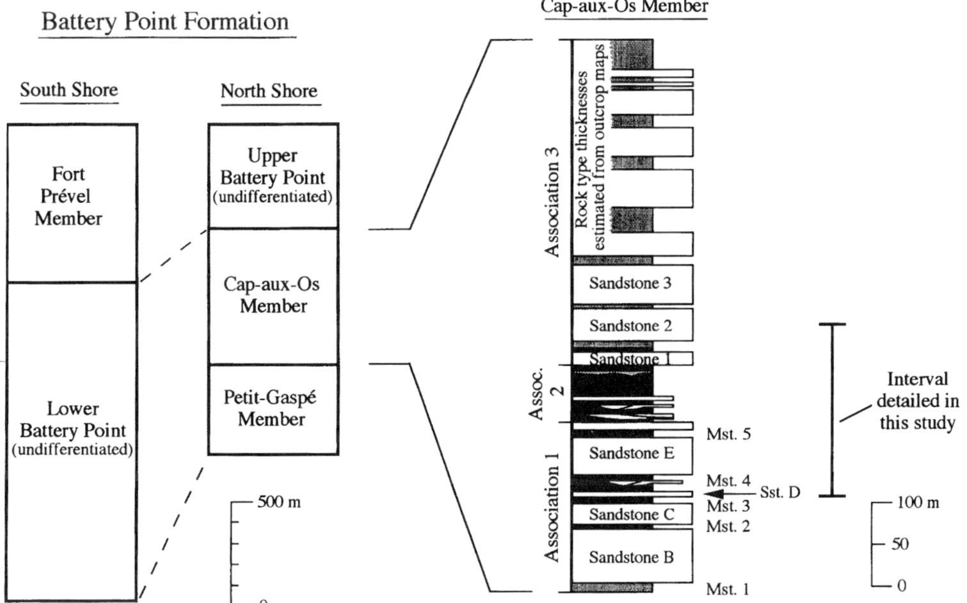

Fig. 2. Stratigraphic subdivision of the Battery Point Formation with expanded stratigraphy of major facies within the Cap-aux-Os Member (right). Black intervals represent dark grey and green–grey mudstone and shale; grey stippled intervals represent red mudstone-rich intervals. (from *Plants Invade the Land*, eds. P. G. Gensel and D. Edwards. ©2001 Columbia University Press. Reprinted by permission of the publisher).

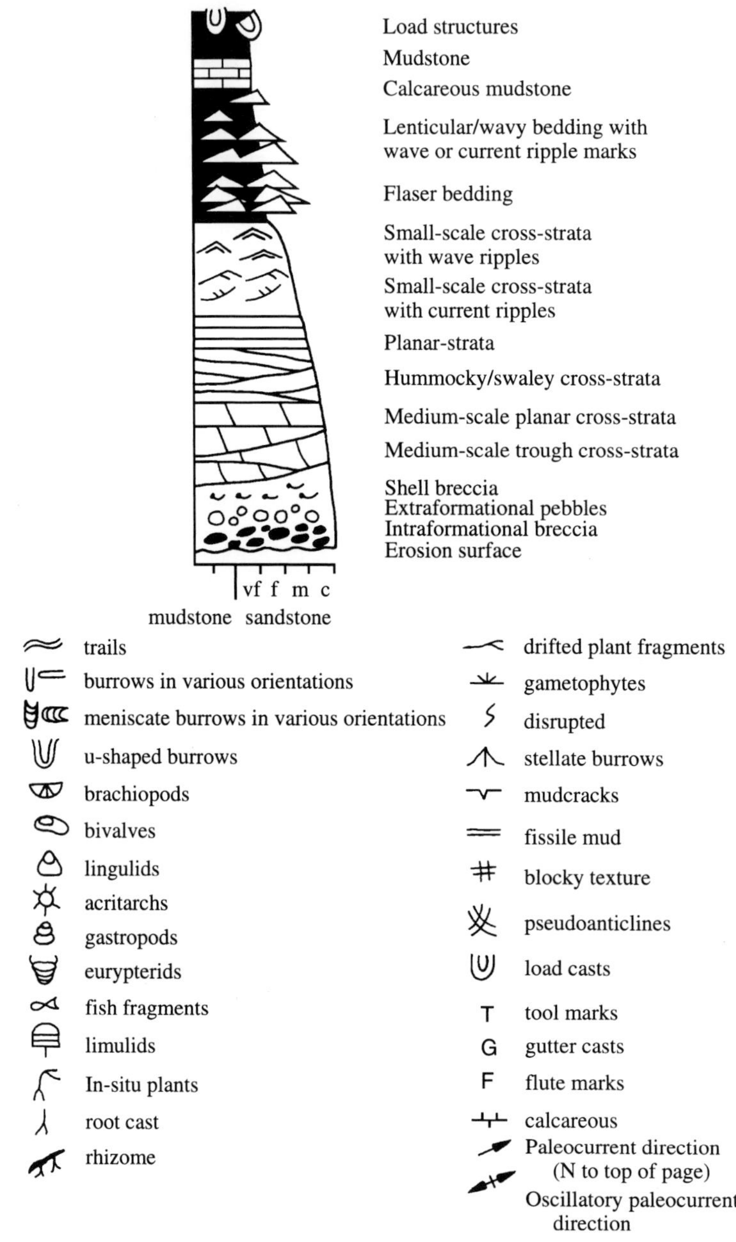

Fig. 3. Sedimentological logs, including correlated logs at SRL and FCW.

Sedimentological description and interpretation

Overview

The strata comprise metres-thick sandstone bodies with minor amounts of mudstone and metres-thick mudstone-dominant intervals that contain relatively thinner (millimetre to decimetre) sandstone strata (Fig. 3). The sandstone bodies are more resistant to weathering and erosion than the mudstones, and therefore form protrusions along the coastline. Sandstones are mainly greenish grey, and range in mean grain

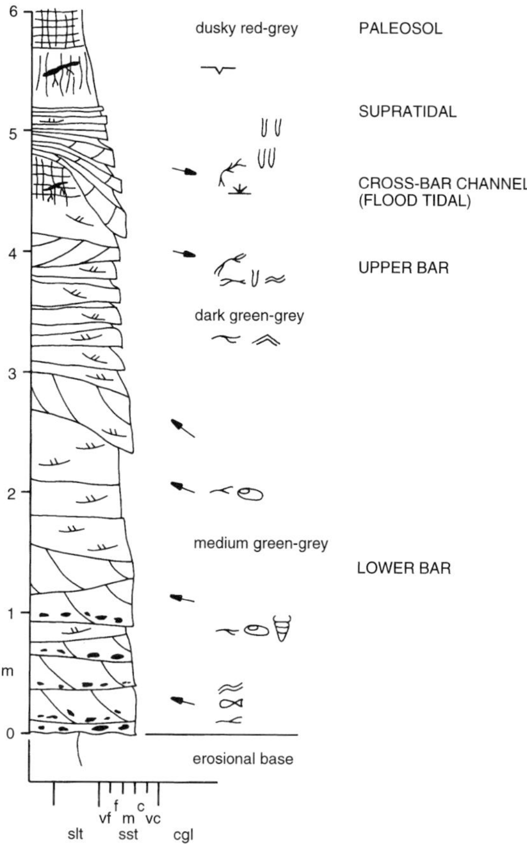

Fig. 4(a).

size from medium to very fine. The thickness, geometry and lithofacies of the sandstone bodies vary greatly (summarized in Fig. 4). Mudstones are dark greenish grey to dusky reddish grey and are mainly siltstones, with relatively rare claystones.

Thick sandstone bodies: description

Thick sandstone bodies are 4–15 m thick. They are made up of one or more sets of large-scale inclined strata (storeys) that are metres thick and tens to hundreds of metres wide (Fig. 5). Storeys have relatively major basal erosion surfaces, commonly overlain by intraformational breccia and uncommon extraformational conglomerate. Large-scale inclined strata range from centimetres to metres thick (generally decimetres), and are bounded by relatively minor erosion surfaces overlain by intraformational breccia (Fig. 4). The large-scale strata are inclined at up to 10° relative to the basal erosion surface of the sets. The largest inclinations generally are seen where palaeocurrents are perpendicular to outcrop faces. Some inclined strata fill channels. Also, relatively small channel-forms (metres to tens of metres wide and decimetres to metres thick) commonly cut through the upper parts of the storeys. Thick sandstone bodies within the mudstone-rich intervals (Lawrence's Association 2) tend to comprise a single storey, whereas those within the sandstone-rich intervals tend to comprise vertically superimposed storeys, the lower ones being truncated by the bases of the upper ones. Untruncated storeys are generally 4–5 m thick. These two types of thick sandstone body (hereafter named Type 1 (single storey) and Type 2 (multi-storey)) can also be distinguished by their internal lithofacies, as described below.

Grain size of the storeys commonly decreases upwards, from medium-grained sand to very fine grained sand (Type 1), or from very coarse or coarse-grained sand to fine- or very fine grained sand (Type 2). Some storeys show little vertical

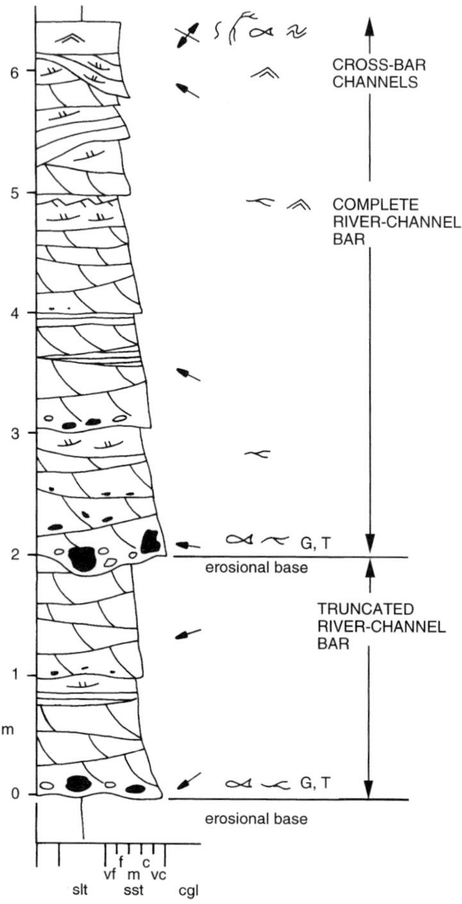

Fig. 4(b).

variation in grain size, or coarsen upwards (Fig. 3). Channel fills generally fine upward: some are dominantly sandstone (coarse-grained channel fills), but others have large proportions of siltstone (fine-grained channel fills). Grain size commonly decreases at the top of individual large-scale strata (e.g. from medium to fine sand or from fine to very fine sand).

Medium-scale trough cross strata (set thickness ranging from 0.03 to 0.5 m, but commonly 0.1–0.3 m) is the dominant internal structure of the Type 2 sandstone bodies, with subordinate small-scale trough cross strata and planar strata. Type 1 sandstone bodies and channel fills tend to have much greater proportions of small-scale trough cross strata. Small-scale cross strata and planar strata tend to occur in the finer-grained parts of the sandstone bodies. Storeys may become finer grained and thinner laterally, with an associated increase in the proportion of small-scale cross strata. Asymmetrical and symmetrical ripple marks occur in the upper, finest-grained parts of the large-scale strata, and flute, gutter and tool marks occur along their erosional bases.

Palaeocurrent directions (corrected for structural dip) derived from medium-scale and small-scale trough cross strata vary vertically and laterally within individual sandstone bodies. Palaeocurrent directions in Type 2 sandstone bodies are generally to the WNW, with most in the range from SW to north (Figs 3, 4 and 6: see also Lawrence & Rust 1988). They vary vertically within a large-scale set by up to 50°. Palaeocurrents vary between superimposed sets by up to 100°. The lower parts of Type 1 sandstone bodies have palaeocurrents similar to those of Type 2 bodies. However, palaeocurrents in the upper parts of Type 1 sandstone bodies are commonly in a near-opposite direction. Rarely, opposing

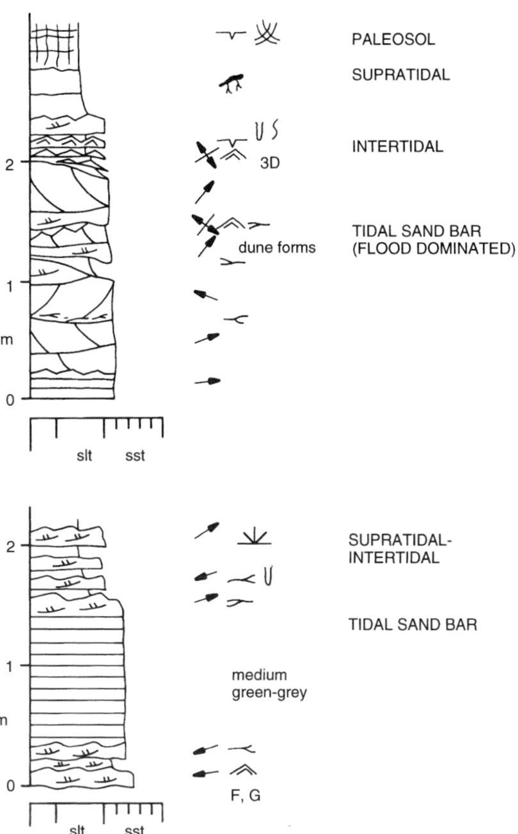

Fig. 4(c).

palaeocurrents are manifested in herringbone cross lamination. Crest-line orientations of symmetrical ripple marks tend to be parallel to unidirectional palaeocurrent indicators.

Evidence of organic activity is more common in Type 1 sandstone bodies than in Type 2, and the organic components differ. Type 1 sandstone bodies have more plant material, and have important occurrences of *in loco* (i.e. minimally transported) and less commonly *in situ* plant assemblages, especially in the upper metre or so. Although aerial axes with attached roots or rhizomes are currently unknown from the Battery Point Formation, probable isolated root tufts or rhizomes have been recognized from *in loco* plant assemblages, and *in situ* root and rhizome traces have been identified in some mudstones (Elick *et al.* 1998; this study). Evidence for *in situ* occurrence or minimal transport (*in loco*) is based primarily on attachment postures of axes, such as regular spacing of toppled axes and orientation of free branches in the palaeoflow direction, suggesting anchorage at one end. Preservation of delicate morphological details on many of these plants also suggests minimal transport. The most common plants in Type 1 sandstones are trimerophytes (*Trimerophyton–Pertica* and *Psilophyton* spp.), rhyniopsids (*Huvenia* sp. nov.), and *Sciadophyton* spp. (probable gametophytes). Zosterophylls such as *Sawdonia ornata* and *Renalia hueberi* are relatively rare. The plants occur in the finer-grained portions of centimetre- to decimetre-thick large-scale strata where mudstone caps sandstone. In places, the plants appear to be anchored in the mudstone and project upwards through the overlying sandstone bed, where they may be bent over in the palaeocurrent direction and display parallel swirling patterns (Fig. 7; see also Elick *et al.* 1998). Aerial axes of these plants range from 0.5 to 1 cm wide (rarely as much as 3 cm in *Drepanophycus*) and from 2 to 10 dm long (Fig. 7). *Sciadophyton* usually occurs *in situ* in the mud-draped troughs of ripple and dune

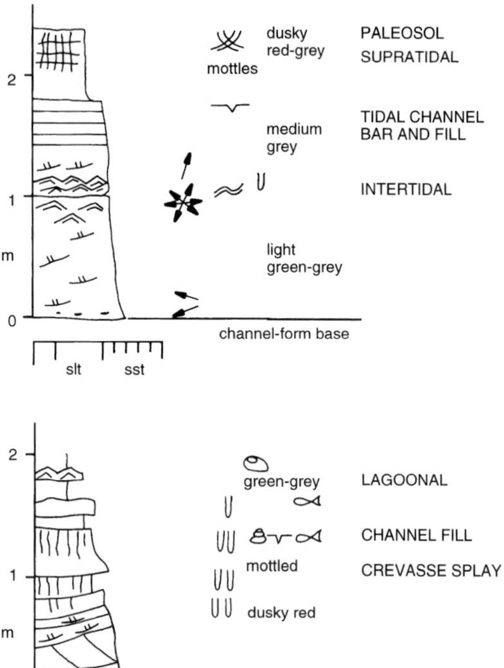

Fig. 4(d).

bed forms. Transported plant material is common, normally as the fragmented axes and sporangia of several species of trimerophytes and zosterophylls. Transported disarticulated modiolopsid bivalves, pterygoid euryterid cuticle fragments and fish-bone fragments (cephalaspid and acanthodian) occur along surfaces of large-scale strata in the lower parts of Type 1 sandstone bodies. Vertical burrows (e.g. *Skolithos*, *Pelecypodichnus*-bivalve escape burrows) and surface trails occur in the tops of large-scale strata in the upper parts of sandstone bodies.

Type 2 sandstone bodies contain virtually no *in situ* plant material and limited trace fossils, and only in their topmost parts (Fig. 4). Fish-bone fragments occur, particularly along basal erosion surfaces. Transported plant assemblages are different from Type 1 sandstone bodies. They consist of mats of *Spongiophyton* (a probable lichen: Stein *et al.* 1993) and, less commonly, variably sized fragments of *Prototaxites* (interpreted as a fungus by Hueber (1996)), ranging from pebble- and cobble-sized fragments to (very rarely) logs up to 1 m wide and of unknown length. Recognizable tracheophyte fragments do occur, but are rare. Transported plant material commonly occurs along medium-scale cross strata or as thin drapes on the top of large-scale strata in the upper parts of sandstone bodies. The largest fragments occur in the coarsest-grained sandstones. Plant-rich strata commonly have yellow limonitic stains, presumably from oxidation of pyrite.

Thick sandstone bodies: interpretation

These sandstone bodies are interpreted as the deposits of the main channels that migrated across the ancient alluvial–coastal plain. Large-scale inclined stratasets (storeys) represent the deposits of migrating channel bars and channel fills. The basal intraformational breccias represent material derived from adjacent cut banks. The vertical and lateral variation in internal structure and grain size within storeys reflect: (1) the spatial distribution of bedforms and surface sediment size on the channel bed during sedimentation events (e.g. floods); (2) the geometry and mode of migration of the bed surface; (3) the outcrop orientation relative to the

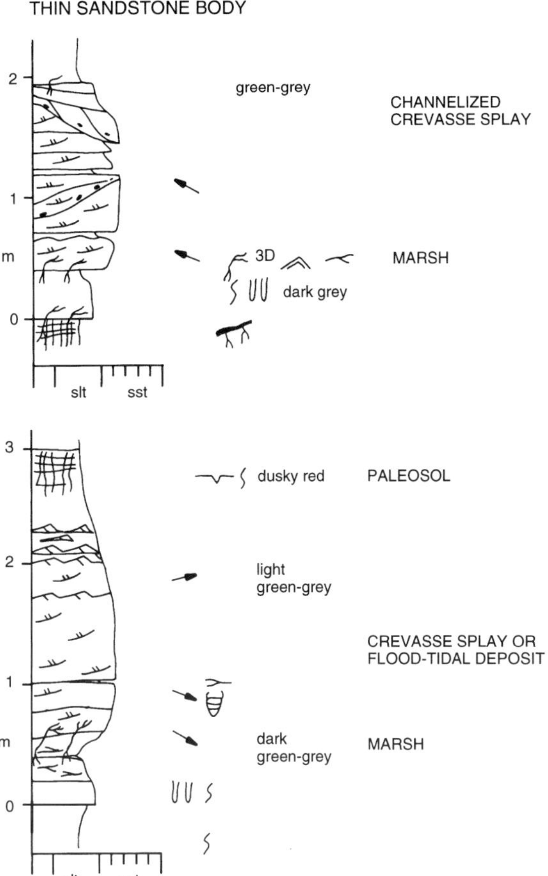

Fig. 4(e).

palaeocurrent orientation (Willis 1989, 1993*a*, *b*; Bridge 1993). Fining-upward storeys indicate preservation of the downstream parts of bars that migrated downstream and laterally. Lateral exposure of these storeys is insufficient to allow a determination of whether or not the channels were divided (braided). Fining-upward channel fills indicate progressive reduction in discharge during filling. Coarsening-upward storeys, and those showing little vertical variation in grain size, represent the upstream parts of bars. These areas can be preserved only if channel bars migrate by expansion as well as downstream translation. The relatively small channel forms cutting through the upper parts of storeys represent cross-bar channels.

The large-scale inclined strata represent the deposits of individual flood events. Vertical changes in internal structure and grain size within large-scale strata represent decreasing flow depth during waning flood stages and decrease or increase in local flow velocity as the water shallowed. The medium-scale cross strata, small-scale cross strata, and planar strata within large-scale strata are associated with deposition from curved-crested dunes and ripples and upper-stage plane beds, respectively. Symmetrical ripple marks on the upper surfaces of large-scale strata indicate the formation of wave ripples in areas of slack water.

Variations in palaeocurrents and lithofacies within Type 1 sandstone bodies indicate the influence of tidal currents (see also Lawrence 1986; Lawrence & Rust 1988). The lowest parts of the channels were dominated by river currents and by ebb-tidal currents, whereas the uppermost parts were dominated by flood-tidal currents. This type of current segregation is typical of the strongly asymmetrical tidal currents expected in channels near the tidal limit of estuaries and tide-influenced deltaic distributaries. The distribution of grain size and sedimentary structures is also

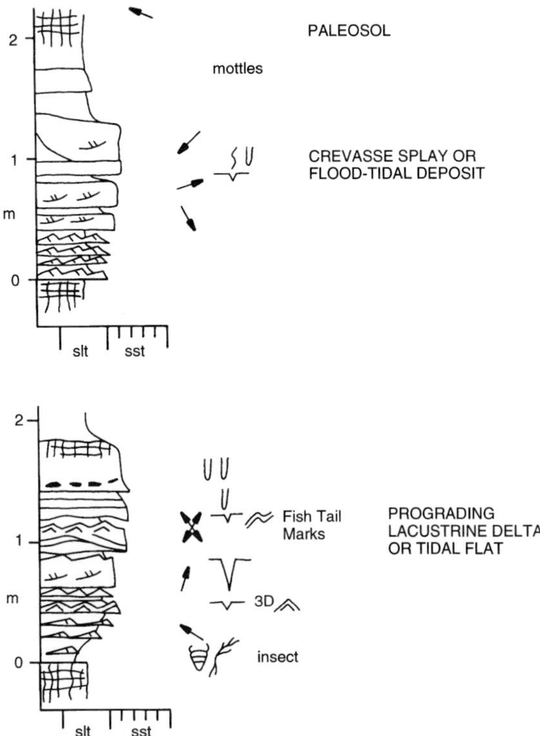

Fig. 4(f).

typical of tidal channel bars near the fluvial–tidal transition (e.g. Barwis 1978; van den Berg 1981; Smith 1988; Allen 1991; Dalrymple *et al.* 1992). Here, medium-scale cross strata are limited to lower (subtidal) parts of bars, whereas there is dominance of small-scale cross strata and relatively high proportions of mud in the upper, intertidal parts. The rarity of tidal bundle sequences may be due to non-depositional subordinate tides and dominant ebb currents that were reinforced by fluvial currents. Type 2 sandstone bodies are relatively coarser grained and do not show evidence of tidal influence. They were therefore upstream of the fluvial–tidal transition. Vertical superimposition of channel bars and fills is uncommon in Type 1 sandstone bodies, indicating isolated channel belts with low deposition rates relative to channel migration rates.

In situ and *in loco* plant assemblages in Type 1 sandstone bodies are due to burial of stands of plants that occupied the upper parts of migrating channel bars, including swales and cross-bar channels (see Fig. 11, below). The sedimentation events were probably associated with seasonal river floods, although flood tides probably also played a role. The plants then re-established themselves on the muddy, falling-stage drapes as water level receded. The morphology and reproductive strategies of these plants were apparently related to this ephemeral environmental setting. The relatively thin cuticle and abundance of fertile specimens suggest that these plants could colonize new sites, grow and reproduce quickly (within weeks or months), before being buried by the next major sedimentation event (Hotton *et al.* 2000). The absence of desiccation cracks in the mudstone drapes indicates that these sediments remained wet. This certainly reflects high water tables expected in near-coastal environments, and probably also reflects frequent flooding during high tides. It is uncertain whether these plants were adapted to brackish water conditions or killed by it. However, the rarity of acritarchs in the *in situ* plant assemblages suggests that the plants did not typically occupy brackish environments. The vertical burrows (*Skolithos* ichnofacies) in Type 1 sandstone bodies are also consistent with a near-coastal setting. It is probable that bivalves were living in the tide-influenced channels, and that their escape burrows were produced during depositional events.

THIN SANDSTONE BODIES

CHANNEL FILL

LACUSTRINE DELTA

2
1
m
0
slt sst

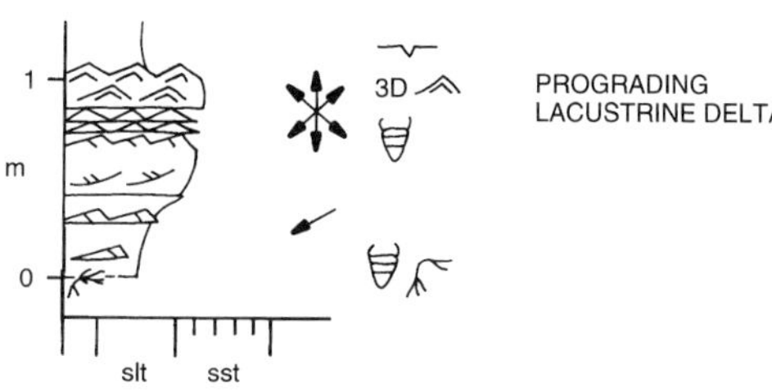

Fig. 4(g).

The fauna and especially the flora of Type 2 sandstone bodies differ strikingly from those of Type 1. *Prototaxites* and *Spongiophyton* are both clearly transported, but probably occupied sites close to the channel or other sources of water, such as abandoned channels and flood-basin lakes. The rarity of identifiable embryophyte material is due at least in part to abrasion and fragmentation during transport, or lack of suitable anaerobic sites for preservation. However, it is also possible that plant diversity and abundance were lower in these fully terrestrial settings.

Quantitative interpretation of thick sandstone bodies

Palaeochannel geometry and hydraulics of some thick sandstone bodies were reconstructed by comparison with a physical model that predicts the sedimentology of single channel-bar deposits. Full description of the technique and examples of the approach have been given by Bridge (1978), Bridge & Diemer (1983), Bridge & Gordon (1985) and Willis (1989, 1993*a*, *b*).

Channel widths, mean bankfull depths, and maximum bankfull depths were approximately 30 m, 2 m and 4–5 m, respectively. These dimensions agree with the empirical regression equation presented by Bridge & Mackey (1993) for the variation of channel width, w, as a function of mean channel depth, d_m:

$$w = 8{\cdot}88 d_m^{1{\cdot}82}$$

Channel-bend wavelength, estimated based on the well-known empirical relationship with bankfull channel width: $L \approx 11w$, is approximately 330 m. Maximum channel sinuosity was about 1.2. This sinuosity is very similar to that calculated from the maximum range of palaeocurrent directions observed for superimposed storeys in sandstone bodies (approximately 100°). The maximum sinuosity is 1.18. Such calculation of sinuosity from palaeocurrent ranges depends on assuming that the channel bends can be represented by sine-generated curves, i.e.

$$sn = 4{\cdot}84/(4{\cdot}84 - \phi^2)$$

where sn is channel sinuosity and ϕ is half of the maximum palaeocurrent range in radians. In most channel sandstone bodies, the observed

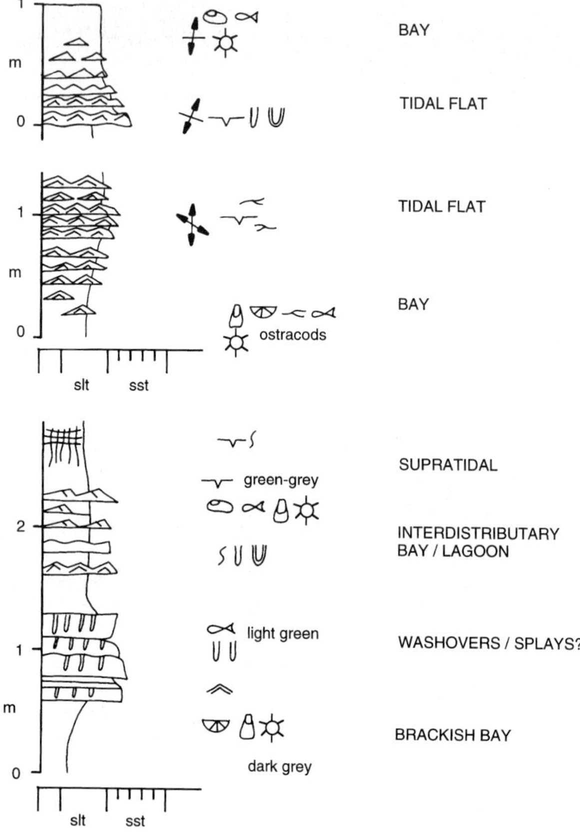

Fig. 4(h).

Fig. 4. Typical types of sandstone bodies and mudstone-rich sequence, with interpretations. Each type is either an actual example from logs of Fig. 3 or idealized based on compilation of several actual examples.

range of palaeocurrents is much smaller than 100°, yielding only estimates of minimum sinuosity. It is expected that channel sinuosity changed in time and space, varying from one (straight) to a maximum of around 1.2.

Bankfull flow velocities and bed shear stresses, averaged over the channel cross-section, are in the ranges 0.4–0.55 m s^{-1} and 2–3.2 Pa, respectively. Mean bankfull water-surface slopes are very difficult to reconstruct accurately, but are probably of order 10^{-4} for Type 2 sandstone bodies, and somewhat less for Type 1 sandstone bodies.

Thin sandstone bodies and mudstones: description and interpretation

Thin sandstone bodies are normally 1–2 m thick, but may thin laterally and pass into mudstone-dominant strata. These sandstone bodies are sheetlike or lenticular (planar base and convex-upward top, or channel filling: Fig. 5). Most sandstone bodies are fine grained to very fine grained, but medium-grained sandstone occurs in places, and interstratified mudstone is common. All sandstone bodies comprise sets of large-scale strata. These strata are centimetres to decimetres thick, generally fine upward with a change of sedimentary structure, and may be inclined by several degrees relative to the base of the sandstone body. Lithofacies sequences within these sandstone bodies are very variable, and some examples are shown in Fig. 4.

Upward-fining sandstone bodies have erosional bases (channel-form in some cases). Large-scale strata with medium-scale cross strata and planar strata tend to be overlain and/or underlain by small-scale cross strata. Dune forms and current-ripple forms are common. Palaeocurrent orientations are normally dominantly in

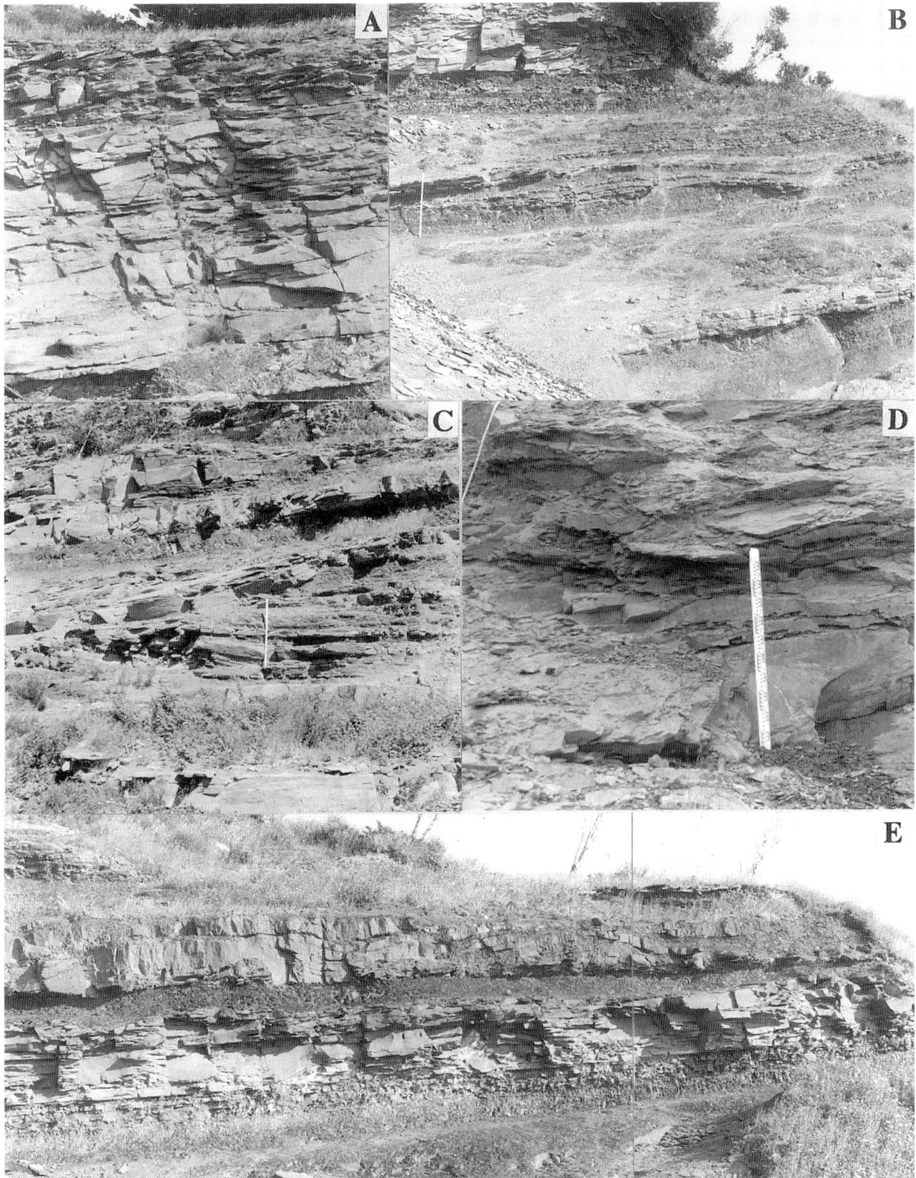

Fig. 5. Photos of sandstone bodies. (**a**) Type 2 multistorey sandstone body of 6 m thickness in lower Association 3 at D'Aiguillon. (**b**) Thin sandstone bodies within grey and red mudstones of uppermost Association 2 at D'Aiguillon (54–62 m). (Note shallow channel fill at centre-right and Type 2 sandstone body of Association 3 in the upper left corner). Two-metre ranging pole for scale. (**c**) Lowermost strata in Association 2 at D'Aiguillon (0–20 m), featuring Type 1 sandstone body (middle foreground, behind 2 m ranging pole), and overlying thin sandstone body (*Sawdonia ornata* type locality. (**d**) Plant-bearing interval of the Type 1 sandstone observed in (**c**). Fine-grained sandstone lenses (dune forms) with mudstone drapes containing *in situ Sciadophyton*. One-meter ranging pole. (**e**) Photomosaic of thin sandstones in Association 2 at D'Aiguillon (45–50 m). The lower sandstone body coarsens upward. The upper sandstone body displays channel forms.

one direction (e.g. westerly or easterly), but with subordinate orientations in a near-opposite direction. Wave-ripple marks and associated cross strata commonly occur, especially in the upper parts of these sandstone bodies, in places superimposed on dune forms. Wave-ripple crest

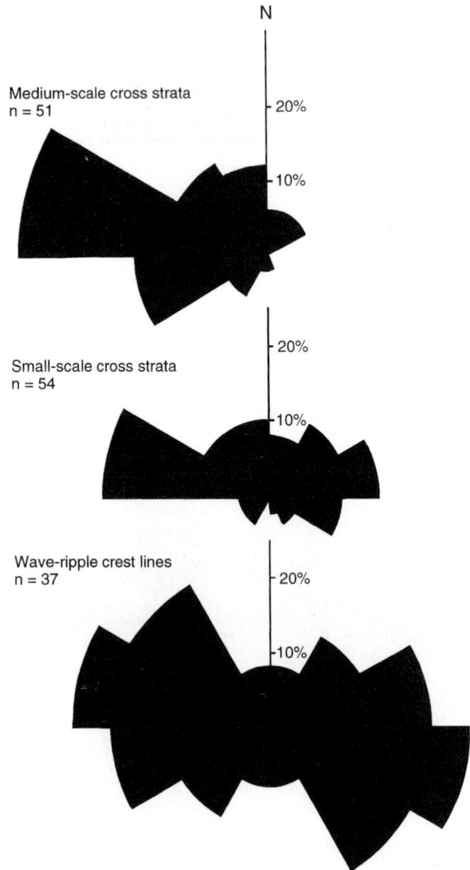

Fig. 6. Palaeocurrent roses for the Cap-aux-Os member.

lines are commonly oriented WNW–ESE, parallel to associated unidirectional current directions: however, crest lines may show a large range of orientations in some cases. Desiccation cracks occur in the upper, muddy parts of sandstone bodies. Transported plant material is common throughout and *in situ* plants occur in the upper parts (e.g. *Trimerophyton/Pertica* spp., *Psilophyton* spp., *Huvenia* sp. nov., *Sciadophyton* spp., *Sawdonia ornata*, *Crenaticaulis verruculosus*, other zosterophylls, *Drepanophycus spinaeformis*). Bioturbation increases upwards within sandstone bodies. Apart from plant roots, the most common form of biotubation is vertical tubes (*Skolithos*). Surface traces also occur on bedding surfaces. These sandstone bodies commonly pass upward into dusky red siltstone with green mottles, blocky fabric, slickensides, pseudoanticlines, root and rhizome casts, and burrows. Green–grey siltstone also has these features but not as well developed. These siltstones are commonly in decimetre-thick, fining-upward units. Topmost parts have limonite staining.

Upward-fining thin sandstone bodies are interpreted as relatively small, tide-influenced channel-bar and channel-fill deposits. They include subtidal, intertidal and supratidal zones. Palaeocurrent orientations generally indicate either flood- or ebb-fluvial dominance. Minor sedimentation on upper-bar surfaces during tidal slack water is represented by wave ripples and mud drapes. Wave-ripple crests are oriented parallel to bar margins in some cases. In others, the large range of orientations is as would be expected on sandy tidal flats. The *Skolithos* ichnofacies is also consistent with sandy tidal flats. Red siltstones are vertic palaeosols and the green–grey siltstones are less well-developed and wetter palaeosols (inceptisols or entisols: Elick *et al.* 1998).

Some thin sandstone bodies coarsen upward, but normally they fine upward at the top (Fig. 4). They appear sheet like to lenticular, with flat base and convex-upward top (Fig. 5), and shallow channels (decimetres deep and metres wide) occur in the upper parts of some of these sandstone bodies. *In situ* plants are common at the base of these sandstone bodies. Plant stands tend to be monospecific. Underlying the sandstones are millimetres- to centimetes-thick, dark green–grey claystone, in places containing accumulations of poorly preserved plant axes. Organic structures interpreted as rhizomes occur in poorly developed soil horizons, generally below plant accumulations but never in close contact. The rhizome-like structures are carbonized, millimetres wide and centimetres long, and boudin like (Fig. 7). Where well preserved, they are regularly spaced and parallel to bedding, and have tiny lateral branches that plunge a short distance into the bedding planes. At one especially well-preserved site (*Sawdonia ornata* type section), *in situ* plant axes are up to 35 cm long and 0.5 cm wide, and are bent over in the palaeocurrent direction (Fig. 7). In places, a given inclined axis bends parallel to bedding and then projects upwards again. Lycophytes (*Sawdonia*, *Crenaticaulus*, undescribed zosterophylls, *Renalia*, *Drepanophycus*) commonly occur in grey–green mudstone, whereas trimerophytes, especially *Trimerophyton–Pertica*, and *Huvenia* occur in the overlying sandstones as *in loco* assemblages. Transported plant material is common, and consists of a random sample of the more common elements of the flora, especially trimerophytes, *Sawdonia* and *Crenaticaulis*, and also commonly includes pterygotid eurypterid fragments and disarticulated small bivalves.

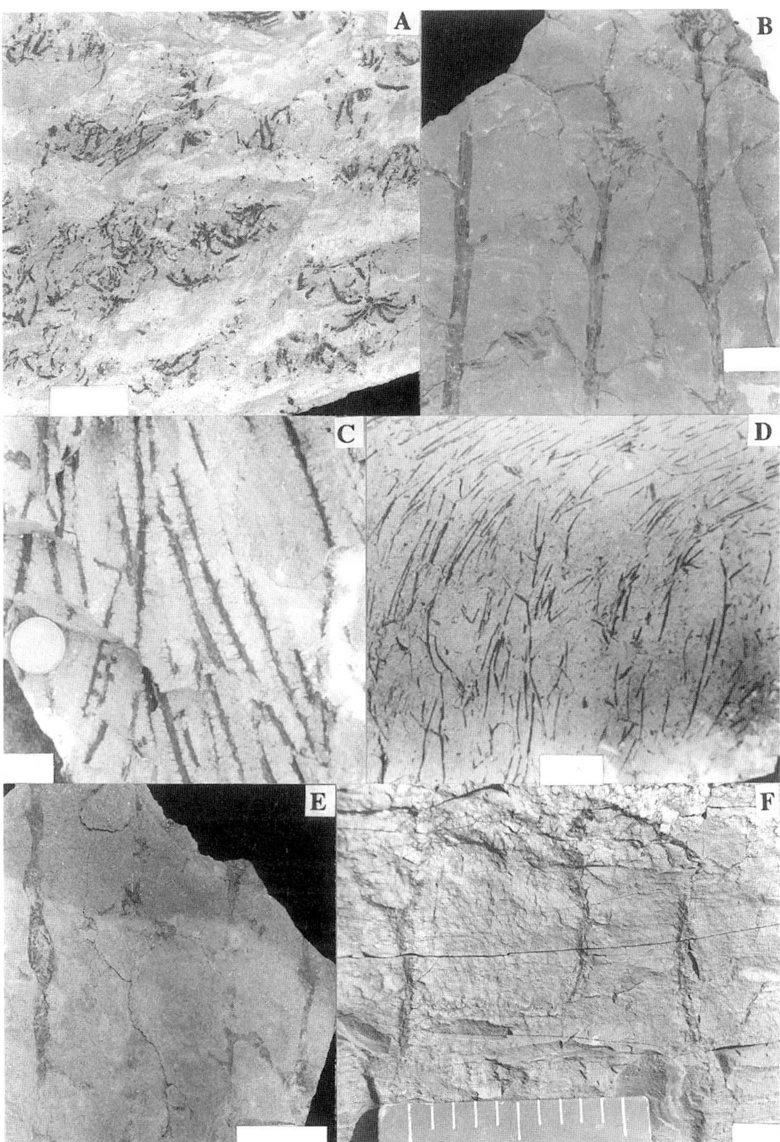

Fig. 7. Plant photographs. White scale bar is 2 cm. (**a**) *Sciadophyton* anchored in a mud drape on the surface of a small dune, uppermost Type 1 sandstone body at Mudstone 4 (First Cove West section 7 m). (**b**) Toppled parallel axes of *Trimerophyton robustius* (type locality, Fort Peninsule). (**c**) Parallel axes of *Sawdonia ornata* protruding upward from a claystone through siltstones and fine sandstones of a coarsening-upward, thin sandstone body. (**d**) Bend of sub-parallel axes of small dichotomous plant (?*Eogaspesia* or ?*Renalia*) suggests current alignment during burial. (**e**) Boudin-like carbon films lying in parallel in lower Association 2 mudstone at D'Aiguillon (8.5 m level), interpreted to be rhizomes. (**f**) Root traces in dusky red–grey mudstone in Mudstone 5 section (16–17 m). Ten-centimetre scale.

Upward-coarsening thin sandstone bodies mostly contain small-scale cross strata associated with current-ripple marks. Palaeocurrents from current ripples and associated cross strata are either westerly or easterly, but widely varying and opposing in some cases. Some sandstone bodies have wave-ripple marks and associated cross strata. These may occur anywhere in the sequence. Wave-ripple crest lines have a large range of orientations. In the upper, coarsest parts of the sandstone bodies, there may be some medium-scale cross strata, planar strata, and,

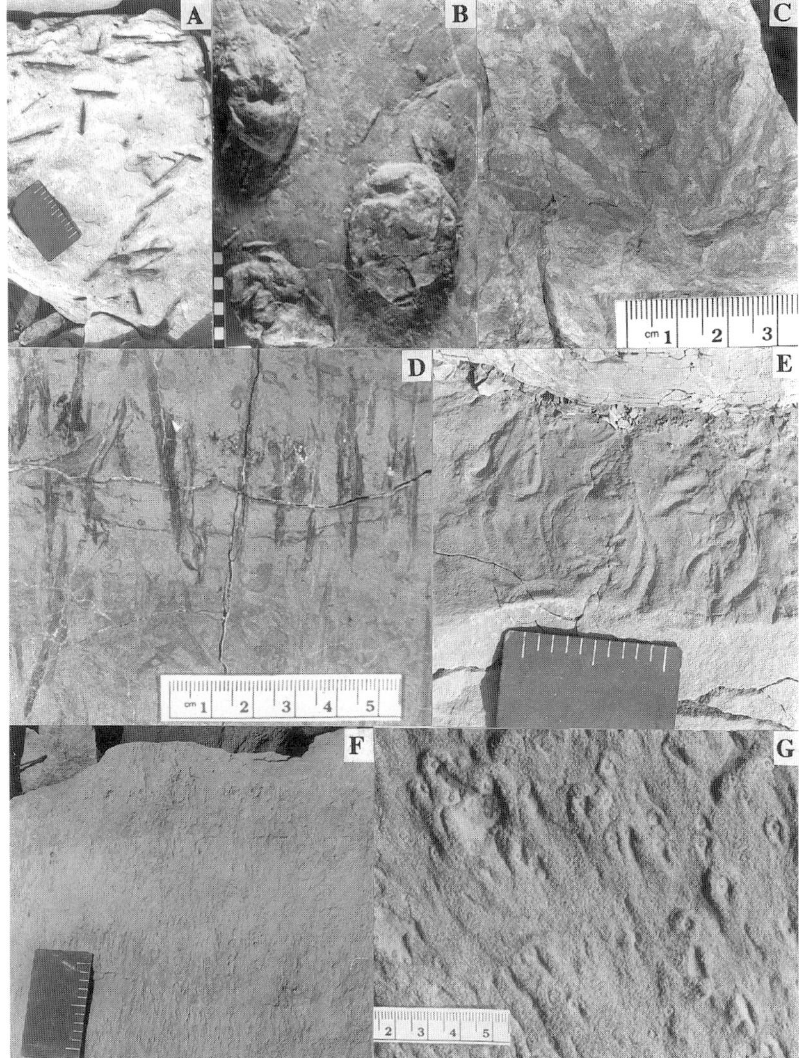

Fig. 8. Trace fossil photographs. All scales are in 1 cm intervals. (**a**) Sandstone in heterolithic strata of First Cove West Section of Mudstone 4 (12–13 m) containing numerous traces of *Diplocraterion*. (**b**) Unidentified hypichnial composite trace associated with *Diplocraterion* in heterolithic beds of Mudstone 4. (**c**) Stellate traces commonly associated with *Skolithos* and *Arenicolites*. (**d**) Deep U-shaped traces with no spreiten, presumably *Arenicolites*. Dense populations of these traces resemble *Skolithos* pipe-rock. (**e**) Fish-tail traces on upper surface of a sandstone bed, Association 2 at D'Aiguillon (34.6 m). (**f**) Light green-grey sandstone with dense concentrations of *Skolithos* ('pipe-rock') in uppermost Mudstone 4 at First Cove West. (**g**) Current crescent scours around *Skolithos* burrows in base of sandstone body that overlies pipe-rock (**f**).

rarely, hummocky cross strata. In general, disruption by desiccation cracks, burrows and roots increases markedly at the top. Typical burrows are vertical tubes (*Skolithos*). Surface traces also occur, including a possible fish-tail trace (Fig. 8). These sandstone bodies commonly pass upwards into dusky red or green–grey palaeosols described above.

Upward-coarsening thin sandstone bodies are interpreted as channelized and unchannelized parts of crevasse splays or minor deltas that prograded into floodplain marshes and lakes or brackish interdistributary bays–lagoons (compare similar deposits described by Coleman (1969), Elliot (1974) Farrell (1987), Tye & Coleman (1989*a*, *b*)). Evidence of waves acting

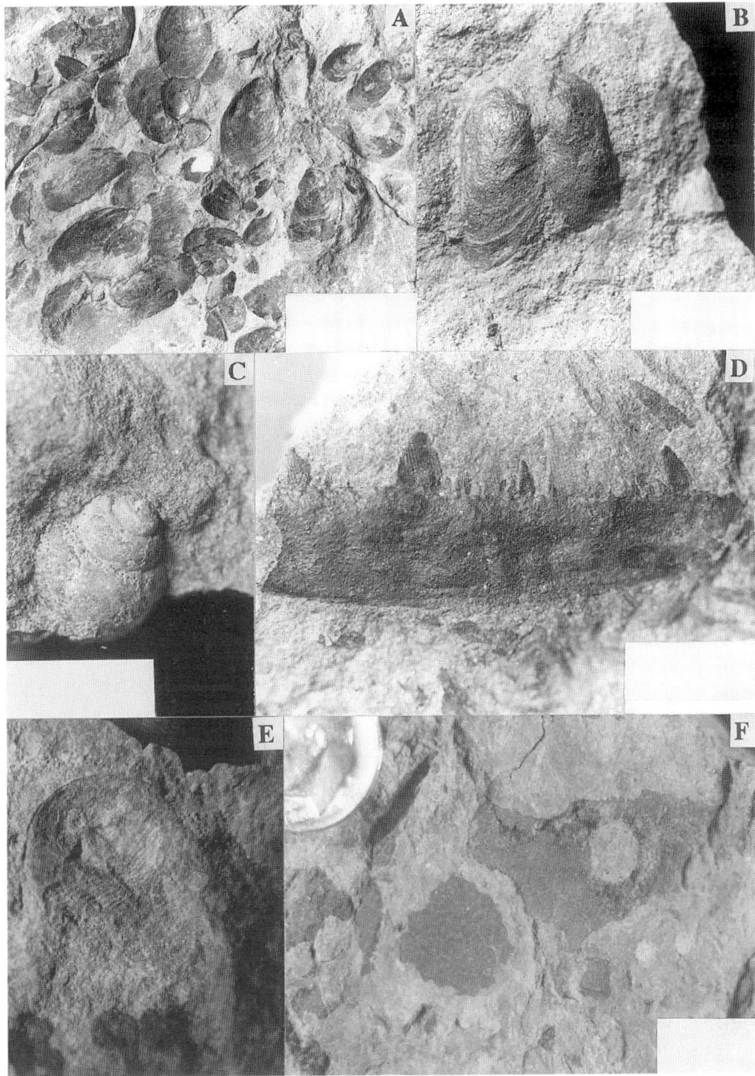

Fig. 9. Other fossil photographs. White bar scales are 1 cm. (**a**) Transported but articulated specimens of *Lingula* sp. From desiccation-cracked mudstones of lower Mudstone 5 Section (14.5 m). (**b**) Articulated shells of mytiloid bivalve from upper Mudstone 5 section. (**c**) internal mold of the gastropod *Holopea*? from calcareous *Skolithos*-bearing mudstone in Association 2 at D'Aiguillon. (**d**) Claw fragment of the eurypterid *Pterygotus gaspesiensis* from upper portion of a Type 1 sandstone body in Association 2 at D'Aiguillon (16.8 m). (**e**) Unidentified limulid from basal mudstone in Association 2 section (*c*. 1 cm diameter). (**f**) Orbit and headshield fragments from a cephalaspid fish is bioturbated siltstone in Association 2 at D'Aiguillon.

in standing water is wave ripples, planar and hummocky cross strata. Easterly directed palaeocurrents suggest that some of these sandstone bodies may be flood-tidal deltas or washovers. Intensity of burrowing increases as deposition rate decreases.

The organic accumulations beneath *in situ* plant assemblages may represent plant litter that accumulates beneath established stands of vegetation (i.e. O horizon of soil). The plants in this setting are generally zosterophylls. These plants have relatively thick cuticle and low sporangial production, both traits that may be associated with stable occupation and vegetative growth. This accords with the inferred environments where they are found, i.e. relatively stable, wet and low-energy habitats, such as floodplain marsh or backswamp (see Fig. 12 below). The bends in the plant axes indicate growth through different episodes of sedimentation, again supporting a degree of longevity. Trimerophytes and rhyniopsids, on the other hand, were probably occupying more ephemeral, riparian sites closer to stream margins, such as levees and cross-bar channels (see more complete discussion by Hotton *et al.* (2001)).

Somewhat thinner (decimetre scale) sequences of very fine sandstone and mudstone have been distinguished in Fig. 4. These sequences may coarsen upward from lenticular to wavy to flaser bedding (with wave-ripple marks) or they may fine upward. Wave-ripple crests tend to be oriented NW–SE or NNE–SSW. The sandstones contain a diversity of trace fossils (e.g. *Skolithos*, *Diplocraterion*, *Arenicolites*, *Planolites*, *Phycodes*, bivalve escape burrows). Muddy parts contain transported articulated lingulid brachiopods, orthid and rhynchonellid brachiopods, disarticulated bivalves (*Cypricardella*, *Modiolopsis*), gastropods, fish-bone fragments, ostracodes and acritarchs. Fine plant fragments ('plant hash') are common throughout these strata. These mudstones may be laminated and contain randomly occurring, centimetre-thick sheets and lenses of very fine sandstone with either current ripples or wave ripples and associated cross strata (Fig. 4). In places, the mudstones are very bioturbated and may contain desiccation cracks.

These mudstone-dominant sequences record transition of sandy–muddy intertidal flats, crevasse splays or minor deltas into muddy, brackish or fully marine bays–lagoons (Reineck & Singh 1973; Terwindt 1981; Coleman & Prior 1982). The sand layers could be due to seasonal floods, tides or storms. These bays were frequently filled with sediment to supratidal level and/or experienced sea-level falls, as evidenced by the upward transition to palaeosols. Although some of the lithofacies variations could be associated with sea-level changes and/or sediment progradation, they could also be associated with seasonal variations in wave or current energy. The sparse fauna, relatively low bioturbation and dominance of *Skolithos* ichnofacies is typical of coastal sands and silts (Howard & Frey 1975; Howard & Reineck 1981). The absence of recognizable plant fragments or whole plants suggests that plants did not occupy brackish or marine environments.

Large-scale vertical variation of the strata

Figure 10 shows the vertical variation of mean sediment size, large-scale stratal thickness, mean palaeocurrent orientation, and the occurrence of specific diagnostic sedimentological and

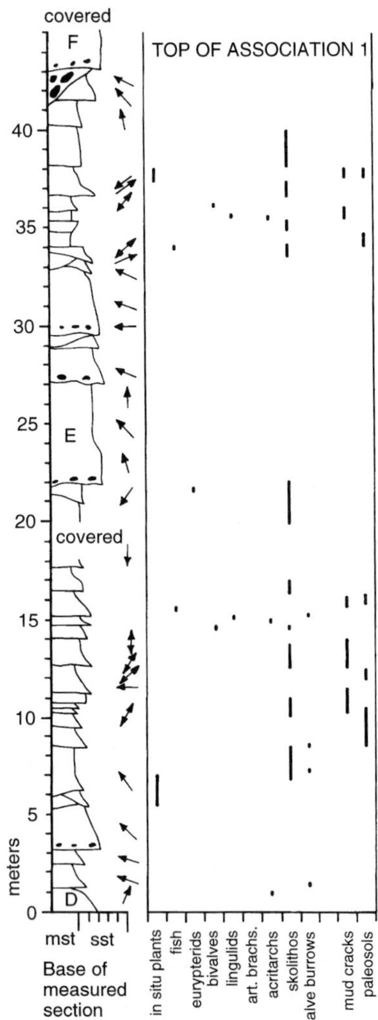

Fig. 10(a).

palaeontological features. There is clear evidence of brackish and marine influence in the upper part of Lawrence's Association 1 and in Association 2, particularly where the proportion of mudstone is relatively high and the thick sandstones bodies are single-storey types. Within the mudstone-dominant parts of this interval, there are abrupt vertical transitions from palaeosols to strata containing trace fossils of the *Skolithos* ichnofacies. Several horizons containing bivalves, inarticulate and articulate brachoipods and acritarchs clearly indicate marine influence. *In situ* plants within the mudstone-dominant strata tend to occur immediately above palaeosols and below strata containing *Skolithos* ichnofacies.

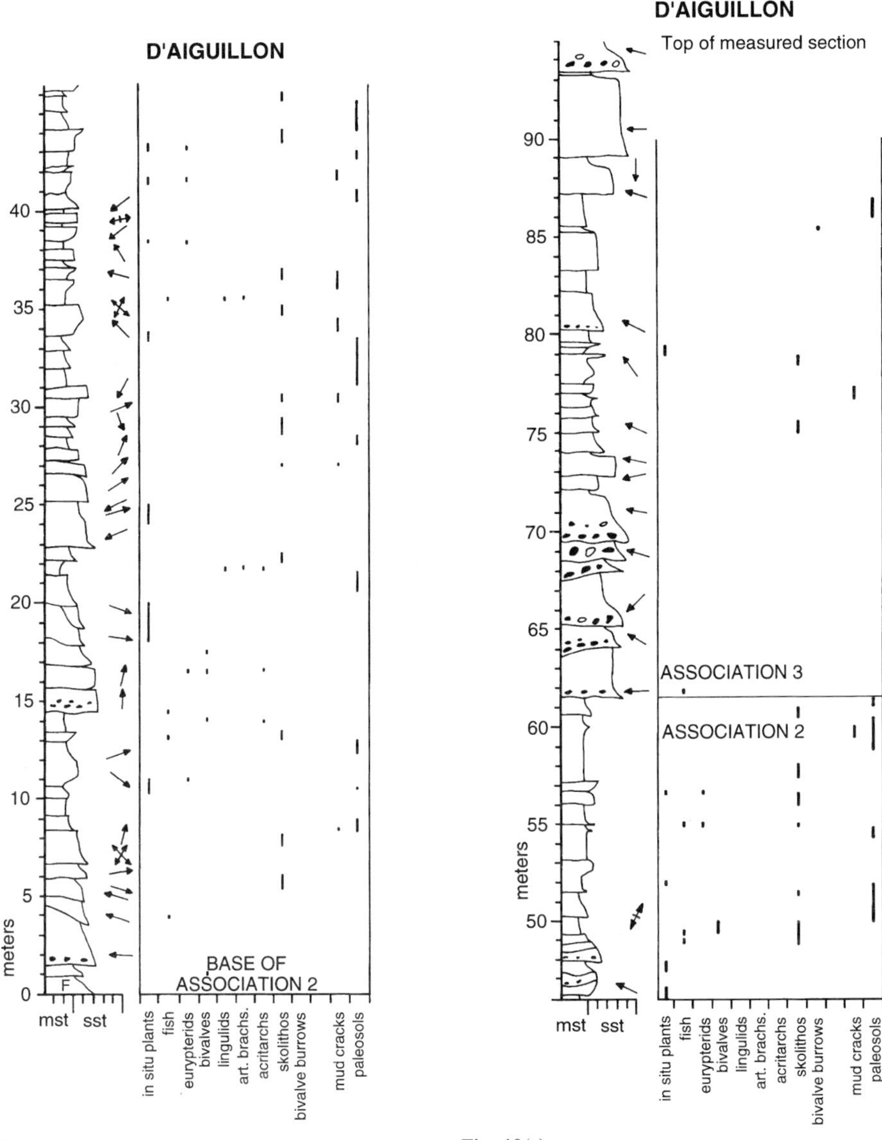

Fig. 10(b). **Fig. 10(c).**

Fig. 10. Vertical variations of mean grain size, palaeocurrents and selected sedimentological and palaeontological features through whole succession.

The abrupt vertical transitions between palaeosols and strata indicative of marine influence indicate frequent changes in relative sea level. It is not possible to establish whether these were of regional or local extent. Some of the marine incursions could possibly have been associated with storm washovers and/or abnormally high tides, and are therefore short lived. The occurrence of the deposits of single-storey, tide-influenced channels within the coastal-plain muds is related here to channel diversions in an overall aggrading setting. Associated deposits, such as crevasse splays and bay-head deltas, are also expected to be bounded by coastal-plain mud. Abandonment of a channel belt and the surrounding coastal plain could easily be followed by incursion of brackish or fully marine water. The preservation of *in situ* plant material is clearly related to rapid, episodic deposition (e.g. associated with crevasse splays, storm washovers, etc.) in near-coastal settings where the water table was high.

McCave (1969) associated mudstone-dominated fluvial sequences from the Catskill Facies of New York State with marine transgressions, based on the assumption that rising sea level would lead to increased rate of floodplain aggradation and higher preservation potential of overbank mudstones. This concept has been used by others more recently (e.g. Shanley & McCabe 1993; Wright & Marriott 1993). Lack of absolute age dating of the Battery Point Formation precludes assessment of relative rates of floodplain deposition. However, there is a clear association of increasing marine influence with a high proportion of mudstone (i.e. association 2). The relatively high proportion of mudstone may be due to broad areas of coastal plain crossed by relatively few distributary channels (e.g. Bridge this volume). It is also possible that the high proportion of mudstone is associated with the muddy bays that are expected on the leeward side of sandy shoals or barriers during relative sea-level rise. More data are needed to test these ideas.

Implications for early land plant ecology

The sedimentological evidence shows clearly that land plants in the Cap-aux-Os Member occupied different portions of the available landscape. Three separate floristic associations can be distinguished: (1) trimerophyte–rhyniopsid–*Sciadophyton*; (2) zosterophyll; (3) *Spongiophyton–Prototaxites*. These associations occur in distinctive sedimentological settings. The first assemblage tends to occur near tops of near-coastal channel bars and channel fills (Fig. 11). These plants tend to occur in a fertile state, and were probably rapidly growing, short lived and dependent on sexual reproduction for dissemination. In contrast, the second association, dominated by zosterophylls, tended to occupy mud-dominated settings such as backswamps

Fig. 11. Reconstructed setting of trimerophyte–rhyniopsid–*Sciadophyton* floristic association, which is near the inner bank of a curved, near-coastal channel. *Sciadophyton* occupies mud drapes in the troughs of dunes and in point-bar swales (foreground). Higher parts of the point bar and banks are inhabited by large trimerophytes (right foreground). The axes of these plants are reoriented and bent over by flood water, and may be buried by sand deposits. The bankfull width of the main channel is about 30 m, whereas the point-bar swale is on the order of a 1 m wide. The dunes (left foreground) are centimetres to decimetres high. (from *Plants Invade the Land*, eds. P. G. Gensel and D. Edwards. ©2001 Columbia University Press. Reprinted by permission of the publisher).

and marshes (Fig. 12). They tended to form laterally extensive, relatively long-lived stands and were dependent primarily on vegetative growth for site occupation. All *in situ* plant stands tend to be dominated by one species, which means that competition among species would have been confined to stand margins. The third type of association (*Spongiophyton–Protataxites*) is transported, but is clearly associated with fully terrestrial fluvial channels. The taxonomic status of both of these taxa remains disputed, but the sedimentological evidence alone clearly indicates that they were fully terrestrial organisms, and renders an algal interpretation of *Protataxites*, in particular, completely untenable. Although *Prototaxites* and *Spongiophyton* did not occupy fluvial channels, their abundance within these channels indicates that they occupied riparian habitats upstream. Furthermore, these two taxa are strikingly absent from the coastal, plant-dominated environments. Why this should be so remains unclear.

In situ plant assemblages lack any of the standard marine indicators (acritarchs, trace fossils, brachiopods), suggesting that they occupied primarily freshwater habitats. Living at the marine–freshwater transition would have exposed plants to occasional brackish conditions, as occurs in extant bayou and coastal swamps. However, it is difficult to determine whether these early land plants tolerated occasional brackish conditions or were killed by them. Their absence in more clearly brackish and marine settings is clear evidence that they were not typically halophytes.

Questions remain as to the ability of early land plants to occupy xeric settings. Initial plant occupation (i.e. successful germination of spores

Fig. 12. Reconstructed setting of zosterophyll floristic association, which is a swampy floodbasin. Monotaxic stands of plants (e.g. *Sawdonia ornata*) are partially buried *in situ* by silts and very fine sands associated with the distal edge of a crevasse splay or levee. These sands also contain fragments of plants transported from near-bank areas. Partially buried plants recover and continue growth between episodes of overbank flooding. Plant density and branching is illustrated as sparser than expected, to demonstrate patterns of vegetative propagation and burial. (from *Plants Invade the Land*, eds. P. G. Gensel and D. Edwards. © 2001 Columbia University Press. Reprinted by permission of the publisher).

and fertilization of gametophytes) probably required at least ephemeral sources of water and a muddy substrate, but these may have been minimal, especially if germination and gametophyte development were rapid. Root traces in reddish mudstones are evidence of occupation of somewhat drier, but still muddy substrates. Both large and small root traces and permineralized plants also occur in fully fluvial channels on the south shore of Gaspé Bay (Elick *et al.* 1998; Hotton *et al.* 2000). Plants were clearly capable of occupying fully terrestrial environments: how xeric these environments may have been remains unclear from the available evidence.

This research was supported by Smithsonian Institution Scholarly Studies Grant no. 1233540G to F. M. Hueber and C. L. Hotton, as well as by grants from the Charles Walcott Fund and Roland Brown Fund. Anne Hull and David Tuttle (Birmingham University) providing drafting and photographic assistance, respectively. The authors also acknowledge J. Dougherty (Geological Survey of Canada), I. Birker (Redpath Museum) and W. Stein (Binghamton University) for access to specimens, S. Braden, R. Harberts and W. Brown for technical help, and J. R. Beerbower, C. C. Labandeira and W. Stein for helpful discussions. The manuscript benefited from the comments of two anonymous referees.

References

ALLEN, G. P. 1991. Sedimentary processes and facies in the Gironde estuary: a recent model for estuarine systems. *In*: SMITH, D. C., REINSON, G. E., ZAITLIN, B. A. & RAHMANI, R. A. (eds) *Clastic Tidal Sedimentology*. Canadian Society of Petroleum Geologists, Memoirs, **16**, 29–39.

ANDREWS, H. N., KASPER, A. E., FORBES, W. H., GENSEL, P. G. & CHALONER, W. G. 1977. Early Devonian flora of the Trout Valley Formation of northern Maine. *Reviews in Palaeobotany and Palynology*, **23**, 255–285.

BARWIS, J. H. 1978. Sedimentology of some South Carolina tidal-creek point bars, and a comparison with their fluvial counterparts. *In*: MIALL, A. D. (ed.) *Fluvial Sedimentology*. Canadian Society of Petroleum Geologists, Memoirs, **5**, 129–160.

BERNER, R. A. 1993. Palaeozoic atmospheric CO_2—importance of solar radiation and plant evolution. *Science*, **261**, 68–70.

—— 1994. GEOCARB II—a revised model of atmospheric CO_2 over Phanerozoic time. *American Journal of Science*, **294**, 56–91.

BOUCOT, A. J., CUMMING, L. M. & JAEGER, H. 1967. Contributions to the age of the Gaspé Sandstone and Gaspé Limestone. *Geological Survey of Canada Paper*, **67–25**, 1–22.

BRIDGE, J. S. 1978. Palaeohydraulic interpretation using mathematical models of contemporary flow and sedimentation in meandering channels. *In*: MIALL, A. D. (ed.) *Fluvial Sedimentology*. Canadian Society of Petroleum Geologists, Memoirs, **5**, 723–742.

—— 1993. The interaction between channel geometry, water flow, sediment transport and deposition in braided rivers. *In*: BEST, J. L. & BRISTOW, C. S. (eds) *Braided Rivers*. Geological Society, London, Special Publications, **75**, 13–71.

—— 2000. The geometry, flow patterns and sedimentary processes of Devonian rivers and coasts. New York and Pennsylvania, USA. This volume.

—— & DIEMER, J. A. 1983. Quantitative interpretation of an evolving ancient river system. *Sedimentology*, **30**, 599–623.

—— & GORDON, E. A. 1985, Quantitative interpretation of ancient river systems in the Oneonta Formation, Catskill Magnafacies. *In*: WOODROW, D. L. & SEVON, W. D. (eds) *The Catskill Delta*. Geological Society of America, Special Papers, **201**, 163–182.

—— & MACKEY, S. D. 1993. A theoretical study of fluvial sandstone body dimensions. *In*: FLINT, S. & BRYANT, I. D. (eds) *Quantitative Description and Modelling of Clastic Hydrocarbon Reservoirs and Outcrop Analogues*. International Association of Sedimentologists, Special Publications, **15**, 213–236.

BRISEBOIS, D. 1981. *Géologie de la région de Gaspé*. Ministere de l'énergie et des résources, direction générale des énergies conventionnelles, rapport intérimaire, **DPV-824**, 1–19.

COLEMAN, J. M. 1969. Brahmaputra River: channel processes and sedimentation. *Sedimentary Geology*, **3**, 129–239.

—— & PRIOR, D. B. 1982. Deltaic environments. *In*: SCHOLLE, P. A. & SPEARING, D. (eds) *Sandstone Depositional Environments*. American Association of Petroleum Geologists, Tulsa, OK, 139–178.

DALRYMPLE, R. W., ZAITLIN, B. A. & BOYD, R. 1992. Estuarine facies models: conceptual basis and stratigraphic implications. *Journal of Sedimentary Petrology*, **62**, 1130–1146.

DAWSON, J. W. 1859. On fossil plants from the Devonian rocks of Canada. *Quarterly Journal of the Geological Society, London*, **15**, 477–488.

—— 1871. *The Fossil Plants of the Devonian and Upper Silurian Formations of Canada*. Geological Survey of Canada, 1–92.

DIMICHELE, W. A. & HOOK, R. *et al.* 1992. Palaeozoic terrestrial ecosystems. *In*: BEHRENSMEYER, A. K., DAMUTH, J. D., DIMICHELE, W. A., POTTS, R., SUES, H. & WING, S. L. (eds) *Terrestrial Ecoystems Through Time*. Chicago, University of Chicago Press, 205–325.

EDWARDS, D. & FANNING, U. 1985. Evolution and environment in the late Silurian-early Devonian: the rise of the pteridophytes. *Philosophical Transactions of the Royal Society of London, Series B*, **309**, 147–165.

ELICK, J. M., DRIESE, S. M. & MORA, C. I. 1998. Very large plant and root traces from the Early to Middle Devonian: implications for early terrestrial ecosystems and atmospheric $p(CO_2)$. *Geology*, **26**, 143–146.

ELLIOTT, T. 1974. Interdistributary bay sequences and their genesis. *Sedimentology*, **21**, 611–622.

FARRELL, K. M. 1987. Sedimentology and facies architecture of overbank deposits of the Mississippi River, False River Region, Louisiana. *In*: ETHRIDGE, F. G., FLORES, R. M. & HARVEY, M. D. (eds) *Recent Developments in Fluvial Sedimentology*. Society of Economic Palaeontologists and Mineralogists, Special Publications, **39**, 111–120.

HOTTON, C. L., HUEBER, F. M., GRIFFING, D. H. & BRIDGE, J. S. 2001. Early land plant environments: an example from the Emsian of Gaspé, Canada. *In*: GENSEL, P. G. & EDWARDS, D. (eds) *Plants Invade the Land: Evolutionary and Environmental Considerations*. Columbia University Press, New York, 179–212.

HOWARD, J. D. & FREY, R. W. 1975. Estuaries of the Georgia coast, U.S.A.: sedimentology and biology. *Senckenbergiana maritima*, **7**, 33–103.

—— & REINECK, H. E. 1981. Depositional facies of high-energy beach-to-offshore sequence: comparison with low-energy sequence. *AAPG Bulletin*, **65**, 807–830.

HUEBER, F. M. 1996. A solution to the enigma of *Prototaxites*. Palaeontological Society Special Publication **8**, Sixth North American Palaeontological Convention, Abstracts of Papers, 183.

LAWRENCE, D. A. 1986. *Sedimentology of the Lower Devonian Battery Point Formation, eastern Gaspé Peninsula*. PhD thesis, University of Bristol.

—— & RUST, B. R. 1988. The Devonian clastic wedge of eastern Gaspé and the Acadian orogeny. *In*: MCMILLAN, N. J., EMBRY, A. F. & GLASS, D. J. (eds) *Devonian of the World*, Vol. II. Canadian Society of Petroleum Geologists, Memoirs, **7**, 53–64.

—— & WILLIAMS, B. P. J. 1987. Evolution of drainage systems in response to Acadian deformation: the Devonian Battery Point Formation, Eastern Canada. *In*: ETHRIDGE, F. G., FLORES, R. M. & HARVEY, M. D. (eds) *Recent Developments in Fluvial Sedimentology*. SEPM Special Publications, **39**, 287–300.

MCCAVE, I. N. 1969. Correlation of marine and nonmarine strata with example from Devonian of New York State. *AAPG Bulletin*, **53**, 155–162.

MCGERRIGLE, H. W. 1950. *The geology of eastern Gaspé*. Québec Department of Mines, Geological Report **15**.

MCGREGOR, D. C. 1973. Lower and Middle Devonian spores of eastern Gaspé, Canada. I. Systematics. *Palaeontographica*, **142B**, 1–77.

—— 1977. Lower and Middle Devonian spores of eastern Gaspé, Canada. II. Biostratigraphy. *Palaeontographica*, **163B**, 111–142.

MORA, C. I., DRIESE, S. G. & COLARUSSO, L. A. 1996. Middle to late Palaeozoic atmospheric CO_2 levels from soil cabonate and organic matter. *Science*, **271**, 1105–1107.

PAGEAU, Y. & PRICHONNET, G. 1976. Interprétation de la paléontologie et de la sédimentologie d'une coupe géologique dans la formation de Battery Point (Dévonien moyen), grès de Gaspé. *Naturaliste Canadien*, **103**, 111–118.

REINECK, H.-E. & SINGH, I. B. 1973. *Depositional Sedimentary Environments*. Springer, Berlin.

RETALLACK, G. J. 1997. Early forest soils and their role in Devonian global change. *Science*, **276**, 583–585.

RICHARDSON, J. B. & MCGREGOR, D. C. 1986. *Silurian and Devonian Spores Zones of the Old Red Sandstone Continent and Adjacent Regions*. Geological Survey of Canada Bulletin, **364**, 1–79.

RUST, B. R. 1981. Alluvial deposits and tectonic style: Devonian and Carboniferous successions in eastern Gaspé. *In*: MIALL, A. D. (ed.) *Sedimentation and Tectonics in Alluvial Basins*. Geological Association of Canada, Special Papers, **23**, 49–75.

—— 1984. Proximal braidplain deposits in the Middle Devonian Malbaie Formation of eastern Gaspé, Québec, Canada. *Sedimentology*, **31**, 675–695.

SCHWEITZER, H. J. 1983. Die Ünterdevonflora des Rheinlandes. *Palaeontographica*, **189B**, 1–138.

SCOTESE, C. R. & MCKERROW, W. S. 1990. Revised world maps and introduction. *In*: MCKERROW, W. S. & SCOTESE, C. R. (eds) *Palaeozoic Palaeogeography and Biogeography*. Geological Society, London, Memoirs, **12**, 1–21.

SHANLEY, K. W. & MCCABE, P. J. 1993. Alluvial architecture in a sequence stratigraphic framework: a case history from the Upper Cretaceous of southern Utah, USA. *In*: FLINT, S. & BRYANT, I. D. (eds) *The Geological Modeling of Hydrocarbon Reservoirs and Outcrop Analogues*. International Association of Sedimentologists, Special Publications, **15**, 21–56.

SMITH, D. G. 1988, Modern point bar deposits analogous to the Athabasca oil sands, Alberta, Canada. *In*: DE BOER, P. L., VAN GELDER, A. & NIO, S. D. (eds) *Tide-influenced Sedimentary Environments and Facies*. D. Reidel, Dordrecht, 417–432.

STEIN, W. E., HARMON, G. E. & HUEBER, F. M. 1993. *Spongiophthon* from the Lower Devonian of North America Reinterpreted as a lichen. *American Journal of Botany Supplement*, **80**, 93.

TERWINDT, J. H. J. 1981. Origin and sequences of sedimentary structures in inshore mesotidal deposits of the North Sea. *In*: NIO, S., SHUTTENHELM, R. T. E. & VAN WEERING, TJ. C. E. (eds) *Holocene Marine Sedimentation in the North Seas*. International Association of Sedimentologists, Special Publications, **5**, 4–26.

TYE, R. S. & COLEMAN, J. M. 1989*a*. Depositional processes and stratigraphy of fluvially dominated lacustrine deltas: Mississippi Delta Plain. *Journal of Sedimentary Petrology*, **59**, 973–996.

——, —— 1989*b*. Evolution of Atchafalaya lacustrine deltas, south–central Louisiana. *Sedimentary Geology*, **65**, 95–112.

VAN DEN BERG, J. H. 1981. Rhythmic seasonal layering in a mesotidal channel fill sequence, Oosterschelde Mouth, the Netherlands. *In*: NIO, S., SHUTTENHELM, R. T. E. & VAN WEERING, TJ. C. E (eds) *Holocene Marine Sedimentation in the North*

Seas. International Association of Sedimentologists, Special Publications, **5**, 147–159.

WILLIS, B. J. 1989. Palaeochannel reconstructions from point bar deposits: a three-dimensional perspective. *Sedimentology*, **36**, 757–766.

—— 1993*a*. Ancient river systems in the Himalayan foredeep, Chinji village area, northern Pakistan. *Sedimentary Geology*, **88**, 1–76.

—— 1993*b*. Interpretation of bedding geometry within ancient point-bar deposits. *In*: MARZO, M. & PUIDEFABREGAS C. (eds) *Alluvial Sedimentation.* International Association of Sedimentologists, Special Publications **17**, 101–114.

WITZKE, B. J. 1990. Palaeoclimatic constraints for Palaeozoic palaeolatitudes of Laurentia and Euramerica. *In*: MCKERROW, W. S. & SCOTESE, C. R. (eds) *Palaeozoic Palaeogeography and Biogeography*. Geological Society, London, Memoirs, **12**, 57–73.

—— & HECKEL, P. J. 1988. Palaeoclimatic indicators and inferred Devonian palaeolatitudes of Euramerica. *In*: MCMILLAN, N. J., EMBRY, A. F. & GLASS, D. J. (eds) *Devonian of the World*, Vol. I. *Canadian Society of Petrolium Geologists, Memoirs,* **7**, 49–63.

WRIGHT, V. P. & MARRIOTT, S. B. 1993. The sequence stratigraphy of fluvial depositional systems: the role of floodplain sediment storage. *Sedimentary Geology*, **86**, 203–210.

ZIEGLER, P. A. 1988. Laurussia—the old red continent. *In*: MCMILLAN, N. J., EMBRY, A. F. & GLASS, D. J. (eds) *Devonian of the World, Vol. I. Canadian Society of Petrolium Geologists, Memoirs,* **7**, 14–48.

The geometry, flow patterns and sedimentary processes of Devonian rivers and coasts, New York and Pennsylvania, USA

JOHN S. BRIDGE

Department of Geological Sciences, Binghamton University, Binghamton, NY 13902-6000, USA

Abstract: The Middle–Upper Devonian strata of New York and Pennsylvania were deposited in a foreland basin adjacent to the Acadian fold–thrust belt. Fluvial (Catskill) strata in the east thin to the west into marine (Chemung) strata. The overall sequence is marine regressive, but there were numerous superimposed marine transgressions. The Catskill river channels changed character with distance from the palaeoshoreline. Sinuous, single-channel rivers (widths tens of metres, maximum bankfull depths 4–5 m, sinuosity 1.1–1.3, mean bankfull flow velocity 0.4–0.7 m s^{-1}) occurred near the coast. With increasing distance from the coast, rivers increased in slope, became wider (up to hundreds of metres) and deeper (up to 15 m), coarser grained, and possibly braided. This suggests that rivers were distributive near the coast. Furthermore, muddy floodplain areas were more extensive near the coast than up-valley, such that the proportion of channel deposits increased up-valley. Coastal areas comprised tide-influenced channels and mouth bars, wave-influenced sandy shoals and washovers, sandy and muddy tidal flats, and muddy interdistributary bays and lakes. Major marine transgressions are thus represented in wholly fluvial successions by a decrease in the proportion of channel sandstone bodies, and a decrease in channel size and grain size. Marine transgressions in near-coastal successions are represented in places by an increase in sandstone proportion associated with progradation of coastal sands over muddy backswamps and interdistributary areas. Small-scale marine transgressions and regressions in coastal deposits (i.e. metres-thick parasequences) are probably associated with switching of coastal channels or delta lobes. There is no evidence for the preservation of incised valleys, nor for any changes in coastal channel patterns associated with these parasequences: therefore, major regional sea-level changes were not responsible. However, large-scale (tens to hundreds of metres) transgressive–regressive sequences were due partly to eustatic sea-level change (e.g. in mid-Givetian time) but also due to changes in the balance between sediment supply and subsidence associated with tectonic uplift of the hinterland and peripheral bulge. Such uplift may have induced some kind of climate change. However, climate changes in the depositional basin are not supported by sedimentological evidence. Further understanding of the origin of these strata will require higher precision in the biostratigraphic correlation of marine and non-marine strata (using miospores), and quantitative interpretation of depositional environments from other regions.

The Middle–Upper Devonian Catskill clastic wedge in northeastern North America is part of a foreland-basin fill (adjacent to the Acadian fold–thrust belt) that has relatively thick non-marine deposits to the east and thinner marine deposits to the west (Fig. 1; review by Woodrow & Sevon (1985)). The overall succession of strata is indicative of marine regression: however, there is ample evidence in coastal and fluvial strata of periodic marine transgressions. There are several superimposed scales of cyclicity in these strata, at least three of which can be readily distinguished.

Asymmetrical, coarsening-upward to fining-upward sequences that are tens to hundreds of metres thick represent on the order of hundreds of thousands to millions of years (based on a long-term average deposition rate for the Middle and Upper Devonian succession). These sequences have been interpreted as recording relative changes in sea level associated with eustasy, tectonically induced changes in sediment supply and subsidence rate, and possibly climate change (e.g. Fletcher 1967; McCave 1969; Dennison & Head 1975; House 1983, 1985; Quinlan & Beaumont 1984; Dennison 1985; Ettensohn 1985; Faill 1985; Johnson *et al.* 1985; Brett & Baird 1986; Van Tassell 1987, 1994*a*, *b*; Willis & Bridge 1988; Dennison & Ettensohn 1994; Prave *et al.* 1996). There is no consensus on which of these factors exerted the

From: Friend, P. F. & Williams, B. P. J. (eds). *New Perspectives on the Old Red Sandstone.* Geological Society, London, Special Publications, **180**, 85–108. 0305-8719/00/$15.00

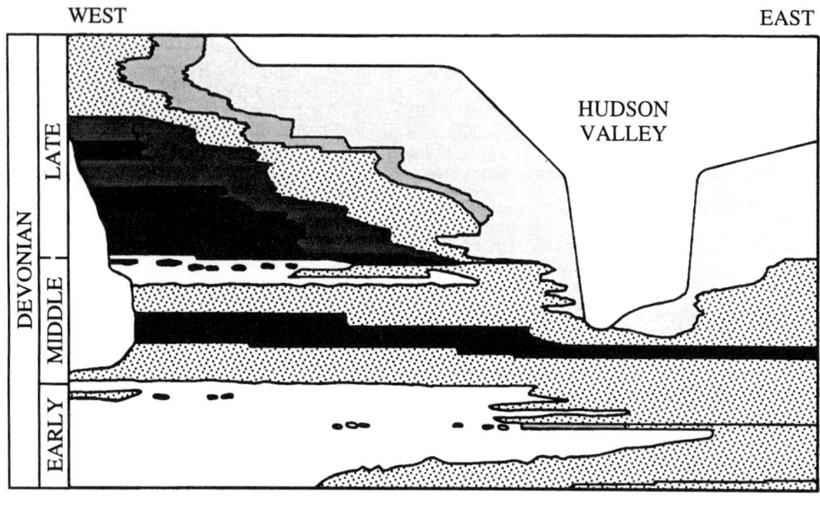

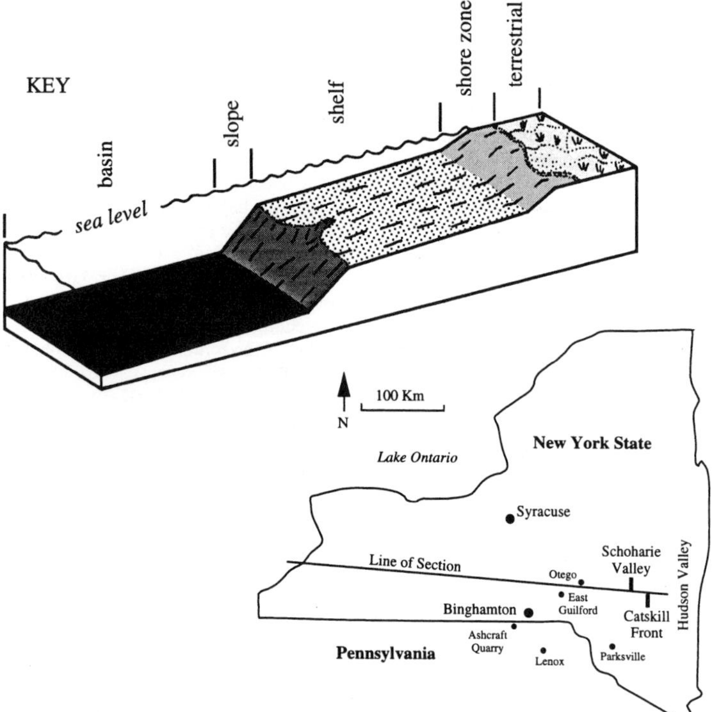

Fig. 1. Summary of Devonian stratigraphy and depositional environments in southern New York State, based on Rickard (1975). The 'slope' in this figure may be more ramp-like (Woodrow, pers. comm.). Location map shows outcrops mentioned in text.

dominant control on these relatively large-scale sequences.

Metres-thick sequences of sandstone and shale that fine upward or coarsen upward represent thousands to tens of thousands of years, and have been interpreted to have been formed by lateral migration, abandonment and filling of channels; progradation and abandonment of

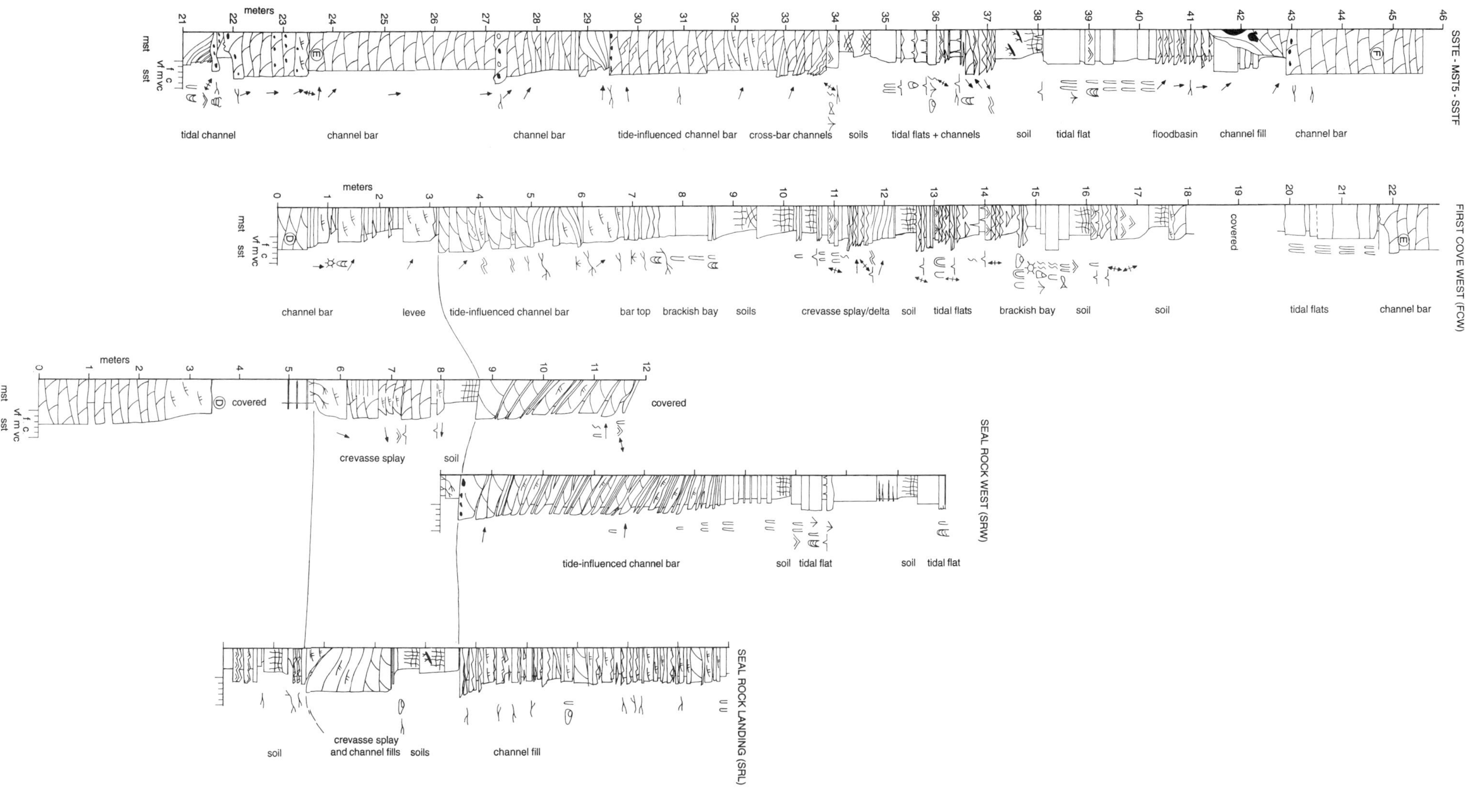

Fig. 3a. Sedimentological logs, including correlated logs at SRL and FCW.

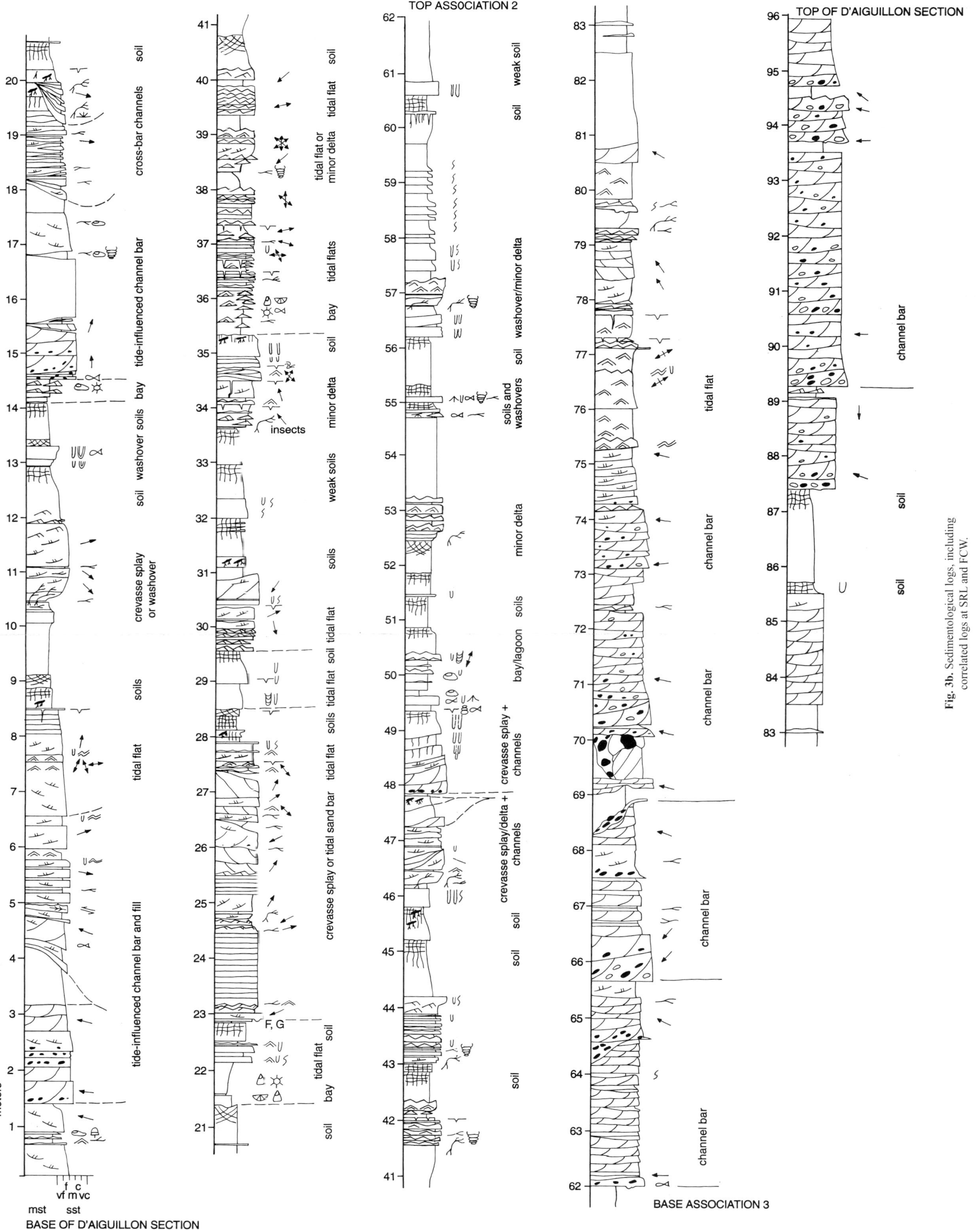

Fig. 3b. Sedimentological logs, including correlated logs at SRL and FCW.

channel-mouth bars, crevasse splays and tidal flats; and filling of coastal bays (e.g. Barrell 1913, 1914; McCave 1968, 1969, 1973; Johnson & Friedman 1969; Gordon & Bridge 1987; Willis & Bridge 1988; Miller & Woodrow 1991; Bridge & Willis 1994; Prave *et al.* 1996; and many others). However, there is disagreement over the plan forms of the palaeochannels. Furthermore, the origin of the coastal sequences (or parasequences) poses many problems, as similar sequences from the Cretaceous Western Interior Seaway have been interpreted in very different ways (discussed by Bridge & Willis (1994)). Although these sequences could have been caused by autocyclic processes such as channel switching (avulsion), allocyclic causes such as eustatic sea-level changes must also be considered (Van Tassell 1987, 1994*a*, *b*; Bridge & Willis 1994; Cotter & Driese 1998). Changes in the nature of these depositional environments over the time scales represented by the thickest cycles described above are poorly known. For example, the relative role of storm waves, tides and river currents in shaping coastal environments during long-term marine regressions and transgressions has not been examined.

Centimetre- to decimetre-thick, sharp-based sandstone stratasets capped by shale are normally interpreted as the deposits of individual floods in rivers and floodplains, of tidal rhythms, or of storm events at sea (e.g. Woodrow & Isley 1983; Craft & Bridge 1987; Gordon & Bridge 1987; Halperin & Bridge 1988; Willis & Bridge 1988; Bridge & Willis 1994; and many others). The origin of these stratasets is much less controversial.

The validity of these interpretations of the different scales of facies sequence hinges on describing them in detail, and on whether or not it is possible to correlate them between outcrops. There is still a critical need for detailed description and quantitative interpretation of these strata: however, this is made difficult by the relatively small, scattered outcrops and the lack of cores. Physical tracing of strata between outcrops is made difficult by the wide spacing of outcrops (and cores), variable dip of strata, lateral changes of facies, lack of distinct marker horizons, and the absence of seismic profiles. Lithostratigraphic correlation of sequences of 10–100 m thickness in marine strata is accomplished using laterally extensive black shales and limestones (Sutton *et al.*, 1962; Sutton 1963; Woodrow & Nugent 1963; McCave 1969, 1973; Rickard 1975, 1989; Brett & Baird 1985, 1986, 1990). However, their correlation with coastal and non-marine rocks remains uncertain (Halperin & Bridge 1988; Bridge & Willis 1994). In coastal and fluvial strata, it is difficult to correlate stratasets of $\leqslant 10$ m thickness for more than on the order of kilometres. Biostratigraphic correlation is very crude because biozones extend for on the order of a million years (100 m of strata) and only recently has it been possible to correlate marine and non-marine rocks using miospores. However, detailed palynostratigraphic work on these strata has barely started. This inability to correlate any but the thickest sequences across the basin makes it very difficult to interpret these deposits in a sequence stratigraphic context, and very difficult to assess whether the controls on their formation are local, regional or global.

The purpose of this paper is to summarize recent work on the Catskill clastic wedge in New York and Pennsylvania that sheds light on some of the problems outlined above. The focus will be on three subject areas: (1) interpretation of fluvial and coastal depositional environments, and especially quantitative interpretation of the geometry, flow patterns and sedimentary processes of the palaeochannels; (2) examination of variations in time and space of depositional environments during marine transgressions and regressions; (3) assessment of the role of climate change, eustasy and tectonism in producing changes in the depositional environments.

Description and interpretation of non-marine strata

Non-marine (Catskill) facies in New York and Pennsylvania have been documented by numerous workers (e.g. Barrell 1913, 1914; Chadwick 1944; Fletcher 1963, 1967; Shepps 1963; Allen & Friend 1968; Johnson & Friedman 1969; Woodrow *et al.* 1973, 1988; Bridge & Gordon 1985; Bridge & Nickelsen 1985; Sevon 1985; Woodrow 1985; Woodrow & Sevon 1985; Bridge *et al.* 1986; Demicco *et al.* 1986; Gordon & Bridge 1987; Gordon 1988; Willis & Bridge 1988). The strata can be divided into grey, metres-thick sandstone bodies (main channel deposits) and grey to red mudstone strata containing relatively thin sandstones (overbank deposits) (Fig. 2).

Main channel deposits: description

Sandstone bodies are sheets of 4–17 m thickness that can be traced laterally for up to 2 km. They consist of single or multiple sets of large-scale inclined strata (storeys), each bounded below by a major erosion surface (Figs 2 and 3). The large-scale strata (bedsets) are inclined at up to about 10° relative to the basal erosion surface of the

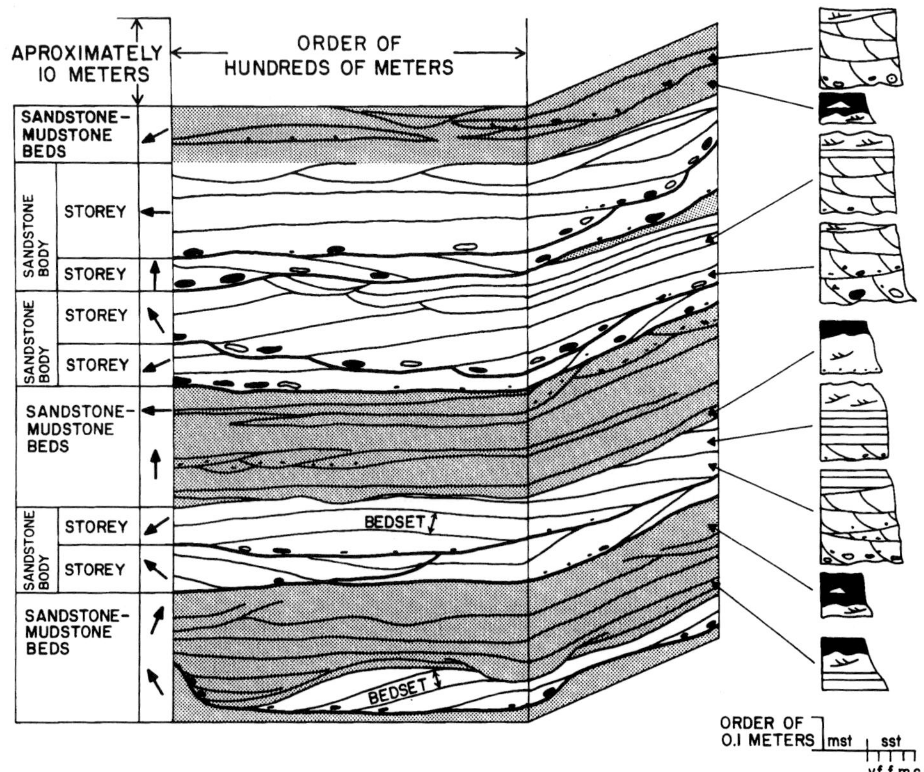

Fig. 2(a). Schematic stratal geometry of the non-marine Catskill facies (modified from Gordon & Bridge (1987)). Thick sandstone bodies (main channel deposits) are unstippled, and mudstone–thin sandstone strata (overbank deposits) are stippled. Stratal geometries lower in the diagram are typical of the finer-grained strata close to the palaeoshoreline, whereas those higher up are characteristic of coarser-grained strata distant from the palaeoshoreline. Symbols explained in (**b**).

storey, and vary in inclination laterally and with outcrop orientation relative to palaeocurrent direction. Large-scale inclined strata are themselves bounded by erosion surfaces, and vary laterally and vertically in composition, grain size and internal structure (Figs 2 and 3). Storeys generally fine upward or vary little in grain size vertically. Most storeys are medium- to fine-grained sandstone, but vary from gravel to very fine sand size. Conglomerates in the lower parts of storeys are both intraformational and extraformational. Medium-scale cross strata (set thickness ranging from 0.03 to 0.5 m, but commonly 0.1–0.3 m) normally dominate the lower, coarser parts of storeys, with more planar strata and small-scale cross strata (set thickness <0.03 m) higher up. The proportion of medium-scale cross strata increases with the mean grain size of the sandstone bodies. Some storeys pass laterally into channel forms that may be filled with sandstone and/or mudstone, and that generally fine upwards. Upper parts of storeys are truncated by relatively small channels (depths up to metres, widths up to tens of metres). Current- and wave-ripple marks also occur in the upper parts of some storeys.

Palaeocurrent directions derived from medium-scale and small-scale trough cross strata vary vertically and laterally within individual sandstone bodies. Palaeocurrent directions are generally to the NW, with most in the range from SW to north. They vary vertically within a storey by up to 50°. Palaeocurrents vary between superimposed storeys by up to about 100°.

Fish fragments, plant remains, and at least two types of burrow are common, and root or

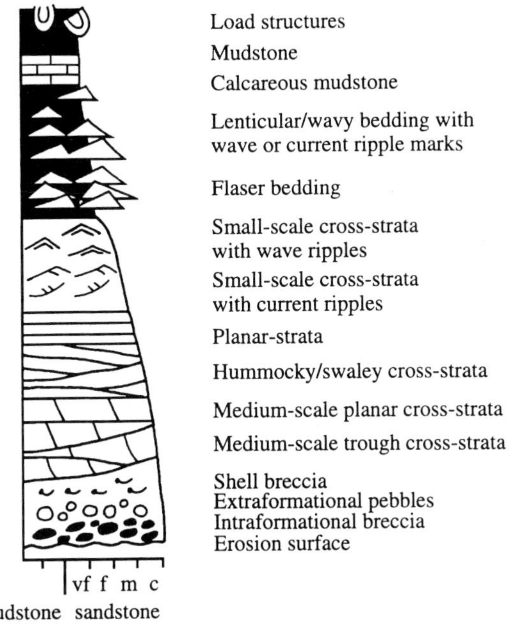

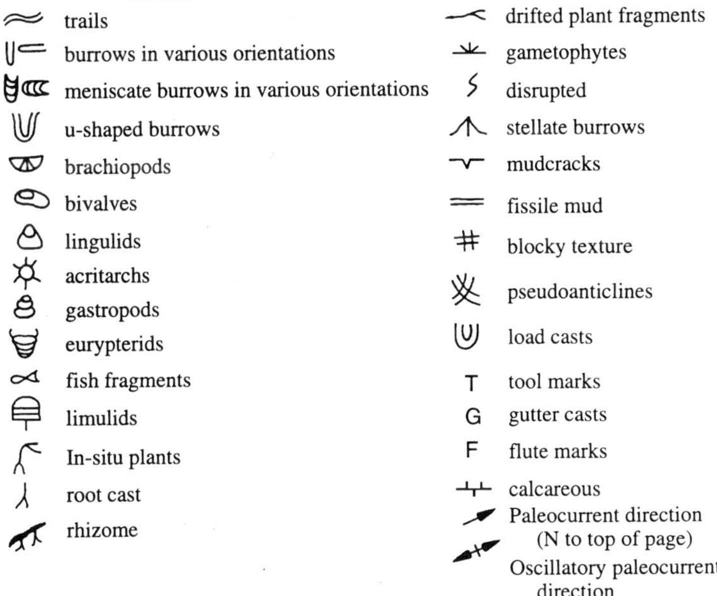

Fig. 2(b). Legend.

rhizome casts penetrate the upper parts of sandstone bodies (Banks *et al.* 1985; Gordon 1988). The freshwater bivalve *Archanodon* and associated burrows also occur in places (e.g. Thoms & Berg 1985; Bridge *et al.* 1986; Gordon 1988).

Main channel deposits: interpretation

Sandstone bodies were interpreted as deposits of predominantly single-channel, sinuous rivers that migrated across vegetated alluvial plains. Large-scale inclined stratasets (storeys) represent

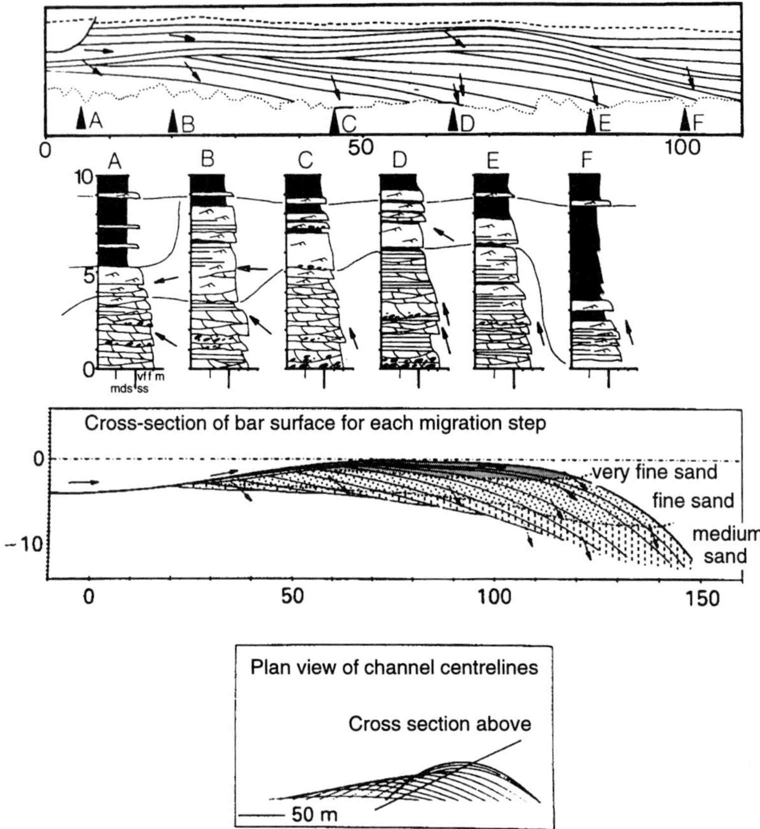

Fig. 3(a).

the deposits of migrating channel bars and channel fills. Vertical and lateral variation in internal structure and grain size within storeys reflect: (1) the spatial distribution of bedforms and surface sediment size on the channel bed during sedimentation events (e.g. floods); (2) the geometry and mode of migration of the bed surface; (3) the outcrop orientation relative to the palaeocurrent orientation (Bridge *et al.* 1986; Willis 1989, 1993*a*, *b*; Bridge 1993). Fining-upward storeys indicate preservation of the downstream parts of bars associated with downstream and lateral migration of bars. Lateral exposure of these sandstone bodies is normally insufficient to allow a determination of whether or not the channels were divided (braided). Fining-upward channel fills indicate progressive reduction in discharge during filling. Coarsening-upward storeys, and storeys showing little vertical variation in grain size, represent the upstream parts of bars. These areas can only be preserved if channel bars migrate by expansion as well as downstream translation. The relatively small channel forms cutting through the upper parts of these sandstone bodies represent cross-bar channels.

The large-scale inclined strata represent the deposits of individual flood events. Vertical changes in internal structure and grain size within large-scale strata represent decreasing flow depth during waning flood stages and decrease or increase in local flow velocity as the water shoaled. The medium-scale cross strata, small-scale cross strata, and planar strata within large-scale strata are associated with deposition from curved-crested dunes and ripples and upper-stage plane beds, respectively. Symmetrical ripple marks on the upper surfaces of large-scale strata indicate the formation of wave ripples in areas of slack water.

Quantitative interpretation of single curved palaeochannels in New York and Pennsylvania

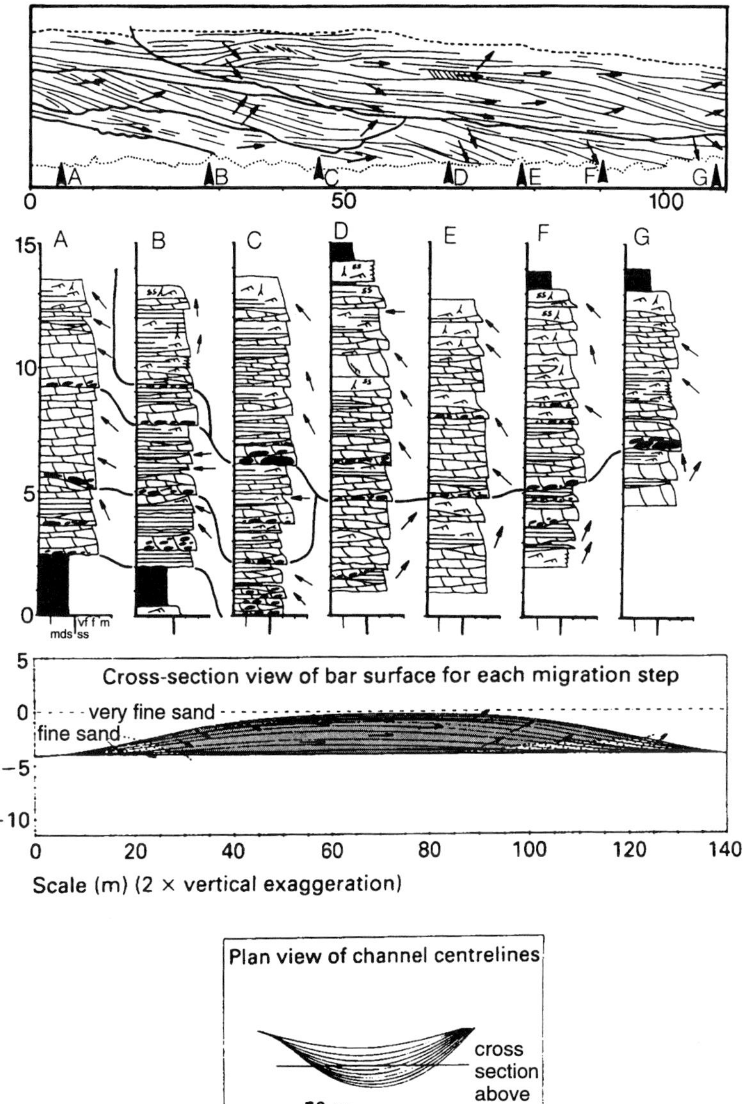

Fig. 3(b).

has been made using 2D static models (Allen 1970), 3D static models (Bridge & Gordon 1985; Gordon & Bridge 1987; Willis & Bridge 1988; Diemer 1992), and 3D dynamic models (Willis 1993*b*: see Fig. 3). River channels were 30–300 m wide, 4–15 m in maximum bankfull depth, and 2–7 m in mean bankfull depth. These dimensions agree with the empirical regression equation presented by Bridge & Mackey (1993) for the variation of channel width, w, as a function of mean channel depth, d_m:

$$w = 8{\cdot}88 d_m^{1{\cdot}82}$$

Channel sinuosity varied up to 1.2 or 1.3 according to the model reconstructions. This sinuosity is very similar to that calculated from the maximum range of palaeocurrent directions

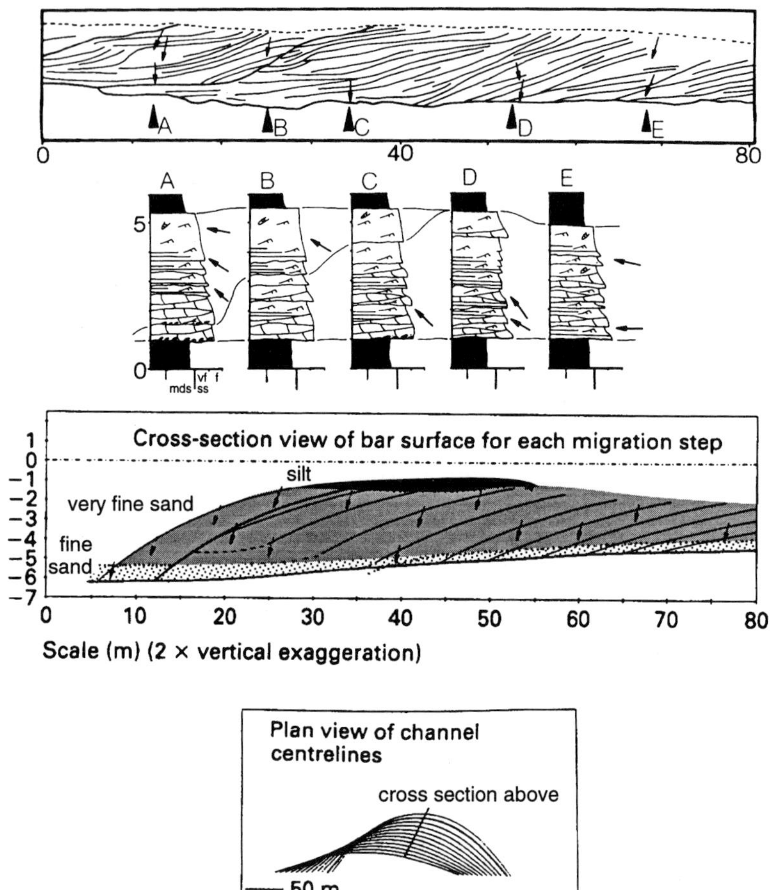

Fig. 3(c).

Fig. 3. Examples of detailed sedimentological studies of sandstone bodies, with quantitative interpretations of paleochannel geometry (redrawn from Willis (1993*b*)). Locations of outcrops shown in Fig. 1.

observed for all of the sandstone bodies (*c.* 100°). The maximum sinuosity is 1.2. Such calculation of sinuosity from palaeocurrent ranges depends on assuming that the channel bends can be represented by sine-generated curves, i.e.

$$sn = 4{\cdot}84/(4{\cdot}84 - \phi^2)$$

where sn is channel sinuosity and ϕ is half of the maximum palaeocurrent range in radians. In most channel sandstone bodies, the observed range of palaeocurrents is much smaller than 100°, yielding only estimates of minimum sinuosity. It is expected that channel sinuosities changed in time and space, varying from unity (straight) to the maximum of around 1.3.

Single channels had mean bankfull flow velocities of 0.4–0.7 m s^{-1}, slopes on the order of 10^{-4}, and bankfull discharge on the order of 10^1–10^3 cumecs. As the full lateral extent of many of the sandstone bodies could not be observed, there is a real possibility that some of them (especially the thicker ones) might show the laterally adjacent channel bars and fills characteristic of braided rivers. There have not yet been any quantitative interpretations of such braided palaeochannels, although it is now theoretically possible (Bridge 1993).

Overbank deposits: description

The mudstone–thin sandstone strata commonly occur in centimetre-thick to decimetre-thick

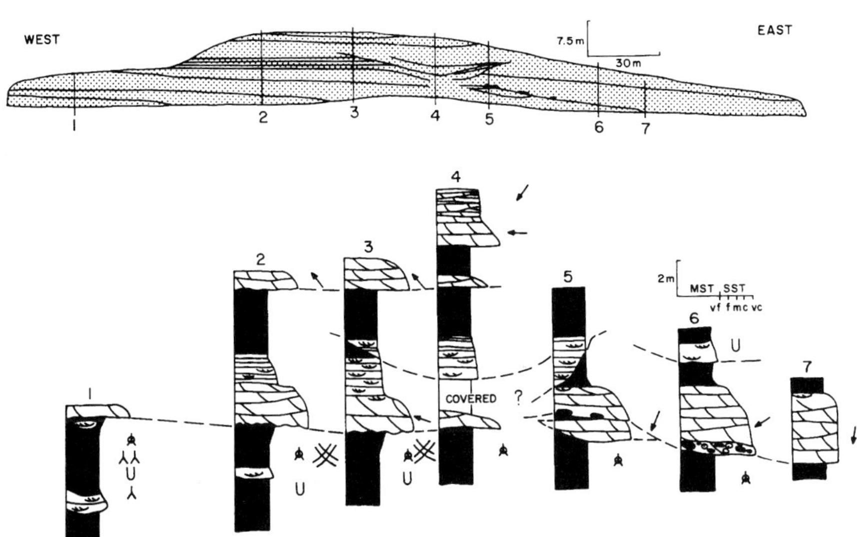

Fig. 4. Examples of overbank deposits from Parksville, New York (redrawn from Gordon & Bridge (1987)). Location of outcrop shown in Fig. 1.

stratasets (hereafter called large-scale strata) that are organized into metres-thick sequences of stratasets (hereafter called large-scale stratasets) (Fig. 4). The large-scale stratasets may be sheets, wedges, lenses, or channel fills. Channel fills commonly fine upwards from sandstone to mudstone. Metres-thick, lenticular and wedge-shaped, large-scale stratasets of sandstone (sandstone bodies) commonly coarsen upwards and may contain relatively small channel fills in their upper parts (Fig. 4). Large-scale sheet-like stratasets commonly fine upwards.

The individual large-scale sandstone strata generally have erosional bases, and fine upward to siltstone. Medium-scale cross strata and planar strata occur near the base of some of these sandstone strata, and dominate the thicker, coarser-grained ones. Small-scale cross strata occur at the top of the thicker sandstone strata and dominate the thinner, finer-grained ones. Wave ripples and current ripples occur on the top surfaces of sandstone stratasets. Most sandstone stratasets have bioturbated tops, and thinner ones are completely bioturbated.

Mudstones are either highly bioturbated (root casts and burrows) and desiccation cracked (giving a blocky texture) or are relatively undisturbed and fissile. Calcareous concretions occur in distinct horizons associated with blocky textured, red, mottled mudstone with slickensided, clay-lined surfaces, pseudoanticlines, burrows and root casts. Some grey–green, bioturbated, blocky mudstones containing roots or rhizomes are overlain by sandstone strata containing casts of *in situ* plants (e.g. Bridge & Willis 1994; Driese *et al.* 1997). Well-preserved, transported plant remains, including spores, occur in grey fissile mudstone (Banks *et al.* 1985). At least four kinds of burrow, and four kinds of trail occur in these mudstone–sandstone strata (e.g. Gordon 1988). Notable is the occurrence of the freshwater bivalve *Archanodon* and associated escape burrows (Bridge *et al.* 1986; Gordon 1988; Friedman & Chamberlain 1995). Ostracodes (Gordon 1988; Friedman & Lundin 1998) and dolomitic carbonate beds (Demicco *et al.* 1987) occur rarely near the palaeoshoreline.

Overbank deposits: interpretation

The mudstone–sandstone strata were deposited on river floodplains. Sandstones were deposited during overbank floods in crevasse and floodplain drainage channels, on crevasse splays, levees and lacustrine deltas, and as overbank sheetflows. Mudstones were deposited in floodbasins and channels, and during waning flow stages in other areas. Blocky textured red mudstones with calcareous concretions, mottles, slickensides and pseudoanticlines indicate repeated wetting and drying and formation of ped structure typical of calcareous palaeosols

CATSKILL SANDSTONE BODIES

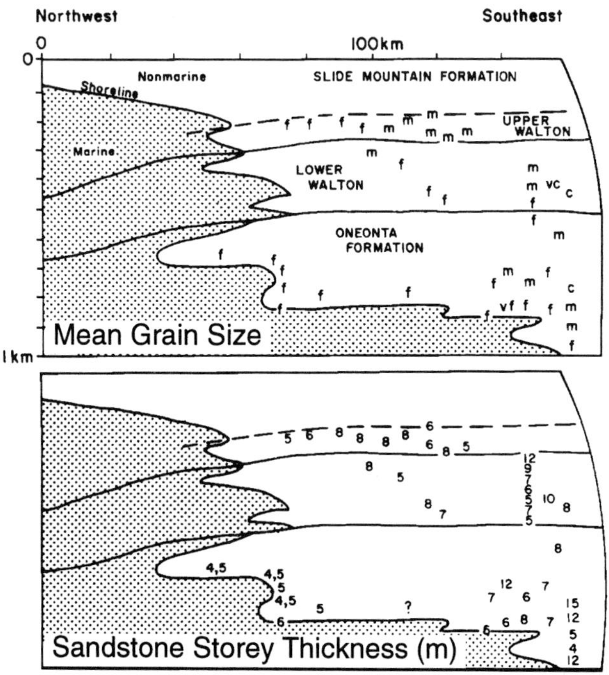

Fig. 5. Variation in the mean grain size and thickness of channel bars (sandstone storeys) in part of the Catskill clastic wedge in southern New York (redrawn from Gordon & Bridge (1987)). It should be noted that fine-grained (f) sandstone bodies occur relatively close to the palaeoshoreline, whereas medium (m) and coarse (c) sandstone bodies occur more distant from the palaeoshoreline. Channel bar thickness, a measure of maximum bankfull channel depth, also tends to increase with distance from the palaeoshoreline.

(vertisols). The grey–green blocky mudstones represent waterlogged (gleyed) palaeosols that are not as well developed as the calcareous vertisols (i.e. inceptisols or entisols: Driese *et al.* 1997). Grey fissile mudstone indicates less disruption and relatively rapid deposition in abandoned channels and lakes on the floodplain. Plant and animal activity was pervasive and diverse.

Variations of non-marine strata in space and time

The proportion, thickness and mean grain size of the main-channel sandstone bodies increase with distance from the palaeoshoreline (Figs 5 and 6). These trends were interpreted to represent an increase in river-channel size, slope and size of the sediment load with distance from the palaeoshoreline, implying a distributive river system (Gordon & Bridge 1987; Willis & Bridge 1988). It is also possible that braided rivers occurred in positions furthest from the palaeoshoreline. As seen below, the tide-influenced river channels were very similar to the near-coastal river channels.

The proportion, thickness and mean grain size of the main-channel sandstone bodies also increase with time at a point over hundreds of metres of strata (Fig. 6). These trends are accompanied by changes in palaeocurrent direction (e.g. from westerly to more northerly: Fig. 6), increase in the proportion of sandstone in overbank deposits, and decrease in the abundance and maturity of calcareous palaeosols. These trends were interpreted by Willis & Bridge (1988) to reflect regionally increased slopes and deposition rates, and progradation of the distributary fluvial systems, during periods of increased rate of source area uplift and sediment supply. Such sediment progradation produced the major, late Givetian marine regression. These uplift-related trends are more marked in proximal alluvium

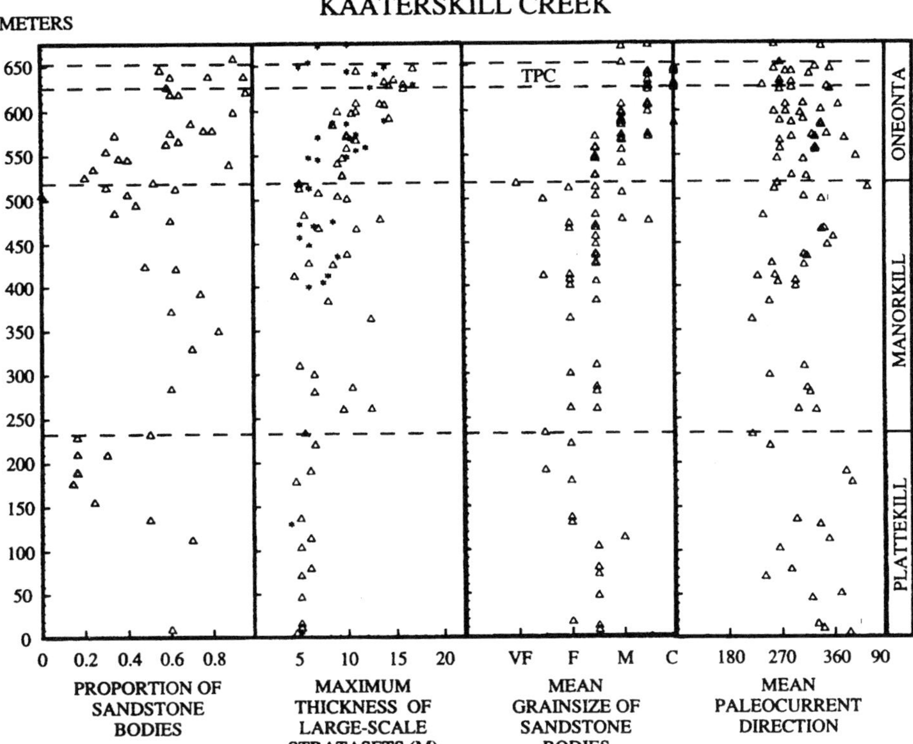

Fig. 6. Stratigraphic variation of proportion of channel sandstone bodies, thickness of large-scale stratasets (storeys), mean grain size of channel sandstone bodies, and mean palaeocurrent direction for the section exposed in Kaaterskill Creek (redrawn from Willis & Bridge (1988)). Kaaterskill Creek is in southern New York at the eastern edge of the Catskill mountains, and the western edge of the Hudson Valley (the Catskill Front: Fig. 1). Formation names given to right.

than that near the palaeoshoreline. Indeed, there is no evidence for changes in channel size and pattern near the coast during relative falls and rises in sea level. There is apparently no clear depositional evidence of climatic change affecting the rivers and floodplains, but a climatic change in the Acadian hinterland cannot be ruled out.

Description and interpretation of coastal strata

The sedimentology of coastal deposits has been studied in New York, Pennsylvania and adjacent areas by Allen & Friend (1968), McCave (1968, 1969, 1973), Johnson & Friedman (1969), Sutton *et al.* (1970), Walker (1971), Walker & Harms (1971, 1975), Bridge & Droser (1985), Woodrow (1985), Slingerland (1986), Boswell & Donaldson (1988), Halperin & Bridge (1988), Slingerland & Loulé (1988), Duke & Prave (1991), Miller & Woodrow (1991), Bridge & Willis (1994), Prave *et al.* (1996), Cotter & Driese (1998) and others. The strata comprise metres-thick sandstone bodies with minor amounts of mudstone and metres-thick mudstone-dominant intervals that contain relatively thinner (millimetre to decimetre) sandstone strata (Figs 7 and 8). Sandstones are mainly greenish grey and range in mean grain size from medium to very fine. There is a large range of variability in the thickness, geometry and lithofacies of the sandstone bodies (summarized in Fig. 7). For this discussion, sandstone bodies are divided into thick (3–6 m) and thin (1–2 m) varieties. Mudstones are dark greenish grey to dusky reddish grey and are mainly siltstones, with relatively rare claystones.

Thick sandstone bodies: description

Some thick sandstone bodies are 4–6 m thick and hundreds of metres wide. They are made up of a single set of large-scale inclined strata (storey) (Figs 7 and 8). Storeys have basal erosion

A
ONSHORE
TIDAL CHANNEL BAR
OFFSHORE
Flood-Oriented Washover
Flood-Dominated or Shoaling Waves
Ebb-Dominated
m
0
1
2
3
4
5
vf f m c vc
mud sand gravel

Fig. 7(a).

B
MAJOR TIDAL CHANNEL BAR TRANSITIONAL TO MOUTH BAR / NEARSHORE SANDS
Tidal Channel Bar
Proximal Channel Mouth Bar
Distal Channel Mouth Bar
m
0
1
2
3
4
5
6
vf f m c vc
mud sand gravel

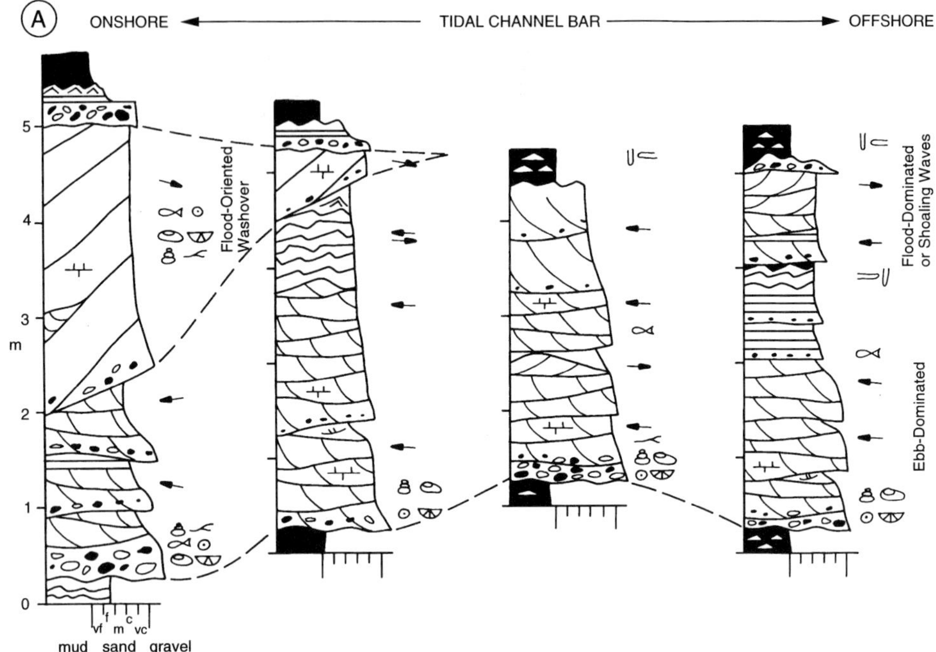

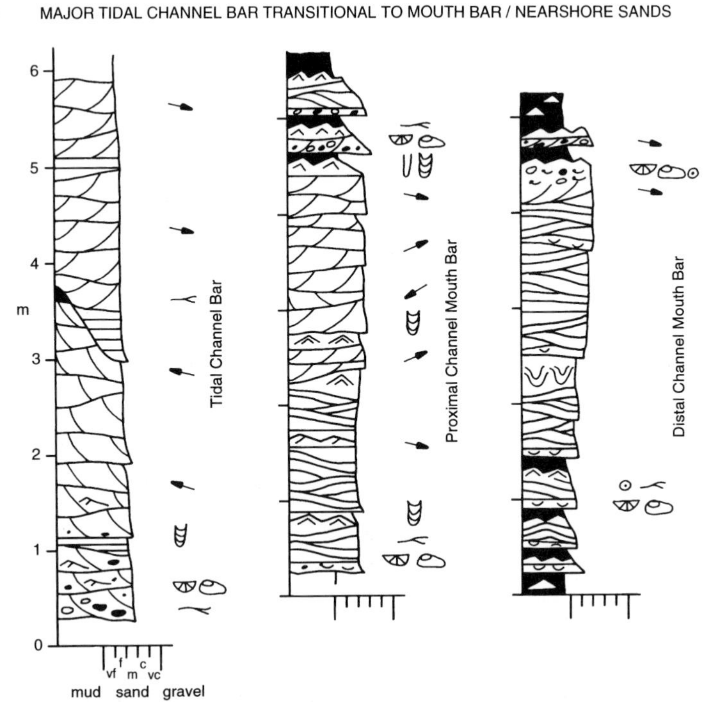

Fig. 7(b).

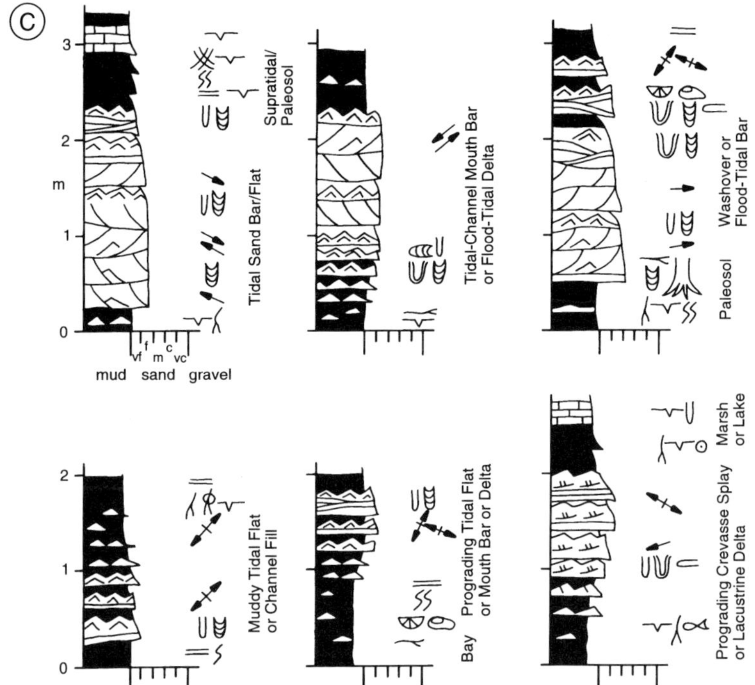

Fig. 7(**c**).

Fig. 7. Examples of metre-scale coastal sandstone bodies and sandstone–mudstone sequences in coastal deposits of the Catskill clastic wedge. (**a**) based on Ashcraft Quarry (Bridge & Droser 1985; Fig. 1). (**b**) and (**c**) based on outcrops in Schoharie Valley (Fig. 1: Bridge & Willis (1994)). Symbols explained in Fig. 2b. Other examples are given by Griffing *et al.* (this volume).

surfaces, commonly overlain by intraformational breccia and extraformational conglomerate. Large-scale inclined strata range from centimetres to metres thick (generally decimetres), and are bounded by relatively minor erosion surfaces overlain by intraformational breccia. The large-scale strata are inclined at up to 10° relative to the basal erosion surface of the sets. The largest inclinations generally occur where palaeocurrents are perpendicular to outcrop faces. Some inclined strata fill channels. Relatively small channel forms (metres to tens of metres wide and decimetres to metres thick) commonly cut through the upper parts of the storeys.

Grain size of the storeys commonly decreases upwards, from medium-grained sand to very fine grained sand. Some storeys show little vertical variation in grain size, or coarsen upward (Figs 7 and 8). Channel fills generally fine upward: some are dominantly sandstone (coarse-grained channel fills), but others have large proportions of siltstone (fine-grained channel fills). Grain size commonly decreases at the top of individual large-scale strata (e.g. from medium to fine sand or from fine to very fine sand).

Medium-scale trough cross strata are the dominant internal structure of the large-scale strata. Medium-scale planar cross strata also occur, especially in the upper parts of storeys. Wave- and current-ripple marks may be superimposed on medium-scale cross strata and their set boundaries, commonly indicating palaeocurrent orientations very different from those of the medium-scale cross strata. Small-scale cross strata and planar strata are normally subordinate to medium-scale cross strata in most of the sandstone bodies, and tend to occur in their finer-grained parts. Small-scale cross strata are mainly in the upper parts of sandstone bodies, whereas planar strata can occur virtually anywhere. The finest-grained sandstone storeys and channel fills tend to have relatively high proportions of small-scale cross strata. Asymmetrical and symmetrical ripple marks occur in the upper parts of the large-scale strata, and flute, gutter and tool marks occur along their erosional bases.

Some sandstone bodies, especially the thinner and finer-grained ones, are dominated by swaley cross strata and planar strata, with subordinate wave-ripple marks in the finer-grained parts of

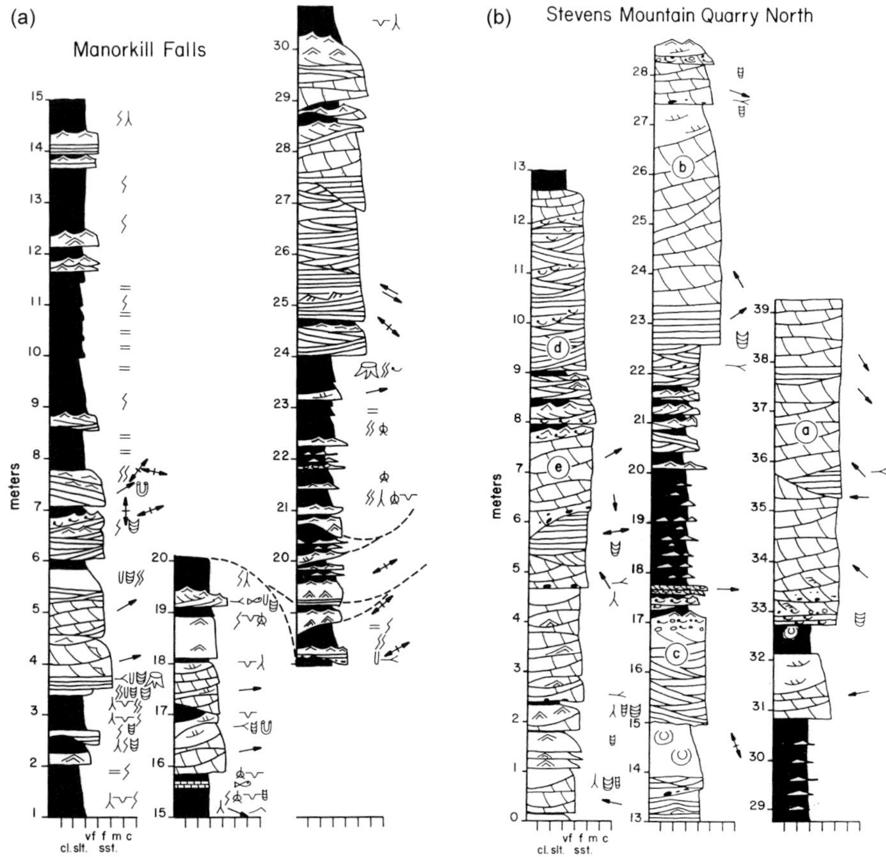

Fig. 8. Sequences of Devonian coastal deposits from the Schoharie Valley in southern New York State (redrawn from Bridge & Willis (1994)). Locations of outcrops shown in Fig. 1.

stratasets. These sandstone bodies have sharp bases, although they may sit above fossiliferous, coarsening-upward sequences containing abundant hummocky cross strata, wave-ripple cross strata, and, in places, soft-sediment deformation structures (load casts). These sandstone bodies may fine upward, very little in grain size, or coarsen upward. The coarsening-upward varieties have conglomeratic, shell-rich strata capping them, in some cases filling small channels. Sandstone bodies dominated by swaley cross strata and planar strata may thicken and coarsen laterally over hundreds of metres and show an increase in the proportion of medium-scale cross strata in the upper part of the sandstone body. With further lateral thickening, the swaley cross strata are replaced completely by medium-scale cross strata.

Palaeocurrent directions derived from medium-scale and small-scale cross strata vary vertically and laterally within individual sandstone bodies. Palaeocurrent directions are generally to the NW, with most in the range from SW to north. Palaeocurrents in the upper parts of sandstone bodies are commonly in the opposite direction (NE–SE). Rarely, opposing palaeocurrents are manifested in herringbone cross stratification. Wave- and current-ripple marks superimposed upon medium-scale cross strata commonly indicate palaeocurrents up or along the lee faces of the host dunes.

Transported disarticulated brachiopods, bivalves, crinoid ossicles, eurypterid cuticle fragments and fish-bone fragments occur along surfaces of large-scale strata in the coarser-grained parts of sandstone bodies. The types of fossils present vary with location of the sandstone body relative to the palaeoshoreline. Robust brachiopod and bivalve shells and crinoid ossicles occur in the sandstone bodies nearest the palaeoshoreline. Burrows such as *Skolithos* and *Arenicolites*, also surface trails, occur in tops of large-scale strata in the upper parts of sandstone bodies. Bivalve escape burrows can occur throughout the sandstone bodies. Transported plant material is very common. Some sandstone bodies may have important occurrences of *in situ* and *in loco* plants, especially in the upper 1 m or so. Evidence for *in situ* occurrence or minimal transport (*in loco*) includes roots or rhizomes, regular spacing of axes that may be bent over in the palaeoflow direction, and complete preservation of delicate detail. *In situ* plants occur in large-scale strata where mudstone caps sandstone. The plants are anchored in the mudstone and project upwards through the overlying sandstone bed, where they are bent over in the palaeoflow direction (Elick *et al.* 1998; Hotton *et al.* 1999; Griffing *et al.* this volume). Most of these plant occurrences tend to be monospecific, although transported taxa may also be present.

Thick sandstone bodies: interpretation

The thickest sandstone bodies are interpreted as the deposits of the main channels that migrated across the ancient coastal plain. Palaeocurrent and lithofacies variations within sandstone bodies indicate the influence of tidal currents. The lower parts of the channels were dominated by river currents and by ebb-tidal currents, whereas the uppermost parts of the channels were dominated by flood-tidal currents. This type of current segregation is typical of the strongly asymmetrical tidal currents expected in the channels of estuaries and tide-influenced deltaic distributaries. In some sandstone bodies, the distribution of grain size and sedimentary structures is typical of tidal channel bars near the fluvial–tidal transition (e.g. Barwis 1978; Van den Berg 1981; Smith 1988; Allen 1991; Dalrymple *et al.* 1992). Here, medium-scale cross strata are limited to the lower (subtidal) parts of bars, whereas there is a dominance of small-scale cross strata and relatively high proportions of mud in the upper, intertidal parts. The rarity of tidal-bundle sequences may be due to non-depositional subordinate tides and dominant ebb currents that were reinforced by fluvial currents. These coastal channel deposits had similar geometry and hydraulics to the smaller fluvial channels described above.

The thick sandstone bodies with swaley cross stratification throughout, or in their lower parts, are interpreted as the deposits of channel-mouth bars and associated nearshore sand accumulations that were prograding into a marine, storm-wave dominated area (Bridge & Willis 1994; Prave *et al.* 1996). Such bars may have been on the seaward side of tidal inlets associated with estuary-mouth shoals or associated with the mouths of tide-influenced deltaic distributaries. The underlying fossiliferous sandstone–mudstone strata represent nearshore marine deposits formed below fairweather wave base. The load casts near the bases of these sandstone bodies represent soft-sediment deformation, a feature common offshore from delta distributaries (Coleman & Prior 1982). The parts of the sandstone bodies with swaley cross strata were deposited under the influence of storm waves above fairweather wave base. The easterly directed palaeocurrents in the upper parts of these sandstone bodies were associated with flood-tidal currents, onshore-directed storm currents, or asymmetrical shoaling wave currents. The coarse shelly strata capping some sandstone bodies may be associated with abandonment of the mouth bar and marine transgression (e.g. Penland *et al.* 1988).

In situ plant occurrences in sandstone bodies are due to burial of stands of plants that were living on the upper parts of migrating channel bars, including in cross-bar channels. The sedimentation events were probably associated with river floods, although flood tides probably also played a role. The plants then became re-established on the muddy, falling-stage drapes as water level receded. The absence of desiccation cracks in the mudstone drapes indicates that these sediments remained wet. This certainly reflects high water tables expected in near-coastal environments, and probably also reflects frequent flooding during high tides. The sparse fauna and low degree of bioturbation (but dominance of the *Skolithos* ichnofacies) are typical of sandy deposits in coastal areas (Howard & Frey 1975; Howard & Reineck 1981). The robust brachiopod–bivalve fauna is typical of Devonian energetic nearshore marine communities (McGhee & Sutton 1985; Sutton & McGhee 1985). The bivalve–eurypterid fauna appears to represent a less energetic and brackish-water setting.

Thin sandstone bodies: description and interpretation

Thin sandstone bodies are normally 1–2 m thick, but may thin laterally and pass into mudstone-dominant strata. These sandstone bodies are sheet-like or lenticular (planar base and convex-upward top or channel filling: Figs 7 and 8). Most sandstone bodies are fine or very fine grained, but medium sandstone occurs in places, and inter-stratified mudstone is common. All sandstone bodies comprise sets of large-scale strata that are centimetres to decimetres thick, generally fine upwards with a change of sedimentary structure, and may be inclined by several degrees relative to the base of the sandstone body. Facies sequences within these sandstone bodies are very variable, and some examples are shown in Figs 7 and 8.

Some thin sandstone bodies have erosional bases (channel-form in some cases) and generally fine upwards. The coarser-grained sandstone bodies commonly have medium-scale cross strata and planar strata overlain by small-scale cross strata, whereas the finer-grained sandstone bodies have dominantly small-scale cross strata. Dune and current-ripple forms are common in these cases. Palaeocurrent orientations are normally dominantly in one direction (e.g. westerly or easterly), but with subordinate orientations in a near-opposite direction. Wave-ripple marks and associated cross strata commonly occur, especially in the upper parts of these sandstone bodies, in places superimposed on dune forms or on medium-scale cross strata. Wave-ripple crest lines are commonly oriented SW–NE: however, crest lines may show a large range of orientations in some cases. Hummocky–swaley cross strata also occur, normally associated with wave-ripple marks, and rarely contain shelly fossils (brachiopods and bivalves).

Desiccation cracks occur in the mudstone layers in the upper parts of sandstone bodies. Transported plant material is common throughout the sandstone bodies, and plant roots or rhizomes occur in the upper parts. In places, sandstone casts of *in situ* tree trunks occur in the base of this type of sandstone body (e.g. the famous Gilboa fossil forest: Banks *et al.* 1985). Trace fossils of the *Skolithos* ichnofacies are common (e.g. *Skolithos*, *Arenicolites*, *Diplocraterion*, *Planolites*, *Meunsteria*: Miller 1979; Miller & Woodrow 1991). Bioturbation increases upwards within sandstone bodies. Surface traces also occur on bedding surfaces. These sandstone bodies commonly pass upward into dusky red to grey siltstones with green mottles, blocky fabric, slickensides, pseudoanticlines, root and rhizome casts and burrows. Green–grey siltstones also have these features but not as well developed. These siltstones are commonly in decimetre-thick, fining-upward units.

This type of thin sandstone body is interpreted as the deposits of prograding sandy tidal flats with relatively small, tide-influenced channel bars and fills (see Bhattacharya & Walker 1992; Dalrymple 1992). They include subtidal and intertidal zones, with the overlying mudstones representing supratidal and floodplain deposits. Palaeocurrent orientations generally indicate either flood- or ebb-fluvial dominance. Minor sedimentation on upper-bar surfaces during tidal slack water is represented by wave ripples and mud drapes. Wave-ripple crests are oriented parallel to bar margins in some cases. In others, the large range of orientations is as would be expected on sandy intertidal flats. Hummocky–swaley cross strata associated with planar strata and wave-ripple marks indicate periodic influence of storm waves. Some sandstone bodies with easterly-directed palaeocurrents (including those that buried tree trunks in place) may represent storm washovers from coastal sand shoals. Red siltstones are vertic palaeosols and green–grey siltstones are less well developed and wetter palaeosols (inceptisols or entisols: Driese *et al.* 1997).

Some thin sandstone bodies coarsen upwards, but normally fine upwards at the top. They are sheet-like to lenticular, with flat base and convex-upward top (Fig. 7), and shallow channels (decimetres deep and metres wide) occur in the upper parts of some of these sandstone bodies. Basal parts are muddy, and are commonly intensely bioturbated, with closely spaced vertical tubes (*Skolithos*). *In situ* plants are common at the base of these sandstone bodies. Immediately below these occurrences, millimetre- to centimetre-thick layers of dark green–grey claystone contain dense accumulations of poorly preserved plant axes, and possibly eurypterid fragments and 'bugs' (Shear *et al.* 1984). Organic structures interpreted as roots or rhizomes occur below *in situ* plant axes (Elick *et al.* 1998; Hotton *et al.* 1999; Griffing *et al.* this volume). The *in situ* plant axes are bent over in the palaeocurrent direction. In places, a given inclined axis is bent parallel to bedding and then projects upwards once again. Transported plant material is also common.

This type of thin sandstone body mainly contains small-scale cross stratification associated with current-ripple marks. Palaeocurrents from current ripples and associated cross strata are either westerly or easterly, but widely varying in direction and opposing in some cases. Some sandstone bodies have wave-ripple marks and associated cross strata. Wave-ripple crest lines have a large range of orientations. In the upper, coarsest parts of the sandstone bodies there may be some medium-scale cross strata, planar strata, and, rarely, hummocky cross strata. In general, disruption in the form of desiccation cracks, burrows and roots increases markedly at the top. Typical burrows are vertical tubes (*Skolithos* ichnofacies). Surface traces are also common. These sandstone bodies commonly pass upward into dusky red or green–grey palaeosols.

These sandstone bodies are interpreted as channelized and unchannelized parts of crevasse splays or minor deltas that prograded into floodplain marshes and lakes or brackish interdistributary bays–lagoons. Similar deposits have been described by Coleman (1969), Elliott (1974), Farrell (1987) and Tye & Coleman (1989*a*, *b*). Evidence of waves acting in standing water is wave ripples, planar strata and hummocky cross strata. Easterly directed palaeocurrents suggest that some of these sandstone bodies may be flood-tidal deltas or washovers. Intensity of burrowing increases as deposition rate decreases.

The thin, plant-rich layers beneath the *in situ* plants may represent the plant litter that accumulates beneath long-established stands of vegetation (i.e. O horizon of soil). The plants in this setting suggest long-term vegetative growth, in accord with a relatively stable, wet and low-energy environment, such as a floodplain marsh or backswamp. The bends in the plant axes indicate growth through different episodes of sedimentation, again supporting a degree of longevity.

Mudstone-dominant strata: description and interpretation

Somewhat thinner (decimetre-scale) sequences of very fine sandstone and mudstone may coarsen upward from lenticular to wavy to flaser bedding (with wave-ripple marks), or they may fine upwards (Figs 7 and 8). Wave ripples may be 2D, 3D or interfering types. Wave-ripple crests have a large range of orientations but tend to be oriented NW–SE or NNE–SSW. The sandstones contain a diversity of trace fossils (e.g. *Skolithos*, *Arenicolites*, *Diplocraterion*, *Planolites*, *Phycodes*). The mudstones may be laminated and contain randomly occurring, centimetre-thick sheets and lenses of bioturbated, very fine sandstone with either current ripples or wave ripples and associated small-scale cross strata (rarely hummocky cross strata). Other mudstones are more disrupted and have desiccation cracks, root casts, burrows and nodular carbonate-rich layers. The distinctive trace fossil *Spirophyton* occurs in isolated horizons of fissile red or grey–green mudstone (Miller & Johnson 1981; Miller 1991). Mudstones may also contain transported articulated lingulid brachiopods, orthid and rhynchonellid brachiopods, disarticulated bivalves, gastropods, fish-bone fragments, ostracods and acritarchs. Transported plant fragments are common throughout.

These mudstone-rich sequences record transition of sandy–muddy tidal flats into muddy, brackish to fully marine interdistributary bays or lagoons (Reineck & Singh 1973; Terwindt 1981; Coleman & Prior 1982). Shallowing or deepening could be due to sediment progradation (e.g. distal edge of a crevasse splay or minor delta) or retrogradation, or could be associated with sea-level change. The facies changes could also be associated with seasonal variations in wave energy. It is difficult to make a distinction in many cases because of lack of lateral exposure. Thin sand layers in mudstone sequences may be due to seasonal floods or storms in coastal embayments. These bays were frequently filled to supratidal level and/or experienced sea-level falls, as evidenced by the upward transition to palaeosols (saltmarsh deposits?). The sparse fauna, relatively low bioturbation and dominance of *Skolithos* ichnofacies are typical of coastal sands and silts (Howard & Frey 1975; Howard & Reineck 1981). *Spirophyton* is apparently typical of muddy, brackish-water settings (Miller & Johnson 1981; Miller 1991). Calcareous mudstones were apparently formed in shallow coastal ponds or supratidal marshes (Demicco *et al.* 1987).

Discussion of metre-scale sandstone–mudstone sequences

The metre-scale sequences of sandstone bounded by mudstone are parasequences using the terminology of Van Wagoner *et al.* (1988, 1990) and Mitchum & Van Wagoner (1991). As average deposition rates for the Middle–Upper Devonian sediments were very approximately 0.1 mm year^{-1}, these sequences represent 10^4–10^5 years on average. The metre-scale sequences in coastal deposits were interpreted by Bridge & Willis (1994) to record changes in relative sea level related to processes such as lateral migration, abandonment and filling of tidal channels; progradation of channel mouth bars and tidal flats; and filling of coastal bays. Interpretations of Givetian and Frasnian coastal and marine deposits in various parts of New York and Pennsylvania have several features in common: (1) sandy, tide-influenced channels; (2) shallow bays and tidal flats where mud and sand were deposited; (3) rarity of beaches; (4) storm-wave domination of the marine shelf (Bridge & Willis 1994; Prave *et al.* 1996). Much of the variability in the deposits across the area could be explained within the context of a wave- and tide-influenced deltaic coastline with a tidal range that varied in time and space (see Slingerland 1986; Erickson *et al.* 1990).

Some metres-thick coastal sequences in the Catskill clastic wedge have been interpreted to reflect eustatic sea-level changes associated with Milankovich climatic cycles (Van Tassell 1987, 1994*a*, *b*; Cotter & Driese 1998). Such an allocyclic control could only be justified if sequences could be traced laterally for distances greater than would be expected for individual delta lobes, and indeed across the whole basin. However, correlation of sequences for more than a few hundred metres has proved difficult in these types of rocks (Duke & Prave 1991; Miller & Woodrow 1991; Prave *et al.* 1996).

Estuaries and sand barriers with muddy bays on their landward sides are expected in coastal areas during relative sea-level rise (e.g. Allen 1991; Dalrymple *et al.* 1992; Reinson 1992). Although there is evidence for washovers from sandy coastal shoals, there is no clear evidence for barrier beaches. Furthermore, no evidence was found for ravinement during regressions and estuarine filling during transgressions. Recently, Cotter & Driese (1998) have interpreted such valley incision and filling for Upper Devonian strata in Pennsylvania: however, the evidence presented is unconvincing. The sparse evidence for tidal bundles in cross-stratified sands indicates that the North Sea tidal flats may not be good analogues for these deposits (*contra* McCave 1968; Johnson & Friedman 1969). The strongly asymmetrical, ebb-dominated currents possibly record the dominance of river flow during deposition.

Sharp-based coastal sandstone bodies similar to those in the Catskill clastic wedge of New York and Pennsylvania, but occurring in rocks of different age in other regions (particularly the Cretaceous Western Interior Seaway), have been ascribed to deposition on prograding strand plains cut by estuaries or tidal inlets, and to wave-formed offshore bars or shoals (references given by Bridge & Willis (1994)). Sandstone bodies with swaley–hummocky cross strata overlain by angle-of-repose cross strata have been interpreted as beach and shoreface deposits, their sharp bases being related to wave-current erosion during relative sea-level fall and so-called forced regression. The coarse-grained strata on the tops of these sandstone bodies were taken as so-called transgressive lags associated with relative rise in sea level. In contrast, Bridge & Willis (1994) explained the sharp bases of some of the sandstone bodies as due to rapid progradation of storm-wave modified channel-mouth sands that were deposited rapidly as a result of channel diversion. A beach face origin of the sandstone bodies was considered unlikely because of absence of: (1) characteristic seaward-dipping planar laminae; (2) alongshore-directed cross strata interbedded with landward-directed cross strata and planar strata typical of ridge and runnel systems; (3) aeolian cross strata. Although beach deposits can be eroded during marine transgressions, and by tidal inlets, universal removal seems unlikely. The apparent absence of beach deposits in both transgressive and regressive sequences along a shoreline that clearly experienced strong storm waves is probably related to high deposition rates of sand and mud near the mouths of a complex of channels. Thus channel switching is a viable mechanism for producing the metre-scale alternations of channel-related sandstone bodies and mudstones. Upward-fining sequences representing filling of coastal bays and/or progradation of tidal flats could be formed in areas away from active channels, perhaps landward of abandoned channel-mouth bars (shoals) that were being reworked by wave currents.

Interpretation of large-scale (tens of metres to 100 m thick) sequences in coastal and fluvial deposits

The tens-of-metres scale regressive–transgressive sequences in coastal deposits can be traced laterally for up to 10 km, and probably represent on the order of 10^5 years of deposition. However, they cannot be correlated with fully marine and fluvial sequences. Palynostratigraphic correlation of thicker sequences of marine, coastal and fluvial strata has been undertaken in few cases (Fig. 9). The correlations are approximate because the resolution of biozones is on the order of a million years at best. McCave (1969) associated mudstone-dominated fluvial sequences with marine transgressions based on the assumption that rising sea level would lead to increased rate of floodplain aggradation and higher preservation potential of overbank mudstones. This concept has been used by others more recently (e.g. Shanley & McCabe 1993; Wright & Marriott 1993). However, floodplain aggradation rate is only one of several controls on proportion of channel-belt deposits in alluvial successions (Mackey & Bridge 1995). Within the 100-metre-scale sequences shown in Fig. 9, the mean grain size and proportion of channel-belt sandstones, and the size of individual channels, increase upward. These trends were interpreted by Willis & Bridge (1988) to record progradation of seaward-fining distributive fluvial systems. Dating of these strata is not accurate enough to

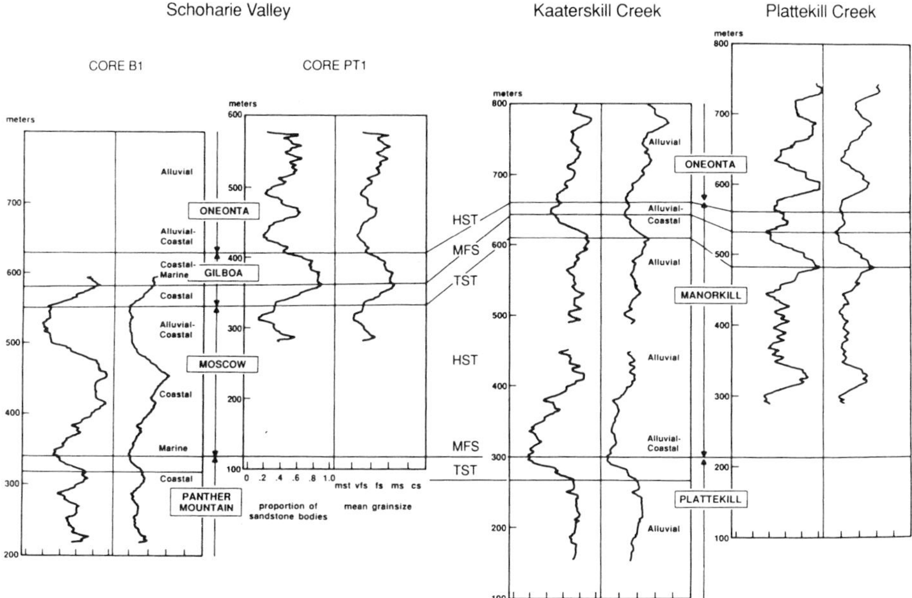

Fig. 9. Correlation of coastal and fluvial deposits of the Schoharie Valley with fluvial deposits of the Catskill Front (Kaaterskill and Plattekill Creeks). Redrawn from Bridge & Willis (1994). Formations indicated in boxes. Location of cores and outcrops is given by Bridge & Willis (1994). MFS, maximum flooding surface; HST, highstand systems tract; and TST, transgressive systems tract.

determine if the upward increase in channel-sandstone proportion is related to decreasing accommodation space and sediment accumulation rate in the way suggested by Shanley & McCabe (1993). However, by analogy with the Miocene Siwalik strata in the Himalayan foredeep, high proportions of channel-belt sandstone bodies are associated with high subsidence and deposition rates, increasing channel sizes, and progradation of megafans (Willis 1993*a*; Khan *et al.* 1997; Zaleha 1997).

Johnson & Friedman (1969) and McCave (1969, 1973) postulated that thin marine limestones (e.g. Tully) were deposited during marine transgressions, because sea-level rise would cause trapping of sand and mud near the coast, starving deeper marine areas of sediment. In contrast, Brett & Baird (1985, 1986, 1990) interpreted such limestones to represent shallowing and sediment starvation-winnowing events following sea-level lowstands. No corresponding erosion surfaces have been recognized in coeval coastal deposits. However, relative sea-level changes are not necessary congruent in different parts of foreland basins, such that cross-basin correlation of regressive or transgressive strata does not necessarily imply coeval events.

Many workers favour eustatic sea-level changes as the main control on the large-scale sequences (see references in introductory parts of this paper, and given by Bridge & Willis (1994)). The transgressive deposits of the well-known Gilboa Formation, New York State, are associated with a widely accepted eustatic sea-level rise that can be recognized world-wide (House 1983; Johnson *et al.* 1985). However, eustatic sea-level changes must be moderated by tectonically induced changes in subsidence and uplift rates and sediment supply (Quinlan & Beaumont 1984; Ettensohn 1985; Faill 1985). To establish whether eustasy or tectonism is the dominant control of sequence development in foreland basins, it is necessary to establish the ages of sequences, and the relative thickness of transgressive and regressive deposits, in different areas of the basin (Jordan & Flemings 1991). It is very difficult to know if rising and falling sea level are coeval or otherwise in different parts of the marine basin, because of dating limitations. Furthermore, it is very difficult to predict the timing and rate of deposition in different parts of the basin in relation to tectonic uplift of the adjacent mountains, and development of the peripheral budge. Jordan and Fleming's (1991) model suggests that presence of the thicker regressive strata relative to transgressive strata in this region points to a tectonic control of some of the large-scale sequences.

Concluding remarks

Further progress in understanding Devonian fluvial and coastal environments will require more detailed description and quantitative interpretation of large outcrops and cores across the region that are as closely spaced as possible. There is a pressing need for high-resolution biostratigraphic correlation of continuous sections of fluvial, coastal and marine strata, primarily using miospores. This work is eminently possible and long overdue.

References

ALLEN, G. P. 1991. Sedimentary processes and facies in the Gironde estuary: a recent model for estuarine systems. *In:* SMITH, D. G., REINSON, G. E., ZAITLIN, B. A. & RAHMANI, R. A. (eds) *Clastic Tidal Sedimentology*. Canadian Society of Petroleum Geologists, Memoirs, **16**, 29–39.

ALLEN, J. R. L. 1970. Studies in fluviatile sedimentation: a comparison of fining-upwards cyclothems, with special reference to coarse-member composition and interpretation. *Journal of Sedimentary Petrology*, **40**, 298–323.

—— & FRIEND, P. F. 1968. Deposition of Catskill facies, Appalachian region: with notes on some other Old Red Sandstone basins. *In:* KLEIN, G. de V. (ed.) *Late Paleozoic and Mesozoic continental sedimentation, northeastern North America*. Geological Society of America, Special Publications, **106**, 21–74.

BANKS, H. P., GRIERSON, J. D. & BONAMO, P. M. 1985. The flora of the Late Devonian Catskill clastic wedge. *In*: WOODROW, D. L. & SEVON, W. D. (eds) *The Catskill Delta*. Geological Society of America, Special Papers, **201**, 125–141.

BARRELL, J. 1913. The Upper Devonian delta of the Appalachian geosyncline. *American Journal of Science, 4th series*, **36**, 429–472.

—— 1914. The Upper Devonian delta of the Appalachian geosyncline. *American Journal of Science, 4th series*, **37**, 87–109.

BARWIS, J. H. 1978. Sedimentology of some South Carolina tidal-creek point bars, and a comparison with their fluvial counterparts. *In*: MIALL, A. D. (ed.) *Fluvial Sedimentology*. Canadian Society of Petroleum Geologists, Memoirs, **5**, 129–160.

BHATTACHARYA, J. P. & WALKER, R. G. 1992. Deltas. *In*: WALKER, R. G. & JAMES, N. P. (eds) *Facies Models: Response to Sea-level Change*. Geological Association of Canada, St. Johns, Newfoundland, 157–177.

BOSWELL, R. M. & DONALDSON, A. C. 1988. Depositional architecture of the Upper Devonian Catskill Delta Complex: central Appalachian Basin. *In*: MCMILLAN, N. J., EMBRY, A. F. & GLASS, D. J. (eds) *Devonian of the World, Vol. II*. Canadian Society of Petroleum Geologists, Memoirs, **14**, 65–84.

BRETT, C. E. & BAIRD, G. C. 1985. Carbonate–shale cycles in the Middle Devonian of New York: an evaluation of models for the origin of limestone in terrigenous shelf sequences. *Geology*, **13**, 324–327.

—— & —— 1986. Symmetrical and upward shallowing cycles in the Middle Devonian of New York and their implications for the punctuated aggradational cycle hypothesis. *Paleoceanography*, **1**, 431–445.

—— & —— 1990. Submarine erosion and condensation in a foreland basin: examples from the Devonian of Erie County, New York. *New York State Geological Association Field Trip Guidebook, 62nd Annual Meeting, Fredonia*, A1–A56.

BRIDGE, J. S. 1985. Paleochannel patterns inferred from alluvial deposits: a critical evaluation. *Journal of Sedimentary Petrology*, **55**, 579–589.

—— 1993. The interaction between channel geometry, water flow, sediment transport and deposition in braided rivers. *In*: BEST, J. L. & BRISTOW, C. S. (eds) *Braided Rivers*. Geological Society, London, Special Publication, **75**, 13–71.

—— & DROSER, M. L. 1985. Unusual marginal-marine lithofacies from the Upper Devonian clastic wedge. *In*: WOODROW, D. L. & SEVON, W. D. (eds) *The Catskill Delta*. Geological Society of America, Special Papers, **201**, 143–161.

—— & GORDON, E. A. 1985. Quantitative interpretation of ancient river systems in the Oneonta Formation, Catskill Magnafacies. *In*: WOODROW, D. L. & SEVON, W. D. (eds) *The Catskill Delta*. Geological Society of America, Special Papers, **201**, 163–182.

—— & MACKEY, S. D. 1993. A theoretical study of fluvial sandstone body dimensions. *In*: FLINT, S. & BRYANT, I. D. (eds) *Quantitative description and modelling of clastic hydrocarbon reservoirs and outcrop analogues*, IAS Special Publication, **15**, 213–236.

—— & NICKELSEN, B. H. 1985. Reanalysis of the Twilight Park Conglomerate, Upper Devonian Catskill Magnafacies, New York State. *Northeastern Geology*, **7**, 181–191.

—— & WILLIS, B. J. 1994. Marine transgressions and regressions recorded in Middle Devonian shore-zone deposits of the Catskill clastic wedge. *Geological Society of America Bulletin*, **106**, 1440–1458.

——, GORDON, E. A. & TITUS, R. C. 1986. Non-marine bivalves and associated burrows in the Catskill magnafacies (Upper Devonian) of New York State. *Palaeogeography, Palaeoclimatology, Palaeoecology*, **55**, 65–88.

CHADWICK, G. H. 1944. *Geology of the Catskill and Kaaterskill Quadrangles, Part 2*. New York State Museum Bulletin, **336**.

COLEMAN, J. M. 1969. Brahmaputra River: channel processes and sedimentation. *Sedimentary Geology*, **3**, 129–239.

—— & PRIOR, D. B. 1982. Deltaic environments. *In*: SCHOLLE, P. A. & SPEARING, D. (eds) *Sandstone Depositional Environments*. American Association of Petroleum Geologists, Tulsa, OK, 139–178.

COTTER, E. & DRIESE, S. G. 1998. Incised-valley fills and other evidence of sea-level fluctuations affecting deposition of the Catskill Formation (Upper Devonian), Appalachian foreland basin, Pennsylvania. *Journal of Sedimentary Research*, **68**, 347–361.

CRAFT, J. H. & BRIDGE, J. S. 1987. Shallow-marine sedimentary processes in the Late Devonian Catskill Sea, New York State. *Geological Society of America Bulletin*, **98**, 338–355.

DALRYMPLE, R. W. 1992. Tidal depositional systems. *In*: WALKER, R. G. & JAMES, N. P. (eds) *Facies Models: Response to Sea-level Change*. Geological Association of Canada, St. Johns, Newfoundland, 195–218.

——, ZAITLIN, B. A. & BOYD, R. 1992. Estuarine facies models: conceptual basis and stratigraphic implications. *Journal of Sedimentary Petrology*, **62**, 1130–1146.

DEMICCO, R. V., BRIDGE, J. S. & CLOYD, K. C. 1987. A unique freshwater carbonate from the upper Devonian Catskill Magnafacies of New York State. *Journal of Sedimentary Petrology*, **57**, 327–334.

DENNISON, J. M. 1985. Catskill Delta shallow marine strata. *In*: WOODROW, D. L. & SEVON, W. D. (eds) *The Catskill Delta*. Geological Society of America, Special Papers, **201**, 91–107.

—— & ETTENSOHN, F. R. (eds) 1994. *Tectonic and Eustatic Controls on Sedimentary Cycles*. SEPM Concepts in Sedimentology and Paleontology, **4**.

—— & HEAD, J. W. 1975. Sea level variation interpreted from the Appalachian Basin Silurian and Devonian. *American Journal of Science*, **275**, 1089–1120.

DIEMER, J. A. 1992. Sedimentology and alluvial stratigraphy of the Upper Devonian Catskill Formation, south–central Pennsylvania, *Northeastern Geology*, **14**, 121–136.

DRIESE, S. G., MORA, C. I. & ELICK, J. M. 1997. Morphology and taphonomy of root and stump casts of the earliest trees (Middle to Late Devonian), Pennsylvania and New York, U.S.A. *Palaios*, **12**, 524–537.

DUKE, W. L. & PRAVE, A. R. 1991. Storm- and tide-influenced prograding shoreline sequences in the Middle Devonian Mahantango Formation, Pennsylvania. *In*: SMITH, D. G., REINSON, G. E., ZAITLIN, B. A. & RAHMANI, R. A. (eds) *Clastic Tidal Sedimentology*. Canadian Society of Petroleum Geologists, Memoirs, **16**, 349–370.

ELICK, J. M., DRIESE, S. G. & MORA, C. I. 1998. Very large plant and root traces from the Early to Middle Devonian: implications for early terrestrial ecosystems and atmospheric $p(CO_2)$. *Geology*, **26**, 143–146.

ELLIOTT, T. 1974. Interdistributary bay sequences and their genesis. *Sedimentology*, **21**, 611–622.

ERICKSON, M. C., MASSON, D. S., SLINGERLAND, R. & SWETLAND, D. W. 1990. Numerical simulation of circulation and sediment transport in the Late Devonian Catskill Sea. *In*: CROSS, T. A. (ed.) *Quantitative Dynamic Stratigraphy*. Prentice–Hall, Englewood Cliffs, NJ, 293–305.

ETTENSOHN, F. R. 1985. Controls on development of Catskill Delta complex basin-facies. *In*: WOODROW, D. L. & SEVON, W. D. (eds) *The Catskill Delta*. Geological Society of America, Special Papers, **201**, 65–75.

FAILL, R. T. 1985. The Acadian Orogeny and the Catskill Delta. *In*: WOODROW, D. L. & SEVON, W. D. (eds) *The Catskill Delta*. Geological Society of America, Special Papers, **201**, 15–38.

FARRELL, K. M. 1987. Sedimentology and facies architecture of overbank deposits of the Mississippi River, False River Region, Louisiana. *In*: ETHRIDGE, F. G., FLORES, R. M. & HARVEY, M. D. (eds) *Recent Developments in Fluvial Sedimentology*. Society of Economic Paleontologists and Mineralogists, Special Publications, **39**, 111–120.

FLETCHER, F. W. 1963. Regional stratigraphy of Middle and Upper Devonian non-marine rocks in southeast New York. *Pennsylvania Geological Survey Bulletin*, **G39**, 25–41.

—— 1967. Middle and Upper Devonian clastics of the Catskill Front, New York. *New York State Geological Association Guidebook 39, New Paltz*, C1–C23.

FRIEDMAN, G. M. & CHAMBERLAIN, J. A. 1995. *Archanodon catskillensis (Vanuxem)*: fresh-water clams from one of the oldest backswamp fluvial facies (Upper Middle Devonian), Catskill Mountains, New York. *Northeastern Geology and Environmental Sciences*, **17**, 431–443.

—— & LUNDIN, R. F. 1998. Freshwater ostracodes from Upper Middle Devonian fluvial facies, Catskill Mountains, New York. *Journal of Paleontology*, **72**, 485–490.

GORDON, E. A. 1988. Body and trace fossils from the Middle–Upper Devonian Catskill magnafacies, southeastern New York. *In*: MCMILLAN, N. J., EMBRY, A. F. & GLASS, D. J. (eds) *Devonian of the World, Vol. II*. Canadian Society of Petroleum Geologists Memoirs, **14**, 139–156.

—— & BRIDGE, J. S. 1987. Evolution of Catskill (Upper Devonian) river systems. *Journal of Sedimentary Petrology*, **57**, 234–249.

GRIFFING, D. H., BRIDGE, J. S. & HOTTON, C. L. 2000. Coastal–fluvial environments and plant palaeoecology of the Early Devonian (Emsian), Gaspé Bay, Québec, Canada. This volume.

HALPERIN, A. & BRIDGE, J. S. 1988. Marine to non-marine transitional deposits in the Frasnian Catskill clastic wedge, south–central New York. *In*: MCMILLAN, N. J., EMBRY, A. F. & GLASS, D. J. (eds) *Devonian of the World, Vol. II*. Canadian Society of Petroleum Geologists Memoirs, **14**, 107–124.

HOTTON, C. L., HUEBER, F. M., GRIFFING, D. H. & BRIDGE, J. S. 2001. Early terrestrial plant environments: an example from the Emsian of Gaspé, Canada. *In*: GENSEL, P. G. & EDWARDS, D. (eds) *Plants Invade the Land*. Columbia University Press, New York.

HOUSE, M. R. 1983. Devonian eustatic events. *Proceeding of the Ussher Society*, **5**, 396–405.

—— 1985. Correlation of mid-Paleozoic ammonoid evolution events with global sedimentary perturbations. *Nature*, **313**, 17–22.

Howard, J. D. & Frey, R. W. 1975. Estuaries of the Georgia coast, U.S.A.: sedimentology and biology. *Senckenbergiana maritima*, **7**, 33–103.

—— & Reineck, H. E. 1981. Depositional facies of high-energy beach-to-offshore sequence: comparison with low-energy sequence. *AAPG Bulletin*, **65**, 807–830.

Johnson, K. G. & Friedman, G. M. 1969. The Tully clastic correlatives (Upper Devonian) of New York State: a model for recognition of alluvial, dune (?), tidal, nearshore (bar and lagoon), and offshore sedimentary environments in a tectonic delta complex. *Journal of Sedimentary Petrology*, **39**, 451–485.

——, Klapper, G. & Sandberg, C. A. 1985. Devonian eustatic fluctuations in Euramerica. *Geological Society of America Bulletin*, **96**, 567–587.

Jordan, T. E. & Flemings, P. B. 1991. Large-scale stratigraphic architecture, eustatic variation, and unsteady tectonism: a theoretical evaluation. *Journal of Geophysical Research*, **96**, 6681–6699.

Khan, I. A., Bridge, J. S., Kappelman, J. & Wilson, R. 1997. Evolution of Miocene fluvial environments, eastern Potwar Plateau, northern Pakistan. *Sedimentology*, **44**, 221–251.

Mackey, S. D. & Bridge, J. S. 1995. Three dimensional model of alluvial stratigraphy: theory and application. *Journal of Sedimentary Research*, **B65**, 7–31.

McCave, I. N. 1968. Shallow and marginal marine sediments associated with the Catskill complex in the Middle Devonian of New York. *In*: Klein, G. de V. (ed.) *Late Paleozoic and Mesozoic Continental Sedimentation, Northeastern North America.* Geological Society of American, Special Papers **106**, 75–108.

—— 1969. Correlation of marine and nonmarine strata with example from Devonian of New York State. *AAPG Bulletin*, **53**, 155–162.

—— 1973. The sedimentology of a transgression: Portland Point and Cooksburg members (Middle Devonian), New York State. *Journal of Sedimentary Petrology*, **43**, 484–504.

McGhee, G. R. & Sutton, R. G. 1985. Late Devonian marine ecosystems of the Lower West Falls Group in New York. *In*: Woodrow, D. L. & Sevon, W. D. (eds) *The Catskill Delta.* Geological Society of America, Special Papers **201**, 199–209.

Miller, M. F. 1979. Paleoenvironmental distribution of trace fossils in the Catskill deltaic complex, New York State. *Palaeogeography, Palaeoclimatology, Palaeoecology*, **28**, 117–141.

—— 1991. Morphology and paleoenvironmental distribution of Paleozoic *Spirophyton* and *Zoophycos*: Implications for the *Zoophycos* ichnofacies. *Palaiois*, **6**, 410–425.

—— & Johnson, K. G. 1981. *Spirophyton* in alluvial-tidal facies of the Catskill deltaic complex: possible biological control of ichnofossil distribution. *Journal of Paleontology*, **55**, 1016–1027.

—— & Woodrow, D. L. 1991. Shoreline deposits of the Catskill deltaic complex, Schoharie Valley, New York. *In*: Landing, E. & Brett, C. E. (eds) *Dynamic Stratigraphy and Depositional Environments of the Hamilton Group (Middle Devonian) in New York State, Part II.* New York State Museum Bulletin, **469**, 153–177.

Mitchum, R. M. & Van Wagoner, J. C. 1991. High-frequency sequences and their stacking patterns: sequence-stratigraphic evidence of high-frequency eustatic cycles. *Sedimentary Geology*, **70**, 131–160.

Penland, S., Boyd, R. & Suter, J. R. 1988. Transgressional depositional systems of the Mississippi delta plain: a model for barrier shoreline and shelf sand development. *Journal of Sedimentary Petrology*, **58**, 932–949.

Prave, A. R., Duke, W. L. & Slattery, W. 1996. A depositional model for storm- and tide-influenced prograding siliciclastic shorelines in the Middle Devonian of the central Appalachian foreland basin. *Sedimentology*, **43**, 611–629.

Quinlan, G. M. & Beaumont, C. 1984. Appalachian thrusting, lithospheric flexure and the Paleozoic stratigraphy of the eastern interior of North America. *Canadian Journal of Earth Science*, **21**, 973–996.

Reineck, H.-E. & Singh, I. B. 1973. *Depositional Sedimentary Environments.* Springer, Berlin.

Reinson, G. E. 1992. Transgressive barrier island and estuarine systems. *In*: Walker, R. G. & James, N. P. (eds) *Facies Models: Response to Sea-level Change.* Geological Association of Canada, St. Johns, Newfoundland, 179–194.

Rickard, L. V. 1975. *Correlation of the Silurian and Devonian Rocks in New York State.* New York State Museum and Science Service, Geological Survey Map and Chart Series, **24**.

—— 1989. *Stratigraphy of the Subsurface Lower and Middle Devonian of New York, Pennsylvania, Ohio and Ontario.* New York State Museum Map and Chart Series, **39**.

Sevon, W. D. 1985. Non-marine facies of the Middle and late Devonian Catskill coastal alluvial plain. *In*: Woodrow, D. L. & Sevon, W. D. (eds) *The Catskill Delta.* Geological Society of America, Special Papers, **201**, 79–90.

Shanley, K. W. & McCabe, P. J. 1993. Alluvial architecture in a sequence stratigraphic framework: a case history from the Upper Cretaceous of southern Utah, USA. *In*: Flint, S. & Bryant, I. D. (eds) *The Geological Modeling of Hydrocarbon Reservoirs and Outcrop Analogues.* International Association of Sedimentologists, Specialo Publications, **15**, 21–56.

Shear, W. A., Bonamo, P. M., Grierson, J. D., Rolfe, W. D. I., Smith, E. L. & Norton, R. A. 1984. Early land animals in North America: evidence from Devonian age arthropods from Gilboa, New York. *Science*, **224**, 492–494.

Shepps, V. C. (ed.) 1963. *Middle and Upper Devonian Stratigraphy of Pennsylvania and Adjacent States.* Pennsylvania Geological Survey Bulletin, **G39**.

Slingerland, R. L. 1986. Numerical computation of co-oscillating paleotides in the Catskill epeiric sea of eastern North America. *Sedimentology*, **33**, 817–829.

—— & LOULÉ, J.-P. 1988. Wind/wave and tidal processes along the Upper Devonian Catskill shoreline in Pennsylvania, U.S.A. *In*: MCMILLAN, N. J., EMBRY, A. F. & GLASS, D. J. (eds) *Devonian of the World, Vol. II*. Canadian Society of Petroleum Geologists, Memoirs, **14**, 125–138.

SMITH, D. G. 1988. Modern point bar deposits analogous to the Athabasca oil sands, Alberta, Canada. *In*: DE BOER, P. L., VAN GELDER, A. & NIO, S. D. (eds) *Tide-influenced Sedimentary Environments and Facies*. D. Reidel, Dordrecht, 417–432.

SUTTON, R. G. 1963. Correlation of Upper Devonian strata in south–central New York. *In*: SHEPPS, V. C. (ed.) *Symposium on Middle and Upper Devonian Stratigraphy of Pennsylvania and Adjacent States*. Pennsylvania Geological Survey Bulletin, **G39**, 87–101.

—— & MCGHEE, G. R. 1985. The evolution of Frasnian marine 'community types' of south–central New York. *In*: Woodrow, D. L. & Sevon, W. D. (eds) *The Catskill Delta*. Geological Society of America, Special Papers, **201**, 211–224.

——, BOWEN, Z. P. & MCALESTER, A. L. 1970. Marine shelf environments of the Upper Devonian Sonyea Group of New York. *Geological Society of American Bulletin*, **81**, 2975–2992.

——, HUMES, E. C., NUGENT, R. C. & WOODROW, D. L. 1962. New stratigraphic nomenclature for Upper Devonian of south–central New York. *AAPG Bulletin*, **46**, 390–393.

TERWINDT, J. H. J. 1981. Origin and sequences of sedimentary structures in inshore mesotidal deposits of the North Sea. *In*: NIO, S., SHUTTENHELM, R. T. E. & VAN WEERING, T. J. C. E. (eds) *Holocene Marine Sedimentation in the North Sea Basin*. International Association of Sedimentologists, Special Publications, **5**, 4–26.

THOMS, R. E. & BERG, T. M. 1985. Interpretation of bivalve trace fossils in fluvial beds of the basal Catskill Formation (Late Devonian), eastern U.S.A. *In*: CURRAN, H. A. (ed.) *Biogenic structures: their use in interpreting depositional environments*, SEPM Special Publication, **35**, 13–20.

TYE, R. S. & COLEMAN, J. M. 1989*a*. Depositional processes and stratigraphy of fluvially dominated lacustrine deltas: Mississippi Delta Plain. *Journal of Sedimentary Petrology*, **59**, 973–996.

—— & —— 1989*b*. Evolution of Atchafalaya lacustrine deltas, south–central Louisiana. *Sedimentary Geology*, **65**, 95–112.

VAN DEN BERG, J. H. 1981. Rhythmic seasonal layering in a mesotidal channel fill sequence, Oosterschelde Mouth, the Netherlands. *In*: NIO, S., SHUTTENHELM, R. T. E. & VAN WEERING, T. J. C. E. (eds) *Holocene Marine Sedimentation in the North Sea Basin*. International Association of Sedimentologists, Special Publications, **5**, 147–159.

VAN TASSELL, J. 1987. Upper Devonian Catskill Delta margin cyclic sedimentation: Brallier, Scherr, and Foreknobs Formations of Virginia and West Virginia. *Geological Society of American Bulletin*, **99**, 414–426.

—— 1994*a*. Evidence of orbitally-driven sedimentary cycles in the Devonian Catskill Delta complex. *In*: DENNISON, J. M. & ETTENSOHN, F. R. (eds) *Tectonic and Eustatic Controls on Sedimentary Cycles*. SEPM Concepts in Sedimentology and Paleontology, **4**, 121–131.

—— 1994*b*. Cyclic deposition in the Devonian Catskill Delta of the Appalachians, U.S.A. *In*: DE BOER, P. L. & SMITH, D. G. (eds) *Orbital Forcing and Cyclic Sequences*. International Association of Sedimentologists, Special Publications, **19**, 395–411.

VAN WAGONER, J. C., MITCHUM, R. M., CAMPION, K. M. & RAHMANIAN, V. D. 1990. *Siliciclastic Sequence Stratigraphy in Well Logs, Cores and Outcrop*. AAPG Methods in Exploration Series, **7**.

——, POSAMENTIER, H. W., MITCHUM, R. M., VAIL, P. R., SARG, J. F., LOUTIT, T. S. & HARDENBOL, J. 1988. An overview of the fundamentals of sequence stratigraphy and key definitions. *In*: WILGUS, C. K., HASTINGS, B. S., KENDALL, C. G. ST C., POSAMENTIER, H. W., ROSS, C. A. & VAN WAGONER, J. C. (eds) *Sea-level Changes: an Integrated Approach*. SEPM, Special Publications, **42**, 39–45.

WALKER, R. G. 1971. Nondeltaic depositional environments in the Catskill clastic wedge (Upper Devonian) of central Pennsylvania. *Geological Society of America Bulletin*, **82**, 1305–1326.

—— & HARMS, J. C. 1971. The 'Catskill Delta': a prograding muddy shoreline in central Pennsylvania. *Journal of Geology*, **79**, 381–399.

—— & —— 1975. Shorelines of weak tidal activity: Upper Devonian Catskill Formation, central Pennsylvania. *In*: GINSBURG, R. N. (ed.) *Tidal Deposits*. Springer, New York, 103–108.

WILLIS, B. J. 1989. Paleochannel reconstructions from point bar deposits: a three-dimensional perspective. *Sedimentology*, **36**, 757–766.

—— 1993*a*. Ancient river systems in the Himalayan foredeep, Chinji village area, northern Pakistan. *Sedimentary Geology*, **88**, 1–76.

—— 1993*b*. Interpretation of bedding geometry within ancient point-bar deposits. *In*: MARZO, M. & PUIDEFABREGAS, C. (eds) *Alluvial Sedimentation*. International Association of Sedimentologists, Special Publications, **17**, 101–114.

—— & BRIDGE, J. S. 1988. Evolution of Catskill River systems, New York State. *In*: MCMILLAN, N. J., EMBRY, A. F. & GLASS, D. J. (eds) *Devonian of the World, Vol. II*. Canadian Society of Petroleum Geologists, Memoirs, **14**, 85–106.

WOODROW, D. L. 1985. Paleogeography, paleoclimate and sedimentary processes of the Late Devonian Catskill Delta. *In*: WOODROW, D. L. & SEVON, W. D. (eds) *The Catskill Delta*. Geological Society of America, Special Papers, **201**, 51–63.

—— & ISLEY, A. M. 1983. Facies, topography and sedimentary processes in the Catskill Sea (Devonian), New York and Pennsylvania. *Geological Society of America Bulletin*, **94**, 459–470.

—— & Nugent, R. C. 1963. Facies and the Rhinestreet Formation in south–central New York. *In*: Coates, D. R. (ed.) *Geology of South–Central New York*. New York State Geological Association, 35th Annual Meeting, Field Trip Guidebook, 59–76.

—— & Sevon, W. D. (eds) 1985. *The Catskill Delta*. Geological Society of America, Special Papers, **201**.

——, Dennison, J. M., Ettensohn, F. R., Sevon, W. D. & Kirchgasser, W. T. 1988. Middle and Upper Devonian stratigraphy and paleogeography of the central and southern Appalachians and eastern Midcontinent, U.S.A. *In*: McMillan, N., J., Embry, A. F. & Glass, D. J. (eds) *Devonian of the World, Vol. I*: Canadian Society of Petroleum Geologists, Memoirs, **14**, 277–301.

——, Fletcher, F. W. & Ahrnsbrak, W. F. 1973. Paleogeography and paleoclimate at the deposition sites of the Devonian Catskill and Old Red Facies. *Geological Society of America Bulletin*, **84**, 3051–3064.

Wright, V. P. & Marriott, S. B. 1993. The sequence stratigraphy of fluvial depositional systems: the role of floodplain sediment storage. *Sedimentary Geology*, **86**, 203–210.

Zaleha, M. J. 1997. Intra- and extrabasinal controls on fluvial deposition in the Miocene Indo-Gangetic foreland basin, northern Pakistan. *Sedimentology*, **44**, 369–390.

Depositional and structural setting of the (?)Lower Old Red Sandstone sediments of Ballymastocker, Co. Donegal

M. McSHERRY[1], J. PARNELL[2], A. G. LESLIE[1] & T. HAGGAN[2]

[1]*Department of Geology, School of Geosciences, The Queen's University of Belfast, Belfast, BT7 1NN, UK (e-mail: M.T.McSherry@qub.ac.uk)*

[2]*Department of Geology & Petroleum Geology, University of Aberdeen, Aberdeen, AB24 2UE, UK*

Abstract: Approximately 250 m of unfossiliferous sediments attributed to the Old Red Sandstone outcrop onshore at Ballymastocker, Donegal in northwest Ireland. Massive sandstones and trough cross-bedded sandstones pass upwards into coarse purple–brown basal conglomerates. The majority of the clast types, which include quartzite, marble, pelite, schist, vein quartz and metadolerite, are derived from the local Dalradian basement. Acid to intermediate porphyry clasts have no local source and may indicate a source to the north-east in the vicinity of the Devonian Lorne Plateau volcanic rocks in Argyll. The rocks are believed to represent an intermontane alluvial fan environment. The sediments occur in an outcrop of 3 km length against the northern flank of Knockalla Mountain, bounded to the southeast by the NE–SW-trending Caledonian Leannan Fault and resting on fractured Argyll Group Dalradian basement. No unconformable contact is observed and all contacts may be tectonic. Coarsening-up cycles within the sediments may reflect syn-sedimentary tectonic activity. The Devonian rocks are transected by a fracture system implying dextral transpressional activity on the Leannan Fault. The Dalradian basement rocks preserve evidence of sinistral strike-slip movement in late Caledonian times. Slickenfibres observed within the sediments and Dalradian rocks provide evidence for oblique dip-slip movement to the northwest. It is considered that these structures represent the last movement of the ORS basin, dropping it below denudation level, which resulted in its preservation. Movement along the Leannan Fault played a major role in the preservation of the offshore basins in the Islay–Donegal Platform region.

Following Geological Survey of Ireland mapping between 1883 and 1887, Hull *et al.* (1891) recorded the occurrence of a 'representative of the Old Red Sandstone' which consisted of red sandstone and conglomerate lying along the northern base of the Knockalla Mountains in Fanad, Co. Donegal (Fig. 1). No fossil evidence has been recorded to date the rocks. However, Hull *et al.* (1891) concluded that they were of Lower Devonian age because of their similarity to sediments of that age in western Scotland. He concluded that the sediments represented the remains of a lake deposit. The southern margin of these sediments is bounded by the Leannan Fault. Cruise (in Hull 1891) reported that the Old Red Sandstone (ORS) was 'brought down against the Knockalla quartzite' by a fault line south of Ballymastocker Bay. The sediments were depicted as lying unconformably on Dalradian basement at the northern margin of the outlier. Cole & Hallissy (1924) suggested that the ORS was brought into position by a thrust plane, but no evidence was presented to support this theory.

Andrew (1951) drew attention to 'sedimentary infiltration' and a reddish stain occurring along joint cavities in the Slieve Tooey Quartzite Formation and, on this basis, suggested that the ORS was deposited against a fault scarp of pre-ORS age. Wilson (1953) documented the presence of well-rounded zircons and angular ilmenite, tourmaline, rutile and authigenic anatase to suggest that the outcrop at Ballymastocker could be correlated with the Lower ORS at Cushendall (Co. Antrim) and in Co. Tyrone and Fermanagh. However, although there is evidence of contemporaneous volcanism at Cushendall, Tyrone and Fermanagh, there is no trace of such igneous activity at Ballymastocker. Inclusions within quartz grains indicated only local derivation from both metamorphic and igneous sources. Powerful and torrential rivers pouring into a basin of deposition were envisaged (Wilson,

From: FRIEND, P. F. & WILLIAMS, B. P. J. (eds). *New Perspectives on the Old Red Sandstone.* Geological Society, London, Special Publications, **180**, 109–122. 0305-8719/00/$15.00

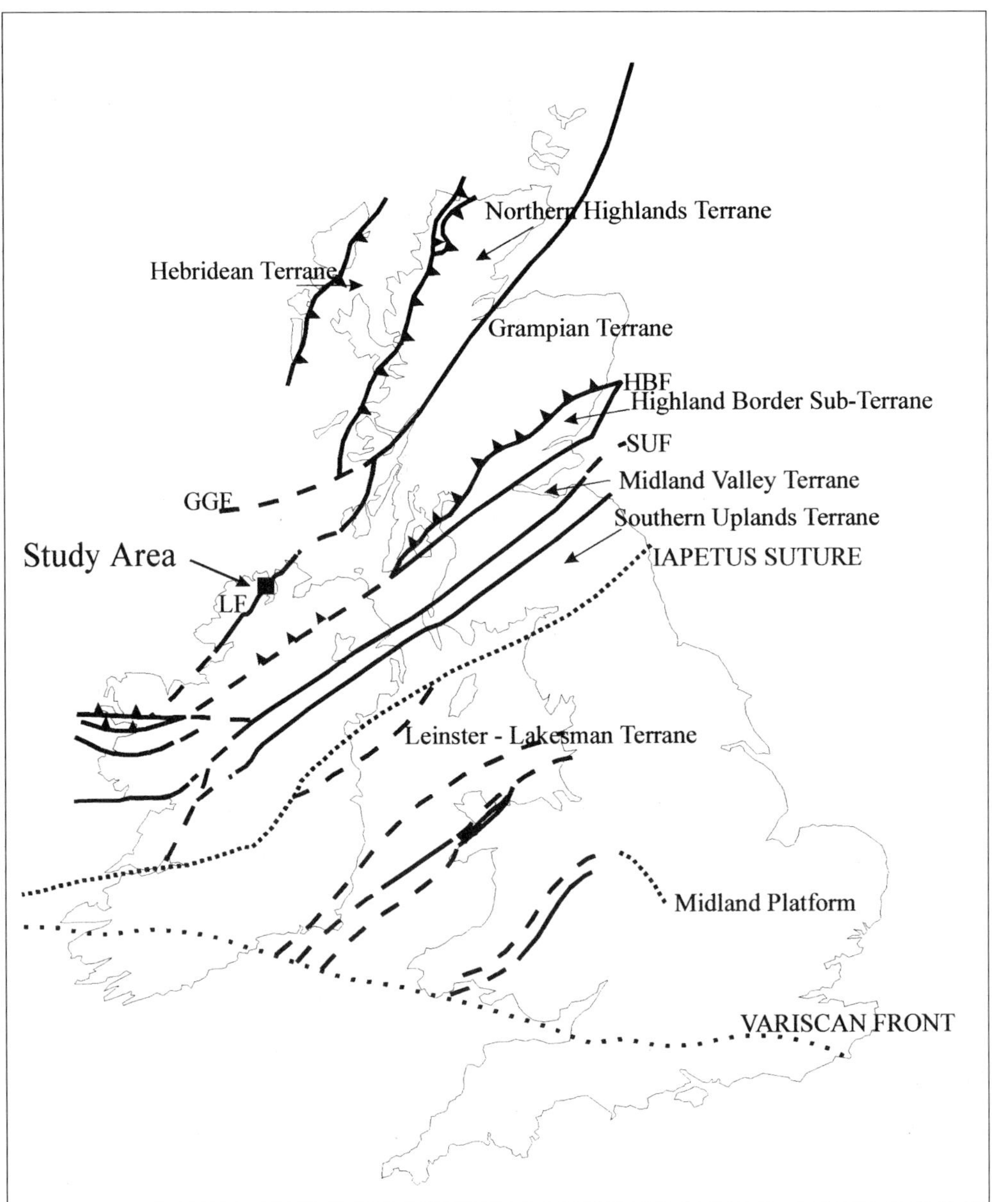

Fig. 1. Location of study area and position of terranes within Britain and Ireland modified from Bluck *et al.* (1992). GGF, Great Glen Fault; HBF, Highland Boundary Fault; LF, Leannan Fault; SUF, Southern Uplands Fault.

1953) as responsible for the formation of the Ballymastocker sediments. Wilson (1972) subsequently postulated that the conglomerate at Ballymastocker was the relic of an isolated valley deposit, whereas those at Tyrone and Cushendall were part of the old Caledonian foreland deposits.

Pitcher *et al.* (1964) considered that a large dip-slip movement best explained the Ballymastocker occurrence on the Leannan Fault system, with downthrow of the ORS to the northwest. These workers also mapped an arcuate splinter fault trending at a gentle angle to the main fault; with the contact between the ORS and basement beds vertical between these two faults. Pitcher *et al.* (1964) were able to constrain the upper age limit of movements on the main strand of the Leannan

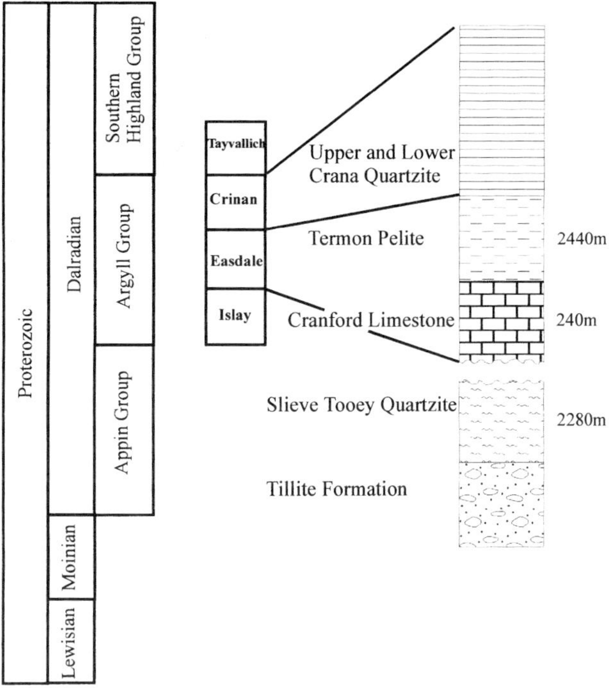

Fig. 2. Dalradian stratigraphy modified from Harris *et al.* (1995) and 1:250 000 geological map of Northern Ireland (GSNI 1997).

Fault, as it did not affect Carboniferous sediments farther to the southwest. Pitcher (1969) documented a zone of cataclasis between the ORS and the underlying Dalradian rocks, concluding that intense faulting had occurred both before and after deposition of the ORS. Long & McConnell (1997) noted that at the centre of the Leannan Fault, there was a subvertical, banded mylonite with subhorizontal sinistral shear indicators. Closely spaced subvertical planes locally cut the mylonite, with strike-slip or down-kinematic indicators. Those workers considered that the ORS sediments at Ballymastocker have been moved several kilometres to the southeast along the Leannan Fault.

We present evidence from new field observations obtained during detailed remapping of the Ballymastocker area, supplemented by fluid inclusion and palaeomagnetic data. The depositional and structural setting of the ORS outlier at Ballymastocker is reassessed on the basis of these data and a reappraisal of published accounts.

The Dalradian basement

The Neoproterozoic Argyll Group Dalradian rocks (Harris *et al.* 1995) typically comprise basin shallowing and basin deepening sequences of cross-bedded orthoquartzites, pebbly siliceous turbidites, graphitic mica schists, garnet–mica schists, metabasic volcanic rocks and limestones, which may often be turbiditic in part (Fig. 2). The base of the Argyll Group is almost always marked by a tillite horizon, the Fanad Boulder Bed, which is the local equivalent of the Port Askaig Tillite Formation. Phillips & Holland (1981) recorded boulder beds with unsorted clasts, up to 2 m in diameter, scattered through a homogeneous matrix of semipelite, psammite and quartzite.

The Fanad Boulder Bed is overlain by pelites with variable dolomite content, which in turn are overlain by the feldspathic Slieve Tooey Quartzite Formation of the Islay Subgroup. This quartzite forms the prominent topographic high of Knockalla Mountain to the southeast of the Ballymastocker outlier and the Leannan fault (Fig. 1), and also crops out some 500 m to the north of the Devonian rocks. Easdale Subgroup rocks are represented in the area by the Cranford Limestone Formation and the Termon Pelite Formation. The basal pelites are often graphitic, whereas psammites and turbiditic pebbly grits mark the upper part of the formation. The Crinan Subgroup Crana Quartzite Formation is a proximal turbiditic sequence of coarse-grained,

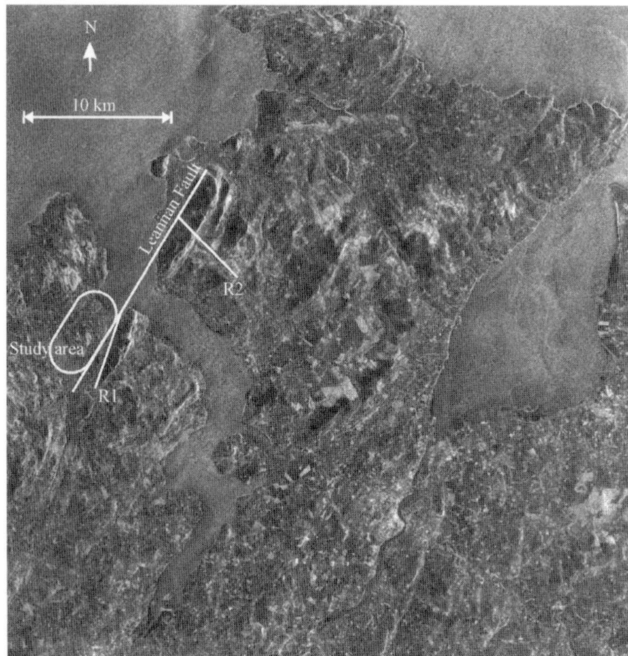

Fig. 3. Radarsat image of study area and northwest Donegal. Data courtesy of Defence Evaluation Research Agency (DERA).

graded, psammitic wackes with pelitic and limestone interbeds.

The Leannan Fault system

The Leannan Fault is a prominent NE–SW-trending fault that cuts across the northwest corner of Ireland. Pitcher *et al.* (1964) described this structure as the most important member of a system of steep strike-slip faults that dissect the Dalradian rocks of Donegal into broadly parallel strips. The Leannan Fault forms a prominent feature on the Radarsat image of northwest Ireland extending northeast from Knockalla Mountain across Lough Swilly to the coast of Inishowen at the southern end of Lennan Bay (see Fig. 3).

Other related NE–SW trending faults and fractures are also present. At Ballymastocker, this fault system comprises steeply inclined fractures, which are clearly visible cutting across Knockalla Mountain and are consistent with a set of synthetic and antithetic Riedel shears in a sinistral strike-slip system (Fig. 3). Leedal & Walker (1954), George & Oswald (1957), Pitcher (1969) and Pitcher & Berger (1972) concurred that sinistral displacements on the Leannan Fault began after emplacement of the main 420–400 Ma Caledonian granites in NW Donegal. Pitcher *et al.* (1964) also noted significant examples of dextral movement on the Leannan Fault between the deposition of ORS beds at Ballymastocker and deposition of Viséan marine sediments in south Donegal.

The Upper Palaeozoic (Lower ORS) sediments

Approximately 250 m of red-bed sediments occur at Ballymastocker, generally dipping between 20° and 55° east to southeast (Fig. 4). Good exposures occur at the eastern end of Port Salon beach (Fig. 5a) and further inland immediately adjacent to the Knockalla Mountain scarp. There is also good exposure in and around the hamlets of Drumfad and Newtown. Two principal red-bed lithofacies were identified during the detailed field mapping. The lower member is a combination of sandstones and trough cross-bedded sandstones; the upper is a coarse purple–brown conglomerate. Clasts can generally be matched with the local Dalradian quartzite, marble, pelite, schist, vein quartz and metadolerite. Some of these clasts are in excess of 1 m in diameter (Fig. 5b). However, there are significant proportions of porphyry clasts in the conglomerate, which do not appear to have a readily identifiable local source.

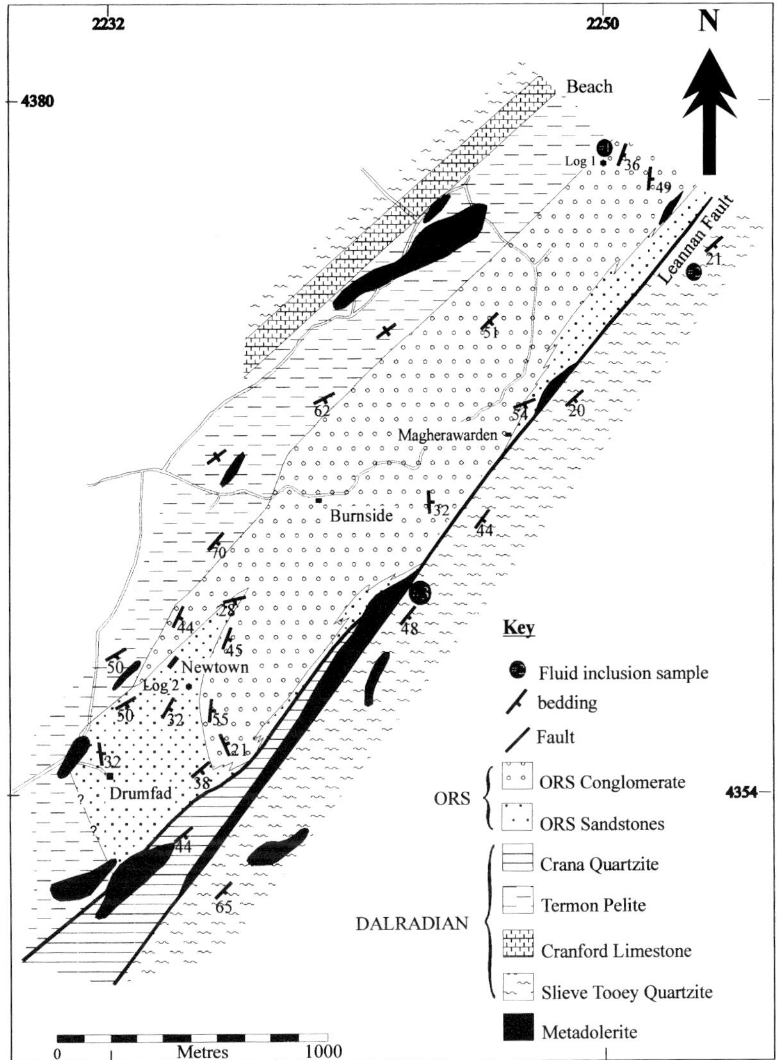

Fig. 4. Geological map of study area.

Figure 6 shows log 1 from within the younger conglomeratic lithofacies recorded at Port Salon beach. It is dominated by poorly sorted conglomeratic beds, which gradually alternate with laterally impersistent thin beds of moderately sorted red to brown sandstones. Maximum clast size ranges from 5 to 56 cm. Towards the top of the section lenses of coarse-grained sandstone in the conglomerates are coincident with a decrease in maximum clast size and an increase in the matrix component. The quartzite component of the clasts rises to a maximum of 25% before reducing again to 10% towards the top of the measured section.

The older lithofacies is illustrated in log 2 (Fig. 7). Coarse-grained, matrix-supported pebble beds with angular clasts occur at the base of the section, followed by a coarse-grained sandstone, which precedes a conglomerate, whose maximum clast size fines upwards. Channelled sandstone (wavelengths of 4–5 m) followed by a fine-grained laminated sandstone occurs some 6 m above the base of the section. Southeasterly palaeocurrents are apparent at this level from measurements on symmetric and asymmetric ripples. Medium- to coarse-grained sandy beds follow with pebble lags at their base and muddy lenses. The upper part of the section

Fig. 5. Field photographs of ORS at Ballymastocker. (**a**) Bedding in conglomeratic ORS facies, Ballymastocker beach, looking NE. Hammer 32 cm long. (**b**) Fractured and displaced pebbles in ORS conglomeratic facies, Ballymastocker beach. Hammer 40 cm long. (**c**) Slickenfabric in sandy ORS facies, caravan park *c.* 10 m northwest of the Leannan Fault.

is dominated by cross-bedded conglomeratic beds, which have an undulatory contact with the underlying sandstones. Erosive contacts are noted between beds whose maximum clast size increases upwards.

The classification schemes of Miall (1978*b*, 1985, 1994) and Rust (1978) were used to divide the Ballymastocker succession into two facies associations. Both of these facies associations are assigned to an alluvial fan environment. Facies Association A represents deposition of crudely bedded conglomerates by proximal rivers with some imbrication, development of coarse sandy lenses and thin beds of medium- to fine-grained laminated sandstones. The older Facies Association B was deposited by distal gravelly rivers. Gravel beds containing trough cross-beds are accompanied by minor gravel beds with planar cross-beds. There are also medium- to coarse-grained sandstones, sometimes pebbly, and minor amounts of fine-grained, horizontally bedded sandstones.

The red–brown colour of these matrix-supported, unfossiliferous conglomerates is indicative of deposition under oxidizing conditions. The gradual increase and subsequent decrease of quartzite clasts suggests unroofing of this Dalradian lithology. The continuous presence of porphyry clasts throughout the sequence reflects the erosion of igneous rocks during the deposition of these beds, possibly originating from the Lower Devonian Lorne Plateau volcanic rocks to the north. There is a notable absence of granite clasts, indicating that the Caledonian granites were not unroofed at the time of ORS deposition. An upward decrease in maximum particle size and an upward increase in the amount of sandstone beds is typical of alluvial fan sequences (Steel & Wilson 1975) and reflects an upward change towards increasing stability. This may represent denudation of the source area (Bluck 1980). At the top of the sequence there is evidence for a renewed sediment supply. An erosive contact between beds followed by a return to the conglomerate facies indicates a reactivated source probably as a result of faulting. The clast provenance at this level is purely local. Movement along the Leannan Fault could have played a role in the deposition of the sediments by renewed faulting but there is no direct evidence for this in the form of facies distribution relative to the fault.

Fault and fracture data

The majority of the Dalradian strata dip toward the southeast. However, some dip towards the northwest (Fig. 8a). Dips vary from near-vertical adjacent to faults, to subhorizontal within the Termon Pelite Formation.

Observed joints and faults in the Slieve Tooey Quartzite on Knockalla Mountain are moderate to steeply dipping and occur in two predominant sets (Fig. 8b and d). The first set (synthetic) are parallel to subparallel (10–20° counter-clockwise) to the Leannan Fault and so broadly trend NE–SW. The second (antithetic) set is orientated perpendicular to that trend. Some of these fracture sets contain a quartz infill (Fig. 8c), whereas others are barren. However, there are also quartz veins that developed before to the

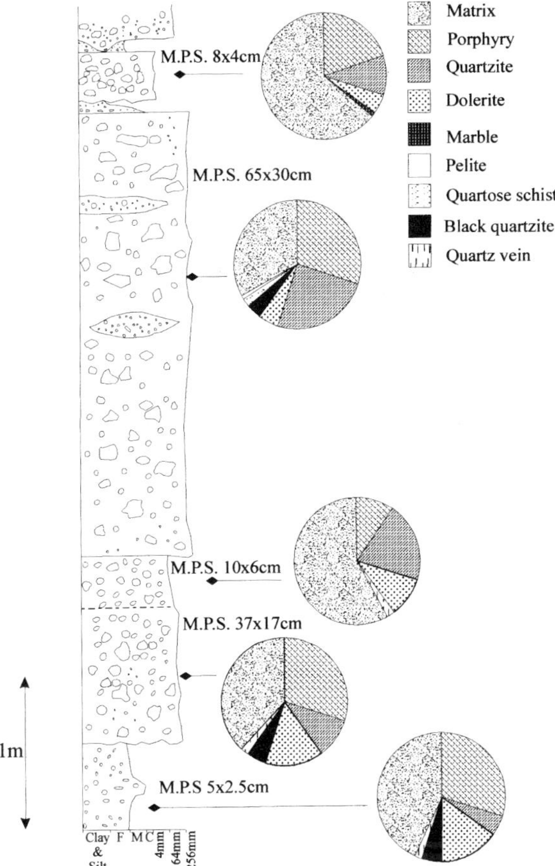

Fig. 6. Log of ORS section exposed at Ballymastocker beach. Pie charts illustrate percentage matrix and clast lithologies within individual beds. MPS, maximum pebble size.

main folding (i.e. they are syn-orogenic) and now appear as intrafolial lenses that trend NE–SW (Fig. 9a). There is some evidence of pervasive infiltration of these quartz veins by red sediment and red staining (Fig. 10a–c) but there was no preferred orientation of open fractures at the time of this event. Vein quartz are brecciated and reddened, which suggests that basement structures were being reactivated and utilized as fluid flow pathways either during or soon after ORS deposition.

The ORS beds dip mostly towards the southeast and east, although some dip towards the northwest (Fig. 11a). The majority of joints and fractures within the ORS are steep to vertical and include some that displace both clasts and matrix (Figs 5b and 11b). Most dip toward either the northeast or the northwest. There are, however, a few joints that dip towards the north. These data are consistent with a system of dextral strike-slip movements along the Leannan Fault, which occurred post-deposition and post-lithification of the ORS rocks. These movements accommodated the infiltration of reddening materials into the active fractures, some of which were reactivated after sinistral movements in late Caledonian or pre-ORS times.

The majority of the mapped boundaries within the outlier are consistent with synthetic Riedel shears in a dextral strike-slip setting (Fig. 4). The fault referred to as an 'arcuate splinter fault' (Pitcher *et al.* 1964) falls into this category. It juxtaposes ORS, metadolerites and Crana Quartzite, all of which are separated from the Slieve Tooey Quartzite by the Leannan Fault (Fig. 4). Dalradian and ORS rocks are commonly steeply dipping in the vicinity of this structure. It is also possible that the arcuate boundary separating the conglomeratic facies and the sandstones east of Newtown could represent a gently to moderately dipping thrust, consequent upon dextral strike-slip. Along a

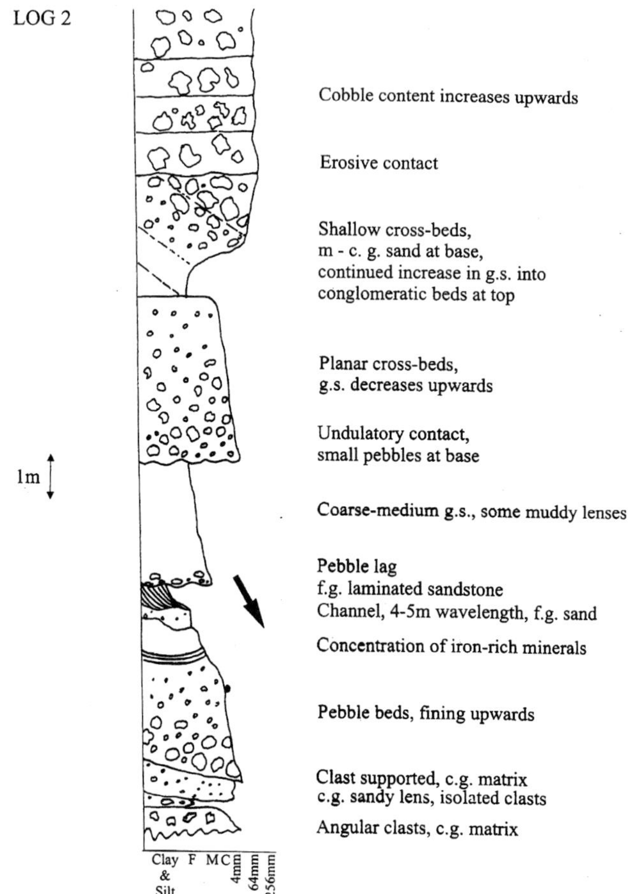

Fig. 7. Log of ORS section at Drumfad. F, M and C, fine, medium and coarse grain sizes; g.s., grain size.

stream section at Newtown, Ballycallan, intensely fractured sandy ORS occurs along a fracture plane that dips 24° towards the southeast. This could reflect overthrusting of conglomeratic facies over sandy facies, but this must remain conjectural.

The majority of preserved fibres on slickensides (Figs 8e and 11c) in both the Dalradian and Devonian rocks are steeply plunging and consistent with oblique dip-slip movements, down to the northwest. This may be the principal reason for preservation of this outlier of ORS red beds.

Palaeomagnetic data

A sample for palaeomagnetic analysis was collected from the ORS sediments at Port Salon beach. The magnetization in this sample was measured by R. D. Elmore using a three-axis cryogenic magnetometer at the University of Oklahoma. Stepwise thermal demagnetization was performed with a Schonstedt thermal demagnetizer and the sample was heated for 30 min at each of 17 steps up to 690°C. The sample was also subjected to stepwise alternating field (AF) demagnetization in 10 mT step intervals up to 120 mT using a 2 G AF demagnetizer.

The magnetization within the Ballymastocker sample is complex and resides predominantly in magnetite (85%), as opposed to hematite. Samples of reddened Dalradian rocks from localities in the Dalradian of Argyll were also subjected to palaeomagnetic data. (Parnell *et al.* 2000). Within these samples, magnetization resides predominantly in hematite, and a consistent late Permian–early Triassic age was found for the reddened Dalradian rocks. Unfortunately, a consistent magnetite direction has not yet been found in the Ballymastocker sample. The magnetite may have lost its stable direction or the magnetite could reside in the large clasts present and retain a direction from the original

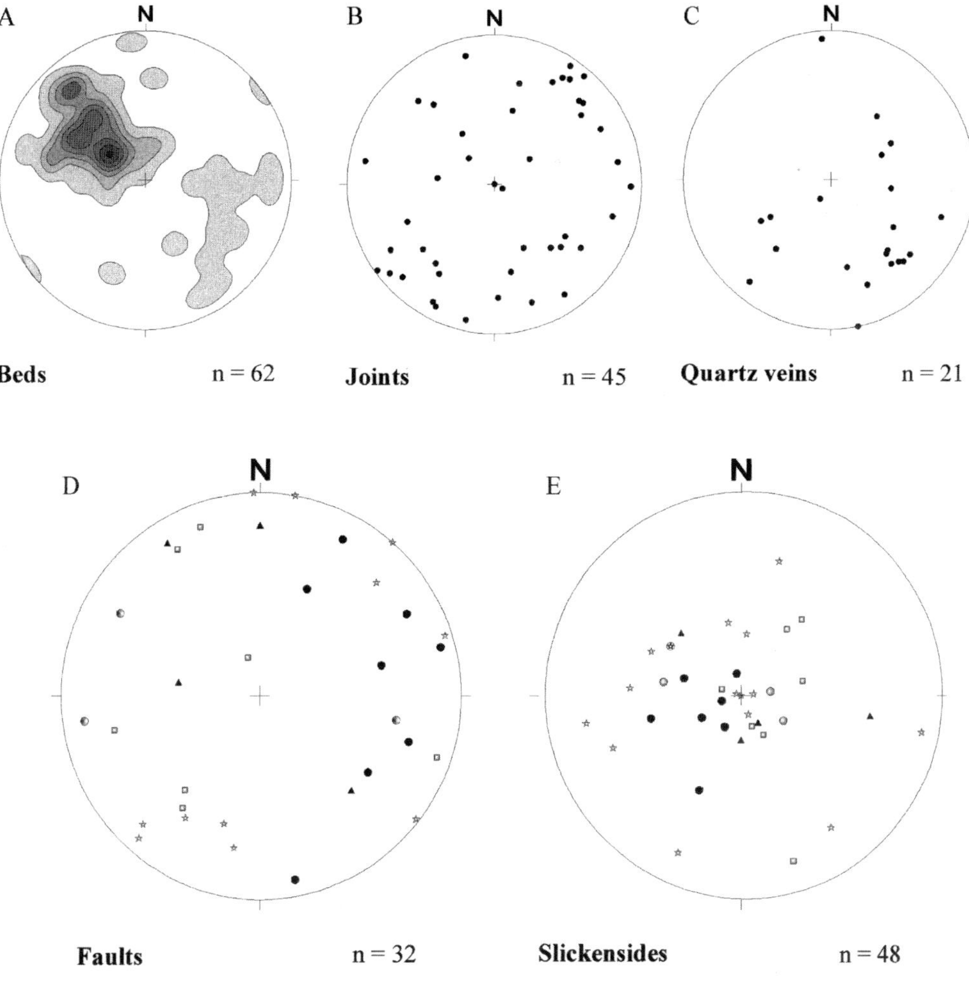

Fig. 8. Equal-area stereonets with Dalradian structural data.

rock. As the clast directions are random, the magnetite direction would also be random (Elmore, pers. comm.). However, above 580°C a hematite component is present and seems to have a late Palaeozoic magnetization. Although not conclusive, the palaeomagnetic data distinguish the Ballymastocker sample from the reddened Dalradian samples in Argyll, in terms of resident magnetization and direction.

Fluid inclusion data

To help constrain the post-depositional history of the basin, measurements of homogenization temperature were made on fluid inclusions in calcite veinlets infilling post-depositional fractures in a conglomerate pebble sampled from the Port Salon beach section. Evidence for late (post-orogenic) behaviour of the basement rocks was

Fig. 9. Field photographs of Dalradian rocks at Ballymastocker. (**a**) Quartz veins in Slieve Tooey Quartzite, Ballymastocker beach, looking northwest. Hammer 32 cm long. (**b**) Slickenfabric on fault plane in Crana Quartzite, Drumfad. Fault plane is reddened and brecciated. Arrow indicates movement direction. Hammer 32 cm long.

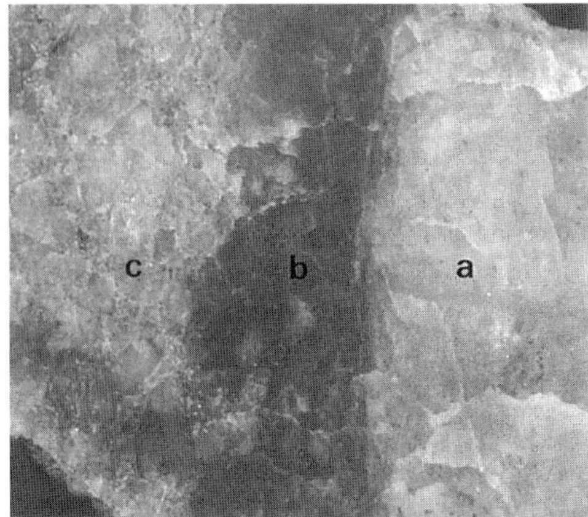

Fig. 10. Section through brecciated/reddened Dalradian rock at basin margin. (**a**) Brecciated and reddened quartzite; (**b**) intensely reddened quartz-rich veinrock; (**c**) brecciated and reddened quartz veinrock. Width of field 1.5 cm.

also sought in the fluid inclusion record of Dalradian-hosted quartz veins (1) parallel with the Leannan Fault at the basin margin, and (2) at a site of intense brecciation and reddening adjacent to the Leannan Fault. Measurements were made on doubly polished wafers using a Linkham THM600 heating–freezing stage.

The homogenization temperatures measured in primary aqueous fluid inclusions in calcite from the beach section ranged from 101.5 to 131.7 °C, with a mean of 115 °C from 20 measurements (Fig. 12a, Table 1). The narrow range of values indicates that the data are reliable. Even without applying a pressure correction, these temperatures suggest burial to several kilometres depth under a normal geothermal gradient.

The fault-parallel quartz yielded a unimodal temperature set (Fig. 12b). The lower temperatures are comparable with those obtained from the calcite, and may reflect the same tectonic event that fractured the pebbles. The brecciated quartz also yields similar temperatures from a late (syn- or post-brecciation) quartz phase associated with reddening. It is likely that a single tectonic episode, involving displacement on the Leannan Fault resulted in fracturing of the basin fill, i.e. the data record post-depositional activity on the fault.

Salinities obtained from freezing (final melting) data for the fluid inclusions within the brecciated quartz and the calcite are low for sedimentary basins, suggesting that the fluids present at the time of mineral precipitation were not influenced by marine or evaporitic strata: this is consistent with a predominantly Devonian clastic infill. The close similarity of the salinities for the pebble-hosted calcite and the brecciated Dalradian-hosted quartz further suggests that they may reflect the same episode of fracturing and fluid entrapment.

Possible scenarios for deep burial are (1) a substantial thickness of Upper Paleozoic (Upper Devonian, Carboniferous) cover, which has subsequently been removed by erosion, and/or (2) similarly thick Mesozoic cover, also since removed by erosion. Predicted patterns of Paleocene lava thickness in the north of Ireland (Preston 1981) suggest that Tertiary burial is not a viable explanation. Regional geological considerations suggest that a thick Upper Palaeozoic cover is more likely than a thick Mesozoic cover; thick Devonian–Carboniferous successions occur in south Donegal and to the east in Northern Ireland and southwest Scotland. Petrographic evidence from the Carboniferous succession of south Donegal and adjacent regions suggests that there is an important episode of uplift and weathering at latest Carboniferous–early Permian times (Parnell 1991), which could also have been responsible for removal of most of the sedimentary cover at Ballymastocker.

Offshore to the north, major faults run parallel to the Leannan Fault. These include

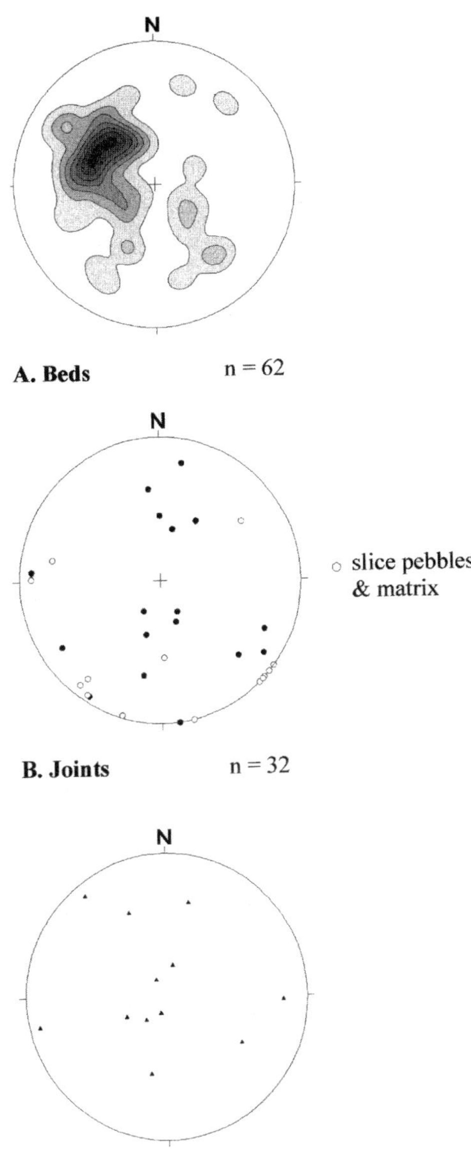

Fig. 11. Equal-area stereonets with ORS structural data.

the Colonsay and Skerryvore faults, which are believed to have controlled Permo-Triassic sedimentation (Evans *et al.* 1980; Dobson & Whittington 1992; Fyfe *et al.* 1993). The extension of the Leannan fault into the Gruinart Fault system at Islay bounds the Loch Indaal basin, which contains Lower Jurassic open marine sediments (Evans *et al.* 1980), indicating movement at a later date. There were clearly numerous episodes when such structures experienced displacement and deformation. No direct evidence allows dating of the movement onshore in Donegal. However, the fluid inclusion temperature data associated with late deformation of the quartz suggest that the last major displacement occurred when the ORS was under several kilometres burial depth and this, as argued above, probably occurred during Palaeozoic time.

Discussion

Structural chronology

The Caledonian Orogeny in the northern British Isles resulted in the formation of southeasterly dipping thrusts, e.g. the Moine Thrust and other steeply inclined, NE–SW-trending fractures. The Radarsat image (Fig. 3) emphasizes the development of fracture systems in the Dalradian rocks. These are likely to reflect sinistral strike-slip movements on the Leannan Fault in late Caledonian times.

Dextral transpressional tectonism on the Leannan Fault is preserved in the ORS sediments with synthetic riedal subvertical shears (R_1), and a prominent fracture set trending 120°C which may represent antithetic riedal shears (R_2). The prominent scarp between the conglomeratic and sandy ORS facies in this dextral strike-slip system may be the result of layer-parallel slip or a thrust boundary.

Movement indicators are preserved as oblique slickenfibres, which represent downthrow of the ORS basin to the northwest.

Depositional framework

Reactivation of the Caledonian NE–SW thrust duplex structures, culminating in strike-slip along the Leannan Fault, would have exerted an influence on the line of the drainage system. A southwesterly flowing river, axial to the Caledonian trend, may have developed, bringing Dalradian clasts including quartzite, marble, pelite and metadolerite along the palaeoslope. Porphyry clasts observed in the ORS conglomerates may be derived by erosion and transport of volcanogenic detritus from the Devonian Lorne Plateau volcanic rocks in Argyll. The absence of granite clasts suggests that deposition occurred before the unroofing of the Caledonian granites. The ORS sediments were deposited in an intermontane basin shortly after orogenic collision. Proximal rivers were responsible for the deposition of Facies A. Distal gravelly rivers produced Facies B. Clast size and roundness indicate that

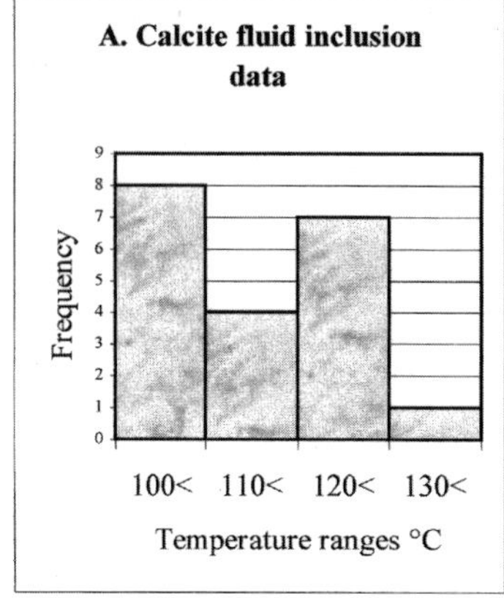

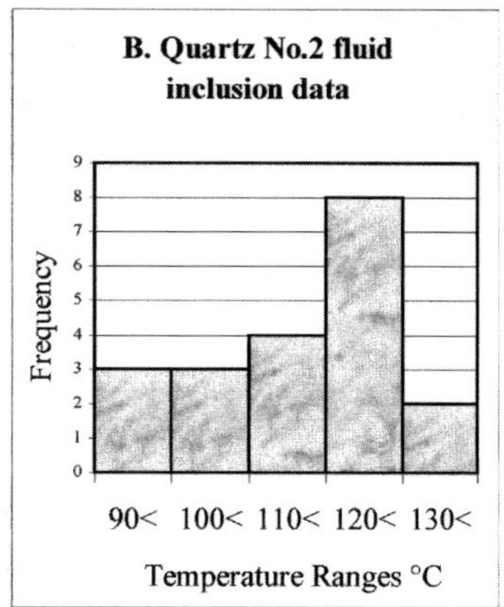

Fig. 12. Histograms displaying (**a**) calcite and (**b**) quartz fluid inclusion data.

Table 1. *Summary of fluid inclusion data*

		Homogenization Temperature (°C)			Freezing Temperature (°C)	
Host	Locality	Min	Min	Max	Mean	Equivalent salinity (wt.% NaCl)
Calcite infilling fractured pebbles in conglomerate	#1	101.5	131.7	115	−2.1*	3.5
Quartz veining at and parallel to basin margin, in Dalradian	#2	130	185		Nd†	Nd†
Brecciated, reddened quarts at basin margin, in Dalradian	#3	90.1	131.9	116	−2.0	3.4

*Omitting one aberrant value of −9.4.
†Nd no data.

these sediments travelled some distance. Few palaeocurrent indicators were ever observed in the field: those measured showed a southeasterly to easterly flow direction. This could represent a local anomaly (tilting of basin by active Leannan Fault?) within the regional southwesterly palaeoflow at the time of deposition.

Fluid inclusion data

The similar homogenization temperatures from primary aqueous inclusions in both the pebble-hosted calcite and brecciated quartz along the Leannan Fault suggest post-depositional movement on the Leannan Fault. Reddening of active fractures within the Dalradian dates movement along the Leannan Fault to either syn- or post-deposition of the ORS. High homogenization temperatures suggest several kilometres of burial. The most likely scenario for deep burial is a substantial thickness of Upper Palaeozoic cover, subsequently removed by erosion at latest Carboniferous–early Permian times (Parnell 1991). The same uplift episode is considered to be responsible for the removal of up to 1.5 km from the present Highland surface of Scotland (Watson 1984).

It is notable that further north along the Leannan–Loch Gruinart Fault system, where the present downthrow is to the east, Devonian rocks are preserved on the easterly side (Fyfe *et al.* 1993). The occurrence of Devonian outcrops on either side of a major strike-slip fault system, according to dip-slip polarity, strongly suggests localized preservation of a more widespread sediment cover.

Conclusions

Studies of the Devonian sediments at Ballymastocker using field relationships, fluid inclusion microthermometry and palaeomagnetic data yield the following conclusions about the Devonian depositional and structural setting:

(1) The ORS sediments were deposited in an alluvial fan environment, before the unroofing of the Caledonian granites.

(2) There is no direct evidence that sedimentation within the basin was controlled by the Leannan Fault. However, it is possible that the Leannan Fault controlled palaeodrainage.

(3) The present distribution of the ORS sediments at Ballymastocker suggests they have been subjected to a stress system with a significant component of dextral shear, followed by downthrow of the basin to the northwest, dropping it below denudation level, and resulting in its preservation.

(4) Limited palaeomagnetic data suggest an Upper Palaeozoic age for the sediments, supporting assumptions in previous published accounts.

(5) Fluid inclusion data are indicative of several kilometres of burial, under a substantial Palaeozoic cover, which has since been removed through uplift and erosion. This is substantiated by a thick Carboniferous succession in south Donegal.

M. McSherry was supported by a PhD grant from Rigel Oil & Gas Ltd. D. Elmore is thanked for his work in the Palaeomagnetism Laboratory in Oklahoma. R. Chapman of DERA in Farnborough is thanked for the provision of Radarsat imagery. A. Rodgers, B. Hartley and G. Alexander also provided skilled technical support.

References

ANDREW, G. 1951. Age of the Port Salon Conglomerate. *Geological Magazine*, **88**, 441–442.

BISHOPP, D. W. 1952. The age of the Port Salon Conglomerate. *Geological Magazine*, **89**.

BLUCK, B. J. 1980. Evolution of a strike-slip fault-controlled basin, Upper Old Red Sandstone, Scotland. International Association of Sedimentologists, Special Publication, **4**, 63–78.

——, GIBBONS, W. & INGHAM, J. K. 1992. Terranes. *In*: COPE, J. C. W. *et al.* (eds) *Atlas of Palaeogeography and Lithofacies*. Geological Society, London, Memoir, **13**, 1–4.

COLE, G. A. J. & HALLISSY, T. 1924. *Handbook of the Geology of Ireland.*

DOBSON, M. R. & WHITTINGTON, R. J. 1992. Aspects of the Geology of the Malin Sea. *In*: PARNELL, J. (ed.) *Basins on the Atlantic Seaboard: Petroleum Geology, Sedimentology and Basin Evolution*. Geological Society, London, Special Publications, **62**, 291–311.

EVANS, D., KENOLTY, N., DOBSON, M. R. & WHITTINGTON, R. J. 1980. *The Geology of the Malin Sea*. Report of the Institute of Geological Sciences, **79/15**.

FYFE, J. A., LONG, D. & EVANS, D. 1993. *The Geology of the Malin–Hebrides Area*. United Kingdom Offshore Regional Report. HMSO, London.

GEORGE, T. N. & OSWALD, D. H. 1957. The Carboniferous rocks of the Donegal Syncline. *Quarterly Journal of the Geological Society, London*, **113**, 137–179.

HARRIS, A. L., HASELOCK, P. J. *et al.* 1995. The Dalradian Supergroup in Scotland, Shetland and Ireland. *In*: GIBBONS, W. & HARRIS, A. L. (eds) *A Revised Correlation of Precambrian Rocks in the British Isles*. Geological Society, London, Special Report, **22**, 33–52.

HULL, E. *et al.* 1891. *Explanatory memoir to accompany Sheets 3, 4, 5, 9, 10, 11, 15 and 16 of the maps of the Geological Survey of Ireland, comprising northwest and central Donegal*. Alexander Thomas, Dublin.

LEEDAL, G. P. & WALKER, G. P. L 1954. Tear faults in the Barnesmore area, Donegal. *Geological Magazine*, **91**, 116–120.

LONG, C. B. & MCCONNELL, B. J. 1997. *Geology of North Donegal. A geological description to accompany the bedrock geology 1; 100 000 scale map series, Sheet 1 and part of Sheet 2, north Donegal*. Geological Survey of Ireland, Dublin.

MIALL, A. D. 1978 Facies types and vertical profile models in braided river deposits: a summary. *In*: MIALL, A. D. (ed.) *Fluvial sedimentology*, Canadian Society of Petroleum Geologists, Memoir, **5**, 597–604.

—— 1985. Architectural-element analysis: a new method of facies analysis applied to fluvial deposits. *Earth Science Reviews*, **22**, 261–308.

—— 1994. Alluvial deposits. *In*: WALKER, R. G. & JAMES, N. P. (eds) *Facies Models*. Geoscience Canada, Reprint Series, **1**, Geological Association of Canada, 119–142

PARNELL, J. 1991. Hydrocarbon potential of Northern Ireland. II. Reservoir potential of the Carboniferous. *Journal of Petroleum Geology*, **14**, 143–160.

——, BARON, M. *et al.* 2000. Dolomitic breccia veins as evidence for extension and fluid flow in the Dalradian of Argyll. *Geological Magazine*.

PHILLIPS, W. E. A. & HOLLAND, C. H. 1981. Late Caledonian deformation. *In*: HOLLAND, C. H. (ed.) *A Geology of Ireland*. Scottish Academic Press, Edinburgh.

PITCHER, W. S. 1969. Northest trending faults of Scotland and Ireland, and chronology of displacements. *In*: KAY, M. (ed.) *North Atlantic Geology and Continental Drift. Memoirs American Association of Petroleum Geologists*, **12**, 724–733.

—— & BERGER, A. R. 1972. *The Geology of Donegal: a Study of Granite Emplacement and Unroofing*. Wiley–Interscience, New York.

——, ELWELL, R. W. D. *et al.* 1964. The Leannan Fault. *Quarterly Journal of the Geological Society, London*, **120**.

PRESTON, J. 1981. Tertiary igneous activity. *In*: HOLLAND, C. H. (ed.) *A Geology of Ireland*. Scottish Academic Press, Edinburgh.

RUST, B. R. 1978. Facies models. *In*: HIALL, A. D. (ed.) *Fluvial sedimentology*. Canada Society of Petroleum Geology, Memoir, **5**, 187–198.

STEEL, R. J. & WILSON, A. C. 1975. Sedimentation and tectonism (?Permo-Triassic) on the margin of the North Minch Basin, Lewis. *Journal of the Geological Society, London*, **131**, 227–233.

WATSON, J. B. 1984. The ending of the Caledonian Orogeny in Scotland. *Journal of the Geological Society, London*, **141**, 193–214.

WILSON, H. E. 1953. The petrography of the Old Red Sandstone rocks of the North of Ireland. *Proceedings of the Royal Irish Academy*, **55**(B), 283–320.

Initiation and early development of the Dingle Basin, SW Ireland, in the context of the closure of the Iapetus Ocean

J. DOUGLAS BOYD[1] & RODERICK J. SLOAN[2]
[1]*BP, Farburn Industrial Estate, Dyce, Aberdeen AB21 7PB, UK*
[2]*C & C Reservoirs Ltd., 93–99 Upper Richmond Road, London SW15 2TG, UK*

Abstract: The Dingle Basin of southwest Ireland lies within 40 km of the present-day trace of the Iapetus Suture. Its late Silurian fill (Dunquin Group and lower Dingle Group) is an important and, in many aspects, unique record of late Caledonian development in the Irish and British sector of the orogen. The upper Wenlock–upper Ludlow Dunquin Group comprises shallow-marine to non-marine siliciclastic and volcanic rocks (acid pyroclastic deposits and predominantly andesitic lavas) deposited on and around a volcanic island(s), whereas the lower, upper Ludlow–Přídolí, part of the overlying Dingle Group comprises sandstones, mudstones and minor conglomerates deposited in lacustrine, lake margin and succeeding fluvial systems. The change from marine Dunquin Group to non-marine Dingle Group (Old Red Sandstone) sedimentation is interpreted to have been tectonically driven. The succession is interpreted in terms of four phases of basin evolution: (1) a phase related to active subduction; (2) a phase related to subduction termination; (3) a post-subduction thermal subsidence phase; (4) a phase of strike-slip fault-controlled subsidence. In the broader, late-Caledonian context, the Dunquin Group volcanic rocks and associated sediments are interpreted as representing localized subduction of a final vestigial portion of Iapetus oceanic crust. The later, inferred, strike-slip influence in the basin is believed to be part of the well-documented regional strike-slip regime affecting this sector of the Caledonides.

Silurian rocks of volcanic and shallow-marine through to fully non-marine (Lower Old Red Sandstone) origin crop out on the Dingle Peninsula in southwest Ireland (Fig. 1) and form part of the Silurian–Devonian fill of the Dingle Basin (Pracht 1996).

At the present day, the Dingle Peninsula lies less than 40 km south of the inferred trace of the 'Iapetus Suture' (Phillips *et al.* 1976; McKerrow & Soper 1989; Fig. 1). The Silurian fill of the Dingle Basin reflects the evolving volcanic and tectonic history of the collisional zone to the north, and the relationship between the basin fill and the nature of that zone forms the subject of this paper. Recent interpretations of the final stages of the Laurentia–Eastern Avalonia segment of the Iapetus Ocean (Fig. 2) fall into two broad groups. The first are those that see subduction of oceanic crust as completed by Late Ordovician time, with subsequent displacement on the plate boundary being strike-slip (e.g. Hutton 1987; Hutton & Murphy 1987). The second group of models infer northwestward subduction of oceanic crust beneath Laurentia to have continued into Silurian time, perhaps developing into continental subduction or underthrusting (e.g. Soper *et al.* 1987; Scotese & McKerrow 1990; Williams *et al.* 1992). Silurian rocks of the Dingle Basin are clearly relevant to this debate by virtue of their age, character and their location on the Eastern Avalonian margin.

The Silurian succession on the Dingle Peninsula comprises the upper Wenlock–upper Ludlow Dunquin Group and the basal, upper Ludlow–Přídolí part of the Dingle Group (Fig. 3). The Dunquin Group was assigned by Holland (1988) to late Wenlock (early Homerian) to early late Ludlow (*leintwardinensis* Biozone or early Ludfordian) time on the basis primarily of its abundant brachiopod faunas. More recently, Benton & Underwood (1994) have discovered a graptolite specimen towards the base of the Group in the Ferriter's Cove Formation (see below), which indicates the *lundgreni* Biozone (early Homerian time), and inferred that the underlying, basal formation of the Dunquin Group, the Foilnamahagh Formation, therefore is of Sheinwoodian or lower Wenlock age. However, as the Foilnamahagh Formation is virtually unfossiliferous, the Sheinwoodian dating is speculative and we

From: FRIEND, P. F. & WILLIAMS, B. P. J. (eds). *New Perspectives on the Old Red Sandstone.* Geological Society, London, Special Publications, **180**, 123–145. 0305-8719/00/$15.00

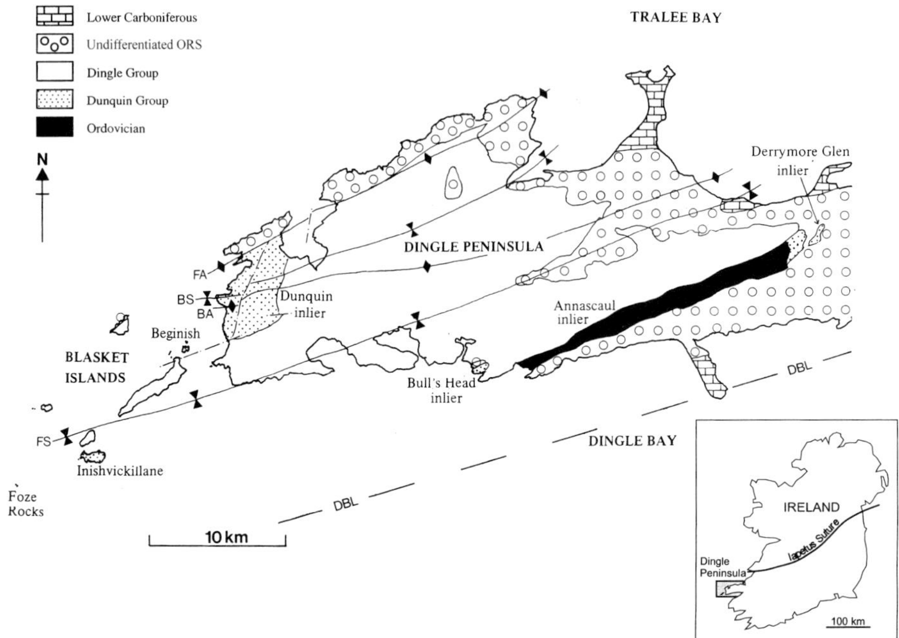

Fig. 1. General geological map of the Dingle Peninsula and Blasket Islands, showing inliers of the Dunquin Group, and major fold axes, and inset showing location of Dingle Peninsula and schematic trace of the Iapetus Suture through Ireland. BA, Ballysitteragh Anticline; BS, Ballyferriter Syncline; DBL, Dingle Bay Lineament; FA, Feohanagh Anticline; FS, Fahan Syncline. Modified after Horne (1974), Todd (1989) and Pracht (1996).

follow the original early Homerian dating of Holland (1988). The Dunquin Group comprises shallow-marine and minor coastal plain siliciclastic sediments and a range of volcaniclastic deposits and lavas (Gardiner & Reynolds 1902; Holland 1969, 1987, 1988; Sloan & Bennett 1990; Sloan 1991). The succeeding Dingle Group consists of non-marine siliciclastic rocks deposited in a range of lacustrine and fluvial environments. Its deposition has been summarized by Todd *et al.* (1988). The base of the Dingle Group is taken as late Ludlow in age, given its locally conformable contact with the underlying fossiliferous Dunquin Group. The youngest date derived from the Dingle Group is early–?mid-Emsian time, for a miospore assemblage from the upper part of the Slea Head Formation (upper part of the Dingle Group succession) (Higgs 1999).

The paper summarizes the key characteristics of the Dunquin Group and lower Dingle Group, based principally on outcrops in the west and southwest of the Dingle Peninsula and the neighbouring Blasket Islands (Boyd 1983; Sloan 1991). The paper outlines the Silurian development of the Dingle Basin in terms of its depositional environments, palaeogeography, igneous petrogenesis and tectonic evolution. A new interpretation of the tectonic evolution of the Dingle Basin is presented that envisages four phases of basin evolution in late Wenlock to late Přídolí time: (1) active subduction, accompanied by episodic volcanism and volcano-tectonically controlled shallow-marine and fan-delta deposition; (2) subduction termination, accompanied by an extension-related hydrothermal event and massive, climactic volcanic eruptions; (3) post-subduction thermal subsidence, accompanied by laterally extensive and relatively homogeneous, shallow-marine deposition; (4) strike-slip fault-controlled subsidence, accompanied by non-marine (Old Red Sandstone) deposition. The mid- to late Ludlow fill of the Dingle Basin records a transition from marine to fully non-marine (Old Red Sandstone) deposition, which is interpreted to have been tectonically driven.

The paper is organized as follows: summary accounts of each formation in the Dunquin Group and lower Dingle Group are presented providing relevant stratigraphical, lithological, sedimentological and geochemical information; the syndepositional tectonic evolution of the basin is then discussed and four phases of development are inferred; lastly, the broader regional context of the Dingle Basin in terms of late Iapetus history is discussed.

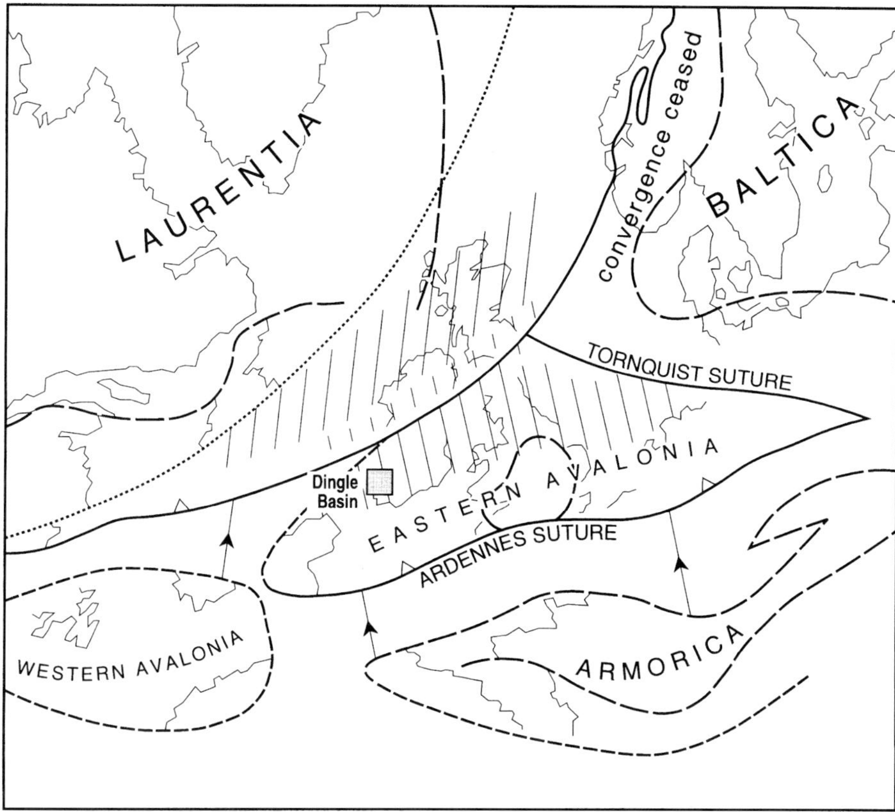

Fig. 2. Early Devonian plate reconstruction (based on Soper (1986)), showing the location of magmatic arcs in the Irish and British regions (hatched). The Dingle Basin lay on the northwestern margin of Eastern Avalonia. It should be noted that in this model, the Iapetus Ocean at this time was still open immediately to the west of the Dingle Basin.

Dunquin Group

Stratigraphy and depositional facies

The Dunquin Group crops out in a number of inliers along the Dingle Peninsula (Horne 1974; Parkin 1976) and in several small islands (the Blasket Islands) off its western end (Figs 1, 4 and 5). The fullest succession is found in the Dunquin inlier in the west of the peninsula, in predominantly coastal sections that expose an interval of *c.* 1500 m thickness of shallow-marine and minor coastal-plain sediments interbedded with a range of volcaniclastic deposits and lavas. The Dunquin Group has been divided into six formations (Holland 1969, 1988; Sloan 1991), which are outlined below in ascending stratigraphical order.

Foilnamahagh Formation. The Foilnamahagh Formation, the lowest unit in the Dunquin Group, is *c.* 150 m thick and crops out in the northwest of the Dunquin inlier (Fig. 4). It has a faulted base and is overlain conformably by the Ferriter's Cove Formation (Fig. 6). The Foilnamahagh Formation is a markedly heterogeneous interval of continental and marine sediments, lavas and pyroclastic deposits. It is dominated by red mudrocks interbedded with sandstones and clast- and matrix-supported conglomerates; minor bioturbated, fossiliferous siltstones occur near the top of the formation. Volcanic facies occur mainly near the base of the formation and include a 40 m succession of thin (<4 m), porphyritic basalt and basaltic andesite lava flows and minor pyroclastic flows. Conglomerate clast lithologies include intraformational lavas, ignimbrites, and sandstones and siltstones similar to the lithologies in the adjacent intervals. The Foilnamahagh Formation is interpreted as a succession of fan-delta sediments shed from volcanic uplands onto

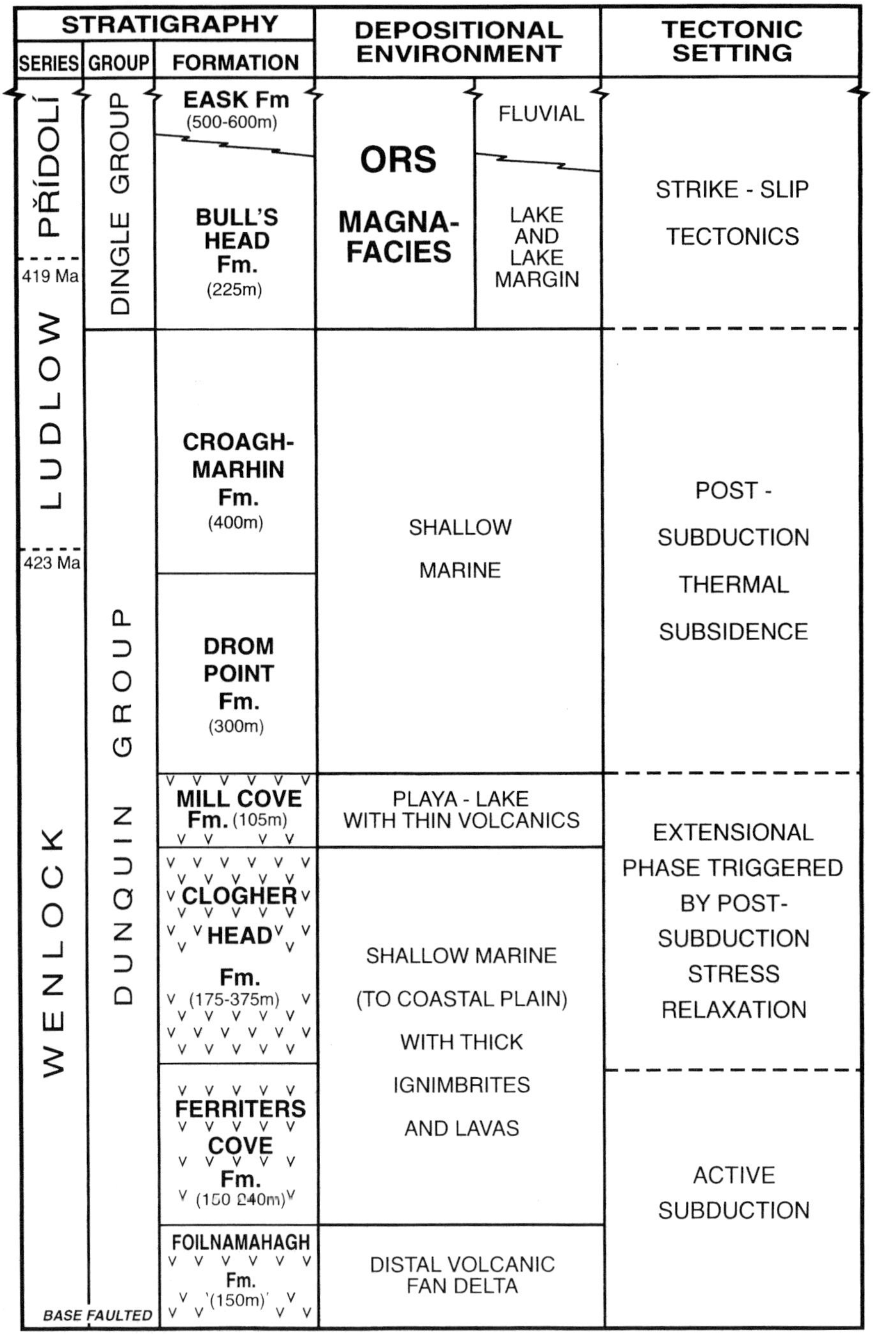

STRATIGRAPHY			DEPOSITIONAL ENVIRONMENT		TECTONIC SETTING
SERIES	GROUP	FORMATION			
PŘÍDOLÍ (419 Ma)	DINGLE GROUP	EASK Fm (500-600m)	ORS MAGNA-FACIES	FLUVIAL	STRIKE - SLIP TECTONICS
		BULL'S HEAD Fm. (225m)		LAKE AND LAKE MARGIN	
LUDLOW (423 Ma)	DUNQUIN GROUP	CROAGH-MARHIN Fm. (400m)	SHALLOW MARINE		POST - SUBDUCTION THERMAL SUBSIDENCE
WENLOCK		DROM POINT Fm. (300m)			
		MILL COVE Fm. (105m)	PLAYA - LAKE WITH THIN VOLCANICS		EXTENSIONAL PHASE TRIGGERED BY POST-SUBDUCTION STRESS RELAXATION
		CLOGHER HEAD Fm. (175-375m)	SHALLOW MARINE (TO COASTAL PLAIN) WITH THICK IGNIMBRITES AND LAVAS		
		FERRITERS COVE Fm. (150 240m)			ACTIVE SUBDUCTION
BASE FAULTED		FOILNAMAHAGH Fm. (150m)	DISTAL VOLCANIC FAN DELTA		

Fig. 3. Summary stratigraphy, depositional environments and phases of basin evolution for Dunquin Group and lower Dingle Group, western Dingle Peninsula. Absolute ages from Gradstein Ogg (1996).

the coastal plain of a volcanic island. The formation fines upward and displays evidence of increasing marine influence, attesting to an overall transgression. This trend was temporarily interrupted by the emplacement of pyroclastic flows near the end of Foilnamahagh Formation time.

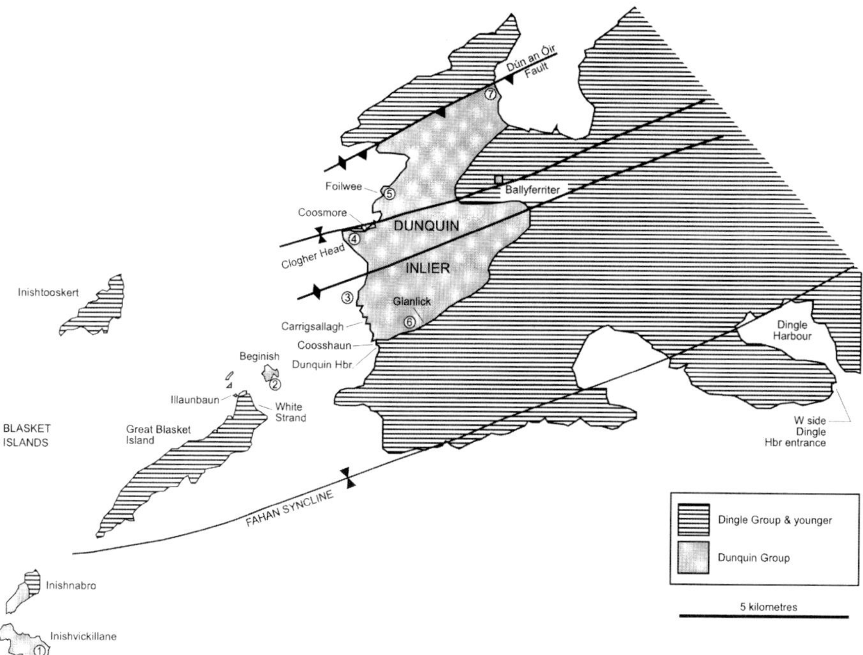

Fig. 4. Outline geological map of the western end of the Dingle Peninsula and the Blasket Islands, showing localities referred to in text. Circled numbers refer to the locations of stratigraphic sections shown in Fig. 5.

Ferriter's Cove Formation. The Ferriter's Cove Formation overlies the Foilnamahagh Formation and crops out in three fault-repeated sections on the west coast of the Dingle Peninsula (Figs 4 and 5). It comprises an interval of 150–240 m thickness of stacked, coarsening-upward parasequences with interstratified pyroclastic deposits (Fig. 6). The lower part of each parasequence is composed of intensely bioturbated siltstones, containing a rich shelly fauna (including brachiopods, corals, bryozoans, crinoids and rare trilobites and graptolites (Watkins 1978; Holland 1987, 1988; Benton & Underwood 1994)). These sediments become sandier and less bioturbated upwards, with wave-rippled siltstones and very fine grained sandstones giving way to hummocky cross-stratified, very fine to fine grained sandstones at the top, with cross-bedded conglomerates containing intraformational clasts also present locally. These facies are interpreted as representing a transition from offshore-marine to wave-dominated shoreface environments (Watkins 1978; Sloan & Williams 1991).

The shoreface successions are overlain by intervals of interlaminated very fine sandstone–siltstone and mudstone, showing common desiccation cracks, interstratified with intervals of intensely bioturbated siltstone, containing *in situ* developments of the coral *Parastriatopora* and rare ostracode-rich beds. These are interpreted as representing tidal-flats and lagoons with subtidal patch reefs. Volcanic deposits are absent in the offshore–shoreface intervals but are invariably associated with the tidal flat and lagoonal deposits. The volcanic intervals vary greatly in thickness (from <1 m to >60 m) and consist predominantly of ignimbrites and pyroclastic fall lapilli-tuffs.

The individual parasequences are of the order of 10–80 m thick and can be correlated between the three exposed sections. In a southward direction the inter-volcanic sedimentary intervals become thinner, whereas the volcanic deposits thicken dramatically. Limited palaeocurrent data suggest a NNE-dipping palaeoslope. The volcanic-rich tidal-flat–lagoonal part of each parasequence is overlain abruptly by the offshore facies of the succeeding parasequence. This apparent abrupt rise in relative sea level and transgression is interpreted as the result of volcanic-related subsidence following eruptive emptying of a nearby magma chamber (Sloan & Williams 1991).

Clogher Head Formation. The Clogher Head Formation overlies the Ferriter's Cove Formation and consists mainly of lavas and pyroclastic flow deposits. It crops out not only on

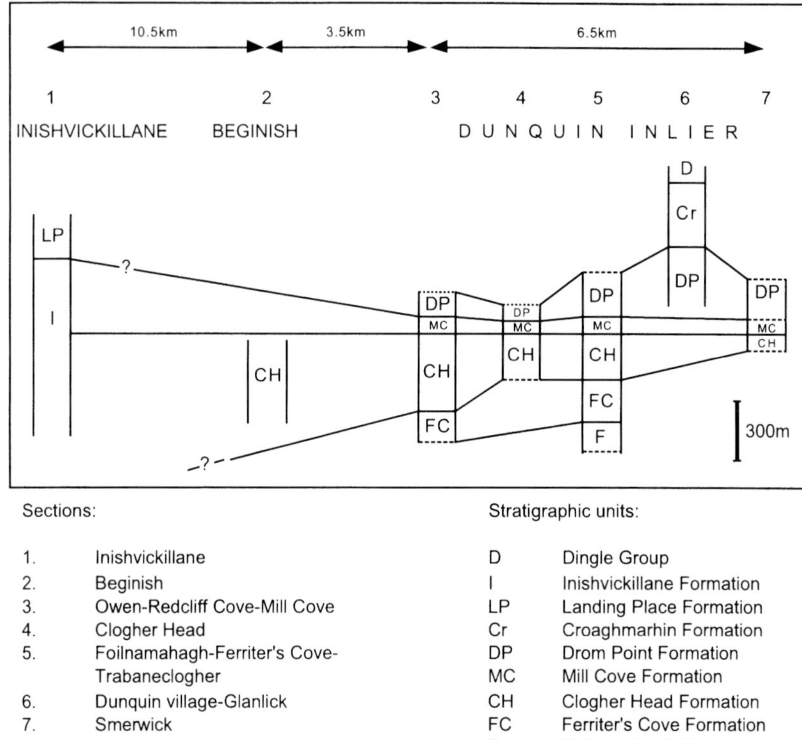

Fig. 5. Summary correlation of Dunquin Group formations exposed on the Blasket Islands and in the Dunquin inlier sections. Locations of sections are shown in Fig. 4.

the west coast of the Dingle Peninsula but also on several of the Blasket Islands, notably Beginish (Fig. 4); partially stratigraphically equivalent volcanic rocks of the Inishvickillane Formation (Parkin 1974) crop out on Inishvickillane, some 10 km from the mainland (Figs 4 and 5). Exposure of the entire Clogher Head Formation is continuous only on the mainland, with fault-repeated sections thickening to the south (from 175 m in the north to 375 m in the south). The formation is dominated by welded ignimbrites and subordinate andesitic–dacitic lava flows, with thin and sporadically developed, co-eruptive pyroclastic fall deposits (Fig. 6). The ignimbrites range in thickness from <1 m to >55 m, but most are 5–10 m thick. Most have a eutaxitic or densely welded texture, with a devitrified matrix, normally graded lithic fragments and inversely graded pumice. Lava-flow thicknesses vary from 30 cm to >45 m, but most are 1–3 m thick. They are generally medium grey to olive green–grey in colour, but many flow tops are reddened, the degree of reddening decreasing downwards, indicating contemporaneous weathering. The Clogher Head Formation, as exposed near Foilwee (Fig. 4), is capped by a lava flow that displays a 5 m-thick weathering profile, which grades downwards from a soft clay (composed largely of chlorite, corrensite, illite and smectite) into a hard basalt with clay-filled fissures (Wright *et al.* 1991). This well-developed weathering profile was formed during a prolonged hiatus before the deposition of the overlying Mill Cove Formation. Sediments in the Clogher Head Formation occur generally as thin (<2 m) intervals of red mudstones, sandstones and pebbly layers that drape ignimbrite and lava-flow surfaces. However, at Carrigsallagh, south of Clogher Head, the volcanic succession is interrupted by a fossiliferous coarsening-upward interval of 40 m thickness similar to the coarsening-upward parasequences in the Ferriter's Cove Formation.

The 400 m succession of densely welded ignimbrites and lava flows on the island of Beginish is correlated with the mainland Clogher Head Formation (Fig. 7) on the basis of their close lithological and geochemical similarity. The >900 m succession of andesite–basalt lavas on the island of Inishvickillane is

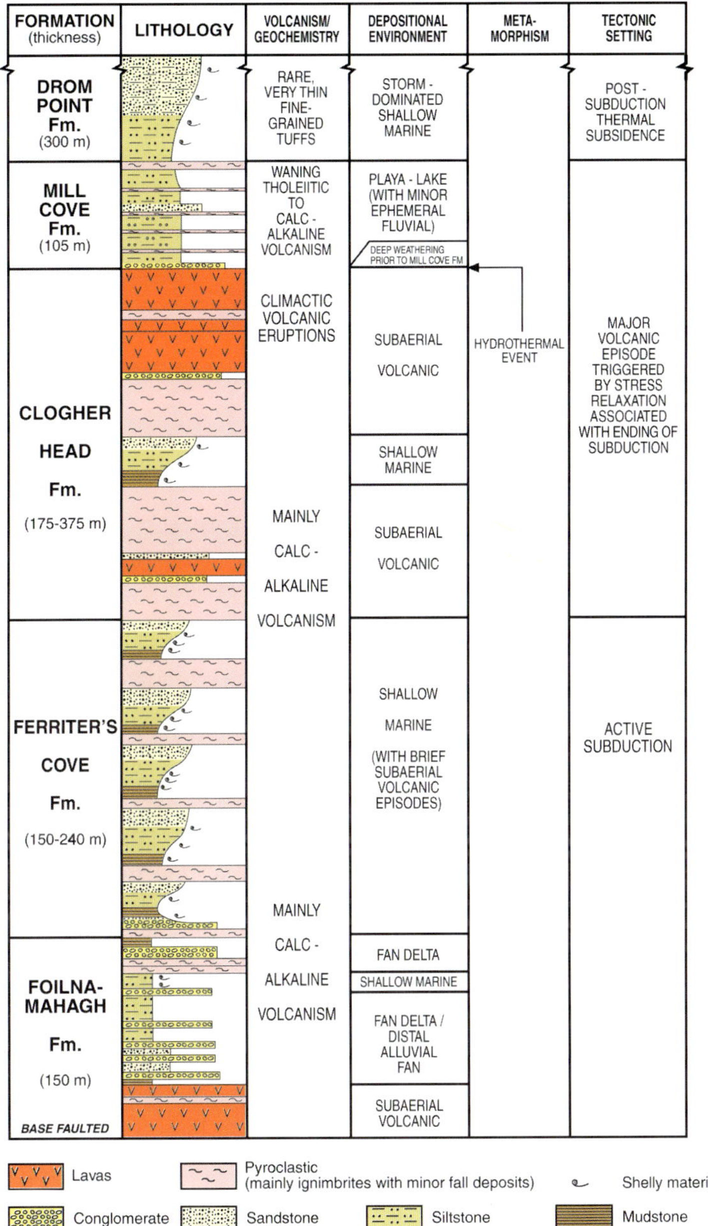

Fig. 6. Lithological, depositional, volcanic and tectonic summary of Dunquin Group (excluding Croaghmarhin Formation), Dunquin inlier, western Dingle Peninsula.

likewise lithologically and geochemically similar to the mainland succession, although welded ignimbrites are rare. The lower *c.* 400 m of this succession is correlated with the Clogher Head Formation, whereas the remaining upper part is the probable correlative of the overlying Mill Cove Formation (Sloan 1991; Fig. 5).

Mill Cove Formation. The overlying Mill Cove Formation is a 105 m interval of purple–red and grey muddy siltstones and mudstones, minor sandstones and conglomerates, and interspersed scoria-flow and pyroclastic fall deposits. The formation is exposed in four fault-repeated sections in the Dunquin inlier, with

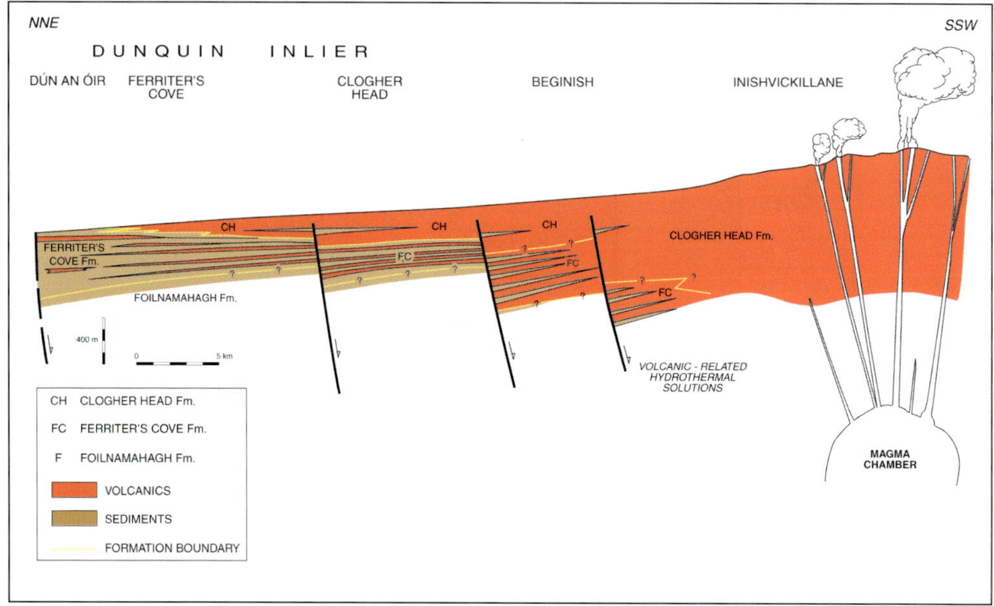

Fig. 7. Schematic depositional and tectonic cross-section of western Dingle Peninsula at late Wenlock time, showing Clogher Head and earlier formations. Subsidence controlled by volcano-tectonism. Volcanic sequences thicken southwards; sedimentary sequences thin southwards, except locally adjacent to faults.

stratigraphical equivalents exposed on the island of Inishvickillane and in small faulted outcrops on Illaunbaun and Great Blasket Island (Fig. 4). A total of nine scoria-flow units have been logged in the Dunquin inlier, each 30–360 cm thick and separated by sedimentary intervals in excess of 1.5 m. The scoria clasts are up to 90 cm across, chlorite rich and generally weakly flattened. Much of the formation is made up of tabular-bedded mudrocks, with typical bed thicknesses of 20–60 cm. They are generally massive or poorly laminated, with local wave-ripple cross-lamination and abundant post-depositional features including complex colour mottling, burrows and desiccation cracks, reflecting deposition in a shallow, ephemeral lacustrine or playa-lake setting. The mottled mudrocks were interpreted as groundwater ferricretes by Wright *et al.* (1992). These rocks were deposited in and around a playa lake (see Motts 1972) situated on a coastal plain built out by emplacement of the thick ignimbrites and lavas of the Clogher Head Formation. The scoria-flow deposits provide evidence of a vulcanian eruptive style (*sensu* Wright *et al.* 1980), characterized by episodic, brief, explosive eruptions.

On Inishvickillane, a deep weathering profile developed in andesite–basalt lavas in the Inishvickillane Formation provides a marker horizon that is correlated with a similar weathering profile at the top of the Clogher Head Formation on the mainland (Fig. 5). The remainder of the Inishvickillane Formation that overlies this weathered horizon (which is correlatable with the mainland Mill Cove Formation) comprises two, *c.* 150 m lava-flow intervals, separated by an 80 m interval of red and grey mudstones, fine sandstones and minor tuffs, similar to the mainland Mill Cove Formation lithologies.

Drom Point Formation. The Drom Point Formation crops out in the Dunquin inlier and on the Great Blasket Island, with a stratigraphical equivalent (Landing Place Formation) occurring on Inishvickillane (Fig. 5). The Drom Point Formation (Fig. 6) consists of a 300 m interval of fossiliferous siltstones, very fine to fine grained sandstones, minor claystones and rare intraclast conglomerates and tuffaceous beds. Abundant developments of *Chondrites* burrows are characteristic of the formation. The base of the formation is marked by an abrupt change from deposition of the playa-lake red beds and scoria-flow units of the Mill Cove Formation to fossiliferous offshore muds, indicating a rapid rise in relative sea level and transgresion. The lower 100 m of the formation comprises an overall coarsening-upward succession,

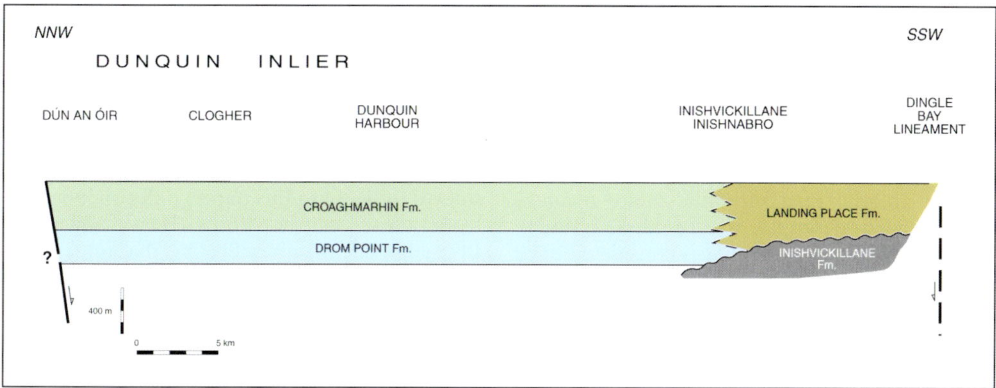

Fig. 8. Schematic depositional and tectonic cross-section of western Dingle Peninsula at late Wenlock–late Ludlow time, showing Drom Point and Croaghmarhin Formations. Shallow shelf environments were developed throughout. Thicknesses from Holland (1987, 1988).

representing a transition from offshore to shoreface environments. The top of the formation was defined by Holland (1969) as the top of decalcified beds rich in the endemic brachiopod *Rhipidium hibernicum*.

Croaghmarhin Formation. The Croaghmarhin Formation overlies the Drom Point Formation and is overlain by the Dingle Group (Figs 3 and 8; see below for details of contact relationships). It is seen only in inland exposures in the Dunquin inlier. The uppermost part of the formation has been assigned to the *leintwardinensis* Biozone (late Ludlow time) by Holland (1988). The formation comprises a 400 m succession of calcareous–decalcified siltstones and impure limestones containing an abundant fauna, dominated by tabulate corals and brachiopods, with rare graptolites. Tuffs are absent. The formation was deposited in a shallow-marine environment. The paucity of sands and the abundance of tabulate corals may reflect deeper-water conditions than in the Drom Point Formation and/or reduced sediment supply from the hinterland.

Igneous petrology and magmatic evolution

The pyroclastic rocks of the Dunquin Group are predominantly rhyolitic in composition, whereas the lavas, although ranging from basaltic to dacitic, are predominantly andesitic. The only exposed intrusion is a fine grained, mafic dyke seen on the west coast of the Dunquin inlier. The geochemical compositions and petrographic characteristics of the pyroclastic rocks, lavas and dyke indicate common derivation from a repeatedly replenished, basaltic magma chamber(s). Plagioclase, Fe–Ti oxides, pyroxene (and olivine) are inferred to have fractionated from parent basaltic magmas, with pyroxene and olivine being retained and accumulating in the upper crust, giving rise to an intense aeromagnetic anomaly to the southwest of Inishvickillane (Lefort & Max 1984). Details of the geochemistry and petrography of these igneous rocks have been given by Sloan & Bennett (1990) and Sloan (1991).

The Dunquin Group volcanic suite is late Wenlock (Homerian) in age (Holland 1988). However, as the base of the Group is faulted, volcanism may have begun in early Wenlock (Sheinwoodian) time or earlier (see Benton & Underwood 1994). The thick ignimbrites and lavas of the Clogher Head Formation represent the climax of volcanism in the Dingle Basin. It is significant that the topmost lavas in the formation are compositionally the most primitive, suggesting that the climactic eruptions had effectively emptied the subvolcanic magma chamber and that deeper levels were being tapped by the end of Clogher Head Formation time. Subsequent volcanic activity is represented by sporadic scoria-flow deposits in the Mill Cove Formation in the Dunquin inlier, although andesite-dominated lavas continued to be erupted farther to the southwest in the vicinity of Inishvickillane. The top of the Clogher Head Formation also marks an overall change in igneous composition from predominantly calc-alkaline to more tholeiitic. The top of the Mill Cove Formation marks the abrupt end of coarse-grained pyroclastic eruptions in the Dingle Basin. Volcaniclastic units in the overlying Drom Point Formation comprise fine grained, dominantly lithic tuffs, possibly deposited during sporadic phreatic–phreatomagmatic eruptions as volcanic activity petered out.

Latest Clogher Head Formation time hydrothermal event

The exposures of the Dunquin Group display localized occurrences of mineral assemblages indicative of prehnite–actinolite facies metamorphism (Liou *et al.* 1985): prehnite–actinolite–epidote–chlorite–quartz–titanite–albite–opaques (for detailed descriptions, see Sloan (1991)). This low-pressure assemblage is considered to be the product primarily of volcanic-related hydrothermal metamorphism $\pm$ burial metamorphism. Evidence of the influence of hydrothermal alteration is well displayed on the southern shore of Beginish, where a 2 m-wide alteration zone, developed along the contact between two lava flows, contains a central assemblage of albite–quartz–epidote–hematite, grading outwards into relatively unaltered lava. Veins of epidote and quartz are widespread in the Clogher Head Formation, but are absent in the overlying units. Abundant detrital grains of epidote (alongside grains of fresh, unaltered plagioclase) have been observed in a sandstone bed in the Mill Cove Formation. Such detrital epidote has not been observed in the Clogher Head Formation sediments or in underlying units, and is thus likely to have crystallized at latest Clogher Head Formation time. These occurrences provide evidence of a hydrothermal event that followed the climactic volcanic eruptions and was probably driven by heat generated from the subjacent magma chamber(s).

Lower Dingle Group stratigraphy and depositional facies

The Dingle Group overlies the Dunquin Group and represents the onset of non-marine, Lower Old Red Sandstone, deposition. This is interpreted to have occurred in late Ludlow time. The Dingle Group crops out extensively across the Dingle Peninsula and its lithostratigraphy and depositional environments have been summarized by Todd *et al.* (1988). In the context of this paper, focusing on the early development of the Dingle Basin, attention is concentrated on the lowest part of the Dingle Group, the Bull's Head and Eask Formations. The main development of these formations is in the south of the Peninsula, in the Fahan Syncline (Fig. 4). Correlation of the Bull's Head and Eask Formations northward from the Fahan Syncline is based on position in succession and models of likely lateral equivalence. Pracht (1996) followed Todd (1989) in correlating them with the Ballyferriter Formation, which crops out around the village of Ballyferriter in the north (Fig. 4), where it is depicted as unconformably overlying the Dunquin Group. This unconformity in the north is considered to be time equivalent to the basal part of the Bull's Head Formation in the Fahan Syncline to the south. Depositional facies in the Ballyferriter Formation (not considered in detail in this paper) differ significantly from those in the Bull's Head and Eask Formations, possibly reflecting the influence of an intervening or underlying intrabasinal high (see below).

Bull's Head Formation

The Bull's Head Formation can be mapped across an area of *c.* 50 km (WSW–ENE) by 10 km (NNW–SSE, the shortening direction) (Pracht 1996), but its main outcrops are on the north and south limbs of the Fahan Syncline, including the Great Blasket Island and Inishnabro (Fig. 4). Its boundary with the underlying Dunquin Group is generally faulted but in two areas a conformable contact is inferred. At Glanlick, near Dunquin village at the western end of the Peninsula (Fig. 4), a stream section and discontinuous hillside exposure can be interpreted in terms of a marine to non-marine transition involving a prograding wave-dominated shoreline (Boyd 1983), and on Inishnabro, Todd (1991) described a gradational succession also interpreted as a shallow-marine to non-marine transition.

The Formation is assumed to be of Late Silurian (late Ludlow–Přídolí) age considering the bounding ages of the Dingle Group as a whole (see above). It is subdivided into four members: the Heterolithic Member, the Boat Cove and Trabane Members, and the Paraconglomerate Member (Fig. 9).

The Heterolithic Member comprises the only lithofacies association of the Bull's Head Formation on the southern limb of the Fahan Syncline and the dominant one on the northern limb. Its thickness in the south is *c.* 220 m, and in the north 100 m. A sedimentological log of a typical interval is shown in Fig. 10. The Heterolithic Member is characterized by purple–grey, very fine sandstone units <1 cm thick, thinly interlaminated with typically much thinner mudstone layers. Sandstones appear structureless internally and show irregular lateral thickness variations. Simple sub-vertical sand-filled burrows of 5 mm diameter and 5 cm length are common, and more pervasive bioturbation is seen at some levels. Bedding planes commonly display straight ripple crests and polygonal desiccation cracks; most measurable crests restore around a NE–SW

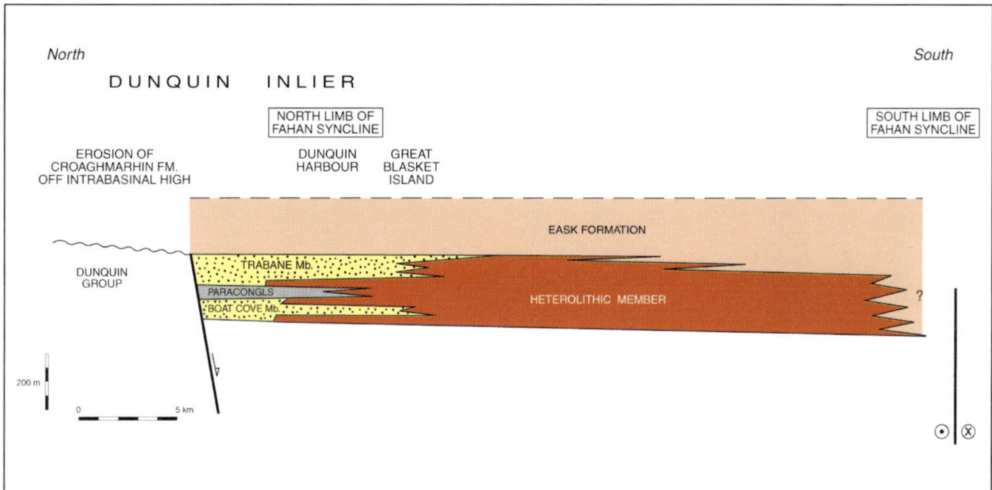

Fig. 9. Schematic depositional and tectonic cross-section of western Dingle Peninsula at late Ludlow–Přídolí time, showing Bull's Head Formation and lower Eask Formation.

orientation. Individual sandstone and mudstone units >1 cm thick also occur, forming subordinate components of the Heterolithic Member; these sands tend to be cleaner and slightly coarser than the thin sand laminae. On the northern limb of the Fahan Syncline these thicker sands and muds are commonly up to 3 cm in thickness, and occur, commonly in association with each other, interspersed through the thinly interlaminated association. On the southern limb thicker sandstone units are more common and typically range up to 10 cm, generally grouped into distinct sandier packages up to 6 m thick.

Sands of the Heterolithic Member are interpreted to have been reworked by repeated wave activity in a lacustrine setting, with the interlaminated muds deposited during intervening calms. Thicker sands and associated muds reflect more intense activity and suspension of fines, whereas discrete sandier packages may represent the influence of lake margin sandflat systems. Although exposure of the lake floor did occur, the persistence of infaunal biogenic activity and the evident wave influence suggest that shallow water cover was the norm. The NE–SW orientation of wave ripple crests is interpreted as the dominant shoreline orientation in a similarly oriented lake. The consistency in character of the Heterolithic Member throughout its thickness points to similar water depth, i.e. a similar balance between sediment supply and relative lake level, throughout its deposition.

The Boat Cove and Trabane Members are two very similar lithofacies associations occurring respectively intercalated with and overlying the Heterolithic Member on the northern limb of the Fahan Syncline, with thicknesses of 21 m and 94 m, respectively (Fig. 9), and best exposed on the Great Blasket Island. They are not developed on the southern limb. These Members are composed principally of very fine to fine, horizontally laminated sandstones, with subordinate current ripple lamination (some climbing) and cross-bedding, and with wave rippling on the tops of some thinner beds. A typical logged section is shown in Fig. 10. Sandstones show sharp to erosive bases with associated mudclast conglomerates in places. Thin siltstone and mudstone units within these Members show evidence variously of desiccation, bioturbation and incipient pedogenesis.

The association of facies comprising the Boat Cove and Trabane Members, and the transitional relationships with the Heterolithic Member, are interpreted as representing deposition by unconfined sheet floods on lake margin sandflats. Palaeocurrent data are consistent with deposition on the northern side of the lake, as is the restriction of these two Members to the northern area.

The Paraconglomerate Member comprises a series of mud matrix conglomerate beds which intercalate with sandy Heterolithic Member facies and possibly also the Boat Cove Member (Fig. 9). It again is restricted to the northern limb of the Fahan Syncline and is exposed only at the cove Coosshaun (Fig. 4), immediately north of Dunquin Harbour. The total exposed, faulted, conglomerate-bearing section is *c.* 10 m thick. Clasts are most commonly 5–30 cm in long axis, but some up to 115 cm are also seen. All clasts

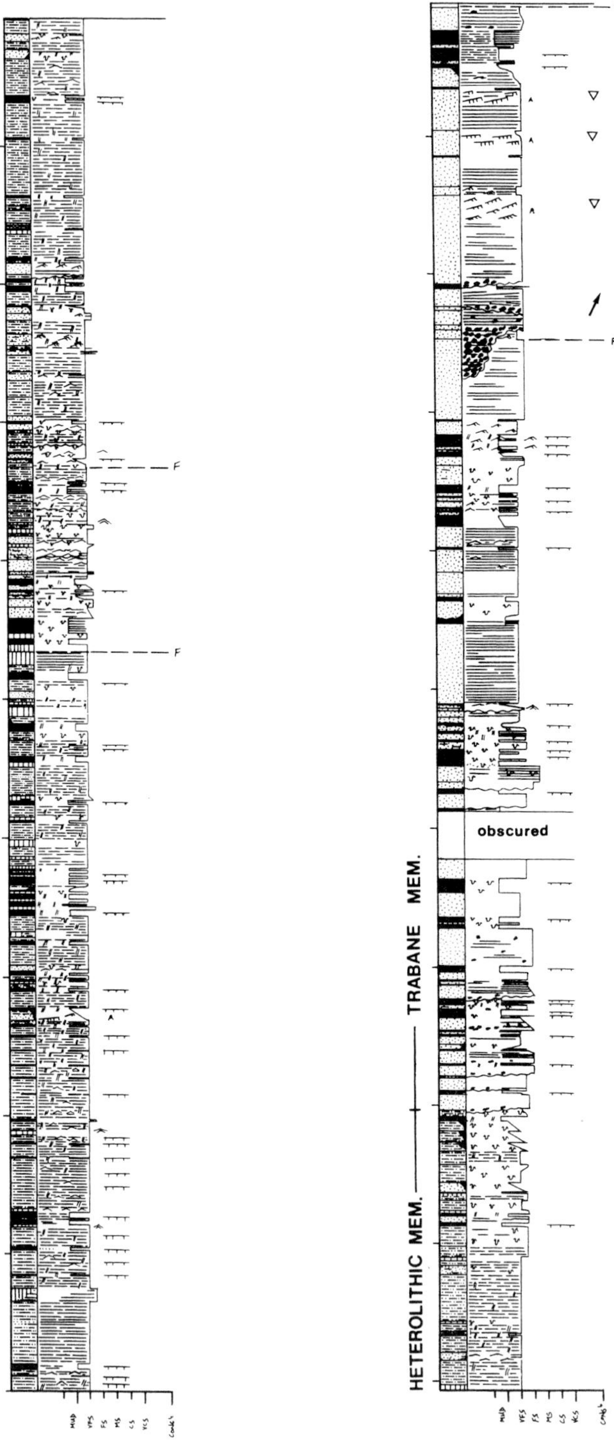

Fig. 10. Detailed sedimentological log through the upper part of the Heterolithic Member and basal Trabane Member, Bull's Head Formation, exposed at White Strand, Great Blasket Island. See Fig. 4 for general location, Fig. 12 for key. Scale increment is 1 m. Base of log at bottom left, top at top right.

are of fine grained (rarely fine–medium) calcite- and quartz-cemented sandstone, which is commonly veined. Lack of consistent orientation suggests that veining occurred before conglomerate deposition. The Paraconglomerate Member is interpreted as representing deposition on a small debris-flow dominated fan delta developed locally on the northern side of the Heterolithic Member lacustrine system.

Eask Formation

The Eask Formation overlies the Heterolithic Member of the Bull's Head Formation on the southern limb of the Fahan Syncline, and overlies the Trabane Member on its northern limb (Fig. 9). The Eask Formation comprises a spectrum of lithofacies (Fig. 11). The coarsest components are relatively thick, composite, fine grained to medium grained sand bodies >1 m thick. The sandstones show parallel and low-angle lamination and cross-bedding, the latter being better developed on the southern limb. The middle of the Eask Formation facies continuum is represented by thinner sandstones, up to fine grained and individually 1–100 cm thick. They mainly show parallel lamination, some current ripple lamination, and minor cross-bedding. These sandstones range from displaying sand-on-sand amalgamation to being interbedded with mudstone. The finest-grained elements of the Eask Formation are mudstone to muddy very fine sandstone intervals, which lack, over intervals of 1 m or more, clean sandstone beds. Internal structures range from horizontal lamination with minor current rippling, through bioturbation, to mudcracks and incipient pedogenic nodules.

The coarser facies decrease in modal and maximum grain size, and as a proportion of the formation thickness, from the southern to the northern limb of the Fahan Syncline. Palaeocurrent data from both limbs show dispersal towards the NNW.

The sandstones are interpreted as the deposits of episodic fluvial flood events on the basis of their tabular geometries and the predominance of upper flow regime parallel lamination. Deposition in shallow channels and unconfined sheets is inferred. The mudstones are seen as having been deposited distal and lateral to the areas of most active flooding, probably derived mainly from the run-out of turbid flood waters. Between flood events these areas were subject to pedogenesis and desiccation. The Eask Formation's geometry, facies and palaeocurrents suggest that it represents a broad alluvial apron comprising laterally coalescing, probably terminal, ephemeral fluvial systems, which paralleled the southern margin of the basin and dispersed sediment towards the NNW.

Eask Formation deposition in the area of the southern limb of the Fahan Syncline was succeeded by the conglomeratic deposits of the Trabeg Conglomerate Formation, also dispersing towards the northwest (Todd *et al.* 1988); uplift of the source area to the south is inferred. On the northern limb of the Fahan Syncline the Eask Formation is overlain by the Coumeenoole Formation, representing a northeast draining braided fluvial system, again reflecting increased runoff from the Dingle Basin catchment. The Coumeenoole Formation is not developed on the southern limb of the Fahan Syncline, where its lateral equivalent is considered to be the upper part of the Eask Formation.

Lower Dingle Group overview

The boundaries of the Dingle Basin during lower Dingle Group time must lie beyond the present-day outcrop area, but are now tectonically removed or concealed by younger geology and can only be inferred. Todd (1989) made the case for the Dingle Bay Lineament being the southern boundary during Trabeg Conglomerate Formation time, and given the continuity of deposition from Eask Formation to Trabeg Conglomerate Formation in the south of the basin it is reasonable to take it as the southern boundary during earlier Dingle Group time also. The eastern and western margins are unconstrained. The Dun an óir Fault (Fig. 4) was taken as the expression of the northern boundary by Todd *et al.* (1988).

Palaeocurrent data from the Heterolithic Member suggest a lacustrine area that was elongate in a WSW–ESE direction. This elongation is interpreted as reflecting the elongation of the overall basin and is also consistent with the general tectonic grain of the collisional margin.

Lower Dingle Group deposition (to top Eask Formation) can be summarized as follows. Bull's Head Formation deposition in the Fahan Syncline area was initiated in late Ludlow time (Fig. 9). A broad shallow lake became established, and persisted to accumulate *c.* 200 m of fairly uniform lacustrine sediment in the southern part of the area. In the north of the area, on the northern side of the lake, a series of lake margin systems developed and prograded lakeward: first the sheetflood dominated sandflats of the Boat Cove Member, then the small debris flow dominated fan delta represented by the Paraconglomerate Member, and finally the renewed sandflat deposition of the Trabane

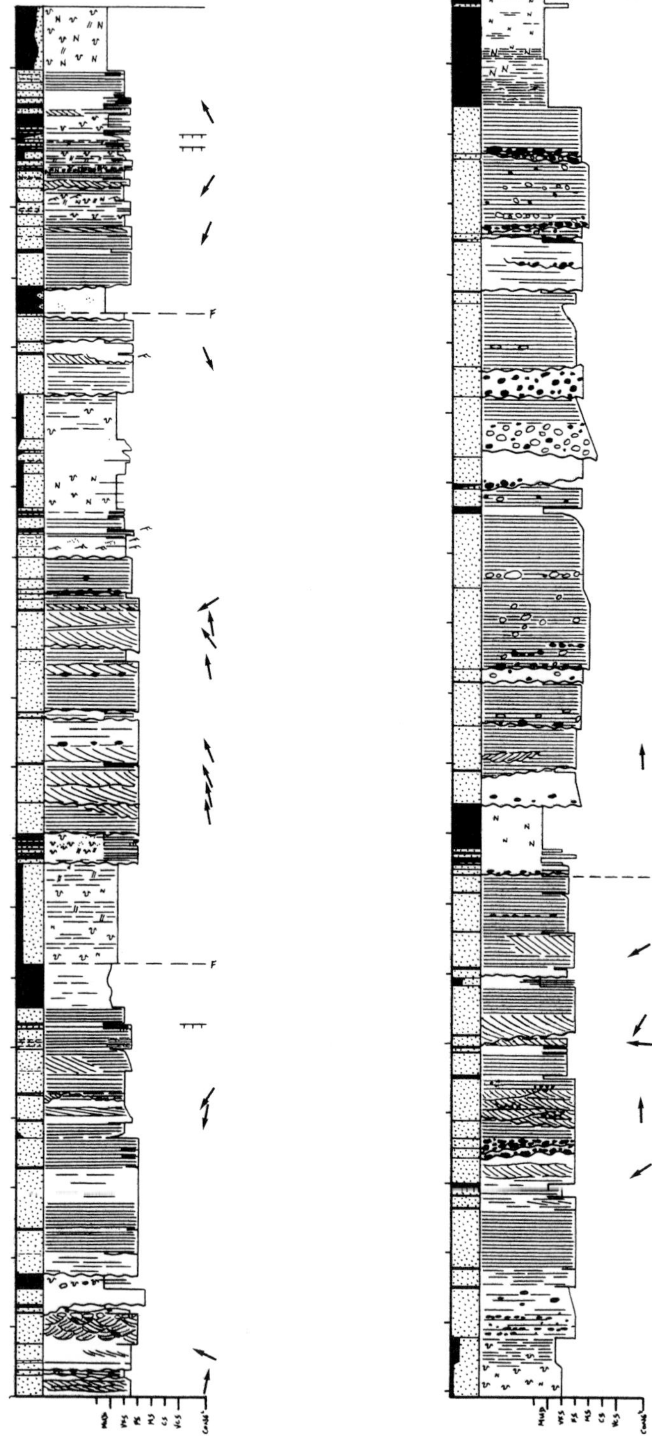

Fig. 11. Detailed sedimentological log showing typical developments of Eask Formation lithofacies, exposed on western shore of Dingle Harbour entrance. See Fig. 4 for general location, Fig. 12 for key. Scale increment is 1 m. Base of log at bottom left, top at top right.

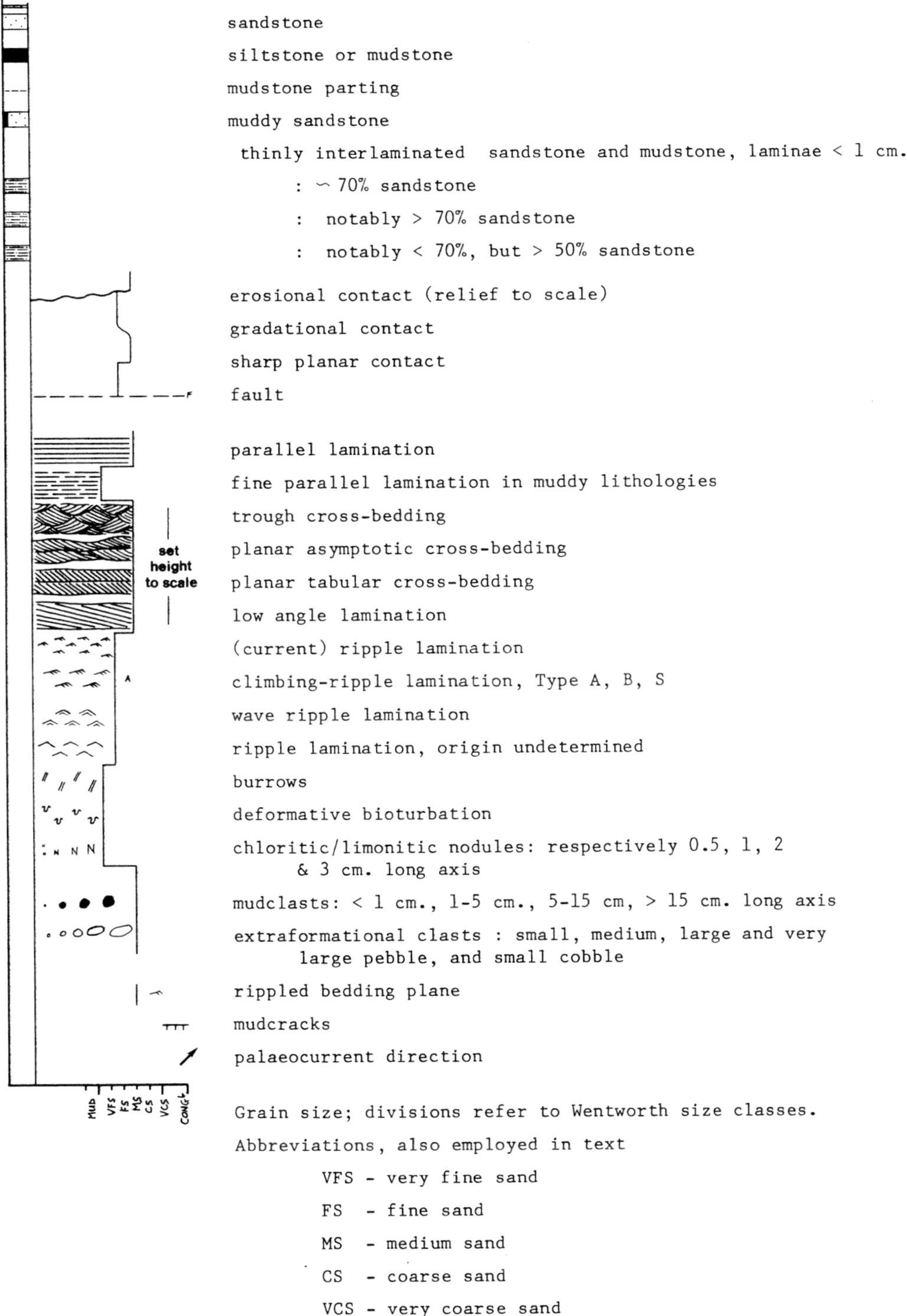

Fig. 12. Key to graphic logs shown in Figs 10 and 11.

Member. The southward prograding Trabane Member met the major NNW advancing fluvial apron of the Eask Formation, thus occluding the lacustrine area (Heterolithic Member) entirely. Bull's Head Formation ended as the Eask Formation continued its northward advance to eventually overlie the Trabane Member.

In the area of the southern limb of the Fahan Syncline, north-northwestward sediment dispersal continued throughout the deposition of the Eask Formation and the overlying Trabeg Conglomerate Formation. In the area of the northern limb, this dispersal pattern was replaced at latest Eask Formation time by a NE-flowing axial braided river system represented by the Coumeenoole Formation.

Basin evolution

Four phases in the evolution of the Dingle Basin are interpreted for the Silurian to Lower Devonian Dunquin Group and lower Dingle Group succession (Table 1): (1) a phase related to active subduction; (2) a phase related to subduction termination; (3) a post-subduction thermal subsidence phase; (4) a phase of strike-slip fault-controlled subsidence. The grounds for these interpretations are set out below, as is the relationship of these phases to the broader regional context of Iapetean closure.

Active subduction phase

The active subduction phase is represented by the Foilnamahagh Formation and Ferriter's Cove Formation, which were deposited during late Wenlock time. At this time the Dingle Basin is interpreted to have lain near the northwest margin of the Eastern Avalonia continental plate (Fig. 2). Although neither detailed mineralogical nor geochemical studies specifically aimed at determining sediment provenance have yet been undertaken, the limited range of pebble clast types (all intraformational) in these formations (and in the Dunquin Group as a whole) suggests sediment derivation from local, predominantly volcanic sources. The presence of two endemic brachiopod species (*Holcospirifer bigugosus* and *Rhipidium hibernicum*) in the Dunquin Group was interpreted by Bassett *et al.* (1976) as indicating isolation of the Dingle Basin from more extensive continental shelf areas to the east (Fig. 2). The Foilnamahagh and Ferriter's Cove Formations (and the succeeding Clogher Head and Mill Cove Formations) of the Dunquin inlier are therefore viewed as representing the preserved fringe of a large volcanic island.

Although the discontinuity of exposure renders the exact configuration of the volcanic centre uncertain, several lines of evidence (including volcanic thickness variations, orientation of cross-beds and slump-folds, and the presence of an intense aeromagnetic anomaly (Lefort & Max 1984)) suggest that a volcanic centre lay to the southwest of Inishvickillane. The lowest-preserved volcanic rocks (lower Foilnamahagh Formation) indicate an early predominantly effusive eruptive style, followed in Ferriter's Cove Formation time by wholly explosive, pyroclastic eruptions (based solely on the evidence of the succession exposed in the Dunquin inlier). That eruptions appear to have been periodic rather than semi-continuous during this phase of basin development is interpreted as the product of repeated filling of the subvolcanic magma chamber(s), which may itself have ultimately reflected periodic generation of magmas above the subducting slab of oceanic crust beneath the Dingle Basin.

The palaeocurrent indicators, and thickness and facies variations suggest that the palaeoslope dipped NNW from the main volcanic centre, with facies belts arranged perpendicular to it. Alluvial fans carrying coarse, eroded volcanic detritus spread onto a coastal plain containing shallow lakes and small braided streams, and fan deltas locally built out across the shoreline. Fluctuation of this shoreline is recorded in the succession of stacked shallow-marine parasequences comprising offshore to wave-dominated shoreface through back-barrier lagoon and tidal-flat facies. These depositional cyclothems were synchronous with volcanic cyclicity. Each eruptive event was accompanied by rapid subsidence, caused by collapse of the subjacent magma chamber, and an abrupt relative rise in sea level. Subsidence occurred along inferred normal faults that dipped towards the volcanic centre.

Subduction termination phase

The Clogher Head Formation is interpreted as representing the terminal phase of subduction beneath the Dingle Basin. The acid ignimbrites that occur in such great thicknesses in the Clogher Head Formation are sporadically present below but are absent above (Fig. 3), suggesting a volcanic episode of basinwide significance. During Clogher Head Formation time, the volcanic rocks became compositionally less evolved (Sloan 1991), suggesting the tapping of deeper, more primitive magmas. The volcanic rocks of the Mill Cove Formation (and its stratigraphical equivalent on Inishvickillane)

Table 1. *Dingle Basin summary: Dunquin Group and lower Dingle Group*

Tectonic phase	Stress regime	Volcanic activity	Igneous composition	Metamorphism	Depositional environments	Controls on sedimentation	Dingle Basin stratigraphy
1, Active subduction	Weak compression	Episodic explosive volcanism from major silicic centre; plus minor effusive activity from minor basic satellite centres	Subduction-related calc-alkaline (with subordinate tholeiitic)		Initially distal fan-delta on volcanic island coastal plain, followed by offshore–storm-dominated shoreface to tidal-flat–lagoonal	Volcano-tectonism, related to repeated filling and eruptive emptying of a nearby magma chamber(s)	Foilnamahagh Fm Ferriter's Cove Fm
2, Ending of subduction	Extension (stress relaxation)	Climactic explosive eruptions triggered by extension; followed by major effusive activity with minor vulcanian eruptions	Initially calc-alkaline, becoming increasingly tholeiitic–less evolved as deeper, more primitive magmas are tapped	Hydrothermal event (widespread prehnite–actinolite facies plus vein mineralization) associated with climactic eruptions and extension	Playa-lake–distal alluvial fan deposition on extensive volcanic coastal plain	Sediment derived by erosion of newly constructed volcanic uplands; more stable depositional setting than previously	Clogher Head Fm Mill Cove Fm
3, Post-subduction	Thermal subsidence	Very minor, residual, phreatic–phreatomagmatic(?) eruptions	Tholeiitic		Offshore to storm-dominated shoreface	Widespread, more-or-less uniform subsidence; very minor fault activity	Drom Point Fm Croaghmarhin Fm
4, Strike-slip tectonism	Sinistral transpression				Shallow impermanent lake, followed by laterally and then axially draining fluvial systems	Intra-basinal faulting	Lower Dingle Group

are, similarly, less evolved and more tholeiitic in composition than the dominantly calc-alkaline Clogher Head Formation succession. A modern analogue is provided by the Plio-Holocene Deschutes Formation of the High Cascades, Oregon, where the tholeiitic composition of its basaltic andesites and andesites was used by Smith *et al.* (1987) to infer an extensional tectonic environment that allowed the tapping of deep, less evolved magmas.

In the Dingle Basin, likewise, the climactic ignimbrite eruptions of the Clogher Head Formation and the post-Clogher Head Formation change to more tholeiitic compositions are inferred as indicating the development of an extensional stress regime. The ending of subduction beneath the Dingle Basin may have been accompanied by stress relaxation, allowing, in late-Clogher Head Formation time, the sudden escape of magmas stored in crustal pools, resulting in catastrophic acid ignimbritic eruptions, and the subsequent tapping of deeper, more tholeiitic magmas. In these circumstances, extensional faulting would also have allowed the circulation of hydrothermal fluids through the Dunquin Group, driven by the magmatic heat source (Fig. 7), forming mineralized veins and widespread prehnite–actinolite facies metamorphism.

The deep soil profile developed in the topmost lava flow of the Clogher Head Formation in the Dunquin inlier indicates a long depositional hiatus following the climactic ignimbrite eruptions. This may reflect the abrupt reduction in sediment supply that would have resulted from blanketing of the landscape by extensive, indurated volcanic units. The Mill Cove Formation was deposited in a relatively quiescent tectonic environment, as suggested by its more or less uniform thickness within the Dunquin inlier. The emplacement of extensive ignimbrites (Clogher Head Formation) built out a wide, low-relief coastal plain where sheetfloods and minor braided streams drained into shallow, ephemeral lakes.

Post-subduction thermal subsidence phase

By the end of Mill Cove Formation time, the supply of subduction-related magmas beneath the Dingle Basin had become exhausted. In the overlying units, volcanic activity is represented by only a few thin, fine grained tuffs in the Drom Point Formation. The Drom Point and Croaghmarhin Formations, consisting of shallow-marine, fossiliferous siltstones and very fine to fine grained sandstones, exhibit laterally relatively uniform facies and thicknesses (Fig. 8), suggestive of a subdued topography and a history of areally uniform subsidence. (Later local erosion of the Croaghmarhin Formation is discussed below.) The accumulation of substantial thicknesses of shallow-marine sediments under such conditions is consistent with basin-wide post-subduction thermal subsidence. Crustal cooling would have resulted from the cut-off of supply of subduction-related magmas to the basin and have led to relatively gradual, but regionally uniform, subsidence. The accommodation space thus created was progressively filled by the shallow-marine shelf sediments of the latest Wenlock to late Ludlow Drom Point and Croaghmarhin Formations. Near latest Drom Point Formation time, the volcanic uplands were onlapped by shallow-marine sediments, as shown on Inishvickillane, where the thick subaerially erupted lava succession is overlain by upper Dunquin Group fossiliferous siltstones.

Strike-slip fault-controlled subsidence phase

Earliest Dingle Group (Bull's Head Formation) time saw a change in the Dingle Basin from relatively uniform subsidence across the depositional area, represented by the outcrops at the western end of the Peninsula, to marked intrabasinal subsidence variations. Activity on discrete faults is inferred, as discussed below. Although this change can be demonstrated, its precise kinematic nature cannot be elucidated from the tectono-sedimentary evidence. Our interpretation is that the change represents initiation of a sinistral strike-slip tectonic regime affecting the Basin, an interpretation based largely on the broader regional context.

The first evidence presented for this fault movement in earliest Dingle Group time is the unconformity at Coosmore (Fig. 4), where Drom Point Formation is erosively overlain by 80 m of fluvial conglomerates assigned to the Dingle Group. No angular discordance is evident, but the Croaghmarhin Formation is absent. The case for these conglomerates being Dingle Group has been made by Horne (1974) and Boyd (1983) and has been followed by Todd *et al.* (1988) and this paper. Doran *et al.* (1973) and Holland (1987), however, interpreted these conglomerates as Upper Old Red Sandstone, mainly on the basis of the Croaghmarhin Formation's absence, in contrast to its conformable relationship with the Dingle Group inland to the south at Glanlick. This point of interpretation is clearly essential to the unconformity's significance in terms of earliest Dingle Group tectonics.

Croaghmarhin Formation is absent at Coosmore but present to the northeast, east and southeast (Holland 1988). Age equivalent sediments are also believed to be represented in the Inishnabro Formation (Todd 1991). Given the lack of angular discordance at the Coosmore contact (the conglomerates' irregular basal surface represents erosion, but bedding above and below is parallel) it could be that the conglomerates are a lateral facies equivalent, such as a fan delta, of the shallow-marine Croaghmarhin Formation. That in itself would suggest intrabasinal fault movement but is considered an unlikely interpretation in the absence of any conglomeratic material within the siltstones and silty fine sandstones of the Croaghmarhin Formation. It is more likely that the Croaghmarhin Formation is absent from Coosmore as a result of early Dingle Group uplift and erosion, based on the interpretation of the Boat Cove and Trabane Members as the erosional product, as discussed below. Newly (re)activated intrabasinal faults separating the uplifted area from the depositional area to the south and east are inferred.

The second line of argument for early Dingle Group intrabasinal fault activity, as intimated above, is the interpreted source of sand for the Boat Cove and Trabane Members. These lake margin sandflat systems were derived from the north and prograded southeastward into the lacustrine system represented by the Heterolithic Member (Fig. 9). They are composed predominantly of very fine to fine sandstone, with striking uniformity of grain size despite the high-energy sheetflood processes evident from the facies. This uniformity is interpreted as reflecting a provenance limited to similar grade sand, such that little or no coarser material was available to be reworked (detailed petrographic study has not been undertaken). The coincidence of this interpretation and the inferred absence of Croaghmarhin Formation from an area immediately to the north is appealing. Recycling of Croaghmarhin Formation sand from an uplifted intrabasinal high to the north could have sourced the Boat Cove and Trabane Member sandy marginal systems on the north side of the Bull's Head lacustrine area.

The third aspect of evidence for a fault-bounded intrabasinal high is the Paraconglomerate Member. Like the Boat Cove and Trabane Members, this is restricted to the northern side of the lacustrine system, i.e. to the northern limb of the Fahan Syncline (Fig. 9). The nature of its provenance is, however, more difficult to account for. Clast lithologies are notably uniform, namely fine or rarely fine–medium sandstone, just as the sandy members (Boat Cove and Trabane Members) show notably uniform grain size. The clasts were cemented and veined before redeposition. It may be that the clasts represent reworking of a particularly coarse facies of the Croaghmarhin Formation, which was prone to early cementation, as is commonly the case in shallow-marine systems. Alternatively, cementation, and also veining, may be fault related, and the associated mud matrix could have been concentrated by clay smearing along faults cutting poorly consolidated sediment before reworking into debris flows.

These three arguments point towards initiation of intrabasinal faulting during early Dingle Group deposition, representing a change in tectonic style. Although several reasons for this could be proposed, and none proven, our preferred explanation on regional grounds is that the Basin at this stage became affected by a strike-slip tectonic regime, which drove the differential subsidence of the basement along suitably oriented, probably pre-existing, intrabasinal and basin margin faults. The general theme of late stage strike-slip activity in this sector of the Caledonides is widely agreed (e.g. Hutton 1987; Bluck 1990). More locally, Todd (1989) has argued for sinistral strike-slip movement on the southern margin of the Dingle Basin during later Dingle Group deposition (Trabeg Conglomerate Formation), and eventual transpressive deformation of the basin fill. We interpret the intrabasinal faulting phase of basin development to have begun at earliest Dingle Group time, and would further argue that the change in tectonics was instrumental in driving the change from marine to non-marine deposition. That gradational depositional change is not readily interpreted as a response solely to simple progradation of non-marine over marine systems, as the initial non-marine system was not fluvial but lacustrine and of itself would not have sustained progradation of a marine shoreline. A more substantial, fault-controlled, rearrangement of the basin floor is suggested.

Tectonic setting

At the present day, the Dingle Peninsula lies less than 40 km south of the inferred trace of the 'Iapetus Suture' (Phillips *et al.* 1976; McKerrow & Soper 1989; Todd *et al.* 1991). That a discrete suture line in the Caledonides exists is now thought unlikely, given subsequent recognition of significant strike-slip tectonics and terrane displacement. However, this lineament remains a fundamental crustal structure within the collisional mosaic, and can be reasonably regarded as an approximate division between crustal fragments originating on the southern, Eastern

Avalonia, and northern, Laurentia, sides of this segment of Iapetus (Fig. 2).

Recent interpretations of the final stages of the Laurentia–Eastern Avalonia segment of Iapetus fall into two broad groups. The first see subduction of oceanic crust as completed by Late Ordovician time, with subsequent displacement on the plate boundary being strike-slip (Hutton 1987; Hutton & Murphy 1987). The second see northwestward subduction of oceanic crust continuing into Silurian time, perhaps developing into continental subduction–underthrusting (Soper *et al.* 1987; Scotese & McKerrow 1990). Williams *et al.* (1992) saw evidence in fresh volcanic detritus within northerly derived Llandovery–lower Wenlock turbidites in north Galway for an early Silurian continental volcanic arc. The Dunquin Group volcanic rocks are clearly very significant to this debate by virtue of their age, character and location, and yet have received very limited consideration in this larger regional context.

The trace element compositions of the Dunquin Group volcanic suite are characteristic of magmatism at a destructive plate margin. The high contents of Ba, La and Th, the depletion of Nb relative to mid-ocean ridge basalt and the enrichment in light rare earth elements (such as La, Ce, Pr and Nd) relative to heavy rare earth elements are particularly diagnostic of subduction-related magmatism. The volcanic suite lacks those geochemical attributes, such as high alkalinity, the presence of ocean-island basalt or bajaites, that may be used to infer a post-subduction tectonic setting (Sloan & Bennett 1990; Sloan 1991). The positive evidence for a subduction-related origin and the absence of evidence in favour of within-plate post-subduction or extension-related origin would appear to argue against the suggestion by Pickering *et al.* (1988) that the genesis of the Dunquin Group volcanic suite could be consistent with bimodal (basaltic–acidic) igneous activity associated with crustal extension.

Elsewhere in the British Isles south of the Iapetus Suture, proximal volcanic rocks of Silurian age occur at Skomer Island, southwest Wales; Tortworth, Gloucestershire; the East Mendips inlier, Somerset; and in a number of boreholes in south–central England and Wales. The Skomer Volcanic Group of Skomer Island, southwest Wales, comprises some 760 m of hawaiite–mugearite lavas, rhyolitic flows and pyroclastic rocks of Llandovery age (Thorpe *et al.* 1989). Their trace element characteristics indicate derivation from a within-plate source with a minor inherited component from subduction that Leat & Thorpe (1989) suggested ended in Caradoc (Late Ordovician) time. Lower Silurian volcanic rocks have been recovered from at least three boreholes in south–central England and southeast Wales, with the thickest succession being penetrated in the Netherton 1 borehole, Worcestershire. The thick andesitic tuffs in this borehole have been dated at 424 ± 8 Ma (late Wenlock time on the time-scale of Gradstein & Ogg (1996)) and have a similar trace-element signature to the Skomer Volcanic Group (Pharoah *et al.* 1991). Likewise, the Llandovery andesitic volcanic rocks of the Tortworth inlier (Gloucestershire) have a mixed within-plate and subduction-related trace element signature (Leat & Thorpe 1989). The East Mendips inlier, Somerset, includes *c.* 500 m of lower Wenlock lava flows and pyroclastic deposits with a predominantly rhyodacitic composition and a subduction-related geochemical signature (van de Kamp 1969; Hancock 1982; Leat & Thorpe 1989). They have thus a similar composition to, but are somewhat older than, the late Wenlock Dunquin Group volcanic suite of the Dingle Basin.

Southeastward subduction of oceanic crust beneath Eastern Avalonia is widely interpreted to be the cause of subalkaline volcanic activity that occurred in Late Ordovician time in the Lake District, northern England, north and mid-Wales, and Wicklow and southeast Ireland. The Dunquin Group volcanic suite was interpreted by Phillips *et al.* (1976) to be part of the same volcanic arc. The timing of the ending of this oceanic subduction and the effective closure of the Iapetus Ocean has been the subject of considerable debate as already noted.

Hutton & Murphy (1987) have argued that the cessation of major regional volcanism in East Avalonia by the end of the Ordovician period indicates that all oceanic crust in this segment of Iapetus had by then been subducted. They contended that by Llandovery time central Ireland was a successor basin overlying the suture and by early Wenlock time northerly derived material had overstepped the entire basin, proving final docking. In this scenario the upper Wenlock Dunquin Group volcanic rocks are too young to relate to Iapetus closure.

Other workers paint a more active impression of the Irish sector of Iapetus during the Silurian period, with Eastern Avalonia continuing to converge on Laurentia. Tectonic, palaeomagnetic, sediment dispersal and igneous arguments have been proposed in support of northwestward continental underthrusting (Soper *et al.* 1987; Soper & Woodcock 1990), and northwestward oceanic subduction (Williams *et al.* 1992). The lower Wenlock overstep succession may have been deposited on a still-underthrusting continental boundary rather than in a successor basin (Soper *et al.* 1987). None the less, there is no evidence apart from Dingle for oceanic subduction continuing as late as late Wenlock time. Three possible explanations for the Dunquin Group volcanic rocks' unique characteristics in terms of location, composition and age present themselves. First, they result from northward subduction under the southern margin of Eastern Avalonia, and may be related to the East Mendip lavas in this respect; if so, the extremely sporadic distribution of volcanic activity requires some explanation. Second, the Dunquin Group volcanic rocks were derived from localized subduction beneath Eastern Avalonia of a vestigial portion of Iapetus oceanic crust caught up in the final collision; strike-slip influence at that stage cannot be inferred directly, nor ruled out. This explanation can be conceived, but the broader regional evidence is no more than permissive. Third, the subduction-related characteristics of the Dunquin Group volcanic rocks reflect post-subduction generation of magmas from lithospheric mantle that had been contaminated during the long period of subduction of Iapetus oceanic crust. However, as stated above, the Dunquin Group lacks those geochemical characteristics that would support such an inherited subduction-related composition. None of these explanations is immediately appealing in that they all require some degree of special pleading, some particular localized circumstances. However, on balance, we see the second option, localized subduction of vestigial Iapetus oceanic crust, as the most straightforward.

This then provides the tectonic setting for the active subduction, subduction termination and post-subduction thermal subsidence phases of Dingle Basin evolution. The last phase, that of intrabasinal faulting, the onset of which drove the onset of Dingle Group deposition, corresponds to the Ludlow–early Devonian period of strike-slip faulting within the Eastern Avalonia–Laurentia collisional zone proposed, for example, by Hutton (1987). This change in the overall tectonic regime of the orogen is believed to have resulted in the increased intrabasinal faulting seen in the Dingle Basin, although the detailed causal linkages between the regional 'tectonic climate' and the basin-specific 'tectonic weather' (Dewey 1982) remain poorly understood.

Conclusions

The Dingle Basin lies less than 40 km from the present-day trace of the Iapetus Suture and its late Silurian fill offers important clues to the late stage development of this sector of the Caledonides by virtue of its character, location and age.

Four phases of basin evolution are inferred from the Wenlock–upper Ludlow Dunquin Group and the upper Ludlow–Přídolí lower Dingle Group. First, a phase of active subduction below the basin, represented by the Wenlock Foilnamahagh Formation (a 150 m heterogeneous succession of continental and marine sediments, lavas and pyroclastic deposits) and Ferriter's Cove Formation (240 m of stacked shallow-marine to shoreline parasequences with interbedded pyroclastic deposits). A subaerial volcanic centre to the southwest is inferred. Subsidence was controlled by normal faults and associated with eruptive phases and magma chamber collapse. This was followed by a subduction termination phase represented by the volcanic rocks of the upper Wenlock Clogher Head Formation (up to 375 m of welded ignimbrites, subordinate andesitic–dacitic lava flows and thin pyroclastic fall deposits) and the overlying coastal plain–playa mudstone-dominated Mill Cove Formation (105 m). The intense volcanic activity is interpreted to reflect rupturing of the magma chamber(s) accompanying stress relaxation as subduction ceased. This also allowed access to deeper, less-evolved, magmas that were erupted during Mill Cove Formation time. A phase of late Wenlock to late Ludlow post-subduction thermal subsidence followed, during which laterally extensive and uniform shallow-marine silts and sands of the Drom Point (300 m) and Croaghmarhin (400 m) Formations accumulated. The final phase inferred is one of strike-slip fault-controlled subsidence, initiated in the late Ludlow time, represented by the lower Dingle Group (lacustrine, lake margin and succeeding fluvial systems). This last change in tectonics probably drove the change from marine to non-marine, Old Red Sandstone, deposition.

The Dunquin Group volcanic rocks and associated sediments are interpreted in the context of late Caledonian development to represent localized subduction of a final vestigial portion of Iapetus oceanic crust beneath Eastern Avalonia. Strike-slip influence during that period cannot be inferred directly, nor ruled out. Stronger evidence for strike-slip activity comes from the final phase of basin evolution, although the detailed pattern of intrabasinal faulting and its linkage to the regional strike-slip regime remains conjectural.

The authors would like to thank the two anonymous reviewers for their comments. We are also grateful to

B. P. J. Williams for his enthusiastic support and to the late M. C. Bennett, to whom this paper is dedicated.

References

BASSETT, M. G., COCKS, L. R. M. & HOLLAND, C. H. 1976. The affinities of two endemic Silurian brachiopods from the Dingle Peninsula, Ireland. *Palaeontology*, **19**, 615–625.

BENTON, M. J. & UNDERWOOD, C. J. 1994. Graptolite evidence for the age of the Dunquin Group (Silurian), Dingle Peninsula, County Kerry. *Irish Journal of Earth Sciences*, **13**, 91–94.

BLUCK, B. J. 1990. Terrane provenance and amalgamation: examples from the Caledonides. *Philosophical Transactions of the Royal Society of London, Series A*, **331**, 599–609.

BOYD, J. D. 1983. *Sedimentology of the lower Dingle Group, southern Dingle Peninsula, southwest Ireland.* PhD thesis, University of Bristol.

DEWEY, J. F. 1982. Plate tectonics and the evolution of the British Isles. *Journal of the Geological Society, London*, **139**, 371–412.

DORAN, R. J. P., HOLLAND, C. H. & JACKSON, A. A. 1973. The sub-Old Red Sandstone surface in southern Ireland. *Proceedings of the Royal Irish Academy*, **73B**, 109–127.

GARDINER, C. I. & REYNOLDS, S. H. 1902. The fossiliferous Silurian beds and associated igneous rocks of the Clogher Head district (Co. Kerry). *Quarterly Journal of the Geological Society, London*, **58**, 226–266.

GRADSTEIN, F. M. & OGG, J. 1996. A Phanerozoic time scale. *Episodes*, **19**, 3–5.

HANCOCK, N. J. 1982. Stratigraphy, palaeogeography and structure of the East Mendips Silurian inlier. *Proceedings of the Geologists' Association*, **93**, 247–261.

HIGGS, K. T. 1999. Early Devonian spore assemblages from the Dingle Group, Co. Kerry, Ireland. *Bollettino della Societa Paleontologica Italiana*, **38**, 187–196.

HOLLAND, C. H. 1969. Irish counterpart of Silurian of Newfoundland. *In*: KAY, M. (ed.) *North Atlantic—Geology and Continental Drift*. Memoirs, American Association of Petroleum Geologists, **12**, 298–308.

—— 1987. Stratigraphical and structural relationships of the Dingle Group (Silurian), County Kerry, Ireland. *Geological Magazine*, **124**, 33–42.

—— 1988. The fossiliferous Silurian rocks of the Dunquin Inlier, Dingle Peninsula, County Kerry, Ireland. *Transactions of the Royal Society of Edinburgh: Earth Sciences*, **79**, 347–360.

HORNE, R. R. 1974. The lithostratigraphy of the Late Silurian to Early Carboniferous of the Dingle Peninsula, Co. Kerry. *Bulletin of the Geological Survey of Ireland*, **1**, 395–428.

HUTTON, D. H. W. 1987. Strike-slip terranes and a model for the evolution of the British and Irish Caledonides. *Geological Magazine*, **5**, 405–425.

—— & MURPHY, F. C. 1987. The Silurian rocks of the Southern Uplands and Ireland as a successor basin to the end-Ordovician closure of Iapetus. *Journal of the Geological Society, London*, **144**, 765–772.

LEAT, P. T. & THORPE, R. S. 1989. Snowdon basalts and the cessation of Caledonian subduction by the Longvillian. *Journal of the Geological Society, London*, **146**, 965–970.

LEFORT, J. P. & MAX, M. D. 1984. Development of the Porcupine Seabight: use of magnetic data to show the direct relationship between early oceanic and continental structures. *Journal of the Geological Society, London*, **141**, 663–674.

LIOU, J. G., MARUYAMA, S. & CHO, M. 1985. Phase equilibria and mineral paragenesis of metabasites in low-grade metamorphism. *Mineralogical Magazine*, **49**, 321–333.

MCKERROW, W. S. & SOPER, N. J. 1989. The Iapetus Suture in the British Isles. *Geological Magazine*, **126**, 1–8.

MOTTS, W. S. 1972. Some hydrologic and geologic processes influencing playa development in the western United States. *In*: REEVES, C. C., Jr (ed.) *Playa Lake Symposium*. International Center for Arid and Semi-Arid Land Studies, **4**, 89–106.

PARKIN, J. 1974. Silurian rocks of Inishvickillane, Blasket Islands, Co. Kerry. *Scientific Proceedings of the Royal Dublin Society*, **A5**, 354–357.

—— 1976. Silurian rocks of the Bull's Head, Annascaul and Derrymore Glen inliers, Dingle Peninsula, Co. Kerry. *Proceedings of the Royal Irish Academy*, **76B**, 577–606.

PHAROAH, T. C., MERRIMAN, R. J., EVANS, J. A., BREWER, T. S., WEBB, P. C. & SMITH, N. J. P. 1991. Early Palaeozoic arc-related volcanism in the concealed Caledonides of southern Britain. *Annales de la Société Géologique de Belgique*, **114**, 63–91.

PHILLIPS, W. E. A., STILLMAN, C. J. & MURPHY, T. 1976. A Caledonian plate tectonic model. *Journal of the Geological Society, London*, **132**, 579–609.

PICKERING, K. T., BASSETT, M. G. & SIVETER, D. J. 1988. Late Ordovician–early Silurian destruction of the Iapetus Ocean: Newfoundland, British Isles and Scandinavia. *Transactions of the Royal Society of Edinburgh: Earth Sciences*, **79**, 361–382.

PRACHT, M. 1996, *Geology of Dingle Bay. A geological description to accompany the bedrock geology 1:100 000 map series, Sheet 20, Dingle Bay.* Geological Survey of Ireland, Dublin.

SCOTESE, C. R. & MCKERROW, W. S. 1990. Revised World maps and introduction. *In*: MCKERROW, W. S. & SCOTESE, C. R. (eds) *Palaeozoic Palaeogeography and Biogeography*. Geological Society, London, Memoir, **12**, 1–21.

SLOAN, R. J. 1991. *Mid-Silurian sedimentation and volcanism in Dingle, southwest Ireland.* PhD thesis, University of Bristol.

—— & BENNETT, M. C. 1990. Geochemical character of Silurian volcanism in SW Ireland. *Journal of the Geological Society, London*, **147**, 1051–1060.

—— & WILLIAMS, B. P. J. 1991. Volcano-tectonic control of offshore to tidal-flat regressive cycles from the mid-Silurian Dunquin Group of southwest Ireland. *In*: MACDONALD, D. I. M. (ed.)

Sedimentation, Tectonics and Eustasy. International Association of Sedimentologists, Special Publication, **12**, 105–119.

SMITH, G. A., SNEE, L. W. & TAYLOR, E. M. 1987. Stratigraphic, sedimentologic, and petrologic record of late Miocene subsidence of the central Oregon High Cascades. *Geology*, **15**, 389–392.

SOPER, N. J. 1986. The Newer Granite problem: a geotectonic view. *Geological Magazine*, **123**, 227–236.

—— & WOODCOCK, N. H. 1990. Silurian collision and sediment dispersal patterns in southern Britain. *Geological Magazine*, **127**, 527–542.

——, WEBB, B. C. & WOODCOCK, N. H. 1987. Late Caledonian (Acadian) transpression in north-west England: timing, geometry and geotectonic significance. *Proceedings of the Yorkshire Geological Society*, **46**, 175–192.

THORPE, R. S., LEAT, P. T., BEVINS, R. E. & HUGHES, D. J. 1989. Late-orogenic alkaline/subalkaline Silurian volcanism of the Skomer Volcanic Group in the Caledonides of south Wales. *Journal of the Geological Society, London*, **146**, 125–132.

TODD, S. P. 1989. *Caledonian tectonics and conglomerate sedimentation in Dingle, SW Ireland*. PhD thesis, University of Bristol.

—— 1991. The Silurian rocks of Inishnabro, Blasket Islands, County Kerry and their regional significance. *Irish Journal of Earth Sciences*, **11**, 91–98.

——, BOYD, J. D. & DODD, C. D. 1988. Old Red Sandstone sedimentation and basin development in the Dingle Peninsula, southwest Ireland. *In*: MCMILLAN, N. J., EMBRY, A. F. & GLASS, D. J. (eds) *The Devonian of the World*. Canadian Society of Petroleum Geologists Memoirs, **14**, 251–268.

——, MURPHY, F. C. & KENNAN, P. S. 1991. On the trace of the Iapetus suture in Ireland and Britain. *Journal of the Geological Society, London*, **148**, 869–880.

VAN DE KAMP, P. C. 1969. The Silurian rocks of the Mendip Hills, Somerset; and the Tortworth area, Gloucestershire, England. *Geological Magazine*, **106**, 542–553.

WATKINS, R. 1978. Silurian marine communities of Dingle, Ireland. *Palaeogeography, Palaeoclimatology, Palaeoecology*, **23**, 79–118.

WILLIAMS, D. M., O'CONNOR, P. D. & MENUGE, J. 1992. Silurian turbidite provenance and the closure of Iapetus. *Journal of the Geological Society, London*, **149**, 349–357.

WRIGHT, J. V., SMITH, A. L. & SELF, S. 1980. A working terminology of pyroclastic deposits. *Journal of Volcanology and Geothermal Research*, **8**, 315–336.

WRIGHT, V. P., SLOAN, R. J., GARVIE, L. A. J. & RAE, J. E. 1991. A polygenetic palaeosol from the Silurian (Wenlock) of southwest Ireland. *Journal of the Geological Society, London*, **148**, 849–859.

——, ——, VALERO GARCES, B. & GARVIE, L. A. J. 1992. Groundwater ferricretes from the Silurian of Ireland and Permian of the Spanish Pyrenees. *Sedimentary Geology*, **77**, 37–49.

A new terrane in the Old Red Sandstone of the Dingle Peninsula, SW Ireland

LORNA K. RICHMOND[1] & BRIAN P. J. WILLIAMS[2]

[1]*Conoco (UK) Limited, Anderson Drive, Aberdeen AB15 6FZ, UK*

[2]*Department of Geology and Petroleum Geology, University of Aberdeen, Aberdeen AB24 3UE, UK*

Abstract: We propose the name Northwest Dingle Domain for the enigmatic Old Red Sandstone terrane whose tectono-sedimentary evolution has perplexed generations of geologists. The Domain remained largely misinterpreted, unappraised or simply disregarded, and its fundamental impact on regional basin dynamics was grossly overlooked. The Northwest Dingle Domain is largely structurally constrained between two ENE-trending Caledonian structures: the North Kerry Lineament and the Fohernamanagh Fault. It comprises four unconformity-bounded Groups: the Lower Devonian Smerwick Group; the Middle Devonian Pointagare Group; and the Upper Devonian Carrigduff and Ballyroe Groups. Their fluvial–aeolian, and locally tidal, sedimentation patterns profile Late Caledonian transpression to Mid–Late Devonian extension. The inherent primary structural control on basin location, development, geometry, sedimentary-fill and preservation is manifest in the Northwest Dingle Domain. The Acadian emplacement of the Smerwick Group set the foundations of the Northwest Dingle Domain. The Smerwick Group documents sandy and gravelliferous ephemeral-fluvial and erg-margin aeolian processes on an ancient terminal fan(s). The Pointagare Group is cogenetic with the Caherbla Group of south Dingle. Together they record the renewed influx of coarse-grained sediment in the form of transverse alluvial fans and axial braidplains in response to increased tectonism followed by overstep of an erg complex. The Pointagare–Caherbla basin model highlights the fundamental structural control on basin topography, palaeodrainage patterns, provenance, palaeowind directions and sedimentation style in tectonically-active extensional basins.

The Dingle Peninsula is the northernmost of three peninsulas, which effectively make-up County Kerry in the southwest corner of Ireland. The Peninsula preserves a unique Ordovician to Carboniferous succession including the most complete ORS magnafacies in SW Ireland (Fig. 1). The oldest part of the ORS sequence was deposited within the Dingle Basin, which formed in a transpressive tectonic regime on convergence between Avalonia and Laurentia. The younger ORS sequences were deposited in extensional basins on the northern margin of the Mid- to Late Devonian Munster Basin of south Kerry and Cork.

A series of fundamental ENE-trending lineaments can be traced across southwest Ireland. The Dingle Peninsula is situated to the south of the putative trace of the Iapetus Suture Zone and to the north of the Dingle Bay Lineament (DBL). The North Kerry Lineament (NKL) can be traced a few kilometres offshore from the present-day coastline of northwest Dingle. These deep crustal lineaments are thought to have acted as important basin-bounding faults. They have to varying degrees controlled the tectonic and sedimentary evolution of southwest Ireland during Late Silurian to Early Devonian Caledonian deformation, Late Devonian extension and Late Carboniferous Variscan deformation (Dodd 1986; Price & Todd 1988; Todd *et al.* 1988*a*; Todd 1989*a*, *b*; Williams *et al.* 1989). The Old Red Sandstone magnafacies is extensive across southwest Ireland and the succession exposed on the Dingle Peninsula ranges in age from Late Silurian to Early Carboniferous (Todd *et al.* 1988*a*; Richmond 1998). The complex tectonic and sedimentary evolution of the area is reflected in the varied collage of Old Red Sandstone sequences emanating in a highly compartmentalized, often controversial, stratigraphy (Figs 1, 2, 3 and 4). The new stratigraphy (Richmond 1998) recognizes the major units outlined in Table 1 together with their revised chronostratigraphy (Figs 2 and 5).

From: FRIEND, P. F. & WILLIAMS, B. P. J. (eds). *New Perspectives on the Old Red Sandstone.* Geological Society, London, Special Publications, **180**, 147–183. 0305-8719/00/$15.00

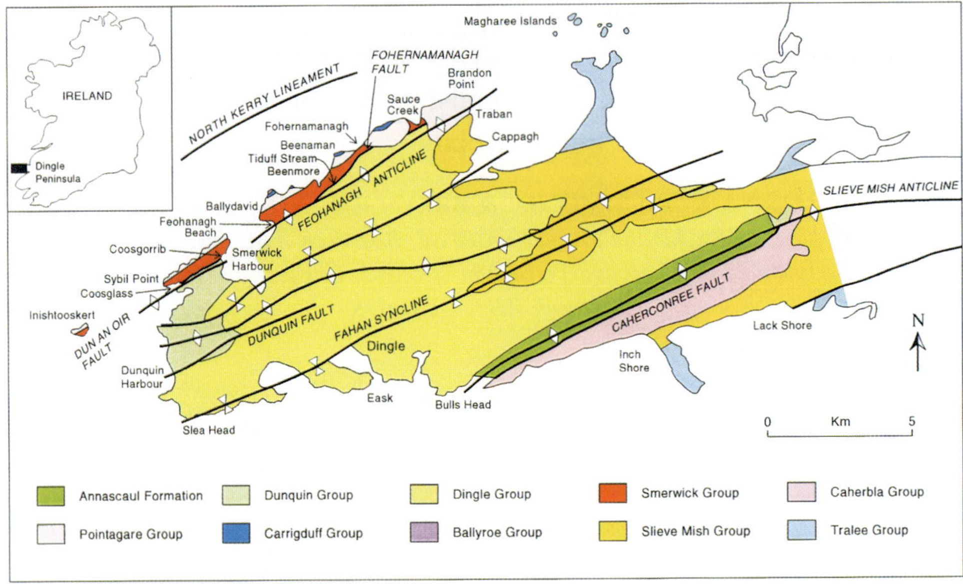

Fig. 1. Geological map of the Dingle Peninsula.

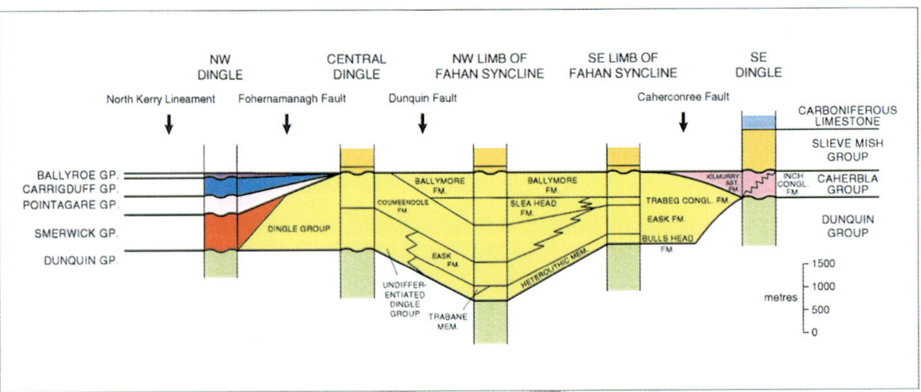

Fig. 2. Lithostratigraphy of the Dingle Peninsula.

Stratigraphic and structural framework

The geology of the Dingle Peninsula was first documented in the Geological Survey of Ireland memoir of 1863, which was produced by the survey officers, J. B. Jukes and G. V. Du Noyer. Jukes & Du Noyer (1863) identified four distinct units, which were mapped across the Peninsula: the Ferriter's Cove, Croaghmarhin, Dingle and unconformable Old Red Sandstone beds. In addition, they very astutely realized that the Smerwick beds, which form Sybil Head and Ballydavid Head, could be differentiated from the Dingle beds. Following the work of Capewell (1951, 1965) on the Old Red Sandstone in the east of the Peninsula and the research of Holland (1969) and Parkin (1974) on the marine Silurian sequence, Gardiner & Horne (1972) and Horne (1974) produced the first detailed lithostratigraphic framework. This was to be an excellent template for future research in the area. Boyd (1983), Dodd (1986), King (1988), Todd (1989*a*) and Sloan (1991) later refined this stratigraphic framework during a series of PhD research studies largely undertaken out of the University of Bristol. The culmination of this work can be seen in the research of Richmond (1998) and the papers of Boyd & Sloan (this volume), and Todd (this volume).

Table 1. *Stratigraphic and tectono-sedimentological summary of the geology of the Dingle Peninsula*

Group or Formation	Age	Basin	Sedimentation
Slieve Mish Group	Late Devonian	Post-rift, extensional half-graben	Perennial braided river–alluvial fan and sinuous stream
Ballyroe Group	Late Devonian	Syn-rift, extensional half-graben	Perennial and ephemeral braided river–alluvial fan, erg-margin, tidal incursions
Carrigduff Group	Late Devonian	Syn-rift, extensional half-graben	Perennial braided river–alluvial fan
Pointagare Group	Mid-Devonian	Post-Caledonian, pre-rift, extensional graben	Perennial and ephemeral braided river–alluvial fan, erg to erg-margin
Caherbla Group	Mid-Devonian	Post-Caledonian, pre-rift, extensional graben	Perennial and ephemeral braided river–alluvial fan, erg to erg-margin
Smerwick Group	Early Devonian	Late Caledonian, pre-rift, strike-slip pull apart or extensional half-graben	Ephemeral-fluvial sinous stream, ephemeral-fluvial terminal fan, erg to erg-margin
Dingle Group	Late Silurian–Early Devonian	Late Caledonian, pre-rift, strike-slip, load-induced foreland basin	Barrier island, ephemeral-lacustrine, ephemeral-fluvial terminal fan, perennial braided river–alluvial fan
Dunquin Group	Early–Late Silurian	Intra-arc basin	Shallow marine, shelfal, ephemeral-fluvial volcanic rocks
Annascaul Formation	Early Ordovician	Early Caledonian, strike-slip basin	Deep-water clastic deposits, volcanic rocks

In the Northwest Dingle Domain, Horne (1974) had encountered correlation difficulties and was forced to define a whole series of members and a formation, to overcome the inherent problems in sequence matching (Fig. 4). Todd *et al.* (1988*b*) questioned the correlation of the Smerwick beds with the Dingle beds, using structural, stratigraphical and preliminary sedimentological analyses. They reverted to the view of Jukes & Du Noyer (1863) suggesting that these two red-bed sequences are indeed separate. This revelation led to the establishment of the Smerwick Group (Fig. 4). Dodd (1986) further revised the stratigraphy in the Northwest Dingle Domain by removing the basal formation and member from the Glengarriff Harbour Group of Horne (1974) and placing them into the Pointagare Group (Fig. 4). Once more, the correlation of strata in the Northwest Dingle Domain with those to the south was thought to be invalid.

The existing stratigraphy provides a broad framework on which to build a new lithostratigraphic scheme for the Northwest Dingle Domain (Figs 1, 2 and 4). The detailed revised lithostratigraphy for the entire Dingle Peninsula and immediately adjacent areas is shown in Figs 2 and 6 and Table 1 (Richmond 1998). In the course of this research, new biostratigraphic, chronostratigraphic, sedimentary, structural and provenance evidence has come to light, resulting in major revisions in the stratigraphy. Herein, the stratigraphic definition of the Smerwick and Pointagare Groups of northwest Dingle are considered in their relationship with the Dingle and Caherbla Groups to the south (Figs 2 and 6).

The Dingle Group

The Dingle Group conformably overlies the Dunquin Group, the age of which is biostratigraphically well constrained. Macrofaunal evidence suggests that Dingle Group sedimentation started at the end of Ludlow time (Holland 1987; Todd *et al.* 1988*a*; Boyd & Sloan this volume). Two Lower Devonian spore assemblages have been identified within the Dingle Group. The oldest was sampled from the Eask Formation at Dunmore Head and is considered to be of Pragian age (Higgs 1999). The youngest was taken from the Slea Head Formation of Todd *et al.* (1988*a*) at Slea Head itself and has been assigned to early to (?)mid-Emsian age by Higgs (1999). These age dates for the Dingle Group are in accordance with the tectonic regime thought to be in operation at the time. Both Late Caledonian deformation and Late Carboniferous Variscan deformation affected the Dingle Group. Evidence elsewhere in the British Isles

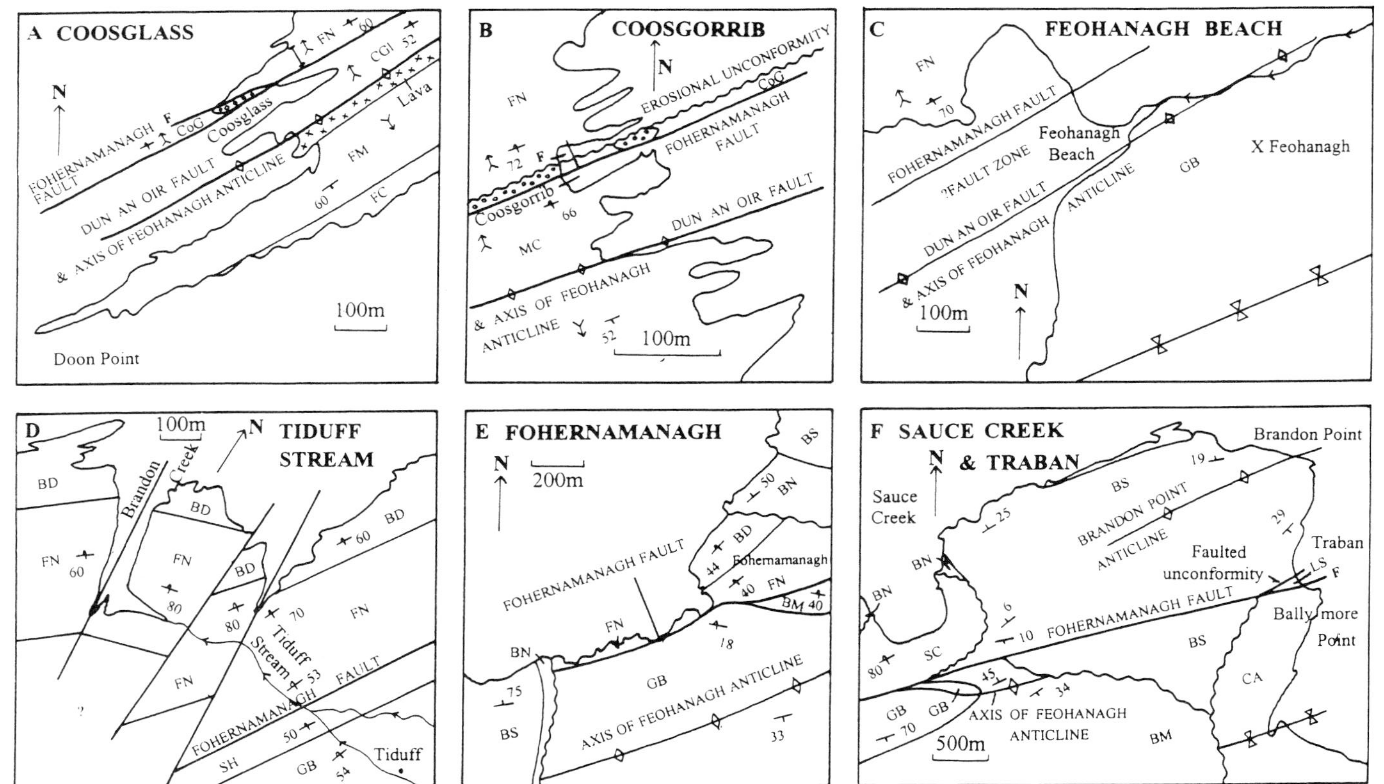

Fig. 3. Structural relationships between the Smerwick Group and adjacent rock sequences (detailed maps of key locations in the NW Dingle Domain depicted along strike from WSW to ENE). Abbreviations used in Figs 3 and 6: A, Ardnaguggen Fm; AF, Annascaul Fm; BaM, Ballymore Fm; BD, Ballydavid Fm; BF, Ballyferriter Fm; BH, Bull's Head Fm; BN, Beenaman Conglomerate Fm; BNa, Ballynane Fm; BS, Beenmore Sandstone Fm; CC, Caherconree Fm; CH, Clogher Head Fm; CM, Clashmelcon Fm; CMo, Coosmore Conglomerate Mbr; Co, Coumeenoole Fm; CoG, Coosgorrib Conglomerate Fm; CP, Cappagh Fm; Cr, Croaghmarhin Fm; CS, Chloritic Sandstone Fm; DG, Dingle Gp (undifferentiated); DP, Drom Point Fm; E, Eask Fm; FC, Ferriter's Cove Fm; FH, Foilnamanagh Fm; FN, Farran Fm; GB, Glashabeg Fm; GD, Glandahalin Fm; GS, Grey Sandstone Fm; IC, Inch Conglomerate Fm; IN, Inishnabro Fm; IV, Inishvickillane Fm; IY, Ishaboy Fm; KM, Kilmore Fm; KY, Kilmurry Fm; LK, Lack Fm; LP, Landing Place Fm; LPS, Lower Purple Sandstone Fm; LS, Lower Limestone Shale; LSC, Lough Slat Conglomerate Fm; MC, Mill Cove Fm; SC, Sauce Creek Fm; SH, Slea Head Fm; TG, Trabeg Conglomerate Fm; UP, Upper Purple Sandstone Fm.

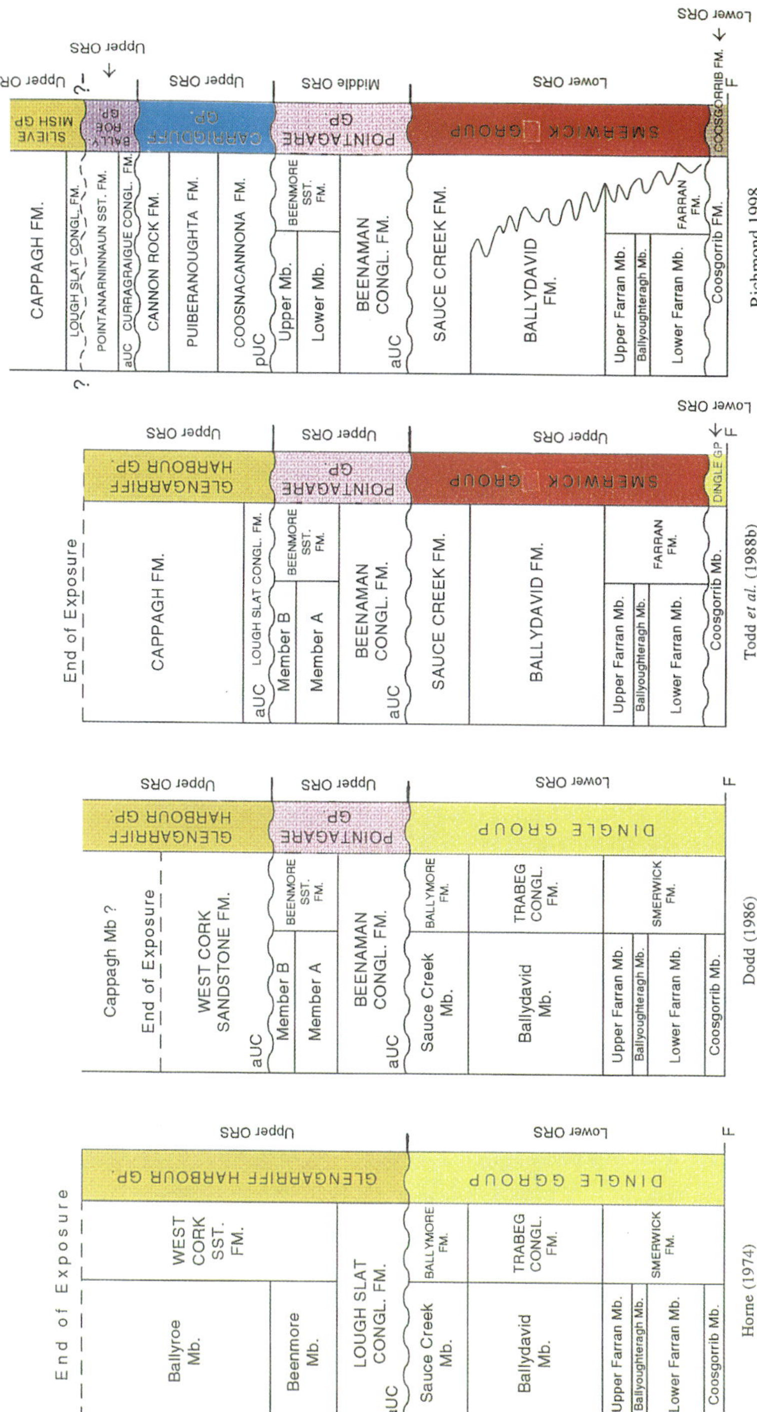

Fig. 4. Lithostratigraphic evolution of the ORS sequences, NW Dingle Peninsula. (F = Fohernamanagh Fault; aUC = Angular unconformity; pUC = Para-unconformity)

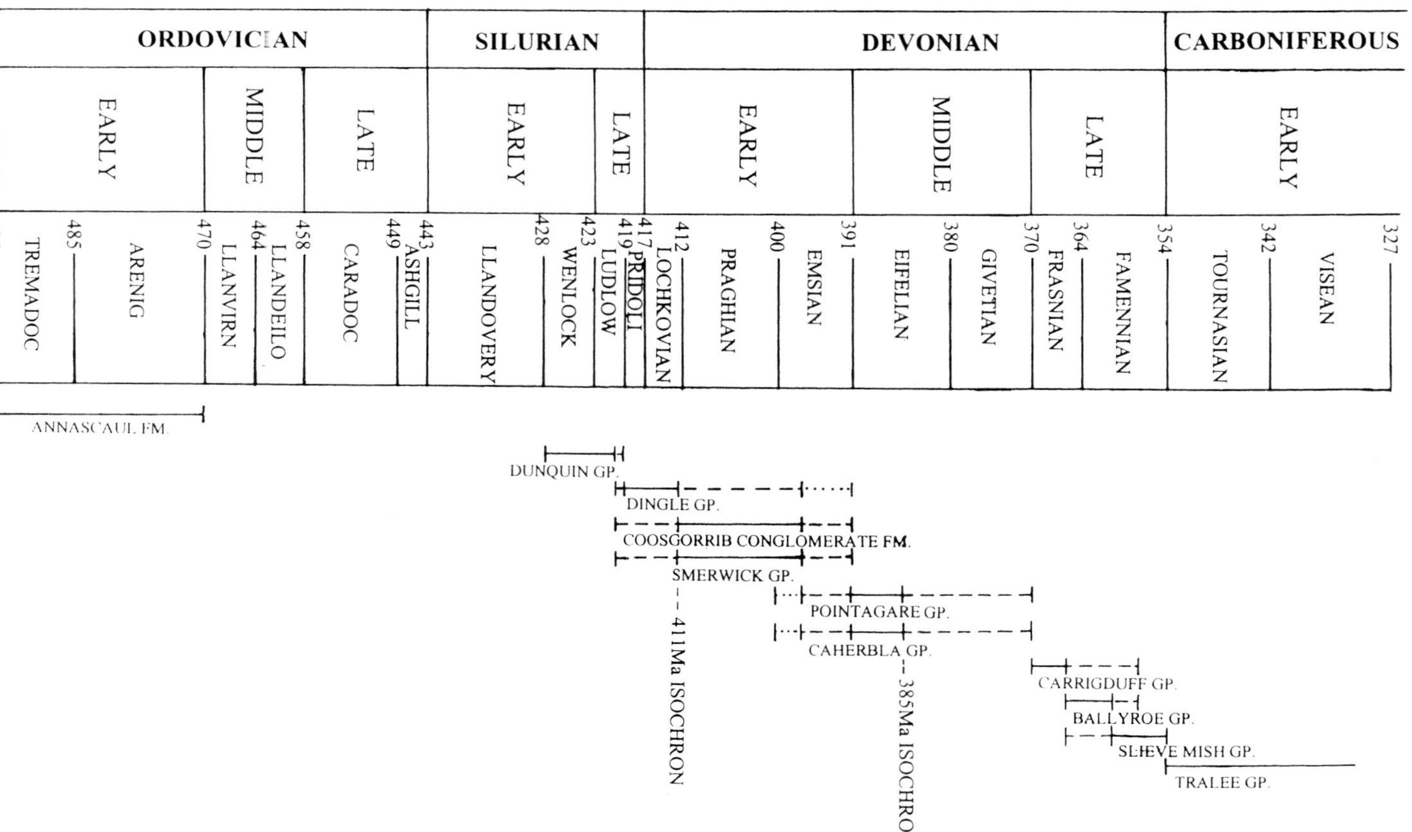

Fig. 5. Revised chronostratigraphy of the Dingle Peninsula (continuous lines, probable age; dashed line, possible age; dotted line, age limit). The time scale is based on Gradstein & Ogg (1996). The 411 Ma isochron is from the Cooscrawn Tuff Bed in the Ballymore Formation of the Dingle Group, and the 385 Ma isochron is from the Enagh Tuff Bed of NW Iveragh (Williams *et al.* 1997, this volume).

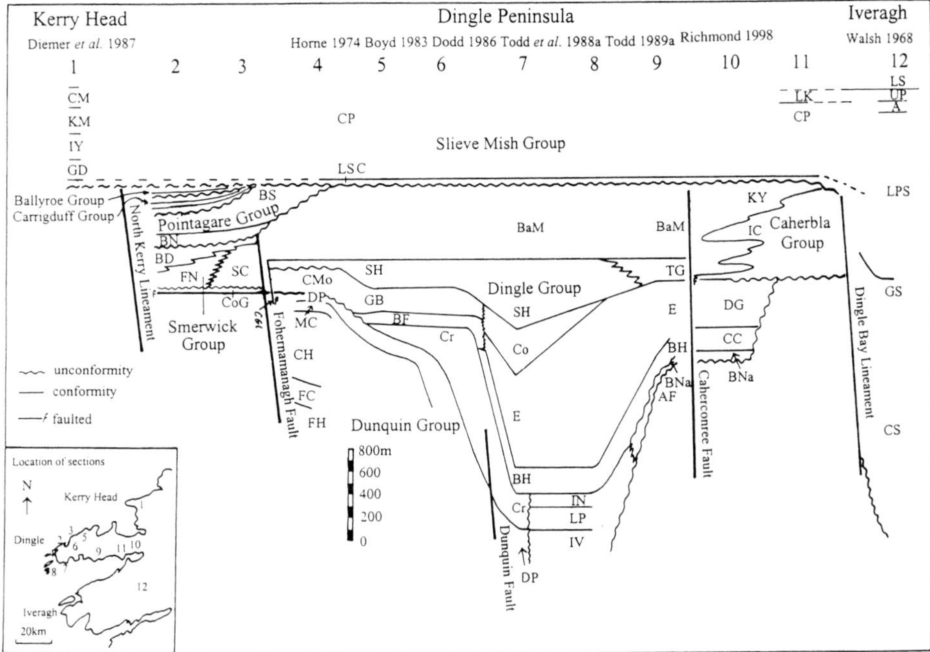

Fig. 6. Lithostratigraphic scheme for the Siluro-Devonian rocks of the Dingle Peninsula (modified mainly after Todd (1989*a*)).

suggests that Late Caledonian deformation took place in Early Devonian time and culminated in Emsian time (Soper *et al.* 1987; McKerrow 1988). The early to (?)mid-Emsian age of the Slea Head Formation confirms that Dingle Group sedimentation continued into Emsian time. Deposition of the younger Ballymore Formation may have extended into mid-Emsian time. Hence using solely biostratigraphical and structural evidence it can be concluded that the Dingle Group more than likely spans from latest Ludlow to early Emsian time and possibly into mid-Emsian time (Fig. 5).

Recent chronometric age dating using zircons extracted from the Cooscrawn Tuff Bed has yielded an age of 411 ± 3 Ma (Williams *et al.* this volume). The Cooscrawn Tuff was first discovered by Horne (1974) on the south side of Parkmore Point in the Ballymore Formation of the Dingle Group. This isochron date places the Ballymore Formation in Ludlow time according to Harland *et al.* (1990). A Ludlovian age date for the upper part of the Dingle Group contradicts the biostratigraphical age dates and is irreconcilable with reasonable Dingle Group sedimentation rates. A more plausible late Lochkovian age is indicated by the Siluro-Devonian time scale established by Gradstein & Ogg (1996), (Fig. 5). A Lochkovian age for the upper part of the Dingle Group is more compatible with sedimentation rates. Reasonable sedimentation rates of *c.* 200 m Ma^{-1} for the Dingle Group (>2.8 km thick) suggest that it would have taken 14 Ma to accumulate (Todd *et al.* 1988*a*). However, this is still in disagreement with the dates obtained from miospore assemblages in the Dingle Group (Higgs 1999). Hence it is apparent that the Siluro-Devonian boundary and the Silurian and Devonian time scales need to be significantly revised and recalibrated (Williams *et al.* this volume). At present, there are two possible maximum and minimum age ranges for the Dingle Group: latest Ludlow to early (to (?)mid-) Emsian time or latest Ludlow to latest Lochkovian time, respectively (Fig. 5). The age range for the Dingle Group cannot be further refined until the Silurian and Devonian time scales have been recalibrated.

The Smerwick Group

The authors uphold the separation of the Smerwick beds from the Dingle Group and the establishment of the Smerwick Group by Todd *et al.* (1988*b*). The internal stratigraphy of the Smerwick Group, which consists of the Farran, Ballydavid and Sauce Creek Formations of Todd *et al.* (1988*b*), is retained (Figs 4 and 6). The subdivision of the Farran Formation (Smerwick Formation of Horne (1974)) by Horne (1974) into the Lower Farran, Ballyoughteragh and Upper Farran Members (Fig. 4) is also unchallenged, although an additional member, the Coosglass Member, is separated out at the base of the Formation (Richmond 1998). The Ballydavid Formation can be divided into Lower, Middle and Upper Members at the type section at Ballydavid Head. Along-strike correlation of their stratigraphic equivalents is limited. The Sauce Creek Formation cannot be subdivided into mappable units of member status, in part, because of lack of exposure.

The Smerwick Group is restricted to the northern, overturned limb of the Feohanagh Anticline (Fig. 3a). The Feohanagh Anticline plunges at between 7 and 25° to the northeast and verges to the northwest. The Dun an Oir Fault is coincident with the axis of the Feohanagh Anticline between Coosglass and Feohanagh where the Anticline is tightest. In the northeast the broad anticlinal structures (e.g. the Brandon Point Anticline, Fig. 3f) are open and their cores are unfaulted. In the absence of biostratigraphical evidence for the age of the Group, structural relationships (Fig. 3) and deformation history are used to structurally infer the Group's age. Unfortunately, in this case, the lower boundary of the Smerwick Group is fraught with controversy. The boundary is exposed at several localities, Coosglass, Coosgorrib and Fohernamanagh, and can be inferred accurately at Feohanagh Beach, Tiduff Stream and Sauce Creek (Fig. 3a–e). It is proposed that the surface expression and extension of this boundary is exposed at Traban, near Brandon Point (Fig. 3f). The revised, extended trace and nature of the contact between the Smerwick and Dingle Groups is of fundamental importance as it represents a major terrane boundary in southwest Ireland.

The preferred stratigraphic model proposes that the Smerwick Group was deposited in an allochthonous terrane coeval with, or more likely shortly after, Dingle Group sedimentation. Hence the Smerwick Group ranges in age anywhere from latest Ludlow to latest Emsian time (Fig. 5). The most likely age range for the Group spans from the latest Lochkovian (using the Cooscrawn Tuff isochron and the time scale of Gradstein & Ogg (1996)), or early to (?)mid-Emsian (using biostratigraphical evidence), to perhaps late Emsian time. The allochthonous terrane model is preferred as it accounts for the high degree of faulting along the Fohernamanagh Fault and the very different facies and structure of the two groups. The provenance of the Glashabeg Formation of the Dingle Group (Todd 1989*a*) indicates that the strike-slip docking of the Smerwick Group would have taken place in post-Glashabeg Formation times (Richmond 1998) and more probably in post-Dingle Group times. Docking is constrained to pre-Pointagare Group deposition because the Pointagare Group oversteps the line of contact between the Dingle and Smerwick Groups. Strike-slip movement on the Fohernamanagh Fault is consistent with the tectonic regime in place at that time. A transpressional tectonic regime is evident from detailed structural analyses (King 1988; Todd 1989*a*). Sinistral strike-slip displacement on the Dingle Bay Lineament in Trabeg Conglomerate Formation times is invoked from fan lithosome geometries and strewing of labile clast types (Todd *et al.* 1988*a*; Todd 1989*a*, *b*).

The Pointagare and Caherbla Groups

Dodd (1986) first defined the Pointagare Group. His internal stratigraphy is retained, such that the basal Beenaman Conglomerate Formation is overlain by the Lower and Upper Members of the Beenmore Sandstone Formation (Members A and B of Dodd (1986)) (Fig. 4). The Pointagare Group is largely restricted to the northern, overturned limb of the Feohanagh Anticline (Figs 1, 2 and 6). At Masatiompan the Beenaman Conglomerate Formation of the Pointagare Group oversteps the postulated trace of the Fohernamanagh Fault. The dip of the Pointagare Group varies from 80° to the NW on the northern limb of the Feohanagh Anticline to 30° to the SE on the southern limb. The Pointagare Group is the only group in the Northwest Dingle Domain to overstep the Fohernamanagh Fault and to occur on the southern limb of the Feohanagh Anticline (Richmond 1998). The Beenmore Sandstone Formation is seen to onlap onto the spectacular angular unconformity on the northeast side of Sauce Creek, where the Smerwick Group beds are at 90° to the Pointagare Group beds (Fig. 7). The Pointagare Group rests with angular unconformity on the overturned Smerwick Group with an angular discordance of between 30 and 90°.

It was noted by Dodd (1986) that the newly defined Pointagare Group occupied a similar stratigraphic position to the Caherbla Group in the southeast of the Peninsula. However, he found no biostratigraphic, provenance or palaeocurrent evidence to suggest that the sequence in the northwest was equivalent to the Caherbla Group. This view is now challenged (Richmond 1998) and it is suggested that the Pointagare Group is at least in part coeval with the Caherbla

Fig. 7. The upper NE wall of Sauce Creek (Figs 1 and 3f). The overturned, subvertical beds of the Sauce Creek Formation of the Smerwick Group are unconformably overlain by sub-horizontal beds of the Beenmore Sandstone Formation of the Pointagare Group. A 90° angular discordance is clear. The portion of cliff face shown is 10 m high and oriented NNW–SSE.

Group on the basis of structure, stratigraphic position, biostratigraphy, facies, petrography, provenance, palaeodispersal patterns, palaeowind directions and clast lithotypes. This is clearly illustrated in Fig. 14 (below), and it could be argued that the Pointagare Group should be abolished and incorporated in the Caherbla Group. However, it is preferred to retain the group status of the Pointagare beds until further biostratigraphic evidence or detailed geochemical analysis of clast types is available to pinpoint the exact relationship between the two groups.

The Caherbla Group is shown to have an unconformable relationship with the Dingle and Dunquin Groups (Horne 1974, 1976; Todd 1989*a*) (Fig. 6). The Caherbla Group was not affected by Late Caledonian deformation, which culminated in Emsian time (Soper *et al.* 1987; McKerrow 1988). The proposed age ranges for the Dingle Group and the late Famennian age of the Slieve Mish Group poorly constrain the Caherbla Group to Mid-Devonian time. The Caherbla Group could range in age from latest Emsian to mid-Famennian time (Fig. 5). The Eifelian U–Pb isotope age date of 385 Ma obtained from the Enagh Tuff Bed of the Old Red Sandstone of the Munster Basin in northwest Iveragh by Williams *et al.* (1997) (Fig. 5) also has important implications on not only the age of the Caherbla Group, but also the relative timing and evolution of the Dingle and Munster Basins. This has been discussed in detail by Richmond (1998).

The base of the Carrigduff Group (introduced by Richmond (1998)) of northwest Dingle has revealed a miospore assemblage considered to be of early to mid-Frasnian age (Higgs, pers. comm.). The underlying Pointagare Group must therefore have been deposited before early to mid-Frasnian time. It is proposed that the Pointagare Group of northwest Dingle is stratigraphically equivalent to the Caherbla Group. The Pointagare and Caherbla Groups potentially range in age from the latest Emsian to the early Frasnian time (Fig. 5).

Towards a new stratigraphic model

The northwest Dingle succession is very largely restricted to the north of the Fohernamanagh Fault and comprises four unconformity-bounded groups and a formation: the Coosgorrib Conglomerate Formation and the Smerwick, Pointagare, Carrigduff and Ballyroe Groups. The Coosgorrib Conglomerate Formation and Smerwick Group belong to the Lower Old Red Sandstone and may have been in part coeval with the Dingle Group. These basal two units are part of a separate, probably allochthonous, terrane. The Fohernamanagh Fault represents the terrane boundary. The Pointagare Group is thought to be stratigraphically equivalent to the Caherbla Group and together they comprise the Middle Old Red Sandstone of the Dingle Peninsula. The Carrigduff and Ballyroe Groups (introduced by Richmond (1998)) are assigned to the Upper Old Red Sandstone (not considered elsewhere in this paper) and pre-date Slieve Mish Group sedimentation. The tectonic and sedimentary evolution of the Northwest Dingle Domain is extremely complex and varied, and substantially different from that of the southern domain.

NW Dingle Domain: the Smerwick Group

The Smerwick Group crops out along the northwest coast of the Dingle Peninsula between the island of Inishtooskert and the embayment of Sauce Creek (Fig. 1). It attains a maximum

Fig. 8. Aeolian dune cross-stratified cosets with planar bounding surfaces and interbedded heterolithic interdune–playa units. The beds dip to the SSE (right) and young NNW (left). Hammer shaft is 36 cm long. Farran Formation, waterfall locality, Ballydavid Head.

thickness of about 1 km on the west side of Smerwick Harbour. The Group is restricted to the northern, overturned limb of the Feohanagh Anticline and is bounded to the south by the Fohernamanagh Fault. At Coosgorrib and north of Coosglass the Coosgorrib Conglomerate Formation is present stratigraphically below the Smerwick Group and to the north of the Fohernamanagh Fault (Figs 3 and 6). The inverted Smerwick Group beds strike WSW–ENE, parallel to the present-day coastline, and dip steeply to the SSE. The Smerwick Group as defined by Todd *et al.* (1988*b*) constitutes three formations: the Farran, Ballydavid and Sauce Creek Formations (Figs 4 and 6). This internal stratigraphy is retained (Richmond 1998).

All three formations composing the Smerwick Group preserve a highly interdigitated sequence of both fluvial and aeolian sediments. In unravelling mixed fluvial–aeolian sequences, the use of cumulative evidence in the positive identification of aeolian deposits cannot be overemphasised (Figs 8 and 9). Fluvial reworking of formerly aeolian sediment and the complete fluidization of sedimentary units can greatly hamper positive identification of aeolian deposits. The presence of wind ripple lamination is the only unequivocal evidence of aeolian sedimentation (Hunter 1977*a*, *b*, 1981; Kocurek & Dott 1981).

Criteria used to aid in the positive identification of aeolian deposits in the Smerwick Group have been documented in detail by Richmond (1998). There is a significant overlap in the features that distinguish aeolian from fluvial deposits, particularly where reworking of formerly aeolian-processed sediment has taken place.

The Farran Formation

The Farran Formation comprises the basal sandstone sequence of the Smerwick Group as defined by Todd *et al.* (1988*b*). The Formation is stratigraphically equivalent to the Smerwick Formation of Horne (1974) (comprising the Ballyoughteragh and Upper and Lower Farran Members). The Coosgorrib Member is excluded from the Smerwick Group and was assigned to attenuated Dingle Group by Todd *et al.* (1988*b*). It was given formation status by Richmond (1998) and considered to be a separate stratigraphic unit (Fig. 4).

The presence of 'oblique lamination on a very extended scale' in the upper part of the Farran Formation was observed by Du Noyer, though its significance was understandably not appreciated at the time (Jukes & Du Noyer 1863). Subsequent sedimentological analysis of the Farran Formation (Farran Member of Horne (1974)) by Horne (1974) led to the interpretation of the 'oblique lamination' as 'very probable aeolian' cross-stratification (Fig. 8). Todd *et al.* (1988*a*, *b*) further elaborated on the sedimentology by dividing the Farran Formation into two deposystems: distal ephemeral-fluvial sheetfloods dispersing to the SE and small, transverse, aeolian dunes.

The orange–red sandstones of the Farran Formation crop out on the northern, overturned limb of the Feohanagh Anticline between Sybil Point and Fohernamanagh (Fig. 1). The inverted beds strike WSW–ENE, parallel to the coastline, and steeply dip at an angle of between 40 and 80° to the SSE. The thickness variations along strike in the Farran Formation can be accounted for by faulting along the basal contact and by the diachronous nature of the upper boundary with the overlying, conglomeratic Ballydavid Formation. The Farran Formation reaches a maximum thickness of 560 m on the west side of Smerwick Harbour.

The nature and significance of the basal contact with the underlying Dingle Group is critical.

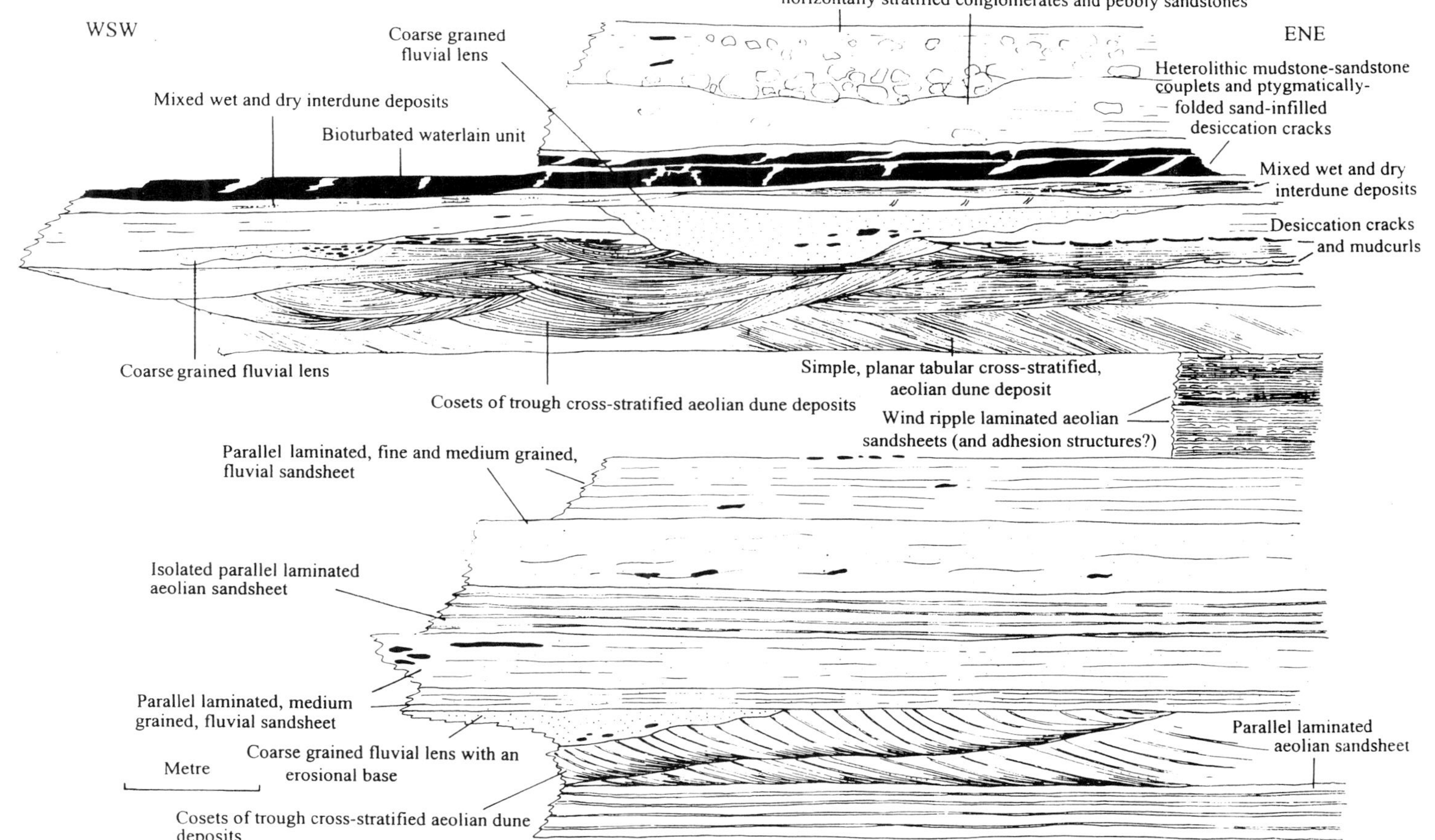

Fig. 9. Lateral profile of fluvial–aeolian architecture, Ballydavid Formation, Ballydavid Head.

This important, unfortunately controversial, contact is exposed only at Fohernamanagh (Fig. 3e). The contact between the Dingle Group and the Smerwick Group is interpreted to be invariably faulted along strike and is demarcated by the Fohernamanagh Fault. The northeastern extent of accessible exposure of the Farran Formation is at Fohernamanagh. Orange–red sandstones of 'Farran Formation affinity' are exposed at Fohernamanagh in direct faulted contact with the Dingle Group. At Fohernamanagh the Smerwick Group is greatly reduced in thickness, probably because of faulting out of the lower part of the Group along the basal contact.

The Farran Formation interfingers with, and is laterally equivalent to, the Ballydavid and Sauce Creek Formations (Figs 10 and 11). The boundary between the Farran Formation and the overlying Ballydavid Formation is transitional and diachronous along strike. The base of the Ballydavid Formation is defined as the horizon above which conglomeratic sedimentation becomes prevalent (i.e. conglomeratic facies packages >20 m thick). Isolated conglomeratic facies packages are present below the base of the Ballydavid Formation (*sensu stricto*). These precursive conglomeratic incursions into the Farran Formation herald the conglomeratic sedimentation of the Ballydavid Formation. They are encountered at stratigraphically lower levels and are of greater thickness on traversing the sequence along strike to the WSW (Fig. 11). The Ballyoughteragh Member defined by Horne (1974) is historically the first, and the most laterally extensive, of these conglomerate horizons to be recognized (Fig. 10). The basal part of the Smerwick Group is not exposed between Fohernamanagh and Sauce Creek. It is proposed that the transition between Farran Formation and the Sauce Creek Formation (*sensu stricto*) takes place between these two locations (over a distance of 3 km) (Figs 3 and 11).

Strike-perpendicular log sections were measured along tectonic strike, at the west side of Smerwick Harbour, Ballydavid Head and Sybil Head (Fig. 1). Ballydavid Head was selected as the type section for the Farran Formation, as the outcrop is readily accessible along its entire length. However, the Ballydavid Head section is of reduced thickness and does not expose the basal part of the Formation in contact with the underlying Dingle Group or Coosgorrib Conglomerate Formation. The reduction in exposed Formation thickness from 560 m to 243 m (a maximum of 350 m) across Smerwick Harbour to Ballydavid Head is more than likely due to faulting out of the basal part of the Farran Formation by the Fohernamanagh Fault and/or Dun an Oir Fault, which run(s) unexposed through the beach at Feohanagh (Fig. 3).

The sedimentological attributes of the Farran Formation are indicative of both ephemeral-fluvial and aeolian sedimentation (Richmond 1998). Taylor (1994) provided a comprehensive review of both modern and ancient ephemeral systems, and defined ephemeral in the rock record as pertaining to high-energy, rapidly varying flow, separated by episodes during which the channel is dry. This is an acceptable definition (Richmond 1998), the characteristics of which certainly apply to the Farran Formation (and Ballydavid and Sauce Creek Formations) fluvial system. In assigning the term ephemeral to a system one has to consider a combination of characteristics that cumulatively classify the system as ephemeral. Every fluvial system is unique and therefore it would be impossible to set out a list of features that always occur in ephemeral deposits. However, the following features recorded from the Farran Formation are cumulative evidence of the ephemerality in this particular fluvial system: predominance of upper flow regime, parallel lamination in poorly channelized to non-channelized sandstone units; a dearth of lower flow regime deposition in the form of ripple laminated caps, mud drapes and heterolithic units; thin flood cycles in overbank–interdune and interdune–playa heterolithic units; desiccation of red mudstone laminae; wind-winnowing of fluvial material to produce deflation lag deposits; abundance of intraclast lags and intraformational conglomerates; abundant scour-and-fill structures, internal and basal, erosional scour surfaces and laminar discontinuities; minor, thin, poorly developed mudstone veneers to beds; massive sandstone units deposited by sand-saturated, hyperconcentrated flows; bioturbation of sheetflood flow tops and interdune units; fluidization, liquefaction and water-escape phenomena; interdigitation with aeolian deposits, silt- and mud-infilled scour and fill structures; scours draped by mud and desiccated before subsequent infill; palaeosols (immature and variable); highly interdigitated facies states and facies associations.

High-energy, short-lived, relatively shallow, non-uniform, unsteady flow in broad sheets or poorly confined channels dominated ephemeral-fluvial sedimentation of the Farran Formation. Channel beds, banks and sand dunes provided abundant unconsolidated sand, which was rapidly removed and incorporated into flows that resulted in rapid channel widening, recession

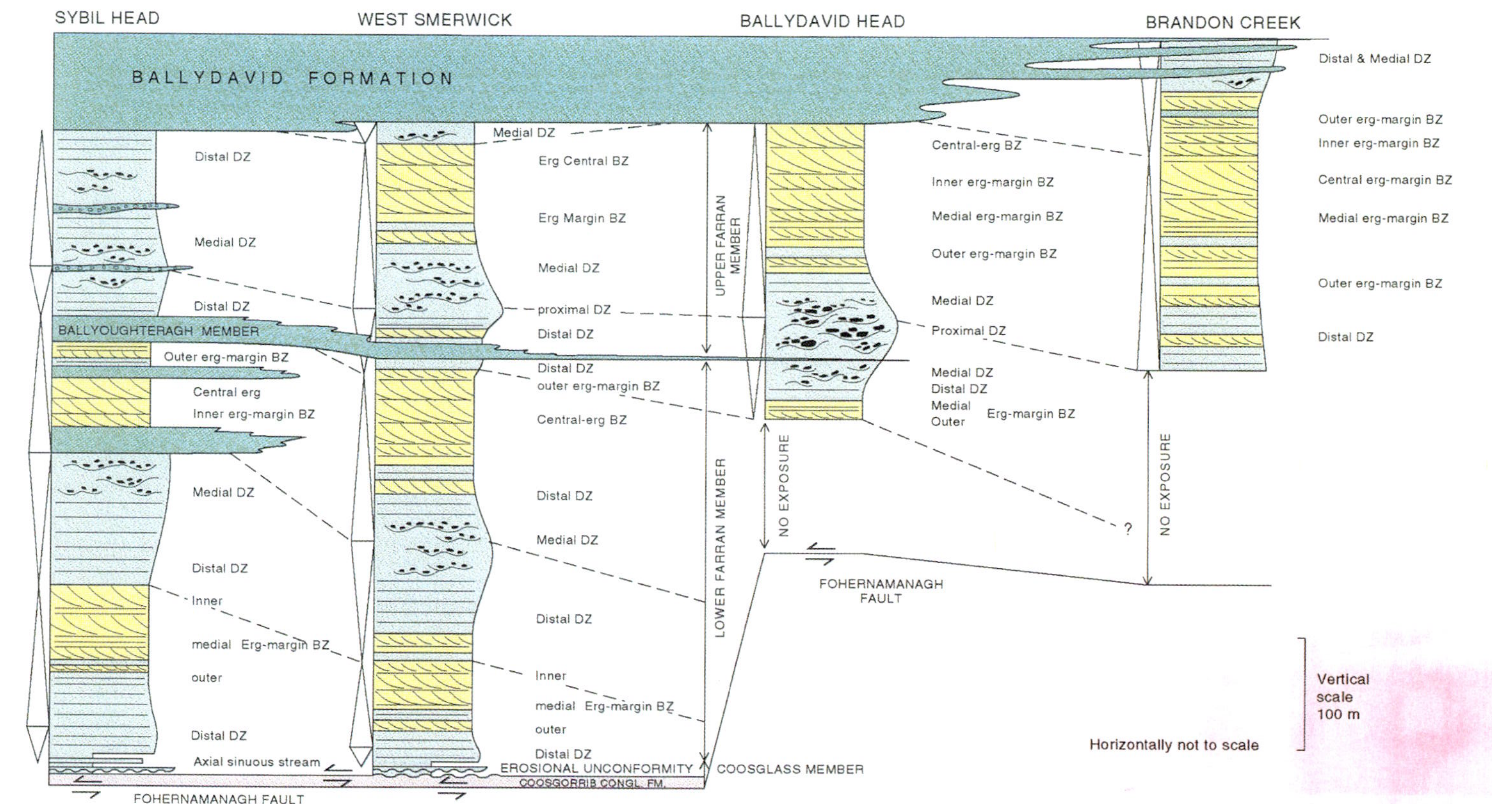

Fig. 10. Lateral facies variation diagram and third-order wetting and drying cycles in the Farran Formation. DZ, distributary zone; BZ, basinal zone.

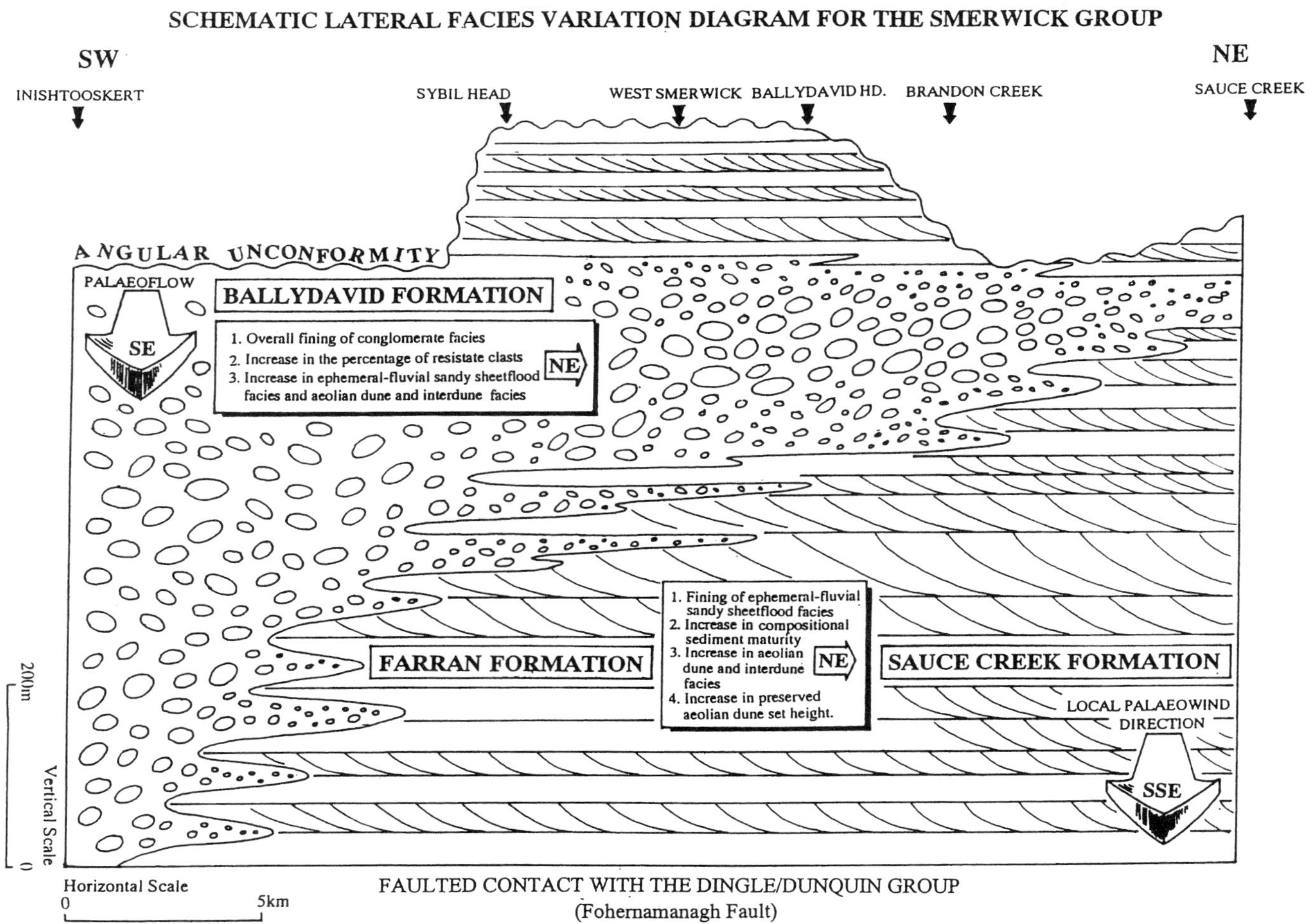

Fig. 11. Schematic lateral facies variation diagram for the Smerwick Group. Sketch perpendicular to palaeoflow.

of flow and high transmission losses. In the Farran Formation fluvial system sand-saturated, hyperconcentrated flows were rapidly deposited, resulting in massive sandstone units. Rapid recession of flows inhibited low stage reworking and the development of fine drapes. The development of bedforms was suppressed by the high sediment concentration of flows. Dramatic falls in water stage on rapid recession of flow, the prevalence of upper flow regime conditions and shallow water depths also inhibited bedform development. Low stage, fine drapes and bedforms may have occasionally formed but were rapidly removed and destroyed by further flood events. Rapid deposition and overburden of sandstone units generated inherent instability of sands and their subsequent liquefaction, fluidization and induced water escape. Rapid fluctuations in flow conditions and pulsating flood events resulted in rapid and abrupt changes in the preserved facies. Scour-and-fill structures, scour surfaces and major, erosional, channelized surfaces mark rapid changes in flow conditions. Where scour and erosion surfaces are draped by mudstone this records rapid recession of flow and sediment by-pass. The flows were in part constrained by the aeolian topography in which channelization of flows increased as a result of funnelling along topographic lows in aeolian sandsheets and interdunes. In the aeolian environment the floods occasionally choked interdune corridors, causing channel avulsion and abandonment. The flow pathways were intermittently dried during subaerial exposure, desiccation, deflation, bioturbation and rare pedogenesis. Channel-bordering aeolian sandstone sheets underwent early lithification during periods of elevated water table. These sheets were subsequently ripped up and incorporated into flows. They also induced terraced scouring at the bases of overlying channelized units.

The facies association divisions of the Farran Formation are mainly derived from the type section at Ballydavid Head. The facies associations reflect the local environment of deposition, and analysis of the hierarchical stacking patterns of the facies associations can be used to illustrate the broader depositional environment and its temporal and spatial evolution (Fig. 10). Within the facies associations, repetitive fining-upward and/or drying–upward or drying-wetting-upward (waterlain facies overlain by aeolian facies, which are in turn capped by a veneer of waterlain facies) facies transitions can be identified (Richmond 1998). The controls on the development of this small-scale cyclicity and the larger-scale (facies association and member) stacking patterns are couched in a hierarchical stratigraphic framework.

The depositional processes recorded within individual facies state units and facies associations themselves are indicative of distributary zone, terminal fan sedimentation and its downstream transition into basinal zone, aeolian deposition (Fig. 10). The facies associations are genetically defined in terms of their environment of deposition, and can be ascribed to proximal, medial and distal distributary zone and outer, medial and inner erg-margin, and central-erg, basinal zone sedimentation. The feeder zone sediments are not preserved.

Terminal fan deposition is characterized by gradual downstream reduction in discharge as a result of infiltration, evaporation and/or low precipitation, such that distributary channels ultimately disappear downstream and surface water flow leaving the system tends to zero (Friend 1978). Hence terminal fans tend to form in semi-arid to arid regions where evaporation and transmission losses are high and precipitation is low. High infiltration rates into high-permeability channel margins result in high transmission losses, which in turn cause a downstream reduction in discharge and water depths. These features are all consistent with the hydraulic processes that can be inferred in Farran Formation times.

Terminal fan deposits are generally relatively thick, of the order of 100–1000 m, and accumulate in small basins adjacent to areas of uplift (Friend 1978; Kelly & Olsen 1993). These features are reconcilable with a terminal fan model for the Farran Formation (Fig. 12a–c). The frequency and dimensions of channelized units decreases downstream and this is accompanied by a decrease in the sediment grain size. The decrease in channel width-to-depth ratios downstream indicates a downstream reduction in discharge, as distributary channels in the proximal zone are replaced by finer-grained sheet-flood deposits in the distal zone. Evidence for this is conspicuous in the Farran Formation. The silt- and mud-sized particles are, however, present only as thin veneers in the distributary zone, and silt–mud accumulations are generally restricted to interdune traps producing interdune–playa and overbank–interdune deposits.

The most complete section exposed through the Farran Formation is that on the west side of Smerwick Harbour (Fig. 1). The Member divisions are established in this section through the Formation and extrapolated along strike to the WSW and ENE (Fig. 10). The boundaries between constituent members are drawn in accordance with Horne (1974) with the addition

Palaeogeographic models for Smerwick Group deposition

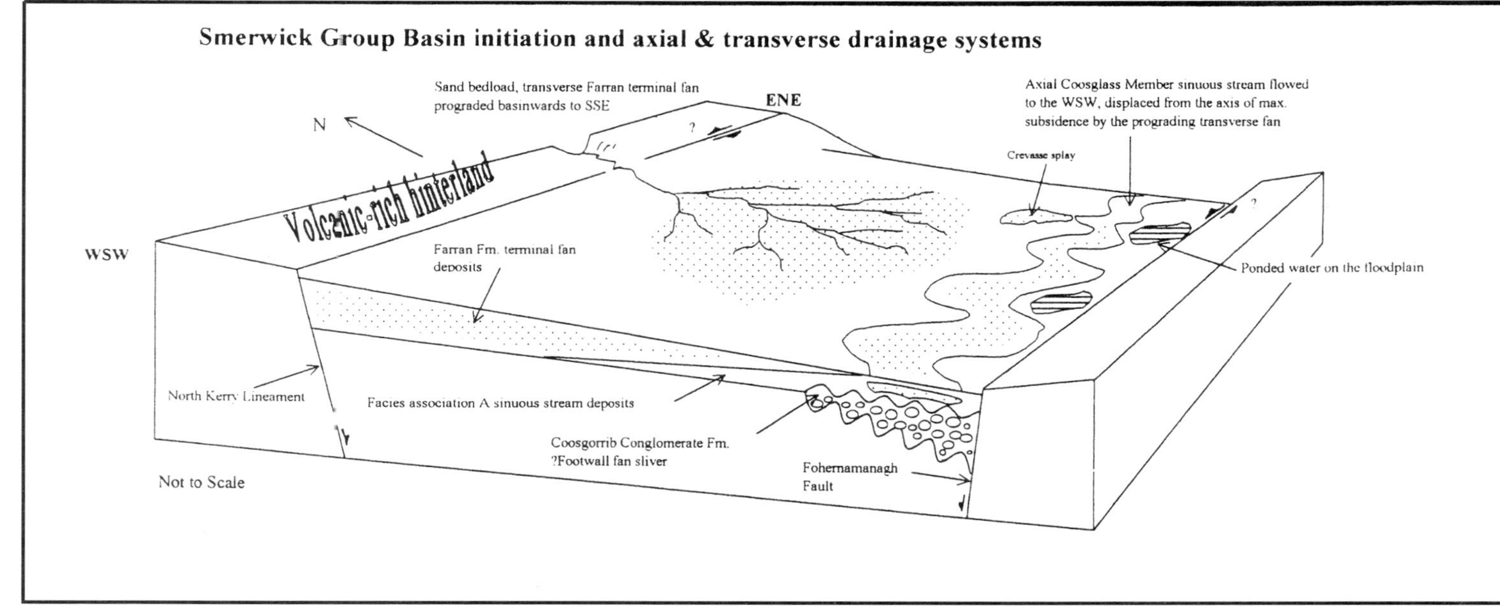

Fig. 12(a).

Smerwick Group Basin development, two separate terminal fan feeders, terminal fan & erg-margin to central erg sedimentation in a simple N-S extensional graben
Volcanic-rich hinterland
N
ENE
WSW
Ballydavid conglomeratic bedload terminal fan feeder
Farran sand bedload terminal fan feeder
Transverse-barchanoid dunes migrated to the S & SSE. Largest dunes formed in the ENE (at Sauce Creek) were the influence of terminal fan activity was reduced
Local winds directed towards the S and SSE
Regional prevailing winds directed towards the NNW
Stable high protected the Smerwick Basin from regional southerly winds
Fohernamanagh Fault
Coosglass Mb.
Coosgorrib Conglomerate Fm. ?Footwall fan sliver
Farran Fm. terminal fan deposits
Ballydavid Fm. Terminal fan deposits
North Kerry Lineament
Not to Scale
Fluvially-flooded interdunes
Ponded interdunes

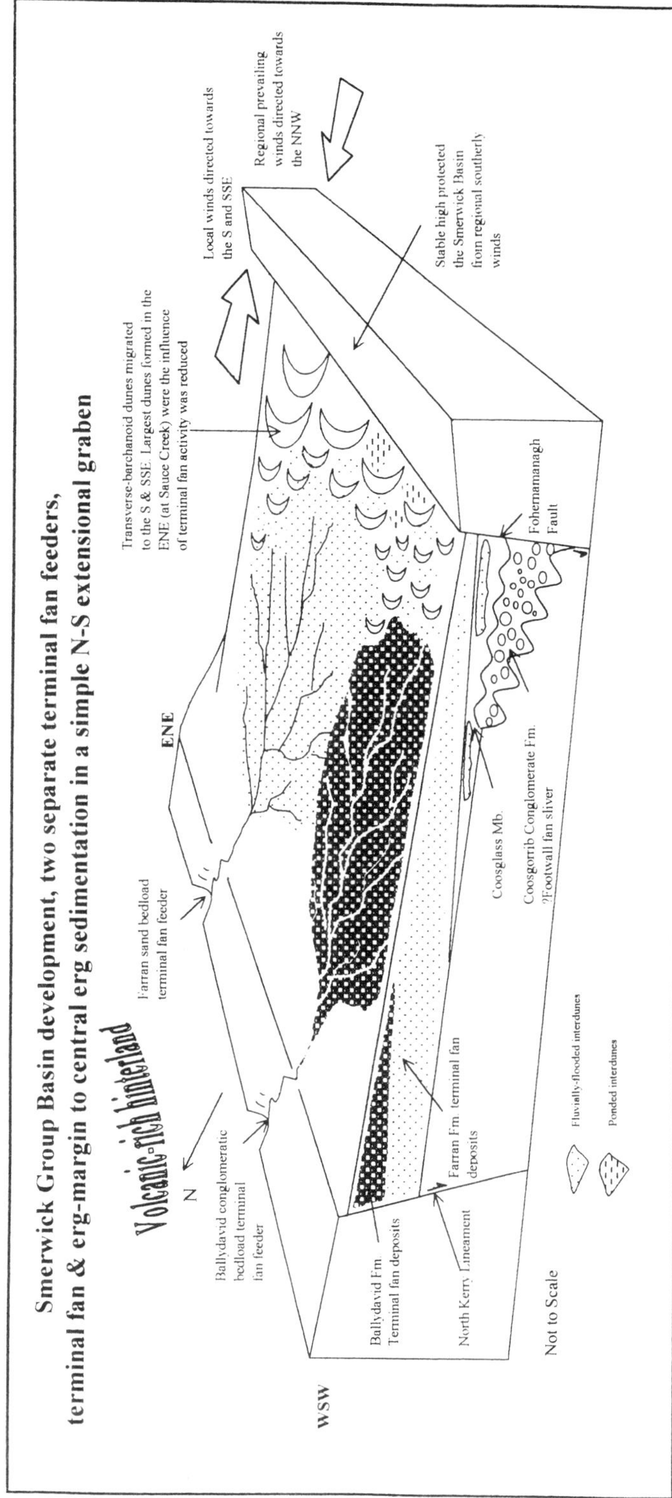

Fig. 12(b).

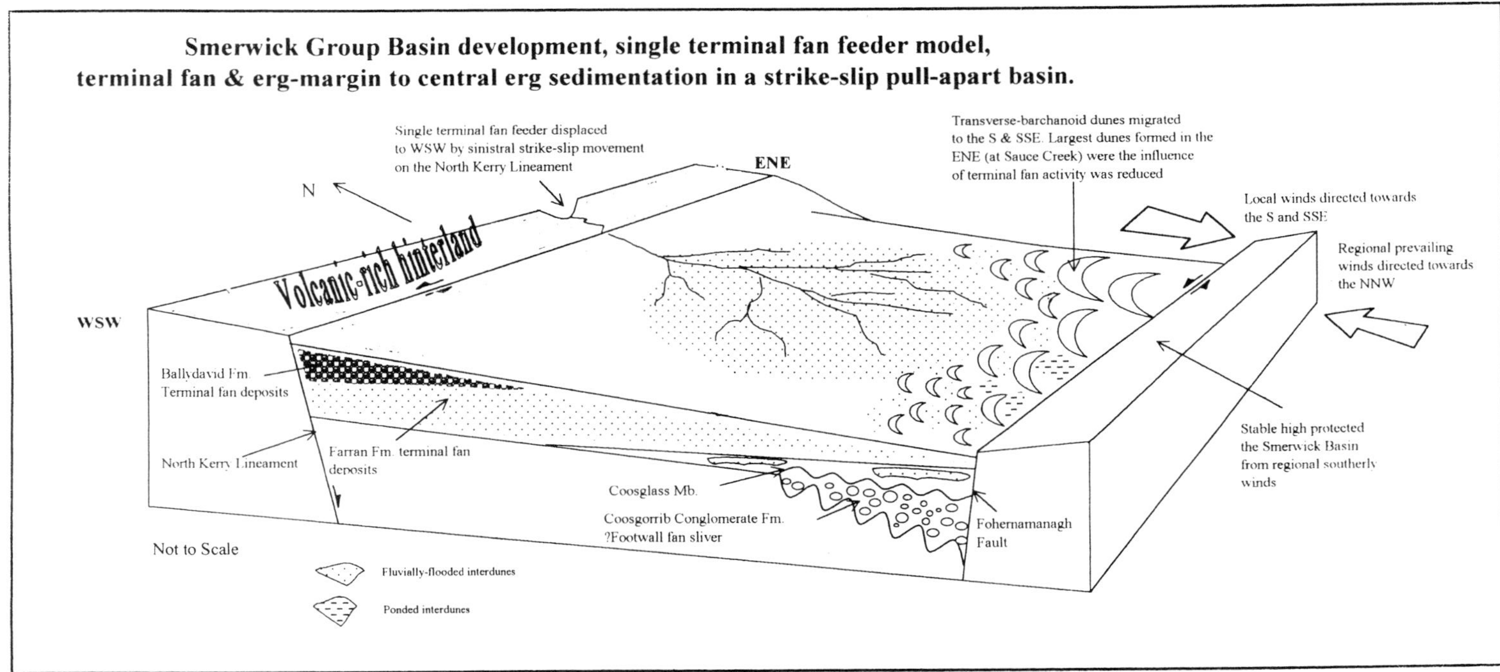

Fig. 12(c).

Fig. 12. Palaeogeographical models for the Smerwick Group. (**a**) Basin initiation, axial and transverse drainage system. (**b**) Basin development with two separate terminal fan feeders; terminal fan and erg-margin to central-erg sedimentation in a simple north–south extensional graben. (**c**) Alternative basin development scenario with single terminal fan feeder; terminal fan and erg-margin to central-erg sedimentation in a strike-slip pull-apart basin.

here of the basal Coosglass Member. Thus four members are encountered in the Formation: the Coosglass, Lower Farran, Ballyoughteragh and Upper Farran Members. The basal Coosglass Member, not recorded by Horne (1974), records initial, axial sinuous-stream deposition, which rapidly gave way to transverse terminal fan deposition (Fig. 12a). The Coosglass Member is best developed to the north of Coosglass (Fig. 3a), where it attains a thickness of 20 m. A 12 m thickness of Coosglass Member is also recorded at Coosgorrib directly overlying the Coosgorrib Conglomerate Formation. The Lower Farran Member extends up to the base of the Ballyoughteragh Member (Fig. 10). This Member records the rapid retreat of the initial terminal fan progradation, with concomitant erg advance, followed by terminal fan progradation and retrogradation, the latter being accompanied once more by erg advance. The Lower Farran Member is interspersed with conglomeratic facies packages of Ballydavid Formation affinity. The Ballyoughteragh Member can be accurately correlated from Sybil Head through to West Smerwick (Fig. 10). Perhaps coincidentally, very poorly developed conglomerate horizons also occur at this level in the stratigraphy in the Ballydavid Head section, such that an extremely tentative correlation can be drawn across Smerwick Harbour. The Ballyoughteragh Member is simply the thickest, most areally extensive of the Ballydavid Formation affinity packages in the Farran Formation (Fig. 11). The Upper Farran Member extends from the top of the Ballyoughteragh Member to the diachronous contact with the overlying Ballydavid Formation. The lower part of this member records erg retreat that was accompanied by the most advanced terminal fan progradation within the Farran Formation. The upper part of the member records subsequent terminal fan retrogradation and erg advance, and in the ENE late stage erg retreat is also recorded.

The Farran Formation was deposited in an active tectonic regime (Fig. 12a–c). Tectonic activity resulted in uplift of the source area and subsidence of the hanging-wall basin-forming region. These two factors determined accommodation space and sediment supply. It is envisaged that tectonism was the primary allocyclic control on deposition during Farran Formation sedimentation. The gross, lower-order progradational phases record incremental uplift of the footwall, followed by equilibration and retrogradation, producing fining-upward trends as sediment supply to the fluvial system was decreased and aeolian reworking took place.

The first-order, basal bounding surface, represented by the locally preserved unconformity at the base of the Farran Formation, records the response to increased tectonism, which caused uplift and erosion of the Coosgorrib Conglomerate Formation on Smerwick Basin initiation. The upper, second-order bounding surface formed as a result of the initiation of the Ballydavid Formation footwall terminal fan, whose depocentre was located along strike to the WSW (Richmond 1998). The onset of Ballydavid Formation terminal fan deposition was almost certainly a product of rejuvenated tectonic activity, resulting in footwall uplift and denudation together with basin subsidence and sediment accumulation. The broad-scale, second-order fan progradation and retrogradation in the Farran Formation resulted from the initial incremental uplift of the footwall in the NNW followed by equilibration of the deposystem. Increased slope gradient, sediment supply and discharge is recorded by the progradational episode, which was superseded by fan retrogradation in response to reduced slope gradient, discharge and sediment supply. The asymmetric profile (Fig. 10) is probably due to a combination of the following: the time-lag in the response of sedimentation to tectonism; the gradual increase in the amount of uplift through time; the differing accumulation rates of distributary zone and basinal zone sediments; the significant spans of geological time not represented and simply the slice through the fluvial–aeolian basin architecture exposed.

Third-order cycles can be identified (Fig. 10 wetting and drying cycles) and these resulted from changes in the ratio of accommodation space to sediment supply. On increasing this ratio, waning fluvial deposition and aeolian reworking are preserved (Richmond 1998); on decreasing it, sediment by-pass occurs and aeolian reworking does not take place. The third-order cycles record the pulsatory retrogradation and progradation of the terminal fan and accompanying erg advance and erg retreat, respectively. The preservational bias towards drying-upward cycles is due to the consistently high ratio of accommodation space to sediment supply. The preservation of wetting-upward components suggests that changes in sedimentation were occurring over a sufficiently protracted length of time to allow for their preservation. This in turn suggests that these changes occurred in response to tectonism rather than climate. The regular-grain size changes are probably indicative of periodic incremental uplift of the source area.

A fourth-order cyclicity is represented by facies transitions that are apparent in the proximal and medial distributary zone sediments of the terminal fan. They formed in response to local, short-term changes on the fan surface determined by subtle changes in climate, such as increased storm activity, initiating channel migration and abandonment. Similar climatically controlled drying–upward-cycles and/or drying-wetting upward cycles of a similar scale in other ancient rock record examples are well documented (e.g. Clemmensen *et al.* 1989; Meadows & Beach 1993; McKie & Garden 1996; Herries 1993; Howell & Mountney 1997). In the Farran Formation their character changes from drying–wetting-upward to drying-upward on third-order terminal fan progradation. This is indicative of a superimposed tectonic control on the preservation of climatically determined processes, or indeed of the possibility that climatic response may be tied to changes in basin configuration linked to tectonism (e.g. Lawrence & Williams 1987). The absence of a well-developed, wetting-upward component in the more proximal deposits could be ascribed either to lack of preservation, such that the wetting-upward element was deposited but subsequently removed, or non-deposition as a result of very rapidly changing conditions.

Thus, the Farran Formation records sand bedload, ephemeral-fluvial, proximal to distal distributary zone terminal fan deposition and erg-margin to central-erg sedimentation. The terminal fan issued from the NNW into the Smerwick Basin at the present-day location of Ballydavid Head (Fig. 12), and inundated and displaced away from the axis of maximum subsidence an initial, 'low-seeking', axial, ephemeral sinuous stream which flowed to the WSW (Richmond 1998; see Leeder & Gawthorpe 1987). The switch from axial to transverse drainage was induced by footwall uplift in the NNW during embryonic Smerwick Basin times. The wind reworked the sand supplied by the terminal fan system into aeolian sandsheets and dunes on the erg margin. The palaeowind was directed towards the SSE. The Farran Formation records sand bedload, second-order, terminal fan advance and retreat in response to incremental footwall uplift, subsequent equilibration and increased aeolian reworking. The Farran Formation terminal fan from time to time coalesced with, and was ultimately engulfed by, the Ballydavid, gravel bedload terminal fan (Fig. 12a–c).

The Ballydavid Formation

The Ballydavid Formation comprises the conglomeratic suite of the Smerwick Group. It stratigraphically overlies the Farran Formation and passes laterally into the Sauce Creek Formation towards the ENE. The Ballydavid Formation was introduced by Todd *et al.* (1988*b*) on the elevation of the Smerwick Formation of Horne (1974) to Group status. The sedimentology of the Ballydavid Formation has not previously been appraised. The conglomerates with finer, sandy interbeds (Horne 1974) were interpreted by Todd *et al.* (1988*b*) to be the product of braided river deposition. Previous to the study by Richmond (1998), aeolian sediments had not been recorded within the Ballydavid Formation.

The conglomerates are interbedded with subordinate orange–red sandstones and crop out on the northern, overturned limb of the Feohanagh Anticline between Sybil Point and Fohernamanagh (Figs 1, 3 and 9). The beds strike WSW–ENE and dip at between 40 and 80° to the SSE. The conglomeratic suite overlies and interfingers with the Farran Formation, both of which are laterally transitional into the Sauce Creek Formation. The contact between the Farran and Ballydavid Formations is diachronous along strike. At Ballydavid Head the contact is sharp and channelized, with an erosive relief of at least 5 m. The base of the Ballydavid Formation is defined as the horizon above which conglomeratic sedimentation is prevalent (i.e. conglomeratic packages >20 m thick) and it ascends the Smerwick Group stratigraphy on traversing along strike from WSW to ENE (Figs 10 and 11). The latter is accompanied by an increased interfingering between upper Farran Formation sandstones and lower Ballydavid Formation conglomerates at the contact. The upper contact with the overlying Sauce Creek Formation is exposed between Sybil Point and Brandon Creek. It also becomes transitional on traversing along strike from WSW to ENE (Fig. 3). This contact at Ballydavid Head is sharp and non-transitional, whereas ENE along strike near Brandon Creek the Sauce Creek Formation and Ballydavid Formation facies packages are interdigitated. Where this upper contact is transitional, it is demarcated by the horizon above which conglomeratic units are subordinate and less than 20 m thick. At the sharp contact at Ballydavid Head, a very marked increase in sandstone compositional maturity is recorded on ascending into the Sauce Creek Formation. The Ballydavid Formation fines and thins along strike to the ENE. It reduces in thickness from 730 m at Inishtooskert to 370 m at Brandon Creek to 90 m of laterally equivalent conglomerates at Sauce Creek (Sauce Creek Formation *sensu stricto*) (Figs 1 and 11). These formation-scale, together with member-scale, facies association scale and facies-scale, spatial variations reflect the proximity to the terminal fan feeder of the Ballydavid Formation. The feeder is shown to be located to the WSW (Richmond 1998) (Fig. 12b and c). Hence on strike-traverse to the ENE the transition into the Ballydavid Formation is more gradual because of the relatively distal setting with respect to the fan feeder. Corroborative evidence for the location of the terminal fan feeder in the form of formation-scale spatial variations in clast lithotypes has been addressed by Richmond (1998).

The type section for the Ballydavid Formation is at Ballydavid Head, where a section of 383 m was measured in detail (Richmond 1998). Along-strike sections through the Ballydavid Formation are of limited access. However, detailed facies package mapping along strike from Inishtooskert to Fohernamanagh (and to Sauce Creek) allows a crude, basin-wide, fluvial–aeolian architecture to be established. The sedimentary processes and their interaction are consistent with ephemeral-fluvial, distributary zone, terminal fan deposition and inner and medial erg-margin, aeolian, basinal zone sedimentation (Richmond 1998), simliar to that of the underlying Farran Formation. The terminal fan issued into the basin from the NNW at, or to the WSW of, the present-day location of Inishtooskert. The palaeowind direction was directed towards the south. Aeolian reworking was restricted by a limited sand supply from an upwind gravel bedload, as opposed to sand bedload, terminal fan. Hence the bedrock geology of the hinterland influenced both the bedload carried by the terminal fan and ultimately aeolian sedimentation (Figs 11 and 12). The Ballydavid Formation records gravel bedload, second-order, terminal fan advance and retreat in response to incremental footwall uplift and subsequent equilibration. Third-order and fourth-order cyclicity similar to that of the Farran Formation is also apparent. The gravel bedload terminal fan coalesced with, and eventually engulfed, the Farran Formation terminal fan.

The Sauce Creek Formation

The Sauce Creek Formation forms the topmost sandstone sequence of the Smerwick Group at Ballydavid Head (and between Sybil Head and Been Dermot; Beenmore & Brandon Creek) and is the lateral equivalent of both the Farran and Ballydavid Formations at the type section at Sauce Creek (Figs 1 and 11). The Formation is stratigraphically equivalent to the Sauce Creek Member of Horne (1974), which was originally assigned to the Ballymore Formation of the Dingle Group (Horne 1974). The Sauce Creek Member was elevated to formation status on recognition of the Smerwick Group by Todd *et al.* (1988*b*).

The Sauce Creek Formation (Sauce Creek Member of Horne (1974)) was identified by Horne (1974) at Ballydavid Head and Sauce Creek. He described the Sauce Creek Formation as comprising laterally persistent, flaggy beds of regular thickness, commonly displaying ripples and shrinkage cracks. Conglomeratic units in the upper part of the Formation at Sauce Creek were also noted. Horne (1974) likened the lithologies encountered in the Sauce Creek Formation to the Bull's Head Formation of the Lower Dingle Group. He took these features to pertain to a possible lacustrine origin. The aim here is to illustrate that the Sauce Creek Formation is emphatically non-lacustrine in origin and is a fluvial–aeolian succession of Smerwick Group affinity. This revelation further supports our removal of the Smerwick beds from the Dingle Group.

The grey–pink sandstones of the Sauce Creek Formation are very distinct from the orange–red sandstones and conglomerates of the Farran and Ballydavid Formations. The Formation crops out over a distance of 20 km on the northern, overturned limb of the Feohanagh Anticline between Sybil Head and Been Dermot, Beenmore and Brandon Creek, Ballydavid Head and at Sauce Creek itself (Fig. 3). The beds are inverted, strike WSW–ENE and dip steeply to the SSE at an angle of between 60 and 88°. The Formation attains a thickness of 200 m at Ballydavid Head and likewise between Sybil Head and Been Dermot. Interdigitated Ballydavid and Sauce Creek Formation facies associations are encountered between Beenmore and Brandon Creek (Fig. 11). The type section for the Sauce Creek Formation is located in the north-east wall of Sauce Creek itself (Fig. 3f). Here the Sauce Creek Formation attains a thickness of 840 m. However, only 700 m of outcrop are exposed, of which only 304 m is accessible.

The nature and implications of the lower contact with the underlying Ballydavid Formation was discussed earlier. The upper contact with the overlying Pointagare Group forms the spectacular angular unconformity (Fig. 7) that can be mapped from Inishtooskert in the WSW to Sauce Creek in the ENE. The unconformity displays, over the 30 km distance along strike, an erosive relief of up to 200 m (Fig. 11).

Unfortunately, because of the lack of exposure, the spatial variations within the Sauce Creek Formation *sensu stricto* are unknown. However, as the Sauce Creek Formation at Sauce

Creek is laterally equivalent to the Farran and Ballydavid Formations along strike to the WSW, comparisons can be drawn between the Formations from a facies to formation scale. The Ballydavid Formation affinity conglomerate unit in the upper part of the Sauce Creek Formation *sensu stricto* is of much reduced thickness compared with the 730 m thick Ballydavid Formation succession at Inishtooskert (30 km along strike). Correlation of third-order, member-scale cycles is unfortunately impossible because of lack of exposure. However, the cycles are of a similar scale and style to those encountered in the Farran Formation. The facies associations are significantly different in terms of their internal make-up, facies states and facies proportions. The fourth-order facies transitions are of a similar style and scale to those encountered in the same association of the Farran Formation. However, wetting-upward caps to drying-upward facies transitions are absent. The sandstones of the Sauce Creek Formation have a higher textural and compositional maturity than the laterally equivalent Farran and Ballydavid Formations (Richmond 1998).

The Sauce Creek Formation at Sauce Creek records more distal, terminal fan deposition relative to both the Farran and Ballydavid terminal fans (Fig. 12). The fan feeders of both the Farran and Ballydavid systems were located to the WSW of Sauce Creek. Erg-margin deposits are better developed at Sauce Creek in the Sauce Creek Formation because of the increased distance from the terminal fan feeders and closer proximity to the erg centre. The sandstones are as a result more texturally and compositionally mature because of increased working by both fluvial and aeolian processes. The conglomerates are also significantly compositionally more mature as a result of increased transportation distance from the terminal fan feeder of the Ballydavid Formation, located to the WSW of the present-day location of Sauce Creek.

The Sauce Creek Formation at Sauce Creek records ephemeral-fluvial sedimentation distal to that of the Farran Formation in the WSW. Accompanying erg-margin and central-erg deposition were more significant in this part of the basin, which was relatively removed from the influence of terminal fan activity (Fig. 12). The erg centre probably lay to the ENE of Sauce Creek. The location of the erg was controlled by the supply of sand provided by the upwind, sand bedload, terminal fan deposits and by the basin topography and wind regime (Fryberger & Ahlbrandt 1979). Back-erg preservation is recorded at the top of the Sauce Creek Formation at Ballydavid Head. The condensed back-erg sequence at Ballydavid Head immediately below the unconformity owes its preservation to rapid basin subsidence and burial (Porter 1986). Back-erg preservation is also likely to have been enhanced in areas more distal to the terminal fan.

Depositional synthesis of the Smerwick Group

The Smerwick Group forms a fluvial–aeolian succession comprising three formations that interfinger laterally on a basin-wide scale. Together with the underlying Coosgorrib Conglomerate Formation the Group is restricted to the Northwest Dingle Domain and is part of a separate terrane bounded to the south by a terrane boundary demarcated by the Fohernamanagh Fault. The Coosgorrib Conglomerate Formation is included within the Smerwick Basin succession because the erosional unconformity between it and the Smerwick Group is interpreted to represent an intra-basinal unconformity (Richmond 1998). This intra-basinal unconformity is locally faulted such that the Coosgorrib Conglomerate Formation forms a thin, fault-bounded, conglomeratic sliver (Fig. 3a and b).

The Smerwick Group displays evidence of both Variscan and weak localized Caledonian deformation. This together with the structural juxtaposition of the Group below the marked angular unconformity with the Middle Devonian Pointagare Group and provenance evidence place the Smerwick Group within the Lower Old Red Sandstone. It ranges in age somewhere between the latest Ludlow and latest Emsian time (Fig. 5). Smerwick Group sedimentation took place at the same time as, or more likely shortly after, the deposition of the Dingle Group. The structural, facies, provenance and palaeocurrent evidence indicate that deposition of the Dingle and Smerwick Groups in the same basin, or immediately adjacent sub-basins, is irreconcilable. However, contrary to Todd *et al.* (1988*b*), the Smerwick Group has by far a greater affinity with the Dingle Group than with the overlying successions in the Northwest Dingle Domain (Richmond 1998).

It is envisaged that the Smerwick Group, together with the Coosgorrib Conglomerate Formation, constitutes an allochthonous terrane that docked certainly in post-Glashabeg and more probably, post-Dingle Group times before Pointagare Group deposition. The docking mechanism for the Smerwick Group was strike-slip.

This is consistent with the strike-slip regime known to have been in operation during Dingle Group times (Todd 1989*a*). The Trabeg Conglomerate Formation fan lithosome displays evidence of sinistral strike-slip on the Dingle Bay Lineament (Todd 1989*a*). Considerable faulting must have taken place along the Fohernamanagh Fault, particularly at Fohernamanagh, where some 620 m of Smerwick Group sequence is missing. This faulting must have been primarily Late Caledonian as the Pointagare Group overlying the terrane boundary is affected only by minor faults at Fohernamanagh. Hence it is proposed that the Smerwick Group docked during Late Caledonian deformation. Faulting along the terrane boundary was variable along strike, resulting in considerable thickness variations in the Smerwick Group. The generation of the Feohanagh Anticline and initial tilting of the Smerwick Group must also have taken place during Late Caledonian deformation. The Anticline was probably later tightened and the Smerwick sequence overturned during Variscan deformation.

The Smerwick Group highlights the interaction and interdependence of fluvial and aeolian processes in a delicately balanced environment. The primary controls on aeolian sedimentation were the wind regime, basin topography, climate, hinterland bedrock geology, accommodation space, sand supply from an upwind terminal fan source, and the frequency and magnitude of flooding and associated elevated water table levels. Additional controls on fluvial deposition were sediment supply, discharge, slope gradient, rates of evaporation and infiltration, precipitation, sediment binding by vegetation, permeability and cohesive properties of the substrate, and the topography of the aeolian landscape. The overriding constraint on both fluvial and aeolian deposition was an allocyclic, most probably tectonic, control which was superimposed upon short-term climatically controlled or storm-related processes. The preservation of aeolian reworking within all orders of cyclicity suggests that ratios of accommodation space to sediment supply were generally high (Richmond 1998). A low suspension load and abundant sand bedload also encouraged aeolian reworking.

All three Formations record concordant palaeodrainage and palaeowind directions. During Farran and Sauce Creek Formation deposition the net sediment transport direction was to the SSE, that is, transverse to the basin axis. There is however a slight swing in wind directions to the south along strike to the ENE in the Sauce Creek Formation system. This may have been in part induced by basin topography. During Ballydavid Formation sedimentation the palaeoflow indicated by Ballydavid Formation conglomeratic facies is towards the south, concordant with the palaeowind direction. The concomitant switch to the south of the palaeowind and palaeodrainage directions suggests a mutual cause, probably induced by minor alterations to basin morphology associated with the initiation of the Ballydavid terminal fan. The Smerwick Group records an overall wind direction towards the SSE and a concordant palaeodrainage. The wind direction recorded in the Smerwick Group is opposed to the regional prevailing wind direction thought to be in operation during Early Devonian times; only minor reversals to the NNW or NW are recorded. The basin must have been largely protected from southerly winds by a topographic barrier which, because of the allochthonous nature of the terrane, remains of unknown geometry, size and orientation.

The model, as presented so far, suggests that two terminal fans issued into the Smerwick Basin, one of which was dominated by a sand bedload and the other by a gravel bedload. However, a second model is also plausible (Fig. 12). The temporal and spatial evolution along strike to the WSW from sand bedload to gravel bedload terminal fan deposition may reflect the unroofing of the footwall accompanying sinistral strike-slip displacement of a single terminal fan feeder that was shifted along strike to the WSW. The sinistral strike-slip invoked is the same sense of slip identified in the Trabeg Conglomerate fan lithosome of the Dingle Group (Todd 1989*a*).

The Smerwick Group accumulated in a small, hydrologically closed, elongate, asymmetric, ENE–WSW-oriented basin developed adjacent to a major fault (NKL) downthrowing to the south as a result of the onset of a simple north–south extensional regime (Fig. 12b) (or a sinistral transtensional regime, Fig. 12c). During embryonic Smerwick Basin times an axial, ephemeral, sinuous stream system flowed along the basin to the WSW. The axial system was rapidly inundated by transverse terminal fan(s) that initiated in response to footwall uplift of the basin margin in the NNW. The terminal fan(s) dispersed to the SSE and terminated downstream and downwind in a basinal zone occupied by an erg. Ephemeral-fluvial and aeolian sedimentation took place under semi-arid or arid conditions. The Group records pulsed terminal fan progradation and retrogradation interspersed with periods of erg retreat and erg advance, respectively. The strike-slip docking of the allochthon along the Fohernamanagh Fault during Late Caledonian deformation emplaced the foundations of the

Northwest Dingle Domain. The Northwest Dingle Domain subsequently continued to behave in separatist manner well into Late Devonian time, resulting in a unique, diverse Old Red Sandstone collage in West Kerry. The Northwest Dingle Domain illustrates the fundamental control of inherent, deep crustal heterogeneities on the structural evolution and ultimately the sedimentary cover of a terrane.

NW Dingle Domain: the Pointagare Group

Dodd (1986) defined the Pointagare Group in the Northwest Dingle Domain to replace the Upper Old Red Sandstone members of Horne (1974) (Fig. 4). The newly defined Pointagare Group was subdivided into the Beenaman Conglomerate and Beenmore Sandstone Formations. The stratigraphy established by Dodd (1986) is retained (Richmond 1998). However, the existence of angular unconformity with the overlying Carrigduff Group (Glengarriff Harbour Group of Dodd (1986)) is refuted. The contact between the two groups is interpreted to be a paraunconformity. The basal beds of the Carrigduff Group have been biostratigraphically dated as early to mid-Frasnian in age (Higgs, pers. comm.).

The Pointagare Group crops out in high sea cliffs along the northwest Dingle coast between the island of Inishtooskert and Brandon Point (Figs 1 and 3), and rests with angular unconformity on the underlying Smerwick Group (Fig. 7). The Group is unique in the Northwest Dingle Domain in that it is both observed to dip to the south on the southern upright limb of an anticline (the Brandon Point Anticline) and oversteps the trace of the Fohernamanagh Fault (at Masatiompan) (Figs 1 and 3). To the WSW of Sauce Creek, the beds dip to the NNW at an angle of between 10 and 80°. Between Ballydavid Head and Beenaman the Pointagare Group attains its maximum exposed thickness of 340 m. The Group displays a pervasive, anastomosing, pressure solution Variscan cleavage. A Caledonian cleavage is absent. It was not deformed by the Late Caledonian deformation that probably culminated in Emsian time (according to Soper *et al.* (1987) and McKerrow (1988)). Structural evidence, the Group's stratigraphic position and the biostratigraphical age date from the base of the overlying Carrigduff Group place the Group largely in the Mid-Devonian time. On these grounds alone, the Pointagare Group probably ranges in age from latest Emsian to early Frasnian time (Fig. 5). Evidence for a more restricted age range from latest Emsian time to 385 Ma from the Enagh Tuff isotopic age date (Williams *et al.* 1997) is persuasive (Richmond 1998) (Fig. 5).

Dodd (1986) stated that there is no biostratigraphic, provenance or palaeocurrent evidence to suggest that the Pointagare beds are equivalent to the Caherbla Group of south Dingle. A more detailed analysis of the Pointagare Group, however, reveals much contrary evidence. Richmond (1998) illustrated that the Pointagare Group is in fact stratigraphically equivalent to the Caherbla Group on the basis of structure, stratigraphical position, biostratigraphy, facies, petrography, provenance, palaeoflow directions, palaeowind directions and clast lithotypes (Figs 13 and 14). The very elegant work of Dodd (1986) on the contiguous Caherbla Group is invaluable. The Pointagare Group is retained as a single entity in the Northwest Dingle Domain for the present. However, it is hoped that further biostratigraphical work and detailed geochemical studies on clast lithotypes will result in the abolition of the Pointagare Group and its inclusion into the Caherbla Group of south Dingle.

The Beenaman Conglomerate and Beenmore Sandstone Formations

The Beenaman Conglomerate Formation overlies the Smerwick Group with angular unconformity and constitutes a distinct, greyish red–purple to grey, conglomeratic suite (Figs 13 and 15). The Formation was described by Dodd (1986) as a fining-upward succession comprising sandy conglomerates, pebbly sandstones and sandstone sheets deposited in a low-sinuosity braided river or alluvial fan that dispersed to the southeast. He noted that the Beenaman Conglomerate Formation on the western flank of Beenaman contains quartz-mica schist clasts. Unfortunately, he did not realize their significance.

The upper boundary with the overlying Beenmore Sandstone Formation is placed at the base of the lowermost aeolian cross-stratal sets. The Beenaman Conglomerate Formation thins along strike to the ENE (Fig. 13). It attains a thickness of 200 m between Sybil Head and Been Dermot, but thins to just 60 m at Deelick Point (Fig. 3). The type section for the Formation was measured at Ballydavid Head (Richmond 1998). Additional small sections were measured along strike in conjunction with clast lithotype quantification and palaeocurrent measurements.

The Beenmore Sandstone Formation (Beenmore Member of Horne (1974)) was first described by Horne (1971), who identified

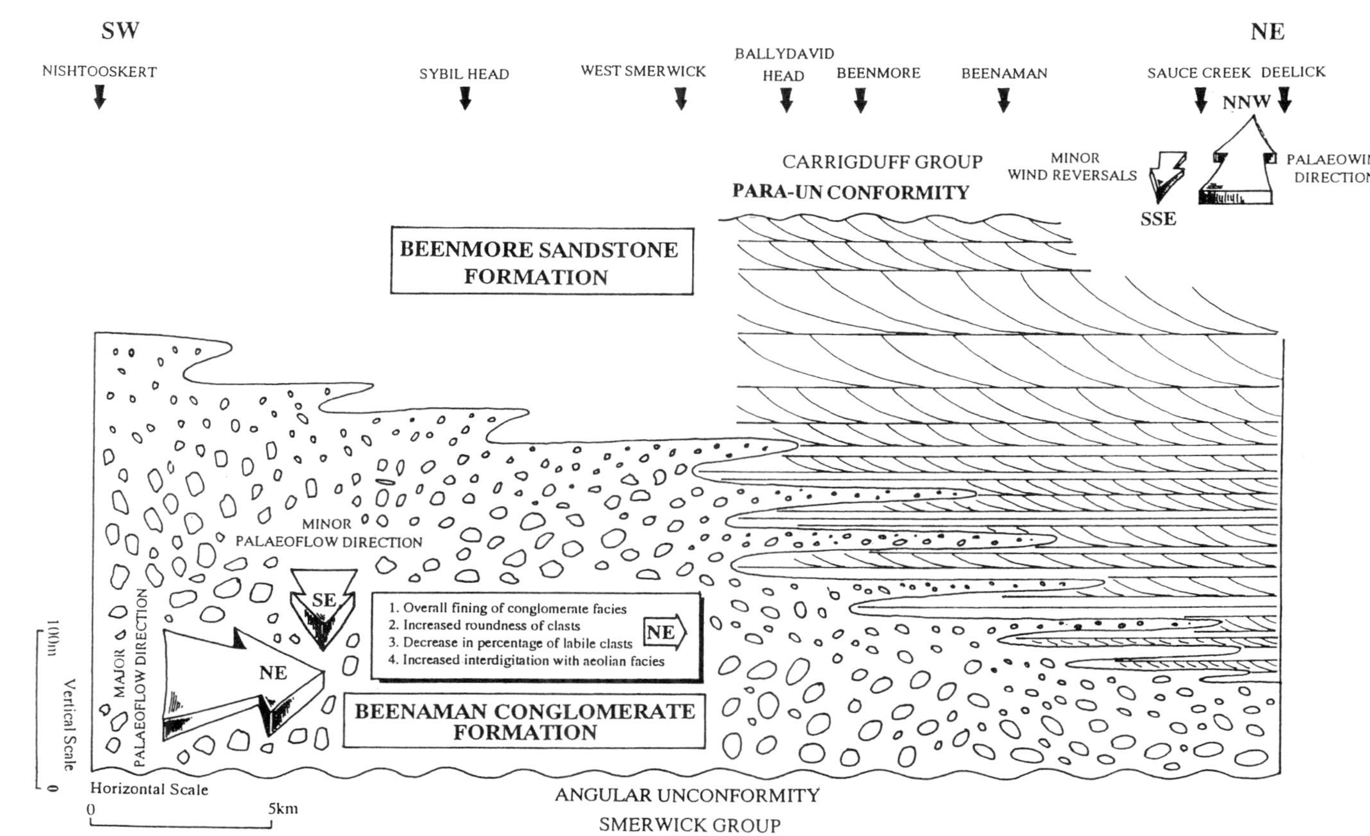

Fig. 13. Schematic lateral facies variation diagram for the Pointagare Group. Sketch parallel to predominant palaeoflow.

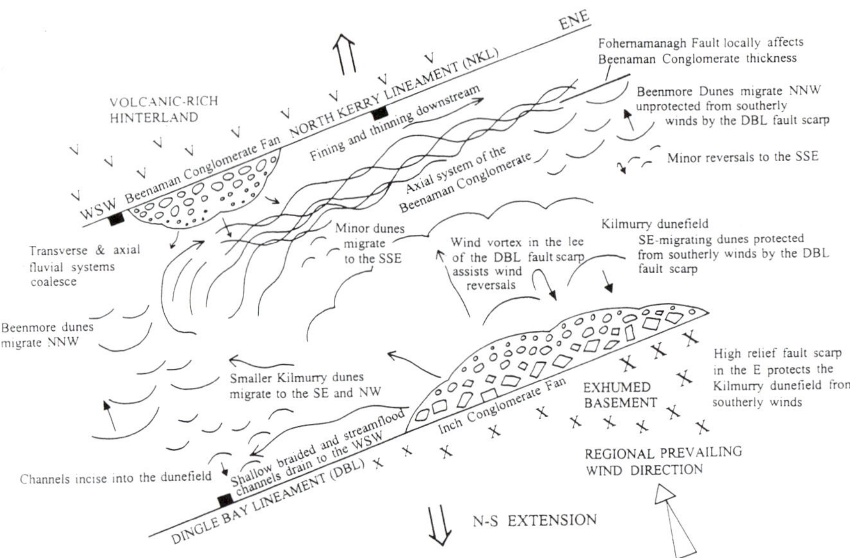

Fig. 14. Palaeogeographical reconstruction for Pointagare and Caherbla Group sedimentation (not to scale).

Fig. 15. Typical petromict conglomerate in the lower part of the Beenaman Conglomerate Formation at Ballydavid Head (Pointagare Group). Clast lithotypes include vein quartz, quartz arenite, metaquartzite, chert, jasper, granite, sandstones–siltstones and volcanic rocks. Rounded to well-rounded quartz-ica schists are conspicuous (right of centre). Lens cap is 6 cm in diameter.

aeolian cross-stratification both in the Beenmore Sandstone Formation exposed at Beenaman (Fig. 16) and the Kilmurry Formation of the Caherbla Group of Horne (1974). A correlation between the two suites of aeolian sandstones was not attempted because of incorrect stratigraphic correlation. Horne (1971) interpreted the thick cross-stratal sets as deposits of aeolian sand dunes. Dodd (1986) further examined the Beenmore Sandstone Formation at Beenaman and

Fig. 16. Aeolian dune cross-bedded set bounded by planar, bedding-parallel surfaces in the Upper Member of the Beenmore Sandstone Formation, Pointagare Group, north side of Sauce Creek (Fig. 3f). The foresets are asymptotic to the basal surface, which truncates underlying wet and dry interdune sheets. Minor modification surfaces are common laterally. The sequence dips gently to the NNW. Hammer shaft is 36 cm long.

found the large-scale, cross-stratified aeolian dune deposits to be interbedded with variably developed, wet and dry interdune deposits and dry sandsheets. The present analysis of the Formation at additional sections along strike and the integration of these results with the underlying Beenaman Conglomerate Formation have led to the development of a new model, which has fundamental basin-wide implications.

The Beenmore Sandstone Formation is exposed between Ballydavid Head and Brandon Point (Figs 3 and 13) and attains a maximum exposed thickness of 256 m at the type section at Beenaman (Fig. 13). Additional sections through the Formation were measured at Beenmore, Ballydavid Head, Sauce Creek and north of Traban (Fig. 1). Gently dipping sections discovered around Sauce Creek and north of Traban lend themselves to lateral profiling of fluvial and aeolian architectures (Fig. 15). The Formation is divided into two members, the Lower and Upper Members, which are equivalent to Members A and B of Dodd (1986) (Fig. 4). The boundary between the Upper and Lower Members is demarcated by the very marked switch from fluvially-dominated to aeolian-dominated sedimentation. Between Ballydavid Head and Beenaman the Carrigduff Group overlies the Upper Member. The decrease in thickness of the Beenaman Conglomerate Formation along strike to the ENE is accompanied by an increase in thickness of the overlying Beenmore Sandstone Formation (Fig. 13). In addition, on the southern limb of the Brandon Point Anticline the Beenmore Sandstone Formation directly overlies the unconformity with the underlying Smerwick Group (Figs 3f and 7). The Beenmore Sandstone Formation onlaps onto the unconformity to the SE as the underlying conglomerate thins and pinches out to the SE.

Temporal and spatial facies variations

The Beenaman Conglomerate Formation above the basal facies association forms a broadly fining-upward succession that locally overlies the unconformity with the underlying Smerwick Group. It records rapid progradation followed by protracted retrogradation of the alluvial–fluvial system almost certainly in response to tectonism. Following uplift and erosion of the source area(s) and associated rapid creation of accommodation space on basin subsidence, the alluvial fan–braided stream(s) prograded basinwards. The alluvial–fluvial system subsequently retreated in response to a decrease in tectonic activity that reduced the slope gradient, sediment supply and discharge. Fining-upward and coarsening-upward facies associations reflect punctuated tectonism. It is envisaged that the primary control on fluvial sedimentation in this active basin-margin setting is allocyclic and tectonic. The Beenaman Conglomerate Formation thins from a maximum thickness of 200 m in the WSW to 60 m in the ENE (Fig. 13). It also loses the basal facies association and broadly fines on traversing along strike to the ENE. In addition, spatial variations in clast lithotypes also take place along strike. These trends are consistent with an axial component to palaeoflow along the basin length.

Latest Ludlow to late Emsian, Late Caledonian, transpressional, intra-suture basin development. Lower ORS, Dingle and Smerwick Group sedimentation.

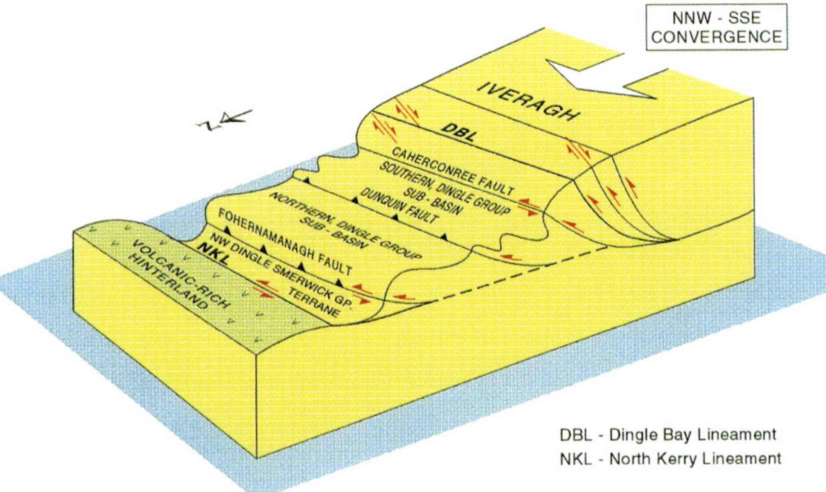

Latest Emsian to ?385 Ma, post - Caledonian, extensional basin development. Middle ORS, Caherbla and Pointagare Group sedimentation.

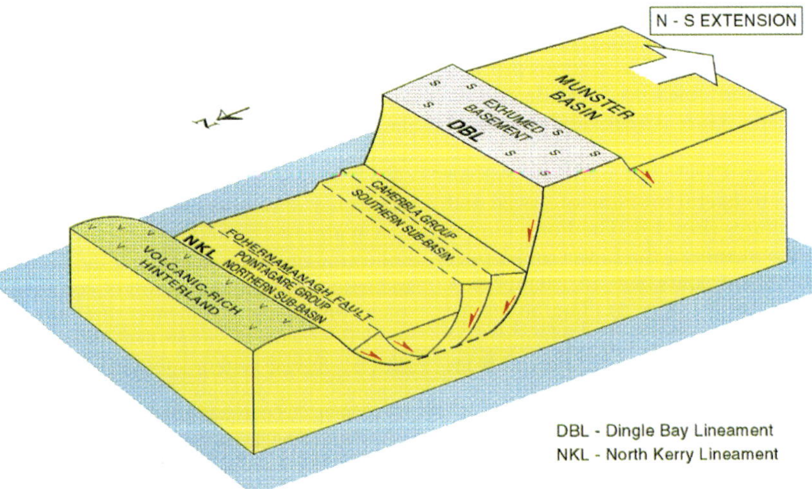

The Beenmore Sandstone Formation forms a broadly drying-upward sequence capped by a wetting-upward sequence. This cyclicity exhibits a strongly asymmetric profile with a drying-upward bias. This formation-scale temporal variation is of second-order frequency and records the abandonment of an alluvial fan–braidplain and subsequent overstep of an aeolian dunefield. This scale of cyclicity records protracted advance of the dunefield followed by its late-stage rapid retreat. These phases of erg advance and erg retreat in the Beenmore Sandstone record either the actual migration of the erg towards and away from the Northwest Dingle Domain or the expansion and contraction of a well-developed erg located to the southwest, respectively.

The Lower Member of the Formation exhibits an increase in the proportion of aeolian to fluvial facies associations on traversing along strike to the ENE over a distance of 33 km (Fig. 13). The aeolian facies also change their internal make-up. The thickness and abundance of interdune and aeolian sandsheet deposits is increased in the ENE. This is accompanied by a reduction in aeolian dune set thicknesses. This is indicative of a decrease in fluvial influx in the ENE (which is consistent with the axial component to the palaeoflow). More erg-marginal facies probably developed in the ENE in response to decreased sediment supply from the fluvial system, whereas to the WSW, larger, although often solitary, simple aeolian dunes formed on channel margins as sediment supply from the fluvial system was greater.

The Upper Member of the Formation preserves an increase in proportion of the central-erg deposits along strike towards the WSW. The most proximal fluvial deposits are also located in the WSW. This fluvial sand was an important sediment source for the aeolian system. Thus the location of the dunefield was probably largely controlled by sediment supply. Hence its maximum development is spatially coincident with the most proximal fluvial deposits. In addition, sand also tended to accumulate in areas already covered by sand.

The palaeoflow and palaeowind directions recorded in the Pointagare Group play an extremely important role in establishing a basin-scale depositional model for the Group. The Beenaman Conglomerate Formation possesses a polymodal palaeocurrent dispersion with a vector mean indicating net drainage to the ESE. There are major flow components to the ESE, ENE and SE, in decreasing order of magnitude. Subsidiary flow is also indicated to the NE and to the NNW. Fluvial facies in the Lower Member of the Beenmore Sandstone indicate a net drainage direction to the ESE. However, the palaeocurrent dispersion is more strongly bimodal than in the underlying Conglomerate Formation. The dominant flow components are to the S–SSE, and ENE. Their resultant palaeoflow vector is directed towards the ESE. The increase in bimodality of palaeoflow within the mixed fluvial–aeolian suite of the Lower Member of the Beenmore Sandstone Formation is attributed to the funnelling of flows along interdune corridors. During Beenmore Sandstone sedimentation the reduction in fluvial influx from the northwest and the onset of aeolian dune construction resulted in an increase in the influence of the aeolian landscape on fluvial drainage patterns. The dunefield relief became capable of diverting the fluvial drainage. This is a phenomenon apparent in many modern dunefields (Glennie 1970; Dodd 1986; Langford 1989; Langford & Chan 1989).

Aeolian facies in the Lower Member of the Beenmore Sandstone Formation indicate a polymodal palaeowind regime. The dominant wind direction recorded has a low dispersion and is directed towards the NNW. There are also significant wind component directions to both the NNE and SSE. The reversals in wind direction to the SSE are recorded in all sections along strike through the Lower Member. The

Fig. 17. Late Silurian to Mid-Devonian basin evolution in the Dingle Peninsula. (**a**) Late Ludlow–?late Emsian, late Caledonian intrasuture basin development. Deposition of the Dingle and Smerwick Groups and the Coosgorrib Conglomerate Formation. Syndepositional, late Caledonian faulting and folding in a sinistral transpressive regime. Northward propagation of thrusts rooted in the Dingle Bay Lineament; development of a positive flower structure. Northward-propagating basin inversion in ?late Emsian time, Acadian deformation and docking of the Smerwick Group. (Substantially modified after Todd (1989a).) (**b**) After basin inversion in ?late Emsian times, the Smerwick and Dingle Basins underwent extensional collapse generating a north–south extensional regime. The post-Caledonian, pre-rift sediments of the Caherbla and Pointagare Groups were deposited between late Emsian time and 385 Ma. The Dingle Bay and North Kerry Lineaments were activated as normal faults. The Dingle Block resembled a broad graben probably disrupted by intrabasinal faults. The Pointagare Group 'stitched' the Smerwick Group on to the Dingle Block. The Dingle Bay Lineament formed a ridge of exhumed basement which separated the Dingle Block from Iveragh to the South, where an embryonic Munster Basin began to form in the very latest Emsian.

Upper Member displays a strong cluster in palaeowind direction to the NNW. Minor, but significant, components are also recorded directed towards the NNE and SSE. Minor reversals in wind direction towards the SSE are recorded in all sections along strike, save at Beenaman and Beenmore.

The clast lithotypes encountered within the Beenaman Conglomerate Formation include vein quartz, quartz arenite, metaquartzite, jasper, sandstone, volcanic rocks, granite or granophyre, chert, quartz–mica schist and gneiss (Fig. 15). The abundance of conspicuous gneiss and quartz–mica schist clasts is a feature exclusive to this particular Formation in the Northwest Dingle Domain. The proportions of the constituent clast lithotypes are seen to vary both temporally and spatially within the Formation. On traversing along strike from WSW to ENE and ascending the Formation, the percentage of volcanic and metamorphic, gneiss and quartz–mica schist clasts broadly decreases (over a distance of 33 km) (Fig. 13). This is accompanied by an increase in the percentage of resistate clasts, namely vein quartz, quartz arenite, metaquartzite and chert. The clasts vary in roundness from angular to well rounded, with the exception of almost invariably rounded to well-rounded quartz–mica schists (Fig. 15) and subrounded to rounded gneisses. All clast lithotypes also display an increase in roundness along strike to the ENE.

Depositional synthesis

The Pointagare Group is inferred to be largely of Mid-Devonian age. It forms a fluvial–aeolian succession that documents the transition from an alluvial fan and braided river environment to a fully developed dunefield. The basal Beenaman Conglomerate Formation records debris flow, streamflood and sheetflood processes in a perennial, medial alluvial fan to distal braided stream environments (Fig. 14). The net ESE palaeodrainage is the result of an ENE-flowing, axial braided stream coalescing with a transverse alluvial fan, which shed erosional detritus to the SE from a hinterland located in the NW. Minor, opposing drainage to the NE and NNW indicates that a small sediment influx also issued from the south. The abundance of quartz–mica schist and gneissic clasts and a major, axial, fluvial trunk system are features unique to this Group in the Northwest Dingle Domain. They are suggestive of a very different source terrane. Quartz–mica schists in particular are very labile clasts. Hence their high degree of roundness. They are first-cycle clasts. An axial component to palaeoflow is consistent with the decrease in labile clast lithotypes and concomitant increase in resistate clast lithotypes and roundness of all clast lithotypes in a downstream direction to the ENE. In addition, the axial component to palaeodrainage is reflected in the thinning and fining of the Formation coupled with the loss of the basal facies association in the downstream direction. The Beenaman Conglomerate Formation records rapid fan progradation (not recorded in the ENE) in response to increased tectonism and uplift of a source terrane(s) (Fig. 14).

The North Kerry Lineament probably represents the northern, basin-bounding fault. It was reactivated during Pointagare Group times as a result of the onset of a north–south extensional regime. The NKL formed an ENE–WSW-trending normal, listric fault downthrowing to the south and generating a graben basin morphology. The active extensional basin margin setting and transverse alluvial fan and axial fluvial drainage coupled with the onset of aeolian deposition in Beenmore Sandstone times are comparable with the rift-basin sedimentation described by Blair (1987) (where aeolian sedimentation takes the place of lacustrine deposition). The cyclic megasequences of Blair (1987) are of the order of hundreds of metres thick. They are attributed to changes in basin subsidence related to tectonism. The third-order, fining- and coarsening-upward cycles in the Pointagare Group are of a similar scale and also record the response of sedimentary architecture to punctuated tectonism. Fourth-order cycles are also present and result from climatically controlled or storm-related channel abandonment and migration processes.

The Beenmore Sandstone Formation records punctuated retrogradation of a distal alluvial fan and/or braided river, coupled with the advance of a dunefield in response to the reduced basin-margin relief that followed basin generation and alluvial fan sedimentation. The thickening of the Formation to the ENE is accompanied by the thinning of the underlying Conglomerate Formation. As the Beenmore Sandstone Formation onlaps onto the unconformity with the underlying Smerwick Group along strike to the ENE, the Beenaman Conglomerate thins to the SSE (Fig. 13). This suggests that the axis of maximum subsidence with respect to the Fohernamanagh Fault lay to the NNW during the acme of fluvial deposition. The dunefield that later advanced from the south covered the area to the SSE, which in the ENE extent of the basin must have been an area of relative uplift during Beenaman Conglomerate times. This area of upwarp was probably related to the Fohernamanagh Fault. Although this fault remained largely inactive during Pointagare Group times (as the Beenaman Conglomerate oversteps it along strike to the WSW), it still influenced sedimentation patterns (Fig. 14). Thus the surface expression of fault activity can have a strong local effect on river aggradation and distribution of fluvial facies.

The palaeodrainage recorded in the Lower Member of the Beenmore Sandstone is strongly bimodal. Palaeoflow components to the south, SSE and ENE can be identified. The increase in bimodality of palaeodrainage within this part of the succession compared with the underlying conglomerate is probably due to a reduction in fluvial influx from the NW coupled with embryonic dunefield development. The axial system and the axis of maximum subsidence were probably displaced away from the footwall located to the NNW during initial footwall uplift. As incremental footwall uplift decreased, the axial system shifted its drainage back towards the footwall (Leeder & Gawthorpe 1987). Hence with decreased fluvial discharge and stream power, the more distal floods were more readily diverted by the aeolian landscape and funnelled along WSW–ENE-oriented interdune corridors.

Subsequently, an aeolian dunefield rapidly encroached from the SSE and buried the abandoned alluvial fan–braidplain: first, with a locally preserved veneer of fore-erg medial erg-margin sediments (Porter 1986); second, with a thick blanket of inner erg-margin and central-erg deposits; third, with a medial erg-margin back-erg cover. The erg-margin sediments owe their preservation to rapid burial and basin subsidence (Porter 1986). Erg-margin sandsheets were in low abundance because of the high sand supply, which favoured rapid dune construction as opposed to sandsheet formation. The dunefield comprised transverse-barchanoid dunes migrating predominantly to the NNW (Fig. 14). With increased sand supply and decreased fluvial influence the dunefield relief increased. First, simple dunes and later compound dunes (draa) marched towards the NNW, resulting in a dunefield which banked-up downwind against the SW-facing, basin-bounding NKL fault scarp in the northwest. Adjacent to the fault scarp maximum local basin subsidence and a drop in sand drift potential resulted in a thick accumulation of aeolian dune sand. The dunefield was best developed in the WSW because of either greater basin subsidence in the WSW or simply a higher supply of sand from the more proximal, alluvial fan and braidplain sediments of the underlying fluvial system in the WSW. The increase in dune size and compound dune bedforms with time suggests that the dunefield became increasingly removed from an upwind sand source. This may have simply been due to the migration of the dunefield to the NNW away from the upwind sand source or else may reflect the southward migration of the sand source (retrogradation of a fluvial system in the south).

The predominant NNW dune migration direction is consistent with the prevailing wind direction thought to be in operation during Mid-Devonian times. Minor palaeowinds directed towards the NNE record the deflection of winds by local basin topography changes or the mountain front in the NNW, or else were produced by the spiralling of winds along interdune corridors. The reversed dunes owe their preservation to rapid burial during periods of high basin subsidence either by aeolian dune or fluvial deposits. Late-stage erg retreat probably heralded rejuvenation of footwall uplift in the NW in response to north–south extension. Rapid basin subsidence to the south associated with north–south extension probably also induced dunefield retreat to the south in the north.

The Pointagare Group is the only sequence of rocks in the Northwest Dingle Domain to overstep the terrane boundary that is demarcated by the Fohernamanagh Fault (Fig. 1). Structural disposition limits its southernmost exposure to the north side of Traban. However, the depositional model suggests that the Pointagare Group was deposited further to the south and hence that here it was subsequently uplifted and eroded. Rapid subsidence of the Northwest Dingle Domain in response to increased north–south extension and the generation of the isolated half-graben in which the Carrigduff Group alluvium accumulated probably facilitated the preservation of the Pointagare Group in the northwest. To the southeast, the Central Dingle Domain became an area of footwall uplift situated to the north of the Dingle Bay Lineament (and in particular north of the Caherconree–Minard Head fault zone) (Fig. 17). Pointagare Group sediments deposited over the Central Dingle Block were removed in Late Devonian times.

Three questions remain unanswered by this model. First, what was the age of the source terrane that provided the high-grade metamorphic clasts? Second, where did the axial system originate? Third, what topographic feature caused the reversals in wind direction? The answers lie in the Caherbla Group of southeast Dingle.

Palaeogeographical reconstruction for the Pointagare and Caherbla Groups

The Caherbla Group crops out in the southeast of the Peninsula in an elongate, NE–SW-trending stretch of ground between Minard Head and Derrymore Glen. The Group was established by Horne (1974), who partitioned the sequence into the Inch Conglomerate Formation and the Kilmurry Formation. The Kilmurry Formation both interfingers with and overlies the Inch Conglomerate Formation. The Caherbla Group thins to the northwest such that only the lower portion of the Group is present to the north of the Caherconree Fault (Horne 1974). The pinch-out of the Group to the NW is the result of uplift and erosion of the Group in the NW during footwall uplift related to Late Devonian extension.

The Inch Conglomerate Formation is characterized by red–purple to grey, predominantly thickly bedded, massive, breccio-conglomerates containing distinctive metamorphic clasts, up to boulder grade, within a red sandy matrix. The clast types include schists and gneisses, together with subordinate vein quartz, chert, jasper, quartz–tourmaline, phyllite (Horne 1970, 1974, 1975), granite, quartzite and pegmatite (Capewell 1951; Dodd 1986). A comprehensive listing of clast lithotypes is documented in Todd (this volume).The clasts were derived from a positive ridge of exhumed basement, in the form of a northward-facing fault scarp (Horne 1970, 1975; Dodd 1986) that was generated by the transpression and uplift along the Dingle Bay Lineament, which accompanied inversion of the Dingle Basin during Late Caledonian deformation (Todd *et al.* 1988*a*; Todd 1989*a*). The ENE–WSW-trending fault scarp lay some 4–5 km to the south of Inch and was coincident with the postulated trace of the Dingle Bay Lineament. Erosion and denudation of the fault scarp spawned one or more alluvial fans that dispersed northwards in the form of streamfloods and sheetfloods recorded by the Inch Conglomerate Formation (Horne 1970, 1975; Dodd 1986; Todd *et al.* 1988*a*). The Inch Conglomerate Formation fines northwestwards away from the metamorphic source terrane (Horne 1975; Dodd 1986). The fining-upward nature of the Inch Conglomerate suggests that subsequent to initial progradation northwards, sediment supply was reduced and fan abandonment took place. The Inch fan was overstepped by the Kilmurry Formation (Dodd 1986; Todd *et al.* 1988*a*).

The Kilmurry Formation is a grey to red–purple, pale yellow–orange, fluvial–aeolian sandstone suite. Large-scale cross-stratification conspicuous in this Formation was first ascribed to an aeolian origin by Horne (1971). In eastern outcrops of the Kilmurry Formation aeolian deposits record compound, transverse-barchanoid dune migration to the SE (Dodd 1986). Western outcrops reveal that dunes migrated both to the SE and the NW. In the west, aeolian dune deposits are interbedded with fluvial units of both shallow braided and streamflood channels, which indicate a palaeoflow to the WSW (Horne 1975; Dodd 1986). It is apparent that in the west channels were capable of incising into the dunefield and flow was funnelled along interdune corridors and/or along the dunefield margin (Dodd 1986). Fluvial incursions were able to breach the dunefield in the west, where it was less well developed and of lower relief than in the east (Dodd 1986; Todd *et al.* 1988*a*, 1990).

The southeasterly migrating dunes of the Kilmurry Formation are at odds with the wind regime indicated by other aeolian deposits of the southern British Isles, which record a regional prevailing wind direction towards the NW. The anomalous wind directions recorded in the Kilmurry Formation can be accounted for by the presence of the postulated fault scarp situated to the south (Dodd 1986). This fault scarp formed an orographic barrier. In the eastern area the fault scarp was sufficiently high to protect the sand dunes from the influence of southerly winds. The Kilmurry dunefield extended at least 3 km north of the fault scarp, suggesting that the scarp was at least 500 m above the basin floor (based on modern analogues in Oman) (Dodd 1986; Todd *et al.* 1988*a*). In addition, a wind vortex in the lee of the scarp during periods of southerly winds may have assisted dune migration towards the south and southeast (Dodd 1986).

The Pointagare and Caherbla Groups share the following characteristics: they both are of Mid-Devonian age and occupy a similar stratigraphic niche; they were only affected by Late Devonian extension and Variscan deformation; they are unconformable on Lower Old Red Sandstone sequences; they both show a marked thinning towards the central Dingle Peninsula, where indeed they are absent; they both contain first-cycle, high-grade, metamorphic basement clasts, in particular quartz–mica schists and gneisses; the sandstones are petrographically similar and both contain mica within aeolian dune, interdune and fluvial sediments; they both record alluvial fan–braided stream retrogradation and subsequent burial by an erg; the style of fluvial sedimentation is similar and is reconcilable with a proximal–distal relationship between the Inch and Beenaman Conglomerate systems; they demonstrate a similar transition between alluvial fan–braided stream sedimentation and central-erg deposition accompanied by the change from perennial to ephemeral-fluvial deposition; a pronounced influence of aeolian topography on fluvial drainage systems is evident in both groups; they both display axial and transverse palaeodrainage systems; the aeolian dune and interdune facies are highly comparable and a similar scale of aeolian dune cross-stratification is observed; they both suggest the presence of a topographic high to the south protecting dunefields in the north.

These shared characteristics are reconcilable with contemporaneous deposition of the Pointagare and Caherbla Groups within adjacent, hydrologically-interconnected sub-basins (Fig. 14). The Beenaman Conglomerate axial system is a distal equivalent to the Inch Conglomerate fan; hence the very marked decrease in first-cycle clasts and their increased rounding in the Beenaman Conglomerate compared with the proximal, breccio-conglomerates of the Inch Conglomerate. A swing in palaeodrainage is invoked from that of the Inch fan, dispersing to the north and WSW, to the axial system in the Northwest Dingle Domain flowing to the ENE. The palaeodrainage was diverted by and directed along the pre-existing, Caledonian ENE–WSW structural grain. The swing from transverse to axial drainage could have been induced by basin tilting towards the ENE or simply structurally determined changes in basin morphology. Small sediment fluxes from the south were recorded by the minor NE and NNW palaeodrainage components. The transverse to axial swing in palaeodrainage bears a close resemblance to that recorded in the Middle Devonian Malbaie Formation of Eastern Canada (Rust 1984). The axial and transverse components of the Malbaie Formation are recorded in specific gravel (transverse) and sand (axial) dominated facies associations, between which there is little mixing. In the Pointagare Group, particularly in Beenaman Conglomerate Formation times (although to a much lesser extent in Beenmore Sandstone Formation times), the axial and transverse systems coalesced with a high degree of mixing.

The dunefield in the Northwest Dingle Domain formed as a result of expansion of the Kilmurry dunefield that rapidly advanced from the south. It was for the most part not protected by the fault scarp to the south generated along the present-day Dingle Bay Lineament and hence the dominant winds were directed towards the NNW. This is consistent with the regional prevailing wind direction during Mid-Devonian times. In the Kilmurry dunefield to the south, particularly in the east where the fault scarp was highest, the dunes were protected from southerly winds and their southeasterly migration was assisted by wind vortices in the lee of the fault scarp. Minor wind reversals to the south and SSE are also recorded in the Northwest Dingle Domain. However, it is not plausible that the fault scarp along the Dingle Bay Lineament to the south was of sufficient relief to protect dunes in this domain from southerly winds. The present-day separation of the two domains is 19 km and taking into account subsequent Variscan shortening (estimated to be 50% by Todd (1989*c*)), their separation during Mid-Devonian times would have been of the order of 38 km. This suggests that the first-cycle clasts survived a transport distance of at least 38 km. The footwall-proximal alluvial fan deposits of the Inch Conglomerate Formation indicate that the relief on the Dingle Bay Lineament fault scarp would have been sufficient to protect the Kilmurry dunefield from southerly winds (Dodd 1986). It is envisaged that minor reversals in wind direction in the Northwest Dingle Domain were generated by a minor local upwarp between the two domains or else convection currents. An upwarp could have periodically protected the Northwest Dingle Domain from southerly winds and wind vortices in the lee of the upwarp may have assisted wind reversals. Such an upwarp may have been coincident with the Fohernamanagh Fault. An area of uplift can also be invoked by the rapid thinning of the Beenaman Conglomerate to the SE.

The Caherbla and Pointagare Groups owe their preservation to rapid basin subsidence and burial adjacent to active normal faults (the Caherconree–Minard Head Fault and the North Kerry Lineament, respectively) (Fig. 1) in response to Late Devonian north–south extension. The Central Dingle Domain formed an uplifted fault block during early Late Devonian times such that the Middle Devonian sequence was removed and is not preserved in this Domain.

The Caherbla and Pointagare Groups illustrate the inherent tectonic control on fluvial drainage patterns, the location of erg accumulation, fluvial–aeolian facies distribution and ultimately sequence preservation. This new extensional basin model for the first time unites stratigraphic units from the Northwest Dingle Domain with those of south Dingle.

Conclusions

The Dingle Peninsula preserves a sedimentary record from Early Ordovician to Carboniferous time that reflects its complex structural evolution, inherited from its intra-arc and intra-suture setting. The sedimentary succession can be subdivided into (1) Early Caledonian, intra-arc (Annascaul Formation and Dunquin Group),

(2) Late Caledonian, intra-suture (Dingle and Smerwick Groups), (3) post-Caledonian, pre-rift (Caherbla and Pointagare Groups), (4) syn-rift (Carrigduff and Ballyroe Groups), and (5) post-rift (Slieve Mish Group) sediments (Table 1). The basin infill-architecture of the Late and immediately post-Caledonian phases only are summarized below.

Late Caledonian, intra-suture sedimentation of the Dingle and Smerwick Groups (Fig. 17a)

Barrier island progradation recorded a conformable marine to non-marine transition and generated a hydrologically closed basin in which Lower Old Red Sandstone sedimentation began in latest Ludlow times (Boyd 1983; Todd *et al.* 1988*a*; Boyd & Sloan this volume). Differential subsidence during the initiation of the Dingle Basin resulted in marked facies and thickness variations across the Peninsula (Todd *et al.* 1988*b*; Todd 1989*a*). Sediment accumulation in an ephemeral lake was superseded by transverse ephemeral-fluvial, sand bedload, terminal fan deposition (Boyd 1983) and subsequent axial braided stream and transverse alluvial fan, sand and gravel bedload sedimentation coeval with sinistral transpression along the DBL (Todd *et al.* 1988*a*; Todd 1989*a*, *b*). Dingle Group sedimentation culminated in deposition in a laterally extensive sandy braided, axial river system, which continued into Lochkovian time according to the 411 Ma isochron date from the Cooscrawn Tuff Bed or at least into early Emsian time using biostratigraphical evidence (Fig. 5).

Smerwick Group and Coosgorrib Conglomerate Formation deposition took place between latest Ludlow and latest Emsian time, coeval with, or more likely shortly after, Dingle Group sedimentation in an isolated basin with its own microclimate. Sedimentation occurred in a small, hydrologically closed, elongate, asymmetric ENE-oriented basin developed adjacent to the NKL (Fig. 17a). The NKL had a component of down-to-the-south displacement, as a result of the onset of a simple north–south extensional regime or a local, sinistral transtensional regime such as in a classic pull-apart. The Coosgorrib Conglomerate Formation represents a sliver of ?footwall fan sediments locally preserved adjacent to the terrane boundary. The Smerwick Group records initial, 'low-seeking', axial ephemeral, sinuous stream sedimentation during a period of relative tectonic quiescence following uplift and erosion of the Coosgorrib Conglomerate. Increased tectonism and footwall uplift to the north of the NKL spawned transverse terminal fans that displaced and engulfed the axial fluvial system. Ephemeral, sand and gravel bedload, terminal fans or fan, in response to punctuated tectonism on the northern basin margin, issued into the basinal zone occupied by an erg. The palaeodrainage and palaeowind directions were concordant, both systems dispersing sediment to the SE. The presence of an orographic barrier on the southern margin of the Basin can be invoked by the reversal in wind direction towards the south and SSE. The barrier protected the Basin from southerly regional winds.

Post-Caledonian, pre-rift sedimentation of the Caherbla and Pointagare Groups (Fig. 17b)

Following uplift and inversion of the Dingle and Smerwick Group Basins and the initiation of a north–south extensional regime Middle Old Red Sandstone sedimentation of the Caherbla and Pointagare Groups took place most probably between latest Emsian time and 385 Ma (Eifelian time) (Fig. 5). The NKL and DBL were reactivated as normal faults with down-to-the-south and down-to-the-north displacements, respectively (Fig. 17b). Footwall uplift on the basin margins and collapse of the Dingle Basin caused hinterland stream incision and generation of prograding alluvial fans, which shed detritus to the south and north respectively into two adjacent sub-basins, hydrologically connected by an axial, braided river system (Fig. 14). First-cycle volcanic rocks were derived from a hinterland in the northwest and high grade metamorphics from the exhumed basement along the DBL (Fig. 17b). Sediment mixing of material derived from separate source terranes took place at the confluence of the transverse and axial river systems. Punctuated tectonism on the basin margins is reflected in the fluvial–aeolian sedimentary fill.

The NKL and DBL formed the basin-bounding faults (Fig. 17b). However, movement on intra-basinal faults would probably be reflected in sediment thickness and facies variations across the Peninsula had the Middle Old Red Sandstone not been largely subsequently removed. Following alluvial fan retrogradation and abandonment, on re-equilibration of the system and cessation of footwall uplift, alluvial fan and braided stream deposits were subjected to aeolian reworking and were ultimately buried by a large, fully developed dunefield with a width of up to 38 km. The dunefield influenced fluvial drainage patterns and was largely sourced from the neighbouring fluvial system highlighted by the presence of conspicuous mica in aeolian deposits. Simple and compound, transverse-barchanoid dunes in the southern sub-basin were largely protected from regional winds by the northerly facing fault scarp along the DBL, whereas in the northern sub-basin dunes were left exposed to the influence of southeasterly, regional prevailing winds. Wind vortices in the lee of the fault scarp along the DBL and perhaps also in the wind-shadow of intra-basinal highs may have enhanced and generated wind reversals to the southeast in the southern and northern sub-basins, respectively. Basin topography, controlled by the underlying inherited Caledonian structure, determined fluvial drainage patterns, erg development and ultimately fluvial–aeolian basin fill architecture and preservation.

The authors would like to thank R. R. Horne (Geological Survey of Ireland) for his interest and encouragement, and for establishing a geological template for Dingle research following his remarkable mapping in the early 1970s. The Dingle ORS research by J. D. Boyd, C. D. Dodd, S. P. Todd and R. J. Sloan provided key insights into the unravelling of the Northwest Dingle Domain. K. T. Higgs (University College Cork) supplied much field discussion and assistance in palynological dating of the sequences. The fieldwork support and research award to L.K.R. was provided by Amerada Hess UK Ltd, and preparation, drafting and manuscript costs were covered by Conoco (UK) Ltd. The people of Dingle are thanked for their kindness and hospitality. P. F. Friend and N. H. Trewin are also thanked for their review of this research and their valuable comments.

References

BLAIR, T. C. 1987. Tectonic and hydrologic controls on cyclic alluvial fan, fluvial, and lacustrine rift-basin sedimentation, Jurassic–lowermost Cretaceous Todos Santos Formation, Chiapas, Mexico. *Journal of Sedimentary Petrology*, **57**, 845–862.

BOYD, J. D. 1983. *Sedimentology of the Lower Dingle Group, southern Dingle Peninsula, southwest Ireland*. PhD thesis, University of Bristol.

—— & SLOAN, R. J. 2000. Initiation and early development of the Dingle Basin, southwest Ireland, in the context of the closure of the Iapetus Ocean. *This volume*.

CAPEWELL, J. G. 1951. The Old Red Sandstone of the Inch and Annascaul District, Co. Kerry. *Proceedings of the Royal Irish Academy*, **54**, 141–168.

—— 1965. The Old Red Sandstone of Slieve Mish, Co. Kerry. *Proceedings of the Royal Irish Academy*, **64B**, 165–175.

CLEMMENSEN, L. B., OLSEN, H. & BLAKEY, R. C. 1989. Erg-margin deposits in the Lower Jurassic Moenave Formation and Wingate Sandstone, southern Utah. *Geological Society of America Bulletin*, **101**, 759–773.

DIEMER, J. A. , BRIDGE, J. S. & SANDERSON, D. J. 1987. Revised geology of Kerry Head, County Kerry. *Irish Journal of Earth Sciences*, **8**, 113–138.

DODD, C. D. 1986. *Sedimentology of fluvio-aeolian interaction with geological examples from the British Isles*. PhD thesis, Plymouth Polytechnic.

FRIEND, P. F. 1978. Distinctive features of some ancient river systems. *In*: MIALL, A. D. (ed.) *Fluvial Sedimentology*. Canadian Society of Petroleum Geologists, Memoirs, **5**, 531–542.

FRYBERGER, S. G. & AHLBRANDT, T. S. 1979. Mechanisms for the formation of eolian sand seas. *Zeitschrift für Geomorphologie, N.F.*, **23**, 440–460.

GARDINER, P. R. R. & HORNE, R. R. 1972. The Devonian and Lower Carboniferous clastic correlatives of southern Ireland. *Geological Survey of Ireland Bulletin*, **1**, 335–366.

GLENNIE, K. W. 1970. *Desert Sedimentary Environments*. Developments in Sedimentology 14. Elsevier, Amsterdam.

GRADSTEIN, F. M. & OGG, J. 1996. The Phanerozoic time scale. *Episodes*, **19**, 3–6.

HARLAND, W. B., ARMSTRONG, R. L., COX, A. V., CRAIG, L. E., SMITH, A. G. & SMITH, D. G. 1990. *A Geological Time Scale 1989*. Cambridge University Press, Cambridge.

HERRIES, R. D. 1993. Contrasting styles of fluvial–aeolian interaction at a downwind erg margin: Jurassic Kayenta–Navajo transition, northeastern Arizona, USA. *In*: NORTH, C. P. & PROSSER, D. J. (eds) *Characterization of Fluvial and Aeolian Reservoirs*. Geological Society, London, Special Publications, **73**, 199–218.

HIGGS, K. T. 1999. Early Devonian spore assemblages from the Dingle Group, County Kerry, Ireland. *Bollettino della Societa Paleontologica Italiana*, **38**(2–3), 187–196.

HOLLAND, C. H. 1969. Irish counterpart of the Silurian of Newfoundland. *In*: KAY, M. (ed.) *North Atlantic Geology and Continental Drift*. American Association of Petroleum Geologists, Memoirs, **12**, 298–308.

—— 1987. Stratigraphical and structural relationships of the Dingle Group (Silurian), County Kerry, Ireland. *Geological Magazine*, **124**, 33–42.

HORNE, R. R. 1970. A preliminary re-interpretation of the Devonian palaeogeography of western County Kerry. *Geological Survey of Ireland Bulletin*, **1**, 53–60.

—— 1971. Aeolian cross-stratification in the Devonian of the Dingle Peninsula, Co. Kerry. *Geological Magazine*, **103**, 151–158.

—— 1974. The lithostratigraphy of the Late Silurian to early Carboniferous of the Dingle Peninsula, Co. Kerry. *Geological Survey of Ireland Bulletin*, **1**, 395–428.

—— 1975. The association of alluvial fan, aeolian and fluviatile facies in the Caherbla Group (Devonian), Dingle Peninsula, Ireland. *Journal of Sedimentary Petrology*, **45**, 535–540.

HOWELL, J. & MOUNTNEY, N. 1997. Climatic cyclicity and accommodation space in arid to semi-arid depositional systems: an example from the Rotliegend Group of the UK southern North Sea. *In*: ZIEGLER, K., TURNER, P. & DAINES, S. R. (eds) *Petroleum Geology of the Southern North Sea: Future Potential.* Geological Society, London, Special Publications, **123**, 63–86.

HUNTER, R. E. 1977*a*. Basic types of stratification in small aeolian dunes. *Sedimentology*, **24**, 361–387.

—— 1977*b*. Terminology of cross-stratified sedimentary layers and climbing-ripple structures. *Journal of Sedimentary Petrology*, **47**, 697–706.

—— 1981. Stratification styles in eolian sandstones: Pennsylvanian to Jurassic examples from the Western Interior U.S.A. *In*: ETHRIDGE, F. G. & FLORES, R. M. (eds) *Recent and Ancient Nonmarine Depositional Environments: Models for Exploration.* Society of Economic Paleontologists and Mineralogists, Special Publications, **31**, 315–329.

JUKES, J. B. & DU NOYER, G. V. 1863. *Explanation of sheets 160, 161, 171 and part of 172.* Memoir Geological Survey of Ireland, Dublin.

KELLY, S. B. & OLSEN, H. 1993. Terminal fans—a review with reference to Devonian examples. *Sedimentology*, **85**, 339–374.

KING, S. C. 1988. *Structural studies in the Dingle Peninsula, Ireland.* PhD thesis, Queens University, Belfast.

KOCUREK, G. & DOTT, R. H. 1981. Distinctions and uses of stratification types in the interpretation of eolian sand. *Journal of Sedimentary Petrology*, **51**, 579–595.

LANGFORD, R. P. 1989. Fluvial–aeolian interactions: Part I, modern systems. *Sedimentology*, **36**, 1023–1035.

—— & CHAN, M. A. 1989. Fluvial–aeolian interactions: Part II, ancient systems. *Sedimentology*, **36**, 1037–1051.

LAWRENCE, D. A. & WILLIAMS, B. P. J. 1987. Evolution of drainage systems in response to Acadian deformation: the Devonian Battery Point Formation, Eastern Canada. *In*: ETHRIDGE, F. G., FLORES, R. M. & HARVEY, M. D. (eds) *Recent Developments in Fluvial Sedimentology*. Society of Economic Paleontologists and Mineralogists, Special Publications, **39**, 243–252.

LEEDER, M. R. & GAWTHORPE, R. L. 1987 Sedimentary models for extensional tilt-block/half-graben basins. *In*: COWARD, M. P., DEWEY, J. F. & HANCOCK, P. L. (eds) *Continental Extensional Tectonics.* Geological Society, London, Special Publications, **28**, 139–152.

MCKERROW, W. S. 1988. Wenlock to Givetian deformation in the British Isles and the Canadian Appalachians. *In*: HARRIS, A. L. & FETTES, D. J. (eds) *The Caledonian–Appalachian Orogen.* Geological Society, London, Special Publications, **38**, 437–448.

MCKIE, T. & GARDEN, I. R. 1996. Hierarchial stratigraphic cycles in the non-marine Clair Group (Devonian) UKCS. *In*: HOWELL, J. A. & AITKEN, J. F. (eds) *High Resolution Sequence Stratigraphy: Innovations and Applications.* Geological Society, London, Special Publications, **104**, 139–157.

MEADOWS, N. S. & BEACH, A. 1993. Structural and climatic controls on facies distribution in a mixed fluvial and aeolian reservoir: the Triassic Sherwood Sandstone in the Irish Sea. *In*: NORTH, C. P. & Prosser, D. J. (eds) *Characterisation of Fluvial and Aeolian Reservoirs.* Geological Society, London, Special Publications, **73**, 247–264.

PARKIN, J. 1974. Silurian rocks of Inishvickillane, Blasket Islands, Co. Kerry. Scientific *Proceedings of the Royal Dublin Society*, **A5**, 277–291.

PORTER, M. L. 1986. Sedimentary record of erg migration. *Geology*, **14**, 497–500.

PRICE, C. A. & TODD, S. P. 1988. A model for the development of the Irish Variscides. *Journal of the Geological Society, London*, **145**, 935–939.

RICHMOND, L. K. 1998. *Fluvial-aeolian interactions and Old Red Sandstone basin evolution, Northwest Dingle Peninsula, Co. Kerry, S.W. Ireland.* PhD thesis, University of Aberdeen.

RUST, B. R. 1984. Proximal braidplain deposits in the Middle Devonian Malbaie Formation of Eastern Gaspé, Quebec, Canada. *Sedimentology*, **31**, 675–695.

SLOAN, R. J. 1991. *Mid-Silurian sedimentation and volcanism in Dingle, Southwest Ireland.* PhD thesis, University of Bristol.

SOPER, N. J., WEBB, B. C. & WOODCOCK, N. H. 1987. Late Caledonian (Acadian) transpression in northwest England: timing, geometry and geotectonic significance. *Proceedings of the Yorkshire Geological Society*, **46**, 175–192.

TAYLOR, K. S. 1994. *Ephemeral–fluvial sediments as potential hydrocarbon reservoirs.* PhD thesis, University of Aberdeen.

TODD, S. P. 1989*a*. *Caledonian tectonics and conglomerate sedimentation in the Dingle Basin, southern Ireland.* PhD thesis, University of Bristol.

—— 1989*b*. Role of the Dingle Bay Lineament in the evolution of the Old Red Sandstone of southwest Ireland. *In*: ARTHURTON, R. S., GUTTERIDGE, P. & NOLAN, S. C. (eds) *Role of Tectonics in Devonian and Carboniferous Sedimentation in the British Isles.* Yorkshire Geological Society Special Publications, 35–54.

—— 1989*c*. Stream-driven, high-density gravelly traction carpets: possible deposits in the Trabeg Conglomerate Formation, SW Ireland and some theoretical considerations of their origin. *Sedimentology*, **36**, 513–530.

—— 2000. Taking the roof off a suture zone: basin setting and provenance of conglomerates in the ORS Dingle Basin of SW Ireland. *This volume.*

——, BOYD, J. D. & DODD, C. D. 1988*a*. Old Red Sandstone sedimentation and basin development in the Dingle Peninsula, southwest Ireland. *In*: MCMILLAN, N. J. EMBRY, A. F. & GLASS, D. J. (eds) *The Devonian of the World, Vol. 2.* Canadian Society of Petroleum Geologists, Memoirs, **14**, 251–268.

——, WILLIAMS, B. P. J. & HANCOCK, P. L. 1988*b*. Lithostratigraphy and structure of the Old Red Sandstone of the northern Dingle Peninsula, Co.

Kerry, southwest Ireland. *Geological Journal*, **23**, 107–120.

——, BOYD, J. D., SLOAN, R. J. & WILLIAMS, B. P. J. 1990. *Sedimentology and tectonic setting of the Siluro-Devonian rocks of the Dingle Peninsula, SW Ireland.* A field excursion guide prepared for the Sediments 1990, 13th International Sedimentological Congress, Nottingham, UK.

WALSH, P. T. 1968. The Old Red Sandstone west of Killarney, Co. Kerry, Ireland. *Proceedings of the Royal Irish Academy*, **66B**, 9–25.

WILLIAMS, E. A., BAMFORD, M. L. F., COOPER, M. A. *et al.* 1989. Tectonic controls and sedimentary response in the Devonian–Carboniferous Munster and South Munster Basins, south west Ireland. *In*: ARTHURTON, R. S., GUTTERIDGE, P. & NOLAN, S. C. (eds) *Role of Tectonics in Devonian and Carboniferous Sedimentation in the British Isles.* Yorkshire Geological Society Special Publication, **6**, 123–141.

——, SERGEEV, S. A., STOSSEL, I. & FORD, M. 1997. An Eifelian U–Pb zircon date for the Enagh Tuff Bed from the Old Red Sandstone of the Munster Basin in NW. Iveragh, south-west Ireland. *Journal of the Geological Society, London*, **154**, 189–193.

——, ——, ——, —— & HIGGS, K. T. 2000. U–Pb zircon geochemistry of silicic tuffs and chronostratigraphy of the earlier Old Red Sandstone in the Munster Basin, SW Ireland. *This volume.*

Taking the roof off a suture zone: basin setting and provenance of conglomerates in the ORS Dingle Basin of SW Ireland

SIMON P. TODD

BP Amoco, Farburn Industrial Estate, Dyce, Aberdeen AB21 0PB, UK
(e-mail: toddsp@bp.com)

Abstract: The Late Silurian to Mid-Devonian Dingle Basin occupies a central position within the Iapetus Suture Zone of SW Ireland. The basin is believed to have formed from late Silurian times onwards as a product of sinistral transpression along several major faults within this suture zone. Conglomeratic sediments were deposited by moderately large gravelly fans shed into the basin from the NW (Glashabeg Formation) and the SE (Trabeg Conglomerate Formation). The systems fed into a large apparently through-going sandy, axial river (Slea Head Formation) that flowed towards the NE. The lateral, basin-margin systems were sourced from two disparate source terranes. To the NW lay a basic volcanic hinterland with some intermediate volcanic rocks and limestones, mudstones, sandstones and chert. During the Early Devonian time the SE drainage basin was underlain by sandstones, quartzites, phyllites and limestones probably intruded by a granite. Some of these lithologies can be found in outcrop in the pre-Dingle Group of the peninsula. Others need to be correlated with rocks of the oceanic terranes in the northern part of the suture zone. The southerly derived clasts have corollaries in the rocks of the Avalonian Leinster Terrane south of the suture. Following partial inversion of the Dingle Basin, the southerly hinterland was apparently further unroofed during mid-Devonian time, when the Inch Conglomerate Formation was deposited by alluvial fans shed northwards from a source area formed along the Dingle Bay Lineament. The Inch conglomerates are characterized by distinctive clasts of schist, gneiss, mylonite, tourmalinite and granite. The general picture of the Early Devonian deformation, intrusion by granite and unroofing of terranes currently partially exposed in central and southern Ireland within the Iapetus Suture Zone is largely consistent with clast lithotypes. However, some exploratory isotopic data indicate at least two possible vagaries in the interpretation. First, model Nd T_{DM} ages of Trabeg sedimentary clasts yield several results older than typical southern Iapetean or Avalonian crustal material. This suggests a complex history of sedimentary mixing of material across the developing Iapetus Ocean. Second, two of three Rb–Sr muscovite–whole-rock dates of Inch metamorphic clasts indicate Silurian ages. These data are similar to Rb–Sr dates derived from the Carnsore and Saltees granites in the Rosslare Terrane, perhaps extending the geographical spread of this Silurian deformation.

Since first proposed by Wilson (1966), the upper Precambrian to upper Devonian rocks of Ireland and Britain have generally been regarded as the products of the opening and closure of an Early Palaeozoic ocean (Dewey 1971). Early models (e.g. Dewey 1971) discussed the Caledonian orogen in terms of tectonic processes effected by orthogonal convergence between the two sides of the Iapetus Ocean. Later models, after recognition of a component of strike-slip within the belt (e.g. Phillips *et al.* 1976; Soper & Hutton 1984), have centred on the movement of tectonostratigraphic terranes within the Caledonides (e.g. Bluck 1985; Hutton 1987; McKerrow 1988).

The Old Red Sandstone (ORS), of upper Silurian to middle Devonian age, of Ireland and Britain was deposited during the final stages of convergence between the northern Laurentian continental margin and the southern Avalonian sub-continent which amalgamated during the Acadian Orogeny (Fig. 1; Soper *et al.* 1987; McKerrow 1988). The later part of the Irish ORS, late Devonian to early Carboniferous in age, records the opening phases of the Variscan cycle. The preserved fills of the lower ORS sedimentary basins developed within the Caledonian orogen may yield data concerning terranes now hidden by strike-slip dispersal, underthrust beneath other blocks, removed by overthrusting and

From: FRIEND, P. F. & WILLIAMS, B. P. J. (eds). *New Perspectives on the Old Red Sandstone.* Geological Society, London, Special Publications, **180**, 185–222. 0305-8719/00/$15.00

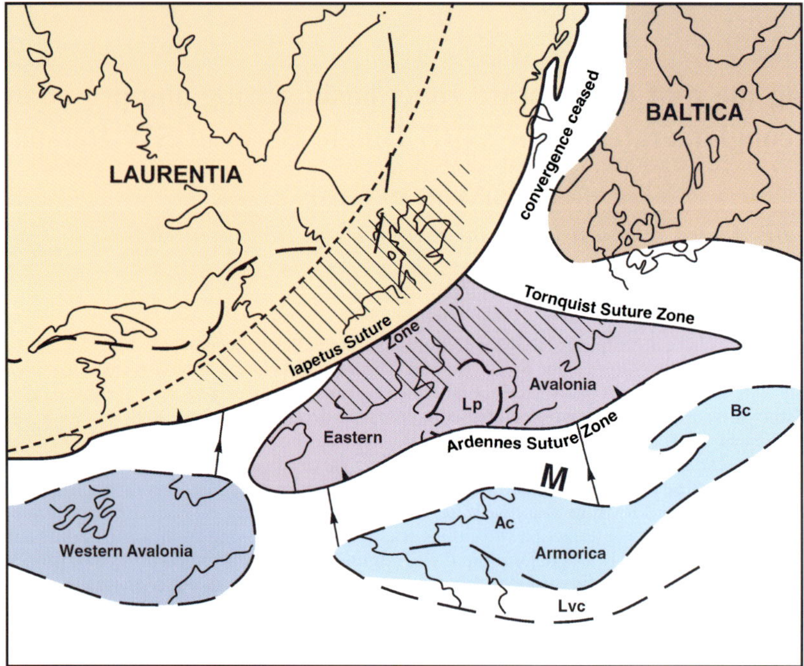

Fig. 1. Sketch of the palaeogeography of the British and Irish Caledonides during Early Devonian time (from Soper & Hutton 1984). Lp, London platform; Mec, Mid-European Caledonides; Bc, Bohemian craton; Ac, Armorican craton; Lvc, Ligurian-Vosgian cordillera.

erosion, or obscured by a cover of post-Caledonian rocks.

The Dingle Basin, which is preserved in the outcrops of the Dingle Peninsula, SW Ireland, was generated within the Iapetus Suture Zone formed by the final convergence between Avalonia and Laurentia (Figs 1 and 2; Todd *et al.* 1991). The ORS rocks that record this phase are the upper Silurian to middle Devonian Dingle and Caherbla Groups. Several lineaments, e.g. the Dingle Bay Lineament (Fig. 2), with a component of strike-slip, influenced sedimentation patterns within an overall transpressive regime (Todd 1989*a*; Todd *et al.* 1988*a*). Following inversion of the Dingle Basin, probably during Emsian time (Todd *et al.* 1988*a*), the conglomerates and sandstones of the Caherbla Group were deposited. Evidence for syn-sedimentary tectonic deformation during Dunquin, Dingle and Caherbla Group times is abundant, and together these groups provide a unique Irish record of the last Caledonian convergence that culminated in the Acadian Orogeny.

The aim of this paper is to describe how the rocks of the Dingle Peninsula, specifically the lower part of the ORS, record the final stages of the Caledonian cycle, the Acadian Orogeny, in the Iapetus Suture Zone. The paper both reviews and summarizes previous work and provides provenance data previously largely unpublished except in a field guide (Todd *et al.* 1990). Summaries of the stratigraphy including a discussion of newly published age data and interpretations for the Dingle ORS structure are first provided. Then the depositional settings and provenance of four conglomeratic formations in the lower part of Dingle's ORS succession are described. Essential to the understanding of the geology of the area is the information that can be gleaned from the provenance of these conglomerates. These data are used to construct the geology of the Early Devonian hinterlands to the Dingle Basin. This reconstruction is placed in the context of a regional tectonic model for the Siluro-Devonian evolution of the Iapetus Suture Zone in Ireland and Britain.

Stratigraphy of the Dingle and Caherbla Groups

Figure 3 depicts a new chronostratigraphic scheme for the Palaeozoic rocks of the area of the Dingle Peninsula, and Fig. 4 illustrates the lithostratigraphy of the rocks of the Dingle Group. Figure 3 includes a schematic

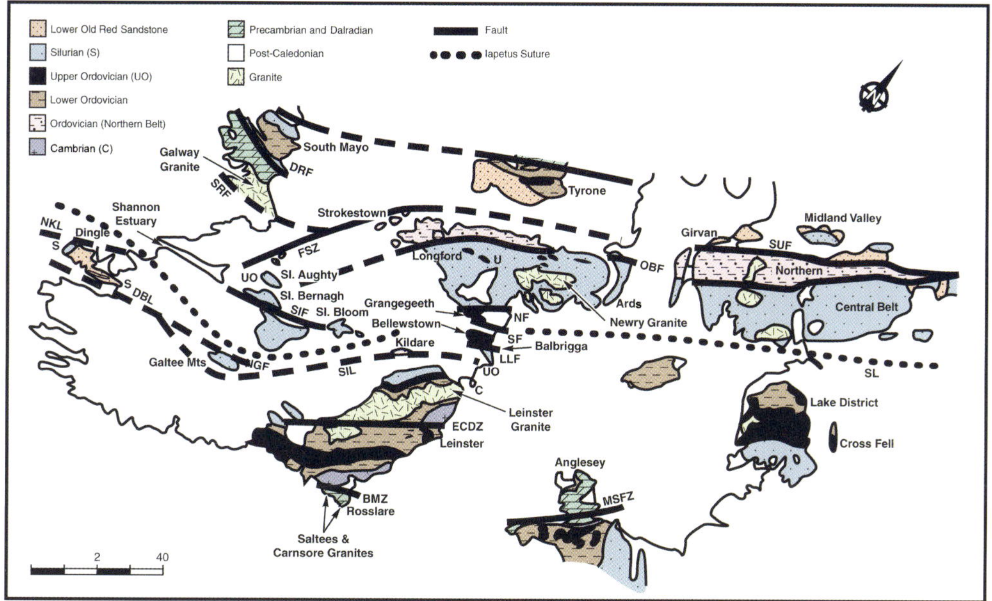

Fig. 2. Geological map of the Iapetus Suture Zone of Ireland and Britain (from Todd *et al.* 1991). BMZ, Ballycogly Mylonite Zone; DBL, Dingle Bay Lineament; DRF, Doon Rock Fault; ECDZ, East Carlow Deformation Zone; FSZ, Fergus Shear Zone; LLF, Lowther Lodge Fault; MSFZ, Menai Straits Zone; NF, Navan Fault; NGF, North Galtees Fault; NKL, North Kerry Lineament; OBF, Orlock Bridge Fault; SF, Slane Fault; SIF, Silvermines Fault, SIL, South Ireland Lineament; SL, Solway Line; SRF, Skird Rocks Fault, SUF, Southern Uplands Fault.

representation of the complex ORS stratigraphic architecture of the Iveragh peninsula region because the age of these rocks has a bearing on the interpretation of the ORS of Dingle (for more comprehensive accounts of the ORS south of Dingle see Williams *et al.* (this volume) and reference therein). The oldest proven rocks of the Dingle Peninsula are the Lower Ordovician shales, sandstones, and mélange of the Annascaul Formation (Todd 1989*c*; Todd *et al.* 2000). These are overlain unconformably by the fossiliferous Wenlock rocks of the Dunquin Group (Parkin 1976*a*; Todd *et al.* 2000). The Dunquin Group in the west of the peninsula is composed of fossiliferous shallow marine sediments and volcanic rocks of Wenlock and Ludlow age (Fig. 3). In at least two localities the Dunquin Group is seen to pass conformably into the red and green non-marine sediments of the Dingle Group (Holland 1987; Todd 1991). In other localities, however, the contact between the two groups is either faulted or unconformable (Todd *et al.* 1988*b*; Todd 1991). The base of the Dingle Group is nevertheless of upper Ludlow in age (Holland 1987). The lower Dingle Group is broadly an upward-coarsening megasequence, culminating in three lithostratigraphically contiguous conglomeratic formations, the Trabeg in the south, the Slea Head in the central area and the Glashabeg Formation in the north. These are overlain by extensive sandstones of the Ballymore Formation (Figs 3 and 4).

The age of the Dingle Group has been the subject of some debate in the past. An upper late Ludlow age for the base of the group is indicated by the conformable transitions from the Dunquin Group in the central Dingle Basin (Todd 1991). Van der Zwan (1980) reported an Emsian age for a spore assemblage collected from the Eask Formation at Dunmore Head (Fig. 5). Several workers, including Holland (1987), Todd *et al.* (1988*a*) and Todd (1989*a*) disputed the validity of the assemblage on the grounds of the surprisingly young age of the medial Dingle Group indicated by the assemblage, the surprisingly low organic maturity and the lack of success of attempts to recover repeat assemblages. Instead, arguments based around sedimentation rates of molasse were used to suggest that the preserved top of the Dingle Group was no younger than Emsian age and probably no younger than the Pragian age (e.g. Todd 1991).

New biostratigraphic evidence for the internal age of the Dingle Group has been recently

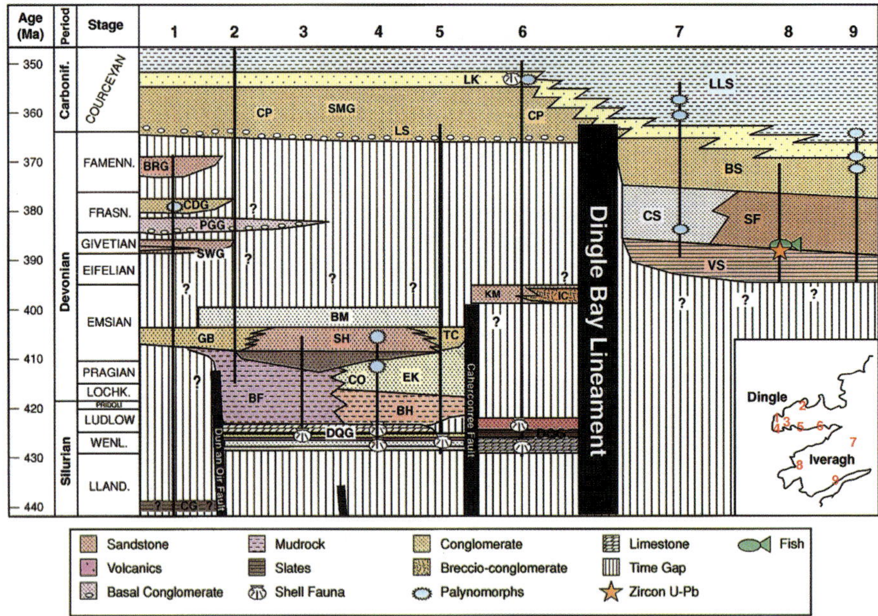

Fig. 3. Chronostratigraphy of the Palaeozoic rocks of the Dingle Peninsula area. Compiled from various sources, by area: 1: Todd *et al.* (1988*a*, *b*), Richmond (pers. comm.), Pracht (1996); 2: Todd *et al.* (1988*b*), Todd (1989*c*); 3: Holland (1988), Todd *et al.* (1988*b*), Todd (1989*c*); 4: Holland (1987, 1988), Todd *et al.* (1988*a*), Todd (1989*a*, *c*, 1991), Higgs (2000); 5: Parkin (1976*a*), Todd (1989*a*, *c*), Todd *et al.* (2000); 6: Horne (1974), Parkin (1976*a*), Higgs *et al.* (1988), Todd (1989*c*), Pracht (1996); 7: Higgs & Russell (1981); Pracht (1996); 8: Russell (1978), Higgs & Russell (1981), Williams *et al.* (1997); 9: Higgs & Russell (1981). BF, Ballyferriter Formation; BH, Bulls Head Formation; BM, Ballymore Formation; BRG, Ballyroe Group; BS, Ballinskelligs Sandstone Formation; CDG, Carrigduff Group; CG, Coosglass slates; CO, Coumeenoole Formation; CP, Cappagh Sandstone Formation; CS, Chloritic Sandstone Formation; DQG, Dunquin Group; EK, Eask Formation; GB, Glashabeg Formation; IC, Inch Conglomerate Formation; KM, Kilmurry Sandstone Formation; LK, Lack Formation; LLS, Lower Limestone Shales; LS, Lough Slat Conglomerate Formation; PGG, Pointagare Group; SF, St Finians Sandstone Formation; SH, Slea Head Formation; SMG, Slieve Mish Group; SWG, Smerwick Group; TC, Trabeg Conglomerate Formation; VS, Valentia Slate Formation. The time scale is from Tucker *et al.* (1998) for the Devonian period and from McKerrow *et al.* (1985) for the Silurian period.

provided by Higgs (2000). Two new spore assemblages have been recovered. The lower one has been collected from a similar stratigraphic and geographic location as the van der Zwan (1980) assemblage; however, this new assemblage has been ascribed an early late Pragian age (Higgs 2000). A mudrock sample in one of the three conglomeratic formations, the Slea Head Formation, has yielded a spore assemblage indicative of an early to ?mid-Emsian age. Hence the Dingle Group extends towards the top of the Lower Devonian from its Silurian base (Fig. 3).

In the north of the Dingle Peninsula, north of the Dún an Oír Fault, the Dingle Group is only represented by a thin interval of Glashabeg Formation (Todd *et al.* 1988*b*). The succeeding ORS succession includes several groups bounded by unconformities. Todd *et al.* (1988*b*) interpreted the oldest of these groups, the Smerwick Group, to be in unconformable contact, sometimes faulted, with the Dingle Group. Richmond & Williams (this volume) argue that the Smerwick Group is the same age as the Dingle Group and that the two groups are juxtaposed as 'terranes' across a significant fault. However, in the absence of definitive age data the previous interpretations of Todd *et al.* (1988*a*, *b*) and Todd (1989*a*) are maintained here: the Smerwick Group is younger than the Dingle Group and is likely to be between Givetian and Frasnian age (Fig. 3).

Similarly, the age of the Pointagare Group is a matter of conjecture without direct evidence. Although Richmond & Williams (this volume) argue that the Pointagare Group is the same age as the Caherbla Group on the basis of similar clast lithotypes, a shared provenance and depositional age is not proven and the original interpretations of a Late Devonian age for the

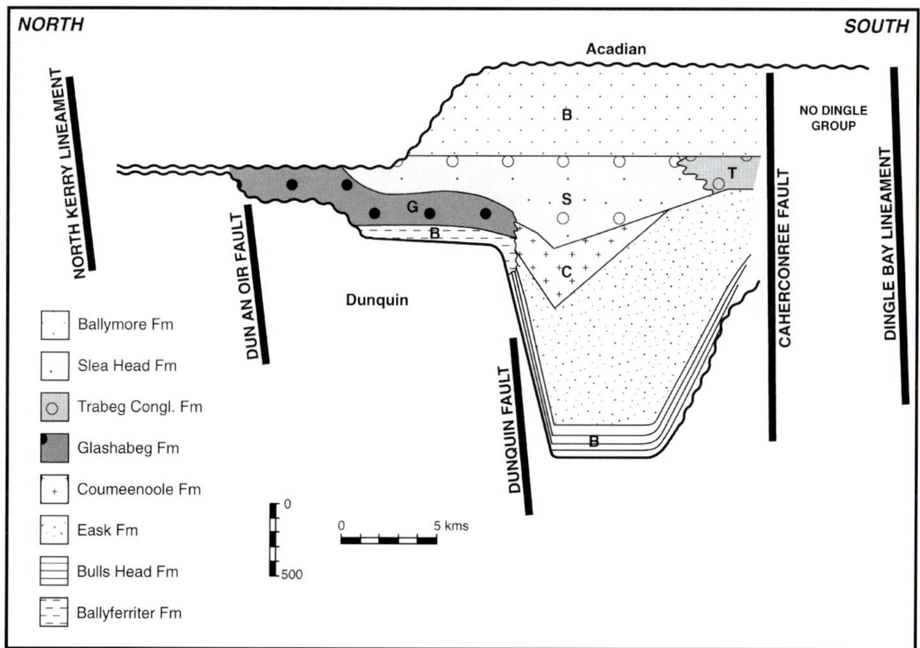

Fig. 4. Lithostratigraphy of the Dingle Group.

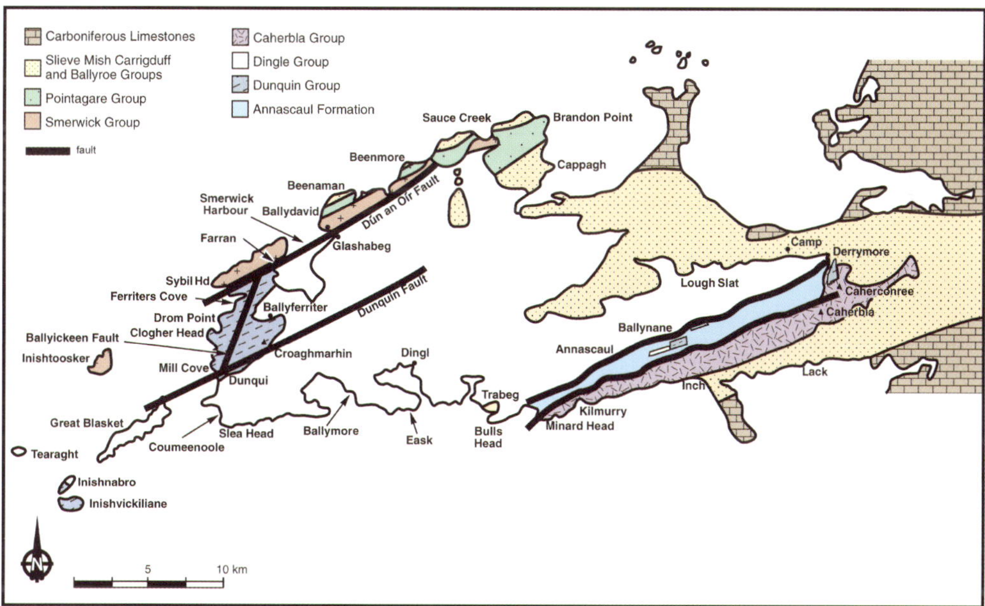

Fig. 5. Geological map of the Dingle Peninsula.

Pointagare Group by Todd *et al.* (1988*a*, *b*) are maintained here (Fig. 3). Two new groups, the Ballyroe and Carrigduff Groups, mapped by Richmond (pers. comm.) in this northern domain stratigraphically between the Pointagare and more ubiquitous Slieve Mish Groups are illustrated in Fig. 3. The older of these Groups, the Carrigduff Group, has yielded spores dated to early–mid-Frasnian time (Richmond, pers. comm.). Todd (1989*a*, p. 51) suggested that these mid- to Late Devonian groups were formed in a rotational half-graben on the northern margin of the Munster Basin and stratigraphically and sedimentologically related to the sandy influxes in the main basin.

In the SE of the Dingle Peninsula, for example around Caherconree, the Dunquin Group is overlain unconformably by the Caherbla Group, with no intervening Dingle Group (Fig. 5). The Caherbla Group is composed of two formations, the aeolian and subordinate sandstones of the Kilmurry Formation and the alluvial breccio-conglomerates and sandstones of the Inch Conglomerate Formation (Horne 1970; Todd *et al.* 1988*a*). The basin setting and provenance of the Inch conglomerates are included in this paper.

Although only ever in faulted contact with the Dingle Group, the Caherbla Group has been considered by all researchers to be younger that the Dingle Group because of the low southerly dips of the Caherbla Group compared with the folded and overturned attitudes of adjacent Dingle Group strata in the southeast Dingle Peninsula (e.g. Todd *et al.* 1988*a*).

Previous workers have attempted to place constraints on the upper age of the Caherbla Group by correlation with the ORS of the Iveragh Peninsula, which forms the main depocentre of the later Devonian Munster Basin (Todd *et al.* 1988*a*; Todd 1989*a*; Williams *et al.* 1989, 1997, this volume; Fig. 3). In the southern Dingle Peninsula the Caherbla Group is overlain by the Slieve Mish Group (Pracht 1996), which records a transition from non-marine fluviatile facies into marine facies. The marginal marine facies (the Lack Formation) yield palynomorphs and macrofossils indicating an Early Carboniferous age (Higgs *et al.* 1988; Todd 1989*c*; Fig. 3). Tighter control on the age of the Caherbla Group may be offered by the presence of metamorphic clasts in conglomerate horizons within the lower Iveragh ORS succession (Capewell 1975). Several workers have suggested that these clasts are second-cycle erosion of the Inch Conglomerate Formation (e.g. Todd *et al.* 1988*a*; Todd 1989*a*; Williams *et al.* 1989, 1997, this volume). If this true, then the Iveragh conglomerates provide an upper bracket on the age of the Caherbla Group through the presence in strata closely adjacent to the conglomerates of fish indicating a late Givetian–early Frasnian age (Russell 1978) and a U–Pb age of 384.9 ± 0.7 Ma (also Givetian on the time scale of Tucker *et al.* (1998)) derived from zircons extracted from the Enagh Tuff Bed (Williams *et al.* 1997). Hence the Caherbla Group is probably pre-late Givetian in age and a late Emsian to early Eifelian age for the group fits best with the currently available data.

Structure

The Caledonian structure of the Dingle Peninsula has been described in detail by Todd (1989*c*). Although there is a strong Variscan overprint and the Caledonian structure has also been modified by Late Devonian extension, it is possible to deduce a Caledonian history for the area. Table 1 summarizes the structural development of the area.

Deep structure

As mentioned above, the Dingle Peninsula lies within the Iapetus Suture Zone of SW Ireland (Fig. 2) (Todd *et al.* 1991). The suture itself is somewhat difficult to resolve as a single trace in Ireland (Todd *et al.* 1991). Nevertheless, the set of N-dipping deep seismic reflectors that are imaged on BIRPS data and interpreted to be the suture (Klemperer 1989) would, if extrapolated to the surface, crop out as an ENE-trending trace about 10 km north of the north coast of the peninsula (Todd *et al.* 1991). This position is in accord with evidence from other geophysical data and stratigraphic, palaeontological, isotopic and structural information (see Todd *et al.* (1991) for a discussion).

Several other lineaments, apparent on gravimetric and aeromagnetic maps of the Dingle area, exerted important influences on the tectonostratigraphic evolution of the area (Todd 1989*a*). To the south of the peninsula is the Dingle Bay Lineament (DBL), which was reactivated several times during Palaeozoic time, for example by forming the southern margin of the Dingle Basin (Todd 1989*a*; Price & Todd 1988) (Fig. 2). Another geophysical lineament, the North Kerry Lineament, is formed by a series of aeromagnetic anomalies along the northern coast of the Dingle Peninsula (Fig. 2) (Todd 1989*a*).

Structure of the Dingle Peninsula

The Palaeozoic rocks of the area were deformed by a series of ENE- to E-trending folds and faults. The folds are generally macroscopic, upright to gently inclined and verge towards the NNW. The major faults of the peninsula trend ENE, but NNE-trending and other faults are also common. Differences in the structural style and major unconformities between the Dunquin Group, the Dingle Group and the Upper ORS indicate that three discrete phases of deformation

Table 1. *Structural history of the Palaeozoic rocks of the Dingle Peninsula*

Tectonic phase	Structural effects	Stratigraphic effects	Age
Variscan inversion and shortening of Munster Basin.	Folding of all Palaeozoic rocks and pervasive pressure solution cleavage F1/S1 folds/cleavage in Slieve Mish Group Main cleavage in Dunquin and Dingle Groups, but F2 folds and S2 cleavage, cutting pre-existing folds and very local fault fabrics F3 folds and second crenulation cleavage (S3) in Annascaul Formation	No post-Westphalian, except for some a minor Cretaceous outlier, preserved in SW Ireland	Late Carboniferous
Extension of Munster Basin; Dingle was footwall margin	Tilting of Caherbla Group and older rocks (Todd 1989*a*)	Unconformity between Caherbla and Slieve Mish Groups (Todd 1989*a*, *c*; Todd *et al.* 1988*b*)	Mid- to Late Devonian
Acadian inversion of Dingle Basin and thrusting–transpression along DBL	Folding and faulting of Dingle Group and older rocks; broadly contiguous with previous phase; no cleavage, except for in fault zones F1 folds in Dingle–Dunquin Group, folds tighter in Dunquin Group; local S1 cleavage in fault zones F2 folds in Annascaul Formation and local S2 cleavage, particularly in fault zones	Unconformity between Slieve Mish–Caherbla and Dingle Groups (Todd 1989*a*); shedding of Inch alluvial fanglomerates from DBL basement ridge	Early Devonian
Acadian transpression along DBL and formation of DBL	Tilting/?folding and faulting of early Dingle Group and Dunquin Group There is no pervasive cleavage of this age in any formation	Variation in Dunquin-Dingle Group transition from conformable marine to non-marine in west Dingle (Todd 1989*a*, 1991; Todd *et al.* 1988*b*) to unconformable in south Dingle (Todd 1989*a*, 1991); shedding of fan/apron sands/conglomerates from DBL high into southern Dingle Basin (Todd 1989*a*)	Late Silurian to Early Devonian
Caledonian arc volcanism, possibly after cessation of subduction	Tilting	Basaltic to rhyolitic lavas and pyroclastic rocks interbedded with marine to non-marine sediments (Sloan & Bennett 1990; Sloan & Williams 1991)	Late Silurian
Caledonian transpression	Folding (F1) and faulting of Annascaul Formation; development of mylonitic fault rocks and crenulation cleavage (S1 in Annascaul Formation)	Faulted ?unconformity between deep-water Annascaul and shallow-marine Ballynane Formations	Within period from Llanvirn to Llandovery
Caledonian extension and possibly backarc volcanism	Bedding-parallel cleavage (S*) in Annascaul Formation, slumps	Deposition of deep-water Annascaul Formation including turbidites with volcanic debris, tuffs and mélange	Early Ordovician

affected the post-Wenlock evolution of the area (Todd 1989*a*; Pracht 1996). The three phases were Acadian (Late Silurian to Early Devonian) transpression, Late Devonian extension, and Variscan (Late Carboniferous) transpression (Todd 1989*a*).

Acadian transpression

Late Caledonian or Acadian oblique convergence was continuing during Dunquin and Dingle Group sedimentation (Todd *et al.* 1988*a*; Todd 1989*a*). Early folding of the Dunquin Group during Dingle Group times is indicated by the tight, overturned WNW-trending folds in the west of the peninsula and the associated attenuated, conglomeratic Dingle Group sequence in the region (Todd *et al.* 1988*a*, *b*). In the SE Dingle Peninsula, in the region of the Annascaul inlier (Fig. 5), the Acadian deformation is more intense and several steep, ENE-trending fault zones deform the lower Dingle Group and older rocks. Shear fabrics within the zones depict a dominantly sinistral sense of strike-slip, which combined with contemporaneous reverse offsets indicate sinistral transpression (Todd 1989*a*). The southerly transition across the Dingle Peninsula from folds and steep reverse faults into steep fault zones with intense sinistral transpressive deformation towards the DBL is interpreted to reflect the geometry of a flower structure centred on the lineament (Todd 1989*a*).

Trabeg Conglomerate Formation

The Trabeg Conglomerate Formation (TCF) occupies a medial position in the Dingle Group on the southern limb of the Fahan Syncline on the south coast of the Dingle Peninsula (Figs 4 and 5). It was deposited along the southern margin of the Dingle Basin (Todd 1989*a*) (Figs 6 and 7). Various aspects of the sedimentology and depositional setting at the southern margin of the Dingle Basin have been previously published by Todd *et al.* (1988*a*), Todd (1989*a*, *b*, 1996) and field guide locality descriptions have been presented by Todd *et al.* (1990). Todd (1989*a*) also used the lateral and vertical distribution of clast types within the TCF lithosome to infer that there was sinistral strike-slip offset of the Trabeg feeder system during deposition, consistent with a sinistral transpressive basin setting. Hence only a brief outline of the depositional facies and basin setting of the TCF is given here; instead, the discussion centres on the size of the Trabeg depositional system and source drainage basin, and the geology of that drainage basin as read from clast lithology.

Sedimentary facies

Sandy and gravelly deposits make up the fluvial channel sediments of the TCF (Todd 1989*a*, *b*, 1996). The sandstone facies dominate the basal and top parts of the formation and are composed of variably complete fining-upward sequences about 1–2 m thick (Fig. 6a). The sequences commence with a planar to irregular erosion surface, with up to 0.5 m relief. A lag of pebble conglomerate often drapes the erosion surface, succeeded by massive (structureless), parallel-laminated or cross-stratified fine sand to coarse sand sandstone. Trough and tabular cross-strata both occur: trough cross-strata sets are 0.19 m thick on average ($n = 24$; mean cross-strata thickness, $XsTh = 0.19$ m; $\sigma = 0.07$ m). Tabular cross-strata tend to be twice as large as the laterally and vertically neighbouring troughs. It should be noted that these measured thicknesses are not adjusted for the substantial tectonic flattening (*c*. 50%) experienced by these upright to overturned strata on the southern limb of the Fahan Syncline (Todd 1989*c*). The sandy fining-upward sequences are usually capped by thin, occasionally calcretized and desiccated siltstone beds (Fig. 6a).

The TCF is typified by sheets of poorly to moderately sorted, clast-supported conglomerate, especially in the medial portion of the formation constituting the top of a first-order coarsening-upward sequence (Todd 1989*a*, *b*, 1996). The conglomerate sheets, which contain clasts up to boulder grade, have erosive planar to irregular bases with relief up to 0.5 m. Ungraded beds are most common, although inversely and normally graded beds also occur. As with the sandstone facies, the conglomerate beds in the gravelly facies also form the basal, but thicker part of small-scale fining-upward sequences. Sandstone beds drape the top of the underlying conglomerate unit (Fig. 6a). The sandstone beds are internally massive, or contain parallel lamination or more rarely small-scale tabular cross-stratification.

A rare facies in the TCF is mud matrix-supported conglomerate, which locally forms beds over 1 m thick (Todd *et al.* 1990).

Depositional environment

The TCF comprises a lithosome constructed by an alluvial apron on the south side of the Dingle Basin (Todd 1989*a*). In vertical succession, a pair of coarsening- and fining-upward, egasequences of *c*. 150–200 m thickness record expansion and then contraction of the fan system, probably caused or enhanced by sinistral strike-slip along

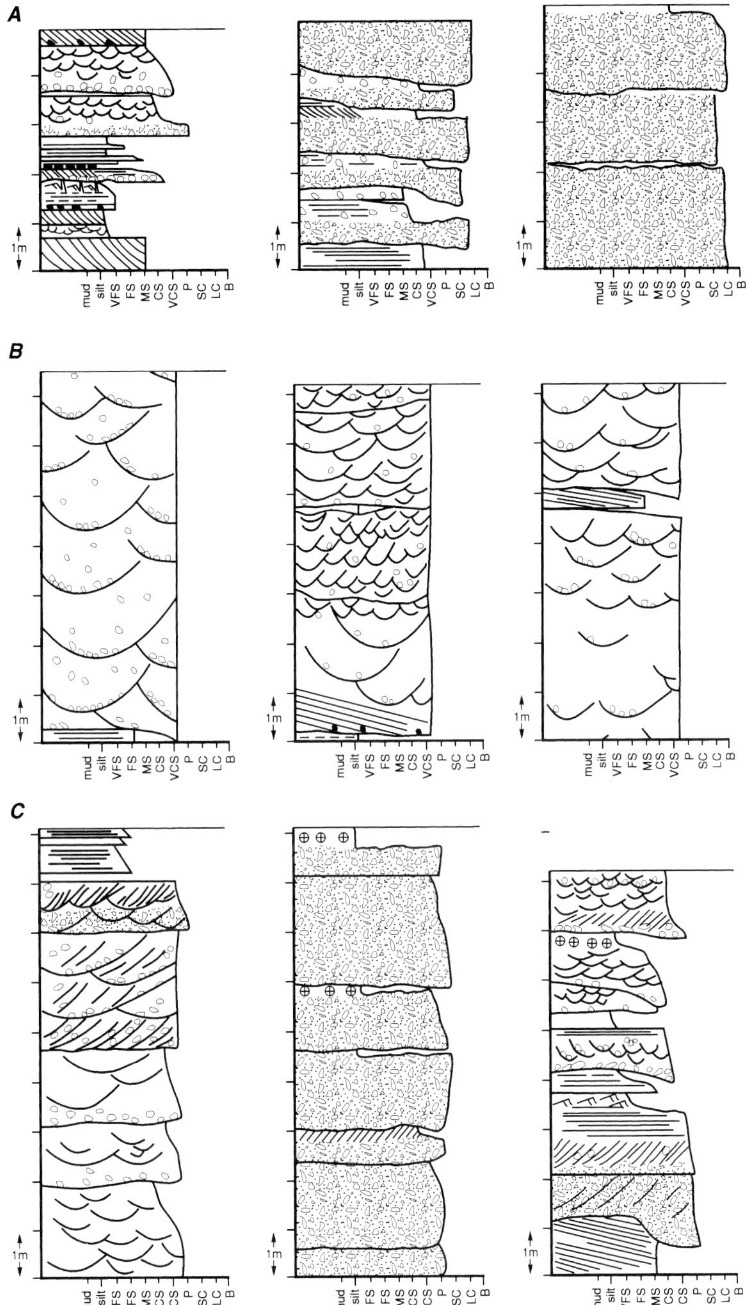

Fig. 6. Sedimentary facies of the Dingle Group conglomerates. (**a**) Trabeg Conglomerate Formation; (**b**) Slea Head Formation; (**c**) Glashabeg Formation.

the fall line of the system (Todd 1989*a*). The sandy facies are interpreted to have been emplaced during stream floods in channels ornamented with transverse bars (tabular cross-strata) and dunes (trough cross-strata) in the deeper reaches (Todd 1989*a*, *b*, 1996). The facies and internal structure of the TCF sandy facies compare well with the deposits and

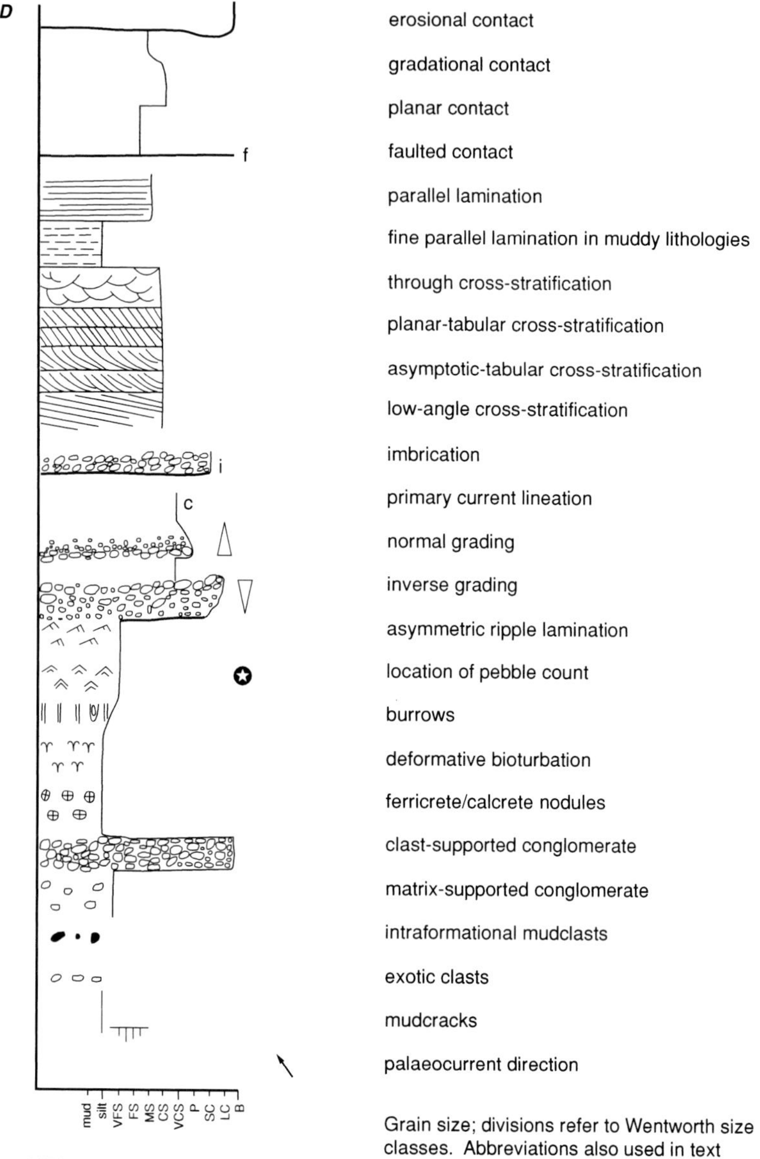

Fig. 6(d).

hydraulic scale of sandy ephemeral streams in Israel (Karcz 1972) and Australia (Williams 1971).

The Trabeg sandy facies give way in vertical succession to the gravelly facies that characterize the TCF as a whole. The gravelly deposits are interpreted to also have been deposited by stream floods, but in more proximal channels of greater discharge in a distributive system (Todd 1989*b*, 1996). Thinner conglomerate sheets (<0.4 m thick) are attributed to a low-density ('normal') bedload origin, whereas thicker, coarser sheets are believed to have been emplaced by thicker, high-density bedload sheets (Todd 1989*b*).

The characteristics of the TCF, including lateral consistent facies, rounded clasts, slow and gradual clast size changes both vertically and laterally, the variety of clast lithologies in any one

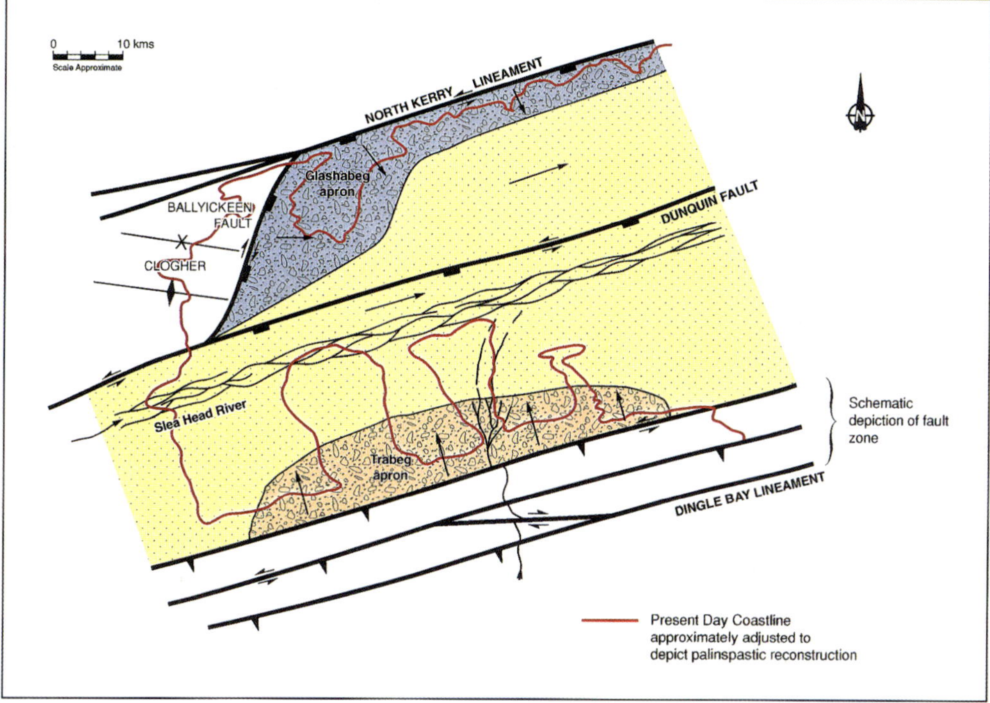

Fig. 7. Palaeogeography of the Dingle Basin during late Pragian to early Emsian time. It should be noted that the present-day coastline has been schematically restored for post-depositional folding and faulting.

conglomerate bed, and a dominance of stream flood processes, are taken to indicate the formation was deposited by a single, moderately large alluvial apron (see Todd (1989*a*) for discussion).

Palaeohydraulic scale

To establish the likely size of the drainage basin that supplied the TCF system, it is worth examining the evidence for the palaeohydraulic scale of the Trabeg rivers. Palaeohydraulic analysis is fraught with difficulties and uncertainties, but an estimate of drainage basin size, accurate to about an order of magnitude, is fit for the purpose of the provenance studies in this paper (see also Miall (1976, p. 476)).

For the sandy TCF facies, bedform height can be used to gain this estimate. The technique adopted here follows to some extent that of Miall (1976) and Turner (1980). The first step is to use the thickness of trough cross-strata as a measure of stream depth. It should be noted that the trough cross-stratal height is a minimum estimate of the formative dune height and that for the TCF the measure is an underestimate because of flattening of the strata. Allen (1968) has shown that the depth of a unidirectional flow is proportional to the height of the dune formed in the flow according to

$$H = 0{\cdot}086 d_{\mathrm{m}}^{1{\cdot}19} \qquad (1)$$

where H is the height of the dune (in metres) and d_{m} is the water depth (m) of the flow. Equation (1) is rearranged to give

$$d_{\mathrm{m}} = (H/0{\cdot}086)^{0{\cdot}84} \qquad (2)$$

It should be noted that there is a $\pm$ 50% error for this equation (Allen 1968), but the method still is fit for the current purpose of order-of-magnitude estimates (see Miall 1976). The next step, and possibly the most tentative, is to estimate the width of the channel from the outcrop if channel margins are preserved, or from empirical relationships between depth and width of modern rivers. Todd (1989*c*) collated such data from Leopold & Maddock (1953), Leopold & Miller (1956) and Leopold & Wolman (1957), and established the following relationship ($n = 197$, $R = 0.78$)

$$w_{\mathrm{a}} = 59 d_{\mathrm{m}}^{0{\cdot}87} \qquad (3)$$

where w_a is the width (m) of the river calculated by this method. An alternative method, proposed by Schumm (1960, 1968), relates the width/depth ratio (F) to the percentage of silt or finer sediment within the channel perimeter (or the sediment load parameter, M, according to

$$F = 255M^{-1\cdot08} \quad (4)$$

Having calculated the cross-sectional area (A_c, in m^2) from the width and depth, a velocity at bankfull discharge is required. The deposits of ephemeral stream floods reported by Williams (1971) contained dunes in sand that had an average height of 0.18 m and that formed in flows with a velocity of 1.2 m s^{-1}. The flash flood deposits in Israel described by Karcz (1972) were contained in dunes that were 0.1 to 0.3 m high and formed in flows with velocities of 0.1–0.8 m s^{-1}. Jackson (1976) found that dunes formed in the sand bed of the meandering Wabash River at flow velocities in the range of 1.2–1.7 m s^{-1}. In the South Saskatchewan River, maximum dune height in the channels between bars is 1.5 m; these dunes form in flows of maximum depths of 3.0 m and velocity 1.75 m s^{-1} (Cant & Walker 1978). From these observations of modern rivers an average flow velocity estimate of 1 m s^{-1} appears reasonable.

If it is assumed that the sandstones were deposited during floods, the mean annual flood discharge, Q_{ma} (in m^3 s^{-1}) can now be calculated from (Leopold *et al.* 1964)

$$Q_{ma} = VA_c \quad (5)$$

The magnitude of flood discharge can also be calculated (Miall 1976), using the following empirical equation developed by Schumm (1968):

$$d_b = 0\cdot09M^{0\cdot35}Q_{ma}^{0\cdot42} \quad (6)$$

where d_b is the bankfull depth in feet, M is the sediment load parameter, and Q_{ma} is the mean annual flood, this time in cubic feet per second. Q_{ma} is related to the area of the drainage basin (A_d; Leopold *et al.* 1964) by

$$Q_{ma} = aA_d^b \quad (7)$$

where a and b are exponents dependent on climate, where a is typically taken as one and b ranges from 0.65 to 0.8, with an average of 0.75 (Leopold *et al.* 1964; Turner 1980). Stream length (L) is related to drainage basin area according to the relationship proposed by Leopold *et al.* (1964):

$$L = 1\cdot4A_d^{0\cdot6} \quad (8)$$

For the purposes of this analysis, it is reiterated that only an order of magnitude is sufficient for the purposes of the provenance study. Table 2 presents a range of estimates for the sandy facies of the TCF. It should be noted that two different approaches give similar results for Q_{ma}. The analysis suggests the TCF sandy facies rivers had a drainage basin that extended some 100 km south of the fall line of depositional system. Also, the sandy TCF rivers are interpreted to represent the distal downstream end of a distributive system on an alluvial fan; hence, again, this more likely to be an underestimate of palaeohydraulic scale of the drainage system. Although no palinspastic correction for Variscan shortening has been applied, the drainage basin of the TCF is therefore likely to have covered an area equivalent to much of the present-day Iveragh peninsula south of Dingle Bay, if not more than that area (Fig. 8).

Clast composition

Twenty-six stations in the TCF were examined quantitatively for clast composition. These data are summarized in Fig. 8 and a representative station is depicted in Fig. 9, but all the data have been given in detail by Todd (1989*a*, *c*). A rich variety of clast types are found in the conglomerates of the TCF, but sedimentary rocks dominate the framework assemblages. Most abundant are grey, pink, yellow to green, very fine sand to medium sand, quartz and sublithic wackes. These greywackes form clasts up to boulder grade and are poorly sorted and often rich in matrix (Figs 9 and 10). Some of these greywacke lithologies have experienced low-grade (contact?) metamorphism. Lithic fragments of micaceous phyllites and fine-grained acidic igneous rocks are dominant in the wackes. Some basic volcanic fragments also occur and detrital opaque minerals are abundant in some clasts. Several of the wacke clasts contain up to 5% white mica, and accessory green tourmaline and zircon are also present. Although affected by the Variscan pressure solution cleavage that cuts the host conglomerates, some of the wacke clasts also show an earlier cleavage.

Rounded pebbles and cobbles of purple, red or black, indurate and rather crystalline arenites, which were counted as 'quartzites' in the field, also occur as clasts in the TCF (Figs 9 and 10h). Microscopically, many of these arenites display

Table 2. *Palaeohydraulic parameters for the sandy channel facies of the Dingle Group*

	H (m)[a]	d_m (m)[b]	w_a (m)[c]	F_a[d]	M[e]	F_b[f]	w_b (m)[g]	V (m s^{-1})[h]	Q_{ma^a} (m^3 s^{-1})[i]	Q_{ma^b} (m^3 s^{-1})[j]	A_d (km^2)[k]	L (km)[l]
Trabeg	0.19	1.9	105	54.1	5	44.8	96	1.0	200	189	1165	97
Glashabeg	0.28	2.7	140	51.8	5	44.8	121	1.0	327	412	2253	144
Slea Head	0.42	3.8	188	49.6	5	44.8	170	1.0	713	924	6369	268

[a]Derived from average XsTh in outcrop, but not that this is a minimum because of tectonic flattening and compaction, particularly for the Trabeg Conglomerate Formation, and the minimum preservation of the cross-strata. [b]Derived from equation (2). [c]Derived from equation (3). [d]Derived from the products of equations (2) and (3). [e]Derived from outcrop estimates. [f]Derived from equation (4). [g]Derived from equations (2) and (4). [h]Taken from analogues described in the text. [i]From estimates of velocity and results of equations (2) and (3). [j]Derived from equation (6). [k] Derived from equation (7) and previous results of Q_{ma}.

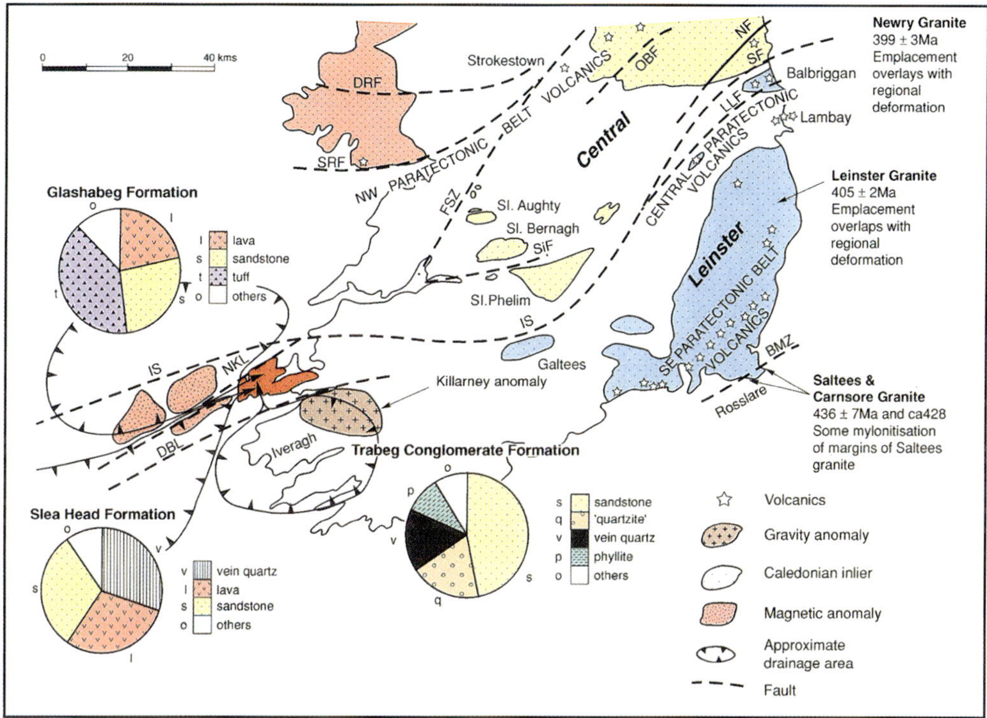

Fig. 8. Drainage basins, provenance data for the Dingle Group conglomerates, and the regional context and potential source types in the Iapetus Suture Zone of Ireland.

straight or curved or interlocking contacts between rounded to sub-rounded quartz grains. Lithic grains, when present, may be fine-grained sedimentary or phyllitic rocks. One of the black sandstones proves in thin section to be a well-sorted, compositionally and texturally mature quartz arenite with an iron ore cement (Fig. 10h). The arenites typically contain abundant quartz veins, suggesting the provenance rocks had been deformed before erosion and inclusion in the Trabeg conglomerates.

Phyllite forms disc- and blade-shaped clasts up to cobble grade. They are typically purple, orange or red, and are often spotted with darker or ringed elliptical spots or segregations up to 2 mm in length (Fig. 11a and b). The phyllites are composed of fine-grained illitic and chloritic micas, aligned parallel to bedding, forming a pervasive slaty cleavage. Porphyroblasts of muscovite overprint this cleavage, as does a later crenulation cleavage, which is conspicuous in some clasts. The spots or segregations

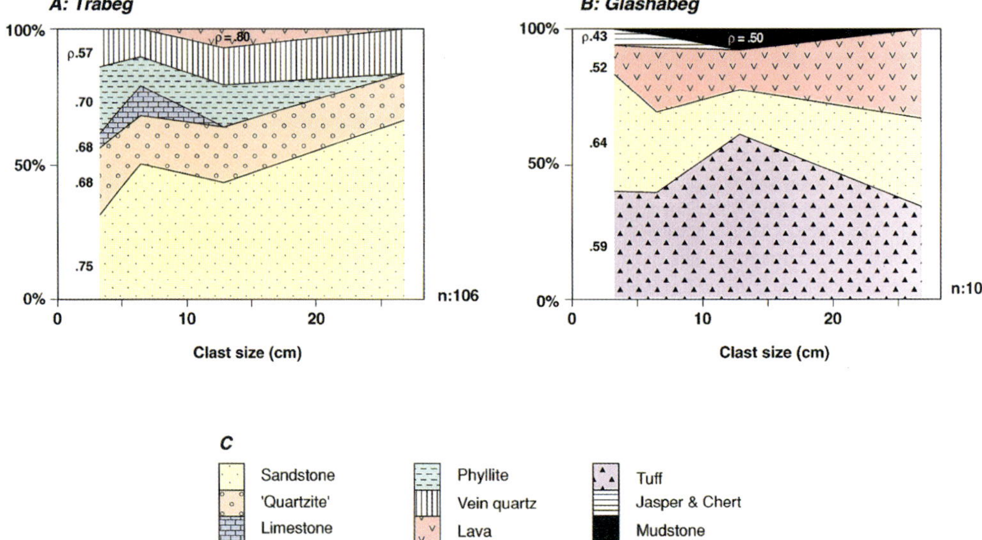

Fig. 9. Clast composition by size and roundness data (ρ) from typical clast counts from beds from the Trabeg Conglomerate Formation (**a**) and Glashabeg Formation (**b**). The complete dataset has been given by Todd (1989*c*). The numbers, e.g. $\rho = 0.80$, refer to the average roundness of the clast types.

observed in hand specimen are deformed by the cleavage and contain a central, elliptical or lath-shaped nucleus of quartz, illite and chlorite, surrounded by haloes of purple and opaque ore. Some of the central nuclei appear to be pseudomorphs, possibly after andalusite or cordierite (Figs 11 and 12).

As first reported by Jukes & Du Noyer (1863), the Trabeg conglomerates also contain clasts of limestone (Figs 9 and 11e). Several varieties are present including tuffaceous pink and red biosparites, creamy white or grey coralliferous biosparites, pale pink and buff micrites, and haematized purple biosparites. The limestones contain an identifiable fauna of crinoid ossicles, corals, brachiopods, bryozoa and algae. Salter, at the request of Du Noyer, first appraised this fauna, and Jukes & Du Noyer (1863) listed the following taxa identified by him: *Strophomena depressa*, *Pentamerus oblongus*, *Stenopora fribrosa*, *Favosites alveolaris*, *Cyclotites lenticulata*. Bassett (pers. comm.) concluded that, at face value, this list with the reported presence of *P. oblongus* and *Palaeocyclus porpita* (a synonym of *C. lenticulata*) would indicate a late Llandovery age. However, efforts so far to corroborate Salter's identifications have failed to uncover these distinctive Llandovery forms. Scrutton (pers. comm.) identified specimens collected by the author as *Palaeofavosites rugosus*, *Haysites* cf. *thomasi* and *Thecia* sp. These corals are Silurian forms that are all found in the Wenlock succession in the Welsh Borderland (Scrutton, pers. comm.). Scrutton and Armstrong (pers. comm.) also processed some of the material for microfauna and found a few fish teeth and scolecodonts, plus one conodont. The conodont is a platform (Pa) element of *Ozarkodina confuens* (γ-morph of Klapper & Murphy (1975)) indicating a Wenlock to Ludlow age. A Wenlock age for the limestone clasts therefore appears most likely.

In addition to vein quartz, which forms clasts up to pebble grade, a variety of other clasts also occur in the TCF in smaller quantities ($<5\%$). These include white or creamy brown chert, lava and granite (an adamellite with biotite, large platy muscovite, garnet and tourmaline, with some deformation of grain boundaries) (Figs 11e and 12b).

The matrix of the Trabeg conglomerates, and indeed the interbedded sandstones in the TCF are composed of VCS- and granule-grade grains of quartz and phyllite, plus plentiful opaque ores, and lava, chlorite, mica, tourmaline and garnet (Fig. 11g). Blue quartz was identified in one sample.

Geochemistry

A reconnaissance of the isotope geochemistry of the clasts of the TCF with an Sm–Nd analysis of four boulders, three greywacke and one phyllite clast was carried out during this research (Table 3). The axiom of this pilot study derives

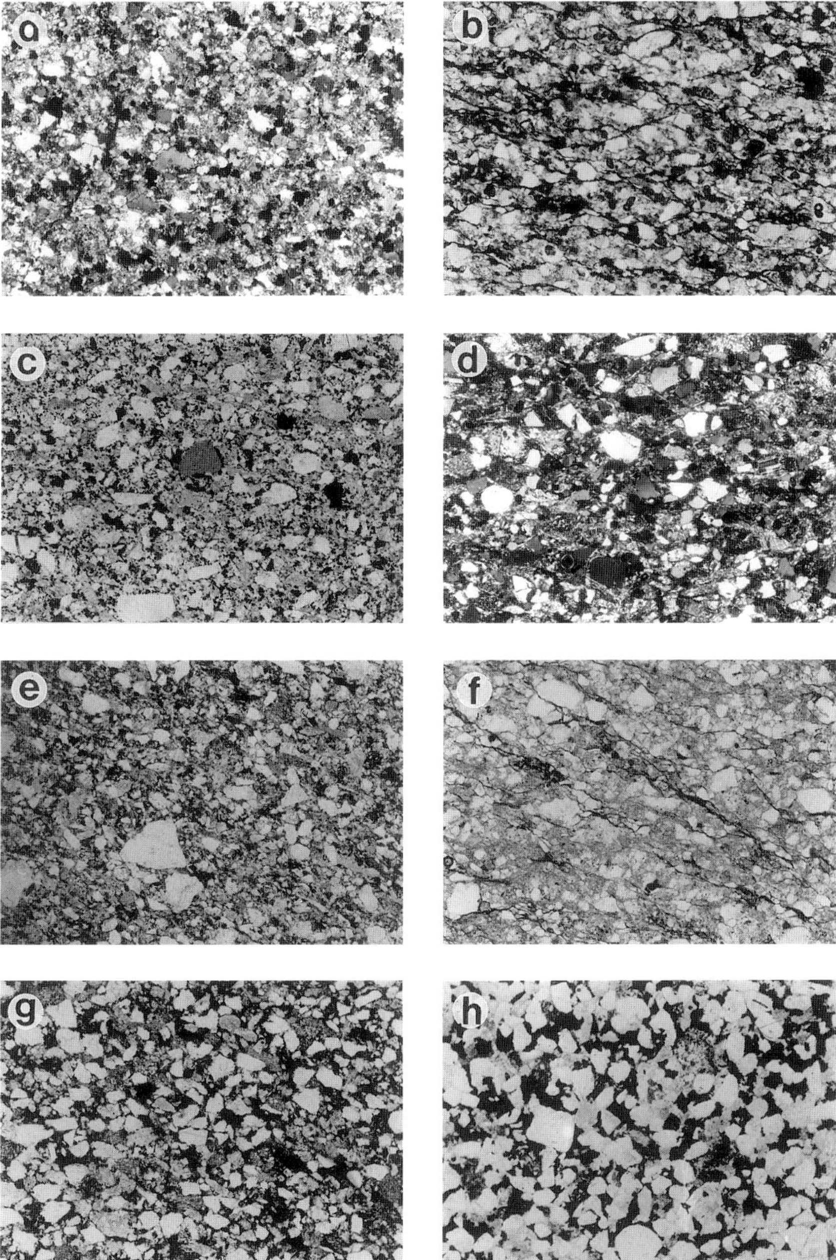

Fig. 10. Photomicrographs of the sandstone clasts of the Trabeg Conglomerate Formation. All photographs shows fields of view about 3 mm wide taken in plane-polarized light unless otherwise stated. (**a**) Recrystallized quartz wacke (sample TBE) (cross-polarized light). (**b**) Lithic wacke (TBF). (**c**) Lithic wacke (TB1E). (**d**) Quartz wacke (TB1D). (**e**) Quartz wacke (ST8646). (**f**) Quartz wacke (ST8827). (**g**) Lithic arenite rich in hematite (PM2B). (**h**) Quartz arenite with a hematite cement (TB1F).

from the fact that Nd analyses of British Precambrian and Lower Palaeozoic sediments, metasediments and granites have shown that there is some difference in the age of the crust of northern Britain and Ireland, from that of southern Britain and Ireland (e.g. Davies *et al.*

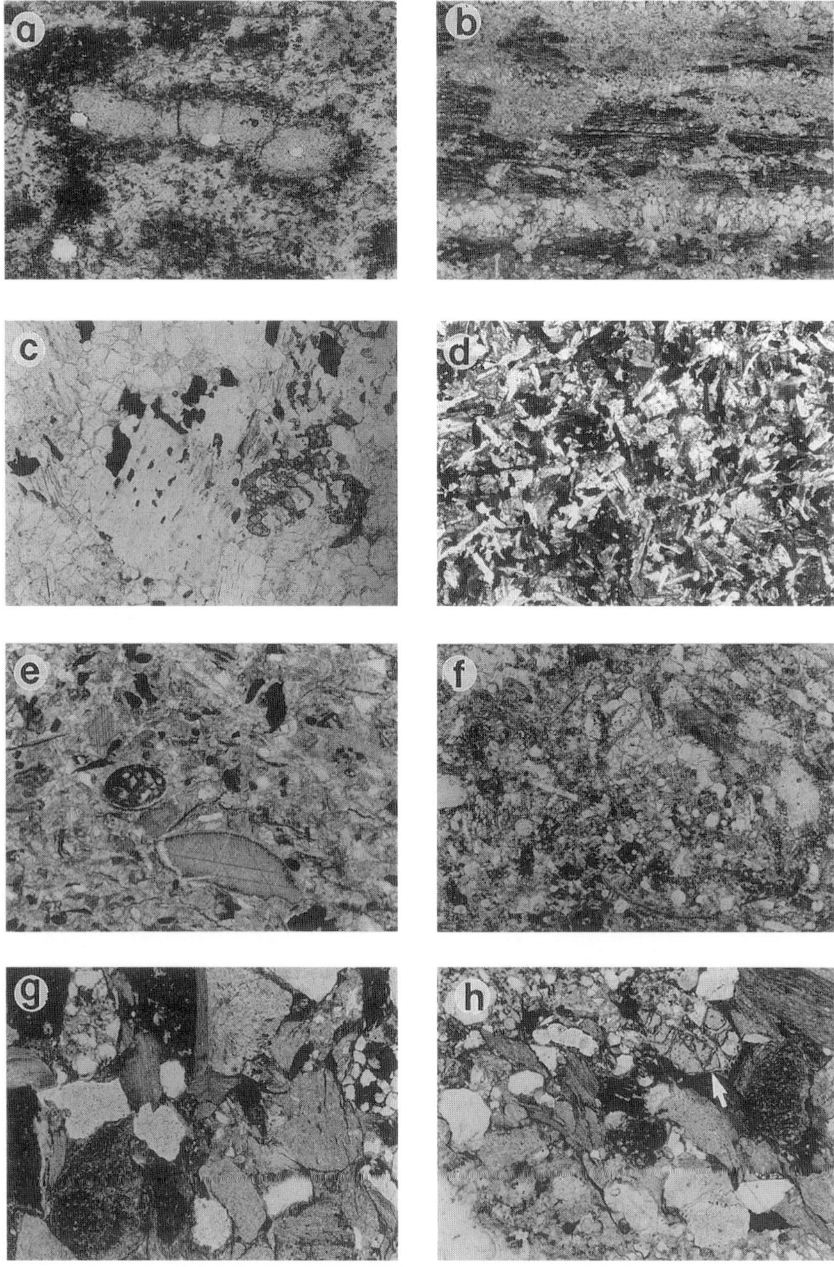

Fig. 11. Photomicrographs of the clasts and matrix of the Trabeg Conglomerate Formation. All photographs shows fields of view about 3 mm wide taken in plane-polarized light unless otherwise stated. (**a**) Phyllite with porphyroblastic laths of illite, chlorite and hematite that possibly represent pseudomorphs after andalusite (sample ST8631). (**b**) Banded phyllite with laminae of recrystallized quartz and spots of hematite, illite and chlorite cut by a bedding-parallel cleavage (ST8624). (**c**) Adamellite with muscovite (centre) and garnet (right) (ST87a). (**d**) Feldspar-phyric basalt (TBA) (cross-polarized light). (**e**) Tuffaceous biosparite (PM2B). (**f**) Tuffaceous biosparite (ST8636). (**g**) Conglomerate matrix with plentiful phyllite fragments, spilite (bottom left) and hematitic arenite (top right) (DH1C). (**h**) Conglomerate matrix with plentiful phyllite fragments, plus quartz, spilite and garnet (arrowed) (DH1A).

Fig. 12. Photographs of two clasts from the Trabeg Conglomerate Formation. (**a**) Pebble of phyllite that contains hematite pseudomorphs possible after andalusite. The pebble is 5 cm wide. (**b**) Pebble of granite cut by vein (which is horizontal in the photograph) of microgranite.

(1985) and references therein). Model Nd ages, calculated from the isotopic ratios of Nd and Sm, reflect the average crustal residence time of the Nd in the rock, or in other words, the average time since the Nd in the sample was extracted from the mantle. Model Nd ages (T_{DM}, calculated according to the depleted mantle model of DePaolo (1981)) for British and Irish sediments and metasediments display regional patterns related to their geographical and hence palaeogeographical context. Of possibly greatest significance to this study is the disparity between the older T_{DM} ages that characterize sediments and metasediments north of the Iapetus Suture from the relatively younger T_{DM} ages from similar rocks south of the suture (Davies *et al.* 1985). Within the Iapetus Suture Zone itself, the distinction is not clear (Todd *et al.* 1991). However, across the British and Irish Caledonides as a whole, rocks with T_{DM} ages greater than 2.0 Ga tend to include some Archaean material (e.g. Lewisian) and this characterizes Laurentian terranes. The data from the Trabeg boulders are therefore interesting in that two of the four samples yield T_{DM} ages greater than 2.0 Ga (Table 3). This is discussed further below after the clast lithologies of the TCF are compared with Lower Palaeozoic outcrops in Ireland.

Provenance

The TCF was probably derived from a drainage basin that covered at least the present area of Iveragh (Fig. 8). This area is now occupied by the Upper Palaeozoic sediments of the Munster Basin. From the typical NE–SW strike of Caledonian rocks in Leinster, the geophysical continuity of southern Ireland from Leinster to Munster (e.g. Max *et al.* 1983; Readman *et al.* 1997), and the lateral continuity of Leinster to the NE into the English Lake District (Murphy 1987*a*) and to the SW into Newfoundland (Kennedy 1979), it is reasonable to assume that a continuation of the Leinster massif extends beneath the Munster Basin and cropped out south of the Dingle Basin during Early Devonian time.

The clasts in the TCF are indeed similar to the lithologies that compose the Lower Palaeozoic succession of the Leinster Massif (see Brück *et al.* (1979) and Holland (1981*a*, *b*) for reviews; compare Allen & Crowley (1983)). The quartz wacke clasts are very similar to Cambrian greywackes of the Bray Group, and also to the

Table 3. *Sm–Nd isotopic data for the sediment clasts in the Trabeg Conglomerate Formation*

No.	Lithology	Sm	Nd	$^{147}Sm/^{144}Nd$	$^{143}Nd/^{144}Nd$	T_{DM} (Ga)
ST8630	Greywacke	5.563	25.74	0.1306	0.512133 ± 19	1.7
ST8633	Greywacke	4.479	17.40	0.1556	0.512115 ± 31	2.4
ST8634	Greywacke	8.052	32.1	0.1516	0.512205 ± 19	2.0
ST8631	Phyllite	5.7	29.66	0.1179	0.512033 ± 16	1.6

Greywacke clasts analysed by C. Elders at Department of Earth Sciences, University of Oxford, using a Micromass 54-E mass spectrometer, using the Sm–Nd techniques of Elders (1987). $^{143}Nd/^{144}Nd$ ratios were determined from spiked samples and are normalized to $^{146}Nd/^{144}Nd = 0.7219$. Quoted errors in the $^{143}Nd/^{144}Nd$ ratios are within-run precisions; reproducibility of the $^{147}Sm/^{144}Nd$ ratios is 0.1%. The phyllite clast was analysed by M. Murphy at University College Dublin using the Sm–Nd techniques described in Table 4.

Silurian Kilcullen Group greywackes (see Brück 1972; Shannon 1978; Hutton & Murphy 1987). The 'quartzite' clasts find close analogues in the quartzites of the Bray Group (see Holland 1981*a*). The spotted phyllite clasts are identical to phyllites occurring in the Cambro-Ordovician Ribband Group of Co. Wexford (Shannon 1977, 1978) and equivalent Lower Ordovician rocks in eastern Ireland (Murphy 1987*a*, *b*, particularly his fig. 5). Some of the phyllite clasts and some of the haematite-rich quartz wackes also could easily have been sourced from the Annascaul Formation of the eastern Dingle Peninsula (Parkin 1976*a*; Todd *et al.* 2000).

The granite clasts invite comparison with the lithologies of the Leinster Granite of SE Ireland. Indeed, identical lithologies occur as marginal igneous facies modified by post-granite emplacement recrystallization to form muscovite–porphyritic adamellite containing tourmaline and garnet (Brück 1968; Stillman 1981, p. 98). The presence of granite in the basement of the Killarney region was proposed by Howard (1975; see also Ford *et al.* 1991) who interpreted a high-amplitude, Bouguer negative anomaly centred around Killarney to be the product of a buried granite at depth (but see Naylor *et al.* (1983), Sanderson (1984), Eimicke (1986) and Price (1989), who interpreted the anomaly to be due to the sedimentary fill of the Munster Basin). If it is a granite body, it could well be an extension of the Leinster Batholith (Readman *et al.* 1997) and the Trabeg granite clasts may be the first record of unroofing of the roof facies of the batholith. There is further discussion of the timing of these events at the end of this paper.

Wenlock limestone clasts in the TCF do not find immediate comparison with the Leinster succession which contains Wenlock greywacke sandstones and mudrocks. However, a local source, consistent with the short distance of transport of the limestone clasts, is available in the Ballynane Formation (Ballynane Member of Parkin 1976*a*; Todd 1989*c*; Pracht 1996; Todd *et al.* 2000) which contains Wenlock limestones and tuffaceous limestones that are closely comparable with the clasts.

The relatively old model T_{DM} ages from the two greywacke clasts are somewhat anomalous when compared with other rocks sampled south of the Iapetus Suture. There are a few samples, representing a small percentage of the available data, from south of the suture with T_{DM} ages greater than 2.0 Ga (see Todd *et al.* 1991, fig. 6). For the Trabeg clasts to have these old ages, and yet be from a southerly provenance relative to the suture, there are three possibilities.

(1) They are Silurian turbidites shed across the Iapetus Suture with a partial Laurentian (e.g. Lewisian) provenance. However, a set of five samples from the central Irish Silurian succession, collected from outcrops either side of the suture, yield T_{DM} ages between 1.3 and 1.7 Ga (Todd *et al.* 1991, fig. 6).

(2) They represent early Palaeozoic sediments partially derived from an old block containing rocks older than 2.5 Ga which has been hitherto unrecognized in the Caledonides south of the Iapetus Suture.

(3) They are Cambrian or older greywackes that as part of the Bray Group were derived from Laurentia before the full opening of Iapetus (e.g. Phillips *et al.* 1976).

Hypothesis (3) appears to be the most likely interpretation at this point. The main hindrance to interpretation is the paucity of data from potential source rocks such as the Bray Group in the Leinster Massif.

In summary, the clasts of the TCF find outcrop correlatives either in local rocks in the southeast of the Dingle Peninsula or in rocks described from the Leinster Massif. This is consistent with the Dingle Basin being located on the northern edge of the Leinster Terrane (Murphy *et al.* 1991; Todd *et al.* 1991, in press). It is also consistent with the model of the Leinster Terrane being deformed, intruded by granite, and uplifted to shed sediment into the Dingle Basin during Early Devonian time.

Glashabeg Formation

The Glashabeg Formation (GF) crops out only in the northwest and north of the Dingle Peninsula. It is best exposed around the shores of Smerwick Harbour (Fig. 5). The Glashabeg Formation represents a depositional system formed along the northern margin of the Dingle Basin and is the lateral equivalent of the SHF and TCF (Fig. 7) (Todd *et al.* 1988*a*). The lithostratigraphy of the three formations, and the discrimination of these Dingle Group conglomerates from the younger rocks of the Smerwick Group, was provided by Todd *et al.* (1988*b*). A detailed interpretation of barform structure, palaeocurrents and the preservation bias of cross-stratification as a result of tectonic tilting was presented by Todd & Went (1991). Field localities in the GF were described by Todd *et al.* (1990) and a full description has been provided by Todd (1989*c*). Therefore only a brief description of sedimentary facies is given here, and instead the focus is placed on evidence of palaeohydraulic scale and conglomerate provenance.

Sedimentary facies

As with the TCF and SHF, the GF forms part of a pair of coarsening- and then fining-upward megasequences, about 400 m thick. Up to 30% of the succession is composed of calcretized mudrocks. Channel deposits form fining-upward sequences, 2–3 m thick, with a basal erosion surface followed by a conglomeratic lag and then cross-stratified, typically pebbly, sandstone, often capped in calcretized mudrock (Fig. 6c). The sandstone beds contain large-scale trough cross-strata, with cosets occasionally over 1 m thick (mean $XsTh = 0.28$ m; $n = 37$; $\sigma = 0.14$ m), which are overstepped by tabular cross-strata commonly 0.6–1.0 m, and up to 1.8 m thick. Conglomerates form the channel deposits in the middle of the GF, at the top of the coarsening-upward megasequence. The conglomerates form poorly to moderately sorted beds, usually a few decimetres thick, which contain crude horizontal stratification (Fig. 6c). Some rarer, thicker conglomerates are massive and unstratified, and two beds of unstratified, mud-matrix-supported conglomerate also occur (Todd *et al.* 1990).

Depositional environment

The GF was deposited in an alluvial apron (Fig. 7), which formed a lithosome that can be mapped out across the NW and north of the Dingle Peninsula. Sandy channel deposits were constructed by slipfaced lateral bars (*sensu* Todd 1996) and by sinuous crested dunes that occupied the deeper reaches of the river (Todd & Went 1991). Flows in the pool region in the lee of the lateral bars must have been at least as deep as the height of the tabular cross-strata. As the GF system prograded, the channel deposits became more gravelly as more proximal channels were superimposed on older ones. Channel incision and the development of terraces shielded from sediment input may have led to the growth of the relatively mature calcretes in the formation (Todd 1996). The mud-matrix-supported conglomerates were deposited by mud flows, suggesting that the fall line was not too distant from the preserved outcrops.

Palaeohydraulic scale

Following a similar approach to that argued for the TCF above, it appears probable that the GF was also deposited by a single, moderately large depositional system sourced from a drainage basin that lay to the west and NW of the Dingle Basin. Table 2 gives a representative estimate of the palaeohydraulic parameters for the GF system. Given the size of the preserved bedforms and barforms in the formation, the size of the GF drainage basin was probably at least as large as the TCF, extending some 100 km or more to the north of the Dingle Peninsula (Fig. 8). If this is correct, then the drainage basin straddles the trace of the Iapetus Suture (Todd *et al.* 1991).

Clast composition

Pebble and cobbles in the Glashabeg Formation are dominated by lithologies of volcanic association. The clast types include lavas, tuffs, jasper, red sandstones, mudrocks, limestone, microgranite and vein quartz (Figs 9b and 13). Purple, black, grey and green lavas are often the most conspicuous in outcrop; in thin section a number of different types have been identified. Spilitic to keratophyric, basic to intermediate lavas rich in plagioclase feldspar are most abundant (Fig. 13). Some lavas are aphyric with large, interlocking laths of feldspar, often displaying trachytic or variolitic textures. The matrix to these laths is formed by opaque ores, presumably derived by alteration of ferromagnesian phases. Other lavas are porphyritic: basalts or andesites with large phenocrysts of feldspar set in a fine-grained groundmass of feldspar, chlorite and epidote. Porphyritic dacites with phenocrysts of quartz and feldspar also occur. Ferromagnesian phenocrysts are not preserved unaltered; in one case, large six-sided crystals pseudomorphed by epidote are interpreted to be the vestige of hornblende in a hornblende andesite. Lava clasts range in size up to cobble grade, and occasionally boulder grade in the Glashabeg conglomerates.

Volcaniclastic rocks are most abundant as clasts in GF conglomerates, including purple welded lapilli-tuffs, and yellow and green dust- and lapilli-tuffs. These clasts are usually of pebble grade, and occasionally of cobble size. Sandstone clasts are lithic arenites containing quartz and rock fragments of fine-grained phyllite and acidic lavas (Fig. 13h).

Limestone clasts, similar to the TCF, have also been found in the GF, but only at one locality (Carrignakeedu, grid reference Q 375 042), where limestone clasts are locally very abundant (about 20% of the framework). Pale pink or creamy micrites, crinoidal biosparites, and coralliferous biosparites are common. Amongst these clasts are some well-preserved macrofauna including crinoid ossicles, bryozoa, corals and brachiopods. One pebble clast is formed by a nearly complete specimen of a large, thick-ribbed brachiopod identified by Holland (pers. comm.) as *Rhipidium* cf. *hibernicum*, a Homerian (Upper Wenlock) species that has so far only been found

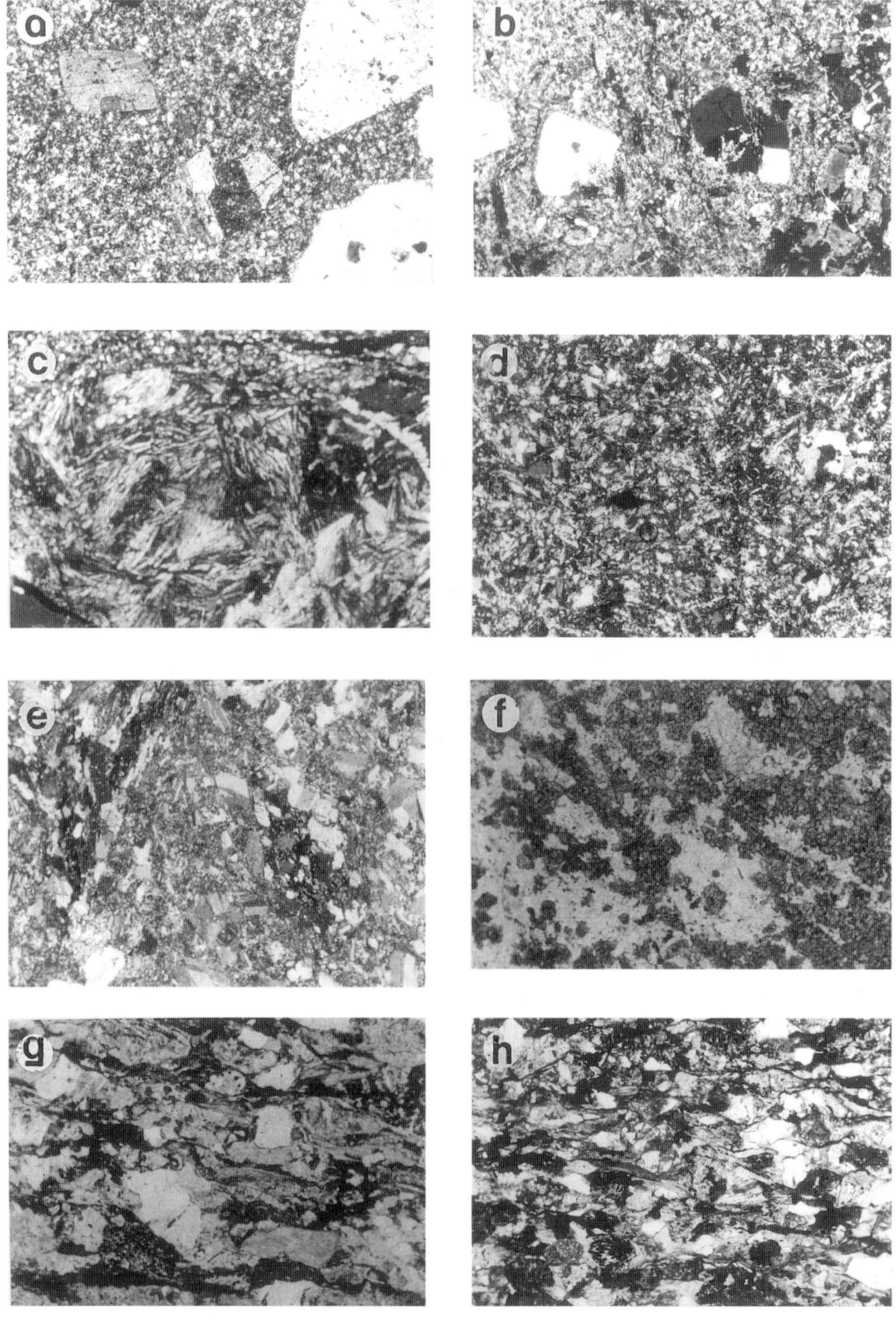
a
b
c
d
e
f
g
h

in the western Dingle Peninsula and neighbouring Great Blasket Island (Bassett *et al.* 1976; Holland 1988).

Conspicuous amongst and characteristic of the GF conglomerates are angular to rounded clasts of red and purple jasper, up to pebble grade typically, and occasionally to cobble and boulder size. The sandy matrix and sandstone interbeds of the GF are formed by volcanic arenites. The sandstones also contain angular to rounded quartz, opaque minerals, limonite, epidote, feldspar, clinopyroxene and chlorite.

Provenance

Palaeocurrents in the GF depict SE, east and NE dispersal directions (Todd *et al.* 1988*b*, 1990; Todd & Went 1991). This indicates a source drainage basin to the NW of the peninsula, currently unexposed, but which lies along Caledonian NE–SW strike from the onshore Lower Palaeozoic inliers such as Longford–Down Massif in the north, and the inliers of eastern (e.g. Grangegeeth) and central (e.g. Slieve Aughty) Ireland to the south (Figs 2 and 8). The speculative reconstruction for the drainage basin of the GF straddles the area that is thought to contain the surface trace of the Iapetus Suture (Phillips *et al.* 1976; Klemperer 1989; Todd *et al.* 1991; Fig. 2).

A hinterland dominated by volcanic rocks is clearly indicated by the petrography of the Glashabeg conglomerates. Some of the clast lithotypes can be found as outcrops in the Silurian volcanic rocks of the Dunquin and Inishvickillane inliers (Gardiner & Reynolds 1902; Holland 1969, 1988; Parkin 1974; Sloan & Bennett 1990; Sloan pers. comm.; author's own observations). Lava flows of fine-grained andesite, some with dispersed, small feldspar laths that occasionally show trachytic textures, which are commonly altered to chlorite- and epidote-rich keratophyres, form part of the Dunquin Group of the western inliers and islands. Other lithologies such as purple welded lapilli-tuffs, purple sandstones and green dust- and lapilli-tuffs also form both Dunquin Group outcrops and form GF clasts. Although there are no outcrops of Wenlock limestones in the western Dingle Peninsula, the endemicity of *R. hibernicum* to the area and the required short transport distances for labile limestone clasts suggest a local source for these clasts.

Some of the lava types in the GF clasts, such as richly feldspar-phyric andesites, do not have local outcrop corollaries (Sloan, pers. comm.). These rocks have possible correlatives with Ordovician volcanic rocks in eastern Ireland. For example, there are extensive Caradocian keratphyres and pyroxene andesites interbedded with volcaniclastic deposits and mudrocks at Lambay, Portrane and Balbriggan (France 1967; Romano 1980*a*, *b*; Stillman 1981; Murphy 1987*a*). Further possible analogues for the volcanic clasts include the Upper Llandeilo–Caradoc basaltic and trachytic and associated volcanic sediments in southern Slieve Aughty (Weir 1973; Figs 2 and 8).

From geophysical evidence, Brown & Williams (1985) considered that these Ordovician volcanic rocks extend across central Ireland beneath the Carboniferous cover in the form of faulted blocks. Similar geophysical evidence to the west of the Dingle Peninsula in the form of paired aeromagnetic and gravimetric anomalies, supports the extension of this subcrop geology into the offshore shelf region (Max *et al.* 1983; Todd 1989*a*) and might represent the remnants of the volcanic hinterland for the GF sediments (Fig. 8).

A clast type that presents something of an enigma is jasper. Jasper, i.e. red or purple chert, is conspicuous in the GF, although virtually absent in the TCF and SHF. The jasper is poorly to moderately rounded (ρ0.46), which for hard, indurate lithology indicates a relatively long transport distance, possibly with several cycles of erosion and deposition (see Mills 1979). Jasper is commonly associated with basic lavas extruded in a submarine setting and is formed as chert with a high trace content of ferrous oxides giving the distinctive colour.

Jasper is not recorded or reported to occur in outcrop south of the putative trace of the Iapetus Suture in Ireland. It does occur in association with pillow lavas within the Precambrian–Cambrian Gwna Mélange in Anglesey and in the Lleyn Peninsula of north Wales (Greenley 1919; Gibbons 1983). The Irish Ordovician

Fig. 13. Photomicrographs of the clasts and matrix of the Glashabeg Formation. All photographs shows fields of view about 3 mm wide taken in plane-polarized light unless otherwise stated. (**a**) Porphyritic quartz–feldspar dacite (sample ST885) (cross-polarized light). (**b**) Porphyritic quartz–feldspar dacite (GB115) (cross-polarized light). (**c**) Variolitic basalt (BR1). (**d**) Aphyric basalt (ST87107). (**e**) Hornblende andesite with amphibole pseudomorphed by feldspar and epidote occupying the centre of the field of view (ST882). (**f**) Epidotized spilite (ST87103). (**g**) Clast of lithic arenite with plentiful clasts of acid volcanic rocks (ST87106). (**h**) Conglomerate matrix with quartz, phyllite, volcanic fragments, epidote, opaque minerals, chlorite and pyroxene.

volcanic successions that were described above from Lambay, Portrane and Balbriggan were developed in submarine settings but contain no jasper (France 1967; Romano 1980*a*, *b*; Murphy 1987*a*, pers. comm.). White chert is interbedded with the lavas of southern Slieve Aughty (Holland 1981*a*). In contrast, jasper occurs in abundance in the lower Ordovician Strokestown Group on the northern margin of the Longford–Down Massif (Morris 1983) and in correlative rocks of the South Connemara Group, south of the Galway Granite (Ffrench & Williams 1984), and in northern Slieve Aughty (Holland 1981*a*; Murphy pers. comm.).

It is very unlikely that the Glashabeg jasper clasts were ultimately derived from the south, particularly as Dingle Group conglomerates (TCF) derived from the south contain virtually no jasper. Given the apparent absence of jasper from the exposed volcanic rocks that form what Stillman (1981) referred to as the Central Paratectonic Volcanic Belt, a major source of jasper south of the Iapetus Suture is unlikely (Figs 2 and 8). The best explanation for the occurrence of jasper (and possibly spilite) clasts in the GF conglomerates is that there were derived, possibly polycyclically, by erosion of the southwestern extrapolation of the jasper-bearing Northwest Paratectonic Volcanic line (Fig. 8) (Stillman 1981). The volcanic line lies within the Northern Belt of the Southern Uplands to Longford–Down zone.

Slea Head Formation

The Slea Head Formation (SHF) crops out in the west and central parts of the Dingle Peninsula and represents the central portion of the fill of the Dingle Basin deposited by a large river system that flowed NE along the axis of the basin (Fig. 5) (Todd *et al.* 1988*a*; Todd 1989*a*). The lithostratigraphic context of the 600 m thick SHF, as the central section of a pair of first an upward-coarsening megasequence (from the Coumeenoole Formation below) and then an upward-fining megasequence (into the Ballymore Formation above) was described by Todd *et al.* (1988*a*, *b*). Field localities in the formation were described by Todd *et al.* (1990) and the deposits of fluviatile compound bars were detailed by Todd (1996). Higgs (2000) has recovered a miospore assemblage of early to ?mid-Emsian age from a level about 200 m from the top of the SHF. As with the TCF and GF, only a brief description of the facies is given here and instead it is the provenance of the SHF that is the prime focus.

Sedimentary facies

The SHF is dominated by thick-bedded coarse to very coarse, pebbly sandstones that comprise multi-storey sheet sandbodies with little interbedded mudrocks (Fig. 6b) (Todd *et al.* 1988*a*, 1990; Todd 1996). By far the dominant sedimentary structure in the SHF sandstones is moderate- to large-scale trough cross-strata (mean $XsTh = 0.41$ m; $n = 29$; $\sigma = 0.22$ m). The trough cross-strata form thick cosets separated from each other by the scoop-shaped and inclined bounding surfaces of the multiple storeys in the sandbodies. Palaeocurrent indicators in the SHF indicate NE palaeoflow (Todd *et al.* 1988*a*; Todd 1989*c*).

Depositional environment

The SHF is interpreted to be the product of deposition from a low-sinuosity, possibly braided river (Todd 1996). The facies compares well with that of the Donjek (Williams & Rust 1969) in that there is a predominance of trough cross-stratification generated by the migration of sinuous-crested dunes (Todd 1996). The dearth of tabular cross-strata probably results from the lack of slip-faced bars in the Slea Head river. Instead, the barforms are envisaged to have been largely slip-faceless compound forms, with both upstream and downstream sides mantled in smaller-scale dunes. The migration, evolution and modification of the compound bars generated the bounding surfaces.

Palaeohydraulic scale

The size of trough cross-strata, sandstone storeys between bounding surfaces, and sandstone sheet thickness all indicate that the SHF was deposited by a relatively major river. For example, by comparison with the sandy channel facies of the TCF, which contain trough cross-strata with a mean thickness of 0.19 m, the SHF trough cross-strata are more than twice as thick. The palaeohydraulic analysis described above if repeated for the SHF, suggests a river with a drainage length of more than 200 km (Table 2). Comparison with the facies and scale of the modern Donjek river would suggest a drainage area greater than 10 000 km^2. Because of the NE-directed palaeoflow of the SHF, and the space constraints imposed by the two other drainage basins that fed sediment to the Dingle Basin (Trabeg and Glashabeg), the Slea Head source is depicted in Fig. 8 as a cone-shaped area mostly off the SW coast of Ireland.

Clast composition

The pebble-grade clasts of the SHF are formed by a wide variety of lithotypes including vein quartz, chert, white and purple quartzite, red and green wackes, arenites (some of which are black hematitic sandstones), tuffs, black aphyric basalt, spilite, brown and white fossiliferous limestones (no fauna apart from crinoid ossicles were identified from these clasts), yellow flow-banded rhyolite and quartz porphyry. These pebbles are set in a matrix formed by lithic arenite that contains an abundance of quartz, phyllitic rock fragments, opaque minerals and lava clasts. Accessory minerals include muscovite, chlorite, calcite, apatite, garnet and tourmaline.

A range of lava types occur as pebbles in the SHF, including aphyric fine-grained or feldspar-phyric, sometimes trachytic, spilite or keratophyre with dispersed feldspar laths set in a matrix of feldspar, epidote, chlorite and opaque minerals; and fine-grained dacites and rhyolites. Arenaceous clasts include quartz arenite (sometimes with a hematite cement), metaquartzite, lithic arenite (itself with framework grains of acidic lavas) and quartz wacke (Fig. 14). Phyllite, with foliated fine-grained mica and chlorite, tends to form only granule- and sand-grade particles. These clast types are identical to clasts in the TCF and in GF. One sample of a lithic arenite sandstone interbed from the SHF is notable for the abundance of garnet in the framework, as well as small pebbles of garnitiferous quartzite (coticule; Kennan 1986) and tourmalinite (Fig. 14).

Provenance

The SHF derives from a very large drainage basin to the southwest of the Dingle Basin. Direct clues to the nature of the pre-Dingle Group bedrock in this area derive from the Silurian outcrops in the west of the Dingle Peninsula and on the Blasket Islands to the SW (Gardiner & Reynolds 1902; Holland 1969, 1988; Parkin 1974, 1976*b*). Coincident with the island of Inishvickillane and the Foze rocks which are composed of Silurian volcanic rocks is an elongate, SW-striking high-amplitude aeromagnetic anomaly that probably reflects the position of a block of Silurian volcanic rocks (Sloan & Bennett 1990). Derivation of some of the detritus in the SHF from this block is supported by the close comparison with the lava-types in the Inishvickillane Formation, which include fine-grained, chloritic keratophyres, some of which contain dispersed feldspar phenocrysts (Parkin 1974; Sloan & Bennett 1990; Sloan, pers. comm.).

The other clasts in the SHF (e.g. the limestone clasts) were derived from a similar source as the TCF, i.e. a southwesterly extrapolation of the Leinster Massif below the Late Palaeozoic Munster Basin and out into the present continental shelf. Especially supportive of this interpretation is the occurrence as clasts of coticule and tourmalinite, which characterize a particular stratigraphic level within the Ribband Group of Leinster (Brindley 1954; Shannon 1977, 1978; Kennan & Kennedy 1983; Murphy 1987*a*, *b*). The other arenaceous and phyllitic clasts in the SHF also have corollaries in outcrops in the Leinster Massif.

Inch Conglomerate Formation

The Inch Conglomerate Formation (ICF), one of two formations in the Caherbla Group, crops out only in the SE of the Dingle Peninsula (Fig. 5). Its formation as an alluvial fan complex shed from a fault scarp composed of metamorphic rocks in the Dingle Bay region was first described by Horne (1975). The stratigraphic position of the Caherbla Group as younger than the late Caledonian folding of the Dingle Group but older than the Late Devonian fill of the Munster Basin was discussed by Horne (1974), Todd *et al.* (1988*a*), Todd (1989*a*), Williams *et al.* (1989, 1997, this volume), and, as summarized above, a late Emsian to Eifelian age for the formation seems most likely. Further description of the facies of breccio-conglomerates and of the interbedded, laterally equivalent and overlying aeolian sandstones of the Kilmurry Sandstone Formation (KSF) was provided by Todd *et al.* (1988*a*) and field localities were described by Todd *et al.* (1990). Only a brief sedimentological description is provided here, and instead palaeohydraulic scale and provenance are the principal concern.

Sedimentary facies

Four main sedimentary bodies or lithosomes composed of the Inch Conglomerate Formation crop out in the southeast of the Dingle Peninsula and are encased in the sandstones of the Kilmurry Formation (Fig. 15). In addition to the main lithosomes, which have been mapped on a 1 : 10 000 scale, some thin bands of fine-grained conglomerate are also intercalated with the sandstones of the Kilmurry Formation (Capewell 1951). The sedimentary facies of the four main lithosomes are similar and can be described in terms of three main facies types.

Disorganized to organized coarse breccio-conglomerate comprises boulder–cobble

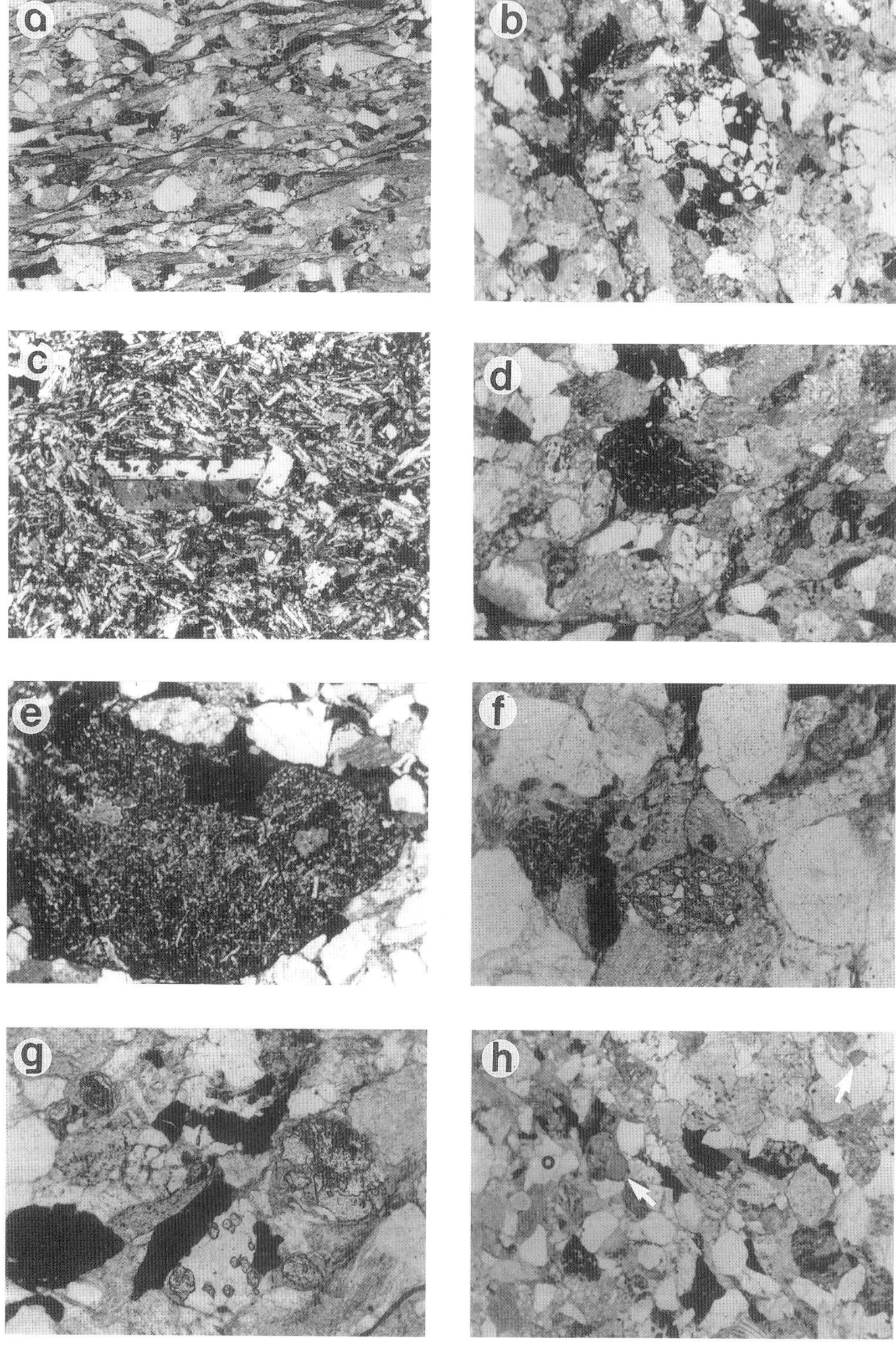
a
b
c
d
e
f
g
h

breccio-conglomerates that are clast to matrix supported, typically disorganized, and display little or no internal stratification, apart from some crude grading. The framework is generally poorly to moderately sorted and a red coarse sand–very coarse sand matrix fills the interstices. The clast fabric tends to be rather chaotic, with many upright clasts, but *a*(t)*b*(i) imbrication, particularly of the larger boulders, and fabrics produced by a preferred bedding-parallel alignment of clast *ab* planes also occur. This facies forms ribbon-like conglomerate-bodies with channelled, concave-up, basal erosion surfaces. The lenticular bodies tend to be *c*. 2–4 m thick and *c*. 10–20 m wide.

Stratified pebble conglomerates and pebbly sandstone facies is composed of stacked stratified sheets of pebbly sandstone and lenticular pebble conglomerates. The different lithologies are interstratified on a scale of 10–50 cm. The pebbly conglomerates are poorly to moderately sorted and composed of mainly spherical clasts. The pebbly sandstones mostly contain horizontal stratification, but tabular and low-angle trough cross-stratification (width to height aspect ratios of the troughs of 15 : 1 or more) also occur. Horizontal stratification in the pebbly sandstones is formed by laminae a few millimetres to several centimetres thick. In several localities, the horizontal strata are inversely graded, with pebble concentrated at the top of the unit. The stratification in the pebble conglomerates and pebbly sandstones is in places disrupted by miniscate *Beaconites* burrows (e.g. Allen & Williams 1981) up to 10 cm wide.

A distinctive cross-stratified sandstone facies is encountered within the main lithosomes of the Inch Conglomerate Formation only at one locality on Inch Shore (see also Horne (1975, fig. 3)). The sandstones are red, clean medium sandstones with well ordered, curved, asymptotic-tabular cross-stratification. The cross-strata are up to 1 m thick and are composed of well-sorted and sharply defined foresets. The 2 m thick package of this sandstone facies is sandwiched between coarse breccio-conglomerate beds.

Depositional environment

The Inch Conglomerate Formation is the deposit of one or more alluvial fans (Fig. 16) (Horne 1975). The fans prograded northwards into an area occupied by an aeolian dune field as represented by the Kilmurry Formation (Horne 1975; Todd *et al.* 1988*a*). The coarse breccio-conglomerates are channel-filling deposits and were probably generated in main trunk channels on the alluvial fan(s). The poor sorting and poor fabric organization attest to rapid deposition from sediment-charged flows. The nature of the facies, particularly in the frequency of normal grading and the *a*(t)*b*(i) imbrication, probably indicates that the beds were deposited from a sediment–water suspension induced by turbulence in a hyperconcentrated or high-density stream flood (*sensu* Todd 1989b). Some of the more organized, sheet-like beds of coarse breccio-conglomerate in the ICF may be the products of depositional lobes at the end of fan channels.

The stratified pebble conglomerates and pebbly sandstone facies is interpreted to have been deposited by lower-energy sheet- and stream floods. Flat stratification is formed in upper flow-regime conditions, whereas the low-amplitude trough cross-strata were formed by the migration of low-relief dunes in shallow and/or sediment-charged flows. The distinctive cross-stratified sandstones which are similar to those that form the bulk of the Kilmurry Formation, are interpreted to the products of small transverse dunes that accumulated and migrated on thin aeolian sand sheets (Horne 1975; Todd *et al.* 1988*a*).

Palaeohydraulic scale

The facies and particle size of the ICF indicate a proximal alluvial fan setting. The fall line is no longer preserved, although a location in the Dingle Bay region, associated with movement on the DBL, seems likely (Todd 1989*a*). The palaeohydraulic estimation methods used for the sandy facies of the TCF, GF and SHF are not appropriate for the ICF. Instead, the fan size is used. Although, because of tectonic dips, the down-fan extent of the fan bodies is not

Fig. 14. Photomicrographs of the pebbly sandstones of the Slea Head Formation. All photographs show fields of view about 3 mm wide taken in plane-polarized light unless otherwise stated. (**a**) Lithic arenite sandstone with plentiful phyllite fragments that are strongly affected by the Variscan cleavage (sample ST8764). (**b**) Lithic arenite including a clast of hematitic arenite (centre) (ST8724). (**c**) Clast of feldspar-phyric andesite (ST8758). (**d**) Lithic arenite sandstone with spilite clast (centre) (ST8758). (**e**) Lithic arenite sandstone with keratophyre–spilite clast occupying most of the field of view (ST8758). (**f**) Lithic arenite including clast of coticule in bottom centre (ST8734) (field of view is about 1.2 mm wide). (**g**) Lithic arenite with conspicuous coticule clasts (e.g. bottom centre) and garnet grains (right and left) (ST8734) (field of view is about 1.2 mm wide). (**h**) Lithic arenite with conspicuous accessory tourmaline (arrowed) (ST8734).

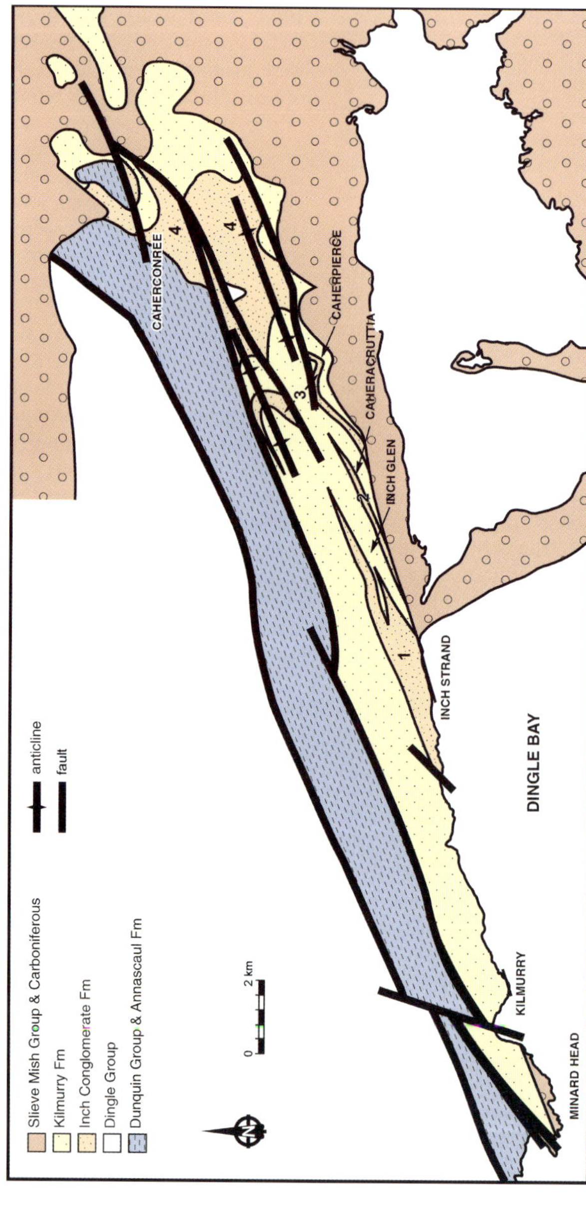

Fig. 15. Geological map of the Caherbla Group, SE Dingle Peninsula. The map is modified after Capewell (1951, 1965) based on geological mapping by the author.

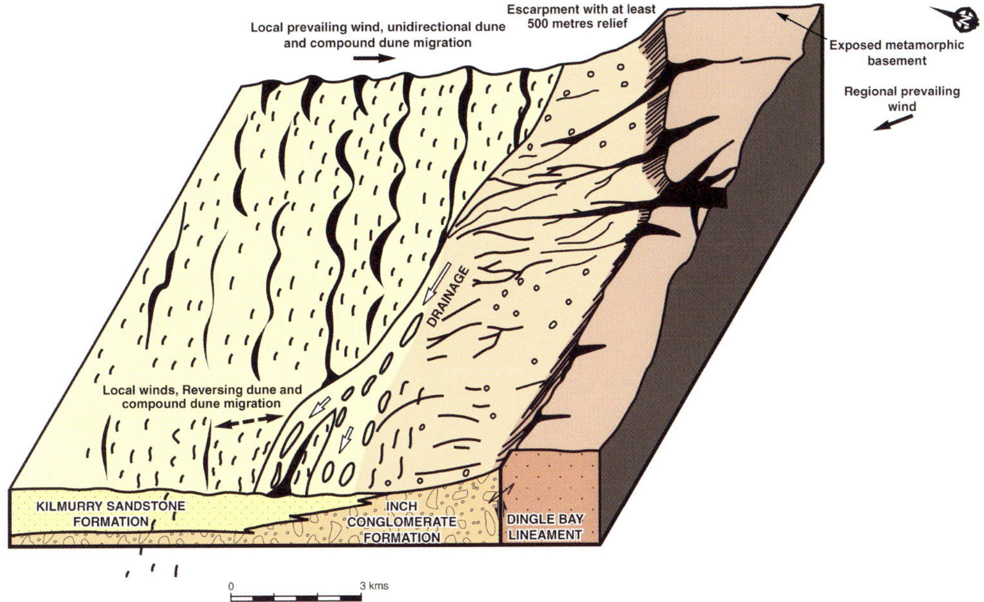

Fig. 16. Palaeogeographical reconstruction of the Caherbla Group from Todd *et al.* (1988*a*).

preserved, the across-fan width of the lithosomes can be mapped and appears to be in the region of 5 km. If this value is taken as a minimum for the fan diameter and assuming a simple semi-circular plan shape, a fan area of about 20 km^2 is estimated.

Empirical relationships between fan area and drainage basin area have been reported by Denny (1965), for example, and typically take the form of

$$A_{\mathrm{f}} = cA_{\mathrm{d}}^{n} \tag{9}$$

where A_{f} is the area of the fan, A_{d} is the area of the drainage basin, n generally varies between 0.8 to 1.0 (Bull 1977), and c is dependent on factors such as lithology, climate slope and tectonic setting, and shows considerable variation. Taking a mean value of n of 0.9 and a range of c from 0.4 to 1.3 (Hooke 1968), a range in size for the Inch drainage areas from 12 to 34 km^2 is estimated. Assuming a simple circular plan shape for the drainage basin, a maximum drainage length of 4–7 km is estimated. A more elongate triangular-shaped drainage basin would suggest a greater possible reach to the south, but still more than 20 km would be unlikely. It is probable that the drainage basins lay in the area of what is now Dingle Bay.

Clast composition

The petrography of the clasts and matrix of the Inch Conglomerate Formation was first described in detail by Capewell (1951). Capewell's admirable account encompasses most of the lithologies encountered in this study, in which more than 80 samples were examined in thin section. Most of these samples were derived from the well-exposed rocks of lithosome 1, although these were supplemented by sampling from the other lithosomes. Several types of lithology are recognized in the clasts (Capewell 1951), and these are in approximate order of abundance: schist; gneiss; mylonite; phyllite; metaquartzite and metagreywacke; vein quartz; tourmalinite and coticule; chert and jasper; granite; serpentinite. Of these, the first three groups make up by far the greatest proportion (>70%) of the clasts, and these are described in detail below. The largest boulders in lithosome 1 are largely composed of gneiss and the cobbles are typically composed of psammitic and semipelitic schists. In lithosome 2, which contains few clasts of small cobble or greater grade, is dominated by clasts of green phyllite, phyllonite, mylonite and cleaved metagreywacke. The clasts of lithosomes 3 and 4 are similar in that they contain abundant psammitic schists and metaquartzites, plus a significant percentage of clasts of deformed granite.

Abundant as clasts in the Inch conglomerates is quartz–muscovite schist which has a mineralogy dominated by quartz and muscovite but also includes feldspar, biotite (typically altered to chlorite), garnet, tourmaline, zircon, apatite and traces of magnetite. Quartz, with interlocking, sutured grain boundaries, comprises 50–80% of these rocks, and large, platy muscovites and biotite make up much of the remainder. The schistosity is formed by the preferred alignment of the micas. Feldspar, when present in small quantities, is usually cloudy and altered. Garnets, when present, occur as small subhedral to anhedral, colourless to buff grains, which are typically chloritized along cracks. These small (<0.5 mm) garnets are often enclosed by quartz, muscovite and tourmaline. Tourmaline occurs as squat, prismatic grains, green or violet in colour and displaying zoning and pleochroism. One sample of micaceous schist also bears pale green actinolite. The quartz–muscovite schists are clearly of metasedimentary origin (Capewell 1951), with protoliths varying from semi-pelitic to psammitic.

Some clasts of fine- to medium-grained schist contain considerably greater quantities of feldspar and chlorite, but are otherwise similar to the quartz muscovite schists described above (Capewell 1951). Indeed, the quartz–muscovite schists may form a gradational series with the quartz–feldspar–muscovite schists. The feldspar is dominated by albite, although smaller quantities of alkali feldspar also occur. In these quartz–feldspar–mica schists, quartz forms 50–70% of the rock. Feldspar makes up about 20%, with mica plus accessory minerals making up the remainder. Both muscovite and biotite occur, although the latter is usually partially or completely altered to chlorite. Garnet, tourmaline, magnetite, zircon and apatite occur as accessories. One sample of a coarse-grained, quartz–feldspar–muscovite schist apparently contains two generations of garnet.

Several more mafic-rich schists occur as pebbles. Amongst these are the serpentenite listed above, which is a schistose rock dominated by stretched and flattened opaque grains of iron ore with subordinate fine-grained fibrous mats of a colourless serpentinite mineral with first-order grey interference colours. Other dark green schist pebbles prove to be in thin section chlorite–sericite schist with subordinate aggregates of quartz.

Foliated pelitic rocks, too fine grained to be termed schists, also occur. Green phyllites have thin bands of quartz set in a foliated matrix of illite and chlorite. A single clast of purple phyllite, rich in hematite, was also found. This clast, with its foliation formed by aligned illite and chlorite and segregations or spots of quartz, sericite and chlorite is identical to a distinctive clast type that occurs in the Trabeg Conglomerate Formation (see above).

Coarser than the schists described above, but gradational with them, are coarse-grained banded to streaky gneisses. Like the schists, these gneisses are also gradational with mylonitic rocks that were derived from gneiss and schist protoliths. A banded to streaky, mesocratic to melanocratic gneiss is a lithology common as clasts in the conglomerates of lithosome 1. Moreover, some of the largest clasts, for example one that measures more than 1 m diameter, are formed of banded gneiss. The gneiss is composed of quartzo-feldspathic bands of coarse-grained colourless quartz, plus mica, chlorite and tourmaline. These bands are enclosed in a dark green, finer-grained matrix of chlorite, muscovite and some quartz and feldspar. The banding is folded into tight to isoclinal folds, which are sometimes rootless owing to extreme thickening of the quartzo-feldspathic layers at the fold hinges. A second schistosity is weakly developed parallel to the axial planes of these small-scale folds. Microscopically, the feldspar proves to be mainly plagioclase, probably albite, with minor amounts of alkali feldspar. Large, porphyroblastic plates of muscovite dominate over smaller laths of greenish brown biotite, which is typically partially or completely replaced by chlorite. Although the main constituent minerals occur in both types of metamorphic layer, it is the concentration of coarse-grained quartz and feldspar into bands that gives the rock its distinctive leucocratic layers.

Fairly common in the Inch conglomerates are clasts with a coarse-grained granitoid mineralogy, but that contain a pervasive foliation. Following Capewell (1951, p. 157), these clasts are labelled granite gneiss. The granite gneiss differs from the gneiss described above in that the granitic lithologies are less mafic and paler in colour in hand specimen. No completely undeformed granitoid rocks were encountered in this study. In fact, the granite gneiss owes its foliation to a sub-mylonitic fabric forming bands of quartz, alkali and plagioclase feldspar. These are intercalated with felted aggregates of sericite, replacing muscovite and feldspar, and chlorite replacing biotite. Relict biotite, muscovite and feldspar persist and the original micas do not have a preferred orientation suggestive of an initial granitic texture of a two-mica adamellite. Garnet also occurs as subhedral grains. It is entirely possible that these granitic lithologies form a gradational series with the gneisses and probable that some of the quartzo-feldspathic mylonites encountered as clasts are more deformed equivalents of these rocks.

One sampled boulder of granite gneiss proved to have a rather unusual lithology, both petrographically and geochemically. In hand specimen, it is pale pink, foliated and contains coarse-grained quartz, feldspar, pink garnet and chlorite. The foliation is produced by banding on a scale of up to 10 mm. The bands comprise subtle alternations of layers rich in albite and quartz with those with quartz, muscovite, garnet and chlorite. The quartz is medium to coarse grained and deformed. Unlike other schist and gneiss lithologies, the quartz in this rock has straight to curved, not sutured or wavy grain boundaries. The plagioclase forms large laths that contain finely repeated lamellar twins. Some of these twins are bent, indicating post-crystallization deformation. Muscovite forms large (sometimes >5 mm long) porphyroblastic plates that lie parallel to the foliation. The large, euhedral to subhedral garnet in this granite gneiss variety also differs from the garnet in other Inch clasts, which is usually small and abraded. Geochemical and isotopic analyses also demonstrate that this leucocratic granite gneiss is rather unusual.

As already noted above, there is a complete gradation, represented in the clast assemblage of the Inch conglomerates, from schist and gneiss to mylonite. The mylonites were clearly derived by high shear-strain deformation of a protolith of schist and gneiss very similar to the lithologies observed in unmylonitized clasts in the Inch conglomerates. The Inch source must have been largely composed of medium-grade schists and gneisses, possibly cut by banded granitoids, traversed by extensive, mylonitic shear zones.

Coarser-grained blastomylonites contain banded aggregates of quartz, sometimes with a small amount of feldspar (depending on the mineralogy of the protolith), set in a finer-grained matrix of sericite, chlorite, magnetite and some larger relict porphyroclasts of feldspar and mica. The quartz ribbons are formed of stretched, interlocking crystals with sutured grain boundaries. The parallelism of the quartz ribbons defines the foliation of the mylonite. The phyllitic bands that were intercalated with the quartz ribbons comprise felted masses of fibrous or less commonly acicular sericite, with chlorite and magnetite. Relict, coarse-grained porphyroclastic grains that float in this phyllitic matrix include abraded and crenulated muscovite, sericitized feldspar and tourmaline. Alteration textures of large mica and feldspar grains with rough, abraded edges passing into sericite are commonly preserved. Some of the larger, relict feldspars form rounded porphyroclastic augen. Garnets are uncommon in the mylonites and tend to reside in the phyllitic matrix as abraded, cracked, broken and chloritized grains, suggesting the mineral did not resist the effects of mylonitization. In contrast, tourmaline survives as euhedral grains. The mylonitic quartz-ribbon foliation of the coarse-grained blastomylonites is commonly contorted by later crenulative microfolds with an accompanying axial planar crenulation cleavage.

Finer-grained ultramylonite usually forms clasts of pebble grade. The rocks are typically green, phyllitic and include abundant quartz veins in addition to ribboned segregations of quartz. The finer-grained mylonites are characterized by much more extensive grain reduction and recovery to produce a more even-grained rock of quartz, sericite, chlorite and abundant opaque grains. Porphyroclastic micas, feldspars and garnets do not survive, although tourmaline is present in some cases. The abundance of opaque grains in both coarse- and fine-grained mylonites, but not in the presumed protoliths, suggests that the movement of ferrous fluids in the shear zones accompanied mylonitization. Like the coarse-grained variety, the phyllonitic foliation in the fine-grained mylonites is often deformed by a later crenulation cleavage. Several clasts in lithosome 2, in Inch Glen, preserve two crenulation cleavages, which deform the mylonitic fabric.

Granitoid lithologies usually form clasts less than cobble grade although several boulders of foliated granite have been encountered. Undeformed granite has not been observed in this study. Adamellite compositions dominate, although one sample contains sufficient plagioclase to be termed a granodiorite. Muscovite is ubiquitous and biotite is represented by pseudomorphs of chlorite and rutile, the latter forming a pseudohexagonal pattern of needles. Quartz grain boundaries commonly exhibit granulitization.

Amongst the other clast types in the Inch conglomerates, not already noted, are quartz wacke, quartz arenite, vein quartz, tourmalinite (quartz–tourmaline rock), garnetiferous quartzite (coticule; Kennan & Kennedy 1983; Kennan 1986) and quartz–feldspar porphyry. Apart from vein quartz, which is usually present, if not abundant, these clast types represent rarities.

The phenoclasts are set in a red sandstone matrix, CS to VCS grade, which is composed of sub-angular to sub-rounded quartz, hematite, chlorite, muscovite, garnet, tourmaline and lithic fragments. Capewell (1951, p. 162) separated out the following assemblage of non-opaque heavy mineral grains: zircon 39%; tourmaline 37%; garnet 16%; rutile 5%; staurolite 3.5%. All these minerals, except staurolite, occur as rock-forming minerals in the clast lithologies and their source is patently obvious. The presence of staurolite, however, possible indicates either the former existence of staurolite-bearing schists in the Inch source that were too soft and/or too distant to provide lithic clasts, or pre-existing sediments with detrital staurolite.

Geochemistry

A reconnaissance study of the isotope geochemistry of the schist and gneiss clasts of the ICF was undertaken as part of this research. An attempt was was made to sample the most important schist and gneiss lithotypes and analyse large boulder clasts from which unspoilt geochemical samples could be specified after trimming of weathered surfaces. The results are listed in Table 4. With the exception of the granitic gneiss sample (ST8646), the clasts yield T_{DM} ages less than 2.0 Ga, in keeping with metasediments sampled from south of the Iapetus Suture (Todd *et al.* 1991). However, the 3.3 Ga T_{DM} age for the granitic gneiss is surprisingly old. If the age for this clast is valid, then an Early Proterozoic (e.g. Lewisian) crustal component must be included. This is apparently inconsistent with an expected Avalonian basement type for this position south of the suture and the phenomenon deserves further sampling and analysis.

The three whole-rock–muscovite (WR–Ms) ages suggest two isotopic events, which, although they need to be substantiated by further sampling and analysis, may be tied to the deformation history observed in the clasts. The two older ages of 429 ± 6 and 435 ± 4 Ma are early Silurian (time scales of McKerrow *et al.* (1985) or Harland *et al.* (1990)) and are interpreted as being related to the retrograde metamorphic event that led to the growth of the large porphyroblastic micas in the schists. The younger 398 ± 5 Ma age is interpreted to represent Early Devonian (late Emsian of the time scale of Tucker *et al.* (1998)) resetting of the isotopic system during the mylonitization that altered the pre-existing schists and gneisses. Given that the ICF is considered to be late Emsian to early Eifelian in age this event must have occurred rapidly and immediately before to deposition of the conglomerates.

Provenance

The Inch clasts therefore indicate a metasedimentary basement source, probably intruded by a granite and cut by mylonitic fault zone(s). This is consistent with the tectonic model of an uplifted block along the Dingle Bay Lineament. The sedimentary cover to this block was probably stripped off the area previously to provide the TCF sediment and other detritus shed into the Dingle Basin from the south (Todd 1989*a*). A simple tectonometamorphic history may be read from the metamorphic fabrics observed in the clasts: (1) sedimentary protolith of wacke sandstones and mudrocks; (2) metamorphism at least to garnet and possibly staurolite grade, gneiss formation; (3) granite intrusion possibly at this stage, but unconstrained; (4) retrograde metamorphism to muscovite grade (micas enclose garnet); (5) further retrograde metamorphism accompanying mylonitization.

The clasts of the Inch Conglomerate Formation are unique in SW Ireland in that they provide an indirect record of the Caledonian or pre-Caledonian basement in an area in which no crystalline basement rocks crop out. However, with regard to the affinity of the Inch metamorphic basement, the petrographic observations and limited isotopic data are inconclusive. As suggested by Capewell (1951), the rocks share some similarities with the Dalradian schists and gneisses of Connemara, which lie some 150 km to the north of the Dingle Peninsula. On the other hand, the Inch clasts have been compared to the basement lithologies of the Rosslare Complex, some 200 km to the ESE of Dingle (Max & Long 1985). Clearly, petrographic observations alone are insufficient in determining the true affinity of the Inch basement. The Sm–Nd isotopic data are largely consistent with an Avalonian-type basement similar to Rosslare (Davies *et al.* 1985). Further similarity between the clasts and the Rosslare block is that the two granites in the Rosslare region yield Silurian ages similar to the WR–Ms ages reported here. The post-metamorphic, relatively undeformed Carnsore granodiorite has been dated (Rb–Sr and U–Pb) by O'Connor *et al.* (1988) at *c.* 428 Ma. It was emplaced contemporaneously with or slightly later than the foliated Saltees granite, which was dated (Rb–Sr) by Max *et al.* (1979) at 436 ± 7 Ma. These shared early Silurian tectonothermal events point towards a linkage between the two areas. Only the very old T_{DM} age from the gneiss (ST8646) is problematic in this interpretation, being much older than anything so far analysed from Rosslare. The further

Table 4. *Sm–Nd and Rb–Sr isotope geochemistry of the Inch clasts*

No.	Lithology	Sm	Nd	$^{147}Sm/^{144}Nd$	$^{143}Nd/^{144}N$	T_{DM} (Ga)	Rb	Sr	$^{87}Rb/^{86}Sr$	$^{87}Sr/^{86}S$	Ms–WR age (Ma)
ST8638	Schist	5.011	26.249	0.1154	0.511940 ± 12	1.701	57.9061	78.7195	2.133554	0.733033	
ST8643	Schist	7.281	36.545	0.1204	0.511945 ± 14	1.784	76.4734	97.4594	2.275428	0.731033	429 ± 6
ST8644	Schist	3.744	19.979	0.1133	0.512133 ± 14	1.376	70.459	180.273	1.132165	0.719876	398 ± 5
ST8645	Mylonite	9.675	48.958	0.1195	0.511922 ± 16	1.804					
ST8646	Gneiss	1.386	4.919	0.1703	0.512123 ± 18	3.338	216.474	111.877	5.729033	0.946517	435 ± 4

All Inch clasts were analysed by M. Murphy using a semi-automated VG Micromass 30 mass spectrometer in the Department of Geology, University College Dublin. Analytical techniques were those described by Menuge (1988) as modified by Menuge & Daly (1990). $^{143}Nd/^{144}Nd$ ratios were determined from spiked samples and are normalized to $^{146}Nd/^{144}Nd = 0.7219$. $^{87}Sr/^{86}Sr$ ratios are corrected for mass fractionation by normalizing to an $^{87}Sr/^{88}Sr$ value of 0.1194. Quoted errors in the $^{143}Nd/^{144}Nd$ ratios are within-run precisions; reproducibility of the $^{147}Sm/^{144}Nd$ and $^{87}Rb/^{86}Sr$ ratios are 0.1% and 1% respectively. All errors are quoted at the $2\sigma_m$ level. Sm–Nd ages were calculated using a decay constant for ^{147}Sm of 6.54×10^{-12} a^{-1}. Uncertainty in arising from analytical errors are typically 40 Ma, and are calculated relative to the depleted mantle model of DePaolo (1981) using presentday CHUR ratios of $^{143}Nd/^{144}Nd = 0.512688$ and $^{147}Sm/^{144}Nd = 0.1966$, renormalized from Jacobsen & Wasserburg (1980). Rb–Sr ages were calculated using a decay constant for ^{87}Rb of 1.42×10^{-11} a^{-1}. Muscovites from sample ST8638 failed to provide a satisfactory analysis and muscovites were not separated from ST8645.

resolution of these problems requires future further isotopic data.

Discussion: regional context

In this section the previous observation and interpretations are drawn together in a discussion of the regional setting of the Dingle Basin and its conglomeratic sedimentary fill during the late Caledonian convergence of Laurentian and Avalonian terranes in the Iapetus Suture Zone.

Basin-forming mechanism

Structural, stratigraphic and sedimentological evidence converge to produce a picture of the Dingle Basin as an elongate, ENE-trending trough, some 30–50 km wide and at least 60 km in length (Todd *et al.* 1988*a*; Todd 1989*a*). The TCF and GF systems were dispersed laterally into the basin from north and south, and the basin centre was occupied by a major river (e.g. SHF; Todd *et al.* 1988*a*). The major Caledonian structure of the region is dominated by the probable surface trace of the Iapetus Suture immediately to the north, and movement on this and related structures may have been responsible for forming the northern margin of the Dingle Basin. To the south lay the Dingle Bay Lineament, which also has a geophysical expression, and this feature probably formed the southern margin of the basin during early Devonian time. Structural data from onshore fault zones parallel to the DBL indicate sinistral transpression (Todd 1989*a*, *c*). Thus the basin was formed by sinistral transpression, with subsidence possibly caused by either crustal loading and/or along-strike (pull-apart) extension (Todd *et al.* 1988*a*; Todd 1989*a*). The main Dingle Basin was inverted before shedding of the ICF fan system from a basement block that was uplifted along the DBL (Horne 1975; Todd *et al.* 1988*a*; Todd 1989*a*).

Early Devonian hinterland geology

The data presented in this paper allow the geology of the region surrounding the Dingle Basin during Early Devonian (Pragian to Early Emsian) time to be reconstructed in terms of the tectonostratigraphic terranes that make up the Lower Palaeozoic geology of Ireland (Hutton 1987; Murphy 1987*a*; Murphy *et al.* 1991). In essence, it appears likely that the current configuration as observed in the various Lower Palaeozoic inliers is largely similar to that deduced from the provenance of the conglomerates.

Within the Iapetus Suture zone of Ireland two main terranes are defined (Murphy *et al.* 1991; Todd *et al.* 1991; Fig. 17). The Central Terrane lies entirely north of the suture and comprises the Longford–Down Massif and related rocks in inliers in central Ireland. The Central Terrane extends northeast into the Southern Uplands of Scotland. South of the suture lies the extensive Leinster Terrane, which is made up of the Lower Palaeozoic rocks of the Leinster Massif and related inliers in SE Ireland. The precise relationship between the Leinster Terrane to the Rosslare Terrane to the south and the smaller Bellewstown and Grangegeeth Terranes to the north is not constrained. Certainly the Leinster Terrane is the Irish along-strike equivalent of the English Lake District.

The TCF had a source area with lithotypes similar to that of Leinster, including one or more high-level granitoid intrusions with an aureole of spotted slates and greywackes and quartzites. Some of the Trabeg clasts have local Dingle provenance, although these rocks are also assigned a Leinster affinity (Todd *et al.* 2000). Thus the Dingle Basin was formed on the Leinster Terrane and received sediment from uplifted Leinster Terrane during Early Devonian time. The ICF metamorphic clasts indicate that part of this source area was completely stripped of cover to expose a basement hinterland in mid-Devonian time.

In contrast, the GF conglomerates were sourced from a largely volcanic hinterland, similar to outcrops in the northern margin of the Leinster Terrane, and the Ordovician succession of the Grangegeeth, Bellewstown and Central Terranes. The jasper in the GF is best explained by provenance from the Northern Belt of the Central Terrane. Hence the drainage area of the GF straddled the Iapetus Suture. The SHF has a provenance that shares aspects of both the TCF and GF.

Siluro-Devonian tectonic history

Figure 17 presents a chronostratigraphy of the regional evolution of the Iapetus Suture Zone. The context of the Early Devonian Dingle Basin lies firmly in the processes of late Caledonian or Acadian convergence of Avalonia and Laurentia (Todd *et al.* 1991). Debate continues on whether collision between these two continents occurred in Ordovician time or later in Silurian time. It is not an objective of this paper to review that debate, rather it is to focus on the Siluro-Devonian part of the history relevant to the Dingle Basin.

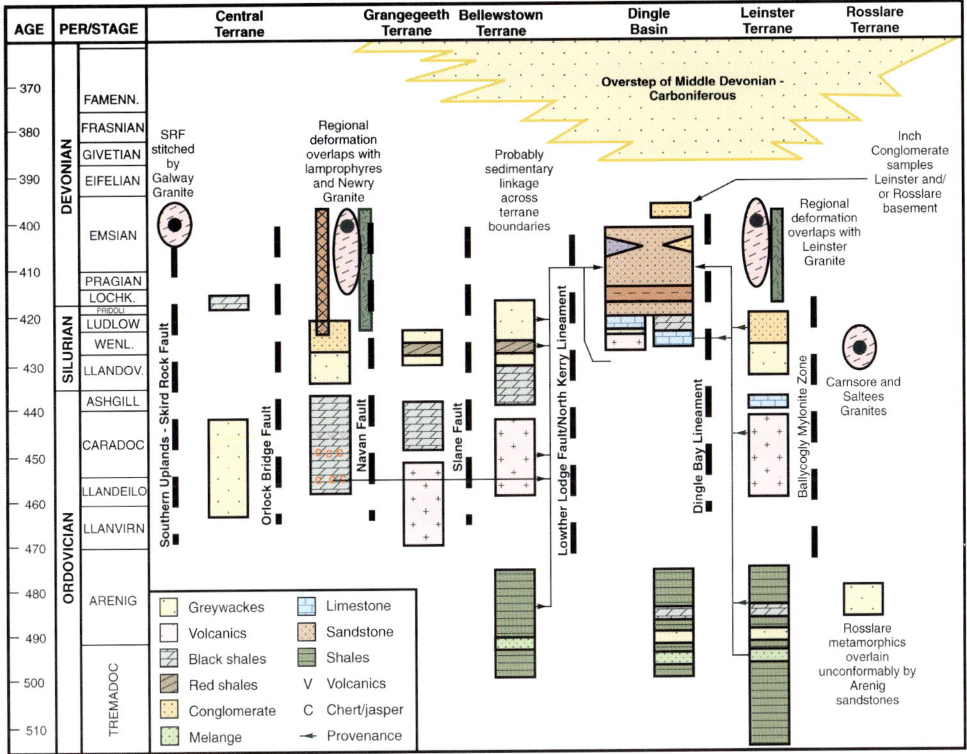

Fig. 17. Regional chronostratigraphy of the Iapetus Suture Zone in Ireland showing the deformation history and linkages between the main terranes or blocks. Data from a variety of sources including Hutton & Murphy (1987), Murphy *et al.* (1991) and other references cited in the text. The time scale is from Tucker *et al.* (1988) for the Devonian period and from McKerrow *et al.* (1985) for the Ordovician and Silurian periods.

During the Llandovery to Wenlock period in central Ireland turbidites were being shed both from the north and south into a basin that straddles the current suture (Hutton & Murphy 1987). The subsidence patterns in the Lake District along-strike reflect foreland basin subsidence as Avalonia was thrust under the Southern Uplands (King 1994). Sediment sources for these mudrocks, sandstones and microconglomerates possibly included reworking of the Ordovician Northern Belt and a now missing volcanic terrane to the north, and the Ordovician rocks of the SE part of the Leinster Terrane to the south. Silurian deformation of the Ordovician and indeed the early Silurian sediments themselves in the Central Terrane is constrained by the structural relationships of two suites of lamprophyre dykes in Co. Down (Anderson & Oliver 1996). The earlier suite post-date the local F1 folds in the Llandovery turbidites but are themselves deformed by D2 (Anderson & Oliver 1996). The younger suite post-date the folding, but are deformed by Caledonian sinistral strike-slip faults; one of this younger suite has been dated as 415 ± 12 Ma, which is in Ludlow time in the time scale of Harland *et al.* (1990) or of Lochkovian time in that of Tucker *et al.* (1998). Thus at least D1 is Silurian in age in the Central Terrane and its equivalent in the Southern Uplands of Scotland (Barnes *et al.* 1989). Early Silurian deformation south of the suture is evidenced by the chronometric ages of the Saltees and Carnsore granites and two WR–Ms ages from Inch clasts (see above).

During Wenlock time in the Dingle region shallow marine to non-marine sediments were deposited on the flanks of acid to intermediate volcanic islands (Sloan & Williams 1991). The geochemical character of the volcanic rocks suggests a subduction-related component (Sloan & Bennett 1990), but regional reconstruction of Iapetus as a diminished seaway during this time suggests that if the subduction was contemporaneous it was occurring at a distant trench, possibly related to the Rheic Ocean to the south (Fig. 1).

Where preserved in Ireland, Ludlow rocks are turbidites in the successor basin of central

Ireland or shallow-marine siliciclastic deposits passing into the non-marine clastic deposits of the Dingle Group in the Dingle Peninsula. Thus the vestigial Iapetean seaway was filled and occluded by Ludlow time. The marine to non-marine transition is not always conformable in the Dingle area, with unconformities being generated along newly defined basin margins during latest Silurian time.

Continued deformation and uplift of the areas surrounding the Dingle Basin during Přídolí to Emsian time is evidenced by the influx of sediment from these sources, bearing reworking Silurian and ?Ordovician rocks. Continued regional deformation in a sinistral transpressive regime within the suture zone during Early Devonian time is dated by an age spread of 418–395 Ma for lamprophyre dykes in the Southern Uplands (Rock *et al.* 1986). These dykes of Ludlow to Pragian age on the time scale of Harland *et al.* (1990) or of Lochkovian to Emsian age in the time scale of Tucker *et al.* (1998). The Leinster and Newry granites were emplaced late in the deformation history of their country rocks and apart from their margins are largely undeformed. With isotopic dates of 405 ± 2 Ma for Leinster (O'Connor *et al.* 1989) and 399 ± 3 Ma (Meighan & Neeson 1979) for Newry, the emplacement of granite in the suture appears to have continued during Early Devonian time and climaxed with uplift, cooling and isotopic closure in late Emsian time (time scale of Tucker *et al.* (1998)). The Dingle Group conglomerates (SHF, TCF and GF) are now dated to early to ?mid-Emsian time and the overlying and youngest preserved sediments in the Dingle Group (Ballymore Formation) are probably of late Emsian age. Hence Dingle Group deposition was largely synchronous with the final stages of granite emplacement in the suture zone. Caherbla Group rocks deposited after (or during?) the inversion of the Dingle Basin are younger than 398 ± 5 Ma (WR–Ms age of clast) or of late Emsian age on the time scale of Tucker *et al.* (1998). Hence the inversion of the Dingle Basin and deposition of the ICF is likely to have followed on quickly from final Dingle Group deposition in the latest Emsian or early Eifelian time.

The oldest proven rocks in the Munster Basin south of the Dingle Basin are Givetian in age (Enagh tuff, 385 ± 0.7 Ma; Williams et al., 1987, this volume). This large late Devonian extensional basin formed on the area of crust that was being uplifted and eroded during Early Devonian time. Thus, as described by Todd (1989*a*), the DBL changed from being an Early Devonian transpressive structure to a Mid–Late Devonian extensional structure, forming the NW margin of the Munster Basin. A period of <15 Ma separates the final stages of the Dingle Basin and emplacement of the Leinster granite in the late Caledonian or Acadian sinistral transpressive regime during Emsian timeand the onset of subsidence in the Munster Basin during ?Eifelian–Givetian time (see Williams *et al.* this volume). This pronounced tectonic change in Mid-Devonian time marks the end of the Caledonian cycle in Ireland, and suturing of Iapetus and the beginning of the Variscan cycle possibly related in some way to the Rheic Ocean to the south.

Conclusions

(1) The Dingle Basin was generated within a regime of late Silurian to Early Devonian sinistral transpressive convergence in the Iapetus Suture Zone (Todd 1989*a*).

(2) The conglomeratic sediments of the Dingle Basin record lateral alluvial apron systems shed from the north and south, and a large axial fluvial system that flowed NE.

(3) The provenance of the Dingle Group conglomerates indicates that the main Caledonian terranes in the Iapetus Suture Zone of Ireland occupied a position during Early Devonian time similar to their current configuration.

(4) The generation, filling and inversion of the Dingle Basin was penecontemporaneous with Late Caledonian deformation of these terranes and emplacement and cooling of the large granite batholiths of the suture zone.

(5) After or during the inversion of the Dingle Basin, probably late Emsian or Eifelian time, the Inch Conglomerate was shed northwards from a uplifted source area of metamorphic rocks, which may form the basement exposed by stripping of the sedimentary cover of same area during the TCF.

(6) The basement sampled by the Inch conglomerates has at least a Silurian metamorphic history and may be Leinster basement and/or be similar to the Rosslare Terrane.

This paper has benefited from the efforts of H. Armstrong, M. G. Bassett, C. H. Holland and C. R. Scrutton for fossil identification, and of M. Murphy and C. Elders for isotopic analysis. The paper forms part of research carried out by the author during a Shell PhD studentship at the University of Bristol and a temporary lectureship at University College Dublin. Although the interpretations expressed are the author's own, the research was enriched by discussion and help from B. P. J. Williams and P. L. Hancock (PhD supervisors) and M. C. Bennett, D. Johnston, P. Kennan, M. J. Kennedy, S. King, F. C. Murphy,

C. P. North, P. Shannon, R. J. Sloan, J. P. Turner and D. Went, and other colleagues in Bristol and Dublin. The manuscript was greatly improved by thorough reviews by J. Graham and an anonymous referee. The paper is dedicated to the memories of M. Bennett, P. Hancock and D. Johnston, who all played an important role in the author's geological education.

References

ALLEN, J. R. L. 1968. *Current Ripples*. North-Holland, Amsterdam.

—— & CROWLEY, S. F. 1983. Lower Old Red Sandstone fluvial dispersal systems in the British Isles. *Transactions of the Royal Society of Edinburgh: Earth Sciences*, **74**, 61–68.

—— & WILLIAMS, B. P. J. 1981. *Beaconites antarticus*: a giant channel associated trace fossil from the Lower Old Red Sandstone of South Wales and the Welsh Borderland. *Geological Journal*, **16**, 255–269.

ANDERSON, T. B. & OLIVER, G. J. H. 1996. Xenoliths of Iapetus Suture mylonites in County Down lamprophyres, Northern Ireland. *Journal of the Geological Society of London*, **153**, 403–407.

BARNES, R. P., LINTERN, B. C. & STONE, P. 1989. Timing and regional implications of deformation in the Southern Uplands of Scotland. *Journal of the Geological Society, London*, **146**, 905–908.

BASSETT, M. G., COCKS, L. R. M. & HOLLAND, C. H. 1976. The affinities of two endemic Silurian brachiopods from the Dingle Peninsula, Co. Kerry, Ireland. *Palaeontology*, **19**, 615–625.

BLUCK, B. J. 1985. The Scottish Paratectonic Caledonides. *Scottish Journal of Geology*, **21**, 437–464.

BRINDLEY, J. C. 1954. The Garnetiferous Beds of the Leinster Granite aureole is south Dublin and their small-scale structures. *Scientific Proceedings of the Royal Dublin Society*, **26**, 245–262.

BROWN, C. & WILLIAMS, B. 1985. A gravity and magnetic interpretation of the structure of the Irish Midlands and its relation to ore genesis. *Journal of the Geological Society, London*, **142**, 1059–1075.

BRÜCK, P. M. 1968. The geology of the Leinster Granite in the Enniskerry–Lough Dan area, Co. Wicklow. *Proceedings of the Royal Irish Academy*, **66B**, 53–70.

—— 1972. Stratigraphy and sedimentology of the Lower Palaeozoic greywacke formations in Counties Kildare and Wicklow. *Proceedings of the Royal Irish Academy*, **72B**, 25–53.

——, COLTHURST, J. R. J., FEELY, M. *et al.* 1979. Southeast Ireland: Lower Palaeozoic stratigraphy and depositional history. *In*: HARRIS, A. L., HOLLAND, C. H. & LEAKE, B. E. (eds) *The Caledonides of the British Isles—Reviewed*. Geological Society, London, Special Publications, **8**, 533–544.

BULL, W. B. 1977. The alluvial fan environment. *Progress in Physical Geography*, **1**, 222–270.

CANT, D. J. & WALKER, R. G. 1978. Fluvial processes and facies sequences in the sandy braided South Saskatchewan River, Canada. *Sedimentology*, **25**, 625–648.

CAPEWELL, J. G. 1951. The Old Red Sandstone of the Inch and Annascaul District, Co. Kerry. *Prodeedings of the Royal Irish Academy*, **58B**, 167–183.

—— 1965. The Old Red Sandstone of Slieve Mish, Co. Kerry *Proceedings of the Royal Irish Academy*, **64B**, 165–174.

—— 1975. The Old Red Sandstone of Slieve Mish, Co. Kerry *Proceedings of the Royal Irish Academy*, **75B**, 155–172.

DAVIES, G. R., GLEDHILL, A. & HAWKESWORTH, C. J. 1985. Upper crustal recycling in southern Britain: evidence from Nd and Sr isotopes. *Earth and Planetary Science Letters*, **75**, 1–12.

DENNY, C. S. 1965. *Alluvial fans in the Death Valley Region, California and Nevada*. US Geological Survey, Professional Papers, **466**.

DEPAOLO, D. J. 1981. Neodymium isotopes in the Colorado Front Range and crust–mantle evolution in the Proterozoic. *Nature, London*, **291**, 193–196.

DEWEY, J. F. 1971. A model for the Lower Palaeozoic evolution of the southern margin of the early Caledonides of Scotland and southern Ireland. *Scottish Journal of Geology*, **7**, 219–240.

EIMICKE, E. A. 1986. An interpretation of the Killarney and Leinster gravity anomalies in the Republic of Ireland. *Irish Journal of Earth Sciences*, **7**, 125–132.

ELDERS, C. F. 1987. *Caledonian tectonics from stratigraphy and isotope geochemistry of Lower Palaeozoic successions*. PhD thesis, University of Oxford.

FFRENCH, G. D. & WILLIAMS, D. M. 1984. The sedimentology of the South Connemara Group—a possible Ordovician trench fill sequence. *Geological Magazine*, **121**, 505–514.

FORD, M., BROWN, C. & READMAN, P. W. 1991. Analysis and tectonic interpretation of gravity data over the Variscides of southwest Ireland. *Journal of the Geological Society, London*, **148**, 137–148.

FRANCE, D. S. 1967. The geology of the Ordovician rocks at Balbriggan, Co. Dublin, Eire. *Geological Journal*, **5**, 291–304.

GARDINER, C. I. & REYNOLDS, S. H. 1902. The fossiliferous Silurian beds and associated igneous rocks of the Clogher Head district (Co. Kerry). *Quarterly Journal of the Geological Society, London*, **58**, 226–266.

GIBBONS, W. 1983. Stratigraphy, subduction and strike-slip faulting in the Mona Complex of north Wales—a review. *Proceedings of the Geologists' Association*, **94**, 147–163.

GREENLEY, E. 1919. *The Geology of Anglesey*. Memoir, Geological Survey of the United Kingdom (2 vols). London.

HARLAND, W. B., ARMSTRONG, R. L., COX, A. V., CRAIG, L. E., SMITH, A. G. & SMITH, D. G. 1990. *A Geologic Time Scale 1989*. Cambridge University Press, Cambridge.

HIGGS, K. 2000. Early Devonian spore assemblages from the Dingle Group, Co. Kerry, Ireland.

Bollettino della societa Paleontologica Italiana, in press.

—— & RUSSELL, K. J. 1981. Upper Devonian faunas and floras from Iveragh, Co. Kerry, Ireland. *Geological Survey of Ireland Bulletin*, **3**, 17–50.

——, CLAYTON, G. & KEEGAN, J. B. 1988. *Stratigraphic and Systematic Palynology of the Tournisian Rocks of Ireland.* Geological Survey of Ireland, Special Papers, **7**.

HOLLAND, C. H. 1969. Irish counterpart of the Silurian of Newfoundland. *In*: KAY, M. (ed.) *North Atlantic Geology and Continental Drift*. Memoirs, American Association of Petroleum Geologists, **12**, 298–308.

—— 1981*a*. Cambrian and Ordovician of the Paratectonic Caledonides. *In*: HOLLAND, C. H. (ed.) *A Geology of Ireland.* Scottish Academic Press, Edinburgh, 41–64.

—— 1981b. Silurian. *In*: HOLLAND, C. H. (ed.) *A Geology of Ireland.* Scottish Academic Press, Edinburgh, 65–82.

—— 1987. Stratigraphical and structural relationships of the Dingle Group (Silurian), County Kerry, Ireland. *Geological Magazine*, **124**, 33–42.

—— 1988. The fossiliferous Silurian rocks of the Dunquin inlier, Dingle Peninsula, County Kerry, Ireland. *Transactions of the Royal Society of Edinburgh: Earth Sciences*, **79**, 347–360.

HOOKE, R. LEB. 1968. Steady state relationships on arid-region alluvial fans in closed basins. *American Journal of Science*, **266**, 609–629.

HORNE, R. R. 1970. A preliminary re-interpretation of the Devonian palaeogeography of western County Kerry. *Geological Survey of Ireland Bulletin*, **1**, 53–60.

—— 1974. The lithostratigraphy of the late Silurian to early Carboniferous of the Dingle Peninsula, Co. Kerry. *Geological Survey of Ireland Bulletin*, **1**, 395–428.

—— 1975. The association of alluvial fan, aeolian and fluviatile facies in the Caherbla Group (Devonian), Dingle Peninsula, Ireland. *Journal of Sedimentary Petrology*, **45**, 535–540.

HOWARD, D. W. 1975. Deep-seated igneous intrusions in Co. Kerry. *Proceedings of the Royal Irish Academy*, **75**, 173–185.

HUTTON, D. H. W. 1987. Strike-slip terranes and a model for the evolution of the British and Irish Caledonides. *Geological Magazine*, **124**, 405–425.

—— & MURPHY, F. C. 1987. The Silurian of the Southern Uplands of Scotland and Ireland as a successor basin to the end-Ordovician closure of Iapetus. *Journal of the Geological Society, London*, **144**, 765–772.

JACKSON, R. G. 1976. Large scale ripples of the Lower Wabash River. *Sedimentology*, **23**, 593–594.

JACOBSEN, S. B. & WASSERBURG, G. J. 1980. Sm–Nd evolution of chondrites. *Earth and Planetary Science Letters*, **50**, 139–155.

JUKES, J. B. & DU NOYER, G. V. 1863. *Explanation of sheets 160, 161, 171 and part of 172. Memoir, Geological Survey of Ireland.*

KARCZ. I. 1972. Sedimentary structures formed by flash floods in southern Israel. *Sedimentary Geology*, **7**, 161–182.

KENNAN, P. S. 1986. The coticule package: a common association of some very distinctive lithologies. *Aardkundige Mededelingen*, **3**, 139–148.

—— & KENNEDY, M. J. 1983. Coticules—a key to correlation along the Appalachian–Caledonian orogen. *In*: SCHENK, P. E. (ed.) *Regional Trends in the Geology of the Appalachian–Caledonian–Hercynian–Mauritanide Orogen.* D. Reidel, Dordrecht, 355–361.

KENNEDY, M. J. 1979. The continuation of the Canadian Appalachians into the Caledonides of Britain and Ireland. *In*: HARRIS, A. L., HOLLAND, C. H. & LEAKE, B. E. (eds) *The Caledonides of the British Isles—Reviewed.* Geological Society, London Special Publications, **8**, 33–62.

KING, L. M. 1994. Subsidence analysis of Eastern Avalonia sequences: implications for Iapetus closure. *Journal of the Geological Society, London*, **151**, 647–657.

KLAPPER, G. & MURPHY, M. A. 1975. Silurian–Lower Devonian conodont sequences in the Roberts Mountains Formation of central Nevada. *University of California Publications, Geological Society*, **111**, 1–62.

KLEMPERER, S. L. 1989. Seismic reflection evidence for the location of the Iapetus suture west of Ireland. *Journal of the Geological Society, London*, **146**, 409–412.

LEOPOLD, L. B. & MADDOCK, T. 1953. The hydraulic geometry of stream channels and some physiographic implications. US Geological Survey, Professional Papers, **252**.

—— & MILLER, J. P. 1956. *Ephemeral Streams—Hydraulic Factors and their Relation to the Drainage Net.* US Geological Survey, Professional Papers, **282-A**.

—— & WOLMAN, M. G. 1957. *River Channel Patterns: Braided, Meandering and Straight.* US Geological Survey, Professional Papers, **202-B**.

——, WOLMAN, M. G. & MILLER, J. P. 1964. *Fluvial Processes in Geomorphology.* Freeman, San Fransisco, CA.

MAX, M. D. & LONG, C. B. 1985. Pre-Caledonian basement in Ireland and its cover relationships. *Geological Journal*, **20**, 341–366.

MAX, M. D., PLOQUIN, A. & SONET, J. 1979. The age of the Saltees Granite in the Rosslare Complex. *In*: HARRIS, A. L., HOLLAND, C. H. & LEAKE, B. E. (eds) *The Caledonides of the British Isles—Reviewed.* Geological Society, London, Special Publications, **8**, 723–725.

——, RYAN, P. D. & INAMDAR, D. D. 1983. A magnetic deep structural interpretation of Ireland. *Tectonics*, **2**, 431–451.

MCKERROW, W. S. 1988. Wenlock to Givetian deformation in the British Isles and Canadian Appalachians. *In*: HARRIS, A. L. & FETTES, D. J. (eds) *The Caledonian–Appalachian Orogen.* Geological Society, London, Special Publications, **38**, 437–448.

——, LAMBERT, R. ST & COCKS, L. R. M. 1985. The Ordovician, Silurian and Devonian periods. *In*: SNELLING, N. J. (ed.). *The Chronology of the Geological Record.* Geological Society, London, Memoirs, **10**, 73–80.

MEIGHAN, I. G. & NEESON, J. C. 1979. The Newry Igneous complex, C. Down. *In*: HARRIS, A. L., HOLLAND, C. H. & LEAKE, B. E. (eds) *The Caledonides of the British Isles—Reviewed.* Geological Society, London, Special Publications, **8**, 717–722.

MENUGE, J. F. 1988. The petrogenesis of massif anorthosites: a Nd and Sr isotopic investigation of the Proterozoic of Rogaland/Vest-agder, SW Norway. *Contributions to Mineralogy and Petrology*, **98**, 363–373.

—— & DALY, J. S. 1990. Proterozoic evolution of the Erris Complex, northwest Mayo, Ireland: neodymium isotope evidence. *In*: GOWER, C., RIVERS, T. & RYAN, B. (eds) *Mid-Proterozoic Laurentia–Baltica.* Geological Association of Canada, Special Papers, **38**, 41–51.

MIALL, A. D. 1976. Palaeocurrent and palaeohydrologic analysis of some vertical profiles through a Creteceous braided stream deposit, Banks Island, Arctic Canada. *Sedimentology*, **23**, 459–483.

MILLS, H. H. 1979. Downstream rounding of pebbles—a quantitative review. *Journal of Sedimentary Petrology*, **49**, 295–302.

MORRIS, J. H. 1983. The stratigraphy of the Lower Palaeozoic rocks in the western end of the Longford–Down inlier, Ireland. *Journal of Earth Sciences, Royal Dublin Society*, **5**, 201–218.

MURPHY, F. C. 1987*a*. Evidence for late Ordovician amalgamation of volcanogenic terranes in the Iapetus suture zone, eastern Ireland. *Transactions of the Royal Society of Edinburgh: Earth Science*, **78**, 153–167.

—— 1987*b*. Late Caledonian granitoids and timing of deformation in the Iapetus Suture Zone of eastern Ireland. *Geological Magazine*, **124**, 135–142.

——, ANDERSON, T. B., DALY, J. S. *et al.* 1991. An appraisal of Caledonian suspect terranes in Ireland. *Irish Journal of Earth Sciences*, **11**, 11–41.

NAYLOR, D., REILLY, T. A. SEVASTOPULO, G. D. & SLEEMAN, A. G. 1983. Stratigraphy and structure in the Irish Variscides. *In*: HANCOCK, P. L. (ed.) *The Variscan Fold Belt in the British Isles.* Hilger, Bristol, 20–46.

O'CONNOR, P. S., AFTALION, M. & KENNAN, P. S. 1989. Isotopic U–Pb ages of zircon and monazite from the Leinster Granite, southeast Ireland. *Geological Magazine*, **126**, 725–728.

——, KENNAN, P. S. & AFTALION, M. 1988. New Rb–Sr and U–Pb ages for the Carnsore Granite and their bearing on the antiquity of the Rosslare Complex, southeastern Ireland. *Geological Magazine*, **125**, 25–29.

PARKIN, J. 1974. Silurian rocks of Inishvickillane, Blasket Islands, Co. Kerry. *Scientific Proceedings of the Royal Dublin Society*, **A5**, 277–291.

—— 1976*a*. Silurian rocks of the Bull's Head, Annascaul and Derrymore Glen inliers, Co. Kerry. *Proceedings of the Royal Irish Academy*, **76B**, 577–606.

—— 1976*b*. The geology of the Foze Rocks, Co. Kerry: a review. *Irish Naturalists Journal*, **18**, 308–309.

PHILLIPS, W. E. A., STILLMAN, C. J. & MURPHY, T. 1976. A Caledonian plate tectonic model. *Journal of the Geological Society, London*, **132**, 579–609.

PRACHT, M. 1996. *Geology of Dingle Bay. A geological description to accompany the bedrock geology 1 : 100 000 series, Sheet 20, Dingle Bay.* Geological Survey of Ireland, Dublin.

PRICE, C. A. 1989. Some thoughts on the subsidence and evolution of the Munster Basin, southern Ireland. *In*: ARTHURTON, R. S., GUTTERIDGE, P. & NOLAN, S. C. (eds) *Role of Tectonics in Devonian and Carboniferous Sedimentation in the British Isles.* Yorkshire Geological Society, Special Publications, **6**, 111–121.

—— & TODD, S. P. 1988. A model for the development of the Irish Variscides. *Journal of the Geological Society, London*, **145**, 935–939.

READMAN, P. W., O'REILLY, B. M. & MURPHY, T. 1997. Gravity gradients and upper crustal tectonic fabrics, Ireland. *Journal of the Geological Society, London*, **154**, 817–828.

ROCL, N. M. S., GASKARTH, J. W. & RUNDLE, C. C. 1986. Lake Caledonian dyke swarms in Scotland: a regional zone of primitive K-rich lamprophyres and associated vents. *Journal of Geology*, **94**, 505–522.

ROMANO, M. 1980*a*. The stratigraphy of the Ordovician rocks between Slane (Co. Meath) and Collon (Co. Louth), Ireland. *Journal of Earth Sciences, Royal Dublin Society*, **5**, 53–79.

—— 1980*b*. The Ordovician rocks around Herbertstown (Co. Meath) and their correlation with the succession at Balbriggan (Co. Dublin), Ireland. *Journal of Earth Sciences, Royal Dublin Society*, **5**, 201–218.

RUSSELL, K. J. 1978. Vertebrate fossils from the Iveragh Peninsula and the age of the Old Red Sandstone. *Journal of Earth Sciences, Royal Dublin Society*, **1**, 151–162.

SANDERSON, D. J. 1984. Structural variation across the northern margin of the Variscides in NW Europe. *In*: HUTTON, D. H. W. & SANDERSON, D. J. (eds) *Variscan Tectonics in the North Atlantic Region.* Geological Society, London, Special Publications, **14**, 149–166.

SCHUMM, S. A. 1960. The shape of alluvial channels in relation to sediment type. *US Geological Survey Professional Paper* **352-B**.

—— 1968. Speculations concerning paleohydrologic controls of terrestrial sedimentation. *Geological Society of America Bulletin*, **79**, 1573–1588.

SHANNON, P. M. 1977. Diagenetic concretions from the Ribband Group sediments of Co. Wexford, Ireland. *Geological Magazine*, **114**, 127–132.

—— 1978. The stratigraphy and sedimentology of the Lower Palaeozoic rocks of southeast Co. Wexford. *Proceedings of the Royal Irish Academy*, **78B**, 247–265.

SLOAN, R. J. & BENNETT, M. C. 1990. Geochemical character of Silurian volcanism in SW Ireland. *Journal of the Geological Society, London*, **147**, 1051–1060.

—— & WILLIAMS, B. P. J. 1991. Volcano-tectonic control of offshore to tidal flat regressive cycles from the Dunquin Group (Silurian) of Dingle, SW Ireland. *In*: MACDONALD, D. I. M. (ed.) *Sea Level Changes at Active Plate Margins: Process and Product*. International Association of Sedimentologists, Special Publications, **12**, 105–119.

SOPER, N. J. & HUTTON, D. H. W. 1984. Late Caledonian sinistral displacements in Britain: implications for a three-plate collision. *Tectonics*, **3**, 781–794.

——, WEBB, B. C. & WOODWOCK, N. H. 1987. Late Caledonian (Acadian) transpression in northwest England: timing, geometry and geotectonic significance. *Proceedings of the Yorkshire Geological Society*, **46**, 175–192.

STILLMAN, C. J. 1981. Caledonian igneous activity. *In*: HOLLAND, C. H. (ed.) *A Geology of Ireland*. Scottish Academic Press, Edinburgh, 83–106.

TODD, S. P. 1989*a*. Role of the Dingle Bay Lineament in the evolution of the Old Red Sandstone of southwest Ireland. *In*: ARTHURTON, R. S., GUTTERIDGE, P. & NOLAN, S. C. (eds.) *Role of Tectonics in Devonian and Carboniferous Sedimentation in the British Isles*. Yorkshire Geological Society, Special Publications, **6**, 35–54.

—— 1989*b*. Stream-driven, high-density gravelly traction carpets: possible deposits in the Trabeg Conglomerate Formation, SW Ireland and some theoretical considerations of their origin. *Sedimentology*, **36**, 513–530.

—— 1989*c*. *Caledonian tectonics and conglomerate sedimentation in the Dingle Basin, southwest Ireland*. PhD thesis, University of Bristol.

—— 1991. The Silurian rocks of Inishnabro, Blaskett Islands, County Kerry and their regional significance. *Irish Journal of Earth Sciences*, **11**, 91–98.

—— 1996. Process deduction from fluvial sedimentary structures. *In*: CARLING, P. A. & DAWSON, M. R. (eds) *Advances in Fluvial Dynamics and Stratigraphy*, Wiley, Chichester, 299–350.

—— & WENT, D. J. 1991. Lateral migration of sand-bed rivers: examples from the Devonian Glashabeg Formation, SW Ireland and the Cambrian Alderney Sandstone Formation, Channel Islands. *Sedimentology*, **38**, 997–1020.

——, BOYD, J. D. & DODD, C. D. 1988*a*. Old Red Sandstone sedimentation and basin development in the Dingle Peninsula, southwest Ireland. *In*: MCMILLAN, N. J., EMBRY, A. F. & GLASS, D. J. (eds.) *The Devonian of the World*. Canadian Society of Petroleum Geologists, Memoirs, **14**(II), 251–268.

——, ——, SLOAN, R. J. & WILLIAMS, B. P. J. 1990. *Sedimentology and Tectonic Setting of the Siluro-Devonian Rocks of the Dingle Peninsula, SW Ireland*. Field Excursion Guide for Sediments 1990, 13th International Sedimentological Congress, Nottingham, UK.

——, CONNERY, C., HIGGS, K. T. & MURPHY, F. C. 2000. An Early Ordovician age for the Annascaul Formation of the SE Dingle Peninsula, SW Ireland. *Journal of the Geological Society, London*, **157**, 823–833.

——, MURPHY, F. C. & KENNAN, P. S. 1991. On the trace of the Iapetus Suture in Ireland and Britain. *Journal of the Geological Society, London*, **148**, 869–880.

——, WILLIAMS, B. P. J. & HANCOCK, P. L. 1988*b*. Lithostratigraphy and structure of the Old Red Sandstone of the northern Dingle Peninsula, Co. Kerry, southwest Ireland. *Geological Journal*, **23**, 107–120.

TUCKER, R. D., BRADLEY, D. C., VER STRAETEN, HARRIS, C. A., EBERT, J. R. & MCCUTCHEON, S. R. 1998. New U–Pb zircon ages and the duration and divison of Devonian time. *Earth and Planetary Science Letters*, **158**, 175–186.

TURNER, B. R. 1980. Palaeohydraulics of an Upper Triassic braided river system in the Karoo Basin, South Africa. *Transactions of the Geological Society of South Africa*, **83**, 425–431.

VAN DER ZWAN, G. J. 1980. Palynological evidence concerning the Devonian age of the Dingle Group, southwest Ireland. *Review of Palaeobotany and Palynology*, **29**, 277–284.

WEIR, J. A. 1973. Lower Palaeozoic graptolitic facies in Ireland and Scotland, review, correlation and palaeogeography. *Scientific Proceedings of the Royal Dublin Society*, **A4**, 439–460.

WILLIAMS, E. A., BAMFORD, M. L. F., COOPER, M. A. *et al.* 1989. Tectonic controls and sedimentary response in the Devonian–Carboniferous Munster and South Munster Basins, south-west Ireland. *In*: ARTHURTON, R. S., GUTTERIDGE, P. & NOLAN, S. C. (eds.) *Role of Tectonics in Devonian and Carboniferous Sedimentation in the British Isles*. Yorkshire Geological Society, Special Publications, **6**, 35–54.

——, SERGEEV, S. A., STOSSEL, I. & FORD, M. 1997. An Eifelian U–Pb zircon date for the Enagh Tuff Bed from the Old Red Sandstone of the Munster Basin in NW Iveragh, SW Ireland. *Journal of the Geological Society, London*, **154**, 189–193.

WILLIAMS, G. E. 1971. Flood deposits of the sand-bed ephemeral streams of central Australia, *Sedimentology*, **17** 1–40.

WILLIAMS, P. F. & RUST, B. R. 1969. The sedimentology of a braided river. *Journal of Sedimentary Petrology*, **39**, 649–679.

WILSON, J. T. 1966. Did the Atlantic close and then re-open? *Nature, London*, **211**, 676–681.

Wide-angle seismic control on the development of the Munster Basin, SW Ireland

N. J. VERMEULEN[1,2], P. M. SHANNON[1], F. MASSON[3] & M. LANDES[4]

[1]*Department of Geology, University College Dublin, Belfield, Dublin 4, Ireland*

[2]*Present address: Chevron Europe, 43–45 Portman Square, London W1H 0AN, UK (e-mail: jvrm@chevron.com)*

[3]*Laboratoire de Géophysique et Tectonique, Université Montpellier 2, CC 060, 4 pl. E. Bataillon, 34095 Montpellier Cedex 05, France*

[4]*Geophysics Section, Dublin Institute for Advanced Studies, 5 Merrion Square, Dublin 2, Ireland*

Abstract: Three wide-angle seismic profiles were acquired from onshore southern Ireland during VARNET-96. These data provide the only subsurface seismic images of a series of predominantly Late Palaeozoic sedimentary basins. The Dingle Basin is a broadly symmetrical rift on the Dingle peninsula, but assumes a half-graben geometry to the east. The southern syn-rift basin margin is defined by the north-dipping Dingle Bay–Galtee Mountains Line. A south-dipping normal fault at Kerry Head marks the northern basin margin in the west, with displacement on this structure decreasing towards the east. The Dingle Basin is separated from the Munster Basin to the south by a basement horst. Seismic data strongly suggest that the Killarney–Mallow Fault Zone, located on the southern margin of this positive feature, is the Munster Basin northern syn-rift marginal fault. The preserved extensional displacement on this detachment decreases from *c.* 2.5 km in northern Iveragh, to *c.* 1 km at Mallow, suggesting that the fault tips out to the east of Mallow. A tilted fault block, possibly cored by shallow granite, is imaged in the northern Munster Basin. This fault terrace is bounded to the south by the Cork–Kenmare Line. The combined Munster Basin and South Munster Basin succession thickens across the Cork–Kenmare Line to a maximum thickness of 8 km on the Beara peninsula. A tectonic thickness of 6 km of sediment is modelled at the Old Head of Kinsale, increasing to 7 km in the southwest.

The surface geology of southern Ireland (Fig. 1) is dominated by a series of inverted and deformed late Silurian to Late Carboniferous basins. The Devonian Munster Basin (Fig. 1) was formally recognized by Capewell (1965), who described the northern basin margin as a buried cliff line at the base of the Devonian succession in the Slieve Mish and Galtee Mountains. This palaeotopographic feature separated a stable or gently subsiding northern region from the rapidly subsiding Munster Basin to the south. Subsequent work has recognized the Dingle Bay–Galtee Mountains Line (DBGML) (Fig. 1), a long-lived Caledonian structural lineament, as the Munster Basin northern syn-rift margin (Naylor & Jones 1967; Gardiner 1975; Gardiner & MacCarthy 1981; Williams *et al.* 1989, 1990, 1998; Williams 1998). This structure approximately coincides with the northern 1 km isopachyte for the basinal Devonian Old Red Sandstone (ORS) succession (Naylor *et al.* 1983). Tectonic reconstructions proposed by Price & Todd (1988) and Meere (1995) favoured the Killarney–Mallow Fault Zone (KMFZ) (Fig. 1), the classic trace of the Variscan thrust front of onshore southern Ireland (Gill 1962), as the northern basin margin fault. The southern margin of the Munster Basin is poorly constrained, but is generally placed just offshore of southern Ireland, following the trend of the present coastline (Gardiner 1975; Gardiner & MacCarthy 1981). Williams *et al.* (1989) interpreted the Dunmanus–Castletown Fault (DCF) (Fig. 1), located in the SW onshore region, as the southern syn-rift basin margin.

The Munster Basin is typically considered to have developed as a half-graben in late Givetian to early Frasnian times. This basin accumulated over 5 km of fluvial, alluvial and aeolian ORS facies sediments (Clayton & Graham 1974;

From: FRIEND, P. F. & WILLIAMS, B. P. J. (eds). *New Perspectives on the Old Red Sandstone.*
Geological Society, London, Special Publications, **180**, 223–237. 0305-8719/00/$15.00

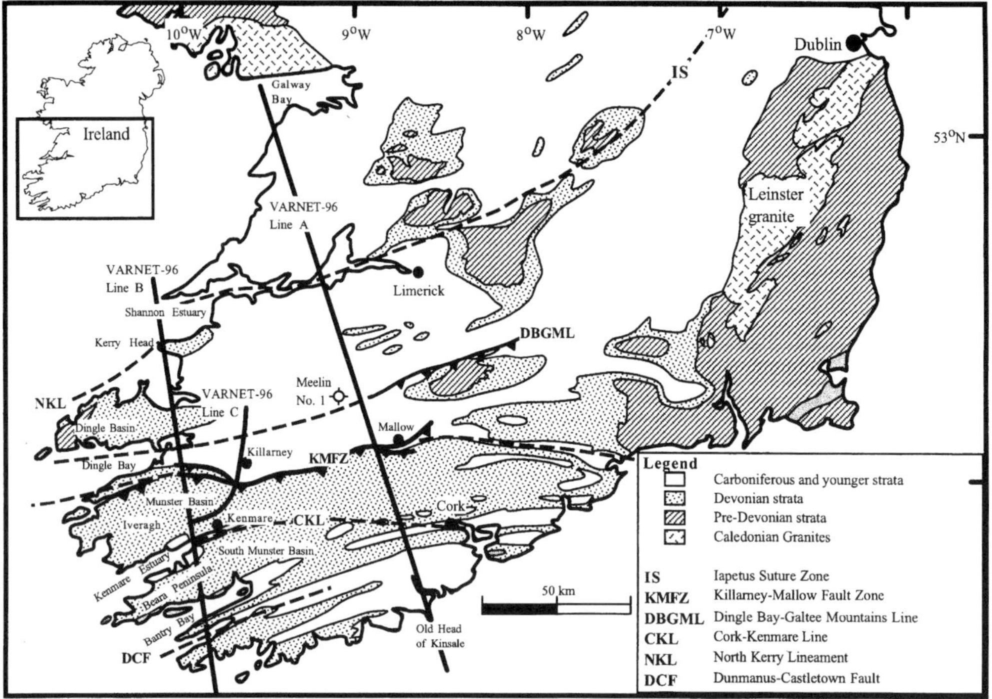

Fig. 1. Geological sketch map of southern Ireland showing the locations of the three VARNET-96 wide-angle seismic profiles and the Ambassador Meelin No. 1 well. The main regional geological elements are also included. Inset map shows the area of detail.

Clayton *et al.* 1980; Williams *et al.* 1989). The most significant syn-rift fault was located on the northern margin, with the basin hinged in the south. Other interpretations suggest that the basin may have evolved as a 150–175 km wide rift, initiated in Early Devonian or early Mid-Devonian times (Williams *et al.* 1997; Williams 1998). The ORS depocentre is generally placed in the vicinity of the Kenmare River estuary (Naylor *et al.* 1983; Williams *et al.* 1989), on Iveragh (Naylor & Jones 1967; Capewell 1975), or on the Beara peninsula (Coe & Selwood 1968).

The South Munster Basin (SMB) (George *et al.* 1976) (Fig. 1), a successor basin to the Munster Basin, developed during rapid subsidence in the central and southern Munster Basin in latest Devonian (Strunian) times (Naylor *et al.* 1989). A pulsed marine transgression strongly influenced deposition of shallow to deep marine siliciclastic strata from Late Famennian to earliest Namurian times (Naylor *et al.* 1974; Naylor 1975; MacCarthy 1987). MacCarthy (1987) considered the SMB to have evolved as an E–W oriented half-graben, although two distinct depocentres were recognized in the southwestern and southern onshore regions by Naylor *et al.* (1989). A facies and sediment thickness divide along the Cork–Kenmare Line (CKL) (Fig. 1) marked the transition from a northern shallow-marine carbonate platform, into the deeper-water basinal succession in the CKL hanging wall (Naylor *et al.* 1989; Williams *et al.* 1989). Although the CKL is recognized as the northern margin of the SMB, the location of the southern basin margin is more speculative. Naylor *et al.* (1989) placed the southern margin in the Celtic Sea, offshore to the south of Ireland.

The Silurian–Devonian Dingle Basin succession (Horne 1970) crops out on the Dingle peninsula, to the north of the Munster Basin and SMB (Fig. 1). The DBGML was recognized as the southern basin margin by Todd (1989), who traced the northern basin margin along a geophysical anomaly extending from north of the Dingle peninsula to Kerry Head (North Kerry lineament (NKL)) (Fig. 1). Shallow- to non-marine siliciclastic deposits, pyroclastic rocks and lavas of the Wenlock–Ludlow Dunquin Group were previously thought to be the oldest beds to crop out in the basin (Holland 1987; Sloan & Bennett 1990). The basinal Silurian strata are conformably and unconformably overlain by the lowest ORS Dingle Group continental clastic rocks of late Silurian to early Devonian age (Holland 1987; Todd 1989). A series of unconformity-bounded groups make up the continental fluvial and alluvial Dingle Basin succession. Although the age of Dingle Basin initiation is fairly well constrained, some disagreement exists regarding the later stages of basin development and ORS deposition in the Dingle region (Todd 1989; Williams & Richmond 1998).

The Munster Basin and SMB were inverted and deformed during the Late Carboniferous–Early Permian Variscan Orogeny (Gill 1962; Cooper *et al.* 1984, 1986; Meere 1995). North-directed Variscan compression imparted an E–W to ENE–WSW tectonic grain on the Munster Basin and SMB. This is reflected in the present-day outcrop pattern. Large-scale folds, with wavelengths of kilometres to tens of kilometres, are recognized in the field and are commonly cut by low-angle thrusts. The greatest amounts of bulk shortening (33–52%) are recorded to the south of the KMFZ (Variscan thrust front of Gill (1962)), decreasing to just 12% to the north of this structure (Cooper *et al.* 1984, 1986; Ford 1987; Meere 1995). The KMFZ clearly represents a significant structural boundary in southern Ireland.

Detailed field studies and geophysical research have resulted in the development of varied tectonostratigraphic models for the Late Palaeozoic basins of southern Ireland. The wide-angle seismic data acquired during the VARNET-96 field campaign provides some added constraints on the subsurface geometry of these basins.

VARNET (VARiscan NETwork)

VARNET was an integrated geological and geophysical project involving researchers from Ireland, Germany and Denmark. The main aim of this research programme was to examine the character of the Variscan fold belt of onshore and offshore southern Ireland. Although this paper is primarily concerned with the evolution and subsurface structure of the Devonian and Carboniferous basins imaged within this study area, some reference is made to the effects of later deformation.

Seismic acquisition

Three VARNET-96 wide-angle seismic profiles were acquired during two seismic deployments onshore in southern Ireland in June 1996 (Masson *et al.* 1998; Vermeulen *et al.* 1998–1999; Landes *et al.* 2000). VARNET-96 Line A (Fig. 1) is a NNW–SSE profile, extending from the Old Head of Kinsale in the south to Galway Bay (Landes *et al.* 2000). One hundred and seventy seismic recording stations were deployed at *c.* 1 km intervals, to give a total line length of 175 km. The maximum range with off-end shots is 200 km. Ten shots were recorded during the first deployment using a conventional in-line geometry (Landes *et al.* 2000). Two shots were fired at intervals along the profile, and this interleaving of shots and stations provides good control on the shallow crustal structure in a number of geologically significant areas. A sparser station grid was left along Line A for the second deployment, to record the 13 fan shots fired during the second acquisition phase.

A total of 109 seismic stations were deployed at *c.* 1 km intervals during the acquisition of Line B (Masson *et al.* 1998), as part of the second seismic deployment. The total length of this NNW–SSE profile (Fig. 1), including off-end shots, is 140 km. Line B crosses a series of peninsulas separated by shallow inlets, making it possible to fire a number of shots at intervals along the line and at both ends of the profile. Again, this interleaving of shots and stations provides good control on shallow crustal structure. Line C (Fig. 1) is a short profile that contains limited regional information and is not included here.

Seismic data processing and velocity modelling

The processing and velocity modelling procedures involved in the production of the final 2D P-wave velocity models have been outlined in detail by Masson *et al.* (1998) and Landes *et al.* (2000). During the data-processing stages, seismic record sections were compiled and plotted using Seismic Handler (Landes *et al.* 2000). The observed phases were correlated on each record section and 1D velocity–depth functions were calculated for each shot using the REFRA program of Sandmeier & Liebhardt (1994). These 1D velocity–depth functions provided the starting point for the 2D velocity models. The 1D results were mapped into preliminary 2D cross-sections and the data were interpreted by 2D forward modelling using the MACRAY ray tracing program (Luetgert 1992) and SEIS83 (Cerveny *et al.* 1977). Travel time inversion (Zelt & Smith 1992) was then carried out on the velocity models primarily to constrain the structure of the lower crust and upper mantle. Finally, synthetic seismograms were calculated in the case of Line A to provide greater control on

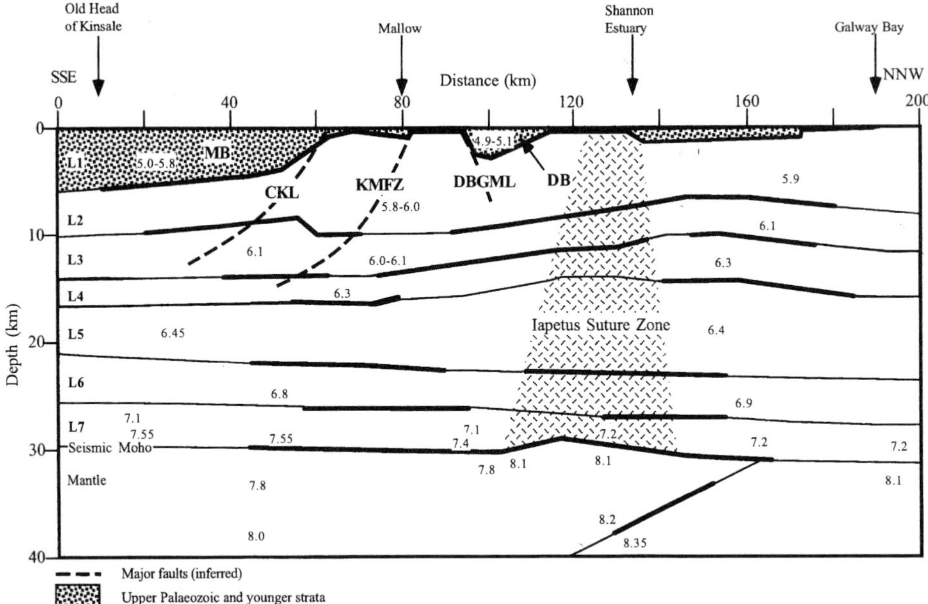

Fig. 2. Tectonic interpretation of the 2D P-wave velocity model along VARNET-96 Line A. All velocities are in kilometres per second (km s^{-1}). MB, Munster Basin; DB, Dingle Basin; CKL, Cork–Kenmare Line; KMFZ, Killarney–Mallow Fault Zone; DBGML, Dingle Bay–Galtee Mountains Line. Bold continuous lines represent crustal reflections.

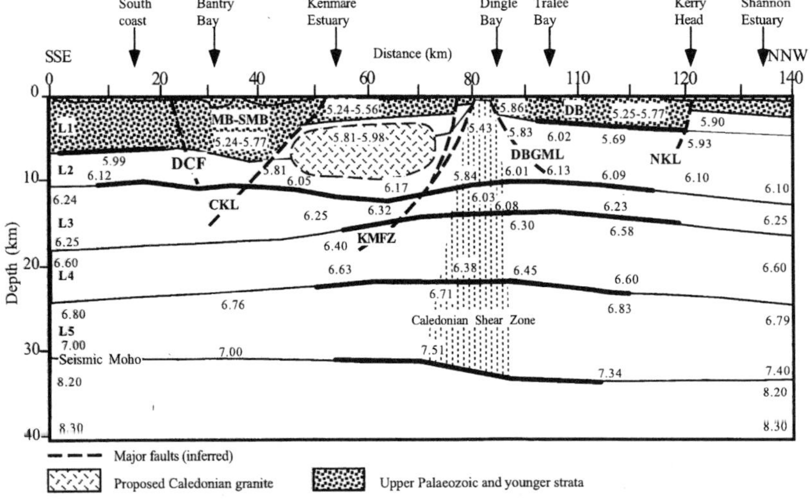

Fig. 3. Tectonic interpretation of the 2D P-wave velocity model along VARNET-96 Line B. All velocities are in kilometres per second (km s^{-1}). MB, Munster Basin; DB, Dingle Basin; KMFZ, Killarney–Mallow Fault Zone; DBGML, Dingle Bay–Galtee Mountains Line; DCF, Dunmanus–Castletown Fault; NKL, North Kerry Lineament. Bold continuous lines represent crustal reflections.

velocity gradients and transition zones in the 2D model. The 2D model was amended until the synthetic seismograms agreed with the observed wave-field. Interfaces shown on the final 2D velocity models represent either direct reflections from the crust (bold lines, Figs 2 and 3), or interpreted crustal discontinuities based on rapid vertical changes in velocity gradients (thin black lines, Figs 2 and 3). Crustal reflections indicate the presence of real rather than interpreted discontinuities and provide direct control on crustal structure. The crustal layers included in the final 2D models are zones of similar velocity characteristics, separated by imaged or inferred velocity boundaries.

Given the large amount of data interpretation involved, 2D wide-angle seismic velocity models clearly are non-unique, and can be described as best-fit solutions designed to honour the velocity data. Source and receiver density is less than in reflection seismic acquisition, and consequently the wide-angle method is a 'broader brush' approach to crustal imaging. The geological models presented here represent what are considered to be the most robust interpretations of the entire VARNET-96 dataset. Uncertainties in depth and velocity do exist on all VARNET-96 profiles. Landes *et al.* (2000) noted that some parts of the 2D model for Line A, in particular at the edges of the profile, are not verified by the observed wave-field. Limitations in the modelling software required that interfaces be extended to the northern and southern ends of the profile. Masson *et al.* (1998) stated that errors in the models are generally cumulative, with greater uncertainty in the deeper parts of the model and at the model edges. For example, the velocity structure of layers 1 and 2 on Line B is well constrained by the observed wave-field (reflections and velocity gradients). However, deep interfaces such as the top of the lower crust and seismic Moho are only well defined only in the central parts of the profile. Seismic resolution is consistently good in the central parts of Line A and Line B to the base of the crust. Resolution at the edges of the profiles is best within the upper crust, but decreases with depth. It is interesting to note that crustal reflections have greatest lateral continuity at shallower levels, but are seen in the central parts of the profiles only at middle- and lower-crustal levels (Figs 2 and 3). This is encouraging for the interpretation of the shallow Late Palaeozoic basins in SW Ireland, as the basins lie in the zone of best data quality (above *c.* 15 km depth).

Geological interpretation of VARNET-96 2D P-wave velocity models

Differentiating Late Palaeozoic basins from Caledonian basement

One of the key points in the following discussion concerns the ability of the wide-angle seismic method to discriminate between Upper Palaeozoic sediments in the Munster Basin, SMB and Dingle Basin, and Caledonian deformed basement to these basins. Clearly no inferences can be made regarding basin geometry and tectonic evolution if the basins cannot be distinguished on the seismic data. A strong upper crustal reflection is imaged on Line A at 6 km depth and 10 km distance (Fig. 2). This reflector marks a sharp change in velocity gradient in the upper crust and separates velocities lower than 5.8 km s^{-1} above the reflector, from velocities faster than 5.8 km s^{-1} below. As the Munster Basin and SMB have suffered severe Late Carboniferous Variscan deformation, which resulted in folding of primary depositional surfaces (Gill 1962), this undeformed south dipping reflection is unlikely to represent a sedimentary interface within the Upper Palaeozoic succession. The reflection may be a south-dipping and north-verging Variscan thrust within the basinal succession, but could also result from an acoustic impedance contrast and change in crustal velocity gradient between Upper Palaeozoic strata and Calaedonian basement (the floor of the Munster Basin). If this reflection is considered to represent an intrabasinal thrust, then the next most likely candidate for the base of the Munster Basin is a deeper reflection imaged at 10 km depth at the south coast. This interpretation requires the Upper Palaeozoic section to be 10 km thick at the south coast (layer 1 and layer 2 in Fig. 2), and up to 6 km thick to the north of the Shannon Estuary. Although a sediment thickness of 10 km may be feasible in the southwest (Ford *et al.* 1991), Lower Palaeozoic basement crops out just to the east of Line A at 80–85 km distance (Fig. 1). This suggests that over 6 km of post-Caledonian strata are not present in the northern region. Therefore, the reflection at the base of layer 1 on Line A is interpreted as the base of the Munster Basin. The lack of apparent deformation on this surface suggests that the basin fill detached from underlying basement during Variscan deformation and the basin floor was undeformed. The range in velocities recorded in layer 1 (4.9–5.8 km s^{-1}) on Line A may indicate the presence of heterogeneities in the sedimentary succession. Velocities in layer 2, interpreted here as pre-Upper Palaeozoic rocks, show a range of only 0.1 km s^{-1}. On this basis, shallow P-wave velocities lower than *c.* 5.8 km s^{-1} are generally considered to represent Upper Palaeozoic and younger strata, whereas velocities higher than 5.8 km s^{-1} are interpreted as pre-Upper Palaeozoic (basement) rocks. Lowe & Jacob (1989) had difficulties in distinguishing Upper Palaeozoic sediments from Caledonian basement on the COOLE I wide-angle seismic profiles, using P-wave velocity alone. However, the source and receiver density used during the VARNET-96 acquisition was significantly greater than that for COOLE I, explaining the improved resolution on the VARNET-96 profiles.

The interpretation of Line B (Fig. 3) is more difficult. A gently north-dipping reflector is imaged at 2–4 km depth beneath the outcrop of the Dingle Basin (84–123 km distance approximately). Given the similarities between this reflector and the interpreted base of the Dingle Basin on Line A (Fig. 2), the Dingle Basin succession is considered to be confined above this reflector. A crustal reflection was imaged at 7 km depth at the south coast, marking the 5.8 km s^{-1} velocity boundary. This velocity interface is again interpreted as the base of the Upper Palaeozoic succession. A sharp step in the basin floor is inferred at 50 km distance in the vicinity of the Kenmare Estuary based on the presence of rapid lateral velocity variations. The interpreted basin floor to the north of this step does not coincide with a crustal reflection. P-wave and S-wave velocity data presented by Rumpel *et al.* (1999) suggest the presence of shallow granite beneath the basinal succession between the CKL and KMFZ. The presence of quartz-rich granite underlying the continental clastic basinal section may explain the absence of an acoustic impedance contrast and associated seismic reflection at the basin floor in this zone.

Slow seismic velocities are recorded in L2 at 80 km distance on Line B, but do not represent Upper Palaeozoic sediments in this case. The depressed velocities coincide with a zone of crust cut by a number of major ENE–WSW faults. The faults, which are perpendicular to the seismic-wave propagation direction (NNW–SSE), slow the P-wave velocities in the faulted zone. Masson *et al.* (1998) suggested that low-velocity granites may also be present in this zone, accounting for the low L2 and L3 velocities at 80 km distance.

In summary, the interpreted Upper Palaeozoic basin geometries are based on the location of well-defined crustal reflections, rapid changes in crustal velocity gradient, crustal composition (Line B, Rumpel *et al.* 1999), and the accepted tectonic evolution of the region. Although the interpretation shown here is one of a number of possible end-members, it will be demonstrated that the proposed geometry of the basins on both profiles can be integrated into a regional tectonic model.

The reader should note that the thickness of sediment imaged on the VARNET-96 profiles is not the true depositional thickness. As the Munster Basin and SMB were intensely deformed during the Late Carboniferous Variscan orogenic episode (Gill 1962; Meere 1995), all sediment thickness values must be treated as post-deformational or tectonic thickness that may not accurately reflect true depositional patterns. Throughout the following discussion, crustal velocity layers are abbreviated as follows: layer 1 as L1, layer 2 as L2.

VARNET-96 Line A

The lateral continuity of the base L1 reflection (Fig. 2), combined with the fact that this interface changes depth at the position of known faults, suggests that this seismic event has regional tectonic significance. As outlined in the previous section, the base of L1 is correlated with the Caledonian unconformity surface. This reflection is considered to mark a change in acoustic impedance between Caledonian deformed rocks and overlying Upper Palaeozoic strata. P-wave velocities above this reflection vary significantly (4.9–5.8 km s^{-1}), perhaps as a result of gross compositional and diagenetic heterogeneities in the Devonian and Carboniferous section. Velocities within L2, interpreted here as Caledonian basement, fall in the range 5.8–6.0 km s^{-1}.

L1 displays a number of dramatic thickness changes along the profile that can be correlated with mapped faults. The western projection of the DBGML (Fig. 1) intersects Line A at 93 km (Fig. 2). Approximately 1 km of Upper Palaeozoic strata (L1) are modelled to the south of the fault, increasing to *c.* 3 km immediately to the north. As the DBGML separates shallow basement to the south from a wedge of sediment to the north, the fault plane must dip and throw to the north. L1 then thins northwards from 3 km in the DBGML hanging wall to 1 km at 115 km, to define a narrow half-graben hinged in the north. This basin lies directly along Caledonian strike from the Dingle Basin in the west (Fig. 1).

Some control on the thickness, age and character of the half-graben succession is provided by the Ambassador Meelin No. 1 well (Fig. 1), sited just west of the half-graben on Line A. The borehole encountered 1690 m of Lower Carboniferous limestones and shales without reaching the base of the Carboniferous section. In addition, the presence of Upper ORS strata in outcrop to the east of the half-graben (Fig. 1) suggests that Devonian sediments may be present beneath the basinal Carboniferous section. Low L1 P-wave velocities (4.9–5.1 km s^{-1}) recorded in the DBGML hanging wall may be caused by karstic dissolution cavities and associated cavings similar to those reported in the Lower Carboniferous succession of the Ambassador

Meelin No. 1 well. Cavities filled with air, fluid or unconsolidated sediment would result in lower P-wave velocities on the wide-angle profile. Seismic data therefore support an interpretation of the DBGML as a north-dipping normal fault, which defines the southern margin of a 22 km wide half-graben on Line A.

The deformed and inverted fill of the Munster Basin and SMB crops out to the south of the KMFZ (78–82 km) on Line A (Fig. 1). The basin floor is placed along the base L1 reflection in this zone. The interpreted thickness of Upper Palaeozoic strata increases southwards to 1–1.5 km in the KMFZ hanging wall, suggesting that only small amounts of extensional displacement are preserved on the fault in this zone. This may reflect low rates of Late Palaeozoic extension on the fault in the east, or subsequent overprinting by Variscan inversion (Meere 1995; Pracht 1997). Although the subsurface trace of the KMFZ on Line A is not imaged by the wide-angle method, the sense of displacement of L1 confirms that the fault downthrows and dips to the south.

L1 is *c.* 1 km thick between 80 and 63 km distance in the KMFZ hanging wall. Rapid thickening of the layer is then imaged between 60 and 50 km, and L1 continues to gradually thicken south to reach 6 km at the Old Head of Kinsale. A second south-dipping and south-throwing Upper Palaeozoic fault is inferred at 63 km distance to account for the modelled thickness increase of the basinal succession in this region (Fig. 2). The only detachment of this scale previously recognized in the Munster Basin is the CKL, which experienced down-to-the-south displacement in Late Devonian and Early Carboniferous times (Naylor *et al.* 1974; Naylor 1975). This E–W oriented lineament is typically shown extending west from Cork Harbour (Fig. 1), some 10–15 km south of the proposed structure on Line A (Naylor *et al.* 1989; Williams *et al.* 1989). As this fault cannot be mapped as a discrete structure in the field, the fault trace is typically inferred from the distribution of Lower Carboniferous facies belts. Lower Carboniferous shallow marine carbonates are confined to the footwall, with deeper marine basinal deposits occurring in the hanging-wall zone (Williams *et al.* 1989). This approach may be flawed given that the SMB succession has been extensively deformed and shortened in Late Carboniferous times. As lateral transport (folding and thrusting) and extensive erosion of the sedimentary succession occurred during deformation, the present distribution of facies belts is not necessarily indicative of the position of basement faults. The observation that the Upper Palaeozoic succession detached from Caledonian basement during Variscan deformation (see above) supports this conclusion. The data presented here suggest that the accepted trace of the CKL (Fig. 1) may need to be reviewed. A step in the Munster Basin floor at 63 km is tentatively correlated with the CKL, placing the fault in the zone between Cork Harbour and the KMFZ. This revised location for the eastern extension of the CKL gives the structure a NE–SW Caledonian orientation similar to the KMFZ and DBGML, rather than the traditional E–W orientation (Williams *et al.* 1989) that cross-cuts the regional tectonic grain.

The VARNET-96 seismic data do not resolve the trace of major Upper Palaeozoic extensional fault planes through the upper and middle crust. The base of the upper crust is not displaced beneath the DBGML on Line A (Fig. 2), suggesting that the amount of offset on this structure decreases with depth. It is therefore difficult to infer the subsurface geometry of this fault, which is shown as a planar north-dipping structure on Line A (Fig. 2). Thickness variations in L1 confirm that the KMFZ and CKL both dip to the south. A step in the base of L2 at 55–60 km may mark the trace of the KMFZ through the upper crust. Meere (1995) suggested that the main faults in the Variscan orogen soled out to the south of Ireland at 18–20 km depth. O'Reilly *et al.* (1991) and Vermeulen *et al.* (1998–1999) recognized a potential crustal detachment surface at 11–14 km depth in the Celtic Sea and onshore in southern Ireland. This surface was considered to define a structural boundary, separating brittle upper crust from more ductile middle crust. Vermeulen *et al.* (1998–1999) proposed that a number of major Variscan and pre-Variscan faults, including the CKL and KMFZ, flatten slightly at this layer before continuing into the middle crust. The subsurface geometry of the CKL and KMFZ remains a topic of debate, although a southerly dip is confirmed by the seismic data. The geometry of the two structures suggests that the CKL may be a footwall shortcut fault that soles into the KMFZ at depth.

The upper upper crustal layer (L2) is attenuated beneath the Munster Basin and SMB (Fig. 2). L2 is *c.* 9–10 km thick in the footwall of the KMFZ (70 km), but thins to 4 km at 50 km distance. The layer retains a thickness of 4 km to the Old Head of Kinsale at the southern end of Line A. L3 also displays a general southward thinning trend beneath the Munster Basin. Southward attenuation is observed at middle crustal levels where L5, the lower middle crust, thins from 9 km at the Shannon River to

4–5 km at the south coast. Crustal attenuation in the southern half of Line A possibly reflects crustal extension during basin development. The most likely candidates are the Devonian–Carboniferous rift episode onshore in southern Ireland or the Mesozoic rift event in the Celtic Sea offshore to the south.

The thickness and velocity of the upper-, middle- and lower-crustal layers on Line A change beneath the Shannon River. A step in the seismic Moho is also imaged. Masson *et al.* (1999) reported a steeply south-dipping mantle velocity interface in this region, extending from the seismic Moho to 110 km depth. This zone of crustal and mantle change is interpreted as the Iapetus Suture Zone, a broad zone of crustal and mantle deformation that marks the site of Caledonian continental collision (Masson *et al.* 1999; Vermeulen *et al.* 1998–1999).

VARNET-96 Line B

L1 on Line B contains velocities higher and lower than 5.8 km s^{-1} suggesting that this is a geologically composite layer, consisting of Upper Palaeozoic strata and Caledonian basement. The basin geometries shown in Fig. 3 incorporate VARNET-96 crustal composition data (Rumpel *et al.* 1999), which were calculated using P-wave and S-wave velocities for Line B. Line B crosses the KMFZ and DBGML between 75 and 85 km approximately. Shallowing of the base L1 reflection to less than 1 km depth is seen in the hanging-wall zone of the faults. This shallow faulted horst separates the outcrop of the Munster Basin and SMB to the south from the Dingle Basin to the north. As mentioned above, low upper- and middle-crustal velocities between 70 and 90 km distance are considered to be due to the presence of crustal anisotropies (faults and fractures) perpendicular to the seismic-wave propagation direction. A step in the seismic Moho is imaged between 70 and 90 km distance, directly beneath this proposed crustal fault zone, indicating that the structures extend to deep levels in the crust. The age of the fault zone cannot be determined directly from the seismic data. However, the strong similarity between this structure and the Iapetus Suture Zone on Line A suggests that it may be a major, deep-going Caledonian shear zone, which had an origin in Late Silurian tectonism.

Line B intersects the trace of the DBGML at 84 km. L1 thickens across the fault from 1 km at 82 km, to 2.5 km at 90 km. Velocities of 5.86 km s^{-1} in the DBGML hanging wall may represent fault slivers of deformed and metamorphosed Silurian rocks, which are seen in outcrop on the Dingle peninsula (Sloan & Bennett 1990). L1 velocities decrease to 5.25–5.77 km s^{-1} at 90 km. A strong reflection is also imaged at the base of L1 between 90 and 120 km distance, implying that the layer consists of the ORS fill of the Dingle Basin and overlying Carboniferous strata in this zone. The interpreted Dingle Basin succession thickens to a maximum of 4 km at 120 km, where a rapid increase in seismic velocity to 5.9 km s^{-1} is observed at the northern termination of the base L1 reflection. This velocity increase suggests a change from Upper Palaeozoic strata in the south to Caledonian basement in the north, and is used to infer the presence of a south-throwing normal fault cutting L1 at 120 km distance. This proposed fault coincides with the trace of the northern syn-rift Dingle Basin margin (NKL) (Fig. 1) of Todd (1989), which intersects Kerry Head at 123 km. Thinning of the Upper Palaeozoic succession to 2 km is predicted in the NKL footwall. The interpreted geometry of the Dingle Basin on Line B is different from that on Line A. The floor of the basin is constrained by a seismic reflection on both profiles. However, in the west the basin has a more symmetrical profile, is considerably wider (35–40 km), and appears to be bounded to the north and south by normal faults.

The thickness of the upper, middle and lower crust does not vary significantly beneath the Dingle Basin, suggesting that little crustal attenuation occurred in this zone in post-Caledonian times. The proximity of the northern half of Line B to the highly deformed Iapetus Suture may have discouraged widespread crustal extension in Late Palaeozoic times. The Dingle Basin appears to have developed as a result of tension on discrete faults rather than by a process of regional crustal extension. This may account for the relatively narrow width and limited sediment thickness within the basin.

The KMFZ crops out at the southern margin of a shallow ridge of L2 at 75–80 km distance. The fill of the Munster Basin is confined to the southern side of the fault on Line B, implying that the fault dips and throws to the south and is the northern syn-rift margin of the basin. P-wave velocities higher than 5.8 km s^{-1} are recorded in the lower levels of L1 in the KMFZ hanging wall, indicating that the layer may contain pre- and post-Caledonian strata. The base of the Munster Basin succession is placed within L1 (*c.* 3–4 km depth) on Iveragh (50–80 km distance). Thickening of the Upper Palaeozoic succession by 2 km across the KMFZ (Fig. 3) represents preserved normal extension on the northern syn-rift Munster Basin margin. The thickness of the

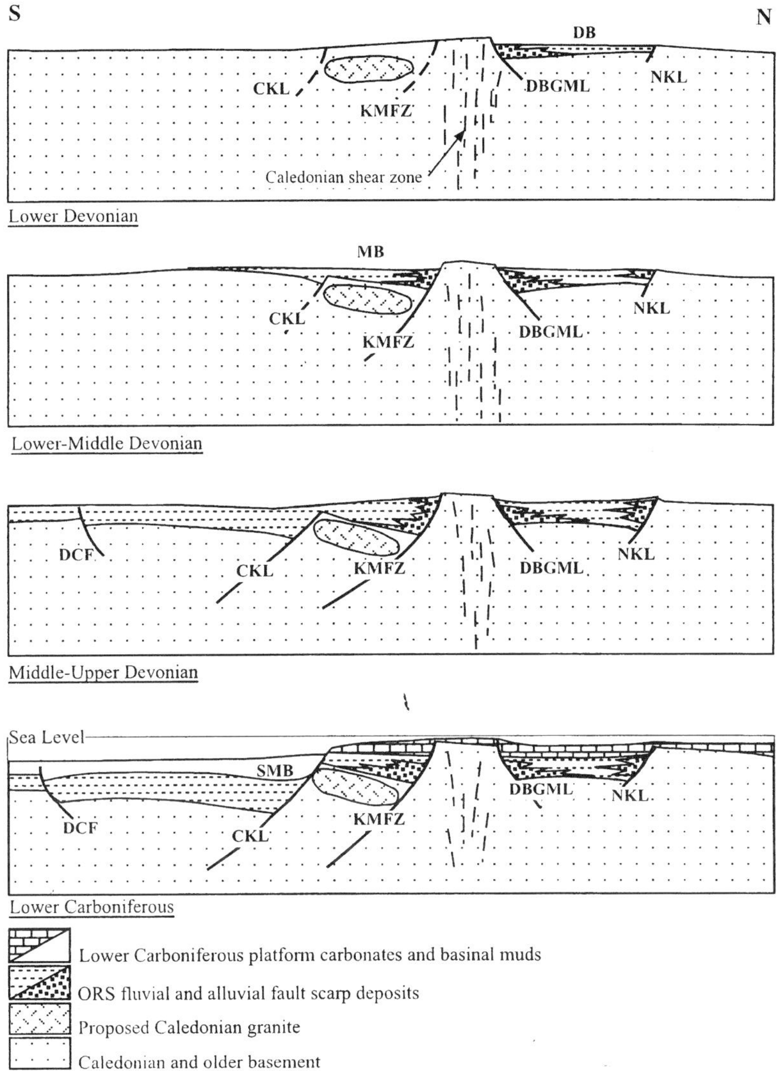

Fig. 4. Schematic tectonic evolutionary model along VARNET-96 Line B (not to scale), from Early Devonian to Early Carboniferous times. The strong control exerted by the pre-existing Caledonian framework on ORS basin development is evident. MB, Munster Basin; DB, Dingle Basin; KMFZ, Killarney–Mallow Fault Zone; DBGML, Dingle Bay–Galtee Mountains Line; DCF, Dunmanus–Castletown Fault; NKL, North Kerry Lineament.

Munster Basin succession in this zone is supported Rumpel *et al.* (1999), who presented a crustal composition model showing shallow granitic intrusions between 3 and 13 km depth and 50 and 80 km distance. Gravity modelling of Line B (Masson *et al.* 1998) also indicated that shallow granites are required to explain the observed gravity field and seismic data.

L1 velocities are all lower than 5.8 km s^{-1} to the south of 45 km, where L1 is considered to consist solely of Munster Basin and SMB strata. The base of L1 also coincides with a crustal reflection to the south of 20 km (Fig. 3), supporting this conclusion. The deformed basinal succession reaches a maximum thickness of 8 km at the Beara peninsula (37 km). Rapid thickening of the Upper Palaeozoic section, from 4 km at 56 km to 8 km at 37 km (Fig. 4), is inferred based on changes in seismic velocity and crustal composition (Rumpel *et al.* 1999). A south-dipping normal fault is shown at the southern margin of the proposed granite intrusions, extending from the surface at 50 km distance. Line B crosses the predicted trace of the CKL at 50–55 km (Fig. 1). Therefore, the inferred thickness change of the Munster Basin and SMB succession on the profile is accounted for by preserved normal displacement on the south-dipping CKL. The basinal succession thins from 8 km at the Beara peninsula to 6 km at Dunmanus Bay (26 km), before thickening to 7 km at the southwest coast. A third major extensional fault is inferred at 26 km distance, to explain the L1 thickness increase and change in slope of the basal reflection in this zone. This inferred structure coincides with the DCF, the Munster Basin southern margin of Williams *et al.* (1989).

The upper crust (L2) is *c.* 10 km thick beneath the northern Munster Basin, but thins to just 3 km at 40 km in the hanging wall of the CKL. L2 is 4–5 km thick between 30 km and the southern end of Line A. Southward thinning of L4, the lower middle crust, is also imaged beneath the southern Munster Basin. The upper- and middle-crustal thickness trends on Line B suggest that the crust beneath southwest Ireland was extended during a regional rift episode.

Regional tectonic synthesis

The southern margin of the Dingle Basin coincides with the ENE–WSW-trending DBGML on both VARNET-96 profiles. In the west, the Dingle Basin is imaged as a symmetrical rift, bounded to the north by the NKL. To the east, the basin assumes a half-graben geometry and narrows from a width of *c.* 40 km on Line B to 22 km on Line A. An eastward decrease in displacement on the NKL is inferred to explain this dramatic change in basin geometry. The preserved Dingle Basin succession also youngs towards the east. Silurian and Devonian strata crop out on the Dingle peninsula, whereas over 1600 m of Carboniferous strata are predicted on Line A, underlain by a thin Devonian ORS succession. Gradual eastward propagation of the DBGML in Devonian and Carboniferous times is considered to have resulted in this younging pattern. Field and seismic studies therefore suggest that subsidence in the Dingle Basin occurred over a prolonged period, as a consequence of Silurian to Carboniferous normal displacement on the north-dipping DBGML (Fig. 4). Whereas the Silurian and Devonian extensional phase can be attributed to post-orogenic collapse and extension of the Caledonian orogen, the later Carboniferous rift episode may be related to the regional basin-forming event that produced the Shannon Trough to the north and the SMB to the south. Strogen (1988) envisaged a regional transtensional tectonic regime during development of the Shannon Trough. A similar tectonic regime to the south may explain the west to east propagation of the DBGML suggested here. The geometry of the upper and middle crust on Line B suggests that crustal thinning did not occur during development of the Dingle Basin. Basin initiation was controlled by localized displacement on discrete faults. Basin development may have been linked to displacement on major Caledonian crustal fault zones which bound the basin to the north (Iapetus Suture) and south (low-velocity zone on Line B).

A shallow ENE–WSW oriented basement ridge, bounded by the DBGML in the north and the KMFZ in the south, separates the Dingle and Munster basins. This horst lies above a crustal low-velocity zone, interpreted here as a deep-going Caledonian shear zone. This fracture zone offsets the seismic Moho beneath the Dingle peninsula (Fig. 3). Elevated lower-crustal P-wave velocities beneath and north of this shear zone indicate the presence of dense mafic rocks and suggest that the lower crust in this region has been intraplated or underplated by mantle melts. The age of the proposed lower-crustal intrusions is uncertain but may be related to Caledonian tectonism or late Caledonian Silurian volcanism on the Dingle peninsula (Sloan & Bennett 1990). The Caledonian shear zone imaged in the crust on Line B may have acted as a focal point for ascent of mantle melts. Rumpel *et al.* (1999) presented modelled compositions for the crust along Line B. These additional seismic data strongly suggested the presence of a lower-crustal mafic body beneath the Dingle peninsula. A silica-rich zone was also imaged in L1 on Line B between 45 and 75 km distance, and 4 and 13 km depth. This silica-rich zone, which has the gross composition of granite (Rumpel *et al.* 1999), coincides exactly with the Killarney Bouguer gravity low (Murphy (1960) and others). Modelling of this geophysical feature indicates the presence of a zone of low-density crust beneath the northern Munster Basin. This has been interpreted as a thick succession of Upper Palaeozoic Munster Basin strata (Eminike 1986; Ford *et al.* 1991), or as a shallow Caledonian granite intrusion (Howard 1975; Cooper *et al.* 1986; Ford *et al.* 1991; Meere 1995; Masson *et al.* 1998). The VARNET-96 seismic data (Rumpel *et al.* 1999) support the buried granite interpretation. Gravity modelling of Line B (Masson *et al.* 1998) also indicates that the observed gravity field cannot be explained by thick Upper

Palaeozoic sediments alone. This inferred granite may be linked to the Caledonian crustal shear zone imaged on Line B. Syn-deformational granite emplacement along shear zones has been inferred for other Irish Caledonian granites (Hutton & Reavy 1992).

Sedimentological evidence exists for a previously exposed, but now buried, granite in SW Ireland. Todd (1989) described two-mica granite clasts from Late Silurian–Early Devonian age strata in the Dingle Basin. Palaeocurrent measurements indicated that the sediments were sourced from an eroding hinterland to the south. As the Munster Basin did not develop until Early or early Mid-Devonian times (Williams *et al.* 1997), the present basin floor would have been a sediment source area during the early development of the Dingle Basin. Capewell (1951, 1975) also recognized granite clasts in the Dingle Basin succession, and Clipstone & Roycroft (1992) recorded granite fragments in the Lower Carboniferous succession in the southwest. The Dingle Basin fill is also rich in tourmaline and tourmaline-rich clasts (Capewell 1951, 1957, 1975; Todd 1998). Tourmaline is also abundant in the metamorphic aureole of the Caledonian Leinster Granite to the east (Kennan 1997–1998).

Seismic, gravity and sedimentary data point to the presence of a shallow Caledonian granite in the southern hinterland of the Dingle Basin. The eroding granite aureole rocks and granite intrusion shed detritus into the developing basin. Increased rates of regional extension in Early Devonian times, in response to collapse of the Caledonian orogen, were accompanied by reactivation of the major shear zone imaged on Line B. Todd (1989) recognized field evidence for sinistral transpression on this Caledonian shear zone in Late Silurian times. Compression on the structural lineament generated a southern provenance area, which sourced a series of ORS alluvial fans in the southern Dingle Basin. The VARNET-96 data suggest that late to post-Caledonian displacement on the shear zone resulted in development of the DBGML on the northern margin of the fracture system. This was followed in Early to Mid-Devonian times by extension on the KMFZ, sited on the southern margin of the shear zone (Fig. 3). The trace of the KMFZ and DBGML to the east of Line A (Fig. 1) reflects the large volume of Leinster Granite in the upper crust in that region. The results of the COOLE I seismic project (Lowe & Jacob 1989) confirm that the marked divergence of the DBGML and KMFZ to the east results from these faults following the northern and southern margins of the granite. The volume and distribution of Caledonian granites and fault structures in the crust clearly influenced Late Palaeozoic extensional patterns in the region.

VARNET-96 confirms the KMFZ as the northern syn-rift margin of the Munster Basin (Figs 2 and 3). Thickening of the Upper Palaeozoic succession into the southern hanging wall of the fault is recorded on Lines A and B. A maximum hanging-wall thickness of 3 km on Line B thins eastwards to *c.* 1 km on Line A, suggesting that crustal extension was focused in the western onshore region. The extensional KMFZ may tip out to the east of Line A. Considerable amounts of Varsican inversion are recognized along the KMFZ, indicating that the present throw on the KMFZ may not represent the true Devono-Carboniferous extensional displacement on this structure (see below). An elevated fault terrace is imaged in the hanging wall of the KMFZ on both profiles. The basinal succession preserved on this terrace reaches a maximum thickness of 4 km on Line B. The upstanding fault block is cored by the inferred granite in the west, perhaps explaining the reduced sediment thickness on the crest of this terrace. Buoyant granite in basement to the northern Munster Basin would not have favoured significant subsidence in the region.

Some difficulties do exist with this model, however. The Munster Basin ORS depocentre has previously been placed on Iveragh (Naylor & Jones 1967; Capewell 1975) and in the vicinity of the Kenmare Estuary (Naylor *et al.* 1983; Williams *et al.* 1989), directly above the buoyant fault terrace imaged on the VARNET-96 data. Price (1989) noted that such a granite in basement to the Munster Basin would have resisted subsidence, and could not underlie the Munster Basin depocentre. However, disagreement does exist regarding the exact location of the basin depocentre. Coe & Selwood (1968) positioned the ORS depocentre on the Beara peninsula, to the south of the proposed granite on Line B. In addition, Williams (1998) suggested that the Munster Basin may have developed as a 150–175 km wide Devonian rift, which was shortened during Variscan compression. A tilted fault-block geometry, similar to that imaged on the VARNET-96 profiles, would be expected in such a wide rift basin.

Ford *et al.* (1991) concluded that a thickness of 10 km of ORS sediment in the northern Munster Basin would explain the observed regional gravity signature. L1 is *c.* 5 km thick at 70 km, but thickens to 8 km at 45 km distance. As mentioned above, seismic velocities, crustal composition and gravity modelling strongly support the presence of a granite beneath 4 km depth in this region. In addition, field studies confirm that normal displacement occurred on the CKL in Late Devonian and Early Carboniferous times, as the SMB developed. A down-to-the-south offset in the base of L1 would be expected across the CKL. No such offset is seen if L1 is considered to consist entirely of Upper Palaeozoic sediments. The granite model introduces the required normal offset across the south-dipping CKL and results in strong similarities in basin geometry between Lines A and B. The location of the CKL on Line B also warrants further discussion. As mentioned above, the location of the KMFZ and DBGML to the east of Line A is controlled by the position and volume of upper-crustal granites. Leeder (1974, 1982) recognized a similar granite control on the Devono-Carboniferous development of a series of basins in Britain. Basin-bounding faults developed on the margins of intrusions, whereas depocentres were typically sited above crust free of low-density plutons. The CKL is located on the southern margin of an inferred granite on Line B. If the interpretation of Coe & Selwood (1968) is correct, then the Munster Basin depocentre is located in the fault hanging wall, to the south of the proposed granite. The preserved basin geometry on Line B resembles the open rift model of Williams (1998), rather than the classic half-graben model (Williams *et al.* 1989).

The VARNET-96 seismic data favour the granite model presented above, which is indirectly supported by sedimentological evidence and gravity modelling. This model cannot be reconciled with all field observations however, perhaps suggesting that a review of the basin evolution may be required in light of the VARNET-96 results.

A dramatic southwards increase in sediment thickness is inferred across the ENE–WSW-trending CKL. L1 reaches a maximum thickness of 8 km beneath the Beara peninsula on Line B. This coincides with the ORS depocentre of Coe & Selwood (1968) and lies within the main SMB depocentre. A thickness of 8 km of Upper Palaeozoic strata on the Beara peninsula thins to 7 km at the southern end of Line B and 6 km at the southern end of Line A. The overall geometry of the Munster Basin and SMB on the VARNET-96 profiles is that of a broad rift extending offshore to the south of Ireland. The development of the northern rift margin, which is characterized by a tilted fault-block geometry, was strongly influenced by Caledonian palaeoplate features.

Sediment thickness values on the seismic profiles represent tectonic thickness and do not necessarily reflect Late Palaeozoic depositional patterns. However, the thickness characteristics of the upper crust, extended during Devono-Carboniferous basin development, may be a more useful indicator of extensional patterns and regional strain distribution. Vermeulen *et al.* (1998–1999) suggested that upper-crustal L1 was not attenuated during Mesozoic extension and basin development, and therefore preserves a pre-Mesozoic geometry. Those workers concluded that Mesozoic extension and crustal thinning was focused beneath a regional sub-horizontal crustal detachment at 12–14 km depth. Therefore, the thickness characteristics of the crust above 12 km depth are considered to reflect the magnitude and location of extension during Devonian and Carboniferous basin development. Interpretation of the VARNET-96 data shows that the upper crust is thinnest (3–4 km) beneath and south of the Beara peninsula (Fig. 3). The greatest amount of cumulative Devonian and Carboniferous extension is therefore believed to have been focused beneath and south of the Beara peninsula. The upper crust thickens to 4 km at the southern end of Line B (Fig. 4), and is *c.* 4 km thick at the Old Head of Kinsale on Line A (Fig. 2). This suggests that the upper crust to the south of the CKL (onshore and offshore) was extended in Devonian and Carboniferous times and that the Late Palaeozoic basin complex extends offshore to the south. No evidence exists on the VARNET-96 profiles to suggest that the southern margin of the Munster Basin or SMB is located onshore in southern Ireland.

Some mention of the influence of post-depositional deformation must be made at this stage. The Munster Basin and SMB were extensively deformed in Late Carboniferous times during the Variscan orogenic episode (Gill (1962) and many others). Northward-migrating deformation folded and thrust the Upper Palaeozoic succession, resulting in development of regional folds and a strong E–W to ENE–WSW tectonic fabric. The KMFZ was recognized as the Variscan thrust front by Gill (1962). Avison (1984) estimated that 2.5 km of Variscan inversion was accommodated by reverse displacement on the KMFZ in the vicinity of Line B. The magnitude of inversion decreases towards the east. Variscan uplift and erosion of the Munster Basin and SMB succession, combined with tectonic thickening as a result of thrusting and folding, have significantly altered the depositional geometry of the basin fill.

The Munster Basin floor lacks any topography and is clearly undeformed in the zone to the south of the CKL. This suggests that kilometre-scale Variscan folds seen at the surface are not cored by Caledonian basement. The basinal succession must have detached from basement to the basin during Variscan compression. Variscan strain was accommodated by folding, cleavage formation and thrusting of the plastic basinal succession. Basement to the basin is considered to have deformed independently of the Upper Palaeozoic cover, with Variscan deformation characterized by inversion of pre-existing extensional faults. Significant thrusting and tectonic thickening of the basement did not occur. The upper crust is attenuated to the south of the CKL and appears to preserve an extensional geometry. In addition, as the KMFZ separates plastic sediments of the Munster Basin from a rigid, basement block, the greatest amount of Variscan inversion would be expected along this structure. This is indeed the case (Gill 1962; Avison 1984). Interpretation of the VARNET-96 profiles therefore suggests that a Variscan décollement separated the Munster Basin from basement to the basin, allowing for independent deformation of the basinal succession and underlying basement.

Conclusions

(1) The Dingle Basin is imaged as a symmetrical, fault-bounded depocentre in the west, but narrows and assumes a pronounced half-graben geometry to the east. The DBGML is confirmed as the southern basin-bounding fault, whereas the northern syn-rift basin margin on Line B coincides approximately with the NKL of Todd (1989). A granite-cored hinterland to the south supplied sediment to the basin.

(2) An ENE–WSW-trending basement horst is imaged, separating the Dingle and Munster Basins. The northern margin of this basement feature coincides with the DBGML, whereas the southern faulted margin is marked by the KMFZ. The initial location of both faults was determined by tensional reactivation of a Caledonian shear zone in the western onshore region. Subsequent fault propagation patterns were strongly influenced by the volume and location of a series of shallow Caledonian granites.

(3) The KMFZ is recognized as the northern syn-rift margin of the Munster Basin. The location of a second major ENE–WSW-trending extensional fault, the CKL, is revealed by a considerable change in sediment thickness on Line A and Line B.

(4) An E–W-trending tilted fault block is imaged in the northern Munster Basin, bounded by the KMFZ in the north and the CKL in the south. This tilted fault block is cored in the west by a proposed Caledonian granite body, explaining the buoyant nature of this terrace. The CKL developed along the southern margin of the inferred granite.

(5) The Upper Palaeozoic Munster Basin–SMB succession reaches a maximum tectonic thickness of 8 km in the hanging wall of the CKL. No evidence exists on the VARNET-96 profiles to suggest that the southern margin of either basin is located in the onshore region. The Munster Basin may have developed as a half-graben initially, controlled by displacement on the KMFZ. As the magnitude of crustal extension increased, subsidence migrated south of the CKL, resulting in a broader rift in Late Devonian and Early Carboniferous times.

VARNET is funded by the EU Human Capital and Mobility Programme under contract number ERBCHRXCT940572 (involving researchers in Ireland, Germany and Denmark).

References

Avison, M. 1984. Contemporaneous faulting, and the eruption and preservation of the Lough Guitane Volcanic Complex, Co. Kerry. *Journal of the Geological Society, London*, **141**, 501–510.

Capewell, J. G. 1951. The Old Red Sandstone of the Inch and Annascaul District, Co. Kerry. *Proceedings of the Royal Irish Academy*, **54**, 141–167.

—— 1957. The stratigraphy and structure of the country around Sneem, Co. Kerry. *Proceedings of the Royal Irish Academy*, **58**, 167–183.

—— 1965. The Old Red Sandstone of Slieve Mish, Co. Kerry. *Proceedings of the Royal Irish Academy*, **64B**, 165–174.

—— 1975. The Old Red Sandstone Group of Iveragh, Co. Kerry. *Proceedings of the Royal Irish Academy*, **75**, 155–171.

Cerveny, V., Molotkov, I. A. & Psencik, I. 1977. *Ray Method in Seismology*. University Karlova, Prague.

Clayton, G. & Graham, J. R. 1974. Miospore assemblages from the Devonian Sherkin Formation of South-west County Cork, Republic of Ireland. *Pollen et Spores*, **16**(4), 565–588.

——, Graham, J. R., Higgs, K., Holland, C. H. & Naylor, D. 1980. Devonian rocks in Ireland: a review. *Journal of Earth Science of the Royal Dublin Society*, **2**, 161–183.

Clipstone, D. & Roycroft, P. 1992. Detrital magmatic muscovite from the Lower Carboniferous of southwest Ireland: buried granites uncovered? *Journal of the Geological Society, London*, **149**, 163–166.

Coe, K. & Selwood, E. B. 1968. The Upper Palaeozoic stratigraphy of West Cork and parts of South Kerry. *Proceedings of the Royal Irish Academy*, **66B**, 113–131.

Cooper, M. A., Collins, D. A., Ford, M., Murphy, F. X. & Trayner, P. M. 1984. Structural style, shortening estimates and the thrust front of the

Irish Variscides. *In*: HUTTON, D. H. W. & SANDERSON, D. J. (eds) *Variscan Tectonics of the North Atlantic Region*. Geological Society, London, Special Publications, **14**, 167–175.

——, ——, ——, ——, —— & O'SULLIVAN, M. 1986. Structural evolution of the Irish Variscides. *Journal of the Geological Society, London*, **143**, 53–61.

EMINIKE, E. A. 1986. An interpretation of the Killarney and Leinster gravity anomalies in the Republic of Ireland. *Irish Journal of Earth Science*, **7**, 125–132.

FORD, M. 1987. Practical application of the sequential balancing technique: an example from the Irish Variscides. *Journal of the Geological Society, London*, **144**, 885–891.

——, BROWN, C. & READMAN, P. 1991. Analysis and tectonic interpretation of gravity data over the Variscides of southwest Ireland. *Journal of the Geological Society, London*, **148**, 137–148.

GARDINER, P. R. R. 1975. Tectonic controls of Devonian and Lower Carboniferous sedimentation in the South of Ireland. *Ninth International Congress of Sedimentology, Nice, Abstracts Volume*, 139–145.

—— & MACCARTHY, I. A. J. 1981. The Late Palaeozoic evolution of southern Ireland in the context of tectonic basins and their transatlantic significance. *In*: KERR, J. W. & FERGUSSON, A. J. (eds) *Geology of the North Atlantic Borderlands*. Canadian Society of Petroleum Geologists, Memoir, **7**, 683–725.

GEORGE, T. N., JOHNSON, G. A. L., MITCHELL, M., PRENTICE, J. E., RAMSBOTTOM, W. H. C., SEVASTOPULO, G. D. & WILSON, R. B. 1976. *A Correlation of Dinantian Rocks in the British Isles*. Geological Society, London, Special Report, **7**.

GILL, W. D. 1962. The Variscan Fold Belt in Ireland. *In*: COE, K. (ed.) *Some Aspects of the Variscan Fold Belt*. Manchester University Press, Manchester.

HOLLAND, C. H. 1987. Stratigraphical and structural relationship of the Dingle Group (Silurian), County Kerry, Ireland. *Geological Magazine*, **124**, 33–42.

HORNE, R. R. 1970. A preliminary reinterpretation of the Devonian palaeogeography of western County Kerry. *Geological Survey of Ireland Bulletin*, **1**, 53–60.

HOWARD, D. W. 1975. Deep-seated igneous intrusions in Co. Kerry. *Proceedings of the Royal Irish Academy*, **75B**, 173–183.

HUTTON, D. H. W. & REAVY, R. J. 1992. Strike-slip tectonics and granite petrogenesis. *Tectonics*, **11**, 960–967.

KENNAN, P. S. 1997–1998. Tourmaline in the aureole of the Leinster Granite, south-east Ireland: towards a model for tourmalinisation around granites? *Irish Journal of Earth Sciences*, **16**, 33–44.

LANDES, M., PRODEHL, C., HAUSER, F., JACOB, A. W. B., VERMEULEN, N. J. & MECHIE, J. 2000. VARNET-96: influence of Variscan and Caledonian orogenies on crustal structure in SW Ireland. *Geophysical Journal International*, **140**, 660–676.

LEEDER, M. R. 1974. The origin of the Northumberland basin. *Scottish Journal of Geology*, **10**, 283–296.

—— 1982. Upper Palaeozoic basins of the British Isles—Caledonide inheritance versus Hercynian plate margin processes. *Journal of the Geological Society, London*, **139**, 479–491.

LOWE, C. & JACOB, A. W. B. 1989. A north–south seismic profile across the Caledonian Suture zone in Ireland. *Tectonophysics*, **168**, 297–318.

LUETGERT, J. H. 1992. *MacRay—interactive two-dimensional seismic raytracing for the Macintosh*. US Geological Survey, Open File Report, **92–356**.

MACCARTHY, I. A. J. 1987. Transgressive facies in the South Munster Basin, Ireland. *Sedimentology*, 34, 389–422.

MASSON, F., HAUSER, F. & JACOB, A. W. B. 1999. The lithospheric trace of the Iapetus Suture in SW Ireland from teleseismic data. *Tectonophysics*, **302**, 83–98.

——, JACOB, A. W. B., PRODEHL, C., READMAN, P. W., SHANNON, P. M., SCHULZE, A. & ENDERLE, U. 1998. A wide-angle seismic traverse through the Variscan of southwest Ireland. *Geophysical Journal International*, **134**, 689–705.

MEERE, P. A. 1995. The structural evolution of the western Irish Variscides: an example of obstacle tectonics? *Tectonophysics*, **246**, 97–112.

MURPHY, T. 1960. *Gravity anomaly map of Ireland, sheet 5-South West*. Dublin Institute for Advanced Studies, Geophysical Bulletin, **18**.

NAYLOR, D. 1975. Upper Devonian–Lower Carboniferous stratigraphy along the south coast of Dunmanus Bay, Co. Cork. *Proceedings of the Royal Irish Academy*, **75B**, 317–337.

—— & JONES, P. C. 1967. Sedimentation and tectonic setting of the Old Red Sandstone of southwest Ireland. *In*: OSWALD, D. H. (ed.) *Proceedings of the International Symposium on the Devonian System, Calgary*, 1089–1099.

——, —— & MATTHEWS, S. C. 1974. Facies relationships in the Lower Carboniferous of southwest Ireland and adjacent regions. *Geological Journal*, **9**, 77–96.

——, REILLY, T. A., SEVASTOPULO, G. D. & SLEEMAN, A. G. 1983. Stratigraphy and structure of the Irish Variscides. *In*: HANCOCK, P. L. (ed.) *The Variscan Fold Belt in the British Isles*, Adam Hilger Ltd, Bristol, 20–46.

——, SEVASTOPULO, G. D. & SLEEMAN, A. G. 1989. Subsidence history of the South Munster Basin, Ireland. *In*: ARTHURTON, R. S., GUTTERIDGE, P. & NOLAN, S. C. (eds) *The Role of Tectonics in Devonian and Carboniferous Sedimentation in the British Isles*. Yorkshire Geological Society, Occasional Publications, **6**, 99–109.

O'REILLY, B. M., SHANNON, P. M. & VOGT, U. 1991. Seismic studies in the North Celtic Sea Basin: implications for basin development. *Journal of the Geological Society, London*, **148**, 191–195.

PRACHT, M. 1997. *The Geology of Kerry–Cork. Sheet 21 Geological Description*. Geological Survey of Ireland, Dublin.

PRICE, C. A. 1989. Some thoughts on the subsidence and evolution of the Munster Basin, southern Ireland. *In*: ARTHURTON, R. S., GUTTERIDGE, P. & NOLAN, S. C. (eds) *The Role of Tectonics in Devonian and Carboniferous Sedimentation in the British Isles*. Yorkshire Geological Society Occasional Publications, **6**, 111–121.

—— & TODD, S. P. 1988. A model for the development of the Irish Variscides. *Journal of the Geological Society, London*, **145**, 935–939.

RUMPEL, H.-M., VERMEULEN, N. J., MASSON, F., JACOB, A. W. B. & PRODEHL, C. 1999. An integrated shear wave study along the VARNET-96 Line B profile in SW Ireland including Vp/Vs ratio. *American Geophysical Union Annual Meeting Extended Book of Abstracts*.

SANDMEIER, J. & LIEBHARDT, G. 1994. *Handbuch für REFRA (refraktionsseismisches Auswerteund Interpretationsprogramm)*. Geophysical Institute of the University of Karlsruhe, internal report.

SLOAN, R. J. & BENNETT, M. C. 1990. Geochemical character of Silurian volcanism in SW Ireland. *Journal of the Geological Society, London*, **147**, 1051–1060.

STROGEN, P. 1988. The Carboniferous lithostratigraphy of southeast County Limerick, Ireland, and the origin of the Shannon Trough. *Geological Journal*, **23**, 121–137.

TODD, S. P. 1989. Role of the Dingle Bay Lineament in the evolution of the Old Red Sandstone of southwest Ireland. *In*: ARTHURTON, R. S., GUTTERIDGE, P. & NOLAN, S. C. (eds) *The Role of Tectonics in Devonian and Carboniferous Sedimentation in the British Isles*. Yorkshire Geological Society Occasional Publications, **6**, 35–54.

—— 1998. Taking the roof off a suture zone: provenance of conglomerates in the ORS Dingle Basin of SW Ireland. *In*: FRIEND, P. & WILLIAMS, B. P. J. (eds) *New Perspectives on the Old Red Sandstone*. Geological Society of London, British Sedimentological Research Group Meeting Book of Abstracts, 24–25.

VERMEULEN, N. J., SHANNON, P. M., LANDES, M., MASSON, F. & THE VARNET GROUP 1998–1999. Seismic evidence for subhorizontal crustal detachments beneath the Irish Variscides. *Irish Journal of Earth Sciences*, **17**, 1–18.

WILLIAMS, B. P. J. & RICHMOND, L. K. 1998. A new Old Red Sandstone terrane in S.W. Ireland. *In*: FRIEND, P. & WILLIAMS, B. P. J. (eds) *New Perspectives on the Old Red Sandstone*. Geological Society of London, British Sedimentological Research Group Meeting Book of Abstracts, 27–28.

WILLIAMS, E. A. 1998. Numerical modelling of Munster Basin extensional subsidence and controls on fluvial system emplacement. *In*: FRIEND, P. & WILLIAMS, B. P. J. (eds) *New Perspectives on the Old Red Sandstone*. Geological Society of London, British Sedimentological Research Group Meeting Book of Abstracts, 29–30.

——, BAMFORD, M. L. F., COOPER, M. A. *et al.* 1989. Tectonic controls and sedimentary response in the Devonian–Carboniferous Munster and South Munster basins, south-west Ireland. *In*: ARTHURTON, R. S., GUTTERIDGE, P. & NOLAN, S. C. (eds) *The Role of Tectonics in Devonian and Carboniferous Sedimentation in the British Isles*. Yorkshire Geological Society Occasional Publications, **6**, 123–141.

——, FORD, M. & EDWARDS, H. E. 1990. Discussion of a model for the development of the Irish Variscides. *Journal of the Geological Society, London*, **147**, 566–571.

——, SERGEEV, S. A., STÖSSEL, I. & FORD, M. 1997. An Eifelian U–Pb zircon date for the Enagh Tuff Bed from the Old Red Sandstone of the Munster Basin in NW Iveragh, SW Ireland. *Journal of the Geological Society, London*, **154**, 189–194.

——, ——, ——, —— & HIGGS, K. 1998. *U–Pb zircon geochronology of Munster Basin silicic tuffs: implications for the chronostratigraphy, cyclicity and sedimentation of the Old Red Sandstone of SW Ireland*. 41st Annual Irish Geological Research Meeting Book of Abstracts, 25.

ZELT, C. A. & SMITH, R. B. 1992. Seismic traveltime inversion for 2D crustal velocity structure. *Geophysical Journal International*, **108**, 16–34.

Flexural cantilever models of extensional subsidence in the Munster Basin (SW Ireland) and Old Red Sandstone fluvial dispersal systems

E. A. WILLIAMS

Department of Geology, University College Cork, Ireland, and Geologisches Institut, ETH-Zentrum, 8092 Zürich, Switzerland

Present address: CRPG-CNRS, B.P. 20, 54501 Vandoeuvre-lès-Nancy Cedex, France (e-mail: williams@crpg.cnrs-nancy.fr)

Abstract: Flexural cantilever (2D) computer modelling of the palinspastically restored Mid–Late Devonian Munster Basin has been used to appraise quantitatively extensional subsidence and Old Red Sandstone (ORS) stratigraphic geometries. One-dimensional decompaction and (Airy isostatic) backstripping were carried out to constrain syn-rift forward models; these specify the Late Palaeozoic rifting history of the region. Forward modelling showed that listric faults and detachments fail to reproduce restored ORS sediment geometries, but instead indicated that multiple planar, upper-crustal faults are necessary to achieve the correct order of syn-rift subsidence across the basin. Modelled (non-unique) sections transverse to the basin bounding fault (Dingle Bay–Galtree Fault Zone) replicated ORS geometries with cumulative extensions of 27 km in the east (stretching factor $\beta = 1.3$) and 59 km in the west ($\beta = 1.48$), with effective elastic thicknesses of 7 and 8 km, respectively, starting from a 40 km thick post-Acadian crust. Resultant peak heat-flow anomalies predict well the location of known syn-rift volcano-magmatic centres. Modelling indicated that significant offshore faults are required to achieve subsidence in the west Cork region, implying that the basin continues offshore. Observed ORS (<0.85 km thick) sections on the regional footwall, considered to be the result of thermal (post-rift) subsidence, are not accounted for by modelling, whereas *c.* 1 km of post-rift ORS is modelled over 5 Ma in the central–southern regions of the basin. These sections buried the principal rift faults during late Famennian time. The ORS of the Munster Basin is dominated by two large-scale transverse fluvial dispersal systems that were largely insensitive to deflection by any extension faults that propagated in the syn-rift fill. A third major system entered in the SW, demarcated by antithetic extension faults south of the depocentre.

The Munster Basin is the largest repository of Old Red Sandstone (ORS) rocks in Ireland (Capewell 1965; Clayton *et al.* 1980; Graham & Clayton 1988), although its exposed record is restricted to the Late (Higgs *et al.* this volume) and Late-Mid to Late Devonian (Williams *et al.* this volume) from palynological and isotopic information available in the west of the basin (Fig. 1). The rocks representing the fill of the Munster Basin were significantly deformed during the Variscan Orogeny into an arcuate slate belt in which several orders of buckle folds are affected by a weakly transecting pressure solution cleavage. The folds are dismembered by generally late, steeply to moderately inclined smooth-trajectory thrusts (*sensu* Cooper & Trayner 1986) and steep cross-faults (Naylor *et al.* 1981; Cooper *et al.* 1986; Ford 1987; Murphy 1990). The Variscan orogen in Ireland comprises several structural zones (Gill 1962; Cooper *et al.* 1986), which overprint the ORS rocks of the Munster Basin. Zone 1, to the south of the Killarney–Mallow Fault and Dingle–Dungarvan Line (Figs 1 and 2), which affects the largest part of the basin, underwent a maximum of *c.* 52% shortening in the centre of the zone (Ford 1987) under epizone facies metamorphism (Blackmore 1995).

The original (pre-Variscan) structure of the basin has been regarded generally (Naylor *et al.* 1981, 1983; Graham 1983; Sanderson 1984; Price & Todd 1988; Price 1989; Williams *et al.* 1989, 1993; MacCarthy 1990) as a half-graben following the original suggestion of Naylor & Jones (1967), based on an apparently asymmetrical sediment prism in a N–S transect through its western part and the presumption of major faulting in Dingle Bay (Fig. 2). Gardiner & MacCarthy (1981), however, considered dispersal patterns consistent with a graben, and Price

From: FRIEND, P. F. & WILLIAMS, B. P. J. (eds). *New Perspectives on the Old Red Sandstone.* Geological Society, London, Special Publications, **180**, 239–268. 0305-8719/00/$15.00

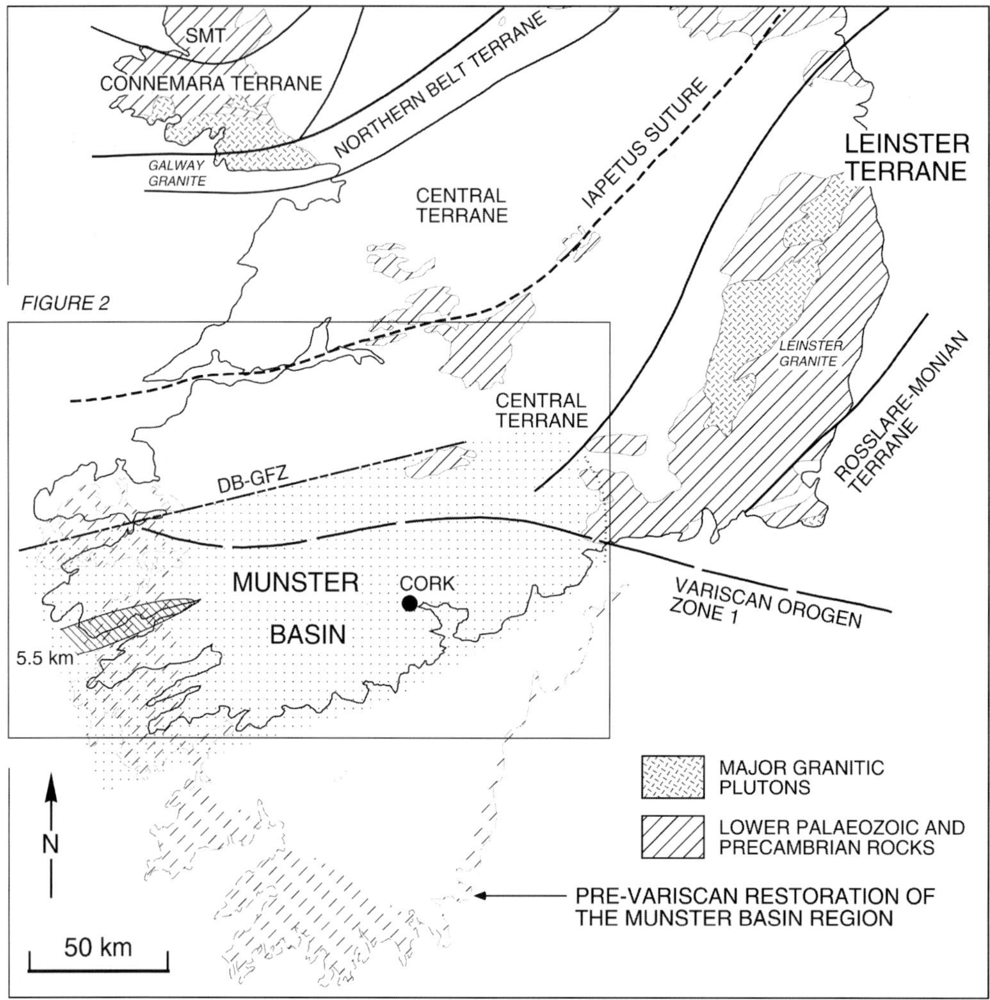

Fig. 1. Tectonic map of southern Ireland, showing the context of the Munster Basin. Caledonian terranes after Hutton (1987). Palinspastic (pre-Variscan) outline of the basin superimposed (present coastline acts as the datum). The northern boundary of Variscan orogen zone 1 (Cooper *et al.* 1986) is referred to as the Dingle–Dungarvan Line, the central part of which is defined by the Killarney–Mallow (thrust) Fault Zone (see Figs 2 and 3). Dotted line is the eastern margin to the Munster Basin. The apparent depocentre in the south Iveragh–north Kenmare River is approximately defined by the 5.5 km isopach.

(1989) suggested that a more symmetrical geometry described the basin in the east (Galtee–Knockmealdown Mountains), although no 2D sections of this part of the basin have been published. The earliest published sections through the western part of the basin showed only one major fault zone (the northern margin in Dingle Bay) and depict a 'hinged basin' with subsidence diminishing at the south coast (Naylor *et al.* 1981). Recent models have incorporated northern marginal faults that are either listric (Price & Todd 1988; Price 1989; MacCarthy 1990) or steep planar faults (Williams *et al.* 1989) linked to detachments, or planar upper-crustal faults (Williams *et al.* 1993).

Despite the common description of the Munster Basin as a half-graben, and by inference its extensional mode of origin, three general criticisms may be levelled at these models. (1) None of the 2D models of the western part of the basin, which normally depict the ORS fill, have been considered palinspastically, except those of Williams *et al.* (1989, 1993). This has important implications for the true width-scale of the basin, and raises the question of the pre-Variscan

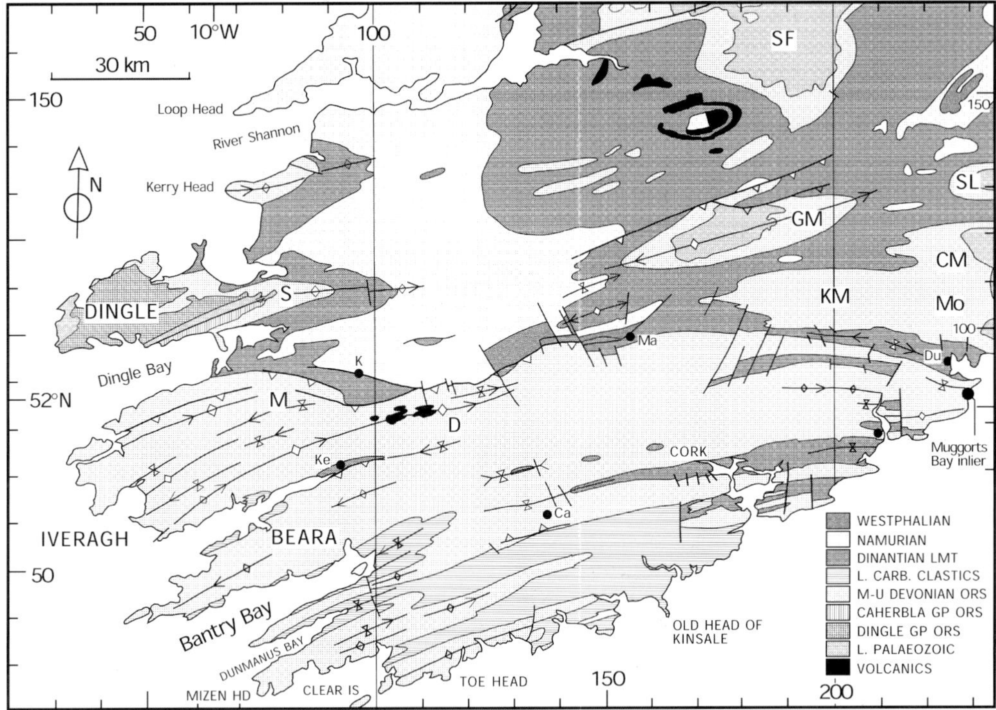

Fig. 2. Geological map of SW Ireland. Abbreviations referred to in text: SF, Slieve Phelim; GM, Galtee Mountains; KM, eastern Kilworth–western Knockmealdown Mountains; CM, Comeragh Mountains; S, Slieve Mish; M, MacGillycuddy's Reeks; Mo, Monavullagh Mountains; D, Derrynasaggart Mountains; SL, Slievenamon; K, Killarney; Ke, Kenmare; Ma, Mallow; Ca, Castletown; Du, Dungarvan; Y, Youghal.

position of major faults (or fault zones), and ORS volcanic or magmatic centres. Moreover, no detailed consideration of the pre-Variscan origin of individual faults has taken place. (2) All of these models, whether 2D (Naylor *et al.* 1981; Price & Todd 1988) or 3D (Williams *et al.* 1989, 1993; MacCarthy 1990) have been qualitative, without consideration of whether subsidence is achievable given the limited number and type of faults employed in the respective models. (3) No analysis has included decompaction or back-stripping of the ORS basin fill, and several workers (Graham & Clayton 1988; Price 1989; Meere 1995*b*) have suggested that such studies are problematical because of the diagenetic and Variscan strain state of the rocks. Despite this, several studies have modelled Late Palaeozoic subsidence in one dimension, employing McKenzie's (1978) uniform extension model (Sanderson 1984; Meere 1995*b*) and Hellinger & Sclater's (1983) two-layer stretching model (Price 1989). These analyses employed somewhat different rifting histories, and predicted stretching factors for the depocentre ranging from $\beta = 2$ to $\beta = 5$.

The uncertainties inherent in all the previous models of the basin raise fundamental questions of: (1) the fill age in relation to rifting, that is, whether the regionally thick and extensive ORS is a syn- or post-rift magnafacies; (2) the general number and nature of faults required to achieve the observed subsidence; (3) the true width-scale of rifting; (4) the governing crustal parameters. In this paper, two sections corrected for Variscan shortening in the east and west of the Munster Basin are forward-modelled using a flexural cantilever model of 2D continental lithospheric extension (Kusznir & Egan 1990; Kusznir *et al.* 1991). Stratigraphic columns along the sections lines are decompacted and backstripped, using established lithological parameters, to provide a more realistic minimum syn-rift subsidence for the forward model experiments, and to characterize the subsidence regime of the region. The flexural forward models provide quantitative estimates of stretching, heat-flow anomaly and maximum horizontal bending stresses, as well as the 2D crustal–basin structure and sediment fill geometry. This paper also addresses the nature and scale of the basin-filling fluvial dispersal systems in the context of the extensional fault arrays modelled, and the gross subsidence history of the region.

Pre-Variscan template

The regional Mid- to Late Devonian tectonic setting of southern Ireland and southern Britain was one of widespread extensional basin subsidence. In SW England, the Gramscatho Basin (south Cornwall), initiated in late Silurian–early Devonian time, contains tholeiitic mid-ocean ridge basalts (MORB) to back-arc basalts and deep-water marine sediments ranging to Late Devonian in age (Sanderson 1984). The Trevone Basin (south and central Devon), evolved from a continental setting, with ?Lochkovian–Pragian aged ORS and volcanic rocks of the Dartmouth Group (south Devon), to a marine basin with a thick Lower to Upper Devonian succession and intra-plate volcanic rocks (Bluck *et al.* 1992). In the north Devon, Bristol Channel and South Wales region, thinner continental ORS and marginal marine facies accumulated more episodically, in an overall rifted settling (Allen 1974, 1979; Sanderson 1984; Tunbridge 1984, 1986).

The exact mode of Variscan deformation in southern Ireland remains a debated point in the literature, as does the quantification of shortening. Dominantly compressional (Cooper *et al.* 1986) and significantly (dextrally) transpressional (Sanderson 1984) deformational models have been proposed, with further subdivision into those that are purely thick skinned (basement involved), suggested by Sanderson (1984), and those that are thin skinned, in turn either basement involved (Ford 1987; Price & Todd 1988; Meere 1995*a*) or detached at the base of the ORS (Cooper *et al.* 1984, 1986; Bamford 1987; Murphy 1990). The transpressive inversion model of Price & Todd (1988) is based on reactivation of a listric extensional detachment and on reuse of planar extension faults in its hanging wall. Filtering of the Bouguer anomaly over the Irish Variscides suggests that basement may core the major folds of the region (Ford *et al.* 1991), indicating that deformation was not significantly detached at least at the base of the ORS basin fill. Folded basement beneath the Munster Basin crops out in the hinge zone of the major Galtees–Ballyhoura Mountains anticline in the eastern orogen transition zone (Fig. 2). Major controversy surrounds the status of the Killarney–Mallow Fault (Fig. 2); in particular, whether it is the reactivated Munster Basin margin fault (Sanderson 1984; Price & Todd 1988; Price 1989; Meere 1995*a*; Vermeulen *et al.* 1998–1999, this volume), a Variscan thrust structure (Cooper *et al.* 1984, 1986; Williams *et al.* 1989, 1990) or a dextral shear zone (Max & Lefort 1984).

The palinspastic restoration utilized in this study (Fig. 1), revised from Williams *et al.* (1989), is based on a compressional deformation mode, but better accounts for the variation of natural strain (e) and shortening direction around the (arcuate) fold belt. The template is modified in the eastern and central parts of the basin. This is a result of restoration of north-directed shortening in the east of the orogen ($e = -0.44$, Cooper *et al.* 1986, fig. 6), and NNW-directed shortening in the region west of Cork City (section 2 of Cooper *et al.* (1984)), where an estimate of e of -0.47 is based on a 10% increase in shortening using the restoration techniques of Cooper & Trayner (1986) and Ford (1987). A conservative value of $e = -0.5$ is used for the western part of the fold belt (see Ford 1987) and a more recent, anomalous, bulk shortening value for this region (33%, Meere 1995*a*) is not adopted. No published balanced section restores deformation from the Munster Basin northern margin; indeed, Cooper *et al.* (1986) estimated *c.* 6 km of unaccounted slip that must continue north of their section, pinned in the eastern orogen transition zone. They also considered that total strain remained underestimated in their balanced sections. For these reasons, the palinspastic template can be regarded as conservative. The Killarney–Mallow Fault is treated as a Variscan age structure in the restored template (see Williams *et al.* 1990), rather than the local basin margin.

The palinspastically restored Munster Basin (Fig. 1) is demarcated by extreme changes in ORS thickness across a NNW margin (widely considered to be fault controlled), and an eastern margin (the eastern Comeragh–Monavullagh Mountains) that is not associated with major faulting (Capewell 1956). A western margin is inferred offshore (Graham 1983) on evidence of the ESE dispersal of the Sherkin Formation (Graham & Reilly 1972) in the Sheep's Head–Toe Head region (Fig. 3). No western source area can be inferred from ORS facies and dispersal in the Iveragh–Beara regions (Fig. 3). Ford *et al.* (1992, fig. 1c), however, suggested an offshore westward closure to the basin based on the nature of the basement from deep seismic and other geophysical data (Makris *et al.* 1988; Conroy & Brock 1989; Klemperer *et al.* 1991). There is no onshore evidence of a southern basin margin (Naylor *et al.* 1983). Proximal facies continental red beds and tuffs of ?Frasnian age (Robinson *et al.* 1981) were drilled *c.* 155 km offshore to the SSE of Toe Head (Fig. 2). This non-palinspastic distance suggests that these

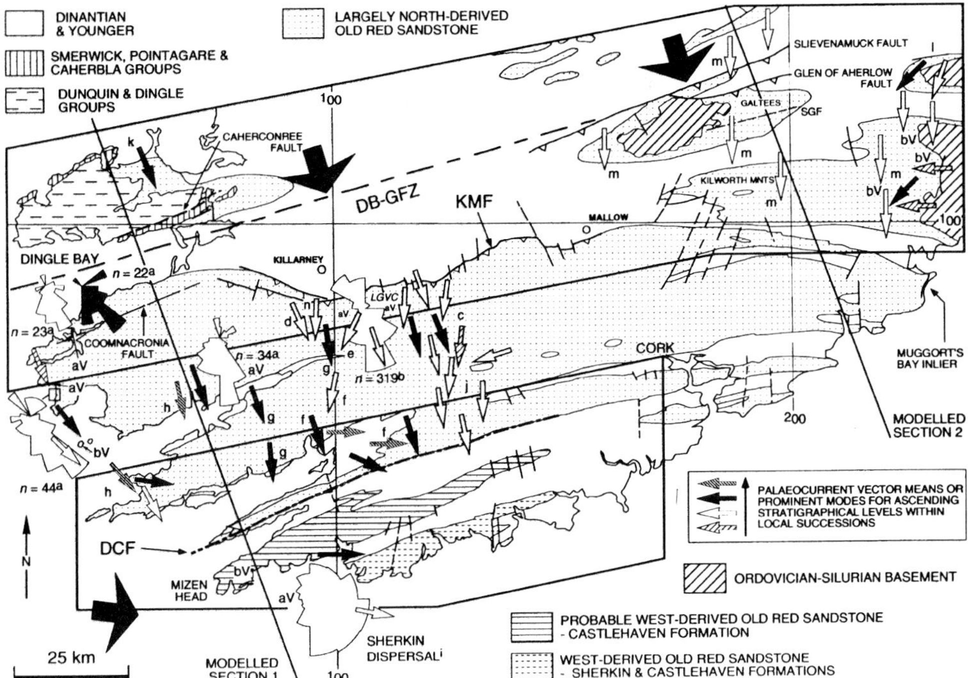

Fig. 3. Simplified geological map highlighting the Dingle Bay–Galtee Fault Zone (DB–GFZ) and the Dunmanus–Castletown Fault (DCF) showing the gross derivation of associated ORS. Also shown are other basin extension faults for which independent evidence exists, distribution of basement, detailed palaeocurrent data and modelled section lines. Sources of palaeocurrent data: a, Russell (1984); b, Avison (1984*a*); c, Williams (1993); d, Wegmann (1993); e, Kelly & Olsen (1993); f, Williams *et al.* (1989); g, Sadler & Kelly (1993); h, Graham *et al.* (1992); i, Graham & Reilly (1972); j, Edwards (pers. comm.); k, Todd (1989); l, Colthurst (1978); m, compilation of Carruthers (1987, fig. 2); n, Wingfield (1968). Large filled arrows indicate the approximate position of stable entry points of the principal river systems that filled the basin. KMF, Killarney–Mallow Fault.

sediments are not closely related to the onshore Munster Basin, but may be a product of the rifted crust of which the basin is a part.

Major faults

The evidence for the principal basin-forming structures used in the forward models is briefly reviewed. The Dingle Bay–Galtee Fault Zone (DB–GFZ; Fig. 3) is considered to be the principal boundary fault of the Munster Basin. In the west it comprises the Caherconree and Dingle Bay Faults, and in the NE, the Glen of Aherlow and Slievenamuck Faults. The Caherconree Fault (Fig. 3), is associated with, and linked to, minor synthetic normal faults in south Dingle (Capewell 1951, 1965), and is partly responsible for the tilting of the pre-late Givetian Caherbla Group (Williams *et al.* this volume) during Munster Basin extension (Horne 1974, 1977; Todd 1989). The main strand of the DB–GFZ in Dingle Bay is defined by geophysical anomalies (Howard 1975), major ORS thickness change and stratigraphical incompatibility (Williams *et al.* 1989), and has a sedimentary signature of steep alluvial fan deposits in north Iveragh. These include the Doulus Conglomerate (Capewell 1975), constructed of pebbly sheetflood and viscous debris flow deposits (Russell 1984), and a suite of pebbly sandstone and pebble conglomerate lithosomes, with high ratios of resistate to high-grade metamorphic clast assemblages, containing lithologies typical of the Caherbla Group (Inch Conglomerate) of south Dingle (Figs 2 and 3). The trace of the DB–GFZ marked the SSE margin of the Siluro-Devonian Dingle Basin, with the opposite sense of downthrow to that operative during Munster Basin time.

The Coomnacronia Fault (Fig. 3), first mapped by Capewell (1975) as a thrust fault, is responsible for major southward stratigraphic thickening based on the correlation of the Keel–Enagh Tuff Bed across the structure (Williams *et al.* this volume; Fig. 4). The fault trace also has very marked Bouguer and filtered gravity signatures (Morris 1980; Ford *et al.* 1991; Readman *et al.* 1997), and has shallow basement

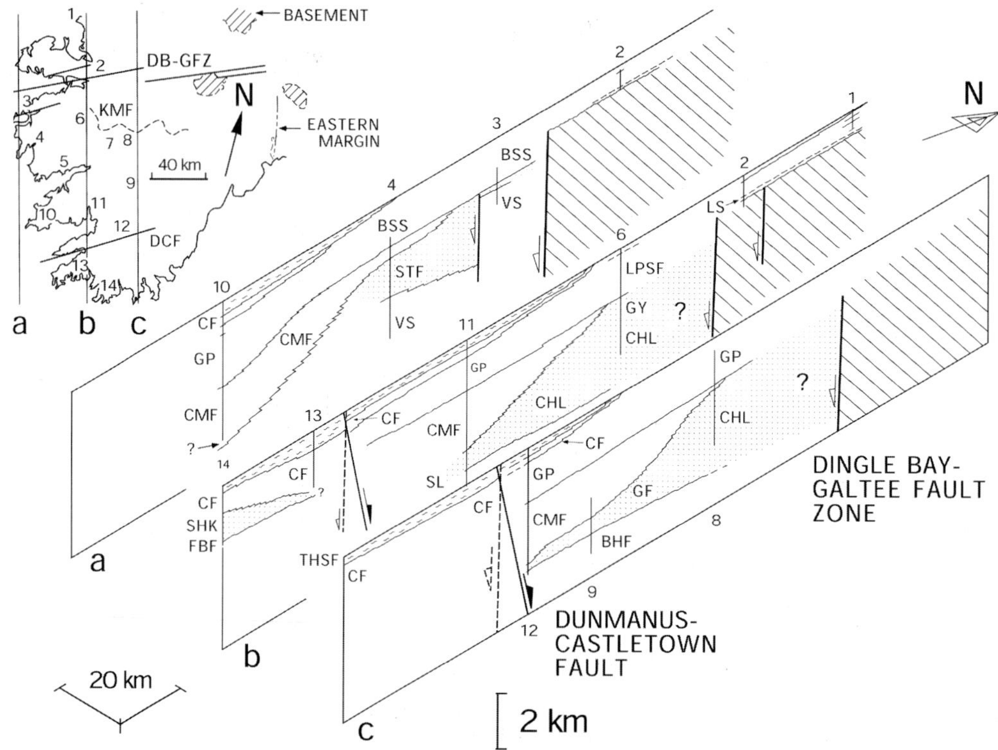

Fig. 4. Isometric serial sections of the ORS stratigraphical architecture in the western part of the Munster Basin (modified from Williams (1993)), forming the database for the modelled section line 1. BSS, Ballinskelligs Sandstone Fm; CMF, Caha Mountain Fm; STF, St Finan's Sandstone Fm; VS, Valentia Slate Fm; GP, Gun Point Fm; SL, Slaheny Fm; CHL, Chloritic Sandstone Fm; GY, Grey Sandstone Fm; LPSF, Lower Purple Sandstone Fm; GF, Gortanimill Fm; BHF; Bird Hill Fm; FBF, Foilcoagh Bay Fm; SHK, Sherkin Fm; CF, Castlehaven Fm; THSF, Toe Head Sandstone Fm. ORS formations of station 2 comprise the Slieve Mish Group, where the basal unit is the Lough Slat Conglomerate Formation (LS). Thickness information: 1, Bridge & Diemer (1983), Diemer *et al.* (1987), Diemer & Bridge (1988); 2, Capewell (1965); 3, Williams *et al.* (1997); 4, 5, Russell (1978, 1984); 6, Walsh (1968); 7, Avison (1984*a*, *b*); 8, 9, 11, 12, Williams *et al.* (1989); 10, James & Graham (1995); 13, Naylor *et al.* (1983); 14, Graham & Reilly (1972). Dashed fault traces (southward downthrow) show the Dunmanus Fault, which was active during South Munster Basin rifting, with Sheep's Head in the footwall (**b**).

in its eastern footwall, imaged on the seismic refraction line of Masson *et al.* (1998). Masson *et al.* (1998) and Vermeulen *et al.* (1998–1999), however, interpreted this basement as a Lower Palaeozoic granite, with a hypothetical intrusive relationship to their upper-crustal basement (L2), without considering the Coomnacronia Fault. In a later paper, Vermeulen *et al.* (this volume) interpreted this shallow basement as a vertical Caledonian shear zone. The Coomnacronia Fault is considered here to be an intrabasinal, south-dipping extension fault, as it preserves a thinned ORS stratigraphy in its footwall, which is at least as old as Givetian time (Williams *et al.* this volume) and is coherent with ORS stratigraphy in Iveragh (Fig. 4).

The closely spaced Glen of Aherlow and Slievenamuck Faults are well-defined palaeo-extensional structures from sedimentary facies, correlation and thickness criteria (Doran *et al.* 1973; Carruthers 1987). However, unlike the NW margin faults, they are inverted (reactivated) structures, as they show local thrust relationships (Shelford 1963) and elevate regional hanging-wall basement (the Lower Palaeozoic Galtees inlier; Figs 2 and 3). South of the axial surface trace of the Galtees anticline, east–west thickness variation in the ORS and the distribution of the Pigeon Rock Formation have been interpreted as due to an extension fault (Sleeman *et al.* 1995; Fig. 3), referred to here as the Southern Galtees Fault (Fig. 3). The basin margin geometry is reconstructed in Fig. 5. Other basin extension faults have been suggested for the eastern part of the Munster Basin, on the northern sides of the Knockmealdown and

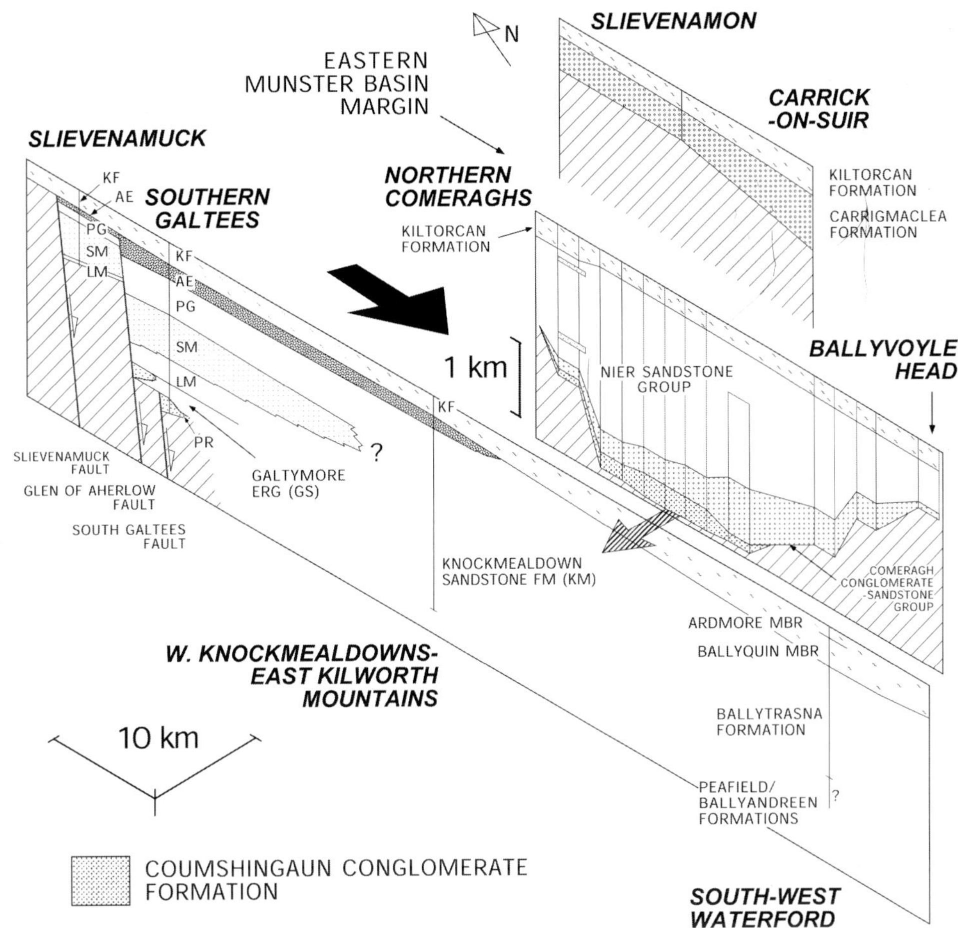

Fig. 5. Isometric (palinspastic) diagram showing the stratigraphical relationships of the ORS in the eastern and northeastern Munster Basin. Stratigraphical data sources: Slievenamuck–Galtees–Knockmealdowns (Carruthers 1987), SW Waterford–east Cork (Trayner 1985; MacCarthy 1990; Sleeman *et al.* 1995), Comeraghs–Ballyvoyle Head (Capewell 1956), Slievenamon–Carrick-on-Suir (Colthurst 1978; Carruthers 1987). PR, Pigeon Rock Conglomerate Formation; GS, Galtymore Sandstone Fm; LM, Lough Muskry Fm; SM, Slievenamuck Conglomerate Fm; PG, Poulgrania Fm; AE, Ardane Conglomerate Fm; KF, Kiltorcan Fm.

Comeragh Mountains (Naylor *et al.* 1981, p. 65, 1983, p. 42; Price 1989, fig. 1 and p. 113). However, specific details of facies and stratigraphic thickness variations have not been presented, hence these faults are not incorporated in Fig. 5.

Except for the mid-Frasnian Carrigduff Group in north Dingle (Richmond 1998; Higgs & Richmond, pers. comm.; Richmond & Williams this volume), there is no known Mid- to early Late Devonian ORS north of the DB–GFZ. The DB–GFZ is therefore regarded as the fundamental extensional fault zone of the northern Munster Basin.

The Dunmanus–Castletown Fault (Fig. 3) was suggested as an antithetic intrabasinal structure by Williams *et al.* (1989) on the basis of integrated facies, palaeodispersal and thickness variation signatures in associated Devonian ORS, latest Devonian coastal plain–shallow marine and Carboniferous marine sequences. Subsequently, gravity modelling (Ford *et al.* 1991) and seismic refraction data across the fault (Masson *et al.* 1998) demonstrate deep crustal changes associated with the structure, compatible with an antithetic offset sense.

Sub-Munster Basin crustal basement

Deep seismic data, 30 km along-strike west of the Munster Basin (Ford *et al.* 1992), image south-dipping reflectors within the mid- to upper crust.

These have been interpreted as due to deformation associated with Early Palaeozoic subduction beneath Eastern Avalonia (Ford *et al.* 1992). Compatibility with structural fabrics and vergence in the Leinster Zone suggests continuity of Leinster Terrane (Fig. 1) psammitic–pelitic metasedimentary and volcanic basement beneath the Munster Basin (Ford *et al.* 1992). Field evidence for this comes from (1) the Muggort's Bay inlier (Fig. 3), a detached wedge of Middle Cambrian Leinster Terrane lithologies overlain by basal ORS gravels (Cooper *et al.* 1986; Brück *et al.* 2000), and more importantly (2) alluvial fan rudites preserved in the Early Devonian Dingle Basin (Horne 1974; Todd 1989), which indicate a supracrustal metasedimentary basement to the Munster Basin beneath Iveragh (Allen & Crowley 1983).

The basement beneath NE Iveragh is controversial because of varying interpretations of a large negative Bouguer anomaly (the Killarney anomaly), taken to indicate either (1) a major granite pluton (Howard 1975; Gardiner & MacCarthy 1981; Murphy 1981; Avison 1984*a*, *b*; MacCarthy 1990, fig. 11; Readman *et al.* 1997) or (2) thick ORS in the depocentre of the Munster Basin (Naylor & Jones 1967; Clayton *et al.* 1980; Sanderson 1984; Emenike 1986; Karner *et al.* 1987; Price 1989) plus additional sediments of a precursor basin (Naylor *et al.* 1983). Ford *et al.* (1991) demonstrated that both scenarios can be successfully gravity modelled using known surface geological data (Fig. 2), but depend on different tectonic interpretations of the Killarney–Mallow Fault and DB–GFZ. Naylor *et al.* (1983), Price (1989, p. 118) and Ford *et al.* (1991) have suggested the improbability of a major hanging-wall granite on the grounds that its low density would have inhibited subsidence. However, minor granite(s) beneath Iveragh is supported by the very rare granitic clasts reported from single stations within the Trabeg and Inch Conglomerates of the Dingle Basin (Capewell 1951; Horne 1974; Dodd 1986; Todd 1989).

Munster Basin subsidence initiated <15 Ma after late Caledonian (Acadian) deformation and plutonism in southern Ireland (Williams *et al.* this volume). It is likely therefore that the post-Acadian (Leinster Terrane) crust beneath the basin was significantly thickened, despite undergoing ?localized (Caherbla Group) erosion, before 385 Ma. Its likely thickness and the probability that it had not thermally re-equilibrated suggests an initially strong lithosphere (Coward 1993), relevant to the flexural model input parameters.

ORS basin fill stratigraphy and fluvial dispersal systems

The internal stratigraphical geometry and nomenclature of the Munster Basin ORS is complex, and it is particularly difficult to link the undated ORS lithostratigraphy of the eastern part of the basin (Capewell 1956; Carruthers 1987; MacCarthy 1990; Sleeman *et al.* 1995; Fig. 5) to that in the west (Fig. 4), where biostratigraphic and isotopic data are available for the older part of the record (e.g. Clayton & Graham 1974; Russell 1978; Higgs & Russell 1981; Williams *et al.* 1997, this volume; Higgs *et al.* this volume). Some of the oldest exposed ORS within the depocentre is now known to be as old as late Givetian in age (*c.* 385 Ma; Williams *et al.* this volume). Although the ORS throughout the basin is conformable with Carboniferous sequences, providing a correlation datum, detailed linkage of eastern and western stratigraphies is not attempted. Instead, the local stratigraphical databases on which the decompaction–backstripping exercises and the flexural forward models rely are collated in two, approximately palinspastic isometric diagrams (Figs 4 and 5), and a summary stressing the component fluvial systems is given.

Integration of the stratigraphy and sedimentology (schemes of Carruthers 1987; Williams *et al.* 1990) of the ORS suggests that the fill of the (onshore) Munster Basin is fundamentally the product of major, low-gradient, fluvial dispersal systems entering the basin at only three entry points (Williams *et al.* 1993; Fig. 3), which were essentially stable for the >23 Ma subsidence history (Williams *et al.* this volume) of the basin. These systems were areally restricted and geometrically complex spreads of relatively coarse-grained alluvium, embedded in a related matrix of fine-grained red-bed clastic deposits (Williams *et al.* 1989). The systems involved major rivers of variable scale or style entering the basin, diversifying downstream and depositing sediment in a complex series of alluvial environments and sub-environments over areas restricted by the location of the main input and the long-term behaviour of the systems. Two of the entry points, which crossed the basin-bounding Dingle Bay–Galtee Fault Zone, gave rise to transverse-draining systems with diffusional lengths of $?>70$ and *c.* 120 km. One, entering in the region of the Galtee Mountains, dominated the eastern part of the basin (the Galtee System). The second, larger system entered ENE of Dingle Bay and dominated the west of the basin, with

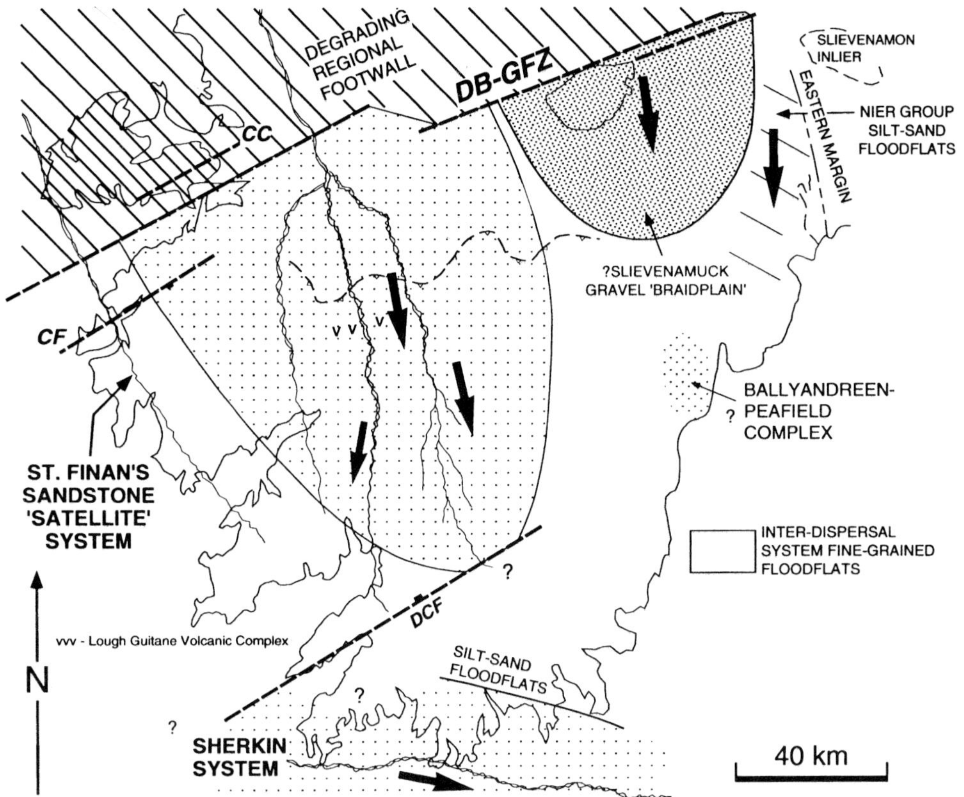

Fig. 6. Palaeogeography of the principal fluvial dispersal systems during the earlier phase of syn-rift subsidence. It is currently not possible to accurately identify coeval laterally dispersing systems; however, the Chloritic Sandstone–Gortanimill–Slaheny System in the W–NW is suggested to have operated at the same time as the Slievenamuck Formation gravel 'braidplain' through the Galtee Mountains (Carruthers 1987). The ?synchronous laterally draining (Carruthers 1987, fig. 2) Nier Sandstone Group (*sensu* Boldy (1982) in Sleeman *et al.* (1995); see also Capewell (1956)) succeeded the sub-axially draining coarse-grained Coumshingaun and Comeragh Conglomerates at the eastern margin of the basin. These units are part of the syn-rift stratigraphy shown in model section 2. CC, Caherconree Fault; CF, Coomnacronia Fault.

exposed proximal zones in the Reeks–Derrynasaggarts (the Iveragh System).

The transverse drainage systems show clear vertical stratigraphical differentiation (Figs 4 and 5), allowing temporally distinct dispersal systems to be defined. In the northwest, the earlier basin history was dominated by the Chloritic Sandstone–Gortanimill–Slaheny fluvial dispersal system (Williams *et al.* 1989; Fig. 6), which was characterized by avulsion-controlled, large sand-bed braided rivers (Williams 1993). The distal end of the system can be traced to the hanging wall of the Dunmanus–Castletown Fault, across which regional stratigraphic elevation changes, concealing evidence of the termination of the system. Palaeocurrents maintain SSE vectors (Fig. 3) up to the fault. This system was coeval with the St Finan's lateral system (Fig. 6), which underwent major syn-rift thickening across the Coomnacronia Fault (Williams *et al.* this volume). It is tentatively suggested that, in the Galtees, coeval sedimentation was represented by the coarse-grained Slievenamuck Formation, which represents a gravel 'braidplain' (Carruthers 1987). The across-basin reach of this system is not known, as correlations are not well established between the Galtees and sections to the south (Fig. 5). The Slievenamuck Formation is a component of south-dispersing alluvium of the NE Munster Basin that succeeded early west and WSW dispersal prevalent until near the top of the Comeragh Conglomerate–Sandstone Group of the Comeragh and Monavullagh Mountains (Fig. 5).

The third major dispersal system operative at this time, the Sherkin System, was confined to the

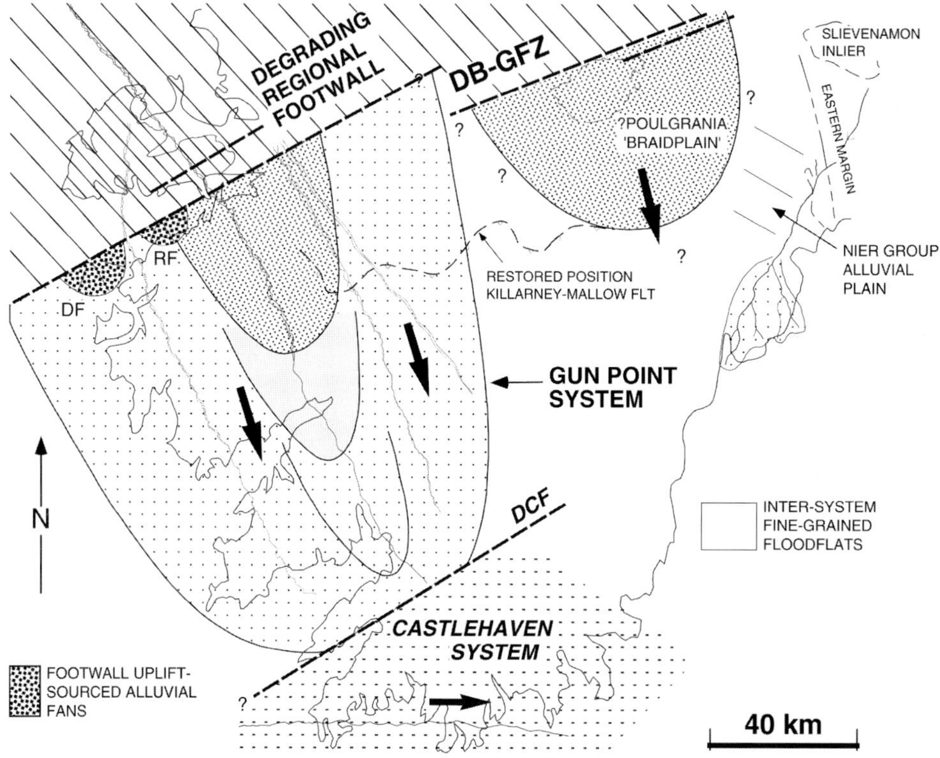

Fig. 7. Palaeogeography of the principal fluvial dispersal systems during the later phase of syn-rift subsidence. The Gun Point System is thought to be coeval with the laterally draining Ballydavid Formation sandy 'braidplain' of Carruthers (1987) in the Galtee Mountains (name redefined to Poulgrania Formation by Sleeman *et al.* (1995)). Dispersal system in SW Waterford–east Cork from MacCarthy (1990, fig. 14). RF, Reacaslagh–Darby's Bridge fan; DF, Doulus fan.

south of the (antithetic) Dunmanus–Castletown Fault, and had an oblique mean dispersal direction (Fig. 3) with respect to the basin trend (Fig. 6). Thick, fine-grained sequences of the younger Castlehaven Formation on Mizen Head (Figs 2–4) may indicate that the Sherkin low-sinuosity rivers (Graham & Reilly 1972) were restricted to the extreme southwest by a cryptic extension fault south of the Mizen peninsula.

Later ORS accumulation in the NW of the basin is represented by the Gun Point System (Fig. 7; Williams *et al.* 1989; Sadler & Kelly 1993). This is a non-radial fluvial dispersal system similar to the earlier ones, with distinct axial–lateral and proximal–distal facies zones, reaching the Dunmanus–Castletown Fault. The largest rivers of the system (low-sinuosity–braided), carrying the coarsest load, were confined to the axial region, and decreased in scale across-basin. Although described as a 'terminal fan' (Sadler & Kelly 1993), it has not yet been established that the Gun Point System terminated at the Dunmanus–Castletown Fault. The Gun Point and Chloritic Sandstone–Gortanimill–Slaheny Systems have different gross stratigraphic geometries (Fig. 4) and alluvial architectures, which indicate that the former was largely aggradational, whereas the latter was strongly progradational–retrogradational. The principal reasons for these differences are thought to be long-period subsidence rate variations convolved with long-term drainage basin evolution (sediment flux and water discharge variations).

South of the Dunmanus–Castletown Fault, the eastwards-draining Castlehaven Formation was coeval with the Gun Point System, and was characterized by silty flood flats, with rare minor river channel belts. In the NE Munster Basin, the sand-dominant Poulgrania 'braidplain' (Fig. 7) is thought to have operated at this time, whereas in the SE small ?terminal rivers flowed southwestward (MacCarthy 1990) draining the Early Palaeozoic Leinster Massif.

Regional footwall to hanging-wall correlation

Thin ORS successions, related to the main fill of the basin, of generally latest Devonian to early Carboniferous age (Holland 1981) are widely developed to the north and east of the basin (Fig. 2). These include successions at Kerry Head (Fig. 2; Diemer *et al.* 1987; Diemer & Bridge 1988) and Slieve Mish–Dingle in the NW (Capewell 1951, 1965), Slieve Phelim (Doran 1970; Naylor *et al.* 1983) and Slievenamon (Colthurst 1978) in the NE, and Carrick-on-Suir (Penney 1980; Carruthers 1987) and Brownstown Head (Ori & Penney 1982) in the east. These form part of the basin periphery successions of MacCarthy (1990), and those to the north of the boundary fault zone are the regional basin footwall successions of this paper (and Williams *et al.* 1989, 1993).

Northwestern Munster Basin. Accurate correlation of the Slieve Mish Group (Dingle peninsula) with the north Iveragh ORS is problematical because of stratigraphic contrasts, and the lack of biostratigraphic data. Although the succession in Iveragh lacks a conglomeratic unit similar to the 122 m thick Lough Slat Conglomerate Formation (Figs 4 and 8), numerous metre-scale conglomerate-bodies have been recorded, of two petrographic types. Type 1, recognized by Capewell (1975), contains <50% of high-grade metamorphic pebbles (quartz schist; acid gneiss), plus high fractions of vein quartz. All occurrences are restricted to within 4 km of the north coast of Iveragh. Capewell (1975, p. 159) recorded several discrete Type 1 units in the upper part of the Valentia Slate Formation, the (30 m) Doulus Conglomerate in NW Iveragh, and the Reacaslagh Conglomerate in NW Iveragh, and the Reacaslagh Conglomerate in the lower part of the Ballinskelligs Sandstone Formation. These pebbly conglomerates are considered to be reworked from the Inch Conglomerate Formation (Capewell 1951) of the Dingle Basin (Williams *et al.* 1989, 1997; Todd 1989). Type 2 conglomerates (and pebbly sandstones) represent resistate assemblage gravels containing vein quartz–jasper–chert–quartzite, and are more frequent and widespread in Iveragh at all stratigraphic levels (Capewell 1957, 1975, p. 159; Husain 1957; Walsh 1968; Wingfield 1968; Russell 1984) across the peninsula. As the (basal) Lough Slat Conglomerate Formation post-dates erosion of the (Dingle Basin) source rocks of Type 1 conglomerates, it is constrained to correlate with a level above the highest (second-cycle) Type 1 gravel facies in the Iveragh succession, currently believed to be the Reacaslagh Conglomerate

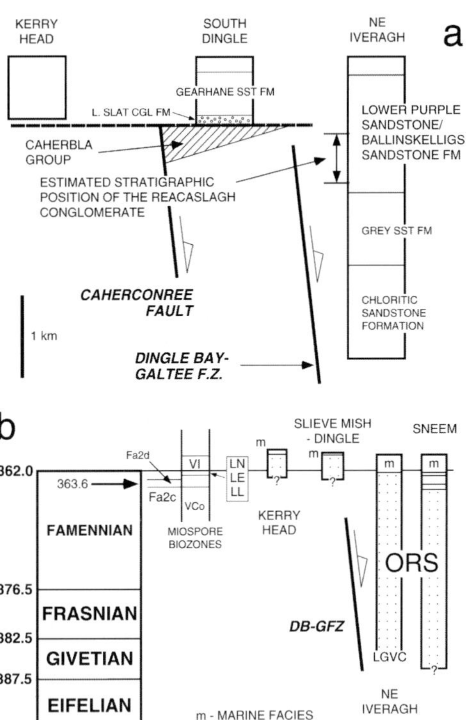

Fig. 8. Section across the northern Munster Basin margin in the Dingle–Iveragh region, showing the suggested (**a**) stratigraphical and (**b**) time relationships between the ORS of the regional footwall (post-rift sequence) and that of the hanging wall depocentre (syn- to post-rift sequences).

(Fig. 8). It is not currently possible to accurately identify this level. However, the correlation implies that the lower Slieve Mish Group (the product of early regional footwall subsidence) is equivalent to the upper part of the Lower Purple and Ballinskelligs Sandstone (and coeval) Formations (Fig. 8) in the basin. The early post-rift basin palaeogeography is shown in Fig. 9.

Northeastern and eastern Munster Basin. Relatively thin (undated) basal conglomeratic sequences (<750 m, Fig. 5) to the east and NE of the basin are pre-LL miospore biozone (latest Famennian) age, based on biostratigraphy in the topmost local ORS (Colthurst 1978; Penney 1980; Ori & Penney 1982). These lack obvious correlatives in the Comeragh–Monvullagh Mountains, and are considered here to be late (post-rift) ORS sequences. Across the basin margin at Slievena-muck and into the basin at Kilworth, the Kiltorcan Formation forms a clear link (Fig. 5). The upper part of this formation at

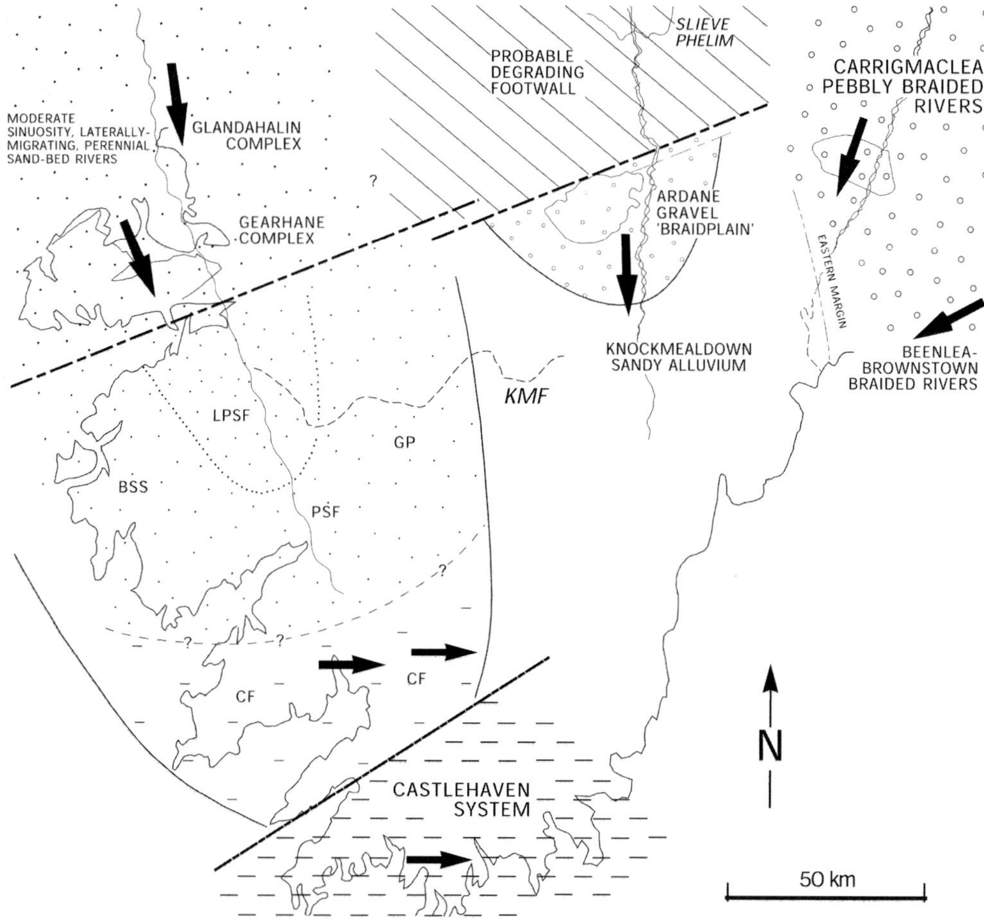

Fig. 9. Suggested palaeogeography of the Munster Basin and its regional footwall and eastern basin periphery during latest Devonian time (pre- to early LL miospore biozone time). Transverse fluvial dispersal systems remained operative in the NW and the Galtee Mountains, although footwall subsidence may have been earlier in the west (Kerry Head–Dingle). Decay and retreat of the Gun Point System led to the deposition of fine-grained ORS (Castlehaven (siltstone) Formation) over the former distal extension of the system. This and the ?upper Castlehaven Formation south of the DCF possibly represent the post-rift subsidence modelled on section 1. Palaeoflow sources as in Fig. 3, except for the Glandahalin Formation at Kerry Head (Bridge *et al.* 1980; Diemer *et al.* 1987), and the Beenlea and Brownstown Head Members (Templetown Conglomerate Formation) at Brownstown Head (calculated vector mean from fig. 4 of Ori & Penney (1982)).

Kilworth is earliest Hastarian in age (VI miospore biozone age; Higgs *et al.* 1988). The underlying Ardane Formation in the Galtees apparently thickens across the Glen of Aherlow Fault (Fig. 5), suggesting late rifting. These relationships suggest that post-rift subsidence in the east and NE occurred somewhat later than in the depocentre and NW (Fig. 9).

Syn-ORS volcanic–magmatic centres

The principal volcanic and magmatic centres are located towards the northern margin of the Munster Basin, mainly in the west (Iveragh) although also in the NE (Comeragh Mountains). The volume and frequency of magmatic products in relation to the basin scale is very low. In the extreme NW, in the footwall of the Coomnacronia Fault, the Valentia Harbour shallow basic intrusive complex (Capewell 1975; Russell 1984; Graham *et al.* 1995) represents pre-385 Ma (Givetian) stretching (Williams *et al.* 1997, this volume). However, the earliest known magmatic product was the 0.3 m thick silicic Puffin Sound airfall tuff (Russell 1984) within the lower Valentia Slate Formation in west Iveragh. The largest volcanic centre in the basin, the tripartite Lough Guitane Volcanic Complex (Avison 1984*b*), comprises up to 300 m of rhyolitic lavas and silicic tuffs located in the hanging wall of the Killarney–Mallow Fault (Fig. 2). The complex was approximately synchronous with the 5–11 m thick sub-plinian Keel–Enagh airfall tuff in west and NW Iveragh (Williams *et al.* this volume). Other, approximately coeval, magmatic rocks are the Beenreagh basalts (Williams *et al.* this volume) and the thin Eskine–Knocknagullion tuff (Russell 1984) within the Chloritic Sandstone Formation (Fig. 4). An undated, <2 m thick epidote-rich tuff bed within fine-grained red beds has been reported on Great Skellig Island (Long, pers. comm.), which may belong to the upper part of the ORS stratigraphy of Iveragh (Fig. 3).

In the east of the Munster Basin, four basic volcanic levels of unknown age occur in the ORS of the Comeragh and Monavullagh Mountains (Penney 1978; Sleeman *et al.* 1995). The lower to mid- (basal) Coumshingaun Conglomerate Formation (Fig. 5), contains 10 m of vesicular hawaiite and 75 m of mugearite lava flows (the Coolnahorna and Carrigduff Volcanic Members, respectively; Penney 1978). To the north of this and up-section, the very thin Moanyarha basic vesicular lava (Penney 1978) occurs in the lower part of the Nier Sandstone Group. To the south, in the Monavullagh Mountains, a vesicular mugearite lava (the Doon Lava Member) occurs in the Ballytrasna Formation (Sleeman *et al.* 1995).

In the SW of the basin only two volcanic occurrences are known. The first, a very thin airfall tuff within the Foilçoagh Bay Formation (on Clear Island), close to the base of the ORS beneath the Sherkin Formation (Graham & Reilly 1972), has an approximate Frasnian age from revised palynology of the section (Higgs *et al.* this volume). Second, within the middle to upper Castlehaven Formation of the Mizen peninsula (Fig. 3), minor (>1.5 m thick) mafic tuffaceous eruptive rocks have been recently identified (Haughton, pers. comm.).

Decompaction–backstripping

Sequential decompaction of five stratigraphic sections along the line of model section 1 and four along model section 2 was undertaken to constrain the syn-rift subsidence produced during forward modelling. Sections were chosen to represent large areas of the cross-sections, where the stratigraphy is relatively complete and well known. Decompacted ORS thicknesses are minimum values in each case, as the basement to the succession is not exposed within the basin, except at Slievenamuck and the Galtees. Moreover, no structural correction to formation thicknesses was applied to account for Variscan strain. As the Variscan deformation sequence (Cooper *et al.* 1986) is likely to have resulted in most cases in thinning of stratigraphy on major fold limbs, where thickness data are gathered, resultant decompaction values are underestimates. Fine-grained (siltstone and finer) sequences are likely to have been most affected. Decompaction was carried out using the computer program of Allen & Allen (1990), which includes an exponential porosity decay with depth. The lithological constants of Sclater & Christie (1980) were used (Table 1), integrated over grouped stratigraphical units, having estimated their lithological composition.

Table 1. *Parameters used in the decompaction and backstripped subsidence analyses*

Lithology	Compaction coefficient c (km^{-1})	Surface porosity ϕ_0 (%)	Sediment density ($kg\ m^{-3}$)
Shale	0.51	63	2.720
Siltstone	0.45	59	2.530
Conglomerate	0.3	50	2.600
Limestone	0.7	50	2.670
Sandstone	0.27	49	2.650
Shaly sandstone	0.39	56	2.680

For the decompaction–backstripping analysis, no addition of thick pre-Permo-Triassic sediments was made to local successions, as has been suggested is necessary based on thermal maturation arguments (Clayton 1989). The effects of this, however, on total and (uncorrected) tectonic subsidence have been tested for the SW Waterford section (see below). The range of Silesian cover indicated by Clayton (1989, 5–7 km) has been computed (Fig. 10a) for a shaly sandstone lithology (Table 1) considered likely to represent Namurian–Westphalian sediments. The results show negligible differences between pre-Silesian tectonic and total subsidence for this range, but depart from the total subsidence derived from the preserved section (Fig. 10a). For the Ballytrasna (Siltstone) Formation there were *c.* 13 and 14% increases in decompacted thickness for hypothetical Silesian cover of 5 and 7 km, respectively (Fig. 10a). The results show predictable increases in mean subsidence rates over the Lower Carboniferous curve for such large additional sections, with downward inflections of the tectonic and total subsidence curves (Fig. 10a). However, known maximum Namurian (1.4–*c.* 2.0 km) and Westphalian (0.35 km)

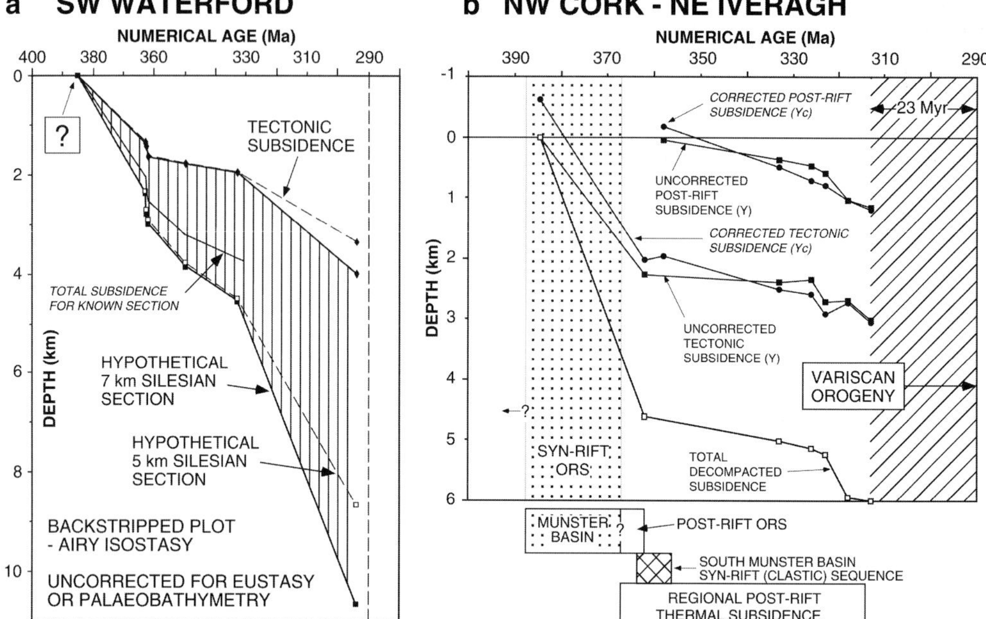

Fig. 10. (**a**) Uncorrected, backstripped subsidence plot of the SW Waterford section (Fig. 12d) incorporating hypothetical 5 and 7 km thick Silesian sections after Clayton (1989). Time scales used are those of Harland *et al.* (1990) for the Carboniferous period and Tucker *et al.* (1998) for the Devonian period. (**b**) Geohistory plot of the composite north Iveragh–NE Cork section (see Fig. 11b), showing eustatically and bathymetrically corrected and uncorrected tectonic subsidence for the complete section and also the post-rift section alone (see text for details). Estimated timing of Munster and South Munster Basin syn-rift phases, Munster Basin thermal subsidence and potential missing section up to the Variscan Orogeny are also shown. The age of the Variscan Orogeny is based on data in Halliday & Mitchell (1983). Late Caledonian (Acadian) deformation culminated at *c*. 400 Ma.

sections in southern Ireland (Sevastopulo 1981; Collinson *et al.* 1991), suggests that $\ll 5$ km of Silesian sediments accumulated (see Clayton 1989, p. 615). With this in mind, realistic plot curves would be unlike those for load-induced flexural subsidence, although few details of the Silesian section are known from south Munster. These considerations imply collectively that the flexural forward models are underestimates in terms of stretching and related parameters.

All sections have been backstripped using the porosity, bulk density and Airy isostatic calculations in the program of Allen & Allen (1990), without correction for sea-level change or palaeobathymetry. The 1D local isostasy used is equivalent to an effective elastic thickness (T_e) of zero in the flexural cantilever model described below. The effects of global eustatic variation and local palaeobathymetry have been investigated for one of the sections (NE Iveragh–North Cork; Fig. 10b). First-order sea-level variation was taken from Vail *et al.* (1977), in conjunction with the time scale of Tucker *et al.* (1998). Palaeobathymetric corrections were 200 m for the Waulsortian carbonate rocks (model of Lees *et al.* (1985)), 250 m for the Clare Shales, and 200 m for Namurian turbidites (Naylor *et al.* 1989). Cyclothemic (paralic) Namurian and Westphalian sediments were taken to be at sea level, as were earliest Carboniferous carbonate successions. Corrected tectonic subsidence curves for the complete section and, more correctly, for the post-rift section alone (Fig. 10b) show minor departures from uncorrected curves. This suggests that the use of uncorrected curves is reasonably valid for describing tectonic subsidence at the other stations.

Results: western Munster Basin

The sections chosen for decompaction represent the syn-rift fill near the basin margin (north Iveragh), in the depocentre (Killarney, south Iveragh and south Beara) and the thinner syn-rift section south of the Dunamanus–Castletown Fault (Cape Clear–Toe Head). The difficulty in identifying accurately the base of the post-rift ORS within the basin (Fig. 8) dictated that the simpler exercise of taking the top of the ORS as the decompaction datum in the flexural forward models be used. Thus, the ORS was decompacted by removal of uppermost Devonian to Upper Carboniferous (Westphalian) sediments. Syn-rift sediment of the South Munster Basin (Fig. 10b) was also taken into account. All sections (Fig. 11) show significant decompaction over the fully compacted case (e.g. Figs 4 and 5). Some 2.37 km, for example, of partially decompacted ORS is present in the footwall block to the Dunmanus–Castletown Fault (Fig. 11e).

All geohistory and backstripped plots from sections in the west of the basin (Fig. 11a–e) show typical curves of extensional subsidence regimes (see McKenzie 1978). The first-order syn-rift and post-rift segments are interpreted as due to Munster Basin rifting (Fig. 10). However, subsidence history of section in south Iveragh, south Beara and Clear Island–Toe Head (Fig. 11c–e) shows evidence for a second pulse of rifting during latest Devonian–earliest Carboniferous time. This is the South Munster Basin rifting phase (Williams *et al.* 1989; Fig. 10b), corresponding initially to the clastic deposits of the coastal Toe Head Formation (Fig. 4) and shallow marine Old Head Sandstone Formation (McAfee 1992), and subsequently to the Kinsale Formation and equivalents (Naylor *et al.* 1989). The relative lack of absolute age control within the ORS prevents a clearer demarcation and analysis of latest ORS subsidence and South Munster Basin rifting. Despite this, the ORS of the Munster Basin is treated as a discrete phase of extensional subsidence, with its latest (?post-rift) history being complicated by South Munster Basin rifting and latest Devonian rising base level (Fig. 10b).

Results: eastern Munster Basin

In contrast, geohistory and backstripped plots from the eastern part of the basin show less clearly defined syn- and post-rift curve segments (Fig. 12). This may be partly a reflection of an incorrect numerical age estimate of the earliest ORS. However, very thick Courceyan and Viséan carbonates accumulated rapidly in the east, before a late Dinantian to early Namurian unconformity affecting the northern decompacted sections (Fig. 12a–c). Only the southernmost section (Fig. 12d) shows evidence of rapid latest Devonian (South Munster Basin) syn-rift subsidence.

Flexural cantilever model

The flexural cantilever model used in this study is particularly advantageous, as it allows graphical 2D models to be made rapidly, which can be compared with stratigraphic templates of the regions under investigation (e.g. Figs 4 and 5). The model has been fully described by Kusznir & Egan (1990) and Kusznir *et al.* (1991). During continental lithospheric extension the brittle, seismogenic upper crust responds by normal faulting (simple shear), whereas the ductile lower crust and lithospheric mantle deforms by pure shear (Fig. 13a). The footwall and hanging wall of the normal fault(s) generated are regarded as interacting flexural cantilevers (Kusznir *et al.* 1991). Mechanical unloading of the footwall block of extensional faults results in rift shoulder (footwall) uplift, at the same time as the down-bending of the hanging wall causes basin subsidence (Fig. 13b). The program models this as instantaneous stretching. The thermal and flexural (regional) isostatic consequences of deformation are accounted for at all stages of the model, as are syn-rift erosion, syn- and post-rift sediment fill and compaction. The model has been successfully applied to a variety of well-constrained extensional settings (Kusznir & Morley 1990; Marsden *et al.* 1991; Roberts & Yielding 1991; Roberts *et al.* 1993, 1995; Kusznir *et al.* 1995).

The model (2D STRETCH, version 5) allows a variety of input parameters related to the fault array used, the sediment fill, bathymetry, erosion and crustal behaviour. Most important are the horizontal coordinates, extensions, dip direction and shape of individual faults in model. The model allows variation in the initial thickness of the crust and the upper-crustal brittle layer (Fig. 13a), as well as the initial height of the crust above base level. Most importantly, the computer model allows variation in the effective elastic thickness (T_e) of the lithosphere, which describes the flexural rigidity of a perfectly elastic lithosphere. In nature T_e is unpredictable, and must be found during modelling by iteration. In all of the models, lithological parameters approximate a sandstone-filled basin (compaction decay coefficient c of 0.27 km^{-1}, sediment 2.65 kg m^{-3}, porosity ϕ_0 of 49%) with an initial zero bathymetry.

Forward models: results

Initial experiments were performed on previous, contrasting, models of the western Munster Basin involving listric (Price & Todd 1988; Price 1989) and planar basin margin faults linked to detachments (Williams *et al.* 1989) to determine whether subsidence is realistic for these fault geometries. Crustal parameters specified by Price (1989) were used in the forward models, except for T_e, which was unspecified;

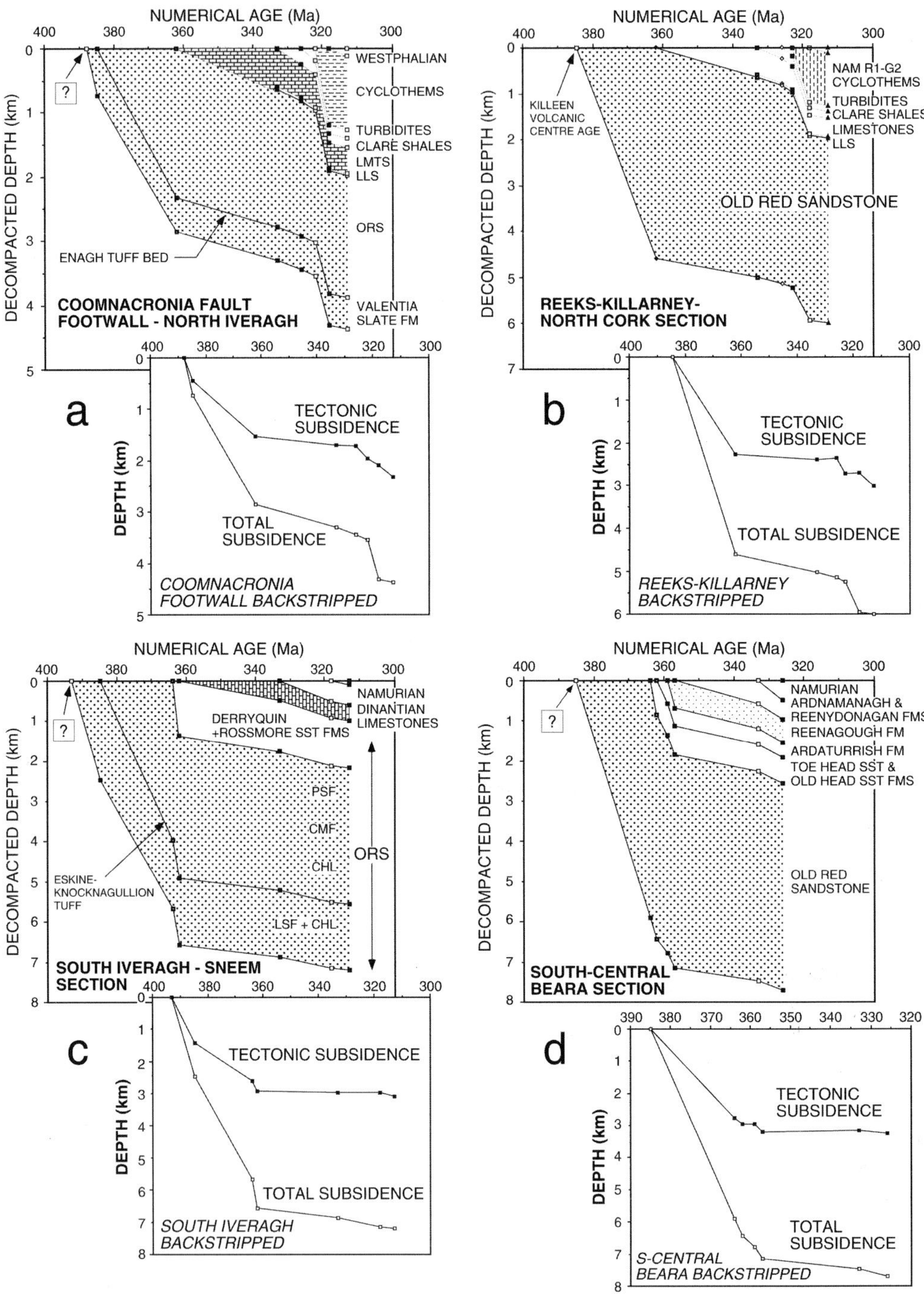

Fig. 11(a–d).

$T_e = 10$ km was used in the models shown (Fig. 14). The listric fault structure (Fig. 14a) detaching at the base of a 30 km thick crust results in a narrow basin, for a very large extension (30 km) on the basin boundary fault implied by Price's fig. 5a (1989). Although not

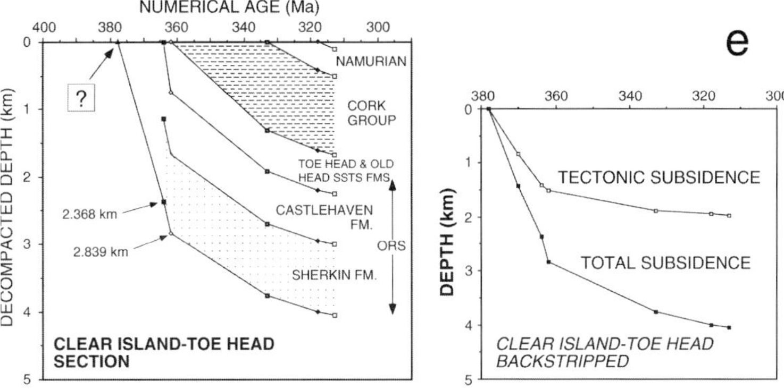

Fig. 11(e).

Fig. 11. Decompacted and backstripped subsidence plots of representative stratigraphical sections along modelled profile 1. Time scales as for Fig. 10. (**a**) NW Iveragh (footwall of Coomnacronia Fault). Data from Williams *et al.* (1997) and (b). (**b**) NE Reeks–Killarney–Castleisland and north Cork. Stratigraphical database from Walsh (1967, 1968), Sevastopulo (1981), Price (1986) and Williams *et al.* (this volume). (**c**) Southern Iveragh–Sneem. Stratigraphical database from Russell (1984) and Naylor & Sevastopulo (1993). Age of base Derryquin Sandstone estimated based on the Famennian 2c age of the VCo (formerly VUs) miospore biozone of Higgs & Russell (1981) and the age of the Piskahegan Group given by Tucker *et al.* (1998). The Eskine–Knocknagullion Tuff (Russell 1984) is assumed to have the same age as the Keel–Enagh Tuff (see Russell (1984) and Williams *et al.* (this volume) for discussion). (**d**) South–central Beara. Data from Williams *et al.* (1989), McAfee (1992) and James & Graham (1995). Boundary ages for the Reenagough Formation are conjectural, approximating to ages within the mid-Courceyan A division (Harland *et al.* 1990). (**e**) Clear Island–Toe Head. Stratigraphical database from Graham & Reilly (1972), Clayton & Graham (1974), Naylor *et al.* (1989) and Williams *et al.* (1989). The base of the Castlehaven Formation (?intra-Famennian) is not known. Age of the base of the ORS (= base of Sherkin Formation) estimated by the palynological data of Higgs *et al.* (this volume) and numerical and biostratigraphical ages of Williams *et al.* (this volume).

specified by Williams *et al.* (1989), a mid-crustal level was used to detach the planar basin margin fault in Fig. 14b. This results in an unrealistic narrow, box-profile basin (see Kusznir *et al.* 1991). Inclusion of an antithetic structure (the Dunmanus–Castletown Fault) has negligible effect on syn-rift subsidence. Both models fail to replicate the lateral extent of Munster Basin ORS subsidence, and require very large heaves on single faults. Both the detachment and listric fault models are inappropriate for these reasons and are unrealistic generally because they incorporate vertical simple shear of a weak hanging wall, an invalid assumption for faulting of essentially similar upper-crustal basement. Planar faults restricted to the brittle layer are therefore used in the following models.

Initial conditions common to both modelled section lines, apart from those described above, are (1) the top of the crust is at base level ($H = 0$), as marine microfossils in the earliest (Frasnian) fill in the southwest of the basin indicate that it was at sea level (Higgs *et al.* this volume), and (2) crustal and brittle layer thicknesses of 40 and 16 km, respectively. The crustal thickness used is considered applicable to post-Acadian conditions. Multiple experiments were performed, varying T_e, extension and crustal thickness, until sufficiently wide and deep 2D sections were produced. Optimum values for T_e were 7–8 km, within an acceptable range for extending lithosphere, and significantly less than values (10 km) used successfully for the Jeanne d'Arc basin (Kusznir *et al.* 1991). The mode of syn-rift filling used is that of 'zero bathymetry', which is an iterative process that terminates when the basin is filled completely to base level, as appropriate to the oversupplied fluvial Munster Basin.

Section 1: northern Dingle–Clear Island

This section (Fig. 15a) incorporates previously discussed, reasonably well-established, bounding and intrabasinal extension faults (see above). The Dingle Bay and Caherconree Faults are given steeper than average dips, reflecting the pre-Munster Basin history of this basement lineament adjacent to the Dingle Basin. The 'North Kerry lineament' (Todd 1989; Richmond & Williams this volume) is assumed to be a small extension fault affecting the Munster Basin footwall, and is located offshore of the north coast of Dingle. A series of synthetic and

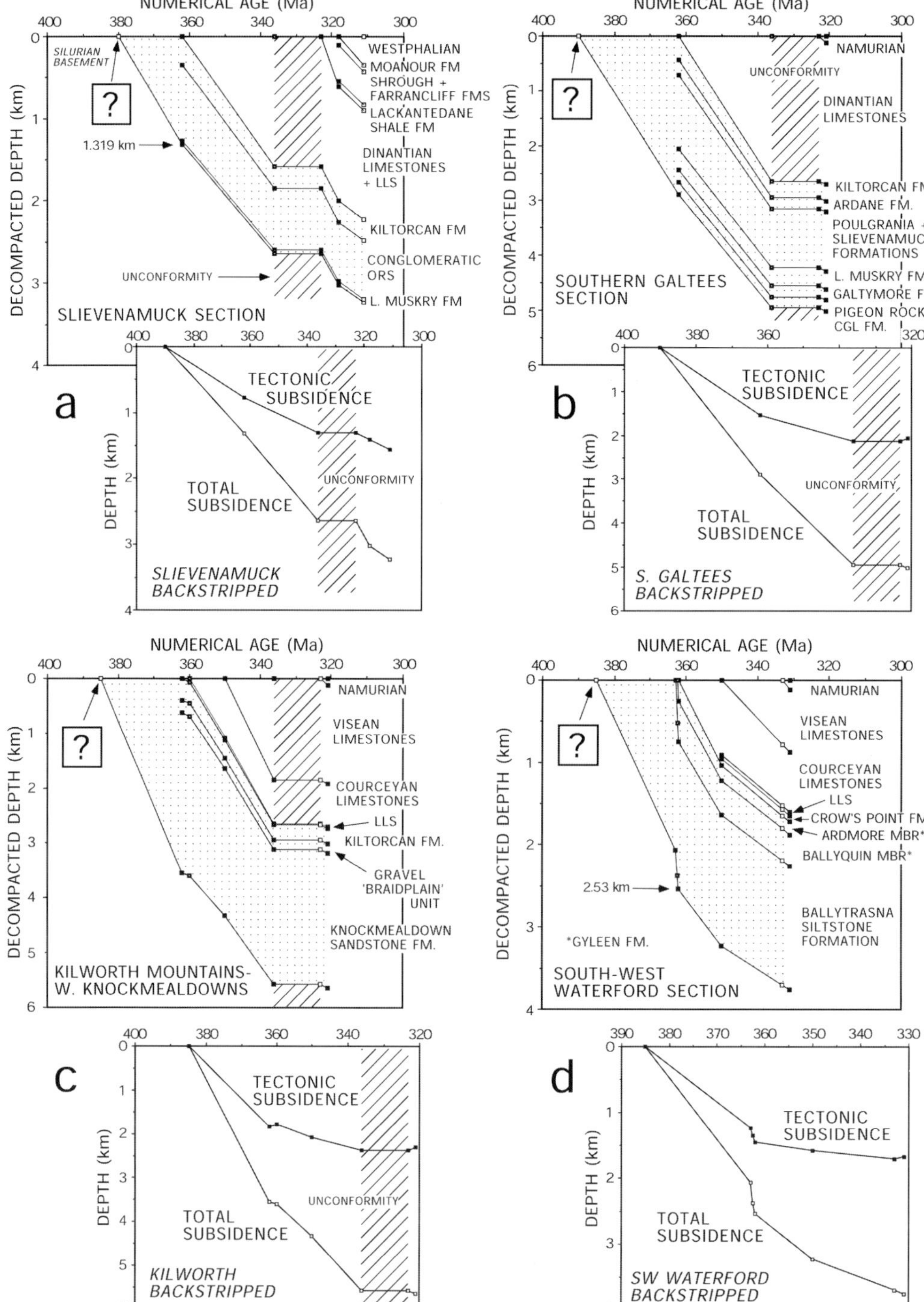

Fig. 12. Decompacted and backstripped subsidence plots for representative stratigraphic sections along model line 2. Time scales as for Fig. 10. (**a**) Slievenamuck. Data from Shelford (1963), Sevastopulo (1981), Carruthers (1987) and Archer *et al.* (1996). (**b**) Southern Galtee Mountains. Data from Carruthers (1987) and Sleeman *et al.* (1995). (**c**) Kilworth Mountains–West Knockmealdown Mountains. Data from Carruthers (1987), Higgs *et al.* (1988) and Sleeman *et al.* (1995). (**d**) SW Waterford. Data from MacCarthy (1990) and Sleeman *et al.* (1994, 1995).

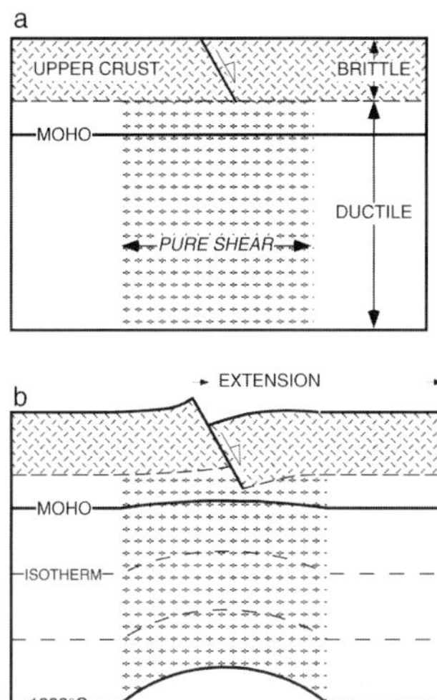

Fig. 13. Definition diagrams of the flexural cantilever extensional model for planar faults utilized in modelled sections in this paper (after Kusznir *et al.* 1991, fig. 2b). (**a**) Undeformed lithosphere. (**b**) Extended lithosphere, with faulting restricted to the brittle upper-crustal layer.

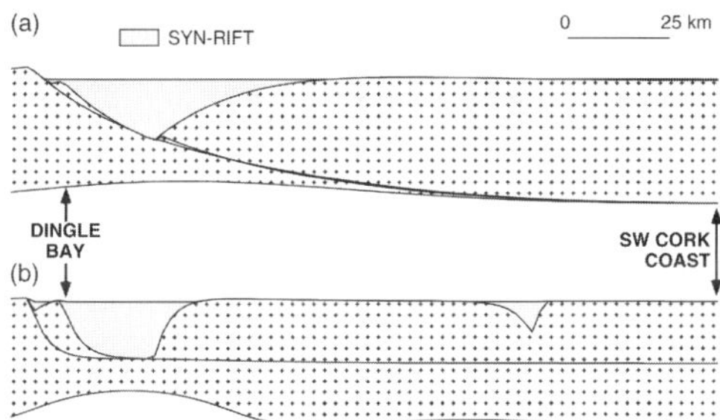

Fig. 14. Flexural cantilever forward models of the western Munster Basin, (**a**) using a low-angle listric basin boundary fault detaching at the base of the crust, based on the model and specified parameters of Price (1989, fig. 5; see also Price & Todd (1988)), and (**b**) using steep, planar upper-crustal basin boundary faults detaching at a mid-crustal level (model of Williams *et al.* (1989, fig. 5)).

antithetic (60° dipping) faults of relatively minor extensions (0.5–3.8 km, mean 2.6 km, SD 1 km), having a mean spacing of 16 km (SD 15 km), have been added to achieve the minimum observed ORS subsidence. Their spacing compares with values of 20 km for thc North Viking Graben, 25 km for the Gulf of Suez, 30–40 km for the Basin and Range (USA) and 40 km in central Greece (Jackson & White 1989). Larger displacement hypothetical offshore faults are necessary to achieve onshore syn-rift subsidence, although different configurations are possible to accomplish this. However, the offshore faults in Fig. 15 were so spaced and their extensions adjusted to minimize footwall uplifts, and to avoid the south coast, as no coarse clastic basement material having a southern provenance is known in the ORS of these sections.

Details of the syn-rift model (Fig. 15b) shows that rotated fault block crests are completely buried by syn-rift (ORS) stratigraphy, and that the basin depocentre and thin succession south of the Dunmanus–Castletown Fault are replicated (see Fig. 11). Major footwall uplift (7.6 km) is produced by cumulative extension on the Caherconree (extension E of 3 km) and Dingle Bay Faults ($E = 5$ km), and to a degree by the Coomnacronia Fault ($E = 2.8$ km). The summed extension for section 1 is 59.1 km. Importantly, thin syn-rift sediment is preserved at the northern end of the section (Fig. 15b) principally by the flexural deformation of the footwall giving a 'rim-type' basin. Such footwall basins are known from the Red Sea region, and are typical of flexural cantilever models employing moderate T_e values (4–10 km). This preservation can be equated with the Frasnian Carrigduff Group of north Dingle (Richmond 1998; Richmond & Williams this volume). The peak (1.65) and mean (lower crustal, 1.48) stretching factors for the model are shown in Fig. 15c. The heat-flow anomaly profile (Fig. 15d) coincides well with the syn-rift magmatic rocks of Iveragh, and a lower peak coincides with the present SW coast, where limited volcanic rocks are known. Bending stresses (Fig. 15e) are within realistic limits, and their high values reflect the relatively high T_e (8 km) used in the model. Modelled erosion of the footwall uplifts and the isostatic response (Fig. 16a and b) shows that fault blocks within the basin remain buried. Subsequent post-rift thermal subsidence (Fig. 16d and e) for a period of 5 Ma, and incorporating a eustatic rise derived from the rate used by Diemer & Bridge (1988) for latest Devonian–early Carboniferous time, gives a thin post-rift fill in the centre of the basin, spanning the Dunmanus–Castletown Fault. The isostatically uplifted basin footwall prevents subsidence across this region. Changing the erosion mechanism in the model to a flat, regionally extensive surface does not help to achieve post-rift subsidence on the Dingle footwall block. Experiments with a constant erosion level at 0 km result in very large isostatic uplift of the basin footwall and marginal hanging wall, and consequent erosion of the syn-rift basin stratigraphy reaching as far as the depocentre. This is clearly invalid, as no such erosion is known for the ORS in the west of the basin. Experiments (not shown) with higher planation surfaces failed to avoid erosion of the basin fill and resulted in

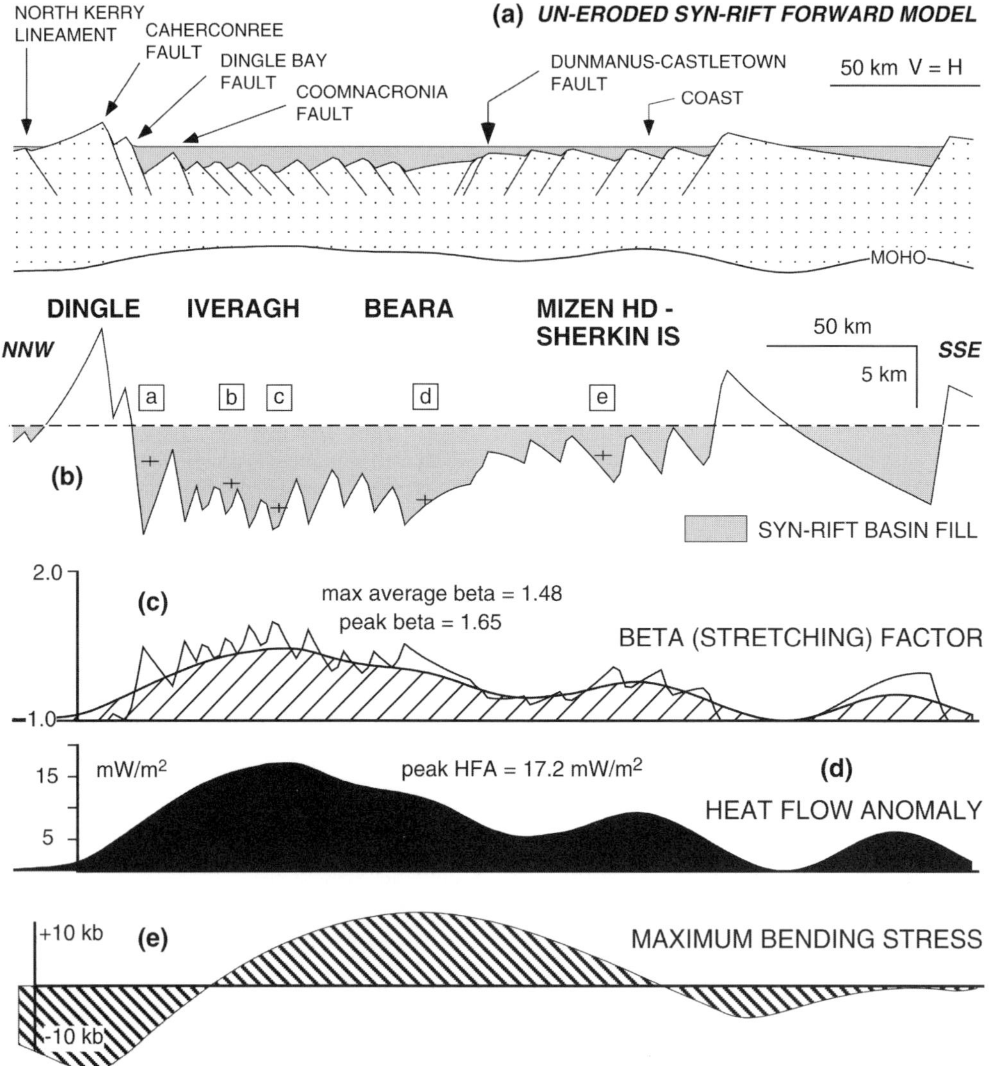

Fig. 15. Flexural cantilever forward model of section 1 through the western Munster Basin. (**a**) Uneroded crustal structure and basin geometry following syn-rift subsidence. (**b**) Detail of syn-rift stratigraphy. Noteworthy features are the footwall uplift of the Dingle block north of the Caherconree Fault, and the preservation of syn-rift ORS in the extreme north of Dingle (possibly represented by the middle Frasnian Carrigduff Group, Richmond (1998); see text for details). Decompacted depths of the ORS at five stations (Fig. 11) are shown by crosses. Forward model data plots of (**c**) the resultant stretching factor (β), (**d**) heat-flow anomaly and (**e**) maximum bending stresses from the flexural cantilever model. The strong asymmetry of both β and heat-flow anomaly towards the northern boundary fault zone should be noted.

remnant footwall block topography that prevented post-rift accumulation.

Section 2: Slieve Phelim–Ardmore

This section contains only the DB–GFZ and South Galtees Fault as well-known basin extension faults. However, the Muggort's Bay Lower Palaeozoic inlier (Fig. 3) has been modelled as a small horst, and projected onto the section line. Minor synthetic and antithetic faults (Fig. 17a) have been randomly located across the section (mean spacing 8 km, SD 3.4 km), with a prevalence of antithetic structures in the southern part. The section obliquely approaches the eastern basin margin to the south, hence the overall

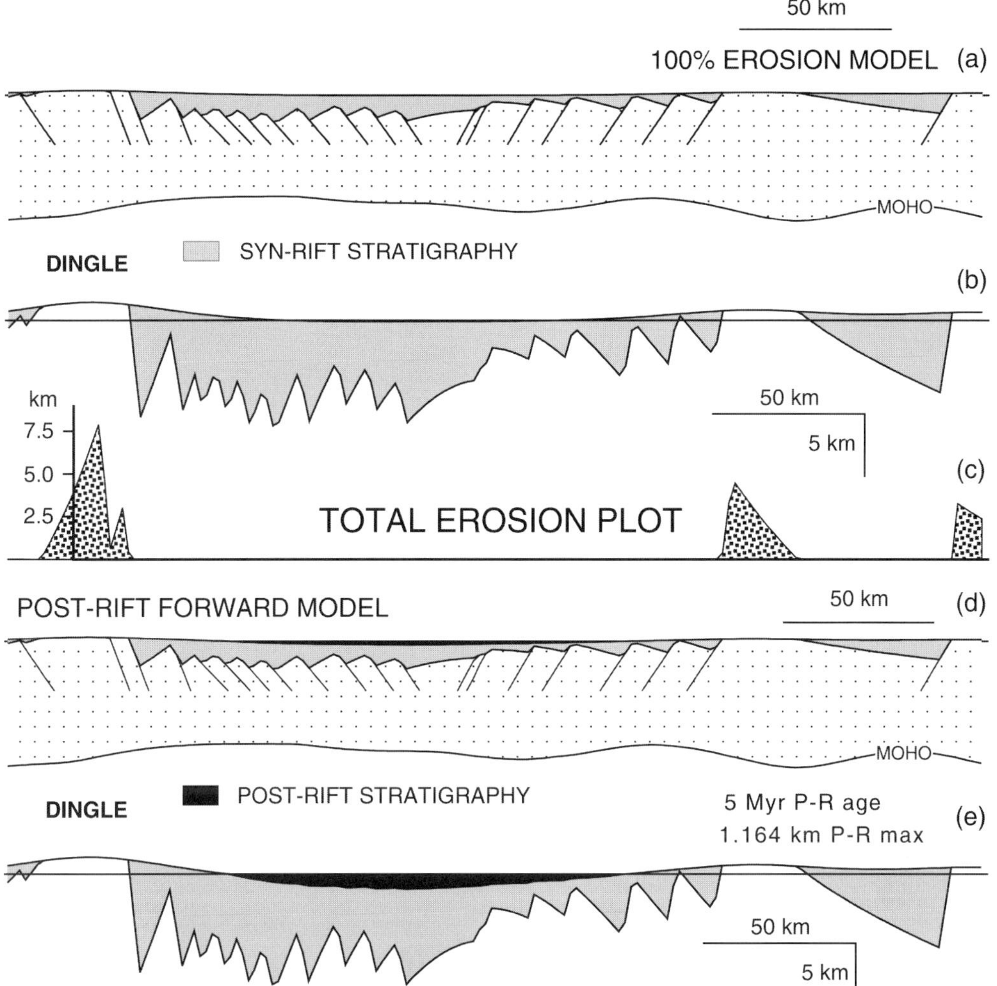

Fig. 16. (**a**) Crustal model of section 1 with 100% erosion of footwall uplifts. The vertically exaggerated section (**b**) shows the remnant topography over the Dingle footwall block, and the plot (**c**) shows the magnitude and distribution of the erosion. It should be noted that none of the ORS syn-rift basin fill is eroded in this model. (**d**) Post-rift forward model, proceeding from eroded model (**a**). Post-rift stratigraphy accumulated over 5 Ma with a 0.1 mm a^{-1} eustatic rise to approximate latest Devonian–early Carboniferous conditions. It should be noted that the model does not replicate known subsidence on the regional footwall (**e**).

length of the model line has been restricted (compare with section 1). Major extension was concentrated on the Glen of Aherlow and Slievenamuck Faults (E 3.75 and 2.0 km, respectively) resulting in significant footwall uplift and subsidence in the hanging wall (Fig. 17b). Minor extensions on the faults (0.5–2.25 km, mean E of 1.34 km) of the synthetic array is, however, necessary to achieve widespread syn-rift subsidence. Total extension for section 2 was 27 km, which resulted in low value (mean 1.3, crustal thickness ratio 1.4) and wide-profile stretching factors (Fig. 17c). The peak heat-flow anomaly is low (12 mW m^{-2}), but its location predicts the rare basic volcanic rocks in the northern Comeraghs (Fig. 17d). Erosion of the footwall region (Fig. 18a) results in a similar outcome to section 1, in that minor isostatic uplift leaves a remnant topography (Fig. 18b), although no erosion of the ORS in the basin (Fig. 18c). Attempts to bury the footwall by post-rift subsidence over the same period as for section 1 (5 Ma), again fail (Fig. 18d). A thin post-rift megasequence is modelled in the basin centre (Fig. 18e).

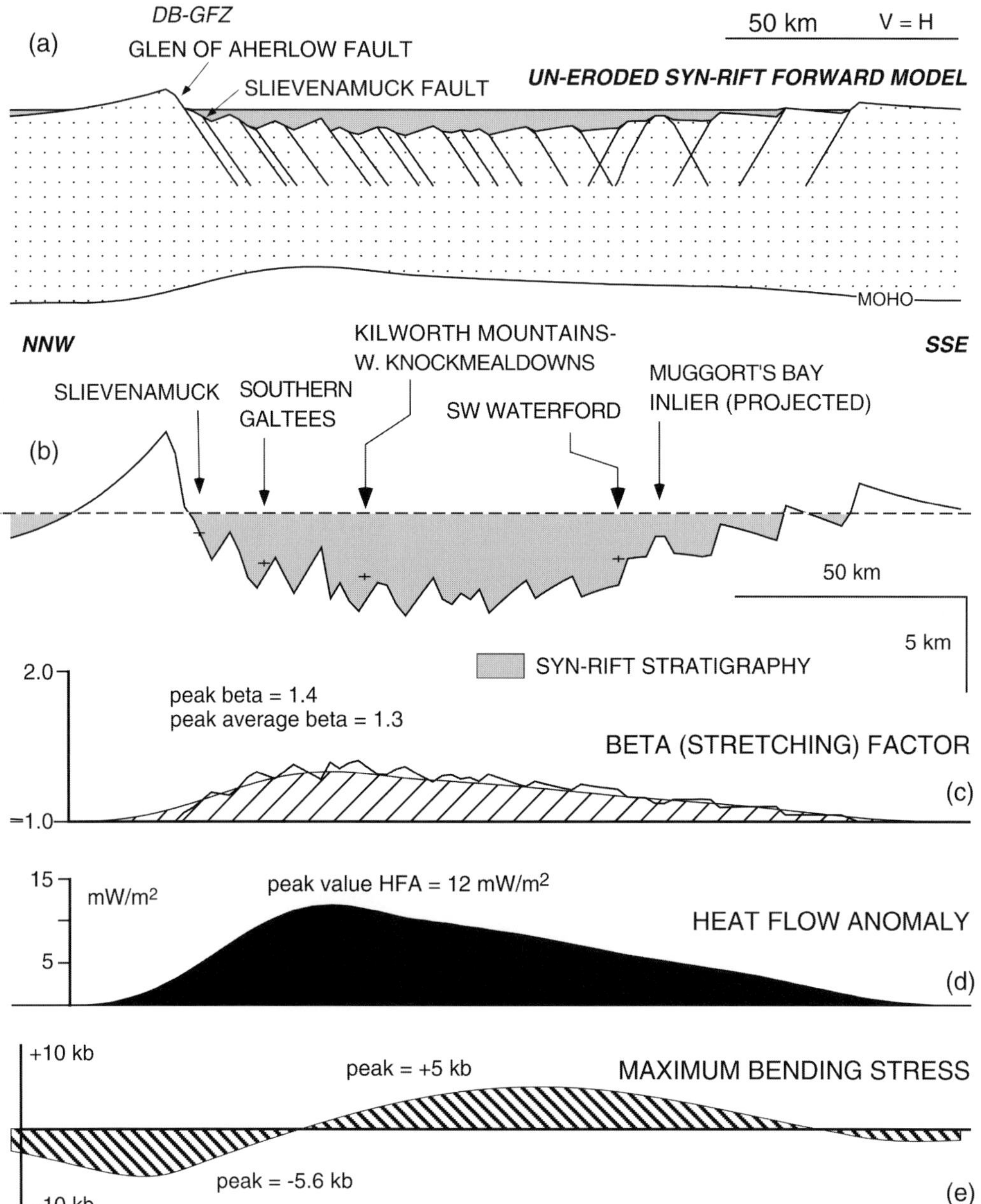

Fig. 17. Flexural cantilever forward model of section 2. (**a**) Crustal structure and basin geometry following syn-rift subsidence, without erosion of footwall uplifts. Crustal and sediment fill parameters as for section 1, except $T_e = 7$ km. Cumulative extension is less than for section 1. (**b**) Enlarged detail of (**a**), showing decompacted depths of ORS (crosses), and significant footwall uplift associated with basin margin DB–GFZ (locally the Slievenamuck and Glen of Aherlow Faults). Plots of (**c**) stretching factor β, (**d**) heat-flow anomaly and (**e**) maximum bending stress predicted for the section.

Discussion

Flexural cantilever forward modelling successfully reproduces, in essential aspects, non-unique syn-rift sediment geometries (Figs 15 and 17) of the palinspastic Munster Basin. It is less successful in producing post-rift (thermal) subsidence on the regional footwall (Figs 16 and 18),

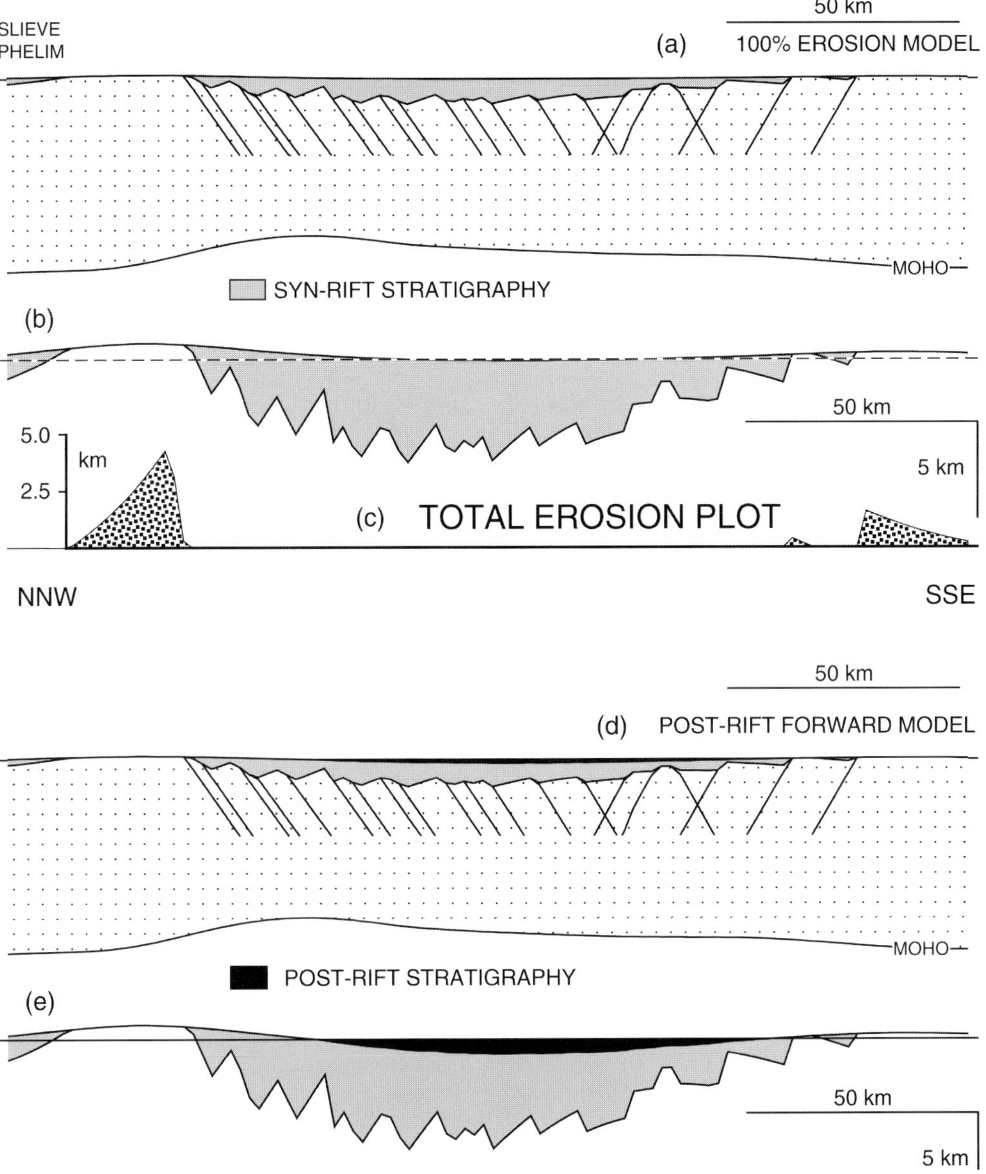

Fig. 18. Forward models of section 2 showing (**a**) 100% erosion of syn-rift footwall uplifts, (**b**) detail of (**a**) illustrating footwall topography predicted after erosion, (**c**) plot of distribution of erosion, (**d**) post-rift forward model showing thermal subsidence in the central region of the section, (**e**) enlargement of (**d**). Crustal parameters and eusatic sea-level rise as for post-rift situation in section 1.

probably of late Famennian age. The forward models produced do not account for heat loss during the syn-rift phase, resulting in over-estimates of footwall uplifts. In reality, for basins with prolonged syn-rift histories (i.e. >20 Ma, Jarvis & McKenzie 1980), heat is lost continuously and a component of thermal subsidence may be superimposed on active rifting. The likely duration of the Munster Basin syn-rift phase (?>18–20 Ma; Fig. 10b) suggests that its behaviour would depart (Jarvis & McKenzie 1980) from instantaneous stretching models of the type used, allowing thermal subsidence to affect the basin footwall in a manner not modelled by STRETCH. Other contributory factors in this case may include overestimated footwall uplift as

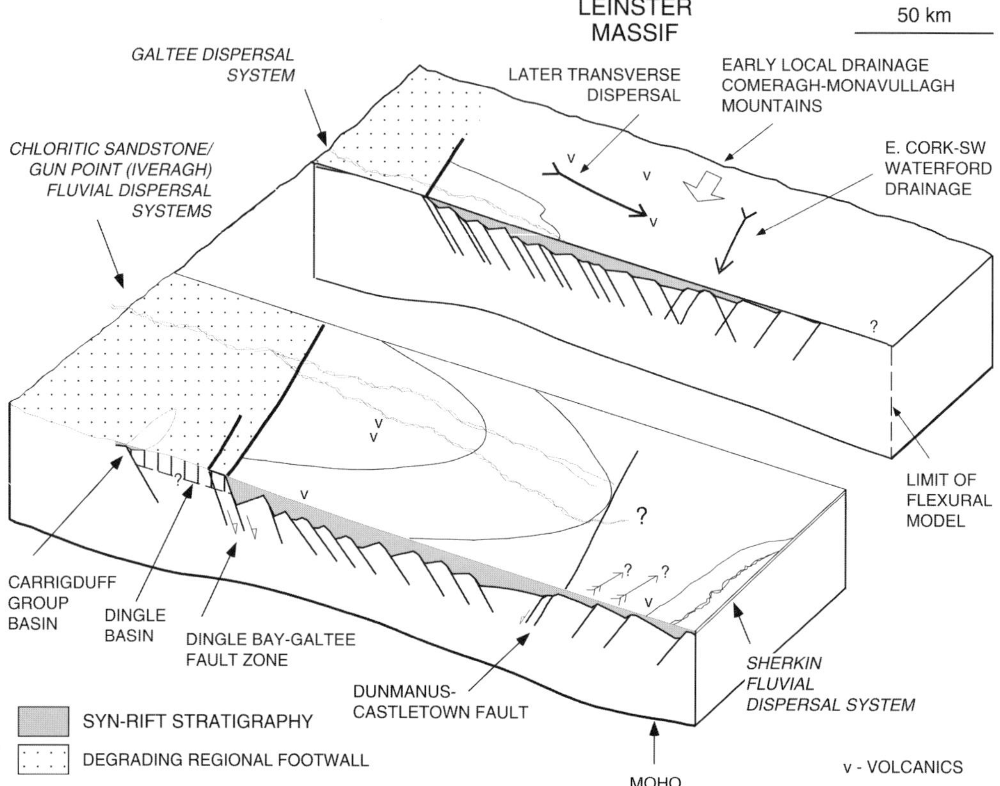

Fig. 19. Simplified block model summarizing the context of ORS accumulation in the onshore part of the Munster Basin. Crustal sections are based on Figs 16a and 18a.

a result of non-recognition of extension faults in the basin margin fault zone, and also the effect of a second phase of (South Munster Basin) rifting during latest Devonian time. Lowering footwall uplift, in particular, would reduce isostatic uplift generated during erosion and thus facilitate conditions for subsidence of the footwall. Experiments with the forward flexural model have shown that the problem of the lack of simple post-rift subsidence of the regional footwall is not solved by utilizing the maximum available time from the end of the Devonian period to the Variscan orogeny (65–70 Ma). Moreover, changing the post-rift T_e to 3 or 4 km, as is reasonable following lithospheric modification by rifting, has no significant effect on subsidence for a post-rift duration of 5 Ma.

The scale and nature of the exposed syn-rift (Figs 6 and 7) alluvial systems filling the basin, and the scale of the basin itself (Fig. 19), are different from widely used facies models of tilt-block half-graben basins (e.g. Leeder & Gawthorpe 1987) and supradetachment basins (Friedmann & Burbank 1995). Small-scale rifts ($\ll$ 50 km wide) with uplifted basement in footwall and hanging wall acting as source areas, are filled by footwall-sourced, steep alluvial fans, hanging-wall-sourced 'alluvial cones' and/or lake or through-going axial fluvial systems (Leeder & Gawthorpe 1987). Supradetachment basins are filled by mass-wasting (steep) alluvial fans, largely of footwall derivation, and lake systems (Friedmann & Burbank 1995). Making the hypothetical assumption that small tilt-block systems may characterize the earliest unexposed fill of the Munster Basin (e.g. Fig. 19), later (visible) dispersal systems show no clear effects on intrabasinal synthetic or antithetic faults propagated into the syn-rift fill, in terms of sediment routing or environments (Figs 3, 6 and 7). Although intrabasinal faults in the syn-rift are exceptionally difficult to detect without a suite of indicators (palaeoflow, offset marker horizons, e.g. airfall tuffs, or major changes in thickness and stratigraphy; Williams *et al.* 1989, this volume), this raises the question of whether the ORS is actually syn-rift, and not very thick post-rift sediment. Experiments were thus performed assuming that the known ORS history (23 Ma) was a thermal (time-dependent) subsidence phase. These experiments generated insufficient post-rift stratigraphy. Thus the preferred model is of a relatively long syn-rift period, which is interrupted by a second rift phase initiating the South Munster Basin (Fig. 10b), and an overlap in thermal subsidence and Munster Basin rifting.

Although the Dingle block footwall uplift (of the rift shoulder) replicates the erosion and tilting of the Caherbla Group (Horne 1974; Todd 1989), based on slip on the Coomnacronia Fault and the main basin fault in Dingle Bay, the overall magnitude of the footwall uplift in the Dingle Basin region (peak value 7.5 km, Fig. 15) is very large. This value does not seem to be in accordance with the sporadic and low volumes of fan gravel preserved in the hanging wall of the Iveragh region (e.g. Fig. 7). However, this relief can be eliminated by continuous erosion over a period of 18 Ma at a rate of 0.4 mm a^{-1}. More extension fault structures in the Dingle block may reduce the high local uplift currently generated in the model by only two major boundary faults. The question of erosion rates is clearly important. The regionally plane surface (Doran *et al.* 1973) underlying the Upper ORS north of the KMF (more strictly the DB–GFZ), must be due to some combination of degradation during (1) post-Acadian to pre-Munster Basin time and (2) syn-Munster Basin time. Taking the average maximum degradation rate in drainage basins of 1 m ka^{-1} (Schumm 1977, p. 34), it is feasible to have fully eroded Acadian uplands in the available time (<15 Ma) from Early Devonian uplift to Munster Basin initiation (?Givetian time, Williams *et al.* this volume). Erosion in pre-Munster Basin time must have been at least sufficient to subdue orographic drainage barriers established by Acadian uplift, as the Munster Basin transverse dispersal systems were sourced from Caledonian terranes to the north (Graham 1983; Figs 1 and 3). Major relief could also have been eliminated during the longer period of Munster Basin history (>23 Ma). Lower erosion rates probably applied during both intervals. By late Famennian time the landscape of the regional footwall to the Munster Basin had been pared down to allow a coastal plain–estuarine tract to advance northwards (Diemer & Bridge 1988). The bulk of this erosion is likely to have occurred during Late Devonian time, when the Munster Basin was supplied from large upland drainage basins.

Significant departures in the stretching factors calculated by the flexural model from previous 1D estimates are evident. Previous modelling (Sanderson 1984, pp. 150–151; Meere 1995*b*) used McKenzie's (1978) stretching model, and interpreted the Late Palaeozoic history of SW Ireland as a single stretching event followed by thermal subsidence. Sanderson (1984) found that *c.* 3.9 km of initial ORS subsidence was followed by *c.* 1.6 km of ORS over 20 Ma by thermal subsidence (to latest Devonian time) for a stretching factor of $\beta = 2$, the suggested value for the onset of volcanicity. This was followed by 1.5 km of sediment to mid-Carboniferous time. Meere (1995*b*) using similar crustal parameters, although different timings, derived (exceptionally high) values of $\beta = 3$–5. Price (1989) used the two-layer stretching model of Hellinger & Sclater (1983), with similar lithospheric parameters to Sanderson (1984), and obtained crustal stretching factors of two and 1.4 for the depocentre and SW of the basin, respectively. Apart from radical differences in timing of basin initiation and duration between these and the current study, other differences are the rifting history and the capability of the forward models to compute flexural rather than Airy isostasy. One-dimensional (Airy) models tend to overestimate β (Roberts *et al.* 1995). Thus, the conservative syn-rift sediment thicknesses modelled here are believed to provide a more accurate assessment of stretching. The lower values of β predicted in the models are in keeping with the low volumes of syn-rift magmatism, although not with recent suggestions of high geothermal gradients during basin formation (Clayton 1989; Blackmore 1995; Meere 1995*b*).

Conclusions

Flexural cantilever forward modelling of palinspastically restored sections across the Munster Basin requires the use of multiple planar upper-crustal extension faults to successfully replicate decompacted ORS sediment geometries. With this configuration, syn-rift subsidence is modelled, over a width scale of >200 km, for two sections that are non-unique in detail, as a result of lack of data on extension faults in the basin. Listic faults and basal extensional detachment models linked to the known basin-bounding faults do not replicate the ORS subsidence of the Munster Basin, and are not thought to have operated.

The overall syn-rift stratigraphical geometry modelled is asymmetrical, replicating the known gross ORS geometry, achieved by major extension on the Dingle Bay–Galtee Fault Zone, *c.* 2–3 km heaves on spaced planar synthetic faults and relatively small antithetic extensional displacement on the previously postulated

Dunmanus–Castletown Fault. This last fault controls the thin ORS succession in its footwall block (the Sherkin and Castlehaven Formations). Subsidence in the footwall of the Dunmanus–Castletown Fault, however, is dependent on offshore, north-downthrowing extension faults. This indicates that previous 'hinged' basin models with a single boundary fault are not viable, and that the Munster Basin continues, at a minimum, for tens of kilometres offshore. Significant footwall uplift of the immediate basin footwall is predicted, in both the northern Galtees and south–central Dingle. In the case of Dingle, this is based on major extensions on the Caherconree Fault and the main basin structure in Dingle Bay. Preservation of the thin Frasnian ORS (Carrigduff Group) on the northern Dingle footwall block is achieved by a combination of 'rim basin'-type subsidence and a minor extension fault located offshore of the north coast of Dingle.

Experiments showed that an effective elastic thickness (T_e) of 7 and 8 km promoted a sufficiently deep and wide syn-rift basin, for summed extensions of 27 to 59 km, respectively. This is consistent with a thickened, cooling post-Acadian lithosphere. The modelled rise in the Moho, and the resultant peak heat-flow anomaly, predicts the palinspastic location of main volcanic centres, which are skewed towards the northern margin. However, the low values of the stretching factor (1.3–1.48) differ significantly from published 1D (Airy) extensional models, and the low heat-flow anomaly predicted contradicts the high geothermal gradients required in recent models of syn-extensional peak metamorphism.

A minor post-rift megasequence across the onshore basin is predicted for a 5 Ma post-rift age, plus a modest positive eustatic component. This is insufficient to bury the regional footwall, and a combination of thermal subsidence overlapping in time with Munster Basin extension, and renewed South Munster Basin (latest Devonian) rifting is suggested to explain Famennian ORS north of the basin margin (Fig. 9).

Sincere thanks are due to N. J. Kusznir (University of Liverpool) for allowing use of his flexural cantilever model (2D STRETCH), and for many discussions on extensional tectonics and modelling. Thanks also to A. M. McAfee for information on the sequences in the South Munster Basin, to H. E. Edwards for ORS palaeocurrent data, to P. D. W. Haughton for permission to quote his discovery of the Castlehaven tuffs, to C. B. Long for data on the Skellig Michael tuff, and to M. Ford for numerous discussions on the geology of the Munster Basin. I am very grateful to an anonymous referee and P. M. Shannon for their thorough and constructive reviews of the typescript.

References

Allen, J. R. L. 1974. The Devonian rocks of Wales and the Welsh Borderland. *In*: Owen T. R. (ed.) *The Upper Palaeozoic and Post-Palaeozoic Rocks of Wales*. University of Wales Press, Cardiff, 47–84.

—— 1979. Old Red Sandstone facies in external basins, with particular reference to southern Britain. *In*: House, M. R., Scrutton, C. T. & Bassett, M. G. (eds) *The Devonian System*. Special Papers in Palaeontology, **23**, 65–80.

—— & Crowley, S. F. 1983. Lower Old Red Sandstone fluvial dispersal systems in the British Isles. *Transactions of the Royal Society of Edinburgh: Earth Sciences*, **74**, 61–68.

Allen, P. A. & Allen, J. R. 1990. *Basin Analysis: Principles and Applications. Blackwell Scientific, Oxford.*

Archer, J. B., Sleeman, A. G., Smith, D. C., Claringbold, K., Stanley, G. & Wright, G. 1996. *A geological description of Tipperary and adjoining parts of Laois, Kilkenny, Offaly, Clare and Limerick, to accompany the bedrock geology 1:100 000 scale map series, Sheet 18, Tipperary.* Geological Survey of Ireland, Dublin.

Avison, M. 1984*a*. Contemporaneous faulting and the eruption and preservation of the Lough Guitane Volcanic Complex, Co. Kerry. *Journal of the Geological Society, London*, **141**, 501–510.

—— 1984*b*. The Lough Guitane Volcanic Complex, County Kerry—a preliminary survey. *Irish Journal of Earth Sciences*, **6**, 127–136.

Bamford, M. L. F. 1987. Discussion of structural evolution of the Irish Variscides. *Journal of the Geological Society, London*, **144**, 997–998.

Blackmore, R. 1995. Low-grade metamorphism in the Upper Palaeozoic Munster Basin, southern Ireland. *Irish Journal of Earth Sciences*, **14**, 115–133.

Bluck, B. J., Cope, J. C. W. & Scrutton, C. T. 1992. Devonian. *In*: Cope, J. C. W., Ingham, J. K. & Rawson, P. F. (eds) *Atlas of Paleogeography and Lithofacies*. Geological Society, London, Memoir, **13**, 57–66.

Bridge, J. S. & Diemer, J. A. 1983. Quantitative interpretation of an evolving ancient river system. *Sedimentology*, **30**, 599–623.

——, van Veen, P. M. & Matten, L. C. 1980. Aspects of the sedimentology, palynology and palaeobotany of the Upper Devonian of southern Kerry Head, Co. Kerry, Ireland. *Geological Journal*, **15**, 143–170.

Brück, P. M., Higgs, K., Mazianne, N. & Vanguestaine, M. 2000. A Middle Cambrian age for the Muggort's Bay Inlier, southwest Co. Waterford. *Abstracts volume, Irish Geological Association Research Meeting, University College Cork*, 9.

Capewell, J. G. 1951. The Old Red Sandstone of the Inch and Annascaul district, Co. Kerry. *Proceedings of the Royal Irish Academy*, **54(B)**, 141–167.

—— 1956. The stratigraphy, structure and sedimentation of the Old Red Snadstone of the Comeragh Mountains and adjacent areas, County Waterford, Ireland. *Quarterly Journal of the Geological Society, London*, **112**, 393–412.

—— 1957. The stratigraphy and structure of the country around Sneem, Co. Kerry. *Proceedings of the Royal Irish Academy*, **58(B)**, 167–183.

—— 1965. The Old Red Sandstone of Slieve Mish, Co. Kerry. *Proceedings of the Royal Irish Academy*, **64(B)**, 165–174.

—— 1975. The Old Red Sandstone Group of Iveragh, Co. Kerry. *Proceedings of the Royal Irish Academy*, **75(B)**, 155–171.

CARRUTHERS, R. A. 1987. Aeolian sedimentation from the Galtymore Formation (Devonian), Ireland. *In*: FROSTICK, L. & REID, I. (eds) *Desert Sediments: Ancient and Modern*. Geological Society, London, Special Publications, **35**, 251–268.

CLAYTON, G. 1989. Vitrinite reflectance data from the Kinsale Harbour–Old Head of Kinsale area, southern Ireland, and its bearing on the interpretation of the Munster Basin. *Journal of the Geological Society, London*, **146**, 611–616.

—— & GRAHAM, J. R. 1974. Miospore assemblages from the Devonian Sherkin Formation of southwest County Cork, Republic of Ireland. *Pollen et Spores*, **16**, 565–588.

——, ——, HIGGS, K., HOLLAND, C. H. & NAYLOR, D. 1980. Devonian rocks in Ireland: a review. *Journal of Earth Sciences, Royal Dublin Society*, **2**, 161–183.

COLLINSON, J. D., MARTINSEN, O., BAKKEN, B. & KLOSTER, A. 1991. Early fill of the Western Irish Namurian Basin: a complex relationship between turbidites and deltas. *Basin Research*, **3**, 223–242.

COLTHURST, J. R. J. 1978. Old Red Sandstone rocks surrounding the Slievenamon inlier, Counties Tipperary and Kilkenny. *Journal of Earth Sciences, Royal Dublin Society*, **1**, 77–103.

CONROY, J. J. & BROCK, A. 1989. Gravity and magnetic studies of crustal structure across the Porcupine basin west of Ireland. *Earth and Planetary Science Letters*, **93**, 371–376.

COOPER, M. A. & TRAYNER, P. M. 1986. Thrust-surface geometry: implications for thrust-belt evolution and section-balancing techniques. *Journal of Structural Geology*, **8**, 305–312.

——, COLLINS, D., FORD, M., MURPHY, F. X. & TRAYNER, P. M. 1984. Structural style, shortening estimates and the thrust front of the Irish Variscides. *In*: HUTTON, D. H. W. & SANDERSON, D. J. (eds) *Variscan Tectonics of the North Atlantic Region*. Geological Society, London, Special Publications, **14**, 167–175.

——, ——, ——, ——, —— & O'SULLIVAN, M. 1986. Structural evolution of the Irish Variscides. *Journal of the Geological Society, London*, **143**, 53–61.

COWARD, M. P. 1993. The effect of Late Caledonian and Variscan continental escape tectonics on basement structure, Paleozoic basin kinematics and subsequent Mesozoic basin development in NW Europe. *In*: PARKER, J. R. (ed.) *Petroleum Geology of Northwest Europe: Proceedings of the 4th Conference*. Geological Society, London, 1095–1108.

DIEMER, J. A. & BRIDGE, J. S. 1988. Transition from alluvial plain to tide-dominated coastal deposits associated with the Tournaisian marine transgression in southwest Ireland. *In*: DE BOER, P. L., VAN GELDER, A. & NIO, S. D. (eds) *Tide-influenced Sedimentary Environments and Facies*. D. Reidel, Dordrecht, 359–388.

——, —— & SANDERSON, D. J. 1987. Revised geology of Kerry Head, County Kerry. *Irish Journal of Earth Sciences*, **8**, 113–138.

DODD, C. 1986. *The sedimentology of fluvio-aeolian interactions, with geological examples from the British Isles*. PhD thesis, Plymouth Polytechnic.

DORAN, R. J. P. Palynological evidence for the age of the Old Red Sandstone near Cappagh White, Co. Tipperary. *Scientific Proceedings of the Royal Dublin Society*, **A3**, 343–350.

——, HOLLAND, C. H. & JACKSON, A. A. 1973. The sub-Old Red Sandstone surface in southern Ireland. *Proceedings of the Royal Irish Academy*, **73**, 109–130.

EMENIKE, E. A. 1986. An interpretation of the Killarney and Leinster gravity anomalies in the Republic of Ireland. *Irish Journal of Earth Sciences*, **7**, 125–132.

FORD, M. 1987. Practical application of the sequential balancing technique: an example from the Irish Variscides. *Journal of the Geological Society, London*, **144**, 885–891.

——, BROWN, C. & READMAN, P. 1991. Analysis and tectonic interpretation of gravity data over the Variscides of southwest Ireland. *Journal of the Geological Society, London*, **148**, 137–148.

——, KLEMPERER, S. L. & RYAN, P. D. 1992. Deep structure of southern Ireland: a new geological synthesis using BIRPS deep reflection profiling. *Journal of the Geological Society, London*, **149**, 915–922.

FRIEDMANN, S. J & BURBANK, D. W. 1995. Rift basins and supradetachment basins: intracontinental extensional end-members. *Basin Research*, **7**, 109–127.

GARDINER, P. R. R. & MACCARTHY, I. A. J. 1981. The Late Paleozoic evolution of southern Ireland in the context of tectonic basins and their transatlantic significance. *In*: KERR, J. W. & FERGUSSON, A. J. (eds) *Geology of the North Atlantic Borderlands*. Canadian Society of Petroleum Geologists, Memoirs, **7**, 683–725.

GILL, W. D. 1962. The Variscan Fold Belt in Ireland. *In*: COE, K. (ed.) *Some Aspects of the Variscan Fold Belt*. Manchester University Press, Manchester, 49–64.

GRAHAM, J. R. 1983. Analysis of the Upper Devonian Munster Basin, an example of a fluvial distributary system. *In*: COLLINSON, J. D. & LEWIN, J. (eds) *Modern and Ancient Fluvial Systems*. International Association of Sedimentologists, Special Publications, **6**, 473–483.

—— & CLAYTON, G. 1988. Devonian rocks in Ireland and their relation to adjacent regions. *In*:

McMillan, N. J., Embry, A. F. & Glass, D. J. (eds) *The Devonian of the World, Vol. 1*. Canadian Society of Petroleum Geologists, Memoirs, **14**, 325–340.

—— & Reilly, T. A. 1972. The Sherkin Formation (Devonian) of south-west County Cork. *Geological Survey of Ireland, Bulletin*, **1**, 281–300.

——, James, A. & Russell, K. J. 1992. Basin history deduced from subtle changes in fluvial style: a study of distal alluvium from the Devonian of southwest Ireland. *Transactions of the Royal Society of Edinburgh: Earth Sciences*, **83**, 655–667.

——, Russell, K. J. & Stillman, C. J. 1995. Late Devonian magmatism in west Kerry and its relationship to the development of the Munster Basin. *Irish Journal of Earth Sciences*, **14**, 7–23.

Halliday, A. N. & Mitchell, J. G. 1983. K–Ar ages of clay concentrates from Irish orebodies and their bearing on the timing of mineralisation. *Transactions of the Royal Society of Edinburgh: Earth Sciences*, **74**, 1–14.

Harland, W. B., Armstrong, R. L. Cox, A. V., Craig, L. E., Smith, A. G. & Smith, D. G. 1990. *A Geologic Time Scale 1989*. Cambridge University Press, Cambridge.

Hellinger, S. J. & Sclater, J. G. 1983. Some comments on two-layer extensional models for the evolution of sedimentary basins. *Journal of Geophysical Research*, **88**, 8251–8269.

Higgs, K. & Russell, K. J. 1981. Upper Devonian microfloras from southeast Iveragh, County Kerry, Ireland. *Geological Survey of Ireland Bulletin*, **3**, 17–50.

——, Clayton, G. & Keegan, J. B. 1988. *Stratigraphic and Systematic Palynology of the Tournaisian Rocks of Ireland*. Geological Survey of Ireland, Special Paper, **7**.

——, MacCarthy, I. A. J. & O'Brien, M. M. 2000. A mid-Frasnian marine incursion into the southern part of the Munster Basin; evidence from the Foilcoagh Bay Beds, Sherkin Formation, southwest County Cork, Ireland. This volume.

Holland, C. H. 1981. Devonian. *In*: Holland, C. H. (ed.) *A Geology of Ireland*. Scottish Academic Press, Edinburgh, 121–146.

Horne, R. R. 1974. The lithostratigraphy of the Late Silurian to Early Carboniferous of the Dingle peninsula, Co. Kerry. *Geological Survey of Ireland Bulletin*, **1**, 395–428.

—— 1977. A tectonic interpretation of a supposed 'Devonian cliff line' in Slieve Mish, County Kerry and the Galty Mountains, Tipperary. *Geological Survey of Ireland Bulletin*, **2**, 85–97.

Howard, D. W. 1975. Deep-seated igneous intrusions in Co. Kerry. *Proceedings of the Royal Irish Academy*, **75**, 173–183.

Husain, S. M. 1957. *The geology of the Kenmare Syncline, Co. Kerry, Ireland*. PhD thesis, University of London.

Hutton, D. H. W. 1987. Strike-slip terranes and model for the evolution of the British and Irish Caledonides. *Geological Magazine*, **124**, 405–425.

Jackson, J. A. & White, N. J. 1989. Normal faulting in the upper continental crust: observations from regions of active extension. *Journal of Structural Geology*, **11**, 15–36.

James, A. & Graham, J. R. 1995. Stratigraphy and structure of Devonian fluvial sediments, western Beara Peninsula, south-west Ireland. *Geological Journal*, **30**, 165–182.

Jarvis, G. T. & McKenzie, D. P. 1980. Sedimentary basin formation with finite extension rates. *Earth and Planetary Science Letters*, **48**, 42–52.

——, Karner, G. D., Weissel, J. K., Dewey, J. F. & Munday, T. J. 1987. Geotectonic imaging of the north-western European continent and shelf. *Marine and Petroleum Geology*, **4**, 94–102.

Kelly, S. B. & Olsen, H. 1993. Terminal fans—a review with reference to Devonian examples. *Sedimentary Geology*, **1–4**, 339–374.

Klemperer, S. L., Ryan, P. D. & Snyder, D. B. 1991. A deep seismic reflection transect across the Irish Caledonides. *Journal of the Geological Society, London*, **148**, 149–164.

Kusznir, N. J. & Egan, S. S. 1990. Simple-shear and pure-shear models of extensional sedimentary basin formation: application to the Jeanne d'Arc Basin, Grand Banks of Newfoundland. *In*: Tankard, A. J. & Balkwill, H. R. (eds) *Extensional Tectonics of the North Atlantic Margins*. Memoirs, American Association of Petroleum Geologists, **46**, 305–322.

—— & Morley, C. 1990. The Lake Tanganyika Rift, East Africa: application of the flexural cantilever model of continental extension. *Geophysical Journal International*, **101**, 275–286.

——, Marsden, G. & Egan, S. S. 1991. A flexural-cantilever simple-shear/pure-shear model of continental lithosphere extension: applications to the Jeanne d'Arc Basin, Grand Banks and Viking Graben, North Sea. *In*: Roberts, A. M., Yielding, G. & Freeman, B. (eds) *The Geometry of Normal Faults*. Geological Society, London, Special Publications, **56**, 41–60.

——, Roberts, A. M. & Morley, C. 1995. Forward and reverse models of rift basin formation. *In*: Lambiase, J. J. (ed.) *Hydrocarbon Habitat in Rift Basins*, Geological Society, London, Special Publications, **80**, 33–56.

Leeder, M. R. & Gawthorpe, R. L. 1987. Sedimentary models for extensional tilt-block/half-graben basins. *In*: Coward, M. P., Dewey, J. F. & Hancock, P. L. (eds) *Continental Extensional Tectonics*. Geological Society, London, Special Publications, **28**, 139–152.

Lees, A., Hallet, V. & Hibo, D. 1985. Facies variation in Waulsortian build-ups, Part 1: a model from Belgium. *Geological Journal*, **20**, 133–158.

MacCarthy, I. A. J. 1990. Alluvial sedimentatioin patterns in the Munster Basin, Ireland. *Sedimentology*, **37**, 685–712.

Makris, J., Egloff, R., Jacob, A. W. B., Mohr, P., Murphy, T. & Ryan, P. 1988. Continental crust under the southern Porcupine Seabight west of Ireland. *Earth and Planetary Science Letters*, **89**, 387–397.

MARSDEN, G., ROBERTS, A. M., YIELDING, G. & KUSZNIR, N. J. 1991. Applications of the flexural cantilever simple-shear/pure-shear model of continental lithosphere extension to the formation of the North Sea Basin. *In*: BLUNDELL, D. J. & GIBBS, A. D. (eds) *Tectonic Evolution of the North Sea Rifts*. Oxford University Press, Oxford, 236–257.

MASSON, F., JACOB, A. W. B., PRODEHL, C., READMAN, P. W., SHANNON, P. M., SCHULZE, A. & ENDERLE, U. 1998. A wide-angle seismic traverse through the Variscan of southwest Ireland. *Geophysical Journal International*, **134**, 689–705.

MAX, M. D. & LEFORT, J. P. 1984. Does the Variscan Front in Ireland follow a dextral shear zone? *In*: HUTTON, D. H. W. & SANDERSON, D. J. (eds) *Variscan Tectonics of the North Atlantic Region*. Geological Society, London, Special Publications, **14**, 177–183.

MCAFEE, A. M. 1992. *Upper Devonian–Lower Carboniferous sedimentation patterns in the western part of the South Munster Basin*. PhD thesis, National University of Ireland.

MCKENZIE, D. 1978. Some remarks on the development of sedimentary basins. *Earth and Planetary Science Letters*, **40**, 25–32.

MEERE, P. A. 1995*a*. The structural evolution of the western Irish Variscides: an example of obstacle tectonics? *Tectonophysics*, **246**, 97–112.

—— 1995*b*. Sub-greenschist facies metamorphism from the Variscides of SW Ireland: an early synextensional peak thermal event. *Journal of the Geological Society, London*, **152**, 511–521.

MORRIS, P. 1980. An analysis of some small scale gravity variations over the Iveragh peninsula, County Kerry, Ireland. *Journal of Earth Sciences, Royal Dublin Society*, **3**, 147–153.

MURPHY, F. X. 1990. The Irish Variscides: a fold belt developed within a major surge zone. *Journal of the Geological Society, London*, **147**, 451–460.

MURPHY, T. 1981. Geophysical evidence. *In*: HOLLAND, C. H. (ed.) *A Geology of Ireland*. Scottish Academic Press, Edinburgh, 225–229.

NAYLOR, D. & JONES, P. C. 1967. Sedimentation and tectonic setting of the Old Red Sandstone of southwest Ireland. *In*: OSWALD, D. H. (ed.) *International Symposium on the Devonian System*. Alberta Society of Petroleum Geologists, Memoirs, **2**, 1089–1099.

—— & SEVASTOPULO, G. D. 1993. The Reenydonagan Formation (Dinantian) of the Bantry and Dunmanus synclines, County Cork. *Irish Journal of Earth Sciences*, **12**, 191–203.

——, REILLY, T. A., SEVASTOPULO, G. D. & SLEEMAN, A. G. 1983. Stratigraphy and structure in the Irish Variscides. *In*: HANCOCK, P. L. (ed.) *The Variscan Foldbelt in the British Isles*. Adam Hilger, Bristol, 20–46.

——, SEVASTOPULO, G. D. & SLEEMAN, A. G. 1989. Subsidence history of the South Munster Basin, Ireland. *In*: ARTHURTON, R. S., GUTTERIDGE, P. & NOLAN, S. C. (eds) *The Role of Tectonics in Devonian and Carboniferous Sedimentation in the British Isles*. Occasional Publications of the Yorkshire Geological Society, **6**, 99–109.

——, ——, —— & REILLY, T. A. 1981. The Variscan fold belt in Ireland. *Geologie en Mijnbouw*, **60**, 49–66.

ORI, G. G. & PENNEY, S. R. 1982. The stratigraphy and sedimentology of the Old Red Sandstone sequence at Dunmore East, County Waterford. *Journal of Earth Sciences, Royal Dublin Society*, **5**, 43–59.

PENNEY, S. R. 1978. Devonian lavas from the Comeragh Mountains, County Waterford. *Journal of Earth Sciences, Royal Dublin Society*, **1**, 71–76.

—— 1980. A new look at the Old Red Sandstone succession of the Comeragh Mountains, County Waterford. *Journal of Earth Sciences, Royal Dublin Society*, **3**, 155–178.

PRICE, C. A. 1986. *The geology of the Iveragh Peninsula, County Kerry, Ireland, incorporating a remote sensing lineament study*. PhD thesis, University of Dublin.

—— 1989. Some thoughts on the subsidence and evolution of the Munster Basin, southern Ireland. *In*: ARTHURTON, R. S., GUTTERIDGE, P. & NOLAN, S. C. (eds) *The Role of Tectonics in Devonian and Carboniferous Sedimentation in the British Isles*. Occasional Publications of the Yorkshire Geological Society, **6**, 111–121.

—— & TODD, S. P. 1988. A model for the development of the Irish Variscides. *Journal of the Geological Society, London*, **145**, 935–939.

READMAN, P. W., O'REILLY, B. M. & MURPHY, T. Gravity gradients and upper-crustal tectonic fabrics, Ireland. *Journal of the Geological Society, London*, **154**, 817–828.

RICHMOND, L. K. 1998. *Fluvial–aeolian interactions and Old Red Sandstone basin evolution, Northwest Dingle Peninsula, County Kerry, Southwest Ireland*. PhD thesis, University of Aberdeen.

—— & WILLIAMS, B. P. J. 2000. A new terrane in the Old Red Sandstone of the Dingle Peninsula SW Ireland. *This volume*.

ROBERTS, A. M. & YIELDING, G. 1991. Deformation around basin-margin faults in the North Sea/mid-Norway rift. *In*: ROBERTS, A. M., YIELDING, G. & FREEMAN, B. (eds) *The Geometry of Normal Faults*. Geological Society, London, Special Publications, **56**, 61–78.

——, ——, KUSZNIR, N. J., WALKER, I. M. & DORN-LOPEZ, D. 1993. Mesozoic extension in the North Sea: constraints from flexural backstripping, forward modelling and fault populations. *In*: PARKER, J. R. (ed.) *Petroleum Geology of Northwest Europe: Proceedings of the 4th Conference*. Geological Society, London, 1123–1136.

——, ——, ——, —— & —— 1995. Quantitative analysis of Triassic extension in the northern Viking Graben. *Journal of the Geological Society, London*, **152**, 15–26.

ROBINSON, K. W., SHANNON, P. M. & YOUNG, D. G. G. 1981. The Fastnet Basin, an integrated analysis. *In*: ILLING, L. V. & HOBSON, G. D. (eds) *Petroleum Geology of the Continental Shelf of North-west Europe*. Heyden, London, 444–454.

RUSSELL, K. J. 1978. Vertebrate fossils from the Iveragh Peninsula and the age of the Old Red Sandstone. *Journal of Earth Sciences, Royal Dublin Society*, **1**, 151–162.

—— 1984. *The sedimentology and palaeogeography of some Devonian sedimentary in rocks in southwest Ireland.* PhD thesis, Plymouth Polytechnic.

SADLER, S. P. & KELLY, S. B. 1993. Fluvial processes and cyclicity in terminal fan deposits: an example from the Late Devonian of southwest Ireland. Sedimentary Geology, **1–4**, 375–386.

SANDERSON, D. J. 1984. Structural variation across the northern margin of the Variscides in NW Europe. *In*: HUTTON, D. H. W. & SANDERSON, D. J. (eds) *Variscan Tectonics of the North Atlantic Region.* Geological Society, London, Special Publications, **14**, 149–165.

SCHUMM, S. A. 1977. *The Fluvial System.* Wiley, New York.

SCLATER, J. G. & CHRISTIE, P. A. F. 1980. Continental stretching: an explanation of the post Mid-Cretaceous subsidence of the central North Sea Basin. *Journal of Geophysical Research*, **85**, 3711–3739.

SEVASTOPULO, G. D. 1981. Upper Carboniferous. *In*: HOLLAND, C. H. (ed.) *A Geology of Ireland.* Scottish Academic Press, Edinburgh, 173–187.

SHELFORD, P. H. 1963. The structure and relationship of the Namurian outcrop between Duntryleague, Co. Limerick and Dromlin, Co. Tipperary. *Proceedings of the Royal Irish Academy*, **62(B)**, 255–266.

SLEEMAN, A. G., MCCONNELL, B., CLARINGBOLD, K., O'CONNOR, P., Warren, W. P. & WRIGHT, G. 1995. *A geological description of east Cork, Waterford and adjoining parts of Tipperary and Limerick to accompany the bedrock geology 1:100 000 scale map series, Sheet 22, East Cork–Waterford.* Geological Survey of Ireland, Dublin.

——, PRACHT, M., DALY, E. P., FLEGG, A. M., O'CONNOR, P. J. & WARREN, W. P. 1994. *A geological description of south Cork and adjoining parts of Waterford to accompany the bedrock geology 1:100 000 scale map series, Sheet 25, South Cork.* Geological Survey of Ireland, Dublin.

TODD, S. P. 1989. Role of the Dingle Bay Lineament in the evolution of the Old Red Sandstone of south-west Ireland. *In*: ARTHURTON, R. S., GUTTERIDGE, P. & NOLAN, S. C. (eds) *The Role of Tectonics in Devonian and Carboniferous Sedimentation in the British Isles.* Occasional Publications of the Yorkshire Geological Society, **6**, 35–54.

TRAYNER, P. M. 1985. *The stratigraphy and structure of parts of east Co. Cork and west Co. Waterford.* PhD thesis, National University of Ireland.

TUCKER, R. D., BRADLEY, D. C., VER STRAETEN, C. A., HARRIS, A. G., EBERT, J. R. & MCCUTCHEON, S. R. 1998. New U–Pb zircon ages and the duration and division of Devonian time. *Earth and Planetary Science Letters*, **158**, 175–186.

TUNBRIDGE, I. P. 1984. Facies model for a sandy ephemeral stream and clay playa complex; the Middle Devonian Trentishoe Formation of North Devon, U.K. *Sedimentology*, **31**, 697–715.

—— 1986. Mid-Devonian tectonics and sedimentation in the Bristol Channel area. *Journal of the Geological Society, London*, **143**, 107–115.

VAIL, P. R., MITCHUM, R. M., JR & THOMPSON, S. 1977. seismic stratigraphy and global changes of sea level, Part 4: global cycles of relative changes in sea level. *In*: PAYTON, C. E. (ed.) *Seismic Stratigraphy—Applications to Hydrocarbon Exploration.* Memoirs, American Association of Petroleum Geologists, **26**, 83–97.

VERMEULEN, N. J., SHANNON, P. M., LANDES, M., MASSON, F. & THE VARNET GROUP. 1998–1999. Seismic evidence for subhorizontal crustal detachments beneath the Irish Variscides. *Irish Journal of Earth Sciences*, **17**, 1–18.

——, ——, MASSON, F. & LANDES, M. 2000. wide-angle seismic control on the development of the Munster Basin, SW Ireland. This volume.

WALSH, P. T. 1967. Notes on the Namurian stratigraphy north of Killarney, Co. Kerry. *Irish Naturalists' Journal*, **15**, 254–258.

—— 1968. The Old Red Sandstone west of Killarney, Co. Kerry, Ireland. *Proceedings of the Royal Irish Academy*, **66(B)**, 9–26.

WEGMANN, M. 1993. *Die geologie des Killarney Nationalpark, County Kerry, SW-Ireland.* Diplomarbeit an der Abteilung XC (Erdwissenschaften) thesis, ETH Zürich.

WILLIAMS, E. A. 1993. *The stratigraphy, fluvial sedimentology and structural geology of the Old Red Sandstone in the Derrynasaggart Mountains, Counties Cork and Kerry.* PhD thesis, National University of Ireland.

——, BAMFORD, M. L. F., COOPER, M. A. *et al.* 1989. Tectonics controls and sedimentary response in the Devonian–Carboniferous Munster and South Munster Basins, south-west Ireland. *In*: ARTHURTON, R. S., GUTTERIDGE, P. & NOLAN, S. C. (eds) *The Role of Tectonics in Devonian and Carboniferous Sedimentation in the British Isles.* Occasional Publications of the Yorkshire Geological Society, **6**, 123–141.

——, FORD, M. & EDWARDS, H. E. 1990. Discussion of a model for the development of the Irish Variscides. *Journal of the Geological Society, London*, **147**, 566–571.

——, ——, —— & O'SULLIVAN, M. J. 1993. An outline of evolution of the Late Devonian Munster Basin, South-West Ireland. *In*: GAYER, R. A., GREILING, R. O. & VOGEL, A. K. (eds) *Rhenohercynian and Subvariscan Fold Belts.* Vieweg, Braunschweig, 131–138.

——, SERGEEV, S. A., STÖSSEL, I. & FORD, M. 1997. An Eifelian U–Pb zircon date for the Enagh Tuff Bed from the Old Red Sandstone of the Munster Basin in NW Iveragh, SW Ireland. *Journal of the Geological Society, London*, **154**, 189–193.

——, ——, ——, —— & HIGGS, K. T. 2000. U–Pb zircon geochronology of silicic tuffs and chronostratigraphy of the earliest Old Red Sandstone in the Munster Basin, SW Ireland. This volume.

WINGFIELD, R. T. R. 1968. *The geology of Kenmare and Killarney.* PhD thesis, University of Dublin.

U–Pb zircon geochronology of silicic tuffs and chronostratigraphy of the earliest Old Red Sandstone in the Munster Basin, SW Ireland

E. A. WILLIAMS[1,5], S. A. SERGEEV[1,2], I. STÖSSEL[1], M. FORD[1,4] & K. T. HIGGS[3]

[1]*Geology Institute and* [2]*Institute of Isotope Geology and Mineral Resources, ETH-Zentrum, 8092 Zürich, Switzerland*

[3]*Department of Geology, University College, Cork, Ireland*

[4]*Present address: École Nationale Supérieure de Géologie, CRPG-CNRS, Rue du Doyen Marcel Roubault, B.P. 40, 54501 Vandoeuvre-lès-Nancy, France*

[5]*Present address: CRPG-CNRS, B.P. 20, 54501 Vandoeuvre-lès-Nancy Cedex, France (e-mail: williams@crpg.cnrs-nancy.fr)*

Abstract: Newly acquired U–Pb magmatic zircon dates from silicic tuffs within the Old Red Sandstone (ORS) magnafacies of the Munster Basin (SW Ireland) are intercalibrated with newly discovered (late Givetian) and reappraised (mid-Frasnian) miospore assemblages to provide the first biostratigraphically constrained numerical ages in the Irish Devonian succession. The weighted mean $^{207}Pb/^{206}Pb$ isotopic age determined for the Keel Tuff Bed (385.0 ± 2.9 Ma) is indistinguishable from that of the previously investigated Enagh Tuff Bed (384.9 ± 0.7 Ma). In conjunction with very similar rare earth element (REE) signatures, this confirms their correlation, placing a minimum age of 384.9 ± 0.4 Ma on the newly discovered Reenagaveen microflora, which is assigned to the late Givetian TCo Oppel zone. The equivalence of the Keel and Enagh Tuffs constrains a vertebrate fauna containing *Bothriolepis* and the Valentia Island tetrapod ichnofauna to pre-date this event. Isotopic dating of thickly bedded subaerial tuffs from the Lough Guitane Volcanic Complex, a major accumulation of rhyolites and silicic volcaniclastic rocks, reveals ages of 384.5 ± 1.0 Ma (Killeen Volcanic Centre), indistinguishable from the Keel–Enagh Tuff Bed, and 378.5 ± 0.2 Ma from the Horses Glen Volcanic Centre, previously considered to be the oldest of the complex. The Horses Glen Centre post-dates the Moll's Gap Quarry microflora, the only current biostratigraphical control on the age of the early ORS in the east of the basin depocentre, thus indicating a minimum age for the (mid-Frasnian) IV Oppel zone, the revised biostratigraphic age of this assemblage. These controls on the early ORS (1) suggest that Munster Basin initiation occurred before late Givetian time and (2) give time-averaged (compacted) accumulation rates of *c*. 0.17–0.25 and 0.18 mm a^{-1} for eastern and western Iveragh, respectively. The minimum basin duration time was *c*. 23 Ma to the end of the Devonian period. The implications of these data for the depocentre stratigraphy, volcanic events, proposed ORS cyclicities and the geohistory of the Munster Basin are examined.

Palinspastic reconstructions of the Munster Basin (SW Ireland) that account for Variscan deformation (Williams *this volume*), reveal the largest onshore Devonian Old Red Sandstone (ORS) basin in Ireland. The basin is widely interpreted as extensional in origin (following Naylor & Jones 1967), controlled by a northern boundary fault zone (the Dingle Bay–Galtee Fault Zone, Fig. 1) and an intrabasinal antithetic fault (the Dunmanus–Castletown Fault) *c*. 120 km to the SSE (Williams *et al*. 1989). The true width of the basin is >180 km as there is no onshore evidence of a southern margin. At least 5.7 km of ORS alluvial magnafacies accumulated in the depocentre (Fig. 1) up to latest Devonian time, when sediments transitional to shallow marine environments marked the end of the Munster Basin as a discrete structure.

The ORS basin fill is widely considered to be of Late Devonian age (Holland 1977; Clayton *et al*. 1980; Graham 1983; MacCarthy 1990), although the age of the oldest exposed ORS is poorly constrained palaeontologically and has only recently been investigated by high-precision U–Pb isotopic techniques (Williams *et al*. 1997). The true initiation age of the basin is unknown, as the base of the ORS is not exposed. Currently,

From: FRIEND, P. F. & WILLIAMS, B. P. J. (eds). *New Perspectives on the Old Red Sandstone.* Geological Society, London, Special Publications, **180**, 269–302. 0305-8719/00/$15.00

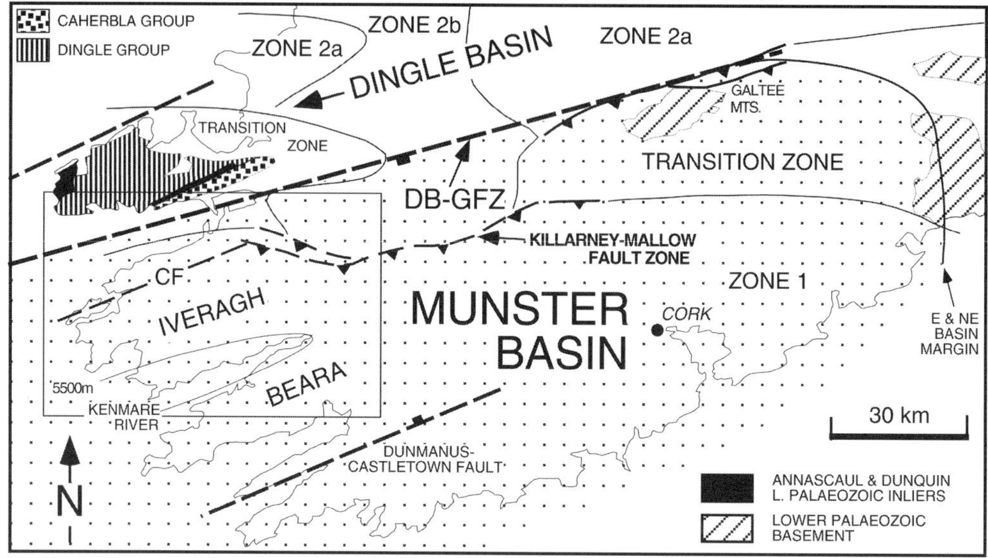

Fig. 1. Location map of the Munster and Dingle basins, showing the basin margins, principal fault zones (DB–GFZ, Dingle Bay–Galtee Fault Zone; CF, Coomnacronia Fault) and boundaries of Variscan structural zones (after Cooper *et al.* (1986); shown as continuous thin lines). Coarse stipple shows the onshore extent of the Munster Basin ORS. Boxed area is that of Fig. 2. The position of the apparent depocentre in southern Iveragh is indicated by the 5500 m isopach.

the chronostratigraphic age of the early ORS in the basin depocentre depends on only two biostratigraphical records. First, the vertebrate *Bothriolepis*, from the west Iveragh succession, has been considered to place a maximum age of late Givetian–early Frasnian on its enclosing sediments (Russell 1978). Second, a miospore assemblage from a section in SE Iveragh (Moll's Gap Quarry) has been assigned slightly conflicting late Givetian–early Frasnian (van Veen & van der Zwan 1980) and early Frasnian (Higgs & Russell 1981) ages. Limitations, however, apply to these records. *Bothriolepis* is long ranging, disallowing precise chronostratigraphic specification and is highly provincial, where its earliest occurrence varies from early Frasnian time in Euramerica–Baltica to possibly Givetian–Eifelian time (Westoll 1979) and Givetian time in China (Pan & Dineley 1988). Apart from the minor difference in chronostratigraphic ages attributed to the Moll's Gap Quarry microflora, its precision is limited because of the coarse nature of biozonal scheme available at the time it was described, and subsequent decisions on boundary stratotypes (Klapper *et al.* 1987, 1993). Finally, although the study of Williams *et al.* (1997) provides a minimum numerical constraint on ORS age (384.9 ± 0.7 Ma), the dated horizon was not then biostratigraphically calibrated, which led to a suggested Eifelian correlation based on a direct comparison with several time scales (summarized by Harland *et al.* (1990, fig. 1.5) and Williams *et al.* (1997, p. 192)).

All of these considerations severely hamper placing the Munster Basin in a realistic time frame in relation to the preceding late Caledonian (Acadian) deformation and plutonism in southern Ireland, and Siluro-Devonian marine and Early to Mid-ORS sedimentation in the Dingle Basin (Fig. 1). Moreover, the absence of quantitative age data from the ORS of the Munster Basin has hindered accurate estimation of subsidence rates, basin duration and the chronology of volcanic events, as well as accurate correlation of basin stratigraphies. The new geochronology and palynology reported in this paper, together with the publication of new isotopic dates and time scales for the Devonian period (e.g. Roden *et al.* 1990; Tucker & McKerrow 1995; Tucker *et al.* 1998; Williams *et al.* this volume) necessitate a reassessment of the time frame of the Munster Basin, in which the above issues are addressed. We characterize selected volcanic rocks from the early ORS in the basin depocentre, in terms of U–Pb numerical ages and geochemical signals, which we relate to a revised mid-Frasnian (IV miospore biozone) chronostratic age for the Moll's Gap Quarry microflora and the newly acquired Reenagaveen microflora (TCo biozone age).

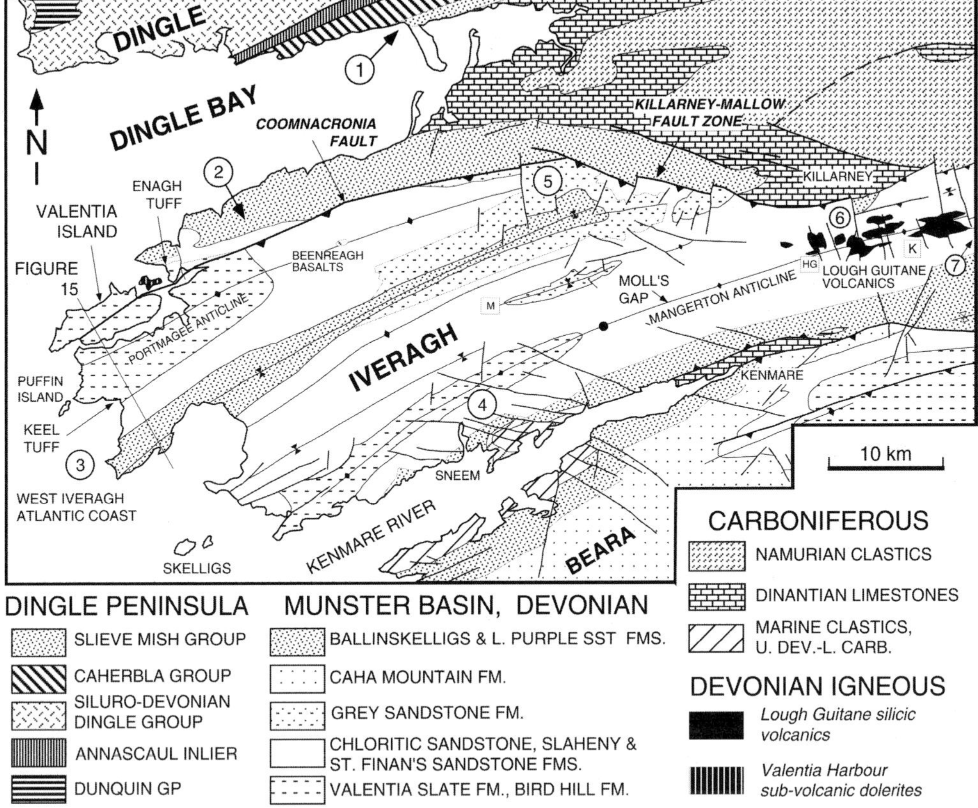

Fig. 2. Geological–structural map of the Iveragh peninsula–Derrynasaggart Mountains region. Sampled localities in the Lough Guitane Volcanic Complex are, Killeen (K) and Horses Glen (HG) Centres. Locations of the Beenreagh igneous centre, Keel and Enagh Tuffs are also indicated. Structures referred to in the text include the Mullaghanattin Syncline (M) and an axial plunge depression of the Mangerton Anticline (indicated by a filled circle). Numbers refer to localities in Fig. 3.

Geological setting

The ORS basin-fill sequence of the Iveragh–Derrynasaggarts region passes conformably upwards to Lower Carboniferous marine strata in the Sneem (Capewell 1957; Higgs & Russell 1981), Kenmare (Husain 1957) and Killarney (Walsh 1968; Higgs *et al.* 1988) areas (Fig. 2). The ORS lithostratigraphy comprises a series of geometrically complex formations of differing lateral and vertical grain size and facies composition, varying both parallel to and transverse to the northern basin margin (Williams *et al.* 1989, fig. 3). This variation is a function of contrasting axial–marginal and proximal–distal environments of two approximately superimposed large-radius fluvial dispersal systems, (1) the Chloritic Sandstone–Gortanimill and (2) Gun Point Systems (Williams *et al.* 1989, fig. 4), which drained transversely across the depocentre. A depositional strike-parallel correlation through Iveragh linking the volcanic deposits and biostratigraphic levels dealt with in this paper was detailed by Williams *et al.* (1989, fig. 3), a modified version of which is shown in Fig. 3. The coarse-grained Chloritic Sandstone Formation of the east Iveragh–Derrynasaggarts region, which contains the Lough Guitane Volcanic Complex in its lower part, is considered the lithostratigraphic equivalent of the very fine grained St Finan's Sandstone Formation of west Iveragh, based on the mapping of Capewell (1975) and Walsh (1968), and regional-scale constraints (Williams *et al.* (1989); but see Graham *et al.* (1992, p. 658) for a contrary view). Dated volcanic rocks and biostratigraphically important horizons in west Iveragh occur within the Valentia Slate and St Finan's Sandstone Formations, and within the Chloritic Sandstone Formation in east Iveragh.

SLIEVE MISH DINGLE (1)	NORTH-WEST IVERAGH (2)	WEST IVERAGH (3)	SNEEM (4)	NORTH EAST IVERAGH (5)	LOUGH GUITANE (6)	NORTHERN (7) DERRYNASAGGARTS
LACK FM. 183m			DERRYQUIN SST FORMATION 661m	UPPER PURPLE SST + ARDNAGLUGGEN FMS		
GEARHANE SST FM. 549m	*top not seen*	*top not seen*			*top not seen*	*top not seen*
LOUGH SLAT CGL. FM. 122m	BALLINSKELLIGS SST FM. >>350m	BALLINSKELLIGS SANDSTONE FORMATION >>350m	PURPLE SANDSTONE FORMATION 1464m	LOWER PURPLE SANDSTONE FORMATION 1525m	GUN POINT FORMATION >>700m	GUN POINT FORMATION >1475m
	Doulus Cgl. Mbr 30m					
	BALLINSKELLIGS SST FM. (70m)					
PRE-MUNSTER BASIN (LOWER PALAEOZOIC & DINGLE BASIN FILL)	ST. FINAN'S SANDSTONE FORMATION 190m	CAHA MOUNTAIN FM. 400m	CAHA MOUNTAIN FM. 470m	GREY SANDSTONE FORMATION 915m	CHLORITIC SANDSTONE FORMATION (2100m)	CHLORITIC SANDSTONE FORMATION >2000m
		ST. FINAN'S SANDSTONE FORMATION 1245m	(315m)			
			Green Sst Mbr 413m	CHLORITIC SANDSTONE FORMATION >1220m	L. Guitane Volc. Complex 200m	*<lowest level>*
		(766m)	CHLORITIC SANDSTONE FM. 1628m		CSF (>800m)	
	ENAGH TUFF	KEEL TUFF		*base not seen*	*base not seen*	
	VALENTIA SLATE FM. >500m	VALENTIA SLATE FORMATION >1561m	LOWER SLATE FORMATION >1458m			
			Eskine Sst Mbr (66m)			
	base not seen	*base not seen*	*base not seen*			

Fig. 3. Table of lithostratigraphical terminology and correlation for the ORS of the Iveragh–Derrynasaggarts region. No vertical or lateral scales implied. Values in parentheses refer to the interval indicated. Terminology, modified for consistency, as in Williams *et al.* (1989, fig. 3). The Slieve Mish–Dingle stratigraphic terminology is slightly modified from Capewell (1965), and collectively referred to as the Slieve Mish Group (Williams 1993; Pracht 1997).

Palaeotectonically, the Iveragh peninsula–Derrynasaggart Mountains region occurs between the apparent depocentre in the Kenmare River–south Iveragh region and the main extensional boundary fault of the Munster Basin located in Dingle Bay (Figs 1 and 2), part of the Dingle Bay–Galtee Fault Zone (DB–GFZ; Williams *et al.* 1989). The Iveragh–Derrynasaggarts region occupies the north-western part of zone 1 of the Irish Variscan orogen (Cooper *et al.* 1986). This region is characterized by arcuate ENE–WSW- to NE–SW-trending first-order macrofolds, at least three orders of congruous smaller-scale folds (Husain 1957), an intensely developed weakly transecting pressure solution cleavage, and variably developed fore- and back-thrusts. The northern limit of zone 1 occurs along the unreactivated main strand of the DB–GFZ, and along the Killarney–Mallow (thrust) Fault bounding the north side of the Derrynasaggart Mountains (Figs 1 and 2).

The metamorphic grade of rocks in orogen zone 1 is low. Illite crystallinity studies of clastic lithologies indicate epizone conditions during Variscan deformation (Blackmore 1995; Meere 1995*a*); metamorphic temperatures, determined by chlorite geothermometry, were 280–315° C (Meere 1995*b*). Peak palaeotemperatures have been also quantified by conodont colour and vitrinite reflectance indices at >250° C (Clayton 1989) and at *c.* 340–350° C (Blackmore 1995, fig. 13). These values, and former estimates of lower greenschist conditions (e.g. Avison 1982, 1984*a*), are well below closure temperatures for U–Pb dating of zircon (>800° C), indicating that their isotopic systems are unlikely to have been disturbed. Variscan deformation has been effectively dated at *c.* 290 Ma, from the mean age of multiple K–Ar determinations of syntectonic vein minerals located largely in the southern part of orogen zone 1 (Halliday & Mitchell 1983).

Stratigraphical-structural context of dated levels

Lough Guitane Volcanic Complex

This group of non-vesicular rhyolitic lavas and associated silicic volcaniclastic tuffs is the largest known accumulation of volcanic rock in the Munster Basin. Despite recrystallization to a quartz–albite–chlorite–phengite assemblage, textural evidence indicates a 1–2% restite content in the magma (Avison 1982). The Lough Guitane Complex consists of three major volcanic centres (Avison 1982, 1984*b*) outcropping for 17 km along the hinge zone of the Mangerton Anticline; from WSW to ENE these are the Horses Glen, Bennaunmore and Killeen volcanic centres (Fig. 2). In the Lough Guitane region the Mangerton Anticline is an ENE-plunging, asymmetrical (NNW-facing) macrofold, located in the

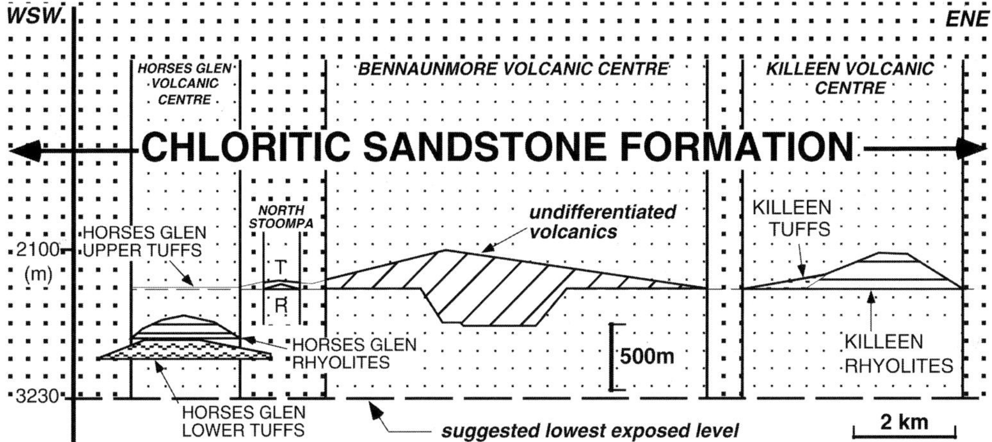

Fig. 4. Simplified stratigraphy of the Lough Guitane Volcanic Complex after Avison (1984*a*, *b*). Samples for isotopic dating were taken from the Horses Glen Lower Tuffs (Horses Glen Centre) and the Killeen Tuffs (Killeen Centre). The complex internal stratigraphy of the Bennaunmore Centre is shown undifferentiated. Abbreviations used for the North Stoompa volcanic rocks: T, tuffs; R, North Stoompa Rhyolite Lava.

hanging wall of the Killarney–Mallow Fault (Fig. 2). The fold is cut by numerous Variscan cross faults, approximately transverse to its axial surface trace (Avison 1984*b*). This first-order fold is traceable from the south–central Iveragh (Sneem) area (Fig. 2), where it is an upright, *c.* 15° ENE-plunging structure referred to as the Kilcrohane Anticline (Capewell 1957).

The individual centres of the Lough Guitane Complex are volcanologically distinct, showing thinning trends, distal facies and syn-volcanic faults that indicate their margins (Avison 1982, 1984*a*, *b*). All of the centres occur entirely within the Chloritic Sandstone Formation (Fig. 4). In the northern Derrynasaggarts–Lough Guitane region, this formation comprises complexly stratified, metre-scale sequences of medium–coarse sandstone and rare pebbly sandstone, which alternate with siltstone and fine-grade sandstone intervals, respectively interpreted as the channel belts of high-discharge braided rivers and related flood plain deposits (Williams *et al.* 1989; Williams 1993). The largest of the volcanic centres, the Bennaunmore Centre, is associated with two sets of syn-volcanic extension faults responsible for complex variations in unit thicknesses and volcanic facies (Avison 1984*a*). The Horses Glen Centre was considered by Avison (1982, 1984*a*, *b*) to be the older of a two-phase volcanic history, in which the time-equivalent Killeen and Bennaunmore centres were erupted following a *c.* 200 m thick interval of Chloritic Sandstone Formation (non-volcanic) alluviation. Because of this, and the structural and volcanic complexity of the Bennaunmore Centre, it was decided to sample the Horses Glen and Killeen volcanic centres for isotopic dating, to bracket the age of the whole complex.

Killeen Tuffs. The Killeen Volcanic Centre is dominated by the Killeen rhyolite lavas, a composite pile of lava flows exceeding 300 m in thickness in the NE of the centre, but on average 200 m thick (Avison 1982; Fig. 4). The Killeen rhyolites are both under- and overlain by rocks of the Chloritic Sandstone Formation, except at the western margin where the Killeen Tuffs intervene. Two separate outcrops are referred to as the Killeen Tuffs (Avison 1984*b*) overlying the main body of the Killeen rhyolites distally, delimiting the northwestern and southwestern margins of the centre (Avison 1982, 1984*b*, fig. 5). The lithosome sampled is that at the south-western limit of the centre. This lithosome is *c.* 120 m thick immediately west of its banking (onlapping) boundary with the Killeen rhyolites, but wedges out westwards over 1 km (Avison 1982). This level was considered by Avison (1984*a*, *b*) to be equivalent to the main deposits of the Bennaunmore Centre. The top of the bedded to massive Killeen tuffs was considered by Avison (1982) to have been subaerially eroded by runoff before burial by Chloritic Sandstone Formation alluvial sedimentation. These relationships suggest that the Killeen tuffs marginally post-date the extrusion of the rhyolites, and that the tuffs were subaerially emplaced, effectively dating the age of the Killeen Centre. Approximately 30 kg of medium- to fine-grained, grey massive tuffaceous rock were collected from a quarry [Irish National Grid reference W08828216], NW of the village of Clonkeen. Weathered and fractured rock was carefully avoided.

Horses Glen Lower Tuffs. The Horses Glen Volcanic Centre (Fig. 2) is deformed by a train of symmetrical, second-order folds within the hinge zone of the Mangerton Anticline. The centre comprises two principal volcano-stratigraphical units (Fig. 4), the basal Horses Glen Lower Tuffs, and the immediately overlying Rhyolite Lavas (Avison 1984*b*). The 10 m thick Horses Glen Upper Tuffs have been estimated by Avison (1984*b*, p. 128) to be 200 m above the top of the rhyolites, and are assumed to be equivalent to the main body of the Bennaunmore Centre (Avison 1982, p. 158, 1984*b*). A strike-parallel fault, however, separates the Upper Tuffs from the main deposits (the Rhyolites and Lower Tuffs) of the Horses Glen Centre (Avison 1984*b*, figs. 3 and 6a). On this correlation, based on fig. 7 of Avison (1984*b*; Fig. 4), there is *c.* 440 m stratigraphic difference between the (sampled) Killeen Tuffs and the middle of the Horses Glen Lower Tuffs.

Over 33 kg of fresh uniform tuffaceous lithology were removed from a thick (metre-scale) package of undulatory to sinusoidally bedded volcaniclastic facies of the Horses Glen Lower Tuffs [V99638186] *c.* 220 m NW of Lough Managh. The section sampled is located on the NNW limb of the (second-order) anticline that crosses the northern part of Lough Managh shown by Avison (1984*b*, fig. 3), and is thus from approximately the mid-part of the Horses Glen Lower Tuffs. The three-dimensional undulatory–sinusoidal bedding prevalent in the sampled section suggests a surging pyroclastic flow origin, and confirms subaerial emplacement of the dated beds.

Moll's Gap Quarry

The horizon of the Moll's Gap Quarry microflora (Higgs & Russell 1981) outcrops in the hinge zone of the Mangerton Anticline [V85937755] 14.25 km to the WSW, along the axial surface trace of the fold from the sample site in the Horses Glen Volcanic Centre (Fig. 2). Higgs & Russell (1981) estimated the stratigraphic position of the microfloral horizon as 800 m down-section of the top of the Green Sandstone Formation of Walsh (1968; the Chloritic Sandstone Formation of this paper) mapped to the north of Moll's Gap in the Mullaghanattin syncline (Walsh 1968, fig. 4).

As formation thicknesses vary with transverse distance from the basin margin as a result of differential subsidence and diachronous boundaries, the stratigraphical separation between Moll's Gap and the Horses Glen Centre is assessed using bedding orientation in the fold hinge, rather than lithostratigraphical thicknesses measured from fold limbs. In the Moll's Gap area the Mangerton Anticline plunges gently WSW (28–246°, Wegmann 1993) to a plunge depression [V800750] located north of Blackwater (Husain 1957, chapter 3; Capewell 1975, p. 168; Fig. 2), which effectively delimits the upright Kilcrohane Anticline. The Mangerton Anticline hinge zone plunges 4–074° in the region WSW of the Horses Glen Centre (Wegmann 1993, p. 39), and plunges consistently ENE in association with the entire Lough Guitane Complex (Avison 1984*b*) and further to the east (Williams 1993). Bedding data from the hinge area and north limb of the anticline compute a fold axis orientation of 1–075° (Wegmann 1993). The location of the axial culmination, which must be located ENE of Moll's Gap, is currently unknown in detail, although preliminary observations suggest that it occurs relatively close to Moll's Gap. Accounting for the topographical difference, and using 1 and 4° ENE fold plunge values, Moll's Gap Quarry is *c.* 430 and 1170 m respectively below the sample horizon in the Horses Glen Centre. This analysis does not account for displacement on faults that intersect the fold axial surface trace. Recent compilation mapping of the region (Pracht 1997) shows at least five cross-faults that potentially affect the fold hinge between Moll's Gap and the Horses Glen Centre. Although displacement data on specific faults are lacking, in general in the east Iveragh and Derrynasaggart Mountains region, dip-slip on cross faults tends to oppose fold axial plunge direction, maintaining structural–stratigraphical elevation along-strike (Walsh 1968, pp. 19–21; Avison 1982, p. 62; Williams 1993, p. 75). A thrust fault obliquely cross-cutting the fold hinge, affecting part of the Horses Glen Lower Tuffs and the Devil's Punch Bowl rhyolite (Avison 1984*b*, fig. 3) would have the opposite effect described for the cross-faults, but detailed data on its throw are not available. The relationship between the fold axial plunge and the cross-faults suggests that the stratigraphical difference between the isotopically and chronostratigraphically dated levels would be minimized, and of the order of $10–10^2$ m.

Keel Tuff Bed

Two major structures affect the isotopically and biostratigraphically dated levels in west and NW Iveragh, the south-dipping Coomnacronia Fault

(Fig. 1) and the Portmagee Anticline located in its hanging wall (Capewell 1975; Fig. 2). Capewell (1975) identified three conformable units (the Valentia Slate, St Finan's Sandstone and Ballinskelligs Sandstone Formations; Fig. 3) in a subtle fine-grained succession cropping out on the SSE limb of the Portmagee Anticline. Structural observations indicate a simple fold limb, lacking significant strike-parallel faults, affected by isolated mesoscopic fold pairs. Russell (1978, 1984) modified slightly the formation boundaries, and identified the Keel Tuff Bed and seven fish beds within this succession (Fig. 5a). This stratigraphy has been generally followed by subsequent workers (e.g. Clayton *et al.* 1980; van Veen & van der Zwan 1980; Gardiner & MacCarthy 1981; Higgs & Russell 1981; Williams *et al.* 1989, 1997), with the exception of Graham *et al.* (1992), who divided the succession into two 'stages', the boundary of which did not coincide with the established formational units. Graham (1983, p. 477) regarded the entire west Iveragh succession as a 'fine-grained fluvial facies' with an overall lithological similarity. Russell (1984), however, quantified the differences between Capewell's (1975) formations by variations in four metre-scale facies associations (Fig. 5a; referred to as 'facies' by Russell (1984)): (1) sandstone-body; (2) rippled and laminated (sandstone and siltstone); (3) sand-laminated siltstone; (4) bioturbated facies associations. This stratigraphy is modified here by the inclusion of the Caha Mountain Formation, replacing the lowest 400 m of Russell's (1984) Ballinskelligs Sandstone Formation. The base of the latter was considered highly gradational within this interval (Russell 1984), which lacks significant reddish purple coarser-grained sandstone-bodies that characterize the higher part of the formation (Capewell 1975), and is notable for markedly red-coloured facies and preferentially developed calcrete horizons compared with the section as a whole. The Caha Mountain Formation is also recognized in this stratigraphical position in the Beara peninsula–west central Cork region, where it comprises red–purple fine-grained sandstones and siltstones (Williams *et al.* 1989). The formation shows marked northward thinning across the basin mainly in relation to the Chloritic Sandstone Formation and its equivalents.

The 5.1 m thick Keel Tuff Bed [V38556845] occurs within the Valentia Slate Formation 766 m below the base of the St Finan's Sandstone Formation, which is defined by the first appearance of significant quantities of cross-stratified and parallel-laminated, grey to grey–green (fluvial channel) sandstone-bodies in the section (Russell 1984; Fig. 5a). The Valentia Slate Formation coarsens upwards because of the increasing incidence of the parallel- and cross-laminated very fine to fine-grained sandstone and coarse siltstone facies association, in a succession otherwise dominated volumetrically by a variably bioturbated, purple and grey–purple coarse siltstone and interbedded fine sandstone association (Russell 1984; Fig. 5a). The base of the Keel Tuff non-erosively overlies a desiccated, moderately bioturbated and very weakly calcretized (0.02 m thick) green siltstone bed. This overlies a 0.5 cm thick medium–coarse grade quartz–feldspar crystal tuff layer, an apparent precursor to the Keel airfall deposit. These beds are part of a 1.2 m thick sand-laminated siltstone facies association (*sensu* Russell 1984) immediately underlying the tuff bed, and follow a thicker interval (>2.7 m) of rippled and laminated sandstones (Fig. 5b). The Keel Tuff shows about eight moderately prominent internal bedding planes spaced at 0.3–0.7 m intervals, across which there is no obvious textural or structural change, excepting the boundary of the lowest 0.61 m thick (coarser-grained) unit. The petrography of this division, and the remainder of the tuff bed, has been described in detail by Graham *et al.* (1995, p. 16). The upper 0.6 m of the Keel Tuff is finely (colour) banded (0.3–0.5 cm) and is terminated by an irregular–sharp, though flat, top surface, suggesting minor erosion of the fall. Approximately 28 kg of fresh rock were collected from the lower to mid-part of the Keel Tuff for isotopic dating, avoiding a set of through-going curved, strike-parallel fractures affecting the middle portion of the bed.

Within the Valentia Slate Formation the Keel Tuff Bed is bracketed up-section by three closely spaced fish beds (Nos 3–5), which did not yield biostratigraphic evidence of their age (Russell 1978), and below by fish beds 1 and 2 (Fig. 5a). Of the two fish beds that reportedly yield diagnostic maximum ages, fish bed 1 (of late Givetian–early Frasnian, Russell 1978) occurs 555 m below the Keel Tuff within the Valentia Slate Formation, and fish bed 6 (of Famennian age, Russell 1978) occurs within the lower part of the St Finan's Sandstone Formation (Fig. 5a). Fish bed 1 is stratigraphically close to a 0.3 m thick silicic tuff, reportedly similar (Russell 1984) to the Keel Tuff, which we term here the Puffin Sound Tuff Bed. Continuing efforts are being made to sample this tuff for isotopic dating.

Reenagaveen microflora

A detailed description of the location of the productive sample is given below. The section is

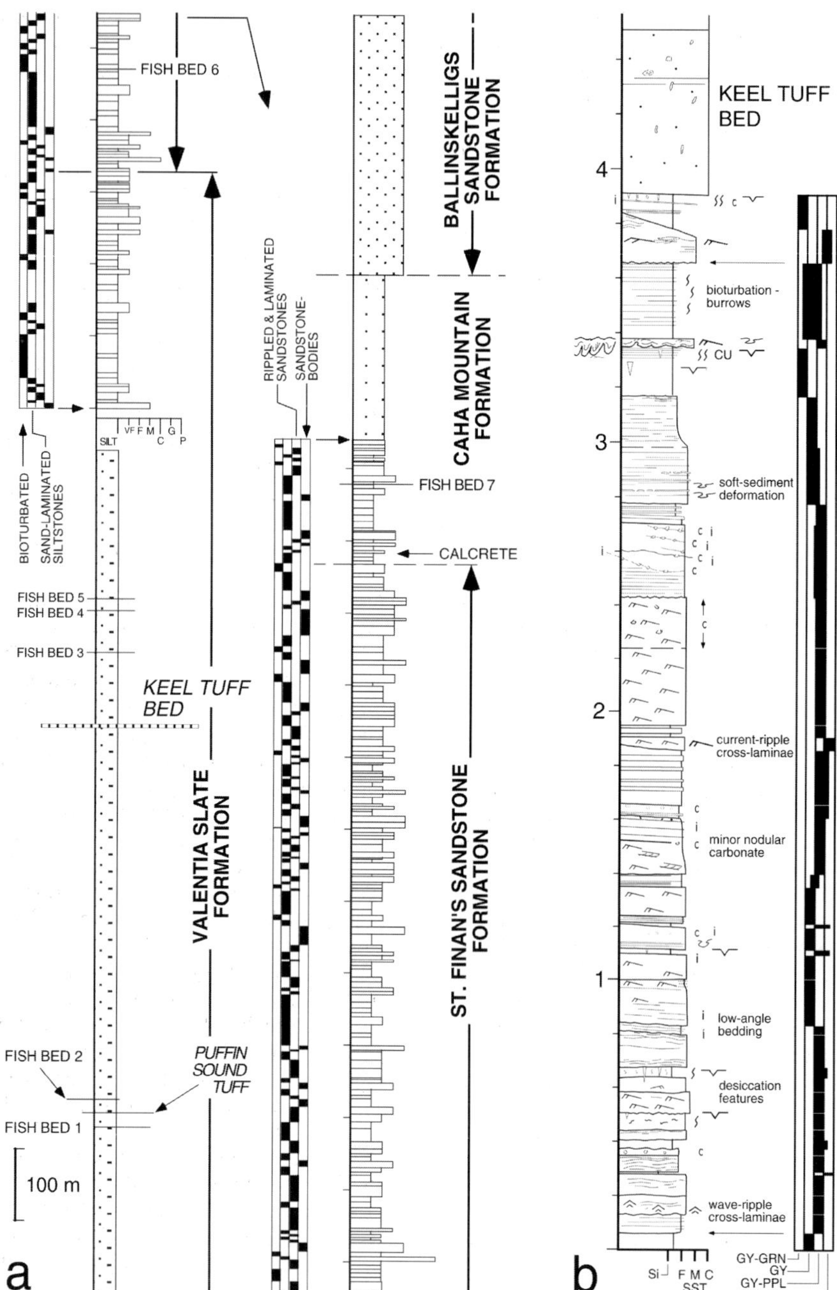

Fig. 5. (**a**) Stratigraphical context of the Keel Tuff Bed within the ORS type section of west Iveragh. Facies association and grain-size data for part of the section are based on Russell (1984). Positions of fish and tuff beds are based on Russell (1984) and Horne & Russell (1980, unpublished field guide). The standard lithostratigraphy is modified by the recognition of the Caha Mountain Formation as the lowest 400 m of the Ballinskelligs Sandstone Formation of Capewell (1975) and Russell (1978), in accordance with facies association data from the section (Russell 1984) and the regional lithostratigraphic position of the Caha Mountain Formation (Williams *et al.* 1989). (**b**) Bed-scale facies data from the Valentia Slate Formation immediately beneath the Keel Tuff Bed (scale in metres). Key to colour column: GY–GRN, grey–green; GY, grey (N5); GY–PPL, grey–purple; PPL–RD, purple–red. i, inclined bedding surfaces.

located on the NE coast of Valentia Island in the immediate hanging wall of the Coomnacronia Fault (Fig. 6a), the trace of which Capewell (1975, fig. 3) inferred west of Reenagaveen Point. In this region, Valentia Slate Formation rocks occupy both the footwall and hanging-wall blocks. The Valentia Harbour–Beginish Island intrusive dolerites–basalts and volcanic breccias (Capewell 1975; Graham *et al.* 1995) are restricted to the Valentia Slate outcrop in the fault footwall (Figs 2 and 6a). The structure of the Reenagaveen coastal section has been mapped by Stössel (1993), and is compatible with the north limb of the regional Portmagee Anticline.

Enagh Tuff Bed

The Enagh Tuff (Russell 1984) crops out in the footwall block of the Coomnacronia Fault in northwest Iveragh (Fig. 6a), where it is 11.6 m thick in the section dated by Williams *et al.* (1997). Its stratigraphical context is summarized in Figs 3 and 6b, following the reassessment given by Williams *et al.* (1997, p. 190), who retained the nomenclature of Capewell (1975) but redefined the position of certain formational boundaries (see Russell (1984) and Graham *et al.* (1995) for differing schemes). At Enagh Point [V41858005] the tuff bed marks the base of the St Finan's Sandstone Formation (Fig. 6b and c), which, in the Coomnacronia Fault footwall, is sandstone-body rich, coarser grained and *c.* 190 m thick compared with west Iveragh (Fig. 5a). Traced inland to the ENE, the base of the St Finan's Sandstone Formation is slightly diachronous, with a younging component in this direction (Stössel 1993). The Enagh Tuff is *c.* 350 m above the horizon of the Valentia Island tetrapod trackway (Stössel 1995) within the Valentia Slate Formation (Williams *et al.* 1997).

Geochemistry

Detailed geochemical analyses (Table 1) were undertaken on the volcanic rocks that were considered for geochronology (the Killeen and Horses Glen Lower Tuffs, Beenreagh basalts, and the Keel and Enagh Tuff Beds) and those that are associated with isotopically and biostratigraphically dated horizons (Reenadrolaun Tuff and the Bealtra Volcanic Breccia Bed), in order to establish stratigraphical correlations (Figs 3 and 6b). The data also allow the evaluation of the tectono-magmatic setting of the Iveragh suite using discrimination diagrams. All samples used for geochronology, as well as associated volcanic and intrusive rocks, are subalkaline from their ($Na_2O + K_2O$)–SiO_2 distributions (Table 1) and low Nb/Y ratios in the plot of Winchester & Floyd (1977; Fig. 7a). On this and a suite of petrological plots using immobile trace elements (e.g. Zr–TiO_2) the rocks range in composition from rhyolite (the Keel and Enagh tuffs) and rhyodacite–dacite (Killeen, Horses Glen L. Tuffs) to basalt (Beenreagh).

On the discrimination diagram applicable to basaltic rocks (Pearce & Norry 1979; Fig. 7b) the tuff and andesite components of the Bealtra Volcanic Breccia show clear within-plate basalt (WPB) affinities, whereas its basalt clasts plot in both the WPB and mid-ocean ridge basalt (MORB) fields. The Reenadrolaun Tuff (Fig. 6b), considered to be a distal equivalent of the Bealtra Breccia (Williams *et al.* 1997), has a clear WPB signature. The sub-volcanic dolerite at Bealtra South, structurally beneath the volcanic breccia, and the Devonian Beginish Island dolerites (Stössel 1993; Fig. 6a) reveal similar WPB affinities (see also Graham *et al.* (1995)). Further discriminant plots (not shown here), including TiO_2 against Y/Nb (Floyd & Winchester 1975), show that components of the Bealtra Breccia, the Beenreagh basalts and the basic–intermediate Reenadrolaun Tuff have mixed continental tholeiite and MORB affinities. Rare earth element (REE) multi-element analysis of the Beenreagh basalts from the Chloritic Sandstone Formation indicate a very similar profile to E-type (anomalous) MORB (<1.4 times enriched in REE), confirmed by the ternary Hf/3–Th–Ta plot of Wood (1980) that indicates an E-type MORB and within-plate tholeiite affinity. The rhyodacitic Lough Guitane samples also plot in the WPB field (Fig. 7b), whereas the more differentiated Keel and Enagh Tuffs (Graham *et al.* 1995) plot outside all fields.

The suite of discrimination diagrams for granitic composition rocks of Pearce *et al.* (1984) is used for the felsic volcanic rocks (Fig. 8a–d). The bivariate diagrams using Ta/Yb and Nb/Y (Fig. 8a and b) show that the silicic volcanic rocks plot in the anomalous ocean ridge field, with an overlap into the volcanic-arc field. The diagrams utilizing Rb (Fig. 8c and d) more clearly show a within-plate affinity. The tectono-magmatic setting for the rocks studied is supported by data from the Bennaunmore Volcanic Centre (Gräfe 1993), which reinforce that the mean positions plot close to the within-plate to volcanic arc boundary (Fig. 8c), and to the anomalous ocean ridge–ocean ridge junction (Fig. 8a).

A multi-element diagram for the target suite (Fig. 9a), using principally large ion lithophile

a
1 km
DINGLE BAY
Cooncrome Harbour
Doulus Head
Enagh Tuff
Enagh Point
Bealtra
Reenadrolaun Tuff & Conglomerate
BEGINISH ISLAND
Laght Point
Cahersiveen
Reenadrolaun Point
Coomnacronia Fault
Reenagaveen microflora
Knight's Town
VALENTIA ISLAND
N
Ballinskelligs Sandstone Fm
St. Finan's Sandstone Fm
Valentia Slate Fm
Tetrapod Trackway
Doulus Conglomerate Mbr
Enagh Tuff Bed
Bealtra Volcanic Breccia Bed
Doleritic Intrusions
b
300 m
BALLINSKELLIGS SANDSTONE FORMATION
CANGLASS POINT / COONANNA HARBOUR - LAGHT POINT
DOULUS CONGLOMERATE MEMBER
ST. FINAN'S SANDSTONE FORMATION
REENADROLAUN POINT VALENTIA ISLAND
ENAGH TUFF BED
Reenadrolaun Conglomerate
Reenadrolaun Tuff
VALENTIA
BEALTRA VOLCANIC BRECCIA
SLATE
FORMATION

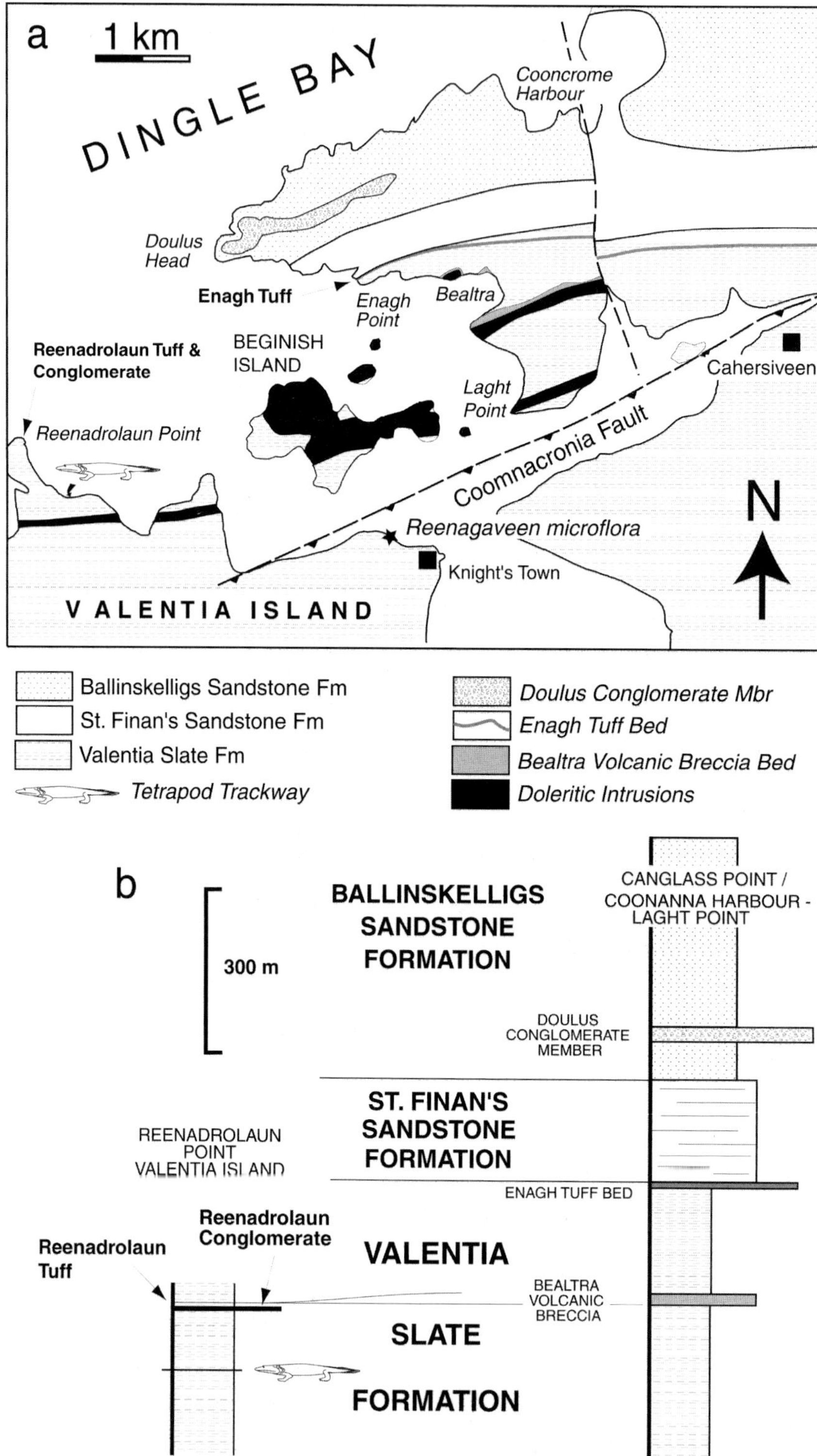

Fig. 6(a & b).

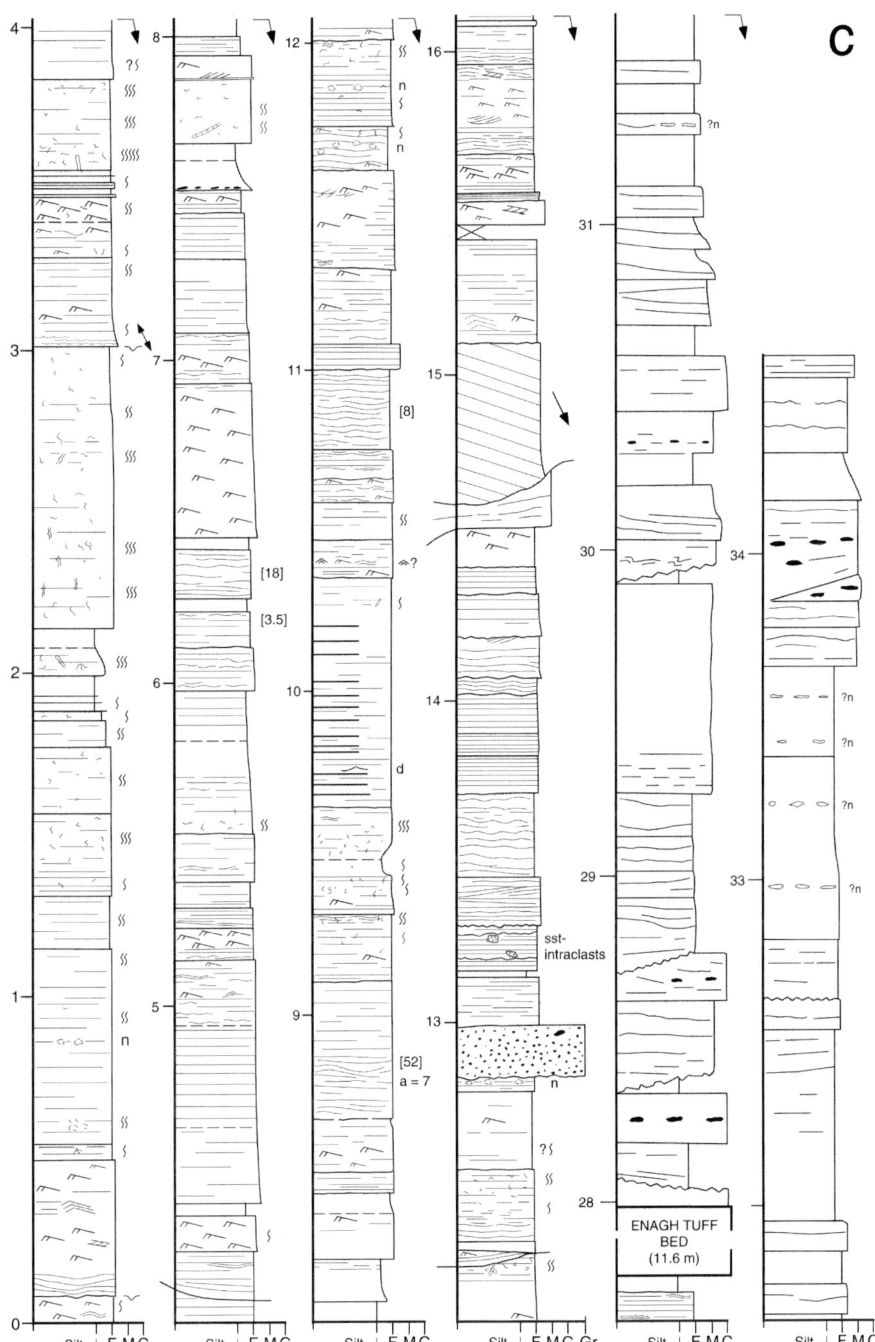

Fig. 6(c).

Fig. 6. (**a**) Map and (**b**) stratigraphic correlation of the ORS in the Valentia Harbour region, showing the locations of the Enagh Tuff Bed, Bealtra Volcanic Breccia bed, Reenadrolaun Tuff, Reenagaveen miospore assemblage and the Valentia Island tetrapod ichnofauna. (**c**) Detailed facies data from the Valentia Slate and St Finan's Sandstone Formations immediately beneath and above the Enagh Tuff Bed, respectively. Key to symbols shown in Fig. 5b. Figures in brackets are wavelengths of in-phase sinuous to wavy lamination in centimetres; amplitudes of these structures are typically 1–3 cm.

Table 1. *Whole-rock geochemical data* (major and trace elements) from sampled lithologies used for dating and correlation*

Sample	Keel	Enagh	Killeen	HGLT	Beenr.	RDPT
SiO_2	72.93	79.25	75.87	78.39	45.57	54.95
Al_2O_3	11.48	11.06	11.13	8.54	15.22	16.81
Fe_2O_3	3.19	2.28	3.66	2.99	11.49	11.43
MnO	0.19	0.02	0.06	0.09	0.2	0.24
MgO	1.94	0.16	0.91	0.67	9.3	5.45
CaO	2.44	0.26	0.66	1.85	9.1	0.5
Na_2O	0.38	5.58	1.78	1.25	3.01	1.19
K_2O	2.64	0.56	3.14	3.21	0.02	2.53
TiO_2	0.24	0.26	0.31	0.31	1.43	1.77
P_2O_5	0.03	0	0.05	0.08	0.24	0.27
Loss	4.13	0.39	2.09	2.51	4	4.62
Total	99.59	99.81	99.65	99.89	99.58	99.76
As	8.77	6.65	1.47	4.7	39.4	35.8
Ba	553	185	609	636	52	431
Be	4.61	5.08	4.19	1.85	0.71	2.1
Bi	0.12	0.24	0.44	0.13	0.01	0.45
Cd	0.31	0.15	0.14	0.45	0.5	0.09
Ce	147.6	204.4	90.41	72.28	19.68	65.96
Co	6.43	10.9	3.43	4.28	41	30.3
Cr	11.1	10	26.3	239	425	246
Cs	4.31	1.45	5.44	2.67	0.376	6.36
Cu	3.4	51.5	2.5	4.9	77.1	12.8
Dy	17.81	16.04	8.75	5.84	4.97	7.21
Er	9.13	7.79	4.25	2.78	2.81	3.76
Eu	2.11	2.32	1.56	1.12	1.26	1.88
Ga	22.1	15.2	21.5	14.7	16.3	22.3
Gd	17.49	15.53	7.93	6.02	4.23	6.84
Ge	1.46	1.19	1.41	1.39	1	1.91
Hf	10.1	9.43	7.64	4.81	2.78	6.44
Ho	3.78	3.61	1.91	1.26	1.21	1.63
In	0.07	0.07	0.1	0.06	0.04	0.06
La	104.8	91.38	43	34.98	7.84	34.52
Lu	1.27	1.25	0.67	0.43	0.45	0.57
Mo	0.12	0.99	0.25	1.55	0.34	0.39
Nb	28.21	28.09	15.68	12.12	5.54	13.5
Nd	95.6	88.48	39.92	33.09	13.32	32.84
Ni	8.02	7.14	8.28	20.1	163	84.8
Pb	9.25	10.7	18.4	5.53	1.8	23.3
Pr	23.05	23.69	10.38	8.1	2.79	8.2
Rb	117	31.28	139.6	113.6	0.711	111.8
Sb	0.4	0.59	0.47	0.64	0.28	2.54
Sm	20.31	17.99	8.52	6.54	3.93	7.3
Sn	7.62	6.17	4.92	3.11	0.69	2.53
Sr	72.4	108	97.2	109	195	55
Ta	2.51	2.66	1.4	0.953	0.436	1.08
Tb	2.85	2.33	1.31	0.91	0.72	1.06
Th	19.44	18.85	14.41	8.76	0.53	6.61
Tm	1.37	1.1	0.62	0.41	0.39	0.54
U	4.86	4.72	3.66	2.39	0.15	2.3
V	17.3	7.6	19.9	30.2	230	206
W	0.87	174	2.82	2.67	0.14	2.03
Y	88.3	76.9	44.5	30.7	28.9	40.9
Yb	9.46	8.49	4.36	2.97	2.98	3.75
Zn	83.2	13	86.9	55.8	80.2	178
Zr	317	314	261	183	110	259

*Major elements (in %) were analysed by inductively coupled plasma-atomic emission spectrometry (ICP-AES), and trace elements (in ppm) by ICP mass spectrometry, at the Service d'Analyses des Roches et des Minéraux, CRPG-CNRS, Vandoeuvre. HGLT, Horses Glen Lower Tuffs; Beenr., Beenreagh basalts; RDPT, Reenadrolaun Tuff.

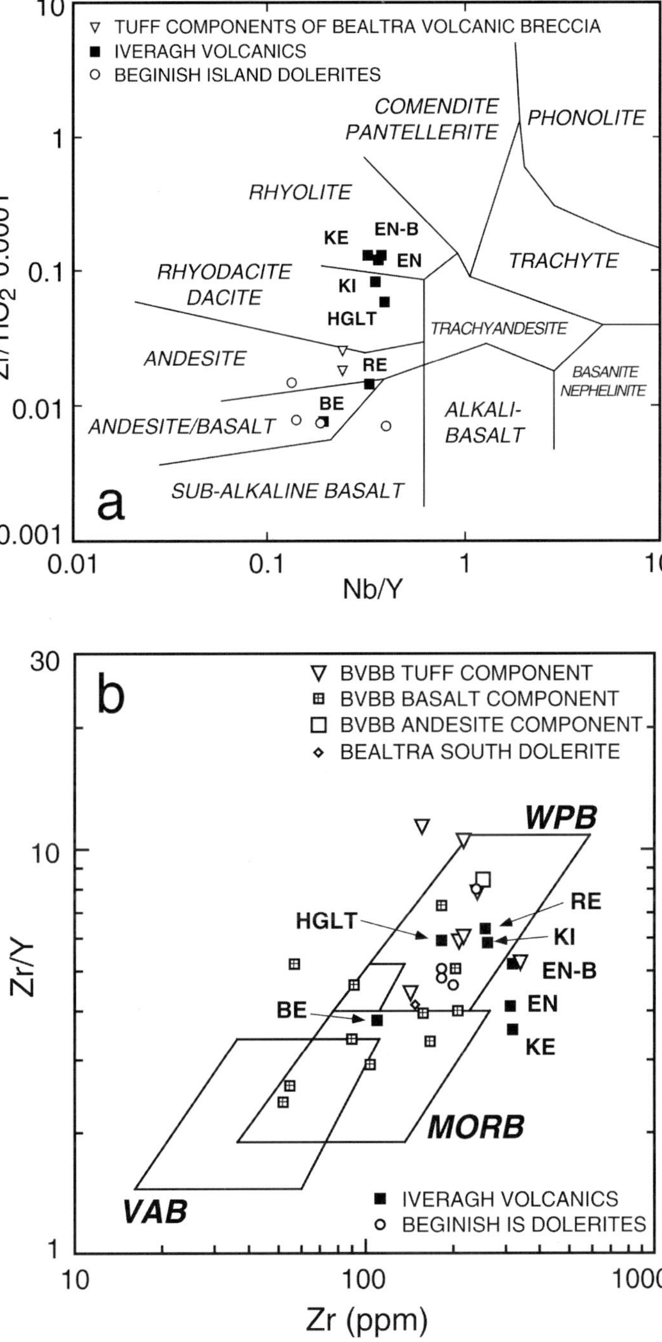

Fig. 7. Plots showing the petrology and tectonic discrimination of sampled and related volcanic rocks. (**a**) Zr/TiO_2 against Nb/Y (Winchester & Floyd 1977, fig. 6), (**b**) Zr/Y against Zr (Pearce & Norry 1979). In (**b**) basalt, tuff and andesite clasts, components of the Bealtra Volcanic Breccia, are plotted along with a closely related sub-volcanic dolerite from Bealtra south (data from Graham *et al.* (1995)), and intrusive dolerites from Beginish Island (data from Stössel (1993)). EN, Enagh Tuff (from Enagh Point); EN-B, Enagh Tuff (from Ballycarbery); KE, Keel Tuff; KI, Killeen Tuffs; HGLT, Horses Glen Lower Tuffs; RE, Reenadrolaun Tuff; BE, Beenreagh basalts; WPB, within-plate basalts; MORB, mid-ocean ridge basalts; VAB, volcanic-arc basalts.

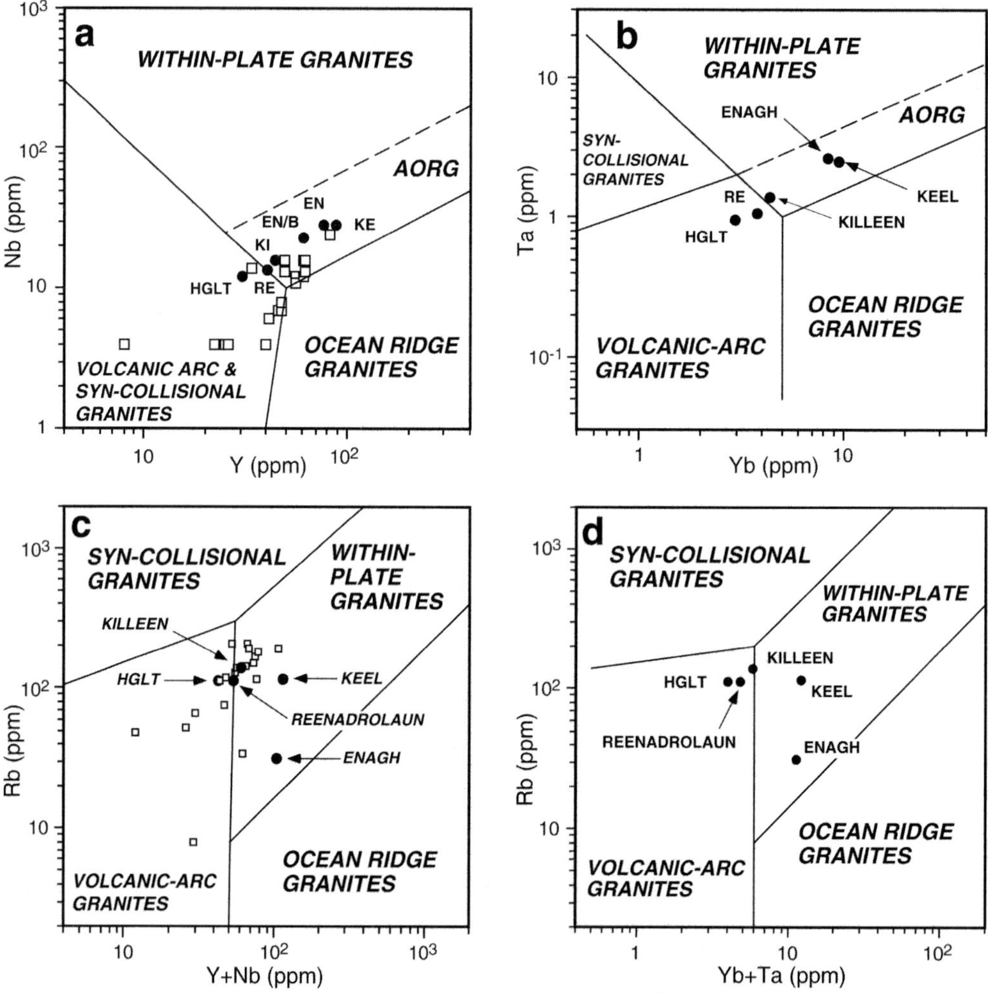

Fig. 8. Tectonic discrimination plots for granitic composition and felsic rocks (Pearce *et al.* 1984). (**a**) Nb against Y, (**b**) Ta against Yb, (**c**) Rb against Y + Nb, (**d**) Rb against Yb + Ta. Open square symbols in (**a**) and (**c**) are silicic volcanic rock data (Gräfe 1993) largely from the Bennaunmore Centre of the Lough Guitane Volcanic Complex, but including five data points from other centres. AORG, anomalous ocean ridge granites.

and high field strength elements, shows close tracking of the Keel and Enagh Tuffs, replicating the pattern found by Graham *et al.* (1995, fig. 9). However, for this range of elements there is no distinction between the Keel–Enagh, Lough Guitane and Reenadrolaun Tuffs. More useful in this respect is a plot using REE normalized against N-type MORB (Fig. 9b), which reveals the very close similarity between the Keel and Enagh Tuffs plus, at the same time, clear differences from the Lough Guitane and Reenadrolaun samples. The very similar patterns exhibited by the Keel and Enagh Tuffs strongly suggest a common genetic origin, from the same magmatic system.

Although sharing a common WPB to MORB signature (Fig. 7b), the proposed direct linkage between the proximal Bealtra Volcanic Breccia and the distal Reenadrolaun Tuff (Fig. 6b) is more stringently tested using MORB-normalized multi-element diagrams (Fig. 10). Both of the principal components of the Bealtra Breccia show close correspondence to the Reenadrolaun Tuff in terms of immobile elements. The mobile elements are significantly depleted with respect to the Reenadrolaun Tuff (although data for Rb are limited), with the exception of Sr, which is modestly enriched in the basalt clast component (Fig. 10a), and more significantly in the tuff component (Fig. 10b).

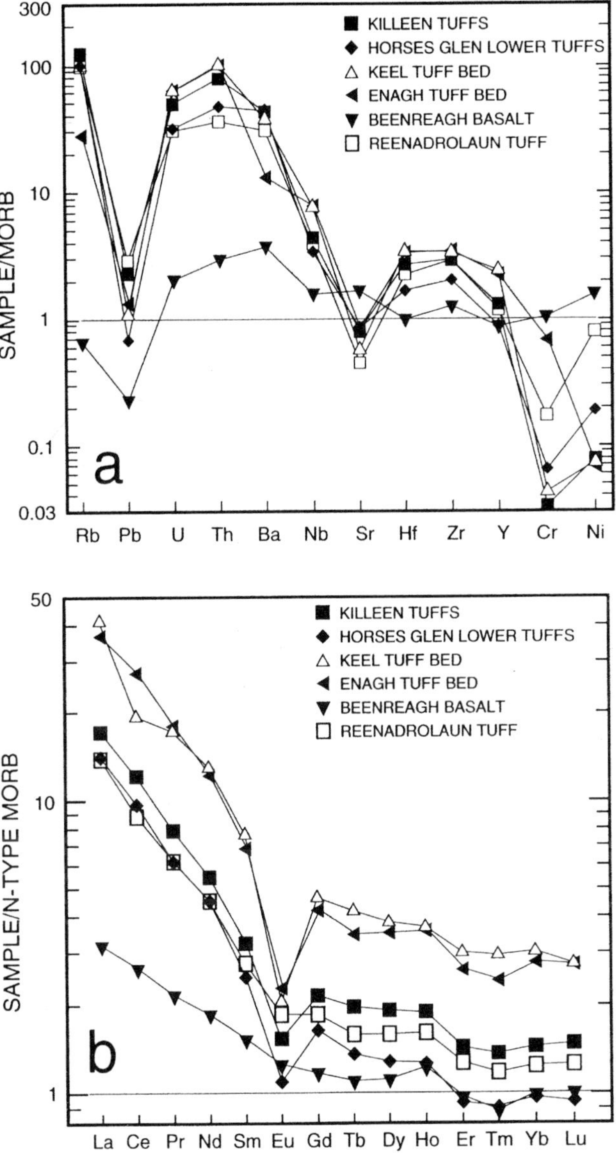

Fig. 9. (**a**) Multi-element diagram (large ion lithophile, high field strength and other trace elements), normalized against a MORB standard, for the isotopically dated and other studied volcanic rocks. (**b**) Rare earth element variation diagram normalized against the N-type MORB standard, for the sample suite in (**a**). Plots created with Minpet software, using its library of normalizing values.

Overall, clast components are relatively depleted in immobile elements (<9 ×), best illustrated by a rock-normalized plot (Fig. 10b, inset). Maximum enrichment in immobile elements is *c*. 2.75 times. Data for REE (La, Ce, Nd) show very narrow spreads and mean correspondence to the Reenadrolaun Tuff (Fig. 10). Comparisons undertaken with the Enagh Tuff, the only other known eruptive rock in the Valentia Harbour area, show poor correspondence. These factors suggest a common origin for the Bealtra Breccia and Reenadrolaun Tuff, thus linking the mainland stratigraphy with that of NE Valentia Island.

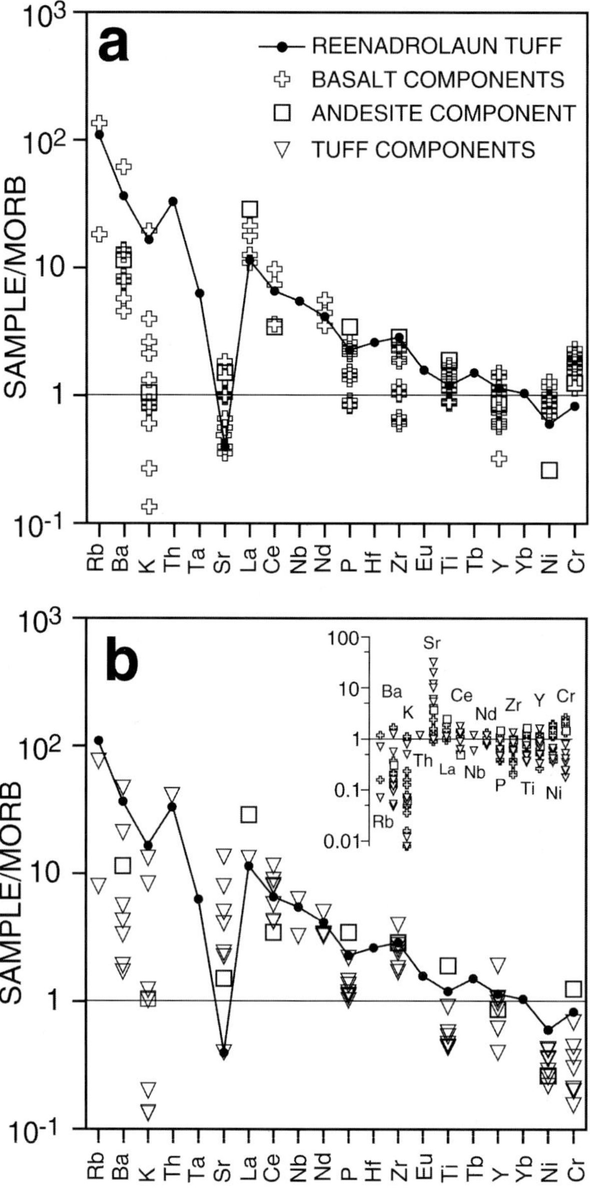

Fig. 10. Multi-element diagrams, normalized against the MORB standard of Bevins *et al.* (1984), that compare (**a**) the basalt clast component and (**b**) the tuff clast component of the Bealtra Volcanic Breccia with the Reenadrolaun Tuff. Determinations from an andesite clast from the Bealtra Volcanic Breccia are plotted in both diagrams. Inset in (**b**) shows a rock-normalized plot, where data from the Reenadrolaun Tuff are used as normalizing values. The Bealtra Breccia data are from Stössel (1993).

Geochronology

Primary igneous zircon crystals, commonly recovered from silicic volcanic tuffs, are generally considered to have crystallized rapidly, immediately before eruption or during cooling following expulsion from the magma chamber (Tucker & McKerrow 1995). The high rate of cooling associated with explosive volcanism ensures that the minerals fall below the closure temperature for the U–Th–Pb isotopic system (>800° C) instantaneously on a geological time scale. Airfall and other explosive tuffs containing magmatic zircons can, therefore, provide accurate chronometric data on the immediate sedimentary successions in which they are incorporated. This is the basis of the following analysis and numerous similar studies (e.g. Roden *et al.* 1990; Tucker *et al.* 1990, 1998; Tucker & McKerrow 1995).

Analytical methods

The zircon concentrates from all three sampled tuff horizons (Killeen Tuffs, Horses Glen Lower Tuffs and Keel Tuff Bed) were separated in a way previously applied to the Enagh Tuff Bed (Williams *et al.* 1997). All of the tuffs are characterized by a low content of accessory zircon, in direct correlation with the Zr content (183–317 ppm) of the host rocks (Table 1). The overwhelming proportion of zircon crystals are small in size and were separated from the fine fraction of the rock matrix (70–120 μm). The single zircon grains that were hand-picked for U–Pb analysis are (1) typical of the dominant igneous population in all tuff samples, and (2) represent the best-quality grains on the basis of transparency, absence of chemical corrosion and minimal amount of solid inclusions. Unfortunately, no one magmatic grain lacking either gaseous or solid inclusions was isolated. The inclusion-rich nature of the crystals, interpreted to be the result of capture during rapid crystallization of volcanic material, is also a feature of zircons from the Enagh Tuff (Williams *et al.* 1997). Because of the abundance of inclusions, forming up to 25% of some grains, individual zircons contain a significant component of non-radiogenic Pb (see below). This effect cannot be reduced by air-abrasion of outer surfaces of the grains. However, with the aim of improving the homogeneity of analysed grains and to focus on pure zircon material, some grains were mechanically fragmented to avoid the most contaminated parts. In addition, before standard chemical treatment, all zircons and zircon fragments were washed ultrasonically in cold 1 : 1 HCl for 1 h to eliminate adherent mineral phases, as well as exposed inclusions.

U–Pb analyses were performed at the IGMR ETH Zürich using chemical and mass-spectrometry procedures fully described by Meier and Oberli in Wiedenbeck *et al.* (1995). During the period of analysis, measured bomb blanks ranged between 5 and 6.6 pg Pb. Analytical and age uncertainties are quoted at 95% confidence limits. Corrections for common Pb content are based on the Stacey & Kramers (1975) model, and decay constants are taken from Steiger & Jäger (1977). The $^{207}Pb/^{206}Pb$ mean ages that we quote are weighted by 1/variance of individual determinations, and uncertainties quoted are external errors.

Zircon morphology

Characteristic zircon grains from the Killeen, Horses Glen and Keel Tuffs are shown in Fig. 11. The accessory magmatic zircons from the Killeen and Keel Tuffs demonstrate very similar features such as simple short-prismatic {100} shapes in combination with {101} pyramids. These zircon grains are transparent and colourless and show oscillatory growth zoning (Fig. 11, 1a and 2a). In the Killeen Tuffs a small number (*c.* 5%) of non-magmatic, presumably incorporated, zircons was found. These grains exhibit two distinct morphological types: (1) complex-faceted, short-prismatic, honey yellow crystals (Fig. 11, 2d) and (2) spheroidal, brownish, semi-transparent grains of distinctly detrital origin (Fig. 11, 2c). The characteristics of both the magmatic and non-magmatic zircons described above are similar to those described from the Enagh Tuff Bed (Williams *et al.* 1997). The zircon population from the Horses Glen Lower Tuffs, in contrast, mainly consists (*c.* 80%) of complex-faceted, euhedral, short- to long-prismatic transparent grains (Fig. 11, 3b) of variable colour. The central interior part of these grains sometimes show slightly turbid domains, surrounded by bubbles, that may indicate the presence of inherited zircon material. Approximately 15% of the bulk population consists of very small, needle-like colourless grains of the same crystallographic type ({100} + {101}) as the magmatic zircons from the Killeen and Keel Tuffs (Fig. 11, 3a). The remaining 5% is represented by ellipsoidal, weakly transparent brown grains with rough scars and corrosional cavities on the grains' surfaces (Fig. 11, 3c). Such grains are thought to be detrital.

U–Pb results

Isotope ratios, elemental concentrations and atomic ratios are presented in Table 2. Analytical data points are shown in Figs 12 and 13.

Killeen Tuffs. To date the age of eruption–crystallization, three samples of single zircon crystals and zircon fragments of magmatic origin were analysed (Nos 2–4, Table 2). All of them are variably discordant (from 87 to 99% of concordancy) but show the same $^{207}Pb/^{206}Pb$ age indicating their attribution to a single generation. The analytical data points combine to give a

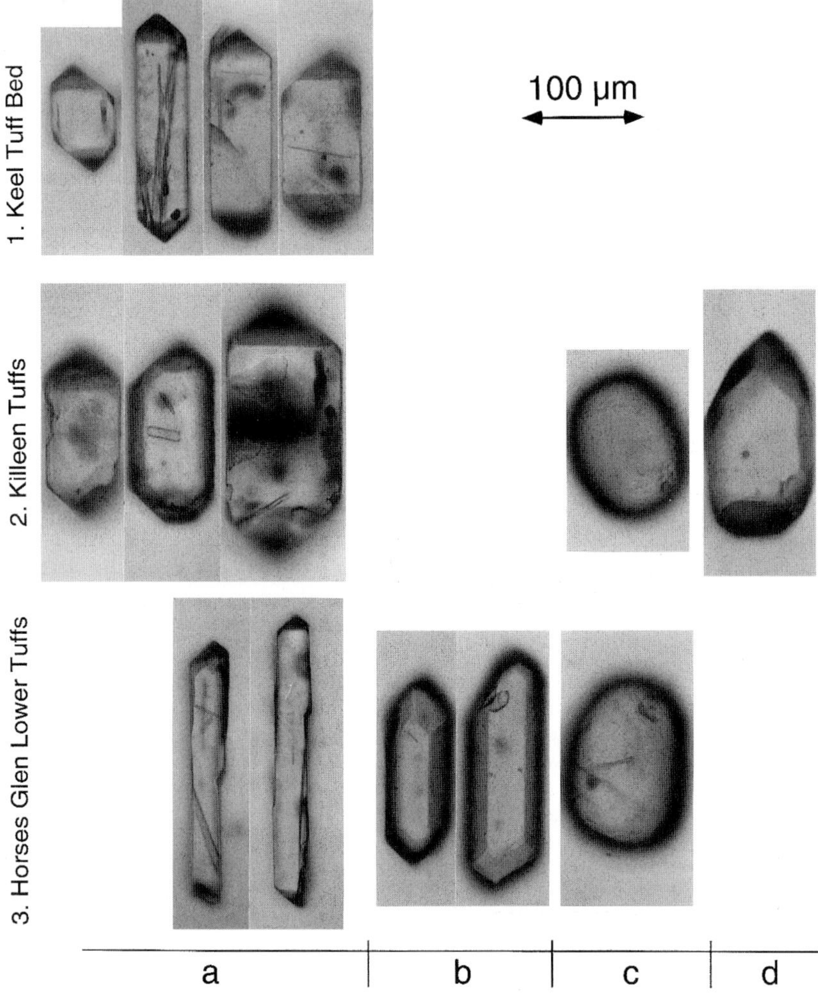

Fig. 11. Photographs of representative zircon types from the Keel, Killeen and Horses Glen Lower tuffs.

discordia line with an upper intercept age of 385.1 +30/−14 Ma (Fig. 12). The lower intercept (*c.* 5 Ma) is not well defined but clearly indicates only one weak alteration of the zircons' U–Pb system in modern time. However, our best estimate for the age of the magmatic event is the weighted mean $^{207}Pb/^{206}Pb$ age of 384.5 ± 1.0 Ma (95% confidence limits (c.l.) external). One complex-faceted single crystal of honey yellow colour (No. 1, Table 2, not shown in Fig. 12) shows clear isotopic differences compared with the other magmatic zircons from this sample, and gave a $^{207}Pb/^{206}Pb$ age of *c.* 1000 Ma. This suggests a (minimum) Late Proterozoic age for hidden basement rocks, material of which was incorporated during magma ascent.

Horses Glen Lower Tuffs. Four analyses were undertaken on single complex-faceted, euhedral, prismatic transparent grains representing the prevailing population of magmatic zircon grains from these deposits (Nos 5–8, Table 2). These relatively large grains (6.7–12.0 μg each) contain small proportions of optically imperceptible inherited material. This effect causes analytical points to define a discordia line (mean square weighted deviation (MSWD) = 0.1) giving a lower intercept age of 378.3 +18/−80 Ma and an upper intercept age of 619 Ma (Fig. 12). The latter value is closely compatible with the age of the main incorporated material obtained for the Enagh Tuff Bed (604 ± 55 Ma, Williams *et al.* 1997) and probably linked to similar zircon seed

Table 2. *U–Pb isotopic data from zircons extracted from the three tuff horizons*

No.	Laboratory typology and colour index[a]	Weight (μg)	Concentration (ppm) U	Pb rad.	Pb com.	Th/U[b]	$^{206}Pb/^{204}Pb$[c]	Isotopic ratio corrected for blank and common Pb[d] $^{208}Pb/^{206}Pb$	$^{207}Pb/^{206}Pb$	$^{207}Pb/^{235}U$	$^{206}Pb/^{238}U$	Rho[e]	Apparent age (Ma) $^{207}Pb/^{206}Pb$	$^{207}Pb/^{235}U$	$^{206}Pb/^{238}U$	DC[f]
Killeen Tuffs																
1	S,sp,y	4.3	231	38.3	0.1	0.32	29869	0.0989 ± 6	0.0733 ± 2	1.6628 ± 70	0.1646 ± 3	0.63	1021.20 ± 6.78	994.45 ± 2.66	982.37 ± 1.64	39.2
2	D,sp,c	4.9	681	39.6	0.3	0.73	8056	0.2267 ± 6	0.0543 ± 2	0.3955 ± 16	0.0528 ± 1	0.63	384.37 ± 7.38	338.35 ± 1.16	331.69 ± 0.54	86.5
3	D,sp,c,F	2.9	404	27.1	3.7	0.73	437	0.2282 ± 14	0.0543 ± 6	0.4554 ± 50	0.0608 ± 1	0.65	384.45 ± 22.54	381.04 ± 3.42	380.48 ± 0.48	99.0
4	D,sp,c	3.6	426	26.2	0.4	0.67	3908	0.2078 ± 10	0.0543 ± 2	0.4243 ± 32	0.0566 ± 1	0.52	385.24 ± 14.58	359.10 ± 2.24	355.07 ± 0.86	92.2
Horses Glen Lower Tuffs																
5	S,lp,c	12.0	346	24.6	0.8	0.73	1706	0.2280 ± 6	0.0552 ± 2	0.4908 ± 28	0.0645 ± 2	0.55	420.29 ± 10.72	405.45 ± 1.90	402.84 ± 1.00	93.8
6	S,sp,y	8.2	132	9.9	2.9	0.74	208	0.2328 ± 24	0.0557 ± 9	0.5157 ± 90	0.0671 ± 2	0.56	441.29 ± 36.34	422.30 ± 6.08	418.83 ± 1.04	90.2
7	S,lp,c	7.7	209	15.1	3.2	0.74	282	0.2324 ± 11	0.0552 ± 3	0.4951 ± 36	0.0650 ± 1	0.67	421.29 ± 14.12	408.38 ± 2.40	406.11 ± 0.58	93.1
8	S,sp,y	6.7	123	9.1	7.6	0.41	91	0.1295 ± 24	0.0568 ± 8	0.5662 ± 88	0.0723 ± 1	0.67	483.54 ± 32.02	455.55 ± 5.70	450.02 ± 0.96	84.0
9	D,lp,c,F	1.3	123	8.0	5.5	0.71	101	0.2205 ± 68	0.0544 ± 30	0.4460 ± 250	0.0594 ± 2	0.72	388.98 ± 120.45	374.44 ± 17.57	372.10 ± 1.42	95.6
10	D,lp,c,F*9	16.3	119	8.2	0.5	0.85	898	0.2660 ± 12	0.0542 ± 4	0.4519 ± 40	0.0605 ± 2	0.49	379.42 ± 17.83	378.63 ± 2.82	378.50 ± 1.06	100.0
Keel Tuff Bed																
11	D,sp,c,F	1.6	103	6.1	4.9	0.56	91	0.1740 ± 76	0.0543 ± 32	0.4167 ± 256	0.0577 ± 2	0.69	382.26 ± 132.02	353.66 ± 18.35	349.32 ± 1.52	91.4
12	D,sp,c,*2	2.1	112	6.7	2.2	0.56	197	0.1737 ± 50	0.0544 ± 22	0.4213 ± 172	0.0562 ± 2	0.69	385.87 ± 87.74	356.98 ± 12.34	352.55 ± 1.06	91.5
13	D,lp,c	3.7	105	6.0	1.6	0.59	239	0.1854 ± 32	0.0544 ± 14	0.4046 ± 108	0.0540 ± 1	0.64	385.76 ± 56.54	344.96 ± 7.77	338.93 ± 0.80	87.8
14	D,sp,c,F	2.6	64	3.8	3.1	0.56	91	0.1739 ± 76	0.0543 ± 32	0.4158 ± 254	0.0556 ± 2	0.65	381.56 ± 130.96	353.06 ± 18.18	348.74 ± 1.62	91.4

[a]D, complex-faceted grains with combination of {100} and {101}; S, complex-faceted grains with combination of {100}, {110} and {101}; sp, short prismatic grains; lp, long prismatic grains; F, zircon fragments; y, yellow; c, colourless; **n*, number of analysed small grains of the same morphology.

[b]Calculated from $^{208}Pb/^{206}Pb$, according to age of the samples.

[c]Corrected for mass fractionation, spike and blank contribution.

[d]Uncertainties (95% confidence level) refer to the last digits of corresponding ratios.

[e]Correlation coefficients of $^{207}Pb/^{235}U$ v. $^{206}Pb/^{238}U$ ratios.

[f]Degree of concordance (%) calculated from (100 × ($^{206}Pb/^{238}U$)age/$^{207}Pb/^{206}Pb$)age) for the magmatic zircons free of inheritance and from (100 × lower concordia intercept age/($^{206}Pb/^{238}U$)age) for the grains bearing an inherited component.

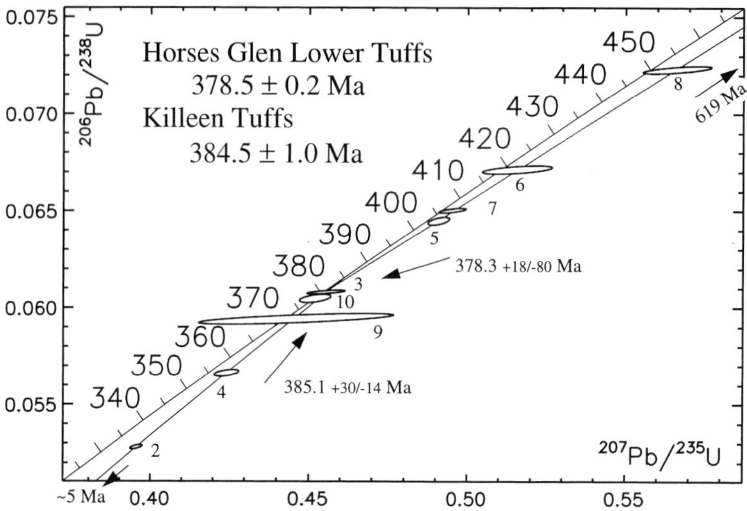

Fig. 12. U–Pb concordia plot of magmatic zircons extracted from the Killeen and the Horses Glen Lower Tuffs.

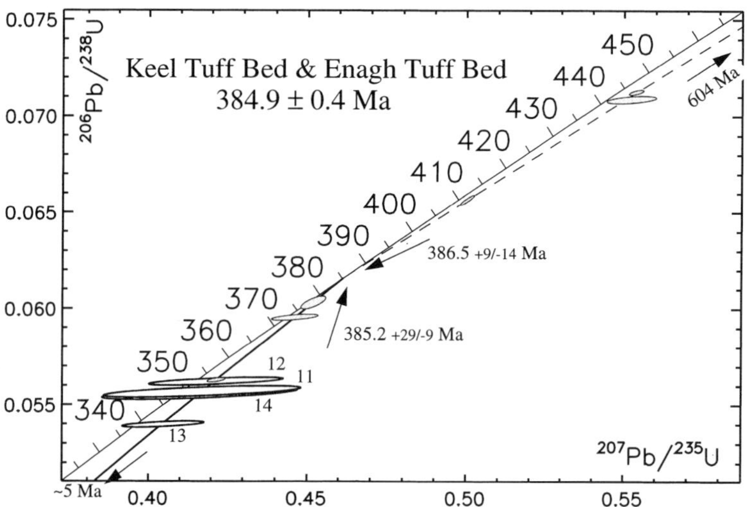

Fig. 13. U–Pb concordia plot of magmatic zircons extracted from the Keel Tuff Bed, including data from the Enagh Tuff Bed (Williams *et al.* 1997, fig. 4).

xenocrysts. To identify the magmatic zircon population lacking inheritance, a clean fragment of a very thin needle-shaped grain (No. 9, Table 2) was analysed. This is nearly concordant (96% of concordancy), although with a large error ellipse, and is interpreted as a magmatic grain lacking inherited material, but having suffered post-magmatic lead loss. A further analytical point (No. 10, Table 2), comprising an aggregate of nine very small needle-like single zircons and zircon fragments of uniform size and form (identical in type to No. 9, Table 2), is absolutely concordant with a $^{206}Pb/^{238}U$–$^{207}Pb/^{235}U$–$^{207}Pb/^{206}Pb$ combined age of 378.5 ± 0.2 Ma. This is in perfect agreement with the lower intercept age of the inheritance-bearing zircons (Fig. 12) and, based on this, we interpret the age of eruption–crystallization, and synchronous emplacement of the sampled horizon as 378.5 ± 0.2 Ma.

Keel Tuff Bed. Four data points were obtained for single magmatic grains and crystal fragments (Nos 11–14, Table 2). The resultant analytical ellipses are large compared with those of the affiliated Enagh Tuff Bed (Williams *et al.* 1997, fig. 4, and see reference points Fig. 13) as a result of the very low $^{206}Pb/^{204}Pb$ ratios of the Keel magmatic zircons (low U and high common Pb content). However, analysed zircons show perfect similarity in U concentrations (64–112 ppm) and have Th/U ratios in the narrow range 0.56–0.59. Because of the very narrow spread of the analytical points (Fig. 13) the upper intercept age carries a large uncertainty, and thus the preferred estimate of the age of the Keel Tuff Bed is the weighted mean $^{207}Pb/^{206}Pb$ age of 385.0 ± 2.9 Ma (95% c.l., external). The Keel and Enagh Tuff samples have slightly contrasting zircon isotopic compositions, which may suggest that they were erupted from different parts of the same magma chamber, on account of their identical ages and their very similar REE signatures (Fig. 9b). The discordia line calculated for magmatic, inheritance-free zircon grains from the Keel and Enagh tuffs yields an upper concordia intercept age of 385.2 $+29/-9$ Ma. However, the preferred eruption–crystallization age for the Keel–Enagh tuff is 384.9 ± 0.4 Ma (95% c.l., external), based on a combined weighted $^{207}Pb/^{206}Pb$ mean age of all seven analysed magmatic zircons.

Biostratigraphy

The samples were processed using standard laboratory techniques. Hydrofluoric acid was used to remove the silicates, and heavy liquid separation (using zinc bromide solution) was applied to remove very fine grained residual minerals. Sample RV 1 contained a very sparse amount of organic residue and the relatively small numbers of sporomorphs present are thermally mature and required extensive oxidation (24 h) in Schultze Solution to render them sufficiently light coloured for identification.

Reenagaveen microflora

Sample RV 1 was collected from the Valentia Slate Formation at Reenagaveen Point, 0.6 km west of Knightstown, along the north coast of Valentia Island (Fig. 6a). The productive lithology is a dark grey–green mudrock cropping out 60 m west of Reenagaveen Point [V422775] near the point where the section becomes inaccessible.

A small but distinctive spore assemblage was recorded, which contained the following spore taxa, *Geminospora lemurata* Balme emend Playford 1993, *Samarisporites triangulatus* Allen 1965, *Chelinospora concinna* Allen 1965, *Retusotriletes pychovii* Naumova 1953, *Rhabdosporites parvulus* Richardson 1965, *Grandispora* cf. *megaformis* (Richardson) McGregor 1973, *Dictyotriletes craticulatus* Clayton & Graham 1974, *Verruciretusispora* sp. and *Punctatisporites* sp. Figure 14(a–i) illustrates examples of the miospore taxa recorded.

In terms of the miospore zonation scheme of Streel *et al.* (1987), this assemblage can be assigned to the *Samarisporites triangulatus–Chelinospora concinna* TCo Oppel Zone based on the presence of the two zonal index species. The TCo biozone ranges from late Givetian to early Frasnian in age and spans earliest Frasnian time (Streel *et al.* 1987). *Samarisporites triangulatus* is a distinctive and widely reported taxon whose first occurrence biohorizon occurs in the Kerpen Formation of the Eifel region of Germany. Loboziak *et al.* (1990) correlate this occurrence with a level within the *ensensis–bipennatus* conodont zone, close to the base of the middle Givetian succession. However, a more precise biostratigraphic determination of the Reenagaveen microflora can be achieved by reference to the stratigraphical inception of *Chelinospora concinna.* The first occurrence biohorizon of this species occurs in the Blacourt Formation of the Boulonnais region of northern France and correlates with the mid- to late *varcus* conodont zone of late Givetian age (Loboziak & Streel 1989; Streel & Loboziak 1996). Therefore, the Reenagaveen spore assemblage can be no older than late Givetian in age.

Moll's Gap Quarry microflora

Higgs & Russell (1981) described a spore assemblage recovered from intraformational grey–green siltstone clasts within an 11.3 m thick sandstone-body of the Green Sandstone Formation (Chloritic Sandstone Formation), exposed in the Moll's Gap Quarry, Macgillycuddy's Reeks (Fig. 2). An early Frasnian age was suggested for the assemblage based on the previously published ranges of taxa such as *Samarisporites triangulatus*, *Geminospora lemurata*, *Grandispora inculta* Allen 1965, *Dictyotriletes perlotus* (Naumova) Mortimer & Chaloner 1971 and *Ancyrospora* simplex Guennel emend Urban 1969. However, since 1981, a considerable amount of new palynostratigraphical information has been published on Devonian spores, including two separate spore zonation schemes (Richardson & McGregor 1986; Streel *et al.* 1987). In the light of the new information, the Moll's Gap material has been re-examined and reappraised. In addition to the spore taxa originally described, two additional species have been identified in the Moll's Gap material; these are *Verrucosisorites bulliferus* Richardson & McGregor 1986 and *Rugospora bricei* Loboziak

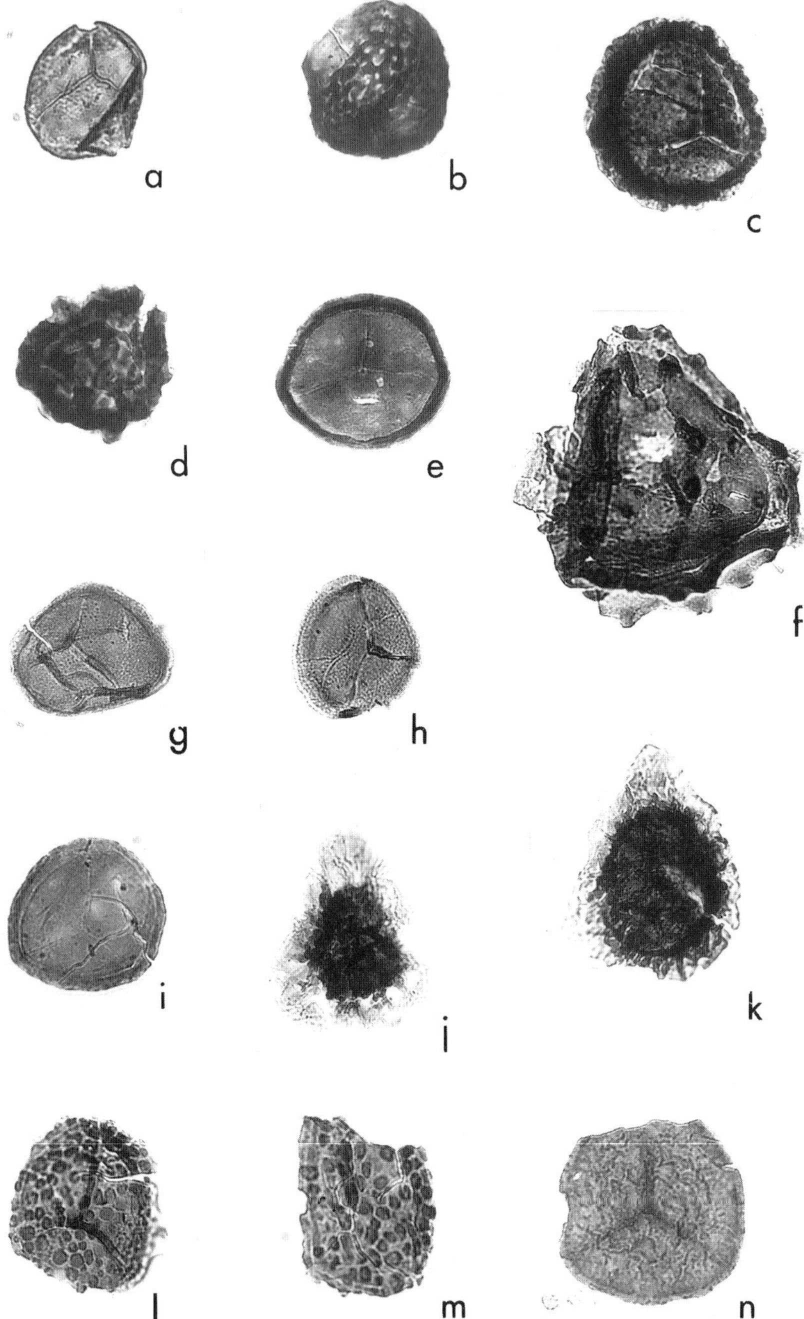

Fig. 14. Photographs of selected spore taxa from the Reenagaveen and Moll's Gap Quarry miospore assemblages. Magnification is × 500 in all cases. Spore taxa from Reenagaveen: (**a**) *Retusotriletes pychovii* Naumova 1953. RV1(A), H231; (**b**) *Dictyotriletes craticulatus* Clayton & Graham 1974. RV1(A), P372; (**c**) *Verruciretusispora* sp. RV1(A), K234; (**d**) *Chelinospora concinna* Allen 1965 RV1(A), G151; (**e**, **g**, **h**, **i**) *Geminospora lemurata* Balme emend Playford 1993. RV1(A), L413, F351, D215, G282; (**f**) *Grandispora* cf. *megaformis* (Richardson) McGregor 1973. RV1(A), P262; (**j**, **k**) *Samarisporites triangulatus* Allen 1965. RV1(A). Spore taxa from Moll's Gap Quarry: (**l**, **m**) *Verrucosisporites bulliferus* Richardson & McGregor 1986. MG1(G), MG1(F); (**n**) *Rugospora bricei* Loboziak & Streel 1989. MG1 (G).

& Streel 1989 (see Fig. 14l–n). These species are particularly important as they are both zonal index species for the Frasnian interval. Re-examination of specimens described as *Geminospora boleta* by Higgs & Russell (1981) in the Moll's Gap microflora revealed a complex of acamerate and variably camerate verrucate forms that could be described as a morphon with *V. bulliferus* comprising the acamerate end member and *G. boleta* the distinctly camerate end member of the morphon. Two specimens regarded as acamerate *V. bulliferus* are shown in Fig. 14l and m.

The first appearance of *Verrucosisorites bulliferus* was used by Richardson & McGregor (1986) to define the *Verrucosisorites bulliferus–Archaoperrisaccus ovalis* Assemblage Zone, and by Streel *et al.* (1987) to define the *Verrucosisorites bulliferus–Lophozonotriletes media* (BM) Oppel Zone of early to mid Frasnian age. The first appearance of *Rugospora bricei* was used by Loboziak & Streel (1989) as a marker species to indicate Zone IV, which overlies the BM Zone. Furthermore, Streel & Loboziak (1996) have shown that the first occurrence biohorizon of *Rugospora bricei* can be correlated within the late *hassi* to *linguiformis* conodont zones, within the mid- to late Frasnian interval. Therefore the presence of *Rugospora bricei* indicates that the Moll's Gap microflora is no more than mid-Frasnian in age.

Derryreag microflora

A third miospore assemblage, associated with isotopically dated rocks detailed in this paper, was located in Derryreag townland [W12888181] south of the Killeen Volcanic Centre (Williams 1993). This was recovered from a green–grey siltstone from a sequence of thick sandstones from the Chloritic Sandstone Formation *c.* 690 m stratigraphically above the Killeen rhyolites. The sample from Derryreag yielded a very limited microfloral assemblage, in which wood fragments were common, but spores were of low abundance. The assemblage comprised the following elements: *Punctatisporities planus* type, *Punctatisporities minutus*, *Geminospora* cf. *lemurata*, *Couverrucosisporities* sp., Spore Type A (apiculate simple), *Auroraspora* cf. *hyalina*, *Retusotriletes simplex* type. The only stratigraphically useful species is *Geminospora* cf. *lemurata*, which suggests a late Givetian–late Famennian age.

Discussion

The existing biostratigraphic data for the basin depocentre (Russell 1978; Higgs & Russell 1981), together with those from the basal ORS in the SW of the basin (Clayton & Graham 1974; Fig. 1), form the basis of several aspects of analysis of the Munster Basin. In one of the most far reaching, Graham (1983, p. 481) used the constraint provided by *Bothriolepis* and contemporary estimates of Devonian geological time to calculate a maximum sedimentation rate of 0.4 mm a^{-1} for the ORS succession of the depocentre. This value has been subsequently widely disseminated (e.g. Graham & Clayton 1988), and has been used to evaluate avulsion-controlled models of ORS sedimentation (Diemer & Bridge 1988) and to equate variously defined types of cyclicity within the ORS with Milankovitch band periodicities (Kelly 1992, 1993; Sadler & Kelly 1993; Kelly & Sadler 1995). The same method to calculate basin duration and subsidence rate has been used in a heterogeneous stretching model of the Munster Basin (Price 1989), with different values (13 Ma and 0.5 mm a^{-1}, respectively) obtained based on the use of revised time scales. Estimates of basin duration vary from 13 Ma (Kelly & Sadler 1995) to 27 Ma (Friend 1985, table 1). These figures have been used to suggest 'exceptionally high sedimentation' rates (Graham 1983; Friend 1985, p. 455; Price 1989, p. 116), which in turn has conditioned views of the fluvial style and evolution of parts of the basin fill (Graham *et al.* 1992) in relation to existing subsidence models (Sanderson 1984; Price 1989). Graham (1983, p. 473) and Graham & Clayton (1988, p. 332) also used the biostratigraphy to suggest that the gross age similarity of the oldest exposed ORS sequences in Iveragh and SW Cork demonstrates the true variation in the basin fill thickness, confirming the half-graben basin geometry (Naylor & Jones 1967). The new geochronological and biostratigraphical results presented in this paper call for a reassessment of these hypotheses (see also Higgs *et al.* this volume).

Correlation of the west Iveragh Old Red Sandstone

Despite showing minor differences in magmatic zircon isotopic composition (Table 2), the identical upper intercept and $^{207}Pb/^{206}Pb$ mean ages determined for the Keel and Enagh Tuffs, together with their very similar REE patterns (Fig. 9b) provide strong evidence that these beds represent the same (sub-plinian) explosive event (Williams *et al.* 1997). On the basis of this chronometric–geochemical correlation, we rename this stratigraphical (event) horizon as the Keel–Enagh Tuff Bed. Accepting this

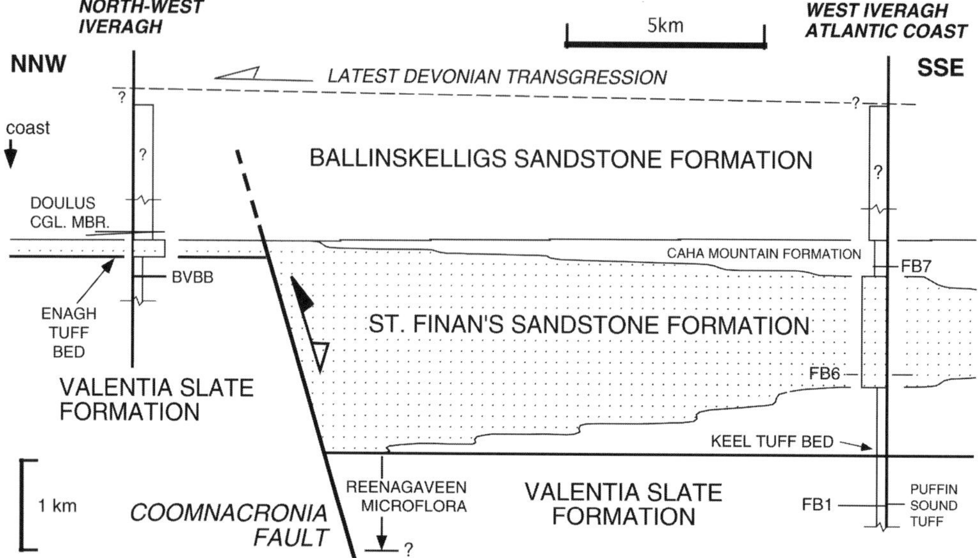

Fig. 15. Stratigraphic architecture of the west Iveragh Atlantic coast–NW Iveragh sections, based on correlation of the Keel and Enagh Tuff Beds. This link and the known stratigraphy (Capewell 1975) indicate a strongly diachronous base to the St Finan's Sandstone Formation, and extensional growth of the Coomnacronia Fault during St Finan's Sandstone Formation time. Stratigraphic positions of fish beds, the Puffin Sound and Keel Tuffs taken from Russell (1978, 1984); details of the Enagh Tuff, Bealtra Volcanic Breccia Bed (BVBB) and Doulus Conglomerate Member from Stössel (1993, 1995). Horizontal distance is approximately palinspastic (Williams *et al.* 1989). Mean palaeoflow was approximately SSE-directed, parallel to the plane of the section.

correlation permits the calculation of a combined geochronological age for the Keel–Enagh Tuff of 384.9 ± 0.4 Ma, refining the estimate of the minimum basin numerical age in the western part of the depocentre (Williams *et al.* 1997).

Although a correlation between the Keel and Enagh Tuffs has been argued previously (Russell 1984; Graham *et al.* 1995; Williams *et al.* 1997), the stratigraphical implications have not been fully explored. The Keel–Enagh Tuff Bed links an ORS lithostratigraphy across the Coomnacronia Fault (Fig. 15) that varies in thickness, alluvial and sequence architecture in several intervals. The provable stratigraphic section in the western part of the footwall block is only *c.* 1.1 km (Stössel 1993; Williams *et al.* 1997); a key unit, the St Finan's Sandstone Formation, is *c.* 190 m thick here, compared with 1245 m (Russell 1984; Fig. 5a) in the type section in the hanging wall. The Keel–Enagh Tuff Bed demarcates the base of the formation in the footwall block, but to the south occurs in the mid-part of the Valentia Slate Formation (Figs 5a and 15). This datum indicates that the St Finan's Sandstone Formation undergoes major thickening to the south, and has a highly diachronous (S-younging) base (Fig. 15). This stratigraphic architecture suggests that the Coomnacronia Fault was a major extensional structure during St Finan's Sandstone time. No definitive evidence yet exists to confirm extensional growth either during pre-Keel–Enagh Tuff (Valentia Slate) time, or during Ballinskelligs Sandstone time. However, apparently thin Upper Palaeozoic crust north of the Coomnacronia Fault (Masson *et al.* 1998) suggests an attenuated section of Valentia Slate in the footwall, and thick Ballinskelligs Sandstone Formation further east in the footwall (Ford, unpublished data) suggests that extensional displacement ceased after the St Finan's Sandstone Formation.

Grain-size contrasts within and between formations associated with the Keel–Enagh Tuff, pointed out by Graham *et al.* (1995), are a combined function of transport distance from the basin margin (the Dingle Bay–Galtee Fault Zone, Fig. 1) and differential subsidence across the Coomnacronia Fault, as well as stratigraphic position. The St Finan's Sandstone Formation in the footwall block of the Coomnacronia Fault has a coarser mean grain size and higher sandstone-body density compared with the west Iveragh type section (Fig. 6c). These sections are proximal and medial parts of a dispersal system from the NW sector of the basin margin, given the SE–SSE palaeoflow for these levels detailed by Russell (1984) and Graham *et al.* (1992). Significant facies differences do not occur in the Valentia Slate Formation across the Coomnacronia Fault. Bulk coarser facies and a higher frequency of discrete conglomerate-bodies occur along the north coast of Valentia Island (Williams *et al.* 1997), suggesting lateral transitions in facies occurred in response to subsidence- and topographic-control by the DB–GFZ in Dingle Bay. The coherent regional stratigraphy across the Coomnacronia Fault linked by the Keel–Enagh Tuff Bed (Fig. 15) indicates that the fault is intrabasinal.

The Keel–Enagh Tuff correlation suggests that the Coomnacronia Fault controlled the northward pinch-out of the Caha Mountain Formation (Fig. 6c), and that of the Valentia Slate Formation above the tuff bed in the hanging wall (Fig. 15). We suggest that this sequence architecture was generated by subsidence-controlled creation of accommodation space in the fault hanging wall, resulting in a thick section of St Finan's Sandstone alluvium. The present outcrop of the Valentia Slate Formation across the fault (Figs 2 and 6a) suggests significant fault inversion, involving an approximate minimum reverse throw of 2.2 km. In turn, this relationship indicates that the Reenagaveen microflora predates the Keel–Enagh Tuff Bed by a currently unknown thickness of Valentia Slate Formation, thus providing a minimum age. The combined $^{207}Pb/^{206}Pb$ weighted mean age for the Keel–Enagh Tuff (384.9 ± 0.4 Ma, 95% c.l., external) correlates with a position in mid-Givetian time on the Devonian time scale of Tucker *et al.* (1998). This chronometric age correlates closely, but not precisely, with the biostratigraphical age indicated by the spore data. However, because of the lack of continuous exposure and the inversion of the Coomnacronia Fault, uncertainties exist in correlating the relative levels of the tuff bed and the Reenagaveen microflora, creating difficulties in refining the Devonian time scale (Tucker *et al.* 1998).

The 384.9 ± 0.4 Ma age for the Keel–Enagh Tuff Bed constrains a minimum age for the occurrence of *Bothriolepis* (Fish Bed 1, Russell 1978) in this section. Given the extended, and uncertain, range of this genus (Westoll 1979; Lelievre & Goujet 1986; Pan & Dineley 1988), its biostratigraphic calibration of the numerical age of the Keel–Enagh Tuff Bed is poor. Fish Bed 2 (Fig. 5a) has a similar minimum numerical age, but provides no biostratigraphical constraint (Russell 1978). However, the correlation provided by Fig. 15 suggests that both fish beds may be of similar age to the Reenagaveen microflora, and thus approximately of late Givetian (*varcus* conodont zone) age. If correct, this would be the first report of *Bothriolepis* of Givetian age in the Euramerican province. The more precise isotopic age determined for the Keel–Enagh Tuff together, the late Givetian control furnished by the Reenagaveen microflora and the geochemical evidence for a correlation between the Reenadrolaun Tuff and the Bealtra Volcanic Breccia reinforce the minimum numerical age of the Valentia Island tetrapod ichnofauna (Stössel 1995; Williams *et al.* 1997), and indicate a pre-late Givetian chronostratigraphic age.

Currently, the only other biostratigraphic constraint on the Keel–Enagh Tuff Bed is Fish Bed 6 (Figs 5a, 15 and 16) from which the vertebrate *Sauripterus* (*sic*) sp. was reported, and attributed a Famennian age (Russell 1978). Assuming an mid-Famennian age for Fish Bed 6 implies a considerable condensation of time from the Keel–Enagh Tuff, and poses problems of across-depocentre correlation (Fig. 16). Even taking an earliest Famennian age for Fish Bed 6, to yield a similar time-averaged accumulation rate to that calibrated from the isotopic dating (Fig. 17), does not resolve the incompatibilities in the section parallel to depositional strike (Fig. 18). Recent re-examination of the material identified as *Sauripterus* sp. from Fish Bed 6 (Russell 1978, figs 7 and 8) indicates that it cannot be used to unequivocally prove the presence of *Sauripteris taylori*, the only certain member of the genus (Jeffery, pers. comm.; Westoll 1979, p. 347). Indeed, the nature of the material (scales) allows only a possible identification at the level of the order Rhizodontida, members of which have ranges from ?late Givetian to late Carboniferous time (Jeffery, pers. comm.). In the light of this, and as the west Iveragh section does not contain clear evidence of time loss or marked condensation (Fig. 5a), we regard the previously employed age designation of Fish Bed 6 as being invalid. The uncertain designation of the material from Fish Bed 6 does not currently permit its use in refining the subsidence history of the section (Fig. 17).

Lough Guitane Volcanic Complex–Moll's Gap correlations

The independent isotopic dates derived for the Lough Guitane Volcanic Complex suggest a different eruption chronology from that of Avison (1982, 1984*a*, *b*). Dates from the Killeen and Horses Glen Lower Tuffs indicate that the Horses Glen Centre is probably the youngest of the complex, although this study has not attempted to date the very complex

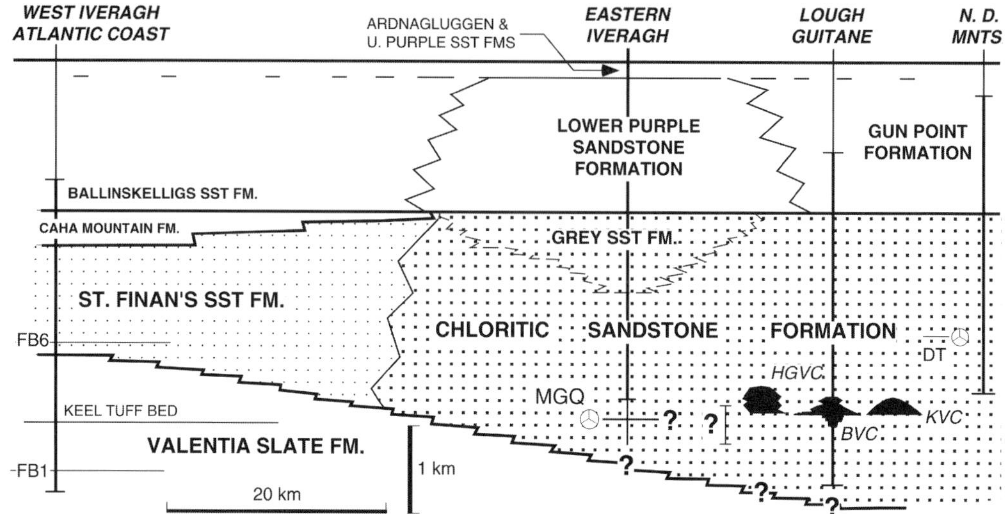

Fig. 16. Stratigraphic correlation of the Lough Guitane–northern Derrynasaggart Mountains (N.D. MNTS) section with the west Iveragh Atlantic coast (modified from Williams (1993, fig. 3.11)). This represents a depositional strike-parallel section through the laterally dispersing fluvial dispersal systems of the western Munster Basin, (1) the Chloritic Sandstone–Gortanimill and (2) Gun Point Systems. The stratigraphic position of the Horses Glen Volcanic Centre has been modified from Avison (1982, 1984*b*; see Fig. 4), although it is not known in detail. Stratigraphic information from Walsh (1968), Russell (1978, 1984), Higgs & Russell (1981) and Avison (1982, 1984*b*). MGQ, Moll's Gap Quarry microflora; DT, Derryreag microflora.

Bennaunmore Volcanic Centre. If the Killeen Tuffs are equivalent to the main products of the Bennaunmore Centre, as suggested by Avison (1982), then these centres may be regarded as the earliest erupted (Fig. 16). The relative ages of the Horses Glen and the Bennaunmore Centre suggest dip-slip displacement on a cross-fault between the centres, either that identified by Avison (1984*b*, fig. 3) to the west of the North Stoompa tuffs and rhyolite (Fig. 4), or one in the unexposed ground between this and the Horses Glen Centre. Detailed information to constrain the stratigraphic position of the Horses Glen Centre is currently lacking, and further mapping of the ORS and evaluation of faults associated with the centre are required to resolve this.

The high-precision age determined for the Horses Glen Lower Tuffs (378.5 ± 0.2 Ma) correlates with a mid-Frasnian age on the time scale of Tucker *et al.* (1998). This is in close accord with the biostratigraphical age indicated by the reappraised Moll's Gap microflora. However, the zircon age from the Killeen Tuffs (384.5 ± 1.0 Ma) correlates with a mid-Givetian age (Tucker *et al.* 1998), significantly older than the age indicated by the Moll's Gap microflora. We infer that the interval containing the Moll's Gap microflora is stratigraphically at a comparable, although probably underlying, level with the Horses Glen Lower Tuffs. In the absence of definitive field evidence, we cannot confidently relate the Killeen Tuffs and the Moll's Gap microflora.

Across-depocentre stratigraphic architecture

The isotopic dates and the new and reappraised miospore assemblages together refine the ORS stratigraphic architecture parallel to the basin margin in the depocentre (Williams *et al.* 1989; Williams 1993; Fig. 16). The ages of the Keel and Killeen Tuffs are consistent with the relative correlation of the sections, and with the new biostratigraphy reported here. The new data support a diachronous lithostratigraphical base to the Chloritic Sandstone–St Finan's Sandstone Formations (see Graham *et al.* 1992, p. 658). The cross-sectional shape of this lithosome (Fig. 16) indicates contemporaneous transversely draining large braided river (Chloritic Sandstone Formation) and ephemeral sand–silt flat (Valentia Slate Formation) fluvial environments earlier in the system history. Later in its history, an internal facies change is evident from a large braided river to a marginal sand flat environment, with an increased proportion of small channel-belts (St Finan's Sandstone Formation). This sequence architecture was primarily influenced by a (probably) fixed entry point of a major river into the basin, supplying relatively coarse (sandy–pebbly)

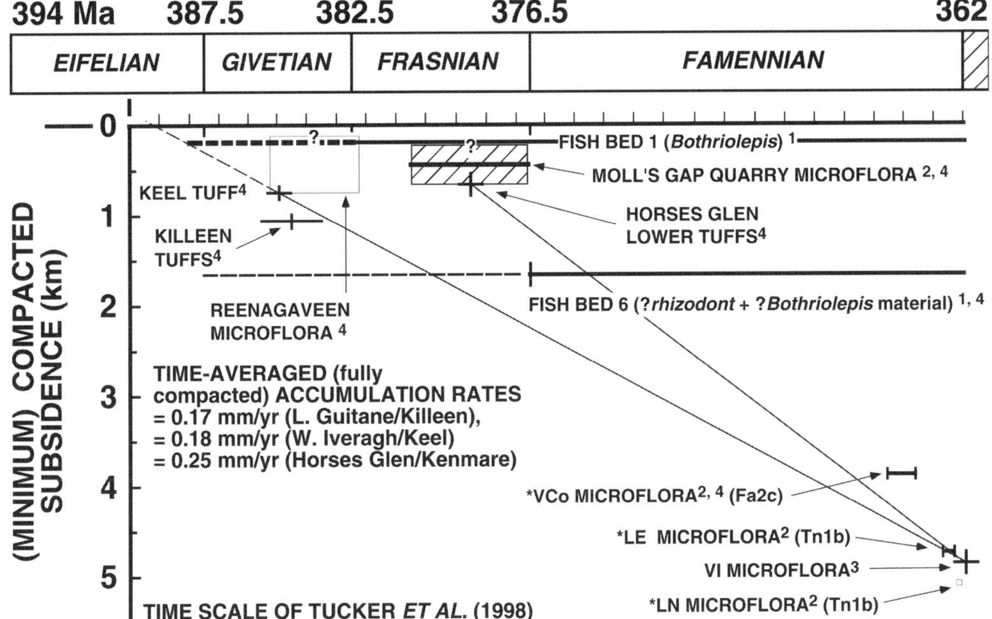

Fig. 17. Composite geohistory plot for the ORS of the Iveragh region of the Munster Basin. Data sources: 1, Russell (1978); 2, Higgs & Russell (1981); 3, Higgs *et al.* (1988); 4, this paper. Fish bed 6 is considered here to contain possible rhizodont vertebrate material, possibly ranging from the ?late Givetian and through the Devonian period (Jeffery, pers. comm.), rather than certain remains of *Sauripterus* (*sic*) (Russell 1978); the former Famennian range based on *Sauripterus* is shown by a continuous line. The maximum age range of late Givetian–early Frasnian age reported by Russell (1978) for Fish bed 1 is shown by a continuous line; its possible extension to satisfy the new data is shown by the dashed line. Uncertainties exist over the positions of the Moll's Gap microflora, the Horses Glen Centre and the Reenagaveen microflora. *Microfloral data from sections in southern Iveragh. The VCo biozone microflora is renamed from the former VU biozone (Higgs & Russell 1981). Miospore biozone durations (VCo, LE, LN and VI) are not known in detail; the VCo biozone (= Fa2c) is shown as older than 363.6 Ma, which is the isotopic age given for Fa2d by Tucker *et al.* (1998).

sediment to the Chloritic Sandstone fluvial dispersal system (Williams *this volume*).

Although there is chronometric evidence for a discrete volcanic interval in the basin depocentre, the Keel–Enagh Tuff Bed and Killeen–Bennaunmore centres, these are not thought to be volcanologically related, i.e. the Keel–Enagh Tuff is not the product of the Lough Guitane Complex. The Lough Guitane volcanism was dominated by the accumulation of very thick rhyolitic lava piles (300 m) with little evidence of vesiculation, and silicic tuffs composed of volcanic ash, non-vesicular lapilli and terrigenous detritus (Avison 1982, 1984*a*, *b*). Avison (1982) concluded that the absence of highly explosive volcanicity and the lack of products of normal airfall processes was due to the large thickness of partially lithified or unconsolidated alluvium through which the magma rose. It is thus unlikely that the Lough Guitane Complex produced a large sub-plinian or plinian eruption of the type characterizing the Keel–Enagh event. Moreover, the respective deposits show dissimilar petrographic, REE and other geochemical signatures (Figs 7a and 9b). These factors and the rapid, apparently southward thinning of the Keel–Enagh Tuff, and the lack of any reported mappable connection with the Lough Guitane Complex, argue against any volcanological link. A 0.04 m thick tuff bed, identified by Russell (1984) 185 m above the base of the Chloritic Sandstone Formation 7.5 km NE of Sneem (Fig. 2), was correlated by him with the Keel Tuff Bed, on the basis of complex facies arguments that are beyond the scope of this discussion. However, based on the stratigraphy detailed here (Figs 15 and 16) and Russell's (1984, p. 226) correlation, Williams (1993, fig. 3.13) showed that the Keel Tuff occurred at approximately the same stratigraphical level as the Lough Guitane volcanic rocks. This is consistent with the chronometric data of this

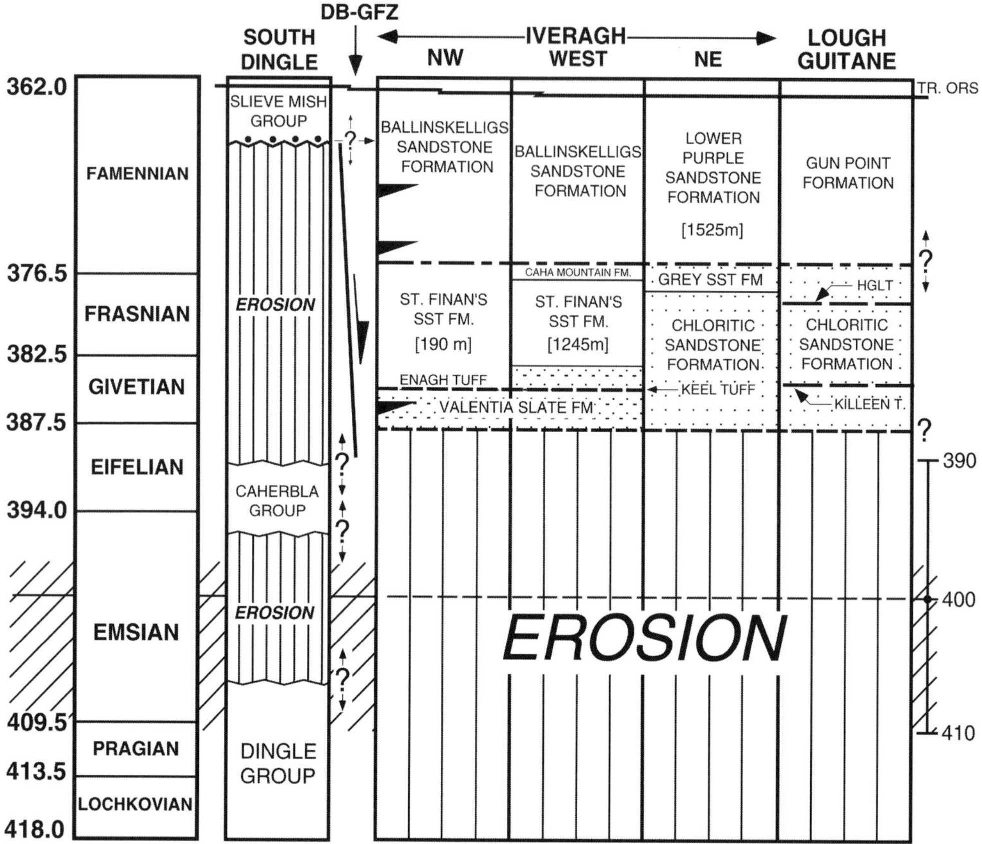

Fig. 18. Suggested chronostratigraphy of the main dated sections in the Munster Basin, and their relationship to the ORS in the south of the Dingle peninsula (the western regional footwall to the Munster Basin). The time scale is that of Tucker *et al.* (1998). The hachured zone shows the suggested timing of late Caledonian (Acadian) deformation based on information given by Murphy (1987) and McKerrow (1988). The 410–390 Ma range bar depicts the time span of syn-Acadian pluton intrusion (Murphy 1987); the median age of *c.* 400 Ma is widely considered to date the peak Acadian (D1) event (e.g. Murphy 1987). The chronostratigraphy of the alluvial fan lithosomes within the Valentia Slate and Ballinskelligs Sandstone Formations of NW Iveragh (filled triangles) is not known in detail. These lithosomes are considered to have been partly sourced from the Caherbla Group of south Dingle, and the deposition of the Slieve Mish Group must post-date the latest of these units (Williams 1993, this volume). Erosion of the north Iveragh area during Dingle and Caherbla Group times was suggested by Allen & Crowley (1983) and Todd (1989). An approximation of the chronostratigraphy of the ORS transitional (TR. ORS) to the latest Devonian marine transgression is based on calibration of Fa2d (see Fig. 17). HGLT, Horses Glen Lower Tuffs; DB–GFZ, Dingle Bay–Galtee Fault Zone.

paper, but the presence of this thin tuff does not constitute a genetic link between the Lough Guitane and the Keel–Enagh volcanic rocks.

Geohistory

The geochronology and biostratigraphy may be used to produce a crude geohistory of the NW basin depocentre (Fig. 17). Using the relatively well-known biostratigraphy of the latest ORS to marine Early Carboniferous period in Iveragh (Walsh 1968; Higgs & Russell 1981; Higgs *et al.* 1988) and the likely age of the Devonian–Carboniferous boundary (362 Ma, Tucker *et al.* 1998; but see Claoué-Long *et al.* 1995) gives a minimum duration time for the basin of *c.* 23 Ma. The west Iveragh (Atlantic coast) section, from the Keel Tuff Bed, gives a time-averaged, fully compacted accumulation rate of 0.18 mm a^{-1}. This derives from the stratigraphy detailed in Fig. 15 and assumes a regionally uniform thickness of 1525 m for the Ballinskelligs Sandstone–Lower Purple Sandstone Formation (see Williams *et al.* 1989), and *c.* 174 m for the transitional ORS to the VI biozone at Killarney (Walsh 1968; Higgs *et al.* 1988; Fig. 3). The Glen Flesk section, associated with the Killeen Tuffs in the Lough Guitane–northern Derrynasaggart Mountains area, gives an accumulation rate of 0.17 mm a^{-1}. This figure derives from stratigraphical data for the Chloritic Sandstone Formation detailed in Fig. 3 (see Avison 1982, 1984*a*, *b*), and the same assumption regarding the thickness of the Gun Point Formation–Lower Purple Sandstone Formation plus transitional ORS. The rate based on the age of the Horses Glen Lower Tuffs and the stratigraphic section from the Mangerton Anticline at Mangerton Mountain [V98038077] to the Kenmare Syncline [V99657275] (Naylor 1978; Ford unpubl. data) is 0.25 mm a^{-1}. Uncertainties, based on the quoted external errors of the weighted mean isotopic ages, are negligible in all cases. These rates are significantly lower than previous estimates of 0.4 mm a^{-1} (Graham 1983; Graham & Clayton 1988), 0.5 mm a^{-1} (Price 1989) and 0.38–0.46 mm a^{-1} (Kelly & Sadler 1995).

Figure 17 shows that there are uncertainties in the stratigraphical positions of the Reenagaveen and Moll's Gap microfloras in relation to the Keel–Enagh and Horses Glen Lower tuffs, respectively. However, the control furnished by the isotopic dating has implications for proposed orbital forcing time scales in the ORS of the depocentre (Kelly 1993; Sadler & Kelly 1993; Kelly & Sadler 1995). These studies have suggested that several Milankovitch band periodicities are represented within the ORS, but on the basis of calibration by the 0.4 mm a^{-1} time-averaged subsidence rate calculated by Graham (1983). Although it is possible that higher subsidence rates applied during intervals of ORS accumulation, the lower isotopically derived rate (0.17 mm a^{-1}) calculated for a principal section used in the study of cyclicity (Glen Flesk–Killeen) introduces uncertainty in ascribing specific, orbitally induced climatic signals to the fluvial sediments. The cyclicities (P1–P3) reported by Sadler & Kelly (1993) and Kelly & Sadler (1995, table 1) have wavelengths ranging from 31 to 175 m; these are optimally converted to orbital eccentricity cycles (111 and 412 ka) using the 0.4 mm a^{-1} mean subsidence rate, where the durations are 78 to 438 ka. However, Kelly & Sadler (1995, table 1) did not evaluate estimated perodicities of their critical cycles for a range of ORS subsidence rates. Using the isotopically derived mean rate for the same section (0.17 mm a^{-1}) gives an overall range of 182–1029 ka. On this basis, cycles P1–P3 (Kelly & Sadler 1995, table 1) have respective periodicities of 182–247, 324–365 and 647–1029 ka. Using the derived 0.25 mm a^{-1} rate from the Horses Glen–Kenmare section gives periodicities for P1–P3 of 124–168, 220–248 and 440–700 ka, respectively. These periodicities deviate significantly from those proposed for P1–P3 by Kelly & Sadler (1995) and attributed to orbital eccentricities. Applying alternative ages to the end of Munster Basin sedimentation, i.e. the beginning of the Carboniferous period (e.g. 353.7 ± 4.2 Ma, Claoué-Long *et al.* 1995) produces lower time-averaged rates, and further deviation from suggested eccentricity cycles. Clearly, there needs to be re-evaluation of possible cyclicity based on the isotopic data presented here, further isotopic data to constrain smaller time slices and also re-evaluation of data based on developments in both fields.

Just as the geohistory plot (Fig. 17) is at present too coarse to calibrate definitely any long-period cyclicity affecting the progradation–retreat of the depocentre fluvial dispersal systems, it is also not possible to define syn- and post-rift megasequences. If the probable age of ORS sedimentation (Slieve Mish Group) on the footwall to the Munster Basin (late Famennian) represents post-rift sedimentation, syn-rift subsidence rates may have remained of the order of the quoted time-averaged value for much of the known basin history (Fig. 17). This suggests a prolonged syn-rift phase, with volcanism being restricted to its earlier part.

The uncertainties in assigning numerical ages to microfloral assemblages in continental sequences, even using the most robust available geological time scales (e.g. Tucker & McKerrow 1995; Tucker *et al.* 1998), remain considerable. Even relating locally associated biostratigraphic and numerical ages presents significant problems in complex fluvial sequences found in the basin depocentre. Thus detailed across-basin correlation of the ORS in the Munster Basin should await further isotopic dating of critical tuffs (e.g. the Clear Island tuff reported by Graham & Reilly (1972)).

Dingle Basin–Munster Basin relationship

A suggested chronostratigraphy for the NW Munster Basin depocentre and its relationship to the Lower and Middle ORS rocks of the immediate footwall region is given in Fig. 18, calibrated against the time scale of Tucker *et al.* (1998). The numerical age of earliest ORS in the Munster Basin remains to be specified, but must be older than the Killeen and Keel–Enagh ages. The Munster Basin must have initiated subsequent to (1) the end of late Caledonian

(Acadian) deformation, probably given by syntectonic granite intrusion at *c.* 400 Ma in the Leinster massif and eastern Ireland (Murphy 1987), and (2) the deposition of the post-Acadian Caherbla Group in the southern part of the Dingle Basin (Fig. 18). The youngest limit of the Acadian Orogeny in eastern Ireland (390 Ma) given by Murphy (1987) is unlikely to have applied to SW Ireland, as little time would be available for accumulation of the Caherbla Group before the onset of Munster Basin subsidence (Fig. 18). The distinctive, high-grade metamorphic conglomerates of the Caherbla Group were reworked by erosion in footwall drainage basins into the NW Munster Basin before Keel–Enagh Tuff emplacement at 384.9 Ma (see Williams *et al.* 1997). Crustal extension to form the Munster Basin thus began <15 Ma after major granite plutonism in the Iapetus Suture zone and the end of Acadian deformation. The new biostratigraphy indicates Mid-Devonian ORS sedimentation and a pre-late Givetian basin initiation, and by extension a pre-late Givetian age for the Caherbla Group of the Dingle Basin.

Conclusions

(1) Three new high-precision U–Pb zircon dates (quoted as weighted mean $^{207}Pb/^{206}Pb$ ages and external errors with 95% confidence limits) have been obtained from subaerial tuffs within the ORS of the NW Munster Basin depocentre, the Keel (385.0 ± 2.9 Ma), Killeen (384.5 ± 1.0 Ma), and the Horses Glen (Lower) Tuffs (378.5 ± 0.2 Ma). These ages allow sequence correlation across the depocentre in Iveragh, and indicate that the earliest ORS in the east and west is older than *c.* 385 Ma.

(2) The newly discovered Reenagaveen microflora, from the Valentia Slate Formation, has been assigned to the TCo miospore biozone, and has been shown to be equivalent to the mid–late *varcus* conodont zone, and is therefore no older than late Givetian time. This microflora predates the Keel–Enagh Tuff Bed (combined age 384.9 ± 0.4 Ma) by a currently unknown thickness of Valentia Slate Formation. The numerical age and geochemical correlations of the Keel and Enagh Tuffs imply probable late Givetian ages for the placoderm *Bothriolepis*, as well as the Valentia Island tetrapod trackway ichnofauna. This would suggest important revisions to views on the evolution of the tetrapods (Ahlberg & Milner 1994; Clack 1997), and the endemism of *Bothriolepis*. An Eifelian chronostratigraphic age is not confirmed for the NW Iveragh ORS (see Williams *et al.* 1997) based on the new biostratigraphy and the Keel–Enagh Tuff correlation.

(3) The biostratigraphic age of the Moll's Gap Quarry microflora is significantly refined, and is assigned to the IV miospore biozone, and can be further equated with the late *hassi* to *linguiformis* conodont zones, and is thus no older than mid-Frasnian age. The microflora probably pre-dates the Horses Glen Lower Tuffs by a few hundred metres of strata. The Killeen Tuffs, from the eastern centre of the Lough Guitane Volcanic Complex, give an older date than the Horses Glen Lower Tuffs, suggesting a reverse eruption sequence from that previously reported for the complex.

(4) The Munster Basin underwent pre-late Givetian initiation, and the oldest dated tuffs constrain a minimum duration of 23 Ma to the end of the Devonian period. The stratigraphy of the isotopically dated sections gives compacted time-averaged subsidence rates of 0.17–0.25 and 0.18 mm a^{-1}, for east Iveragh–northern Derrynasaggarts and west Iveragh, respectively. These rates are significantly different from those used to calibrate orbitally forced cycles proposed for the ORS of the depocentre, as well as crustal stretching models.

(5) Geochemical analysis of the dated silicic and selected basic volcanic rocks shows that they are sub-alkaline, and that the basic rocks have continental tholeiite to MORB affinities. Discrimination diagrams using different immobile trace elements indicate a within-plate tectonic environment with an anomalous (MORB) geochemical signature, compatible with the inferred continental stretching origin of the Munster Basin. Statistically identical ages of inherited zircon material from the Enagh and Horses Glen Lower Tuffs suggest an homogeneous source crust beneath the Iveragh region of the depocentre.

(6) Previous time scales (McKerrow *et al.* 1985; Harland *et al.* 1990) applied to the Devonian ORS of the Munster Basin (e.g. Williams *et al.* 1997) have been superseded by that of Tucker *et al.* (1998), which is broadly supported by the new geochronology and biostratigraphy reported. Using the technique of Tucker *et al.* (1998, fig. 2) for time-scale calibration, an error box for the Horses Glen age intersects their best-fit time line, but that of the Keel–Enagh Tuff plots close to the line, suggesting modification of the Givetian stage boundary.

We thank the Schweizerische Akademie der Naturwissenschaften and the ETH Zürich for supporting the fieldwork. We are grateful to R. H. Steiger (ETH Zürich) for his support of the project, and to H. Derksen for her expert supervision of the mineral separation. Thanks are also due to K. J. Russell for data and discussions on the west Iveragh type stratigraphy, to R. D. Tucker for pre-publication information on Devonian geochronology and to D. L. Dineley, J. A. Clack and, in particular, J. Jeffery for information on the vertebrates. Thanks are due to J. A. Evans for a painstaking review of the isotope geology, as well as to S. B. Kelly and S. P. Todd for their constructive critiques.

References

AHLBERG, P. E. & MILNER, A. R. 1994. The origin and early diversification of tetrapods. *Nature*, **368**, 507–514.

ALLEN, J. R. L. & CROWLEY, S. F. 1983. Lower Old Red Sandstone fluvial dispersal systems in the British Isles. *Transactions of the Royal Society of Edinburgh: Earth Sciences*, **74**, 61–68.

AVISON, M. M. 1982. *The geology of the Lough Guitane Volcanic Complex and associated sediments, County Kerry, Ireland.* PhD thesis, University of Keele.

—— 1984*a*. Contemporaneous faulting and the eruption and preservation of the Lough Guitane Volcanic Complex, Co. Kerry. *Journal of the Geological Society, London*, **141**, 501–510.

—— 1984*b*. The Lough Guitane Volcanic Complex, County Kerry—a preliminary survey. *Irish Journal of Earth Sciences*, **6**, 127–136.

BEVINS, R. E., KOKELAAR, B. P. & DUNKLEY, P. N. 1984. Petrology and geochemistry of lower to middle Ordovician igneous rocks in Wales: a volcanic arc to marginal basin transition. *Proceedings of the Geologists' Association*, **95**, 337–347.

BLACKMORE, R. 1995. Low-grade metamorphism in the Upper Palaeozoic Munster Basin, southern Ireland. *Irish Journal of Earth Sciences*, **14**, 115–133.

CAPEWELL, J. G. 1957. The stratigraphy and structure of the country around Sneem, Co. Kerry. *Proceedings of the Royal Irish Academy*, **58(B)**, 167–183.

—— 1965. The Old Red Sandstone of Slieve Mish, Co. Kerry. *Proceedings of the Royal Irish Academy*, **64(B)**, 165–174.

—— 1975. The Old Red Sandstone Group of Iveragh, Co. Kerry. *Proceedings of the Royal Irish Academy*, **75(B)**, 155–171.

CLACK, J. A. 1997. Devonian tetrapod trackways and trackmakers; a review of the fossils and footprints. *Palaeogeography, Palaeoclimatology, Palaeoecology*, **130**, 227–250.

CLAOUÉ-LONG, J. C., COMPSTON, W., ROBERTS, J. & FANNING, C. M. 1995. Two Carboniferous ages: a comparison of SHRIMP zircon dating with conventional zircon ages and $^{40}Ar/^{39}Ar$ analysis. *In*: BERGGREN, W. A., KENT, D. V., AUBRY, M.-P. & HARDENBOL, J. (eds) *Geochronology, Time Scales and Global Stratigraphic Correlation.* Society of Economic Palaeontologists and Mineralogists, Special Publications, **54**, 3–21.

CLAYTON, G. 1989. Vitrinite reflectance data from the Kinsale Harbour–Old Head of Kinsale area, southern Ireland, and its bearing on the interpretation of the Munster Basin. *Journal of the Geological Society, London*, **146**, 611–616.

—— & Graham, J. R. 1974. Miospore assemblages from the Devonian Sherkin Formation of southwest County Cork, Republic of Ireland. *Pollen et Spores*, **16**, 565–588.

——, ——, Higgs, K., Holland, C. H. & Naylor, D. 1980. Devonian rocks in Ireland: a review. *Journal of Earth Sciences, Royal Dublin Society*, **2**, 161–183.

COOPER, M. A., COLLINS, D. A., FORD, M., MURPHY, F. X., TRAYNER, P. M. & O'SULLIVAN, M. 1986. Structural evolution of the Irish Variscides. *Journal of the Geological Society, London*, **143**, 53–61.

DIEMER, J. A. & BRIDGE, J. S. 1988. Transition from alluvial plain to tide-dominated coastal deposits associated with the Tournaisian marine transgression in southwest Ireland. *In*: DE BOER, P. L., VAN GELDER, A. & NIO, S. D. (eds) *Tide-influenced Sedimentary Environments and Facies*. D. Reidel, Dordrecht, 359–388.

FLOYD, P. A. & WINCHESTER, J. A. 1975. Magma-type and tectonic setting discrimination using immobile elements. *Earth and Planetary Science Letters*, **27**, 211–218.

FRIEND, P. F. 1985. Molasse basins of Europe: a tectonic assessment. *Transactions of the Royal Society of Edinburgh: Earth Sciences*, **76**, 451–462.

GARDINER, P. R. R. & MACCARTHY, I. A. J. 1981. The Late Paleozoic evolution of southern Ireland in the context of tectonic basins and their transatlantic significance. *In*: KERR, J. W. & FERGUSSON, A. J. (eds) *Geology of the North Atlantic Borderlands*. Canadian Society of Petroleum Geologists, Memoirs, **7**, 683–725.

GRÄFE, K. 1993. *Geologische und petrographische untersuchungen des Bennaunmore Volcanic Centre (Südwest Irland)*. Diplomarbeit an der Abteilung XC (Erdwissenschaften) thesis, ETH Zürich.

GRAHAM, J. R. 1983. Analysis of the Upper Devonian Munster Basin, an example of a fluvial distributary system. *In*: COLLINSON, J. D. & LEWIN, J. (eds) *Modern and Ancient Fluvial Systems*. International Association of Sedimentologists, Special Publications, **6**, 473–483.

—— & CLAYTON, G. 1988. Devonian rocks in Ireland and their relation to adjacent regions. *In*: MCMILLAN, N. J., EMBRY, A. F. & GLASS, D. J. (eds) *The Devonian of the World, Vol. 1*. Canadian Society of Petroleum Geologists, Memoirs, **14**, 325–340.

—— & REILLY, T. A. 1972. The Sherkin Formation (Devonian) of south-west County Cork. *Geological Survey of Ireland, Bulletin*, **1**, 281–300.

——, JAMES, A. & RUSSELL, K. J. 1992. Basin history deduced from subtle changes in fluvial style: a study of distal alluvium from the Devonian of southwest Ireland. *Transactions of the Royal*

Society of Edinburgh: Earth Sciences, **83**, 655–667.

——, RUSSELL, K. J. & STILLMAN, C. J. 1995. Late Devonian magmatism in west Kerry and its relationship to the development of the Munster Basin. *Irish Journal of Earth Sciences*, **14**, 7–23.

HALLIDAY, A. N. & MITCHELL, J. G. 1983. K–Ar ages of clay concentrates from Irish orebodies and their bearing on the timing of mineralisation. *Transactions of the Royal Society of Edinburgh: Earth Sciences*, **74**, 1–14.

HARLAND, W. B., ARMSTRONG, R. L., COX, A. V., CRAIG, L. E., SMITH, A. G. & SMITH, D. G. 1990. *A Geologic Time Scale 1989*. Cambridge University Press, Cambridge.

HIGGS, K. & RUSSELL, K. J. 1981. Upper Devonian microfloras from southeast Iveragh, County Kerry, Ireland. *Geological Survey of Ireland Bulletin*, **3**, 17–50.

——, CLAYTON, G. & KEEGAN, J. B. 1988. Stratigraphic and systematic palynology of the Tournaisian rocks of Ireland. *Geological Survey of Ireland*, Special Paper, **7**.

——, MACCARTHY, I. A. J. & O'BRIEN, M. M. 2000. A mid-Frasnian marine incursion into the southern part of the Munster Basin; evidence from the Foilcoagh Bay Beds, Sherkin Formation, southwest County Cork, Ireland. This volume.

HOLLAND, C. H. 1977. Ireland. *In*: HOUSE, M. R., RICHARDSON, J. B., CHALONER, W. G., ALLEN, J. R. L., HOLLAND, C. H. & WESTOLL, T. S. (eds) *A Correlation of the Devonian Rocks in the British Isles*. Geological Society of London, Special Report, **7**, 54–66.

HUSAIN, S. M. 1957. *The geology of the Kenmare Syncline, Co. Kerry, Ireland*. PhD thesis, University of London.

KELLY, S. B. 1992. Milankovitch cyclicity recorded from Devonian non-marine sediments. *Terra Nova*, **4**, 578–584.

—— 1993. Cyclical discharge variation recorded in alluvial sediments: an example from the Devonian of southwest Ireland. *In*: NORTH, C. P. & PROSSER, D. J. (eds) *Characterization of Fluvial and Aeolian Reservoirs*. Geological Society of London, Special Publications, **73**, 157–166.

—— & SADLER, S. P. 1995. Equilibrium and response to climatic and tectonic forcing: a study of alluvial sequences in the Devonian Munster Basin, Ireland. *In*: HOUSE, M. R. & GALE, A. S. (eds) *Orbital Forcing Timescales and Cyclostratigraphy*. Geological Society of London, Special Publications, **85**, 19–36.

KLAPPER, G., FEIST, R., BECKER, R. T. & HOUSE, M. R. 1993. Definition of the Frasnian/Famennian Stage boundary. *Episodes*, **16**, 433-441.

——, —— & HOUSE, M. R. 1987. Decision on the boundary stratotype for the Middle/Upper Devonian Series boundary. *Episodes*, **10**, 97–101.

LELIEVRE, H. & GOUJET, D. 1986. Biostratigraphic significance of some uppermost Devonian Placoderms. *Annales de la Société Géologique de Belgique*, **109**, 55–59.

LOBOZIAK, S. & STREEL, M. 1989. Middle–Upper Devonian miospores from the Ghadamis Basin (Tunisia–Libya): systematics and stratigraphy. *Review of Palaeobotany and Palynology*, **58**, 173–196.

——, —— & WEDDIGE, K. 1990. Miospores, the *lemurata* and *triangulatus* levels and their faunal indices near the Eifelian/Givetian boundary in the Eifel (F.R.G.). *Annales de la Société Géologique de Belgique*, **T113**, 299–313.

MACCARTHY, I. A. J. 1990. Alluvial sedimentation patterns in the Munster Basin, Ireland. *Sedimentology*, **37**, 685–712.

MASSON, F., JACOB, A. W. B., PRODEHL, C., READMAN, P. W., SHANNON, P. M., SCHULZE, A. & ENDERLE, U. 1998. A wide-angle seismic traverse through the Variscan of southwest Ireland. *Geophysical Journal International*, **134**, 689–705.

MCKERROW, W. S. 1988. Wenlock to Givetian deformation in the British Isles and the Canadian Appalachians. *In*: HARRIS, A. L. & FETTES, D. J. (eds) *The Caledonian–Appalachian Orogen*. Geological Society of London, Special Publications, **38**, 437–448.

——, LAMBERT, R. S. J. & COCKS, L. R. M. 1985. The Ordovician, Silurian and Devonian Periods. *In*: SNELLING, N. J. (ed.) *The Chronology of the Geological Record*. Geological Society of London, Memoirs, **10**, 73–80.

MEERE, P. A. 1995*a*. The structural evolution of the western Irish Variscides: an example of obstacle tectonics? *Tectonophysics*, **246**, 97–112.

—— 1995*b*. Sub-greenschist facies metamorphism from the Variscides of SW Ireland: an early synextensional peak thermal event. *Journal of the Geological Society, London*, **152**, 511–521.

MURPHY, F. C. 1987. Late Caledonian granitoids and the timing of deformation in the Iapetus suture zone of eastern Ireland. *Geological Magazine*, **124**, 135–142.

NAYLOR, D. 1978. A structural section across the Variscan fold belt, southwest Ireland. *Journal of Earth Sciences, Royal Dublin Society*, **1**, 63–70.

—— & JONES, P. C. 1967. Sedimentation and tectonic setting of the Old Red Sandstone of southwest Ireland. *In*: OSWALD, D. H. (ed.) *International Symposium on the Devonian System*. Alberta Society of Petroleum Geologists, Calgary, **2**, 1089–1099.

PAN, J. & DINELEY, D. L. 1988. *Proceedings of the Royal Society of London, Series B*, **235**, 29–61.

PEARCE, J. A. & NORRY, M. J. 1979. Petrogenetic implications of Ti, Zr, Y and Nb variations in volcanic rocks. *Contributions to Mineralogy and Petrology*, **69**, 33–47.

——, HARRIS, N. B. W. & TINDLE, A. G. 1984. Trace element discrimination diagrams for the tectonic interpretation of granitic rocks. *Journal of Petrology*, **25**, 956–983.

PRACHT, M. 1997. Geology of Kerry–Cork. *A geological description to accompany the bedrock geology 1:100 000 scale map series, Sheet 21,*

Kerry–Cork. Geological Survey of Ireland, Dublin.

PRICE, C. A. 1989. Some thoughts on the subsidence and evolution of the Munster Basin, southern Ireland. *In*: ARTHURTON, R. S., GUTTERIDGE, P. & NOLAN, S. C. (eds) *The Role of Tectonics in Devonian and Carboniferous Sedimentation in the British Isles*. Occasional Publications, of the Yorkshire Geological Society, **6**, 111–121.

RICHARDSON, J. B. & MCGREGOR, D. C. 1986. Silurian and Devonian spore zones of the Old Red Sandstone continent and adjacent regions. *Bulletin Geological Survey of Canada*, **364**, 1–79.

RODEN, M. K., PARRISH, R. R. & MILLER, D. S. 1990. The absolute age of the Eifelian Tioga ash bed, Pennsylvania. *Journal of Geology*, **98**, 282–285.

RUSSELL, K. J. 1978. Vertebrate fossils from the Iveragh Peninsula and the age of the Old Red Sandstone. *Journal of Earth Sciences Royal Dublin Society*, **1**, 151–162.

—— 1984. *The sedimentology and palaeogeography of some Devonian sedimentary rocks in southwest Ireland*. PhD thesis, Plymouth Polytechnic.

SADLER, S. P. & KELLY, S. B. 1993. Fluvial processes and cyclicity in terminal fan deposits: an example from the Late Devonian of southwest Ireland. *Sedimentary Geology*. **1–4**, 375–386.

SANDERSON, D. J. 1984. Structural variation across the northern margin of the Variscides in NW Europe. *In*: HUTTON, D. H. W. & SANDERSON, D. J. (eds) *Variscan Tectonics of the North Atlantic Region*, Geological Society, London, Special Publications, **14**, 149–165.

STACEY, J. S. & KRAMERS, J. D. 1975. Approximation of terrestrial lead isotope evolution by a two-stage model. *Earth and Planetary Sciences Letters*, **26**, 207–221.

STEIGER, R. H. & JÄGER, E. 1977. Subcommission on geochronology: convention on the use of decay constants in geo- and cosmochronology. *Earth and Planetary Sciences Letters*, **36**, 359–362.

STÖSSEL, I. 1993. *Geologie der Umgebung von Cahersiveen, Co. Kerry, Südwestirland*. Diplomarbeit an der Abteilung XC (Erdwissenschaften) thesis, ETH Zürich.

—— 1995. The discovery of a new Devonian tetrapod trackway in SW Ireland. *Journal of the Geological Society, London*, **152**, 407–413.

STREEL, M, & LOBOZIAK, S. 1996. Middle and Upper Devonian miospores. *In*: JANSONIUS, J. & MCGREGOR, D. C. (eds) *Palynology: Principles and Applications*. American Association of Stratigraphic Palynologists Foundation, **2**, 575–587.

——, HIGGS, K., LOBOZIAK, S., RIEGEL, W. & STEEMANS, P. 1987. Spore stratigraphy and correlation with faunas and floras in the type marine Devonian of the Ardenne Rhenish regions. *Review of Palaeobotany and Palynology*, **50**, 211–229.

TODD, S. P. 1989. Role of the Dingle Bay Lineament in the evolution of the Old Red Sandstone of southwest Ireland. *In*: ARTHURTON, R. S., GUTTERIDGE, P. & NOLAN, S. C. (eds) *The Role of Tectonics in Devonian and Carboniferous Sedimentation in the British Isles*. Occasional Publications of the Yorkshire Geological Society, **6**, 35–54.

TUCKER, R. D. & MCKERROW, W. S. 1995. Early Paleozoic chronology: a review in light of new U–Pb zircon ages from Newfoundland and Britain. *Canadian Journal of Earth Sciences*, **32**, 368–379.

——, BRADLEY, D. C., VER STRAETEN, C. A., HARRIS, A. G., EBERT, J. R. & MCCUTCHEON, S. R. 1998. New U–Pb zircon ages and the duration and division of Devonian time. *Earth and Planetary Science Letters*, **158**, 175–186.

——, KROGH, T. E., ROSS, R. J., JR & WILLIAMS, S. H. 1990. Time-scale calibration by high-precision U–Pb zircon dating of interstratified volcanic ashes in the Ordovician and Lower Silurian stratotypes of Britain. *Earth and Planetary Science Letters*, **100**, 51–58.

VAN VEEN, P. M. & VAN DER ZWAN, C. J. 1980. Palynological evidence concerning the middle/late Devonian age of the Green Sandstone Formation, McGillycuddy's Reeks, southwest Ireland. *Geologie en Mijnbouw*, **59**, 405–408.

WALSH, P. T. 1968. The Old Red Sandstone west of Killarney, Co. Kerry, Ireland. *Proceedings of the Royal Irish Academy*, **66(B)**, 9–26.

WEGMANN, M. 1993. *Die Geologie des Killarney Nationalpark, County Kerry, SW-Irland*. Diplomarbeit an der Abteilung XC (Erdwissenschaften) thesis, ETH Zürich.

WESTOLL, T. S. 1979. Devonian fish biostratigraphy. *In*: HOUSE, M. R., SCRUTTON, C. T. & BASSETT, M. G. (eds) *The Devonian System*. Special Papers in Palaeontology, **23**, 341–353.

WIEDENBECK, M., ALLE, P., CORFU, F. *et al.* 1995. Three natural zircon standards for U–Th–Pb, Lu–Hf, trace element and REE analyses. *Geostandards Newsletter*, **19**, 1–23.

WILLIAMS, E. A. 1993. *The stratigraphy, fluvial sedimentology and structural geology of the Old Red Sandstone in the Derrynasaggart Mountains, Counties Cork and Kerry*. PhD thesis, National University of Ireland.

——, BAMFORD, M. L. F., COOPER, M. A. *et al.* 1989. Tectonic controls and sedimentary response in the Devonian–Carboniferous Munster and South Munster Basins, south-west Ireland. *In*: ARTHURTON, R. S., GUTTERIDGE, P. & NOLAN, S. C. (eds) *The Role of Tectonics in Devonian and Carboniferous Sedimentation in the British Isles*. Occasional Publications of the Yorkshire Geological Society, **6**, 123–141.

——, FRIEND, P. F. & WILLIAMS, B. P. J. 2000. A review of Devonian timescales: databases, construction and new data. *This volume*.

——, SERGEEV, S. A., STÖSSEL, I. & FORD, M. 1997. An Eifelian U–Pb zircon date for the Enagh Tuff Bed from the Old Red Sandstone of the Munster Basin in NW Iveragh, SW Ireland. *Journal of the Geological Society, London*, **154**, 189–193.

WINCHESTER, J. A. & FLOYD, P. A. 1977. Geochemical discrimination of different magma series and their differentiation products using immobile elements. *Chemical Geology*, **20**, 325–343.

WOOD, D. A. 1980. The application of a Th–Hf–Ta diagram to problems of tectonomagmatic classification and to establishing the nature of crustal contamination of basaltic lavas of the British Tertiary volcanic province. *Earth and Planetary Science Letters*, **50**, 11–30.

Controls on magmatism in the Munster Basin, SW Ireland

M. PRACHT
Geological Survey of Ireland, Beggars Bush, Dublin 4, Ireland
(e-mail: prachtm@tec.irlgov.ie)

Abstract: The following paper is intended as a review of the magmatism in the Munster Basin and an initial attempt to link it with the basin's history. During the initiation, evolution and inversion of the Devono-Carboniferous Munster Basin magmatic activity was widespread, although small in volume. The episodic occurrence and diversity of the magmatism gives insights into the basement involvement and structural controls relating to the basin's history. The differences in composition, location and structural relationship of the magmatic bodies are used as indicators for the timing of their emplacement and their relationship to the basin's evolution. On the western Beara Peninsula in SW Ireland alone, more than 160 sheet-like intrusions, a wide variety of tuff bands and a deep-seated pipe-like structure of lamprophyric affinity occur. Other centres of magmatism in the basin are the Lough Guitane area, where rhyolitic lava flows and acidic pyroclastic rocks are associated with contemporaneous basin faults, and the Valentia Harbour area, where doleritic sills are associated with a volcanic breccia. A rhyodacitic tuff can be traced as far as St Finan's Bay on the western shore of the Iveragh Peninsula. The chemical composition of the extrusive and intrusive magmatic bodies ranges from tholeiitic dolerites (Valentia Harbour) and subalkaline rhyolites (Lough Guitane) in Mid-Devonian time to subalkaline basalts and tuffs (e.g. Beara Peninsula) in Late Devonian time. In Late Carboniferous and possibly Permian time alkali basalts, trachytes and phonolites (e.g. Beara Peninsula) occur. The igneous activity in the Munster Basin is linked to the basin's history by the interaction between active faults and fractures opening up during various stages of stress imposed on the basin and exploitation of these faults and fractures by rising magma.

This paper is an attempt to synthesize our current knowledge of the magmatism in SW Ireland with special emphasis on the Munster Basin. In trying to link the magmatism to the Munster Basin development, the author is aware of the paper's shortcomings, that is, the sparsity of age data for many of the igneous rocks and the larger part of the sedimentary infill and the still unsatisfactory understanding of the structural evolution and the role of the basement. There exists also a clear bias towards the better documented magmatic occurrences (i.e. those of Lough Guitane, Valentia Harbour and the Beara Peninsula). The Munster Basin, as referred to, comprises a region of continued sedimentation from the Devonian ORS to Carboniferous marine deposition in SW Ireland. No differentiation has been made between the Munster Basin and the temporal sub-basin, the South Munster Basin, or indeed further sub-basins thereof (see below). Because very few absolute age dates exist in the notoriously difficult to date 'Old Red Sandstone' facies and only sporadic detailed studies have been published, the review has to be preliminary and somewhat arbitrary, and it is hoped that future work will improve our understanding of the relationship between basin evolution and magmatism. A broad range of magmatic rocks of Devono-Carboniferous age occur across Ireland (Fig. 1), but only the major magmatic centres in the immediate vicinity of, and thought to be relevant to the understanding of the development of the Munster Basin, are discussed briefly.

The southern region of Ireland is part of the Rhenohercynian structural zone or Zone 1 of Gill (1962) and Cooper *et al.* (1986). This region is separated from an area to the north, which equates with Zone 2 of Gill (1962) and the Transition Zone (Zone 2 of Cooper *et al.* (1986)) by a major thrust, which is conventionally named the 'Variscan Front' (*sensu* Cooper *et al.* 1984). To the south of this thrust a thick ORS megasequence crops out and the region is dominated by Variscan structures. The metamorphic grade reaches upper anchizone to lower epizone grade (Clayton 1989; Blackmore 1995; Meere 1995).

Despite their small volume, Upper Palaeozoic igneous rocks in Ireland are widespread and

From: FRIEND, P. F. & WILLIAMS, B. P. J. (eds). *New Perspectives on the Old Red Sandstone.* Geological Society, London, Special Publications, **180**, 303–317. 0305-8719/00/$15.00

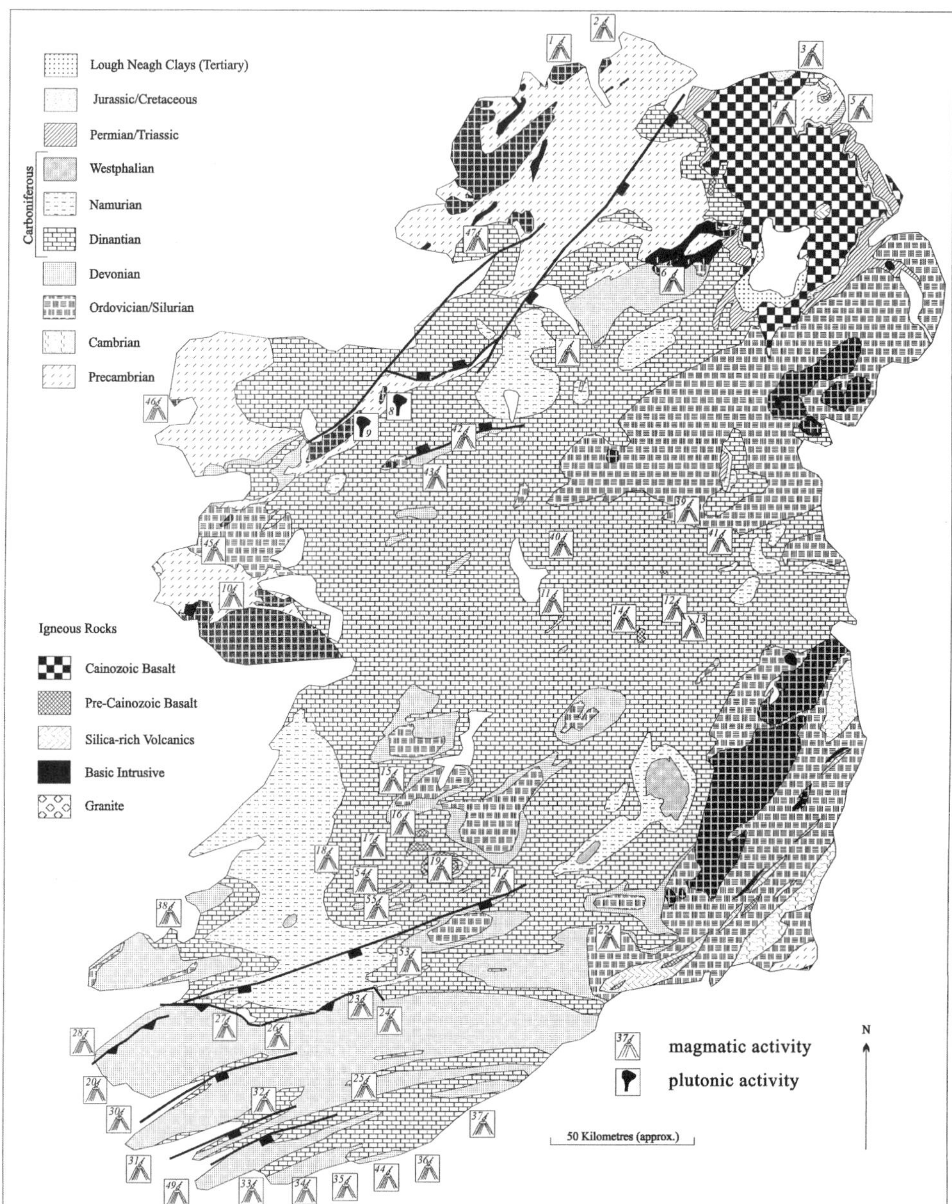

Fig. 1. Occurrences of Upper Palaeozoic igneous rocks in Ireland. Numbers refer to Fig. 2.

occur in a variety of styles. Extrusive and intrusive igneous rocks, including widespread tuff bands, can be found from Co. Cork in the south to the Inishowen Peninsula, Co. Donegal in the north (e.g. Murthy 1948; Coe 1969; Mitchell & Mohr 1987; Brück *et al.* 1988; Figs 1 and 2).

Although less well developed in Ireland than in Britain (Floyd 1982*a*, *b*; Francis 1988), tholeiitic volcanism in Mid-Devonian (Williams *et al.* 1997) times associated with continental ORS facies can be recognized within the Munster Basin (Graham *et al.* 1995). During Late

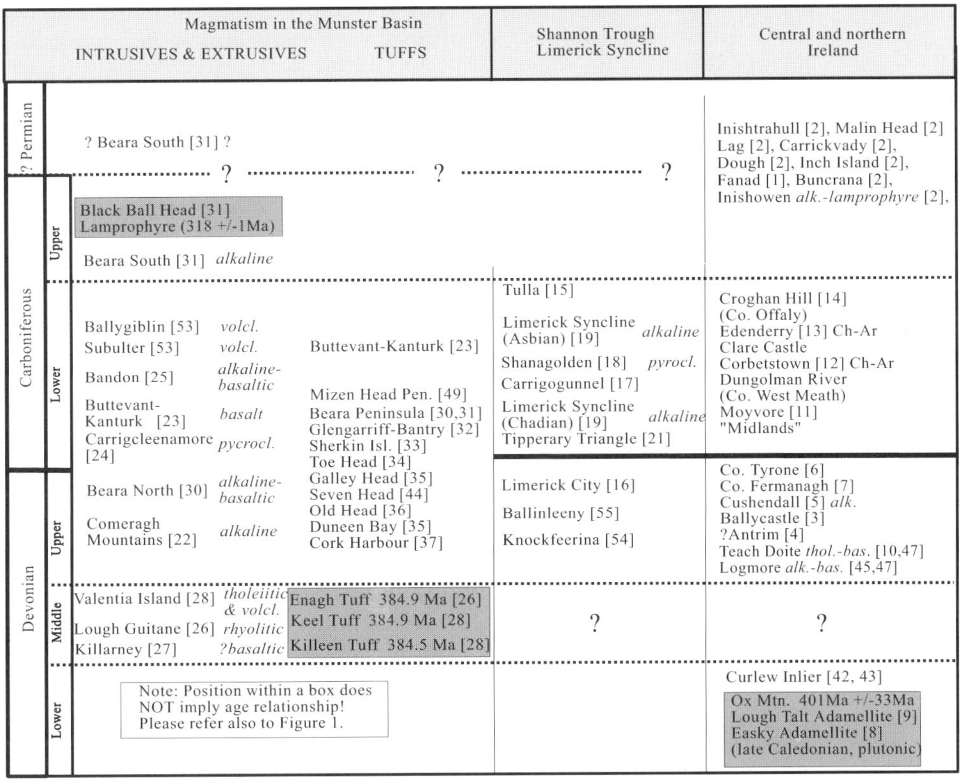

Period		Magmatism in the Munster Basin: INTRUSIVES & EXTRUSIVES		Magmatism in the Munster Basin: TUFFS	Shannon Trough Limerick Syncline	Central and northern Ireland
? Permian		? Beara South [31] ?				Inishtrahull [2], Malin Head [2] Lag [2], Carrickvady [2], Dough [2], Inch Island [2], Fanad [1], Buncrana [2], Inishowen *alk.-lamprophyre* [2],
		?		?	?	
Carboniferous	Upper	Black Ball Head [31] Lamprophyre (318 +/-1Ma)				
		Beara South [31]	*alkaline*			
	Lower	Ballygiblin [53] Subulter [53] Bandon [25] Buttevant-Kanturk [23] Carrigcleenamore [24]	*volcl.* *volcl.* *alkaline-basaltic* *basalt* *pycrocl.*	Buttevant-Kanturk [23] Mizen Head Pen. [49] Beara Peninsula [30,31] Glengarriff-Bantry [32] Sherkin Isl. [33] Toe Head [34]	Tulla [15] Limerick Syncline (Asbian) [19] *alkaline* Shanagolden [18] *pyrocl.* Carrigogunnel [17] Limerick Syncline (Chadian) [19] *alkaline* Tipperary Triangle [21]	Croghan Hill [14] (Co. Offaly) Edenderry [13] Ch-Ar Clare Castle Corbetstown [12] Ch-Ar Dungolman River (Co. West Meath) Moyvore [11] "Midlands"
Devonian	Upper	Beara North [30] Comeragh Mountains [22]	*alkaline-basaltic* *alkaline*	Galley Head [35] Seven Head [44] Old Head [36] Duneen Bay [35] Cork Harbour [37]	Limerick City [16] Ballinleeny [55] Knockfeerina [54]	Co. Tyrone [6] Co. Fermanagh [7] Cushendall [5] *alk.* Ballycastle [3] ?Antrim [4] Teach Doite *thol.-bas.* [10,47] Logmore *alk.-bas.* [45,47]
	Middle	Valentia Island [28] Lough Guitane [26] Killarney [27]	*tholeiitic & volcl.* *rhyolitic* *?basaltic*	Enagh Tuff 384.9 Ma [26] Keel Tuff 384.9 Ma [28] Killeen Tuff 384.5 Ma [28]	?	?
	Lower	Note: Position within a box does NOT imply age relationship! Please refer also to Figure 1.				Curlew Inlier [42, 43] Ox Mtn. 401Ma +/-33Ma Lough Talt Adamellite [9] Easky Adamellite [8] (late Caledonian, plutonic)

References

[1, 46] LEUTWEIN, F. 1970. [2] MURTHY, M.V.N. 1948. **[3, 4, 5, 6, 7]** *Geological Survey of Northern Ireland. 1977.* **[8, 9, 42, 43]** MacDERMOT, C.V.et al. 1996. **[10, 45]** MITCHELL, J.G.. & MOHR, P. 1987.**[11, 12, 13]** Geological Survey of Ireland. **[14]** STROGEN, P. 1974. **[15]** SCHULTZ, R.W. & SEVASTOPULO, G.D 1965. **[16, 18, 20]** SLEEMAN, A.G. & PRACHT, M. (1998). **[17]** STROGEN, P. 1973. **[19]** STROGEN, P. et al. 1996. **[21]** SWEENEY, E. 1998. **[22]** PENNEY, S.R. 1978. **[23, 24, 27]** PRACHT, M. 1997. **[23, 32, 35, 36]** PRACHT, M. & BATCHELOR, R.A. 1998/9. **[25]** BRUEK, P. et al. 1988. **[26]** AVISON, M. 1984a. **[26]** WILLIAMS, E.A. et al. 1997. **[28]** GRAHAM, J.R. et al. 1995. **[28]** WILLIAMS, E.A. et al. 1998. **[30, 31]** PRACHT, M. & KINNAIRD, J.A 1997. **[31]** PRACHT, M. & KINNAIRD, J.A. 1995. **[33]** GRAHAM, J.R. & REILLY, T.A. 1972. **[34, 35, 44]** GRAHAM, J.R. & REILLY, T.A. 1976. **[34]** CLIPSTONE, D. 1992. **[37, 48]** MacCARTHY, I.A.J. 1990. **[46]** SUTTON, J.S. 1970. **[47]** GALLAGHER S. & ELSDON, R. 1990. **[49]** NAYLOR, D. 1975. **[53,54,55]** SLEEMAN, A.G. & PRACHT, M. (1998).

Fig. 2. Localities and source cross-references of Devonian to Permo-Carboniferous magmatism in Ireland.

Devonian and Dinantian time, alkali basaltic intrusions were emplaced predominantly as sills and dykes, with volcanic ash being deposited as tuff layers within Devonian to Lower Carboniferous successions. In Late Carboniferous and possibly Permian times, at the time of the main deformation phase (Coller 1984), volcanism culminated during the inversion and deformation of the basin-fill caused by the onset of the Variscan deformation of the Munster Basin with the intrusion of extensive alkaline sills and dykes (Pracht & Kinnaird 1997). Recent research indicates that magmatism on the Beara Peninsula continued after the Variscan deformation. Preliminary whole-rock analysis (K–Ar) on two separate intrusions from the southern magmatic province indicates a Permian age of some of the intrusions. To verify the significance and validity of the dates (K-loss?), an Ar–Ar step-heating study will follow (M. Timmerman, pers. comm.). A faulted Tertiary dyke on the westernmost part of the Beara Peninsula was studied in detail by Pracht (1994).

The area to the north, Zone 2a of Cooper *et al.* (1986), the Limerick area of the Shannon Trough, was the focal point of igneous activity, which peaked in Early (Chadian to Asbian) Carboniferous times (Strogen 1988). However, it is important to bear in mind that although Devonian magmatism in this region was widespread, no detailed studies have yet been carried out on these rocks.

In Ireland each structural zone and individual areas within each zone display temporal and spatial differences in the nature, i.e. composition and volume, of their magmatic history. Intermittent volcanic activity of the Munster Basin and the region to the north, as far north as Donegal, persisted for a long period of time and does not appear to have been directly related to any deformation phase of the Variscan Orogeny. Instead, a relationship between extensional (basin development phase) and compressional (basin inversion phase) stress regimes of the Munster Basin and the Variscan Orogeny and the magmatic inventory is sought.

Metamorphic studies (e.g. Blackmore 1995; Meere 1995) have so far failed to find any relationship between the intrusions and geothermal temperatures within the area of intrusive bodies. This is not surprising considering that the area underwent lower greenschist facies metamorphism and that, based on their small volume and despite their high intrusion temperature (likely to be >800 °C), the intrusions had little heat capacity. Thin thermal contact zones are locally present. Potential geothermal trends may also have been obliterated by the effects of mineralizing fluids passing through the host rock (e.g. Allihies Copper Mines). The role of the intrusive bodies in the thermal history of the Munster Basin therefore remains unquantified.

The Munster Basin

The Munster Basin in SW Ireland, which crops out over an area of $>12\,500$ km^2, was initiated as a major intracratonic depositional area in Early to Mid-Devonian times (Williams *et al.* 1997, 1998) by extension on probably pre-existing Caledonian faults (Gall 1991). The basin developed as a half-graben (Naylor & Jones 1967; Naylor *et al.* 1981) and inherited its ENE–WSW trend from the controlling influence of bordering Caledonian lineaments in the Lower Palaeozoic basement (Gardiner & MacCarthy 1981; Cooper *et al.* 1986; Williams *et al.* 1989). More than 6 km of non-marine 'Old Red Sandstone' sediment (Cooper *et al.* 1986; MacCarthy 1990) accumulated in the depocentre at Kenmare River, to the north of the Beara Peninsula (Fig. 4). The basin's northern margin is ill defined and covered by a thick sequence of Namurian rocks.

During Dinantian time, the area to the north of a line drawn from approximately Cork City to Kenmare developed as a shelf area, the site of a thick sequence of Carboniferous limestones. Rapid clastic sedimentation continued in the basin, to the south of the Cork–Kenmare Line (Naylor & Sevastopulo 1979) during early to mid-Courceyan time with only thin equivalent sequences on the shelf. During Viséan time the distinction between the basin and the shelf was maintained. Little sediment was deposited during Viséan times in the basin, however, whereas sedimentation on the shelf kept pace with subsidence (Naylor *et al.* 1989).

Compression of the Munster Basin towards the end of the Carboniferous period resulted in basin inversion, with reactivation of original extensional faults and minor thrusts (Meere 1992; Price & Todd 1988). The basin fill was deformed by layer-parallel shortening, folding and faulting above an interpreted reactivated basal detachment (Cooper *et al.* 1986; Price & Todd 1988).

During latest Devonian and early Carboniferous time several centres of magmatism developed within the Munster Basin (the shelf area being part of the Munster Basin), which are discussed in their basin evolutionary context below (Fig. 3). Minor occurrences are listed in Fig. 2, but are not discussed here because of lack of information.

Magmatism in the Munster Basin

Lough Guitane

At Lough Guitane quartz keratophyres, originally rhyolitic lava flows, and associated pyroclastic rocks have been described by Avison (1984*a*, *b*) as being associated with contemporaneous faulting. The volcanic rocks are interstratified with the Glenflesk Chloritic Sandstone Formation (Pracht 1997). Three separate volcanic centres were identified, the largest of which is part of a graben structure *c.* 1.5 km wide. The thickest lava flow exceeds 350 m. On the basis of the relationship between the fluviatile sediment and the lava flows two phases of extrusion were recognized by Avison (1984*a*, *b*). An early phase produced the westerly Horses Glen Volcanic Centre and a later phase the Killeen and Bennaunmore Volcanic Centres (Avison 1984*a*, *b*).

Valentia Harbour

In the Valentia Harbour area, Co. Kerry, doleritic sills of tholeiitic affinities are contemporaneous with a spatially associated volcanic breccia and pre-date a rhyodacitic tuff, the Enagh Tuff Bed (Capewell 1975; Graham *et al.* 1995). The magma here is thought to have exploited major basin faults during a period of stretching for its ascent (Graham *et al.* 1995). Compared with the Beara basalts the dolerites are

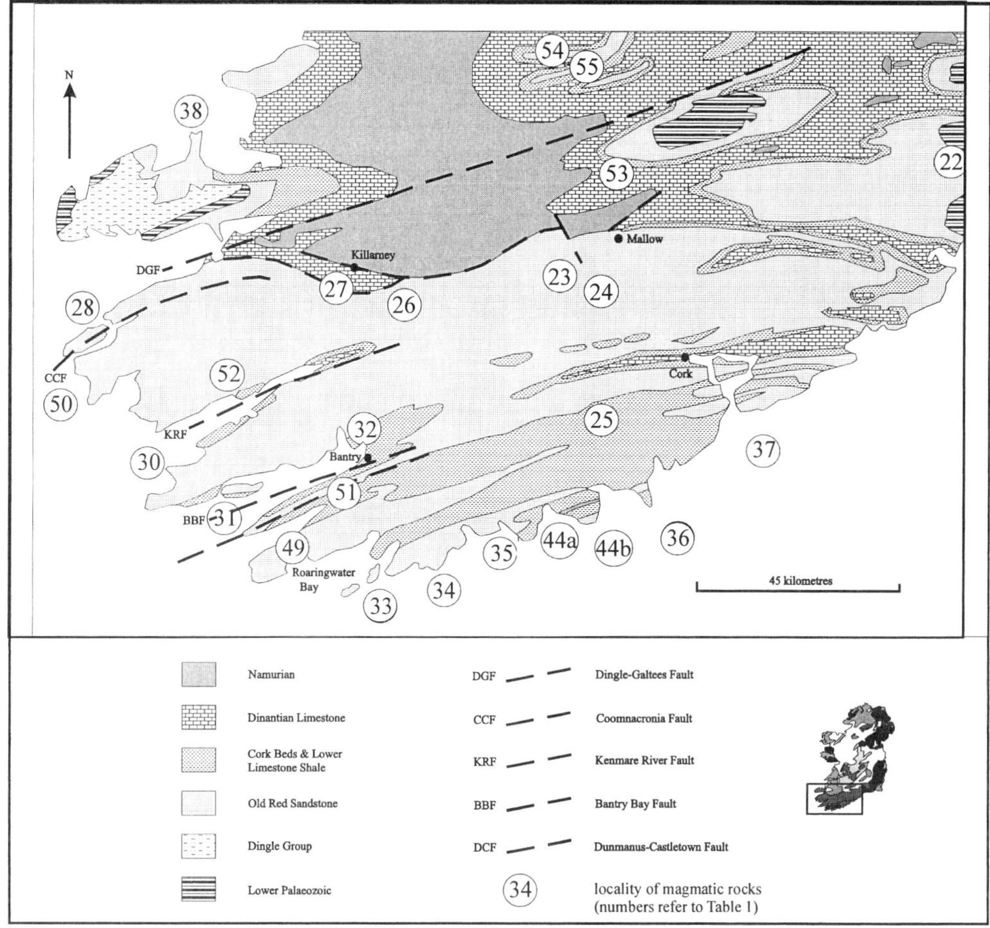

Fig. 3. Occurrences of Upper Palaeozoic igneous rocks in the Munster Basin.

characterized by low Nb/Y (0.2–0.3) and low Zr/Y ratios (3.79–4.13), and they plot in the midocean ridge basalt (MORB) field of the Zr v. Zr/Y discrimination diagram of Pearce & Norry (1979) and the basalt to andesite field of Winchester & Floyd (1977; see Fig. 5, below).

Beara Peninsula

The main centre of magmatism in the Munster Basin during Late Devonian and Carboniferous times was on the western extremity of the Beara Peninsula, which is characterized by a broad ENE-trending antiform that plunges gently to the SW (Coe and Selwood 1963, 1968; Fig. 4). Deformation resulted in three broad structural zones on the peninsula, a northern and southern zone representing the northern and southern limbs of the main structure and a central zone representing the core of the antiform, the 'transition zone' of Coe & Selwood (1963) being broadly part of the central fold zone (James & Graham 1995). On the basis of the change in fold style and termination of stratigraphic units, major faults between all fold zones have been inferred (Coe & Selwood 1963; Pracht 1994; James & Graham 1995). A broad range of basic and felsic alkaline rocks were emplaced into consolidated Upper Devonian and Lower Carboniferous sediments on the Beara Peninsula, close to the depocentre of the Munster Basin (Fig. 4).

Sheet-like intrusions, sills and dykes. The sheet-like intrusions of the Beara Peninsula (Coe 1966) can be divided principally into basic and felsic rocks. One hundred and sixty-four sills and dykes have been mapped, which are all sub-alkaline to alkaline in nature (Pracht & Kinnaird 1997). The sills have a strike length ranging from

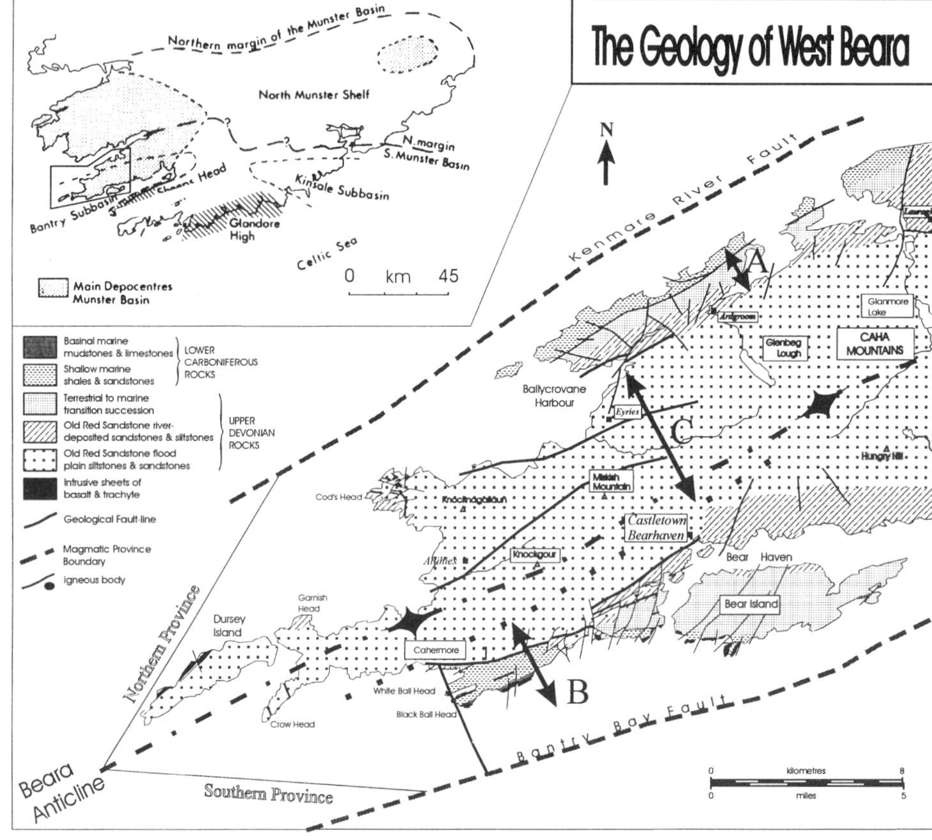

1 Cahermore- Castletown Bearhaven Fault

Fold Zones according to James & Graham (1995)
A Northern Fold Zone
C Central Fold Zone
B Soutern Fold Zone

Fig. 4. Simplified geological map of the western part of the Beara Peninsula.

a few metres to 1.9 km. The thickness of the sills and dykes varies from <10 cm to >80 m and thickness variations along strike of individual sills are common. Transgressive sills can be observed occasionally (e.g. at Black Ball Head). The dykes are generally much shorter in length, although some (e.g. on Bere Island) are >1 km long. All sills and dykes appear to have intruded lithified sediments that are locally baked, and against which the intrusions are chilled. The rocks are in general feldspar-phyric with phenocrysts up to 1.5 cm. Most of the felsic samples show a trachytic texture and feldspar phenocrysts constitute up to 30% of the rock, often forming clusters of over 10 cm in diameter. A refractory cleavage can be clearly distinguished in most sheet-like intrusive bodies, and folding of the dykes is common. Often quartz veins, some of them sheared, can be seen in sills and dykes, and sigmoidal tension gashes indicate intrusion-parallel dextral shear movement. Some sheeted intrusions show boudinage with joints perpendicular to the bedding of the country rock. The variable structural relationship between individual intrusions suggests a broadly contemporaneous intrusion time (Pracht 1994).

On the basis of detailed modal and geochemical analyses two separate suites have been identified (Pracht & Kinnaird 1997). The modal and geochemical data used were carefully screened, and in a parallel study the effects of alteration were monitored (Pracht 1994). On the basis of a wide range of discriminant diagrams (Pracht & Kinnaird 1997), the geochemical

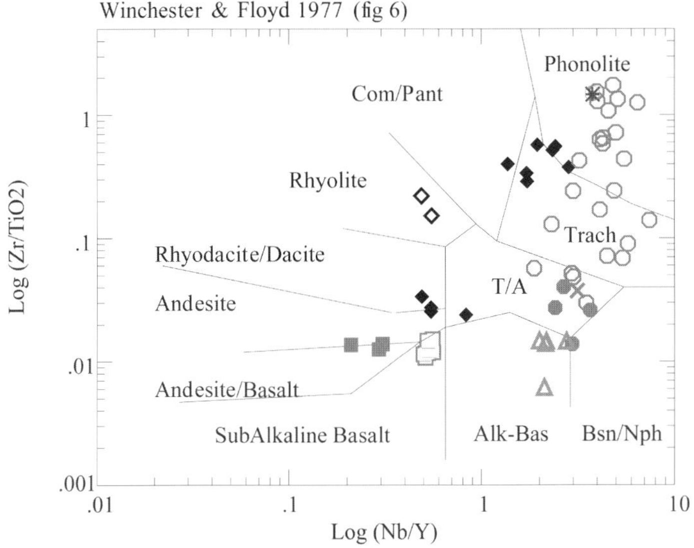

Fig. 5. Magmatic rocks of the Munster Basin and one sample from a Tertiary intrusive rock from Donegal in the Zr/TiO_2–Nb/Y diagram of Winchester & Floyd (1977). Sources: please refer to Fig. 2.

distinction between the two suites can be best demonstrated using the Zr/TiO_2–Nb/Y diagram of Winchester & Floyd (1977), which shows that, using Nb/Y as a discriminant, both subalkaline and alkaline basalts occur (Fig. 5). Alkaline basalts with $Nb/Y > 1$ and more evolved rocks are confined to the south of the area whereas the subalkaline basalts with $Nb/Y < 1$ are restricted to the north. Discriminant diagrams that plot compatible against incompatible trace elements or that use pairs of incompatible trace elements further emphasize the differences in geochemistry between the two suites (Pracht & Kinnaird 1997).

The northern suite of sheet-like intrusions comprises subalkaline basalts of Cod's Head and Dursey Island, which are intruded into Devonian red beds. They are restricted to the central fold zone of the Beara Peninsula. Analytical data cluster closely on discriminant diagrams. The basalts have a range in silica values of 44–49% and have much lower Nb/Y ratios than the southern suite. Although Cr data are limited, Cr and Ni values are well below typical primary melt compositions (> 400 ppm Ni and >1000 ppm Cr; Frey & Prinz (1978)). However, the range of Cr values suggests fractionation of related basalts. All the basalts have a Zr/Y ratio between 5.08 and 7.94, constraining them to within-plate basalts using the criteria of Pearce & Norry (1979) (Pracht & Kinnaird 1997).

The southern suite of sheet-like intrusions comprises alkali basalts, trachytes and phonolites, which crop out along 9 km of the south coast of the Beara Peninsula. These are intruded into Devonian Old Red Sandstone beds and marine Lower Carboniferous strata and occur within the southern fold zone of the peninsula. These intrusions have a much broader compositional spectrum than the northern suite, and have higher Ba, La, Nb, Th, Ce and Zr, and lower V, Sc and Ni contents. They are characterized by higher Nb/Y (2.42–3.68) and higher Zr/Y ratios (7.66–11.53) and they plot in the within-plate basalt field of the Zr v. Zr/Y discrimination diagram of Pearce & Norry (1979). The associated felsic rocks fall on a fractionation trend from the basalts but are unrelated to the basalts of the northern province (Pracht & Kinnaird 1997).

Pipe-like intrusion with lamprophyric affinities. In addition to the sheet-like intrusions, a pyroclastic (Fisher & Schmincke 1984) pipe of lamprophyric affinities (*sensu* Bergman 1987; Rock 1987) occurs at Black Ball Head, extending to White Ball Head and Cahermore (Coe 1966, 1969; Pracht & Kinnaird 1995). This pipe is closely associated with the sheet-like intrusions of the southern suite, both cross-cutting and being cut by sills and dykes. The Black Ball Head pipe has a megacryst assemblage of kaersutite, phlogopite and Ti-magnetite, accompanied by nodules of amphibole pyroxenites set in a completely altered matrix. A mantle origin has been established for the mineral assemblage of the pipe, with a crystallization depth of >75 km (Pracht & Kinnaird 1995). On the basis of field observation,

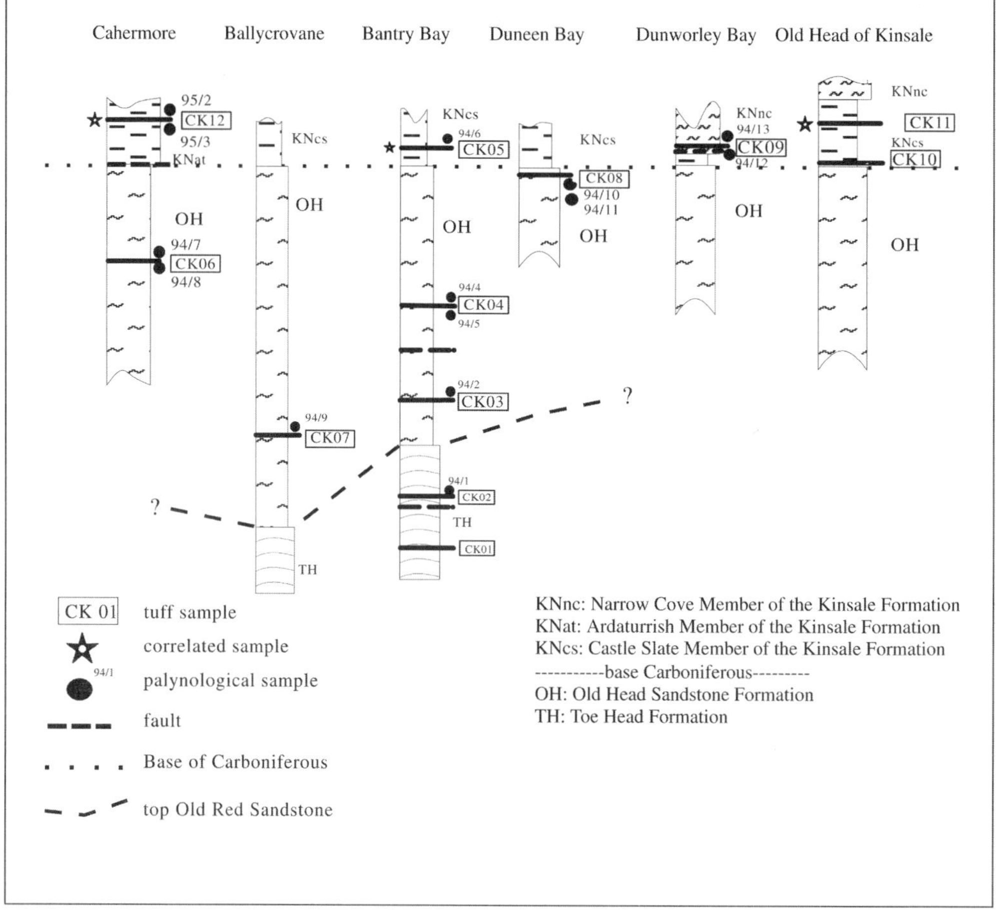

Fig. 6. Stratigraphic position of studied South Munster Basin tuff layers at selected locations and their correlation.

a structural relationship between the location of the Black Ball Head Pipe and a NNW-trending tensional shear fault is invoked. The fault is interpreted as the possible result of strike-slip movement along the Cahermore–Castletownbere Fault (between the central and southern fold zone). Recent (1999) Ar–Ar step heating dating of kaersutite megacrysts from the pipe at Black Ball Head yielded a *c.* 318 Ma plateau age (Namurian) (M. Timmerman, pers. comm.). The pipe is located within the southern igneous suite and is cut by a trachytic dyke belonging to this suite.

Tuff layers in the Munster Basin. Volcanic activity is marked by several tuff layers within the Munster Basin (Fig. 6). They occur in middle Devonian to lower Carboniferous strata. A tuff unit from the Killeen Volcanic Centre of the Lough Guitane area has been dated by the U/Pb method on zircon crystals as 384.5 ± 1.0 Ma (Eifelian) (Williams *et al.* 1998). The Enagh Tuff bed from the Valentia Harbour area, which is closely associated with the Bealtra igneous rocks, is about 4–10 m thick and has been dated by the Pb–Pb method on zircons by Williams *et al.* (1997) as 384.9 ± 0.7 Ma (Eifelian). A preliminary study of some tuff layers further south (Beara Peninsula to Old Head of Kinsale) was carried out by Pracht & Batchelor (1998/99).

None of the southern tuffs has been radiometrically dated but chronostratigraphical control is based on miospore assemblages obtained from above and below each tuff layer (Higgs in Pracht & Batchelor 1998/99; Fig. 6). On the basis of heavy mineral content and major and trace element analysis, the southern tuff layers of the Munster Basin show evidence of bimodal volcanism, represented by mafic and evolved alkaline compositions. Bulk-rock geochemical data allow the evolved alkaline southern tuff layers to be subdivided into two distinct chemical groups. Analysis of zircon crystals extracted from the southern tuff layers confirms two chemical associations within the evolved alkaline group, and one within the mafic group (Pracht & Batchelor 1998/99).

There are at least six distinct volcanic events preserved as tuff layers in the Upper Devonian to Lower Carboniferous successions between Ballycrovane Harbour on the north side of the Beara Peninsula and the Old Head of Kinsale, west of Cork Harbour (Fig. 6). The data suggest a temporal progression from the basic alkali extrusive rocks of a mafic group M1 (CK01, CK02) in the Toe Head Formation to the evolved alkali groups A1 (CK03, CK04, CK07) and A3 (CK06) in the Old Head Sandstone Formation, and from a mafic group M2 (CK08) in the upper part of the Old Head Sandstone Formation to the highly evolved group A2 (CK05, CK11, CK12) in the Castle Slate Member–Ardaturrish Member of the Kinsale Formation. The mafic group M3 (CK09) indicates yet another event of volcanic activity in the Ardaturrish Member. Four sets of these tuff layers can be correlated in the Munster Basin. Two mafic groups M1 and M2 and two evolved groups A1 and A2 are differentiated. The recognition of these groups provides a tool for the correlation of stratigraphic units and allows verification of biostratigraphical data (Pracht & Batchelor 1998/99).

Discussion

Relationship between magmatism and basin evolution

So far no attempt has been made to create a concept for magmatic evolution of the Munster Basin in Devono-Carboniferous times. However, the increase of research carried out over the past decade or so provides us with enough data to attempt a preliminary review.

In Eastern Canada, the geochemical evolution of Devonian to Carboniferous igneous rocks of the Magdalen Basin has been described by Pe-Piper & Piper (1998). The Magdalen Basin is the largest sub-basin of the Maritimes Basin and covers an area of approximately 25 000 km^2. The basin is situated in the northern Appalachians and is a major basin of Mid-Devonian to Early Permian age. The basin fill consists of *c.* 7 km of clastic fluviatile to lacustrine sediments overlain by additional clastic deposits, evaporites, carbonates and mafic volcanic rocks (LaFlèche *et al.* 1998). Although the Magdalen Basin is about double the size of the Munster Basin, its age, palaeogeographical position, sedimentary history and magmatic inventory make a comparison worth while.

Five phases of magmatic evolution of the Magdalen basin have been distinguished by Pe-Piper & Piper (1998): (1) Mid- to Late Devonian partial melting of lithospheric mantle, producing tholeiites and minor alkalic basalt; (2) anhydrous base-of-crust melting producing A-type granites; (3) in latest Devonian time, the development of a large fractionating magma chamber at the base of the crust with uniform Nd isotopes and high Ti in the mafic magma; (4) intrusion of late Tournaisian moderately well-fractionated dykes; (5) local Viséan to Westphalian alkalic magmas, which ascend along crustal-scale faults (Pe-Piper & Piper 1998).

It is also interesting to note here that some samples from basalts of the Magdalen Islands contain a very similar mineral assemblage to the pipe at Black Ball Head, consisting of clinopyroxene (Ti-rich salite), kaersutite and biotite with accessory apatite, quartz, titanomagnetite and pyrite (LaFlèche *et al.* 1998).

As discussed by Piper *et al.* (1993, 1995), faulting and thrusting to create space for the intrusion of magmas were important processes in the Magdalen Basin. The same applies for the Munster Basin. In the following section, characteristics and features of the major faults recognized in the Munster Basin are described briefly.

At Lough Guitane magmatism, probably of Mid-Devonian age (Williams *et al.* 1998), is interpreted to be associated with contemporaneous basin faulting (Avison 1984*a*). In the Valentia Harbour area the magmatism is thought to be associated with major basin faults during a period of stretching (Graham *et al.* 1995), similar to the Coomnacronia Fault, a major basinal fault extending for >35 km in an ENE–WSW direction. Likewise, on the Beara Peninsula, the sheet-like intrusions are thought to be controlled by strike-slip on major basin faults and resulting cross-cutting shear faults. Three major faults have been recognized (Coe & Selwood 1963, 1968; Naylor 1978; James & Graham 1995; Pracht & Kinnaird 1997), the Cahermore–Castletownbere Fault, Kenmare River Fault and Bantry Bay Fault. All of these faults extend in an approximately ENE–WSW direction.

The Cahermore–Castletownbere Fault can be mapped on land. The fault corresponds approximately to the fold zone boundaries between the central and southern fold zone of James & Graham (1995). The fault offsets an uppermost Devonian to lower Carboniferous broadly NE–SW-striking sedimentary sequence to the south against ENE–WSW-striking continental red-bed facies to the north. The stratigraphic units to the south cannot be traced across the fault to the north. As the fault itself is not exposed a quantification of fault movement is not possible, but it is estimated to be several hundreds of metres.

Two important south-dipping high angle reverse faults in the Bantry Bay area were mapped by Naylor (1978). One fault runs through the north side of the Sheeps Head Peninsula and the other to the north of Whiddy Island, with a structural low to the north (Naylor 1978). The existence of a major structural lineament in Bantry Bay is strongly supported by a geophysical survey and bottom sampling by divers (Caston & Max 1978), which has shown that the igneous rocks of the Beara Peninsula extend into Bantry Bay with an approximate ENE–WSW strike.

The tracing of a fault through Kenmare River is more circumstantial and to a large extent hypothetical. However, there exists an important major thrust on land on the southern side of the Kenmare Syncline, the Kenmare Thrust. This displaces the Caha Mountain Formation against the Dinantian succession in the core of the Kenmare Syncline (Husain 1957; Pracht 1997).

In addition to the discussed major faults, the existence of the lamprophyric pipe at Black Ball Head, with its upper-mantle xenolith assemblage, shows that very deep-seated basement-involved zones of structural weaknesses existed in Late Carboniferous times.

Magma composition in continental basins changes during progressive lithospheric thinning (McKenzie & Bickle 1988). In general, small batches of magma generated by a crustal stretching factor of $b > 2$ are alkaline, with a change to tholeiitic basalts as the amount of stretching ($b > 4$) and melting increases. However, at stretching factors larger than two, the basic principle of subsidence is significantly changed by the presence of partial melt in the underlying asthenosphere, limiting subsidence to *c.* 2.5 km, which in general is shallower than the level of emplacement of 'oceanic crust' (i.e. tholeiitic magma) (Foucher *et al.* 1982). For the Munster Basin this might be the cause for the intrusive character of the MORB-type dolerites of the Valentia Harbour area.

As already suggested by Leeder (1988), the Mid- to Late Devonian and succeeding Carboniferous basins throughout Ireland have a dominantly extensional character; basin subsidence occurred in response to N–S-oriented tensional stress, related to a subduction zone south of the British Isles.

As suggested by Pe-Piper & Piper (1998) for the Magdalen Basin, the magmatism in the Munster Basin is interpreted to have been induced by extension or lithospheric thinning. The magmatism was controlled partly by pre-existing zones of weakness in the Caledonide crust and partly by fracture zones that developed perpendicular to the main extensional stress regime that dominated during the development of the Munster Basin as it subsided (Leeder 1988; Price & Todd 1988; Todd 1989; Graham *et al.* 1995). Graham *et al.* (1995) suggested that the tholeiitic nature of the Valentia dolerite sills indicates substantial stretching and lithospheric thinning which had taken place by early Mid-Devonian times (Williams *et al.* 1997), with the formation of a large enough melt fraction at the base of the lithosphere to produce the magma (Foucher *et al.* 1982). The extension was probably localized adjacent to major faults similar to the Coomnacronia Fault, which formed during the extension of the Munster Basin and probably represent the reactivation of pre-existing zones of weakness in the underlying crust.

A similar process was suggested by LaFlèche *et al.* (1998) for the evolution of the magmatism in the Magdalen Basin (phase (1)). Here, the magmatism was caused by passive upwelling of the mantle as a result of lithospheric extension following the Acadian Orogeny. Likewise, those workers explained the geochemical heterogeneity within the group of tholeiites by variable degrees of partial melting of a slightly enriched asthenospheric mantle and mixing between alkalic and tholeiitic mafic melts (LaFlèche *et al.* 1998).

Following the argument that alkaline melts form during limited amount of lithospheric streching factors, the subalkaline nature of the sills on the north side of the Beara Peninsula suggests that extension decreased by the time of their emplacement. Evidence for the timing of emplacement of the northern suite is circumstantial in that no age date exists so far (discussed by Pracht & Kinnaird (1997)). On the basis of the available data, the basic rocks from the Valentia area and the basalts from the northern suite of the Beara Peninsula are geochemically distinct (Graham *et al.* 1995; Pracht & Kinnaird 1997; Fig. 5). However, as discussed by Pracht & Kinnaird (1997), the emplacement of the northern suite is interpreted as having taken place before the emplacement of the southern suite, i.e. sometime during late Devonian to early Carboniferous time. This, in turn, is interpreted to reflect a change from an initially dominantly extensional stress regime in mid- to late Devonian times to a subsidence-controlled stress regime in Late Devonian to early Carboniferous times.

Assuming a consistant and homogeneous asthenosphere during Mid-Devonian to Carboniferous times beneath the Munster Basin, the stretching factor for the lithosphere underneath the Beara Peninsula in Carboniferous time was never large enough to allow for the necessary degree of decompression of the mantle source to lead to the generation and extrusion of tholeiitic magmas as on the Iveragh Peninsula in Mid-Devonian time.

The geochemical composition of the uppermost Devonian and Lower Carboniferous syn-depositional alkaline tuffs in the Munster Basin seems to indicate that basin subsidence and small-scale partial melting of the upper mantle, or tapping of a reservoir thereof, continued into Dinantian times. However, it is not clear at this stage to what extent crustal contamination of the magma occurred and if the reservoir was situated in the upper mantle or lower crust. The data from the tuff layers also suggest a temporal progression from basic alkali to evolved alkali compositions. The repeated occurrences of less evolved to progressively more evolved alkaline tuff layers in the stratigraphic succession might indicate pulses of subsidence within the basin leading to repeated tapping of a magma reservoir at the base of the lithosphere (phases 3 and 4 in the Magdalen Basin development).

The magmatic evolution of the Munster Basin continued with the intrusion of the alkaline sills and dykes (all syn- to post-diagenetic and most pre-deformational; phase 5 of the Magdalen Basin development). Assuming a Late Carboniferous Variscan deformation phase and knowing the age of the kaesutite megacrysts of the Black Ball Head Pipe and its structural relationship to the sills and dykes interacting with it, a late Carboniferous age for the southern suite seems reasonable. This would indicate a switch from a subsidence phase to a compressional stress regime and is a possible indication of the beginning of basin inversion. A deeper source for this later magmatism in the southern province of the Beara Peninsula could reflect stress release and associated 'opening up' of the faults (Coe 1966; Matthews *et al.* 1983) that controlled basin formation. Alternatively, the pipe could also be the result of fracture propagation as a result of the presence of a low-viscosity volatile component at the depth of formation and fluidization processes closer to the surface (Anderson 1979). In each case, a zone of structural weakness in the penetrated crust is envisaged.

Overall, the earlier suite (of Mid- to Late Devonian age) of the Valentia Harbour sub-volcanic rocks on the Iveragh Peninsula resulted either from extensive partial melting as a result of greater stretching or equilibration at the base of the crust whereas later alkaline magmatism may have resulted from a smaller degree of partial melting at greater depth (closer to the depocentre?).

With respect to the Beara Peninsula, on the basis of field observation and geochemical data it is interpreted that, as the *focus et tempus* of magmatism shifted southwards, the intrusions became more alkaline in nature and fractionated to give related felsic differentiates. Thus, the different compositions in their occurrence might be explained by different time, place, and depth of magma generation.

However, for some of the intrusions on the Beara Peninsula late Variscan (?Permian) stress relief causing opening of fractures, and potential conduits cannot be ruled out (P. Meere pers. comm.). Permian magmatic activity is indicated by preliminary K–Ar studies (see above).

Conclusions

The data available to date suggest that the alkaline magmatism in the Munster Basin was the result of decompression fractional melting of the mantle induced by lithospheric extension, controlled partly by pre-existing zones of weakness in the Caledonide crust and partly by fracture zones that developed parallel to the Munster Basin margin as it subsided. The magmatism is linked to major fault systems with a general ENE–WSW strike. The age and detailed history of these faults, however, remains unclear.

In this paper it was attempted to link the magmatism in the Munster Basin to extension, subsidence and inversion of the basin. It is concluded that the magmatism is related to zones of structural weaknesses in the crust, represented by faults and thrusts and that the various magmatic events took place at certain stages of the basin evolution. The dolerites of the Valentia Harbour area were intruded during an extensional phase. On the Beara Peninsula, the earlier northern suite is interpreted to have intruded during a subsidence phase of the Munster Basin development, close to the depocentre. The southern suite, including the Black Ball Head lamprophyre, intruded during a longer period of time possibly beginning at the subsidence–dominated phase but terminating the latest, during the Variscan deformation. The change in geochemical composition we see from the Valentia Harbour dolerites to the Beara trachytes is consistent with studies of magmatism in continental basins, which show that magma composition changes during progressive lithospheric thinning

(McKenzie & Bickle 1988). Small batches of magma generated by a crustal stretching factor of $b > 2$ are alkaline, with a change to tholeiitic basalts as the amount of stretching ($b > 4$) and melting increases.

As pointed out above, a deeper source for the later magmatism (Black Ball Head lamprophyre) could reflect stress release and associated opening up of previously extensional basin faults. This southward shift of magmatism from the Valentia Harbour area to the depocentre near the Beara Peninsula is consistent with basin development which began in the Mid-Devonian (Williams *et al.* 1997), to the north of Kenmare, and transferred southward to form the Munster Basin, south of the Cork–Kenmare Line, in early Carboniferous time (Williams *et al.* 1989).

The nature and composition of the magmatism in the Munster Basin and beyond demonstrate that: (1) partial melting of the upper mantle took place; (2) structural weaknesses existed to facilitate the ascent of these magmas from considerable depth; (3) these fault systems were activated repeatedly; (4) the change in geochemical composition can be explained by the difference in *loci et tempi* of the magmatism.

The occurrence of tholeiitic basalts in the Valentia Harbour region, the subalkaline composition of the northern province on the Beara Peninsula and the alkaline magmas of the southern province indicate a complex history of intrusions still to be resolved. Much work remains to be done, especially with respect to the many Devonian igneous rocks occurring to the north of the 'Variscan Front', but also in assigning absolute age dates to many of the igneous occurrences within the Munster Basin and the detailed study of the structural inventory of the region.

I wish to thank the referees and editors for their constructive comments on an earlier version of this paper. Thanks also go to the Geological Survey of Ireland. Special thanks are due to M. Timmerman for allowing me to use unpublished data on the Beara Intrusives.

References

Anderson, O. L. 1979. Diatremes and carbonatites. The role of fracture dynamics in kimberlite pipe fromation. *In*: Boyd, F. R. & Meyer, H. O. A. (eds) *Kimberlites, Diatremes and Diamonds: Their Geology, Petrology and Geochemistry. Proceedings of the Second International Kimberlite Conference*, **2**, 344–353.

Avison, M. 1984*a*. Contemporaneous faulting and the eruption and preservation of the Lough Guitane Volcanic Complex, Co. Kerry. *Journal of the Geological Society of London*, **141**, 501–510.

—— 1984*b*. The Lough Guitane Volcanic Complex, County Kerry—a preliminary survey. *Irish Journal of Earth Sciences*, **6**, 127–136.

Bergman, S. C. 1987. Lamproites and other potassium-rich igneous rocks: a review of their occurrence, mineralogy and geochemistry. *In*: Fitton, J. G. & Upton, B. G. J. (eds) *Alkaline Igneous Rocks*. Geological Society, London, Special Publications, **30**, 103–190.

Blackmore, R. 1995. Low-grade metamorphism in the Upper Palaeozoic Munster Basin, southern Ireland. *Irish Journal of Earth Sciences*, **15**, 115–133.

Brück, P. M., Beese, A. P., Sweetman, T. M. & Wheildon, J. 1988. Alkali gabbro sills in the Lower Carboniferous near Bandon, County Cork. *Irish Journal of Earth Sciences*, **9**, 23–37.

Capewell, J. G. 1975. The Old Red Sandstone Group of Iveragh, Co. Kerry. *Proceedings of the Royal Irish Academy*, **75B**, 155–171.

Caston, G. F. & Max, M. D. 1978. Rock outcrops and sediments, Northwest Bantry Bay, Ireland. *Geological Survey of Ireland Bulletin*, **2**, 205–213.

Clayton, G. 1989. Vitrinite reflectance data from the Kinsale Harbour–Old Head of Kinsale area, southern Ireland, and its bearing on the interpretation of the Munster Basin. *Journal of the Geological Society, London*, **146**, 611–616.

Clipstone, D. 1992. *Biostratigraphy and lithostratigraphy of the Dinantian of the Kilmaclenine area, North County Cork*. PhD thesis, National University of Ireland.

Coe, K. 1966. Intrusive tuffs of West Cork, Ireland. *Journal of the Geological Society, London*, **122**, 1–28.

—— 1969. The geology of the minor intrusions of West Cork, Ireland. *Proceedings of the Geological Association*, **80**(4), 441–457.

—— & Selwood, E. B. 1963. The stratigraphy and structure of part of the Beara Peninsula, Co. Cork. *Proceedings of the Royal Irish Academy*, **63**, 33–59.

—— & —— 1968. The Upper Palaeozoic stratigraphy of West Cork and parts of South Kerry. *Proceedings of the Royal Irish Academy*, **66**, 113–131.

Coller, D. W. 1984. Variscan structures in the Upper Palaeozoic rocks of west central Ireland. *In*: Hutton, D. H. W. & Sanderson, D. J. (eds) *Variscan Tectonics of the North Atlantic Region*. Geological Society, London, Special Publications, **14**, 185–194.

Cooper, M. A., Collins, D. A., Ford, M., Murphy, F. X., Trayner, P. M. 1984. Structural style, shortening estimates and thrust front of the Irish Variscides. *In*: Hutton, D. H. W. & Sanderson, D. J. (eds) *Variscan Tectonics of the North Atlantic Region. Geological Society, London, Special Publications*, **14**, 167–176.

——, ——, ——, ——, —— & O'Sullivan, M. 1986. Structural evolution of the Irish Variscides. *Journal of the Geological Society, London*, **143**, 53–61.

DALY, J. S. 1998. *Mid-late Devonian magmatism in Connemara.* 41st Annual Irish Geological Research Meeting, Galway, Abstract, 13.

FISHER, R. V. & SCHMINKE, H.-U. 1984. Pyroclastic rocks.

FLOYD, P. A. 1982*a*. Introduction: geological setting of upper Palaeozoic magmatism. *In*: SUTHERLAND, D. S. (ed.) *Igneous Rocks of the British Isles.* Wiley, Chichester, 217–225.

—— 1982*b*. The Hercynian Trough: Devonian and Carboniferous volcanism in south-western Britain. *In*: SUTHERLAND, D. S. (ed.) *Igneous Rocks of the British Isles.* Wiley, Chichester, 227–242.

FOUCHER, J.-P., LE PICHON, X. & SIBUET, J.-C. 1982. The ocean-continent transition in the uniform lithospheric stretching model: role of partial melting in the mantle. *In*: KENT, P., BOTT, M. H. P., MCKENZIE, D. P. & WILLIAMS, C. A. (eds) *The Evolution of Sedimentary Basins.* Philosophical Transactions of the Royal Society of London, Series A, **305**, 27–40.

FRANCIS, E. H. 1988. Mid-Devonian to early Permian volcanism: Old World. *In*: HARRIS, A. L. & FETTES, D. J. (eds) *The Caledonian–Appalachian Orogen.* Geological Society, London, Special Publications, **38**, 573–584.

FREY, F. A. & PRINZ, M. 1978. Ultramafic inclusions from San Carlos, Arizona: petrologic and geochemical data bearing on their petrogenesis. *Earth and Planetary Science Letters*, **38**, 129–176.

GALL, LE B. 1991. Crustal evolutionary model for the Variscides in Ireland and Wales from SWAT seismic data. *Journal of the Geological Society, London*, **148**, 759–774.

GALLAGHER, S. & ELSDON, R. 1990. Spinel lherzolite and other xenoliths from a dolerite dyke in southwest Donegal. *Geological Magazine*, **127**(2), 177–180.

GARDINER, P. R. R. & MACCARTHY, I. A. J. 1981. The Late Palaeozoic evolution of Southern Ireland in the context of tectonic basins and their transatlantic significance. *In*: KERR, J. W. & FERGUSSON, A. J. (eds) *Geology of the North Atlantic Borderlands.* Memoirs, Canadian Society for Petroleum Geologists, **7**, 683–725.

Geological Survey of Northern Ireland 1977. *Geological map of Northern Ireland*, 1 : 250 000 scale. Geological Survey of Northern Ireland.

GILL, W. D. 1962. The Variscan fold belt in Ireland. *In*: COE, K. (ed.) *Some Aspects of the Variscan Fold Belt.* Manchester University Press, Manchester, 49–64.

GRAHAM, J. R. & REILLY, T. A. 1972. The Sherkin Formation (Devonian) of south-west County Cork. *Bulletin of the Geological Survey of Ireland*, **1**, 355–366.

—— & —— 1976. The stratigraphy of the area around Clonakilty Bay, South County Cork. *Proceedings of the Royal Irish Academy*, **76**, 379–391.

——, RUSSELL, K. J. & STILLMAN, C. J. 1995. Late Devonian magmatism in West Kerry and its relationship to the development of the Munster Basin. *Irish Journal of Earth Sciences*, **14**, 7–23.

HUSAIN, S. M. 1957. *The geology of the Kenmare Syncline, Co. Kerry, Ireland.* PhD thesis University of London.

JAMES, A. & GRAHAM, J. R. 1995. Stratigraphy and structure of Devonian fluvial sediments, western Beara Peninsula, south-west Ireland. *Geological Journal*, **30**, 165–182.

LAFLÈCHE, M. R., CAMIRÉ, G. & JENNER, G. A. 1998. Geochemistry of post-Acadian, Carboniferous continental intraplate basalts from the Maritimes Basin, Magdalen Islands, Québec, Canada. *Chemical Geology*, **148**, 115–136.

LEEDER, M. R. 1988. Recent developments in Carboniferous geology: a critical review with implications for the British Isles and NW Europe. *Proceedings of the Geologists Association*, **99**, 73–100.

LEUTWEIN, F. 1970. Preliminary remarks on some geochronological analyses of Irish granites and gneisses. *Irish Naturalist Journal*, **16**, 306–308.

MACCARTHY, I. A. J. 1990. Alluvial sedimentation patterns in the Munster Basin, Ireland. *Sedimentology*, **37**, 685–712.

MACDERMOT, C. V., LONG, C. B. & HARNEY, S. J. 1996. *A geological description of Sligo, Leitrim, and adjoining parts of Cavan, Fermanagh, Mayo and Roscommon to accompany the bedrock geology 1 : 100 000 scale map series, sheet 7, Sligo–Leitrim.* Geological Survey of Ireland, Dublin.

MATTHEWS, S. C., NAYLOR, D. & SEVASTOPULO, G. D. 1983. Palaeozoic sedimentary sequences as a reflection of deep structure in southwest Ireland. *Sedimentary Geology*, **34**, 83–95.

MCKENZIE, D. P. & BICKLE, M. J. 1988. The volume and composition of melt generated by extension of the lithosphere. *Journal of Petrology*, **29**, 625–679.

MEERE, P. 1992. *Structural and metamorphic studies of the Irish Variscides from the Killarney–Ballydehob transect, southwest Ireland.* PhD thesis, National University of Ireland.

—— 1995. Sub-greenschist facies metamorphism from the Variscides of SW Ireland: an early synextensional peak thermal event. *Journal of the Geological Society, London*, **152**, 511–521.

MITCHELL, J. G. & MOHR, P. 1987. Carboniferous dikes of West Connacht, Ireland. *Transactions of the Royal Society of Edinburgh, Earth Sciences*, **78**, 133–151.

MURTHY, M. V. N. 1948. Camptonitic dyke rocks from Inishowen, County Donegal, Ireland. *Transaction of the Geoolgical Society of Glasgow*, **21**, 205–206.

NAYLOR, D. 1975. Upper Devonian–Lower Carboniferous stratigraphy along the south coast of Dunmanus Bay, County Cork. *Proceedings of the Royal Irish Academy*, **75B**, 317–337.

—— 1978. A structural section across the Variscan fold belt, southwest Ireland. *Journal of Earth Sciences, Royal Dublin Society*, **1**, 63–70.

—— & JONES, P. C. 1967. Sedimentation and tectonic setting of the Old Red Sandstone of southwest Ireland. *In*: OSWALD, D. (ed.) *International*

Symposium on the Devonian System (Calgary). Publication of the Society of Petrology and Geology, **2**, 1089–1099.

—— & Sevastopulo, G. D. 1979. The Hercynian 'Front' in Ireland. *Krystallinikum*, **14**, 77–90.

——, —— & Sleeman, A. G. 1989. Subsidence history of the South Munster Basin. *In*: Athurton, R. S., Gutteridge, P. & Nolan, S. C. (eds) *Role of Tectonics in Devonian and Caboniferous Sedimentation in the British Isles*. Occassional Publications of the Yorkshire Geological Society, **6**, 99–109.

——, ——, —— & Reilly, T. A. 1981. The Variscan fold beld in Ireland. *In*: Zwart, H. J. & Dornsiepen, U. F. (eds) *The Variscan Orogen in Europe*. *Geologie en Mijnbouev*, **60**, 49–66.

Pearce, J. A. & Norry, M. J. 1979. Petrogenetic implications of the Ti, Zr, Y and Nb variations in volcanic rocks. *Contributions to Mineralogy and Petrology*, **69**, 33–47.

Penney, S. R. 1978. Devonian lavas from the Comeragh Mountains, County Waterford. *Journal of Earth Sciences Royal Dublin Society*, **1**, 71–76.

Pe-Piper, G. & Piper, D. J. W. 1998. Geochemical evolution of Devonian–Carboniferous igneous rocks of the Magdalen basin, Eastern Canada: Pb- and Nd-isotope evidence for mantle and lower crustal sources. *Canadian Journal of Earth Sciences*, **35**, 201–221.

Piper, D. J. W., Pe-Piper, G. & Loncarevic, B. D. 1993. Devonian–Carboniferous igneous intrusions and their deformation, Cobequid Highlands, Nova Scotia. *Atlantic Geology*, **29**, 219–232.

——, —— & Pass, D. J. 1995. The stratigraphy and geochemistry of late Devonian to early Carboniferous volcanic rocks of the northern Chignecto peninsula, Cobequid Highlands, Nova Scotia. *Atlantic Geology*, **32**, 39–52.

Pracht, M. 1994. *The geology of the Beara Peninsula, Ireland*. PhD thesis, National University of Ireland.

—— 1997. *A geological description to accompany the bedrock geology 1 : 100 000 scale map series, sheet 21, Kerry–Cork*. Geological Survey of Ireland, Dublin.

—— & Batchelor, R. A. 1998/99. A geochemical study of late Devonian and early Carboniferous tuffs from the South Munster Basin, Ireland. *Irish Journal of Earth Sciences*, **17**, 25–38.

—— & Kinnaird, J. A. 1995. Mineral chemistry of megacrysts and ultramafic nodules from an undersaturated pipe at Black Ball Head, County Cork. *Irish Journal of Earth Sciences*, **14**, 47–58.

—— & —— 1997. Carboniferous subvolcanic activity on the Beara Peninsula, SW Ireland. *Geological Journal*, **32**, 297–312.

Price, C. A. & Todd, S. P. 1988. A model for the development of the Irish Variscides. *Journal of the Geological Society, London*, **145**, 935–939.

Rock, N. M. S. 1987. The nature and origin of lamprophyres: an overview. *In*: Fitton, J. G & Upton, B. G. J. (eds) *Alkaline Igneous Rocks*. Geological Society, London, Special Publications, **30**, 191–226.

Schultz, R. W. & Sevastopulo, G. D 1965. Lower Carboniferous volcanic rocks near Tulla, Co. Clare, Ireland. *Scientific Proceeding of the Royal Dublin Society, Series A*, **2**(3), 153–162.

Sleeman, A. G. & Pracht, M. 1999. *Geology of the Shannon Estuary. A geological description of the Shannon Estuary region including parts of Clare, Limerick and Kerry, to accompany the bedrock geology 1 : 100 000 Scale Map Series, Sheet 17, Shannon Estuary*. Geological Survey of Ireland, Dublin.

Strogen, P. 1973. The volcanic rocks of the Carrigogunnel area, Co. Limerick. *Scientific Proceedings of the Royal Dublin Society, Series A*, **5**, 1–26.

—— 1974. The sub-Palaeozoic basement in central Ireland. *Nature*, **250**, 562–563.

—— 1983. *The geology of the volcanic rocks of southeast County Limerick*. PhD thesis, National University of Ireland.

—— 1988. The Carboniferous lithostratigraphy of SE County Limerick, Ireland, and the origin of the Shannon Trough. *Geological Journal*, **23**, 121–137.

——, Sommerville, I. D., Pickard, N. A. H., Jones, G. LL. & Fleming, M. 1996. Controls on ramp, platform and basinal sedimentation in the Dinantian of the Dublin Basin and Shannon Trough, Ireland. *In*: Strogen, P., Somerville, I. D. & Jones, G. L. (eds) *Recent Advances in Lower Carboniferous Geology*. Geological Society, London, Special Publications, **48**, 263–279.

Sutton, J. S. 1970. The Termon Granite and associated minor intrusives, Co. Mayo, Ireland. *Scientific Proceedings of the Royal Dublin Society*, **3A**, 293–302.

Sweeney, E. 1998. *Petrology, volcanology, geochemistry and age of Irish Carboniferous volcanic rocks and their significance in the evolution of sedimentary basins*. Unpublished Final Technical Report **SC/95/311**, University College Dublin.

Todd, S. P. 1989. Role of the Dingle Bay Lineament in the evolution of the Old Red Sandstone of south-west Ireland. *In*: Arthurton, R. S., Gutteridge, P. & Nolan, S. C. (eds) *Role of Tectonics in Devonian and Carboniferous Sedimentation in the British Isles*. Occasional Publications of the Yorkshire Geological Society, **6**, 39–54.

Williams, E. A., Bamford, M. L. F., Cooper, M. A. *et al.* 1989. Tectonic controls and sedimentary response in the Devonian-Carboniferous Munster and South Munster Basins, southwest Ireland. *In*: Arthurton, R. S., Gutteridge, P. & Nolan, S. C. (eds) *The Role of Tectonics in Devonian and Carboniferous Sedimentation in the British Isles*. Occasional Publication of the Yorkshire Geological Society, **6**, 123–141.

——, Sergeev, S. A., Stössel, I. & Ford, M. 1997. An Eifelian U–Pb zircon date for the Enagh Tuff Bed from the Old Red Sandstone of the Munster

Basin, SW Ireland. *Journal of the Geological Society, London*, **154**, 189–193.

——, ——, —— & Higgs, K. T. 1998. U–Pb zircon geochronology of Munster Basin silicic tuffs: implications for the chronostratigraphy, cyclicity and sedimentation of the Old Red Sandstone of SW Ireland. *Abstracts of the Irish Geological Research Meeting, University of Galway.*

Winchester, J. A. & Floyd, P. A. 1977. Geochemical discrimination of different magma series and their differentiation products using immobile elements. *Chemical Geology*, **20**, 325–343.

A mid-Frasnian marine incursion into the southern part of the Munster Basin: evidence from the Foilcoagh Bay Beds, Sherkin Formation, SW County Cork, Ireland

K. T. HIGGS, I. A. J. MACCARTHY & M. M. O'BRIEN

Department of Geology, University College Cork, Cork, Ireland

Abstract: The Munster Basin of southern Ireland contains a thick (7 km +) succession of Old Red Sandstone sediments interpreted as the product of various alluvial processes. The present study presents a preliminary sedimentological and palynological analysis of a grey succession informally known as the Foilcoagh Bay Beds, which is the lowest unit of the Sherkin Formation exposed in the southern part of the basin. Sedimentological analysis of the succession suggests that it is the product of sinuous distributary channels, flanked by permanently flooded overbank areas that endured occasional crevasse splay floods. These conditions evolved into a protected lagoon or lake that received periodic high energy floodings from an adjacent marine environment. Palynological study has refined the age of the Foilcoagh Bay Beds as mid-Frasnian time. Palynofacies analysis has provided direct evidence of marine influence as revealed by the presence of marine microfossils and abundant amorphous organic matter at some dark grey mudrock levels. This suggests deposition in a well-established lacustrine or lagoonal environment in which anoxic conditions prevailed at intervals and which was subjected to a period of marine incursion. The recognition of a marine influence in the Munster Basin at an early stage in its history has several important implications, including the following (1) previous models for the basin that suggested an enclosed centripetally draining entirely non-marine system have to be re-evaluated; (2) the drainage pattern and direction of the marine incursion were probably controlled by localized subsidence along an east–east direction; initiation of subsidence associated with the development of the east–west-trending South Munster Basin may have commenced much earlier than previously considered; (3) the occurrence of marine conditions has been tentatively correlated with the Rhinestreet mid-Frasnian sea-level highstand; (4) the recognition of marine conditions early in the known history of the basin provides a pathway for fish to have entered the basin; (5) the basin may have had a marine connection throughout much of Late Devonian time, opening the possibility for base-level control of the alluvial systems within the basin by external eustatic factors.

The Munster Basin of southern Ireland contains a thick (7 km +) succession of continental Old Red Sandstone sediments interpreted as the product of various alluvial processes. The basin has been interpreted as a wide (>150 km) half-graben. The northern margin is defined by the Dingle Bay–Galtee Fault Zone, which was responsible for the major southward thickening of the Old Red Sandstone succession (Williams *et al.* 1989) with the depocentre of the basin located in the MacGillycuddy Reeks of Iveragh (see Fig. 1). The basin fill ranges from late Mid-Devonian to Late Devonian in age, with the Valentia Slate Formation, the oldest strata exposed in the Munster Basin, being geochronometrically dated at 384.9 ± 0.7 Ma by Williams *et al.* (1997), which correlates with a late Mid-Devonian (mid-Givetian) age in terms of the Devonian time scale of Tucker *et al.* (1998).

The Old Red Sandstone succession is considerably thinner (2 km) towards the southern margin of the basin, and in southwest County Cork is composed of three formations. These are, in ascending stratigraphical order, the Sherkin Formation, the Castlehaven Formation and the Toe Head Formation (see Fig. 2). This paper presents a sedimentological and palynological analysis of the Foilcoagh Bay Beds, the lowest stratigraphical unit of the Sherkin Formation (Fig. 2), and hence the oldest exposed strata in the southern part of the Munster Basin. The aim of the study was to refine the biostratigraphical age and to determine the depositional environment of the Foilcoagh Bay Beds.

From: FRIEND, P. F. & WILLIAMS, B. P. J. (eds). *New Perspectives on the Old Red Sandstone.* Geological Society, London, Special Publications, **180**, 319–332. 0305-8719/00/$15.00

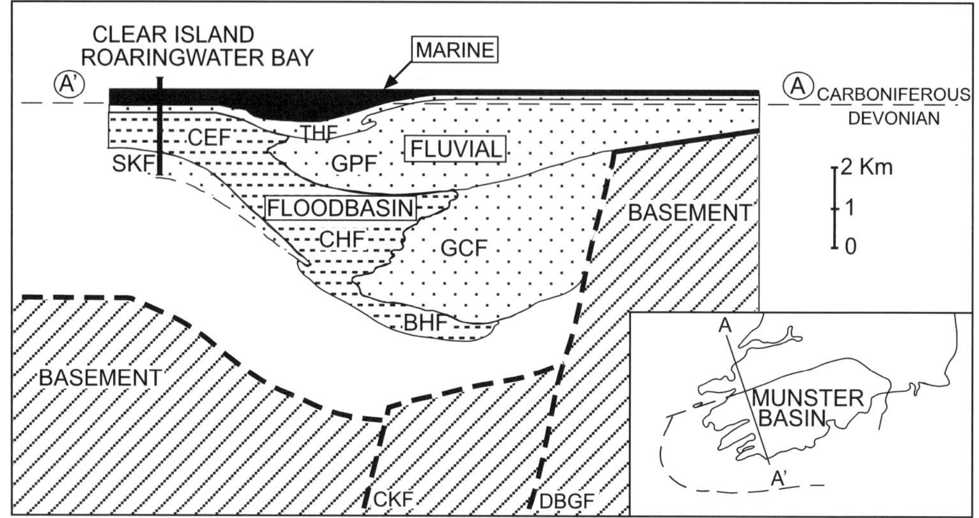

Fig. 1. Stratigraphical cross-section across the western part of the Munster Basin. The vertical bar on the left side of the section shows the stratigraphical location of the Sherkin Sandstone Formation (SKF) (see Fig. 2 for details of this section). BHF—Bird Hill Fm., GCF—Glenflesk Chloritic Sandstone Fm., CHF—Caha Mountain Fm., CEF—Castlehaven Fm., GPF—Gun Point Fm., THF—Toe Head Sandstone Fm., DBGF—Dingle Bay–Galtee Fault, CKF—Cork–Kenmare Fault. Position of basement inferred.

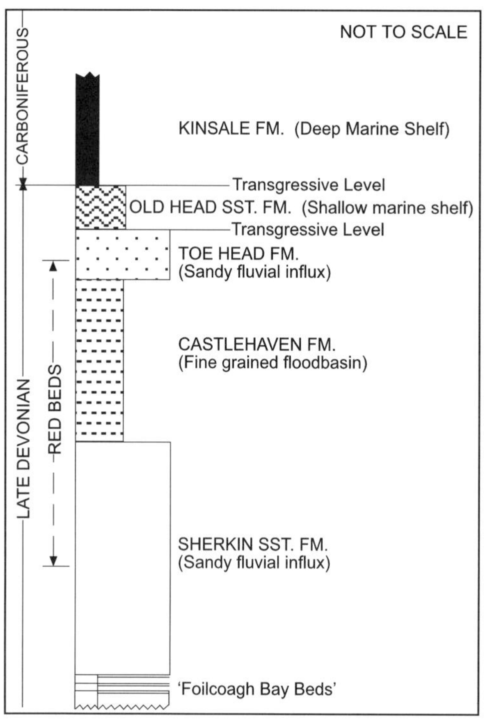

Fig. 2. Stratigraphy of the Clear Island–Roaringwater Bay area showing the location of the studied sections (Foilcoagh Bay Beds).

Stratigraphy

The Sherkin Formation is the lowest lithostratigraphical division of the Old Red Sandstone magnafacies in the southern part of the Munster Basin. Its outcrop is restricted to southwest County Cork, where it forms the core of the Rosscarbery Anticline and can be traced for over 40 km from Clear Island in the west to Rosscarbery in the east (Graham & Reilly 1972). It comprises a 1050 m sequence of fine to medium grey and grey–green sandstones interbedded with subordinate grey–green and purple siltstones and mudrocks. The dominant colour is grey, and purple rocks are restricted to the upper part of the formation. The basal 60 m of the formation is distinguished by intervals of dark grey to black mudrocks. The basal sequence, termed the Foilcoagh Bay Beds is exposed in and around Foilcoagh Bay on the northern coast of Clear Island (Graham & Reilly 1972) (Fig. 3). The basal part of this unit crops out in the cliffs around the bay and on a wave-cut platform immediately to the southwest of the bay (Fig. 3). The middle part of the succession is not fully accessible and consequently a continuous log of the entire succession is not yet possible. Nevertheless, two preliminary logs have been made (Fig. 4), one covering the lower part of the Foilcoagh Bay Beds (Section A) and the other

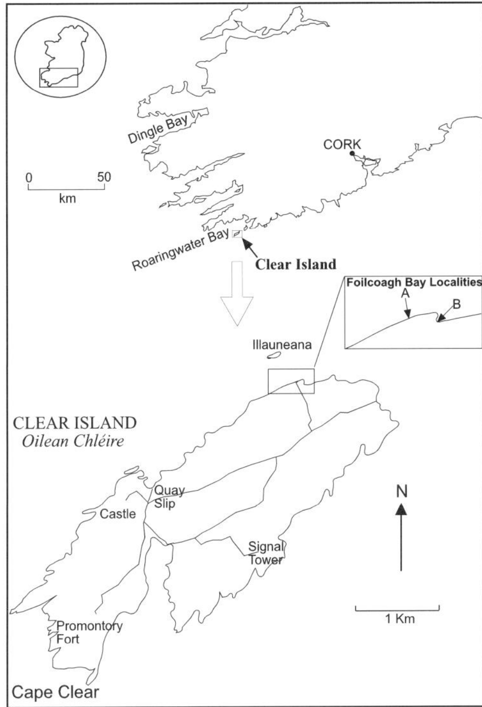

Fig. 3. Geographical location of Foilcoagh Bay in Clear Island in SW Ireland.

covering the middle to upper part of the succession (Section B). A suite of palynology samples was collected from each section, together with one spot sample (CC4) collected from a dark grey mudrock horizon that crops out just west of the inaccessible headland between the two logged sections.

Previous palaeoenvironmental interpretation of the Sherkin Formation

Coe & Selwood (1968) considered the Sherkin Formation to be a lateral equivalent of part of the much thicker and more purple-coloured Caha Mountain Formation to the north. They also considered the Sherkin Formation to represent a marine incursion into the basin. This marine interpretation was probably based on the overall grey colour of the formation and the occasional presence of trace fossils identified by Coe & Selwood (1968) as *Chondrites*.

Graham & Reilly (1972) made a detailed sedimentological study of the Sherkin Formation and concluded that the most likely environment of deposition was terrestrial under fluvial conditions. In particular, they interpreted the sandstones to have been deposited in a braided river environment and the mudrocks as alluvial floodplain deposits. They also considered the Foilcoagh Bay Beds to be of continental origin. They suggested that the dark grey colour may be indicative of more continuous water coverage of the sedimentary surface, and so coastal plain or limnic conditions were suggested. However, Graham & Reilly (1972, p. 297) believed that there was certainly no sedimentological evidence that the Foilcoagh Bay Beds were ever marine, and the marine trace fossils reported by Coe & Selwood (1968) were better interpreted as freshwater burrows. The palynological study of Clayton & Graham (1974) confirmed non-marine palaeoenvironments similar to those described by Graham & Reilly (1972). More recently, Kelly & Olsen (1993) have interpreted the Sherkin Formation to represent a fluvial distributive system in a terminal fan setting, with the Foilcoagh Bay Beds interpreted as lacustrine deposits. Those workers suggested that a possible modern-day analogue for the Foilcoagh Bay Beds is the 'lagoon' deposits of the Lake Eyre Basin of Australia, a type of playa that develops in topographic depressions fed by major rivers.

Lithofacies of the Foilcoagh Bay Beds

The bulk of Section A consists of extensive good-quality exposure, which extends laterally for more than 100 m. However, the lower 9 m of the section lies within the intertidal zone and sedimentological details here are difficult to observe. Section B consists of a narrow outcrop at the back of Foilcoagh Bay immediately south of the slip such that lateral control is limited to less than about 3 m. In addition, this section is structurally complex so that sedimentological analysis of it is particularly difficult and interpretations can only be tentative. A preliminary analysis of the two logged sections has allowed the recognition of five lithofacies (Fig. 4).

Lithofacies 1

This facies is confined to Section A and occurs at two levels. The lower occurrence consists of a 9.8 m multistorey medium-grained sandstone complex but detailed structures are difficult to observe because of the poor exposure. The upper occurrence is a sandstone-dominant erosively based unit 4.2 m thickness. This commences with a 1 m large-scale trough cross-stratified unit indicating a south-southeasterly directed palaeocurrent. This is overlain by a 0.6 m set of large-scale low-angle planar cross-laminated sandstones. The upper part of the facies consists of an epsilon cross-stratified unit of 2.2 m thickness

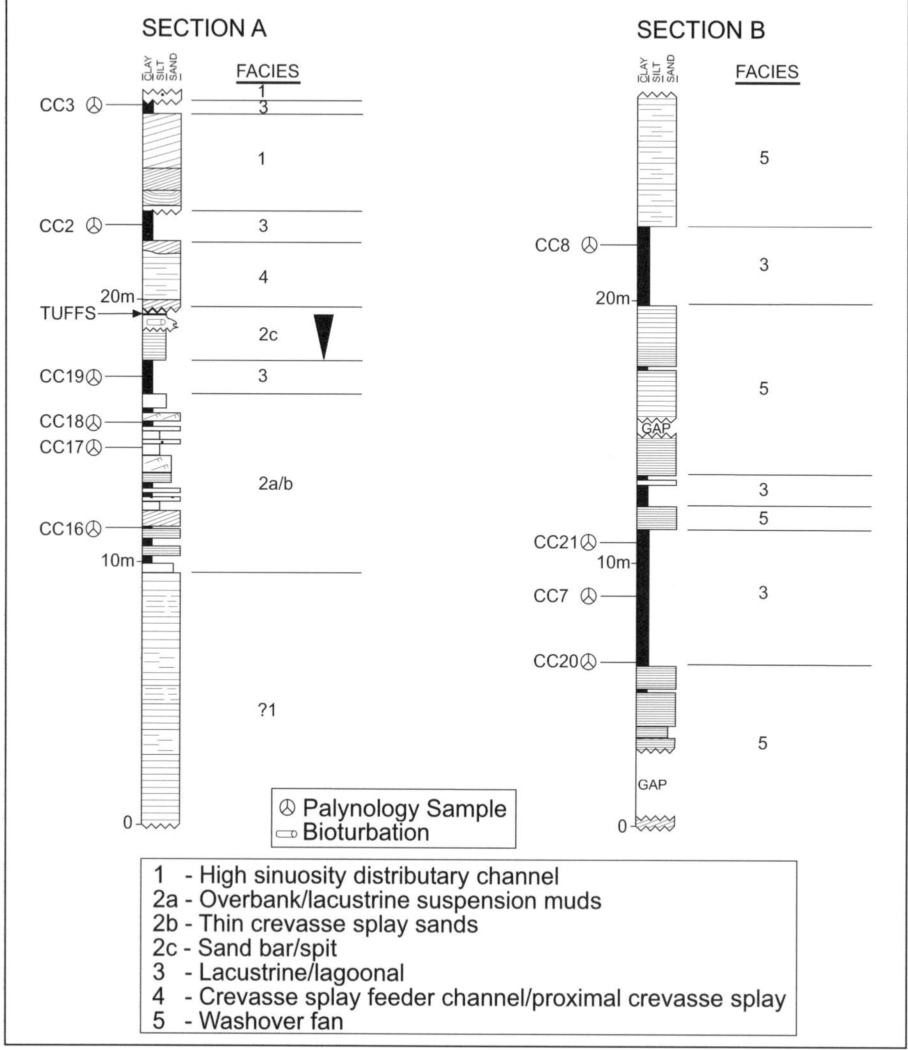

Fig. 4. Logs of Sections A and B showing the locations of palynological samples, the subdivision of the successions into lithofacies and summary sedimentological interpretations.

(Allen 1963) in which there are abundant low-angle depositional and erosional surfaces, often overlain by intraformational clasts, and parallel lamination is common. The upper metre of the facies contains a number of mudrocks ranging to 0.4 m thick, which are interleaved with the epsilon foresets, resulting in a configuration comparable with inclined heterolithic stratification (IHS) described by Thomas *et al.* (1987).

Lithofacies 2

This facies is recognized at only one level in Section A. It consists of three interbedded subfacies: grey mudrocks or siltstones (Facies 2a); fine- to medium-grained grey sandstones (Facies 2b); brown weathering heterolithic siltstones and sandstones (Facies 2c).

The mudrocks and siltstones of Facies 2a range from 0.1 to 0.2 m in thickness and are generally massive with occasional thin sandy or silty laminae. In places, incipient linsen bedding with poor lithological contrast is developed. Such levels are often cross-laminated.

Facies 2b consists of fine- to medium-grained grey sandstones. The facies consists of sheet or lenticular units of 0.1–0.65 m thickness, which have a sharp or erosive base. The sandstones are often apparently massive or contain small-scale current-generated cross-lamination occasionally arranged as climbing sets. Some sandstones have large-scale low-angle cross-lamination.

Facies 2c consists of a brown weathering, coarsening-up laminated siltstone unit of 0.95 m thickness with incipient heterolithic character although with poor lithological separation. The unit is strongly burrowed and is capped by a thin sandstone containing small-scale wave-generated ripple bedforms with a wavelength of 3 cm. Facies 2c is overlain by a 0.65 m interval containing two thin tuffaceous horizons.

Lithofacies 3

This lithofacies consists of massive or laminated dark grey to black mudrocks which range from 1.3 to 4.5 m in thickness. Linsen bedding is often present, with up to 10% of the facies being composed of thin millimetre thick sandy or silty laminae. Minor small-scale cross-laminated levels are often developed in coarser levels. At one level (23 m in section A, Fig. 4), detached sandstone ball and pillow structures are suspended within the mudrock.

Lithofacies 4

There is only one occurrence of this facies and this consists of an erosively based sandstone unit ranging from 1.7 to 2.7 m in thickness. The facies commences with a single set (0.2–1.0 m thick) of large-scale low-angle cross-laminated sandstones showing an easterly directed palaeocurrent. This is overlain by a thin (0.1 m) mudrock drape. The bulk of the facies consists of apparently massive or parallel-laminated fine-grained brownish weathering sandstones. The upper part of the facies consists of a large-scale trough cross-laminated sandstone unit indicating south-southeasterly directed currents. The facies is characterized by numerous internal erosional surfaces.

Lithofacies 5

This consists of apparently parallel-sided sheet grey sandstones ranging from 0.5 to 3.9 m in thickness characterized by multistorey parallel lamination. Occasional thin mudrock intervals and intraformational clasts are developed at some levels. The facies is interbedded with Facies 3 and is confined to Section B.

Interpretation of lithofacies

The predominance of grey colours and absence of red coloration and of any evidence of subaerial exposure strongly suggest that this sequence accumulated in an area where there was a relatively permanent cover of water or at least high water tables for prolonged periods. There are no obvious sedimentary structures visible in the field that would indicate unequivocal marine conditions. These observations, allied with the predominance of fine sediment grades, point to a depositional setting located in the very distal reaches of a drainage network. The palynofacies evidence presented below indicates that Facies 3 shows evidence of marine influence.

Lithofacies 1

The lower cross-stratified part of Facies 1 is interpreted as the product of currents that produced large-scale bedforms and thin low-amplitude sand waves. The epsilon cross-stratified unit in the upper part of the facies is suggestive of a lateral accretion unit similar to that which could have been produced by lateral point bar migration (Allen 1964, 1965). The geometry and variable lithology of this unit and extensive development of erosive surfaces are consistent with variable hydrodynamic conditions. These could have been produced in a point bar setting of a high-sinuosity channel that experienced either variable discharge or fluctuating base level. The occurrence of lenticular suspension deposited mudrock in the upper part of point bars (IHS) is suggestive of deposition in the lower reaches of a very low energy high-sinuosity stream (Nanson 1980; Jackson 1981; Smith 1987; Jordan & Prior 1992). The presence of IHS may also suggest tidal influence (Smith 1987; Miall 1996). Such a setting would have produced the fluctuating base level required to deposit mudrock drapes on point bar surfaces. A distal high-sinuosity distributary channel setting with a possible open connection to a marine environment is therefore suggested as the most likely interpretation for this facies.

Lithofacies 2

The massive or laminated mudrocks and siltstones of this facies (2a) are clearly the product of low-energy suspension conditions in a permanent water body such as a lake or lagoon. The coarser-grained interbeds (2b) represent mostly single event influxes into this environment. The influxes were, however, consistently in the lower flow regime and probably were sourced from breaching of a levee of a nearby permanent fluvial channel such as that represented by Facies 1. The influxes probably represent the distal part of crevasse splays that discharged into a lake or lagoon. The brown coarsening-up burrowed unit capped by wave-ripple bedforms (Facies 2c) indicates a shallowing sequence. It strongly resembles minor sand spits that result from wave reworking of crevasse sediment described by Elliott (1974).

Lithofacies 3

These mudrocks are interpreted as having accumulated in a relatively permanent water body under low-energy suspension conditions. The thickness of the facies suggests that these were substantial water bodies, which persisted in spite of any seasonal variations in sediment flux or water-table fluctuations. The most likely environment in which these conditions would apply are a lacustrine or a protected lagoonal environment. The new palynofacies evidence suggests that there was marine influence in the upper three occurrences of this facies and this indicates that such a lagoon must have been largely protected from marine influence for most of the time, apart probably from washover floodings, which would have accompanied major storm events or periodic sea-level rises.

Lithofacies 4

The range of structures in this facies indicate that it accumulated mainly under upper flow regime conditions and that there was extensive erosion and sediment reworking. The basal large-scale planar cross-laminated unit indicates an initial phase of bar or sandwave migration. This was followed by extensive upper flow regime conditions, which resulted in erosive surfaces and parallel lamination. The presence of occasional mudrock drapes indicates that high-energy influxes were punctuated by low-energy phases of suspension sedimentation. The large-scale trough cross-laminated unit at the top of the facies indicates a return to lower flow regime sandwaves or sinuous megaripples. This succession testifies to rapid waxing and waning of the flow regime for the facies. These processes could have been produced either within a distributary channel or in an overbank setting. The distal setting of the studied section within the basin, the predominance of extensive upper flow regime conditions in the facies and the evidence for highly variable discharge are suggestive of deposition in an overbank rather than an in-channel situation. If this is correct, the facies may have been the product of deposition in a feeder channel that led to crevasse splays (Jordan & Prior 1992) or may have accumulated in the proximal reaches of a crevasse splay. In such a situation, repeated high-energy influxes during periods of overbank flooding would have been accompanied by extensive erosion and reworking, and would have been punctuated by low-energy phases of non-deposition or mudrock deposition. The limited palaeocurrent information is also consistent with this interpretation.

Lithofacies 5

The predominance of parallel lamination throughout this facies testifies to repeated phases of upper flow regime conditions. The facies strongly resembles sheet-flood deposits described by Tunbridge (1981). Each occurrence of the facies was introduced as a high-energy influx. Deposition was rapid and there was no time for reworking or the development of lower flow regime structures. The facies also shows a strong resemblance to deposits of washover fans that build into lagoonal environments (Schwartz 1975, 1982). As the facies is interbedded with possible lagoonal sediments (Facies 3) bearing some marine influence on palynological grounds (see below), this interpretation is regarded as being the most probable for this facies.

Summary

The sedimentological characteristics of this succession are consistent with deposition in a distal setting in the drainage network. In Section A, Facies 1 occurs twice and is interpreted as the product of a high-sinuosity channel that was probably influenced by tidal action and hence had a marine connection. The upward gradation to overbank conditions (Facies 2), which shows a bulk fining-upward trend, suggests lateral migration of the major channel but the absence of subaerial indicators provides evidence for sustained high water tables even in the overbank areas, which must have been permanently flooded. The overbank areas were essentially shallow lacustrine environments. Where these conditions persisted, the water bodies became more extensive and deeper, resulting in the typically anoxic conditions recorded in Facies 3 (see below). This applies particularly to occurrences of Facies 3 in Section B. The marine influence within the upper three sampled occurrences of Facies 3 points to the establishment of a protected coastal lagoonal environment that frequently experienced high-energy washovers (Facies 5). Such floodings of the lagoonal areas could have been initiated by storm surge events or overtopping of barrier systems during periodic highstands of sea level.

Palynology

Previous palynological work

Clayton & Graham (1974) described two spore assemblages from the Sherkin Formation on Clear Island. One of these samples was collected from Foilcoagh Bay Beds *c.* 30 m above the base of the beds, whereas the other sample was from Coosadouglas on the northeast coast, *c.* 375 m above the base of the formation. The spore assemblages suggested a late Mid-Devonian (Givetian) or early Late Devonian (Frasnian) age. In terms of palaeoenvironment, Clayton & Graham (1974) found no evidence of marine influence in the microflora.

Present palynological study

Ten grey to dark grey mudrock samples were collected from the two logged sections of the Foilcoagh Bay Beds, together with one spot sample. The stratigraphical positions of the palynology samples and the lithofacies from which they were sourced are shown in Fig. 4. The samples were processed using standard palynological laboratory techniques and the productive samples yielded moderate to abundant amounts of organic residue. The organic material is thermally mature with a thermal alteration index TAI 4.5–5 on the Staplin (1969) scale. The residues therefore required extensive oxidation in Schultze Solution to render the palynomorphs light enough for identification. Material studied for palynofacies analysis was studied in both unsieved and sieved (at 10 μm) state and in both unoxidized and oxidized preparations. Moderately well-preserved palynomorph assemblages were recovered from most of the samples, although fracturing of palynomorphs was commonly seen as a result of the strong penetrative cleavage developed in the rocks. Pyrite pitting of palynomorphs is present in a few of the dark grey samples from Lithofacies 3. Spores are by far the most abundant type of palynomorph present in the assemblages. However, a small number of acritarchs, prasinophyte phycoma and scolecodonts were also recorded in several samples. Although numerically less significant than the spores, their presence is of palaeoenvironmental importance and will be discussed below.

Stratigraphical palynology

A detailed account of the systematics and stratigraphical distribution of the spores, acritarchs and prasinophyte phycoma occurring in the samples will be presented in a future paper. A brief summary of the microflora is given here.

The following spore taxa have been identified in the samples: *Ancyrospora simplex* Guennel emend Urban 1969, *Chelinospora concinna* Allen 1965, *Converrucosisporites liratus* Clayton & Graham 1974, *Diaphanospora reticulata* Guennel 1963, *Dictyotriletes craticulatus* Clayton & Graham 1974, *Dictyotriletes perlotus* (Naumova) Mortimer & Chaloner 1971, *Geminospora lemurata* Balme emend Playford 1993, *Geminospora plicata* Clayton & Graham 1974, *Grandispora tomentosa* (Naumova) Taugourdeau-Lantz 1967, *Grandispora inculta* Allen 1965, *Grandispora saetosa* Clayton & Graham 1974, *Hystricosporites delectabilis* McGregor 1960, *Lophozonotriletes media* Taugourdeau-Lantz 1967, *Retusotriletes pychovii* Naumova 1953, *Rhabdosporites parvulus* Richardson 1965, *Rugospora bricei* Loboziak & Streel 1989, *Samarisporites triangulatus* Allen 1965, *Verrucosisporites bulliferus* Richardson & McGregor 1986, *Videospora glabrimarginata* (Owens) Higgs & Russell 1981.

The microflora recorded contains most of the spore taxa described by Clayton & Graham (1974); however, several additional species have been identified in the present study, which allow a more precise biostratigraphical age determination for the Foilcoagh Bay Beds. The additional species include the stratigraphically significant taxa *Verrucosisorites bulliferus*, *Lophozonotriletes media* and *Rugospora bricei*. These three species are important zonal index species for the early Late Devonian (Frasnian) interval and allow correlation with the Devonian spore zonation scheme of Streel *et al.* (1987). The first appearance of *V. bulliferus* and *L. media* defines the *Verrucosisorites bulliferus–Lophozonotriletes media* (BM) Oppel Zone of early to mid-Frasnian age, and the first appearance of *Rugospora bricei* is used as a marker species for defining Zone IV, which overlies the BM Zone (Loboziak & Streel 1989). Furthermore, Streel & Loboziak (1996) have shown that the first occurrence biohorizon of *Rugospora bricei* can be correlated within the late *hassi* to *linguiformis* Conodont Zones, which correspond to the mid- to late-Frasnian interval. Consequently, the presence of *R. bricei* in all of the samples allows an assignment to Zone IV of Streel *et al.* (1987), which indicates an age no older than mid-Frasnian time for the Foilcoagh Bay Beds microflora.

Palynofacies analysis

The nature and distribution of the particulate organic matter in each sample was analysed by making counts of the relative abundances of the following components: (1) palynomorphs (spores, acritarchs, prasinophtye phycoma, scolecodonts); (2) phytoclasts (structured wood fragments, structured opaque fragments, cuticle, tubes); (3) amorphous organic matter (AOM; fine structureless organic fragments). Counts of 500 particles were made from a slide of each sample. The count data for the particulate matter in each sample are plotted graphically on the two section logs (see Fig. 5).

Section A. Samples CC16 and CC18, taken from Lithofacies 2 show high scores of phytoclasts in terms of both structured wood and opaque fragments, and low scores of palynomorphs and AOM. Sample CC19, taken from the stratigraphically lowest interval of Lithofacies 3, shows a similar distribution of particles to that seen in CC18 with a perhaps small increase in AOM and a slight decrease in palynomorphs.

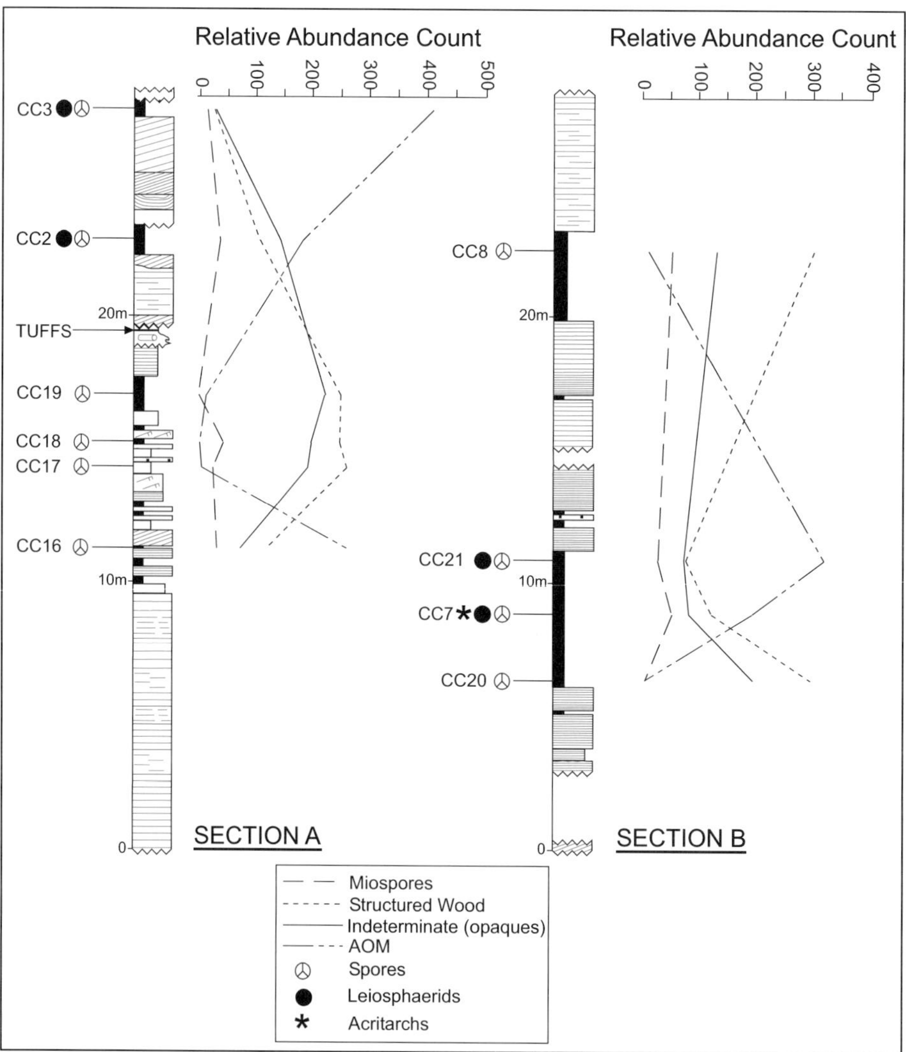

Fig. 5. Logs of Sections A and B showing variations in Relative Abundance Counts of Miospores, Structured Wood, Indeterminate Opaques, Amorphous Organic Matter, Leiosphaerids and Acritarchs.

Sample CC2, taken from the succeeding interval of Lithofacies 3, shows a marked reduction in phytoclasts, a significant increase in AOM, low numbers of palynomorphs and the first record of prasinophyte phycoma (*Leiosphaeridia*). Sample CC3, from Lithofacies 3, shows very low numbers of phytoclasts and palynomorphs (including *Leiosphaeridia*). and abundant amounts of AOM.

Section B. Sample CC20, taken from the base of the dark mudrock interval of Lithofacies 3, shows high scores of phytoclasts in terms of both structured wood and opaque fragments, and low scores of palynomorphs and AOM, similar to the distribution seen in samples from Lithofacies 2 in Section A. Sample CC7, taken from just above the middle of the same dark mudrock interval, shows a significant increase in AOM and a marked reduction in phytoclasts. The palynomorphs increase slightly in this sample and there is an important occurrence of acritarchs (*Gorgonisphaeridium* and *Veryhacium*), prasinophytes (*Leiosphaeridia* and *Cymatiosphaera*) and scolecodonts. Sample CC21, taken from near the top of the mudrock unit, contains abundant amounts of AOM, pyritized spores and prasinophyte phycoma. Sample CC8, taken from a grey mudrock near the top of the logged section, shows high scores of phytoclast fragments and low scores of palynomorphs and AOM, similar to the distribution seen in sample CC20 (Facies 2).

A spot sample CC4, taken from a dark grey mudrock interval stratigraphically between the two logged sections, shows a very similar palynofacies to sample CC7, with high AOM and small numbers of acritarchs (*Gorgonisphaeridium*, *Micrhystridium* and *Veryhacium*) and prasinophyte phycoma (*Leiosphaeridia* and *Cymatiosphaera*).

Discussion of palynofacies results

The palynofacies of the samples analysed from the fluvially influenced Lithofacies 2 are characterized by high amounts of terrestrially derived phytoclast particles, relatively low levels of AOM, and low to moderate amounts of palynomorphs composed exclusively of plant spores.

The palynofacies of samples collected from Lithofacies 3 show a trend towards moderate to large amounts of AOM and significantly reduced levels of terrestrial phytoclasts. Most of the samples analysed from the dark grey mudrock intervals of Lithofacies 3 yielded prasinophtyte phycoma suggestive of a brackish to marginal marine influence; furthermore, in two samples acritarchs and scolecodonts have been recorded from approximately the middle part of each respective dark grey mudrock unit, indicating the development of maximum marine conditions.

Palaeoenvironmental implications

Evidence for a marine influence

The palynological study of the dark grey mudrock intervals of Lithofacies 3 has provided direct evidence for a marine influence in the Foilcoagh Bay Beds. This evidence is summarized below.

(1) Acritarchs: these are considered to be the resting cysts of marine phytoplankton (Strother 1996). The presence of acritarch genera such as *Gorgonisphaeridium*, *Veryhacium* and *Micrhystridium* in samples CC 4 and CC 7 of Lithofacies 3 indicates a marine influence in the sediments.

(2) Prasinophyte phycoma: these represent the non-motile stage of green algae. In the present study, forms such as *Leiosphaeridia* and *Cymatiosphaera* have been recorded from most intervals of Lithofacies 3. Although living prasinophyceae are not exclusively found in marine environments, fossil prasinophyceae are rarely associated with fresh water (Tappan 1980). Their presence in basin margin sequences is generally taken to indicate marine or brackish marine conditions, or short-lived marine incursions (Batten 1996), and they are most common in organic-rich laminated sediments that accumulated under conditions of reduced oxygen concentration (Tappan 1980)

(3) Scolecodonts: these are the chitinous jaw parts of marine polychaete annelid worms and are widely accepted as having been restricted to marine environments. Several scolecodont elements were recorded from Sample CC7 in Lithofacies 3 in Foilcoagh Bay

(4) Amorphous organic matter (AOM): high amounts of AOM indicate increased anoxia in the aquatic environment, particularly in a marine realm (Tyson 1987), where this condition is normally associated with little or significantly reduced terrestrial input. However, conditions of dysoxia–anoxia may also occur in non-marine environments, and in these situations associated floral or faunal evidence is needed to confirm the specific environment

(5) Occurrence of pyrite: the formation of pyrite is controlled by three factors: concentration of organic carbon, dissolved sulphate and detrital iron minerals (Berner & Raiswell 1983). Anoxic marine environments rich in AOM satisfy all three criteria and so pyrite commonly forms in this marine situation. However, in

freshwater anoxic environments the formation of pyrite is limited by low sulphate concentrations. Pyrite occurs in some of the dark grey mudrocks (Samples CC4, CC7) of Lithofacies 3 and pyrite pitting of palynomorph exines has been observed in these samples (see Fig. 6n).

In conclusion, the documented palynomorph, palynofacies and pyrite evidence suggests that the dark grey mudrock of Lithofacies 3 represents deposition in a well-established lacustrine or lagoonal environment in which anoxic conditions prevailed at certain intervals and which was subjected to one or more periods of marine incursion.

Palaeogeographical model for the southern part of the Munster Basin in mid-Frasnian times

The Munster Basin is widely depicted as an intracontinental sedimentary basin located towards the southern margin of the Old Red Sandstone Continent, with sediment being supplied from major source areas located to the west, northwest, north and northeast. There is no evidence for southerly derived sediment at any stage in the basin history, and a low-relief southern basin margin lying just to the south of the present Irish coastline has been suggested (MacCarthy 1990). MacCarthy (1990) also suggested that the drainage pattern within the basin consisted of fluvial inputs from the northeast, north, northwest and west and that these merged distally into an eastward draining system which ultimately led to a marine environment (Old Head Sandstone Formation) during Late Devonian (Famennian) time.

The marine-influenced Foilcoagh Bay Beds appear to pass upwards gradationally into a progressively more fluvially influenced succession as represented by the bulk of the Sherkin Sandstone Formation. This eventually culminated in a fully red-bed succession (Fig. 2). Extensive palaeocurrent evidence from the Sherkin Sandstone Formation (Graham & Reilly 1972) indicates that the drainage direction was generally towards the east in this part of the basin. The drainage direction and pattern of the marine incursion, which has been identified in this paper, must have been influenced by the structural framework of the basin, which would have had an approximate east–west trend. This would imply that the marine incursion advanced from the east. It follows that there must have been an east–west-oriented structurally positive area lying along the southern side of the Munster Basin (Fig. 7). Palynological evidence indicates that part of the Glenflesk Chloritic Sandstone Formation in the northwestern part of the basin was deposited at about the same time as the Sherkin Sandstone Formation (Higgs & Russell 1981; Williams *et al.* this volume). Its drainage direction was towards the southeast, and this must have merged with the Sherkin Sandstone fluvial system so that there was a notable swing in the drainage direction as it was traced through the basin (Fig. 7). This is a remarkably similar pattern to that which occurred in late Late Devonian time (MacCarthy 1987, 1990) when subsidence associated with movement on the Cork–Kenmare basement inferred fault led to the development of the South Munster Basin (MacCarthy & Gardiner 1987; Naylor *et al.* 1989). The development of the latter influenced both the fluvial drainage pattern and the direction of the late Late Devonian transgression into the basin (MacCarthy 1987, 1990). The similar drainage pattern and axial direction of the inferred marine incursion identified in the lower part of the Sherkin Sandstone Formation (Fig. 7) suggest that the influence of basement control associated with the development of the South Munster Basin may have commenced much earlier than has previously been considered. In this context, it is interesting to note that an easterly directed drainage persisted throughout the entire exposed Upper Devonian succession in the southwestern part of the basin (Graham & Reilly 1972; Graham 1975; Naylor 1975; Reilly & Graham 1976; Cotter & Graham 1991). This strongly suggests that the initiation of the South Munster Basin may have taken place earlier than late Late Devonian time.

The evidence presented in this study shows that in early Late Devonian times the southern part of Munster Basin was affected by a marine incursion which may represent a highstand of sea level at this time. House (1983) has documented a series of global sea-level fluctuations during the Devonian period. Of these, the highest sea levels occurred in the early Late Devonian, with perhaps the greatest being the Rhinestreet highstand, which occurred in mid-Frasnian time, as noted by House & Kirchgasser (1993). Those workers have shown that the mid-Frasnian sea-level rise in New York State was made up of several small-scale rhythms. The marine incursion documented in the Foilcoagh Bay Beds can tentatively be attributed to this Rhinestreet flooding event. The observation that marine microfossils occur only at the mid-point of the dark grey mudrock units suggests a slow progression to maximum marine influence rather than a rapid transgressive flooding event.

Fig. 6. (**a–c**) *Rugospora bricei* Loboziak & Streel 1989. (**d**) *Samarisporites triangulatus* Allen 1965. (**e**) *Chelinospora concinna* Allen 1965. (**f**) *Lophozonotriletes media* Taugourdeau-Lantz 1967. (**g**) *Geminospora lemurata* Balme emend Playford 1993. (**h**) *Gorgonisphaeridium* sp. (**i**) *Cymatiosphaera* sp. (**j**) Scolecodont. (**k**) *Gorgonisphaeridium* sp. (**l**) *Leiosphaeridia* sp. (**m**) Scolecodont. (**n**) Pyritized spore. (**o**) *Leiosphaeridia* sp. For all parts of figure, magnification ×750.

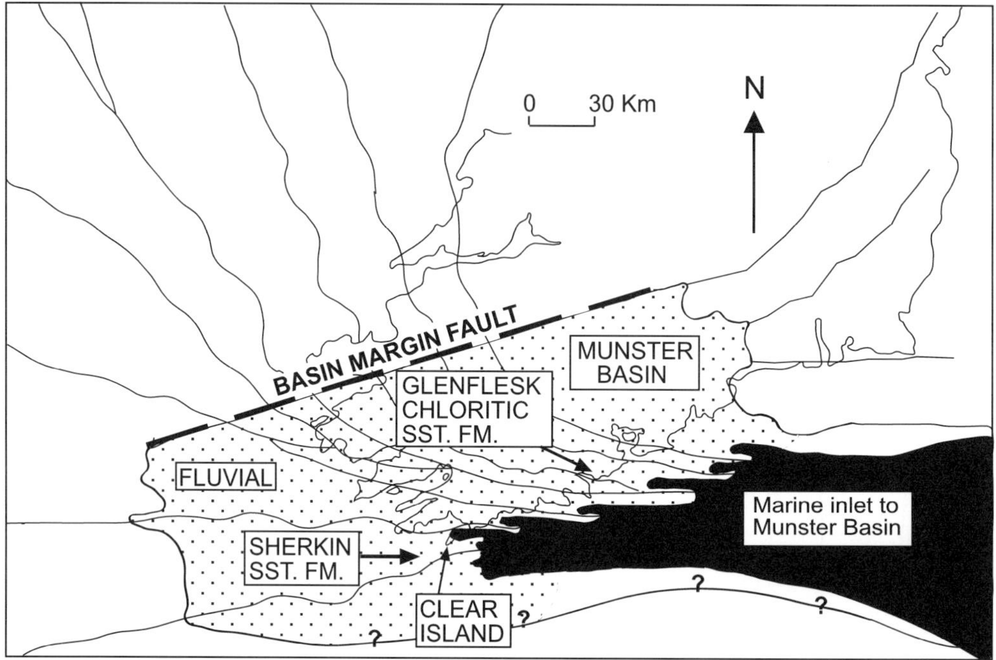

Fig. 7. Palaeogeographical reconstruction of the Munster Basin in the early Upper Devonian (modified from MacCarthy 1990). Discharge directions of the Glenflesk Chloritic Sandstone and Sherkin Formations are shown by arrows. No account has been made for tectonic shortening which is approximately 50% in a north–south direction.

The Rheic Ocean lay to the southeast of Ireland as evidenced by the Frasnian marine sediments of the Morte Slates of north Devon in southwest England (Bluck *et al.* 1989). In speculating about the most likely route of marine waters into Munster Basin, one possible route may have been via a seaway to the southeast (Fig. 7). As a result of the mid-Frasnian sea-level maxima, the sea would have flooded northwestwards and transgressed the southern margin of the Munster Basin. The evidence from the Foilcoagh Bay Beds of fluvial, brackish and marine conditions suggest that Clear Island was close to the shoreline. Eventually sea levels dropped or the shoreline prograded and the southern part of the Munster Basin returned to exclusively non-marine sedimentation.

A similar picture of marine incursions has emerged in other areas of the Old Red Sandstone Continent. Marshall (1996) has described two incursions in late Givetian and early Frasnian time of the Old Red Sandstone in the Orcadian Basin of Scotland. Also, Cloutier *et al.* (1996) reported a marine influence in the lower part of the middle Frasnian Escuminac Formation of eastern Quebec, Canada. This latter case is particularly interesting, as this area would have been palaeogeographically situated on the southwest part of the Old Red Sandstone Continent and therefore may have been affected by the same mid-Frasnian sea-level maxima as the Munster Basin.

In terms of biogeography, the late Mid- and early Late-Devonian successions in eastern Quebec and in southwest Ireland both contain similar Placoderm (antiarch) fish faunas. Russell (1978) has reported *Bothriolepis* from the Valentia Slate Formation, in the western part of the Munster Basin, and Cloutier *et al.* (1996) has described the distribution of a much more diverse antiarch fish fauna from the Escuminac Formation of eastern Quebec. In the latter study, it was demonstrated that *Bothriolepsis* lived in both brackish and marine waters as well as fresh water. Therefore, the evidence of marine influence in the southern part of the Munster Basin shows that marine connections would have been present to possibly allow fish such as *Bothriolepis* to enter the basin and migrate to freshwater habitats.

The recognition of a marine connection to the Munster Basin at an early stage in its known history and the return of marine conditions in the Upper Devonian (Famennian) Old Head Sandstone Formation (Figs 1 and 2) suggests that base-level control of the progradation and abandonment of alluvial systems within the basin may have been influenced by eustatic fluctuations in addition to any influence from local tectonic factors and sediment flux. The position of the shoreline probably oscillated, and the connection between the basin and the marine environment may well have been maintained throughout Late Devonian time.

References

Allen, J. R. L. 1963. The classification of cross-stratified units, with notes on their origin. *Sedimentology*, **2**, 93–114.

—— 1964. Studies in fluviatile sedimentation: six cyclothems from the Lower Old Red Sandstone, Anglo-Welsh Basin. *Sedimentology*, **3**, 163–198.

—— 1965. A review of the origin and characteristics of recent alluvial sediments. *Sedimentology*, **5**, 89–191.

Batten, D. J. 1996. Palynofacies and palaeoenvironmental interpretations. *In*: Jansonius, J. & McGregor, D. C. (eds) *Palynology: Principles and Applications*. American Association of Stratigraphic Palynologists Foundation, **3**, Publishers Press, Salt Lake City, 1011–1064.

Berner, R. A. & Raiswell, R. 1983. A new method for distinguishing freshwater from marine sedimentary environments. *Geology* **12**, 365–368.

Bluck, B. J., Haughton, P. D. W., House, M. R., Selwood, E. B. & Tunbridge, J. P. 1989. Devonian of England, Wales and Scotland. *In*: McMillian, N. J., Embry, A. F., Glass, D. J. *et al.* (eds) *Devonian of the World, Vol. 1*. Canadian Society of Petroleum Geologists, Memoirs, **14**, 305–324.

Clayton, G. & Graham, J. R. 1974. Miospore assemblages from the Devonian Sherkin Formation of south-west County Cork, Republic of Ireland. *Pollen et Spores*, **16**, 565–588.

Cloutier, R., Loboziak, S., Chandilier, A. M. & Bliek, A. 1996 Biostratigraphy of the Upper Devonian Escuminac Formation, eastern Quebec, Canada: based on miospores and fishes. *Review of Palaeobotany and Palynology*, **93**, 191–215.

Coe, K. & Selwood, E. B. 1968. The Upper Palaeozoic stratigraphy of west Cork and parts of south Kerry. *Proceedings of the Royal Irish Academy*, **66B**, 113–131.

Cotter, E. & Graham, J. R. 1991. Coastal plain sedimentation in the late Devonian of southern Ireland; hummocky cross-stratification in fluvial deposits. *Sedimentary Geology*, **72**, 201–224.

Elliott, T. 1974. Interdistributary bay sequences and their genesis. *Sedimentology*, **21**, 611–622.

Graham, J. R. 1975. Deposits of a near-coastal fluvial plain: the Toe Head Formation (Upper Devonian) of south-west Cork, Eire. *Sedimentary Geology*, **14**, 45–61.

—— 1983. Analysis of the Upper Devonian Munster Basin, an example of a fluvial distributary system. *In*: Collinson, J. D. & Lewin, J. (eds) *Modern and Ancient Fluvial Systems*. International Association of Sedimentologists, Special Publications, **6**, 473–483.

—— & Reilly, T. A. 1972. The Sherkin Formation (Devonian) of south-west County Cork. *Geological Survey Ireland Bulletin*, **1**, 281–300.

Higgs, K. & Russell, K. J. 1981. Upper Devonian microfloras from southeast Iveragh, County Kerry, Ireland. *Geological Survey of Ireland Bulletin*, **3**, 17–50.

House, M. R. 1983. Devonian eustatic cycles. *Proceedings of the Ussher Society*, **5**, 395–405.

—— & Kirchgasser, W. T. 1993. Devonian goniatite biostratigraphy and timing of facies movements in the Frasnian of eastern North America. *In*: Hailwood, E. A. & Kidd, R. B. (eds) *High Resolution Stratigraphy*. Geological Society, London, Special Publications, **70**, 267–292.

Jackson R. G., II 1981. Sedimentology of muddy fine-grained channel deposits in meandering streams of the American Middle West. *Journal of Sedimentary Petrology*, **51**, 1169–1192.

Jordan, D. W. & Prior, W. A. 1992. Hierarchical levels of heterogeneity in a Mississippi River meander belt and application to reservoir systems. *AAPG Bulletin*, **76**, 1601–1624.

Kelly, S. B. & Olsen, H. 1993. Terminal fans—a review with reference to Devonian examples. *Sedimentary Geology*, **85**, 339–374.

Loboziak, S. & Streel, M. 1989. Middle–Upper Devonian miospores from the Ghadamis Basin (Tunisia–Libya): systematics and stratigraphy. *Review of Palaeobotany and Palynology*, **58**, 173–196.

MacCarthy, I. A. J. 1987. Transgressive facies in the South Munster Basin, Ireland. *Sedimentology*, **34**, 389–422.

—— 1990. Alluvial sedimentation patterns in the Munster Basin Ireland. *Sedimentology*, **37**, 685–712.

—— & Gardiner, P. R. R. (1987). Dinantian cyclicity: a case study from the Munster Basin of southern Ireland. *In*: Miller, J., Adams, A. E. & Wright, V. P. (eds) *European Dinantian Environments*. Wiley, Chichester, 199–238.

Marshall, J. E. A. 1996 Devonian marine incursions into the Orcadian Basin, Scotland. *Journal of the Geological Society, London*, **153**, 451–466.

Miall, A. D. 1996. *The Geology of Fluvial Deposits*. Springer, Berlin.

Nanson, G. C. 1980. Point bar and floodplain formation of the meandering Beatton Formation, northeastern British Columbia, Canada. *Sedimentology*, **27**, 3–30.

Naylor, D. 1975. Upper Devonian–Lower Carboniferous stratigraphy along the south coast of Dunmanus Bay, Co. Cork. *Proceedings of the Royal Irish Academy*, **75**, 317–337.

——, Sevastopulo, G. D. & Sleeman, A. G. 1989. Subsidence history of the South Munster Basin, Ireland. *In*: Arthurton, R. S., Gutteridge, P. & Nolan, S. C. (eds) *The Role of Tectonics in Devonian and Carboniferous Sedimentation in the*

British Isles. Occasional Publications of the Yorkshire Geological Society, **6**, 99–110.

Reilly, T. A. & Graham, J. R. 1976. The stratigraphy of the Roaringwater Bay area of south-west County Cork. *Geological Survey of Ireland Bulletin*, **2**, 1–13.

Russell, K. J. 1978. Vertebrate fossils from the Iveragh Peninsula and the age of the Old Red Sandstone. *Irish Journal of Earth Science*, **1**, 151–162.

Schwartz, R. K. 1975. *Nature and Genesis of Some Storm Washover Deposits*. US Army Corps of Engineers, Coastal Engineering Research Centre, Technical Memoir, **61**.

—— 1982. Bedform and stratification characteristics of some modern small-scale washover sand bodies. *Sedimentology*, **29**, 835–849.

Smith, D. G. 1987. Meandering river point bar lithofacies models: modern and ancient examples compared. *In*: Ethridge, E. G., Flores, R. M. & Harvey, M. D. (eds) *Recent Developments in Fluvial Sedimentology*. Society of Economic Paleontologists and Mineralogists, Special Publications, **39**, 83–91.

Staplin, F. L. 1969. Sedimentary organic matter, organic metamorphism and oil and gas occurrence. *Bulletin of Canadian Petroleum Geology*. **17**, 46–66

Streel, M. & Loboziak, S. 1996. Middle and Upper Devonian miospores. *In*: Jansonius, J. & McGregor, D. C. (eds) *Palynology: Principles and Applications*. American Association of Stratigraphic Palynologists Foundation, **2**, 575–587.

——, Higgs, K., Loboziak, S., Riegel, W. & Steemans, P. 1987. Spore stratigraphy and correlation with faunas and floras in the type marine Devonian of the Ardenne–Rhenish regions. *Review of Palaeobotany and Palynology*, **50**, 211–229.

Strother, 1996. Acritarchs. *In*: Jansonius, J. & McGregor, D. C. (eds) *Palynology: Principles and Applications*. American Association of Stratigraphic Palynologists Foundation, **1**, Publishers Press, Salt Lake City, 81–106.

Tappan, H. N. 1980. *Paleobiology of Plant Protists*. Freeman, San Francisco.

Thomas, R. G., Smith, D. G., Wood, J. M., Visser, J., Calverley-Range, E. A. & Koster, E. H. 1987. Inclined heterolithic stratification—terminology, description, interpretation and significance. *Sedimentary Geology*, **53**, 123–179.

Tucker, R. D., Bradley, D. C., Ver Straeten, C. A., Harris, A. G., Ebert, J. R. & McCutheon, S. R. 1998. New U–Pb zircon ages and the duration and division of Devonian time. *Earth and Planetary Science Letters*, **158**, 175–186.

Tunbridge, I. P. 1981. Sandy high energy flood sediments—some criteria for their recognition, with an example from the Devonian of S.W. England. *Sedimentary Geology*, **28**, 79–95.

Tyson, R. V. 1987. The genesis and palynofacies characteristics of marine petroleum source rocks. *In*: Brooks, J. & Fleet, A. J. (eds) *Marine Petroleum Source Rocks*. Geological Society, London Special Publications, **26**, 47–67.

Williams, E. A., Bamford, M. L. F., Cooper, M. A. *et al.* 1989. Tectonic controls and sedimentary response in the Devonian–Carboniferous Munster and South Munster Basins, south-west Ireland. *In*: Arthurton, R. S., Gutteridge, P. & Nolan, S. C. (eds) *The Role of Tectonics in Devonian and Carboniferous Sedimentation in the British Isles*. Occasional Publications of the Yorkshire Geological Society, **6**, 123–141.

——, Sergeev, S. A., Stössel, I. & Ford, M. 1997. An Eifelian U–Pb zircon date for the Enagh Tuff Bed from the Old Red Sandstone of the Munster Basin in NW Iveragh, SW Ireland. *Journal of the Geological Society, London*, **154**, 189–193.

——, ——, ——, —— & Higgs, K. T. 2000. U–Pb zircon geochronology of silicic tuffs and chronostratigraphy of the earliest Old Red Sandstone in the Munster Basin, SW Ireland. This volume.

Palaeoenvironment of the plant bearing horizons of the Devonian–Carboniferous Kiltorcan Formation, Kiltorcan Hill, Co. Kilkenny, Ireland

D. E. JARVIS

Department of Geology, National University of Ireland, Cork, Ireland
(e-mail: ejarvis@ucc.ie)

Abstract: The environment of deposition of the plant-bearing horizons of the Devonian–Carboniferous Kiltorcan Formation in Co. Kilkenny is poorly known. Previous studies suggested a possible lacustrine palaeoenvironment with plant material being swept in and fluvial conditions. Reinvestigation of the plant-rich localities on Kiltorcan Hill using both sedimentological facies and palynofacies suggests that there were high-sinuosity streams carrying fine-grained sediments with the plant fossils preserved on point bars, and marginal back-swamp conditions with plant material deposited close to the growth site.

The Kiltorcan Formation forms the uppermost part of the Old Red Sandstone and is a distinctive Upper Devonian–Lower Carboniferous unit in Southern Ireland. It comprises predominantly non-red lithologies, being characterised by green mudstones, siltstones and fine sandstones and yellow sandstones. The type locality is at Kiltorcan Hill, near Ballyhale in County Kilkenny where an internationally renowned fossil assemblage containing *Cyclostigma* and *Archaeopteris* has been recorded. The arborescent lycopod *Cyclostigma kiltorkense* appears to be of biostratigraphic importance. It is an index species of a subzone of Banks' floral assemblage VII (Banks 1980) and may be restricted to the 'Strunian' part of the uppermost Devonian succession. Other elements of the assemblage have evolutionary importance, such as the progymnosperm *Archaeopteris hibernica* (Beck 1981). Despite the extensive research undertaken on the fossil material little work has been done on the environment of deposition of the Kiltorcan Formation at its type locality near Ballyhale, Co. Kilkenny (Fig. 1). This study aims to provide an accurate determination of the palaeoenvironments in which the plants grew, died and were subsequently deposited in the sediments. Both sedimentological facies and palynofacies analyses were employed to establish the fullest reconstruction possible.

The Kiltorcan fossil assemblage

The Kiltorcan assemblage is famous for its striking preservation and evolutionary importance. Much work was undertaken on the assemblage in the middle of the 19th century (Forbes 1853; Haughton 1855, 1859; Griffith & Brongniart 1857, 1858; Baily 1859, 1861, 1869, 1874) and during the early part of the 20th Century (Johnson 1912*a*, *b*, 1913*a*, *b*, 1914*a*, *b*); this work concentrated on the description of the assemblage and suggestions as to the age of the material. No work was carried from the 1920s until the late 1960s, when new studies took place (Chaloner 1968; Chaloner *et al.* 1977). Most of this work focused on the cones and seeds of the assemblage. Chaloner *et al.* (1977) redescribed the earliest platyspermic seed *Spermolithus devonicus* Johnson from the Old Plant Quarry on Kiltorcan Hill (Fig. 1), indicating that the platyspermic and radiospermic seed habits were synchronous in their origin and represent two separate lineages (Chaloner *et al.* 1977).

Stratigraphic distribution of the flora

The principle outcrops of the Kiltorcan Formation on Kiltorcan Hill occur in three quarries: the New Quarry, the Old Plant Quarry and the Roadstone Quarry (Fig. 1).

In the New Quarry *C. kiltorkense* has been found at the base of the exposed section on Kiltorcan Hill (Jarvis 1990).

Chaloner & Lacey (pers. comm.) have described the 'Classic' assemblage from the Old Plant Quarry (Fig. 1) as including *Archaeopteris hibernica* (Forbes) Dawson, *Cyclostigma kiltorkense* Haughton, *Sphenopteris hookeri* Baily and *Sphenopteris* sp. as well as the seed *Spermolithus*

From: FRIEND, P. F. & WILLIAMS, B. P. J. (eds). *New Perspectives on the Old Red Sandstone.* Geological Society, London, Special Publications, **180**, 333–341. 0305-8719/00/$15.00

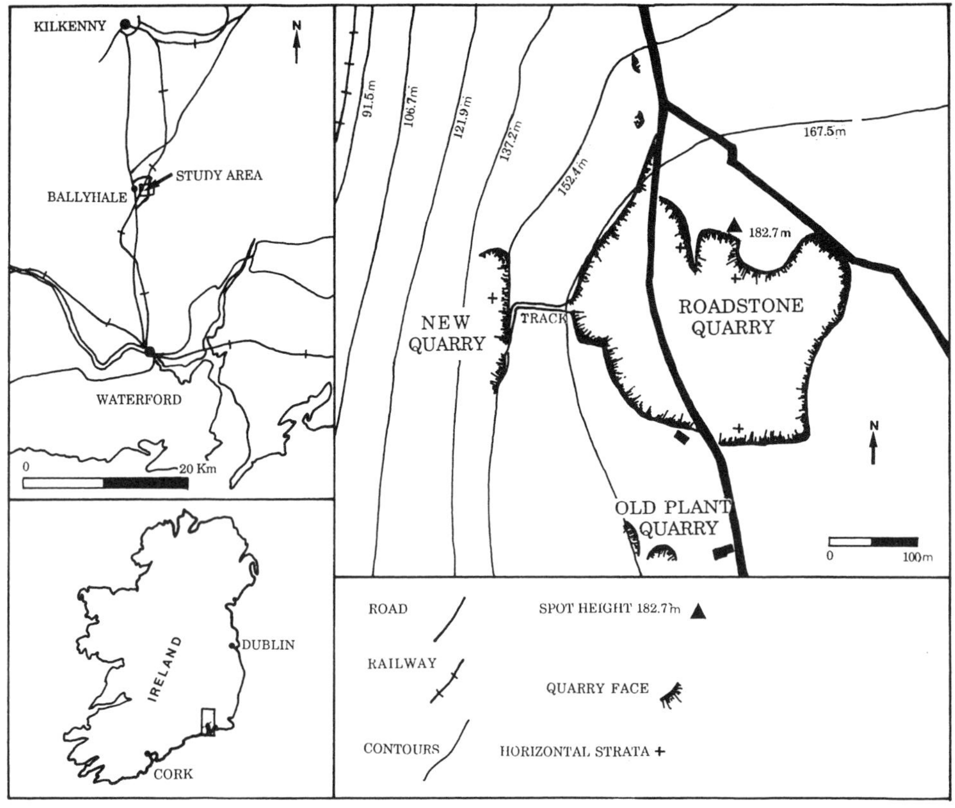

Fig. 1. Map of Kiltorcan Hill, Co. Kilkenny, displaying the position of the New Quarry, the Old Plant Quarry and the Roadstone Quarry.

devonicus Johnson. Chaloner & Lacey (pers. comm.) also noted differences in the leaf scar attachment to be the main distinguishing factor between the lycopod species *C. kiltorkense* from the Old Plant Quarry and *Lepidodendropsis* aff. *L. hirmeri* from the Roadstone Quarry.

The stratigraphically highest plant bed on Kiltorcan Hill in the Roadstone Quarry (Figs 1 and 2) has yielded *Lepidodendropsis* sp., aff. *L. hirmeri* Lutz, cf. *Rhacopteris* sp. (fertile pinna) and *Bythotrephis* sp. (an alga) but none of the plants of the 'Classic' Kiltorcan Assemblage (Lacey, pers. comm.).

Jarvis (1990) extracted palynological residues from the New Quarry and the Roadstone Quarry on Kiltorcan Hill (Figs 1 and 2). The New Quarry yielded an Upper Devonian LE Biozone miospore assemblage. The Roadstone Quarry yielded a lowermost Carboniferous VI Miospore Biozone assemblage. This proved that the Classic Kiltorcan flora of *C. kiltorkense* and *A. hibernica* is in part and possibly entirely latest Devonian LE to ?LN Miospore Biozone in age, whereas the upper flora from the Roadstone Quarry is earliest Carboniferous, VI Miospore Biozone in age.

Sedimentology

Colthurst (1977, 1978) made a regional study of the Old Red Sandstone rocks of the district including a somewhat more detailed sedimentological reinvestigation of the Kiltorcan Hill sequence. He suggested that the plant fossils at the Old Plant Quarry were probably deposited on a bar-tail in a meandering river channel whereas the fossiliferous beds at the higher Roadstone Quarry were interpreted as the result of deposition in a lacustrine environment.

The use of sedimentological facies analysis combined with palynofacies studies and the location of additional outcrops on Kiltorcan Hill has allowed a much more detailed and accurate palaeoenvironmental reconstruction to be made during the present study. This has greatly aided the understanding of the environment of deposition of the plant material.

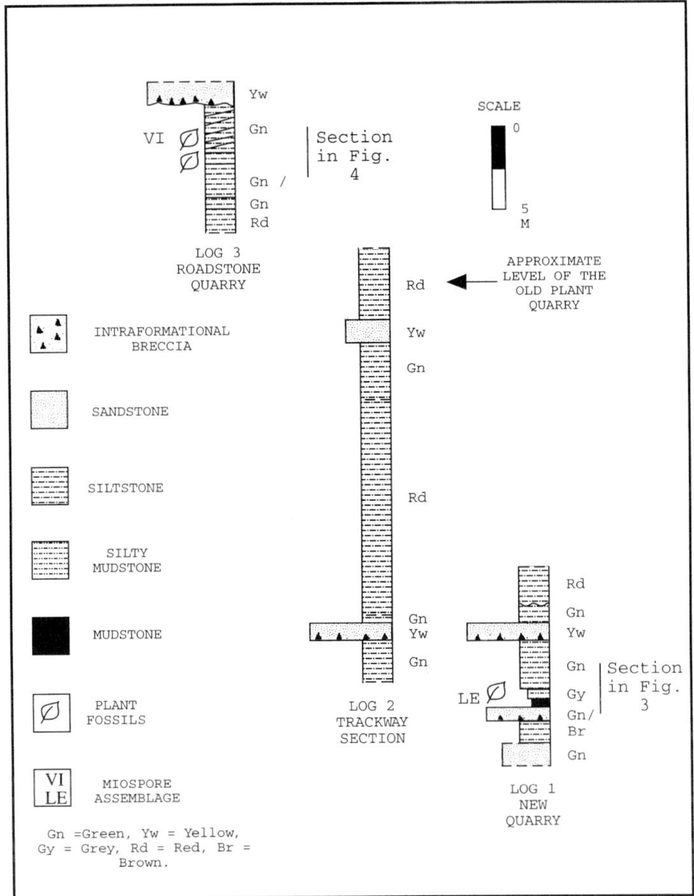

Fig. 2. Summary stratigraphic logs of the main locations studied on Kiltorcan Hill.

Unfortunately, the Old Plant Quarry is now mostly filled in and sedimentological analysis of this level is now impossible. The main emphasis was therefore placed on a study of the New Quarry and Roadstone Quarry sections (Fig. 2). The plant fossil material within each of the quarries is preserved in different facies types and represents preservation in disparate environments (Figs 3 and 4).

New Quarry

In the New Quarry a 200 m strike section exposes a north to south gradational sequence. The nature of the quarrying produces oblique outcrop making the production of longitudinal cross-sections difficult. At the south end the sequence (Fig. 3) consists of green massive siltstones with brown and black rootlet-like features (1–3 mm thick, up to 7 cm long). This is overlain by grey siltstones with abundant coalified and permineralized large stems of *C. kiltorkense* up to 14 cm long and 7 cm wide with the long axes of the stems arranged chaotically on individual bedding planes. Further north the sequence passes laterally into green 'flaggy' bedded coarse siltstone–fine sandstone with a very high mica content. These latter beds contain little or no plant material. The flaggy green siltstone passes laterally into flaggy red micaeous siltstone.

The plant bed exposed in the southern part of the quarry is considered to be of significance. Although the fossils of *C. kiltorkense* found at this locality are not seen *in situ* it appears likely, from their large size and abundance, that they were growing close to the site of sedimentation. The horizon below this plant bed is a green siltstone with brown and grey rootlet-like vertical traces and convex–concave slip surfaces which are very similar to what Gray & Nickelsen (1989) have described from North America as pedogenic slickensides. These pedogenic slickensides

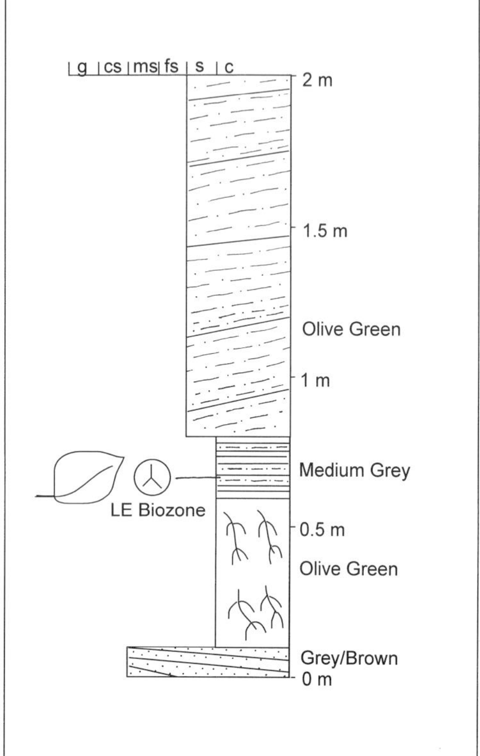

Fig. 3. Log of the plant-bearing horizon in the New Quarry horizon.

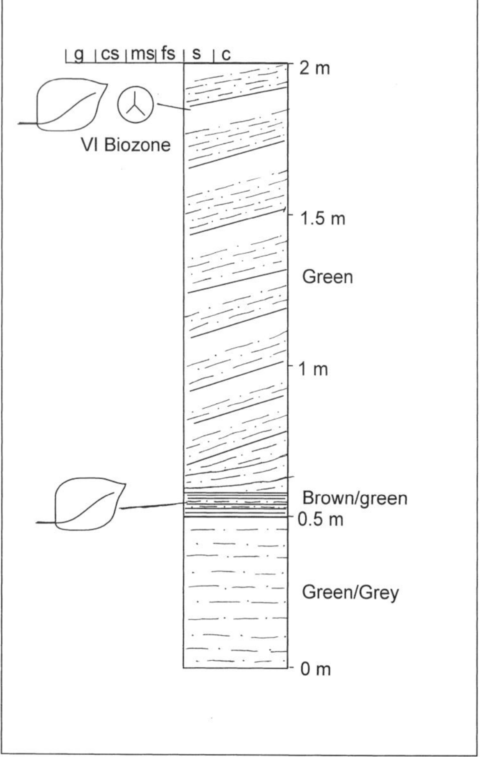

Fig. 4. Log of the plant-bearing horizon in the Roadstone Quarry horizon.

are found in modern soils that experience repeated cycles of drying and wetting (Gray & Nickelsen 1989). The presence of rootlets and pedogenic slickensides suggests that this horizon represents a palaeosol. The high carbonaceous content of the plant bed suggests that conditions were normally waterlogged, with only minor phases of drying out. The plant bed is therefore interpreted as having been deposited in a water-saturated swamp-type environment with a normally high water table. The 'flaggy' green and red siltstones to the north are possibly the deposits of a levee, behind which the swamp was developed.

Roadstone Quarry

In the north of this quarry, a 3 m thick unit of interlaminated mudstone and green siltstone with low-angle cross stratification and plant fossils is exposed (Fig. 4). The cross beds are dipping gently to the northeast (Fig. 5). A plentiful, chlorite-preserved fossil flora is seen in the green mudstone units; here plant material is seen to be concentrated along the bedding planes and laminae directly below siltstones. The best-preserved fossil material is found in the lower parts of the foresets on the top of mudstones. The upper part of the foresets and the top of siltstones display much less fossil plant material. In the present investigation only fragmentary plant material was observed. However, more complete plant remains have been extracted in the past: Colthurst (1977) listed a floral assemblage for this location including *Lepidodendropsis* sp. aff. *L. hirmeri* Lutz, cf. *Rhacopteris* sp. (fertile pinna) and *Bythotrephis* sp. During the present study numerous stems of *Lepidodendropsis* sp. were observed ranging up to 10 cm long and 4 cm wide. The plant stems are current orientated, suggesting they were deposited in a moderate-energy environment. A small scale alternation between medium siltstone and fine siltstone and mudstone with plant material concentrated in the finer lithologies suggests that the hydrodynamic energy of the environment was fluctuating. The low-angle cross-bedded units are seen accreting towards the north, in a fine grained point bar deposit (Fig. 5).

Bridge *et al.* (1980) have noted that the most abundant plant material in plant-bearing

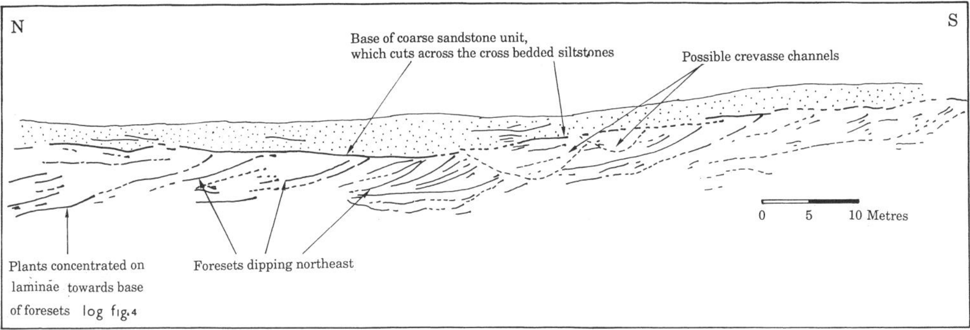

Fig. 5. Cross-section of the plant-bearing level of the Roadstone Quarry (looking east; vertical scale same as horizontal scale).

sandstones at Kerry Head, County Kerry appears to be restricted to facies interpreted as levee or crevasse deposits in areas marginal to channels. In such environments plants would be bent over and broken by floodwaters and then buried very quickly. Bridge *et al.* (1980) concluded that the restriction of major gymnosperm plants to areas near channels may be due to the locally high water table of perennial streams. It appears that the plants preserved in the Roadstone Quarry were also growing marginal to a river, from which they were incorporated in point bar deposits. The mudstones to siltstone alternations represent fluctuating energy conditions on the point bar.

Palynofacies

Palynological samples were extracted from the plant beds in the New and Roadstone Quarries. These samples yielded miospore assemblages that were used to date the plant horizons (Jarvis 1990); this work proved that the strata on Kiltorcan Hill straddle the Devonian–Carboniferous boundary.

More recently, additional work has been undertaken on these samples to determine their palynofacies. Modern classification of palynological matter such as the Amsterdam Palynological Organic Matter Classification (APOMC) has allowed a better understanding of how different types of palynological organic matter (POM) behave in sedimentary environments. To date, little palynofacies work has been undertaken on non-marine environments, a notable exception being the work of Fisher (1980), who has investigated Mesozoic non-marine palynofacies. In the present study the APOMC is used throughout and deductions are based on Fisher (1980) and Tyson (1993) with addition of my own interpretations.

New Quarry plant bed

Description

Preservation of the POM is good. Short oxidation times of only 30 min to 1 h in Schultze solution were required for miospore identification. The size of the organic material ranges up to 250 μm for woody structured debris (w.s.d.) fragments, with an average size of 60–70 μm. Most spores were 50–60 μm in size, although larger spores such as *Retispora macroreticulata* are present. The sample displays a high absolute abundance of spores (*c.* 30 000 g^{-1}). Approximately 30 miospore taxa were recorded and a high (50%) relative abundance of spores to other major groups of POM (see Fig. 6). Woody structured debris (w.s.d.) and oxidized structured debris (o.s.d.) are both present but only make up a small relative abundance (*c.* 20–25% each) of the POM as a result of the high abundance of spores. The ratio for non-oxidized structured debris to o.s.d. is around 1 : 1. Cuticle makes up 5% of the POM.

In the relative percentages of miospores, *Retispora lepidophyta* is seen to be dominate, making up 68% of the miospore population. *Diducites* spp. and *Indotriradites explanatus* all make up consistent but small parts of the assemblage (see Fig. 6 for details).

Interpretation

The near-equal proportions of o.s.d. and w.s.d. suggests the environment was not severely oxidizing. The o.s.d. content represents a low level of oxidation of the palynological material, which may have taken place during transportation or *in situ* under a fluctuating water table. The good preservation also reflects deposition in an environment with only a low oxidation level of the sediment. The low percentage of cuticle and

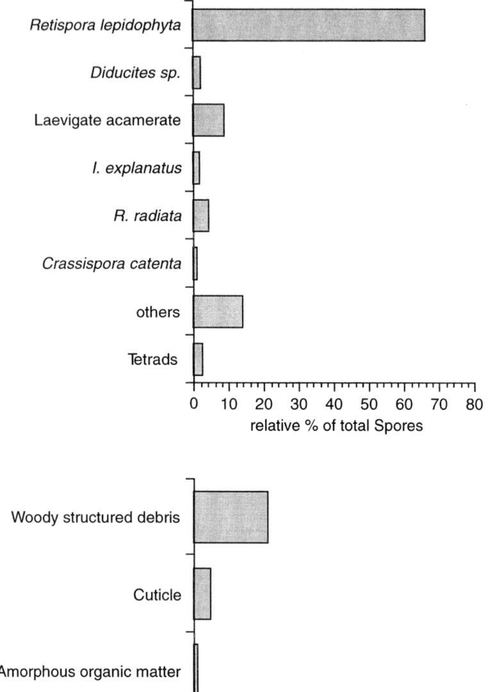

Fig. 6. Relative percentages of important elements of the palynofacies assemblage recorded from sample 88/1, Kiltorcan New Quarry.

lack of very large fragments of w.s.d. suggest a lack of mechanical breakdown of the plant stems and/or the possible transportation of most of the medium-sized woody and buoyant cuticle. The high absolute abundance and relative percentage to w.s.d. of miospores suggests deposition in a quiet environment away from active fluvial deposition. The presence of megaspores, some large spores and tetrads suggests deposition close to vegetation source. The sedimentary facies here has been interpreted as representing a near-swamp depositional site (see above). The high percentage of *R. lepidophyta* also suggests deposition was not within the swamp, because of its possible origin from swamp margin arborescent lycopods such as *C. kiltorkense* (Jarvis 1992). A quiet to moderate-energy, low-oxidizing swamp margin is therefore suggested by the palynofacies present.

Roadstone Quarry plant bed

Description

This sample, CQ-1, yielded a small absolute abundance of miospores of *c.* 3000–4000 spores per gram of sediment, but the spores are well preserved with 15 species present. The relative abundance of both cuticle and spores to total POM was *c.* 20% (see Fig. 7); w.s.d. made up nearly 40% of the total POM and was more abundant than o.s.d. Amorphous organic matter (AOM) of degraded wood origin makes up a very small percentage of the POM.

Most of the spores observed are simple leavigate acamerate taxa (*c.* 60% of total spores). *Spelaeotriletes* spp., *Verrucosisporites nitidus* and thick-walled spores make up only small percentages of the miospore assemblages (less than

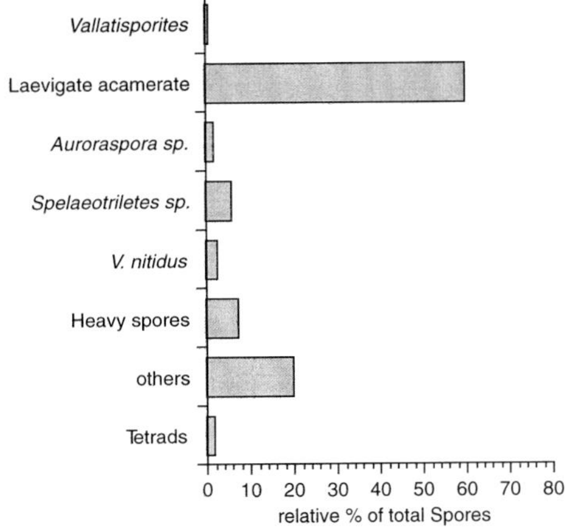

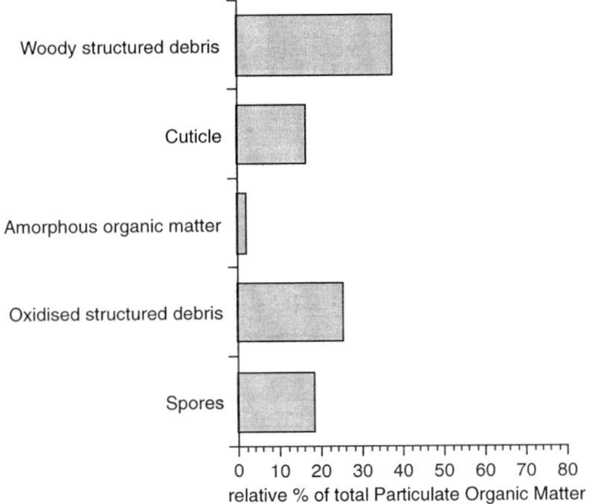

Fig. 7. Relative percentages of important elements of the palynofacies assemblage recorded from sample CQ-1, Kiltorcan Roadstone Quarry.

10%) (see Fig. 7 for details). No large spores were noted.

Interpretation

The high proportion of w.s.d., the presence of AOM and excellent preservation of the material suggest very little oxidation of the POM, both during transportation and within the sediment. The lack of large miospores and w.s.d. suggests some current sorting. The low proportion of spores to other POM indicates hydrodynamic sorting and dilution by structured debris. The moderately high proportion of well-preserved cuticle suggests that the energy of transportation was not high and transportation time was not prolonged.

A moderate- to low-energy reducing environment with rapid deposition of plant material is therefore deduced for this palynofacies. This would be consistent with a point bar palaeoenvironment in a stream carrying fine-grained sediment.

Floristics of the plant beds

Scheckler (1986), in a palaeoecological analysis of some late Devonian coal swamps from Appalachian Laurentia (USA), has noted that the back-swamp environments in non-coastal plain upland areas are dominated by the pre-fern *Rhacophyton* and a near-swamp community consisting of tree lycopods. Scheckler (1986) noted that the drier flood plain environment was represented by trees such as *Archaeopteris–Callixylon* of progymnosperm affinity.

The abundance of the arborescent lycopod *C. kiltorkense* and the corresponding absence of the progymnosperm *A. hibernica* from the New Quarry plant horizon seems to fit into Scheckler's (1986) back-swamp environmental model. This further reinforces the sedimentological and palynofacies analyses made here.

Furthermore, the presence of well preserved and abundant fronds of *Archaeopteris* found in the Old Plant Quarry may suggest deposition near a drier part of the flood plain according to Scheckler's (1986) model. The inclusion of tree lycopod material in this assemblage as well would indicate some transportation of the lycopods from wetter parts of the flood plain. This would be consistent with a bar-tail environment for the Old Plant Quarry fossil horizon (Colthurst 1978).

The lack of progymnosperms in the Roadstone Quarry may be due solely to stratigraphic reasons; there are no definite records of *A. hibernica* from VI Miospore Biozone (Jarvis 1992). The presence of lycopod fragments would suggest erosion of swampy parts of the alluvial plain by the stream carrying fine-grained sediment interpreted here.

Conclusions

Analysis of sedimentological, palynofacies and paleofloristical data suggests that the plant beds of the Kiltorcan Formation at Kiltorcan Hill were deposited on point bars and in back-swamp environments. The lowest exposed fossil horizon in the New Quarry represents a near-swamp palaeoenvironment. The Old Plant Quarry fossils were deposited on a bar tail of a stream flowing through a dry part of the alluvial plain. The Roadstone Quarry plant beds were laid down in a fine-grained high-sinuosity river running through a wet area of the alluvial plain. No evidence for a lacustrine palaeoenvironment at Kiltorcan was noted.

I would like to thank K. T. Higgs (UCC) for guidance throughout this work, which formed part of a PhD study, and P. M. Brück (UCC) for the use of departmental facilities. I would like to thank the late W. Lacey, University of Wales at Bangor, for information on the floras. This work was carried out under a University College Cork postgraduate studentship.

References

Baily, W. H. 1859. On the fructification of *Cyclopteris hibernica* (Forbes), from the Upper Devonian or Lower Carboniferous strata at Kiltorkan Hill, Co. Kilkenny. *British Association for the Advancement Science, Report* (28th meeting), **28**, 75–76.

—— 1861. Palaeontological notes. *In*: Jukes, J. B., DuNoyer, G. V., Baily, W. H. & Kinahan, G. H. (eds) *Explanations to accompany sheets 147 and 157 of the maps of the Geological Survey of Ireland, illustrating the parts of Co. Kilkenny, Carlow and Wexford.* Memoir, Geological Survey of Ireland.

—— 1869. On fossils obtained at Kiltorkan Quarry, Co. Kilkenny. *British Association for the Advancement Science, Report*, **1869**, 72–75.

—— 1874. On the fossils from the Upper Old Red Sandstone of Kiltorcan Hill, in the County of Kilkenny. *Proceedings of the Royal Irish Academy Report*, **2**, 44–48.

Banks, H. P. 1980. Floral assemblages in the Siluro-Devonian. *In*: Dilcher, D. L. & Taylon, T. N. (eds) *Biostratigraphy of Fossil Plants.* Dowden, Hutchinson and Ross, Stroudsburg, PA.

Beck, C. B. 1981. *Archaeopteris* and its role in vascular plant evolution. *In*: Niklas, K. J. (ed.) *Palaeobotany and Palaeoecology and Evolution. Vol. 1.* Praeger, New York, 193–230.

Bridge, J. S., Van Veen, P. M. & Matten, L. C. 1980. Aspects of the sedimentology, palynology and palaeobotany of the Upper Devonian of southern Kerry Head, Co. Kerry, Ireland. *Geological Journal*, **15**, 143–170.

Chaloner, W. G. 1968. The cone of *Cyclostigma kiltorcense* Haughton from the Upper Devonian of Ireland. *Journal of the Linnean Society (Botany)*, **61**(384), 25–36.

——, Hill, A. J. & Lacey, W. S. 1977. First Devonian platyspermic seed and its implications in gymnosperm evolution. *Nature*, **256**, 233–235.

Colthurst, J. R. J. 1977. *The Geology of the Lower Palaeozoic and Old Red Sandstone rocks of the Slievenamon inlier.* PhD thesis, University of Dublin.

—— 1978. Old Red Sandstone rocks surrounding the Slievenamon Inlier, Counties Tipperary and Kilkenny, *Journal of Earth Sciences Royal Dublin Society*, **1**, 77–103.

Fisher, M. J. 1980. Kerogen distribution and depositional environments in the Middle Jurassic of Yorkshire, U.K. *In*: *Proceedings, International Palynological Conference, Lucknow, 1976–1977*, **2**, 574–580.

Forbes, E. 1853. On the fossils of the Yellow Sandstone of the south of Ireland. *British Association for the Advancement Science, Report* (22nd meeting), **22**, 43.

Gray, M. B. & Nickleson, R. P. 1989. Pedogenic Slickensides, indicators of strain and deformation processes in bed sequences of the Appalachian foreland. *Geology*, **17**, 72–75.

Griffith, R. & Brongniart, A. 1857. On the fossil plants which have been discovered in the rocks at the base of the Carboniferous system in Ireland. *Journal of the Geological Society, Dublin*, **7**, 287–293.

—— & —— 1858. On the fossil plants which have been discovered in the rocks at the base of the

Carboniferous system in Ireland. *Journal of the Royal Dublin Society*, **1**, 313.

HAUGHTON, S. 1855. On the evidence afforded by fossil plants, as to the boundary line between the Devonian and Carboniferous rocks. *Journal of the Geological Society, Dublin*, **6**, 227–241.

—— 1859. On *Cyclostigma* at Ballyhale, Co. Kilkenny. *Natural History Review*, **7**, 209.

JARVIS, D. E. 1990. New palynological data on the age of the Kiltorcan flora of Co. Kilkenny, Ireland. *Journal of Micropalaeontology*, **9**, 87–94.

—— 1992. *The stratigraphic palynology, palynofacies and sedimentology of the Devonian–Carboniferous Kiltorcan Formation of Southern Ireland.* PhD thesis, National University of Ireland, UCC.

JOHNSON, T. 1912*a*. Is *Archaeopteris* a pteridosperm. *Scientific Proceedings of the Royal Dublin Society*, **13**, 114–136.

—— 1912*b*. *Forbesia cancellata*, gen. et sp. nov. (*Sphenopteris* sp. Baily). *Scientific Proceedings of the Royal Dublin Society*, **13**, 177–183.

—— 1913*a*. The occurrence of *Archaeopteris tschermaki*, stur, and other species of *Archaeopteris* in Ireland. *Scientific Proceedings of the Royal Dublin Society*, **13**, 137–141.

—— 1913*b*. On *Bothrodendron* (*Cyclostigma*) *kiltorkense*, Haughton. *Scientific Proceedings of the Royal Dublin Society*, **13**, 500–525.

—— 1914*a*. *Ginkgophyllum kiltorkense* sp. nov. *Scientific Proceedings of the Royal Dublin Society*, **14**, 169–178.

—— 1914*b*. *Bothrodendron kiltorkense*, Haughton sp. its stigma and cone. *Scientific Proceedings of the Royal Dublin Society*, **14**, 211–214.

SCHECKLER, S. E. 1986. Geology, floristics and palaeoecology of Late Devonian coal swamps from Appalachian Laurentia (U.S.A.). *Annales de la Société Géologique de Belgique*, **109**, 209–222.

TYSON, R. V. 1993. Palynofacies analysis. *In*: JENKINS, D. G. (ed.). *Applied Micropalaeontology*, Kluwer Academic, Dordrecht, 53–191.

Silurian marginal marine sedimentation and the anatomy of the marine–Old Red Sandstone transition in Pembrokeshire, SW Wales

ROBERT D. HILLIER

Department of Geology and Petroleum Geology, Meston Building, Kings College, Aberdeen University, Aberdeen AB24 3UE, UK

Present address: Ysgol Greenhill, Heywood Lane, Tenby, Pembrokeshire SA70 8BN, UK

Abstract: Deposition of the Silurian (Wenlock) siliciclastic Gray Sandstone Group of southwest Pembrokeshire took place within littoral environments close to the palaeogeographical shelf-margin of the Welsh Basin. Sedimentation described a northerly, basinward progradation across an earlier Late Ordovician? to Aeronian rift basin. Changes in relative sea level (rsl) had profound effects on depositional environments, and five depositional sequences are recognized. During highstands of rsl, the area was influenced by wave-dominated, shallow-marine conditions. During lowstands of rsl, shelf incision and sediment bypass occurred. Associated valley fills vary in nature from high-sinuosity estuarine channels, tidal flats and tidally-influenced, high-sinuosity fluvial channels. The last of these predominate within the youngest sequence, with abundant subaerial emergence indicators heralding the onset of true continental deposition, and the conformable transition into the overlying Old Red Sandstone Red Cliff Formation within the Marloes Peninsula. A renewal of tectonic activity ensued within the Lower Old Red Sandstone, with pebbly low-sinuosity alluvium of the Albion Sands Formation, and fanglomerates of the Lindsway Bay Formation reflecting reactivation of earlier rift-margin faults, probably within a transtensional tectonic regime.

The interpretation of the transition between the marine Silurian deposits and the overlying continental deposits of the Lower Old Red Sandstone in southern Pembrokeshire has been a problematic one. In particular, the argument as to whether the relationship within the Marloes Peninsula to the north of Milford Haven is conformable (Sanzen-Baker 1972; Walmsley & Bassett 1976; Hurst *et al.* 1978) or unconformable (Allen & Williams 1978; Allen 1985*a*) remains unresolved. In this paper, the environments of deposition of the youngest marine Silurian sediments, the Gray Sandstone Group (of Wenlock age, Walmsley & Bassett 1976) are described. The recognition of incised valleys containing tidal-influenced deposits within the Gray Sandstone Group has shed new light upon depositional environments within Wenlock time, providing evidence for a conformable transition into the Lower Old Red Sandstone. The relationship between tectonic activity and sedimentologywithin the Lower Old Red Sandstone is also addressed.

Regional setting

The Welsh Basin lay on the northwestern margin of an important Lower Palaeozoic microplate, Eastern Avalonia (Bluck *et al.* 1992). Geodetically, the basin probably lay within subtropical latitudes approximately between 13 and 17° S, a position it had reached through northward drift by late Llandovery time (Channell *et al.* 1993; Torsvik *et al.* 1993). The Silurian Welsh Basin had a clearly defined structural control on sedimentation (Fig. 1), with deep-marine sediments of the basin proper being separated from their correlative shallow-marine and continental deposits by an anastomosing array of long-lived faults stretching from Pembrokeshire into the English Midlands (Woodcock & Gibbons 1988; Bassett *et al.* 1992).

The Gray Sandstone Group crops out within the Variscan Foldbelt (Strahan *et al.* 1914; Cantrill *et al.* 1916; Dunne 1983). In Pembrokeshire this zone of structural deformation is typified by meso- and macro-scale folding (Hancock *et al.* 1982, 1983). These folds are related to the positive inversion of long-lived, pre-existing normal faults (Powell 1987, 1989; Hayward & Graham 1989), namely the Wenall, Benton and Ritec Faults, and their associated footwall shortcuts (the Johnson and Musselwick Thrusts, Fig. 2). It is likely that these normal faults were active during latest Ordovician to

From: FRIEND, P. F. & WILLIAMS, B. P. J. (eds). *New Perspectives on the Old Red Sandstone.* Geological Society, London, Special Publications, **180**, 343–354. 0305-8719/00/$15.00

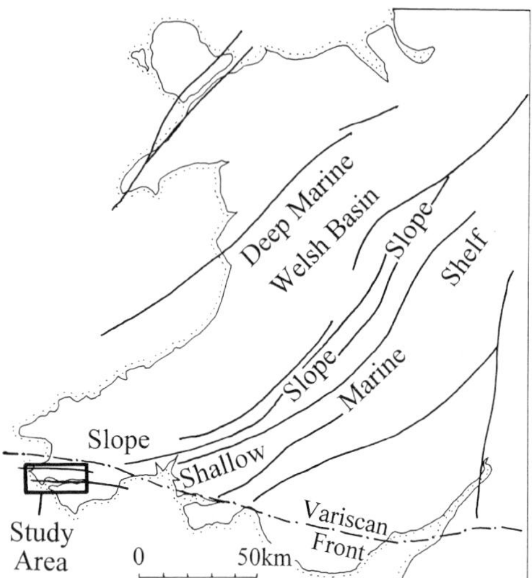

Fig. 1. Structural control on bathymetry of the Silurian Welsh Basin.

Llandovery times, defining a series of half-grabens here termed the Skomer Basin. The Skomer Basin accommodated over 1000 m of basaltic and rhyolitic lavas, pyroclastic deposits and associated continental to shallow-marine sediments (Ziegler *et al.* 1969; Thorpe *et al.* 1989). Such rifting on the shelf platform was synchronous with other intra-shelf basin development to the east, notably the Woolhope and Usk Basins (Butler *et al.* 1997). It is uncertain whether rifting was related to the development of the Rheic Ocean to the south (Pharaoh *et al.* 1991), or to transtensional stresses associated with the docking of Eastern Avalonia with Laurentia (Soper & Woodcock 1990).

The Late Llandovery to Wenlock rift-related subsidence was episodic, being punctuated by local uplift and erosion; for example, the unconformity developed during latest Aeronian to earliest Telychian time, at the top of the Skomer Volcanic Group. Within this episodic subsidence phase, accommodation space was developed that allowed the deposition of a thick, northward prograding siliciclastic wedge containing the late Llandovery to Wenlock Coralliferous and Gray Sandstone Groups, and their correlative fluvial feeder systems, the embryonic Old Red Sandstone.

Environments of deposition

Deposition of the Gray Sandstone Group described a northerly, basinward progradation of a thick (*c.* 500 m maximum thickness in the Marloes Peninsula), predominantly shallow-marine siliciclastic wedge across the Skomer Basin. Changes in relative sea level (rsl) had profound effects on the depositional environment and sequence architecture within this wedge, with five sequences being identified. During rsl highstands, the area was influenced by wave- or storm-dominated shallow-marine conditions, with the shoreline having an approximate east–west orientation, and northerly strand plain progradation being fed by a ready supply of siliciclastic material from the southerly hinterland Pretannia (Cope & Bassett 1987). Falls in rsl resulted in subaerial exposure and valley incision of the shelf (Fig. 3), with sediment bypassing to the northerly, deeper parts of the Welsh Basin. It is the description of estuarine deposits within the incised valleys of the Marloes Peninsula, and the relationship with the overlying Old Red Sandstone, that is the focus of the remainder of this paper.

Sedimentology of valley fills

Tidal-dominated conditions are recognized within the Gray Sandstone Group predominantly by the recognition of mud couplets or

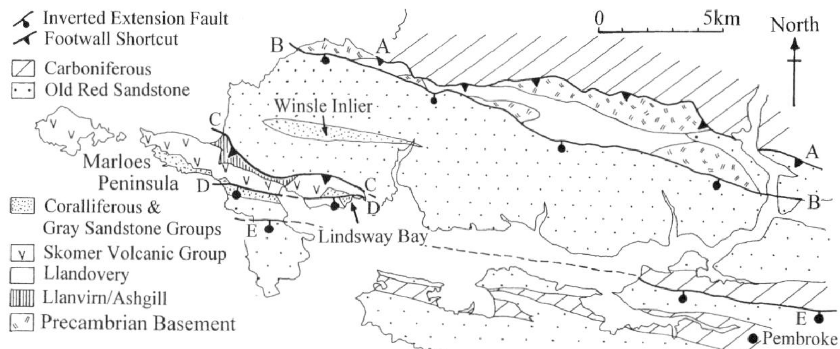

Fig. 2. Simplified geological map of southwest Pembrokeshire. A–A, Johnson Thrust; B–B, Benton Fault; C–C, Musselwick Thrust; D–D, Wenall Fault; E–E, Ritec Fault.

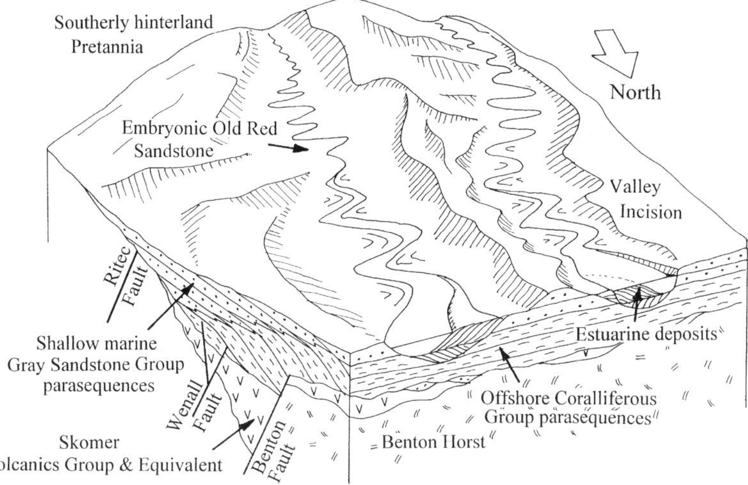

Fig. 3. Schematic sequence development and palaeogeography during deposition of the Gray Sandstone Group.

Fig. 4. Mud couplets from megaripple toesets (SM 7698 0775). (Note mud couplet enclosing a thin sand layer of the subordinate current stage (**A**), separating thicker sand layers of the dominant current stage (**B**).) Coin is 3 cm in diameter.

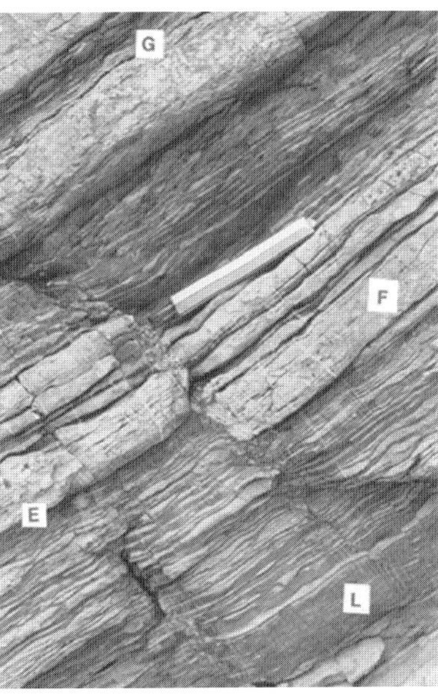

Fig. 5. Seasonal flaser-bedded sandstones (**F**), and linsen-bedded mudstones (**L**) (SM 7634 0815). (Note erosive (**E**) and gradational contacts (**G**).) Scale bar is 25 cm long.

bundles (Fig. 4). Here, two mudstone laminae encapsulate a thin sandstone layer deposited by the subordinate tidal current, with each couplet being separated by a thicker sandstone layer deposited during the dominant current stage (Nio & Yang 1991). The mudstone drapes record deposition from suspension during slack tide intervals. Also observed are sequential changes within bundle thickness, manifesting themselves as thick–thin alternations. These are interpreted as deposition from spring and neap tide cyclicity. Because of the thin nature of the laminae, a detailed mathematical analysis similar to that of Visser (1980) has not been undertaken.

Other rhythmically bedded heterolithic deposits are frequently observed within the Gray Sandstone Group. Commonly, these beds are of centimetre to decimetre scale and comprise sandstone-rich–mudstone-rich rhythmic bedding (Fig. 5). Sandstone units are principally flaser-bedded whereas the mudstone-rich beds comprise a myriad of linsen bedding. In reality, these complex heterolithic units record a continuum of

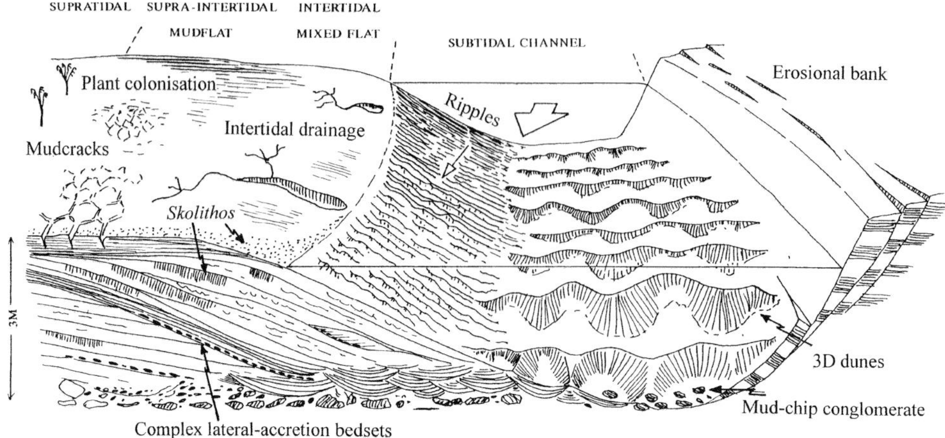

Fig. 6. Idealized sedimentary association of Gray Sandstone Group tidal channel and associated mudflats.

deposits related to the sandstone–mudstone ratio (*sensu* Reineck & Wunderlich 1968). A gradual transition from flaser-bedded, fine-grained sandstone beds is observed into the overlying, interbedded mudstone units, whereas the base of the flaser-bedded units is often mildly erosive (Fig. 5) Successions of rhythmic layers are commonly seen to be laterally extensive over tens of metres. Similar, repeated, rhythmic cycles have been recorded from recent sediments in a mesotidal channel fill sequence from the Dutch Oosterschelde (Van den Berg 1981), where a seasonal origin for this distinctive bedding type has been advocated. In winter periods, currents within the tidal system would be elevated as a result of increased freshwater runoff, together with the effects of higher storm frequency (and associated fluid power). Thus the sharp-based flaser-bedded sandstones are interpreted as a result of highly dynamic processes limiting the amount of flocculated mud fallout. During the summer season, however, current velocities would be expected to diminish as a result of decreased freshwater runoff and storm activity. Here, flocculated mud would accumulate at the sediment–water interface more readily. Reduced current velocities would also have the effect of reducing the sand content in the system, this being manifested as thin discontinuous sandstone linsen.

Sequentially, sedimentary structures and grain-size trends within sedimentary rocks interpreted as being deposited by the action of tidal currents comprise a non-random vertical distribution. In simple terms, this distribution is that of upward fining and upward thinning over the scale of metres, representing the deposits of laterally accreted tidal channels and associated mud flats. The following description documents an 'ideal' cyclothem generated through numerous field observations within the Gray Sandstone Group (Fig. 6).

The base to each channel succession is marked by a clearly defined erosion surface. This surface typically has a relief of decimetre scale or less, more often than not being associated with a lag of clast-supported mud-chip conglomerate (Figs 6 and 7). These mud chips are angular, and commonly of pebble to cobble grade. The conglomerates have an associated matrix of fine-grained quartz wacke. In rare instances, reworked phosphate nodules and vein-quartz pebbles are incorporated within the basal lag,

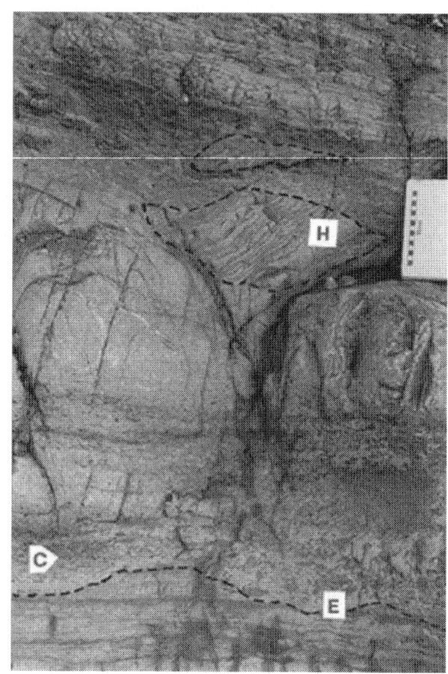

Fig. 7. Tidal channel base (SM 7647 0799). (Note erosive base (**E**), mud-chip conglomerate (**C**) and rotated heterolithic blocks (**H**).) Scale in centimetres.

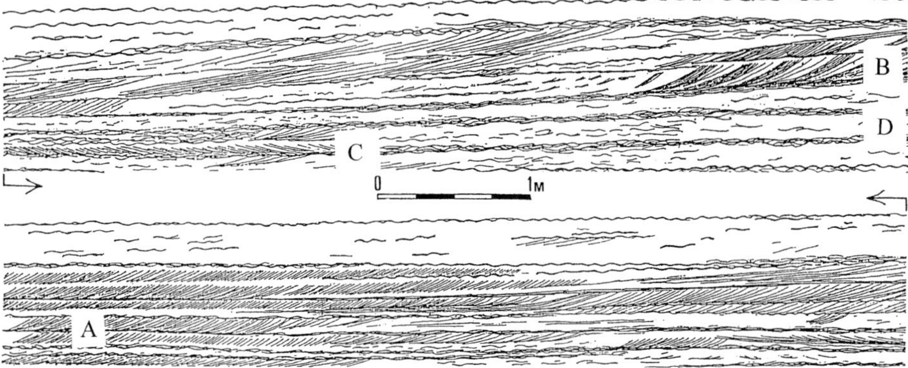

Fig. 8. Structures within estuarine, laterally accreted bedsets (sense of accretion to the left) (SM 7643 0810). Dominant structures seen are megaripple toesets, each defined by thin muddy drapes (**A**). The megaripples possibly exhibit spring–neap cyclicity (**B**), and rare herringbone cross-stratification (**C**). Possible annual couplets of mudstone-rich–sandstone-rich heterolithic units are also present (**D**). Reactivation surfaces are also prevalent.

with the exceptional occurrence of eroded, reworked *lingulid* debris at SM 7703 0777. Clasts occasionally are larger than cobble grade, with instances of rotated heterolithic blocks (Figs 6 and 7), probably recording events related to channel-bank failure through erosive meander migration. The basal lag is typically massive, although subtle clast imbrication is rarely observed. The thickness of the basal lag varies from decimetres to centimetres. Where the lag is thin or absent, the basal erosion surface is defined in two ways. First, the sediments overlying the channel base (medium- to fine-grained quartz wacke), are trough cross-bedded, implying the migration of sinuous-crested, 3D megaripples along the channel floor. These megaripples exhibit evidence of pulsed flow with reactivation surfaces being commonly developed. Unidirectional reactivation (Nio & Yang 1991) is probably associated with spring–neap current variability, whereas bidirectional reactivation (Nio & Yang 1991) is associated with the daily ebbing and flooding of the tide. Laminae are typically draped with thin mudstone flasers (Fig. 8). Set thicknesses vary from centimetre to decimetre scale. Similar bedforms and internal geometries are commonly described from modern intertidal (and subtidal) environments (e.g. Dalrymple *et al.* 1990).

Vertically above the trough cross-stratified sandstones, and in areas where these large-scale bedforms are absent, is found a distinctive suite of heterolithic fine-grained sandstones and mudstones. These heterolithic deposits typically comprise complex associations of flaser- and lenticular-bedded sedimentary rocks with parallel to sub-parallel set boundaries that possess dips of up to 20° (Figs 6, 8 and 9). These 'master-bedding' planes comprise the bulk of the tidal channel association. Bed geometry falls into Allen's (1963) epsilon cross-stratification category, with decreasing bottom and topset dips. Internally, these inclined master bedsets are complex, with a myriad of small-scale tidal-induced structures being present. These include flaser- or linsen-bedded heterolithic deposits, spring–neap bundles, annual(?) cycles, megaripple foresets defined by mudstone drapes and numerous reactivation or erosion surfaces. The master bedding surfaces have a bed-normal separation in the centimetre to decimetre range (Fig. 5). Small, 2D, and more commonly linguoidal current ripples are present on the surfaces of the inclined master beds, with palaeocurrent orientations at high angles to the bed dip direction. It is proposed that the master-bedding surfaces constitute lateral-accretion structures (the inclined heterolithic stratification of Thomas

Fig. 9. Laterally accreted estuarine tidal bedsets (**L**), overlying channel base (**B**) (SM 7698 0775). (Note preservation of tidal flat topsets (**T**).) Scale bar is 1 m long.

Fig. 10. Centimetre–decimetre-bedded muddy heterolithic deposits typical of tidal flat deposits (SM 7646 0801). Coin is 3 cm in diameter.

et al. (1987)), formed by meander migration of tidal point bars, with the inclined bedding surfaces dipping toward the channel thalweg (Reineck 1967; De Mowbray 1983). Interestingly, evidence of bimodal palaeocurrent flow indicators is rare within the observed channel bodies, probably reflecting the fact that, as in many modern estuaries, discrete channels were dominated by either flood- or ebb-dominant tidal currents.

In some instances, topsets of the epsilon, laterally accreted bedsets have been preserved (Fig. 9). These bedsets are typified by centimetre- to decimetre-bedded sandstone–mudstone heterolithic deposits (Fig. 10). These topsets are subparallel with respect to their respective channel bases, reflecting a low-gradient primary depositional surface. The sandstones are typically fine-grained centimetre scale thin–flat lenticles, suggesting starved sand-grade bedload transport. The high mudstone content to these beds indicates the predominance of deposition from suspension of mud–silt grade aggregates. The interbedded nature of the sandstone–mudstone lithologies reflects the changing nature of bottom currents during tidal cycles, together with probable seasonal changes in sediment load and competence. Sedimentary structures within these thinly bedded heterolithic deposits indicate regular subaerial exposure at the time of deposition. Most commonly, this is suggested by the presence of sandstone-filled desiccation cracks on a variety of scales dependent on the mudstone bed thickness (Fig. 11). Current ripples are seen to have flat, planed-off crestal portions (Fig. 12). This is consistent with subaerial exposure soon after deposition, with ripple crests being reworked into a lower flow regime flat-bed by small

Fig. 11. Large sandstone-filled desiccation cracks in tidal flat heterolithic deposits (SM 7703 0777). Hammer is 30 cm long.

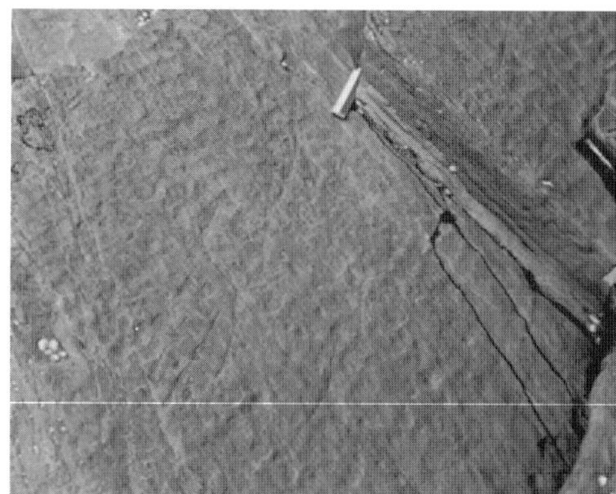

Fig. 12. Flat-topped current ripples indicating emergence of tidal flat at low tide (SM 7702 0777). Scale bar is 25 cm long.

wind-driven currents, producing a characteristic 'washed-out' morphology (De Vries Klein 1977).

In close association with these deposits are small-scale depressions and ridges on the upper surface of sandstone layers (Fig. 13). These structures are similar to the 'wrinkle' marks described from intertidal sandy coasts by Allen (1985*b*). These structures form immediately after emergence of the sandy substrate at low tide, and are believed to represent aseismic soft-sediment loading of small-scale sand pillows into underlying muds. Loading structures are also present

Fig. 13. Wrinkle marks: intertidal, small-scale aseismic soft-sediment deformation structures (SM 7648 0800). Scale bar is 25 cm long.

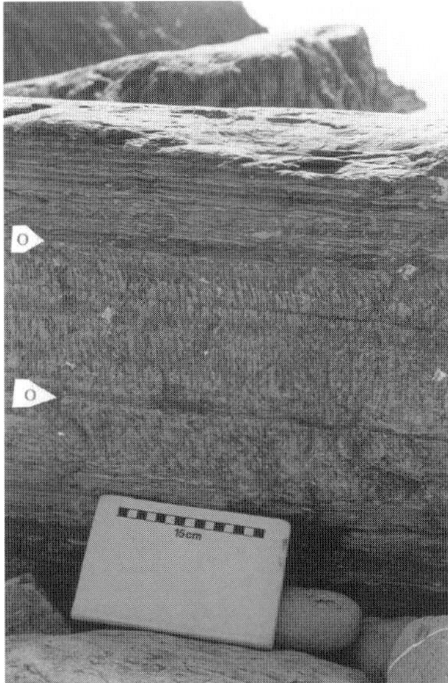

Fig. 14. Dense colonization of intertidal heterolithic deposits by mudstone-lined, sandstone-filled *Skolithos* burrows. (Note separation of discrete colonization events by marked omission surfaces (**O**).) Fallen block, SM 7647 0802.

at a larger scale within these mudstone-rich heterolithic deposits. Such deformation is a response to differential water content and density differences associated with the interbedded heterolithic deposits, caused by rapid emplacement of sands onto water-saturated muds. Such a phenomenon is common from many modern intertidal settings (De Vries Klein 1977). The existence of subaerial exposure from sedimentary rocks lying above laterally accreted tidal channel bedsets is consistent with an interpretation of mixed mudstone–sandstone intertidal flats for these deposits. Similar intertidal mixed-flat deposits have been described in the modern intertidal environments, for example, by Evans (1975), Yeo & Risk (1981) and Frey *et al.* (1989).

The intertidal deposits exhibit a characteristic trace fossil assemblage. Most diagnostic is the presence of mudstone-lined, sandstone-filled vertical tubes attributed to the genus *Skolithos* (Fig. 14). These traces form distinct, dense, monospecific colonies. Each colonization event is characterized by a single size population. Individual burrows are typically 5–10 cm long, and less than 1 cm in diameter. Burrows are never seen to crosscut each other. Specific colonization 'cycles' are bounded by omission surfaces (Fig. 14). Sinuous grazing trails similar to that developed by grazing gastropods on modern tidal flats are also common, being easily recognized on bedding-plane surfaces. More complex, paired, indented trails, though much rarer, are also present. Sandstone-filled horizontal burrows of the genus *Planolites* are also numerous within tidal channel complexes.

Spatial distribution of Gray Sandstone Group incised valleys

The mapped distribution of facies associations and depositional sequences within the Marloes Peninsula is portrayed in Fig. 15, with Fig. 16 showing a schematic west-to-east correlation within the Gray Sandstone Group across the Peninsula. The base of each valley fill is marked by sequence boundaries separating mid- to upper-shoreface deposits of the underlying highstand of rsl, from the overlying estuarine deposits of the lowstand and transgressive systems tract(s). The top to each valley fill is typified by a planar erosion surface with abundant nodular phosphogenesis and open-marine indicators (sedimentary structures and faunas) indicating marine flooding and ravinement. At Pitting Gales Point (SM 7615 0834), the valley fill (VF1, Figs 15 and 16) comprises 5 m of tidal, laterally accreted bedsets, whereas to the southeast of Marloes Sands (SM 7885 0704) a correlative 10.6 m thick heterolithic unit is observed, probably representing deposits

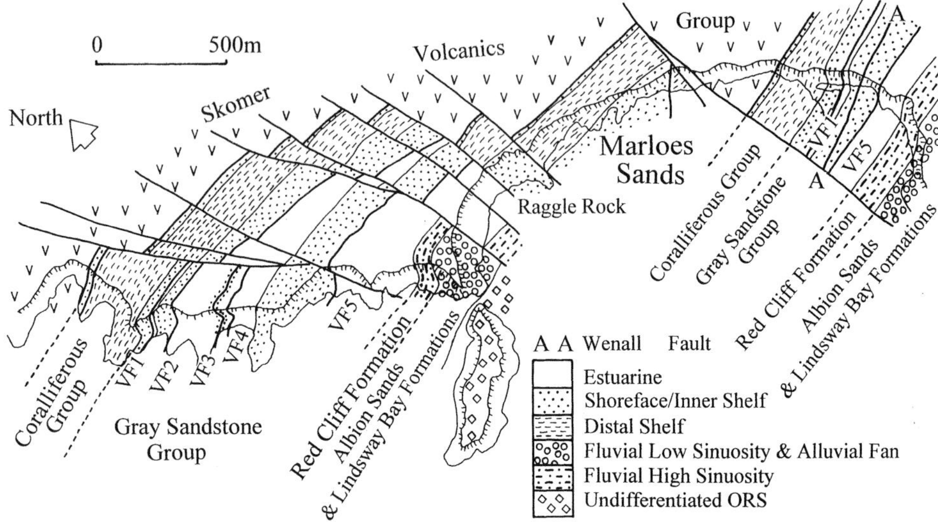

Fig. 15. Simplified depositional sequence–facies association map of the Marloes Peninsula. VF, valley fill.

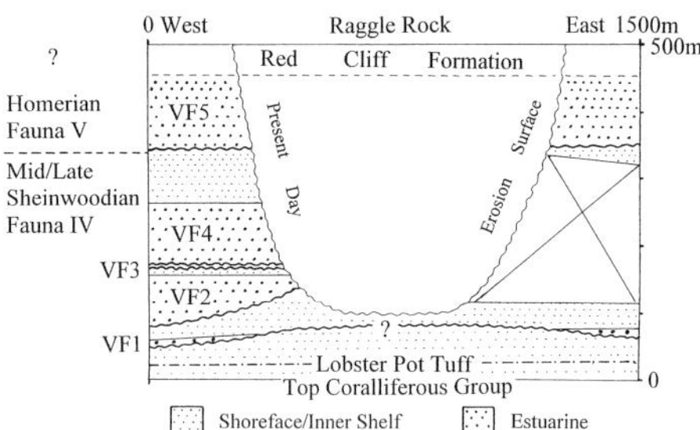

Fig. 16. East–west correlation of Gray Sandstone group facies associations across the Marloes Peninsula. VF, valley fill.

of tidal flat origin. Although rare, palaeocurrent data from current ripples within the valley fill suggests northward (fluvial or ebb-tide dominated?) flow. Correlation between the two sections is aided by the recognition of a distinctive, but thin semi-regional crystal-lithic tuff band found close to the base of the Gray Sandstone Group, herein called the Lobster Pot Tuff. The spatial distribution of valley fill as mapped between Pitting Gales Point and Marloes Sands is intriguing. Although present at both locations, within the area between Raggle Rocks (SM 777 076) and Matthews Slade (SM 784 075), estuarine facies are absent at the correct stratigraphic horizon above the Lobster Pot Tuff (Figs 15 and 16). Instead, this section is characterized by stacked shoreface parasequences. This fact probably reflects primary topographic variations along the extent of sequence boundary at the base of the valley fill of the order of 10 m or so.

The base of valley fill 2 (VF2, Fig. 15) crops out at Three Doors (SM 7622 0827). The valley fill comprises *c.* 90 m of stacked laterally accreted tidal heterolithic deposits. The tidal channel bedsets are typically 2 m thick, with preservation of laterally accreted bottomsets being the norm. Rarely, preserved bedset thicknesses exceed 4 m. Bounding surfaces between stacked channel bedsets are planar features, with little or no depositional topography. The basal bounding surface is often marked by a thin conglomerate lag, or small in-channel megaripples. The internal stratification of the laterally accreted bedsets is complex, although a dominance of flaser- and linsen-bedded current ripples is noted. Sandstone content is greater than 50% of the preserved rock volume. The preserved tidal channels are identical so that no inferences with regard to increasing freshwater or marine influences can be ascertained. This suggests probable vertical aggradation of the facies belt, with sediment supply counterbalancing the effects of accommodation space development. Palaeocurrent analysis (from current ripples) suggests that flow was bimodal, with a component to the northeast and another to the southwest. These observations are consistent with flood–ebb-oriented flow directions. A dominance of one orientation over the other was not noted. A break in section at SM 7638 0812 probably covers the top of VF2, as 10 m to the southeast a 4 m thick sandstone-dominated interval with hummocky cross-stratification and wave ripples attests to open-marine conditions returning to the area during a period of high rsl. Tidal-dominated conditions are quickly re-established within the section, as over the next 10 m, two valley fills crop out (VF3 and VF4, Figs 15 and 16). Valley fill 3 comprises 5.7 m of stacked estuarine laterally accreted bedsets, with palaeocurrent data suggesting flow to the east-southeast. Valley fill 4 crops out 1 m above the top of Valley fill 3, continuing southward to SM 765 079 where the 83 m thick estuarine facies association is terminated by a planar erosion surface. Within VF4, tidal channels and tidal flat deposits each approximate to 50% of the preserved stratigraphy. Palaeocurrent measurements, although sparse in VF4, still indicate flood–ebb tidal flow in a SW–NE direction, respectively.

Valley fills 1–4 fall within the Fauna IV of Walmsley & Bassett (1976), indicating a mid- to late Sheinwoodian age (immediate post-*riccartonensis*). This date allows correlation (in part) with a deep-marine equivalent within the Welsh Basin, namely the Penstrowed Grits Formation (Woodcock *et al.* 1996). It seems likely that sequence boundaries within the Gray Sandstone Group reflect pulses of regional tectonism forcing drops in base level, basinward shift in sedimentary facies belts, and rejuvenating sediment source areas to the south. In this respect, the incised valleys were conduits of sediment transport, supplying the deep-marine sediment sinks to the north. The correlation of VF2–VF4 from the west of the Marloes Peninsula to the equivalent section along the southeast of Marloes Sands is not possible (Fig. 16), as the Wenall Fault has a normal throw of *c.* 200 m, which cuts out this part of the section.

Valley fill 5 and its relationship with Lower Old Red Sandstone in the Marloes area

Valley fill 5 documents the final stage of Gray Sandstone Group development within the Marloes Peninsula. To the southeast of Marloes Sands, VF5 is continuously exposed, constituting a measured stratigraphic thickness of 120 m (Figs 15 and 16). The section comprises interbedded tidal lateral-accretion and tidal flat deposits yielding rare brachiopod debris dated by Walmsley & Bassett (1976) as Homerian in age. The sequence is repeated to the north of Albion Sands, where 71.5 m of section are accessible from SM 7696 077 to SM 7707 0771. Here, interpreted mixed–mud flats comprise nearly 40% of the valley fill, the remainder being composed of tidal lateral-accretion bedsets. The tidal flat association has abundant evidence of emergence, with desiccation cracks, planed-off current ripples and shallow erosive channels interpreted as emergence runoff creeks being commonplace. There is also a high proportion of convolutely laminated fine-grained sandstones associated with this section. Such deformation takes the form of small-scale (centimetre) slumped or slide-like features on inclined laterally accreted bedsets, or larger (metre) scale convolutions with abrupt basal and upper surfaces. The origin(s) of such features is problematic, but include wave-induced liquefaction (Dalrymple 1979), pore-pressure changes brought on by changing levels of tide-water (De Boer 1979), or earthquake-induced deformation associated with passing seismic waves (Barsch-Winkler & Schmoll 1984).

The tidal flat association exhibits colour and textural changes particular to this valley fill. Rarely, the tops of such deposits are seen to change colour from the dark grey so characteristic of the Gray Sandstone Group, to pinks and reds more akin to the overlying Old Red Sandstone. A close association with desiccation cracks (e.g. at SM 7700 0774, 84 m beneath the base of the Lower Old Red Sandstone) hints at prolonged subaerial emergence, with the red coloration probably reflecting associated iron oxidation. Such reddening also occurs within lateral-accretion deposits at the top of the Gray Sandstone Group, magnificently exposed to the north of Red Cliff (SM 7898 0676). Here the basal grey–buff portions of lateral-accretion sets contain rare *Planolites* and mudstone-lined sandstone-filled *Skolithos* burrows. The latter are exclusive to the tidal channel association described earlier, and as such possibly indicate a tidal influence to these channels. A gradual transition upward is seen into pink–red mottled desiccated mudflat deposits (Figs 17 and 18)

Fig. 17. Interbedded grey–buff upward-fining laterally accreted bedsets with red, desiccated mudflat deposits, north of Red Cliff (SM 7897 0676). Channel bases at B1, B2 and B3.

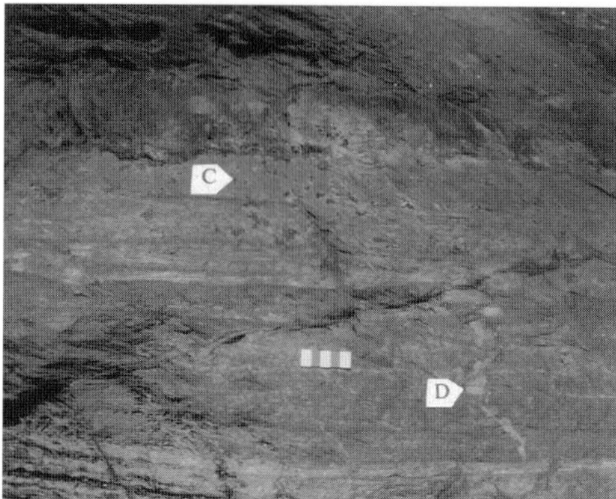

Fig. 18. Desiccated mudflat deposits (**D**), with immature calcrete nodules (**C**) to north of Red Cliff (SM 7897 0676). Scale bar in centimetres.

defining the tops to each channel cycle. The mudflat association also have small (centimetre scale) spheroidal calcite-cemented nodules in place (Fig. 18), which are interpreted as immature calcretes implying prolonged terrestrial exposure. Thus, in conjunction, it seems highly probable that these channel deposits at the top of VF5 represent tidal-influenced fluvial channels (or vice versa).

The recognition of tidal-influenced sedimentation, the close association with emergence indicators, colour changes, and the lack of an observable change in sedimentary facies within the basal beds of the Lower Old Red Sandstone strongly indicate a gradual, conformable transition between the Gray Sandstone Group and the overlying Red Cliff Formation at Albion Sands, to the north of Red Cliff, at Dale Roads (SM 8100 0646) and at Lindsway Bay. The red staining of the topmost Gray Sandstone Group (together with inferred facies changes), led Allen & Williams (1978) to conclude, however, that this transition in the Marloes Block was unconformable, being similar in nature to the clearly unconformable contacts observed to the south of Milford Haven at Freshwater West and East and at the top of the Coralliferous Group within the Winsle Inlier (Allen *et al.* 1976). The concept of a conformable contact within the Gray Sandstone Group of the Marloes Peninsula is not new, being advocated by Cantrill *et al.* (1916), Sanzen-Baker (1972), Walmsley & Bassett (1976) and Hurst *et al.* (1978). Dating of this transitional contact is problematic because of the absence of macrofaunas. Walmsley & Bassett (1976) put the age as late Wenlock or at the earliest, early Ludlow time. This age was agreed upon by Hurst *et al.* (1979). Allen & Williams (1978), however,

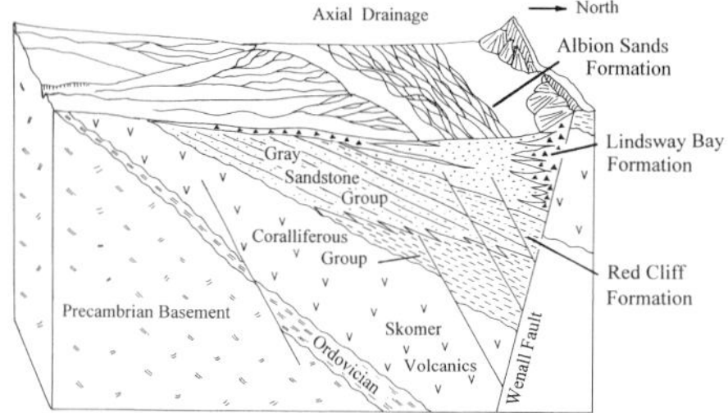

Fig. 19. Schematic diagram of tectonic framework during deposition of the Albion Sands and Lindsway Bay Formations.

for reasons mentioned above, placed the basal beds of the Old Red Sandstone in the Marloes Block at a late Silurian, possibly Downtonian age. Palaeocurrent data from VF5 indicate NE–SW bimodal tidal transportation, with data from the Red Cliff Formation demonstrating fluvial transport from the south (Allen *et al.* 1981; Allen 1985*b*), consistent with a NE–SW trending valley.

By late Ludlow times, there occurred a widespread reduction to the sea that occupied the Welsh Basin (Allen 1985*a*), with shoaling of the basin demonstrated by shallow-marine deposition as sedimentation rates exceeded accommodation space generation (Bassett *et al.* 1992). Soper & Woodcock (1990) inferred that final 'docking' of the Eastern Avalonia microplate with Laurentia occurred at this time, probably resulting in significant strike-slip activity. It seems possible that the renewal in tectonic activity resulted in the observed change of sedimentation in the Old Red Sandstone of the study area: pebbly low-sinuosity alluvium of the Albion Sands Formation (derived from the west and southwest: Allen & Williams 1978; Allen *et al.* 1981; Allen 1985*a*), and alluvial fanglomerates of the Lindsway Bay Formation (derived from the south and southeast: Allen & Williams 1978; Allen *et al.* 1981), being deposited within the accommodation space developed by transtensional reactivation of the basin margin faults (Fig. 19). Large-scale syn-sedimentary extension faults exposed within the Albion Sands Formation at Horse's Neck (SM 777 076) reflect this tectonic activity. Felsic volcanic debris from the Lindsway Bay Formation is similar to that described from the Skomer Volcanic Group (Allen & Williams 1978), this implying that the older basin fill was being recycled, probably along uplifted highs or at the basin margins. This certainly explains the observed contact in the Winsle Inlier, where Lindsway Bay Formation unconformably rests directly atop the Coralliferous Group (Allen *et al.* 1976).

Conclusions

The Gray Sandstone Group within the Marloes Peninsula is seen to be a continuation of sedimentation from the underlying Coralliferous Group. As a whole, this continuum describes a basinward prograding clastic wedge deposited during a phase of thermal subsidence over the Skomer Basin. Fluctuations in rsl produced changes in sedimentation on varying scales. Minor fluctuations (fourth order?), produced periods of marine flooding and subsequent strand-plain progradation, implying a ready supply of clastic material from the south. Progradation of such parasequences and parasequence sets, however, are punctuated by periods of falling rsl on a larger scale (third order?). Here, basinward shifts in depositional environment are interpreted, with valley incision of the shelf resulting from reduced base level. Such incised valleys presumably were highly focused sediment transport pathways to the northern, deeper parts of the Welsh Basin, with active submarine fan development within the lowstand systems tracts. With rising rsl, these valleys constrained the tidal prism to such an effect that tidal-dominated (estuarine) environments prevailed. Terrestrial conditions were established southwards towards the hinterland, with deposition of the embryonic Old Red Sandstone. Five such episodes of incision and valley fill are recognized in the Marloes Peninsula, with younger valley fills displaying an ever-increasing 'proximal' or indeed fluvial signature. The youngest valley fill heralds deposition into the overlying Lower Old Red Sandstone, with a conformable contact being recognized into the Red Cliff Formation. Renewed transtensional tectonic activity within the area resulted in tilt-fault block formation, and axial deposition of low-sinuosity fluvial channels of the Albion Sands Formation. Erosion of uplifted fault blocks led to the deposition of the Lindsway Bay Formation fanglomerates. The transtensional nature of the regional tectonic regime dominated basin development and associated sediment fill of the Lower Old Red Sandstone.

This work was carried out under the tenure of an NERC funded research post at the University of Aberdeen. The project was the brainchild of B. P. J. Williams, whose enthusiasm, guidance and support is wholeheartedly appreciated. The paper benefited from constructive reviews by R. J. Sloan, B. P. J. Williams and N. H. Woodcock, to whom the author is gratefully indebted.

References

ALLEN, J. R. L. 1963. The classification of cross-stratified units, with notes on the origin. *Sedimentology*, **2**, 93–114.

—— 1985*a*. Marine to fresh water: the sedimentology of the interrupted environmental transition (Ludlow–Siegenian) in the Anglo-Welsh region. *Philosophical Transactions of the Royal Society, London, Series B*, **309**, 85–104.

—— 1985*b*. Wrinkle marks: an intertidal sedimentary structure due to aseismic soft-sediment loading. *Sedimentary Geology*, **41**, 75–95.

—— & WILLIAMS, B. P. J. 1978. The sequence of the earlier Lower Old Red Sandstone (Siluro-Devonian), north of Milford Haven, southwest Dyfed (Wales). *Geological Journal*, **13**, 113–136.

——, BASSETT, M. G., HANCOCK, P. L., WALMSLEY, V. G. & WILLIAMS, B. P. J. 1976. Stratigraphy and structure of the Winsle Inlier, southwest Dyfed, Wales. *Proceedings of the Geologists' Association*, **87**, 221–229.

——, ELLIOT, T. & WILLIAMS, B. P. J. 1981. Old Red Sandstone and Carboniferous fluvial sediments in South Wales. *In*: ELLIOT, T. (ed.) *Field Guides to Modern and Ancient Fluvial Systems in Britain and Spain*. Proceedings of the 3rd International Symposium on Fluvial Sedimentology, Keele, 1–39.

BARTSCH-WINKLER, S. & SCHMOLL, H. R. 1984. Bedding types in Holocene tidal channel sequences, Knik Arm, Upper Cook Inlet, Alaska. *Journal of Sedimentary Petrology*, **54**, 1239–1250.

BASSETT, M. G., BLUCK, B. J., CAVE, R., HOLLAND, C. H. & LAWSON, J. D. 1992. Silurian. *In*: COPE, J. C. W., INGHAM, J. K. & RAWSON, P. F. (eds) *Atlas of Palaeogeography and Lithofacies*. Geological Society, London, Memoir, **13**, 37–56.

BLUCK, B. J., GIBBONS, W. & INGHAM, J. K. 1992. Terranes. *In*: COPE, J. W. C., INGHAM, J. K. & RAWSON, P. F. (eds) *Atlas of Palaeogeography and Lithofacies*. Geological Society, London, Memoir, **13**, 1–4.

BUTLER, A. J., WOODCOCK, N. H. & STEWART, D. M. 1997. The Woolhope and Usk basins: Silurian rift basins revealed by subsurface mapping in the southern Welsh Borderland. *Journal of the Geological Society, London*, **154**, 209–223.

CANTRILL, T. C., DIXON, E. E. L., THOMAS, H. H, & JONES, O. T. 1916. *The geology of the South Wales Coalfield. Part XII: The country around Milford.* Memoir of the Geological Survey, UK.

CHANNELL, J. E. T., MCCABE, C. & WOODCOCK, N. H. 1993. Palaeomagnetic study of Llandovery (Lower Silurian) red beds in northwest England. *Geophysical Journal International*, **115**, 1085–1094.

COPE, J. C. W. & BASSETT, M. G. 1987. Sediment sources and Palaeozoic history of the Bristol channel area. *Proceedings of the Geologists' Association*, **98**, 315–330.

DALRYMPLE, R. W. 1979. Wave-induced liquefaction: a modern example from the Bay of Fundy. *Sedimentology*, **26**, 835–844.

——, KNIGHT, R. J., ZAITLIN, B. A. & MIDDLETON, G. V. 1990. Dynamics and facies model of a macrotidal sand-bar complex, Cobequid Bay–Salmon River Estuary (Bay of Fundy). *Sedimentology*, **37**, 577–612.

De Boer, P. L. 1979. Convolute lamination in modern sands of the estuary of the Oosterschelde, the Netherlands, formed by the result of entrapped air. *Sedimentology*, **26**, 283–294.

De Mowbray, T. 1983. The genesis of lateral accretion deposits in recent mudflat channels, Solway Firth, Scotland. *Sedimentology*, **30**, 425–435.

De Vries Klein, G. 1977. *Clastic Tidal Facies*. CEPCO, Illinois.

Dunne, W. M. 1983. Tectonic evolution of SW Wales during the Upper Palaeozoic. *Journal of the Geological Society, London*, **140**, 257–265.

Evans, G. 1975. Intertidal flat deposits of the Wash, western margin of the North Sea. *In*: Ginsburg, R. (ed.) *Tidal Deposits: a Casebook of Recent Examples and Fossil Counterparts*. Springer, Berlin, 13–20.

Frey, R. W., Howard, J. D., Han, S.-J. & Park, B.-K. 1989. Sediments and sedimentary sequences on a modern macrotidal flat, Inchon, Korea. *Journal of Sedimentary Petrology*, **59**, 28–44.

Hancock, P. L., Dunne, W. M. & Tringham, M. E. 1982. Variscan structures in south Pembrokeshire. *In*: Bassett, M. G. (ed.) *Geological Excursions in Dyfed, Southwest Wales*. University of Wales Press, Cardiff, 215–248.

——, Dunne, W. M. & Tringham, M. E. 1983. Variscan deformation in southwest Wales. *In*: Hancock, P. L. (ed.) *The Variscan Foldbelt in the British Isles*. Hilger, Bristol, 47–73.

Hayward, A. B. & Graham, R. H. 1989. Some geometrical characteristics of inversion. *In*: Cooper, M. A. & Williams, G. D. (eds) *Inversion Tectonics*. Geological Society, London, Special Publications, **44**, 17–39.

Hurst, J. M., Hancock, N. J. & McKerrow, W. S. 1978. Wenlock stratigraphy and palaeogeography of Wales and the Welsh Borderland. *Proceedings of the Geologists' Association*, **89**, 197–226.

Nio, S. D. & Yang, C. S. 1991. *Recognition of Tidally-influenced Facies and Environments*. International Geoservices Short Course Notes 1.

Pharaoh, T. C., Merriman, R. J., Evans, J. A., Brewer, T. S., Webb, P. C. & Smith, N. J. P. 1991. Palaeozoic arc-related volcanism in the concealed Caledonides of Southern Britain. *Annales de la Société Géologique de Belgique*, **114**, 63–91.

Powell, C. M. 1987. Inversion tectonics in southwest Dyfed. *Proceedings of the Geologists' Association*, **98**, 193–203.

—— 1989. Structural controls on Palaeozoic basin evolution and inversion in southwest Wales. *Journal of the Geological Society, London*, **146**, 439–446.

Reineck, H. E. 1967. Layered sediments of tidal flats, beaches and shelf bottoms of the North Sea. *In*: Lauff, G. D. (ed.), *Estuaries*. American Association for the Advancement of Science, Washington, DC, 191–206.

—— & Wunderlich, F. 1968. Classification and origin of flaser and lenticular bedding. *Sedimentology*, **11**, 99–104.

Sanzen-Baker, I. 1972. Stratigraphic relationships and sedimentary environments of the Silurian–early Old Red Sandstone of Pembrokshire. *Proceedings of the Geologists' Association*, **83**, 139–164.

Soper, N. J. & Woodcock, N. H. 1990. Silurian collision and sediment dispersal patterns in southern Britain. *Geological Magazine*, **127**, 527–542.

Strahan, A., Cantrill, T. C., Dixon, E. E. L., Thomas, H. H. & Jones, O. T. 1914. *The geology of the South Wales Coalfield. Part XI: The country around Haverfordwest*. Memoir of the Geological Survey, UK.

Thomas, R. G., Smith, D. G., Wood, J. M., Visser, J., Calverey-Range, E. A. & Koster, E. H. 1987. Inclined heterolithic stratification—terminology, interpretation and significance. *Sedimentary Geology*, **53**, 123–179.

Thorpe, R. S., Leat, P. T., Bevins, R. E. & Hughes, D. J. 1989. Late-orogenic alkaline/subalkaline Silurian volcanism of the Skomer Volcanic Group in the Caledonides of south Wales. *Journal of the Geological Society, London*, **146**, 125–132.

Torsvik, T. H., Trench, A., Svensson, I. & Walderhaug, H. J. 1993. Palaeogeographic significance of mid-Silurian palaeomagnetic results from southern Britain—major revision of the apparent polar wandering path for Eastern Avalonia. *Geophysical Journal International*, **113**, 651–668.

Van den Berg, J. H. 1981. Rhythmic seasonal layering in a mesotidal channel fill sequence, Oosterschelde Mouth, the Netherlands. *In*: Nio, S. D., Schuttenhelm, R. T. E. & van Weering, T. C. E. (eds) *Holocene Marine Sedimentation in the North Sea Basin*, International Association of Sedimentologists, Special Publications, **5**, 147–159.

Visser, M. J. 1980. Neap–spring cycles reflected in Holocene subtidal large-scale bedform deposits: a preliminary note. *Geology*, **8**, 543–546.

Walmsley, V. G. & Bassett, M. G. 1976. Biostratigraphy and correlation of the Coralliferous Group and Gray Sandstone Group (Silurian) of Pembrokeshire, Wales. *Proceedings of the Geologists' Association*, **87**, 191–220.

Woodcock, N. H. & Gibbons, W. 1988. Is the Welsh Borderland Fault System a terrane boundary? *Journal of the Geological Society, London*, **145**, 915–923.

——, Butler, A. J., Davies, J. R. & Waters, R. A. 1996. Sequence stratigraphical analysis of late Ordovician and early Silurian depositional systems in the Welsh Basin: a critical assessment. *In*: Hesselbo, S. P. & Parkinson, D. N. (eds) *Sequence Stratigraphy in British Geology*. Geological Society, London, Special Publications, **103**, 197–208.

Yeo, R. K. & Risk, R. J. 1981. The sedimentology, stratigraphy and preservation of intertidal deposits in the Minas Basin system, Bay of Fundy. *Journal of Sedimentary Petrology*, **51**, 245–260.

Ziegler, A. M., McKerrow, W. S., Burne, R. V. & Baker, P. E. 1969. Correlation and environmental setting of the Skomer Volcanic Group, Pembrokeshire. *Proceedings of the Geologists' Association*, **80**, 409–439.

Progress in reconstructing vegetation on the Old Red Sandstone Continent: two *Emphanisporites* producers from the Lochkovian sequence of the Welsh Borderland

DIANNE EDWARDS[1] & JOHN B. RICHARDSON[2]

[1]*Department of Earth Sciences, Cardiff University, PO Box 914, Cardiff CF10 3YE, UK*

[2]*Department of Palaeontology, Natural History Museum, Cromwell Road, London SW7 5BD, UK*

Abstract: Small coalified fossils (mesofossils) have yielded new insights into vegetation of the Old Red Sandstone Continent in early Devonian times. Particularly useful are those containing spores that can be placed in dispersed spore taxa, although patinate and emphanoid spores have not hitherto been found *in situ*. *Emphanisporites* cf. *micrornatus* Richardson & Lister is described in a bifurcating cylindrical sporangium preserved as a cuticular sheath. A terminal dehiscence feature is compared with that in *Horneophyton*. The sporangium is encased in amorphous detritus with some tubular fragments. Similar associations occur on other sporangia, e.g. *Tortilicaulis* and axes at this North Brown Clee Hill locality, and they are interpreted as remains of a microbial or fungal film. Fragmentary cuticles, interpreted as isolated sporangial valves, bear an undescribed species of *Emphanisporites* with fine interdigitating proximal muri and laevigate distal surfaces referred to *Emphanisporites* sp. A Richardson & Lister. Analysis of dispersed spore assemblages from the locality and others in the Welsh Borderland indicate that the two emphanoid taxa were not common components of the spore 'rain'. This evidence, coupled with the dearth of mesofossils of the producers, suggests that the plants grew at the upper reaches of the drainage basin of the river that deposited the sediment, although the paucity of sporangia may also be attributed to their low fossilization potential.

The Lower Old Red Sandstone in southern Britain has provided important evidence in elucidating the early history of vascular plants on land (Edwards 1979). Of particular importance is the more or less continuous sequence of plant assemblages from a relatively localized area (South Wales and the Welsh Borderland). These have been recovered from rocks spanning the transition from marine Upper Silurian into fluvial Lower Devonian rocks (*inter alia* Allen & Tarlo 1963; Allen 1974) and were documented in the classic papers of Lang (1937) and Croft and Lang (1942). Subsequent studies have added new taxa (e.g. *Deheubarthia*, Edwards *et al.* 1989; *Thrinkophyton*, Kenrick & Edwards 1988*b*), revisions of existing taxa (e.g. *Zosterophyllum llanoveranum*, Edwards 1969; *Uskiella spargens*, Shute & Edwards 1989) and yielded new localities (e.g. Hassan 1982; Kenrick 1988).

Although these studies have fleshed out the 'traditional' overview of plant diversification on land (e.g. Edwards & Davies 1990), they also have necessitated radical reappraisal of the nature and affinities of early land vegetation. Examples include the following:

(1) Re-collecting at Lang's 1937 localities has confirmed that the architecture of most Late Silurian–earliest Devonian plants was indeed an axial dichotomously branching system with terminal sporangia, but that there are numerous taxa, with wide variety of morphologies when compared with the cooksonias Lang described (e.g. Fanning *et al.* 1990, 1991, 1992). In the majority of cases, it has been impossible to extract the anatomical information from their coalified axes that would allow assignation to the tracheophytes and so these fossils are informally termed 'rhyniophytoid' (*sensu* Edwards & Edwards 1986) or 'cooksonioid' (Taylor 1982).

(2) The discovery of a new locality, Allt Ddu, in the Brecon Beacons, containing a number of zosterophylls with both smooth and spiny axes has extended the initial major diversification in Laurussia of that group back into the late Gedinnian time (BZ spore zone roughly equivalent to latest Lochkovian time) (Hassan

From: FRIEND, P. F. & WILLIAMS, B. P. J. (eds). *New Perspectives on the Old Red Sandstone*. Geological Society, London, Special Publications, **180**, 355–370. 0305-8719/00/$15.00

1982; Richardson *et al.* 1982; Kenrick 1988; Edwards 1990; Wellman *et al.* 1998*c*).

(3) Application of mineralogical techniques to Pragian pyrite permineralizations has contributed detailed information on the construction of tracheids (e.g. *Gosslingia*, Kenrick & Edwards 1988*a*; *Sennicaulis*, Kenrick *et al.* 1991), with implications both for the delimitation of major lineages of early land plants (Kenrick & Crane 1991; Kenrick *et al.* 1991) and for taphonomic processes involving lignified tissues.

(4) Studies of minute coalified fossils (mesofossils) with preserved 3D anatomy isolated from grey relatively unconsolidated siltstones have demonstrated a further facet of early Devonian vegetation: a short 'turf' of axial plants of diverse affinity (Edwards 1996). The demonstration of tracheids and stomata in *Cooksonia pertoni* Lang so preserved provided unequivocal evidence for its vascular status and homoiohydric life-style (Raven 1984; Edwards *et al.* 1992). Axial plants, with terminal sporangia containing tetrads and dyads similar to those found dispersed in Ordovician and Silurian rocks, provided the first body fossil evidence for these pioneering land plants, albeit from relict populations (Edwards *et al.* 1995, 1999; Wellman *et al.* 1998*b*).

(5) Ultrastructural studies of spores involving transmission electron microscopy (TEM), scanning electron microscopy (SEM) and light microscopy have been used with some effect to demonstrate diversity and relationships between the producers, but less successfully, their general affinities (e.g. Fanning *et al.* 1988; Rogerson *et al.* 1993; Wellman *et al.* 1998*a*). Such *in situ* spore studies (that also encompass presumed tracheophytes) in these coalified mesofossils and compressions from elsewhere provide a link with the coeval dispersed spore assemblages (Edwards & Richardson 1996) that are widely used in correlation (Richardson & Lister 1969; Richardson & McGregor 1986).

(6) More recent developments in dispersed spore studies include the subdivision of the MN spore zone, based on excellent spore preservation in the Welsh Borderland (Richardson *et al.* 1984), and the distinction of the *breconensis–zavallatus* (BZ) spore zone in the uppermost Gedinnian succession of the Brecon Beacons (Richardson *et al.* 1982). These two spore zones are also recorded in the Belgian Ardennes (Streel *et al.* 1987), where spores occur in similar sedimentary associations with vertebrates throughout. The latter allow precise correlation with Rhenish Gedinnian and Siegenian sequences. Lochkovian and Pragian Stages are defined in marine sequences from Bohemia and contain few identifiable land-derived plant microfossils. They thus can only be approximately correlated with the Old Red Sandstone facies (Richardson 1984). Hence we are reluctant to use the standard stage terminology for the Welsh sequences that in particular straddle the Gedinnian–Siegenian boundary, especially as on current evidence this boundary cannot be directly correlated with the Lochkovian–Pragian boundary, although the two are thought to be approximately coeval (Fig. 1).

These dispersed spore studies in the transition between late Gedinnian and early Siegenian times have demonstrated a major change in the composition of spore assemblages in South Wales, presumably reflecting major changes in land vegetation, although increases in size (and possibly composition) may merely reflect local changes in depositional environments (namely, from silts to sands). The increase in diversity in spores cannot solely be attributed to the radiation of the zosterophylls (see above) as the latter, as far as is known, are characterized by laevigate retusoid spores that belong to relatively few taxa. However, the appearance of larger retusoid spores with apiculate exospore may herald the appearance of trimerophytes in local vegetation, first represented in the megafossil record by *Dawsonites* sporangia in the Lower Siegenian (Pragian) deposits of the Brecon Beacons. The majority of dispersed spores still cannot be assigned to megafossil taxa. In this paper we use information on dispersed and *in situ Emphanisporites* to demonstrate how integrated studies between palaeobotanist and palynologist can shed light on understanding the vegetation of the Old Red Continent in early Devonian times.

The genus *Emphanisporites* McGregor 1961

Emphanisporites, first described from the Emsian succession of eastern Canada, is characterized by distinctive interradial proximal sculpture in the form of radiating muri, specific variation being based on the degree of expression and shape of the muri. The distal surface may be laevigate, micrograbulate, conate or spinose, or have annulate thickenings. The genus is typically associated with the Lower Devonian succession, but the earliest records (*E. protophanus* Richardson & Ioannides) occur in the Wenlock deposits (Homerian: *ludensis* Graptolite Zone) of subsurface Libya (Richardson & Ioannides 1973) and in Wenlock–Gorstian units of the boreholes and outcrops of South Wales (Burgess & Richardson 1995). Thus it appears at approximately the same time in Gondwana and Laurussia. Thereafter Ludfordian and Přídolí

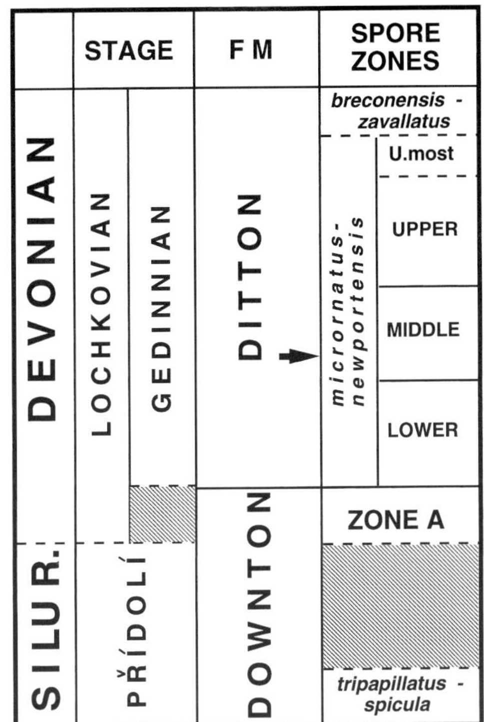

Fig. 1. Generalized stratigraphical succession for North Brown Clee Hill, Shropshire, showing the approximate position of the plant locality (arrow), and correlation with spore zones, subzones, stage and system boundaries. Zones based on Richardson (1974), Richardson *et al.* (1982, 1984, 2000) and Richardson (unpublished data).

records appear to show divergence although the poor representation of the emphanoid features in the Ludfordian type area is probably greatly influenced by facies, because spores are often badly preserved and recovery is poor in both numbers and variety when compared with Libya (e.g. 10 v. 26 taxa: Richardson & Lister 1969; Richardson & Ioannides 1979). Particularly notable is the appearance of *E. rotatus* McGregor in the Ludfordian, and possibly upper Gorstian deposits of North Africa, a species first recorded in the Lochkovian sequences of Laurussia. In addition, *Emphanisporites* with a distal annulus (*E. splendens* Richardson & Ioannides) is found in Přídolí rocks of Gondwana (Richardson & Ioannides 1973) but this feature appears in the Emsian succession of Laurussia (e.g. *E. annulatus*, McGregor 1961). Species with weak proximal muri are common in Lochkovian deposits, but are replaced by more robust ribbed forms in the overlying Pragian–Siegenian and Emsian sequences of Europe, North America (Canada and USA), North Africa and South America (see references given by Richardson & McGregor (1986)). The distal sculpture also shows a distinctive trend through time. The earliest forms are laevigate and then become successively micrograngulate, conate to spinose. This trend parallels the azonate retusoid trend in sculpture (Richardson 1996*a*; Richardson & Burgess 2000), but does not show the verrucate–murornate stage seen in equatorially crassitate and distally patinate structural forms.

The highly distinctive nature of the proximal radial muri raises the possibility that all the species derived from the same clade and, using the number of species as a measure of the diversity of the producers, its maximum diversification occurred in Emsian times (Richardson & McGregor 1986). Post-Emsian records extending into the Lower Carboniferous succession are rarer and must be scrutinized carefully, because many of these spores closely resemble Lower Devonian species, and occur in offlapping sedimentary sequences that potentially represent reworking events. The nature of the spores makes them both easily recognizable and relatively resistant to decay, especially in the Siegenian–Emsian forms with coarse ornament. Thus, more than any other spore taxa, they are likely to survive and be identified. Studies in the Famennian units of New York State have revealed spore assemblages with a large amount of reworked material (Richardson & Ahmed 1988). Basal Famennian strata contain well-known Middle Devonian miospore species, and progressively higher in the Famennian sequence occur annulate *Emphanisporites* typical of Emsian strata and then robust-ribbed forms lacking an annulus, which are possibly derived from Siegenian deposits. This succession of *Emphanisporites* taxa mirrors the erosive events resulting from progressive degradation of older Devonian rocks in the Appalachians.

The rapid successive morphological changes in species of *Emphanisporites* and their wide geographical spread both in Laurussia and Gondwana have led to their use as zonal indices in the Lower Devonian succession, namely, *E. micrornatus* (MN zone: early (but not earliest)–early late Gedinnian time), *E. zavallatus* (BZ zone: late Gedinnian time) and *E. annulatus* AS zone: early and early late Emsian time (Richardson & McGregor 1986; Fig. 1).

Preparation procedures

The fossils were recovered from Lochkovian rocks (Fig. 1) exposed in the banks of a stream on the north side of Brown Clee Hill, Shropshire (Edwards 1996). Grey siltstone with flecks of coaly material were disaggregated by soaking and gently shaking in water. The residues were filtered through 250 μm mesh, and individual coalified mesofossils picked out using a brush or needle. Selected specimens were treated with 40% HF to remove adhering quartz grains, and washed thoroughly in water. Some were oxidized in concentrated nitric acid until translucent and washed. After air drying, fertile axes were mounted on carbon films on stubs for initial screening using a Cambridge SEM360. Following photography and archiving (Wellman *et al.* 1996), promising specimens were broken open and re-examined.

Descriptions of mesofossils and spores

cf. Horneophyton *sp. (Figs 2, 3a and 5a–d and g)*

This specimen was picked out during a project to investigate conducting cells because it superficially appeared to be a bifurcating naked axis (Fig. 2a and b). On mounting on the stub, it fell apart revealing spores and hence its true status. The entire specimen is thought to be spore bearing, but the proximal end is irregularly fractured. Its length is *c.* 1.5 mm. There is no pronounced widening before the bifurcation that occurs *c.* 0.33 mm below the tips. The more intact daughter segment is 0.24 mm wide. The most obvious features are the internal reticulum of raised ridges, the very disorganized, irregular superficial covering and the distal smooth rims of the daughter segments. Although it seems likely that there was just one central cavity proximal to the superficial bifurcation, the curvatures of the edges of the fragments after fracture do not preclude a longitudinal partitioning into two (Fig. 2b).

The limiting layer is interpreted as a cuticle, with the ridges marking the positions of the junctions of adjacent cells in the sporangial wall. The latter are longitudinally elongate with overlapping ends but show no regular pattern. The ridges are rounded (Fig. 2c); the inner surface is smooth, except at very high magnification, when it shows wrinkling and bears very small granules. In very restricted areas, the remains of the internal periclinal wall of the superficial cells are preserved and emphasize the 'shallow' nature of this layer (Fig. 2c and d). The inner wall is slightly thinner than the outer. The cells may be filled with pyrite (Fig. 2c).

In the vicinity of the apex, the cells in the sporangial wall became less elongate and under the recurved 'collar' less shallow. Figure 2e shows these distal cells in section, revealing their thick outer periclinal and anticlinal walls, and greater height. Inner periclinal walls are incomplete, suggesting that it was their collapse that caused the inward curvature of the sporangial tip. The cuticle is completely exposed in this region and apart from a few small 'craters' lacks the microbial film (see Fig. 5a, below). In addition to shallow depressions, the more complete tip shows two elliptical depressed areas with central similarly shaped area enclosing disorganized coalified material (Fig. 2f). These superficially resemble stomata, but there are no indications of two guard cells. The outer surface of the bifurcating structure is almost completely covered by a disorganized layer of varying appearance and thickness (Figs 2a, b and d, 5a and g). Where absent, particularly at the tips, the surface below is more or less smooth (Fig. 5a). The chaotic debris may be in the form of irregular sheets with adhering irregular masses of strands, or an odd plate with marginal extensions. Also present are 'craters' (possibly broken eruptions of cuticle) of varying size (Fig. 2d), or well-defined lengths of tubes, either smooth (Fig. 5c) or with internal annular thickenings (Fig. 5a, b and d) of varying thickness. Other associations of smaller tubes may be more fragmented. Such features are interpreted as possible fungal components of a microbial film (see p. 365).

Emphanisporites *cf.* micrornatus Richardson & Lister 1969

The following descriptions are based solely on SEM studies. The spores occur in small clusters, occasionally with intact tetrads, adhering to the inside of the cuticle (Fig. 2c and d). They show a circular to subcircular amb and were originally subspherical with flattened proximal pole, but are frequently preserved in oblique compression with invaginated proximal face (Fig. 2h and i). Curvaturae perfectae are distinct and slightly proximal from the equator (Figs 2h and 3a); the contact areas are slightly less than the spore radius, with 7–9 radial muri in each interradial area. The muri are ± straight, nearly reaching the spore apex, and they are evenly tapered, being equatorially 4–5 μm wide, and 1 to <1 μm near the apex. Distal sculpture is of microconi and microbaculae with elements 1–2 μm apart and *c.* 0.5 μm wide and 0.5–1 μm high (Fig. 2g). The

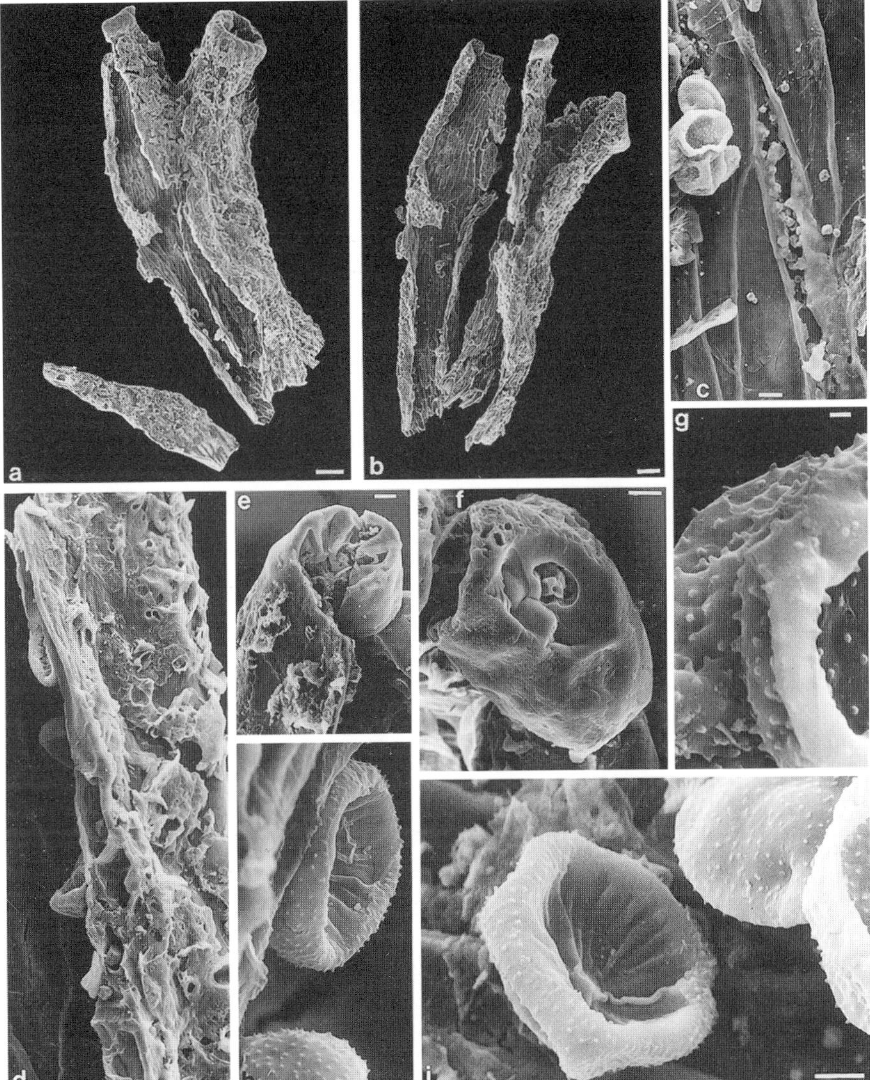

Fig. 2. Scanning electron micrographs of cf. *Horneophyton* sp. North Brown Clee Hill (NBCH), Shropshire, England. Lochkovian. NMW 99.8G.1. (Scale bars represent 10 µm, unless otherwise noted.) (**a**, **b**) Two views of entire specimen. Scale bar represents 100 µm. (**c**) Internal surface of part of sporangia) cuticle showing one cell with internal periclinal wall preserved and adhering cluster of spores. (**d**) Longitudinally fractured wall with single layer of shallow cells and encrusting film. (**e**) Recurved cuticle around pore. (**f**) Superficial feature on rim. (**g**) Ornament on distal surface of spore. Scale bar represents 1 µm. (**h**, **i**) *In situ* spores assigned to *Emphanisporites* cf. *micrornatus*. Scale bars represent 5 µm.

trilete mark is in the form of an apertural ridge (*sensu* Rogerson *et al.* 1993) and accompanied by folds nearly equal to the spore radius. As measured from scanning electron micrographs, the spores are 24–27 µm in diameter.

Remarks and comparisons. Diagnostic features for comparison are the straight to slightly sinuous, proximal muri that taper polewards from a narrow, 1 µm curvatural ridge, and always extend to the curvatural ridge, and spaced distal sculpture of micrograna and microconi. They are thus closely similar to *Emphanisporites micrornatus* Richardson & Lister 1969 in sculpture and size, but the zonal index fossil has spatulate muri that do not reach the curvatural

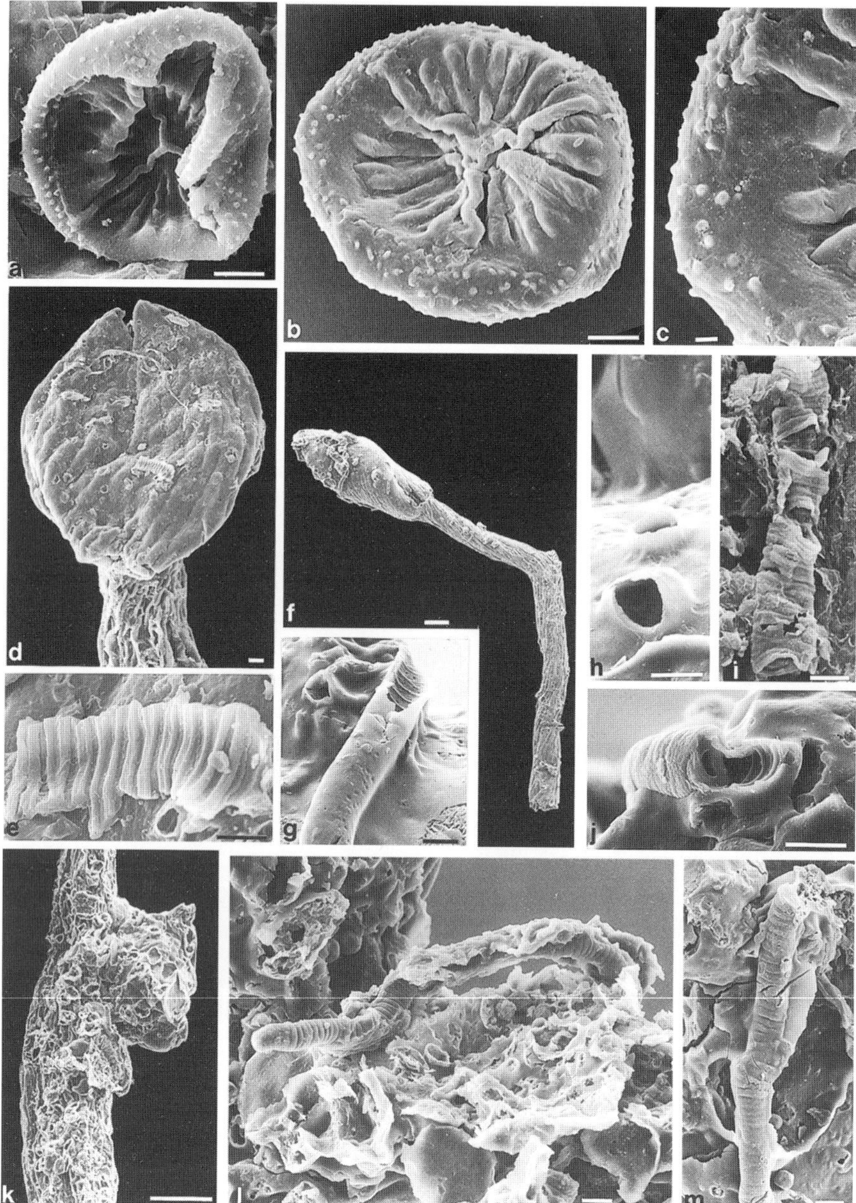

Fig. 3. Scanning electron micrographs of Lochkovian specimens from the Welsh Borderland. (Scale bars represent 10 μm unless otherwise noted.) (**a**) *In situ E.* cf. *micrornatus* with well-defined apertural fold. From specimen illustrated in Fig. 2a–i. Scale bar represents 5 μm. (**b, c**) *Emphanisporites micrornatus* in dispersed spore assemblage from same locality. BM113415. (**b**) Scale bar represents 5 μm. (**c**) Scale bar represents 1 μm. (**d**) Unidentified sporangium with adhering banded tube and 'cratered' cuticle. NBCH. NMW 99.8G.2. (**e**) Banded tube in (**d**) magnified. (**f–h, m**) *Tortilicaulis offaeus*. NBCH. NMW 96.5G.9. (**f**) Entire specimen. Scale bar represents 100 μm. (**g**) Tube with internal thickening coated with amorphous film on surface of sporangium. (Note 'crazing' above scale bar.) (**h**) 'Craters' in sporangial cuticle. (**i**) Poorly preserved banded tube and further debris on a sterile naked axis. NBCH. NMW 99.8G.3. (**j, k**) Encrusted axis with lateral outgrowth. (Note extensive cratering of cuticle.) NMW 99.8G.4. (**j**) Adhering tube with internal ridging enclosing a further tube of uniform wall thickness in TS. NMW 99.8G.4. (**k**) Scale bar represents 100 μm. (**l**) Part of (**k**) enlarged showing cast of tube with internal ridging and remains of encrusted tube wall. (**m**) Tube with external surface exposed adhering to specimen in (**f**).

ridge (Fig. 3b and c). Comparisons are hampered because the dispersed taxon was described optically from strewmounts on slides, whereas the very small numbers of spores in the sporangia have limited our present study to SEM observations and do not allow assessment of intrasporangial variation. A scanning electron micrograph of dispersed *E. micrornatus*, from the same locality, is thus included for direct comparison (Fig. 3b and c). It is not impossible that the zonal index spore derives from the same taxon, but requires more material both dispersed and *in situ* to substantiate this possibility.

Discussion. This spore-containing coalified mesofossil, identified as cf. *Horneophyton* sp., although fragmentary, does not appear to be represented in the compression fossil record, where it might well be identified as a minute *Hostinella*, a form genus for naked axes isotomously branching. As such it would probably have been overlooked in a study (as in Targrove locality; Fanning *et al.* 1992) where any potential sporangia were film-pulled for spores. Truncated apices characterize the sporangia of the Upper Silurian–Lower Devonian *Steganotheca*, where there is sometimes a lenticular terminal thickening of coalified material (Edwards 1970; Edwards & Rogerson 1979), but branching has not been recorded in the presumed terminal sporangium (spores have not been demonstrated). Truncated bifurcating structures from the Upper Silurian sequence of Bolivia are slightly swollen and thus distinct from the axial shape recorded here (Edwards *et al.* 1999). It is possible that the truncated tips here are a post-dehiscence maturation feature, and that mature, but intact, sporangia had differently shaped apices, perhaps similar to those in *Salopella* (Fanning *et al.* 1992). However, gross sporangial shape in the latter taxon is different: the sporangia do not have extensive parallel sides, split longitudinally into two valves and, in the Lower Devonian sequence, contain spores of *Aneurospora* type (Edwards *et al.* 1994). Comparison of conventional compression fossils lacking cellular detail with silicified permineralizations from the Rhynie Cherts always presents identification problems, resulting in doubts that the Scottish Siegenian–Pragian plants do occur elsewhere in the fossil record (see, e.g. Ishchenko (1975) and comments by Powell *et al.* (2000)). In this instance, the Welsh Borderland specimen resembles the cylindrical, lobed sporangia sometimes described as 'with broad flat tips', of *Horneophyton lignieri* (Kidston & Lang 1920). Particularly relevant are Bhutta's findings (Bhutta 1973) that although the majority of *in situ* spores may be assigned to *Apiculiretusispora*, smaller examples (44–56 μm) were related to *Emphanisporites decoratus* Allen, a species with distal spines rather than the micrograna on the spores described here. Recognition of dispersed spora taxa in the silicified sporangia is particularly difficult, with spores showing varying degrees of preservation and hence appearance even within the same sporangium (see, e.g. Powell *et al.* 2000). Eggert (1974) illustrated scanning electron micrographs of isolated spores that he compared with *Retusotriletes* or *Apiculiretusispora* in their possession of curvaturae perfectae, although the roughened surfaces may well be an artefact associated with silicification. The permineralized sporangia have walls comprising parenchymatous tissue, often showing poor cellular preservation, limited by a conspicious cuticle with inwardly directed flanges. Further decay of such wall tissues would produce the cuticular skeleton seen in the coalified specimen. The small areas of cuticle exposed in the lateral walls show no evidence of the distinctive stomatal complexes that characterize the Rhynie taxon, although cell outlines are similar (Hass 1991). The columellae of *Horneophyton* are also absent, but would not be expected to survive taphonomic processes. Although the Rhynie Chert sporangia have been extensively studied (Kidston & Lang 1920; Bhutta 1972, 1973; Eggert 1974; El-Saadawy & Lacey 1979) there is limited consensus on their apical features and dehiscence mechanisms. The original researchers thought they were indehiscent; recent workers have mentioned distal slits or pores, with the most elaborate mechanism involving a central dome shaped area (Bhutta 1972). A number of figured specimens show increased thickening of outer periclinal and radial walls in the vicinity of the tips and, where dehiscence has occurred, a recurving of the cuticle as noted in the specimen here. This led to Eggert's suggestion of a small central stomium (Eggert 1974). However, whereas in some of the silicified material there is a central depressed area of displaced (apparently isolated) tissue that might represent the remains of the central dome (see Bhutta 1972, 1973; El-Saadawy & Lacey 1979), this feature might escape preservation in the compression fossil. Such differences, and lack of unequivocal identity of the spores of the Rhynie plant make it impossible to conclude that the new specimen belongs to *Horneophyton*, but influenced by gross morphology (including branching) and distal features it seems not unlikely.

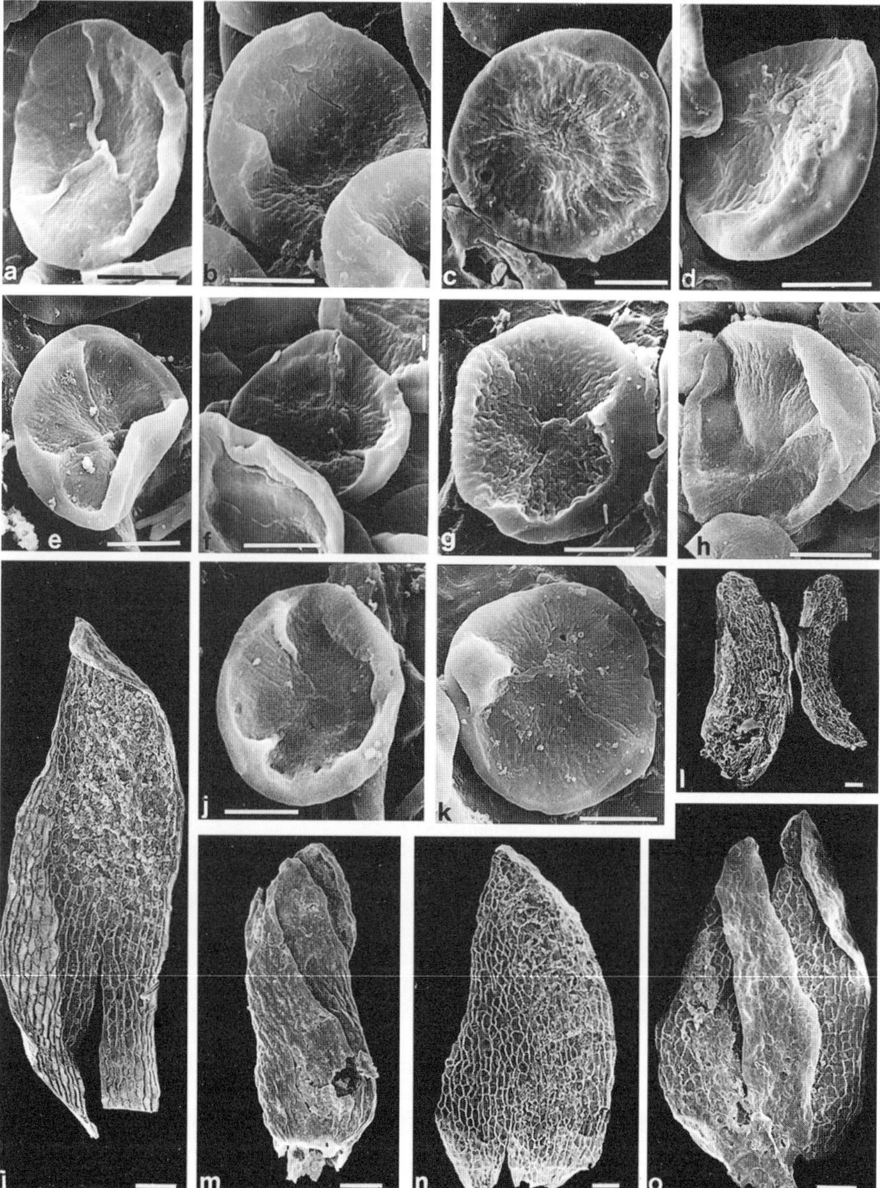

Fig. 4. Scanning electron micrographs of spores and sporangial cuticles. North Brown Clee Hill, Shropshire, England. Lochkovian. (**a–h**, **j**, **k**) *Emphanisporites* sp. A Richardson & Lister 1969 adhering to internal surfaces of sporangial cuticles and showing variations in proximal features. (For fuller explanation see text.) Scale bars represent 5 μm. (**a**) NMW 99.8G.5. (**b**) NMW 99.8G.6. (**c**, **d**) NMW 99.8G.7. (**e**, **j**, **k**) NMW 99.8G.8. (**f**) NMW 99.8G.9. (**g**) NMW99.8G.10. (**h**) NMW99.8G.11. (**i**) ± intact valve of sporangium (adhering spores figured in (**e**), (**j**) and (**k**)). Scale bar represents 110 μm. (**l**) Cuticle with bluntly rounded apex, with part of second valve to right. NMW 99.8G.12. Scale bar represents 100 μm. (**m**) External view of almost intact sporangial cuticles, showing distal split into two valves and partial twisting (adhering spore figured in (**h**)). Scale bar represents 100 μm. (**n**) Internal surface of sporangial cuticle incomplete on left-hand margin. NMW 99.9G.13. Scale bar represents 100 μm. (**o**) Dehisced bivalved sporangium represented by two partially overlapping cuticles. NMW99.8G.14. Scale bar represents 100 μm.

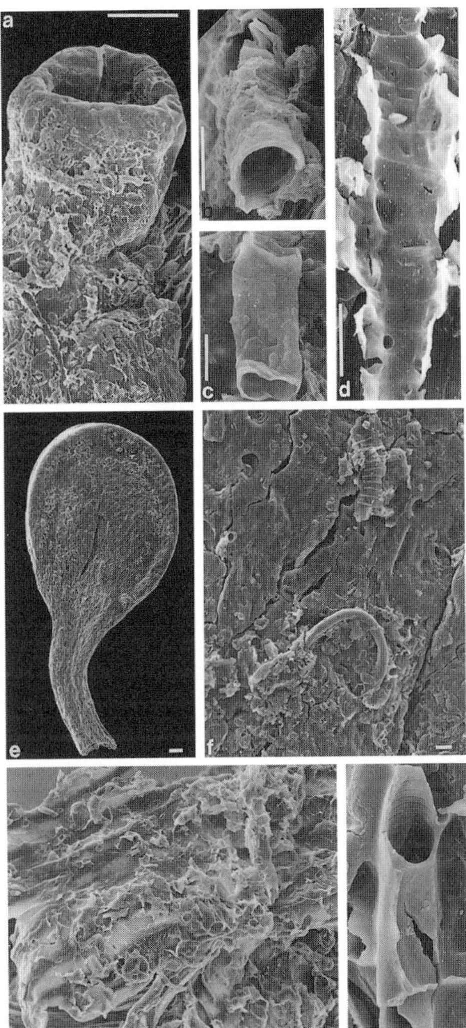

Fig. 5. Scanning electron micrographs of Lochkovian specimens from North Brown Clee Hill, Shropshire, England. (Scale bars represent 10 μm, unless otherwise noted.) (**a**–**d**, **g**) Aspects of the encrusting material on cf. *Horneophyton* sp. NMW 99.8G.1 (see also Figs 2a–i and 3a). (**a**) General view below sporangium apex. Arrowed tube magnified in (**d**). (**b**) Fragment of tube with internal thickening. (**c**) Fragment of smooth tube. (**d**) Fractured tube with low, widely spaced, narrow ridges, (**e**) Unidentified sporangium with traces of encrusting material. NMW 96.11G.5. Scale bar represents 100 μm. (**f**) Part of surface of (**e**) with fragments of tubes. (**g**) General view of external surface of sporangial cuticle of cf. *Horneophyton* sp. (**h**) Part of a banded tube that appears to penetrate tissues of axis. NMW 99.8G.15.

Sporangial cuticles (Figs 4 and 5)

The commonest fossils associated with *Emphanisporites* are pieces of cuticle of various shapes and sizes with isolated spores or clusters of spores attached to the surface with ridges. Rare examples appear to be remains of complete collapsed sporangia that have disintegrated during preparation (Fig. 4l, m and o). Smooth, sometimes recurved margins characterize those with well-defined outlines and completely preserved peripheral 'cells' (Fig. 4i). In contrast, irregularly shaped fragments tend to have minutely ragged edges. Figure 4i shows a possibly complete example that is truncated at one end and $\pm$ parallel sided before tapering into an obtusely rounded tip. That the 'cells' appear complete at the margins suggests they bordered a predisposed longitudinal split. The cuticle is thus interpreted as the limiting layer of the valve of a $\pm$ fusiform, bivalved sporangium. The limiting wall layer, as is reflected by the ridges, is longitudinally elongate, but never more than twice as long as width and in $\pm$ longitudinal files with truncated and not extensively overlapping ends. There is no regular patterning; cells become shorter and of more variable shape distally. Some examples show a cleft at the proximal end suggestive of flattening of an originally curved (?spoon-shaped) structure (Fig. 4i and n). Smaller possibly intact specimens have rounded ends. Examples of variation in dimensions and shape are given in Table 1. Most examples suggest that the sporangium comprised two valves. In one there is evidence of twisting.

Emphanisporites *sp. A Richardson & Lister 1969*

Spores have a circular to subcircular amb and were generally subspherical with a flattened proximal pole (Fig. 4a–h, j and k). They are frequently preserved in oblique compression with slightly invaginated proximal face. Curvaturae perfectae are distinct and slightly proximal from the equator; contact areas are thus slightly less than the spore radius. Numerous radial muri occur in each interradial area. Muri are $\pm$ straight and evenly tapered, anostomosing or interdigitating and reach the labra at the spore apex. They are equatorially <0.5 μm wide. The trilete mark, represented by a very narrow apertural ridge, is accompanied by folds that extend nearly 4/5 spore radius, and the labra usually taper towards the spore apex (Fig. 4e and j). The distal hemisphere is laevigate. Equatorial diameter varies between specimens (Table 1) but is in the

Table 1. *Distribution of sporangial cuticles and adhering spores assigned to* Emphanisporites *sp. A Richardson & Lister 1969*

Number (NMW 99)	Valve Length (*L*, mm)	Valve Width (*W*, mm)	Spore (mean, μm)	Range (μm)	*n*	Comments on specimens
8G.7	1.1	0.5	16	14–17	16	Well-defined muri (Fig. 4c and d)
8G.13	1.24	0.63	13	12–14	7	Same as 8G.8 (Fig. 4n)
8G.19	0.96	0.6	14	12–15	6	Rugulate ornament, but rare cfd with above Trilete rarely developed
8G.14	–	–	16	14–17	5	Two valves (Fig. 4o)
8G.10	–	–	–	–	–	Two valves (Fig. 4g)
8G.8	1.26	0.42	–	–	–	(Fig. 4e and i–k)
8G.5	1.45	0.35	15.5	14–17	3	Good trilete (Fig. 4a)
8G.6	1.63	0.58	–	15–17	–	(Fig. 4b)
8G.18	1.54	0.4	11.5	10.5–13	9	Twisting in sporangium
8G.11	0.59	0.24	13	11–15	4	Note labra and muri in spores (Fig. 4h and m)
8G.22	–	–	17.5	16–19	5	Sporangium poorly preserved
8G.20	–	–	–	15.5, 11	2	Rounded tip to sporangium
8G.23	–	–	16	15–17	3	Fragment
8G.21	1.2	0.1/0.28	14.5	13–16	15	Two long fragments Very faint muri
8G.17	1.0	0.42	16	14–18	9	Two valves Well-defined apertural fold Very faint muri
8G.9	0.63	0.3	13.5	13–15	8	Rounded tip Very delicate apertural fold Regular muri (Fig. 4f)
8G.16	1.33	0.49	–	13	3	Rounded tip to sporangium
8G.12	1.2	0.43	17	16–18	3	Faint muri, good folds (Fig. 4l)

range 11–18 μm. There is also variation in the degree of expression of the proximal ornament (inter- and intra-sporangia) and the nature of the trilete (apertural fold). Some specimens show much better resolution of the interdigitating muri than others (compare Fig. 4a, c and k), and there may also be differences in their length (compare Fig. 4c, d and h). On one cuticular specimen only (Fig. 4g) the spores show sculpture more reminiscent of the irregular radial ornament of *Scylaspora* Burgess & Richardson. Occasional spores in a sporangium lack a trilete mark (Fig. 4b). In one cuticular specimen, the lips of the attached spores become wider or are parallel sided towards the spore apex (Fig. 4a).

Comparison and remarks. The diagnostic features of the spores are the proximal, radial, very narrow anastomising–interdigitating muri and smooth distal surfaces. Although the features of the proximal surface may appear to be the result of compression, the intricate pattern of disjointed and anastomosing muri in our view precludes this possibility. The spores are almost identical with *Emphanisporites* sp. A in Richardson & Lister 1969 except that the latter (40–50 μm) are much larger. Such a discrepancy in size between spores is much greater than for taxa we have previously studied. The consistency of *in situ* spore size between specimens convinces us that this is not an artefact, but we have no explanation for the differences. One spore (Fig. 4g) has coarser sculpture and more irregular muri to rugulae. This type of sculpture is found in the genus *Scylaspora*, where the proximal surface is ornamented with 'irregularly oriented muri, rugulae and verrucae' (Wellman 1999). That our specimens show indications of radial alignment is perhaps indicative of some intergrading between the two taxa. Indeed, *Scylaspora* is common in the upper Homerian succession, where a species of *Emphanisporites* and *Artemapyra* (a hilate cryptospore) with short proximal equatorial radial muri also occur. Whether or not there are close relationships between these dispersed taxa may well be resolved by TEM studies.

Discussion. The more complete examples suggest that the cuticles are part of the valves of a flattened ± ovoid to fusiform or naviculate sporangium. In compression they would thus be similar to certain species of *Salopella* (e.g. *S. marcencis*, Fanning *et al.* 1992), but the latter can be distinguished because they contain either retusoid or crassitate spores of *Aneurospora* type (Edwards & Richardson 1996). Reniform sporangium cuticles assigned to *Resilitheca* contain *Retusotriletes* sp. (Edwards *et al.* 1995). We thus conclude that the *Emphanisporites*-containing forms should be placed in a new taxon to emphasize the distinctive nature of the spores and the construction of the sporangial wall, such that it is preserved only as its cuticle. Both the ornament of the spores and sporangial dehiscence mechanism are different from the single specimen named here as cf. *Horneophyton* sp., although the preservation as cuticles is similar. This and the absence of evidence indicating similarities in the spore ultrastructure make us reluctant to assign the specimens to the genus *Horneophyton*. Furthermore, our uncertainties in establishing gross sporangial morphology preclude giving these sporangial cuticles a name. This is an example of where, given sufficient material, ultrastructural studies using TEM would be particularly useful.

Isolated cuticles with similarly shaped cell outlines have been recorded from the lower Downtonian sequence of Gloucestershire (Gorsley Common: Burgess 1987) but lack details of spores. The earliest examples from the Upper Wenlock Cae Castell Formation (Rumney Borehole, South Wales) that are coeval with the earliest *Cooksonia* fossils in Ireland (Edwards *et al.* 1983) are more fragmentary with less regular cell patterning (Burgess 1987; Edwards & Wellman 1996).

Sporangial and axial encrustations (Figs 2a and d, 4d–m and 5)

In searching for critical evidence for the vascular nature of Silurian and early Devonian fossils there is a tendency to select the most pristine specimens for investigation. Those with highly irregular surfaces are discarded as apparently disintegrating, but here we show that such surface irregularities may reflect the presence of a microbial or fungal coating. It is unfortunately impossible to distinguish organisms or well-organized tissues in the covering. The disorganized coalified residues may represent amorphous material, e.g. mucilage or remains of aggregations of filaments or hyphae. Larger tubes, some with internal thickenings, are the only discrete entities, but they too are incomplete. Exotic spores sometimes adhere to the surfaces (Fig. 5g). However, such features together with the crater-like structures that may represent remnants of eruptions on the surface of the cuticle are consistently found on other fossils at this locality. Typical examples are described here.

Edwards *et al.* (1996) illustrated a sporangium belonging to *Tortilicaulis offaeus* (Fanning *et al.* 1992) on which a small number of well-preserved tubes with internal thickenings were partially embedded in an amorphous layer, resembling a consolidated slime, that sometimes has a slightly crazed appearance in patches (Fig. 2f–h and m). Similar associations are present on the surface of the subtending axes (Fig. 3m). Comparable associations of hyphae, amorphous material and modified, crazed cuticle are seen today during the initial stages of colonization by germinating spores of certain fungi, e.g. *Colletrichum* and *Uromyces*, raising the possibility that the fossil forms represent parasites or saprotrophs.

An unidentified sporangium (Fig. 5e), shows small patches of disorganized adhering tissue again with banded tubes (Fig. 5f), whereas in others (e.g. Fig. 3d and e) only fragments of tube and eruptions of the cuticle are seen. Indeed, in the past we assumed that isolated occurrences of tubes as seen in the latter were the results of fortuitous superpositioning. Such examples are not confined to sporangia, and appearances of axes may be even more variable. Figure 3i shows a small portion of a vertically aligned tube on the surface of a heavily degraded unidentifiable axis. This appears merely adpressed to the axis (hence fortuitous), but in other examples they may be partially buried, possibly indicating invasion of the underlying tissue (Fig. 5h). The cuticle may be radically disrupted as in an axial specimen with a lateral swelling (Fig. 3k and l) and again is associated with tubes (Fig. 3j). One specimen is preserved unusually as an internal cast, with a possible tip, which has a coating of indistinguishable material possibly representing the remains of the ensheathing smaller tubes (Fig. 3l). The lateral swelling has not yielded any anatomical information and may be an area distended by pyrite, whereas the rest of the axis has shrivelled.

Discussion. The consistent association of disorganized 'flaky' material with occasional banded tubes or smooth larger tubes intimately associated with plant cuticles showing crater-like structures, abrasion or corrosion, and areas of amorphous material suggests that the plant surface was colonized by an organism, colony or integrated association of organisms capable of interaction with the cuticle. Both biological and trophic relation of this encrustation remain highly conjectural. It is possible that it derived no nutritional benefit from the plant tissues, but used them as an inert substrate, although the cuticular reaction and occasional superficial penetration by tubes are more suggestive of parasitism or saprotrophism. In relation to the first possibility, we compared the unevenly thickened tubes with the sheaths of cyanobacteria, although their size is closer to the cells–filaments of a eukaryotic alga, and they are less frequent than those that usually occur in such microbial mats. The similarity of such tubes to those associated with complexes of small tubes in *Nematothallus*, and to a lesser extent, *Nematosketum*, is striking (Burgess & Edwards 1991). The example in Fig. 31 shows adhering remnants of possible small tubes reminiscent of similarly preserved isolated fragments of the latter. The affinities of the nematophytes remain obscure. In the original description, Lang (1937) considered them neither algal nor higher plant and created a new higher taxon, the Nematophytales. More recently, *Prototaxites*, also placed in the nematophytes, has been assigned to the fungi (Hueber 1996). Our findings appear to support such a relationship inasmuch as the complex encrusts plant tissues and appears to interact with them, although in their dimensions and architecture the thickened tubes have no extant counterparts.

General discussion on palaeoecology and taphonomy

The mesofossils associated with *Emphanisporites* provide some evidence to explain why some spore producers are under-represented in the meso- or megafossil record. In previous discussions we have postulated that: (1) the plants lived outside the catchment area of the rivers that delivered the sediments; (2) their sporangia were of low fossilization potential; (3) the plants occupied restricted ecological niches and thus were rare in the local vegetation.

Attempts to distinguish between these, or combinations of these possibilities, may be assisted by quantitative evaluation of the dispersed spore assemblage at the same horizons and in neighbouring areas (Tables 2–4). Thus in relation to points (1)–(3) above: (1) spores may be represented in the dispersed spore rain, but would be swamped by that from local plants; (2) spores would be present in quantities dependent on local numbers of individuals and spore-producing capacity; (3) there would be limited or even sporadic to no representation of spores from assemblages in local geographical areas.

Low fossilization potential of intact sporangia holds here where both types are represented by cuticles, softer tissues having been destroyed. Further, in the case of the *Emphanisporites* sp. A sporangia, dehiscence into two valves at maturity has destroyed the integrity of the sporangium. The lack of any evidence for a subtending axis suggests that its tissues (and particularly a stereome) were not endowed with any recalcitrant polymers that enhanced fossilization. Such cuticles are not rare, so it is possible that the parent plants grew in local vegetation but were not particularly common, although an individual sporangium might have contained a large number of small spores (see, e.g. Fig. 4i). Surprisingly this species of *Emphanisporites* sp. A is not recorded at this locality, although it is present (but not recorded in every spore count of 250) at the youngest horizon in the same local sequence. Likewise, it is not recorded in strata from the MN zone in Herefordshire and the Ammons Hill section. Thus we tentatively conclude that the plants grew palaeogeographically further to the north of the Brown Clee Hill area, i.e. closer to the source areas of the rivers.

The single specimen of cf. *Horneophyton* sp. probably survived taphonomic processes because it was encrusted by some kind of microbial or fungal growth. As mentioned above, its gross similarity to a *Hostinella* might have resulted in its not being recognized as a sporangium in our screening procedures, and had it fragmented, we probably would not have paid attention to very small irregular pieces of cuticle. With these provisos, it remains possible that the plant was either very rare or grew some distance from the depositional environment. In the dispersed taxa, *E.* cf. *micrornatus* occurs once in a count of 250 spores from the mesofossil-bearing horizon, whereas *E. micrornatus* itself is the most abundant representative of the genus both at 14 m below the plant horizon and within it (Table 2). Combining the two taxa, they comprise 5.6% of the total spores at the plant-bearing horizon. In more southern parts of the Anglo-Welsh Basin (Herefordshire: Table 3) where the MN zone has been divided into three parts (see Fig. 1), in the lower, *E.* cf. *micrornatus* occurs before *E. micrornatus*, and occurs with it in the middle part, the latter occurrence being at approximately the same stratigraphic level as the mesofossil locality, but neither are found in the more coastal (i.e. occasionally marine influenced) lowermost MN zone rocks at Ammons Hill (Table 4). Such analyses indicate that the producers (possibly a complex of closely related species or a single taxon showing infraspecific variation) were widespread but not very common constituents of

Table 2. *Distribution of* Emphanisporites *spp. in the North Clee Hills sequence*

	lower MN		mid-MN		
	PLR	+35 m	+46 m	+60 m	+*c*. 200 m
E. epicautus	–	–	1	1	
E. cf. *neglectus* complex	–	3	1	4	
E. cf. *micrornatus* A	–	–	–	1	
E. micrornatus	–	–	9	11	P
E. sp. A	–	–	–	–	
E. sp. B	–	–	–	P	
E. sp. A *sensu* R&L 1969	–	–	–	–	P
Emphanoid cryptospores *Cymbohilates variabilis* var. *variabilis*	–	–	1	P	
C. variabilis var. A	–	–	–	P	

Mesofossils occur at the + 60 m horizon, i.e. 60 m above the *Psammosteus* Limestone Horison (PLR). The + 200 m locality is in a stream nearby. Counts based on 250 spores. P indicates presence in occasional strewmounts but not consistently when 250 spores counted (Richardson, unpublished data).

Table 3. *Distribution of the genus* Emphanisporites *in the southern part of the Anglo-Welsh Basin in Herefordshire*

			lower MN		mid-MN		
	−39 m	−33 m	+2 m	+41 m	+97 m	+109 m	+168 m
E. neglectus 1	X	X	X	X	–	X	X
E. neglectus 2	X	X	X	X	X	X	X
E. neglectus 4	X	–	–	–	–	–	–
E. neglectus 5	X	–	–	–	–	–	–
E. sp. 2	X	X	X	–	X	X	X
E. sp. A	–	–	X	X	X	X	–
E. cf. *epicautus*	–	–	X	X	X	X	–
cf. *E. micrornatus*	–	–	X	–	X	X	–
E. sp. B	–	–	–	X	–	–	–
E. cf. *micrornatus*	–	–	X	X	X	–	–
E. rotatus?	–	–	–	X	X	–	–
E. micrornatus	–	–	–	–	X	X	X

Horizons samples above (+m) and below (−m) the *Psammosteus* Limestone Horizon (Richardson, unpublished data).

Table 4. *Distribution of* Emphanisporites *spp. in the Ammons Hill section, St Maughan's Formation*

	lower MN					mid-MN
	25239–40	25242	25245–7	25248–9	25256	25252
E. neglectus 4	X	–	–	–	X	–
E. neglectus 5	X	–	–	–	–	–
E. cf. *epicautus*	X	–	–	–	–	–
E. neglectus 2	–	–	X	–	X	–
E. epicautus	–	–	X	–	–	–
E. neglectus 1	–	–	–	–	X	–
E. sp. C	–	–	–	–	X	–
E. sp. B	–	–	–	–	–	X

The *Psammosteus* Limestone is faulted out in this section but all of the samples in the table are from the lowermost subzone of the *micrornatus–newportensis* Spore Assemblage Zone. Numbers referred to BGS samples: 25242 is a horizon with evidence of marine influence.

proximal flood-plain vegetation in the present Clee Hills and Herefordshire areas, possibly occupying more specialized ecological niches some distance away from the riverside habitats. Our uncertainty as to its precise relationship with the Rhynie Chert taxon *Horneophyton lignieri* (Kidston & Lang) Darrah makes any conclusions relating to taphonomic or ecophysiological issues of the Rhynie Chert plants premature (see Powell *et al.* 2000). However, should further evidence from the spores of the Rhynie plants indicate that *Horneophyton* is indeed the first Rhynie taxon to be recorded with confidence in clastic facies, then a preservational explanation for the lack of Rhynie taxa elsewhere in the Lower Devonian sequence gains support, although tolerance of water-stressed environments cannot be discounted!

This paper provides results of a limited integrated investigation of *in situ* and dispersed spores in the Lochkovian succession. Preliminary general conclusions indicate that no one spore taxon dominates spore assemblages, but the commonest (namely, *Streelispora–Aneurospora* complex) at this Brown Clee Hill locality derive from *Cooksonia pertoni* and *Salopella* spp. (Edwards 1996). The former is the commonest mesofossil at this locality. Also abundant is the genus *Ambitisporites*, whose source is more controversial, although it is possible that some examples derive from *Tortilicaulis*, another common mesofossil at the locality. This plant possesses spores with a microgranulate ornament, which may have sloughed off after dispersal or may be too small to be observed by light microscopy (Edwards *et al.* 1995). Although *Ambitisporites* (Hoffmeister) has been recorded in Silurian cooksonias, Lochkovian examples contain either *Aneurospora* or *Streelispora newportensis* (Chaloner & Streel) Richardson & Lister. There remain a number of mesofossils at the locality from which spores have not yet been recorded (e.g. plate III in fig. 2 of Edwards (1996)) and some major spore taxa have still not been found *in situ*. Important here are trilete distally patinate, proximally laevigate and thin-walled spores, and also emphanoid hilate cryptospores. The latter, thought to derive from dyads, are particularly abundant and diverse in dispersed spore assemblages from the Clee Hills (Richardson 1996*b*).

We thank L. Axe and C. Wellman for assistance in the preparation of the material, and D.E. is grateful to the NERC for continued support (Grants GR9/02903 and GR9/1441).

References

ALLEN, J. R. L. 1974. Sedimentology of the Old Red Sandstone (Silurian–Devonian) in the Clee Hills Area, Shropshire, England. *Sedimentary Geology*, **12**, 73–167.

—— & TARLO, L. B. 1963. The Downtonian and Dittonian facies of the Welsh Borderland. *Geological Magazine*, **100**, 129–155.

BHUTTA, A. A. 1972. Observations on the sporangia of *Horneophyton lignieri* (Kidston and Lang) Barghoorn and Darrah 1938. *Pakistan Journal of Botany*, **4**, 27–34.

—— 1973. On the spores (including germinating spores) of *Horneophyton* (*Hornea*) *lignieri* (Kidston and Lang) Barghorn and Darrah (1938). *Pakistan Journal of Botany*, **5**, 45–55.

BURGESS, N. D. 1987. *Micro- and megafossils of land plants from the Silurian and Lower Devonian of the Anglo-Welsh Basin*. PhD thesis, University of Wales, Cardiff

—— & EDWARDS, D. 1991. Classification of uppermost Ordovician to Lower Devonian tubular and filamentous macerals from the Anglo-Welsh Basin. *Botanical Journal of the Linnean Society*, **106**, 41–66.

—— & RICHARDSON, J. B. 1995. Late Wenlock to early Prídolí cryptospores and miospores from south and southwest Wales, Great Britain. *Palaeontographica, B*, **236**, 1–44.

CROFT, W. N. & LANG, W. H. 1942. The Lower Devonian flora of the Senni Beds of Monmouthshire and Breconshire. *Philosophical Transactions of the Royal Society of London, Series B*, **231**, 131–163.

EDWARDS, D. 1969. Further observations on *Zosterophyllum llanoveranum* from the Lower Devonian of South Wales. *American Journal of Botany*, **56**, 201–210.

—— 1970. Fertile Rhyniophytina from the Lower Devonian of Britain. *Palaeontology*, **13**, 451–461.

—— 1979. A late Silurian flora from the Lower Old Red Sandstone of south-west Dyfed. *Palaeontology*, **22**, 23–52.

—— 1990. Constraints on Silurian and Early Devonian phytogeographic analysis based on megafossils. *In*: MCKERROW, W. S. & SCOTESE, C. R. (eds) *Palaeozoic Palaeogeography and Biogeography*. Geological Society, London, Memoir, **12**, 233–242.

—— 1996. New insights into early land ecosystems: a glimpse of a Lilliputian world. *Review of Palaeobotany and Palynology*, **90**, 159–174.

—— & DAVIES, M. 1990. Interpretations of early land plant radiations: 'facile adaptationist guesswork' or reasoned speculation? *In*: TAYLOR, P. D. & LARWOOD, G. P. (eds) *Major Evolutionary Radiations*. Systematics Association Special Volume, **42**, 351–376.

—— & EDWARDS, D. S. 1986. A reconsideration of the Rhyniophytina, Banks. *In*: SPICER, R. A. & THOMAS, B. A. (eds) *Systematic and Taxonomic Approaches in Palaeobotany*. Systematics Association Special Volume, **31**, 199–220.

—— & RICHARDSON, J. B. 1996. Review of *in situ* spores in early land plants. *In*: JANSONIUS, J. & MCGREGOR, D. C. (eds) *Palynology: Principles and Applications, Volume 1, Principles*. American Association of Stratigraphic Palynologists Foundation. Publishers Press, Salt Lake City, UT, 391–407.

—— & ROGERSON, E. C. W. 1979. New records of fertile Rhyniophytina from the late Silurian of Wales. *Geological Magazine*, **116**, 93–98.

—— & WELLMAN, C. H. 1996. Older plant macerals (excluding spores). *In*: JANSONIUS, J. & MCGREGOR, D. C. (eds) *Palynology: Principles and Applications, Volume 1, Principles*. American Association of Stratigraphic Palynologists Foundation. Publishers Press, Salt Lake City, UT, 383–387.

——, ABBOTT, G. D. & RAVEN, J. A. 1996. Cuticles of early land plants: a palaeoecophysiological evaluation. *In*: KERSTEINS, G. (ed.) *Plant Cuticles—an Integrated Functional Approach*. Bios, Oxford, 1–31.

——, DAVIES, K. L. & AXE, L. 1992. A vascular conducting strand in the early land plant *Cooksonia*. *Nature*, **357**, 683–685.

——, DUCKETT, J. G. & RICHARDSON, J. B. 1995. Hepatic characters in the earliest land plants. *Nature*, **374**, 635–636.

——, FANNING, U., DAVIES, K. L., AXE, L. & RICHARDSON, J. B. 1995. Exceptional preservation in Lower Devonian coalified fossils from the Welsh Borderland: a new genus based on reniform sporangia lacking thickened borders. *Botanical Journal of the Linnean Society*, **117**, 233–254.

——, —— & RICHARDSON, J. B. 1994. Lower Devonian coalified sporangia from Shropshire: *Salopella* Edwards & Richardson and *Tortilicaulis* Edwards. *Botanical Journal of the Linnean Society*, **116**, 89–110.

——, FEEHAN, J. & SMITH, D. G. 1983. A late Wenlock flora from Co. Tipperary, Ireland. *Botanical Journal of the Linnean Society*, **86**, 19–36.

——, KENRICK, P. & CARLUCCIO, L. M. 1989. A reconsideration of cf. *Psilophyton princeps* (Croft and Lang, 1942), a zosterophyll widespread in the Lower Old Red Sandstone of South Wales. *Botanical Journal of the Linnean Society*, **100**, 293–318.

——, WELLMAN, C. H. & AXE, L. 1999. Tetrads in sporangia and spore masses from the Upper Silurian and Lower Devonian of the Welsh Borderland. *Botanical Journal of the Linnean Society*, **130**, 111–155.

EGGERT, D. A. 1974. The sporangium of *Horneophyton lignieri* (Rhyniophytina). *American Journal of Botany*, **61**, 405–413.

EL-SAADAWY, W. EL-S. & LACEY, W. S. 1979. The sporangia of *Horneophyton lignieri* (Kidston and Lang) Barghoorn and Darrah. *Review of Palaeobotany and Palynology*, **28**, 137–144.

FANNING, U., EDWARDS, D. & RICHARDSON, J. B. 1990. Further evidence for diversity in late Silurian land vegetation. *Journal of the Geological Society, London*, **147**, 725–728.

——, —— & —— 1991. A new rhyniophytoid from the late Silurian of the Welsh Borderland. *Neues Jahrbuch für Geologie und Paläontologie, Abhandlungen*, **183**, 37–47.

——, —— & —— 1992. A diverse assemblage of early land plants from the Lower Devonian of the Welsh Borderland. *Botanical Journal of the Linnean Society*, **109**, 161–188.

——, RICHARDSON, J. B. & EDWARDS, D. 1988. Cryptic evolution in an early land plant. *Evolutionary Trends in Plants*, **2**, 13–24.

HASS, H. 1991. Die Epidermis von *Horneophyton lignieri* (Kidston & Lang) Barghorn & Darrah. *Neues Jahrbuch für Geologie und Paläontologie, Abhandlungen*, **183**, 61–85.

HASSAN, A. M. 1982. *Palynology, stratigraphy and provenance of the Lower Old Red Sandstone of the Brecon Beacons (Powys) and the Black Mountains (Gwent and Powys) South Wales*. PhD thesis, King's College, University of London.

HUEBER, F. M. 1996. *A Solution to the Enigma of* Prototaxites. Palaeontological Society Special Publication, **8**. Sixth North American Paleontological Convention Abstracts.

ISHCHENKO, T. A. 1975. *The Late Silurian Flora of Podolia*. Institute of Geological Science, Academy of Science of the Ukrainian SSR, Kiev, 1–80.

KENRICK, P. 1988. *Studies on Lower Devonian plants from South Wales*. PhD thesis, University of Wales, Cardiff.

—— & CRANE, P. R. 1991. Water-conducting cells in early land plants: implications for the early evolution of tracheophytes. *Botanical Gazette*, **152**, 335–356.

—— & EDWARDS, D. 1988*a*. The anatomy of Lower Devonian *Gosslingia breconensis* Heard based on pyritized axes, with some comments on the permineralization process. *Botanical Journal of the Linnean Society*, **97**, 95–123.

—— & —— 1988*b*. A new zosterophyll from a recently discovered exposure of the Lower Devonian Senni Beds in Dyfed, Wales. *Botanical Journal of the Linnean Society*, **98**, 97–115.

——, —— & DALES, R. C. 1991. Novel ultrastructure in water-conducting cells of the Lower Devonian plant *Sennicaulis hippocrepiformis*. *Palaeontology*, **34**, 751–766.

——, REMY, W. & CRANE, P. R. 1991. The structure of the water-conducting cells in the enigmatic early land plants *Stockmansella langii* Fairon-Demaret, *Huvenia kleui* Hass et Remy and *Sciadophyton* sp. Remy *et al.* 1980. *Argumenta Palaeobotanica*, **8**, 179–191.

KIDSTON, R. & LANG, W. H. 1920. Ibid. Part II. Additional notes on *Rhynia gwynne-vaughani*, Kidston and Lang; with descriptions of *Rhynia major*, n.sp. and *Hornea lignieri*, n.g., n.sp. *Transactions of the Royal Society of Edinburgh*, **52**, 603–627.

LANG, W. H. 1937. On the plant-remains from the Downtonian of England and Wales. *Philosophical*

Transactions of the Royal Society of London, Series B, **227**, 245–291.

McGregor, D. C. 1961. Spores with proximal radial pattern from the Devonian of Canada. *Geological Survey of Canada, Bulletin*, **76**, 1–11.

Powell, C., Trewin, N. & Edwards, D. 2000. Palaeoecology and plant succession in a borehole through the Rhynie Cherts, Lower Old Red Sandstone, Scotland. *This volume.*

Raven, J. A. 1984. Physiological correlates of the morphology of early vascular plants. *Botanical Journal of the Linnean Society*, **88**, 105–126.

Richardson, J. B. 1974. The stratigraphic utilization of some Silurian and Devonian miospore species in the Northern Hemisphere: an attempt at a synthesis. *International Symposium on Belgian Micropalaeontological Limits*, **9**, 1–13.

—— 1984. Mid Palaeozoic palynology, facies and correlation. *Proceedings of the 27th International Geological Congress, Moscow*, **1** *(Stratigraphy)*. VNU Science Press, Utrecht, 341–365.

—— 1996*a*. Lower and Middle Palaeozoic records of terrestrial palynomorphs. *In*: Jansonius, J. & McGregor, D. C. (eds) *Palynology: Principles and Applications, Volume 2, Principles.* American Association of Stratigraphic Palynologists Foundation. Publishers Press, Salt Lake City, UT, 555–574.

—— 1996*b*. Taxonomy and classification of some new Early Devonian cryptospores from England. *Special Papers in Palaeontology*, **55**, 7–40.

—— & Ahmed, S. 1988. Miospores, zonation and correlation of Upper Devonian sequences from western New York State and Pennsylvania. *In*: McMillan, N. J., Embry, A. F. & Glass, D. J. (eds), *Devonian of the World. Proceedings of the Second International Symposium on the Devonian System. Calgary, Canada. 1988, Volume III.* Canadian Society of Petroleum Geologists, Calgary, AL, 541–558.

—— & Burgess, N. D. 2000. Sporomorph evolution in the Anglo-Welsh Basin: tempo and parallelism. *In*: Kurmann, M. H. & Hemsley, A. R. (eds) *The Evolution of Plant Architecture.* Royal Botanic Gardens, Kew.

—— & Ioannides, N. 1973. Silurian palynomorphs from the Tanezzuff and Acacus Formations, Tripolitania, North Africa. *Micropalaeontology*, **19**, 257–307.

—— & Lister, T. R. 1969. Upper Silurian and Lower Devonian spore assemblages from the Welsh Borderland and South Wales. *Palaeontology*, **12**, 201–252.

—— & McGregor, D. C. 1986. Silurian and Devonian spore zones of the Old Red Sandstone Continent and adjacent regions. *Geological Survey of Canada, Bulletin*, **364**, 1–79.

——, Ford, J. H. & Parker, F. 1984. Miospores, correlation and age of some Scottish Lower Old Red Sandstone sediments from the Strathmore region (Fife and Angus). *Journal of Micropalaeontology*, **3**, 109–124.

——, Rodriquez, R. M. & Sutherland, S. J. E. 2000. Palynology and recognition of the Silurian/Devonian boundary in some British terrestrial sediments by correlation with Cantabrian and other European sequences—a progress report. *Courier Forschungsinstitut Senckenberg*, **220**, 1–7.

——, Streel, M., Hassan, A. & Steemans, Ph. 1982. A new spore assemblage to correlate between the Breconian (British Isles) and the Gedinnian (Belgium). *Annales de la Société Geologique de Belgique*, **105**, 135–143.

Rogerson, E. C. W., Edwards, D., Davies, K. L. & Richardson, J. B. 1993. Identification of *in situ* spores in a Silurian *Cooksonia* from the Welsh Borderland. *In*: Collinson, M. E. & Scott, A. C. (eds) *Studies in Palaeobotany and Palynology in honour of Professor W. G. Chaloner, F.R.S.* Special Papers in Palaeontology, **49**, 17–30.

Shute, C. H. & Edwards, D. 1989. A new rhyniopsid with novel sporangium organization from the Lower Devonian of South Wales. *Botanical Journal of the Linnean Society*, **100**, 111–137.

Streel, M., Higgs, K., Loboziak, S., Riegel, W. & Steemans, P. 1987. Spore stratigraphy and correlation with faunas and ferras in the type marine Devonian of the Ardenne–Rhenish regions. *Review of Palaeobotany and Palynology*, **50**, 211–229.

Taylor, T. N. 1982. The origin of land plants: a paleobotanical perspective. *Taxon*, **31**, 155–177.

Wellman, C. H. 1999. Sporangia containing *Scylaspora* from the Lower Devonian of the Welsh Borderland. *Palaeontology*, **42**, 67–81.

——, Edwards, D. & Axe, L. 1996. Curation of exceptionally preserved early land plant fossils: problems and solutions. *Curator*, **39**, 208–216.

——, —— & —— 1998*a*. Permanent dyads in sporangia and spore masses from the Lower Devonian of the Welsh Borderland. *Botanical Journal of the Linnean Society*, **127**, 117–147.

——, —— & —— 1998*b*. Ultrastructure of laevigate hilate cryptospores in sporangia and spore masses from the Upper Silurian and Lower Devonian of the Welsh Borderland. *Philosophical Transactions of the Royal Society, London*, **353**, B, 1983–2004.

——, Thomas, R. G., Edwards, D. & Kenrick, P. 1998*c*. The Cosheston Group (Lower Old Red Sandstone) in southwest Wales: age, correlation and palaeobotanical significance. *Geological Magazine*, **135**, 397–412.

Sedimentology, cyclicity and floodplain architecture in the Lower Old Red Sandstone of SW Wales

SUSAN E. LOVE[1,2] & BRIAN P. J. WILLIAMS[1]

[1]*Department of Geology and Petroleum Geology, University of Aberdeen, Meston Building, Aberdeen AB24 3UE, UK (e-mail: b.williams@abdn.ac.uk)*

[2]*Present address: ExxonMobil International Ltd, Mobil Court, 3, Clements Inn, London WC2A 2EB, UK (e-mail: sue_love@email.mobil.com)*

Abstract: The high-quality, laterally continuous coastal exposures of the Moor Cliffs Formation have allowed a highly detailed 2D reconstruction of the floodplain sediments and their contained pedogenic horizons to be made. The thick siltstone packages were actively deposited as finely laminated and rippled sheets, or as intraformational clasts forming larger bedforms. It is proposed that the unusual sediment geometries preserved are intimately related to the timing of land plant colonization of the Old Red Sandstone continent. Channels were extremely broad with low relief and flow over interfluvial areas was common. Evidence for ephemerality and regular desiccation is also closely related to the lack of rooted vegetation and not to palaeoclimate, which is postulated to be warm and seasonally wet. The low net sandstone (<10%) fluvial sediments are the product of deposition by shallow, high width to depth fluvial 'channels' flanked by broad, low-relief silty plains on which Vertisols formed. The reconstructions of this fluvial system reveal distinctive and systematic vertical and lateral variations in floodplain architecture and palaeosol development. Pedogenic maturity consistently increases with distance both vertically and laterally from channels. The vertical patterns of palaeosol development and maturity suggest that autocyclic processes of aggradation and avulsion predominated.

This paper presents the detailed observations made on a 2.5 km by 70 m coastal exposure of Old Red Sandstone (ORS) in the South Pembrokeshire area of the Anglo-Welsh Basin (Fig. 1). The sedimentary system is low net-to-gross and the sandstones are separated by thick siltstone packages containing palaeosols. The detailed description and correlation of these palaeosols has revealed consistent and cyclic variations both laterally and vertically in the development of the soils. The strong vertical cyclicity has implications for the dominant sedimentary processes in existence at the time of deposition. The incorporation of the pedogenic styles and other features in the overbank sediments has led to a reinterpretation of the climate prevalent at the time.

Regional sedimentology

Introduction and past research

The first detailed description of the ORS succession in Dyfed was made by Dixon (1921). Detailed sedimentological study of the fluvial sequences, stratigraphy and palaeogeography of the Lower ORS successions followed much later (e.g. Allen 1963, 1964, 1974*a*, *b*; Allen & Williams 1978, 1979, 1981*b*, 1982; Williams *et al.* 1982). The calcretes and associated pseudo-anticlinal fracture systems were described and interpreted by Allen (1973, 1974*c*, 1986), who recognized that these features form palaeosol components and that their widespread occurrence in the varied Lower ORS fluvial systems has implications for depositional environment, climate and fluvial architecture.

Although there is a large volume of published research on the Lower ORS in general, the coastal outcrops of the Moor Cliffs Formation between Freshwater East and Old Castle Head have been the subject of more limited study (Dixon 1921, 1933; King 1933; Leach 1933; Edwards 1979; Williams *et al.* 1982; Love 1993; Love *et al.* 1993; Marriott & Wright 1993).

Old Red Sandstone stratigraphy and sedimentology

The sediments comprising the Lower ORS lack well-preserved flora and fauna. Consequently, the stratigraphy and position of the

From: FRIEND, P. F. & WILLIAMS, B. P. J. (eds). *New Perspectives on the Old Red Sandstone.* Geological Society, London, Special Publications, **180**, 371–388. 0305-8719/00/$15.00

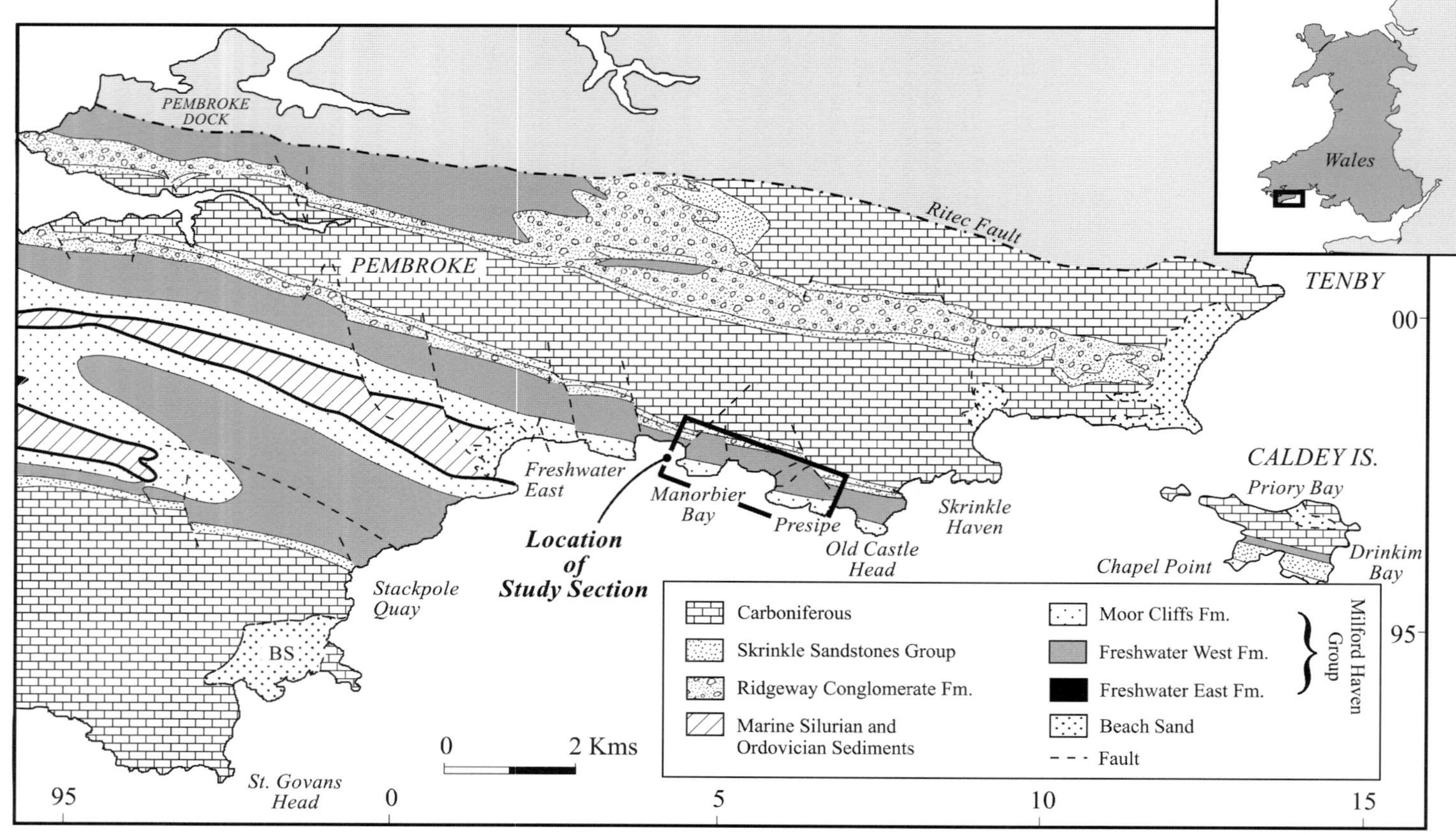

Fig. 1. Map showing location of study section and the Old Red Sandstone outcrop of South Pembrokeshire, south of the Ritec Fault.

Siluro-Devonian boundary has been the subject of speculation and debate (Cocks *et al.* 1971; Walmsey 1974; Ziegler *et al.* 1974; Allen & Williams 1981*b*; Allen & Crowley 1983). The recognition of regional marker beds such as the *Psammosteus* Limestone, renamed locally as the Chapel Point Calcrete (Williams *et al.* 1982), and regionally extensive tuff falls, such as the Townsend Tuff Bed (TTB), form significant stratigraphic marker beds within the Lower ORS. In SW Dyfed the Lower ORS comprises a 1–4.5 km thick succession of red beds, which has been divided into Groups, Formations and Members using a combination of lithofacies changes and laterally extensive marker beds (Allen & Williams 1978, 1982; Williams *et al.* 1982) rather than by the sparse occurrence of fossil remains (Fig. 2).

The cliff sections herein described comprise the lower part of the Moor Cliffs Formation, underlain by sequences containing Silurian-aged invertebrate and vertebrate fauna and overlain by Lower Devonian sediments containing vertebrate faunas. The Townsend Tuff Bed is taken locally as the Silurian–Devonian boundary (Allen & Williams 1981*b*; Allen & Crowley 1983). The studied section lies *c.* 90 m below the TTB and is therefore assigned to the Přídolían stage of the Upper Silurian succession (Fig. 2). These Lower ORS sediments were deposited on a large alluvial plain bordering a sea to the south (Allen 1985). Marine influence is restricted to lower Sandy Haven and upper Freshwater East Formations, which comprise coastal mudflats with intertidal channels (Allen & Williams 1978, 1981*b*; Williams *et al.* 1982; Allen 1986). There is no evidence for marine influence in the Moor Cliffs Formation at the study location (Love 1993; Marriott & Wright 1993; Jenkins 1998).

The Moor Cliffs Formation, and laterally equivalent Sandy Haven Formation, is dominated by siltstones with over 75% of the formations' thickness comprising red–brown siltstones. Intraformational conglomerates, subordinate sandstones, tuff horizons and rare extraformational conglomerates make up the remainder of the succession (Allen & Williams 1978; Friend & Williams 1978; Allen *et al.* 1982; Williams *et al.* 1982). The siltstones host conspicuous and well-developed palaeosols dominated by carbonate nodules with columnar morphologies and by slickensided fracture planes, which form pseudoanticlinal fracture systems variously filled with carbonate cement or fine-grained sediment. The Sandy Haven and Moor Cliffs Formations represent more proximal and distal components, respectively, of the same fluvial system. Rivers flowed south and southeast with sediment source areas to the north and east (Allen & Williams 1982; Allen & Crowley 1983). The Chapel Point Calcrete marks the top of the Moor Cliffs Formation.

Structural and tectonic setting

The sedimentary succession preserved in the Dyfed area represents the distal part of an extensive alluvial system. Volcanic activity from an inferred igneous centre to the east or west (Allen & Williams 1981*b*) periodically showered the alluvial plains with ash. Several major basement lineaments trending E–W, such as the Benton and Musselwick Faults to the north of Milford Haven and the Ritec Fault to the south, are thought to have periodically influenced ORS deposition in the area (Sanzen-Baker 1972; Allen & Williams 1978, 1982; Allen & Crowley 1983; Powell 1989; Jenkins 1998).

The Chapel Point Calcrete, a distinctive and thick sequence of superimposed pedogenic carbonates (Allen 1974*c*, 1986; Williams *et al.* 1982), occurs at the top of the Moor Cliffs Formation (Fig. 2) and is characterized by high maturities and erosional top surfaces. These calcretes form a conspicuous stratigraphic marker horizon south and east of the Ritec Fault and its extension, the Llandyfaelog Fault, and represent a prolonged period of landscape stability punctuated by episodes of mild erosion and renewed deposition. North of the Ritec Fault in equivalent sediments, the Chapel Point Calcrete is apparently absent (Allen & Williams 1978). The E–W-trending extensional Ritec and Benton Faults probably controlled subsidence rates and thus earliest sedimentation in the area (Allen & Williams 1978, 1982; Powell 1989; Jenkins 1998) and continued to have an episodic influence on deposition. A structurally controlled, elevated, sediment-starved terrace is interpreted to have existed to the south of the Ritec Fault upon which the mature Chapel Point palaeosols formed. It is unclear how dominantly south-draining fluvial systems responded to this structurally controlled terrace. Within the coastal sections between Freshwater East and Old Castle Head, (Fig. 1) no evidence for contemporaneous sedimentation and tectonic activity was observed.

The Sandy Haven Formation north of Milford Haven, and the partially equivalent Freshwater East and Moor Cliffs Formations have very similar facies distributions. However, the Sandy Haven Formation north of the Ritec Fault is much thicker (850–900 m) than the Freshwater East and Moor Cliffs Formations (140–415 m) to the south (Allen & Williams 1978; Williams *et al.* 1982). The much thicker

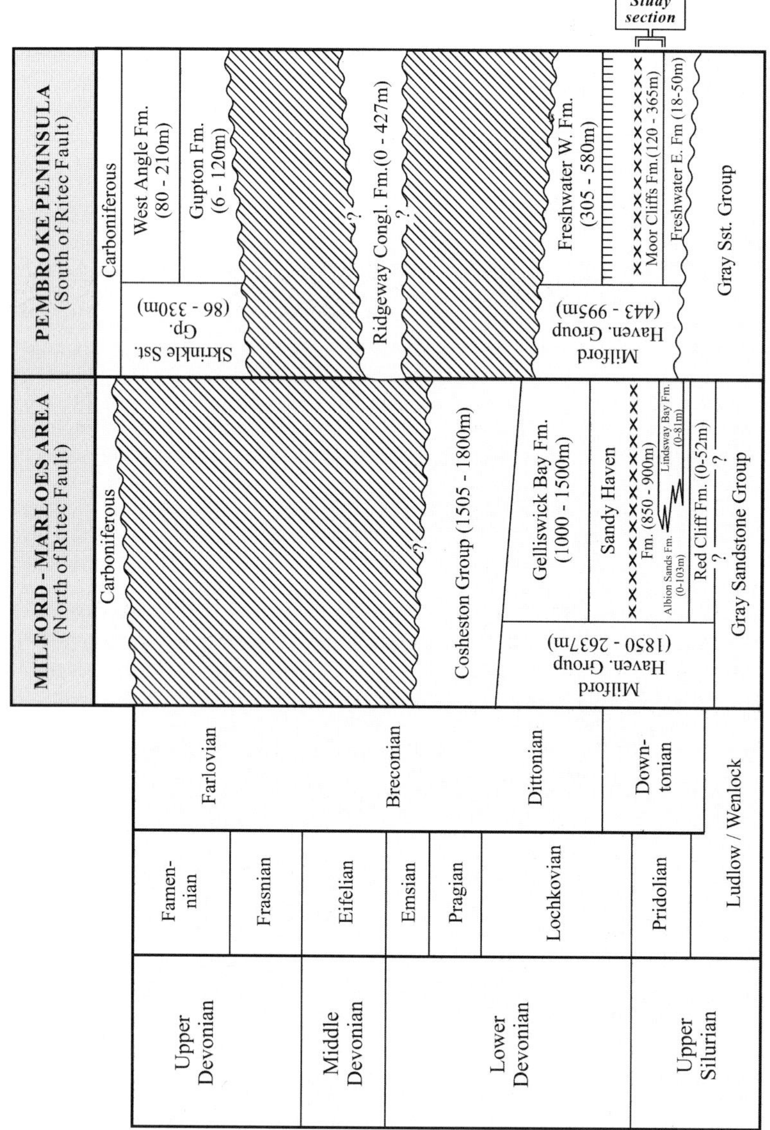

Fig. 2. The ORS stratigraphy of southwest Dyfed (modified after Williams *et al.* (1982) and Wellman *et al.* (1998).

and coarser-grained succession north of the Ritec Fault suggests that this fault was a controlling factor on sedimentation in the area, perhaps periodically partitioning the basin (Powell 1987, 1989).

Detailed sedimentology and depositional environment, Moor Cliffs Formation

The cliff sections between Freshwater East and Old Castle Head provide excellent high-quality exposures. The near-vertical succession is well lithified and subtly weathered, and delicate structures within the fine-grained units are preserved. The outcrop described and discussed here comprises a 50–70 m vertical thickness of mudstone-dominated sediment over a 2.5 km lateral distance. Two headlands of East Moor Cliff and Priests Nose were studied in detail and a distinctive tuff horizon, the Rooks Cave Tuff (RCT) (first recognized by Williams *et al.* (1982)), was used as a marker bed to correlate from one headland to another (see Fig. 3). A detailed 2D lateral profile of the coastal exposures was constructed by measuring a series of vertical facies logs and linking them by lateral profiling (Figs 3 and 4). Using this method, important vertical and lateral changes in depositional geometries, facies and palaeosol development were recorded.

Conglomerates

Two types of conglomerate are present in the Moor Cliffs Formation; intraformational, comprising reworked carbonate and siltstone material, and extraformational, which occur in association with sandbodies or as discrete isolated lenses. Both types of conglomerates are a volumetrically minor component of the fluvial package (Allen & Williams 1979, fig. 1).

Extraformational conglomerates are rare, comprising quartz grains in addition to reworked siltstone and carbonate material. Two examples of this conglomerate were identified in the study section: on East Moor Cliff, where it forms the basal part of a relatively thick sandbody (Fig. 4, log EM1, 43 m above the RCT), and on the Priests Nose section, where the conglomerate forms an isolated 30 m wide lens (Fig. 3, 40 m east of log MN3). In both cases, the conglomerates erosively overlie a siltstone.

The most common conglomerates found in the Moor Cliffs Formation are intraformational and comprise varying proportions of carbonate and siltstone clasts that have a siltstone matrix (Fig. 5a). The conglomerates often overlie laterally extensive (minimum 500 m) erosion surfaces, and occur in multi-storey lenses or as isolated lenses encased in siltstones and are not related to the sandbodies (Allen & Williams 1979). Although the intraformational conglomerates do not always overlie carbonate-bearing palaeosols, examples of truncated Vertisols in the section provide good evidence for clast source (Fig. 5b, c). The conglomerates grade from clast-supported to pebbly siltstone, and low-angle cross-bedding, parallel-bedding and lenticular structures are common. Where clasts are elongate, they are commonly imbricated. Intraformational conglomerates comprising rip-up clasts occur at the base of sandbodies, at the bases of individual sandstone units and along foresets. The clasts of red or grey–green siltstone occasionally preserve original lamination and are either elongate with long axes parallel to bedding or rounded.

One prominent and laterally extensive conglomerate occurs directly above a moderately mature carbonate-bearing palaeosol, which has had its top horizons removed by erosion (Fig. 3, 15 m below the RCT, Fig. 5b). In some places the conglomerate is only a few clasts thick, or has had its fabric partially obliterated by continued pedogenesis, incorporating the conglomeratic material into the developing soil. Siltstones containing desiccation cracks and burrows intercalated with the conglomerate (e.g. Fig. 3, logs MN3 and MN7, 15 m below the RCT) provide evidence for pulsed, episodic deposition.

Sandstones

The sandstones comprise distinct fining-upward units, which total a maximum of 10% of the succession thickness. The sandstones are highly micaceous, quartz rich and well sorted, ranging from 0.25 to 4.5 m thick, averaging 1–1.5 m thick. The sandbodies are generally medium grained, grey–green with increasing red–brown colour towards their siltier tops. They comprise broad sheets with apparent width-to-thickness ratios in excess of 250:1. Palaeocurrent directions were variable, with a predominantly southwards drainage direction. A conglomerate lag or, more commonly, sands hosting siltstone rip-up clasts overlie planar, mildly erosive bases with occasional shallow scour structures (Fig. 5d).

The sandbodies are invariably well structured and often fine upwards to very fine sandstones and siltstones. Sedimentary structures preserve an upward progression from planar and trough cross-bedding to climbing ripples, reflecting the decrease in grain size. All sandbodies in the cliff sections are multi-storey with internal erosion

Fig. 5. (**a**) Intraformational conglomerate comprising reworked carbonate and siltstone clasts. (**b**) East side of Rooks Cave, Moor Cliffs Formation below the Rooks Cave Tuff. Younging from right to left. (**c**) Carbonate-bearing Vertisol (maturity III–IV) with rotated columnar carbonate nodules, erosively overlain by intraformational conglomerates. Younging from right to left. (**d**) Base of sandstone with mild erosion, overlying red–brown siltstone. (**e**) Wedge-shaped ped forms coated with clay slickensides and high volume of fanned columnar carbonate nodules. Younging from right to left. (**f**) Drab haloes surrounding cores of red–brown siltstone.

surfaces, fine-grained drapes and rip-up clasts. Some depositional packages have burrowed or desiccated tops, indicating pauses in deposition and subaerial exposure between successive depositional events (e.g. Fig. 3, either side of log MN3, 5 m below the RCT) or have sandstone units separated by rippled and desiccated siltstone, indicating that the channels were episodically active and subject to ephemeral flow.

Volcanic ash

Tuffs occur at intervals throughout this sedimentary succession; they are generally fine grained with pale grey and purple banding. The thicker tuffs are often highly cleaved and preserve a variety of sedimentary structures such as ripples and soft sedimentary deformation.

The Rooks Cave Tuff varies in thickness (up to 2.5 m) and preserves original floodplain topography. The sudden influx of volcanic ash blanketed the floodplain surface and preserves many delicate trace fossils that are not found elsewhere in the section. These trace fossils include arthropod tracks, funnel-shaped vertical burrows and asymmetrical pathways with pellets in the base (Allen & Williams 1981*b*; Williams *et al.* 1982). The minor tuffs are not always laterally continuous and on East Moor Cliff (Fig. 4, 5 m above the RCT, *c.* 50 m east of log EM1) a thin tuff pinches out in the middle of a palaeosol.

Fine-grained sediments

Siltstone comprises 65–95% of the outcrop thickness and is generally micaceous and poorly sorted. Red–brown coloration is dominant, although grey–green coloration below sandstones is common. The siltstones contain many primary and secondary features including palaeosols. The most abundant structures are low-angle ripples, parallel lamination and parallel bedding, desiccation cracks, soft-sediment deformation, and a variety of trace fossils including small horizontal, vertical and U-shaped burrows and trackways. Less abundant features include climbing ripple sets, low-angle accretion surfaces, primary current lineations, adhesion ripples and warts, rain pits and carbonate nodules.

Burrows are commonly U-shaped (*Diplocraterion*) or horizontal and between 2 and 8 mm in diameter. Only one convincing example of *Beaconites antarcticus* was observed on the Priests Nose section beneath the Rooks Cave Tuff 30 m east of log MN3 (Fig. 3) although some large (3.5 cm diameter) U-shaped burrows occur elsewhere in the siltstones. Other trace fossils include asymmetrical locomotion traces with clusters of pellets in the base, vertical funnel-shaped burrows, and arthropod tracks and resting traces (Trewin, pers. comm.). The burrows often have an associated colour change, with burrow-fills and the sediment immediately surrounding the burrow either partially or totally green–grey, or drab purple–grey. Desiccation cracks and desiccation polygons of varying sizes are common throughout the section, cracks being generally less than 2 cm in width and filled with silt or sand, some having multiple layers of sediment infill, suggesting that the cracks reopened several times.

Carbonate nodules with a parallel-to-bedding form grow around and engulf host sediment, and in some well-exposed examples primary ripple laminations are preserved within the nodules. Carbonate also preferentially precipitates in and around burrows. The nodules are not associated with palaeosol profiles and occur in discrete bands in siltstones that contain well-preserved ripples, burrows, adhesion ripples and desiccation cracks. Perpendicular-to-bedding nodules occur either in massive siltstones or at the bases of palaeosol profiles (e.g. Fig. 4, log EM3, 15 m above the RCT; Fig. 3, log MN4, 12 m below the RCT).

Pedogenic horizons

The pedogenic structures described below are characteristic of modern Vertisols (Fig. 6) and the Moor Cliffs Formation palaeosols are categorized as Vertisols. The most visually conspicuous features of the thick red–brown siltstones are palaeosols that are dominated by pale carbonate nodules and bowl-shaped fracture systems (Fig. 5c). Pedogenic features include desiccation cracks, drab haloes, prismatic soil structures, ped textures, carbonate nodules, wedge-shaped ped forms and large curved fracture planes (Fig. 5b, c, e and f). The palaeosols have developed in siltstones and very fine silty sandstones, and have not developed in cleaner sandy parent material. The pseudoanticlinal and wedge-shaped features are attributed to soil formation rather than to structural processes that affected the Moor Cliffs Formation during Variscan deformation. Pseudoanticlinal fractures often have slickenside coatings. Slickenside fabrics produced during structural movements are usually unidirectional or show a limited number of overprinted unidirectional phases. The clay and carbonate slickensides found coating pseudoanticlinal fractures in the Moor Cliffs Formation siltstones and mudstones have radial patterns, reinforcing the interpretation that the fracture systems are pedogenic in origin rather than the product of subsequent burial and structural movements.

Although most of the siltstones are red–brown, tinges of orange and purple are also present and reflect the proportions of various constituents of silt, mud and tuffaceous material in the rock. Fine-grained siltstones and mudstones have a paler red–orange colour whereas tuffaceous units are purple or grey. The predominant colour of the palaeosols is red–brown, as a result of hematite grain coatings and disseminated hematite in the groundmass. The present vivid red–brown palaeosols may not reflect the original colour of the soil because pigments dehydrate through time and on burial (e.g. Wright 1989; Retallack 1990). The original sediment may have contained goethite and therefore had a more drab hue. The dominant red–brown coloration of these Vertisols, in

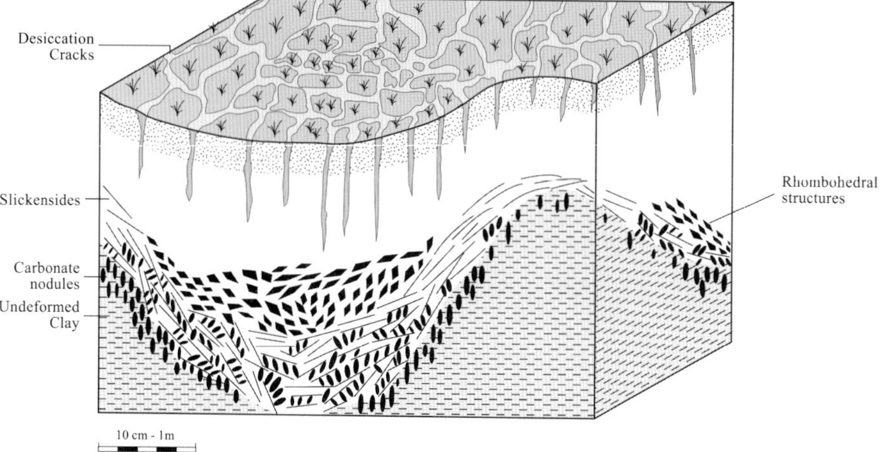

Fig. 6. Section through a typical modern Vertisol illustrating surface relief (gilgai), carbonate nodules and pseudoanticlines at depth (modified after Dudal & Eswaran (1988)).

contrast to their more usual black colour, indicates that these soils formed in alluvial sediments with reasonable drainage and in a humid climate (Fitz Patrick 1980; Duchaufour 1982; Ahmad 1983).

Localized colour variations comprise 'drab haloes' and reduction spots. Colour mottling is very limited in the succession and is restricted to tuffaceous horizons and units immediately adjacent to the tuffs. Drab haloes are common in rippled and burrowed siltstone units and in palaeosols where the fabric is not dominated by carbonate nodules. The haloes are purple–grey and tend to be elongate perpendicular to bedding. They occur around fine pedogenic fractures or burrows, or as discrete features with cores of reddened silt (Fig. 5f). In cross-section the haloes are generally circular and some bifurcate down. Reduction spots are rare, but where present are blue–green and tend to occur in distinct bands, in coarser desiccation crack fills, or associated with burrows.

Many siltstone units have a distinctive textured fabric, which can range from fine and hackly to coarser and more blocky. The texture does not vary consistently with grain size and only occurs in certain siltstone horizons in association with palaeosols. The ped fabric is not present in palaeosol horizons characterized by high amounts of carbonate accumulation or pseudoanticlines, and is thought to be a primary pedogenic texture of the upper soil horizon rather than a tectonic fabric.

Carbonate precipitation in various forms is very common in siltstone units that have experienced pedogenesis. The size and morphology of the nodules and the volume of the unit that they occupy is highly variable. Carbonate morphologies range from fine filaments or sheets trending both parallel and horizontal to bedding, to small subspherical nodules and large columnar nodules typically 30 cm long, which trend perpendicular to bedding, or which fan in response to pseudoanticline development (Fig. 7). Where carbonate dominates the sediment and makes up over 75% of the unit, the perpendicular-to-bedding nodules are sometimes capped with sub-horizontal planar to wavy sheets of carbonate (e.g. Fig. 3, 30 m east of log MN3 6 m below the RCT). This is the most mature carbonate morphology observed in the study section.

Large angular wedges of siltstone within palaeosols are the products of intersecting fracture planes that are covered in claystone slickensides (Fig. 5c and e). Pseudoanticlinal and synclinal fractures of various thickness are widespread and are lined with clay slickensides, filled with silt, or more rarely with carbonate cement. Some fractures are open with no evidence of any original infilling material. The host sediment is invariably reddened siltstone that has no surviving primary sedimentary features.

A typical pedogenic profile is between 2 and 10 m thick and consists of massive red–brown siltstone with ped texture, drab haloes trending sub-perpendicular to bedding and, more rarely, burrows and desiccation cracks. Below this zone is a package of curved pseudoanticlinal fracture systems with variable amounts of carbonate nodule precipitation and drab haloed fractures (see Love 1993; Wright & Marriott 1996). The nodules are commonly fanned, thus trending perpendicular to the pseudoanticline surface

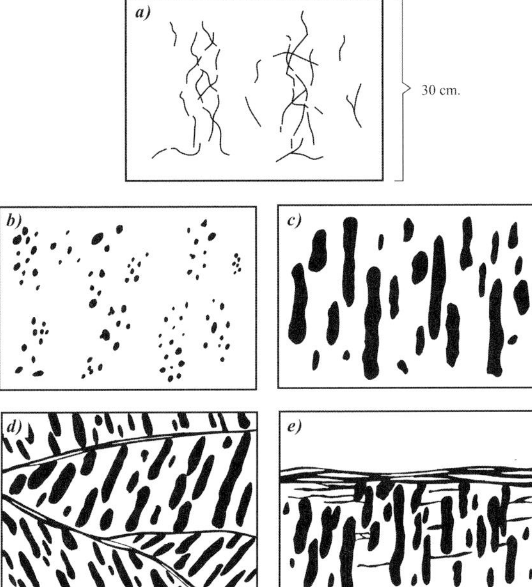

Fig. 7. Pedogenic carbonate development hosted in Moor Cliffs Formation siltstones, becoming progressively more mature from (**a**) to (**e**). (**a**) Filamentous; (**b**) sub-spherical; (**c**) columnar; (**d**) fanned columnar with slickensides; (**e**) columnar capped with laminar sheets.

(e.g. Fig. 4 log EM2, 30 m above the RCT). The wavelength and amplitude of pseudoanticlines, the subsurface equivalent of surface gilgai, indicate that the mounds were typically 0.5–1 m high and 2–15 m across. Composite soil profiles were traced for distances of up to 400 m. Limited by the coastal exposures, this represents the minimum lateral extent of these palaeosols.

Climate and vegetation

A semi-arid climate for the ORS sediments has been suggested by several workers (e.g. Allen 1973, 1974*b*, *c*, 1979, 1986; Allen & Williams 1981*b*, 1982; Marriott & Wright 1993), whereas Woodrow *et al.* (1973) interpreted a savannah, wet and dry climate with high temperatures and rainfall. Plate reconstruction places Britain in tropical to sub-tropical latitudes on the southern part of the ORS continental landmass at *c.* 5–15° S (Woodrow *et al.* 1973; Drewry *et al.* 1974; Ziegler *et al.* 1977; Van der Voo 1983; Channell *et al.* 1992). At present, these latitudes are characterized by dense, lush vegetation and tropical soils with high amounts of organic litter and deeply weathered horizons (Birkland 1984; Goudie 1984). There is clearly divergence in the interpretations of the climate prevalent during deposition of the ORS sediments and the reconstructed palaeolatitudes. This divergence has probably been caused by interpreting sediments with clear evidence for ephemeral flow, in the context of primitive vegetation.

By Late Silurian times, the first vascular land plants, comprising bacteria, fungi, algae and lichens, had diversified and become established across many continental areas including Scotland and the Anglo-Welsh Basin (e.g. Edwards 1970, 1979, 1980; Edwards & Richardson 1974; Gray & Boucot 1977; Chaloner & Sheerin 1979; Stewart 1983; Gray 1985; Shute & Edwards 1989). Plant fossils in the Lower ORS of the Anglo-Welsh Basin show that the land was colonized by dense stands of small primitive vascular plants, and fragments of well-preserved *Cooksonia* have been recovered from ORS exposures at Freshwater East (Edwards 1979). This vegetation was low growing with limited anchoring roots and would have been vulnerable to destruction during heavy rainfall and flooding.

Vertisol formation requires a seasonal climate ranging from humid to semi-arid. Subtle oscillations from wetter rainy seasons to drier dry seasons do not greatly affect the resulting soils, which are not sensitive indicators of climate variation (e.g. Oakes & Thorp 1967; Duchaufour 1982). The Vertisols in the Moor Cliffs Formation are unlikely to record humid–arid cyclicity and there is no evidence for such changes in climate. Vegetation is an important component of soil formation (e.g. Wilding *et al.* 1983*a*, *b*; Birkland 1984; Retallack 1990) and fungal, bacterial and algal biomass can contribute significant proportions of organic matter to the soil (Ugolini & Edmonds 1983). Present-day Vertisols are characterized by very low percentages of organic matter (typically less than 2%), their formation being dependent more on a high clay content and wetting and drying cycles (e.g. Duchaufour 1982). This may explain the dominant Vertic nature of the ORS palaeosols, which developed in the absence of dense vegetation.

Vegetation acts as a baffle between rainfall, runoff and evaporation (Goudie 1984), which is of critical importance to fluvial systems. Without sufficient plant cover and biomass in ORS times, despite high amounts of rainfall, wet–dry cyclicity would have been extreme. Upstream areas did not have sufficient vegetative cover to retain water and runoff rates were high, resulting in short-duration, high-discharge floods followed by rapid drying out and desiccation. Seasonal desiccation within this interpreted monsoonal, sub-tropical climate would have presented a challenging environment to the pioneer terrestrial vegetation. Although the land may have looked fairly green close to perennial water sources, plants were incapable of binding substrate together, stabilizing riverbanks and levees, or of resisting erosion by floodwaters.

The lack of extensive root systems to stabilize substrate has an important effect on river and flood behaviour, and ultimately on fluvial archiecture (e.g. Schumm 1968). In a landscape containing primitive vegetation, drainage channels would be broad and shallow, lacking cohesive retaining banks. These fluvial systems would quickly flood, and mass runoff from upland and interfluvial areas caused large-volume, high peak discharge floods of short duration, giving the appearance of a more arid regime. Transport of weathered material from upland areas would have been rapid and sedimentation on broad low-relief plains predominant. This scenario is envisaged for the Lower ORS alluvial system.

Fluvial facies association

The fluvial facies association describes discrete sandbodies preserved in the section that are interpreted to be the main fluvial channel systems that drained across the area. The sandbodies are multi-storey and several have large, low-angle accretion surfaces indicating that the broad shallow channels contained low-angle bar forms. It is unclear from outcrop whether these are downstream or laterally accreting. The sandstones are separated by thick siltstone packages containing palaeosols and although connectivity within each multi-storey sandbody is very good, no connectivity between the sandbodies was observed either vertically or laterally.

A single channel margin is preserved in the study section at Priests Nose (Fig. 3, west of log MN3, 3 m above the RCT). Over a distance of 220 m towards log MN7, the sandbody progressively fines to rippled, burrowed and desiccated very fine sandstone. The channel margins did not comprise sharp or steep banks and were not resistant to erosion. Evidence for downcutting and entrenchment was not observed.

Coarse, quartz-rich sediment forms the basal parts of thicker sandstones or rarely occurs as thin, lenticular beds encased in siltstone (Fig. 3, between log MN4 and Rooks Cave, 1–3 m below the RCT). The single observed example of a lenticular unit of conglomerate not associated with a sandstone is overlain by rippled and parallel-laminated siltstones and very fine sandstones interpreted to have been deposited in proximal settings to the main fluvial channels. The conglomerate may represent a lag to a fine-grained crevasse deposit.

The siltstone clasts occurring within the sandstones were most probably derived from floodplain silts that were eroded and entrained in out-of-channel flow (Allen & Williams 1979). Vertisols with upper smectitic horizons broken up by desiccation cracks had a high source potential for intraformational clasts. Once entrained in the flow, these clasts were transported as cohesive lumps and thus became rounded rather than breaking up and disaggregating. Lower soil horizons comprising large wedge-shaped peds, and which had time after flooding and wetting to swell, would have been resistant to erosion, thus limiting the depth of scouring either by interfluvial runoff or by newly established fluvial systems. Pedogenic aggregates have been documented in modern formations by Rust & Nanson (1989) and from the Ridgeway Conglomerate (Ekes 1993) and Freshwater West Formations (Marriott & Wright 1996) (Fig. 2).

The channel system supplied conglomerates and sand to a broad floodplain, which had a relief of a few metres or less. The channels carried sand grade material as bed load and deposited broad bar forms and low-relief dunes. Frequent flooding caused water to flow onto interfluvial areas and high-intensity storms produced runoff sourced from interfluvial areas. The channel systems frequently dried out and were ephemeral in nature. Extensive root systems were not present and vegetation was not capable of stabilizing channel banks. This produced wide, shallow fluvial channels that rapidly filled and flooded during high-discharge floods.

The thick floodplain packages deposited between the sandstones and the broad morphology of the sandbodies suggest that the fluvial channels avulsed considerable distances (several kilometres) to isolate areas of the alluvial plain from the depositional influence of fluvial channels. Flow directions were variable with a predominantly southwards drainage direction, suggesting lateral avulsion in an easterly or westerly direction. Although younger sequences in the Moor Cliffs Formation contain evidence for shallow palaeovalleys (Allen & Williams 1982), incised valley margins were not identified in the study section. Lateral correlations between the East Moor Cliff and Priests Nose sections indicate that channel sandbodies, moderately mature and mature Vertisols are juxtaposed over distances of 1 km (compare Figs 3 and 4), suggesting that the influence of these fluvial systems was laterally limited, and that any out-of-channel flow did not result in deposition of significant thicknesses of sediment, or prolonged submergence sufficient to kill the developing soil.

Floodplain facies association

The floodplain facies association embraces siltstones, conglomerates and tuffs that were subject to alteration and pedogenesis during prolonged periods of non-deposition and floodplain stability (Love 1993; Wright & Marriott 1996). The palaeosols record variations in relative rates of aggradation, minimum thicknesses of sediment removed by erosion and the climatic conditions that prevailed over the Lower ORS alluvial plains. The floodplain experienced periods dominated by subaqueous deposition interspersed with times of pedogenesis in subaerially exposed sediment. There is no evidence in the sedimentary sequence for marine conditions or periodic marine influence and all sedimentary features are consistent with a freshwater environment of deposition (Love 1993; Jenkins 1998).

Moderately mature palaeosols are often erosively overlain by intraformational conglomerates, indicating that the floodplain was subject to prolonged periods of landscape stability followed by shorter-lived instability, erosion and reworking. Laterally extensive bands of intraformational conglomerate containing a high percentage of carbonate nodules erosively overlie truncated Vertisols that lack their top structural zones (e.g. Fig. 3, log MN4, 17 m below the RCT; Fig. 5c). Clasts are often rounded and have a complex polygenetic history, indicating that the carbonate debris has been transported over varying distances before deposition. Identification of these conglomerates is biased in favour of those containing carbonate clasts because the aggregate texture in siltstones is destroyed by compaction at shallow depths (Rust & Nanson 1989; Marriott & Wright 1996).

Where pedogenesis has not obliterated previous sedimentary structures, the siltstones and very fine silty sandstones contain ample evidence for active deposition by flowing water. The sub-tropical climate with a combination of high temperatures and abundant rainfall would have led to rapid weathering of source areas, producing large quantities of silt and mud. Ripples from over 1 cm high to only 4 mm from crest to crest are hosted in fine-grained siltstones and these units are commonly finely laminated. Rare examples of fine sandstones grading laterally into structured siltstones (e.g. Fig. 4, logs EM1 to EM2 21 m above the RCT) indicate that these sediments are closely related to channel processes and are probably low-angle 'levee' and avulsion deposits resulting from over the low-relief floodplain that travelled for a minimum of several hundred metres.

Some siltstone packages contain low-angle accretion surfaces (Fig. 3, west of log MN7, directly below the RCT; Fig. 4, 70 m east of log EM3, 15 m above the RCT), suggesting that these silts were actively deposited as cohesive silt aggregates and that these silt packages comprise bar forms composed entirely of silt-grade material. Individual depositional units pinch out against major bedding planes, and coarser stringers preserve planar and ripple lamination. Pebbly siltstones are commonly cross- or parallel-bedded and contain intraformational carbonate clasts.

The depositional processes responsible for both conglomerate and siltstone packages are distinct from those in fluvial channels and their lateral equivalents, which carried sand into the system. The intraformational clasts were sourced from interfluvial areas where they were entrained and transported by sheet runoff generated within interfluvial lowlands by an ephemeral drainage network active for short times following intense rainfall. In some cases, these drainage networks eroded the top surfaces of the underlying palaeosols.

The Rooks Cave Tuff's variable thickness preserves original topographic relief on the floodplain that was a minimum of 2.5 m (see Fig. 3, log MN3). Allen & Williams (1982) suggested that the topographic lows were either channels or ponds, whereas Marriott & Wright (1993, 1996) interpreted the thickness variations as due to gilgai relief on the top surface of a Vertisol; however, the substrate immediately beneath the tuff does not contain a palaeosol. The preservation of delicate trace fossils, which occur only in the topographic lows, favours the pond interpretation, because active flow would have rapidly destroyed such delicate features. It is not possible to establish the 3D morphology of the ponds from outcrop and it remains conjectural whether these areas were isolated bowl-shaped depressions or linked interfluvial channel or drainage systems.

Rippled, deformed and parallel lamination together with burrows signify a subaqueous influence, whereas wrinkle marks and adhesion ripples signify damp surfaces with desiccation cracks indicating fully dried sediment. Pale grey–green and blue reduction spots are not common, tending to occur in weakly pedogenically altered siltstone and at contacts between deposition units, suggesting that the floodplain was not subjected to prolonged submergence. The reduction spots indicate that the substrate was periodically saturated, although the highly localized nature of the spots and a predominance of oxidized, red–brown coloration show that prolonged reduction of iron pigments did not take place.

The siltstones preserve a whole range of bioturbation phenomena, from slightly to completely churned, indicating that residence time in the top 50 cm of sediment, and therefore sedimentation rates, was variable. The sedimentation rate in these cases was an order of magnitude faster and conditions were wetter than when palaeosols were developing.

Drab haloes mostly occur around long thin siltstone-filled tubes, which occasionally bifurcate down suggesting an organic origin. They were also found to surround or permeate burrows and line some fractures and nodules in palaeosols (Fig. 5f). The purple–grey coloration of the haloes is not a product of reduction and removal of iron pigments, but forms independently of reduction spots and mottles. The haloes occur in both bioturbated and unbioturbated substrate. As primitive vascular plants that colonized the Lower ORS landscape did not have extensive root systems (Stewart 1983), communities of bacteria and algae that were abundant at this time are the most likely originator of these features.

Several stages of nodule growth are present in the study section, from thin isolated filaments to well-developed concretions several centimetres in diameter and in excess of 30 cm in length. In some cases carbonate accumulation volumetrically exceeds that of the host siltstone. The Vertisols are dominantly red–brown (coloured by hematite), and gleying (reduced zones) is rare and if present is usually associated with sandbodies. Colour mottling is rare and there are no rust-coloured concentrations of iron and/or manganese, which are indicative of water-table levels fluctuating through the soil (Gerrard 1981). The lack of evidence for prolonged water saturation of the soils indicates that water-table levels remained, for the most part, several metres below the landsurface surface. However, water-table levels would have fluctuated in response to seasonal changes in rainfall. Carbonate accumulations are not restricted to palaeosols; some have developed in waterlain and otherwise non-pedogenically altered sediment, supporting the hypothesis that floodwaters provided a significant source of calcium carbonate to the Moor Cliffs Formation sediments.

Modern Vertisols comprise a distinctive series of structural zones (Fig. 6) with a desiccated top, peds coarsening and becoming more angular with depth into a zone of slickenside development and finally a zone of carbonate precipitation. Soil, or ped texture, is distinguished from cleavage fabrics because the texture is not present throughout the siltstone succession, is not grain size dependent and occurs only in the upper horizons of Vertisols and in poorly developed profiles. The ped texture is a primary soil structure, obliterated by pseudoanticline development and carbonate nodule growth.

The Moor Cliffs palaeosols contain all the features that characterize Vertisols, but do not follow the clear soil zones that define Vertisols. Different features overlap and interact, the most abundant example of this being pseudoanticlines and carbonate nodules, which coexist in the same siltstone units. The nodules are fanned and deformed as a consequence of continued movement on pseudoanticline fractures. Several development sequences were presented by Love (1993) and Marriott & Wright (1993, 1996), explaining the succession of events necessary to produce the combination of pedogenic features observed in the Moor Cliffs Formation Vertisols.

A significant number of moderately mature to mature Vertisols have had their top surfaces erosively removed. Fanned nodules are more common below these pronounced erosion surfaces (e.g. Fig. 4 between logs EM3 and EM2, 35–40 m above the RCT). The zones of carbonate precipitation and pseudoanticline movement are not necessarily mutually exclusive and zones within the soil may experience both nodule growth and pseudoanticline development (Ahmad 1983). As the soil shrinks and swells, sediment between the planes of movement is squashed and deforms, moving the nodules within the host sediment.

Vertisols form rapidly, and features such as pseudoanticlines and gilgai take of the order of 550 years or less to form (Parsons *et al.* 1973). Carbonate nodules take much longer to grow and are therefore a good indicator of relative maturity variations within the palaeosols, assuming delivery of carbonate to the floodplain remains constant. Carbonate maturity ranges were assigned stages from I to IV after the morphogenetic sequence derived by Gile *et al.* (1966). Stages I–III are common whereas stage IV maturity is rare and only encountered in association with tuffs or as a truncated profile on Priests Nose 5 m below the RCT. The most mature carbonate accumulations are dominated by vertical nodules with thin laminar upper surfaces; they are assigned maturity stage IV, and represent the lower parts of mature Vertisols that have had their upper zones removed by erosion (Fig. 5b and c). These stage IV palaeosols formed during prolonged pauses in sedimentation and some estimates from other field studies suggest that this pause was of the order of 2000–100 000 years (Yaalon 1983; Birkland 1984; Retallack 1990).

Allen (1986) proposed a maximum of 30 000 years for palaeosol development, calculated by considering the duration of the Lower ORS, the number of calcretes (carbonate-bearing soil horizons) contained in the succession and the amount of time required to deposit the sediment. The abundance of carbonate nodules in pedogenic and non-pedogenic host sediment indicates that calcium carbonate was in abundant supply, favouring the shorter rather than the longer time span. Reworked carbonate conglomerate and frequent tuff falls provided additional concentrated sources of carbonate to developing soils, thus preferentially enhancing this maturity indicator in some soils.

The range of depositional and pedogenic features in the thick siltstones implies that the floodplain did not aggrade in a steady and uniform manner. The palaeosol data suggest that there were pauses in sedimentation of the order of 500–30 000 years. During development of the Vertisols, aggradation continued and discrete erosive events affected the system. During times of higher and more continuous sedimentation with associated increased floodplain submergence, palaeosols did not have the opportunity to form. Within this dominantly aggradational system, sedimentation rates slowed or stopped and the floodplain frequently dried out when sedimentation ceased. The seasonal cycles of wetting and drying rapidly led to the formation of hummocky topography comprising mounds and hollows of varying breadth and depth and distinctive Vertisol structural zones (see Fig. 6). The Vertisols were flooded and saturated in the rainy season, but became dried out and desiccated in the dry months. These Vertisols show a range of maturities and morphologies that depart from modern Vertisol profiles. Such differences reflect rates of aggradation and degradation, and the order in which these events took place (Love 1993; Marriott & Wright 1993, 1996).

Interfluvial areas in the wet season were characterized by flooded areas with shallow standing water, providing a favourable habitat for a variety of vertebrate and invertebrate fauna including fish and arthropods (e.g. Allen & Williams 1981*a*, *b*; Walker 1985). Emergent areas would have been colonized by dense stands of primitive vascular plants, lichens and algae (e.g. Edwards 1979, 1980; Gray 1985). In the dry season, the alluvial plains dried out and oxidizing conditions prevailed. Desiccation polygons developed preferentially in the hollows, where plants may have survived the dry season. Vertisols are not sensitive to climate change and subtle oscillations would not have been recorded. The dominant red–brown coloration of the Vertisols, in contrast to their more usual black colour, indicates that these soils formed in alluvial sediments with reasonable drainage and in a humid climate (Duchaufour 1982; Ahmad 1983).

Variations in palaeosol development

Variations in palaeosol maturity occur on two scales: local variations on a scale of a few metres to tens of metres, and larger-scale variations revealed by correlating logs along the 2.5 km section.

Small-scale variations in palaeosol maturity were caused by topographic relief and substrate composition. In some sedimentary packages, Vertisol maturity does not increase consistently in any one direction. The degree of carbonate development varies over distances of 15–30 m and generally increases towards topographic lows. Modern Vertisols favour lowlying, poorly drained conditions (e.g. Ahmad 1983), which enhance seasonal wet–dry cycles and therefore soil development. The relationship between topographic lows and Vertisol development explains variations in soil maturity over short lateral distances in the section (Fig. 3, westwards from log MN3, 5 m below the RCT).

Development of Vertisols is considerably enhanced immediately above tuff horizons. The

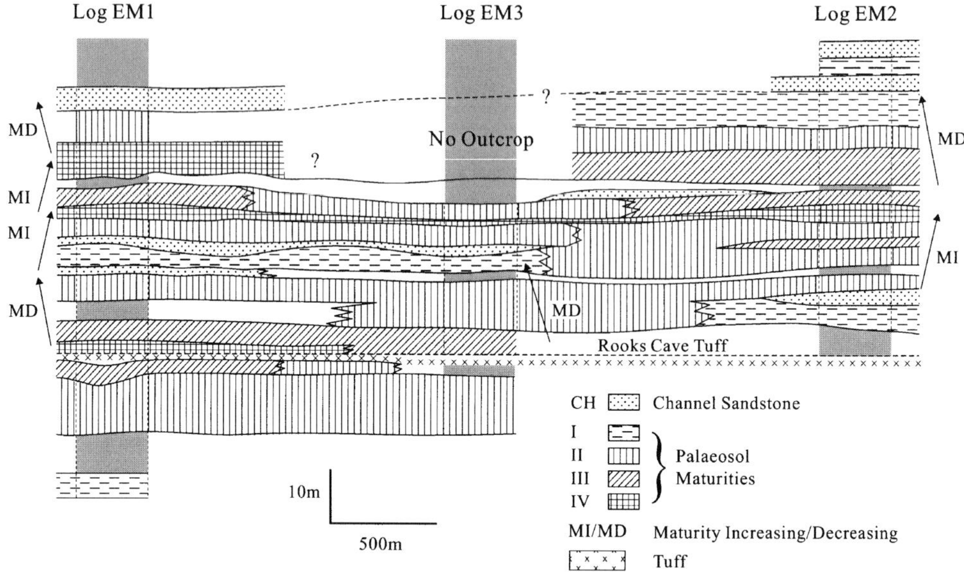

Fig. 8. Strong vertical cyclicity preserved in the Moor Cliffs Formation at East Moor Cliff (see Fig. 3 for location) interpreted to be pedofacies sequences.

Vertisol profiles have a much higher percentage of carbonate and pseudoanticline fractures are more pronounced, wider and commonly filled with carbonate (e.g. Fig. 4, log EM1, immediately above the RCT). Rapid soil development in parent materials containing volcanic ash has been studied by other workers (e.g. Harris 1971; Jahn *et al.* 1985), who concluded that ash rapidly weathers, releasing compounds from which many soil-forming clay minerals can form. The enhanced palaeosol development in the Moor Cliffs Formation is localized and affects between 0.1 and 1.5 m of sediment above the Rooks Cave Tuff fall.

The lateral profile from East Moor Cliff to Priests Nose illustrates that, in contrast to the initial sheet-like appearance, few units are continuous over the 2.5 km section. The sandbodies comprise broad multi-storey systems, only occasionally continuous over distances greater than 1 km. The tuffs provide important correlation horizons, but thinner tuff horizons can be incorporated into developing Vertisols and obliterated (Fig. 4, 50 m east of log EM1, 6 m above the RCT), as well as being discontinuous as a result of non-deposition or removal by erosion. In several instances, sandbodies in one section correspond to Vertisols of varying maturities in laterally equivalent sedimentary packages. For example, the prominent sandbodies on Priests Nose just above the Rooks Cave Tuff give way westwards to moderately mature Vertisols in the East Moor Cliff section.

Maturity values of I–IV were assigned to the palaeosols, allowing patterns in palaeosol development to be identified. The most mature palaeosols were found to consistently occur in positions furthest from the channels in both a lateral and vertical sense. Closer to sandbodies Vertisols are less well developed, with maturity values of less than III, and contain a low percentage of carbonate nodules. At more distal locations, Vertisols occur with maturity values of III or IV, and carbonate accumulations are abundant, comprising fanned columnar nodules. The lateral changes in palaeosol maturity developed in the floodplain siltstones are consistent with the 'pedofacies relationship' first described by Bown & Kraus (1987), where soil maturity consistently increases with increased lateral distance from active channels.

In addition to the lateral pattern of pedogenic maturity a strong vertical cyclicity was identified in all the logged sections within the Moor Cliffs Formation. Channel sandstones are overlain by siltstones with no pedogenic modification, which in turn are overlain by siltstones hosting palaeosols of increasing maturity to a maximum stage III or IV. This pattern is then reversed, with decreasing palaeosol maturity before another sandbody is encountered. This is especially well developed on the East Moor Cliff exposures,

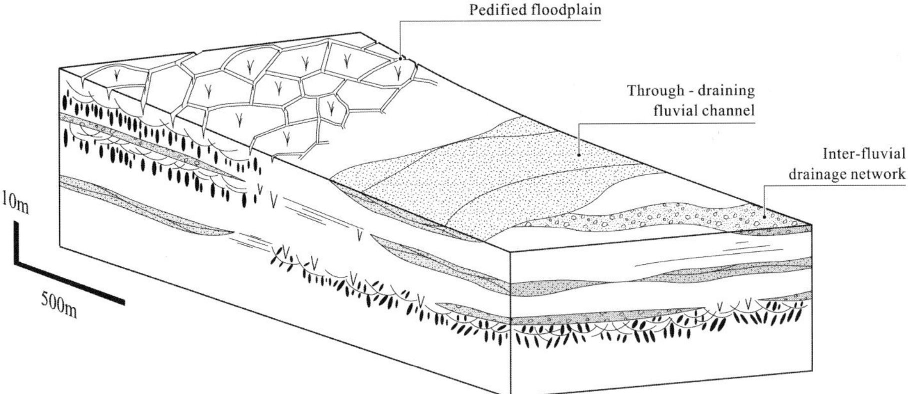

Fig. 9. Schematic reconstruction of the Moor Cliffs Formation floodplain.

where patterns of maturity can be correlated for 500 m along-strike (Fig. 8). This type of cyclicity, called 'pedofacies sequences' by Kraus (1987), is produced by autocyclic fluvial processes of avulsion and aggradation.

Volcanic ash falls and erosive events interrupt the strongly cyclical pattern. The erosive events interrupt maturity progressions and compartmentalize the section. The erosion surfaces overprint the autocyclic signature of the pedofacies sequences, suggesting that processes other than channel avulsion and aggradation affected the Moor Cliffs Formation sedimentary system.

Periodic erosion of previously stable land surfaces indicates that there was a change in the climate or that there was a relative change in base level (e.g. Allen 1974*c*, 1986). However, there is no evidence in the Moor Cliffs Formation palaeosols for climate change. The palaeoshoreline is estimated to follow the present-day Bristol Channel (Simon & Bluck 1982; Allen 1985; Jenkins 1998), close enough to the Dyfed area that changes in sea level could have affected the fluvial system (e.g. Salter 1993), and the Temeside Formation contains calcretized intertidal sediments (Allen 1985), implying that pedogenesis of shore zone sediments took place during times of relative lowered sea level. However, the Moor Cliffs Formation studied preserves no evidence for marine influence and only local, small-scale entrenched fluvial systems and palaeovalleys. The proximity to large and active fault systems favours small base-level changes related to fault movements.

Conclusions

As a result of this detailed study of the Moor Cliffs Formation, with particular emphasis on the palaeosols, a re-evaluation of the depositional system has been made, which significantly departs from previously published models. From an understanding of the interplay between vegetation, fluvial style and pedogenic development presented in this paper, it is postulated that a sub-tropical, monsoonal climate prevailed during deposition of the ORS in the Anglo-Welsh Basin. The climate during deposition of the Moor Cliffs Formation was therefore characterized by high temperatures, high annual rainfall and marked wet and dry seasons. The climate coupled with the primitive vegetation produced rapid weathering of upland areas, and weathered products were swiftly transferred into the drainage network to be transported downstream. The broad alluvial plains in Dyfed are interpreted to be the distal component of this extensive fluvial system.

The sedimentary architecture of the Moor Cliffs Formation is envisaged to be the product of deposition by a combination of through-flowing high width to depth sandy channel systems flanked by broad low-relief siltstone plains, which had a minimum floodplain topography of 2.5 m on which Vertisols formed (Fig. 9). River systems had highly variable discharge rates and ephemeral flow. There is little evidence for deposition by passive suspension fall-out from floodwaters. Deposition of siltstones took two forms: first, where silt was deposited in ripple forms and finely laminated sheets deposited in association with through-draining channels; second, where cohesive lumps of silt and other intraformational of debris were transported as coarse sand and conglomerate grade bed-load and deposited in bar forms within interfluvial areas. Palaeosol maturities vary in a consistent pattern attesting to a fluvial regime controlled, for the most part, by random channel avulsion and aggradational conditions. Interfluvial areas contained a separate network of broad shallow channels, which periodically carried runoff and eroded floodplain material. The interfluvial channels acted as tributaries to the through-draining channels, introducing siltstone and carbonate clasts into the channels.

We acknowledge The Petroleum Science and Technology Institute (PSTI), particularly R. Johnston, for funding the PhD component of this study, V. P. Wright, S. B. Marriott, G. Jenkins and R. D. Hillier for thoughtful discussions in the field, Presentational Graphics, especially for production of the lateral panels, ExxonMobil for funding colour production, N. H. Trewin and P. F. Friend for review and helpful comments, and the staff at the Fourcroft Hotel, Tenby for their warm hospitality.

References

AHMAD, N. 1983. Vertisols. *In*: WILDING, L. P., SMECK, N. E. & HALL, G. F. (eds) *Pedogenesis and Soil Taxonomy: II. The Soil Orders*. Elsevier, Amsterdam, 91–123.

ALLEN, J. R. L. 1963. Depositional features of Dittonian rocks: Pembrokeshire compared with the Welsh Borderland. *Geological Magazine*, **100**, 385–400.

—— 1964. Studies in fluviatile sedimentation: six cyclothems from the Lower Old Red Sandstone, Anglo-Welsh Basin. *Sedimentology*, **3**, 163–198.

—— 1973. Compressional structures (patterned ground) in Devonian pedogenic limestones. *Nature: Physical Science*, **243**, 84–86.

—— 1974*a*. The Devonian rocks of Wales and the Welsh borderland. *In*: OWEN, T. R. (ed.) *The Upper Palaeozoic and Post-Palaeozoic Rocks of South Wales*. University of Wales Press, Cardiff, 47–84.

—— 1974*b*. Geomorphology of Siluro-Devonian alluvial plains. *Nature*, **249**, 644–645.

—— 1974*c*. Studies in fluviatile sedimentation: implications of pedogenic carbonate units, Lower Old Red Sandstone, Anglo-Welsh outcrop. *Geological Journal*, **9**, 181–207.

—— 1979. Old Red Sandstone facies in external basins, with particular reference to Southern Britain. *In*: HOUSE, M. R., SCRUTTON, C. T. & BASSETT, M. G. (eds) *The Devonian System*. Palaeontological Association, Special Papers in Palaeontology, **23**, 65–80.

—— 1985. Marine to fresh water: the sedimentology of the interrupted environmental transition (Ludlow–Siegenian) in the Anglo-Welsh region. *Philosophical Transactions of the Royal Society of London*, **309**, 85–104.

—— 1986. Pedogenic calcretes in the Old Red Sandstone facies (Late Silurian–Early Carboniferous) of the Anglo-Welsh area, southern Britain. *In*: WRIGHT, V. P. (ed.) *Paleosols: Their Recognition and Interpretation*. Blackwell Scientific, Oxford, 58–86.

—— & CROWLEY, S. F. 1983. Lower Old Red Sandstone fluvial dispersal systems in the British Isles. *Transactions of the Royal Society of Edinburgh*, **74**, 61–68.

—— & WILLIAMS, B. P. J. 1978. The sequence of the earlier Lower Old Red Sandstone (Siluro-Devonian), north of Milford Haven, southwest Dyfed (Wales). *Geological Journal*, **13**, 113–136.

—— & —— 1979. Interfluvial drainage on Siluro-Devonian alluvial plains in Wales and the Welsh Borders. *Journal of the Geological Society, London*, **136**, 361–366.

—— & —— 1981*a*. *Beaconites antarcticus*: a giant channel-associated trace fossil from the Lower Old Red Sandstone of South Wales and the Welsh Borders. *Geological Journal*, **16**, 255–269.

—— & —— 1981*b*. Sedimentology and stratigraphy of the Townsend Tuff Bed (Lower Old Red Sandstone) in South Wales and the Welsh Borders. *Journal of the Geological Society, London*, **138**, 15–29.

—— & —— 1982. The architecture of an alluvial suite: rocks between the Townsend Tuff and Pickard Bay Tuff Beds (Early Devonian), Southwest Wales. *Philosophical Transactions of the Royal Society of London, Series B*, **297**, 51–89.

——, THOMAS, R. G. & WILLIAMS, B. P. J. 1982. The Old Red Sandstone north of Milford Haven. *In*: BASSETT, M. G. (ed.) *Geological Excursions in Dyfed, south-west Wales*. National Museum of Wales, Cardiff, 123–149.

BIRKLAND, P. W. 1984. *Soils and Geomorphology*. Oxford University Press, New York.

BOWN, T. M. & KRAUS, M. J. 1987. Integration of channel and floodplain suites, I. Developmental sequence and lateral relations of alluvial paleosols. *Journal of Sedimentary Petrology*, **57**, 587–601.

CHALONER, W. G. & SHEERIN, A. 1979. Devonian macrofloras. *In*: HOUSE, M. R., SCRUTTON, C. T. & BASSETT, M. G. (eds) *The Devonian System*. Palaeontology Association, Special Papers in Palaeontology, **23**, 145–161.

CHANNELL, J. E. T., MCCABE, C. & WOODCOCK, N. H. 1992. Early Devonian (pre-Acadian) magnetization directions in Lower Old Red Sandstone of south Wales (UK). *Geophysical Journal International*, **108**, 883–894.

COCKS, L. R. M., HOLLAND, C. H., RICKARDS, R. B. & STRACHAN, I. 1971. A correlation of Silurian rocks in the British Isles. *Geological Society, London*, Special Report **1**, 103–136.

DIXON, E. E. L. 1921. *The geology of the South Wales Coalfield. Part XIII. The country around Pembroke and Tenby*. Memoir of the Geological Survey of England and Wales.

—— 1933. Notes on the geological succession in South Pembrokeshire. *Proceedings of the Geologists' Association*, **44**, 402–411.

DREWRY, G. E., RAMSAY, A. T. S. & GILBERT SMITH, A. 1974. Climatically controlled sediments, the geomagnetic field, and trade wind belts in Phanerozoic time. *Journal of Geology*, **82**, 531–553.

DUCHAUFOUR, P. 1982. *Pedology*. George Allen & Unwin, London.

DUDAL, R. & ESWARAN, H. 1988. Distribution, properties and classification of Vertisols. *In*; WILDING, L. P. & PUENTES, R. (eds) *Vertisols: their Distribution, Properties, Classification and Management*. Soil Management Support Services. Texas A&M University, Technical Monograph **18**, 1–22.

EDWARDS, D. 1970. Fertile Rhyniophytina from the Lower Devonian of Britain. *Palaeontology*, **13**, 451–461.

—— 1979. A Late Silurian flora from the Lower Old Red Sandstone of south-west Dyfed. *Palaeontology*, **22**, 23–52.

—— 1980. Early land floras. *In*: PANCHEN, A. L. (ed.) *The Terrestrial Environment and the Origin of Land Vertebrates*. Systematics Association, Special Volume, **15**, 55–85.

—— & RICHARDSON, J. B. 1974. Lower Devonian (Dittonian) plants from the Welsh borderland. *Palaeontology*, **17**, 311–324.

EKES, C. 1993. Bedload-transported pedogenic mud aggregates in the Lower Old Red Sandstone in southwest Wales. *Journal of the Geological Society, London*, **150**, 469–472.

FITZPATRICK, E. A. 1980. *Soils: their Formation, Classification and Distribution*. Longman, Harlow.

FRIEND, P. F. & WILLIAMS, B. P. J. (eds) 1978. *A Field Guide to Selected Outcrop Areas of the Devonian of Scotland, the Welsh Borderland and South Wales*. Palaeontological Association, London.

GERRARD, A. J. 1981. *Soils and Landforms: an Integration of Geomorphology and Pedology*. George Allen & Unwin, London.

GILE, L. H., PETERSON, F. F. & GROSSMAN, R. B. 1966. Morphological and genetic sequences of carbonate accumulation in desert soils. *Soil Science*, **101**, 347–360.

GOUDIE, A. S. 1984. *The Nature of the Environment: an Advanced Physical Geography*. Basil Blackwell, Oxford.

GRAY, J. 1985. The microfossil record of early land plants: advances in understanding of early terrestrialization, 1970–1984. *Philosophical Transactions of the Royal Society of London*, **309**, 167–195.

—— & BOUCOT, A. J. 1977. Early vascular plants: proof and conjecture. *Lethaia*, **10**, 145–174.

HARRIS, S. A. 1971. Podsol development on volcanic ash deposits in the Talamanca Range, Costa Rica. *In*: YAALON, D. H. (ed.) *Paleopedology: Origin, Nature and Dating of Paleosols*. Israel Universities Press, Jerusalem, 191–209.

JAHN, R., GUDMUNDSSON, T. & STAHR, K. 1985. Carbonatisation as a soil forming process on soils from basic pyroclastic fall deposits on the island of Lanzarote, Spain. *In*: CALDAS, E. F. & YAALON, D. H. (eds) *Volcanic Soils: Weathering and Landscape Relationships of Soils on Tephra and Basalt*. Catena Supplement, **7**, 87–97.

JENKINS, G. 1998. *An investigation of marine influence during deposition of the Lower Old Red Sandstone, Anglo-Welsh Basin, UK*. PhD thesis, University of Wales, Cardiff.

KING, W. W. 1933. The Downtonian and Dittonian strata of Great Britain and north-western Europe. *Quarterly Journal of the Geological Society, London*, **90**, 526–570.

KRAUS, M. J. 1987. Integration of channel and floodplain suites, II. Vertical relations of alluvial paleosols. *Journal of Sedimentary Petrology*, **57**, 602–612.

LEACH, A. L. 1933. The geology and scenery of Tenby and the South Pembrokeshire coast. *Proceedings of the Geologists' Association*, **44**, 187–216.

LOVE, S. E. 1993. *Floodplain deposits as indicators of sandbody geometry and reservoir architecture*. PhD thesis, Aberdeen University.

——, WILLIAMS, B. P. J. & DAVIES, D. K. 1993. Fluvial channel reservoirs: prediction and location using floodplain paleosols. *American Association of Petroleum Geologists Abstracts*, 201.

MARRIOTT, S. B. & WRIGHT, V. P. 1993. Palaeosols as indicators of geomorphic stability in two Old Red Sandstone alluvial suites, South Wales. *Journal of the Geological Society, London*, **150**, 1109–1120.

—— & —— 1996. Sediment recycling on Siluro-Devonian floodplains. *Journal of the Geological Society, London*, **153**, 661–664.

OAKES, H. & THORP, J. 1967. Dark-clay soils of warm regions variously called Rendzina, Black Cotton Soils, Regur, and Tirs. *In*: DREW, J. V. (ed.) *Selected Papers in Soil Formation and Classification*. Soil Science Society of America, Special Publication Series, **1**, 136–149.

PARSONS, R. B., MONCHAROAN, L. & KNOX, E. G. 1973. Geomorphic occurrence of Pelloxerets, Willamette Valley, Oregon. *Soil Science Society of America: Proceedings*, **37**, 924–927.

POWELL, C. M. 1987. Inversion tectonics in S.W. Dyfed. *Proceedings of the Geologists' Association*, **98**, 193–203.

—— 1989. Structural controls on Palaeozoic basin evolution and inversion in southwest Wales. *Journal of the Geological Society, London*, **146**, 439–446.

RETALLACK, G. J. 1990. *Soils of the Past: an Introduction to Paleopedology*. Unwin Hyman, Boston, MA.

RUST, B. R. & NANSON, G. C. 1989. Bedload transport of mud as pedogenic aggregates in modern and ancient rivers. *Sedimentology*, **36**, 291–306.

SALTER, T. 1993. Fluvial scour and incision; models for their influence on the development of realistic reservoir geometries. *In*: NORTH, C. P. & PROSSER, D. J. (eds) *Characterization of Fluvial and Aeolian Reservoirs*. Geological Society, London, Special Publications, **73**, 33–52.

SANZEN-BAKER, I. 1972. Stratigraphical relationships and sedimentary environments of the Silurian–Early Old Red Sandstone of Pembrokeshire. *Proceedings of the Geologists' Association*, **83**, 139–164.

SCHUMM, S. A. 1968. Speculations concerning paleohydrologic controls of terrestrial sedimentation. *Geological Society of America Bulletin*, **79**, 1573–1588.

SHUTE, C. H. & EDWARDS, D. 1989. A new rhyniopsid with novel sporangium organization from the Lower Devonian of South Wales. *Botanical Journal of the Linnean Society*, **100**, 111–137.

SIMON, J. B. & BLUCK, B. J. 1982. Palaeodrainage of the southern margin of the Caledonian mountain chain in the northern British Isles. *Transactions of the Royal Society of Edinburgh*, **73**, 11–15.

STEWART, W. N. 1983. *Paleobotany and the Evolution of Plants*. Cambridge University Press, Cambridge.

UGOLINI, F. C. & EDMONDS, R. L. 1983. Soil biology. *In*: WILDING, L. P., SMECK, N. E. & HALL, G. F. (eds) *Pedogenesis and Soil Taxonomy: I. Concepts and Interactions*. Elsevier, Amsterdam, 193–231.

VAN DER VOO. 1983. Paleomagnetic constraints on the assembly of the Old Red continent. *Tectonophysics*, **91**, 271–283.

WALKER, E. F. 1985. Arthropod ichnofauna of the Old Red Sandstone at Dunure and Montrose, Scotland. *Transactions of the Royal Society of Edinburgh*, **76**, 287–297.

WALMSEY, V. G. 1974. The base of the Upper Palaeozoic. *In*: OWEN, T. R. (ed.) *The Upper Palaeozoic and Post-Palaeozoic Rocks of South Wales*. University of Wales Press, Cardiff, 31–46.

WELLMAN, C. H., THOMAS, R. G., EDWARDS, D. & KENRICKS, P. 1998. The Cosheston Group (Lower Old Red Sandstone) in southwest Wales: age, correlation and palaeobotanical significance. *Geological Magazine*, **135**(3), 397–412.

WILDING, L. P., SMECK, N. E. & HALL, G. F. (eds) 1983*a*. *Pedogenesis and Soil Taxonomy: I. Concepts and Interactions*. Elsevier, Amsterdam.

——, —— & —— (eds) 1983*b*. *Pedogenesis and Soil Taxonomy: II. The Soil Orders*. Elsevier, Amsterdam.

WILLIAMS, B. P. J., ALLEN, J. R. L. & MARSHALL, J. D. 1982. Old Red Sandstone facies of the Pembroke peninsula, south of the Ritec fault. *In*: BASSETT, M. G. (ed.) *Geological Excursions in Dyfed, South-west Wales*. National Museum of Wales, Cardiff, 151–174.

WOODROW, D. L., FLETCHER, F. W. & AHRNSBRAK, W. F. 1973. Paleogeography and paleoclimate at the deposition sites of the Devonian Catskill and the Old Red facies. *Geological Society of America Bulletin*, **84**, 3051–3064.

WRIGHT, V. P. 1989. Paleosol recognition. *In*: ALLEN, J. R. L. & WRIGHT, V. P. (eds) *Paleosols in Siliciclastic Sequences*. Postgraduate Research Institute for Sedimentology, University of Reading, 1–25.

—— & MARRIOTT, S. B. 1996. A quantitative approach to soil occurrence in alluvial deposits and its application to the Old Red Sandstone of Britain. *Journal of the Geological Society, London*, **153**, 907–913.

YAALON, D. H. 1983. Climate, time and soil development. *In*: WILDING, L. P., SMECK, N. E. & HALL, G. F. (eds) *Pedogenesis and Soil Taxonomy: I. Concepts and Interactions*. Elsevier, Amsterdam, 233–251.

ZIEGLER, A. M., HANSEN, K. S., JOHNSON, M. E., KELLY, M. A., SCOTESE, C. R. & VAN DER VOO, R. 1977. Silurian continental distributions, paleogeography, climatology and biogeography. *Tectonophysics*, **40**, 13–51.

——, RICKARDS, R. B. & MCKERROW, W. S. (eds) 1974. *Correlation of the Silurian Rocks of the British Isles*. Geological Society of America, Special Papers, **154**.

Depositional setting of the Lower Old Red Sandstone at Pantymaes Quarry, central South Wales: new perspectives on the significance and occurrence of 'Senni Beds' facies

GERAINT OWEN[1] & DUNCAN HAWLEY[2]

[1]*Department of Geography, University of Wales Swansea, Singleton Park, Swansea SA2 8PP, UK (e-mail: g.owen@swansea.ac.uk)*

[2]*Department of Education, University of Wales Swansea, Hendrefoelan, Swansea SA2 7NB, UK*

Abstract: The 200 m long face of Pantymaes Quarry, central South Wales, exposes part of the upper (Dittonian) Red Marl Group (Lower Old Red Sandstone). A lower Sandstone Facies Association (up to 15 m thick) comprises interbedded grey–green micaceous sandstones, pebbly intraformational conglomerates, and grey siltstones. Major bounding surfaces define stacked channel complexes, representing the deposits of braided river systems. An erosively overlying Mudstone Facies Association (up to 15 m thick) comprises tabular-bedded, bioturbated and calcretized red siltstones and fine sandstones representing overbank deposits, possibly from a meandering system. The change in fluvial style is interpreted as a local response to pulses of oblique-slip movement on nearby fault systems. The period of fault activity resulted in the braided Sandstone Facies Association, the Mudstone Facies Association recording a gradual return to a meandering system. Individual pulses of activity produced the stacked channels within the Sandstone Facies Association. The Sandstone Facies Association is similar in character to the overlying Senni Beds, suggesting that 'Senni Beds' may describe a recurring, possibly tectonically controlled, braided channel facies in the Lower Old Red Sandstone.

Pantymaes Quarry lies 3 km south of Sennibridge in Powys, some 14 km WSW of Brecon, and 6 km north of the main Brecon Beacons escarpment, at an altitude of about 300 m (grid reference SN 913 264, Fig. 1). Access is gained via a public bridleway west of the A4067 road. The main face extends over 250 m from north to south, facing east (Fig. 2). Additional exposures occur on a 15 m long west–east face at the north end and lower north–south faces on the east side of the quarry. The beds dip gently south, and the main face exposes an approximately constant stratigraphical interval about 25 m thick through Lower Old Red Sandstone rocks. The southern end of the quarry is bounded by a NE–SW-trending fault, which downthrows to the NW.

The lower part of the quarry face exposes a sandstone-dominated interval up to 15 m thick, with subordinate intraformational conglomerate and grey siltstone: this is here termed the Sandstone Facies Association. It is separated by an erosion surface from an overlying mudstone interval up to 15 m thick, termed the Mudstone Facies Association, which is dominated by red, massive mudstone with bedded fine-grained siltstone and sandstone at its base and top.

The quarry worked the lower, sandstone unit. Opened in the early 20th Century, large-scale

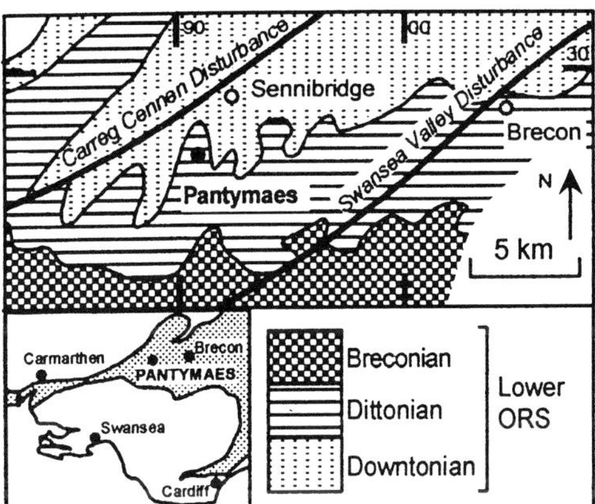

Fig. 1. Location and geological setting of Pantymaes Quarry. Inset map shows South Wales outcrop of Old Red Sandstone.

From: FRIEND, P. F. & WILLIAMS, B. P. J. (eds). *New Perspectives on the Old Red Sandstone.* Geological Society, London, Special Publications, **180**, 389–400. 0305-8719/00/$15.00

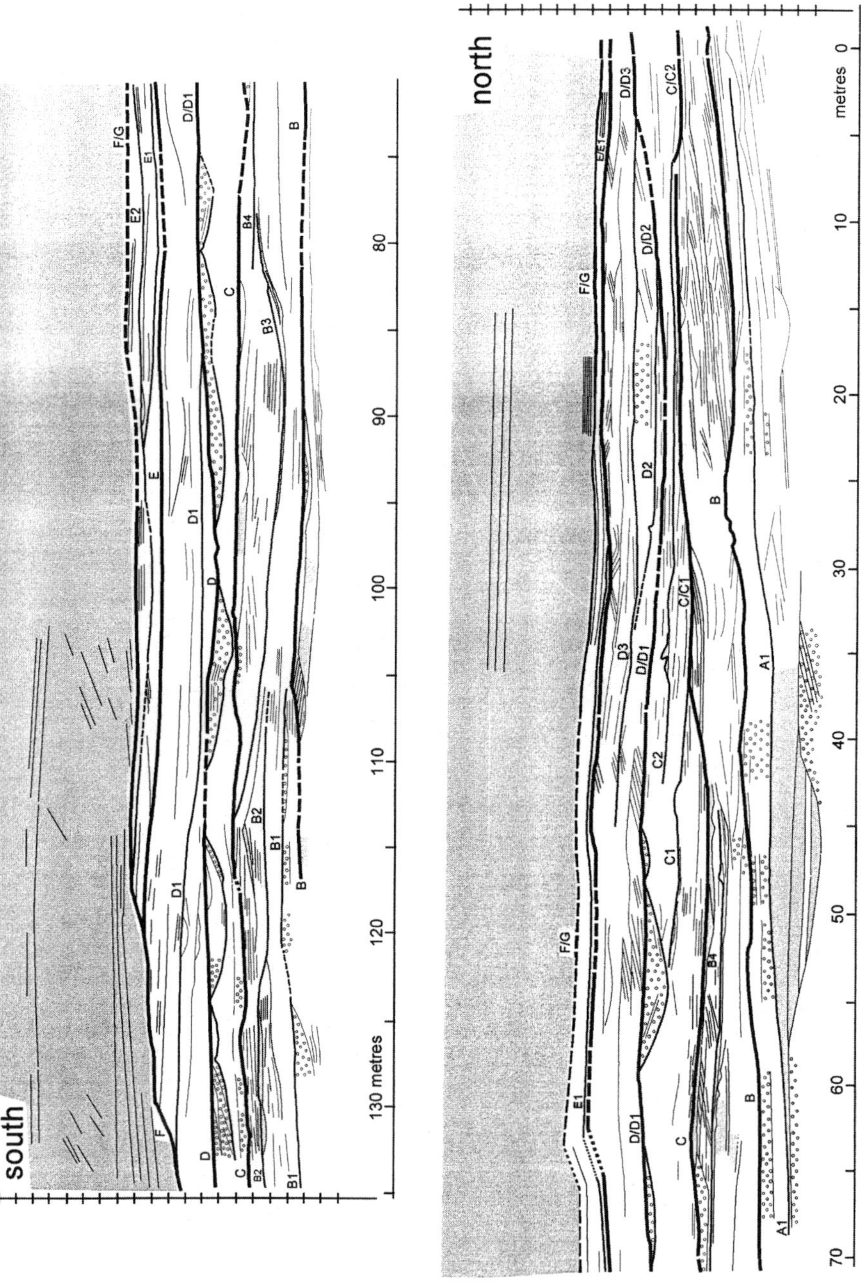

Fig. 2. Detailed diagram of main north–south face at Pantymaes Quarry, compiled from photographs annotated in the field. Horizontal and vertical scales are equal. Shaded areas represent mudstone. The upper part of the face is the red Mudstone Facies Association, other areas of mudstone are grey–green. Open circles are intraformational conglomerate. Unornamented areas are mainly grey–green sandstone, although in some parts of the face the lithology cannot be determined because of difficulty of access. Bounding surfaces are labelled A, A1, B, etc. and are described in the text.

operations by the Harpur family began in 1929, to supply sandstone for roadstone and bridge masonry in Mid- and South Wales. The quarry was closed in 1947 when the thickness of the red mudstone overburden made quarrying uneconomic, and operations were shifted to a new quarry at Heol Senni (Harpur, pers. comm.). Descriptions of the Heol Senni quarry have been given by Edwards *et al.* (1978) and Loeffler & Thomas (1980).

The detailed sedimentology of Pantymaes Quarry has not previously been published. Hassan (1982) compiled a representative sedimentological log to support a study of miospores and palynomorphs from the site. Arthropod trackways collected in 1986 and 1989 are housed in the Department of Geology at the National Museum of Wales in Cardiff. They are currently under investigation at the Department of Earth Sciences, University of Bristol, and an unpublished report outlining the sedimentology was produced by S. Marriott in 1989 to document the context of the trackways (Briggs, pers. comm.).

This paper presents results from a sedimentological survey of Pantymaes Quarry undertaken in the winter of 1998–1999. The aims are to provide new perspectives on the palaeoenvironments of the upper Red Marls in central South Wales and to consider the relationship between the Red Marls and the overlying Senni Beds. Data were collected by logging accessible parts of the quarry faces and mapping onto photomontages. Data for the upper parts of the succession were collected from the fault gully at the southern end of the quarry, from loose blocks on the quarry floor, and by logging a section along an abseil descent of the main quarry face.

Geological setting

The stratigraphical position of Pantymaes Quarry within the Lower Old Red Sandstone has not been accurately determined, but Hassan (1982) attributed the strata to the upper Red Marl Group on evidence from miospores and palynomorphs. Because of a paucity of datable material, internationally defined stage names are not easily applied to the Old Red Sandstone in central South Wales, and the use of local stage names persists (see Fig. 1). The upper Red Marl Group is placed within the 'Dittonian stage' of the Lower Old Red Sandstone of South Wales and the Welsh Borderland (House *et al.* 1977), lithostratigraphically equivalent to the St Maughans Formation in the region to the east (Barclay 1989), and the Llanddeusant Formation to the west (Almond *et al.* 1993). The Red Marl Group comprises red siltstones with minor sandstones and well-developed calcretes, interpreted as deposits of meandering streams on an extensive coastal mud-flat (Allen 1979, 1985; Bluck *et al.* 1992). The Dittonian Red Marls are overlain by the Breconian Senni Beds, which are dominated by green sandstones with pebbly intraformational conglomerates and subordinate siltstones and immature calcretes (Allen 1974, 1979; Edwards *et al.* 1978; Loeffler & Thomas 1980; Owen 1995). The thickness of the red mudstone unit exposed at Pantymaes (10–15 m) is consistent with Hassan's (1982) interpretation that the strata belong to the upper (Dittonian) Red Marl Group.

Chronostratigraphically the upper Red Marl Group is placed in the Lochkovian (upper Gedinnian stage), in the upper part of the *Micrornatus newportensis* zone (Richardson, pers. comm.). Hassan (1982) considered the uppermost part of the upper Red Marl Group barren of palynomorphs. The occurrence of the enigmatic plant *Parka decipiens* at Pantymaes Quarry is also consistent with this stratigraphical position (Hemsley 1990).

Pantymaes Quarry lies about 2 km south of the Caledonoid Church Stretton Lineament, represented in central South Wales by the Carreg Cennen Disturbance (Fig. 1). This is the southernmost element of the NE–SW-trending Welsh Borderland Fault System, which controlled the character and extent of Acadian (mid-Devonian) deformation (Woodcock & Bassett 1993) and was probably active in early to mid-Devonian time (Woodcock 1988). The Swansea Valley Disturbance, a narrow fault belt sub-parallel to the Welsh Borderland Fault System, is located about 6 km south of Pantymaes Quarry (Fig. 1), and the Heol Senni fault, a sub-parallel peripheral of the Swansea Valley Disturbance, crosses the intervening ground (Weaver 1975). Although the main displacements of the Swansea Valley Disturbance are Variscan in age, the faults are probably nucleated on pre-Acadian basement weaknesses (Woodcock & Bassett 1993) and there is evidence that activity along the Swansea Valley Disturbance, and the Vale of Neath Disturbance a further 10 km to the SE, influenced Upper Palaeozoic sedimentation (Weaver 1976; Ramsay 1989). It therefore seems likely that central South Wales was tectonically active during the early Devonian. Local subsidence patterns would have been controlled by one or more local faults with strikes trending across the regional drainage system.

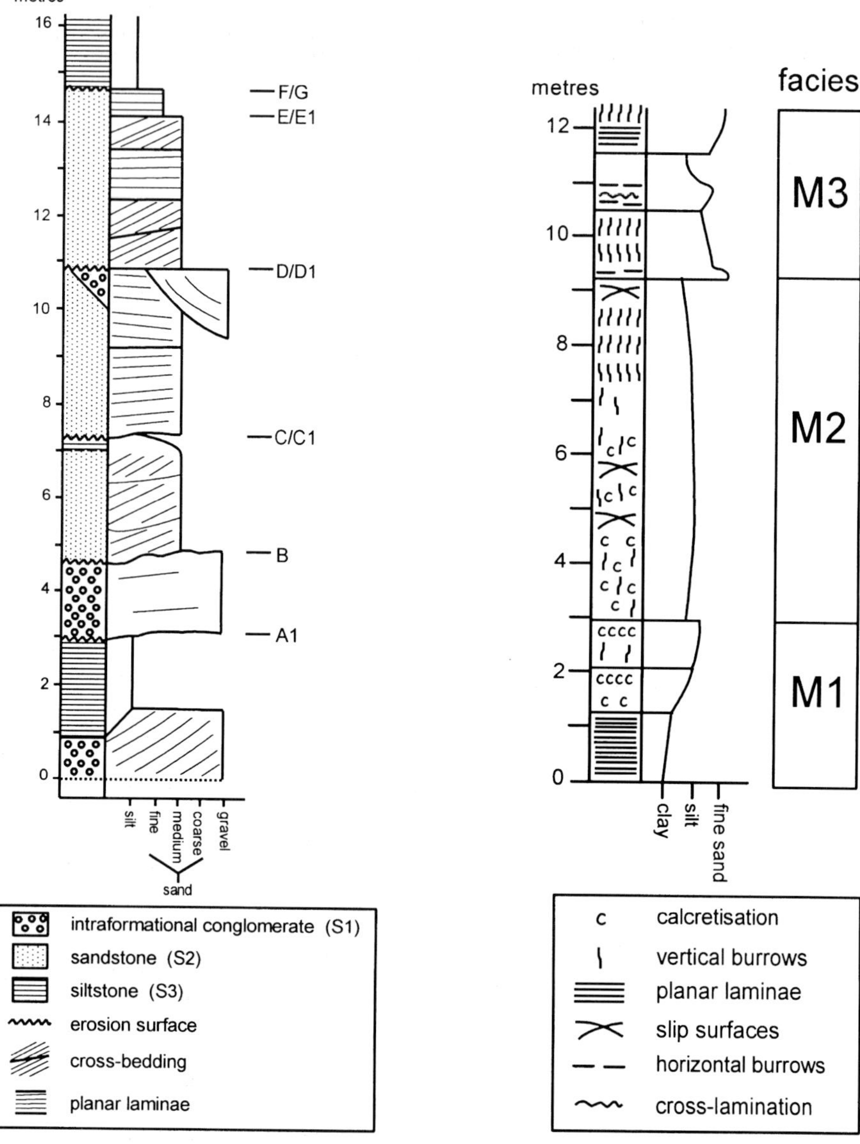

Fig. 3. Representative graphic log through the Sandstone Facies Association at a point *c.* 35 m from the north end of the main face. The main bounding surfaces are identified to the right of the log (A1–F/G).

Fig. 4. Graphic log through the Mudstone Facies Association, compiled on an abseil descent *c.* 125 m from the north end of the main face.

Sedimentary facies (see Figs 2–4)

Facies S1: intraformational conglomerate

This facies is characterized by centimetre-sized, sub-rounded to rounded pebbles of reworked calcrete, with subordinate cobble-sized angular clasts of grey siltstone, in a matrix of green medium- to coarse-grained sandstone. Fragmented plant debris is common, and one small vertebrate fragment has been found. The facies occurs as lenticular or laterally persistent beds up to 2 m thick, and as lag horizons at the base of, or within, parallel laminated sandstone units (see Facies S2). Some thicker units are cross-bedded, indicating palaeocurrent directions ranging from SE to SW (Fig. 2). The facies is interpreted to represent high-energy channel-fills and bars (see Miall 1992).

Facies S2: grey–green sandstone

This facies dominates the Sandstone Facies Association. It comprises grey to green, micaceous fine- to coarse-grained flaggy sandstone. The dominant sedimentary structure is planar, horizontal to gently inclined parallel lamination, with some trough cross-bedding, particularly near the north end of the quarry face. In fallen blocks, trough cross-lamination is picked out by matrix-rich laminae, but this is rare in accessible parts of the face. Some sandstone beds fine upwards, with a lag of intraformational conglomerate at the base and cross-lamination near the top. Plant material, found in units of flaggy micaceous sandstone and at the bases of upward-fining units, includes well-preserved *Parka decipiens* and stems and spines, some over 10 cm in length, as well as fragmented debris. Arthropod tracks have been collected from finer sandstones in this facies (Marriott, pers. comm.). The bounding surfaces of the sandstone units define a complex of nested channels (Fig. 2), described more fully below. This facies is interpreted to represent the deposits of channelized aqueous flows. The dominance of planar lamination over cross-bedding suggests a high-energy, perhaps flashy discharge regime (Tunbridge 1981).

Facies S3: grey siltstone

Grey siltstone occurs as clasts in intraformational conglomerate, and as beds up to 1.5 m thick interbedded with sandstone and intraformational conglomerate. No desiccation features have been observed, but possible burrow traces are present. Some horizons are rich in plant debris. The siltstones are interpreted as overbank deposits that accumulated on a wet floodplain to allow preservation of organic debris.

Facies M1: red mudstone with calcretized horizons

This facies comprises finely laminated red mudstone (Fig. 4) with bioturbation of vertical burrow traces (ichnofabric index 2 *sensu* Droser & Bottjer (1986)) grading up into a series of repeated horizons of isolated but closely packed calcareous nodules 3–9 cm in diameter interbedded with red mudstone (well-developed Stage II calcrete of Machette (1985)). Fallen blocks of similar lithology show desiccation cracks but these have not been found *in situ*. This facies represents overbank deposits. Its occurrence above the erosion surface that separates the Sandstone Facies Association from the Mudstone Facies Association (see below) suggests that the deposits accumulated as the infilling of an abandoned channel.

Facies M2: mudstone with calcrete nodules

This facies is characterized by red, purple or blue mudstone containing scattered calcareous nodules 0.5–3 cm in diameter (Fig. 4). The mudstone has a prismatic to angular blocky ped structure, giving it a friable character, with occasional shallow curved slickensided slip surfaces. The mudstone displays *Skolithos*-type vertical burrow bioturbation (ichnofabric index 4). This facies represents flood basin overbank deposits that received no coarse sediment from active channels.

Facies M3: upward-fining red siltstone units

Facies M3 is a succession of upward-fining red micaceous siltstone units, each about 55–70 cm thick, with sharp bases (Fig. 4). The lowest 25–35 cm of each unit comprises a centimetre-thick basal fine sandstone interfingering with overlying coarse red siltstone. The siltstone has weakly developed cross-lamination and is bioturbated throughout with *Skolithos*-type vertical burrows (ichnofabric index 3), and less commonly occurring ovate, slightly sinuous, horizontal endichnial burrows packed with pellets (ichnofabric index 2). Parting surfaces reveal arthropod trackways. The top of this horizon contains lobate sub-horizontal burrows. The upper 25–35 cm of each unit comprises upward-fining coarse to fine siltstone or mudstone, with thin blue–grey vein-like features and is intensely bioturbated with vertical burrows (ichnofabric index 4–5), of a variety of diameters and between 25 and 35 cm in length. Desiccation cracks occur at the top of this horizon. Fallen blocks with fine-grained lithologies and desiccation cracks similar to the top of Facies M3 show arthropod tracks, but these have not been found *in situ*. This facies represents flood basin overbank deposits receiving regular sediment input from an active channel belt.

Architecture of the Sandstone Facies Association

The Sandstone Facies Association comprises a complex interbedding of facies S1, S2 and S3 with a maximum exposed thickness of about 15 m (Figs 2 and 3). The base is not exposed. The most extensive of the many internal erosion surfaces can be traced laterally for over 100 m and represent major (first-order, *sensu* Allen (1983)) bounding surfaces that allow subdivision of the Association into distinct 'channel complexes'. The bounding surfaces are labelled B–F in Fig. 2, defining the bases of channel complexes B–E. The basal surface of channel complex A is not exposed, and surface F is overlain by the Mudstone Facies Association. Less extensive erosion surfaces are present within channel complexes A–E, and in places second-order surfaces cut down to form the base of a channel complex (see channel complexes C, D and E in Fig. 2). Few palaeocurrent indicators are accessible, but most indicate a strong component to the south or SW, almost parallel to the main quarry face.

Channel complex A

The lowest exposed channel complex is at least 5 m thick and contains one second-order bounding surface (A1). Cross-cutting minor erosion surfaces separate interfingering units of flaggy micaceous sandstone, intraformational conglomerate and grey siltstone (Fig. 5) and define concave-upward channel features 5–15 m wide. At the base of the exposed section (between distances 40 and 60 m on Fig. 2) the top of a unit of cross-bedded intraformational conglomerate defines a trough between two bars that is filled with grey siltstone. Plant material is common, particularly in a thin bed of ferruginous and micaceous sandstone near the base of the face between distances 6 and 25 m. No desiccation cracks were discovered.

The preservation of plant material and lack of desiccation features suggest an environment that was permanently wet. The complex cross-cutting channels and the close interbedding of conglomerate, sandstone and siltstone, however, suggest a highly variable discharge. The environment may have been one of fluctuating, but permanent flow within a network of probably braided channels of the order of 10 m wide. Deposition occurred on the slip faces of cross-bedded gravel bars, and as planar laminae of sand, possibly on bar-tops. Bar surfaces were draped with mud during low stages. Plants probably grew on the channel margins and were preserved within channel deposits by bank erosion. The abundance of calcrete nodules in the intraformational conglomerates points to active pedogenesis in the fluvial basin, and suggests that the channel system may have been entrenched, possibly within a terraced flood basin. The complex cross-cutting of the channels and preservation of lenses of fines are similar to features described from the Senni Beds by Owen (1995).

Fig. 5. Cross-cutting channel surfaces in channel complex A, from the northern end of the subsidiary north–south face (north to left). Erosion surfaces are numbered successively from 1 (oldest) to 3 (youngest). These surfaces are approximately equivalent to erosion surfaces between 0 and 20 m on the main face below surface A1 (see Fig. 2).

Channel complex B

Channel complex B consists of 2–5 m of planar laminated and cross-bedded sandstone. At its northern end it consists of up to 4 m of compound cross-bedded sandstone with some preservation of bar slip faces. Traced to the south, the bar complex passes into a complex of sub-channels filled by planar laminated sandstone. Discontinuous beds of siltstone are preserved between lower-order erosion surfaces, some of which are concave-upward and define the bases of channels. Cross-cutting relationships between second-order erosion surfaces B1–B4 indicate that younger surfaces occur successively to the north (Fig. 2). Surface B3 is well preserved and defines an asymmetrical channel with a steeper northern margin, which is some 35 m wide by a maximum of 3 m deep. The channel is filled with planar laminated sandstone and capped by less than 1 m of grey siltstone. Similar upward-fining channel-fills occur above surfaces B1, B2 and B4. At its northern margin, surface B3 is underlain by more steeply inclined beds of alternating sandstone and grey siltstone, which

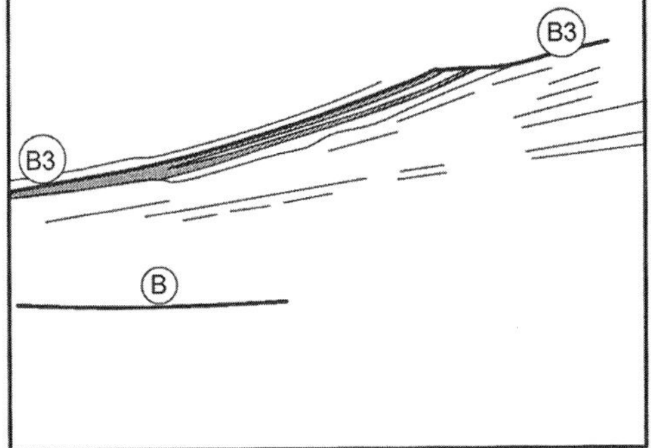

Fig. 6. Inclined, alternating beds of sandstone and siltstone below surface B3 at *c.* 85 m from north end of the main face. Mudstone is shaded on the interpretive diagram. Cross-bedded sandstone passes laterally into cross-stratified interbedded sandstone and mudstone that is directly overlain by erosion surface B3, forming the base of a largely planar laminated channel-fill sandstone body. Similar features occur within channel complex B at *c.* 35 m, 45 m (below B4) and, on a smaller scale, at 115 m.

are concordant with large-scale cross-bedding in the underlying sandstone (Fig. 6). Some of the inclined sandstone beds wedge out down-dip. Similar features are present beneath the northern margin of surface B4.

Channel complex B differs from A in having less intraformational conglomerate, and in its internal organization of compound cross-bedding and asymmetrical upward-fining channel-fills. The architecture records the cutting of channel surface B followed by the migration of compound bars several metres in height in a braided channel system. The inclined beds of alternating siltstone and sandstone (Fig. 6) record the abandonment of the bars with decreasing discharge. Bar troughs were subsequently occupied by a network of smaller channels preserved as the asymmetrical, upward-fining channel-fills. The organization of this channel complex suggests the gradual abandonment of an active and permanent braided channel system.

Channel complex C

Channel complex C is between 1 and 3.5 m in preserved thickness. From south to north its base is formed by surfaces C, C1 and C2. Planar laminated sandstone fines upwards to siltstone, preserved as isolated erosional remnants. The upper part of the complex is characterized by symmetrical lenses of intraformational conglomerate that are truncated at the top of the complex (Figs 2 and 3). The lenses are typically 10 m wide by 2 m deep and some are cross-bedded.

Channel complex C records a renewed phase of erosion that spread from south to north. High-energy, shallow, possibly sheet flows were succeeded by confined, channelized flows bringing coarse intraformational material. The architecture suggests an increasing availability of coarse intraformational material, possibly recording the migration into the area of a channel belt. Given the south to north migration of the basal erosion surface, this phase could have been initiated by tectonic movements, perhaps centred on the Carreg Cennen Disturbance to the north (see Discussion). Alternatively, the conglomerate-plugged channels may have formed in response to a major storm event on the floodplain (see Allen & Williams 1979).

Channel complex D

Channel complexes D and E are inaccessible from the quarry floor. The base of complex D is formed successively from south to north by erosion surfaces D, D1, D2 and D3. Surface D2 defines a symmetrical channel some 30 m wide by 2 m deep. These surfaces are overlain by up to 4 m of planar laminated and cross-bedded sandstone with occasional beds of intraformational conglomerate. The compound cross-bedding and concave-upward channel surfaces are similar to channel complex B, and a return to a similar style of deposition within a sandy braided system is envisaged, although the vertical zonation in complex B is replaced by a lateral back-stepping of erosion surfaces from south to north in complex D.

Channel complex E

This is the highest channel complex preserved in the Sandstone Facies Association. It is up to 2 m thick and characterized by laterally persistent, tabular sandstone beds with some cross-bedding and planar laminae. Several gently inclined internal erosion surfaces are present, and surface E1 replaces E northwards as the base of the complex. Although the lithology of this complex associates it with the Sandstone Facies Association, the style of bedding is similar to parts of the Mudstone Facies Association (see below), which suggests a gradual change in fluvial style.

Architecture of the Mudstone Facies Association

The Mudstone Facies Association makes up the upper third to half of the succession in Pantymaes Quarry and is at least 15 m thick, although its top is not seen. The facies are arranged as laterally extensive beds, extending across the length of the exposed face, except where confined by channel margins.

The contact between the Sandstone Facies Association and the Mudstone Facies Association (surface F, Fig. 2) marks a change in colour from grey to red and a change in dominant lithology from sandstone to mudstone. Surface F is parallel to the bedding in channel complex E in the north, but becomes strongly erosional in the south, cutting down close to the base of channel complex D (Fig. 7), and forming a concave-upward channel feature about 80 m wide and 3 m deep, filled by red mudstones of facies M1 (channel-fill F). A higher erosion surface (surface G, Fig. 7) cuts into these mudstones, defining a channel some 35 m wide, and overlain by mudstones of facies M2 that extend across the whole exposed extent of the Sandstone Facies Association. Both channels are asymmetrical in cross-section, with steeper margins to the north.

Mudstones of facies M1 occur only in channel-fill F, between surfaces F and G. Laminated mudstone in the lower part of the channel-fill is overlain by a compound calcretized profile (*sensu* Wright & Marriott 1996) with three thin nodular horizons interbedded with red mudstone. Surface G is overlain by a thick interval (7.5 m) of mudstone facies M2 that passes up into alternating units of facies M2 and M3 (Fig. 7).

The Mudstone Facies Association represents a major change in depositional style and palaeoenvironment from the Sandstone Facies Association. Channel-fills F and G represent infilling of abandoned channels cut by floodwaters from an adjacent major channel.

Channel-fill F was subject to periodic ponding. The lack of clear evidence of pedogenic development in the laminated sediments of the lowest part of this channel suggests that depositional events were frequent, in the order of 30+ events per year (Wright & Marriott 1996). The compound arrangement of the nodular calcrete horizons at the top of channel-fill F indicates long periods of abandonment leading to pedogenesis, followed by erosive events that removed most of the vertic palaeosol features.

Another phase of major channel erosion and abandonment (rejuvenation) led to the cutting of the upper channel (G). This was filled with overbank deposits followed by a period of slow aggradation across the flood basin. The thick

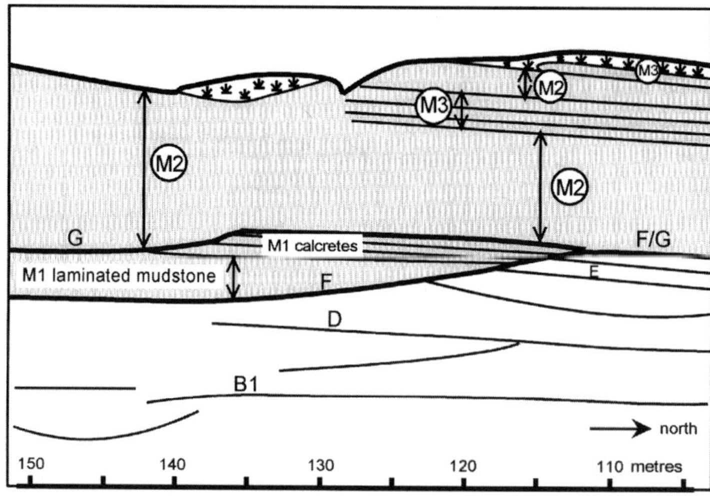

Fig. 7. Erosion surfaces F (between the Sandstone Facies Association and Mudstone Facies Association) and G (within the lower part of the Mudstone Facies Association). Red mudstones of the Mudstone Facies Association are stippled. Labels show mudstone facies M1, M2 and M3, and identify bounding surfaces B1, D and E within Sandstone Facies Association. Scale at the base is as in Fig. 2.

cumulate (Stage 2) palaeosol of facies M2 suggests small increments of sediment input with a low depositional event frequency in the range of 0.3–10 per year (Wright & Marriott 1996). Further evidence for intermittent deposition in facies M2 is given by the intense bioturbation and low ichnodiversity, which imply rapid, widespread and repeated burrowing into the sediments through opportunistic colonization by organisms.

In facies M3 there is no evidence of soil development, suggesting frequent depositional events, in the order of 100+ per year. The greater ichnodiversity in this facies and the preservation of arthropod trackways on the parting surfaces of the laminated component support this interpretation. The coarser sediments and the higher rates of deposition suggest deposition from overbank flows in a near-channel-belt floodbasin setting (Walling *et al.* 1992). The fining-upwards and cyclic arrangement of this facies is interpreted as having been produced by the gradual and increasingly distant lateral migration of a meandering channel, eventually resulting in a brief cessation of sediment input sufficient for the development of desiccation cracks, followed by a reoccupation of the old river channel location at the start of the next cycle. The change in facies from M2 to M3 is interpreted as a fluctuation in depositional rate, produced by a change of channel position within the flood basin resulting from avulsion.

Discussion

The Sandstone Facies Association records deposition within sandy braided river channels with variable discharge. The preservation throughout of plant material suggests that the environment remained sufficiently wet to allow plant growth and preservation, and prevent desiccation. The abundance of calcrete debris records the dominance of a semi-arid climate, allowing pedogenesis on the floodplain. The Mudstone Facies Association records much drier ground conditions, resulting in an abundance of *in situ* calcrete material, the development of desiccation cracks, lack of plant fossil preservation, and less influence of channel processes.

The interpretation of the Sandstone Facies Association has demonstrated major differences in fluvial style between successive channel complexes. This suggests that they represent distinct phases of fluvial development, rather than the punctuated filling of a single fluvial channel: the Sandstone Facies Association represents a complex history of changing fluvial conditions, not a single multi-storey channel sandbody. Extending this argument to the Mudstone Facies Association suggests that this represents another distinct phase of fluvial activity, and not simply overbank deposits to channel deposits of the Sandstone Facies Association. Therefore, despite the exposed succession being only about 25 m thick, we offer an interpretation based on changes in fluvial style influenced by external, allocyclic controls, rather than shifting environments on a single, essentially unchanging floodplain.

The changing pattern of fluvial palaeoenvironments represented at Pantymaes can be summarized thus. An initial development of small, flashy, braided gravelly channels (channel complex A) was followed by the migration into the area of a more stable, sandy braided channel system with large, compound cross-bedded bars (lower part of channel complex B). This system was gradually abandoned (upper part of channel complex B) and after another episode of incision was replaced by a much more flashy regime, culminating in the steep-sided conglomerate-plugged channels of channel complex C. This was succeeded by a return to sandy braided stream conditions (channel complexes D and E). The Mudstone Facies Association is interpreted as overbank sediments deposited in the floodbasin of a meandering river system. At this stage, river channels appear to have been more stable in position, allowing floodplain sediments to dry out. The extensive bioturbation, however, suggests a plentiful supply of floodplain moisture, and the calcrete palaeosols indicate a seasonally wet, semi-arid climate.

The changes in colour (grey–green to red) and dominant grain size (sandstone and conglomerate to mudstone and fine sandstone) between the Sandstone and Mudstone Facies Associations occur abruptly at the level of bounding surface F, but the change in bedding style (channelized to tabular) is more gradual: surfaces F and G preserve channels that are comparable in size and shape with those much lower within the Sandstone Facies Association (Figs 2 and 7). This suggests that the change from a braided channel system to a drier overbank environment occurred gradually, possibly in response to local tectonic or subsidence effects, or to the availability of sediment controlled by more distant tectonic effects. A climatic cause is not favoured, for the following reasons: (1) although *in situ* calcretes occur only in the Mudstone Facies Association, there was a plentiful supply of calcrete material throughout the Sandstone Facies Association, suggesting that calcretes were forming during both periods; (2) arthropod tracks occur in both the Sandstone Facies

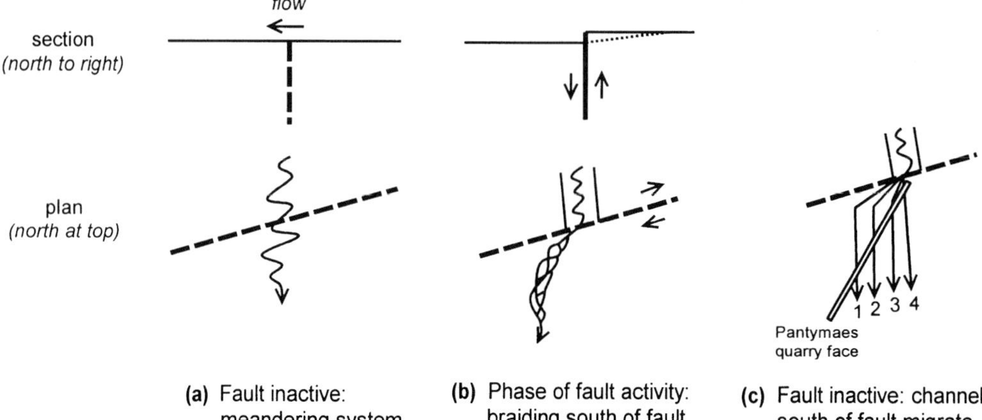

Fig. 8. Model for fluvial development at Pantymaes. (**a**) Carreg Cennen Disturbance inactive: meandering system corresponds to Mudstone Facies Association and 'normal' Red Marl Group sedimentation. (**b**) Phase of fault activity represented by oblique slip movement with uplift to north and dextral strike-slip: channels north of fault become entrenched; increased gradient and coarse intraformational bedload encourage braiding south of fault, corresponding to Sandstone Facies Association. (**c**) Fault activity ceases: channels south of fault migrate consistently eastwards, seen on main quarry face as consistent northward stepping of successive erosion surfaces.

Association and the Mudstone Facies Association; (3) there is evidence for major fluctuations in discharge in both the Sandstone Facies Association and the Mudstone Facies Association. We therefore favour an external, tectonic trigger for the change from the Sandstone Facies Association to the Mudstone Facies Association.

Although it is difficult to reconstruct accurately the geometry and dimensions of the palaeochannels in the Sandstone Facies Association because of the slight angular difference between the general palaeocurrent direction and the strike of the main quarry face, there is a consistent northward stepping of successive erosion surfaces in several channel complexes (Fig. 2). This further suggests a tectonic control on fluvial development, each channel complex representing a response of the fluvial system to a pulse of movement on a nearby fault (Todd & Went 1991; López-Gómez & Arche 1993; Mack & Leeder 1999).

The most likely source of active fault movement in the area is the Carreg Cennen Disturbance (see Fig. 1). Woodcock (1988) and Woodcock & Gibbons (1988) have presented evidence for Devonian activity along this lineament, which probably took the form of small-scale strike-slip and reverse dip-slip displacements. The general pattern of mid-Gedinnian fluvial dispersal systems in Wales is from north to south (Allen & Crowley 1983; Almond *et al.* 1993), such that major river systems would have crossed the Carreg Cennen lineament rather than developing flow parallel to it (Fig. 8a; see Alexander & Leeder 1987; Mack & Leeder 1999).

A tentative model for tectonic control on fluvial behaviour at Pantymaes involves pulses of movement along the Carreg Cennen Disturbance, each resulting in slight uplift to the north (Fig. 8b; see López-Gómez & Arche 1993). Each pulse resulted in a temporary increase in gradient across the fault zone and an input of coarse intraformational bedload to the river system, both factors favouring the development of a braided system with renewed erosive energy (see Leopold & Wolman 1957). A component of dextral strike-slip movement would have displaced the block south of the fault to the SW and allowed consistent migration of successive channels towards the north, as seen within channel complexes B, C, D and E (Fig. 8c; see Todd & Went 1991).

Our model (Fig. 8) therefore envisages each renewed phase of erosion within the Sandstone Facies Association, represented by the bases of the channel complexes, as a response of the river system to activity along the Carreg Cennen Disturbance to the north (see López-Gómez & Arche 1993). We further suggest that the Sandstone Facies Association itself developed because of the onset of local fault activity. Cessation of movements on the Carreg Cennen Disturbance led to a return to more regional tectonic conditions and restoration of the sedimentary conditions and fluvial styles more typical of mid-Gedinnian sediments in the central South Wales region, represented by the Mudstone Facies Association (Fig. 8). The Sandstone Facies Association therefore represents a locally developed tectonic facies in response to a period of active faulting within the sedimentary basin. Its characteristics depend significantly on the fault strike trending at a high angle to the direction of regional drainage.

The characteristics and sedimentological interpretation of the Sandstone Facies Association at Pantymaes are similar to published accounts of the overlying Breconian Senni Beds (Allen 1974; Edwards *et al.* 1978; Loeffler & Thomas 1980; Owen 1995). This raises the possibility that sedimentation of Senni Beds style may represent a recurring facies type, deposited by large, braided, flashy rivers, that is not confined to a particular stratigraphical interval but developed during episodes of intra-basinal tectonics. In areas of limited exposure, such as central South Wales, the interpretation of the mapped boundary between Red Marls and Senni Beds must be treated with care.

Conclusions; new perspectives

(1) Exposures at Pantymaes Quarry lie stratigraphically within the upper (Dittonian) Red Marl Group. A grey–green Sandstone Facies Association is erosively overlain by a red Mudstone Facies Association.

(2) The Sandstone Facies Association is characterized by laterally extensive first-order bounding surfaces that subdivide it into distinct channel complexes, representing the deposits of large, permanent but flashy, braided river systems. Successive channel complexes record different styles of fluvial behaviour.

(3) The Mudstone Facies Association represents overbank deposits, possibly from a meandering river system. Although a channelized, erosional contact separates the Sandstone and Mudstone Facies Associations, the changes across the contact occurred gradually.

(4) The varying fluvial styles within the Sandstone Facies Association and the gradual change to the Mudstone Facies Association suggest that the accumulation of fluvial deposits at Pantymaes Quarry was tectonically controlled, probably in response to movements on nearby Caledonoid fault systems.

(5) The braided channel system at Pantymaes is envisaged as a local response to oblique-slip displacement along a regional fault, which affected the equilibrium of the dominant fluvial system draining across the fault.

(6) The characteristics of the Sandstone Facies Association are similar to those of the overlying Senni Beds. This style of deposition may therefore represent a recurring, tectonically controlled facies within the middle part of the Lower Old Red Sandstone.

Work at Pantymaes Quarry is continuing, with the aim of clarifying the controls on the depositional systems and the palaeoecological significance of the trace fossils.

We acknowledge helpful discussions with D. Briggs, D. Harpur, S. Marriott and J. Richardson. The original manuscript was greatly improved by the constructive comments of J. Almond and an anonymous referee. This work would not have been carried out without the enthusiasm, encouragement and tolerance of the conference organizers P. Friend and B. Williams.

References

Alexander, J. & Leeder, M. R. 1987. Active tectonic control on alluvial architecture. *In*: Ethridge, F. G., Flores, R. M. & Harvey, M. D. (eds) *Recent Developments in Fluvial Sedimentology*. Society of Economic Paleontologists and Mineralogists Special Publications, **39**, 243–252.

Allen, J. R. L. 1974. The Devonian rocks of Wales and the Welsh Borderland. *In*: Owen, T. R. (ed.) *The Upper Palaeozoic and Post Palaeozoic Rocks of Wales*. University of Wales Press, Cardiff, 47–84.

—— 1979. Old Red Sandstone facies in external basins, with particular reference to southern Britain. *In*: *The Devonian System*. Palaeontological Association, Special Papers in Palaeontology, **23**, 65–80.

—— 1983. Studies in fluviatile sedimentation: bars, bar complexes and sandstone sheets (low-sinuosity braided streams) in the Brownstones (L. Devonian), Welsh Borders. *Sedimentary Geology*, **33**, 237–293.

—— 1985. Marine to fresh water: the sedimentology of the interrupted environmental transition (Ludlow–Siegenian) in the Anglo-Welsh region. *Philosophical Transactions of the Royal Society of London, Series B*, **309**, 85–104.

—— & Crowley, S. F. 1983. Lower Old Red Sandstone fluvial dispersal systems in the British Isles. *Transactions of the Royal Society of Edinburgh: Earth Sciences*, **74**, 61–68.

—— & Williams, B. P. J. 1979. Interfluvial drainage on Siluro-Devonian alluvial plains in Wales and the Welsh Borders. *Journal of the Geological Society, London*, **136**, 361–366.

Almond, J., Williams, B. P. J. & Woodcock, N. H. 1993. The Old Red Sandstone of the Brecon Beacons to Black Mountain area. *In*: Woodcock, N. H. & Bassett, M. G. (eds) *Geological Excursions in Powys, Central Wales*. University of Wales Press, National Museum of Wales, Cardiff, 311–330.

Barclay, W. J. 1989. *Geology of the South Wales Coalfield, Part II, the country around Abergavenny*. Memoir of the British Geological Survey, Sheet 232 (3rd edn). HMSO, London.

Bluck, B. J., Cope, J. C. W. & Scrutton, C. T. 1992. Devonian. *In*: Cope, J. C. W., Ingham, J. K. & Rawson, P. F. (eds) *Atlas of Palaeogeography and Lithofacies*. Geological Society, London, Memoir, **13**, 57–66.

Droser, M. L. & Bottjer, D. J. 1986. A semi-quantitative field classification of ichnofabric. *Journal of Sedimentary Petrology*, **56**, 558–559.

Edwards, D., Richardson, J. B. & Thomas, R. G. 1978. Locality B9: Heol Senni Quarry, Powys. *In*: Friend, P. F. & Williams, B. P. J. (eds) *A Field Guide to Selected Outcrop Areas of the Devonian of Scotland, the Welsh Borderland and South Wales. International Symposium on the Devonian System (PADS 78), September 1978*. Palaeontological Association, London, 77–78.

Hassan, A. 1982. *Palynology, stratigraphy, and provenance of the Lower Old Red Sandstone of the Brecon Beacons (Powys) and the Black Mountains (Gwent and Powys)*. PhD thesis, University of London.

Hemsley, A. R. 1990. *Parka decipiens* and land plant spore evolution. *Historical Biology*, **4**, 39–50.

House, M. R., Richardson, J. B., Chaloner, W. G., Allen, J. R. L., Holland, C. H. & Westoll, T. S. 1977. *A Correlation of Devonian Rocks in the British Isles*. Geological Society, London, Special Report, **8**.

Leopold, L. B. & Wolman, M. G. 1957. *River Channel Patterns; Braided, Meandering, and Straight*. US Geological Survey Professional Paper, **282-B**.

Loeffler, E. J. & Thomas, R. G. 1980. A new pteraspidid ostracoderm from the Devonian Senni Beds Formation of South Wales and its stratigraphic significance. *Palaeontology*, **23**, 287–296.

López-Gómez, J. & Arche, A. 1993. Architecture of the Cañizar fluvial sheet sandstones, Early Triassic, Iberian Ranges, eastern Spain. *In*: Marzo, M. & Puigdefábregas, C. (eds) *Alluvial Sedimentation*. International Association of Sedimentologists, Special Publications, **17**, 363–381.

Machette, M. N. 1985. Calcic soils of the southwestern United States. *In*: Weide, D. L. (ed.) *Soils and Quanternary Geology of the South West United States*. Geological Society of America, Special Papers, **203**, 1–21.

Mack, G. H. & Leeder, M. R. 1999. Climatic and tectonic controls on alluvial-fan and axial-fluvial sedimentation in the Plio-Pleistocene Palomas half graben, southern Rio Grande rift. *Journal of Sedimentary Research*, **69**, 635–652.

Miall, A. D. 1992. Alluvial deposits. *In*: Walker, R. G. & James, N. P. (eds) *Facies Models: Response to Sea Level Change*. Geological Association of Canada, Toronto, Ont., 119–142.

Owen, G. 1995. Senni Beds of the Devonian Old Red Sandstone, Dyfed, Wales: anatomy of a semi-arid floodplain. *Sedimentary Geology*, **95**, 221–235.

Ramsay, A. T. S. 1989. Tectonics and sedimentation of late Dinantian limestones in South Wales. *In*: Arthurton, R. S., Gutteridge, P. & Nolan, S. C. (eds) *The Role of Tectonics in Devonian and Carboniferous Sedimentation in the British Isles*. Yorkshire Geological Society Occasional Publication, **6**, 225–241.

Todd, S. P. & Went, D. J. 1991. Lateral migration of sand-bed rivers: examples from the Devonian Glashabeg Formation, SW Ireland and the Cambrian Alderney Sandstone Formation, Channel Islands. *Sedimentology*, **38**, 997–1020.

Tunbridge, I. P. 1981. Sandy high-energy flood sediments—some criteria for their recognition, with an example from the Devonian of S.W. England. *Sedimentary Geology*, **28**, 79–95.

Walling, D. E., Quine, T. A. & He, Q. 1992. Investigating contemporary rates of flood plain sedimentation. *In*: Carling, P. A. & Petts, G. E. (eds) *Lowland Floodplain Rivers: Geomorphological Perspectives*. Wiley, Chichester, 165–184.

Weaver, J. D. 1975. The structure of the Swansea Valley Disturbance between Clydach and Hay-on-Wye, South Wales. *Geological Journal*, **10**, 75–86.

—— 1976. Seismically-induced load structures in the basal Coal Measures, South Wales. *Geological Magazine*, **113**, 535–543.

Woodcock, N. H. 1988. Strike-slip faulting along the Church Stretton Lineament, Old Radnor Inlier, Wales. *Journal of the Geological Society, London*, **145**, 925–933.

—— & Bassett, M. G. (eds) 1993. *Geological Excursions in Powys, Central Wales*. University of Wales Press, Cardiff.

—— & Gibbons, W. 1988. Is the Welsh Borderland Fault System a tearrane boundary? *Journal of the Geological Society, London*, **145**, 915–923.

Wright, V. P. & Marriott, S. B. 1996. A quantitative approach to soil occurrence in alluvial deposits and its application to the Old Red Sandstone of Britain. *Journal of the Geological Society, London*, **153**, 907–913.

Fault-bounded basin fill: fluvial response to tectonic controls in the Skrinkle Sandstones of SW Pembrokeshire, Wales

J. D. MARSHALL

Shell Research and Technical Services (EPT-AE), Postbus 60, 2280 AB, Rijswijk, The Netherlands (e-mail: j.d.marshall@siep.shell.com)

Abstract: The Upper Devonian to Lower Carboniferous Skrinkle Sandstones of the Pembroke Peninsula are predominantly continental deposits from the post-Caledonian syn-rift succession at the southern margin of the Late Palaeozoic Welsh Landmass. The Sandstones record deposition in the 30 km × 10 km Tenby–Angle fault block, the southernmost of a series of fault-bounded depositional basins in SW Dyfed. Activity on the bounding faults strongly influenced sedimentation through Lower Palaeozoic time. The Skrinkle Sandstones are conventionally assigned to a phase of relative fault inactivity, passive transgression of the area and southward drainage off the landmass. The Ritec Fault at the northern block boundary defined a temporary shoreline during final submergence. In contrast to this, it is argued that the lower half of the Skrinkle Sandstones represent a separate structural configuration, where SE-directed palaeocurrents and the high textural maturity of two superimposed basin-fill sequences indicate axial basin fill and potential closure to the south. The upper half records an influx of immature clastic deposits as fluvial sediments that disperse to the southwest, indicating relative uplift of either the Ritec or a more northerly fault. This phase records the true transgression, during which a thick barrier–lagoon coastline is preserved against the footwall ramp of the Ritec Fault.

The Upper Devonian to Lower Carboniferous Skrinkle Sandstones of the Pembroke Peninsula in southwest Wales record the end of Devonian red-bed deposition as a transition through grey–red interbeds into grey deposits of more Carboniferous affinity. They have been extensively studied for determination of the Devonian–Carboniferous boundary (Dolby 1971; Bassett & Jenkins 1977; Higgs *et al.* 1988) and are considered to represent a response to progressive subsidence and probable sea-level rise (Clayton *et al.* 1986) at the margin of the Rhenohercynian Trough.

Seen from a Devonian perspective, the Skrinkle Sandstones are part of a post-Caledonian syn-rift succession at the southern margin of the Late Palaeozoic Welsh Landmass (Fig. 1). The succession accumulated in fault-bounded basins as alluvial fan, alluvial plain and lacustrine deposits whose thicknesses and grain-size trends are thought to reflect growth of the bounding faults with time (Powell 1987, 1989). The Upper Devonian sequence is preserved only in the southerly Tenby–Angle basin, whose northern margin was defined by the active Ritec Fault (Fig. 2) and whose southern margin lay at least 10 km to the south, in the present-day Bristol Channel. The fault activity in the basin allowed unusually complete preservation of continental margin sedimentation during the Carboniferous transgression. The resultant southward-thickening sediment wedge, the Skrinkle Sandstones, has been considered to record a relatively subdued phase of fault activity before a final pulse later in Early Carboniferous time (Powell 1989).

This paper proposes three phases of structural growth and sedimentary response in Late Devonian time, and indicates the probable continued presence of a basin boundary to the south in the first phase. Use is made of contrasting groups of facies associations and their inferred environments of deposition, palaeocurrent directions and their inferred palaeo-drainage directions, and lithological compositions with their inferred provenances.

Geological setting

The Skrinkle Sandstones were defined by Dixon (1921) to encompass the uppermost Devonian red-bed succession in the Pembroke Peninsula, lying unconformably on the presumed Middle–Lower Devonian Ridgeway Conglomerate and passing up gradationally into the Tournaisian Lower Limestone Shales (Fig. 3). Dixon (1921)

From: Friend, P. F. & Williams, B. P. J. (eds). *New Perspectives on the Old Red Sandstone.* Geological Society, London, Special Publications, **180**, 401–416. 0305-8719/00/$15.00

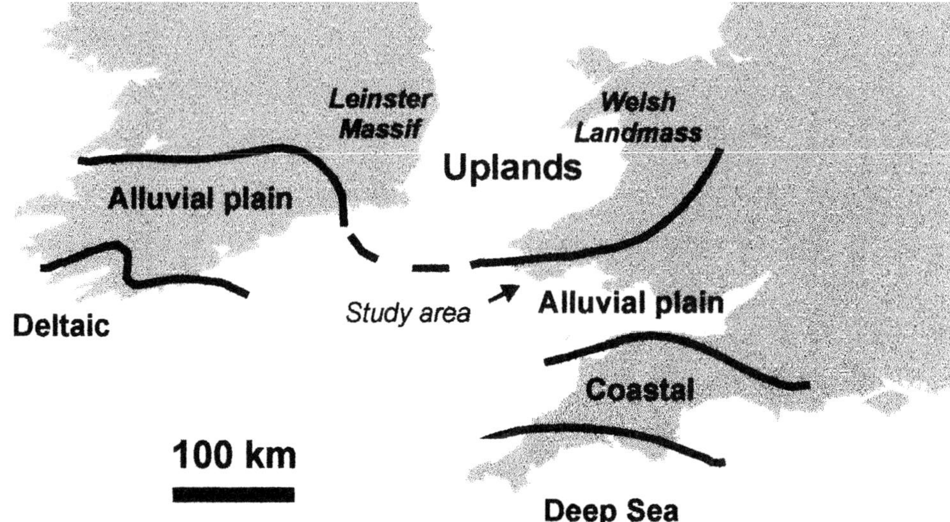

Fig. 1. Late Famennian palaeogeography, SW Britain and southern Ireland. Adapted from Webby (1966), MacCarthy (1990) and Cope *et al.* (1992).

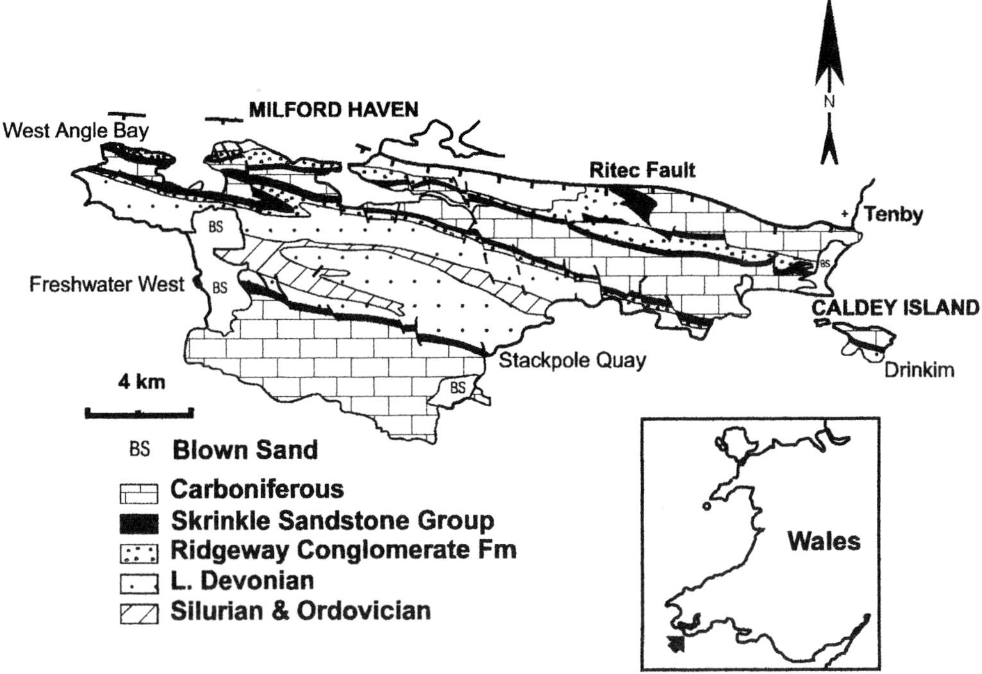

Fig. 2. Geological map of the Pembroke Peninsula, Wales.

established a late Devonian age for the base of the Skrinkle Sandstones from scales of *Holoptychius* sp., a crossopterygian fish of Frasnian–Famennian age. The succession was subsequently interpreted as a large-scale fining-upwards system, starting with braided stream sandstones and ending in barrier coastline heterolithics deposits (Allen 1965, 1974; Hassan 1966). The topmost heterolithic units can be dated as belonging to the Early Carboniferous Courceyan

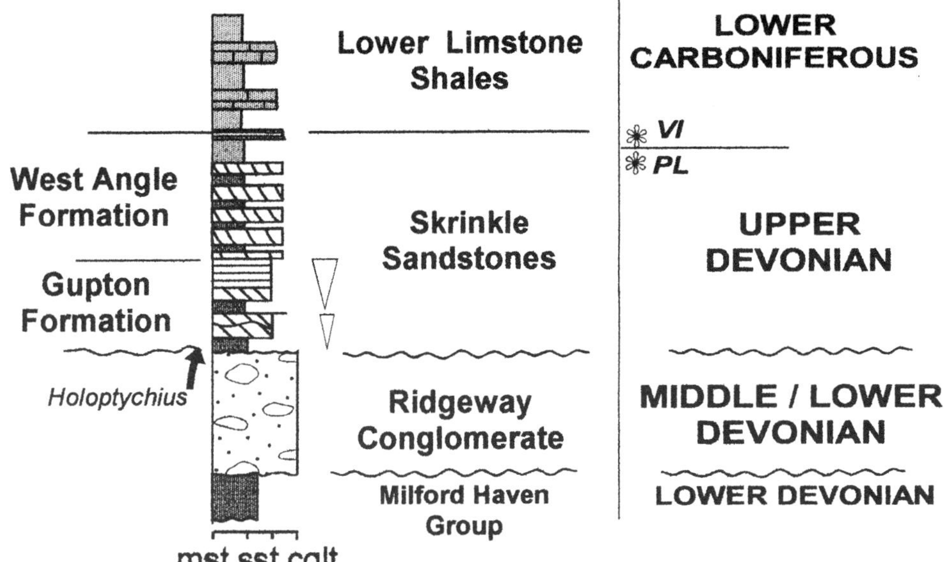

Fig. 3. Stratigraphic summary of the Skrinkle Sandstones. Positions of key spore zone samples indicated in relation to Devonian–Carboniferous boundary and Formation boundary. Position of *Holoptychius* sample similarly indicated.

Stage on the basis of spore data (Bassett & Jenkins 1977). The succession thickens southward from 100 m close to the Ritec Fault to 330 m at Freshwater West, the southern limit being lost below the Bristol Channel. During Famennian time the area is inferred to have been at latitude 5–8° S (Ziegler 1990). The Skrinkle Sandstones are informally divided into the Gupton Formation and the overlying West Angle Formation on the basis of compositional, palaeocurrent and facies contrasts (Marshall 1977). More recent publications are limited to field guides with some details of logged sections (Williams *et al.* 1982).

Sedimentological and structural research during the 1970s–1980s in southern Pembrokeshire has developed a picture for pre-Variscan sedimentation in which a series of long-lived and intermittently active extensional faults defined structural blocks with distinct fills. From north to south these were the Benton, Wenall and Ritec Faults, defining between them the Broad Haven, Winsle and Tenby–Angle fault blocks (Fig. 4; Dunne 1983; Powell 1987, 1989). Associated thrust faults in the area were shown to be later features resultant from Variscan tectonics. Recognition of the Townsend Tuff ash-fall sequence in the Lower Devonian succession allowed the establishment of a datum that extended across the blocks (Allen & Williams 1981). The next stratigraphically continuous horizon across the fault blocks was the Lower Carboniferous Lower Limestone Shales. With the help of facies distributions and cross-fault thickness changes, different phases of tectonically driven depositional activity were identified through Silurian to Early Carboniferous time, with all three faults acting as intermittent sources for basin-margin alluvial fans or as limits to deposition (Fig. 5 and Powell 1987, 1989).

A second stream of activity had established the presence of a Devonian positive area in what is now the Bristol Channel, on the basis of facies and palaeocurrent trends, source rock studies and geophysical information (Mechie & Brooks 1984; Tunbridge 1986; Cope & Bassett 1987). This 'Bristol Channel Landmass' is delineated in part by the modern Bristol Channel Fault Zone (BCFZ), which extends south of the study area and westward and into the Celtic Sea (Fig. 6). Seismic lines across the BCFZ indicate a number of synthetic faults running parallel to the main southward-dipping fault (Kamerling 1979). As with onshore equivalents, some of these faults are younger Variscan or Tertiary features, but others may be reactivated Devonian and older structures. One of these faults could represent the southern margin of the Tenby Angle fault block, which has been postulated at an offshore position since the recognition of a southerly source for the

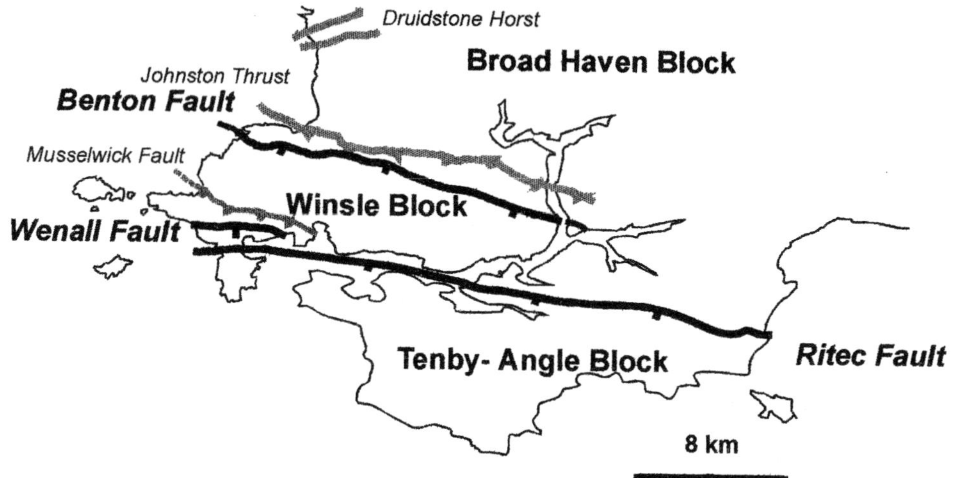

Fig. 4. Fault blocks of SW Dyfed. Based on Powell (1989). Post Devonian (Variscan) Musselwick, Johnston and Druidstone Faults in grey.

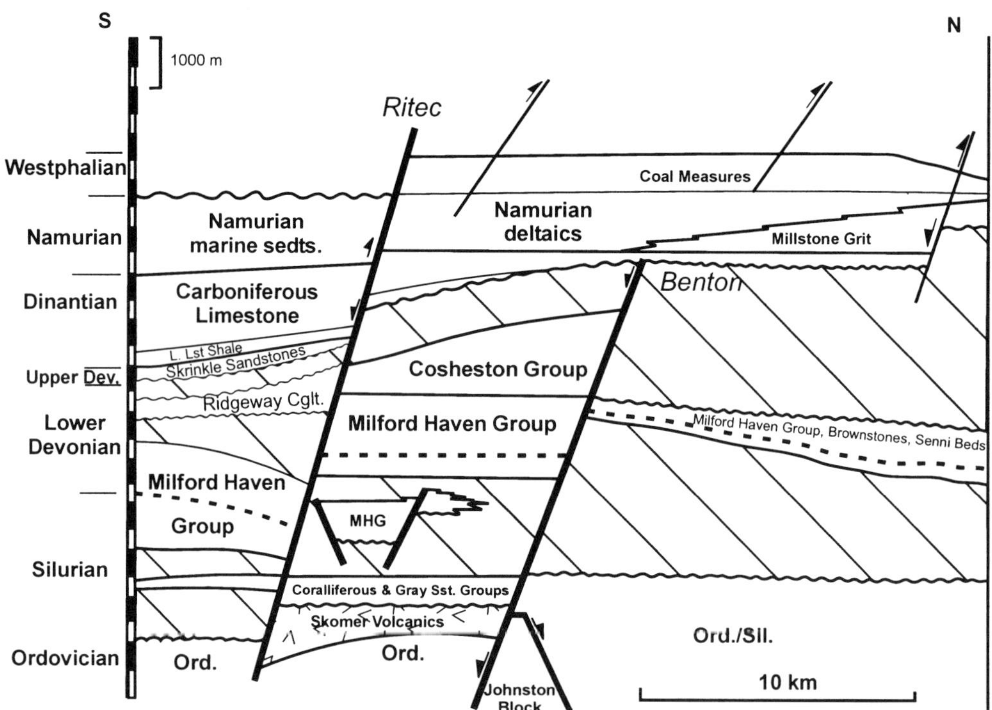

Fig. 5. Facies–tectonism chart of the Pembrokeshire Palaeozoic succession to demonstrate the relationship of stratigraphy within fault blocks to the major bounding faults with time. Diagonal shading indicates hiatus, dashed line in the Milford Haven Group indicates the Townsend Tuff Bed. Adapted from Powell (1987, 1989).

Lower–Middle Devonian Ridgeway Conglomerate (Williams 1964; Tunbridge 1986).

Deformation in the late Carboniferous Variscan orogeny compressed the southern Pembrokeshire area into east–west-trending folds and reactivated the Ritec Fault and other faults into reverse or thrust features (Powell 1989). In consequence, the rocks studied are steeply dipping to overturned and crop out almost exclusively in dip-oriented coastal

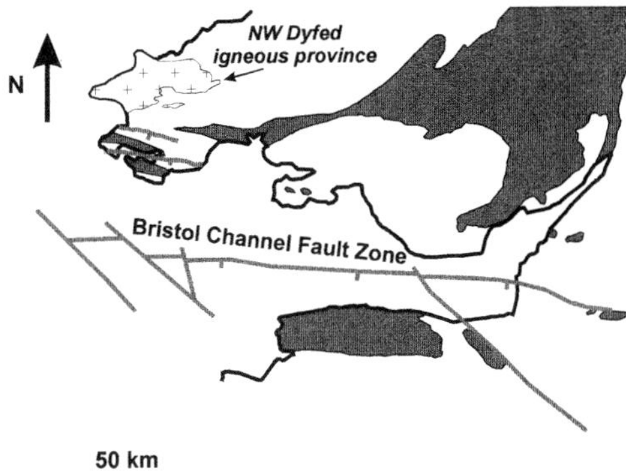

Fig. 6. The Bristol Channel Fault Zone. Based on Kamerling (1979) and Tunbridge (1986). Devonian outcrop shaded, northwest Pembrokeshire igneous province patterned.

sections. It is also likely that a 40° clockwise rotation was imposed on the area south of the Benton Fault during the Variscan event (McClelland Brown 1983). Where palaeocurrents are represented here they have been corrected for structural dip and a *c.* 10° eastward plunge, but not for the clockwise block rotation. However, the conclusions drawn from the palaeocurrent data would not be significantly altered as the faults and the depositional basins were rotated together. The readings are from trough cross-bed laminae, grouped in cosets.

The first phase: axial basin fill

The Gupton Formation rests unconformably on the Lower–Middle Devonian Ridgeway Conglomerate, overstepping it to the east to lie directly on the older, Upper Silurian to Lower Devonian Milford Haven Group. Above the unconformity, the formation comprises two successive progradational sequences, each starting in red mudstones and coarsening upward to clean well-sorted sandstones (Fig. 7). The first sequence, the Lower Sandstone Member, is exposed only in the southernmost outcrop at Freshwater West. It is 55 m thick and shows an upward change through three facies associations: from small multi-storey sandstones set in a background of mudstone and siltstone, to thicker single-storey sandstones and eventually stacked pebbly sandstone-based fining-upward units (Fig. 8). The three associations have been interpreted as small sheetflood systems set in lacustrine or floodplain mudstones, overlain by isolated channel fills and finally by thick meandering channel deposits (Marshall 1977). The coarsening and thickening-upward nature of the succession is suggestive of a distal to proximal change and may represent a prograding terminal fan system. Palaeocurrents indicate transport to the southeast, into the topographic low of the Tenby–Angle fault block.

The second sequence, the Stackpole Sandstone Member, oversteps both the Lower Sandstone Member and the Ridgeway Conglomerate to extend across the entire South Pembroke Peninsula. It is 6 m thick close to the Ritec Fault, thickening to 68 m at Freshwater West (Fig. 9), and is composed of a lower mudstone-rich heterolithic facies association and an upper trough cross-bedded and parallel-laminated sandstone facies association. Vertical trends and sedimentary structures have led to an interpretation as a lacustrine interval that was eventually infilled by a high-energy sandy braidplain (Fig. 10 and Marshall 2000). Palaeocurrents again indicate southeasterly transport directions.

These two sequences indicate successive phases of basin fill, both with transport directions subparallel to the regional east–west fault trend. They also share a similar petrology. The sandstones in both are mature sublitharenites with sparse admixtures of devitrified acid volcanic quartz (Fig. 11b), suggestive of long transport paths rather than local sourcing from a nearby fall line defined by the Ritec or Benton Fault. The lacustrine deposits suggest fault or dip slope closure to the east and south, such that the Bristol Channel fault system could have acted as a topographic limit at the time although not as a major sediment source.

The second phase: transverse basin fill

The onset of the West Angle Formation is marked by a significant textural and compositional change, to red sandstones and conglomerates rich in lithic debris including both acid igneous clasts and lithified sandstones and phyllites. The formation has an erosional base in northern exposures and a 2–5 m vertical transition in the south suggestive that there was little time gap from the underlying Gupton Formation. The sedimentary style also changes, to more 'classical' Old Red Sandstone coarse–fine couplets with abundant calcretes in the fine members (Fig. 12). These dominate the lower half of the formation, termed the Conglomerate Member. Conglomeratic channel fills and sheets are particularly abundant near the Ritec Fault, but thick conglomeratic beds are still recorded at Freshwater West, set in a higher percentage of thick sandstones and muddy siltstones. Pebbles near the Ritec Fault are

Fig. 7. Cliff section of the Gupton Formation at Freshwater West, beds younging to the right. (**a**) The Lower Sandstone Member. R, Ridgeway Conglomerate; MS, multi-storey sandstone association; LS, lenticular sandstone association; TS, tabular sandstone association; Sl, Stackpole Sandstone Member, siltstone association. Cliff height 30 m, width of field of view 80 m. (**b**) Continuation of section, boundaries arrowed. Stackpole Sandstone Member: Sl, siltstone association; SS, sandstone association. C, Conglomerate Member. Width of field of view 65 m.

sub-angular, with roundness increasing southwards. The recorded palaeocurrent directions are south and southwest directed, with considerable variability between successive sandstones and between outcrops.

Fine examples of lateral accretion sets have been recorded from both conglomerate-based and sandstone-based sequences (Fig. 13), indicating point-bar development and meandering channels. Thin beds of intraformational

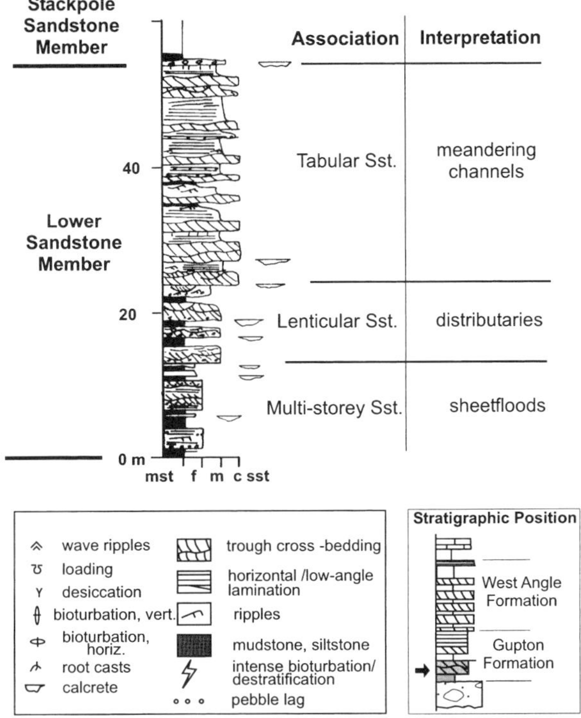

Fig. 8. Schematic log of the Lower Sandstone Member, Gupton Formation, at Freshwater West.

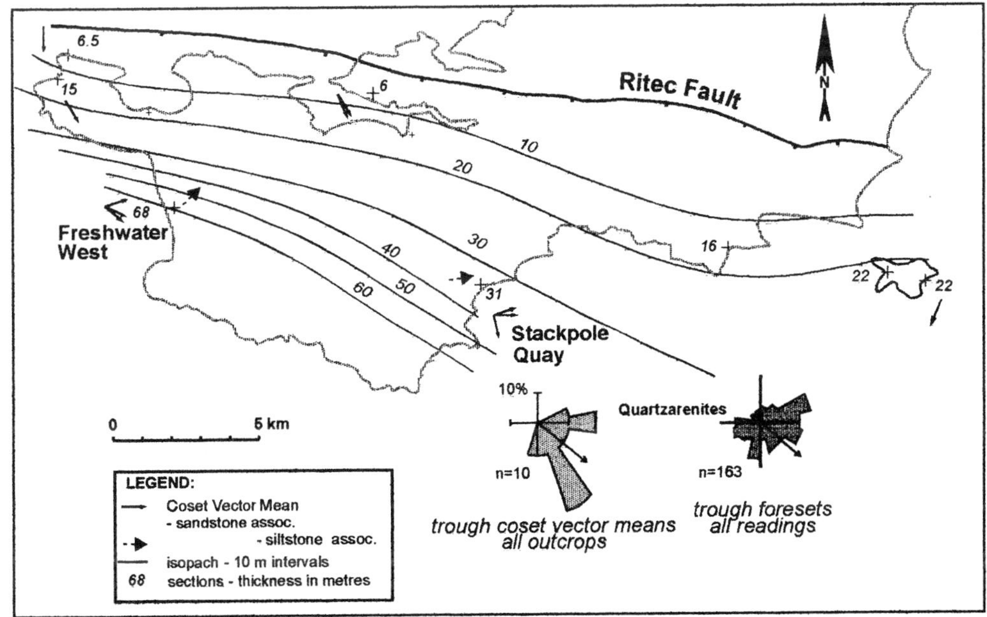

Fig. 9. Isopachs and palaeocurrents of the Stackpole Sandstone Member, Gupton Formation.

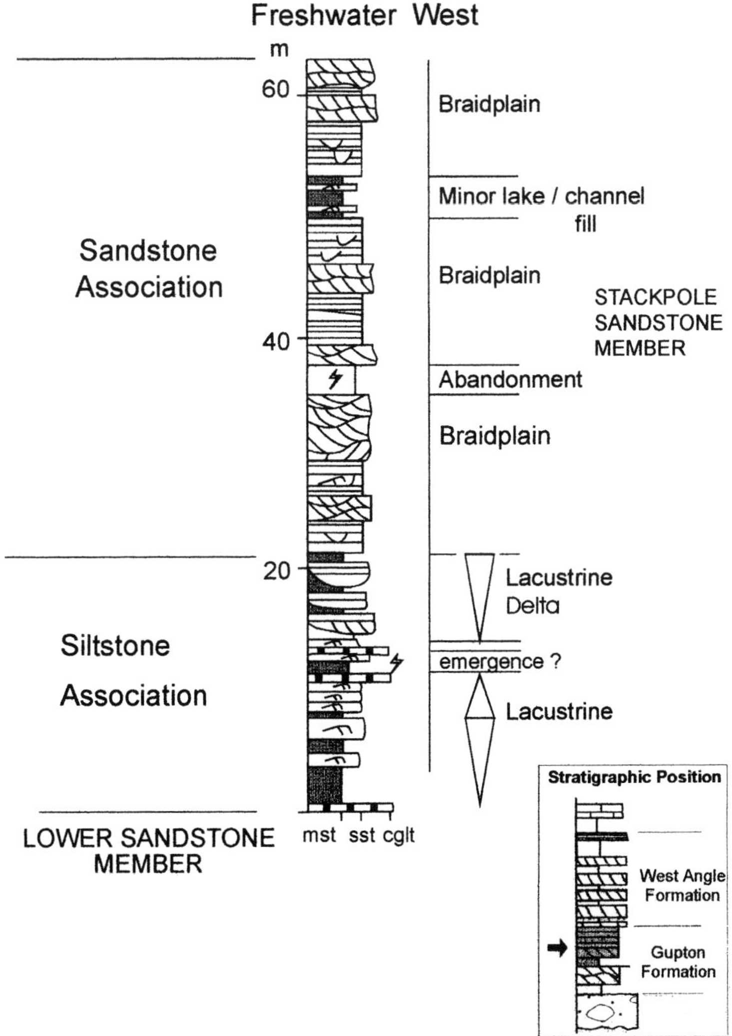

Fig. 10. Stackpole Sandstone Member, schematic log from Freshwater West.

Table 1. *West Angle Formation clasts*

Igneous	Metamorphic	Sedimentary and other
rhyolites, felsites, trachytes, keratophyres (spilitised trachytes), crystal and vitric tuffs, microgranophyre	quartz–mica tectonites, muscovite–chlorite–quartz phyllites, schists	sub-arkosic arenites, vein quartz, jasper

granule- to pebble-grade clasts overlain by rippled sandstones occur in the mudstone-dominated intervals of the formation and are suggestive of floodplain sheetfloods (Fig. 12, e.g. at 10 m). At West Angle Bay a number of tabular, sharp-based and topped extraformational conglomerate beds suggests that gravel-grade sheetfloods could have contributed to deposition close to the basin margin (Fig. 12, 40 m). There is, however, no indication of true alluvial fan development (e.g. mudflow deposits, radial palaeocurrents and downstream reductions in maximum clast size). The deposits are interpreted as the products of meandering

Fig. 11. Gupton Formation, (**a**) *Holoptychius* sp., a weathered cycloid scale from the base of the Lower Sandstone Member. IGS specimen 1812. Width of specimen 2 cm. (**b**) Thin section showing dominantly sub-rounded monocrystalline quartz grains with sericitic surrounds; scale bar represents 1 mm; cross-polarized light.

channel systems in a semi-arid floodplain that spread southwards from the Ritec or Benton Fault, with accommodation space developing in the Tenby–Angle fault block. Close to the fault the depositional style shows the influence of sheetflooding.

The pebbles are dominated by igneous clasts (Fig. 14), with subordinate metamorphic and sedimentary types (Table 1). The igneous pebble suite is dominated by acid extrusive rocks and is comparable with the Precambrian Pebidian volcanic complex of the St David's Bay area (Baker 1982) and the associated Ordovician volcanic complexes between St David's and Fishguard on the northern Pembrokeshire coast (Bevins & Roach 1980) (Fig. 6). Similar pebble assemblages are also known from the Lower Devonian Cosheston Group north of the Ritec Fault (Allen *et al.* 1982). It is therefore likely that the Skrinkle pebbles are derived from the North Pembrokeshire igneous complexes, either directly or by recycling of the Cosheston Group clasts.

The third phase: submergence

The upper half of the West Angle Formation is here termed the Red–Grey Member and records the onset of the Carboniferous transgression. It is marked first by the appearance of occasional grey–green sandstones and pebbly sandstones with preserved plant debris. These occur at the

Table 2. *Faunal assemblages of the Carboniferous transgression*

Name	West Angle Formation	Lower Limestone Shales	Assemblage
Brachiopods			
Cyrtospirifer aff. *verneuili*	+	–	'Devonian'
Leiorhynchus laticosa	+	–	
Macropotamorhynchus sp.	+	+	Devonian and Carboniferous
Lingula sp.	+	+	
Orbiculoidea nitida	+	+	
Spirifer tornacensis	–	+	'Carboniferous'
Rugosochonetes hardrensis	–	+	
Cleiothyridina royssii	–	+	
Leptagonia analoga	–	+	
Schizopora resupinata	–	+	
Bivalves			
Dolbara (Cucculaea) unilateralis	+	–	'Devonian'
Prothyris contorta	+	–	
Sanguinolites minimus	+	–	
Ptychopteria damnoniensis	+	–	
Cypricardinia sp.	+	–	
Actinopteria sp.	+	–	
Myophoria deltoidea	+	+	'Carboniferous'
Aviculopecten sp.	+	+	
Ctenodonta sp.	+	+	
Modiola lata	+	+	

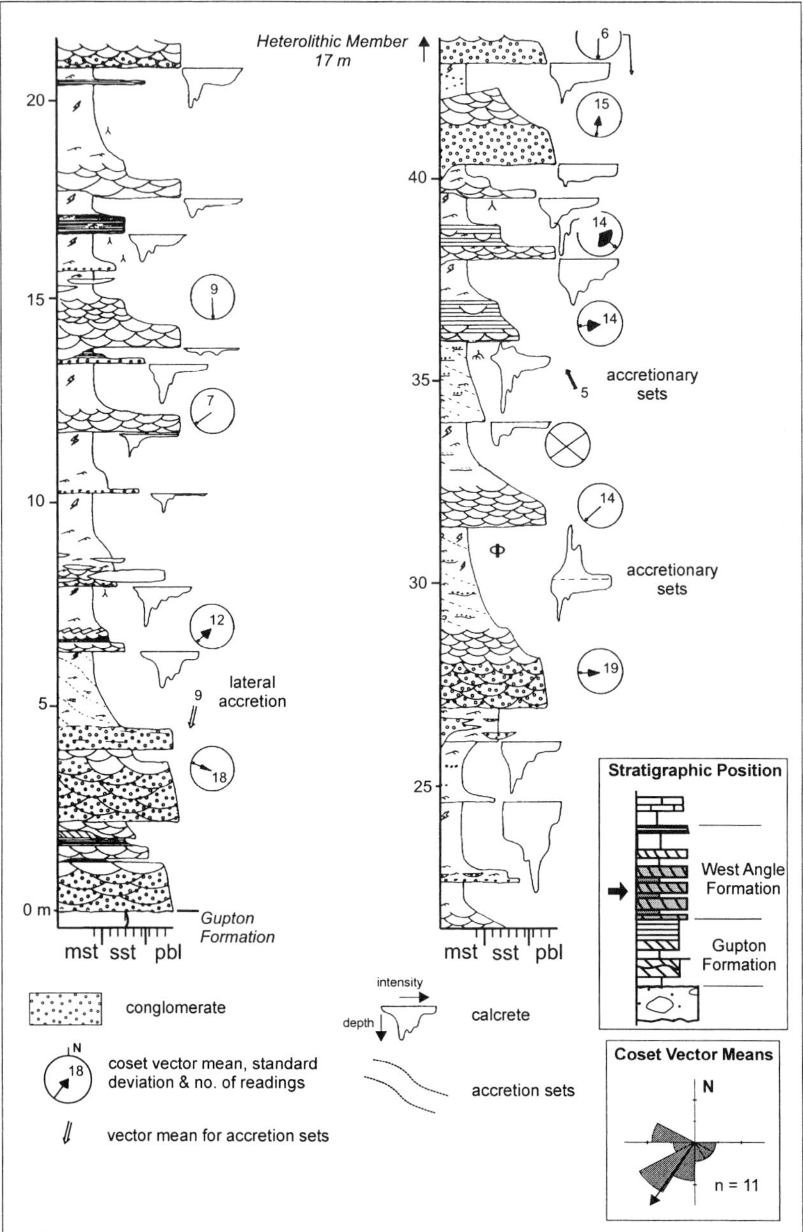

Fig. 12. Lithological log of the Conglomerate Member, West Angle Formation, West Angle Bay (north). Coset vector means plot (inset) shows variability of transport direction between fining-upwards sequences at this outcrop.

bases of dominantly red-bed fining-upwards sequences, with mudstone and calcrete tops reminiscent of the underlying fluvial deposits of the formation. A comparable depositional environment is likely, but with a higher water table allowing the preservation of plant material.

Progressively, grey mudstones appear, with freshwater microfossils, coaly plant material and lingulid brachiopods (Fig. 15), suggesting local lake development. Occasional sheets with phosphatized pebbles, bryozoa and shark's teeth indicate sporadic input from the open sea.

Fig. 13. Outcrop photographs of the Conglomerate Member. (**a**) Fining-upwards units with lateral accretion sets, Stackpole Quay. Section youngs to left; metre ruler lower centre for scale. (**b**) Conglomerate with capping heterolithic lateral accretion sets, West Angle Bay. Scale bar 4 m; section youngs to right.

Fig. 14. Thin-section photographs of the West Angle Formation. (**a**) Trachytic rock fragment, with large feldspar phenocryst in microcrystalline background, scale bar represents 1 mm; cross-polarized light. (**b**) Vitric tuff with residual curved glass-shard textures, scale bar represents 1 mm; plane-polarized light.

Near the top of the formation, a very varied assemblage of sediments includes red- and grey-coloured sandstone and mudstone interbeds with a fauna having similarities to those of clastic intervals at the Devonian–Carboniferous boundary in North Devon (Bassett & Jenkins 1977). In most outcrops the formation is terminated by a mature calcareous sandstone with well-rounded quartz pebbles and sparse marine brachiopods and bryozoa (Fig. 16). Above this sandstone the carbonates and mudstones of the Lower Limestone Shales develop, containing a markedly different faunal assemblage (Table 2).

The calcareous sandstone at the top of the Red–Grey Member represents a laterally extensive sand body separating faunal assemblages and depositional styles (wherever marine indicators such as crinoid ossicles, orthoceratids and shark's teeth are found below the sandstone, they are in erosively based beds, distinct from surrounding sediments, and containing well-rounded quartz pebbles: Fig. 17). The calcareous sandstone provides evidence of palaeocurrents indicating northerly directed transport. These features lead to its interpretation as a coastal

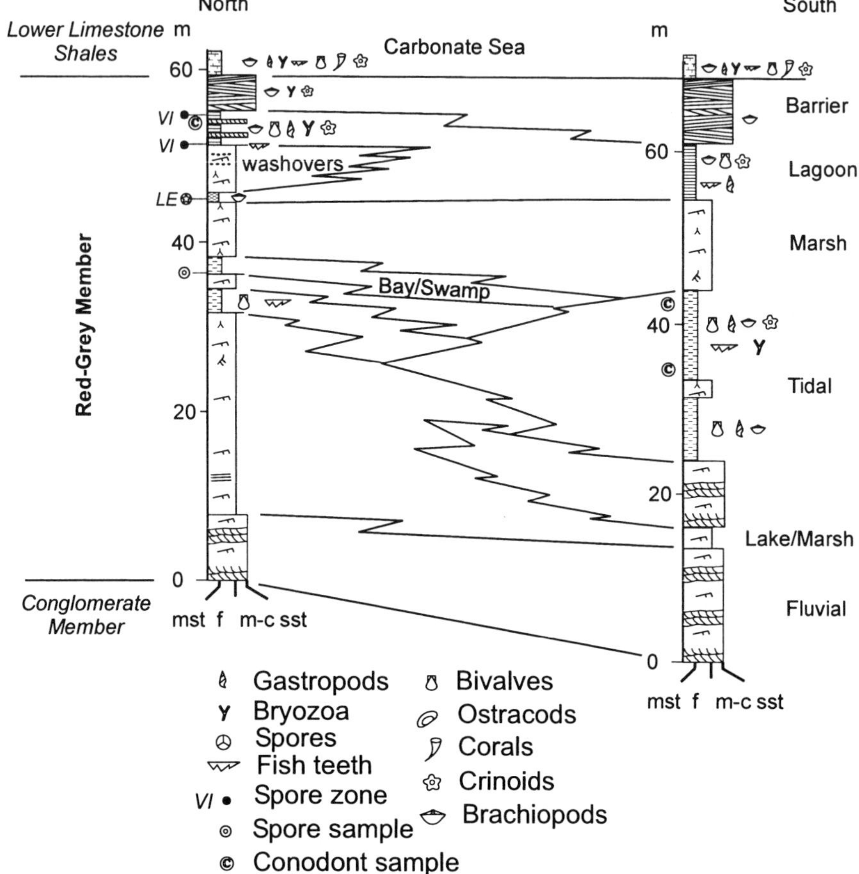

Fig. 15. The Red–Grey Member of the West Angle Formation: schematic section and interpretation of depositional environments across West Angle Bay.

barrier sand, which supplied sediment landwards into protected lagoonal areas as storm washovers. The diversity of sediments underlying the sandstone represent lagoonal and washover fan deposits and (at West Angle South) tidal-inlet and tidal-flat sands (summarized in Fig. 15). The faunal assemblage is lagoonal, subsequently replaced by the more open-marine fauna of the Lower Limestone Shales. Both the barrier sand and one of the underlying washover sands indicate northerly (onshore) directed palaeocurrents, in contrast to the southerly directed fluvial palaeocurrents indicated in the member.

In terms of stratigraphic age, the spore data of Dolby (1971) and Bassett & Jenkins (1977) at West Angle Bay identify the lower, possibly tidal, heterolithic units to be of the Devonian PL zone (sample D1 of Fig. 14, now LE zone, Higgs *et al.* 1988) but the topmost lagoonal interval to be of the Carboniferous VI zone (sample B of Fig. 17). Clayton & Higgs (1979) have demonstrated that these spore zones are independent of the sedimentary facies in which they are found, supporting their use as time markers. Transgression of the area therefore spans these zones and was not completed until after the onset of the Carboniferous time.

Throughout the Red–Grey Member, the sandstone composition is sublitharenitic and comparable with that of the underlying fluvial succession. Conglomerate beds continue to occur and to contain sub-rounded clasts. The alluvial palaeocurrent directions are southerly directed as before, implying the presence of a continued palaeoslope southwards from the Ritec Fault. It is notable that as the Carboniferous transgression moved northwards, the alluvial plain did not, there being no alluvial sediments preserved north of the Ritec Fault. This indicates a basin margin at the Ritec or Benton Fault that was

(a) (c)

(b) (d)

Fig. 16. Photographs of the Red–Grey Member (**a**) Barrier–lagoon section from West Angle Bay south. Red bioturbated flood–plain siltstones at base (N) are sharply overlain by grey bioturbated lagoonal heterolithic units (L) and then by parallel-laminated and cross-bedded barrier sandstones (B). Scale bar is 1 m long. (**b**) Barrier–lagoon section at West Angle Bay north from 50 m–60 m in Fig. 17; section youngs to left. Red floodplain siltstones to right (R); grey shelly lagoonal heterolithic deposits (L) centre; thick grey shelly sandstones of barrier (B). Scale bar is 4 m long. (**c**) Dolomitic algal laminite from West Angle south lagoonal sequence. (**d**) Cross-bedded and low-angle cross-laminated calcareous sandstones from West Angle north barrier sequence, at 56 m in Fig. 17. Scale rule is 0.5 m long.

only later overstepped, in later Courceyan time. The preservation of a thick (60–70 m) barrier–lagoon sequence in the hanging wall of the Ritec Fault testifies to differential subsidence at the time that was not matched in the more northerly Winsle Block, where the Courceyan Lower Limestone Shales have an erosional base on Lower Devonian rocks.

Synthesis: sedimentation and tectonics

A summary of the inferred interplay between tectonics and sedimentation is presented in Fig. 18. After a period of uplift and erosion during Middle- to Late Devonian time, the Gupton Formation was deposited in the Tenby–Angle Block. The formation was produced by an eastward-directed axial depositional system, with the Ritec Fault determining the northern margin of the depositional area. A synthetic or antithetic fault located in the Bristol Channel probably defined a southern margin. The general northward onlap against the Ritec Fault and the southward increase in thickness argues for differential subsidence across the block. Axial drainage via connected sub-basins is a common phenomenon (e.g. Mack & Seager 1990) and can result in sediment transport paths of hundreds of kilometres. This would imply a system that left the mountain front somewhere in the present Celtic Sea, or in southern Ireland. Drainage in southern Ireland at the time was dominated by subsidence in the Munster Basin and stable granitic plutons to the southwest and northeast of the basin (MacCarthy (1990) and Fig. 1). These factors deflected drainage in the Munster Basin to the east at certain intervals of Late Devonian time, and provide a possible source for the Gupton Formation. The north-easterly pluton, the Leinster Massif, is closer and

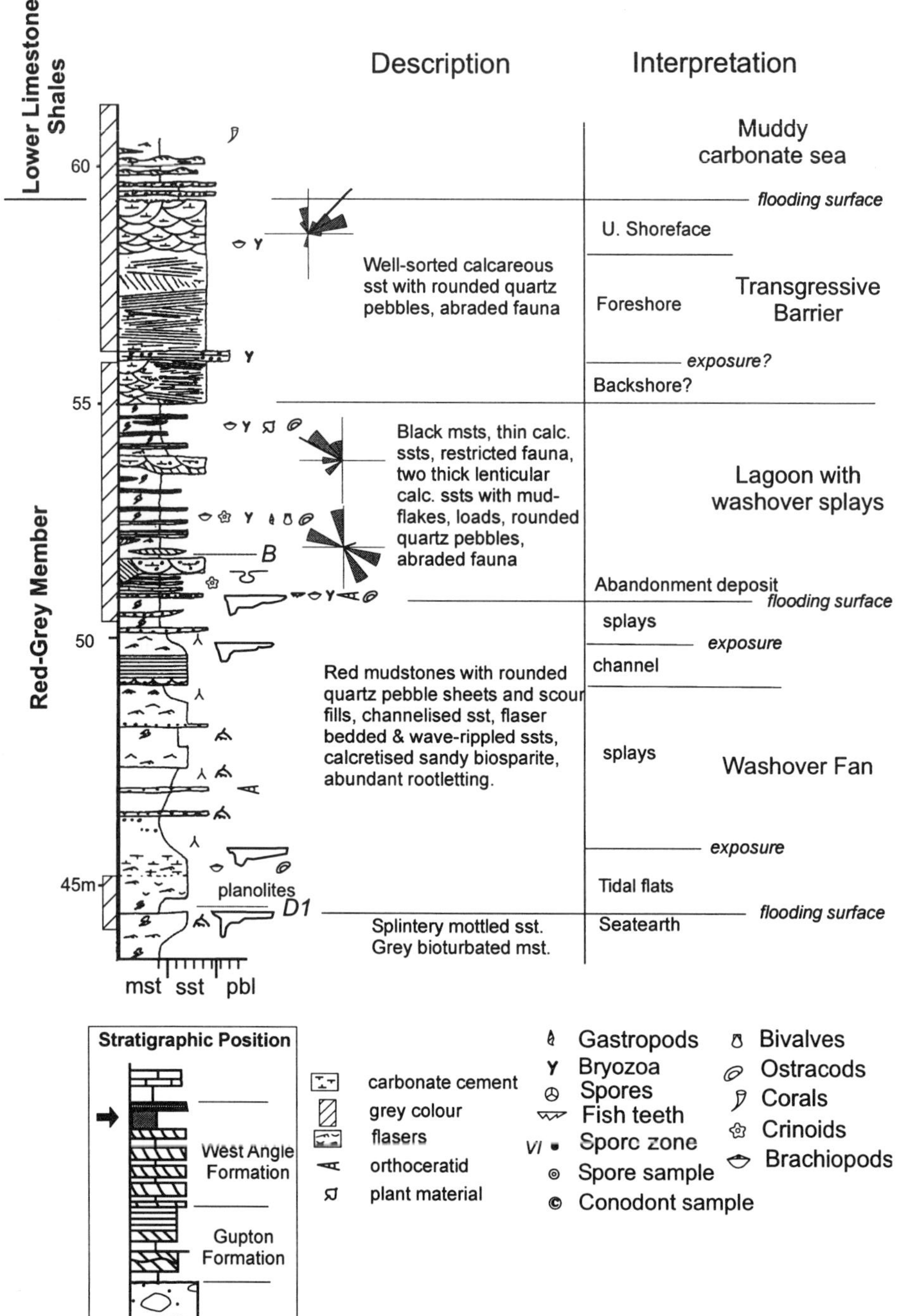

Fig. 17. The Red–Grey Member at West Angle north. Sedimentary log and interpretation of depositional environment of top half of member. Depths quoted are metres from base of member. Spore sample D1 is sample 1 of Dolby (1971); B is sample B of Bassett & Jenkins (1977).

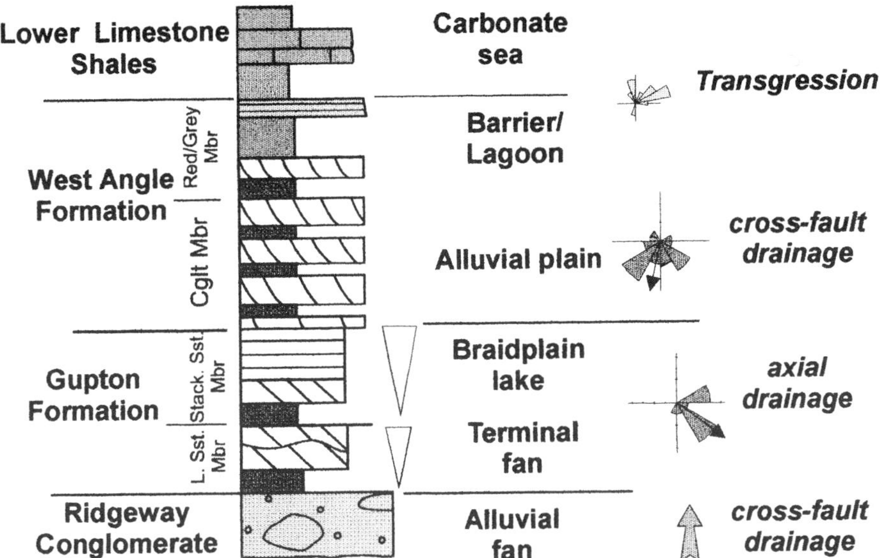

Fig. 18. Summary of inferred environments of deposition, drainage directions as exemplified by representative palaeocurrent roses and relations with basin margin faults.

could itself have given rise to drainage to the south and east into the Celtic Sea area and towards Pembrokeshire.

At the onset of deposition of the West Angle Formation, footwall uplift is postulated for the blocks to the north of the Ritec Fault to allow the development of a coarse-grained transverse drainage system. This either reworked coarse material from the Cosheston Group in the Winsle Block or allowed supply of material from Precambrian and Lower Palaeozoic acid igneous rock in northwest Dyfed (Fig. 4). By this time there is no suggestion of drainage obstruction by a Bristol Channel fault barrier.

It is possible that the Benton Fault acted as the true fall line, but the absence of Upper Devonian sediments in the Winsle Block today indicates that at most only a thinner accumulation was likely and that the main differential movement was along the Ritec Fault. Any Upper Devonian section in the Winsle Block would need to have been eroded during the Carboniferous transgression, and there is no variation of the Lower Limestone Shales across the Ritec Fault that could testify to significant footwall uplift and erosion at that time.

The Carboniferous transgression is a regional event resulting from probable eustatic rises in sea level coupled with gradual subsidence of the Old Red Sandstone continent (Clayton *et al.* 1986). Local tectonics interacted with these to determine the preservation of transgressive deposits. In south Pembrokeshire, subsidence in the hanging wall of the Ritec Fault allowed preservation of a complete barrier–lagoon complex during transgression, as opposed to the more normal ravinement process, which reworks the coastal deposits and carries them offshore sometimes leaving an erosively based lag deposit. The absence of any significant sand in the Lower Limestone Shales of the Winsle Block suggests that the barrier sand at the top of the Skrinkle Sandstones was drowned *in situ* and that non-deposition or ravinement characterized the northerly fault blocks during the continued transgression.

This research forms part of a PhD thesis conducted at Bristol University from 1974 to 1977. I am grateful to B. P. J. Williams and D. L. Dineley for their supervision and guidance, and along with J. R. L. Allen, for introducing me to such an excellent field area. I would also like to thank R. W. W. Lovell, R. G. Thomas and I. P. Tunbridge for stimulating discussions and ideas. And I would like to thank S. J. Marshall, my wife, for giving the encouragement to update and capture a long overdue story. The work was carried out during the tenure of a Shell International Petroleum Student Scholarship, which is gratefully acknowledged.

References

Allen, J. R. L. 1965. Upper Old Red Sandstone (Farlovian) palaeogeography in South Wales and the Welsh Borderland. *Journal of Sedimentary Petrology*, **35**, 167–195.

—— 1974. The Devonian rocks of Wales and the Welsh Borderland. *In*: Owen, T. R. (ed.) *The*

Upper Palaeozoic and Post-Palaeozoic rocks of Wales. University of Wales Press, Cardiff, 47–84.

—— & Williams, B. P. J. W. 1981. Sedimentology and stratigraphy of the Townsend Tuff Bed (Lower Old Red Sandstone) in South Wales and the Welsh Borders. *Journal of the Geological Society, London*, **138**, 15–29.

——, Thomas, R. G. & Williams, B. P. J. 1982. The Old Red Sandstone north of Milford Haven. *In*: Bassett, M. G. (ed.) *Geological Excursions in Dyfed, South-West Wales*. National Museum of Wales, Cardiff, 123–149.

Baker, J. W. 1982. The Precambrian of south-west Dyfed. *In*: Bassett, M. G. (ed.) *Geological Excursions in Dyfed, South-West Wales*. National Museum of Wales, Cardiff, 15–25.

Bassett, M. G. & Jenkins, T. B. H. 1977. Tournaisian conodont and spore data from the uppermost Skrinkle Sandstones of Pembrokeshire, South Wales. *Geologica et Palaeontologica*, **11**, 121–134.

Bevins, R. E. & Roach, R. A. 1980. Early Ordovician volcanism in Dyfed, SW Wales. *In*: Harris, A. L., Holland, C. H. & Leake, B. E. (eds) *The Caledonides of the British Isles—Reviewed*. Geological Society, London, Special Publublications, **8**, 603–609.

Clayton, G. & Higgs, K. 1979. The Tournaisian marine transgression in Ireland. *Journal of Earth Sciences, Royal Dublin Society*, **2**, 1–10.

——, Graham, J. R., Higgs, K., Sevastopulo, G. D. & Welsh, A. 1986. Late Devonian and Early Carboniferous palaeogeography of southern Ireland and southwest Britain. *Annales de la Société Géologique de Belgique*, **109**, 103–111.

Cope, J. C. W & Bassett, M. G. 1987. Sediment sources and Palaeozoic history of the Bristol Channel area. *Proceedings of the Geologists' Association*, **98**, 315–330.

——, Ingham, J. K. & Rawson, P. F. (eds) 1992. *Atlas of Palaeogeography and Lithofacies*. Geological Society, London, Memoir, **13**.

Dixon, E. E. L. 1921. *The Geology of the South Wales Coalfield Part 13. The Country around Pembroke and Tenby*. Memoir of the Geological Survey of the UK, i–vi.

Dolby, G. 1971. Spore assemblages from the Devonian–Carboniferous transition measures in southwest Britain and southern Eire. *In*: *Colloque sur la stratigraphie du Carbonifère*. Congrès et colloques de l'Université de Liège, **55**, 267–274.

Dunne, W. M. 1983. Tectonic evolution of S. W. Wales during the Upper Palaeozoic. *Journal of the Geological Society, London*, **140**, 257–265.

Hassan, T. H. 1966. *The petrology of the Skrinkle Sandstone and contiguous deposits of South Pembrokeshire*. PhD. thesis, Chelsea College of Science and Technology, London.

Higgs, K., Clayton, G. & Keegan, J. B. 1988. *Stratigraphic and Systematic Palynology of the Tournaisian Rocks of Ireland*. Geological Survey of Ireland Special Paper, **7**.

Kamerling, P. 1979. The geology and hydrocarbon habitat of the Bristol Channel Basin. *Journal of Petroleum Geology*, **2**, 75–93.

Mack, G. H. & Seager, W. R. 1990. Tectonic controls on facies distribution of the Camp Rice and Palomas Formations (Pliocene–Pleistocene) in the southern Rio Grande rift. *Geological Society of America Bulletin*, **102**, 45–53.

Marshall, J. D. 1977. *Sedimentology of the Skrinkle Sandstones Group (Devonian–Carboniferous), south-west Dyfed*. PhD thesis, University of Bristol.

—— 2000. Sedimentology of a Devonian fault-bounded braidplain and lacustrine fill in the lower part of the Skrinkle Sandstones, Dyfed, Wales. *Sedimentology*, **47**, 325–342.

MacCarthy, I. A. J. 1990. Alluvial sedimentation patterns in the Munster Basin, Ireland. *Sedimentology*, **37**, 685–712.

McClelland Brown, E. 1983. Palaeomagnetic studies of fold development and propagation in the Pembrokeshire Old Red Sandstone. *Tectonophysics*, **98**, 131–149.

Mechie, J. & Brooks, M. 1984. A seismic study of deep geological structures in the Bristol Channel area. *Geophysical Journal of the Royal Astronomical Society*, **78**, 661–689.

Powell, C. M. 1987. Inversion tectonics in S. W. Dyfed. *Proceedings of the Geologists' Association, London*, **98**, 193–203.

—— 1989. Structural controls on Palaeozoic basin evolution and inversion in southwest Wales. *Journal of the Geological Society, London*, **146**, 439–446.

Tunbridge, I. P. 1986. Mid-Devonian tectonics and sedimentation in the Bristol Channel area. *Journal of the Geological Society, London*, **143**, 107–115.

Webby, B. D. 1966. Middle-upper Devonian palaeogeography of north Devon and west Somerset, England. *Palaeogeography, Palaeoclimatology, Palaeoecology*, **2**, 27–46.

Williams, B. P. J. W. 1964. *The stratigraphy, petrology and sedimentation of the Ridgeway conglomerate and associated formations in south Pembrokeshire*. PhD thesis, University of Wales, Swansea.

——, Allen, J. R. L. & Marshall, J. D. 1982. Old Red Sandstone facies of the Pembroke Peninsula south of the Ritec Fault. *In*: Bassett, M. G. (ed.) *Geological Excursions in Dyfed, South-West Wales*. National Museum of Wales, Cardiff, 151–174.

Ziegler, P. A. 1990. *Geological Atlas of Western and Central Europe*, 2nd edn. Shell Internationaal Petroleum Maatschappij, The Hague.

Old Red Sandstone basins and alluvial systems of Midland Scotland

B. J. BLUCK

Department of Geography and Topographic Science, University of Glasgow, Glasgow G12 8QQ, UK (e-mail: BBLUCK@geog.gla.ac.uk)

Abstract: Old Red Sandstone rocks of the Midland Valley of Scotland record the amalgamation history of blocks widely separated from each other in Ordovician times. As a response to the lateral juxtaposition of the Midland Valley against the Highland block, the Midland Valley, weakened during its long history of igneous activity, was subjected to transtension and transpression, which opened and closed basins resulting in much recycling of sediment and the renewed development of intermittent and sometimes prolific volcanic activity. The Old Red Sandstone comprises two cycles of basin fill: the older (the Lower Old Red Sandstone) is separated by a major unconformity from the younger (the Upper Old Red Sandstone). Each cycle begins with conglomeratic sedimentation in pull-apart basins then fines and petrographically matures upward. The cycles, initiated by faulting, record the decline in tectonic influence and the concomitant reduction in source relief. The sediment source to the Old Red Sandstone is enigmatic. The Dalradian basement to the north and Southern Uplands to the south were, in many places, extensively eroded by Late Silurian–Early Devonian times. The thickest sequence of coarse sediment in the UK, if not in Europe, therefore has no obvious nearby uplift to provide both the sediment and the persistent, substantial slopes required to deliver sediment of that calibre to the basin. To the northeast, the Greenland–Baltica collision had created a major Silurian–Carboniferous (Scandian) uplift. Major river systems draining this mountain belt entered the Lower and Upper Old Red Sandstone basins at a late stage in their development when relief was lowered sufficiently to allow access to them.

The Old Red Sandstone of the Midland Valley is, in many ways, one of the most paradoxical sedimentary sequences in the whole Caledonian collision event. It has an abundance of coarse conglomerate and immature sandstone deposited in thick sequences yet the flanking, potential source blocks appear not to have made a significant contribution to the sediment. There is strong evidence that both the Dalradian basement to the north and the Southern Uplands to the south were substantially eroded by Devonian times and could not therefore supply a great deal of the sediment required (Bluck 1984).

The Old Red Sandstone has been traditionally divided into a lower unit, which ranges from Wenlock–Early Llandovery to Emsian age, and an upper unit, which extends to Carboniferous in age from a base of uncertain age. The upper unit rests unconformably on lower, and each comprises an upward-fining, upward-maturing sedimentary sequence.

The Dalradian, Midland Valley and Southern Uplands blocks were widely separated during part of Early Palaeozoic time but the Midland Valley–Dalradian blocks had probably become attached to each other during Late Ordovician or Llandovery time (Bluck & Leake 1986). It is probable that all three blocks were completing their final amalgamation during the time that the Lower and Upper Old Red Sandstone accumulated.

At the same time, to the northeast, the Laurentian plate was otherthrusting the Baltic plate. This resulted in a major orogen, which extended along western Scandinavia, Eastern Greenland and the Canadian Arctic (Ziegler 1988). By Early Devonian times, in parts of western Norway, regional eclogite facies metamorphic rocks were cooling from an uplift that had already brought them up from some depth (Cuthbert *et al.* 1983; Andersen *et al.* 1998). Throughout this extensive orogenic belt rocks have cooling ages that span Late Silurian–Devonian (Nordgullen *et al.* 1993; Dallmeyer *et al.* 1994; Northrup 1997).

These cooling ages provide evidence for the former presence of one of the world's great orogenic belts, which was to have considerable influence on the dispersal of sediment over northern Europe. Old Red Sandstone rocks of

From: FRIEND, P. F. & WILLIAMS, B. P. J. (eds). *New Perspectives on the Old Red Sandstone.* Geological Society, London, Special Publications, **180**, 417–437. 0305-8719/00/$15.00

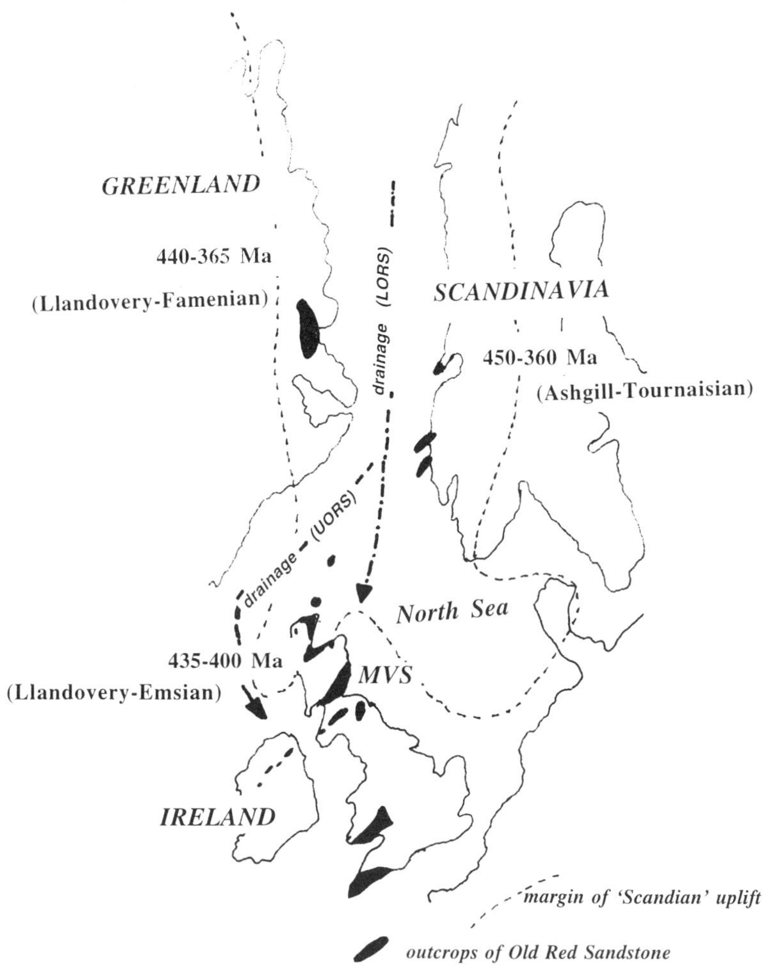

Fig. 1. Map of the North Atlantic with the outcrops of Old Red Sandstone and with the ranges of uplift ages for Greenland and Scandinavia obtained from Anderson *et al.* (1992), Dallmeyer *et al.* (1994), Hames & Andresen (1996) and Northrup (1997) and Andersen *et al.* (1998). The inferred major drainage (dashed line) out of this orogen is based on river scales determined from sediments in the Midland Valley of Scotland (MVS, and see text). Stratigraphical age equivalents are based on the time scale of Gradstein & Ogg (1996) and Tucker *et al.* (1998).

the Midland Valley were deposited under the influence of both these regimes of early strike-slip basin generation during the final throes of terrane amalgamation and later, of this major, distant sediment source (Fig. 1).

The Lower Old Red Sandstone sedimentary rocks, mainly exposed along the margins of the Midland Valley in Scotland, extend southwestward, along strike, into Ireland. On the north margin of the Scottish Midland Valley they are seen either to rest unconformably on the Highland Border Complex (Campbell 1913; Hutchison 1928; Bluck 1992) or to lie in fault contact with either it or the Dalradian basement. Here they make up the asymmetrical Strathmore syncline, which extends from the North Sea to the Firth of Clyde and beyond (Fig. 2). Its axis lies roughly parallel to the Highland Boundary Fault and on its northern margin Old Red Sandstone beds are, in places, overturned.

The southern outcrops are less continuous than those of the northern Midland Valley. Their stratigraphy is less well known but appears to be simpler (Fig. 2). Here the Lower Old Red Sandstone may rest with apparent conformity on the Silurian rocks that form a number of inliers in this region of the Midland Valley. The top of the Lower Old Red Sandstone sequence is undated but is seen to be unconformably overlain by rocks ascribed to the Upper Old Red Sandstone.

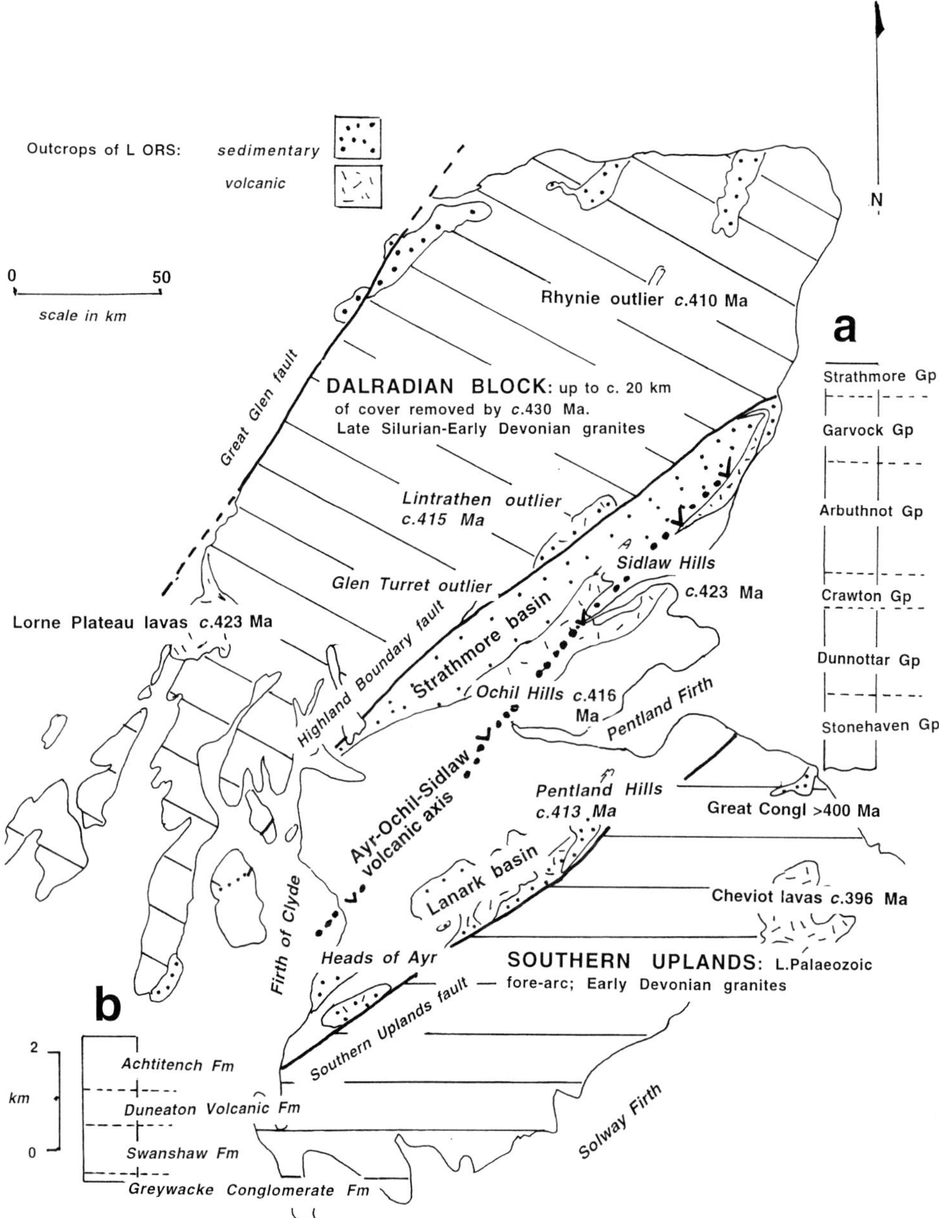

Fig. 2. Map showing the distribution and absolute ages of the Old Red Sandstone outliers on the flanking blocks to the Midland Valley together with ages of dated rocks within the Midland Valley (after Thirlwall 1988). Stratigraphic columns for (**a**) Strathmore basin from Armstrong & Paterson (1970), (**b**) Lanark basin from Phillips *et al.* (1997) on the same thickness scale.

In contrast to the Lower Old Red Sandstone of the Strathmore syncline these strata are thinner and interstratify with lavas, which are far more abundant in these southern outcrops.

Running axially through the centre of the Midland Valley, dividing the northerly and southerly outcrops, lies a thick sequence of basaltic and andesitic lavas with their interleaved volcanogenic sediments (Francis *et al.* 1970). Although volcanic activity was widespread throughout the Midland Valley and on the flanking basements, in this central region it was dominant, forming a series of major volcanic centres aligned along a lineament of Caledonian trend and probably on the site of an older, Ordovician–Silurian magmatic arc (Bluck 1983, 1984). This Ayr–Ochil–Sidlaw volcanic axis formed a topographic high in the Devonian Midland Valley and separated the Strathmore basin to the north from the Lanark basin to its south (Figs 2 and 3).

The 'rift' shoulders to the Midland Valley

Both stratigraphical and radiometric dating have greatly aided the understanding of the relationship between the flanking blocks to the Midland Valley, i.e. the Southern Highlands and Southern Uplands. The Dalradian (Southern Highland) block began its cooling–uplift in Late Cambrian times (*c.* 515 Ma, Dempster 1985; Dempster *et al.* 1995; compare Evans *et al.* 1997) and continued to *c.* 430 Ma(Late Llandovery times). Parts of the block were eroded to a plain surface before *c.* 420 Ma (Late Ludlow time) as both the sub-horizontal Lorne Plateau lavas (dated at *c.* 423 ± 6 Ma (Thirlwall 1988)) and the underlying, *c.* 250 m of coarse alluvial sediments rest on a near-flat, eroded Dalradian basement (Morton 1979).

More critically in relation to the Midland Valley, in the outliers of Lower Old Red Sandstone at Glen Turret and Lintrathen (Fig. 2), both north of the Highland Boundary Fault, the southern edge of the Dalradian block is also seen to be partly planed by Late Silurian time. Here conglomerates and breccias with angular–subrounded clasts of local Dalradian rock are interbedded with lavas and sandstones that are all either contained in broad valley systems or rest on a widespread planation surface cut in Dalradian basement. These sediments were dispersed towards the southeast (Figs 3 and 4).

The 'Lintrathen porphyry', reinterpreted as an ignimbrite, is found resting both on the sediment–lava sequence of the Lintrathen outlier as well as directly on the Dalradian basement near Dunkeld (Paterson & Harris 1969). With the Lintrathen porphyry being dated at 415 ± 6 Ma(Pragian; Thirlwall 1988) and in a sequence that is estimated to be *c.* 1000 m thick (Armstrong & Paterson 1970) there is clearly evidence for a Lower Old Red Sandstone basin, immediately north of the Highland Boundary Fault, which had already accumulated a substantial sequence of first-cycle sediment from a Dalradian source.

Further to the north, however, there is evidence in the radiometric ages of the basement for areas of Devonian cooling and uplift, particularly around the Great Glen Fault and regions to the northwest of it (Dewey & Pankhurst 1970; Kelley 1988; Bluck *et al.* 1988; Fig. 2). This suggests that there were some sediment sources within the Highland blocks and these sources, often rich in quartzite, may have contributed to nearby Lower Old Red Sandstone basins. Although the primary contribution from local basement appears to be limited in some cases, the ubiquitous well-rounded quartzite clasts (which also characterize much of the sediment within the Midland Valley) may be the mature residue of prolonged cycles of uplift and erosion, which were finally responsible for the planation of this basement and which may have continued on into Devonian time.

The Southern Uplands have a similar erosional history. Lavas in the Cheviot Hills that rest directly on folded Silurian strata have ages of 396 ± 4 Ma (Thirlwall 1988). In the northeast margin of the Southern Uplands, near Berwick, both the Great Conglomerate and St Abbs volcanic rocks, cut by a dyke with an age of 400 ± 9 Ma (Rock & Rundle 1986), rest on folded Lower Silurian rocks. It therefore seems probable that, as with the Southern Highlands, the folded Palaeozoic rocks of the Southern Uplands had completed most of their uplift and were already substantially eroded by Early Devonian times.

With both the flanking blocks receiving sediment by Early Devonian times, it is most unlikely that they could also be major contributors to the thick conglomerates that occur within the Midland Valley. Haughton *et al.* (1990), recognizing the problem of source and the paucity of first-cycle Dalradian clasts south of the Highland Boundary Fault, suggested that there could have been thick conglomerates overlying the basement, which were then recycled into the northeastern Midland Valley basins.

This suggestion is unlikely to apply further to the southwest, where the palaeoflow in Lower Old Red Sandstone sediments is sometimes towards the Dalradian block (Fig. 3). In addition, the Lower Old Red Sandstone outliers (particularly those immediately to the north of the Highland Boundary Fault), which rest unconformably on the Dalradian basement and contain Dalradian basement clasts, clearly indicate that basement was, in places at least, exposed at the surface at *c.* 415 Ma (Lochkovian time) and devoid of a thick conglomerate cover (Bluck 1984; Fig. 3). It is clear that flank uplift as the source and as a source of transport energy for thick boulder-bearing conglomerates is questionable.

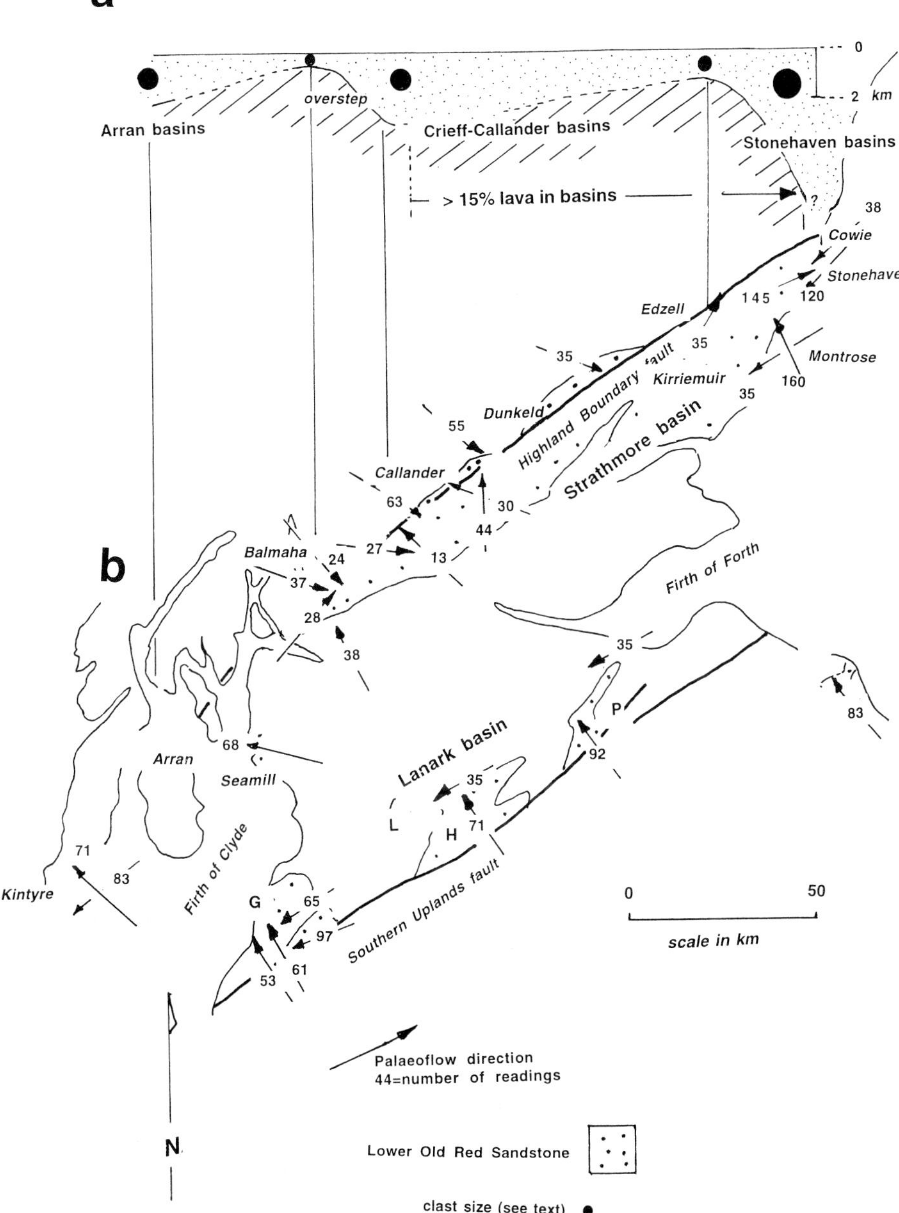

Fig. 3. Palaeoflow directions and thickness variations for Lower Old Red Sandstone. (**a**) Section along north limb of Strathmore syncline (vertical lines tie into localities in (**b**) shows variations in thickness of preserved, combined lower and upper divisions of the Lower Old Red Sandstone. The ground between each culmination (Edzell and Balmaha) is thought to comprise a stack of superimposed basins rather than a single one: three basin stacks are recognized: the Stonehaven basin, Crieff–Callander basin and Arran basins. Filled circles indicate clast size (see text). (**b**) Palaeoflow data for the lower division of the Lower Old Red Sandstone from Friend & MacDonald (1968), Wilson (1980), Bluck (1984), Haughton (1989), Syba (1989) and new data presented in this paper. Silurian inliers: G, Girvan; L, Lesmahagow; H, Hagshaw; P, Pentland Hills.

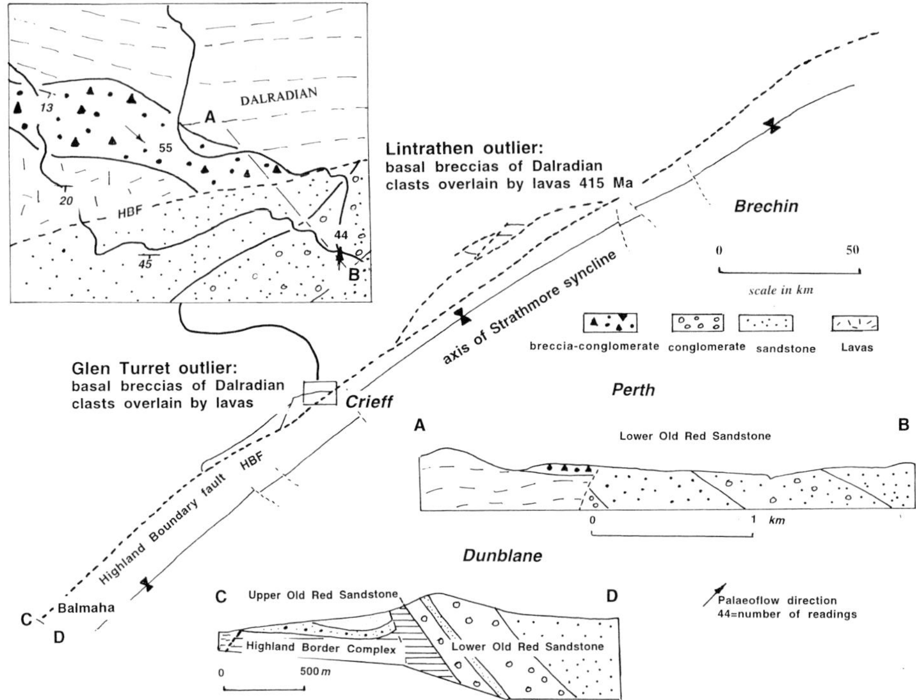

Fig. 4. Sections through the north limb of the Strathmore syncline near Crieff (Glen Turret outlier, after Allen (1940)) and Balmaha. Inset map (after Allen 1940) showing converging palaeoflows for the conglomerates on either side of the Highland Boundary Fault. The conglomerates–breccias with first-cycle Dalradian clasts are dispersed southeast towards the Strathmore basin on the southern side of the Highland Boundary Fault; the conglomerates in the Strathmore basin are dispersed to the north or northeast and contain no certain first-cycle clasts. These observations are repeated in the 'Lintrathen outlier', where an age of 415 Ma has been obtained on the 'Lintrathen ignimbrite (porphyry)', which rests on the basal Dalradian breccias. The Lintrathen and Glen Turret outliers are thought to belong to the northeast margin of a basin that opened to the southeast and is at present truncated at the Highland Boundary Fault.

Lower Old Red Sandstone of the Midland Valley

The Lower Old Red Sandstone rocks, ranging from Late Wenlock (Marshall 1991; Wellman 1993; Wellman & Richardson 1993) to Emsian in age (Richardson 1967; Scott *et al.* 1976; Richardson *et al.* 1984), occur in three NE–SW groups of outcrops in the Midland Valley: (1) along the northern margin as the long, continuous Strathmore syncline; (2) to the south of it, the Ochil–Sidlaw anticline, comprising mainly volcanic rocks yielding ages of 424 ± 6 Ma, 416 ± 6 Ma and 411 ± 6 Ma (Ludlow–Pragian time, Thirlwall 1988); and (3) along the southern Midland Valley, as discontinuous outcrops with interstratified lavas of 413 ± 6 Ma (Pragian time).

The Strathmore syncline

The Lower Old Red Sandstone of the Strathmore basin is a major conglomerate–sandstone–lava sequence, which generally thins to the southwest. A lower part of the Lower Old Red Sandstone sequence is dominated by conglomerates and lava flows (here referred to as the lower division) and the upper by sandstones and mudstones (the upper division). For sequences along the north limb of the Strathmore syncline, the boundary between the two divisions is at the base of the Strathmore Group, but for the axis of the syncline, distal to the Highland Boundary Fault, a major break in style of sedimentation is recognized at the base of the Garvock Group (Fig. 2).

Where seen along the north limb of the Strathmore syncline and in the coastal northeastern outcrops, the southwest thinning of the succession is partly achieved by overlap onto a floor that comprises various rocks of the Highland Border Complex, and partly by deposition in a series of basins containing thinner sequences when traced in this direction (Fig. 3). In the northeast, where they are magnificently exposed along the coast from Cowie to Montrose, the lower division is possibly up to 5 km thick (Marshall *et al.* 1994), but thins dramatically when traced towards the southwest (Armstrong & Paterson 1970; Bluck 1978; Fig. 3). This thinning is accompanied by a reduction in both the thickness and number of individual conglomerate units within the division (Fig. 3). Within the Cowie to Montrose section there are at least two conglomerates >750 m thick and four >250 m thick. In the Edzell region, by contrast, the total sequence has thinned to *c.* 1500 m, with conglomerate beds <250 m thick (although in this instance the total thickness change may be due to the basal part of the sequence being cut out by faulting).

The lower division is relatively thick, possibly up to 4000 m, in sections southwest from Kirriemuir to Callander (Fig. 3) and the individual conglomerate units are often >500 m thick. However, once again, the total lower sequence is reduced to *c.* 500 m at Balmaha, where two individual conglomerate units are <150 m thick. This culmination at Balmaha is reasonably well founded as the sections northeast as far as Callander are seen, in places, to lie unconformably on the Highland Border Complex and can be traced out as a complex, southwest-overstepping sequence of rocks. There is, therefore, little if any of the succession structurally cut out at the base of the sequence and the thinning is stratigraphical.

Further to the southwest, out into Arran and Kintyre, the conglomerate-bearing sequence thickens and repetition of thick conglomerate units once again characterizes the sequences. On Arran, the conglomerate sequence is 1200 m thick with three conglomerate–breccia units >250 m thick (Friend *et al.* 1963). Conglomerates are generally thinner on Kintyre (Friend & MacDonald 1968; Morton 1979).

These southwest changes in thickness of the sequence and the contained conglomerate units along the north limb of the Strathmore syncline are accompanied by parallel changes in both clast sizes and the number of interstratified lava flows (Fig. 3a). Lavas are absent from both the Edzell and Balmaha regions, where the clast sizes tend to be finer (generally <200 mm) and both the sequences and their contained conglomerate units are thinner. In the thick sequences, with thick conglomerates along the northeast coast, clasts are often >300 mm and may reach >1 m. Here there are also many lavas interbedded in the sequence. Clast sizes coarsen in the Crieff–Callander area, where they are often >250 mm and associated with lavas, and also in Arran, where there is also a thin lava flow interstratified with the conglomerates.

Rocks of the lower division, cropping out along the northern edge of the Strathmore basin, represent a section through a number of basins, which are probably aligned roughly parallel to the present trace of the Highland Boundary Fault (Fig. 3). Haughton & Bluck (1988) proposed that there was a stack of superimposed basins in the coastal areas south of Stonehaven and that these Stonehaven basins may have terminated near Edzell. The basins that characterize the outcrop from Edzell to Balmaha include the Crieff–Callander basins, and the basins southwest of Balmaha, the Arran basins (Fig. 3a). The status of the ground between Edzell and Dunkeld is uncertain.

The association of thick conglomerates with thick sedimentary sequences and both with increased clast size and lava flows strongly suggest that crustal thinning produced small, rapidly filled basins (see Crowell 1974; Hempton & Dunne 1984). Bluck (1978, 1992), Haughton & Bluck (1988) and Haughton (1989) have suggested that these sediments accumulated in basins that were controlled by strike-slip faulting. The evidence for a strike-slip regime comes mainly from the clear southwest-overlapping relationship suggesting the progressive opening of the basin floor; internal unconformities (Du Toit 1905; Robertson 1987; Phillips *et al.* 1997) and soft-sediment deformation (Robertson 1987). These characteristics, in conjunction with the relatively small sizes of the basins and their coarse conglomeratic fills, suggest rather than prove a strike-slip control.

The repetition of conglomerates and vertical variations in grain size record source-wide, repeated rejuvenation, which resulted in base-level changes and renewed phases of gravel deposition in the basin. If, as is likely, this repetition was related to fault activity, then faulting was more common and of greater magnitude in the northeast, where the conglomerates are thickest, coarsest and are associated with thick lavas. Rapid subsidence–uplift, induced by faulting, increases slope and stream power so that large clasts occur in thick gravels, which themselves are found in thick sequences.

The presence of broken and re-rounded clasts in many of the conglomerates indicates much recycling. In addition, there is a maturity contrast, in some conglomerates, between very well rounded quartzite and sometimes less well rounded other, softer clast lithologies. This implies much recycling to produce a residue of mature clasts, which are then available for mixing with the more labile, first-cycle clasts derived from elsewhere (mainly lavas within either the basin or the immediate source block). Repeated closure and opening of basins may have provided the means by which these maturity contrasts, and clear recycling of resistant clasts, were achieved.

The Lower Old Red Sandstone of the north limb of the Strathmore syncline is usually steeply dipping or slightly overturned. It rests either unconformably on or is faulted against the Highland Border Complex, a thin sliver of ophiolitic and other rocks lying along the Highland Boundary Fault. Where there is a fault contact, but little evidence for major displacement, this boundary is likely to be a faulted unconformity. The current dispute about the affinities of the Complex has a considerable bearing on the interpretation of the position of the Highlands with respect to the Old Red Sandstone basin. One view is that the Complex stratigraphically overlies the Dalradian sequence (Tanner 1995); the other view is that it is a discrete group of rocks in fault contact with the Dalradian rocks (Bluck 1984; Bluck & Ingham 1997). In the former interpretation, the Old Red Sandstone effectively rests on Dalradian rocks within the Midland Valley; the other is free of that constraint and allows the Dalradian sequence to be near but not juxtaposed with the Midland Valley during Devonian times (i.e. allows for a break between the Dalradian block and rocks with stratigraphic contininuity into those of the Strathmore syncline as discussed below).

The floor to the Strathmore basin is at least partly made up of Highland Border Complex, which is thought to extend some unknown distance south beneath the syncline. By the same reasoning the northern margin of the basin is thought to extend, with its basement of Highland Border Complex, some distance to the north (see Figs 4 and 5). This point is evident when the sub-vertical edge of the Strathmore basin is untilted: the Highland Border Complex, which now forms its floor, extends to the north.

However, the basin could not have extended directly onto the Dalradian block as the composition of its fill, being mainly meta-quartzite and volcanic clasts, is incompatible with that of the outliers of Old Red Sandstone at Glen Turret and Lintrathen, now seen resting on Dalradian rocks and juxtaposed against the northern limb of the Strathmore syncline. Conglomerates and breccias in both outliers contain first-cycle Dalradian clasts, which are dispersed towards the southeast yet are not common in these early sediments of the Strathmore basin. Basins north and south of the Highland Boundary Fault have broadly similar ages but were, for a time at least, filled with sediment dispersed in radically different directions and from radically different sources (see Fig. 5).

It follows from this discussion that the Strathmore basin in which the Lower Old Red Sandstone accumulated originally extended over ground now occupied by the Dalradian block, and this block has to be moved to the northwest relative to the Midland Valley to accommodate this northerly extension of the Strathmore basin and retain its floor of Highland Border Complex.

The Dalradian block, either during or after deposition of the Lower Old Red Sandstone, has moved southwards over this ground to generate the steep northern limb of the Strathmore syncline and effect the removal of the northern margin of the Strathmore basin. There is confirmation for this view, both in the xenoliths of rocks of Highland Border aspect contained in a lamprophyre dyke of probable Devonian age some 10 km north of the present southerly edge of the Dalradian block (Dempster & Bluck 1991) and in the geophysical profiling (Dentith *et al.* 1992), where rocks with Highland Border characteristics were evidenced beneath the present Dalradian outcrop.

It is therefore postulated that, along the entire length of the northern limb of the Strathmore syncline, there was a northerly extension of the Strathmore basin. The northern most margin to this basin may have been the present southern edge of the Dalradian block, with its outliers of Old Red Sandstone carrying first-cycle Dalradian rock (see Figs 4 and 5). The presence of an extended basin to the north of the present outcrop goes some way to solve the problem of the provenance of the bulk of the Old Red Sandstone sediments.

The provenance of the Lower Old Red Sandstone conglomerates in the Strathmore syncline has presented difficulties for some time. All along the northern outcrop of the Strathmore basin the sediments are dominated by meta-quartzites and igneous clasts. The meta-quartzites are polycyclic, almost always being very well rounded and some showing broken and re-rounded clasts. The igneous rocks are dominated by volcanic clasts bearing a resemblance in chemistry and mineralogy to the lavas that are currently found in the basin. Clasts with a clear provenance in the adjacent Dalradian block are rare in the early sediments of the Strathmore basin.

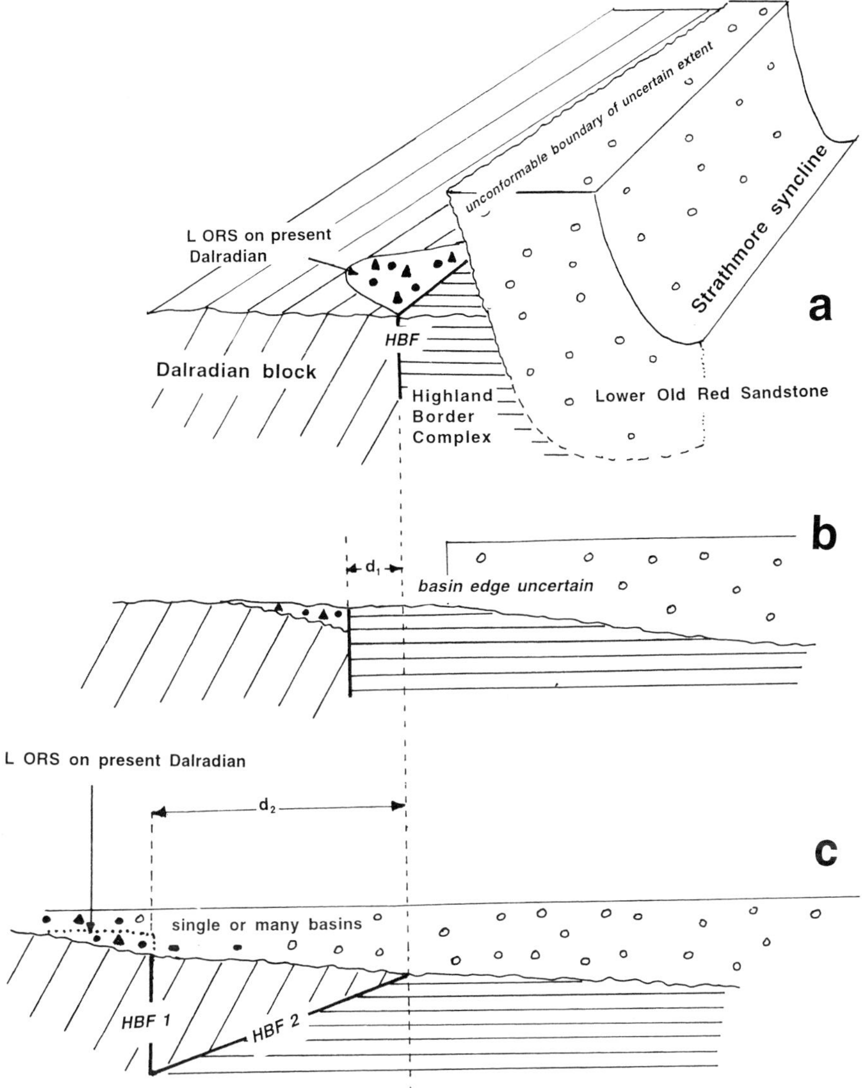

Fig. 5. (**a**) A diagrammatic section through the Highland Border showing the nature of the Strathmore syncline, the basin resting on the Dalradian block with first-cycle Dalradian clasts (see Fig. 4) and the extension of the Highland Border Complex beneath the basin. In this interpretation, the Dalradian block has to be moved relative to the Midland Valley to allow the basin to extend in that direction. HBF, Highland Boundary Fault. (**b**) Unfolding the northern limb of the Strathmore syncline extends the Old Red Sandstone basin into ground now occupied by the Dalradian block by a distance d_1. (**c**) As the sediments in the Strathmore syncline are often dispersed towards the Dalradian block (see Figs 3 and 4) and do not contain abundant first-cycle Dalradian clasts, the basin edge must be some distance further to the north (d_2). The subsequent closing of that basin by the southeast thrusting of the Dalradian block possibly rotated the HBF from its original (HBF 1) to HBF 2. At about this time the Strathmore syncline developed.

In the northeastern coastal sequence there are a variety of additional clasts that include granites with affinities with those to the north in Aberdeenshire (Haughton *et al.* 1990); porphyries with an uncertain provenance; gabbros, spilites and lithic arenites that may have had a source in the Highland Border Complex; psammitic clasts that resemble rocks now seen in the local Dalradian basement; and others. With the exception of the lava clasts nearly all are very well rounded and suggest a multi-cycle provenance. Clasts of Highland Border rock are found in the basal conglomerates further southwest, but are so infrequent as to indicate that the Complex formed no great relief and was probably confined to the floor to the basin in this area.

The Dalradian block lay near to the Strathmore basin by Late Silurian–Early Devonian times (Bluck 1984; Trench & Haughton 1988). Despite this, the Dalradian block contributed little to the early sediments in this basin. This is consistent with the fact that the main uplift of the Dalradian block was complete by 440 Ma (Llandovery time) and that by *c.* 420–410 Ma (Late Silurian–Early Devonian time) there were major unconformities developed over it. Moreover, the development of a basin between the Dalradian block and the present outcrops of Old Red Sandstone would have been a barrier to the inclusion of first-cycle clasts into the Strathmore basin.

Haughton & Farrow (1989) recorded in the detrital garnets from sandstones collected along the north limb of the Strathmore syncline a lateral compositional change, which matches the lateral change in garnet compositions now found in the Dalradian block. This change suggests that the finer Dalradian sediments were able to extend across the basin thought to occupy the intermediate ground between the Dalradian block and the Strathmore basin. This evidence also puts a limit on the amount of lateral displacement undergone between the Dalradian block and Strathmore basin during and after the deposition of the Lower Old Red Sandstone.

In the later sediments of the Strathmore basin, clasts resembling the local Dalradian rocks appear in the conglomerates at Strathfinella Hill (Campbell 1913; Haughton & Bluck 1988) and on Arran (Friend *et al.* 1963), and possibly elsewhere. These deposits are very distinctive fan-like features, which have built out onto finer sediments of the Strathmore Group. This suggests local uplift in the Dalradian block, convergence onto the Midland Valley bringing the Dalradian block nearer, or the uplift and bypassing of the intervening basin.

The deposition of conglomerates continued throughout the Lower Old Red Sandstone. However, in the Garvock Group, there is a record of increasing sandstone deposition and towards the top of this group a calcrete developed (Armstrong & Paterson 1970). This calcrete extends intermittently along the outcrop from Brechin in the northeast to Dunblane in the southwest, a strike distance >100 km. Sparse petrographic data on the sandstones are highly variable within formations, but suggest a slight maturing of the sandstones in the Garvock Group (see Fig. 11, below).

The Garvock Group, with its considerable stratigraphical continuity, relative abundance of sandstones and development of a fairly widespread caliche, is seen to record a time when fault-controlled sedimentation was drawing to a close. Conglomerates, however, still provided a substantial contribution along the north limb of the syncline, where they persisted as discrete fans well into the succeeding Strathmore Group.

The general fining of the sedimentary sequence seen in the Garvock Group was further developed in the Strathmore Group, where there are mudstone units >1000 m thick and sandstones that show a relatively high proportion of metamorphic clasts (Fig. 11). This Garvock–Strathmore Group interval represents a significant change in sedimentation within the Midland Valley. Thick and laterally extensive sandstone bodies associated with thick mudstone deposits provide evidence for very large-scale drainage systems. An example of the internal structure of one of the sandstone bodies is given below.

Evidence for large-scale drainage: the bar at Crossgates–Burnside

A road cut in gently dipping sandstones at Crossgates–Burnside, southwest of Perth, has revealed the internal structure of some very large sediment bars that existed in the Garvock Group (Pragian time, *c.* 414–410 Ma, Tucker *et al.* 1998) of the southeastern limb of the Strathmore syncline. From this, and exposures elsewhere in the Old Red Sandstone, it can be established that alluvial sediment bars of this scale generally show a transition from coarse sandstone and occasional thin, fine conglomerates in planar cross strata at their upstream end (head) to medium sand in trough cross strata and finer sand with mud drapes in large-scale cross strata at their downstream end (tail; Figs 6 and 7).

The road cut is a 800 m, strike section in lithic arenites, which dip a few degrees to the northwest. The basal unit, the top of which is at road level, is a breccia comprising mudstone clasts (C1, Fig. 6) followed by lithic arenites, which, at the west end of the outcrop, comprise a single set of cross strata >12 m thick (C2 and C3, Fig. 6). These large foresets are in sandstone and siltstone with fine mudstone drapes. They interfinger with and are largely replaced laterally (towards the east) by large-scale trough cross strata, some of which are >30 m wide and >2 m thick and are in medium to coarse sandstone (C4,

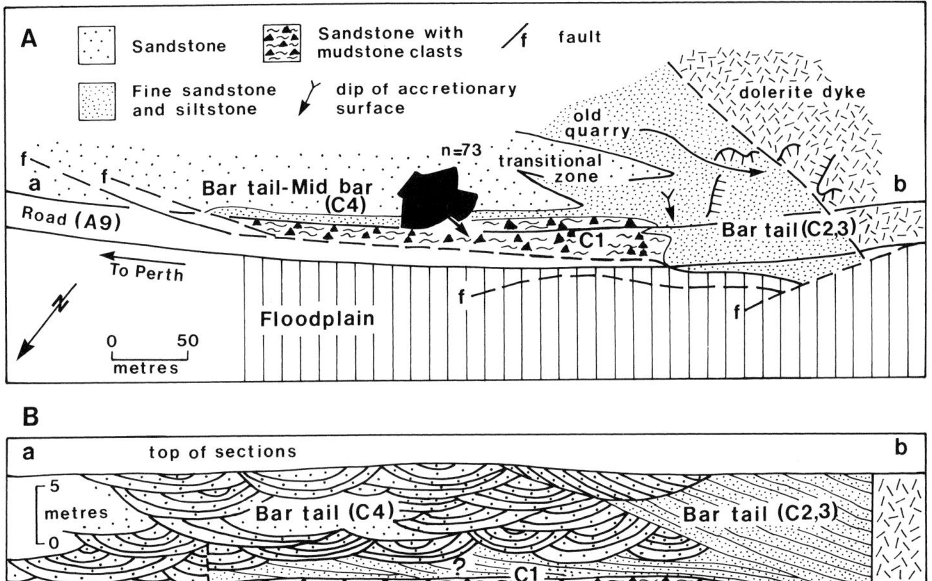

Fig. 6. A, plan and B cross-section (a–b) of the exposed bar *c.* 7 km southwest of Perth (see Fig. 4) between Crossgates (NGR NN 049 209) and near Burnside (NGR NN042 206). Bar tail and head explained in text. C1, breccias comprising sandstone with claystone clasts up to 2.7 m in diameter; C2 and C3, large-scale cross strata; C4, large trough cross strata; n, number of observations; arrow is vector mean.

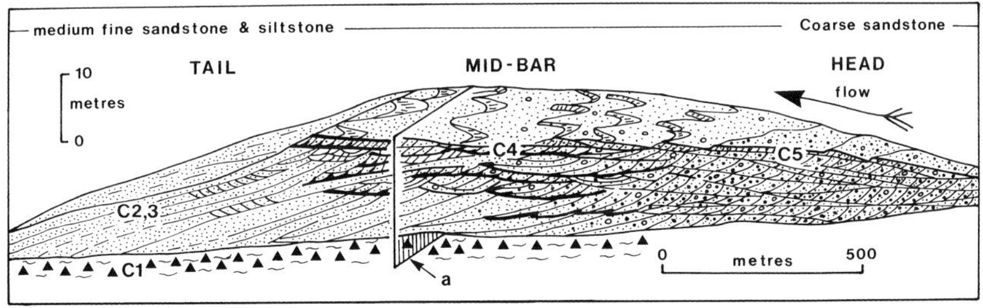

Fig. 7. Explanation of the section at Crossgates–Burnside based partly on sections at various orientations seen elsewhere in the Old Red Sandstone. (**a**) is the section given in Fig. 6, where abbreviations are explained.

Fig. 6). These cross-stratified beds are thought to be a lens within the unit of large-scale planar strata (Fig. 6b), the latter continuing immediately east above the breccias and small exposures high on the outcrop.

This section lies oblique to the palaeoflow. The following interpretation is based on the general lateral relationships in lithologies, the cross-strata orientation within each lithological unit, and on the relationships seen in other exposures (Fig. 7). The large-scale cross strata (C2 and C3) built into a pool in which a basal bed of mud-clast conglomerate and breccia (C1) had accumulated. At low flow-stage, suspended mud draped over these large foresets, and upstream of them coarser sediment accumulated in large trough-producing bed forms of the mid-bar (C4). The sediments of the mid-bar interfingered with the large cross strata of the tail as the bar migrated into the pool, which itself migrated downstream ahead of it (Fig. 7).

The minimum depth of the pool is determined by the thickness of these tail cross strata (C2 and C3), and the minimum depth of the river by the thickness of the entire bar. The large scale cross strata (C2 and C3) are not seen to be affected by the low flow-stage erosion and would have formed in a channel that would have had banks extending well above these bar deposits.

It is not possible to determine the total thickness of the sediment bar, because of lack of vertical exposure. A minimum bar thickness of

12 m is given by the maximum measurable thickness of the large cross strata of the tail. The total river depth was probably 15–20 m, so the river that deposited the bar was substantially deeper than those that had been responsible for sediment accumulation in the Midland Valley up to this time. Haughton & Bluck (1988) and Bluck *et al.* (1992) have suggested that these sediments were deposited by a major river system that drained the coeval, uplifting Scandinavian block far to the northeast (see Fig. 1).

The Lanark basin

The Lower Old Red Sandstone of the Lanark basin rests on rocks of Silurian age (Wellman & Richardson 1993), which are exposed in a series of well-studied inliers (possibly representing a series of discrete sedimentary basins), initially marine but gradually filled with fluvial sediment. Sedimentation in these basins culminated in a widespread conglomerate, the Greywacke Conglomerate. This conglomerate is taken as the base of the Old Red Sandstone in the Lanark basin, and, although in the inliers at Girvan and the Pentland Hills there is an angular unconformity between the Greywacke Conglomerate and the underlying rocks, there is no angular break in both the Hagshaw and Lesmahagow inliers.

The Lower Old Red Sandstone sequence in the Lanark basin is thinner, has more sandstone and has finer-grained conglomerates than found in the Strathmore basin, but both basins have interstratified andesitic–basaltic lavas and an abundance of lava clasts in many of the conglomerates. As with the Strathmore basin, the involvement of the flanking block as a source for the sediments fill is both equivocal and paradoxical.

Smith (1995) and Phillips *et al.* (1997) have subdivided the Lower Old Red Sandstone of the Lanark basin into four formations (see Fig. 2). The basal Greywacke Conglomerate is dominated by greywacke clasts (but still with some clasts of fine-grained lava) and, in places, comprises wedges thinning to the northwest and with a palaeoflow in that direction (McGiven 1967; Bluck 1978). In addition, Syba (1989) identified a major flow component from the northeast and this flow continues into the Girvan region (Fig. 3).

The petrographic compositions of the clasts in the Greywacke Conglomerate have been studied in detail by McGiven (1967), Bluck (1983, 1984), Syba (1989) and Smith (1995), all of whom have failed to match them with rocks now exposed in the bordering Southern Uplands. However, preliminary studies have suggested that the clasts do have a petrographic similarity to those now seen in the Great Conglomerate, which is seen to unconformably overlie the greywackes of the northeast Southern Uplands (Figs 2 and 3).

The conglomerates associated with the Swanshaw, Duneaton and Auchtitench Formations (see Fig. 2) are dominated by igneous clasts that closely resemble the local lavas with which they are interstratified. The palaeoflow measurements so far obtained suggest dispersal from the south (Syba 1989; Smith 1995), but palaeoflow indicators from the associated sandstones clearly show the southwest palaeoflow seen in the underlying Greywacke Conglomerate (Bluck 1978). Exposures of these axial sandstones are generally limited, but those that are large enough show some cross strata >6 m thick. The sandstones may have formed in river systems as large as those postulated for the Strathmore basin and therefore shared in this drainage.

Syba (1989), Smith (1995) and Phillips *et al.* (1997) concluded that the Greywacke Conglomerate was deposited in a series of strike-slip generated basins, with a dominant source to the northeast. These basins could be the continuation of the earlier (Silurian) strike-slip basins postulated by Phillips *et al.* (1997). The role of the Southern Uplands as a source at this time is doubtful.

The shared composition and palaeoflow orientation of the Greywacke and Great Conglomerates suggests their lateral correlation. Absolute age considerations would allow this: the Great Conglomerate is older than *c.* 400 Ma (Emsian time, Rock & Rundle 1986) and the Greywacke Conglomerate underlies lavas dated at 412 Ma (Emsian–Pragian time, Thirlwall 1988). The location of the source for the greywacke clasts is problematic. An allochthonous sheet of greywacke of unknown provenance, at one time discordantly covering southern parts of the present Southern Uplands or its extension to the northeast, is clearly possible. However, this sheet would have to be removed before the extrusion of the Cheviot lavas at *c.* 396 Ma, which rest directly on sediments of the Southern Uplands. Alternatively, a source block could have existed to the northeast in what is now the North Sea.

As discussed above (see also Fig. 2), with reference to the age and position of the Great Conglomerate and the Cheviot lavas, the Southern Uplands were substantially eroded at this time. Along with the Midland Valley, their northern margin could have been part of the floor to the contemporary Lower Old Red Sandstone basin. Alternatively, the Lanark basin could have extended southwards into the area

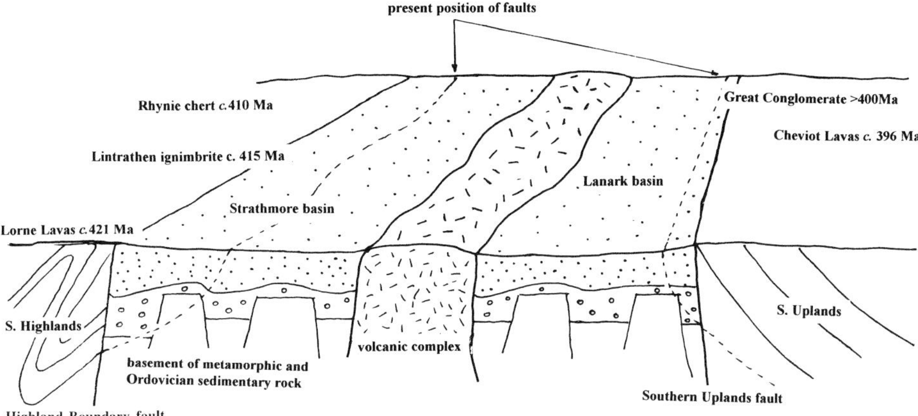

Fig. 8. Block diagram showing the evidence for the planation of the blocks flanking the Midland Valley and the probable Mid-Devonian shortening of the Midland Valley as a result of convergence and the thrusting of the flanking blocks.

now occupied by the Southern Uplands and, as with the Strathmore basin, the southernmost edge of this extended basin would then be the Old Red Sandstone (Great Conglomerate) at present sited on the Southern Uplands block (see Fig. 8).

The role of the Southern Uplands Fault in this case is difficult to establish. Whatever Devonian sediments may have been deposited on the Southern Uplands, they were certainly removed from large tracts of it before Carboniferous times, as rocks of this age are seen to lie unconformably on Lower Palaeozoic rocks (e.g. at Sanquhar). As with the case proposed for the Highlands, the emplacement of the Southern Uplands along the Southern Uplands Fault may have been late, and possibly later than Lower Devonian times. Although fault activity continued into post-Carboniferous times, with the truncation of Carboniferous rocks, Balin (1993), in recording a similarity in Upper Old Red Sandstone sequences north and south of the fault, suggested this later movement at least, to be relatively minor.

The Ayr–Ochil–Sidlaw volcanic axis

The central region of the Midland Valley was occupied by a series of major volcanic centres seen to extend southwest into Ireland and northeastwards to the edge of the North Sea. They are *c.* 2 km thick in the Pentland Hills (Mykura 1960) and up to 3 km thick in the Ochils (Francis *et al.* 1970). These centres have been studied in some detail, and they are seen to thin away from both regions. Lavas are also common elsewhere in the Midland Valley, and are often local in extent and probably related to rapid opening of the fault-controlled sedimentary basins (see Crowell 1974; Hempton & Dunne 1984).

These lavas probably formed on the site of a volcanic–plutonic province that existed intermittently since Ordovician times (Bluck 1984). Clasts of plutonic rocks derived from the north are as young as *c.* 440 Ma (Llandovery time, Longman 1980) in the southwest Midland Valley at Girvan and those from the south are dated at *c.* 420 Ma (Ludlow time) in the NE Midland Valley (Haughton 1988), suggesting this continuity in igneous activity.

Some general considerations

As suggested by George (1960), the Old Red Sandstone basins of the Midland Valley were not originally confined by the major faults that now bound it. There is evidence that, along the northern margin, a basin or even a series of basins intervened between the Strathmore basin and the Dalradian block. The northern edge of one of those basins is preserved in the outliers at Lintrathen and Glen Turret. On the southern margin, it is possible that the Lower Old Red Sandstone basin either extended onto the present Southern Uplands or, as with the Strathmore basin, the southern edge extended partly into ground now occupied by the Southern Uplands (Fig. 8).

In both these scenarios, the Lower Old Red Sandstone history has a significant role in the understanding of the later movement of the bounding faults, and this later history has to be unravelled before any attempt is made to understand the geometry and earlier history of the fractures. The incoming of first-cycle Dalradian clasts into the Strathmore basin during deposition of the upper part of the Lower Old Red Sandstone sequence suggest that there was a possible convergence of the Dalradian block onto the Midland Valley. The age of this convergence is bracketed by 415 Ma (age of the 'Lintrathen porphyry' and the basin sited on Dalradian rocks) and the stratigraphical age of these conglomerates in the Strathmore basin (probably of Emsian time >394 Ma). A regional compression of roughly this age has been suggested by Soper *et al.* (1987).

Early basin evolution was mainly under the control of strike-slip faulting and it is likely that there was much opening and closing of these basins, so that slight changes in the configuration of the stress pattern turned basins of deposition into sources for younger basins. Later in the history of the basin, the dominant influence was the large river system that is believed to have had its source in Scandinavia–Greenland (Fig. 1).

The Upper Old Red Sandstone

The Upper Old Red Sandstone rests unconformably on the Lower and has an uncertain age span. However, the top is gradational into the Carboniferous sequence (Paterson & Hall 1986), and in Fife both fish and miospores are assigned to Late Famennian time (see Browne 1980). The base of the Upper Old Red Sandstone has an unknown age, but Eifelian and Givetian rocks at least, are thought to be absent, indicating a regional uplift somewhere in Mid-Devonian time and a time-gap of >20 Ma.

In contrast to the Lower, the rocks of Upper Old Red Sandstone are thinner, finer grained, and both sandstones and conglomerates are, in general, petrographically more mature. Lavas, with one possible exception, are not present, although lavas are seen to make a contribution to some of the conglomerates and sandstones. Reaching its maximum thickness of >2 km in the west, around the Firth of Clyde, it thins east and south to <300 m in places (Bluck 1980*a*, Paterson & Hall 1986). The eastward thinning direction is also the direction of regional palaeoflow and sediment fining (Bluck, 1978, 1980*a*).

Almost throughout the Midland Valley the Upper Old Red Sandstone sequence fines and matures upward, often terminating with thick deposits of caliche (Balin, this volume). It therefore comprises a major upward-fining cycle, similar to, but more complete than the cycle that characterizes the Lower (see Fig. 11, below). As with the Lower, the early sediments are thought to have accumulated in a transtensional basin, the Clyde basin, centred in the region of the Firth of Clyde and associated with a releasing bend in the Highland Boundary Fault (Bluck 1980*a*). The later fill involved the incoming of large river systems, which on the basis of existing evidence, were a little smaller than those that fed sediment into the Strathmore basin.

The upward-fining sequence in the Firth of Clyde

Although in its general evolution the Upper Old Red Sandstone basin resembles that of the Lower, it differs in both scale and in the relationship between basin and palaeoflow. In the early, fault-controlled basins the palaeoflow was almost always from the southwest, and this flow direction persists through most of the sequence. An exception is seen in the uppermost Old Red Sandstone in the Clyde area, where a late uplift along the Highland Boundary Fault locally shed first-cycle Dalradian clasts southeastwards into the basin (Bluck 1980*a*).

Fluvial sediment bars have been determined from various parts of the Upper Old Red Sandstone sequence (Bluck 1980*b*, 1986). From these findings, both the nature of the stream channels depositing the bars as well as the general nature of the fluvial regimes have been deduced. The thicknesses of the bars have been particularly useful as they determine the minimum depth of the river.

Data collected for the well-exposed Upper Old Red Sandstone rocks around the Firth of Clyde demonstrate that both the structure of the alluvium and the thickness of bars change vertically in the sequence (Fig. 9). The lower part of the sequence is characterized by braided stream deposits with thin bars and shallow streams; these deposits are replaced gradually by deposits of deeper streams, culminating in a deep (>10 m depth) fluvial system replete with flood basins, flood basin deltas and aeolian dunes (Bluck 1992).

Coastal exposures along the Clyde estuary have the advantage of revealing the interrelationships between these large-scale bars and the overbank matrix in which they are found. Many sections record a downstream transition from pebbly sandstones in planar cross strata to medium–coarse sandstones in trough cross strata to medium–fine sandstones with mud drapes in large-scale cross strata. This transition can be traced in single outcrops and is typical of recent alluvial bars as well as other sediment bars in the Old Red Sandstone (Bluck 1986; and, e.g. Fig. 10). Detailed measurements of cross-strata

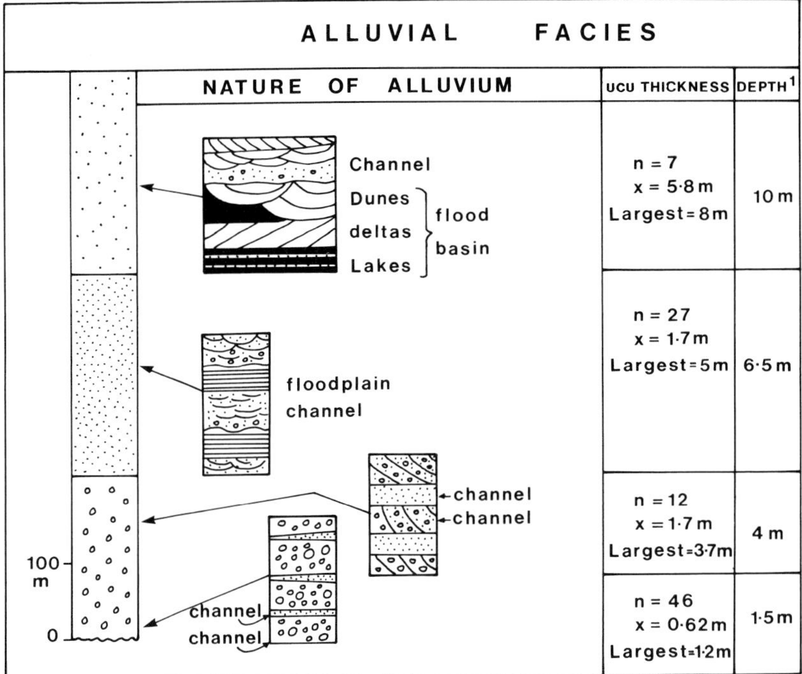

Fig. 9. Vertical changes in alluvial style, the thickness of the upward-coarsening units (UCU) and the suggested channel depth. Data from analysed sections in the Upper Old Red Sandstone of the Clyde region (see Fig. 4 for legend; black indicates mudstone and siltstone). n, number of data; x, mean; UCU, upward-coarsening units; [1], depth is maximum depth of river, estimated from thickness of UCU, a proportion of thickness of floodplain and channel deposits. Lithological legend as for Fig. 11.

dips in relation to lithofacies and to position on the bar demonstrate the complexity of palaeoflow over these structures and permit their reconstruction as set out in detail in Fig. 10.

The Alluvial mega-sequences of the Midland valley

Both Lower and Upper Old Red Sandstone mega-sequences fine and petrographically mature upward, and begin with small-scale fluvial drainage (with local sediment sources) later replaced by a regional drainage. There are significant differences between the Lower and Upper Old Red Sandstone cycles, which illustrate the different controls on their development.

The Lower Old Red Sandstone sequence, with a regional unconformity at its top, is less complete, not as well defined and in many ways more complex than the Upper (see Fig. 11 for comparisons). The Lower Old Red Sandstone cycle begins with deposition in a complex series of basins characterized by recycling of sediment. This pattern suggests rapid basin opening and closing, the continual rejuvenation of the source blocks to yield repetitions of coarse, thick conglomerates, and sufficiently rapid extension to yield lavas coeval with sedimentation. But, as with the Upper Old Red Sandstones in the Clyde basin, the Lower basins show a progressive thinning in one direction, which, in the case of the Lower Old Red Sandstone, was to the southwest.

The sequence terminates upward in a thick, still fairly immature sandstone, mudstone and a single poorly developed calcrete. Basal sandstones within the conglomeratic divisions are characterized by lithic arenites whereas at the top of the sequence the sandstones include both lithic arenites and sublithic arenites. Within this upper division there appears to be a sudden increase in the thickness of alluvial bars and, by inference, in the depth of the rivers that generated them. As with the Upper Old Red Sandstone, these rivers are thought to have drained the contemporary mountainous regions of Scandinavia and Greenland.

In contrast, the Upper Old Red Sandstone cycle is thinner, less conglomeratic, overall finer grained, has little evidence of contemporaneous

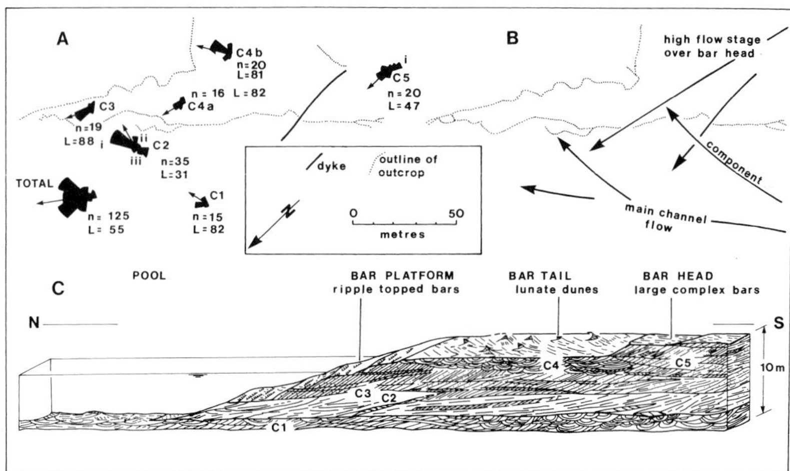

Fig. 10. Showing the Upper Old Red Sandstone alluvial bar exposed at Seamill (NGR NS 2015 4696) on the Clyde coast (see Fig. 3 for location). The exposure is about mid-way up section (i.e. in the middle division of Fig. 11) and is reconstructed from plane-table maps. (**A**) palaeoflow determination for each of the lithological units shown in (**C**). n, number of observations; L, vector magnitude; i, ii, etc. refer to modes of cross strata discussed below. (**B**) interpretation of the palaeoflow measurements based on scales of cross strata and relationships between cross strata units. (**C**) an interpretation of the outcrop (see Bluck 1992), where a major bar–sand body migrates into a pool to yield an upward-coarsening sedimentary unit attaching to a bank to the south. C1, pebbly sandstone with mudstone clasts and trough cross strata (channel deposits); C2, large-scale (> 1 m) cross strata in medium sand with clay drapes (channel pool infill, with a minor mode of cross strata AC2ii representing cross flow in pool, and mode AC2 iii representing counterflow); C3, planar foresets thickening down flow direction (tiers of cross-stratified sand sheets infilling pool); C4, trough cross strata of the mid-bar showing the maximum effects of change in palaeocurrent orientation on falling stage flows; C5, planar cross-stratified pebbly sand (bar head), see Figs 6 and 7).

vulcanicity and is generally characterized by a far simpler vertical sequence. The coarse basal conglomerates show a more simple upward fining, which is not repeated within a basin, but is repeated within basins that are progressively initiated towards the southeast (Bluck 1980*a*). Conglomerates are replaced upward by sandstones and thin conglomerates with a gradual upward increase in alluvial bar thickness (and implied river depth), and the sequence terminates in regional stability with extensive and repeated caliche formation. This vertical change is accompanied by a clear change in sandstone petrography from basal lithic and sublithic arenites to sublithic arenites and quartz arenites towards the top (Fig. 11).

It is clear that the Upper Old Red Sandstone cycle is far more compositionally mature than the Lower and this is partly related to the Lower providing a source for the Upper. In addition, the initial phase of fault-controlled basin development produced restricted drainage basins which, through time, expanded, although possibly still remaining within the confines of the Northern British Isles and its immediate surroundings. As shown to occur in parts of the east African rift system (Soreghan & Cohen 1993), drainage basin expansion is accompanied by increased mineralogical maturity of the sediment. In contrast, the Lower Old Red Sandstone cycle suggests that throughout the sequence there is high-magnitude rejuvenation of the drainage systems by repeated faulting accompanied by much igneous activity, although the rate at which this rejuvenation occurs is reduced through time.

Upward-fining, upward-maturing sequences similar to those described here have been widely recorded. They characterize basins generated by extension or transtension. Examples are seen in the Triassic rocks surrounding the present Atlantic, where they are produced during continental break-up; the Neoproterozoic basins of the North Atlantic that preceded the Caledonian cycle; the Permo-Triassic sediments in Central Spain; and Tertiary sediments in the extensional–transtensional Guadix basin, Spain (e.g. Viseras 1991; Soria *et al.* 1998).

Upward-fining cycles contain a great deal of information that can be translated into

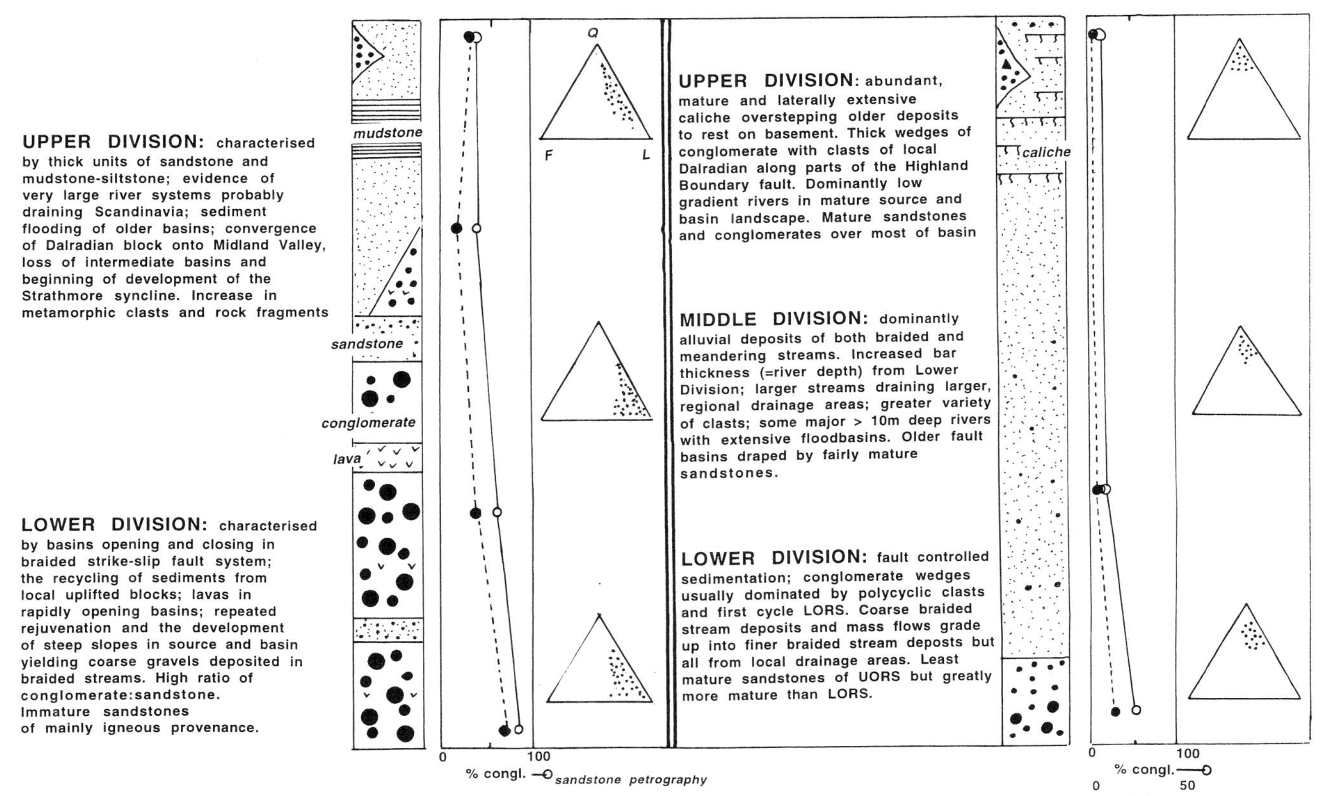

Fig. 11. Upward-fining sequences for both Lower and Upper Old Red Sandstones illustrating the difference in the nature of the two basins. Vertical changes in clast size (MCS, maximum clast size, averaged); changes in per cent of conglomerates and sandstone petrography. Q, quartz; F, feldspar, L, lithic fragments. Summary of facts and interpretations.

source-basin behaviour. They have a key role to play in uniting the geomorphological development of a source with the sedimentary record of a basin. The difference between the Upper and Lower Old Red Sandstone illustrates that point.

Conclusions

The Old Red Sandstone of the Midland Valley is a remarkably good record of the final stages of terrane amalgamation. During Silurian and Early Devonian times, the Midland Valley was a complex of faults, along which volcanic activity was concentrated and basins opened and closed as blocks shifted into place. Immediate sources and basins were mainly within the Midland Valley, and highly mature, recycled sediments were mixed with first-cycle labile volcanic clasts to yield conglomerates with unusual compositions.

Terranes on either side of the Midland Valley had lost most of their cover by Devonian times so that these partly peneplained blocks failed to provide a great deal of sediment to the basins sited there. However, the minor contributions from the lateral blocks are invaluable in their documentation of terrane boundary activity. There is ample evidence of convergence of the Dalradian block onto the Midland Valley as faults, initially involved in probable lateral translation, may have been converted to thrusts. Basins separating the Dalradian block from the Strathmore basin were destroyed and the converging Dalradian block, both at the end of Lower and Upper Old Red Sandstone deposition, yielded first-cycle, Dalradian sediment directly to the Strathmore basin. The convergence recorded in the upper parts of the Lower Old Red Sandstone culminated in the development of the Strathmore syncline in Mid-Devonian time (Kennedy 1958).

Lower and Upper Old Red Sandstone basins are characterized by upward-fining sedimentary sequences. In both units, the lower parts of the sequence develop under the influence of strike-slip faulting, and in the Lower, and possibly in the Upper Old Red Sandstone, these basins are flooded by finer sediments in which there is strong evidence for deposition from large-scale river systems with distant drainage basins.

An external river system would come from the contemporaneous uplifts taking place in the Scandinavian Caledonides. It would seem that these large-scale river systems, draining the Scandian orogen, were ready to enter the Midland Valley whenever there was space available to fill with sediment. In Lower Old Red Sandstone times the route was direct from the northeast; in the Upper it would have come in via western Scotland and dispersed sediment towards the northeast.

The vertical sequence records the changing nature of both basin and source landscape. In the initial stage, the thick units of coarse conglomerates were deposited by streams flowing on steep slopes and the upward-fining and -maturing of the sediment was a response to a more subdued landscape where sediment was stored for long periods before entering the basin. Short-headed streams were replaced by deeper drainage deriving sediment from a wider source area. As the new or rejuvenated drainage network expanded, so the rivers deepened, bar thickness increased, the source yielded more mature sediment and the basin reverted to the development of thick and laterally extensive caliches.

T. Dempster, G. Rogers, P. Haughton and E. Syba are amongst the many who have discussed the problems of provenance of the Lower Old Red Sandstone at length and in these discussions my own views have been focused. Thanks are also due to I. Frazer, whose hospitality over the decades has made all our stays at The Burn a great pleasure. I was fortunate to have two reviewers, N. Woodcock and D. Balin, who read the manuscript with great care and gave constructive advice.

References

Allen, D. A. 1940. The geology of the Highland Border from Glen Almond to Glen Artney. *Transactions of the Royal Society of Edinburgh*, **60**, 171–193.

Andersen, T. B., Berry, H. N. IV, Lux, D. R. & Andresen, A. 1998. The tectonic significance of pre-Scandian $^{40}Ar/^{39}Ar$ phengite cooling ages in the Caledonides of western Norway. *Journal of the Geological Society, London*, **155**, 297–309.

Anderson, W. M., Barker, A. J., Bennett, D. G. & Dallmeyer, R. D. 1992. A tectonic model for Scandian terrrane accretion in the Northern Scandinavian Caledonides. *Journal of the Geological Society, London*, **149**, 727–741.

Armstrong, M. & Paterson, I. B. 1970. *The Lower Old Red Sandstone of the Strathmore Region*. Institute of Geological Sciences Report, **70/12**.

Baliin, D. F. 1993. *Upper Old Red Sandstone sedimentation in the eastern Midland Valley area, Scotland*. PhD thesis, University of Cambridge.

Balin, D. F. 2000. Calcrete morphology and karst development in the Upper Old Red Sandstone at Milton Ness, Scotland. This volume.

Bluck, B. J. 1978. Sedimentation in a late orogenic basin: the Old Red Sandstone of the Midland Valley of Scotland. *In*: Bowes, D. R. & Leake, B. E. (eds) *Crustal Evolution in NW Britain and Adjacent Regions*. Seel House Press, Liverpool, 249–278.

——1980*a*. Evolution of a strike-slip fault-controlled basin, Upper Old Red Sandstone, Scotland. *In*: BALLANCE, P. F & READING, H. G. (eds) *Sedimentation in Oblique-Slip Mobile Zones*. International Association of Sedimentologists, Special Publication, **4**, 63–78.

——1980*b*. Structure, generation and preservation of upward fining, braided stream cycles in the Old Red Sandstone of Scotland. *Transactions of the Royal Society of Edinburgh: Earth Sciences*, **71**, 29–46.

——1983. The role of the Midland valley in the Caledonian Orogeny. *Transactions of the Royal Society of Edinburgh: Earth Sciences*, **74**, 119–139.

——1984. Pre-Carboniferous history of the Midland Valley of Scotland. *Transactions of the Royal Society of Edinburgh: Earth Sciences*, **75**, 275–295.

——1986. Upward coarsening sedimentation units and facies lineages, Old Red Sandstone, Scotland. *Transactions of the Royal Society of Edinburgh: Earth Sciences*, **77**, 251–264.

——1992. Upper Old Red Sandstone of the Firth of Clyde. *In*: LAWSON, J. D. & WEEDON, D. S. (eds) *Geological Excursions around Glasgow and Girvan*. Geological Society of Glasgow, 200–229.

—— & INGHAM, J. K. 1997. The Highland Border Controversy: a discussion of 'New evidence that the Lower Cambrian Leny Limestone at Callander, Perthshire, belongs to the Dalradian Supergroup, and a reassessment of the "exotic" status of the Highland Border Complex'. *Geological Magazine*, **134**, 563–570.

—— & LEAKE, B. E. 1986. Late Ordovician to Early Silurian amalgamation of the Dalradian and adjacent Ordovician rocks in the British Isles. *Geology*, **14**, 917–919.

——, COPE, J. C. W. & SCRUTTON, C. T. 1992. Devonian, *In*: COPE, J. C. W., INGHAM, J. K. & RAWSON, P. F. (eds) *Atlas of Palaeogeography and Lithofacies*. Geological Society of London, Memoir, **13**, 57–66.

——, HAUGHTON, P. D. W., HOUSE, M. R., SELWOOD, E. B. & TUNBRIDGE, I. P. 1988. Devonian of England, Wales and Scotland. *In*: MCMILLAN, N. J., EMBRY, A. F. & GLASS, D. J. (eds) *Devonian of the World*. Canadian Society of Petroleum Geologists, Memoir, **14**, 305–324.

BROWNE, M. A. E. 1980. *The Upper Devonian and Lower Carboniferous (Dinantian) of the Firth of Tay, Scotland*. Institute of Geological Sciences Report, **80/9**.

CAMPBELL, R. 1913. The Geology of South-Eastern Kincardineshire. *Transactions of the Royal Society of Edinburgh*, **48**, 923–960.

CROWELL, J. C. 1974. Origin of late Cenozoic basins in Southern California. *In*: DICKINSON, W. R. (ed.) *Tectonics and Sedimentation*. Society of Economic Paleontologists and Minerologists, Special Publications, **22**, 190–204.

CUTHBERT, S. J., HARVEY, M. A. & CARSWELL, D, A. 1983. A tectonic model for the metamorphic evolution of the Basal Gneiss Complex, Western South Norway. *Journal of Metamorphic Geology*, **1**, 63–90.

DALLMEYER, R. D., STRACHAN, R. A. & HENRIKSEN, N. 1994. ^{40}Ar/^{39}Ar mineral age record in NE Greenland: implications for tectonic evolution of North Atlantic Caledonides. *Journal of the Geological Society, London*, **151**, 615–628.

DEMPSTER, T. J. 1985. Uplift patterns and orogenic evolution in the Scottish Dalradian. *Journal of the Geological Society, London*, **142**, 111–128.

—— & BLUCK, B. J. 1991. Xenoliths in the lamprophyre dykes of Lomondside: constraints on the nature of the crust beneath the southern Dalradian. *Scottish Journal of Geology*, **27**, 157–165.

——, HUDSON, N. F. C. & ROGERS, G. 1995. Metamorphism and cooling of the NE Dalradian. *Journal of the Geological Society, London*, **152**, 383–390.

DENTITH, M. C., TRENCH, A. & BLUCK, B. J. 1992. Geophysical constraints on the nature of the Highland Boundary fault zone in western Scotland. *Geological Magazine*, **129**, 411–419.

DEWEY, J. F. & PANKHURST, R. J. 1970. The evolution of the Scottish Caledonides in relation to their radiometric age patterns. *Transactions of the Royal Society of Edinburgh*, **68**, 361–389.

DU TOIT, A. L. 1905. The Lower Old Red Sandstone rocks of the Balmaha–Aberfoyle region. *Transactions of the Edinburgh Geological Society*, **8**, 315–325.

EVANS, J. A., SOPER, N. J. DEMPSTER, T. J., HUDSON, N. F. C. & ROGERS, G. 1997. Discussion on metamorphism and cooling of the NE Dalradian. *Journal of the Geological Society of London*, **154**, 357–360.

FRANCIS, E. H., FORSYTH, I. H. READ, W. A. & ARMSTRONG, M. 1970. *The Geology of the Stirling District*. Memoirs of the Geological Survey, UK.

FRIEND, P. F. & MACDONALD, R. 1968. Volcanic sediments, stratigraphy and tectonic background of the Old Red Sandstone of Kintyre, W. Scotland. *Scottish Journal of Geology*, **4**, 265–283.

——, HARLAND, W. B. & HUDSON, J. D. 1963. The Old Red Sandstone and Highland Boundary in Arran, Scotland. *Transactions of the Edinburgh Geological Society*, **19**, 363–425.

GEORGE, T. N. 1960. The stratigraphical evolution of the Midland Valley. *Transactions of the Geological Society of Glasgow*, **25**, 32–107.

GRADSTEIN, F. M. & OGG, J. 1996. A Phanerozoic time scale. *Episodes*, **19**, 3–5.

HAMES, W. E & ANDRESEN, A. 1996. Timing of Palaeozoic orogeny and extension in the continental shelf of north-central Norway as indicated by laser ^{40}Ar/^{39}Ar muscovite dating. *Geology*, **24**, 1005–1008.

HAUGHTON, P. D. W. 1988. A cryptic Caledonian flysch terrane in Scotland. *Journal of the Geological Society, London*, **145**, 685–703.

——1989. Structure of some Lower Old Red Sandstone conglomerates, Kincardineshire, Scotland: deposition from late-orogenic antecedent stream? *Journal of the Geological Society, London*, **146**, 509–525.

—— & Bluck, B. J. 1988. Diverse alluvial sequences from the Lower Old Red Sandstone of the Strathmore region, Scotland—implications for the relationship between Late Caledonian tectonics and sedimentation. *In*: McMillan, N. J., Embry, A. F. & Glass, D. J. (eds) *Devonian of the World.* Canadian Society of Petroleum Geologists, Memoir, **14**, 269–293.

—— & Farrow, C. M. 1989. Compositional variations in Lower Old Red Sandstone detrital garnets from the Midland Valley of Scotland and the Anglo-Welsh basin. *Geological Magazine*, **126**, 373–396.

——, Rogers, G. & Halliday, A. N. 1990. Provenance of Lower Old Red Sandstone conglomerates, SE Kincardineshire: evidence for the timing of Caledonian terrane accretion in central Scotland. *Journal of the Geological Society, London*, **147**, 105–120.

Hempton, M. R. & Dunne, L. A. 1984. Sedimentation in pull-apart basins: active examples in Eastern Turkey. *Journal of Geology*, **92**, 513–530.

Hutchinson, A. D. 1928. A lava flow at the base of the Kincardineshire Downtonian. *Geological Society of Edinburgh Transactions*, **12**, 69–73.

Kelly, S. 1988. The relation between K–Ar mineral ages, mica grain size and movement on the Moine Thrust Zone, NW Highlands, Scotland. *Journal of the Geological Society, London*, **145**, 1–10.

Kennedy, W. Q. 1958. The tectonic evolution of the Midland Valley. *Transactions of the Geological Society of Glasgow*, **23**, 106–133.

Longman, C. D. 1980. *Age and affinity of granitic detritus in Lower Palaeozoic conglomerates, S. W. Scotland: implications for Caledonian evolution.* PhD thesis, University of Glasgow.

Marshall, J. E. A. 1991. Palynology of the Stonehaven Group: evidence for a Mid-Silurian age and its geological implications. *Geological Magazine*, **128**, 283–286.

——, Haughton, P. D. W. & Hillier, S. J. 1994. Vitrinite reflectivity and the structure and burial history of the Old Red Sandstone of the Midland Valley of Scotland. *Journal of the Geological Society, London*, **151**, 425–438.

McGiven, A. 1967. *Sedimentation and provenance of the post Valentian conglomerates up to and including the basal conglomerate of the Old Red Sandstone in the southern part of the Midland Valley of Scotland.* PhD thesis, University of Glasgow.

Morton, D. J. 1979. Palaeogeographical evolution of the Lower Old Red Sandstone basin in the Western Midland Valley. *Scottish Journal of Geology*, **15**, 97–116.

Mykura, W. 1960. The Lower Old Red Sandstone rocks of the Pentland Hills. *Bulletin of the Geological Survey of Great Britain*, **16**, 131–155.

Nordgullen, Ø. Bickford, M. E., Nissen, A. L. & Worrtman, G. L. 1993. U–Pb ages from the Bindal Batholith, and the tectonic history of the Helgeland Nappe Complex, Scandinavian Caledonides. *Journal of the Geological Society, London*, **150**, 771–783.

Northrup, C. J. 1997. Timing structural assembly, metamorphism, and cooling of Caledonian nappes in the Ofoten–Efjorden area, North Norway: tectonic insights from U–Pb and $^{40}Ar/^{39}Ar$ geochronology. *Journal of Geology*, **105**, 565–582.

Paterson, I. B. & Hall, I. H. S. 1986. *Lithostratigraphy of the Late Devonian and Early Carboniferous Rocks of the Midland Valley of Scotland.* Report of the British Geological Survey, **18/3**.

—— & Harris, A. L. 1969. *The Lower Old Red Sandstone Ignimbrites from Dunkeld, Perthshire.* Institute of Geological Sciences Report, **69/7**.

Phillips, E. R., Smith, R. A. & Carroll, S. 1997. Strike-slip terrane accretion and pre-Carboniferous evolution of the Midland Valley of Scotland. *Transactions of the Royal Society of Edinburgh: Earth Sciences*, **89**, 209–224.

Richardson, J. B. 1967. Some British Lower Devonian spore assemblages and their stratigraphic significance. *Reviews of Palaeobotany and Palynology*, **1**, 111–129.

——, Ford, J. H. & Parker, F. 1984. Miospores, correlation and age of some Scottish Lower Old Red Sandstone sediment from the Strathmore region (Fife and Angus). *Journal of Micropalaeontology*, **3**, 109–124.

Robertson, S. 1987. Early sinistral transpression in the Lower Old Red Sandstone of Kincardineshire, Scotland. *Scottish Journal of Geology*, **23**, 261–268.

Rock, N. M. S. & Rundle, C. C. 1986. Lower Devonian age for the 'Great (basal) Conglomerate', Scottish Borders. *Scottish Journal of Geology*, **22**, 285–288.

Scott, A. C., Edwards, D. & Rolfe, W. D. I. 1976. Fossiliferous Lower Old Red Sandstone near Cardross, Dunbartonshire. *Proceedings of the Geological Society of Glasgow*, **117**, 4–5.

Smith, R. A.1995. The Siluro-Devonian evolution of the Southern Midland Valley of Scotland. *Geological Magazine*, **132**, 503–513.

Soper, N. J, Webb, B. C. & Woodcock, N. H. 1987. Late Caledonian (Acadian) transpression in north-west England: timing geometry and geotectonic significance. *Proceedings of the Yorkshire Geological Society*, **46**, 175–192.

Soreghan, M. J. & Cohen, A. S. 1993. The effects of basin asymmetry on sand composition: examples from Lake Tanganyika, Africa. *In*: Johnsson, M. J. & Basu, A, (eds) *Processes Controlling the Composition of Clastic Sediments.* Geological Society of America, Special Papers, **284**, 285–301.

Soria, J. M. Viseras, C. & Fernandez, J. 1998. Late Miocene–Pliocene tectono-sedimentary evolution and subsidence history of the Central Betic Cordillera (Spain): a case study in the Guadix intermontane basin, *Geological Magazine*, **135**, 565–574.

Syba, E. 1989. *The sedimentation and provenance of the Greywacke Conglomerate, southern Midland Valley, Scotland.* PhD thesis, University of Glasgow.

TANNER, P. W. G. 1995. New evidence that the Lower Cambrian Leny Limestone at Callander, Perthshire, belongs to the Dalradian Supergroup, and a reassessment of the 'exotic' status of the Highland Border Complex. *Geological Magazine*, **132**, 956–483.

THIRLWALL, M. F. 1988. Geochronology of Late Caledonian magmatism in northern Britain. *Journal of the Geological Scoiety, London*, **145**, 951–967.

TRENCH, A. & HAUGHTON, P. D. W. 1988. Palaeomagnetic study of the 'Lintrathen porphyry' Highland Border region, Central Scotland: terrane linkage of the Grampian and Midland Valley blocks in Late Silurian time? *Geophysical Journal*, **93**, 554.

TUCKER, R. D., BRADLEY, D. C., VER STRAETEN, C. A., HARRIS, A. G., EBERT, J. R. & MCCUTCHEON, S. R. 1998. New U–Pb zircon ages and the division and duration of Devonian time. *Earth and Planetary Science Letters*, **158**, 175–186.

VISERAS, C. 1991. *Estratigrafia y sedimentalogia del relleno alluvial de la Cuenca de Guadix* (*Cordilleras Beticas*). Secretariado de Publicaciones de la Universidad de Granada, Granada.

WELLMAN, C. H. 1993. A land plant microfossil assemblage of mid-Silurian age from Silurian inliers from the Midland Valley of Scotland. *Journal of Micropalaeontology*, **12**, 47–66.

—— & RICHARDSON, J. B. 1993. Terrestrial plant microfossils from the Silurian inliers of the Midland Valley of Scotland. *Palaeontology*, **36**, 155–193.

WILSON, A. C. 1980. The Devonian sedimentation and tectonism of a rapidly subsiding, semi-arid, fluvial basin in the Midland Valley of Scotland. *Scottish Journal of Geology*, **16**, 291–313.

ZIEGLER, P. T. 1988. Laurussia—the Old Red Continent. *In*. McMillan, N. J., Embry, A. F. & Glass, D. J. (eds) *Devonian of the World*. Canadian Society of Petroleum Geologists, Memoir, **14**, 15–48.

Palaeoecology and plant succession in a borehole through the Rhynie cherts, Lower Old Red Sandstone, Scotland

CLARE L. POWELL[1], NIGEL H. TREWIN[1] & DIANNE EDWARDS[2]

[1]*Department of Geology and Petroleum Geology, Aberdeen University, Aberdeen, AB24 3UE, UK*

[2]*Department of Earth Sciences, Cardiff University, PO Box 914, Cardiff CF1 3YE, UK*

Abstract: A cored borehole through the Early Devonian Rhynie cherts at Rhynie, Aberdeenshire, NE Scotland, has revealed 53 chert beds in 35.41 m of core. The cherts originated as sinters deposited by hot-spring activity. Chert comprises 4.20 m of the cored succession, with the thickest bed, representing a single silicification event, being 0.31 m thick and the thickest composite chert (comprising six beds) 0.76 m thick. Average chert bed thickness is 80 mm. Forty-five plant-bearing chert beds are interbedded with sandstones, mudstones and shales. The sediments were deposited on an alluvial plain with local lakes, the area being periodically affected by hot-spring activity. Plants initially colonized both subaerial sand and sinter surfaces. *Rhynia gwynne-vaughanii* and *Horneophyton lignieri* commonly form the basal parts of the profiles with subsequent colonization by other genera. *Rhynia* is commonly found in life position above originally sandy substrates, and *Horneophyton* above sinter surfaces. The composition of the Rhynie vegetation is compared with coeval assemblages and, on the basis of current knowledge, it is concluded that there is no unequivocal evidence that the plants were adapted to life in the stressed environments in the immediate vicinity of hot springs.

The Rhynie cherts occur in the northern part of the Rhynie outlier of Lower Old Red Sandstone in Aberdeenshire, within a sequence of fluvial, overbank and lacustrine sediments, the 'Shales with thin sandstones' of Trewin & Rice (1992). These rocks were interpreted to overlie tuffaceous sandstones, a lava and pre-lava sandstones in the local succession (Trewin & Rice 1992; Fig. 1 and Fig. 3, below), but more recent drilling in 1997 has revealed that the cherts overlie over 100 m (estimated stratigraphic thickness) of lacustrine and fluvial facies, necessitating revision of local stratigraphy and structure. The cherts represent the surface expression of a hydrothermal system in the Rhynie area (Rice *et al.* 1995), which resulted in extensive alteration and partial mineralization of sediments, lava and tuffaceous deposits. Quartz and chert veining, faulting and brecciation are further effects of the hydrothermal activity in the area.

Petrographic examination of the cherts shows a variety of textural types that can be matched with textures found in modern hot-spring sinters (Trewin 1994, 1996), and they represent the surface deposits of a precious metal-bearing hot-spring system (Rice & Trewin 1988). The cherts have been radiometrically dated to 396 ± 12 Ma by the $^{40}Ar/^{39}Ar$ method (Rice *et al.* 1995). The palynology of the cherts implies a Pragian–Emsian age (Richardson 1967), refined to a Pragian or possibly earliest Emsian age by Batten (cited by Rice *et al.* 1995). Thus, the Rhynie deposit is the earliest reported subaerial expression of a hot-spring system. This recent work confirms the opinions of Mackie (1913) and Kidston & Lang (1917) that the chert was of hot-spring origin.

Plants were frequently preserved in growth position with the lower parts of aerial axes still upright, as litter lying on the substrate, or as organic matter in thin soils. Siliceous fluids migrated downwards from surface flooding, and also permeated the deposit from the underlying source in the vicinity of hydrothermal vents. The flooding of small pools by silica-rich waters preserved algae, cyanobacteria, and aquatic arthropods in silica. The essentially instantaneous nature of the aqueous preservation process provides snapshots of the biota and surface material (both organic and mineral) at times of hot-spring eruption, and their description forms the main subject of this paper.

From: FRIEND, P. F. & WILLIAMS, B. P. J. (eds). *New Perspectives on the Old Red Sandstone.* Geological Society, London, Special Publications, **180**, 439–457. 0305-8719/00/$15.00

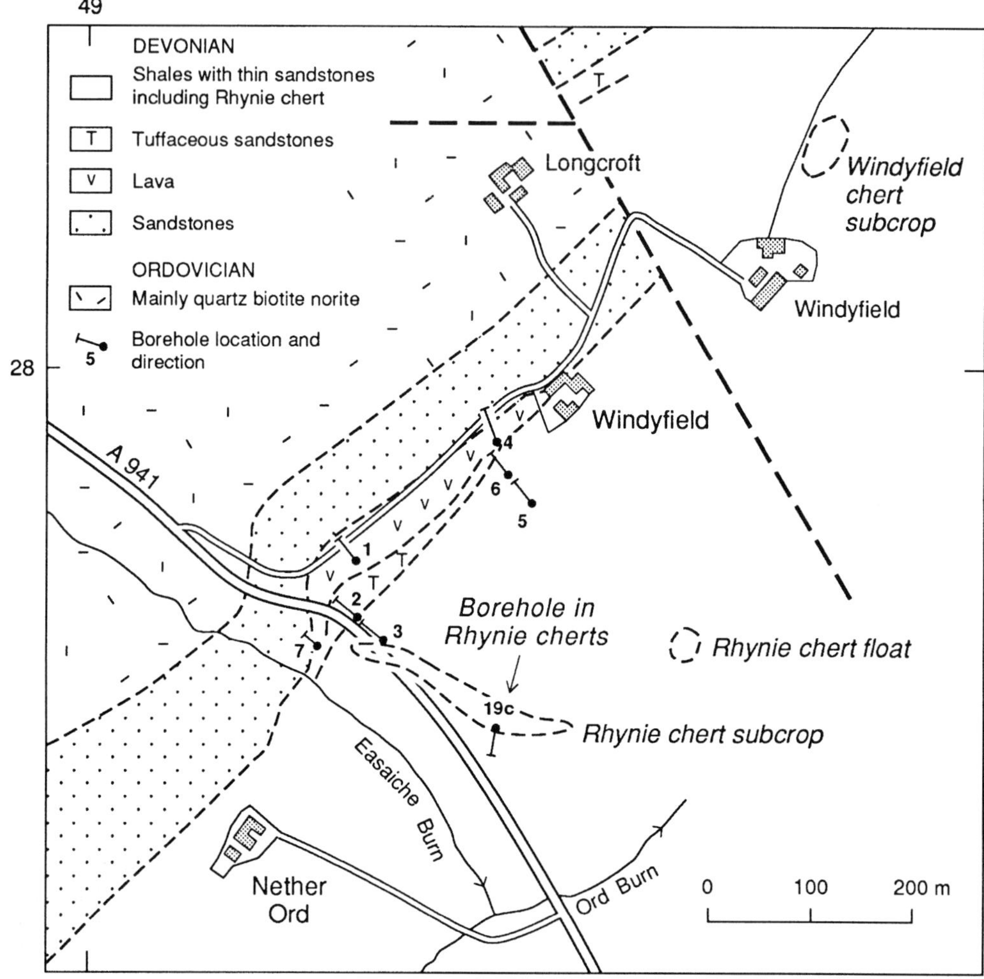

Fig. 1. Map of the immediate area of the Rhynie cherts locality, Rhynie, Aberdeenshire, to show location of borehole 19c (modified from Trewin & Rice (1992)).

Whereas previous studies (e.g. Kidston & Lang 1921, Tasch 1957) have concentrated on loose rocks or material obtained by trenching the weathered zone, a cored borehole through the deposit allows, for the first time, analysis and interpretation of a sequence of plant-bearing cherts from this world-renowned locality. A summary of the Rhynie flora was given by Cleal & Thomas (1995), and of the fauna by Rolfe (1980). Recent additions have been a lichen (Taylor *et al.* 1995*a*), and new information on fungi (Taylor *et al.* 1992, 1995*b*), and on gametophytes (Remy & Hass 1991*a*, *b*, 1996). General considerations of early terrestrial ecosystems and terrestrial colonization have been addressed by Selden & Edwards (1989) and Edwards & Selden (1993).

Material

The material under study comes from the cored borehole 19c, supplemented by loose blocks of chert. It is one of the eight drilled in 1988 reported by Trewin & Rice (1992), who presented a summary log (Trewin & Rice 1992, fig. 7). The borehole was located (Fig. 1) to intercept the known subcrop of the cherts by drilling at an angle of 49° to vertical in direction 206°. A detailed log of the borehole is given in Fig. 2.

The dip of the strata intersected in the borehole is highly variable and several faults and veins are present. Depths are measured along the borehole (maximum 35.41 m) to avoid confusion with sample positions. The borehole was continuously cored, with good core recovery below

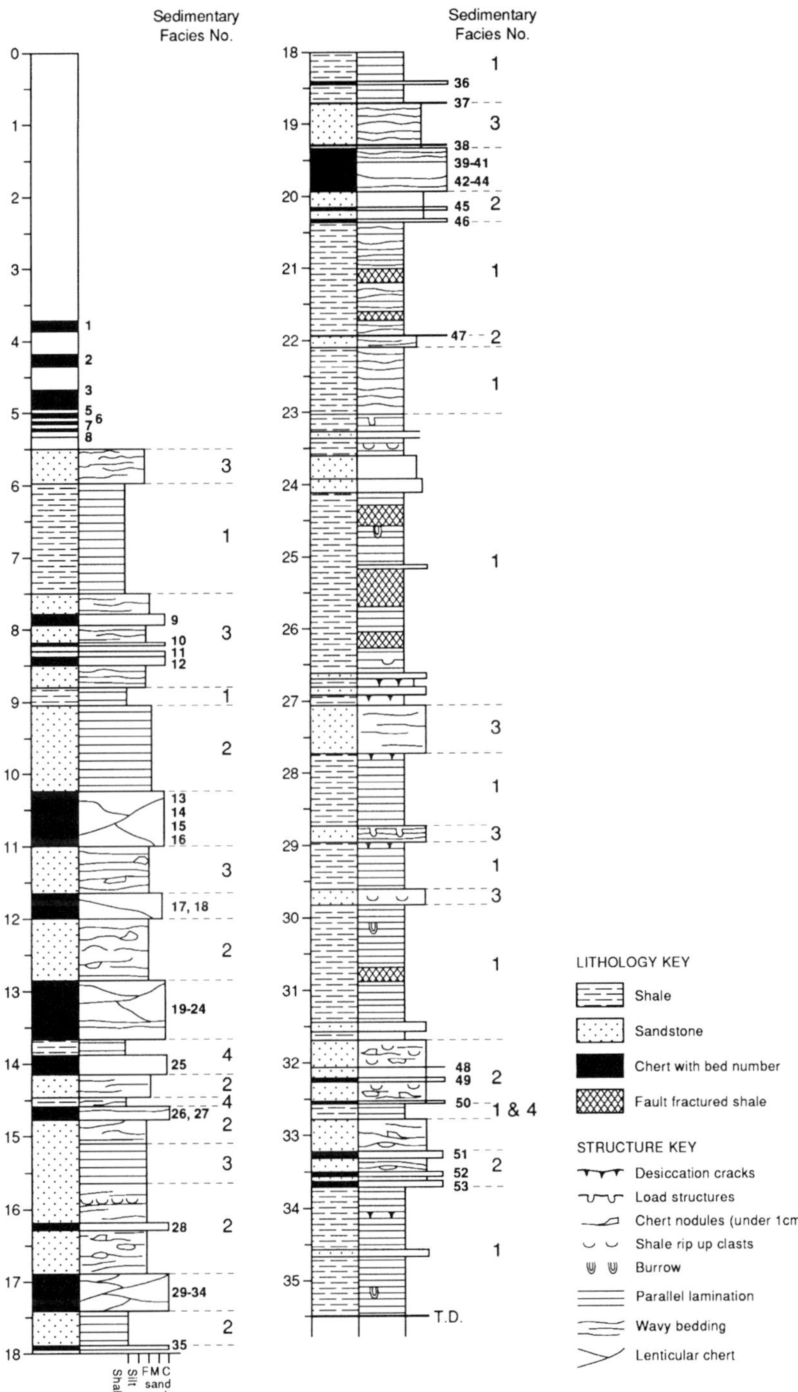

Fig. 2. Sedimentary log of borehole 19c, through the 'Shales with thin sandstones' unit including the Rhynie cherts. See Trewin & Rice (1992) for further stratigraphic details. Depths refer to down-hole measurements. Facies 1: laminated, illite-rich, grey–green shale; Facies 2: carbonaceous, micaceous, chert-cemented sandstone; Facies 3: laminated, carbonaceous, micaceous buff to blue-coloured sandstones; Facies 4: structureless, very pale grey 'porcellaneous' beds.

the weathered zone. The 54 mm diameter core contains soil and drift down to 2.80 m overlying fragmented rock. The topmost rock unequivocally *in situ* occurs at 5.49 m; weathering is apparent down to 11.5 m. The chert beds are numbered from the top (bed 1) to the base of the core (bed 53). A reference slice of the core was embedded in resin, and thin sections and polished blocks from the remaining core material formed the basis for this study. Loose blocks of chert from surface float in the area were also studied. All material is held in the Department of Geology and Petroleum Geology, University of Aberdeen (Powell 1994).

Previous palaeoecological interpretations

The brief assessments of the succession of cherts by Kidston & Lang (1917, 1921), Tasch (1957) and El-Saadawy (1966) were based on trench material (Table 1). Kidston & Lang (1917) recorded *c.* 2.5 m of cherts in a sequence *c.* 3.7 m thick. They observed cherts dominated by monotypic stands of plants immediately succeeding each other, in addition to rarer cherts containing more diverse assemblages. They noted the composite nature of some beds, each with distinctive character, and recognized wide lateral variation in the sequence. Kidston & Lang (1921) developed Mackie's (1913) idea that hot-spring activity had permineralized the plants. El-Saadawy (1966) used material from trenches dug as close as possible to those of Kidston & Lang, and in recording a different sequence of plants, suggested this resulted from great lateral variation in both lithology and plant assemblages. All these workers believed the chert to represent some kind of peat deposit. Knoll (1985) applied Leo & Barghoorn's (1976) model for wood silicification to silicified peats, including the Rhynie chert, and considered the depositional environment to have been a 'swamp'.

Lithologies

The lithologies present in the borehole have been previously described (Trewin & Rice 1992; Powell 1994; Trewin 1994) and are reviewed briefly to provide palaeoenvironmental support for consideration of plant ecology and succession.

Cherts

Chert is present as single or composite beds, almost exclusively bounded by chert-cemented fine-grained sandstone. It also occurs as isolated and composite nodules up to 5 cm in thickness within sandstones (Fig. 3). Chert textures have been described by Trewin (1994) as massive to vuggy, lenticular, laminated, nodular and brecciated. Individual cherts may display more than one character; for example, laminated chert may be brecciated. Plant material occurs in cherts of all textures but is generally best preserved in massive to vuggy cherts and is least common in laminated chert. Detailed textures in the cherts compare closely with those illustrated for

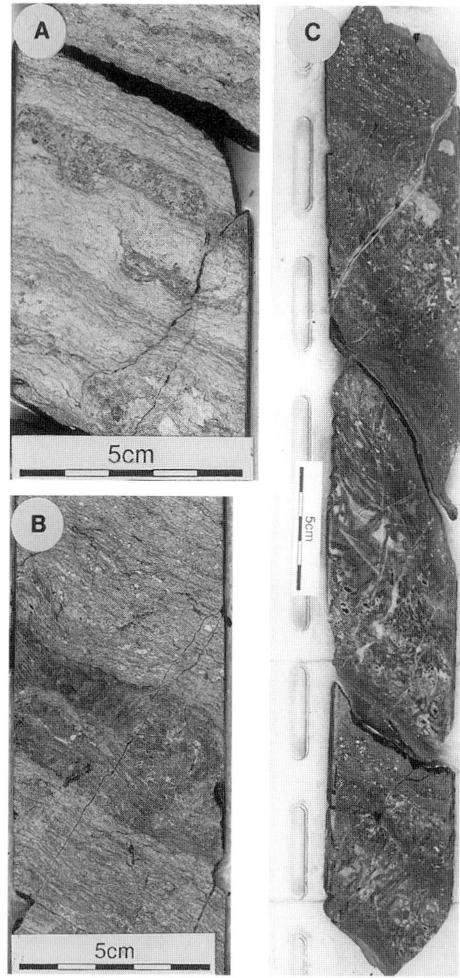

Fig. 3. Chert bed morphologies encountered in borehole 19c. (**a**) Nodular chert: small chert nodules, including individually silicified plant axes, are present in facies 2 sandstone (carbonaceous, chert-cemented sandstone); 19c, 19.3 m. (**b**) Thin chert bed in facies 2 sandstone, representing a single silicification event; 19c, 16.2 m. (**c**) Composite chert: sequence of several chert beds with 3D plant remains, separated by more compacted carbonaceous chert; 19c, base at 13.20 m.

Table 1. *Summary (above) of the palaeontology of Kidston & Lang's trench (1917, 1921), and (below) El-Saadawy's trench 2A (1966)*

K & L's Horizons	Thickness cm (approx.)	Lithology	N	As	R	Ag	H	Other Groups
O	13	cherty sst						
N	15	cherty sst					•	F
						•	•	F
M	23	chert	•			•	•	F
L	30	chert				•		F
K	8	chert				•		
I	8	chert				•		
H	3	chert				•		
G	8	chert				•		
F	13	chert		•		•		A/F/Cy
E	33	chert				•		
D	8	cherty sst				•		F
C	8	cherty sst						
B		cherty sst		•		•	•	
B		chert		•			•	
B	23	chert					•	
B		chert		•			•	
B		chert				•	•	
A′	15	cherty sst						
A″		cherty sst						
A″		cherty sst		•				
A″	38	chert		•			•	
A″		chert				•	•	
A″		chert					•	F
A″		chert			•	•		A/F/Ar/C/Cy/N

El-S's Horizons	Thickness cm (approx.)	Lithology	N	As	R	Ag	H
A	76	cherty sst	•	•	•	•	•
B	91	chert			•		
C	18	sandy shale					
D	20	sst					
E	10	laminated clay					
F	30	chert			•	•	
G	10	shale					
H	25	sst					
I	38	sandy clay					
J	23	cherty sst					
K	28	chert		•	•	•	
L	46	sst					
M	28	chert				•	
N	41	cherty sst					
O	15	cherty sst					
P	8	cherty sst					

Key: N = *Nothia*; As = *Asteroxylon*; R = *Rhynia*; Ag = *Aglaophyton; H = Horneophyton*; F = fungi; Ar = arthropods; C = crustaceans; A = algae; Cy = cyanophytes; N = nematophytes; sst = sandstone.

modern siliceous sinters in Yellowstone National Park by Cady & Farmer (1996).

Interbedded lithologies

The interbedded lithologies (Fig. 2) have been divided into lithofacies 1–4 (Powell 1994).

Lithofacies 1. Grey–green shale and thin sandstones dominate the core below a depth of 20 m. One chert bed is present in this facies. The shales are illite rich and display parallel or wavy laminations up to 3 mm thick and contain mud and sand-filled desiccation cracks. Sandstone and siltstone beds with current ripples and normal grading are up to 60 mm thick.

Lithofacies 2. Carbonaceous, micaceous, chert-cemented sandstone dominates the core above 20 m. This fine-grained sandstone is dominated by detrital quartz and mica; beds are up to 30 mm thick. Thin (to 3 mm) shales and desiccation cracks are present. This facies encloses over 90% of cherts in the borehole. The upper contacts of chert beds with these sandstones are commonly sharp, and the lower contacts gradational over a few millimetres. Plants in the sandstone are preserved as compacted and stylolitized carbonaceous streaks.

Lithofacies 3: Laminated, carbonaceous, micaceous, buff to blue sandstones form units up to 400 mm thick. Beds are up to 80 mm thick, displaying wavy or parallel bedding, and contain shale and siltstone laminae to 5 mm, and shale rip-up clasts are also present.

Lithofacies 4. Structureless, very pale grey 'porcellaneous' clay beds, composed of an illite or illite and smectite–chlorite intergrowth form beds a maximum of 30 mm thick that are associated with nodular cherts. This facies is a minor component, comprising only 320 mm of the core.

Lithological association of chert beds

Of the 45 cherts that occur *in situ* below 5.49 m in the borehole, 27 cherts are underlain by sandstone (60%) and 16 cherts are underlain by chert in composite beds (36%). One bed is underlain by porcellaneous clay and one by shale. Thus, there is a strong association between chert beds and sandstones of lithofacies 2 and 3.

Depositional environments

The cherts at Rhynie display textures typical of modern subaerial sinters (Trewin 1994, 1996) deposited in the immediate vicinity of hot-spring vents. In particular, the laminated cherts are typical of deposition on sinter terraces, and small domed structures resemble the abiogenic stromatolites of Walter (1976). A single loose block has been found with a texture typical of sinters deposited in the splash zone around geyser vents (Trewin 1994). Some massive and lenticular cherts apparently accumulated in water-filled hollows and preserved a freshwater biota (*Palaeonitella* and *Lepidocaris*), but the majority of plants were engulfed where they grew on subaerial surfaces. Sandy substrates and old sinter surfaces were favoured sites for plant growth.

The interbedded lithofacies 1–4 in the core (Fig. 2) are characteristic of shallow-water to emergent conditions. Lithofacies 1 (grey–green shales) was deposited in shallow muddy pools of an ephemeral nature as shown by desiccation features; periodic floods introduced thin graded silty and sandy laminae and beds as well as ripple laminated sands. The sandstones of lithofacies 2 and 3 show evidence of rapid deposition with parallel or wavy lamination and shale rip-up clasts. Desiccation cracks indicate post-depositional exposure, and plant colonization resulted in carbonaceous sandstones underlying plant-bearing cherts.

The general environment was an alluvial plain with lakes, bordering a fluvial system draining north to the present area of the Moray Firth. Periodic overbank flooding resulted in deposition of the observed lithologies. Hot-spring activity probably disrupted local drainage and resulted in the formation of pools and low sinter mounds. Surface drainage and subsurface fluid flow clearly would have been complex. The general setting of the Rhynie locality in a subsiding depositional half-graben basin accounts for the preservation of the hot-spring sinters and associated alluvial plain deposits in an interbedded deposit.

General observations on silicification and plant accumulation

Within the cherts, plants, algae, fungi and bacterial filaments are three-dimensionally preserved by a combination of silica coating on organic surfaces and permineralization of tissue. The state of preservation depends on timing and rapidity of silicification, and the decay state of the organic material before permineralization. The plants in the cherts display a gradation of preservation states, sometimes within one bed, from perfect cellular preservation to unidentifiable remains (Trewin 1994, fig. 5c). Intermediate stages in the silicification of plant remains were illustrated by Trewin (1996, fig. 2).

Decay of modern plants aids precipitation of silica in hot springs (Ferris *et al.* 1986); the initial silica coating on plant material provides support. Hence, sinters with plant debris are generally highly porous and have loose packing. Further silica deposition may lead to preservation of this texture, as seen in the massive texture cherts at Rhynie. However, the highly porous sinter is prone to compaction and brecciation if the silicification process ceases at an early stage. Massive cherts commonly contain aerial parts of plants in growth positions above a litter layer, which can also contain *in situ* rhizomes. Thus, a single silicification event can result in the preservation of a complete cross-section through the living plant community and litter present above a primitive soil traversed by rhizomes.

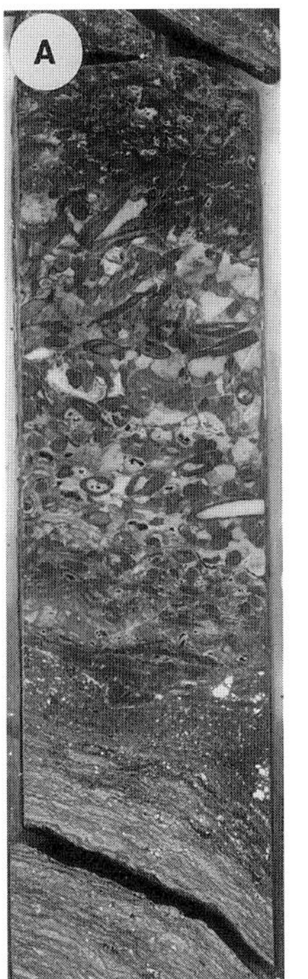

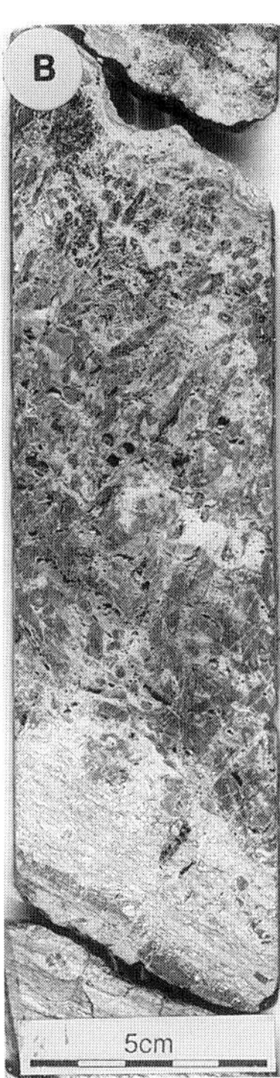

Fig. 4. Sections of plant-rich chert beds. (**a**) Plant-bearing chert overlying compacted chert-cemented sandstone (lithofacies 2). Gradation in compaction of plant debris is symmetrical about the central part of the bed in which plant axes are preserved in very loose packing, with chert infill between the axes; 19c, 17.3 m. (**b**) Plant-bearing chert overlying chert-cemented sandstone. Many plant axes are preserved in growth position. The spaces between the axes are only partly filled by chert, leaving a porous vuggy texture; 19c, 19.80 m.

That silicification was sometimes rapid is indicated by the fact that plants in some beds are in growth positions (axes upright; rhizomes horizontal) and internal parts show no sign of decay. The maintenance of upright axes, and the uncompressed nature of the permineralized fossils suggests that silicification fronts moved inwards preserving the cells, and silica deposition on the outsides of plant axes filled the spaces between axes (see Figs 3c and 4). In many cases silicification was incomplete, and remaining organic matter in the stem axes decayed to leave a hollow cylinder. The degree of silicification probably related to factors such as silica content of the water, temperature, rate of supply of silica-rich water and time. The plants clearly were submerged during silicification.

There is evidence in the cherts for both *in situ* (autochthonous) and transported (allochthonous) plant accumulation. Unicellular rhizoids still attached to the undersides of rhizomes that cut through a litter or soil are taken as evidence of *in situ* growth. It is possible that the litter itself could be allochthonous, representing plant debris accumulated after flooding episodes, and possibly wind transport. Geopetal infill of vugs demonstrating that aerial axes were upright and rhizomes prostrate at the time of silicification provides evidence for *in situ* vegetation. Some cherts contain plant axes that appear to have been bent over by water currents immediately before silicification. This is paralleled at Yellowstone, USA, where plants growing in little-used runoff channels from Giantess Geyser were killed and bent over in the flow direction by hot water from a major eruption, and remained bent after flow had ceased (Edwards & Trewin, pers. obs. Aug. 1999). At Rhynie, in beds where plant material has been compressed before silicification, this has been taken to indicate a predominantly autochthonous litter. However, the degree of compression will be affected by the amount of sediment present and the state of decay of the plants entering the litter from the standing vegetation. The abundance and diversity of fungal hyphae and cysts indicate that profiles were probably not continually waterlogged. Many horizons consist of autochthonous material, but others contain both transported and non-transported material. All the organic material is considered to have been of local origin. The variations in compression of the plant debris relate both to very early (less than 1 m burial) pre-silification compaction and deeper burial compaction associated with cementation.

Regions of chert with cuticular 'straws', spore masses, pieces of xylem and some siliciclastic detritus typical of a young soil lack any indication of the bioturbation (e.g. from worms) that might be anticipated in similar modern-day soils, or as is commonly seen as burrows in palaeosols. Bioturbation is present in sandstones in the core, but is not recognized in the plant-bearing cherts. It is concluded that both the allochthonous and autochthonous elements of the plant-rich beds accumulated over very short periods of time, seldom exceeding a few years, between hot-spring eruptions or clastic depositional events. Mature soil profiles did not have time to develop, and use of the term 'peat' (Kidston & Lang 1917; El-Saadawy 1966; Knoll 1985) for the organic-rich beds is possibly inappropriate given that only one bed exceeds 30 cm, and most contain significant quantities of sediment. The organic-rich beds represent silification of surface plant litter plus growing plants, rather than preservation of mature soil profiles.

The nature of the contribution of standing crop to underlying litter and subsequent humus in Early Devonian soils is conjectural as little is known of the longevity and type of senescence of aerial parts. Exceptions would be abscissed organs such as sporangia and branches of *Rhynia gwynne-vaughanii,* or organs physically dislodged by wind or flooding, which anatomically would probably vary little from the living state. The essentially parenchymatous construction of the Rhynie plants suggests they may have behaved rather like the leaves and stems of monocot bulbs (e.g. *Muscaria*), where on completion of flowering the aerial parts shrivel up and become essentially cuticular envelopes (straws) that sometimes contain the remains of xylem. Such decay states commonly are seen in the litter in the chert, although whether the decay occurred pre- or post-transport or even during silicification frequently cannot be determined (see above). The predominantly parenchymatous Rhynie plants thus contrast with the senescent state of plants with more woody cortical tissues typical of plants found preserved in sandstones and shales for which a modern analogue might be *Psilotum*. In *Psilotum*, after sporangial dehiscence, stems turn first yellow then brown and shrivel, eventually becoming brittle and falling off. More detailed studies of senescence v. decay states of individual fossil taxa are clearly required.

Physical and biotic environments in the environs of hydrothermal systems

Geothermal areas are dynamic. Most modern hot springs are surrounded by prolific plant growth, although not on the active, hot areas of sinter terraces. A zonation of vegetation is present in the immediate vicinity of hot-spring vents and geysers. This zonation is largely controlled by soil temperature and also by soil acidity and chemistry (Burns 1997).

Plants growing at the margins and overhanging pools or steam vents become coated by sinter and draw siliceous water into their roots (preserving plants *in situ*); or fall into the water, and are transported via outflow channels and incorporated into sinter (allochthonous deposits). In modern geothermal areas, plants are not encrusted where there is no standing water, but can be incorporated into sinter, where they are usually preserved as plant moulds (Walter *et al.* 1996).

Autochthonous deposits are preserved as a result of change in the hydrodynamics of the system (White *et al.* 1989; Trewin 1994). Water conduits frequently choke with sinter deposition and the site of activity will relocate, resulting in silicification of plants growing in that area. The previously hot pool will cool, perhaps drain, and be colonized subsequently by a new succession of plants. In contrast, Jones *et al.* (1998) described a scenario leading to silicification in flat-lying, marshy areas. Here development of rhizoliths around the roots of trees, shrubs and grasses, some showing cellular permineralization, occurred after an increase in the height of the water table after flooding with a mixture of hydrothermal fluids and meteoric water. Silicification was thought to result from permeation of the tissues followed by silica nucleation and precipitation on templates produced on walls of decomposing cells. Jones *et al.* provided an additional or alternative scenario where siliceous fluids were drawn into the still-living plants via the transpiration stream, and after death by evaporation and capillary movements with eventual precipitation within and around the tissues. Evidence for this comes from silicification of plants growing at sites away from hot pools, where high-temperature fluids are not present.

The environments envisaged for sinter accumulation at Rhynie (floodplain with ephemeral and hot-spring pools) may have provided many potential habitats for plant colonization, although these sites varied in terms of water availability, water chemistry including concentrations of heavy metals, pH and amount of humus. Plants would have grown either on the sandy sediments (Fig. 4) or on sinter surfaces (Fig. 5), where water from rivers, rain or ground water was readily available and the substrate relatively well drained. These areas periodically were invaded and flooded by hot-spring waters, for sufficient time to deposit sinter and preserve the biota. When little or no clastic material was deposited between flooding events, composite cherts were formed (Fig. 6). Modern hot-spring areas are characterized by sinter terraces that are frequently spectacular and extensive. Terraces have not yet been encountered in boreholes in the Rhynie area, possibly because hot-spring activity was in a depositional area, and sinter deposition was frequently interrupted by clastic deposition. It is also possible that individual hot springs did not flow for sufficient time to build terraces. However, sinter mounds with splash textures were present at hot-spring vents (Trewin 1994). Modern sinter terraces are generally devoid of clastic supply, being fed direct from hot springs with continuous flow.

Hot-spring eruptions within a low-lying alluvial plain might have assisted in maintaining

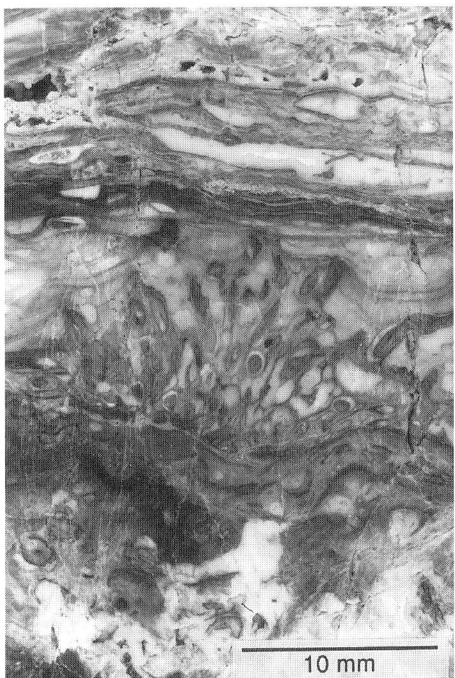

Fig. 5. Part of laminated sinter bed illustrating colonization of sinter surface by plant colony, and subsequent resumption of laminated sinter deposition.

a locally high water table. Although sinter itself is porous and permeable, it would be poorly drained in a low-lying depositional area such as that at Rhynie. The hygroscopic nature of sinter would also help retain moisture. The absence of red-beds and preservation of carbon in the chert-bearing part of the section implies the presence of a high water table, possibly at the surface. Hence the hot spring waters could have flowed into, and flooded, low-lying marshy hollows. Accumulation of plant litter would enhance water retention and also encourage marshy conditions.

Pools in the terrestrial vegetated landscape would have been colonized by cyanobacteria, algae and invertebrates, depending on the temperature and chemistry. However, the aquatic fauna is represented only by *Lepidocaris*, suggesting that either environmental conditions were unsuitable for most invertebrates or that small, isolated pools on the floodplain resulted in logistic problems for dispersal of the aquatic fauna as discussed by Tasch (1957), who speculated that *Lepidocaris* eggs might have been wind transported. Preservation of freshwater biota is to be expected only in very small pools (*c.* 1 m diameter), which are frequent on modern sinter surfaces with a high water table,

Fig. 6. Section through composite chert bed with summary of sequence of events. Three episodes of flooding and silicification by hot-spring waters are recorded in the profile. (**a**) Compacted chert-cemented carbonaceous sandstone. (**b**) First period of flooding. Chert with *Asteroxylon* and ?*Rhynia* at base, some laminated sinter, open vugs. (**c**) Erosion surface colonized by *Rhynia*. (**d**) Second flooding period. Chert beds with sharp irregular base and containing *Rhynia* partly in upright position. Cavities with geopetal floors. (**e**) Detrital influx covers sinter surface. (**f**) Plants recolonize surface and are preserved by third flooding period. Plants poorly preserved, and axis moulds filled with chert–quartz or remaining as empty tubes. (**g**) Sharp sinter top overlain by clastic deposits which have filtered down into the porosity of the sinter. (**h**) Irregular cracks in sinter of all three units are invaded by clastic deposits (darker), indicating existence of extensive porosity at the time of infiltration.

because the temperature and chemistry of the small water volume can be changed rapidly by an influx of siliceous hot-spring water. In lakes the dilution effect is greater, and sinter is not usually deposited subaqueously where hot-spring waters flow into the lake from land, but is found where hot springs erupt through the floor of a lake and a steep temperature gradient is established.

The extensive specialized microbial ecosystems associated with the hotter regions of modern hot springs (e.g. Jones *et al.* 1998), elements of which have been recognized in Devonian hot-spring deposits of Australia (Walter *et al.* 1996), are generally absent in the borehole under discussion because the chert beds represent the cooler end of the system. Material obtained by recent (1997) drilling at Rhynie does contain microbial material such as bacterial stromatolites and mats, which are characteristic of the hot end of the system (Walter *et al.* 1972). This material is the subject of current investigations.

The alluvial plain was occasionally flooded by the local river system and sediment covered the colonized surface, resulting in a break in colonization and an accumulation of sandstone and shale between the chert beds. That plants also grew on the floodplain is indicated by the amount of plant material preserved as coalified compressions in the sandstone partings, and by the diverse spore assemblages (Richardson 1967; Rice *et al.* 1995). The frequent presence of plants in growth position at the base of chert beds indicates their colonization of a sandy substrate onto which spores would have been blown or washed. The influx of sand, which contained locally derived volcanic debris, would initially have produced a damp substrate of high nutrient status. Such substrates would have been accessible to colonizers that would have taken advantage of the available resource and possibly completed the sexual cycle before the substrate became drier and leached. As organic matter content increased in the substrate, the water and nutrient retention capacity presumably also rose, the latter as a result of the provision of more cation–anion exchange sites. The presence of more water and humus in the substrate had two effects: an increase in microbial activity and a subsequent increase in nutrient recycling.

Evidence of bacterial decay is difficult to substantiate and quantify, but Taylor *et al.* (1992, 1995*b*) illustrated a rich mycological flora, including examples of mycoparasitism and, more important to this account, of saprotrophs. In addition, the host and parasite or symbiont display a level of preference, suggesting mycorrhizal activity (Taylor *et al.* 1992). This existence of mycorrhizal fungi in the Rhynie chert plants confirms their ancient origin and possible essential role in plant terrestrialization (Pirozynski & Malloch 1975; Hass *et al.* 1994). In this study, fungi were observed in 39 chert beds (74%). The fungi have thin-walled filamentous and spherical, thick-walled morphologies, common in the mycorrhizal and saprophytic modes of life suggested by Taylor *et al.* (1995*b*). Fungal morphologies observed in the cherts are recorded in Table 2.

Different chert beds preserve different plant genera (Table 3), thus indicating the sampling of different parts of a strongly partitioned ecosystem, seasonality of plant growth or opportunist-dominated systems. The same sequence of plants is not repeated in beds throughout the borehole. Although some plant combinations are more frequent than others (Table 4) no genera are mutually exclusive. This again may suggest slightly different environmental requirements for plant growth in a variety of ecological niches, differing responses to competition of the taxa, or stochastic colonization dynamics among plants with substantially different resource requirements. When the ephemeral and dynamic nature of modern, active geothermal areas is considered, it is not difficult to envisage a wide range of habitats for the Rhynie plants. However, given the short transport distances implied for macroplants in the hot-spring area, all plants probably grew in close proximity. Seasonality in plant growth is a strong possibility but cannot be directly demonstrated from the Rhynie section. However, slightly older (Lochkovian) lacustrine sediments in the Midland Valley display laminites with clastic–carbonate–organic triplets interpreted as indicating a seasonal climate (Trewin & Davidson 1996). Dry summers alternated with cooler and/or wetter winters in a position about 30° south of the Devonian equator. The presence at Rhynie of chert beds with or without spores, and the abundance of decayed plants in some beds but not others, could be taken as evidence that plant growth, spore production and decay were seasonal. Snapshots of the ecosystem at different times of year could have been preserved by random eruptions of hot springs. This provides further evidence that many chert beds record short (maybe one season) periods of plant colonization, and helps explain the surprising absence of spores from some beds (Table 3).

Surface water availability resulted at Rhynie from two sources: rainfall and runoff from hot springs. The rain source was either direct, via outwash from river flooding (indicated by the interpretation of the associated clastic deposits) or through ground water. Geothermal waters contaminated fresh water as shown by the presence of heavy metals (Rice *et al.* 1995) in the Rhynie fluids, hence raising the possibility of phytotoxicity. In addition, an alkaline pH as observed in many modern silica-depositing

Table 2. *Details of spores, fungi, algae and arthropods encountered in the 53 chert beds of borehole 19c; the top of certain bedrock lies between cherts 8 and 9 at 5.49 m (see Fig. 2)*

Chert band no. Borehole 19c	Depth in metres (down hole)	Spores				Fungi			Algae	Arthropods
		Tetrad	Colour	Occ.	Ab.	Morphology	Occ.	Ab.	Occ.	Occ.
1	3.76–3.78		Br	C	2	S/U	Co	5		•
2	4.16–4.32		B/Br	C	4	S/U	C	2		
3	4.65–4.76				1			1		
4	4.76–4.90		B	C	5			1		
5	5.00–5.03				1	S/U	C	2		
6	5.09–5.11		Br	C	3			1		
7	5.18–5.21		B	C	2	S/U&F/M	C/Co	4	•	
8	5.28–5.34		B	C	2	S/F, U&M	Co/C	4		
9	7.81–7.92	•	B/Br	S/C	4			1		•
10	8.27–8.29	•	Br	C	4	S/U&F/M	Co/C	4		•
11	8.35–8.41		Br	C	2	S/U	X	3		
12	8.41–8.50	•	Br	?S/C	3			1		•
13	10.31–10.40				1	F/M	Co	4		•
14	10.40–10.51		Br	C	2	F/M	Co/C	3		
15	10.51–10.72		B	C	2			1		
16	10.72–11.03	•	B	C	2	S/U	C	2		
17	11.70–11.75				1	S/U&F/M	Co/X	5		•
18	11.75–12.00				1	S/U	Co/C	5		
19	12.81–13.09		Br	C	3	S/U&F/M	Co/C	4		•
20	13.09–13.17		B	C	3	S/U	C	4		
21	13.17–13.23				1	S/U	Co	3		
22	13.23–13.31				1	S/U&F/M	Co/C	4		
23	13.31–13.49	•	B	C	2	S/U&F/M	Co	4		
24	13.49–13.57		B	C	3	S/U&F/M	Co/C	3		
25	13.88–14.13		B/Br	C	5	S/U&F/M	Co/C	5		
26	14.63–14.68		Br	C	2			1		
27	14.68–14.74	•	Br	C	2	S/U	Co/C	4		
28	16.19–16.22				1			1		
29	16.84–16.94	•	B	C	5	S/U	Co/C	4		
30	16.94–17.00		B	C	4	S/U&M	Co/C	3		
31	17.00–17.09		B	C	3	S/U	Co/C	2		
32	17.09–17.13		B/Br	S/C	5	S/U	Co/C	5		
33	17.13–17.22	•	Br	C	4	FS/U&M	Co/C	4		
34	17.22–17.33		Br	C	4	S/U&S/M	Co/C	4		
35	17.87–17.89				1	S/U&F/U	Co/C	4		
36	18.31–18.35	•	B/Br	C	5	S/U&F/U	Co	3		
37	18.57–18.58	•	Br	C	4	S/M	Co/C/X	4		
38	19.30–19.31		B/Br	C	4	S/U&F/M	Co	3		
39	19.35–19.36		B	S/C	3	S/U&F/U	Co	4		
40	19.37–19.38		B/Br	C	2	S/U&F/M	Co/C	2		
41	19.41–19.43		B/Br	C	4	S/U&F/U	Co/C	3		
42	19.59–19.65		B/Br	C	3	S/U&F/M	Co	3		
43	19.65–19.79		Br	C	4	S/U&FS	Co/C	3		
44	19.82–19.86		Br	S	3			1		
45	20.09–20.11		B/Br	C	3	S/U&F/M	Co/C	4		•
46	20.28–20.30		Br	S/C	2			1		•
47	21.85–21.90		Br	C	4	S/U&FS	Co/C	5		
48	32.04–32.07		Br	C	5			1		
49	32.21–32.24				1			1		
50	32.51–32.52		Br	C	2	S/U	C	2		
51	33.21–33.26				1			1		
52	33.45–33.49		Br	C	3	S/U	C	3		
53	33.52–33.60				1			1		

Key: 5 = abundant; 4 = common; 3 = rare; 2 = very rare; 1 = absent; B = Black; Br = Brown; S = in sporangium; C = loose in chert; Co = in cortex; X = in xylem; F = Filamentous; FS = Flask Shaped; S = Spherical; U = unicellular; M = multicellular; Occ. = Occurrence; Ab. = Abundance.

Table 3. *Details of plant genera encountered in life position and in plant litter in the 45 plant-bearing chert beds of borehole 19c (only the plant-bearing beds are listed)*

Bed no.	Depth (m)	Life	Axis/rhizome/ sporangium	Litter	Axis/rh/sp	Preservation
1	3.76–3.78			As	ax	C
3	4.65–4.76			Rgv	ax	B
4	4.76–4.90			Rgv	ax	C/E
6	5.09–5.11			Ag	ax	E
7	5.18–5.21			Rgv	ax	D
8	5.28–5.34	As	rh	N//H	ax//ax	D
9	7.81–7.92	Ag	ax/sp	As	ax	D
10	8.27–8.29	As	ax/rh	Ag//H	ax//ax	C/D
11	8.35–8.41	As	rh	N//Ag	ax//ax	D
12	8.41–8.50	As above	ax/rh	Ag//Rgv	ax//ax	D
		Rgv below	ax	N//Rgv//Tri	ax//ax//ax	E
13	10.31–10.40			Rgv//Ag	ax//ax	C/E
14	10.40–10.51			N//Rgv	ax//ax	E
15	10.52–10.72			Agv//Ag	ax//ax	D/E
16	10.72–11.03			Rgv	ax	E
17	11.70–11.75	Rgv		Rgv//N//Ag	ax//ax//ax	D
18	11.75–12.00	Ag//H	ax/rh//rh	H//Ag	ax//ax/rh	B
19	12.81–13.09	H	ax/rh	H	ax	B/C
20	13.09–13.17			Rgv	ax/sp	C
21	13.17–13.23			Ag//Rgv//N	ax//ax//ax	B
22	13.23–13.31	Rgv	ax	Rgv	ax	C
23	13.31–13.49	H above	ax			C/D
		Rgv below	ax/rh	Rgv	ax	C/D
24	13.49–13.57	Rgv above	ax			D/E
		H below	rh	Tri	ax	D/E
25	13.88–14.13	N//Tri//H above	ax//ax//ax/sp	N//Tri/Rgv//H	ax//ax//ax//ax	B/C
		As below	rh	Ag	ax/rh/sp	B/C
27	14.68–14.74	Rgv	ax	Rgv	ax	C/D
28	16.19–16.22			Ag//N//As	sp//ax//ax	B/C
29	16.84–16.94	As above	rh/ax	Rgv//H	ax//ax	A/B
		H below	rh/ax	Rgv//As//H	rh/ax//rh/ax//rh	A/B
30	16.94–17.00	Rgv above	rh/ax	N//H//Rgv	ax//ax//ax	B/C
		H below	rh/ax	N	ax	B/C
31	17.00–17.09	Ag	rh/ax	Rgv//Ag	ax//ax	B/C
32	17.09–17.13	As//N//Ag	ax/rh/sp//ax/sp//ax/rh/sp	Ag//As//N	ax//ax/rh//ax	B
33	17.13–17.22	Ag	ax	Rgv//N	ax//ax	B
34	17.22–17.33	Rgv above	ax/sp			A/B
		As//Ag below	rh/ax//ax/sp	Rgv//Ag//N	ax//ax//ax	A/B
35	17.87–17.89	Rgv//As//N	ax//ax//ax	Rgv//As//N	ax//ax//ax	D/E
36	18.31–18.35			Rgv//N	ax//ax	E
37	18.57–18.58	As	ax/rh/	Rgv//Ag//N//H	ax//ax//ax//ax/sp	D
38	19.30–19.31			H//N	ax//ax	C/D
39	19.35–19.36	H	ax//sp	H	ax	C/D
40	19.37–19.38	N//H	ax//ax	H//N	ax//ax	B/D
41	19.41–19.43	H	ax	H	ax	B/D
42	19.59–19.65	As//N	ax//ax	As//N	ax//ax	D
43	19.65–19.79	Rgv	rh//ax	Rgv//Ag//N	ax//ax//ax	A/B
44	19.82–19.86			H//As	sp/ax//ax	D
45	20.09–20.11			Ag//Rgv//N	ax/sp//ax//ax	B/C
46	20.28–20.30	As	ax/sp	As	ax	B
47	21.85–21.90	Ag	ax	H//As	ax//ax	C
52	33.45–33.49	Rgv	ax/rh	Rgv	ax	B

Preservation: A = full preservation; B = cortical breakdown; C = cuticle & stele; D = unidentifiable cells; E = hollow straw; Note: bed numbers 2, 5, 26, 48, 49, 50, 51, 53 lack macro-plants and are not recorded here.

Table 4. *Summary of occurrence of plant genera in growth position and as litter in the cored sequence of 39* in situ *chert beds (beds 9–53, of which six are not plant bearing)*

	Plants in Life Position			Plants in Litter			
	Rhizomes	Aerial Axes	Total of Beds	Rhizomes	Aerial Axes	Total of Beds	Total No. of Beds with Genus
Rhynia	4	11	11	1	23	23	24
Horneophyton	4	8	10	1	12	13	15
Aglaophyton	3	7	7	2	15	15	18
Asteroxylon	8	9	11	2	9	9	15
Nothia	–	5	5	–	19	19	19
Trichopherophyton	–	1	1	–	3	3	3

streams (e.g. Whakarewarewa, New Zealand, pH 8.1–9.6 at 20 °C at surface (Lloyd 1975)) would have affected availability of mineral nutrients. Thus abundant available Ca would have repercussions for availability of Mg and PO_4, and high pH would also decrease solubility of Fe and Mn, although Al would be available as the aluminate anion. The presence of salts would also affect osmotic water uptake as well as plant growth. Unfortunately, in the fossil situation it is not possible to determine the mix of hot-spring and surface water and hence to have a complete picture of the environmental conditions in which the plants grew. We know the plants were killed and preserved by hot-spring waters, but it is not possible to be certain that they lived in such conditions.

Plant occurrences and successions in the cherts

There is no one vertical succession of plants that can be regarded as typical of a chert bed and thus a generalized cycle of colonization cannot be identified on current evidence. *Rhynia* and *Horneophyton* are commonly present as monotypic stands in the basal parts of beds, with subsequent colonization by other genera. A maximum of three taxa preserved in life position, either as aerial axes and/or rhizomes has been noted in a single bed, a close spatial relationship being obvious from the small area (23 cm^2) of the core. Chert beds containing plant axes in life position are usually underlain by fragmentary plants with compressed or collapsed axes at the base; a maximum of five taxa are recorded as litter in bed 25. The creeping nature of the rhizomes means that different species may be recorded as rhizomes, and as *in situ* vertical aerial axes in the same thin section. Not all possible combinations of taxa occur together in the autochthonous beds but this may be a consequence of small sample size. However, any combination of taxa can occur in the allochthonous plant debris. The total diversity of plants in growth position and in litter only exceeds three in four of the beds.

On a sandy substrate and also on litter, the pioneering colonizer was *Rhynia* (beds 12, 17, 22, 23, 27, 30, 34, 35, 43 and 52). On sinter it was *Horneophyton* as in the beds of a composite unit; beds 30, 29, 23 and 19. However, *Horneophyton* also forms the basal parts of beds 41, 40, 39, 24 and 18 rooted in a sandy substrate.

The occurrence of plants in the chert beds that are undoubtedly *in situ* in the borehole (chert beds 9–53) is summarized in Table 4. The main features of interest are that *Rhynia* is the most common genus, being present in 24 of the 39 *in situ* beds encountered. The common occurrence of *Rhynia* in litter is possibly due to the abscission of adventitious branches during the lifespan of the plant. *Aglaophyton* and *Nothia* are also more than twice as frequent as litter than in life position. On the other hand, *Horneophyton* and *Asteroxylon* are represented more frequently by material in growth position. This may be an artefact of the relative ease of identification, or may indicate that these plants decayed rapidly in autochthonous litter and did not survive transport. *Horneophyton*, *Aglaophyton*, *Asteroxylon* and *Nothia* have a similar rate of occurrence on a bed-by-bed basis and *Trichopherophyton* is rare. Fungi are present in 36 plant-bearing beds (Table 2). Of the chert beds that contain common or abundant fungi, over half (55%) are part of composite units. There are ten chert beds that contain plant litter with fungi, but no plants in life position (1, 7, 13, 14, 16, 20, 21, 36, 38 and 45); eight of these beds contain *Rhynia*.

Whereas the plants in the borehole are commonly in low-diversity stands, the gametophytes recorded by Remy *et al.* (1993) were observed in more mature stands of wider diversity, suggesting that pressures determining the composition of vegetation acted mainly on the later phases of sporophytic growth.

Spore occurrence in the chert

Spores, in the form of dispersed, diverse assemblages or spore masses, are generally common in plant-bearing chert beds but are absent from 12 beds (see Table 2). They might be expected to have been ubiquitous in the air above the vegetation, although there is some evidence that dispersal by air was inefficient in early land plants. That they are absent in cherts with plants in life position (e.g. beds 17, 18, 22 and 35) suggests that the plants were in a vegetative state when silicification occurred, or that sporangia had not shed spores. This might apply particularly to plants with terminal sporangia (e.g. *Aglaophyton*) where sporangial maturity and dehiscence might have been synchronous for all aerial axes in a clone, but is less likely where individual plants exhibited prolonged periods of spore production (e.g. *Nothia*, *Asteroxylon*) and possibly controlled dispersal under favourable conditions (e.g. *Trichopherophyton*).

The complete absence of spores in litter (e.g. beds 4, 13, 21 and 28), where they might be expected to have accumulated via air or water with transported plant debris, is puzzling, even allowing for different transport dynamics. It appears that spore-free litter must have accumulated in a single growing period, and the plants were killed by a silicification event before spore production. Although it may seem unlikely, bearing in mind the resilience of sporopollenin, that the plants were selectively preserved relative to spores, there are examples of single sporangia showing a range in spore appearances, including 'ghosts', that demonstrate local variability in preservation.

Favoured conditions of the Rhynie plants

From the position of the taxa within chert from the core beds and their associations (Tables 3 and 4), and bearing in mind that the area sampled in each horizon is very small, only tentative conclusions can be drawn regarding the typical growth conditions for the Rhynie plants. It is clear that there is no regular overall plant sequence through the chert beds (Tables 1 and 3), and the sequence of events and palaeoecological implications discussed by Tasch (1957) based on consideration of the early literature (particularly Kidston & Lang (1917, 1921)) cannot be substantiated.

Rhynia gwynne-vaughanii

This is the most common plant, both numerically and in terms of ground cover, observed in the borehole, being present in 27 beds (Table 3). In life position it usually forms monotypic stands, either directly above a sandy substrate or with its rhizomes traversing litter and transported debris of other genera and mineral detritus (chert beds 12, 17, 22, 23, 24, 27, 30, 34, 43 and 52). *Rhynia* is thus interpreted as a primary colonizer of sandy, and occasionally muddy, surfaces. This initial colonization was followed by a different sequence of plants in different cherts. This suggests that the gametophytes of *Rhynia gwynne-vaughanii* were opportunists rapidly colonizing newly exposed transiently nutrient-rich sandy, or occasionally muddy, substrates. The sporophytic phase was able to colonize sandy surfaces generally lacking an accumulation of humus.

However, *Rhynia* is found in association with all other genera and is present throughout the borehole, suggesting that it was a generalist tolerant of a wide range of environmental conditions and disturbance, and was capable of withstanding interspecific competition. In the short trench sequence reported by Kidston & Lang (1921), *Rhynia* was confined to the basal bed, leading Tasch (1957) to speculate that it was killed off by a flooding event. Later work has shown the extensive distribution of *Rhynia* in the sequence and Tasch's flooding event cannot be substantiated.

Horneophyton lignieri

Horneophyton is present in 18 beds, with rhizomes and aerial axes commonly in life position, and as monotypic stands (Table 3). *Horneophyton* and *Rhynia* occur together in life position only in three beds, suggesting that they had different optimum conditions for growth (see beds 23, 24 and 30; Table 3). That *Horneophyton* is present as monotypic stands initially colonizing sinter surfaces indicates that it may have tolerated conditions unfavourable for other taxa, or was a more vigorous competitor, or indeed a combination of both.

Horneophyton is present *in situ* at the base of chert beds within composite units, most commonly present as rhizomes cutting through an *in situ* litter of different genera. It was anchored on a porous sinter surface, which probably had good water-retention properties as a result of the hygroscopic nature of sinter. This suggests that *Horneophyton* thrived in wetter conditions, perhaps because of the presence of an intermittent film of water, than the other Rhynie plants and reinforces the view of Remy & Hass (1991*a*) in noting *Horneophyton* was often associated with Chytridomycetes (damp or aquatic fungus) activity. That *Horneophyton* and *Rhynia* stands alternate in composite cherts possibly indicates changes in environmental conditions related in part to flooding events associated with silicification. Wetter conditions might have led to the decline of *Rhynia*, subsequent recolonization by *Horneophyton*, and the process repeated.

Aglaophyton major

Aglaophyton occurs in 20 chert beds, in life position in seven beds and as monotypic stands in four beds, and is associated with *Nothia*, *Horneophyton* and *Asteroxylon* in other beds. In loose blocks it has also been recorded with *Rhynia*. *Aglaophyton* generally grew on a substrate characterised by abundant organic material of diverse origin. It is recorded at a number of horizons in composite cherts, but never colonized sinter surfaces (compare *Horneophyton*), although it occurs once in the core in life position with *Horneophyton*. Where it coexists with other taxa it is the only fertile one.

Remy & Hass (1996) found that germinating spores were preserved in wet or shallow aquatic environments based on their association with algae and aquatic fungi, and speculated that their synchronous early development was rapid to escape colonization by the latter. In the sporophytic phase they considered that the arrested apices and buds on rhizomes were adaptations to resume growth after periods of flooding, having noted that stands of growing *Aglaophyton* sporophytes were frequently associated with abundant *Palaeonitella*, a green aquatic alga. Well-preserved apices and fragmentary aerial axes of *Aglaophyton* infested with aquatic chytrids are often recorded, suggesting that preservation occurred soon after the aerial axes were selectively killed by such flooding events, the absence of fungi in co-occurring rhizome apices and buds indicating their ability to survive and regenerate under such conditions. The homoiohydric features in both gametophytes and sporophytes, including a complex cuticle and stomata that show adaptations to minimize water loss, suggest that the plants were also capable of surviving in drying environments or tolerated physiological drought. However, the presence of stomata in rhizomes among rhizoids suggest that the substrate itself was dry, leading to an inference that *Aglaophyton* was not an aquatic plant but one adapted to periodic flooding and waterlogging, and one exploiting ephemeral very wet conditions during germination.

Nothia aphylla

This plant is present in 21 beds, occurring in both allochthonous and autochthonous litter throughout the borehole (Table 3) usually as fragments in nodular cherts, and in chert-cemented sandstones (lithofacies 2). It appears to have been environmentally versatile, as the taxon is persistent in autochthonous litter across the boundaries of composite chert beds, which see the loss of other genera. *Nothia* is most frequently observed as litter in association with *Rhynia*, *Aglaophyton* and *Asteroxylon*, and instances of its growth with each of these genera have been noted, but it does not occur on its own either as litter or in growth position in the borehole. *Nothia* may have colonized areas different from those colonized by *Rhynia*, or have succeeded *Rhynia* with other genera, but it did not colonize the sinters favoured by *Horneophyton*. The amount of organic material in the substrate may also be significant, as it has not been recorded in mineral soils. All other genera occur in growth position with *Nothia*. This suggests that it needed stable conditions and was not a primary colonizer, although such interpretation is hampered by lack of information on the rhizomes of this plant.

Asteroxylon mackiei

Asteroxylon is found in 17 beds (Table 3), usually as rhizomes traversing a diverse assemblage of plants present as fragments in accumulations of plant debris, and where other genera are established (beds 10, 11, 12, 25, 29, 32, 34, 35 and 37). *Asteroxylon* has the most extensive 'rooting' system of the Rhynie plants and thus could exploit larger volumes for water and mineral nutrients. This, combined with its large size compared with other Rhynie taxa, is here postulated to have conferred major advantages in competition for resources. In ten out of 12 occurrences of *Asteroxylon* in life position it is associated with two or more other genera in the litter. *Asteroxylon* seems to have been part of a plant community, rather than a monotypic colonizer of sediment or sinter surfaces. However, this is in contrast to certain extant herbaceous lycophytes, e.g. *Lycopodium cernuum*, typical of stressed environments including hot springs, where they flourish in mineral 'soils' in the vicinity of steam vents (Burns 1997).

Trichopherophyton teuchansii

This plant has limited occurrence in the borehole, being present in only three beds (12, 24, and 25; Table 3), and in association with a diverse flora including *Rhynia*, suggestive of a late colonizer of humus-rich substrates.

Postscript

The core through the cherts provided an opportunity to document plant ecosystems preserved in cherts in the vicinity of a dynamic hot-springs complex through time, but the limited spatial coverage in a clearly heterogeneous environment does not allow confident conclusions on the palaeoecology of the plants to be drawn. However, the basic record of plant distribution and associations provides valuable data to combine with anatomical studies of the plants. Detailed studies on the growth habits or strategies of Rhynie chert taxa as carried out by the late Professor Remy and his research team (e.g. *Nothia* and *Aglaophyton*) are essential to testing the hypotheses formulated here.

Are the Rhynie plants typical of Early Devonian assemblages?

Detailed cataloguing of the plants in the core, coupled with information on their anatomy and general organisation (e.g. Remy & Hass 1996; Edwards *et al.* 1998) permits a reconsideration of an old controversy relating to the Rhynie plants: were they typical of Pragian wetland–mire vegetation, or were they more specifically adapted to upland inland valleys where preservation is rare but in this case was facilitated by hot-spring activity? Alternatively, were they adapted to the habitats associated with hot springs, and specialized, occupying ecological niches that were unlikely to have been challenged? Evidence for specialization include the folliowing: (1) the composition of the assemblage: no Rhynie taxa have been unequivocally recorded from coeval localities elsewhere; (2) limited palynological data allow correlation with Pragian assemblages, but reveal discrepancies in terms of diversity (Batten, cited by Rice *et al.* (1995)); (3) substrates in the immediate vicinity of hot springs in modern analogues are stressed in terms of water chemistry, pH and temperature (Burns 1997) and thus provide habitats for a specialized flora; this is exemplified by halophytic taxa, such as *Triglochin maritima* on sinter terraces at Yellowstone.

Comparisons of species lists are hampered by taphonomic problems. The low fossilization potential of the predominantly parenchymatous construction of the ground tissues of the Rhynie chert plants mitigates against their common preservation in the generally oxidizing clastic depositional environments elsewhere on the Old Red Continent. Coalified fossils in the fluvial Pragian and older deposits in southern Britain are generally the remains of plants, e.g. *Gosslingia breconensis*, characterized by a thick-walled stereome.

It could be argued that parenchymatous organization and the structural role of turgid tissues are related to colonization of substrates where water is readily available (quite apart from the specialized situation here where flooding by mineral-rich water occurs). However, a recent palaeoecophysiological analysis based on anatomical features associated with stomata suggested sophisticated anatomical adaptations to enhance water use efficiency despite growth on such presumed wet substrates (Edwards *et al.* 1998). Of interest to our conclusions on succession above are the almost identical substomatal cavities in *Rhynia gwynne-vaughanii*, postulated to have colonized well-drained sandy surfaces, and in *Aglaophyton*, which is associated with humus-rich beds that probably had higher water retention capacities. Of course, conclusions based on single environmental factors are dangerous; such adaptations could relate to seasonal atmospheric dryness as in a monsoonal climate, and we need information on the longevity of the aerial parts of the plant. However, we have little evidence to suggest that the Rhynie plants were aquatic or semi-aquatic as described for *Nothia* (Kenrick & Crane 1991) based on absence of tracheidal thickenings and cuticular and stomatal characters. Indeed, the abundance of rhizoids is an indication that the plants were not growing in waterlogged soils.

Comparisons of dispersed spore assemblages in the clastic fluvial and lacustrine sediments between the cherts with those from the cherts themselves (Wellman, unpublished data) should at least reveal any differences between the hot-springs flora and those of the flood plains, and between the latter and clastic assemblages elsewhere, although Rhynie was situated in an inland intermontane basin rather than on a coastal plain. Analysis here is hampered by details of the spore taxa in the Rhynie sporangia: preliminary data from Bhutta (1973*a*, *b*) need enhancement and verification.

Comparisons with vegetation of modern analogues are disappointingly uninformative, and of course, surrounding mature vegetation is usually angiosperm dominated. There appear to be very few floristic surveys (but see Burns

(1997)) and a dearth of ecophysiological studies. Our personal comments are anecdotal and based on general observation. In New Zealand, they confirm that mature vegetation is typical of the region. Detailed analysis of vegetation in the vicinity of steam vents shows different associations (see also Burns (1997)), related in part to thermal gradients but also pH and soil chemistry. However, unlike the alkalinity frequently associated with siliceous hot springs, in these modern New Zealand examples pH may be very low. Personal observations (Edwards) on the steam vents and environs at Craters of the Moon, New Zealand, indicate that recently cooled areas are not vegetated by higher plants. The primary colonizers are mosses, and most surprisingly the tops of steam vents are colonized by *Lycopodium cernuum*, which appears to thrive in the humid atmosphere and is clearly tolerant of high temperatures. In contrast, the records of herbaceous lycopods in the mid-Palaeozoic hot-spring cherts in Queensland, Australia, are interpreted as members of the marshland communities growing at ambient temperatures around geysers and springs (Walter *et al.* 1996). There are limited occurrences where mature vegetation overhanging 'new' vents is encrusted in silica. In the hot-spring localities themselves, young sinter terraces are not vegetated, typical local vegetation occurring on cool 'mature' terraces. The pioneering vegetation here is a sparse covering of grasses of low diversity. There appear to be no autecological studies on these plants, to determine whether or not they are physiologically adapted, nor analysis of the physical characteristics of the substrate. At Yellowstone, hot and dry sinter aprons are dominated by bryophytes and grasses, and relatively hot and wet aprons by *Triglochin*, a salt-marsh plant, frequently rooted in bacterial mats. The marsh plants *Juncus* and *Eleocharis* colonize detrital reworked sinter at about ambient temperature. Mature sinters are colonized by the arborescent weed *Pinus contorta*. Inundation of *P. contorta* by fluids (of thermal or meteoric origin) leads to death, with flooding by siliceous waters producing silicified plants by similar processes to those described by Jones *et al.* (1998).

Knoll (1985) revived the suggestion that the Rhynie plants were relictual, being more primitive and displaced from the flood plain by a 'physiologically more advanced flora'. He envisaged the Rhynie flora retreating to a restricted environment where competition was limited. Alternatively, escape from competition might have been achieved because the plants had specialized physiologies (e.g. to combat stress, as in Yellowstone halophytes) that would have reduced their competitive effectiveness in 'normal' habitats (i.e. stress tolerators). Such hypotheses are difficult to refute or support on physiological grounds, but at the anatomical level the Rhynie plants were sophisticated (Edwards 1993; Edwards *et al.* 1998). At the morphological level, the chert flora is representative of most of the disparity exhibited by coeval plants, although possibly of lower diversity. With the exception of trimerophytes, all the major lineages are represented. *Asteroxylon*, for example, is at a similar morphological and anatomical grade to *Drepanophycus*. Many taxa show growth strategies that enhance prolonged spore production. In addition, the phenomenal preservation reveals anatomical disparity (e.g. in *Aglaophyton*) not recorded elsewhere, a possible reflection of the specialized environment in which they grew that allowed their preservation, rather than a specialized nature enabling them to grow in such an environment. The absence of trimerophytes may reflect the fact that this period saw the beginnings of that group, for example, as *Dawsonites* sp. is very rare in the coeval Brecon Beacons assemblages. The latter are more diverse than the Rhynie chert assemblage, in total 12 taxa, but this number is an aggregate of transported forms: those that were autochthonous are preserved as monospecific stands (e.g. *Gosslingia* and *Tarella*). The diverse assemblages reflect flooding of a far larger catchment area and presumably a variety of habitats, but it is worth reiterating that none of the 'clastic Old Red Sandstone' taxa are recognized in the Rhynie chert. Whether or not Rhynie plants grew in these areas, but were not preserved in such clastic sedimentary successions, might be resolved by detailed scrutiny of dispersed spore assemblages, but this would require more precise information on *in situ* species than is currently available.

In the gametophyte generations Remy *et al.* (1993) concluded that Rhynie examples displayed the same level of organization (terminal gametangiophores) as in other Early Devonian representatives; for example, *Calyculiphyton* (of Emsian age) and *Sciadophyton* (of Pragian–Emsian age). A detailed analysis of their occurrences in relation to the range of substrates described in this study would be rewarding. On the basis of extant ferns, it might be anticipated that they are capable of surviving in a wider variety of stressed environments (e.g. in soil chemistry) than their sporophytes (Young, pers. com.) and would have enjoyed the further advantages of homoiohydry as regards water relations and infection (Remy *et al.* 1993).

Conclusions

The preservation of the Rhynie cherts was due to their accumulation in a subsiding half-graben basin. The host rocks are dominantly fine-grained sandstones, mudstones and shales that were deposited on an alluvial plain with small lakes and ponds. The plants colonized subaerial substrates and autochthonous plant litter accumulated to provide humus. Local, short distance, transport of plant debris resulted in beds of allochthonous litter. The water table was generally close to the surface and small pools existed for long enough to be colonized by aquatic algae and *Lepidocaris*. Hot-spring activity resulted in flooding by hot silica-rich waters of communities of growing plants, litter accumulations and small ponds. The biota was preserved in silica with varying degrees of perfection, resulting in individual chert beds that provide snapshots of the flora and fauna inhabiting surface environments in the vicinity of the hot springs. Silicification of sediments in the borehole described was mainly due to surface flooding with some downward percolation of silica-rich fluid. Studies of loose blocks of chert show that in other areas, near vents, fluids permeated the sediment from below. With burial the original amorphous silica was converted to chert. The resulting chert textures are typical of modern sinters from hot springs.

The frequency of hot-spring eruptions cannot be directly determined, but the observations here of variability between event beds in spore abundance, plant maturity and state of decay indicate that eruptions may have preserved communities at different seasons. There was generally a sufficient gap between eruptions to allow normal sediment accumulation to resume and colonization of cooled sinter surfaces to take place. The lack of pervasive mineralization underlying the cherts indicates that the site of the borehole was not generally one of hot ground or unusual chemistry. There is no direct evidence to believe that conditions between eruptions were anything other than normal for the local alluvial plain. However, we remain uncertain as to whether the vegetation was capable of surviving, or was adapted to, flooding by siliceous waters.

Thus, the snapshots preserved by hot-spring activity are at the cool end of the system. Cherts lacking plants could have resulted when hot waters covered areas lacking plant growth, rather than conditions being too hot to allow plant growth. Thicker, laminated cherts recently discovered at Rhynie probably represent the hotter parts of the system and contain only filamentous bacteria, sometimes in stromatolitic form.

The cherts do not preserve any typical plant successions, but record that in the small area of a borehole penetration of a chert bed the plant diversity is generally low. *Rhynia* appears to have been an initial colonizer of well-drained clastic substrates, and *Horneophyton* occurs as an initial colonizer of sinter substrates. *Asteroxylon* and *Aglaophyton* favoured a situation with accumulations of litter and may have required conditions provided by the developing humus.

This work was carried out while C.L.P. was in receipt of an NERC studentship and a Ciba (now Novartis) Foundation bursary. The borehole was funded by Aberdeen University, and subsequent studies by NERC grant GR3/7048 (C. M. Rice principal investigator). The Royal Society, Carnegie Trust for the Universities of Scotland and the Leverhulme Trust supported N.H.T. in field investigations at Rhynie and in New Zealand and the USA relevant to this paper.

References

BHUTTA, A. A. 1973*a*. On the spores (including germinating spores) of *Horneophyton* (*Hornea*) *lignieri* (Kidston and Lang) Barghoorn and Darrah, 1938. *Pakistan Journal of Botany*, **5**, 45–55.

—— 1973*b*. On the spores (including germinating spores) of *Rhynia major* Kidston and Lang. *Biologia*, **19**, 47–57.

BURNS, B. 1997. Vegetation change along a geothermal stress gradient at the Te Kopia steamfield. *Journal of the Royal Society of New Zealand*, **27**, 279–294.

CADY, S. L. & FARMER, J. D. 1996. Fossilization processes in siliceous thermal springs: trends in preservation along the thermal gradient. *In*: BOCK, G. R. & GOODE, J. (eds) *Evolution of Hydrothermal Ecosystems on Earth (and Mars?)*. Ciba Foundation Symposium, **202**, 150–173.

CLEAL, C. J. & THOMAS, B. A. 1995. *Palaeozoic Palaeobotany of Great Britain*. Chapman and Hall, London.

EDWARDS, D. 1993. Tansley review no. 53: Cells and tissues in the vegetative sporophytes of early land plants. *New Phytologist*, **125**, 225–247.

—— & SELDEN, P. 1993. The development of early terrestrial ecosystems. *Botanical Journal of Scotland*, **46**, 337–366.

——, KERP, H. & HASS, H. 1998. Stomata in early land plants: an anatomical and ecophysiological approach. *Journal of Experimental Botany*, **49**, 255–278.

EL-SAADAWY, W. EL-S. 1966. *Studies on the flora of the Rhynie chert*. PhD thesis, University College of Wales, Aberystwyth.

FERRIS, F. G., BEVERIDGE, T. J. & FYFE, W. S. 1986. Iron silica crystallite nucleation by bacteria in a geothermal sediment. *Nature*, **320**, 609–611.

HASS, H., TAYLOR, T. N. & REMY, W. 1994. Fungi from the Lower Devonian Rhynie Chert: mycoparasitism. *American Journal of Botany*, **8**, 29–37.

JONES, B., RENAULT, R. W., ROSEN, M. R. & KLYEN, L. 1998. Primary siliceous rhizoliths from Loop Road hot springs, North Island, New Zealand. *Journal of Sedimentary Research, A*, **68**, 115–123.

KENRICK, P. & CRANE, P. R. 1991. Water-conducting cells in early fossil land plants: implications for the early evolution of tracheophytes. *Botanical Gazette*, **152**, 335–356.

KIDSTON, R. & LANG, W. H. 1917. On Old Red Sandstone plants showing structure, from the Rhynie chert bed, Aberdeenshire. Part 1: *Rhynia gwynne-vaughanii. Transactions of the Royal Society of Edinburgh*, **51**, 761–784.

—— & —— 1921. On Old Red Sandstone plants showing structure, from the Rhynie chert bed, Aberdeenshire. Part 5: The thallophyta occurring in the peat-bed; the succession of the plants throughout a vertical section of the bed, and the conditions of accumulation and preservation of the deposit. *Transactions of the Royal Society of Edinburgh*, **52**, 852–902.

KNOLL, A. H. 1985. Exceptional preservation of photosynthetic organisms in silicified carbonates and silicified peats. *Philosophical Transactions of the Royal Society of London, Series B*, **311**, 111–122.

LEO, R. F. & BARGHOORN, E. S. 1976. Silicification of wood. *Botanical Museum Leaflets, Harvard University*, **25**(1), 1–47.

LLOYD, E. F. 1975. *Geology of Whakarewarewa hot springs*. NZ Department of Industrial and Scientific Research, Information Series, **111**, 1–24.

MACKIE, W. 1913. The rock series of Craigbeg and Ord Hill, Rhynie, Aberdeenshire. *Transactions of the Edinburgh Geological Society*, **10**, 205–237.

PIROZYNSKI, K. A. & MALLOCH, D. W. 1975. The origin of land plants; a matter of mycotrophism. *Bio Systems*, **6**, 153–164.

POWELL, C. L. 1994. *The palaeoenvironments of the Rhynie Cherts*. PhD thesis, Aberdeen University.

REMY, W. & HASS, H. 1991*a*. *Langiophyton mackiei* nov. gen., nov. spec., ein Gametophyt mit Archegoniophoren aus dem Chert von Rhynie (Unterdevon, Schottland). *Argumenta Palaeobotanica*, **8**, 69–117.

—— & —— 1991*b*. *Kidstonophyton discoides* nov. gen., nov. spec., ein Gametophyt aus dem Chert von Rhynie (Unterdevon, Schottland). *Argumenta Palaeobotanica*, **8**, 29–45.

—— & —— 1996. New information on gametophytes and sporophytes of *Aglaophyton major* and inferences about possible environmental adaptations. *Review of Palaeobotany and Palynology*, **90**, 175–193.

——, GENSEL, P. G. & HASS, H. 1993. The gametophyte generation of some early Devonian land plants. *International Journal of Plant Science*, **154**, 35–58.

RICE, C. M. & TREWIN, N. H. 1988. A Lower Devonian gold-bearing hot-spring system, Rhynie, Scotland. *Transactions of the Institution of Mining and Metallurgy (Section B, Applied Earth Sciences)*, **97**, 141–144.

——, ASHCROFT, W. A., BATTEN, D. J. *et al.* 1995. A Devonian auriferous hot-spring system, Rhynie, Scotland. *Journal of the Geological Society, London*, **152**, 229–250.

RICHARDSON, J. B. 1967. Some British Lower Devonian spore assemblages and their stratigraphic significance. *Review of Palaobotany and Palynology*, **1**, 111–129.

ROLFE, W. D. I. 1980. Early invertebrate terrestrial faunas. *In*: PANCHEN, A. L. (ed.) *The Terrestrial Environment and the Origin of Land Vertebrates.* Academic Press, London, 117–157.

SELDEN, P. A. & EDWARDS, D. 1989. Colonisation of the land. *In*: ALLEN, K. & BRIGGS, D. E. G. (eds) *Evolution and the fossil record.* Bellhaven, London, 122–152.

TASCH, P. 1957. Flora and fauna of the Rhynie Chert: a palaeoecological reevaluation of published evidence. *University of Witchita Bulletin*, **36**, 1–24.

TAYLOR, T. N., HASS, H. & REMY, W. 1992. Devonian fungi: interactions with the green alga *Palaeonitella. Mycologia*, **84**, 901–910.

——, ——, —— & KERP, H. 1995*a*. The oldest fossil lichen. *Nature*, **378**, 244.

——, REMY, W., HASS, H. & KERP, H. 1995*b*. Fossil arbuscular mycorrhizae from the Early Devonian. *Mycologia*, **87**, 560–573.

TREWIN, N. H. 1994. Depositional environment and preservation of biota in the Lower Devonian hot-springs of Rhynie, Aberdeenshire, Scotland. *Transactions of the Royal Society of Edinburgh: Earth Sciences*, **84**, 433–442.

—— 1996. The Rhynie Cherts; an Early Devonian ecosystem preserved by hydrothermal activity. *In*: BOCK, G. R. & GOODE, J. (eds) *Evolution of Hydrothermal Ecosystems on Earth (and Mars?).* Ciba Foundation Symposium, *202*, 131–149.

—— & DAVIDSON, R. G. 1996. An Early Devonian lake and its associated biota in the Midland Valley of Scotland. *Transactions of the Royal Society of Edinburgh: Earth Sciences*, **86**, 233–246.

—— & RICE, C. M. 1992. Stratigraphy and sedimentology of the Rhynie Chert locality. *Scottish Journal of Geology*, **28**, 37–47.

WALTER, M. R. 1976. Geyserites of Yellowstone National Park: an example of abiogenic stromatolites. *In*: WALTER, M. R. (ed.) *Stromatolites.* Developments in Sedimentology, **20**, 87–112.

——, BAULD, J. & BROCK, T. D. 1972. Siliceous algal and bacterial stromatolites in hot-spring and geyser effluents of Yellowstone National Park. *Science*, **178**, 402–405.

——, DESMARAIS, D., FARMER, J. D. & HINMAN, N. W. 1996. Lithofacies and biofacies of mid-Paleozoic thermal spring deposits in the Drummond Basin, Queensland, Australia. *Palaios*, **11**, 497–518.

WHITE, N. C., WOOD, D. G. & LEE, M. C. 1989. Epithermal sinters of Palaeozoic age in north Queensland, Australia. *Geology*, **17**, 718–722.

Age and provenance of limestone clasts in Lower Old Red Sandstone conglomerates: implications for the geological history of the Midland Valley Terrane

HOWARD A. ARMSTRONG[1] & ALAN W. OWEN[2]

[1]*Department of Geological Sciences, University of Durham, South Road, Durham DH1 3LE, UK (e-mail: h.a.armstrong@durham.ac.uk)*

[2]*Division of Earth Sciences, Department of Geography, University of Glasgow, Gregory Building, Lilybank Gardens, Glasgow G12 8QQ, UK*

Abstract: Conodont-bearing limestone clasts in Lower Old Red Sandstone conglomerates in the Lanark and Strathmore basins and the Pentland Hills Inlier, Midland Valley, Scotland, indicate a source in a cryptic arc terrane with a mid-Ordovician (*P. serra–P. anserinus* Biozone) limestone cover. Simpson coefficients of similarity indicate that the faunas from the limestone clasts are closer to conodont faunas from the Holy Cross Mountains, Poland, and the Wrae Limestone in the Northern Belt of the Southern Uplands, than to those in coeval strata from the Laurentian margin including Girvan. Conodont colour alteration index values indicate separate thermal histories for the limestone clasts and coeval strata in the Girvan Inlier. The cryptic arc was located to north of the Northern Belt of the Southern Uplands during Ashgill time and to south of the Midland Valley in Late Silurian–Early Devonian time and clearly had a complex tectonic history.

The Midland Valley Terrane is bounded by the Highland Boundary and Southern Upland faults, both major tectonic boundaries within the northern British Caledonides (Fig. 1; Bluck *et al.* 1992). Elucidation of the Early Palaeozoic and older geology of the Midland Valley Terrane (MVT) has proved difficult because of the extensive middle and Upper Palaeozoic cover. Exposure is largely restricted to the Ordovician–lower Silurian succession of the Girvan Inlier and the Silurian succession of the Central Inliers (Lesmahagow, Pentlands–North Esk and Hagshaw Hills). Geological studies (e.g. Bluck 1985; Hutton 1987; Haughton 1988) have suggested that the basement of the Midland Valley comprises a collage of terranes. Indirect evidence for a complex history is found in suites of clasts within Lower Old Red Sandstone (LORS) conglomerates derived from basement and cryptic cover sequences (Bluck 1978, 1983, 1985; Houghton 1988; Bluck *et al.* 1992).

New finds in LORS conglomerates of limestone clasts of mid-Ordovician age in the Lanark and Strathmore basins, and the Pentland Hills Inlier are documented herein and shed new light on conglomerate provenance. Conodont dating shows these clasts to be all of mid-Ordovician, *P. serra–P. anserinus* Biozone age. We also review clast age and provenance within the regional stratigraphical framework, both north and south of the Southern Upland Fault.

Fossiliferous limestone clasts

LORS conglomerates bearing rare limestone clasts are known principally from the Strathmore Basin and Lanark Basin (including the Pentland Sub-basin), located along the northern and southern margins respectively of the Midland Valley (Fig. 1).

Strathmore Basin

The LORS of the Strathmore succession (Fig. 2) at the NE end of the outcrop has been studied in detail by Haughton (1988). Here the Stonehaven Group overlies the Highland Border Complex (Armstrong & Patterson 1970) and the Cowie Formation at the base of the group yields palynomorphs that indicate a late Wenlock to earliest Ludlow age (Marshall 1991; Wellman 1993). Haughton (1988, 1993; Haughton & Bluck 1988) concluded that the younger Dunnottar (Ludlow) and Crawton (Ludlow–Přídolí) groups were deposited in the so-called Crawton Sub-basin. This developed in a

From: FRIEND, P. F. & WILLIAMS, B. P. J. (eds). *New Perspectives on the Old Red Sandstone.* Geological Society, London, Special Publications, **180**, 459–471. 0305-8719/00/$15.00

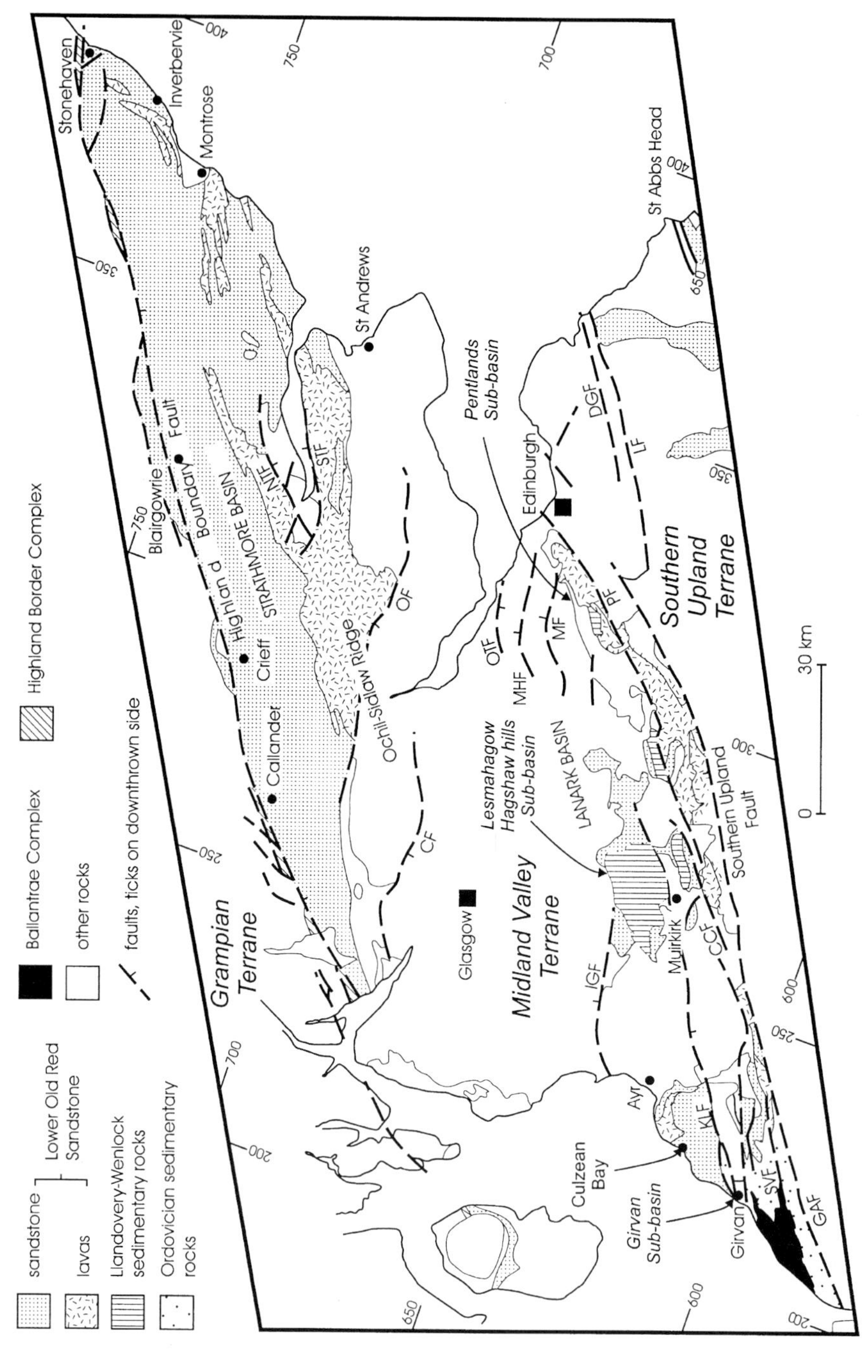

Fig. 1. General geological map of the Midland Valley of Scotland showing the distribution of the Ordovician, Silurian and Lower Old Red Sandstone sedimentary successions. Lower Old Red Sandstone volcanic rocks and the two principal sedimentary basins and sub-basins are marked (redrawn after Phillips *et al.* (1997)). CCF, Carmacoup Fault; CF, Campsie Fault; DGF, Dunbar Gifford fault; GAF, Glen App Fault; IGF, Inchgotrid Fault; KLF, Kerse Loch Fault; LF, Lammemuir Fault; MF, Murieston Fault; MHF, Middleton Fault; NTF, North Tay Fault; OF, Ochil Fault; OTF, Ochiltree Fault; PF, Pentland Fault; STF, South Tay Fault; SVF, Stinchar Valley Fault.

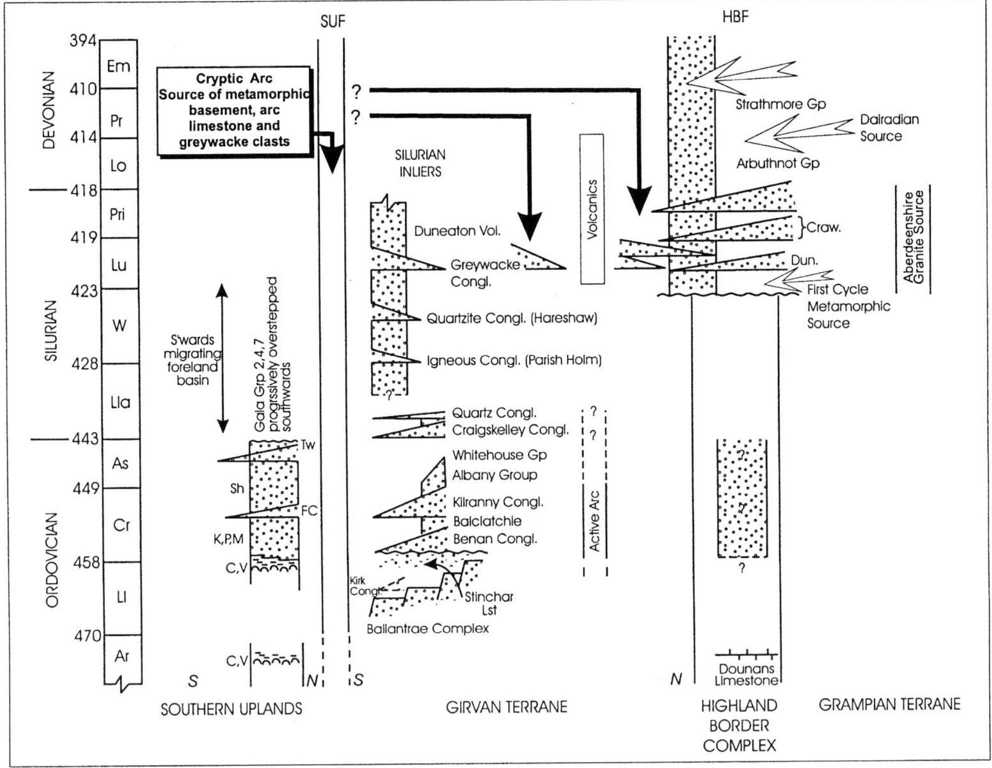

Fig. 2. Chrono- and lithostratigraphy of Ordovician to Devonian successions of the Southern Uplands (composite section), Midland Valley and Grampian terranes. Data synthesized from various sources, principally Smith (1995), Floyd (1996) and Phillips *et al.* (1997). Ordovician and Lower Silurian time-scale based on Harland *et al.* (1990); Upper Silurian and Devonian time-scale based on Tucker *et al.* (1998). SUF, Southern Uplands Fault; HBF, Highland Boundary Fault; C, V, chert and volcanic rocks; Craw, Crawton Group; Dun, Dunnottar Group; K, Kirkcolm Formation; Kirk, Kirkland Conglomerate; M, Marchburn Formation; P, Portpatrick Formation; Sh, Shinnel Formation; Tw, Tweeddale Member.

strike-slip setting and was fed with coarse sediment that had sources in gravel deposits mantling or lying within small basins on a metamorphic, probably Dalradian surface. Consequently, Phillips *et al.* (1997) suggested that the Dalradian and Highland Border Complex occupied broadly comparable positions and were at a similar erosion level to those seen at the present day. The Crawton Group also contains southerly derived conglomerates, which yield a suite of foliated and non-foliated calc-alkaline granite boulders, first-cycle low-grade metamorphic clasts including metagreywacke, psammite and limestones (Haughton 1988; Haughton & Bluck 1988; Haughton *et al.* 1990). Haughton (1988) considered a source for the meta-sedimentary clasts (including limestone) within a cryptic terrane that lay to the south and contained a shallow-marine carbonate, and greywacke succession intruded by high-level plutons (Haughton & Halliday 1991).

A limestone clast from the Crawton Group at Inverbervie yielded a silicified brachiopod, provisionally assigned by Ingham *et al.* (1985, fig. 2b,c), to the Lower Ordovician genus *Archaeorthis*. The preservation of this specimen is extremely poor and its identification is probably best considered as no more than an indeterminate orthidine brachiopod (Harper, pers. comm.).

The present study has isolated a poorly preserved conodont faunule from a limestone boulder at Inverbervie, comprising *Periodon aculeatus* and *Polonodus* sp. (Fig. 3j). The former is abundant in the middle Ordovician succession of southern Scotland (Bergström & Orchard 1985; Armstrong 1997). *Polonodus* is only known from the Wrae Limestone conglomerate in Tweeddale in the Southern Uplands (Bergström

& Orchard 1985) and indicates a probable mid-Ordovician, *Pygodus anserinus* Biozone, i.e. Llandeilo to Aurelucian, latest Llanvirn–earliest Caradoc age.

Lanark Basin

Upper Silurian–Lower Devonian (LORS) strata of the Lanark Basin are well exposed on the Ayrshire coast; on the northern side of Culzean Bay (NS 245130) the strata dip gently towards the northwest and are overlain by volcanics rocks. Here the sedimentary rocks of the Lanark Group consist predominantly of reddish purple or brown micaceous and feldspathic sandstones with thick pebble to boulder conglomerates. The conglomerates contain clasts mainly of greywacke, chert, jasper, felsitic porphyry and rare limestone. Fossiliferous sparry limestone cobbles first recognized by Smith (1895) are more common towards the top of the succession. Preliminary palaeocurrent analysis in Culzean Bay suggests that the conglomerates here were derived from the southeast.

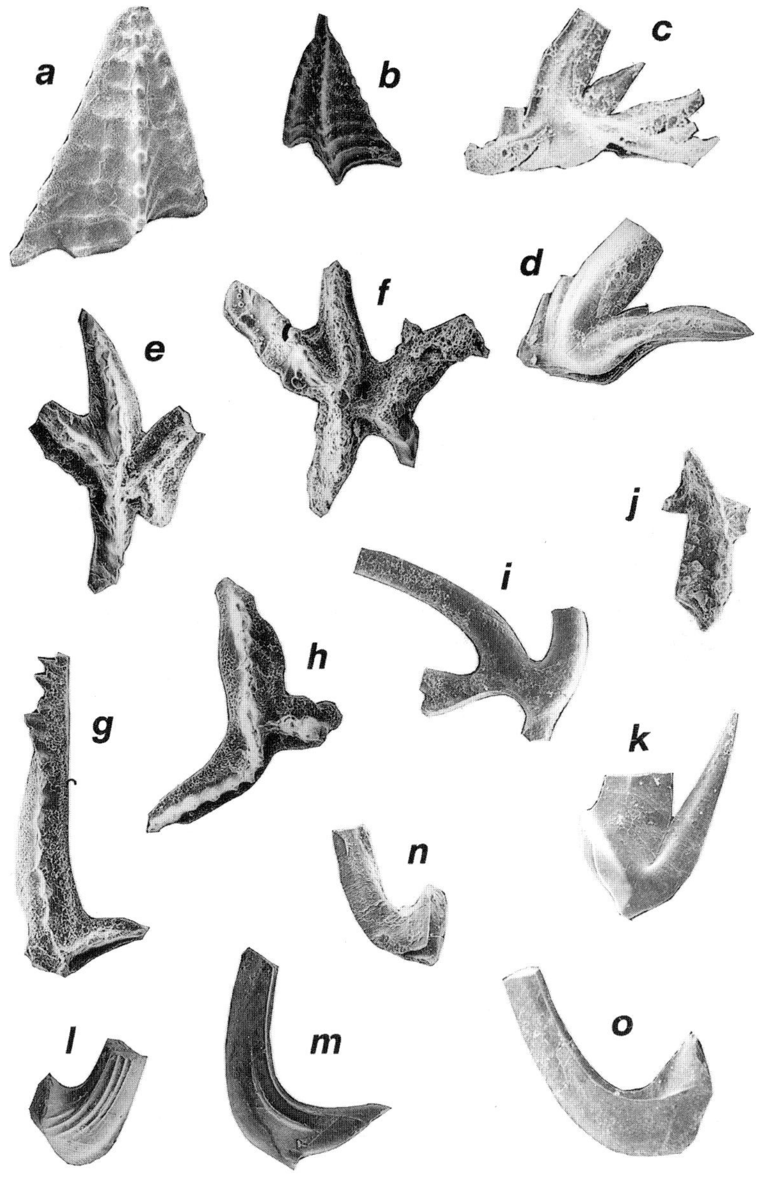

The sparry limestone clasts have yielded an abundant and diverse conodont fauna illustrated in Figs 3 and 4 and listed in Fig. 5. It includes *Pygodus anserinus* (Fig. 3a), *Pygodus serra* (Fig. 3b), *Strachanognathus parvus* (Fig. 3k), *Eoplacognathus lindstroemi* (Fig. 3e–h), *Spinodus spinatus* (Fig. 3i), *Periodon aculeatus* (Fig. 3c and d) and coniform species (Fig. 4).

The limestone clasts also contain a rich and well-preserved shelly fauna, including trilobites such as *Sphaerexochus* sp., *Pseudosphaerexochus*? sp., *Decoroproetus*? sp. and unidentifiable illaenids; strophomenide brachiopods, orthide brachiopods, a conical phosphatic non-articulated brachiopod species, lingulate brachiopods, stick bryozoans, ostracodes, gastropods and crinoids.

Pentlands Sub-basin

Limestone blocks from the Greywacke Conglomerate in Logan Burn (NT 183 619) were discovered by Henderson (1874), who suggested (Henderson 1880) that they were derived from bands of Silurian limestone similar to those cropping out in the Gutterford Burn in the nearby North Esk Inlier. Peach & Horne (1899, p. 606) recorded a diverse fauna including corals, trilobites, brachiopods and orthocones (see also Peach *et al.* 1910; Mitchell & Mykura 1962; Mykura 1986), supporting Henderson's assertion that the clasts were Silurian in age. Armstrong *et al.* (2000) isolated a conodont fauna attributed to the latest Llanvirn to early Caradoc, *P. anserinus* Biozone. A diverse macrofauna including brachiopods, trilobites and corals of mid-Ordovician aspect is currently being described.

Biostratigraphy

Armstrong *et al.* (1996) reviewed the biostratigraphy of mid-Ordovician conodont faunas in southern Scotland and indicated that correlations could be made across the Southern Upland Fault. Bergström (1990) documented the stratigraphical distribution of conodont species in the Barr Group at Girvan. In this section *Pygodus serra* first appears in the lower Stinchar Limestone (Bergström 1990, fig. 3) and the *P. serra–P. anserinus* Biozone boundary lies at a level in the middle of the Stinchar Limestone.

Bergström (1973, 1986) tied the base of the *P. serra* Biozone within the mid-*D. murchisoni* graptolite biozone (Abereiddian) and the top of the *P. anserinus* Biozone to a level within the mid-*N. gracilis* graptolite Biozone (Aurelucian). The top of the *P. serra* Biozone is the first appearance of *P. anserinus* within the upper part of the *H. teretiusculus* graptolite Biozone (Fortey *et al.*, 2000, fig. 34).

The presence of the eponymous biozonal species indicates a late Llanvirn–early Caradoc, Llandeilo to Aurelucian age (*sensu* Fortey *et al.* 1995) for the limestone clasts in the LORS of the Midland Valley.

Faunal similarity analysis

The similarity between two related faunas can be measured in terms of the Simpson coefficient of similarity, *S*, where *S* is the number of species in common between the two faunas divided by the total number of species in the smallest fauna, expressed as a percentage. Coefficients of similarity for broadly coeval conodont faunas from Europe and the eastern USA, are listed in Fig. 5,

Fig. 3. Selected elements of biostratigraphically useful conodont species referred to in the text and listed in Fig. 6. Figured specimens prefixed GLAHM are housed in the palaeontological collections of the Hunterian Museum; other sample numbers are those in the micropalaeontological collections, Department of Geological Sciences, University of Durham. Specimens are from Culzean Bay, sample D842 unless otherwise stated. Element orientation and locational nomenclature as described by Armstrong (1997). Armstrong (2000) has provided a full systematic description of the fauna. **(a)** *Pygodus anserinus* (Lamont & Lindström 1957), specimen GLAHM 109237, sample D841, oral view of Pa element, ×70. **(b)** *Pygodus serra* (Hadding 1913), specimen GLAHM 109230, oral view of Pa element, ×50. **(c)** *Periodon aculeatus* (Hadding 1913), specimen GLAHM 109243, inner lateral view of Pa element, ×50. **(d)** *Periodon aculeatus* (Hadding 1913), specimen GLAHM 109247, inner lateral view of M element, ×50. **(e)** *Eoplacognathus lindstroemi*, specimen GLAHM 109226, oral view of Pa element, ×70. **(f)** *Eoplacognathus lindstroemi* (Hamar 1966), specimen GLAHM 109227, oral view of Pa element, ×70. **(g)** *Eoplacognathus lindstroemi* (Hamar 1966), specimen GLAHM 109228, oral view of ?Pc element, ×70. **(h)** *Eoplacognathus lindstroemi* (Hamar 1966), specimen GLAHM 109229, oral view of ?Pb element, ×70. **(i)** *Spinodus spinatus* (Hadding 1913), specimen GLAHM 109261, lateral view of Pa element, ×70. **(j)** *Polonodus* sp., specimen 683.38, Inverbervie sample D817, oblique oral view of P element, ×70. **(k)** *Strachanognathus parvus* (Rhodes 1955), specimen GLAHM 109215, inner lateral view of qg element, ×70. **(l)** *Drepanodus robustus* (Hadding 1913), specimen GLAHM 109167, inner lateral view of pf element, ×40. **(m)** *Protopanderodus graeai*, specimen GLAHM 109165, inner lateral view of qt element, ×40. **(n)** *Cornuodus* cf. *C. longibasis*, specimen GLAHM 109139, inner view of q element, ×70. **(o)** *Protopanderodus graeai* (Hamar 1966), specimen GLAHM 109151, inner lateral view of ?pt element, ×80.

and are plotted as minimal difference trees in Fig. 6.

The clustering and hierachy indicated by tree A (Fig. 6) shows that faunas from the same palaeo-continents have the greatest number of species in common. For example, 80% of the species present at Girvan also occur in coeval strata in Alabama and Tennessee. Both areas lay on the southern margin of Laurentia in mid-Ordovician time. Faunas from the clasts from Culzean and the Wrae Limestone ('Tweed') are 75% similar to each other, but are only 50% and 56% similar similar to those at Girvan. These data indicate the limestone clasts at Culzean and the Wrae Limestone were not derived from the Stinchar Limestone as typified by present exposures.

The clustering and hierarchy indicated by tree B shows that the faunas from the limestone clasts are 85% similar to those from the Holy Cross Mountains, Poland, an area thought to have lain close to the edge of Avalonia within the Tornquist Ocean (Dzik 1976; Lewandowski 1993). Many of the taxa present in Poland and the clasts

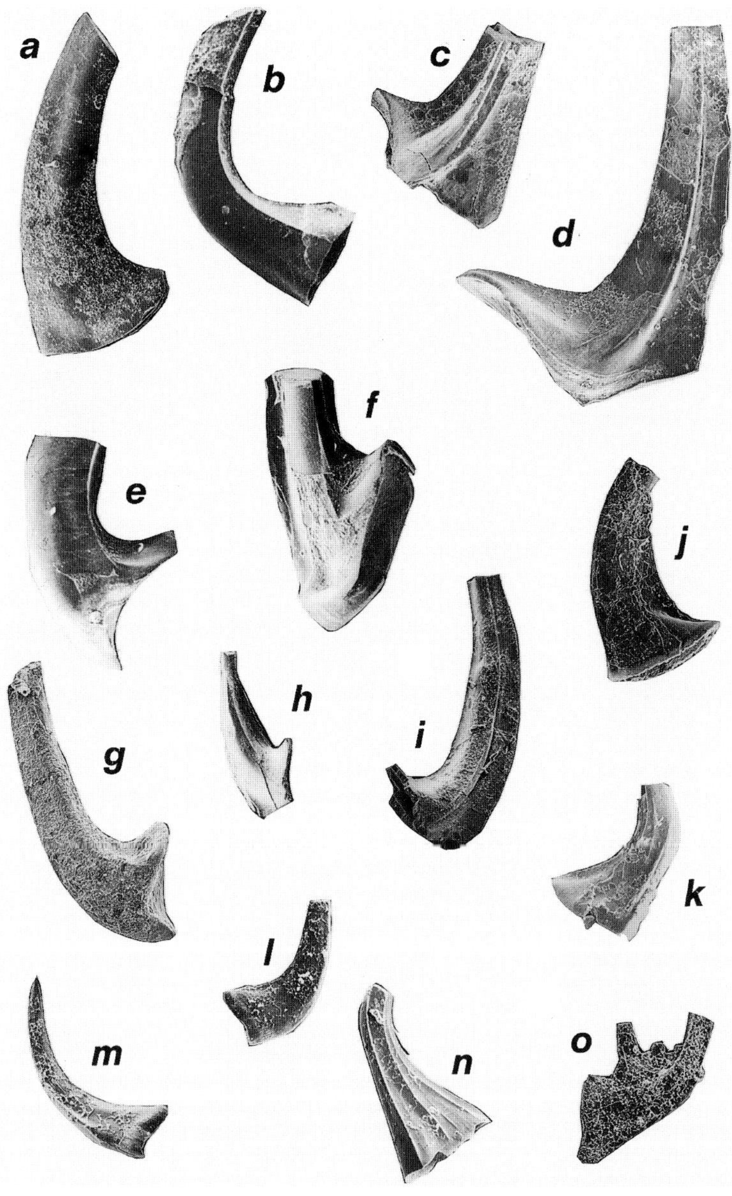

at Culzean have been attributed to the *Periodon–Pygodus* restricted species association (RSA *sensu* Bergström & Carnes 1976). This has traditionally been thought to be indicative of the North Atlantic Realm but is now considered to be a long-lived, deep-water, cosmopolitan fauna within the Iapetus Ocean and neighbouring platform-margin areas (Armstrong 1997; Rasmussen 1998), a view supported by the similarity analysis. The occurrence of this conodont fauna in pure carbonates, basinal muds and cherts suggests either that the constituent taxa were nektonic or that palaeo-oceanographic conditions on the outer shelf of Laurentia and in the deeper marginal basins were similar.

The associated trilobite fauna in the LORS clasts probably represents a pure carbonate biofacies, the Illaeinid–Cheirurid Association. Similar faunal associations, at generic level, occur in the Stinchar Limestone (of latest Llanvirn–earliest Caradoc age; Ingham & Tripp 1991) and the Craighead Limestone (of mid-Caradoc age; Tripp 1980), Girvan district. Trilobites from the Wrae Limestone in the Southern Uplands are characteristic of the margins of the deeper–cooler water Nileid Association (Owen *et al.* 1996).

Conodont colour alteration index (CAI)

Conodonts from Culzean Bay have a CAI of three indicative of heating to 100–200 °C (Epstein *et al.* 1977). Conodonts from the Pentland Hills and Inverbervie have a CAI of seven, indicating heating to 480–610 °C (Rejebian *et al.* 1987). Such a range in CAI values indicates localized heating of the source area consistent with igneous intrusion, rather than a widespread regional metamorphism.

CAI data can be used to constrain the provenance of the limestone clasts. Bergström (1980) noted the marked difference between CAI values to the north and south of the Southern Upland Fault. The Stinchar Limestone, to the north of the Southern Upland Fault is characterized by low CAI values of 1.5–2 (Bergström 1980). These values are comparable with those of the lowermost Carboniferous sequence and indicate heating to no higher than 80 °C and burial to depths of *c.* 2.5 km (Dean & Turner 1995). Conodonts from the Wrae Limestone (within the Ashgill Shinnel Formation) have a CAI of five (Bergström 1980; Armstrong 1997). At an average continental geothermal gradient of 30 °C km^{-1} the recorded temperatures would suggest burial to 2–4.5 km to the north of the Southern Upland Fault and 10–12 km to the south. These data indicate that limestone clasts from the Lanark and Strathmore basins were not derived from the Stinchar Limestone at Girvan.

The metamorphic event had a variable effect in the source of the limestone clasts and can be placed between the age of the conodonts and the deposition of the Shinnel Formation, i.e. between early Caradoc and mid-Ashgill time, 458 to *c.* 445 Ma. Haughton & Halliday (1991, p. 1476) noted hornfels clasts and xenoliths in igneous boulders suggesting intrusion through a pre-existing Lower Ordovician flysch cover succession. They also reported regional thermal events, as recorded by granite clast ages in southerly-derived conglomerates in the Strathmore Basin, at 443 ± 6.4 Ma and between 433 and 417 Ma. The earliest event is consistent with a heating event inferred from the CAI data, during early Ashgill time.

The thermal history of the Girvan basement can be inferred from derived granite clasts and detrital micas in the Northern Belt of the Southern Uplands. Longman *et al.* (1979) reported the age of a granite clast within the Benan Conglomerate at Girvan as 459 ± 10 Ma.

Fig. 4. Selected elements of biostratigraphically useful species referred to in the text and listed in Fig. 5. Specimens described and curated as in Fig. 3. (**a**) *Drepanodus robustus* (Hadding 1913), specimen GLAHM 109145, inner lateral view of ?pf element, ×50. (**b**) *Protopanderodus graeai* (Hamar 1966), specimen GLAHM 109153, inner lateral view of ?pt element, ×40. (**c**) *Protopanderodus varicostatus* (Sweet & Bergström 1966), specimen GLAHM 109157, Culzean Bay sample D843, inner lateral view of pf element, ×50. (**d**) *Protopanderodus varicostatus* (Sweet & Bergström, 1966), specimen GLAHM 1091600, Culzean Bay sample D843, inner lateral view of ?pt element, ×50. (**e**) *Drepanodus arcuatus* (Pander 1856), specimen GLAHM 109179, inner lateral view of pf element, ×50. (**f**) *Drepanodus arcuatus* (Pander 1856), specimen GLAHM 109181, oblique inner lateral view of pt element, ×50. (**g**) *Drepanodus arcuatus* (Pander 1856), specimen GLAHM 109172, inner lateral view of pf element, ×50. (**h**) *Protopanderodus* cf. *P. varicostatus*, specimen GLAHM 109192, inner lateral view of pf element, ×40. (**i**) *Protopanderodus* cf. *P. varicostatus*, specimen GLAHM 109191, inner lateral view of pf element, ×50. (**j**) *Walliserodus costatus* (Dzik 1976), specimen GLAHM 109195, inner lateral view of pt element, ×50. (**k**) *Walliserodus costatus* (Dzik 1976), specimen GLAHM 109201, ae element, ×50. (**l**) *Panderodus sulcatus* (Fåhřaeus 1966), specimen GLAHM 109205, inner lateral view of qg element, ×70. (**m**) *Panderodus sulcatus* (Fåhřaeus 1966), specimen GLAHM 109209, inner lateral view of pf element, ×70. (**n**) *Walliserodus nakholmensis* (Hamar 1966), specimen GLAHM 109204, ae element, ×70. (**o**) ?*Belodina* sp. indet Sweet (1979), specimen 672/29, rastrate element, ×70.

Species presence-absence data

species	Poland	Girvan	S. Wales	W. Border	Tweed	Sweden	Alabama	Culzean	Pentlands	Inverbervie
Amorphognathus ineaequalis Rhodes			x							
A. tvaerensis Bergström	x			x		x		?		
Ansella nevadensis (Eth & Sch)		x	x			x	x			
"*Acontiodus*" *nevadensis* Eth & Sch		x				?	x			
Baltoniodus prevariabilis (Fahraeus)			x		x	x				
Baltoniodus variabilis (Bergström)	x		x		x	x	x	x		
Belodina confluens Sweet		x						x	x	
Belodina monitorensis (Eth & Sch)		x					x			
Cahabagnathus sweeti (Bergström)		x				x	x			
Chirognathus sp.			x							
"*Coelocerodontus*" *digonius* S & B		x					x			
Cornuodus longibasis (Lindström)	x				x			x		
Dapsilodus mutatus (Branson & Mehl)	x			x	x	x	x	x	x	
Drepanodus arcuatus Pander	x							x		
Drepanodus robustus Hadding	x							x		
Drepanodus santacrucensis Dzik	x									
Drepanoistodus spp.	x	x	x	x	x	x	x			
Eoplacognathus lindstroemi (Hamar)	x		x			x		x		
Eoplacognathus reclinatus (Fahraeus)						x	x			
Erraticodon balticus Dzik		x				x	x			
Icriodella cf. *I. praecox* Lindström *et al.*			x	?						
Leptochirognathus sp.							x			
Oistodus cf. *O. venustus* Stauffer		x	x		x	x	x			
"*Ozarkodina*" *joachimensis* Andrews							x			
Panderodus spp.	x	x	x	x	x	x	x	x	x	
Periodon aculeatus Hadding	x	x			x	x	x	x	x	x
Phragmodus flexuosus Moskalenko							x			
"*Plectodina*" *bergstroemi* Šav & Bass			?	x						
Plectodina flexa (Rhodes)			x	x						
Polonodus sp.					x					x
Prattognathus rutriformis (S & B)							x			
Protopanderodus varicostatus (S & B)	x	x			x	x	x	x		
Protopanderodus graeai (Hamar)	x							x		
Protopanderodus graduatus Serpagli	x									
Pygodus anserinus Lamont & Lindström	x	x			x	x	x	x	x	
Pygodus serra (Hadding)	x	x			x	x	x	x		
Rhodesgnathus sp.			x	?						
Spinodus spinatus (Hadding)	x				x	x		x	x	
Strachanognathus parvus Rhodes	x	x			x	x		x	x	
Walliserodus curvatus Cooper	x				x			x	x	
Walliserodus ethingtoni (Fahraeus)		x			x	x		x		
Walliserodus nakholmensis (Hamar)	x							x		

Simpson Coefficient of Similarity

	No. spp	Poland	Girvan	S. Wales	W. Border	Tweed	Sweden	Alabama	Culzean
Poland	20	-	44	33	66	75	63	40	85
Girvan	16		-	33	33	56	75	81	50
S. Wales	12			-	50	42	58	42	25
W. Border	6				-	50	67	50	50
Tweed	16					-	75	56	75
Sweden	19						-	68	64
Alabama	20							-	37
Culzean	19								-

Fig. 5. Species presence–absence data and Simpson coefficients of similarity for localities mentioned in the text. Presence–absence data for localities other than Culzean, Pentlands and Inverbervie are taken from Bergström (1990, table 1). 'Tweed' is the fauna from the Shinnel Formation, Wrae Limestone in Tweedale, Scotland (Armstrong 1997).

Kelley & Bluck (1989) reported the ages of detrital micas in the Southern Uplands to range from 502 to 458 Ma and inferred a source to the north in a metamorphic block that was cooling during Ordovician time. They also noted that the same source had a substantial volume of volcanic and plutonic detritus derived from a magmatic arc, including hornblende-bearing granite clasts in the Corsewall Conglomerate dated to 465 Ma, and contain xenoliths of garnet–mica schist and staurolite schist (Longman 1980). These plutons were considered to be the roots of a middle Ordovician arc in which the time between arc formation, uplift and deposition was as little as 10 Ma (Kelley & Bluck 1989).

The absence of clasts recording pre-458 Ma magmatism, in the southerly-derived conglomerates in the Midland Valley, reinforces the

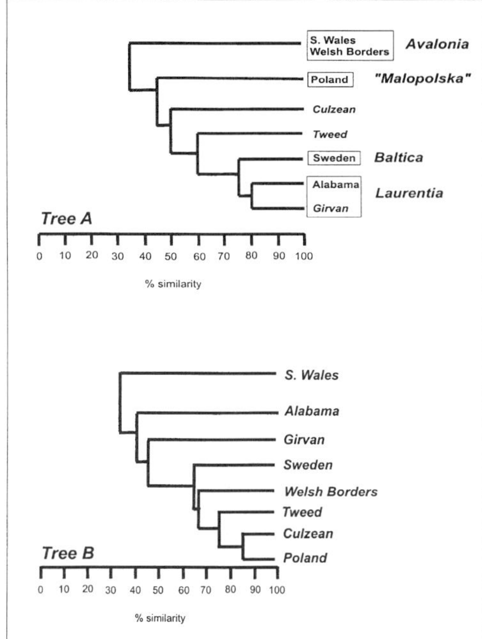

Fig. 6. Minimal difference trees based on the Simpson coefficient of similarity matrix in Fig. 5. Tree A illustrates the clustering and hierachy when comparing localities with Girvan. Tree B illustrates the clustering and hierarchy when comparing localities with the Holy Cross Mountains section in Poland.

view that the clasts were not derived from the Girvan arc and their source had a substantially different early thermal history. The onset of magmatism and regional heating was later (up to *c.* 40 Ma) in the source of the clasts than in the Girvan arc.

Source terrane

Limestone clasts in the LORS conglomerates of the Midland Valley are latest Llanvirn to earliest Caradoc in age. This is identical to the age of the Stinchar Limestone in Girvan (Ince 1984; Bergström 1990) and to Wrae Limestone boulders in the Shinnel Formation, Northern Belt of the Southern Uplands (Owen *et al.* 1996; Armstrong 1997). Palaeogeographical analysis indicates that the conodont faunas in the clasts are distinct from those in the Stinchar Limestone, probably representing a deep shelf–oceanic biofacies. The limestone clasts in southerly-derived conglomerates were not derived from the Girvan area.

The source of southerly-derived conglomerates along the southern margin of the MVT

The lithostratigraphy and provenance of southerly-derived conglomerates that crop out in the Lesmahagow, Pentlands–North Esk, Carmichael and Hagshaw Hills Inliers have been described by McGiven (1967), Bluck (1983, 1984), Williams & Harper (1988) and Smith (1995). Rolfe & Fritz (1966) recorded limestone clasts from the Igneous Conglomerate in the Parish Holm Burn that contained stromatoporoids and bryozoa of Wenlock age. The present workers were unable to find additional clasts.

Basin analysis (Bluck 1983) and geochemical studies (Heinz & Loeschke 1988) indicate that the Southern Uplands were not a source for the Igneous Conglomerate, and Williams & Harper (1988) suggested a source in an Ordovician igneous and metamorphic terrane to the south-east of the MVT. Similarly, the greywacke clasts in the Quartzite and Greywacke conglomerates are not derived from the Southern Uplands (Bluck 1983; Syba 1989). Bluck (1983) suggested a derivation of the Quartzite Conglomerate in a drainage basin up to 60 km to the SE of the present outcrop, and he considered the present outcrops of the Greywacke Conglomerate to be proximal to the source area. This placed the source area beneath the Southern Uplands and consequently Bluck (1985) proposed that the Southern Uplands had been thrust northwards to obscure the source area. The presence of middle Ordovician limestone clasts also suggests a similar source for the Greywacke Conglomerate at Culzean Bay and in the Pentland Hills. Phillips *et al.* (1997, p. 222) thought this source was a probable Llandovery, greywacke sandstone basin to the east. We are now confining the age of the clasts to the Ordovician period.

We follow Williams & Harper (1988) and Syba (1989) in considering the source of the southerly-derived conglomerates along the southern margin of the Midland Valley to be a cryptic metamorphic and volcanic arc terrane with a cover of middle Ordovician limestone and flysch. This terrane lay to the south of the MVT by late Silurian times.

The source of southerly-derived conglomerates in the Strathmore Basin

Three possible hypotheses to explain the source of limestone clasts in the southerly-derived conglomerates in the Strathmore Basin include the following.

(1) A source in the Highland Border Complex or a southerly extension of the complex within the Midland Valley. Limestones are known at two levels in the Highland Border Complex. The Dounans Limestone yields North American faunas of mid-Arenig age (Ingham *et al.* 1985; Ethington & Austin 1991) and is significantly older than those documented herein. The Margie Limestone Formation has yielded dark (but not black) chitinozoans of mid- to late Caradoc age (Burton *et al.* 1984), but despite several attempts, no conodonts. The available evidence suggests that the limestones in the Highland Border Complex are the wrong age to be a source of the limestone clasts in the southerly-derived conglomerates in the Strathmore Basin and this hypothesis is rejected.

(2) A source in a displaced fore-arc sliver(s) within the Midland Valley basement (Haughton 1988). Distinct isotopic signatures of the Midland Valley greywacke clasts in southerly-derived conglomerates in the Strathmore Basin led Haughton (1988) to suggest that the MVT basement included a sliver of a displaced fore-arc that had received sediment from a Ketilidian hinterland. The nearest along-strike outcrops of middle Ordovician carbonate rocks are found in Newfoundland (see Hiscott (1984) for a review), an area located hundreds of kilometres to the west of the Scotto-Irish Promontory (*sensu* Williams *et al.* 1997). Any terrane derived from Newfoundland would have had to tectonically bypass this promontory. We therefore reject this hypothesis. Alternately, the proposed fore-arc sliver could be part of an autochthonous succession as exposed in the Girvan Inlier. In this hypothesis the conodont fauna of the cryptic source and Girvan area should be identical. They are not and the hypothesis is rejected.

(3) A source in the cryptic arc terrane. In this hypothesis the contained clasts of all southerly-derived conglomerates would be the same. The palaeontology of the limestone clasts and the generally similar nature of the other clast types support this hypothesis. However, further work is necessary to establish the contemporaneous origin and provenance of the diverse lithoclasts within these conglomerates

Terrane analysis

The recognition of a cryptic terrane to the north of the basin in which the Southern Uplands successions were deposited has ramifications (Armstrong *et al.* 2000). Lower Caradoc greywackes of the Kirkcolm Formation in the Northern Belt of the Southern Uplands were transported axially and derived from a mature continental source to the northeast whereas the middle Caradoc fossiliferous conglomerate was sourced from the Pomeroy shelf (Scrutton *et al.* 1998). Ashgill conglomerates in the Northern Belt of the Southern Uplands include the Tweeddale Member (Wrae Limestone) of the Shinnel Formation, a distinctive group of mudstones and fine-grained pebbly greywackes restricted to ground between the Fradingmullach and Glen Fumart faults (Leggett 1980; Floyd 1996). Evans *et al.* (1991) postulated a northwesterly derivation within the Midland Valley for the greywackes of the Shinnel Formation. The limestone clasts in the Tweeddale Member (Wrae Limestone) contain an outer shelf macrofauna of trilobites (marginal Nileid Association, Owen *et al.* 1996), brachiopods (Owen *et al.* 1996), gastropods, crinoids and conodonts indicative of a late Llanvirn–early Caradoc, *P. anserinus* Biozone age (Armstrong 1997). Conodont faunas in the Wrae Limestone have a similarity coefficient of 75% to those in the limestone clasts in the LORS at Culzean (Fig. 5). The absence of abundant granite clasts in the Wrae conglomerates and the absence of coarse conglomerates at the same level in the Girvan succession preclude a direct link between the Girvan shelf and Northern Belt Basin at this time. The only alternative explanation is that the Wrae Limestone clasts were derived from the cover succession of the cryptic arc, postulated as a source for the later southerly-derived conglomerates in the Midland Valley including those of the LORS. Thus the cryptic arc would have been displaced northeastwards during Ashgill to Ludlow–Přídolí time before the final emplacement of the Southern Uplands over it.

Conclusions

Micropalaeontological analysis of limestone clasts in the Ordovician conglomerates of the northern Southern Uplands and southerly-derived LORS conglomerates of the Midland Valley has shown them all to be of the same *P. serra* to *P. anserinus* Biozone (late Llanvirn to early Caradoc) age. Simpson coefficients of similarity indicate that the conodont faunas from these limestone clasts have a close similarity to coeval faunas from the Holy Cross Mountains, Poland, and the Wrae Limestone and are significantly different from those from coeval strata in lower latitudes around Laurentia, Avalonia and Baltica. It is most likely that the taxa present in the clasts represent an outer shelf–ocean margin biofacies and not necessarily the close proximity of these various palaeocontinents. CAI data and granite radiometric ages indicate that a regional thermal event in the clast source area occurred in early Ashgill time.

The southerly-derived conglomerates in the Lanark Basin were derived from a cryptic metamorphic–volcanic arc with a cover that included middle Ordovician limestone and flysch. This lay between the MVT and the Northern Belt of the Southern Uplands. Similarly, the southerly-derived conglomerates in the Strathmore Basin could have been derived from this terrane. Magmatism in the cryptic arc caused a regional heating event during early Ashgill time, as magmatism in the Girvan area was waning. By Late Silurian–Early Devonian time the cryptic arc terrane lay south of the MVT. This terrane is now concealed beneath the allochthonous Southern Uplands. Clast provenance studies in the LORS have enormous potential in unravelling the Early Palaeozoic history of the northern British Caledonides.

The authors are indebted to B. Bluck (Glasgow) and P. Haughton (Dublin) for helpful discussions on the geology of the Midland Valley and substantive reviews of an earlier draft of this manuscript. T. Morse processed the conodont samples and developed the photographs. K. Atkinson kindly drew the diagrams. This work was funded by NERC grant GR9/02834.

References

Armstrong, H. A. 1997. Conodonts from the Shinnel Formation, Tweeddale Member (middle Ordovician), Southern Uplands, Scotland. *Palaeontology*, **40**, 763–799.

—— 2000. Conodont micropalaeontology of mid-Ordovician aged limestone clasts from LORS conglomerates, Lanark and Strathmore basins, Midland Valley, Scotland. *Journal of Micropalaeontology*, **19**, 45–59

——, Owen, A. W. & Clarkson, E. N. K. 2000. Ordovician limestone clasts in the Lower Old Red Sandstone, Pentland Hills, southern Midland Valley Terrane. *Scottish Journal of Geology*, **36**, 33–37

——, ——, Scrutton, C. T., Clarkson, E. N. K. & Taylor, C. M. 1996. Evolution of the Northern Belt, Southern Uplands: implications for the Southern Uplands controversy. *Journal of the Geological Society, London*, **153**, 197–205.

Armstrong, M. & Patterson, I. B. 1970. *The Lower Old Red Sandstone of the Strathmore Region*. Institute of Geological Sciences, Report **70/12**.

Bergström, S. M. 1973. Biostratigraphy and facies relations in the lower Middle Ordovician of easternmost Tennessee. *American Journal of Science*, **273-A**, 261–293.

—— 1980. Conodonts as paleotemperature tools in Ordovician rocks of the Caledonides and adjacent areas in Scandinavia and the British Isles. *Geologiska Föreningens i Stockholm Förhandlingar*, **102**, 377–392.

—— 1986. Biostratigraphic intergration of Ordovician graptolite and conodont zones—a regional review. *In*: Hughes, C. P. & Rickards, R. B. (eds) *Palaeoecology and Biostratigraphy of Graptolites*. Geological Society, London, Special Publications, **20**, 61–78.

—— 1990. Biostratigraphic significance of Middle and Upper Ordovician conodonts in the Girvan Succession, south-west Scotland. *Courier Forschungsinstitut Senckenberg*, **118**, 1–43.

—— & Carnes, J. B. 1976. Conodont biostratigraphy and paleoecology of the Holston Formation (Middle Ordovician) and associated strata in eastern Tennessee. *Geological Association of Canada, Special Papers*, **15**, 27–57.

—— & Orchard, M. J. 1985. Conodonts of the Cambrian and Ordovician Systems from the British Isles. *In*: Higgins, A. C. & Austin, R. L. (eds) *A Stratigraphical Index of Conodonts*. Ellis Horwood, Chichester, 32–67.

Bluck, B. J. 1978. Sedimentation in a late orogenic basin: the Old Red Sandstone of the Midland Valley of Scotland. *Geological Journal Special Issue*, **10**, 249–279.

—— 1983. Role of the Midland Valley of Scotland in the Caledonian Orogeny. *Transactions of the Royal Society of Edinburgh: Earth Sciences*, **73**, 119–136.

—— 1984. Pre-Carboniferous history of the Midland Valley of Scotland. *Transactions of the Royal Society of Edinburgh: Earth Sciences*, **75**, 275–295.

—— 1985. The Scottish paratectonic Caledonides. *Scottish Journal of Geology*, **21**, 437–464

——, Gibbons, W. A. & Ingham, J. K. 1992. Terranes. *In*: Cope, J. W. C., Ingham, I. J. K. & Rawson, P. F. (eds) *Atlas of Palaeogeography and Lithofacies*. Geological Society, London, Memoirs, **13**, 1–4.

Burton, C. J., Hocken, C., MacCallum, D. & Young, M. E. 1984. Chitinozoa and the age of the Margie Limestone of the North Esk. *Proceedings of the Geological Society of Glasgow, Sessions*, **124/125**, 27–32.

Dean, M. T. & Turner, N. 1995. Conodont colour alteration index (CAI) values for the Carboniferous of Scotland. *Transactions of the Royal Society of Edinburgh: Earth Sciences*, **85**, 211–220.

Dzik, J. 1976. Remarks on the evolution of Ordovician conodonts. *Acta Palaeontologica Polonica*, **21**, 395–455.

Epstein, A. G., Epstein, A. G. & Harris, L. D. 1977. *Conodont Color Alteration—an Index to Organic Metamorphism*. US Geological Survey, Professional Papers, **995**.

Ethington, R. L. & Austin, R. L. 1991. Conodonts of the Dounans Limestone, Highland Border Complex, Scotland. *Journal of Micropalaeontology*, **10**, 51–56.

Evans, J. A., Stone, P. & Floyd, J. D. 1991. Isotopic characteristics of Ordovician greywacke provenance in the Southern Uplands, Scotland. *In*: MORTON, A. C., TODD, S. P. & HAUGHTON, P. D. W. (eds) *Developments in Sedimentary Provenance Studies*. Geological Society, London, Special Publications, **57**, 161–172.

FLOYD, J. D. 1996. Lithostratigraphy of the Ordovician rocks in the Southern Uplands: Crawford Group, Moffat Shale Group, Leadhills Supergroup. *Transactions of the Royal Society of Edinburgh: Earth Sciences*, **86**, 153–165.

FORTEY, R. A., HARPER, D. A. T., INGHAM, J. K., OWEN, A. W. & RUSHTON, A. W. A. 1995. A revision of the Ordovician Series and Stages in the historical type area. *Geological Magazine*, **132**, 15–30.

——, ——, ——, ——, PARKES, M. A., RUSHTON, A. W. A. & WOODCOCK, N. H. 2000. *A revised correlation of Ordovician rocks in the British Isles*. Geological Society, London, Special Report, **24**.

HARLAND, W. B., ARMSTRONG, R. L., COX, A. V., CRAIG, L. E., SMITH, A. G. & SMITH, D. G. 1990. *A Geologic Time Scale 1989*. Cambridge University Press, Cambridge.

HAUGHTON, P. D. W. 1988. A cryptic Caledonian flysch terrane in Scotland. *Journal of the Geological Society, London*, **145**, 685–703.

—— 1993. Simultaneous dispersal of volcaniclastic and non-volcaniclastic sediment in fluvial basins: examples from the Lower Old Red Sandstone, east–central Scotland. *In*: MARZO, M. & PUIGDEFABREGAS, C. (eds) *Alluvial Sedimentation*. Blackwell Scientific, Oxford, 451–471.

—— & BLUCK, B. J. 1988. Diverse alluvial sequences from the Lower Old Red Sandstone of the Strathmore region, Scotland—implications for the relationship between the late Caledonian tectonics and sedimentation. *In*: MCMILLAN, N. J., EMBRY, A. F. & GLASS D. J. (eds) *Devonian of the World*. Canadian Society of Petroleum Geologists, Memoirs, **14**, 269–293.

—— & HALLIDAY, A. N. 1991. Significance of late Caledonian igneous complex revealed in the Lower Old Red Sandstone conglomerates, central Scotland. *Geological Society of America Bulletin*, **103**, 1476–1492.

——, RODGERS, G. & HALLIDAY, A. N. 1990. Provenance of Lower Old Red Sandstone conglomerates, SE Kincardineshire: evidence for the timing of Caledonian terrane accretion in central Scotland. *Journal of the Geological Society, London*, **147**, 105–120.

HEINZ, W. & LOESCHKE, J. 1988. Volcanic clasts in Silurian conglomerates of the Midland Valley (Hagshaw Hills inlier) Scotland, and their meaning for Caledonian plate tectonics. *Geologische Rundschau*, **77**(2), 453–466.

HENDERSON, J. 1874. Notice of some fossils from the conglomerate at Habbie's Howe, Logan Burn, near Edinburgh. *Transactions of the Edinburgh Geological Society*, **2**, 389–390.

—— 1880. On some recently discovered fossiliferous beds in the Silurian rocks of the Pentland Hills. *Transactions of the Edinburgh Geological Society*, **3**, 353–356.

HISCOTT, R. N. 1984 Ophiolitic source rocks for Taconic-aged flysch: trace element evidence. *Geological Society of America Bulletin*, **95**, 1261–1267.

HUTTON, D. H. W. 1987. Strike-slip terranes and a model for the evolution of the British and Irish Caledonides. *Geological Magazine*, **124**, 405–425.

INCE, D. 1984. Sedimentation and tectonism in the Middle Ordovician of the Girvan district, SW Scotland. *Transactions of the Royal Society of Edinburgh: Earth Sciences*, **75**, 225–237.

INGHAM, J. K. & TRIPP, R. P. 1991. The trilobite fauna of the Middle Ordovician Doularg Formation of the Girvan district, Scotland, and its palaeoenvironmental significance. *Transactions of the Royal Society of Edinburgh: Earth Sciences*, **82**, 27–54.

——, CURRY, G. B. & WILLIAMS, A. 1985. Early Ordovician Dounans Limestone fauna, Highland Border Complex, Scotland. *Transactions of the Royal Society of Edinburgh: Earth Sciences*, **76**, 481–513.

KELLEY, S. & BLUCK, B. J. 1989. Detrital mineral ages from the Southern Uplands using ^{40}Ar–^{39}Ar laser probe. *Journal of the Geological Society, London*, **146**, 401–403.

LEGGETT, J. K. 1980. The sedimentological evolution of a Lower Palaeozoic accretionary fore-arc in the Southern Uplands of Scotland. *Sedimentology*, **27**, 401–417.

LEWANDOWSKI, M. 1993. *Palaeomagnetism of the Palaeozoic rocks of the Holy Cross Mountains and the Origin of the Variscan Orogen*. Publication of the Institute of Geophysics Polish Academy of Science, **A-2**.

LONGMAN, C. D. 1980. *Age and affinity of granitic detritus in Lower Palaeozoic conglomerates, S.W. Scotland: implications for Caledonian evolution*. PhD thesis, University of Glasgow.

——, BLUCK, B. J. & VAN BREEMEN, O. 1979. Ordovician conglomerates and the evolution of the Midland Valley. *Nature*, **280**, 578–581.

MARSHALL, J. E. 1991. Palynology of the Stonehaven Group, Scotland: evidence for a Mid Silurian age and its geological implications. *Geological Magazine*, **128**, 283–286.

MCGIVEN, A. 1967. *Sedimentation and provenance of some post-Valentian conglomerates, Midland Valley, Scotland*. PhD thesis, University of Glasgow.

MITCHELL, G. H. & MYKURA, W. 1962. *The geology of the neighbourhood of Edinburgh (explanation of one-inch sheet 32)*. Memoir of the Geological Survey, Scotland, 3rd edn.

MYKURA, W. 1986. Pentland Hills. *In*: MCADAM, A. D. & CLARKSON, E. N. K. (eds) *Lothian Geology, an Excursion Guide*. Edinburgh Geological Society, 161–174.

——, HARPER, D. A. T. & CLARKSON, E. N. K. 1996. The trilobites and brachiopods of the Wrae Limestone, an Ordovician limestone conglomerate in the Southern Uplands. *Scottish Journal of Geology*, **32**, 133–149.

PEACH, B. N. & HORNE, J. 1899. *The Silurian rocks of Britain, 1: Scotland.* Memoir of the Geological Survey of the United Kingdom.

——, CLOUGH, C. T., HINXMAN, L. W., GRANT WILSON, J. S., CRAMPTON, C. D., MAUFE, H. B. & BAILEY, E. B. 1910. *The geology of the neighbourhood of Edinburgh (explanation of sheet 32 with part of sheet 31).* Memoir of the Geological Survey of Scotland, 2nd edn.

PHILLIPS, E. R., SMITH, R. A. & CARROLL, S. 1997. Strike-slip, terrane accretion and pre-Carboniferous evolution of the Midland Valley of Scotland. *Transactions of the Royal Society of Edinburgh: Earth Sciences*, **89**, 209–224.

RASMUSSEN, J. A. 1998. A reinterpretation of the conodont Atlantic Realm in the late Early Ordovician (early Llanvirn). *Palaeontologica Polonica*, **58**, 67–77.

REJEBIAN, V. A., HARRIS, A. G. & HUEBNER, J. S. 1987. Conodont color and texture alteration: an index to regional metamorphism, contact metamorphism and hydrothermal alteration. *Geological Society of America Bulletin*, **99**, 471–479.

ROLFE, W. D. I. & FRITZ, M. A. 1966. Recent evidence for the age of the Hagshaw Hills inlier, Lanarkshire. *Scottish Journal of Geology*, **2**, 159–164.

SCRUTTON, C. T. S., JERAM, A. J. & ARMSTRONG, H. A. 1998. Kilbuchophyllid corals from the Ordovician (Caradoc) of Pomeroy, Co. Tyrone: implications for coral phylogeny and for movement on the Southern Uplands Fault. *Transactions of the Royal Society of Edinburgh: Earth Sciences*, **88**, 117–126.

SMITH, J. 1895. From the Doon to the Girvan Water, along the Carrick shore. *Transactions of the Geological Society Glasgow*, **10**, 1.

SMITH, R. A. 1995. The Siluro-Devonian evolution of the Southern Midland Valley of Scotland. *Geological Magazine*, **132**, 503–513.

SYBA, E. 1989. *The sedimentation and provenance of the Old Red Sandstone Greywacke Conglomerate southern Midland Valley, Scotland.* PhD thesis, University of Glasgow.

THIRLWALL, M. F. 1988. Geochronology of British Late Caledonian magmatism in northern Britain. *Journal of the Geological Society, London*, **145**, 951–967.

TRIPP, R. P. 1980. Trilobites from the Ordovician Ardwell Group of the Craighead Inlier, Girvan district, Scotland. *Transactions of the Royal Society of Edinburgh: Earth Sciences*, **71**, 123–145.

TUCKER, R. D., BRADLEY, D. C., VERSTRAETEN, C. A., HARRIS, A. G., EBERT, J. R. & MCCUTCHEON, S. R. 1998. New U–Pb zircon ages and the duration and division of Devonian time. *Earth and Planetary Science Letters*, **158**, 175–186.

WELLMAN, C. H. 1993. A land plant microfossil assemblage of Mid Silurian age from the Stonehaven Group, Scotland. *Journal of Micropalaeontology*, **12**, 47–66.

WILLIAMS, D. M. & HARPER, D. A. T. 1988. A basin model for the Silurian of the Midland Valley of Scotland and Ireland. *Journal of the Geological Society, London*, **145**, 741–748.

——, HARKIN, J. & RICE, A. H. N. 1997. Umbers, ocean crust and the Irish Caledonides: terrane transpression and the morphology of the Laurentian margin. *Journal of the Geological Society, London*, **154**, 829–838.

Devonian (Givetian) miospores from the Walls Group, Shetland

J. E. A. MARSHALL
School of Ocean and Earth Science, University of Southampton, Southampton Oceanography Centre, European Way, Southampton, SO14 3ZH, UK
(e-mail: jeam@mail.soc.soton.ac.uk)

Abstract: Miospores demonstrate that the Walls Group of West Shetland is of early and possibly in part late Givetian age (Late Mid-Devonian age). This is both substantially younger in age and of a shorter duration than previously estimated. The correlative rocks of the Walls Group are those of southeast Shetland, Fair Isle, the Eday Group of Orkney and the John O'Groats Sandstone Group of Caithness. The two formations of the Walls Group are, at least in part, time equivalent rather than a stacked sequence. The total thickness of sediment is likely to be much less than the generally cited 12 km, given the short time duration represented by the single miospore assemblage identified throughout the succession. The low diversity of the miospore assemblage and its fluctuating dominant species is interpreted as reflecting a local vegetation source in a low preservation environment. The age and thermal maturity contrast between the Walls Group and the sedimentary successions to the west (Melby, Papa Stour and Foula) provide further constraints on the movement history along the St Magnus Bay Fault system.

Within the Shetland Islands (Fig. 1) there are three distinct, fault-bounded, successions of Old Red Sandstone (ORS) sedimentary rocks. West of the St Magnus Bay Fault (SMBF) system occur the isolated exposures of Melby, Eshaness and Foula. The central succession is the Walls Group, which lies between the St Magnus Bay and Walls Boundary Faults. To the east of both the Walls Boundary and Nesting Faults is the southeast Shetland succession. Out of these three successions it is the Walls Group that is the least understood, lacking both a secure lithostratigraphy and any accurate biostratigraphical (Mykura & Phemister 1976) or geochronological (Flinn *et al.* 1968) dates. However, despite this lack of basic stratigraphical information, the Walls Group has been attributed some significance within geotectonic models of the development of the ORS basins. It has also been regarded as distinctly different from the main Caithness and Orkney development of ORS in having both a complex structural history and a very significant thickness (12 km) of both Lower and Middle Devonian sediment. This paper reports the results of a palynological study of the Walls Group. This allows it, for the first time, to be accurately biostratigraphically dated. In addition, this date has significant implications for the accepted stratigraphy of the Walls Group.

The geology of the Walls Group (Fig. 2) has been detailed by Mykura & Phemister (1976), who subdivided it into the Sandness and Walls Formations, separated by the Sulma Water Fault. The Sandness Formation (1.4–3 km thick) has an unconformable basement contact and largely consists of sandstones with subordinate conglomerates and siltstones of fluvial origin. In the upper part of the formation there is a series of intrusive and extrusive rocks known as the Clousta Volcanic series. The Walls Formation (9 km thick), is presumed to overlie the Sandness Formation and is composed of finer-grained dark-coloured sandstone interpreted as lacustrine in origin. The southern part of the Walls Formation was intruded by the Sandsting Granite Complex in approximately Late Devonian times. However, the published geochronological dates (K/Ar 360 ± 11, 369 ± 10 Ma; Snelling, cited by Mykura & Phemister (1976)) should now be regarded with caution given the recent advances (Tucker *et al.* 1998) in Devonian time-scale calibration using more accurate isotope methods. The Walls Group is truncated to the east by the Walls Boundary Fault. Its contact to the west is the Melby Fault, part of the St Magnus Bay Fault system. This has been interpreted as a transcurrent system by Mykura (1975) and Donovan *et al.* (1976), and as a reverse fault with no discernible lateral

From: FRIEND, P. F. & WILLIAMS, B. P. J. (eds). *New Perspectives on the Old Red Sandstone.* Geological Society, London, Special Publications, **180**, 473–483. 0305-8719/00/$15.00

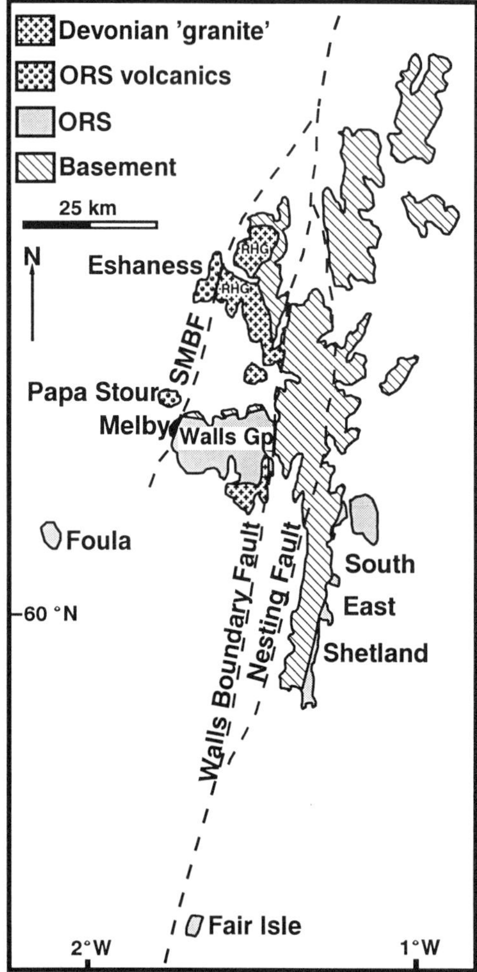

Fig. 1. Map of the Shetland Islands showing location of the Walls Group, the major faults, southeast Shetland, Fair Isle, and the westernmost Devonian sequences of Eshaness, Papa Stour, Melby and Foula. RHG, Ronas Hill Granite; SMBF, St Magnus Bay Fault. Map largely after Mykura (1976).

displacement by Flinn (1977). The Melby Formation immediately to the west has two lacustrine incursions in a sand-rich sequence (Mykura & Phemister 1976). Fish fossils (Watson 1934) and megaspores (Fletcher 1976) show these lakes to be correlatives of the well known Achanarras–Sandwick Fish Bed(s), which are of Eifelian age (Marshall 1996). Subsequently, Melvin (1985) has further interpreted the sedimentary environments of the coastal part of the Walls Formation and largely attributed them to a fluvial origin. Other contributions to the geology of the Walls Group include structural interpretations (Coward & Enfield 1987; Coward *et al.* 1989; Seranne 1992; Coward 1993) where the structural deformation has been related to an episode of basin inversion caused by Mid-Devonian sinistral movement along the Walls Boundary Fault, i.e. evidence for active Devonian strike-slip tectonics (but contrast Rogers *et al.* (1989)).

In addition the Walls Group has been comprehensively remapped by Astin (1982) who dealt with its stratigraphy, sedimentology and structure, and made significant reinterpretations of the geology of the Walls Group.

Previous palaeontological studies on the Walls Group

A summary of the existing palaeontological age evidence for the Walls Group has been given by Mykura & Phemister (1976; see also the correlation table of Mykura (1976)). The Sandness Formation (Chaloner, cited by Mykura & Phemister, 1976) was dated to Early or Mid-Devonian age using a sparse and poorly preserved macroflora found at two localities within sedimentary intercalations in the Clousta Volcanic series. In the Walls Formation, Miles (cited by Mykura & Phemister (1976)) identified fish fossils that indicated a 'Middle Old Red Sandstone age' although the plants (Chaloner, cited by Mykura & Phemister (1976)) proved to be of no stratigraphic value. This evidence was summarized as giving an ?Early to Mid-Devonian age for the Walls Group. However, the Sandness Formation is usually attributed (e.g. Mykura & Phemister 1976) an Early Devonian age as it was placed as the basal formation to the 12 km thick Walls Group, which, in its upper part, is of proven Mid-Devonian age.

Palynological studies on the Walls Group have previously been attempted by Fletcher (pers. comm.) and Owens (cited by Mykura & Phemister (1976)) but neither found material suitable for identification. However, Batten (pers. comm., cited by Melvin (1976, 1985)) recovered better preserved material including miospores with grapnel tip appendages. These led him to suggest a Devonian age, no older than Emsian time (Late Early Devonian), for part of the Walls Formation.

Material and methods

In view of the strong degree of deformation and metamorphism (greenschist facies) present in the Walls Group and the poor results from previous palynological studies, an intensive field sampling programme was undertaken in the area. Much of the coastline was traversed together with the inland lake shore

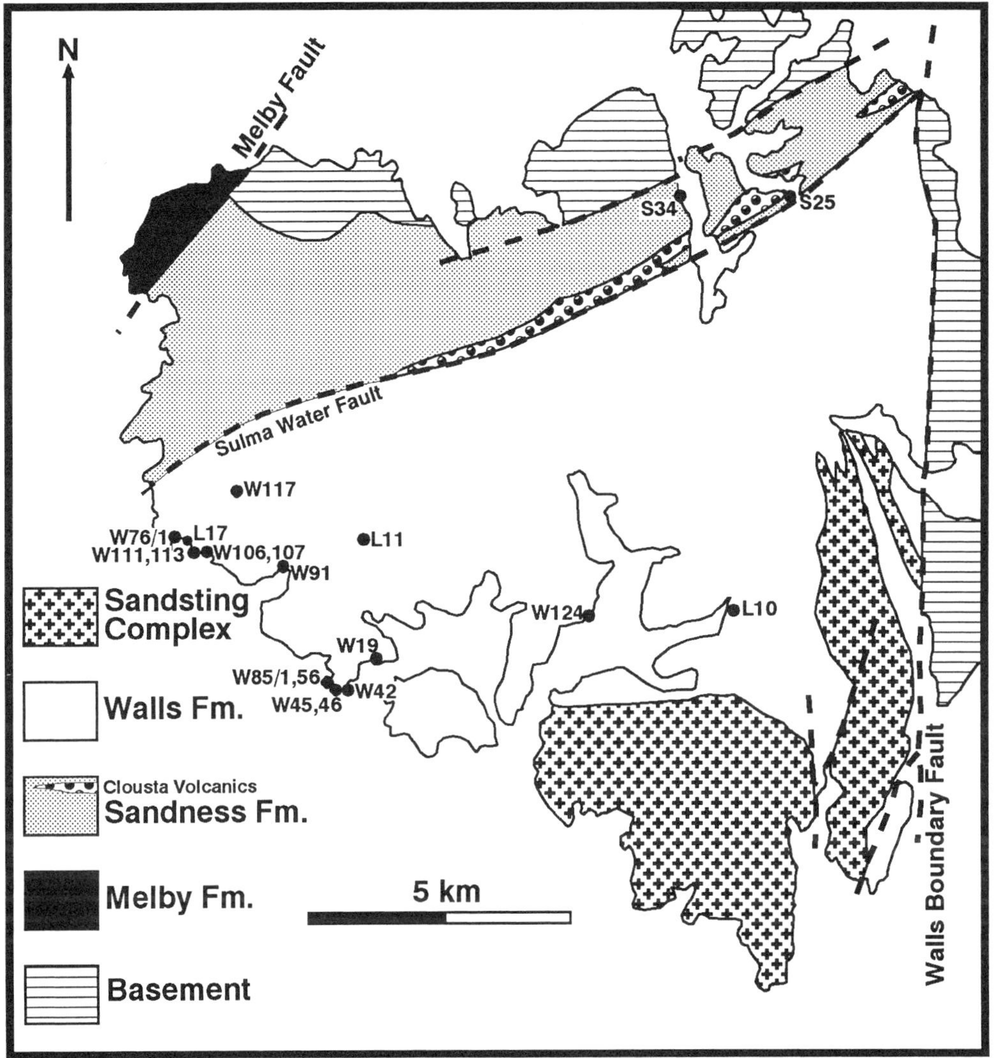

Fig. 2. Map of West Shetland showing the Sandness and Walls Formations of the Walls Group. Location of palynologically productive samples indicated including two from the Sandness Formation. (Note the structural juxtaposition of the Melby Formation against the Sandness Formation and pre-Devonian metamorphic basement.) Largely after Mykura & Phemister (1976).

exposures. This resulted in a collection of 205 samples. Unlike most palynological samples the colour of these fine-grained sediments from the Walls Group is not a helpful guide for predicting the presence of miospore-rich kerogens. Those sediments that are darkest in colour (black) do have the highest organic contents but this is almost entirely composed of amorphous organic matter (AOM) and contains only rare poorly preserved miospores. Preferential selection of dark green siltstones, which elsewhere in the Orcadian Basin contain miospore-rich kerogens, was difficult within the Walls Group, as most sediments are green hued through the growth of diagenetic low-grade metamorphic minerals. Therefore, to expedite palynological processing, the samples were screened by first measuring their TOC (total organic carbon content) using a 'Girdel Rock Eval Oil Shows Analyzer' (results summarized by Marshall *et al.* 1985). The samples were then palynologically processed in order of decreasing organic richness, with those containing less than 0.1% TOC being eventually rejected. All 49 Sandness Formation samples were processed, as the recovery of miospores from this part of the sequence was regarded as particularly important. Palynological processing was by standard techniques as outlined by Marshall & Allen (1982). From a total of 112 samples processed,

Table 1. *Location of palynologically productive samples from the Walls Group*

Sample no.	Locality	Grid reference	Formation	*Rv* (%)	*n*	SD
S25	Clousta	HU 30265696	Sandness	7.1	45	1.2
S34	Shungalong Point	HU 28475707	Sandness			
L10	Seli Voe	HU 29384842	Walls	7.4	33	1.8
L11	Hill of Elvister	HU 22934980	Walls	6.4	34	1.3
L17	Wick of Watsness	HU 17705052	Walls	6.4	40	1.4
W19	Lunga Taing	HU 21864736	Walls			
W42	Rusna Stacks	HU 20874704	Walls	8.9	30	1.5
W45	Black Head	HU 20804709	Walls			
W46	Black Head	HU 20804709	Walls			
W56	Uski Geo	HU 20884718	Walls			
W76/1	Wick of Watsness	HU 175520	Walls	5.0	24	0.85
W91	Footaborough	HU 19924954	Walls			
W106	Rams Geo	HU 18114966	Walls	6.8	29	1.8
W107	Rams Geo	HU 18114966	Walls			
W111	Rams Geo	HU 18114966	Walls	6.9	20	1.1
W113	Gorsendi Geo	HU 17885010	Walls	7	36	1.3
W117	Turdale	HU 19735077	Walls			
W124	Gruting Voe	HU 26294813	Walls			
W85/1	Black Head	HU 20804709	Walls	6.8	47	2.5

All sample locations are shown on Fig. 2. The vitrinite reflectivity determinations were made under oil (RI = 1.515) on polished thin sections using a Zeiss UMSP 50 microscope equipped with a ×40 oil immersion objective. *Rv* is the mean random reflectivity value, *n* the number of measurements and SD the standard deviation. Details of method, calibration, precision and vitrinite selection have been given by Hillier & Marshall (1992).

only 17 from 11 distinct localities contained reasonable palynofloras, of which nine could be regarded as having an 'abundant' palynomorph content (over 250 miospores from some 30–60 g of palynologically processed rock). The location of all productive miospore samples is shown in Fig. 2. Two samples come from the Sandness Formation and the remainder are from the Walls Formation. Details of these sample sites are given in Table 1. The spores recovered were all black, highly carbonized and difficult to oxidatively clear to a translucency level suitable for light microscopy. Vitrinite reflectivity measurements (as mean random reflectance) were made on ten of the kerogen isolates and the values obtained (Table 1) range from 5.0 to 8.9%. These are very high, comparable with values from Fair Isle (Hillier & Marshall 1992), and indicate maximum formation temperatures in excess of 350–425 °C (Barker & Goldstein 1990), which are almost certainly related to the intrusion of the Sandsting Complex into the Walls Group. These highly carbonized miospores assemblages were successfully oxidized for transmitted light microscopy using fuming nitric acid (24 ml) diluted with water (16 ml) with potassium chlorate (2.5 g). Typical oxidation times were between 15 and 45 min. They are probably the highest thermal maturity palynomorphs to be successfully studied using such oxidation methods. However, in common with similar high-rank material from elsewhere in Shetland (Marshall 1980), the miospores redarken to opacity in a matter of several hours. Hence multiple oxidations were made of each sample, immediately before observation, and the cleared residue was mounted in water under a coverslip sealed with nail varnish to prevent evaporation. Careful control of the amount of water gives a preparation with the coverslip held down by surface tension and suitable for observation with high-power (×100) oil-immersion objectives. The use of temporary mounts means that no slides remain for curation. Therefore the unoxidized kerogen concentrates and original samples have been retained and are available for further study. They are stored in the School of Ocean and Earth Science, University of Southampton. As can be seen from the illustrated miospores (Fig. 3), the oxidized miospores do not figure well as the preservation is poor, the colour still somewhat dark and genera such as *Ancyrospora* lack most of their diagnostic spines. Fortunately, all the taxa from the Walls Group were previously encountered in the much better preserved and abundant material of southeast Shetland (Allen & Marshall 1981) and Fair Isle (Marshall & Allen 1982), which considerably facilitated identification.

The miospore assemblage

Table 2 lists the miospore taxa identified from the Walls Group; some of which are illustrated on Fig. 3. Table 3 shows the proportion of the more abundant taxa to indicate assemblage composition. The most striking feature of the Walls Group microflora in comparison with other Orcadian Basin assemblages of the same age is the low diversity and low absolute abundance of spores in the rock. In addition, the species composition is erratic, with local variation between closely spaced samples being as great

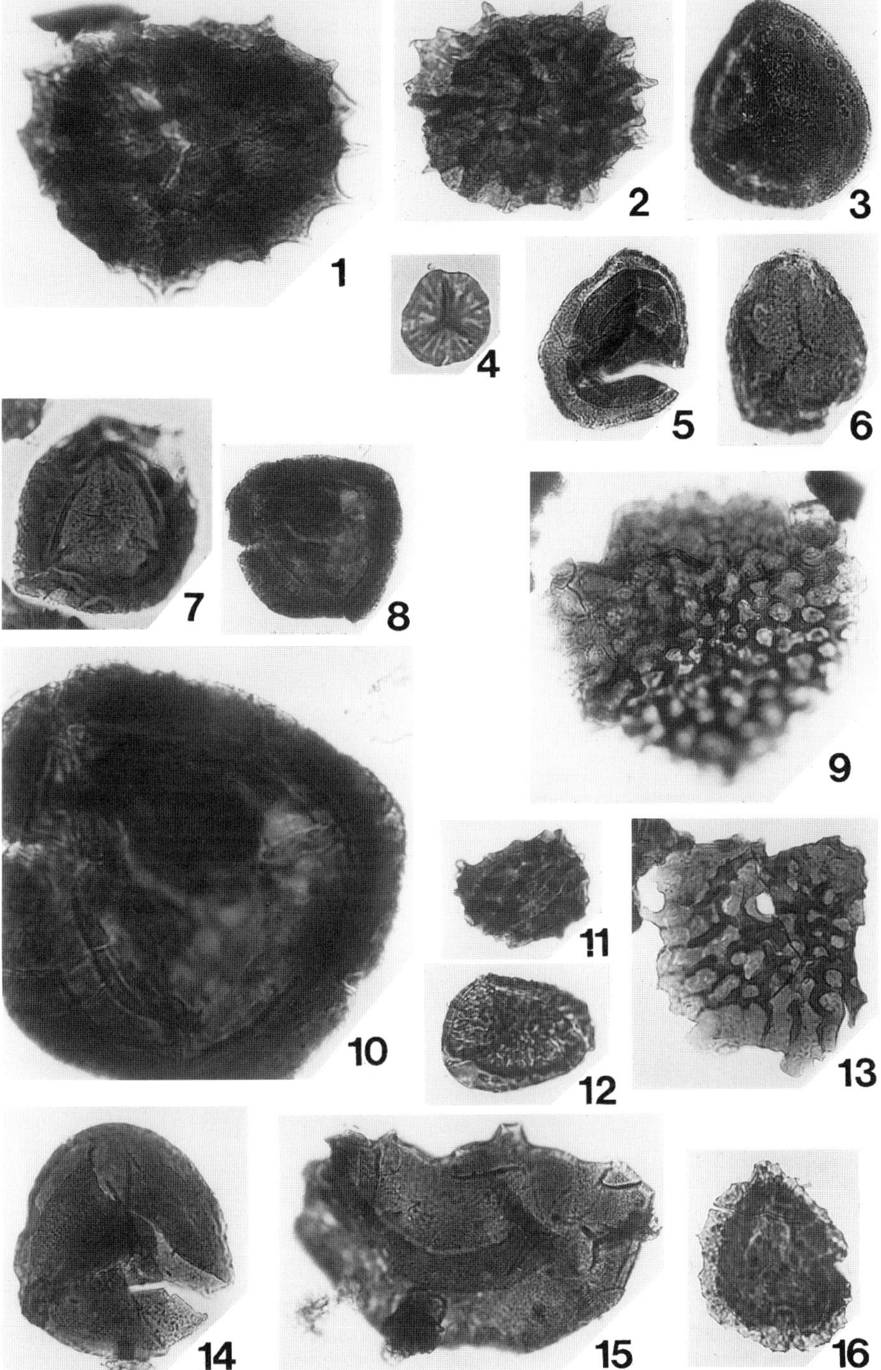

Fig. 3. Miospores from the Walls Group. (All ×400 unless otherwise stated.) (**1**) *Ancyrospora ancyrea*, S25. (**2**) *Ancyrospora ancyrea*, S25. (**3**) *Geminospora lemurata*, L17. (**4**) *Emphanisporites rotatus*, W106. (**5**) *Geminospora lemurata*, L10. (**6**) *Geminospora lemurata*, S25. (**7**) *Geminospora* sp. A, W106. (**8**) *Geminospora* sp. A, W113. (**9**) '*Cirratriradites*' *monogrammos*, L17. (**10**) *Geminospora* sp. A, W113, ×1000. (**11**) *Verrucosisporites premnus*, W113. (**12**) *Retusotriletes rugulatus*, S25. (**13**) '*Cirratriradites*' *monogrammos*, S25. (**14**) *Insculptospora confossa*, S25. (**15**) *Contagisporites optivus* var. *optivus*, W76/1. (**16**) *Densosporites concinnus*, S25.

Table 2. *Miospore taxa from the Walls Group*

Ancyrospora ancyrea (Eisenack) Richardson 1962
Ancyrospora ancyrea cf. var. *brevispinosa* of Marshall & Allen (1982)
Auroraspora macromanifestus (Hacquebard) Richardson 1960
Auroraspora micromanifestus (Hacquebard) Richardson 1960
Calamospora atava (Naumova) McGregor 1973
'*Cirratriradites*' *avius* Allen 1965
Contagisporites optivus (Chibrikova) Owens 1971 var. *optivus* Owens 1971
Cristatisporites mediconus Richardson 1960
Densosporites concinnus (Owens) McGregor & Camfield 1982
Densosporites devonicus Richardson 1960
Emphanisporites rotatus (McGregor) McGregor 1973
Geminospora lemurata (Balme) Playford 1983
Geminospora sp A of Marshall & Allen 1982
'*Cirratriradites*' *monogrammos* (Arkhangelskaya) Arkhangelskaya 1985, synonymous with *Cirratriradites* sp. A of Marshall & Allen 1982
Insculptospora confossa (Richardson) Marshall 1985
Retusotriletes distinctus Richardson 1965
Retusotriletes rotundus (Streel) Lele & Streel 1969
Retusotriletes rugulatus Riegel 1973
Rhabdosporites langii (Eisenack) Richardson 1960
Trileites langii Richardson 1965
Verruciretusispora dubia (Eisenack) Richardson & Rasul 1978
Verrucosisporites premnus Richardson 1965
Verrucosisporites scurrus (Naumova) McGregor & Camfield 1982

All taxonomic citations not listed in the references have been given by McGregor & Camfield (1982) or Marshall (1996).

Table 3. *Percentage abundance determined from the better-preserved miospore samples from the Walls Group*

	S25	L10	L11	L17	W42	W106	W113	W117	W76/1
Ancyrospora ancyrea	45	3	39	32	2	+	17	5	18
A. ancyrea cf. var. *brevispinosa*	+		+	+			+		+
Calamospora atava	5	+		+	76	1	6	60	6
'*Cirratriradites*' *monogrammos*	2	6	+	3		+	4	2	3
Contagisporites optivus									+
Cristatisporites mediconus	+		2	4	+	+	4		3
Densosporites concinnus	3	13	4	14	+	+	10		3
Densosporites devonicus	5	1	14	3	+	8	4	2	5
Geminospora lemurata	+	17	5	1				+	11
Geminospora sp. A	2	+		2		12	1	+	1
Geminospora spp.		3	3				+		
Insculptospora confossa	25	48	19	23	1	58	23	4	18
Rhabdosporites langii	2	4	3	7	1	6	8	11	8
Others	11	5	11	11	20	15	23	16	24
Total miospore count	261	280	362	273	97	158	210	250	202

+, indicates that the abundance is <1%. Note the variable total miospore count caused by low abundances within the sediment. Different samples that are often closely placed geographically and stratigraphically (e.g. W76/1 and W106) can show very different proportions of a few dominant miospores.

as total variation within all the samples. All the samples can only be attributed to one miospore assemblage and a restricted age range is implicit. Significantly, the miospore assemblage in the Sandness Formation is the same as that from the Walls Formation. One unusual feature of the microflora is the dominance of any one species in a single sample. These dominant taxa (*Ancyrospora ancyrea*, *Insculptospora confossa* and *Calamospora atava*) fluctuate between samples, often achieving a high proportion (e.g. 50–70%), which is uncommon in other Orcadian Basin microfloras.

In comparison with miospore zonal schemes based on conodont dated Euramerican assemblages (Richardson & McGregor 1986; Streel *et al.* 1987; Avkhimovitch *et al.* 1993) the Walls Group miospore assemblage is clearly Mid-Devonian in age. Key diagnostic taxa include *Ancyrospora ancyrea*, '*Cirratriradites*' *monogrammos*, *Densosporites concinnus*, *Densosporites devonicus* and *Rhabdosporites langii*, which all have Eifelian first occurrences. The presence of *Geminospora lemurata* in most samples, and its abundance in three, is significant, as this zonal taxon has an inception (Loboziak *et al.* 1991; Streel & Loboziak 1994) in the earliest part of the Givetian stage. However, its absence in other samples cannot be given any age significance, as its distribution is erratic and not confined to stratigraphically higher levels in the Walls Group succession. Determining the upper age limit of the Walls Group is more difficult. One potentially significant taxon is *Contagisporites optivus* which has previously been given zonal significance with a late Givetian inception (Richardson & McGregor 1986). However, the timing of its first occurrence varies widely across the ORS continent (McGregor 1981) and within the Orkney succession it has an inception (Marshall 1996) close to that of *Geminospora lemurata*. Hence its stratigraphic significance is discounted here. Absent from the Walls Group, but occurring in both southeast Shetland and Orkney (Allen & Marshall 1981; Marshall 1996), are a number of miospores such as *Chelinospora concinna* and *Cristatisporites triangulatus*, which have been used to define mid- and late Givetian subzones (Avkhimovitch *et al.* 1993; Streel *et al.* 1987; Turnau 1996). The absence of these younger Givetian miospore taxa from the Walls Group assemblage suggests that the succession pre-dates their inception and is entirely of early Givetian age. However, given the limitations of the Walls Group assemblage in terms of miospore abundance and diversity and sample distribution it is best to regard the succession as Givetian in age, definitely early Givetian but possibly also of younger Givetian age.

In comparison with other Orcadian Basin microfloras (Fig. 4), the Walls Group assemblage is clearly a correlative of those described from southeast Shetland (Allen & Marshall 1981) and Fair Isle (Marshall & Allen 1982). It is also a correlative of the Lower Eday Sandstone and Eday Flagstone Formations (both Eday Group) of Orkney (Marshall 1996). However, it is clearly distinctly different from and younger than the Lower and Upper Stromness Flagstone Formations of Orkney. These flagstone formations are the major interval of lacustrine sediments in Orkney and are of Eifelian age. This Eifelian part of the lacustrine sequence contains the important basin-wide lacustrine flooding event commonly referred to as the Sandwick (Orkney) or Achanarras (Caithness) Fish Bed. Significantly, this level can be clearly identified in the successions west of the Melby Fault at Melby (Fletcher 1976; Allen, pers. comm.), Papa Stour (Marshall 1988) and on the island of Foula (Blackbourn & Marshall 1985).

Palynofacies and miospore palaeoecology

Two general palynofacies types occur in the Walls Group kerogens. One is dominated by AOM which, from its restriction to dark-coloured lacustrine facies and through comparison with lower thermal maturity kerogens from elsewhere (Marshall *et al.* 1985) in the Orcadian Basin, has formed diagenetically from lacustrine algae. The other palynofacies is the phytoclast dominated material of land plant origin which can be relatively spore rich and occurs largely within lake margin sediments. As noted earlier (Table 2) the miospore-'rich' assemblages are rare, and of both low diversity and low absolute abundance with dominant taxa fluctuating significantly between stratigraphically close samples. This contrasts with other Orcadian Basin microfloras of the same age, which are more diverse, more uniform and richer in terms of absolute abundance. The Walls Group microfloras are interpreted as more strongly reflecting the local vegetation pattern. The sedimentary facies show the depositional environments to be dominated by fluvial processes with episodic lacustrine incursions. Such sedimentary environments have a low potential for the preservation of both micro- and macrofossil plant remains, although plants certainly did occur, as shown by the presence of vertically aligned poorly preserved stems (Astin, pers. comm.). Mudstones (in fact contact altered to pelites), which do contain microfloras, are not particularly common and have a low organic content (i.e. low TOC). This, together with the low miospore diversity is believed to indicate a sparse vegetation cover of a few common plants occurring locally in monospecific stands, and thus giving palynofloras characterized by fluctuating dominants. This does not seem to be the situation in other areas of the Orcadian Basin where microfloras have commonly undergone significant transport and homogenization by lake and fluvial processes so that the effects of the parent plants' immediate environments are removed.

One notable occurrence in the Walls Group microflora is the high proportion of *Ancyrospora ancyrea* in minor mudstone lithologies within sandstone-rich intervals (e.g. W46, W56, Black Head, HU 208471). This miospore is of unknown

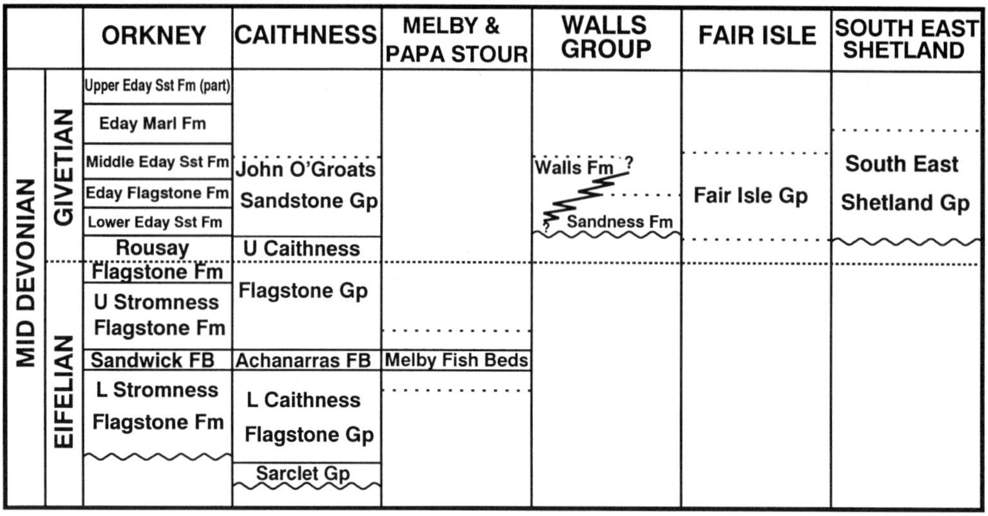

		ORKNEY	CAITHNESS	MELBY & PAPA STOUR	WALLS GROUP	FAIR ISLE	SOUTH EAST SHETLAND
MID DEVONIAN	GIVETIAN	Upper Eday Sst Fm (part)					
		Eday Marl Fm					
		Middle Eday Sst Fm	John O'Groats Sandstone Gp		Walls Fm ?	Fair Isle Gp	South East Shetland Gp
		Eday Flagstone Fm					
		Lower Eday Sst Fm			? Sandness Fm		
		Rousay Flagstone Fm	U Caithness Flagstone Gp				
	EIFELIAN	U Stromness Flagstone Fm					
		Sandwick FB	Achanarras FB	Melby Fish Beds			
		L Stromness Flagstone Fm	L Caithness Flagstone Gp				
			Sarclet Gp				

Fig. 4. Correlation table showing the age of the Walls Group relative to other Orcadian Basin successions. The successions in Orkney and Caithness have a long interval of lacustrine sedimentation of Eifelian and earliest Givetian age, which contains the important basin-wide lacustrine flooding event of the Sandwick Fish Bed and correlatives. This initial lacustrine sedimentation is interrupted by an episode of basin extension with uplift followed by an interval of fluvial sedimentation (the Lower Eday Sandstone Formation). The lacustrine environment is then re-established (the Eday Flagstone Formation) although variable in both lateral distribution and duration. The Eday Marl Formation was deposited in a muddy sabkha plain intermittently flooded by the sea. The Upper Eday Sandstone Formation marks the return to fluvial sedimentation but within an open basin with through drainage. The successions of the Walls Group, Fair Isle and southeast Shetland are all younger than the main Eifelian episode of lacustrine sedimentation in being entirely of Givetian age. They all contain lacustrine intervals that are equivalent to the Eday Flagstone Formation. (Note the presence of the Sandwick Fish Bed equivalent at Melby and Papa Stour adjacent to the Walls Group and separated by the Melby Fault.) The Walls Group and southeast Shetland successions have exposed basal unconformities where the oldest sediments are Givetian in age. Hence sedimentation within these sub-basins is presumed to be initiated following the pre-Lower Eday Sandstone Formation (early Givetian) episode of basin extension. However, the possibility that the unconformity represents basin onlap onto an earlier and now concealed Eifelian succession should always be considered. The partial age equivalence of the Walls and Sandness Formations should be noted. The stratigraphical nomenclature is both inconsistent and in part unpublished. The wider-spaced hatched lines within the regional lithostratigraphic columns are where section is absent through erosion or not exposed. The relative durations of the Eifelian and Givetian ages are from Tucker *et al.* (1998). The relative durations of the Eifelian lacustrine formations are based on both thickness and the number of lake cycles. The Eday Group formations have estimated relative time durations.

macroplant affinity despite being a very common element of the Orcadian Basin floras. It has been suggested (e.g. Allen 1980) that this lack of a known macroplant is due to it being from an 'upland flora'. However, if these Walls Group microfloras do reflect local vegetation patterns it would seem more likely to be a plant from a low preservation environment such as a sand-rich fluvial system. Also significant is the co-association of a very high abundance of *A. ancyrea* with the presumed lycopod *Thursophyton* at various localities elsewhere in the Orcadian Basin (Foula, Blackbourn & Marshall 1985; southeast Shetland, Marshall, pers. obs.) and Canning Land, East Greenland (Marshall, pers. obs.).

Significance of the Givetian age of the Walls Group microflora

The results of the palynological analysis show that the entire Walls Group can be constrained to a Late Mid-Devonian (Givetian) age. This is younger than the major Eifelian development of lacustrine facies in Orkney and Caithness but a correlative of the ORS of southeast Shetland and Fair Isle. The Sandness Formation is thus clearly not Early Devonian in age (Mykura & Phemister 1976). In addition, it contains a palynoflora that is identical to that of the Walls Formation. Therefore these two units are at least, in part, age equivalents. The recognition that the same time-restricted palynological assemblage occurs in both the Walls and Sandness Formations also has implications for the total thickness of the Walls Group as estimated by Mykura & Phemister (1976). Quite simply, the time duration as represented by the single palynological assemblage is much shorter that the time duration within which a 12 km sequence of sediment could accumulate. These issues of sequence thickness, lithostratigraphical succession and equivalence of the Walls and Sandness Formations have been comprehensively discussed by Astin (1982).

The recognition of a Givetian age for the Walls Group confirms the correlations of Mykura & Young (1969), Mykura (1972*a*, *b*) and Rogers *et al.* (1989), who lithostratigraphically compared this succession with that of Fair Isle across the Walls Boundary Fault. Through this correlation they determined a post-Devonian dextral transcurrent movement along the fault of some 60–95 km (but see Flinn (1992) for an alternative view). The age of the Walls Group is also significant in the debate over the sense of movement on the Melby Fault. Mykura (1975), Donovan *et al.* (1976) and Seranne (1992) interpreted the Melby Fault as transcurrent with a significant dextral displacement which has juxtaposed rock sequences from the southern part of the Orcadian Basin (Melby, Papa Stour, Eshaness) against the dissimilar Walls Group. However, Flinn (1977, 1985, 1992; but see also Ritchie *et al.* (1993)) from field observations, has interpreted the Melby Fault (St Magnus Bay Fault, SMBF) as a reverse fault dipping to the east with no evidence for transcurrent movement, the Walls Group being faulted up-dip relative to the Melby Formation. Flinn (1985) also proposed that the Melby Formation was deposited unconformably over the Walls Group implying that the latter was deformed, intruded by the Sandsting Complex and then eroded before the deposition of the Eifelian Melby Fish Bed. This latter suggestion is clearly untenable now that it has been shown that the Walls Group is younger than the successions of Melby, Papa Stour, Eshaness and Foula.

That the Melby Fault is an important discontinuity is clear from the disparate ages of the two successions. There is also a strong contrast in thermal maturity; the Walls Group kerogens have very high vitrinite reflectivity levels (5–8.9%), which almost certainly result from intrusion of the Sandsting Complex. In comparison, kerogens from Melby (1%) and Papa Stour (0.75%) have very low thermal maturity levels, being unaffected by anything other than normal burial. This distinction is continued at Eshaness, where the largely volcanic succession has a faulted contact against the Ronas Hill Granite (RHG, Fig. 1). This granite, like the Sandsting Complex, is an intrusion of approximately Late Devonian age (Rb/Sr 358 ± 8 Ma, Miller & Flinn 1966) and is therefore younger than the Eshaness succession. However, its lack of any discernible contact metamorphism in the sediments immediately west of the St Magnus Bay Fault (Finlay 1930) shows that this intrusion cannot have been emplaced whilst in its present location relative to the Eshaness succession. There must therefore be, at least, some element of post-ORS strike-slip movement on this fault (Rogers *et al.* 1989) although the most recent observed displacement is reverse movement.

Financial support is gratefully acknowledged from the University of Newcastle upon Tyne (Research Grant 1101, fieldwork) and Gearhart Geo-Consultants, Aberdeen (pyrolysis facilities) Tim Astin (PRIS, University of Reading) provided invaluable discussion and information on the Walls Group.

References

ALLEN, K. C. 1980. A review of *in situ* Late Silurian and Devonian spores. *Review of Palaeobotany and Palynology*, **29**, 253–270.

ALLEN, P. A. & MARSHALL, J. E. A. 1981. Depositional environments and palynology of the Devonian South-east Shetland Basin. *Scottish Journal of Geology*, **17**, 257–273.

ARKHANGELSKAYA, A. D. 1985. Zonal spore assemblages and stratigraphy of the Lower and Middle Devonian in the Russian Plate. *In*: MENNER, V. V. & BYVSHEVA, T. V. (eds) *Atlas of Spore and Pollen of Phanerozoic Oil- and Gas-bearing Strata of the Russian and Turanian Plates*. Trudy Vsesoiuznogo Nauchno-Issledovatel'skogo Geologorazvedochnogo Neftianogo Instituta (VNIGNI), **253**, 5–14, 32–80 (in Russian).

ASTIN, T. R. 1982. *The Devonian geology of the Walls Peninsula, Shetland*. PhD thesis, University of Cambridge.

AVKHIMOVITCH, V. I., TCHIBRIKOVA, E. V., OBUKHOVSKAYA, T. G. *et al.* 1993. Middle and Upper Devonian miospore zonation of Eastern Europe. *Bulletin du Centre Recherches Elf Exploration Production*, **17**, 79–147.

BARKER, C. E. & GOLDSTEIN, R. H. 1990. Fluid-inclusion technique for determining maximum temperature in calcite and its comparison to the vitrinite geothermometer. *Geology*, **18**, 1003–1006.

BLACKBOURN, G. A. & MARSHALL, J. E. A. 1985. The Geology of Foula, Shetland. *In*: BLACKBOURN, G. A. (ed.) *Geological Field Guide to Foula, Shetland*. Britoil, Glasgow, 1–46.

COWARD, M. P. 1993. The effect of Late Caledonian and Variscan continental escape tectonics on basement structure, Palaeozoic basin kinematics and subsequent Mesozoic basin development in NW Europe. *In*: PARKER, J. R. (ed.) *Petroleum Geology of Northwest Europe: Proceedings of the*

4th Conference. Geological Society, London, 1095–1108.

—— & ENFIELD, M. A. 1987. The structure of the West Orkney and adjacent basins. *In*: BROOKS, J. & GLENNIE, K. W. (eds) *Petroleum Geology of North West Europe*. Graham and Trotman, London, 687–696.

——, —— & FISCHER, M. W. 1989. Devonian basins of Northern Scotland: extension and inversion related to Late Caledonian–Variscan tectonics. *In*: COOPER, M. A. & WILLIAMS, G. D. (eds) *Inversion Tectonics*. Geological Society, London, Special Publications, **44**, 275–308.

DONOVAN, R. N., ARCHER, R., TURNER, P. & TARLING, D. H. 1976. Devonian palaeogeography of the Orcadian Basin and the Great Glen Fault. *Nature, London*, **259**, 550–551.

FINLAY, T. M. 1930. The Old Red Sandstone of Shetland. Part II. North-western area. *Transactions of the Royal Society of Edinburgh*, **56**, 671–694.

FLETCHER, A. E. J. 1976. *Investigations on Devonian megaspores*. PhD thesis, University of Bristol.

FLINN, D. 1977. Transcurrent faults and associated cataclasis in Shetland. *Journal of the Geological Society, London*, **133**, 231–248.

—— 1985. The Caledonides of Shetland. *In*: GEE, D. G. & STURT, B. A. (eds), *The Caledonides Orogen-Scandinavia and Related Areas*. Wiley, Chichester, 1159–1172.

—— 1992. The history of the Walls Boundary Fault, Shetland: the northward continuation of the Great Glen Fault from Scotland. *Journal of the Geological Society, London*, **149**, 721–726.

——, MILLER, J. A., EVANS, A. L. & PRINGLE, I. R. 1968. On the age of the sediments and contemporaneous volcanic rocks of western Shetland. *Scottish Journal of Geology*, **4**, 10–19.

HILLIER, S. J. & MARSHALL, J. E. A. 1992. Organic maturation, thermal history and hydrocarbon generation in the Orcadian Basin, Scotland. *Journal of the Geological Society, London*, **149**, 491–502.

LOBOZIAK, S., STREEL, M. & WEDDIGE, K., 1991. Miospores, the *lemurata* and *triangulatus* levels and their faunal indices near the Eifelian/Givetian boundary in the Eifel (F.R.G.). *Annales de la Société Géologique de Belgique*, **113**, 299–313.

MARSHALL, J. E. A. 1980. A method for the successful oxidation and subsequent stabilisation of high rank, poorly preserved spore assemblages. *Review of Palaeobotany and Palynology*, **29**, 313–319.

—— 1985. *Insculptospora*, a new genus of Devonian camerate spore with a sculptured intexine. *Pollen et Spores*, **27**, 453–470.

—— 1988. Devonian miospores from Papa Stour, Shetland. *Transactions of the Royal Society of Edinburgh: Earth Science*, **79**, 13–18.

—— 1996. *Rhabdosporites langii*, *Geminospora lemurata* and *Contagisporites optivus*: an origin for heterospory within the Progymnosperms. *Review of Palaeobotany and Palynology*, **93**, 159–189.

—— & ALLEN, K. C. 1982. Devonian miospore assemblages from Fair Isle, Shetland. *Palaeontology*, **25**, 277–312.

——, BROWN, J. F. & HINDMARSH, S. 1985. Hydrocarbon source rock potential of the Devonian rocks of the Orcadian Basin. *Scottish Journal of Geology*, **21**, 301–320.

MCGREGOR, D. C. 1981. Spores and the Middle–Upper Devonian boundary. *Review of Palaeobotany and Palynology*, **34**, 25–47.

—— & CAMFIELD, M. 1982. Middle Devonian miospores from the Cape de Bray, Weatherall, and Hecla Bay Formations of north-eastern Melville Island, Canadian Arctic. *Bulletin of the Geological Survey of Canada*, **348**, 1–105.

MELVIN, J. 1976. *Sedimentological studies in Upper Palaeozoic sandstones near Bude, Cornwall and Walls, Shetland.* PhD thesis, University of Edinburgh.

—— 1985. Walls Formation, Western Shetland: distal alluvial plain deposits within a tectonically active Devonian basin. *Scottish Journal of Geology*, **21**, 23–40.

MILLER, J. A. & FLINN, D. 1966. A survey of the age relations of Shetland rocks. *Geological Journal*, **5**, 95–116.

MYKURA, W. 1972*a*. The Old Red Sandstone sediments of Fair Isle, Shetland Islands. *Bulletin of the Geological Survey of Great Britain*, **41**, 1–31.

—— 1972*b*. Igneous intrusions and mineralisation in Fair Isle, Shetland Islands. *Bulletin of the Geological Survey of Great Britain*, **41** 33–53.

—— 1975. Possible large scale sinistral displacement along the Great Glen Fault in Scotland. *Geological Magazine*, **112**, 91–93.

—— 1976. *Orkney and Shetland.* British Regional Geology. HMSO, Edinburgh.

—— & PHEMISTER, J. 1976. *The Geology of Western Shetland.* Memoirs of the Geological Survey of Great Britain, Scotland, Sheet 127 and parts of 125, 126 and 128.

—— & Young, B. R. 1969. *Sodic Scapolite (Dipyre) in the Shetland Islands.* Institute of Geological Sciences, Report, **69/4**, 1–8.

RICHARDSON, J. B. & MCGREGOR, D. C. 1986. Silurian and Devonian spore zones of the Old Red Sandstone continent and adjacent regions. *Bulletin of the Geological Survey of Canada*, **364**, 1–79.

RITCHIE, J. D., HITCHEN, K., UNDERHILL, J. R. & FLINN, D. 1993. Discussion on the location and history of the Walls Boundary fault and Moine thrust north and south of Shetland. *Journal of the Geological Society, London*, **150**, 1003–1008.

ROGERS, D. A., MARSHALL, J. E. A. & ASTIN, T. R. 1989. Devonian and later movements on the Great Glen fault system. *Journal of the Geological Society, London*, **146**, 369–372.

SERANNE, M. 1992. Devonian extensional tectonics versus Carboniferous inversion in the northern Orcadian Basin. *Journal of the Geological Society, London*, **149**, 27–37.

STREEL, M. & LOBOZIAK, S. 1994. Observations on the establishment of a Devonian and Lower

Carboniferous high-resolution miospore biostratigraphy. *Review of Palaeobotany and Palynology*, **83**, 261–273.

——, HIGGS, K., LOBOZIAK, S., RIEGEL, W. & STEEMANS, P. 1987. Spore stratigraphy and correlation with faunas and floras in the type marine Devonian of the Ardenne–Rhenish regions. *Review of Palaeobotany and Palynology*, **50**, 211–229.

TUCKER, R. D., BRADLEY, D. C., VER STRAETEN, C. A., HARRIS, A. G., EBERT, J. R. & MCCUTCHEON, S. R. 1998. New U–Pb zircon ages and the duration and division of Devonian time. *Earth and Planetary Science Letters*, **158**, 175–186.

TURNAU, E. 1996. Miospore stratigraphy of the Middle Devonian deposits from Western Pomerania. *Review of Palaeobotany and Palynology*, **93**, 107–125.

WATSON, D. M. S. 1934. Report of fossil fish from Sandness, Shetland. *Memoirs of the Geological Survey of Great Britain, Summary of Progress for 1933*, **1**, 74–76.

Calcrete morphology and karst development in the Upper Old Red Sandstone at Milton Ness, Scotland

DONNA F. BALIN

University of Cambridge, Department of Earth Sciences, Downing Street, Cambridge CB2 3EQ, UK

Present address: Balin & Associates, 127 Claywell Drive, San Antonio, TX 78209, USA (e-mail: Balin@alumni.utexas.net)

Abstract: The Upper Old Red Sandstone at Milton Ness, Scotland, is notable for its excellent preservation of calcrete textures, which are comparable with some of the best Quaternary examples. It is also significant for the implications that can be drawn from the association between karst and calcrete, with this example interpreted to have formed entirely within a semi-arid environment. Karst cavities were developed in a mature hardpan calcrete, generated in sandy fluvial sediments with associated aeolian deposits. Subsequent to karst cavity generation, clasts derived from the subaerially exposed hardpan were locally transported and deposited as a laterally traceable bed connecting the tops of all the cavities. Both this bed and the karst infills were subsequently recalcretized in the final phase of the profile's evolution. Although calcrete–karst associations often are interpreted as the alternation between semi-arid and humid climates, respectively, this example is interpreted to be a result of water accumulating on the nearly impervious hardpan surface under fairly constant semi-arid conditions, evidenced by the recalcretization of both the karst infill and the calcrete-derived breccia ('boulder calcrete'). Additional substrate modification also has taken place by plant roots; the remarkable development of rhizoliths in these Old Red Sandstone sediments should emphasize the need to consider plant influence on other non-marine rocks of post-Silurian age.

Calcretes have proven to be useful tools in evaluating palaeoclimates and sedimentation rates as well as in stratigraphic correlation. Many excellent examples of mature Quaternary calcretes have been described in the literature and, in more recent years, emphasis has been given to their identification and description in older rocks (Wright 1986). This study from the Upper Old Red Sandstone (ORS) of Scotland describes some exceptionally well-preserved calcretes with many features commonly cited in Quaternary examples. It also identifies some features that are unusual in calcretes of any age. The calcrete is associated with karst solution cavities, a combination that is rare in calcretes developed within a detrital host. Although calcrete–karst associations are often interpreted as the alternation between semi-arid and humid climates, respectively, the association at Milton Ness is explained in terms of the evolution of the calcrete profile under relatively uniform semi-arid conditions. Additionally, the excellent development of rhizoliths is surely one of the finest examples in Palaeozoic rocks. The Upper ORS here has not been described before in detail, although the calcrete and associated section were briefly referenced by Trewin (1980, 1987) and Parnell (1981, 1983*a*, *b*).

This study focused on Upper ORS exposures at Milton Ness, located 9 km NNE of the town of Montrose in eastern Scotland (Fig. 1). It is located within the Midland Valley, a major structural province of the British Isles. The Upper ORS crops out as a wave-cut cliff along the North Sea, in places reaching up to about 10 m in height. The bedding is nearly horizontal and extends for *c*. 500 m, thus giving excellent lateral exposure. Locally, a northeast-trending fault separates the block of Upper ORS from Lower ORS conglomerates and volcanic rocks to the west.

The Upper ORS section at Milton Ness correlates lithostratigraphically with the calcrete-bearing Kinnesswood Formation of Fife, which is assigned a Late Devonian (Famennian) or Early Carboniferous (Tournaisian) age

From: Friend, P. F. & Williams, B. P. J. (eds). *New Perspectives on the Old Red Sandstone.* Geological Society, London, Special Publications, **180**, 485–501. 0305-8719/00/$15.00

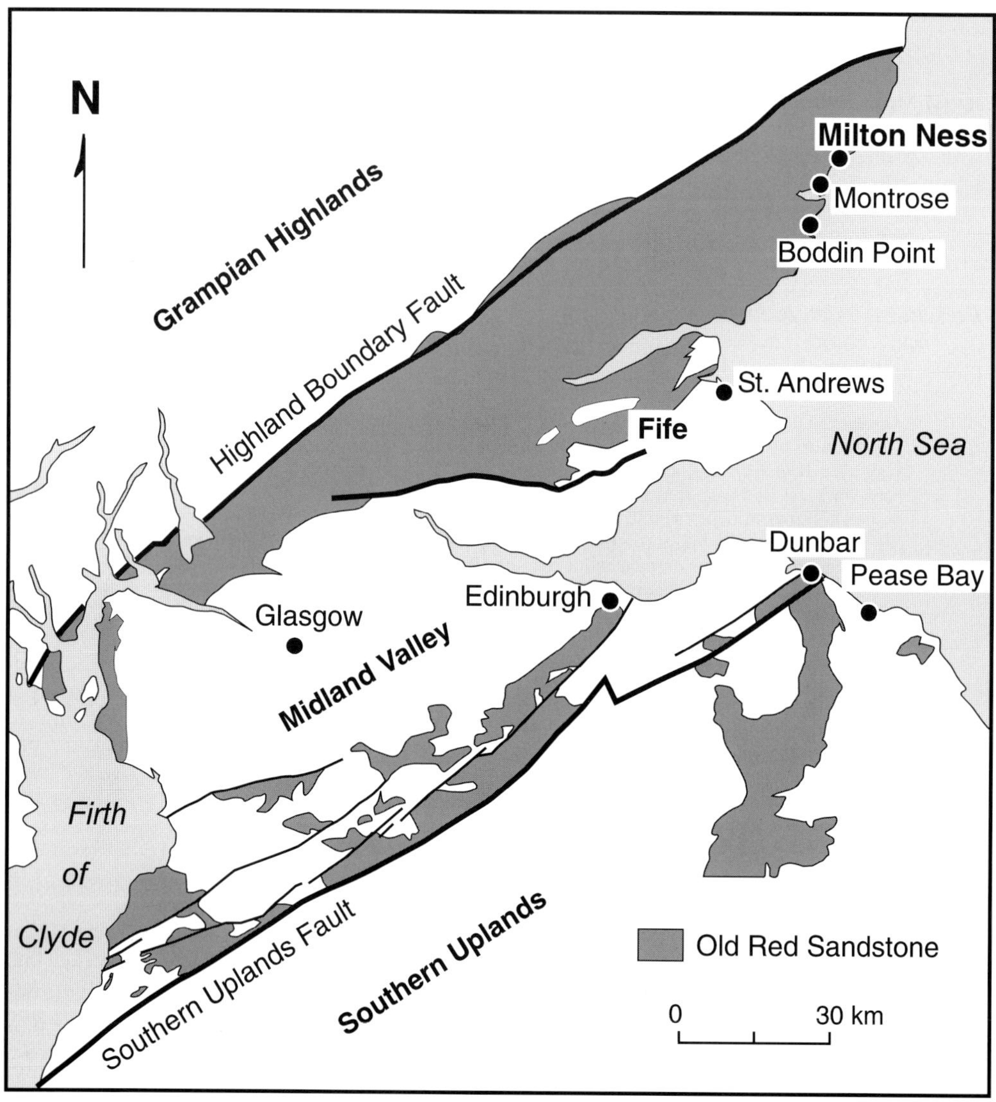

Fig. 1. Index map also showing the distribution of the Old Red Sandstone in the Midland Valley area. National grid references for localities cited in the text include: Milton Ness (NO 770 647), Dunbar (NT 6755 7930), Pease Bay (NT 7915 7108) and Boddin Point (NO 715 536).

(Chisholm & Dean 1974; Paterson & Hall 1986; Wright *et al.* 1993). The Milton Ness outcrop also correlates with the calcrete-bearing part of the Upper ORS exposed in the southern Midland Valley area, specifically at Dunbar and north of Pease Bay (Fig. 1). The calcrete-bearing Upper ORS of these southern outcrops has been assigned to the Cockburnspath Formation by Balin (1993).

Regionally, the Midland Valley of Scotland forms a low-lying, NE–SW-trending belt confined to the north by the Grampian Highlands and to the south by the Southern Uplands. It is defined structurally by its position between the Highland Boundary Fault and the Southern Uplands Fault zones. Devonian and Carboniferous rocks make up most of the Midland Valley's surface geology, in contrast to the late Precambrian–Cambrian rocks of the Grampian Highlands that underwent subsequent metamorphism (Dalradian Supergroup) and the Ordovician and Silurian rocks of the Southern Uplands. Global palaeogeographical reconstructions for Late Devonian time place the British

Isles 5–10° south of the equator (e.g. Woodrow *et al.* 1973; Livermore *et al.* 1985; Tarling 1985).

Much of the Upper Old Red Sandstone in the eastern Midland Valley area represents the distal part of a major river system that flowed eastwards through the Midland Valley, with the thicker and coarser Upper ORS of the Clyde area being a more proximal part of the system. Bars with an estimated thickness of 9 m are present in the Upper ORS of the western Midland Valley (Seamill Fm, Bluck 1990), indicating deposition by a major alluvial system, which ultimately may have been derived from the Scandinavia–Greenland foldbelt to the north (Scandian orogen; Bluck *et al.* 1988). Palaeocurrents obtained from the fluvial facies at Milton Ness and the nearby Upper ORS outcrop at Boddin Point, however, are southerly in direction, implying that these deposits are unlikely to have been the result of the eastward-flowing axial system evident further south in the Midland Valley (Balin 1993). Instead, they probably were deposited by a tributary to that system, in which case, the relatively small catchment would have contributed to the low rates of sedimentation necessary for calcrete generation.

Milton Ness calcrete profile

This study is the first detailed presentation of the calcretes at Milton Ness, and the first to recognize the existence here of karst cavities, reworked hardpan ('boulder calcrete'), and rhizoliths, with the interpretation being derived directly from the author's own observational data. At this locality, there is significant lateral variation in amount and type of pedogenic alteration. However, in places, calcrete profile development shows an idealized vertical zonation of pedogenic carbonate types: (1) an upper compact crust or hardpan (70 cm); (2) platy, sheet-like carbonate (70 cm); (3) carbonate developed as clusters of nodules (120 cm); (4) uniformly distributed, chalky carbonate (55 cm); (5) unaltered host material at the base (Fig. 2). The thicknesses given are from a measured section and do not reflect lateral variation along the outcrop. The marked vertical zonation indicates genesis as a true pedogenic carbonate.

At the eastern end of the outcrop, karst cavities and boulder calcrete represent additional stages in profile development. Both the boulder calcrete and the karst infills were subsequently recalcretized in the final phase of the profile's evolution.

The following sections will first describe the mature hardpan and its associated micromorphological fabrics. A discussion of textural features associated with subsequent karst and boulder calcrete generation will be followed by a description of the recalcification of the karst infills. The processes and paleoclimate implications responsible for karstification and recalcification also will be evaluated. The final section will discuss the morphology of rhizoliths within the Milton Ness section and at nearby Boddin Point.

Hardpan calcrete

The Milton Ness hardpan forms a well-indurated, highly resistant bed measuring up to 1.5 m in thickness. This hardpan is developed at the eastern end of the outcrop, where it weathers as a resistant ledge extending laterally for nearly 150 m, forming the upper part of a 4 m calcrete profile.

Megascopically, the hardpan is characterized by non-tectonic brecciation, irregular subhorizontal carbonate layers, chert laminae, and relicts of the original sandstone host. Generally, the carbonate content of the hardpan tends to increase slightly upwards, and, in places, calcrete development completely obliterates all host rock structures (Fig. 3). Microscopically, many of the features cited in Quaternary calcretes are preserved, including clotted micrite, silicified pisolites, microspar fringes around detrital grains, and floating grain texture.

Floating grain texture and brecciated fabrics: evidence for displacive and replacive calcite

Floating grain texture has been cited by many workers as characteristic of calcretes (Gardner 1972; Steel 1974; Goudie 1983; Solomon & Walkden 1985) and is an extremely common feature of the Milton Ness calcretes (Fig. 4). Although many of the quartz grains show evidence of corrosion at their edges as a result of replacement by calcite, there is considerable evidence that displacive growth was a major, if not predominant, process. A typical example is shown in Fig. 5a where domal calcite growth has shattered the overlying chert laminae. The jigsaw fit of the brecciated chert layer is evidence for the displacive growth of the calcite, and the calcite dome itself has an expanded internal fabric. Calcite growth also has disrupted the underlying chert laminae, with the domal features including fragments of the 'exploded' chert layer. The irregular edges of the chert fragments suggest that replacement also was operative.

The force of crystallization is known to cause displacive growth (e.g. Rothrock 1925; Schuiling

Fig. 2. Calcrete profile showing an idealized vertical zonation of pedogenic carbonate morphologies (B–F). A bed of recalcretized clasts derived from the hardpan caps the sequence ('boulder calcrete'; Balin 1993). Hammer 28 cm long for scale.

Fig. 3. Hardpan layers represent the most mature type of soil profile development. Evaporation of soil water leads to carbonate being precipitated predominantly as concentrated horizontal laminae distinctive of the hardpan. Extensive calcretization has resulted in near-obliteration of all original textures in the sandstone host material. Hammer 28 cm long for scale.

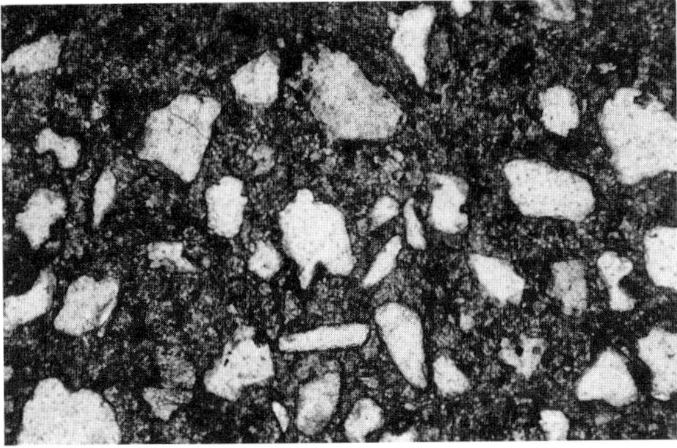

Fig. 4. Floating grain texture in the hardpan calcrete. Some of the quartz grains also show evidence of corrosion by calcite crystallization along the grain boundaries. Horizontal field of view 2 mm.

& Wensink 1962). Watts (1978) discussed the conditions for displacive calcite growth in calcretes, and suggested that precipitation from supersaturated solutions is involved. Additionally, a lower average effective normal stress facilitates displacive growth at lower supersaturation levels. Weyl (1959) stated that with the pressure coefficient under hydrostatic conditions equal to the stress coefficient of the solution film, calcite precipitating from a 1% supersaturated solution will grow displacively up to normal stresses of 10 atm, indicating that calcite needs only mildly supersaturated solutions to grow displacively in the near-surface pedogenic environment. Displacive calcite growth has been documented along fractures in quartz grains (Saigal & Walton 1988), and similar evidence is found in the Milton Ness calcretes, where quartz grains appear to have been split apart by the force of calcite crystallization.

Silica laminations with pisolites

The hardpan also is characterized by horizontal, wavy to irregular layers of silica. Some vertically oriented silica additionally is present but is considerably less abundant. Megascopically, these layers are buff to yellow to pink, ranging up to *c.* 1 cm in thickness, and are composed of chert and drusy amorphous silica. The silica layers are internally laminated and typically are brecciated, with the fractures being infilled by calcite that often can be shown to have grown displacively. Additionally, some silica layers contain pisolites, with characteristic spherical to sub-spherical concentric internal structures, ranging from 0.1 to 2.5 mm in diameter (Fig. 5b). The layers also contain vugs filled with chalcedony, megaquartz and calcite spar. Because permeability would have been progressively reduced as the hardpan matured, water would have been held within vugs and fractures for longer time periods; therefore, crystallization would have proceeded more slowly, resulting in the larger grain sizes of quartz and calcite (Knox 1977). It is likely that the silica in the laminae was derived locally from quartz grain corrosion. Corroded quartz grains are clearly evident at Milton Ness, demonstrating that silica must have been removed in solution (Fig. 4). Summerfield (1983) stated that quartz replacement, especially by carbonates, could provide an abundant source of silica in favourable environments, and Steel (1974) has supported this process to generate

Fig. 5(a).

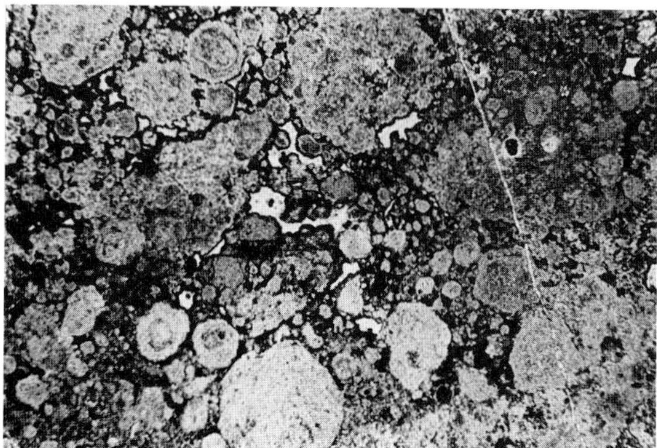

Fig. 5(b).

Fig. 5. (**a**) Displacive calcite growth has shattered the siliceous laminae within the hardpan. The jigsaw fit of the brecciated chert laminae is evidence for the displacive growth of the calcite (arrow). In the lower half of the photograph, fragments of an 'exploded' chert layer have been incorporated into domal growth of the calcite. Expanded fabrics are characteristic of displacive growth of calcite in calcretes. Crossed nicols. Vertical field of view 11.3 mm. (**b**) Pisolites present within siliceous laminae in the hardpan. Transmitted light. Horizontal field of view 4 mm.

silica laminae in New Red Sandstone calcretes of western Scotland.

Clotted micrite

Clotted micrite is a frequently reported feature of calcretes (e.g. 'agglomeratic fabric' of Brewer 1964; 'clotted texture' of Hay & Wiggins 1980; Esteban & Klappa 1983). In the Milton Ness hardpan, micrite frequently appears in irregular patches giving the rock a clotted microfabric; an example of this texture is shown in Fig. 6a. The dark micritic patches typically grade into zones of lighter-coloured micrite or microspar, often with diffuse boundaries. Fenestrae composed of calcite spar also may transect these features, and a few grains of sand or silt commonly 'float' within the clotted fabric. The very fine grain size of the clotted micrite is attributed to rapid precipitation rates.

Microspar fringes around detrital grains

Within the micritic groundmass, zones of more coarsely crystalline calcite surround, or partially surround, floating quartz grains (Fig. 6b). The grain size of the carbonate fringes (4–20 μm) is within the size range of microspar (Folk 1959). The microspar zones, although frequently forming a halo around detrital grains, often are very irregular. 'Fingers' of microspar may extend outward and coalesce with microspar patches in the micritic groundmass. Calcite also forms isopachous fringes around some detrital grains, which commonly suggests phreatic-zone cementation; however, isopachous fringes also may precipitate in the vadose zone if pore spaces between detrital grains are small (Klappa 1983), as interpreted at Milton Ness.

Although microspar fringes have been attributed to shrinkage of the calcrete around the detrital grain surface and the subsequent infill by microspar, Goudie (1973) maintained that both the hydration and linear expansion coefficient of calcite is very low and unlikely to be of significance in calcrete. This implies that calcite dehydration and thermal contraction are similarly insignificant, although contraction may be aided by dehydration of interstitial clays or salts (Assereto & Kendall 1977). If shrinkage had resulted in the surrounding micritic matrix pulling away from the detrital grain surface, the grain would have dropped slightly to the base of the 'cavity' and the spar infill would have developed preferentially above the detrital grain. However, the spar does not show this type of preferred distribution and often completely surrounds and parallels the grain boundaries (Fig. 6b).

Although many of the detrital grains have corroded edges, a large number do not. The microspar rims, therefore, cannot be explained solely as the corroded edges of detrital grains that have been replaced by microspar.

One possibility is that displacive, and occasionally replacive, growth of the micritic matrix may have been followed by selective recrystallization to microspar adjacent to the quartz grains.

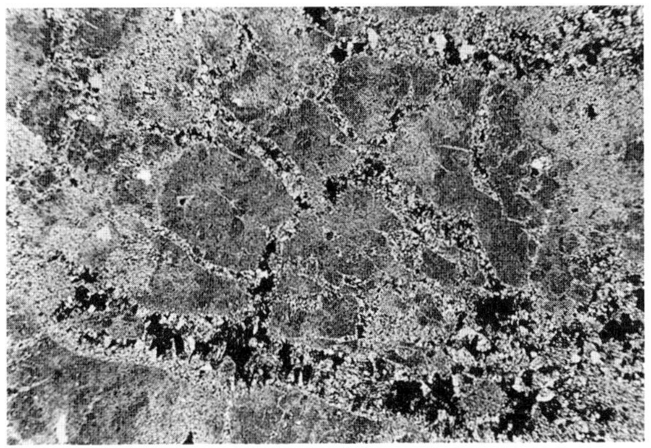

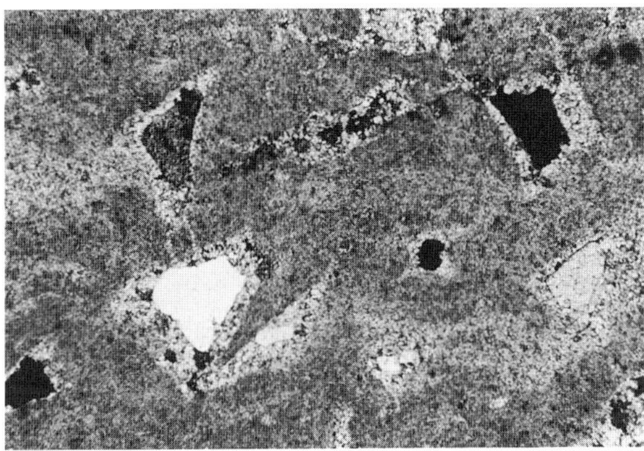

Fig. 6. (**a**) Clotted micrite (dark patches) with diffuse lighter zones of microspar. Spar-filled fenestrae transect the finer-grained matrix. Floating grains are very sparsely distributed. From karst infill. Crossed nicols. Horizontal field of view 8.2 mm. (**b**) Microspar fringes around floating detrital grains. The fringes are not always regular and commonly merge with diffuse microspar zones in the matrix (e.g. grain on lower right). From karst infill. Horizontal field of view 2 mm.

Micrite will recrystallize to microspar when acted upon by calcium-rich waters (Ca:Mg greater than about 3:1; Folk 1974). A thin solution film between the detrital grain and the micrite could promote recrystallization adjacent to the grain. This idea was advanced by Folk (1959), who explained microspar fringes by the initial recrystallization of micrite around allochems or quartz grains, and the outward advancement of the recrystallization front. The observed features would appear to be consistent with this hypothesis.

Karstification and reworking of the hardpan surface

Karst cavities

Several solution cavities penetrate the hardpan layer at the easternmost end of Milton Ness. The karst features are exposed two-dimensionally along the cliff face for *c.* 50 m. At least ten solution cavities are present, with the largest measuring 1.2 m in depth and also 1.2 m in width at its upper edge (Fig. 7a). The largest cavity is located centrally with respect to the other karst features. The smallest solution cavities have dimensions of only a few centimetres, and generally have width-to-depth ratios *c.* 1:1. The cavities extend from the top of the hardpan surface, abruptly truncating the horizontal laminations within the hardpan, and have smooth rounded bases that may be partially lined by carbonate rinds. The cavities tend to be contained wholly within the hardpan, with only one example penetrating its undersurface.

The hardpan's top surface appears to be slightly lower in the area where the largest cavity is developed. It may reflect the original topography of the calcrete rather than later structural

Fig. 7. (**a**) Large karst cavity and sediment infill developed in hardpan calcrete. This is the largest of at least ten similar solution cavities developed in the hardpan, measuring 1.2 m in depth. The section is overlain here by recent glacial till. (**b**) A second solution cavity developed in the hardpan (arrow on L). Sediment infill is composed predominantly of carbonate clasts of reworked calcrete. (Note the boulder calcrete bed, which forms a laterally traceable unit between cavities (arrow on R).) Hammer 28 cm long for scale.

modification. This downwarp is extremely slight, but may have been sufficient to locally concentrate runoff.

Solution hollows are not uncommon in well-developed Quaternary calcretes, although older examples are less frequently described, perhaps because of lack of recognition. They are usually generated in calcretes with carbonate substrates (e.g. Monroe 1966; Panos & Stelcl 1968; Sweeting 1969; Walkden & Davies 1983; Mustard & Donaldson 1990), but they are also known from mature calcretes with detrital hosts (e.g. Judson 1950; Partridge & Brink 1967; Netterberg 1980; Allen 1986; Balin 1993).

Partridge & Brink (1967) suggested that wedging of deep-rooting vegetation may be significant in initiating or enlarging joints within calcretes, thereby facilitating solution along these lines of weakness. The influence of vegetation also has been noted in the development of late Dinantian palaeokarst pits of England and Wales; Vanstone (1998) suggested that rainwater intercepted by the crowns of trees became concentrated beneath the trunks ('stem-flow drainage'), initiating dissolution of limestones underlying the soil horizon. This process also was invoked by Herwitz (1993) to explain karst genesis in Pleistocene limestones of Bermuda.

Plant growth is certainly evident within the Milton Ness calcretes, evinced by abundant rhizoliths in some parts of the section, although in this case it cannot be demonstrated that plants helped to control the distribution of the karstic features. The palaeokarst noted by Vanstone (1998) differs from Milton Ness in the uniform diameter of both deep and shallow pits, which is interpreted to reflect their formation around a particular species of tree. In contrast, the diameters of the solution cavities at Milton Ness are distinctly nonuniform, varying from a few centimetres to 1.2 m in width. Additionally, the isolated karst features at Milton Ness, with their width-to-depth ratios *c.* 1:1, lack the narrower cylindrical morphology that would more easily compare with the isolated, non-amalgamated solution pits of Vanstone (1998) and Herwitz (1993). It is recognized that neither of these factors would rule out the influence of stem-flow drainage as the initiator of karst development at Milton Ness, as, for example, the width of cavities may have been related to different tree species rather than a single type, and the more equilateral shape may be attributed to continued dissolution subsequent to a stem-flow drainage phase of karst generation. The slight topographic downwarp of the nearly impermeable hardpan also may have served to concentrate water in the subsurface, facilitating growth of overlying vegetation. Accordingly, this process remains as a possible mechanism for karst initiation at Milton Ness.

The karst cavities represent a very late-phase modification of the calcrete profile. It is notable that the hardpan in the vicinity of the karst features is composed almost entirely of carbonate, which would have hindered the percolation of water through the subsurface. As the hardpan matured and became increasingly impervious, rainwater probably would have become increasingly confined to the uppermost portions of the profile, initiating dissolution where the water became concentrated. Whether the karst at Milton Ness was initiated by stem-flow drainage or was generated as a result of the slight topographic downwarp concentrating runoff either at the surface or in the subsurface, it is clear that at some point the hardpan became subaerially exposed. The next section describes evidence that the hardpan was reworked at the surface.

Boulder calcrete: an unusual example of recalcretized fabric

There is evidence that the hardpan surface was subaerially exposed and reworked. Erosion produced clasts from the sandstone relicts in the hardpan, as well as carbonate and chert clasts. These clasts were locally transported and deposited as a laterally traceable bed connecting the tops of all the karst cavities (Fig. 7b). Small abraded chert clasts indicate some degree of transport and not simply *in situ* modification of the underlying calcrete. Significantly, this bed shows downward flexure at the edge of some of the karst features, demonstrating that karst generation had already begun at the time of its deposition.

There also are clear examples of displacive micrite growth brecciating clasts of the relict sandstone, evidenced by the jigsaw fit of sandstone-rich areas separated by zones of carbonate. Some of the calcite also has precipitated as passive pore-filling within the sandstone zones. Both the carbonate and sandstone-rich zones in this bed have a noticeable bias for columnar structure, and carbonate zones also have a secondary tendency for horizontal alignment.

Within the carbonate-rich zones, some unusual circular features are present (Fig. 8). These structures range from *c.* 0.25 to 1.80 mm in diameter, and may be oblong to slightly irregular in shape. They commonly have a concentric outer rim of micrite separated from the inner zone by a thin coating of hematite. Hematite may also form a thin coating on the outer rim. Occasionally bridges are formed between the circular features. These structures have very weak cathodoluminescence characteristics. The outer micritic rim luminesces only a slightly brighter orange than the inner zone of duller luminescing micrite.

Following the terminology of Netterberg (1980), the unit described above will be referred to as 'boulder calcrete', denoting clasts derived from a degenerative state of a hardpan during weathering. Extensive recalcretization of the bed does not allow determination of original clast sizes. Therefore, the term as used here reflects only the mode of origin, and not clast dimension.

The columnar fabric of many of the sandy zones makes it unlikely that they comprised individual clasts of sandstone. Instead, the sand-rich zones may have been part of considerably larger clasts that were forced apart by the displacive crystallization of the calcite. It is likely that calcium carbonate from the calcrete clasts was remobilized and reprecipitated during phases of wetting and drying, with calcite commonly being reprecipitated as displacive micrite. Young (1964) reported the splitting of sandstone cobbles in terrace gravels as calcite exploited pore spaces and small cracks during crystallization. Floating quartz grains within the carbonate-rich areas may be supportive of this suggestion. The repeated phases of dissolution and reprecipitation resulted in the distinctive fabric of the recalcified boulder calcrete.

The circular features described within the carbonate zones may be calcite pisolites that lack concentric internal laminations; they are too large to reflect dissolution of detrital quartz grains and replacement by micrite. Alternatively, similar structures have been attributed to the biological activity of roots in calcretes (Pye, pers.

Fig. 8. Unusual circular features in carbonate portions of the boulder calcrete, which may have a biological origin. Thin hematite rinds delimit the outer concentric rims of these structures. Transmitted light. Horizontal field of view 4.2 mm.

comm.). Netterberg (1980) has noted that roots, even trees, often permeate the soil-filled gaps between boulder calcrete clasts. Esteban & Klappa (1983) have described how a reworked and brecciated hardpan, which remains subaerially exposed, may be modified by the biological activity of lower plants, such as algae and fungi. Their action can form a protosoil between the weathered calcrete clasts which is subsequently colonized by higher plants. Comparable structures also have been attributed to lichen hyphae that penetrate limestone (Curtis *et al.* 1976). A biological origin is plausible, as processes of pedoturbation, accompanied by carbonate dissolution and reprecipitation, are known to form reworked, recemented, breccia–conglomeratic fabrics in hardpans (Esteban & Klappa 1983).

Chapman (1971) described a recent lag gravel from Saudi Arabia consisting of rubble derived from the underlying duricrust. The thick desert pavement, which bestowed 'a bizarre aspect to the scenery', is similar in origin to the boulder calcrete of this study, although apparently lacking a recalcretized fabric. A more ancient example comes from the Lower Old Red Sandstone of Wales, where Allen & Williams (1979) and Allen (1986) noted a calcrete conglomerate overlying a thick calcrete profile. The conglomerate had a patchy distribution and ranged from a single clast to several centimetres in thickness.

Gardner (1972) described a 15 cm 'fragmental' bed of weathered calcrete clasts overlying a mature Quaternary calcrete profile. Here, the irregular calcrete clasts were set in an uncemented sand matrix. The top of the underlying calcrete profile had undergone brecciation as a result of physical weathering and had been recemented *in situ*. Although few textural details were provided, the fabric of the recemented calcrete breccia resembles that of the Milton Ness boulder calcrete.

Although in Gardner's example it is the upper part of the duricrust rather than the 'fragmental' bed that is recalcified, other conditions could make cementation of the rubble bed possible. For instance, the formation of the karst cavities at Milton Ness suggests that the hardpan was nearly impermeable by the time the boulder calcrete bed was generated. As rainwater could not easily infiltrate this horizon, much of the water would have been held within the pore spaces of the overlying boulder calcrete promoting calcite solution from the clasts. The calcite would have been reprecipitated as rainfall was followed by evaporation. Furthermore, as the boulder bed became cemented, it would have had the reinforcing effect of shielding the hardpan surface from further weathering. Chapman (1971) noted that the lag gravel of his study provided a 'protective armor' for the underlying topography.

Bretz & Horberg (1949) recorded the phenomenon described above in calcretes of southeastern New Mexico, USA. In these Pliocene calcretes, weathered slabs from the duricrust became brecciated, and the pieces rotated and recemented to form a continuous rock body, which Bretz & Horberg called 'Rock House structure' (after a nearby town). They described repeated cycles of brecciation, dissolution and recementation by growth of 'laminar wrappings' of the calcite. Displacive growth of the calcite is evident by the support of caliche fragments and siliceous pebbles in the recalcified matrix. Netterberg (1980) considered this fabric to be relatively rare and confined to calcrete hardpans, boulders and cobbles. Dissolution of the calcrete fragments was not as severe as in those of Milton Ness, where the fabric retains sandstone and siliceous clasts, but none of the original carbonate clasts.

Recalcification of karst infill

The Milton Ness karst cavities have been infilled and recalcified. The infill is a mixture of carbonate and terrigenous clastic materials in mottled shades of red, purple, white, yellow and orange. It was previously noted that the boulder calcrete bed in some cases shows downward flexure at the edge of the karst features, demonstrating that karst had already generated at the time of boulder calcrete formation. Some of these clasts eroded from the hardpan undoubtedly infilled the solution hollows. Figure 7b shows one such cavity, in which the boulder calcrete bed drapes into the solution hollow, and where the morphology of the original carbonate clasts is still evident. In most of the cavities, however, the infill has undergone more extreme dissolution and recalcification, obscuring the original fabric. The karst infills show floating grain texture, microspar fringes around detrital grains, clotted micrite and spar-filled fenestrae, demonstrating substantial recalcretization of the infill (Fig. 6a and b).

As in the boulder calcrete, pink chert clasts (<2 cm) are scattered throughout the matrix. Some show pisolitic structures within them, thus supporting derivation from the hardpan. Solution has affected some of the clasts' boundaries.

Partridge & Brink (1967) recognized solution hollow or 'makondo' infills that were recalcified in calcretes of the lower Vaal River basin, South Africa; calcite cementing the alluvial sediments was removed by solution and frequently reprecipitated along the makondos. Knox (1977) also noted solution hollows developed in calcretes near Saldanha Bay, South Africa, which were infilled by aeolian sediments and recalcified. Solution cavities infilled with calcrete clasts also have been reported from the High Plains, USA (Bretz & Horberg 1949). The calcrete fragments were mixed with red silt from the overlying deposits and the infill was recalcretized.

Significantly, some of the karst features at Milton Ness are clearly infilled to the base of their cavities with clasts derived from the hardpan (Fig. 7b). This observation suggests that at least some of the karst cavities were not soil filled before subaerial exposure and reworking of the hardpan surface. If all of the karst cavities had been generated under a soil cover, then soil should have been retained in the karst depressions once they became subaerially exposed, and the clasts derived from hardpan would have been deposited on top of a soil infill. Therefore, at least some of the karst cavities probably were generated subaerially.

Karst and calcrete association: implications for genesis

At Milton Ness, recalcretization of the karst infill argues against change to a humid climate subsequent to hardpan generation, as does recalcretization of the boulder calcrete bed. Many previous studies have described karst development in humid climatic regimes (e.g. Corbel & Muxart 1970; Jennings 1972), whereas other studies have described karst genesis in semi-arid and arid climates (e.g. Jennings 1983; Pelechaty *et al.* 1991). Karst generation has been attributed to climate changes during Early Carboniferous time, with karstification associated with a more humid environment (Wright 1980; Smith & Dorobek 1993). As previously noted, the Upper Old Red Sandstone at Milton Ness may be Late Devonian in age, so any climatic fluctuations during Early Carboniferous time are not necessarily relevant to this example. Although a brief humid period is possible, the association between the hardpan calcrete, karst and boulder calcrete at Milton Ness may also be explained in terms of the natural evolution of the landscape without invoking external factors for which there is no additional evidence.

Walkden & Davies (1983) similarly preferred an internal mechanism to explain the common association between karst and calcrete in northern Wales (Upper Dinantian succession). In their study, alluvial channels were outbuilding onto a limestone substrate. Those workers suggested a model where overbank flooding promoted water accumulation and solution. Calcretization took place simultaneously away from static pools or following channel abandonment.

At Milton Ness, the primary cause of karst generation is interpreted to be decreasing permeability associated with hardpan maturation, resulting in concentration of water above the hardpan surface. The water would have been supplied by precipitation, and, as in the Walkden & Davies model, by possible overbank flooding from laterally accreting channels. The calcrete described here from Milton Ness is developed in sandy fluvial sediments and associated aeolian deposits. The fluvial facies display well-defined, fining-upward grain sizes characteristic of the classic meandering stream model of Allen (1970). Aeolian deposits are present at the base of the section, characterized by bidirectional tabular–planar and wedge–planar cross-bedding. The presence of aeolian deposits in association with the calcretes is further supportive of low depositional rates at Milton Ness.

Time and calcrete accumulation

Radiocarbon dates on mature hardpans have yielded minimum times of *c.* 10 ka for their formation (Ruhe 1965; Leeder 1975). However, there is a significant magnitude of error in these calculations, as a result, for instance, of the input of younger aeolian dust or possibly older reworked calcrete (Goudie 1983).

Some immature forms of calcrete take only a few years to develop. For example, Cohen (1982) noted that rubbish left by excavation crews had developed carbonate rinds within 12 years. On the other hand, the most mature calcrete types may take hundreds of thousands, or even millions of years to develop (Gardner 1972). Rates of accumulation may also vary considerably. Radiocarbon dates from Cenozoic calcretes indicate accumulation rates of 5–50 cm ka^{-1} (Netterberg 1978).

The most mature calcrete type at Milton Ness is the hardpan. With its thickness of 1.5 m and the textural evidence for its maturity, it would have taken an absolute minimum of 10 ka years to develop (Leeder 1975). However, a considerably longer time period seems probable based purely on subjective comparison with other Cenozoic examples (Gardner 1972; Netterberg 1978).

At Milton Ness, the hiatus represented by the hardpan must also include the duration of karst development. Because limestone dissolution in the meteoric carbonate environment is dependent on rainfall regime, temperature, distribution of soil cover, biological activity and lithology of the carbonate substrate (Vanstone 1998), attempts to compare palaeokarst relief with rates attributed to modern examples may be prone to appreciable error. With this caution in mind, Sweeting (1973) has listed rates of limestone corrosion in modern karstified areas. Estimates are based on factors that are not measurable in ancient examples, such as fluid discharge and mean total hardness of the water. A wide range of values is evident from 15.1 to 333.0 mm ka^{-1}. As these figures relate to overall lowering of an area, and dissolution is not uniform, locally solution may be considerably greater than the overall figures suggest. The highest dissolution rates have tended to be in colder climates (Corbel 1959), although these high rates cannot be strictly allied to low temperatures (Sweeting 1973).

Given the semi-arid climate and the presumably low discharge at Milton Ness, carbonate dissolution rates would have fallen towards the lower end of the spectrum. Thus, it would appear that the Milton Ness karst cavities, with depths up to 1.2 m, would have taken several thousand years at a minimum to generate. Considering the subsequent recalcification of the karst infill and the boulder calcrete, the surface stabilization at this horizon must have persisted for several tens of thousands of years or, more probably, hundreds of thousands of years.

Rhizolith development at Milton Ness

Rhizoliths are exceptionally well developed in the Upper ORS outcrop at Milton Ness and are a useful tool in the identification of pedogenically altered substrates. The term 'rhizolith' was proposed by Klappa (1980) to identify 'accumulation and/or cementation around, cementation within, or replacement of, higher plant roots by mineral matter'. This term includes structures that preserve the actual root anatomical features (rhizocretions, root petrifactions). It also includes structures that preserve the general root morphology but that lack root components (root moulds, root casts and root tubules).

Root tubules, calcite-cemented cylinders around root moulds (Klappa 1980), are especially abundant at Milton Ness. These structures are generated by the uptake of CO_2 and water by roots for use during photosynthesis. The removal of CO_2 and water results in calcite precipitation adjacent to the root. Although promoted by plant physiological activities, the calcite is precipitated by inorganic processes (Schneider 1977). The root material may eventually decay leaving a void in the central area of the tubule.

Figure 9 shows examples of root tubules at Milton Ness. As the surrounding sandstone is uncemented or only weakly cemented by hematite, it has been more easily eroded away, leading to preferential preservation of the cemented cylinders. Tubules range from less than 1 cm in diameter to 11 cm in diameter and over 50 cm in length.

Many of the Milton Ness tubules bifurcate and show reduced diameters on the second-order branches (Fig. 9b). These features distinguish rhizoliths from burrows, which are usually unbranched with uniform diameters (Klappa 1980). Some tubules are vertically-oriented although the larger ones have a greater tendency for horizontal ramifying networks.

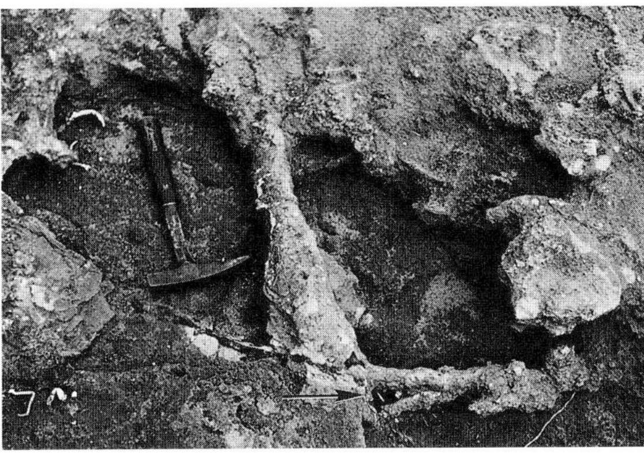

Fig. 9. Root tubules at Milton Ness. (**a**) A small part of the calcareous cylinder has been dislodged to reveal the inner cavity. Hand lens for scale. (**b**) Bifurcation reflecting branching of the original root material (arrow). A decrease in the diameter of the lower-order branches helps to distinguish roots from burrow systems, which tend to have uniform diameters. Hammer 28 cm long for scale.

The position of the tubules within the stratigraphic framework reinforces their origin as rhizoliths. They are located within the sandy portions of fluvial fining-upward cycles (Fig. 10). The basal conglomerates lack rhizoliths because of their unfavourable origin as channel lag deposits. The overlying red siltstones also tend to lack rhizoliths although some vertically oriented zones of carbonate are present. The floodplain may have been less hospitable to plant growth, perhaps because of increased distance from the laterally migrating channel limiting the source of water. Alternatively, the relatively impermeable sediment type may not have been as conducive to plant propagation. A combination of these factors is the preferred hypothesis. The development of rhizoliths would be expected in sandy point-bar soils adjacent to a channel, and their channel-side position also explains the predominant development of horizontally oriented ramifying networks rather than vertical tap roots among the larger rhizoliths. Water availability would preclude the need for vertical tap roots in this environment (Weaver 1958; Cohen 1982).

The dimensions of the Milton Ness root tubules correspond to those given by Esteban & Klappa (1983) for usual rhizolith development; that is, millimetres to centimetres in diameter and centimetres to metres in length. Retallack (1983) reported tubules that were 1–3 cm in diameter, each with a central calcite-filled vug measuring 4–5 mm in diameter. Because of their size, he interpreted these as the root traces of trees or large shrubs. Many of the Milton Ness rhizoliths are considerably larger than Retallack's examples, and thus may also be root traces of relatively large plants. The fossil record of spores suggests that several major lineages of vascular plants had already evolved by mid-Silurian time

Fig. 10. Vertical section showing horizon with large, calcareous rhizoliths (arrow) capping the sandy portion of a fluviatile fining-upward cycle. Vertical field of view *c.* 2.5 m.

Fig. 11. Extremely large root tubules from Boddin Point. Both vertical and horizontally oriented tubules are present. Length-wise cross-sections across the horizontal tubules demonstrate that the central vug, where the root was originally positioned, extends along its entire length. Hammer 28 cm long for scale.

(Kenrick & Crane 1997), and in the Devonian period, land plants proceeded through three major evolutionary steps (homospory, heterospory and, near the close of the Devonian period, seed formation; Chaloner & MacDonald 1980). Plant development was such that by Late Devonian time, land plants were well advanced and included large tree-like forms. Thus, relatively abundant growth of large plants was certainly possible during Upper ORS deposition at Milton Ness.

Boddin Point rhizoliths

Extremely large Upper ORS root tubules also are preserved at Boddin Point, 12.5 km south of the Milton Ness outcrop, where several dozen rhizoliths are present (Figs 1 and 11). The rhizoliths are exposed on the northeast side of the point, near the faulted contact with the Lower ORS. The Boddin Point tubules exceed the maximum size of the Milton Ness examples; the largest rhizolith has a 13 cm central vug with the entire tubule measuring 30 cm in diameter. Smaller tubules measure 11 cm in diameter with a 3 cm central vug. The tubules show a definite concentric pattern around the vug, which is identifiable megascopically. Lengthwise cross-sections through the tubules, shown in Fig. 11, demonstrate that the central vugs extend along the entire length of the tubules. The Boddin Point rhizoliths show a preferential vertical orientation, although some horizontally oriented rhizoliths also are present. These rhizoliths are associated with a hardpan containing pink siliceous laminae; however, the calcrete is not as well developed as the hardpan at Milton Ness, and it is not implied that the hardpans at Boddin Point and Milton Ness necessarily represent the same stratigraphic horizon.

The Boddin Point rhizoliths, with their preferred vertical orientation, suggest that they may have represented large tap roots that descended relatively deeply into the subsurface to reach the water table. Plants that utilize this mechanism to exploit the local water resources are called 'phreatophytes' and have been described in association with calcretes in the Middle Triassic Otter Sandstone of southwest England (Purvis & Wright 1991). The part of the Boddin Point outcrop containing these rhizoliths occurs low on the shoreface along the North Sea and precludes an estimate of the vertical range of the individual tubules.

Summary

The Upper Old Red Sandstone at Milton Ness, with its well-developed calcrete types and abundant rhizoliths, gives clear evidence for subaerial exposure. At the eastern end of the outcrop, a 4 m calcrete profile shows idealized vertical zonation of carbonate morphologies. The profile is capped by a mature hardpan with well-preserved textures commonly noted in Quaternary calcretes. These textures include: (1) irregular, subhorizontal carbonate layers; (2) clotted micrite; (3) floating grains; (4) silicified pisolites; (5) microspar fringes around detrital grains; (6) siliceous laminae; (7) non-tectonic brecciation; (8) corroded detrital grains. The floating grain texture and the jigsaw fit of fragments within the hardpan are highly suggestive of displacive growth. Corroded edges on some detrital grains indicate that replacement by calcite also was operative.

Calcrete maturation was followed by a degenerative phase with karst generation and reworking of the hardpan surface. The reworking produced a breccia that covered the hardpan surface and infilled many of the karst cavities. Small abraded chert clasts indicate some degree of transport and not simply *in situ* modification of the underlying calcrete. Both the breccia and the karst infills were subsequently recalcretized in the final phase of the profile's development.

Although calcrete–karst associations are often interpreted as the alternation between semi-arid and humid climates, respectively, this example at Milton Ness is interpreted as the result of water accumulating on the nearly impervious hardpan under fairly constant semi-arid conditions. The recalcretization of the karst infill and the calcrete breccia argues against change to a humid climate. Although a brief humid period is possible, the association is explained in terms of the natural evolution of the landscape without invoking external factors for which there is no additional evidence. Water would have been supplied by precipitation and possibly by overbank flooding from laterally accreting channels.

Substrate modification by plant roots has also taken place. Many of the rhizoliths, especially those that form horizontal ramifying networks, were generated in point-bar sandstones at Milton Ness. The channel-side position of the roots allowed them to exploit the water resources within the semi-arid landscape. At Boddin Point, vertical root tubules suggest a possible origin as phreatophytes, exploiting a deeper water table at that locality. The remarkable development of rhizoliths in these Old Red Sandstone sediments should emphasize the need to consider plant influence on other non-marine rocks of post-Silurian age.

This work was carried out as part of the author's PhD dissertation at the University of Cambridge as part of a larger study of Upper Old Red Sandstone sedimentation in the eastern Midland Valley of Scotland. The study was made possible through funding by a National Science Foundation Graduate Fellowship. Other financial assistance provided through an Overseas Research Student Award, Cambridge Chancellor's Bursary Award and the American Friends of Cambridge University is gratefully acknowledged. I am thankful to my supervisor, P. F. Friend, for his helpful comments and suggestions throughout the duration of the study. S. K. Tandon provided much useful advice in the initial stages of the work on the palaeosols, and this paper benefited from insightful reviews by V. P. Wright and S. Love. Additionally, I wish to thank R. Lee and K. Harvey for photographic assistance, B. Harris for help with thin sections, and A. R. Scott for assistance with figures and editing. S. Birnbaum provided beneficial comments on an earlier version of this paper.

References

Allen, J. R. L. 1970. Studies in fluviatile sedimentation: a comparison of fining-upwards cyclothems, with special reference to coarse-member composition and interpretation. *Journal of Sedimentary Petrology*, **40**, 298–323.

—— 1986. Pedogenic calcretes in the Old Red Sandstone facies (Late Silurian–Early Carboniferous) of the Anglo-Welsh area, southern Britain. *In*: Wright, V. P. (ed.) *Paleosols: their Recognition and Interpretation.* Blackwell Scientific, Oxford, 58–86.

—— & Williams, B. P. J. 1979. Interfluvial drainage on Siluro-Devonian alluvial plains in Wales and the Welsh Borders. *Journal of the Geological Society, London*, **136**, 361–366.

Assereto, R. L. A. M. & Kendall, C. G. St. C. 1977. Nature, Origin and Classification of Peritidal Tepee Structures and Related Breccias. *Sedimentology*, **24**, 153–210.

BALIN, D. F. 1993. *Upper Old Red Sandstone sedimentation in the eastern Midland Valley area, Scotland.* PhD thesis, University of Cambridge.

BLUCK, B. J. 1990. Alluvial Upper Old Red Sandstone deposits, Firth of Clyde. *In*: BLUCK, B. J., HAUGHTON, P. D. W., FRIEND, P. F. & BALIN, D. F. (eds) *Old Red Sandstone of Scotland: a Record of Late-Stage Caledonian Terrane Amalgamation.* International Association of Sedimentologists Guidebook, International Association of Sedimentologists Congress, Nottingham, UK, August 1990, 36–42.

——, HAUGHTON, P. D. W., HOUSE, M. R., SELLWOOD, E. B. & TUNBRIDGE, I. P. 1988. Devonian of England, Wales, and Scotland. *In*: MCMILLAN, N. J., EMBRY, A. F. & GLASS, D. J. (eds) *Devonian of the World.* Canadian Society of Petroleum Geologists, Memoirs, **14**, 305–324.

BRETZ, J. H. & HORBERG, L. 1949. Caliche in southeastern New Mexico. *Journal of Geology*, **57**, 491–511.

BREWER, R. 1964. *Fabric and Mineral Analysis of Soils.* Wiley, New York.

CHALONER, W. G. & MACDONALD, P. 1980. *Plants Invade the Land.* Royal Scottish Museum. HMSO, Edinburgh.

CHAPMAN, R. W. 1971. Climatic changes and the evolution of landforms in the Eastern Province of Saudi Arabia. *Geological Society of America Bulletin*, **82**, 2713–2728.

CHISHOLM, J. I. & DEAN, J. M. 1974. The Upper Old Red Sandstone of Fife and Kinross: a fluviatile sequence with evidence of marine incursion. *Scottish Journal of Geology*, **10**, 1–30.

COHEN, A. S. 1982. Paleoenvironments of root casts from the Koobi Formation, Kenya. Journal of *Sedimentary Petrology*, **52**(2), 401–414.

CORBEL, J. 1959. Erosion en terrain calcaire: vitesse d'érosion et morphologie. *Annales de Geographie*, **68**, 97–120.

—— & MUXART, R. 1970. Karsts des zones tropicales humides. *Zeitschrift für Geomorphologie*, **14**(4) 411–474.

CURTIS, L. F., COURTNEY, F. M. & TRUDGILL, S. 1976. *Soils in the British Isles.* Longman, Harlow.

ESTEBAN, M. & KLAPPA, C. F. 1983. Subaerial exposure. *In*: SCHOLLE, P. A., BEBOUT, D. G. & MOORE, C. H. (eds) *Carbonate Depositional Environments.* American Association of Petroleum Geologists, Memoirs, **33**, 1–54.

FOLK, R. L., 1959. Practical petrographic classification of limestones. *AAPG Bulletin*, **43**, 1–38.

—— 1974. The natural history of crystalline calcium carbonate: effect of magnesium content and salinity. *Journal of Sedimentary Petrology*, **44**(1), 40–53.

GARDNER, L. R. 1972. Origin of the Mormon Mesa caliche, Clark County, Nevada. *Geological Society of America Bulletin*, **83**, 143–156

GOUDIE, A. S. 1973. *Duricrusts in Tropical and Subtropical Landscapes.* Clarendon Press, Oxford.

—— 1983. Calcrete. *In*: GOUDIE, A. S. & PYE, K. (eds) *Chemical Sediments and Geomorphology.* Academic Press, London, 93–131.

HAY, R. L. & WIGGINS, B. 1980. Pellets, ooids, sepiolite and silica in three calcretes of the southwestern United States. *Sedimentology*, **27**, 559–576.

HERWITZ, S. R. 1993. Stem-flow influences on the formation of solution pipes in Bermuda eolianite. *Geomorphology*, **6**, 253–271.

JENNINGS, J. N. 1972. The character of tropical humid karst. *Zeitschrift für Geomorphologie*, **16**(3), 336–341.

—— 1983. The disregarded karst of the arid and semiarid domain. *Karstologia*, **1**, 61–73.

JUDSON, S. 1950. Depressions of the northern portion of the southern High Plains of eastern New Mexico. *Geological Society of America Bulletin*, **61**, 253–274.

KENRICK, P. & CRANE, P. R. 1997. The origin and early evolution of plants on land. *Nature London*, **389**(6646), 33–39.

KLAPPA, C. F. 1980. Rhizoliths in terrestrial carbonates: classification, recognition, genesis and significance. *Sedimentology*, **27**, 613–629.

—— 1983. A process–response model for the formation of pedogenic calcretes. *In*: WILSON, R. C. L. (ed.) *Residual Deposits: Surface Related Weathering Processes and Materials.* Geological Society, London, Special Publications, **11**, 211–220.

KNOX, G. J. 1977. Caliche profile formation, Saldanha Bay (South Africa). *Sedimentology*, **24**, 657–674.

LEEDER, M. R. 1975. Pedogenic carbonates and flood sediment accumulation rates: a quantitative model for alluvial arid-zone lithofacies. *Geological Magazine*, **112**(3), 257–270.

LIVERMORE, R. A., SMITH, A. G. & BRIDEN, J. C. 1985. Palaeomagnetic constraints on the distribution of continents in the late Silurian and early Devonian. *In*: CHALONER, W. G. & LAWSON, J. D. (eds) *Evolution and Environment in the Late Silurian and Early Devonian.* Royal Society of London, Transactions, **B309**, 29–56.

MONROE, W. H. 1966. Formation of tropical karst topography by limestone solution and precipitation. *Caribbean Journal of Science*, **6**(1–2), 1–7.

MUSTARD, P. S. & DONALDSON, J. A. 1990. Paleokarst breccias, calcretes, silcretes and fault talus breccias at the base of the upper Proterozoic 'Windermere' strata, northern Canadian Cordillera. *Journal of Sedimentary Petrology*, **60**(4), 525–539.

NETTERBERG, F. 1978. Dating and correlation of calcretes and other pedocretes. *Transactions of the Geological Society of South Africa*, **81**, 379–391.

—— 1980. Geology of southern African calcretes: 1. Terminology, description, macrofeatures, and classification. *Transactions of the Geological Society of South Africa*, **83**, 255–283.

PANOS, V. & STELCL, O. 1968. Physiographic and geologic control in the development of Cuban mogotes. *Zeitschift für Geomorphologie*, **12**(2), 117–164.

PARNELL, J. 1981. *Post-depositional processes in the Old Red Sandstone of the Orcadian Basin, Scotland.* PhD thesis, Imperial College of Science and Technology, London.

—— 1983*a*. The Cothall Limestone. *Scottish Journal of Geology*, **19**, 215–218.

—— 1983*b*. Ancient duricrusts and related rocks in perspective: a contribution from the Old Red Sandstone. *In*: WILSON, R. C. L. (ed.) *Residual Deposits: Surface Related Weathering Processes and Materials*. Geological Society, London, Special Publications, **11**, 197–209.

PARTRIDGE, T. C. & BRINK, A. B. A. 1967. Gravels and terraces of the lower Vaal River basin. *South African Geographical Journal*, **49**, 21–38.

PATERSON, I. B. & HALL, I. H. S. 1986. Lithostratigraphy of the Late Devonian and Early Carboniferous Rocks in the Midland Valley of Scotland. British Geological Survey Report **18**(3).

PELECHATY, S. M., JAMES, N. P., KERANS, C. & GROTZINGER, J. P. 1991. A middle Proterozoic palaeokarst unconformity and associated sedimentary rocks, Elu Basin, northwest Canada. *Sedimentology*, **38**, 775–797.

PURVIS, K. & WRIGHT, V. P. 1991. Calcretes related to phreatophytic vegetation from the Middle Triassic Otter Sandstone of South West England. *Sedimentology*, **38**, 539–551.

RETALLACK, G. J. 1983. *Late Eocene and Oligocene paleosols from Badlands National Park, South Dakota*. Geological Society of America, Special Papers, **193**.

ROTHROCK, E. P. 1925. On the force of crystallisation of calcite. *Journal of Geology*, **33**, 80–82.

RUHE, R. V. 1965. Quaternary palaeopedology. *In*: WRIGHT, H. E. & FREY, D. G. (eds) *The Quaternary of the United States*. Princeton University Press, Princeton, NJ, 755–764.

SAIGAL, G. C. & WALTON, E. K. 1988. On the occurrence of displacive calcite in Lower Old Red Sandstone of Carnoustie, eastern Scotland. *Journal of Sedimentary Petrology*, **58**, 131–135.

SCHNEIDER, J. 1977. Carbonate construction and decomposition by epilithic and endolithic microorganisms in salt and freshwater. *In*: FLUGEL, E. (ed.) *Fossil Algae*. Springer, Berlin, 248–260.

SCHUILING, R. D. & WENSINK, H. 1962. Porphyroblastic and poikiloblastic textures: the growth of large crystals in a solid medium. *Neues Jahrbuch für Mineralogie, Monatschefte*, **11–12**, 247–254.

SMITH, T. M. & DOROBEK, S. L. 1993. Stable isotopic composition of meteoric calcites; evidence for Early Mississippian climate change in the Mission Canyon Formation, Montana. *Tectonophysics*, **222**(3–4), 317–331.

SOLOMON, S. T. & WALKDEN, G. M. 1985. The application of cathodoluminescence to interpreting the diagenesis of an ancient calcrete profile. *Sedimentology*, **32**, 877–896.

STEEL, R. J. 1974. Cornstone (fossil caliche)—its origin, stratigraphic, and sedimentological importance in the New Red Sandstone, western Scotland. *Journal of Geology*, **82**, 351–369.

SUMMERFIELD, M. A. 1983. Silcrete. *In*: GOUDIE, A. S. &. PYE, K. (eds) *Chemical Sediments and Geomorphology*. Academic Press, London, 59–91.

SWEETING, M. M. 1969. Karstic morphology in Yucatan. *In*: *Expedition to British Honduras–Yucatan 1966*. University of Edinburgh Report, Section 4, 37–40.

—— 1973. *Karst Landforms*. Macmillan, London.

TARLING, D. H. 1985. Siluro-Devonian palaeogeographies based on palaeomagnetic observations. *In*: CHALONER, W. G. & LAWSON, J. D. (eds) *Evolution and Environment in the Late Silurian and Early Devonian*. Royal Society of London, Transactions, **B309**, 81–83.

TREWIN, N. H. 1980. Geology. *In*: MARREN, P. (ed.) *The Natural History of St. Cyrus*. Nature Conservancy Council, Edinburgh, 6–15.

—— 1987. Devonian of St. Cyrus and Milton Ness. *In*: TREWIN, N. H., KNELLER, B. C. & GILLEN, C. (eds) *Excursion Guide to the Geology of the Aberdeen Area*. Scottish Academic Press, Edinburgh, 251–258.

VANSTONE, S. D. 1998. Late Dinantian palaeokarst of England and Wales: implications for exposure surface development. *Sedimentology*, **45**, 19–37.

WALKDEN, G. & DAVIES, J. 1983. Polyphase erosion of subaerial omission surfaces in the Late Dinantian of Anglesey, North Wales. *Sedimentology*, **30**, 861–878.

WATTS, N. L. 1978. Displacive calcite: evidence from recent and ancient calcretes. *Geology*, **6**, 699–703.

WEAVER, J. E. 1958. Classification of root systems of forbs of grassland and a consideration of their significance. *Ecology*, **39**, 393–401.

WEYL, P. K. 1959. Pressure solution and the force of crystallization—a phenomenological theory. *Journal of Geophysical Research*, **64**(11), 2001–2025.

WOODROW, D. L., FLETCHER, F. W. & AHRNSBRAK, W. F. 1973. Paleogeography and paleoclimate at the deposition site of the Devonian Catskill and Old Red Facies. *Geological Society of America Bulletin*, **84**, 3051–3064.

WRIGHT, V. P. 1980. Climatic fluctuation in the Lower Carboniferous. *Naturwissenschaften*, **67**, 252–253.

—— (ed.) 1986. *Paleosols: their Recognition and Interpretation*. Blackwell Scientific, Oxford.

——, TURNER, M. S., ANDREWS, J. E. & SPIRO, B. 1993. Morphology and significance of supermature calcretes from the Upper Old Red Sandstone of Scotland. *Journal of the Geological Society, London*, **150**, 871–883.

YOUNG, R. G. 1964. Fracturing of sandstone cobbles in caliche cemented terrace gravels. *Journal of Sedimentary Petrology*, **34**, 886–889.

Architecture of the Middle Devonian Kvamshesten Group, western Norway: sedimentary response to deformation above a ramp-flat extensional fault

P. T. OSMUNDSEN[1,2], B. BAKKE[1], A. K. SVENDBY[1] & T. B. ANDERSEN[1]

[1]*Department of Geology, University of Oslo, Pb 1047 Blindern, 0316 Oslo, Norway*

[2]*Present address: Geological Survey of Norway, 7491 Trondheim, Norway*

Abstract: The Mid-Devonian Kvamshesten basin in western Norway formed during late- to post-Caledonian extensional and strike-slip tectonics. The basin is entirely continental in origin, and was probably a closed basin through most of its history. It was filled by alluvial–fluvial deposits belonging to two main terminal fan systems, one sourced in the hanging wall, one sourced in the footwall of the basin-controlling fault(s). Footwall-directed migration of the hanging-wall-sourced fluvial system probably took place during phases of relatively rapid, fault-controlled subsidence. This gave rise to retrogradational stacking of marginal conglomerates and to the footwall-directed migration of the main depocentre(s). During periods of increased sediment influx from footwall catchments and/or decreased subsidence rates, coarse, footwall-sourced material prograded far into the basin. The intercalation of material derived from footwall- and hanging-wall source areas, respectively, gave rise to a pronounced, kilometre-scale rhythmicity that is inferred to reflect fault-block rotation and footwall erosion. The basin fill also displays a pronounced rhythmic organization on the 80–150 m scale. The Kvamshesten basin was deformed during deposition as a result of its position in the hanging wall of a ramp-flat extensional fault and because of extension-parallel shortening. Thus, the pattern of subsidence was non-uniform and the sediment preservation potential varied across the basin. Deposition and preservation of floodbasin fine sediments was enhanced in areas close to the breakaway of the basin-bounding fault and in the hanging-wall syncline that developed above the ramp. The crest of the rollover anticline was mainly occupied by high-energy braided fluvial systems represented by amalgamated sandstone complexes with abundant channel scour surfaces. The migration of the rollover anticline towards the breakaway of the principal basin-bounding fault resulted in the footwall-directed migration of facies belts upwards in the stratigraphy. In our interpretation, a kilometres-thick, footwall-skewed amalgamated sandstone complex was constructed that marks the trace of the migrating rollover anticline through the basin stratigraphy. On the basis of stacking patterns of marginal facies, large-scale grain-size variations, the geometry of stratigraphic units and the overall lithofacies distribution, we present a sequence stratigraphic model for the basin fill. Within individual systems tracts, lithofacies distribution as well as inferred A/S ratios vary laterally because of the effects of differential subsidence and multidirectional sediment transport. Inferences of A/S ratio from vertical sections in one particular part of the basin thus cannot be used to characterize a systems tract as such. Grain-size turnarounds may, however, be used in the definition of systems tracts and sequences if the above relationships are well constrained and if turnarounds occur in stratigraphic units that are easy to correlate across the basin.

The Mid-Devonian basins of western Norway are classic study areas for tectonically controlled continental sedimentation (Kolderup 1921; Jarvik 1949; Bryhni 1964*a*, *b*; Nilsen 1968; Bryhni & Skjerlie 1975; Steel *et al.* 1977; Steel & Gloppen 1980). The rapid facies transitions and the rhythmic organization of the basin fill made the Hornelen basin (Fig. 1) a testing ground for tectono-sedimentary models in the late 1970s and early 1980s (Steel *et al.* 1977, 1985; Steel & Gloppen 1980; Steel 1988). Our understanding of Devonian tectonics in western Norway has, however, changed dramatically through the last 15 years (Bjørlykke 1983; Roberts 1983; Hossack 1984; Norton 1986, 1987; Torsvik *et al.* 1986, 1988; Séranne & Seguret 1987; Andersen & Jamtveit 1990; Fossen 1992; Krabbendam & Dewey 2000). The new tectonic framework that emerged from these studies (see review by Andersen (1998)) has

From: Friend, P. F. & Williams, B. P. J. (eds). *New Perspectives on the Old Red Sandstone.* Geological Society, London, Special Publications, **180**, 503–535. 0305-8719/00/$15.00

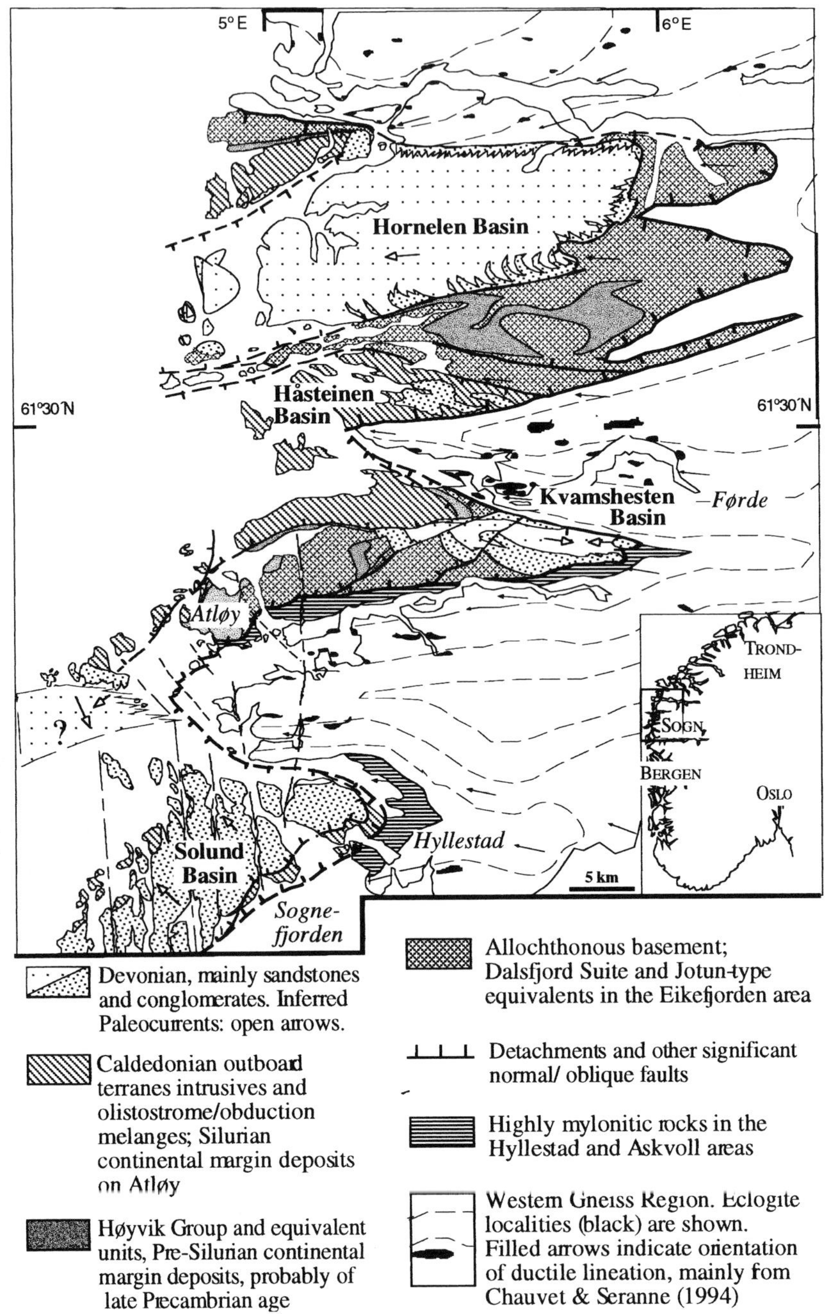

Fig. 1. Overview map of the Nordfjord–Sogn region with Devonian basins and substrate (modifed from Kildal (1970) and from Osmundsen & Andersen (1994)).

called for a re-investigation of the Devonian basins in terms of contact relationships, structure and sedimentary architecture.

Recent developments in sedimentology and basin analysis have provided new approaches to the study of sedimentary rocks. Studies of processes acting in catchment areas have improved our understanding of the links between tectonics, climate and sedimentation (e.g. Leeder & Jackson 1993; Burbank *et al.* 1996; Allen &

Hovius 1998; Leeder *et al.* 1998). Sequence stratigraphy of fluvial successions that can be correlated directly with marine deposits provides a new framework (Posamentier & Vail 1988; Shanley & McKabe 1994; Olsen *et al.* 1995; Burns *et al.* 1997; Currie 1997). Work on deposits that are entirely continental has attempted to establish an independent framework and terminology for successions that cannot be correlated with conventional sequences in offshore areas (e.g. Legarreta *et al.* 1993; Currie 1997). Few basins have, however, been analysed in this way and more work is clearly needed to test concepts in a variety of basin settings.

In many published models, the subdivision of continental rocks into sequences and systems tracts appears to depend strongly on vertical successions of lithofacies. In the original definition, however, a depositional systems tract (Brown & Fischer 1977) is a unit that links contemporaneous depositional systems. In a tectonically controlled basin, however, subsidence rates as well as lithofacies distribution vary across the basin (e.g. Leeder & Gawthorpe 1987; Prosser 1993). In the present paper, we discuss concepts of continental sequence stratigraphy in a Devonian continental extensional basin where stratigraphic units have been mapped laterally towards the basin margins. Although the Devonian basins of western Norway represent a particular type of continental basin, we believe that much of the discussion below has relevance to tectonically controlled basins elsewhere.

Geological setting

The Devonian basins of western Norway

In western Norway, Middle Devonian sediments were deposited in fault-bounded basins above the extensional Nordfjord–Sogn Detachment Zone (NSDZ, Norton 1986, 1987). The NSDZ, which separates the Precambrian Western Gneiss Region (WGR) from the basins and their depositional substrate of Caledonian nappe units (Fig. 1), developed during late- to post-orogenic extension of the western Norwegian Caledonides (Norton 1986, 1987; Séranne & Seguret 1987; Séranne *et al.* 1989; Andersen & Jamtveit 1990; Fossen 1992). The WGR experienced Late Silurian eclogite-facies metamorphism (Griffin *et al.* 1985; Kullerud *et al.* 1986) related to subduction of western Baltica during the final stages of the Caledonian orogeny. The tectonometamorphic break observed across the NSDZ in the Sunnfjord area thus corresponds to excision of *c.* 40 km of crust and the inferred normal, top-to-the west displacement is 40–100 km (Andersen & Jamtveit 1990). Most of this displacement took place in early to mid-Devonian time as evidenced by Ar–Ar geochronology on the rocks in the footwall (Chauvet & Dallmeyer 1992; Berry *et al.* 1995; Eide *et al.* 1997). The brittle faults that separate the Devonian basins from the footwall rocks did, however, experience slip events in Permian and Late Jurassic–Early Cretaceous times (Torsvik *et al.* 1992; Eide *et al.* 1997). As no clasts derived from the footwall of the detachment zone have been unambiguously identified in the basins, the Devonian sedimentary rocks probably had their provenance in the Caledonian nappe units (Cuthbert 1991; Wilks & Cuthbert 1994; Osmundsen *et al.* 1998). The structures that directly controlled sedimentation in the basins are thought to be normal and oblique faults that cut the Caledonian nappe-stack but were rooted in the detachment zone (Cuthbert 1991; Chauvet & Séranne 1994; Osmundsen & Andersen 1994; Wilks & Cuthbert 1994; Osmundsen *et al.* 1998).

The Devonian basins probably formed under a combination of extensional and strike-slip tectonics where the component of sinistral strike-slip deformation became more pronounced to the north (Chauvet & Seranne 1994; Osmundsen *et al.* 1998; Krabbendam & Dewey 1998). The principal elongation direction as inferred from stretching lineations in the footwall of the NSDZ is progressively rotated from a NW–SE trend south of the Solund basin to a WSW–ESE trend to the north of the Hornelen basin (Chauvet & Seranne 1994; Krabbendam & Dewey 1998; Fig. 1). In the array of basins, this appears to be reflected by a change in orientation of syndepositional structures as well as by facies architecture and palaeocurrent directions (Osmundsen & Andersen in press). Whereas the geometry of the Solund basin is compatible with that of a SE-tilted half-graben (Nilsen 1968; Steel 1976, 1988), sedimentological and structural data from the Kvamshesten and Hornelen basins indicate that they were more strongly affected by a component of regional, sinistral strike-slip deformation (Osmundsen & Andersen in press). The regional, syndepositional strain field may have been one of bulk transtension (Osmundsen *et al.* 1998; Krabbendam & Dewey 1998); alternatively, sinistral strike-slip may have followed NW-directed extension during basin formation (Osmundsen & Andersen in press).

The Kvamshesten basin

In the present configuration, the Kvamshesten basin (Fig. 2) constitutes a rotated half-graben in

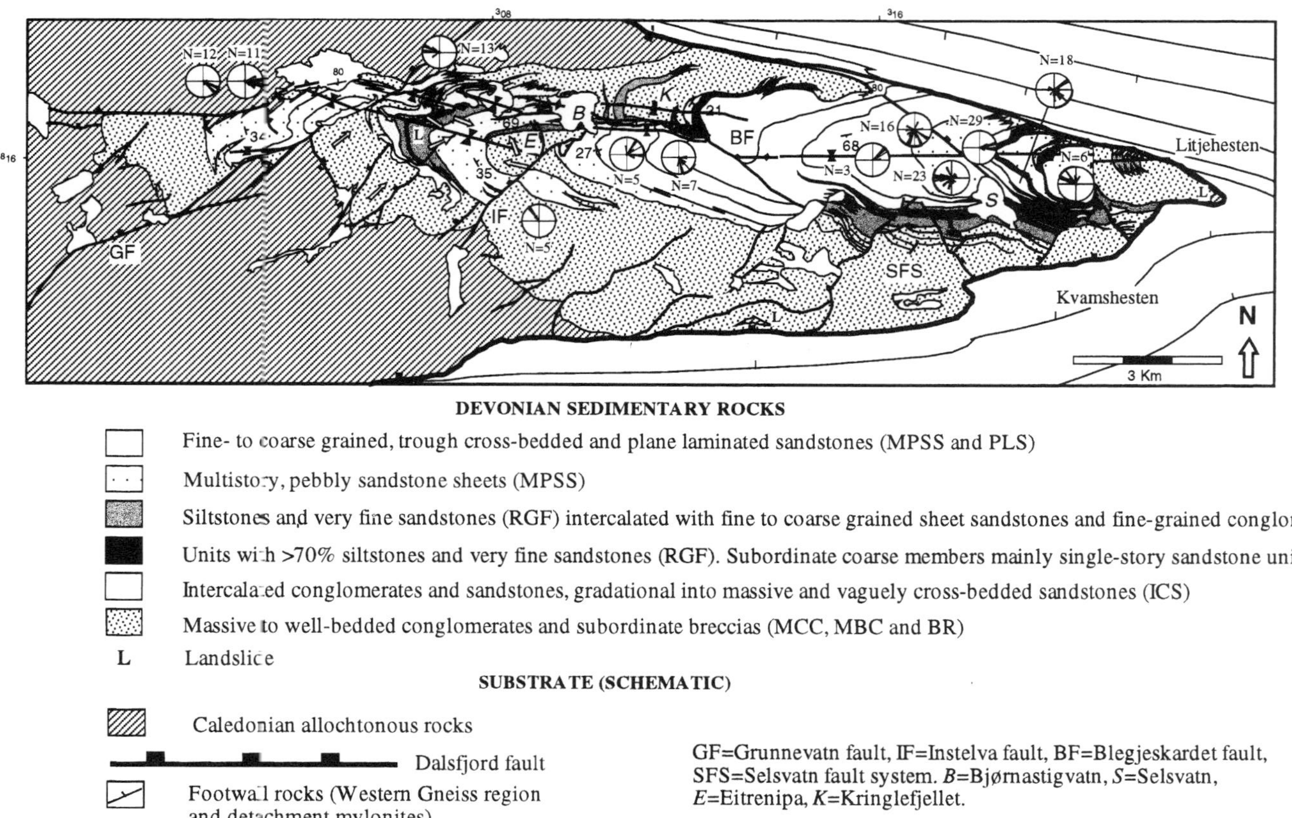

Fig. 2. Geological map of the Kvamshesten basin. (Note asymmetrical distribution of sedimentary units (in particular at low stratigraphic levels), large-scale retrogradational and progradational stacking patterns and sediment dispersal directions (mainly from trough cross-bedding).) Arrows are inferred palaeocurrent directions from Asphaug (1975).

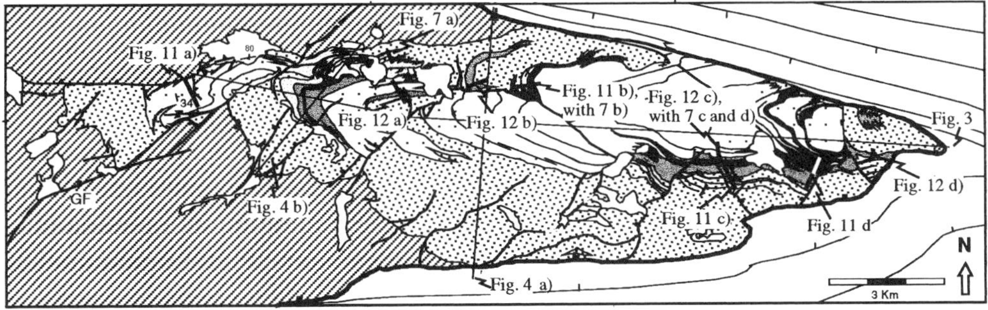

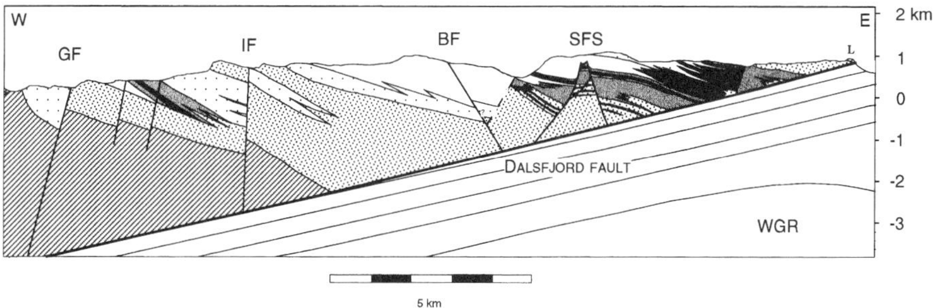

Fig. 3. (**a**) Location map for profiles and logs. (**b**) East–west cross-section through the Kvamshesten basin (i.e. parallel to principal extension direction). The decrease in apparent dip values from intermediate to high stratigraphic levels indicates *c.* 20° of syndepositional rotation. (See (**a**) for location.)

east–west section (Fig. 3), whereas in a section normal to the principal direction of extension, contractional structures are apparent (Fig. 4a and b). The basin displays a pronounced asymmetry with respect to the configuration of main sedimentary units, and the oldest deposits preserved in the basin are located along the present southern basin margin (Osmundsen 1996; Osmundsen *et al.* 1998). At low stratigraphic levels in the western parts of the basin, sedimentary units onlap basement towards the NE (Fig. 4b) and east; east of the Instelva fault, however, onlap is towards the (N)W as reflected by the westward-tapering wedges of conglomerate that overlie the hanging-wall cutoff of basement (Figs 2 and 3). Thus, the Devonian strata were onlapping a syndepositional basement high, probably the crest of a rollover anticline (Osmundsen *et al.* 1998).

Along parts of the northern basin margin, the Devonian succession is cut by a thrust that places pre-Devonian metamorphic rocks upon the basin strata (Figs 2 and 4a). Thus, the basin may have extended for a considerable distance towards the northwest before thrusting and later erosion. The brittle top of the detachment zone (i.e. the Dalsfjord fault, Figs 2 and 3) cuts the deformed basin and cannot be taken to represent the original, basin-bounding fault (Torsvik *et al.* 1986, 1992; Osmundsen *et al.* 1998).

Sedimentary architecture of the Kvamshesten Group

The Middle Devonian sedimentary rocks in the Kvamshesten basin (Fig. 2) comprise a central belt of sandstones bordered by conglomerates (Kolderup 1921), collectively named the Kvamshesten Group (Skjerlie 1971). The Kvamshesten Group is entirely continental in origin and consists mainly of alluvial and fluvial strata deposited under semi-arid conditions (Bryhni & Skjerlie 1975; Steel *et al.* 1985). Osmundsen *et al.* (1998) subdivided the basin fill into five main depositional units based on contact relationships, grain size and sedimentary structures. Here, a further subdivision is made, based on lithofacies associations (Tables 1 and 2) that constitute fundamental stratigraphic building blocks and that have been mapped through the basin fill. The main depositional units are subdivided into numerous coarsening- to fining-upward units (CUFU cycles, Steel *et al.* 1977) on a variety of scales (Fig. 5). Units ranging from some tens of metres up to *c.* 150 m in thickness are mappable as stratigraphic markers through

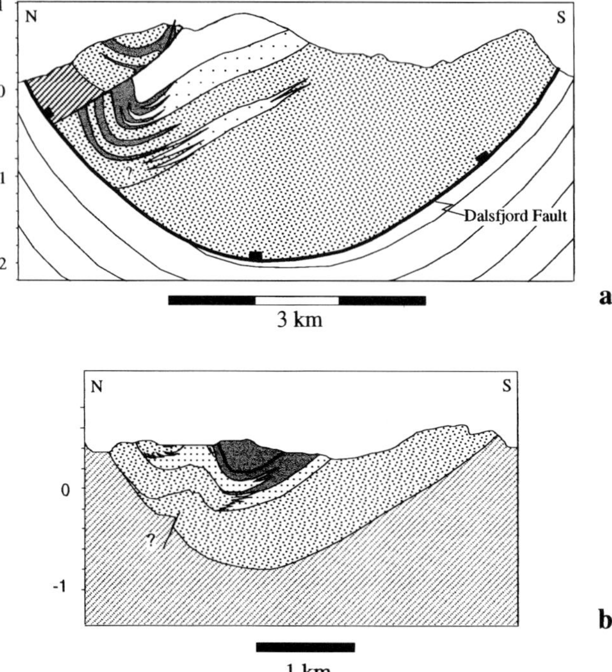

Fig. 4. North–south and NE–SW cross-sections through the Kvamshesten basin (i.e. normal to principal extension direction). In (**a**), Devonian strata are cut by the scoop-shaped Dalsfjord Fault, whereas (**b**) shows onlap relationship against the unconformable northern basin margin. Large-scale contractional structures and asymmetry in the configuration of main sedimentary units should be noted. In particular, the inferred pinchout of the southern margin fan complex (**b**) induces an important element of asymmetry. It follows from this interpretation that sedimentation in the northern margin fan complex started later than in the SMFC. (See Fig. 3a for location.)

large parts of the basin. Within each CUFU unit, the assemblage of lithofacies changes laterally as well as vertically, generally becoming more fine grained towards the central basin area.

The marginal fan complexes

The conglomeratic fan complexes exposed along the present northern and southern basin margins show distinct variations with respect to sedimentary architecture as well as to the timing of initial sedimentation (Osmundsen *et al.* 1998). The lower parts of the southern margin fan complex (SMFC) constitute the oldest deposits encountered in the Kvamshesten basin as shown by the NE-wards onlap onto basement displayed by the Devonian strata. The lowermost parts of the northern margin fan complex (NMFC) interfinger with sandstones that overlie the SMFC in the southwestern parts of the basin (Fig. 2).

Massive and crudely bedded, clast-supported conglomerates (MCC). Both the marginal fan complexes contain several hundred metres of clast-supported conglomerate (Gcm, Gh, Gci, Tables 1 and 2). The more massive of the MCC deposits lack well-developed sedimentary structures, although lithofacies Gt and Gp are observed at high stratigraphic levels in the SMFC in the westernmost basin area. At higher levels in the SMFC, fan segmentation becomes prominent, and the MCC rocks are organized in CUFU units separated by pebbly sandstones. At this stratigraphic level, lithofacies Gci is important, whereas Gt and Gp are observed towards the top of individual fan segments.

In the NMFC, thick MCC deposits are observed above the hanging-wall cutoff of basement. Towards the west and SW, however, the organization of the conglomerates becomes more complex, and involves a variety of lithofacies associations organized in well-defined CUFU units (Fig. 5a). Thus, the uneven distribution of MCC between the SMFC and NMFC contributes significantly to the asymmetry of the preserved basin.

Table 1. *Lithofacies codes of Miall (1996) used in this paper.*

Lithofacies code	Facies	Sedimentary structures	Interpretation
Gmm	Matrix-supported, massive gravel	Weak grading	Plastic debris flow (high strength, viscous)
Gcm	Clast-supported massive gravel	(None)	Pseudoplastic debris flow (inertial bedload, turbulent flow)
Gci	Clast-supported gravel	Inverse grading	Clast-rich debris flow (high strength or pseudoplastic debris flow (low strength)
Gmg	Matrix-supported gravel	Inverse to normal grading	Pseudoplastic debris flows (low strength, viscous)
Gh	Clast-supported, crudely bedded gravel	Horizontal bedding imbrication	Longitudinal bedloads, lag deposits, sieve deposits
Gt	Gravel, stratified	Trough cross-beds	Minor channel fills
Gp	Gravel, stratified	Planar cross-beds	Transverse bedforms, deltaic growth from older bar remnants
St	Sand, fine to very coarse, may be pebbly	Solitary or grouped trough cross-beds	Sinous-crested and linguoid (3D0 dunes
Sp	Sand, fine to very coarse, may be pebbly	Solitary or grouped planar cross-beds	Transverse and linguoid bedforms (2D-dunes)
Sh	Sand, very fine to coarse, may be pebbly	Horizontal lamination, parting or streaming lineation	Plane-bed flow (critical flow)
Sl	Sand, very fine to coarse, may be pebbly	Low angle (<15°) cross-beds	Scour fills, humpback or washed-out dunes, antidunes
Sr	Sand, very fine to coarse, may be pebbly	Ripple cross-lamination	Ripples (lower flow regime)
Ss	As above	Broad, shallow scours	Scour fill
Sm	Sand, fine to coarse	Massive, or faint lamination	Sediment-gravity flow deposits
Fl	Sand, silt, mud	Fine lamination, very small ripples	Overbank, abandoned channel, or waning flood deposits
P	Paleosol carbonate (calcite, siderite)	Pedogenic features: nodules, filaments	Soil with chemical precipitation

Table 2. *Summary of lithofacies associations defined in the text.*

Lithofacies association	Abbreviation	Lithofacies
Massive and crudely bedded, clast-supported conglomerates	MCC	Gcm, Gh, Gci, Gt, Gp
Massive and bedded, clast-to matrix supported conglomerates	MBC	Gci, Gmm, Gh, Sm, Sh
Breccias	BR	Gcm, Gmm, Gmg, Gci, Sm
Intercalated conglomerate and sandstone	ICS	Gh, Gt, Gp, St, Sp, Sl, Sh
Multi-storey, pebbly sandstone sheets	MPSS	St, Ss, Sp, Sl, Sh, Gt, Fl
Fine- to coarse-grained, plane-laminated and cross-bedded sandstones	PCS	Sh, Sl, Sp, St, Fl
Single-storey channel sandstones	SSCH	St, Ss, Sp, Sr, Sl, Gt, Gp
Red and green very fine-grained sandstone and siltstone	RGF	Fl, Sh, Sl, Sr, St, Sp, P

Massive and bedded, clast to matrix-supported conglomerates (MBC). The massive and poorly bedded conglomerates are clast- to matrix supported and comprise lithofacies Gcm, Gmm, Gh, Sm and Sl (Tables 1 and 2). The matrix comprises medium to very coarse sandstone, containing vague plane laminae and occasional red, very fine grained sandstone stringers up to a few centimetres thick. Clasts are subrounded to well rounded and occasionally imbricated.

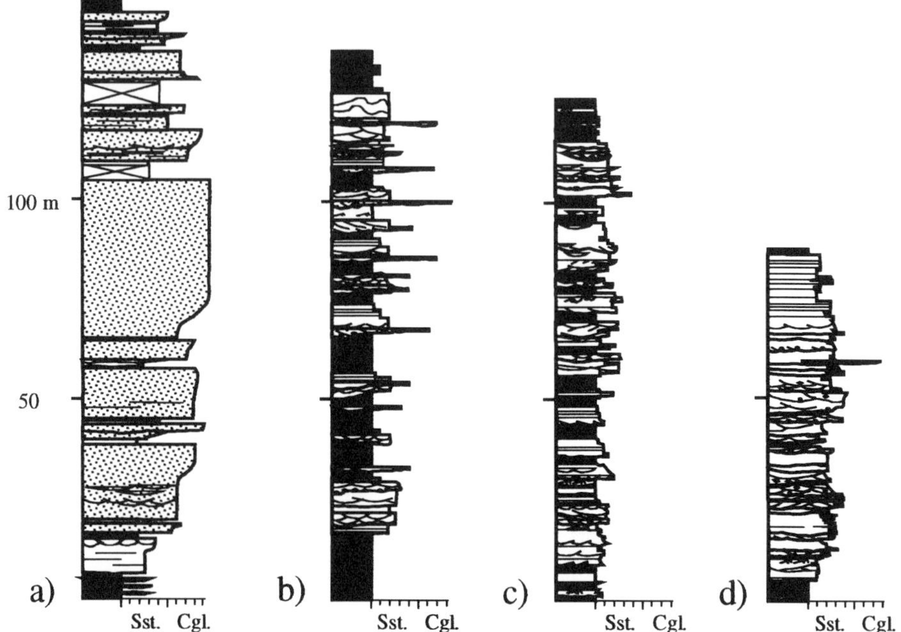

Fig. 5. Medium-scale coarsening- to fining-upward units from various parts of the Kvamshesten basin (see Fig. 3a for locations). The stacked fanglomerates in (**a**) typify medium-scale rhythmicity in the coarse-grained deposits of the northern basin margin; (**b**) displays single- and multi-storey channel units encased in overbank fines. The CUFU unit in (**b**) occurs in a fine-grained interval that caps marginal fanglomerate units (see Figs 2 and 5) and that represents several kilometres of margin-directed shift of the conglomerate–sandstone boundary. The CUFU units in (**c**) and (**d**) were recorded at high stratigraphic levels in the thick sandstone complex that occupies much of the central basin area. (Note multi-storey character and smaller thickness of (**d**).) The architectural variations displayed by the CUFU units are interpreted to reflect their position with respect to the basin margins, and with respect to the larger-scale rhythmicity (see text).

Breccias (BR). The sedimentary breccias in the Kvamshesten basin (Bryhni & Skjerlie 1975) are found intermittently along the basin's northern margin. The breccias typically occur as sheets above the basal unconformity, as infill of palaeotopographic lows, and as lenticular or fan-like bodies that intercalate with other parts of the basin fill. Breccias are either massive and clast supported or matrix supported (Table 2) with angular fragments dispersed in a red, very fine-grained sandstone or siltstone matrix (Osmundsen *et al.* 1998). The fragments commonly consist of quartzite, gneiss, quartz–mica schists or greenschists; these are rock types that make up large parts of the Caledonian nappe-stack below the unconformity (Bryhni & Skjerlie 1975; Osmundsen *et al.* 1998). Discontinuous beds of red, fine-grained rock a few centimetres thick occur intermittently in some of the breccias.

Intercalated conglomerate and sandstone (ICS). Both the marginal fan complexes contain significant proportions of conglomerate and sandstone intercalated on a scale less than 1 m (Fig. 6). In both the marginal fan complexes, ICS rocks are found close to the top of CUFU units, in particular in the areas where the fan complexes grade laterally into the central belt of fluvial sandstones. ICS rocks also constitute the lower parts of conglomeratic coarsening-upwards units in the NMFC. Organization of the ICS units is generally small scale, with discontinuous bedforms of conglomerate and sandstone intercalating on a decimetre scale (Fig. 6). Conglomerates occur as lithofacies Gh, Gt and Gp, notably in discontinuous conglomerate sheets up to 50 cm thick and as pebble clusters a few clasts thick. Clasts are generally within the pebble fraction, but occasionally, cobble-sized clasts are found. Sandstone occurs as lithofacies St and Sp, with bedforms usually small scale, and with lamina that wrap around pebble clusters. The rocks grouped as ICS here grade from conglomerate dominated to sandstone dominated, so that the mean grain size in the distal parts of the ICS units is close to that of the sandstones in the

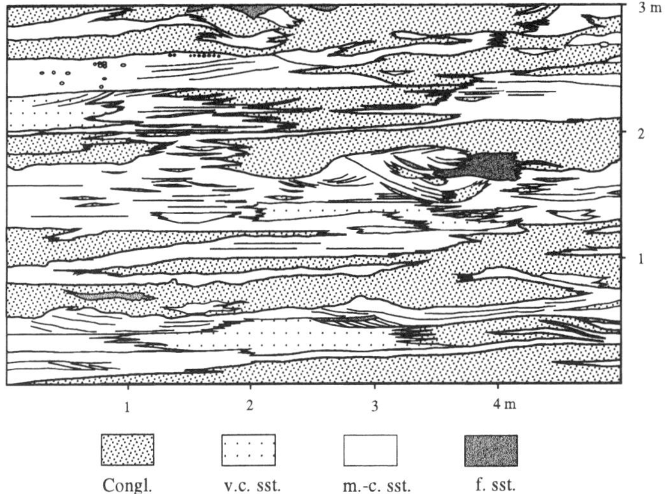

Fig. 6. Intercalated conglomerates and sandstones (ICS), northern margin fan complex. The ICS usually displays small-scale trough cross bedding and abundant plane lamination. ICS comprises both conglomerate- and sandstone-dominated units, the first being gradational into the latter. Towards the basin margins, the ICS grade into coarser conglomeratic units and probably represent downfan deposition or reworking in a braided network of small-scale channels.

central fluvial belt. The fine-grained ICS rocks are, however, separated from the latter by their internal organization as massive, vaguely laminated or small-scale trough cross-bedded units. Palaeocurrent directions in the ICS rocks appear to have been away from the marginal fan complexes towards the central basin area as inferred from the gradual transition from MCC and BR into ICS, and eventually MPSS. A number of readings on imbricate clasts from one locality close to the northern margin (Fig. 2) do, however, indicate west-northwestwards transport.

The central fluvial sandstones

The sandstones in the central basin area are clearly fluvial (Steel *et al.* 1985; Osmundsen *et al.* 1998) and comprise a number of lithofacies associations that record pronounced variations in fluvial style. Their organization is markedly cyclic, and the CUFU motif recognized in the marginal fanglomerates is reflected in laterally correlative sandy units (Fig. 5). On a smaller scale, however, sandstone units may display both coarsening- and fining-upward motifs. The palaeocurrent directions inferred from readings of trough cross bedding are variable (Fig. 2), but appear to reflect a domination of palaeocurrents that were either east- or west-flowing, parallel to the present basin (fold) axis.

Multi-storey, pebbly sandstone sheets (MPSS). The pebbly sandstones of the central basin area are typically organized in 2–40 m thick multi-storey units separated by beds of red, very fine-grained sandstone and siltstone (Fig. 7). The bases of individual multi-storey units are generally semi-conformable with small-scale (<1 m) erosional topography displayed on the outcrop scale. The MPSS commonly display a high density of erosional bounding surfaces. The topography of erosional surfaces is usually low to moderate, and the ratio between height and width of the bounding-surface topography is typically low (Fig. 7). Locally, however, erosional surfaces within units have a pronounced topography with height–width ratios that are similar to, or greater than those displayed by the base of the unit (middle right in Fig. 7). Clasts occur as pebble lags as well as scattered within individual sets; the amount of scattered pebbles generally increases towards the marginal fan complexes. Typically, the upper parts of MPSS units are conglomeratic for a considerable distance away from the marginal fan complexes (Fig. 2). Pebble sizes range from very fine to very coarse. Locally, cobble-sized clasts occur. Clast material comprises gneissic, metasedimentary and metavolcanic lithologies as well as intraformational, silt and very fine grained sandstone. The mean grain size in the pebbly sandstones ranges from medium to very coarse. Lithofacies St, Sp, Sl and Sh (Miall 1996) typically constitute close to 100% of individual units (Fig. 7), with St and Sp as the dominating lithofacies. The thickness of individual cross-bedded sets varies,

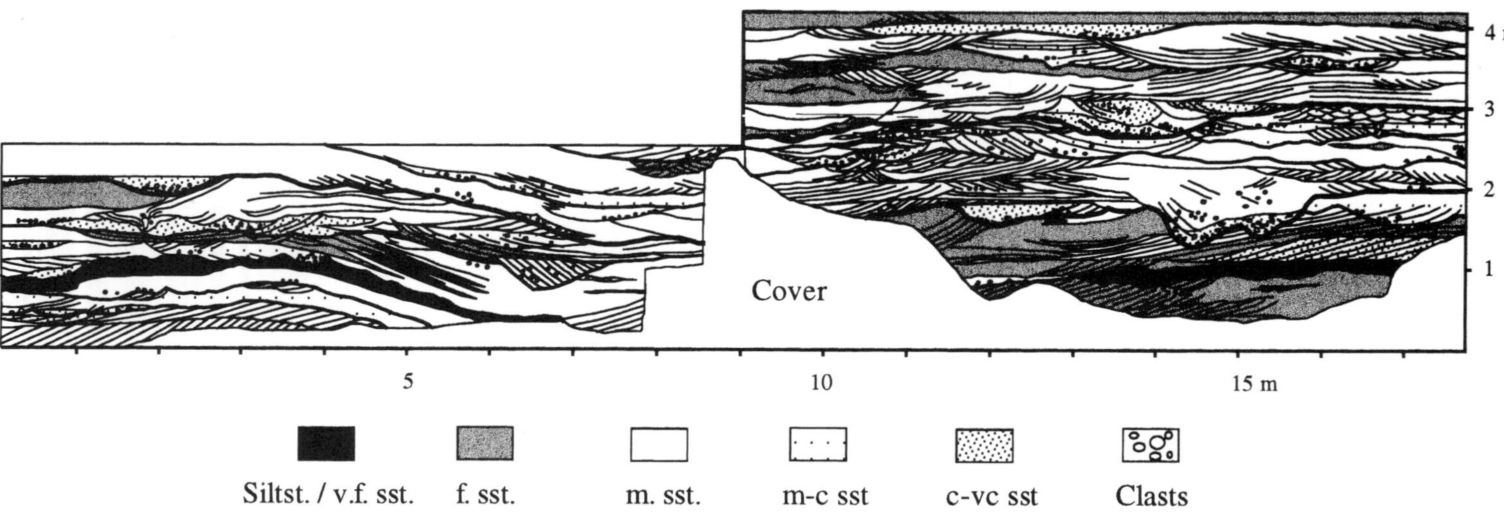

Fig. 7. Part of multi-storey, pebbly sandstone sheet (MPSS), western basin area. MPSS is the dominant lithofacies association in large parts of the central fluvial sandstones and commonly occurs as units several metres or tens of metres thick separated by red, fine-grained strata. Channel-base scours (enhanced), locally with associated pebble lags, ad abundant trough and planar cross bedding should be noted. Scours generally have low depth-to-width ratios, but more localized incision is observed (middle right). Planar cross-bedded, lenticular sandstone bodies are interpreted to represent bar deposits (e.g. lower right). (Note erosion of siltstone unit in lower part of section.) The relatively coarse grain sizes, the frequent channel scour surfaces and the overall sheetlike geometry of the MPSS units indicate deposition in broad and shallow braided systems characterized by frequent channel erosion.

but is generally within the range of 0.2–0.7 m. In the MPSS, entire channel sequences are seldom preserved; the number of erosional surfaces and the abundance of rip-up intraformational clasts suggests that the MPSS represent a relatively low sediment preservation potential.

Fine- to coarse-grained, plane-laminated and cross-bedded sandstones (PCS). At medium to high stratigraphic levels, MPSS units are intercalated with units that are characterized by finer mean grain sizes and a higher abundance of plane lamination and low-angle, plane tangential cross-bedding (e.g. fig. 7b of Osmundsen *et al.* 1998). Horizons of very fine-grained green, brown and grey sandstone and siltstone are present in variable proportions. Individual sandstone units usually have semi-conformable or erosional bases with minor topography. Grain size is usually fine, medium or coarse, with some very fine grained intervals. Compared with the pebbly sandstones, plane lamination and low-angle planar cross bedding constitute a significantly higher proportion of the stratigraphic column. Both trough- and planar cross-bedding are relatively common, but the density of erosional bounding surfaces is considerably less than in the pebbly sandstones. Some beds, commonly 0.02–0.3 m thick, lack sedimentary structures apart from vague grading.

Single-storey channel sandstones (SSCH). Single-storey channel sandstones (Fig. 8) make up a subordinate part of the basin fill. The scoured bases of individual units present a topography that in some cases makes up more than half of the unit thickness. The single-storey channel deposits commonly have most of the channel succession preserved, including both large-scale and small-scale trough cross-bedded sets, followed by rippled and plane laminated strata (Fig. 5b). Thus, the SSCH units constitute a pronounced contrast to the poor preservation of the channel units that build up the MPSS units. The SSCH units are interbedded with thick, red fine-grained rocks (RGF).

Red and green very fine sandstone, siltstone and mudstone (RGF). The fine-grained rocks of the Kvamshesten basin comprise mainly plane laminated red (most abundant), pink and green siltstone and very fine sandstone. RGF units are commonly plane laminated, variably displaying wavy, rippled, folded or contorted lamination. Some of the RGF units contain thin (2–10 cm) stringers of rippled grey, fine-grained sandstone. Raindrop markings and desiccation cracks have been observed on RGF bedding surfaces at several localities (see also Bryhni & Skjerlie (1975)). RGF rocks in the Kvamshesten basin are typically found as units up to a few metres thick that separate thicker ICS, MPSS or PCS units. The amount of red fine-grained rocks v. intercalated coarse members varies systematically from the central basin area towards the basin margins, where the RGF units usually wedge out between conglomeratic CUFU units (Fig. 2). There is a systematic increase in the number and thickness of RGF units from base to top of fining-upwards successions several hundred metres thick (Fig. 9). In some areas, notably at intermediate and high stratigraphic levels, units tens to hundred of metres in thickness with >70% RGF rocks occur. In these areas, intercalated coarse members comprise MPSS, PCS and SSCH units. Parts of exposed RGF sections often contain calcite nodules, commonly 1–3 cm in diameter, that are commonly weathered out in exposed sections. Intervals containing calcite nodules range in thickness from a few centimetres up to 1.5 m, but are commonly 10–75 cm thick.

Rhythmicity

The largest-scale rhythmic signal identified in the Kvamshesten basin is defined by the occurrence of thick conglomerate successions in the central basin area at three stratigraphic intervals, and by the separation of these by sandstone intervals. Thus, on a large scale, the basin can be separated into coarsening (CUP)- and fining-upwards (FUP) units (Figs 9 & 10). The large-scale rhythmicity is generally defined by variations in depositional style as exemplified by the various lithofacies associations. Whereas the lower parts of the FUP intervals usually contain the transition from conglomerate to sandstone, the upper parts of FUP units are commonly characterized by an up-section increase in the amount and thickness of RGF rocks (Fig. 9). The top of the FUP units generally consist of a thick (several tens of metres to >100 m) unit dominated by RGF.

The lower parts of the large-scale CUP units are generally dominated by sheet sandstones composed of MPSS or PCS that intercalate with abundant RGF units. The number and thickness of the RGF units decrease up-section and the large-scale CUP units contain thick successions of amalgamated channel sandstones (MPSS) where the medium-scale CUFU successions on a scale of 80–130 m are readily identified (Fig. 10a and c). The upper parts of the large-scale CUFU units are usually conglomeratic, exemplified by the several hundred metres thick

Fig. 8. Single-storey, trough cross-bedded and plane laminated channel unit (SSCH) encased in floodbasin fine sediments (RGF), high stratigraphic levels west of Kvamshesten. The lateral termination of individual units cannot always be observed in the field as a result of faulting or cover, but the single-storey units are interpreted as isolated channel deposits. They commonly comprise the coarse members in the thickest RGF successions encountered in the basin.

conglomerates (mainly MCC, BR and ICS) that occur at intermediate and high stratigraphic levels (Fig. 10a, b and d).

The CUFU unit on the scale of *c.* 80–150 m (Fig. 5) is a fundamental building block in the Kvamshesten basin. As in the Hornelen basin (Steel *et al.* 1977), this grain-size motif can be recognized in the marginal conglomerates as well as in the sandstones of the central basin area. A subordinate rhythmicity is identified at approximately half the scale of the 80–150 m unit (Fig. 5). In the central basin area, this is reflected by the 20–50 m thick sandstone units that are separated by a few metres of red siltstone. In the marginal fanglomerates, small-scale CUFU units on the scale of a few metres are common. The small-scale rhythmicity becomes less pronounced in the central basin area, and individual channel units a few metres thick are commonly fining upwards. On the 80–150 m scale, the anatomy of individual CUFU units is strongly variable (Fig. 5). Partly, this reflects the transition from conglomerate to sandstone away from the basin margins (Fig. 5; see Gloppen & Steel (1981)), but variations are pronounced also within the central fluvial sandstones (Fig. 5b–d). A 130 m thick CUFU unit shown in Fig. 5b comprises multi-storey as well as single-storey channel units separated by red siltstones. The CUFU unit in Fig. 5d consists of a multi-storey sandstone unit with numerous channel scours. This unit is almost 50 m thinner than that in Fig. 5b. The variable thickness and architecture of CUFU units can be interpreted to reflect variations in sediment preservation potential. Thus, the position of CUFU units in the basin and with respect to larger-scale rhythmicity is critical to evaluate the importance of the anatomical variations described above.

The large-scale (i.e. kilometre) and medium-scale (80–150 m) rhythmic variations that are mappable through large parts of the basin must reflect major controls on basin sedimentation. An important question is whether there is an element of self-similarity between large- and medium-scale rhythmicity. That is, do the large-scale grain-size motifs reflect large-scale CUFU units similar to the medium-scale ones? In the present case, this is a difficult point to address; first, progressively higher stratigraphic levels crop out eastwards in the basin, so true vertical logs can be constructed only within

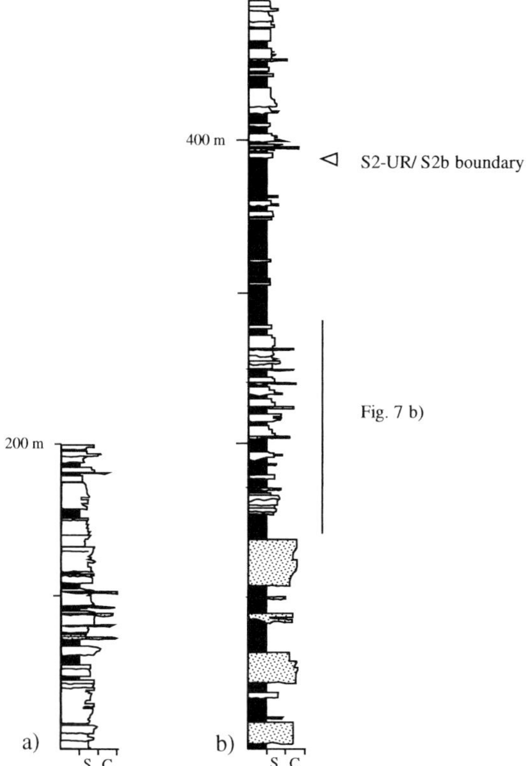

Fig. 9(a & b).

restricted intervals; second, onlap and fanning wedge relationships displayed by the Devonian strata indicate that basin formation was associated with development of considerable intrabasinal topography (Osmundsen *et al.* (1998) and below), thus the thickness of any part of a large-scale unit is dependent on position in the basin, so that the symmetry of a large-scale unit is skewed in one part of the basin compared with another. Thus, self-similarity is not obvious between the large- and medium-scale rhythmites.

Unconformities

Unconformable relationships between Devonian strata and the depositional substrate are exposed for >15 km along the southwestern and northern basin margins (Fig. 2). The Devonian strata onlap the basal unconformity along the northern basin margin (Bryhni & Skjerlie 1975). The direction of onlap is generally towards the northeast, in accordance with the diachroneity between the lowermost deposits in the SMFC and NMFC, respectively. The wedge shape of the SMFC east of the Instelva fault indicates that the Devonian sequence onlaps basement towards the northwest in this area (Osmundsen *et al.* 1998). Intrabasinal unconformities include an onlap or interfingering relationship between floodplain units of the central basin area and the top of the SMFC west of the Instelva fault, onlap onto the top of the SMFC south of the Kringlefjellet fault, erosional truncation of a floodplain unit in the NW part of the basin, and onlap of a heterolithic succession onto the flanks of an anticline at high stratigraphic levels (Fig. 11). Onlap onto the basal unconformity is also observed in the hanging wall of a reverse fault that is exposed in the westernmost part of the basin (Fig. 2). In the southeastern parts of the basin, an angular relationship exists between rotated fault blocks in the SMFC and overlying heterolithic units. In much of the basin, however, angular relationships are characterized by low-angle fanning wedge relationships where unconformities are difficult to identify in the field. Complex intrabasinal faulting, and to some degree Quaternary talus and vegetation, complicates the identification of the pinchout of individual units. Common to the unconformities that have been identified is that they cannot be traced for considerable distances in the basin. In

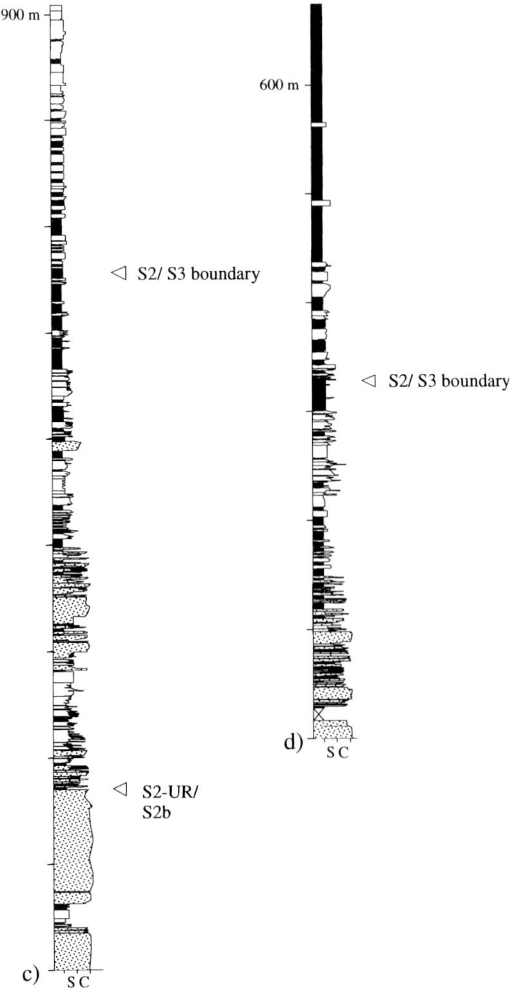

Fig. 9(c & d).

Fig. 9. Large-scale fining-upward (FUP) units (see Fig. 3a for locations; note reduced scale in (**c**) and (**d**)). The log in (**a**) records the fining-upward motif of the MPSS succession in the westernmost basin area. In (**b**)–(**d**) the fining upward from fanglomerates into floodplain fine sediments (RGF) is recorded. Common to these is a rapid transition from fanglomerates into successions dominated by floodbasin deposits. The upper parts of the FUP units are dominated by homogeneous floodbasin deposits. The transition into the sandstone complex dominated by channel and channel-mouth splay deposits is included at the top of the successions in (**b**) and (**c**).

summary, unconformities were apparently produced by several mechanisms, and the potential of unconformities and correlative conformities as markers in the basin stratigraphy is generally low.

Discussion

Interpretation of lithofacies associations and general facies architecture

The large clast sizes encountered in the MCC units and the relations between MCC and other parts of the basin fill justify an interpretation as relatively proximal alluvial fan deposits (Steel *et al.* 1985; Osmundsen *et al.* 1998). The relative scarceness of sedimentary structures in the MCC does, however, hamper a detailed interpretation of depositional processes. The subrounded to well-rounded clasts that dominate the MCC suggest transport aided by fluvial processes. Much of the rounding may, however, have taken place in the drainage basin before the clasts entered the fan. Upon entering the basin, coarse, clast-supported lags were probably produced in stream

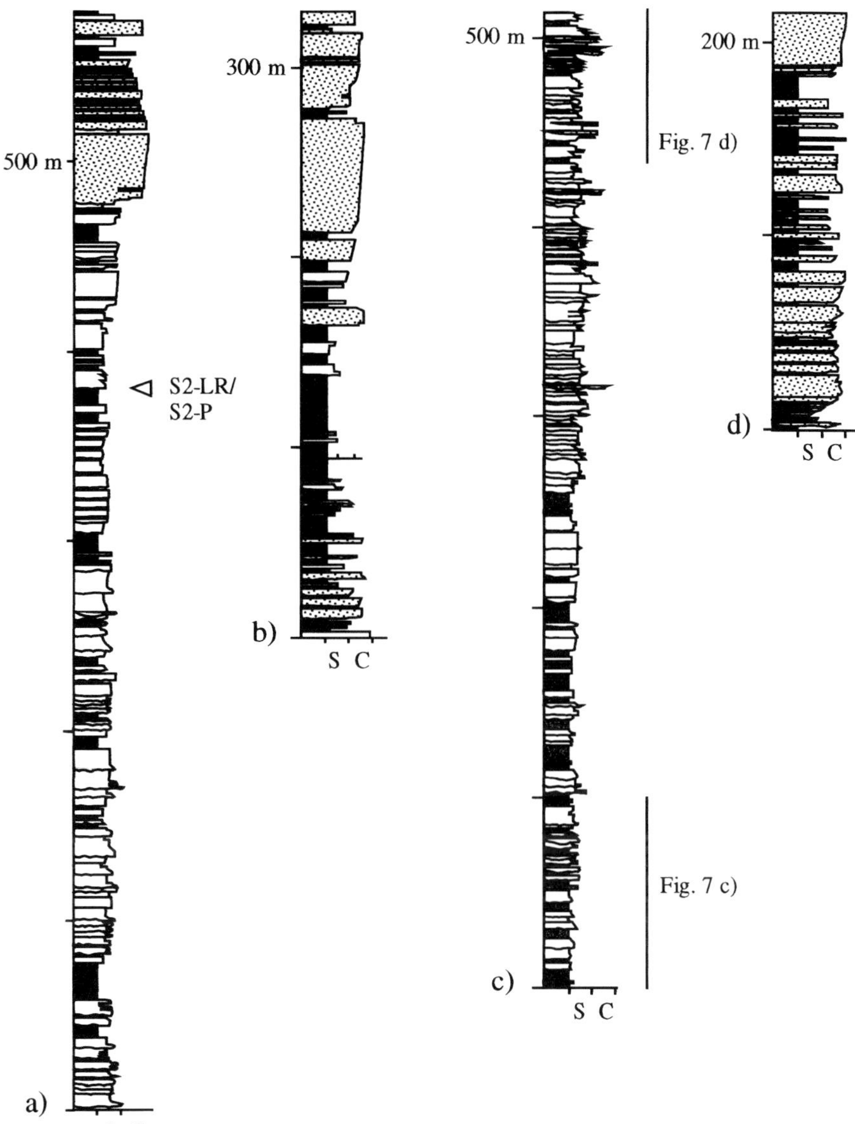

Fig. 10. Large-scale coarsening-upward successions (see Fig. 3a for locations). The succession in (**a**) records the lower two-thirds of the CUFU unit that is based in the heterolithic deposits of the central basin area and has its coarsest part represented by the conglomerates in Eitrenipa. (**b**) Records the upward-coarsening motif in units that are correlated with the Eitrenipa conglomerate across the Kringlefjellet reverse fault. In (**c**), a large-scale CUFU motif is recognized in the sandstone complex of the central basin area, and (**d**) records upward coarsening into the conglomerates in Litjehesten.

channels close to the fan apex. The identification of landslides in the SMFC (Osmundsen *et al.* 1998; Fig. 2) suggests that to some degree, mass-wasting processes were active on the fan surface. The breccias (BR) that occur abundantly in the NMFC were probably laid down by depositional processes involving rock avalanches, rockfall and debris flows. The scarcity of matrix-supported conglomerates and breccias excludes a number of debris-flow processes as major agents of fan construction in the SMFC, but pseudoplastic debris flows may theoretically produce massive, clast-supported conglomerates of the type grouped as MCC here (Miall 1996).

In the ICS rocks, evidence of stream activity is abundant and associated with downfan

Fig. 11. Fanning wedge–onlap relationship onto an east–west-trending anticline at high stratigraphic levels west of Litjehesten (right half of photograph); view towards the east.

reduction of grain size from conglomerate to sandstone. Lithofacies St and Gt (Fig. 6) give evidence of fluvial channel activity. Lenticular, convex-upward conglomerate bodies that display low-angle, plane tangential cross-stratification where they intercalate with sandstones laterally (Fig. 6) are interpreted as gravel bars in a braided channel network. The ICS contain significant volumes of plane laminated medium to very coarse sandstones. These may represent sheet-flood deposits. Sheet-floods may contribute significantly to fan construction in semi-arid climates, in particular on the lower midfan below the intersection between the feeder channel and the fan surface (Blair & McPherson 1994*a*, *b*). Flash flooding of valleys incised on the fan surface may leave coarse-grained lags inside the valley whereas sheetflood deposition takes place downfan (Blair & McPherson 1994*a*, *b*).

In summary, both marginal fan complexes were constructed with the aid of several depositional mechanisms. The volumetric significance of stream-laid deposits versus deposits laid down by other mechanisms is hard to quantify at the present stage, but fluvial processes were probably important in the construction of the MCC units and most certainly in ICS. The NMFC is characterized by a relatively higher occurrence of rocks deposited by gravity flow–fluid gravity flow mechanisms than the SMFC (see Steel *et al.* 1985; Osmundsen *et al.* 1998). The difference between the SMFC and NMFC with respect to architecture, inferred sediment transport directions and onset of deposition indicates that through parts of the basin history, the NMFC and SMFC represent depositional systems that were partly separated in time (Osmundsen *et al.* 1998). At several stratigraphic levels it can be demonstrated that ICS rocks pass directly into MPSS units characterized by eastward-flowing palaeocurrents. Thus both the NMFC and the SMFC represent terminal fan complexes that interacted with a central fluvial channel belt as well as with finer-grained, low-energy floodbasin environments.

The coarse mean grain size, the abundance of pebble-sized material and the multi-storey architecture of the MPSS units indicate that they represent a high-energy braided stream environment. Flow was probably confined to broad, shallow channels (see Fig. 7). The high density of erosional scours and abundant rip-up siltstone clasts encountered in many of the multi-storey units indicate high rates of channel migration and thus a relatively low sediment preservation potential. The difference between MPSS and

PCS lies mainly in the finer average grain size, the relative scarcity of pebbles and the higher abundance of plane and very low angle lamination in the PCS units. The PCS units probably represent a system characterized by less confined flow and, probably, abundant sheetflood activity. The abundance of low-angle plane stratification and plane lamination, the occurrence of trough crossbeds, and the close association with MPSS deposits point towards an interpretation of PCS as channel mouth bars or splays. Similar rocks are common in the Hornelen basin, and were interpreted (Steel & Aasheim 1978; Folkestad 1995; Folkestad & Steel in press) to represent channel mouth splay deposits. In the single-storey channel units, much of the channel sequence is often preserved, including high-level rippled and plane laminated sets. The SSCH units and the siltstones with which they intercalate thus represent a situation characterized by a higher sediment preservation potential than the MPSS units. The grain size and the sedimentary structures encountered in the RGF units indicate that they mainly represent suspension-load material deposited in floodplain–floodbasin and, possibly lacustrine settings (see also Steel *et al.* (1985)). The central basin area experienced numerous widespread flood events, as shown by the many, laterally continuous RGF units that separate MPSS and PCS units at a variety of stratigraphic levels. In several areas, RGF units a few metres thick can be traced into thick fine-grained intervals dominated by RGF (Fig. 2). This situation is excellently illustrated at intermediate as well as high stratigraphic levels. The thick heterolithic units dominated by RGF are typically associated with kilometre-scale eastward translation of the boundary between the central fluvial sandstones and the marginal fan complexes. Thus, they are important markers with respect to the eastward migration of depocentre previously inferred for the Devonian basins in western Norway (Bryhni 1964*a*, *b*; Steel & Gloppen 1980). In the RGF units, intervals rich in calcite nodules probably represent early diagenetic processes, notably soil-forming processes. The formation of soils in floodplain environments takes place between flood events and marks periods of non-deposition. Raindrop markings and desiccation cracks provide further evidence of dry periods. The lack of laterally continuous calcrete beds indicates that sedimentation rates were relatively high. Lacustrine deposition may tentatively have taken place in the Kvamshesten basin (Steel *et al.* 1985; Osmundsen 1996), but critical observations that indicate the presence of permanent, deep bodies of standing water are lacking.

The MPSS display a variety of palaeocurrent directions. In particular, both SE–E- and NW–W-flowing palaeocurrents appear to have dominated in the central basin area. This indicates that the channel deposits of the central basin area cannot be ascribed to one depositional system. Probably, the sandstones represent deposits from two major depositional systems; the other sourced in the footwall of the basin-controlling fault (i.e. to the (S)E), one sourced in the hanging wall (i.e. to the (N)W). We interpret the central fluvial sandstones dominated by MPSS and PCS to represent two terminal fans or river systems (e.g. Friend 1978; Kelly & Olsen 1993; Sadler & Kelly 1993) that drained into a floodbasin represented mainly by RGF rocks. Both river systems were dominated by transport directions that were mainly transverse with respect to the syndepositional tilt direction. The frequent occurrence of PCS rocks in the upper half of the stratigraphy probably reflects increasing distance from the hanging-wall source area. That is, in our interpretation, the hanging-wall-sourced river system was fringed by a belt of channel mouth bars or splays that formed a transition zone between the main braided channel network and the floodbasin. The lateral association of thick RGF dominated units with retrogradational stacking patterns in the marginal fanglomerates indicates that much of the RGF material was transported by eastward flowing rivers.

Among the terminal river deposits, the MPSS represents the lowest sediment preservation potential. The preservation of SSCH units within thick RGF successions at various stratigraphic levels indicates that they represent a higher sediment preservation potential than MPSS. The Kvamshesten basin received sediment from all sides and was apparently a closed basin through much of its history. Thus, the RGF units do not represent the floodplain of an axial river system in the usual sense. Rather, the RGF deposits were associated with sediment dispersal that was transverse with respect to the tilt direction and was trapped and preserved in depositional sinks located in the proximal and central basin areas.

Progradation and retreat of marginal facies; tectonics v. climate as causes of rhythmicity

Tectonics and climate are considered the fundamental controls on sedimentary architecture in continental basins (e.g. Leeder *et al.* 1996). Both factors are able to produce a long-term unsteadiness in the sediment supply and thus a rhythmic signal in the basin fill (e.g. Leeder *et al.* 1998). Cyclic variation in climate is commonly interpreted to reflect changes in astronomical factors such as eccentricity, obliquity and precession, and slip events on large faults are believed to occur at semi-regularly spaced intervals as a result of rhythmic build-up and release of stresses along the fault plane (e.g. Crowell 1974; Steel & Gloppen 1980; Olsen 1990, Schwarzacher 1993). The distinction between tectonic and climatic signals is often not obvious; climatic fluctuations may produce patterns of fan retreat and progradation similar to those produced by tectonism (e.g. Leeder *et al.* 1996).

The Milankovitch time-band. Many basins lack stratigraphic resolution in the order of 10^4–10^5 years (time-band for eccentricity, obliquity and precession). Thus, workers on cyclicity have compared the ratio between the duration of the various astronomical cycles with the ratio between the thicknesses of rhythmites observed at various scales in the stratigraphic succession (e.g. Olsen 1990).

The rhythmites displayed in Fig. 5 contain several orders of smaller-scale CUFU units. A transient fining-upward signal occurs below the middle part of each unit and defines a subordinate rhythmicity. A rough estimate of the ratio of subrhythmite v. rhythmite thickness averages at *c.* 2.3. Similar relations can be seen in several of the medium-scale rhythmites in Figs 9 and 10. A comparison with Kelly's (1992) data from terminal fan deposits in the Munster basin shows a similar ratio (2.36) between the 130 m and 55 m signal. Kelly (1992) explained this as a 'component' of the 110 ka eccentricity cycle. The 36 m signal identified by Kelly (1992) and interpreted to represent the 110 ka eccentricity cycle is less accentuated in our data and may be confused with higher-frequency signals. Thus, from the present dataset, an interpretation of the medium-scale rhythmites as products of orbitally forced climatic fluctuations is not obvious. CUFU units down to metre and even decimetre scale occur frequently along the basin margins. Thus a whole range of small-scale CUFU rhythmites can be defined that may be unrelated to orbital forcing. The Milankowitch time-band appears insufficient to explain all orders of cyclicity in the Kvamshesten basin.

Tectonic and tectono-climatic signals. The most rapid response to a phase of tectonically induced tilt is thought to be the migration of fluvial systems towards the area(s) of maximum subsidence (Alexander & Leeder 1987; Leeder & Gawthorpe 1987; Blair & Bilodeau 1988; Alexander *et al.* 1994). Large parts of the basin floor will be covered by hanging-wall-derived material whereas footwall-derived debris will be trapped in the area proximal to the basin-controlling fault (see sources referred to above and Whipple & Trayler (1996)). If the sediment supply from the footwall is relatively constant, fan progradation distance will be controlled mainly by the subsidence rate (Gordon & Heller 1993; Whipple & Trayler 1993). In the footwall catchments, however, a period of high slip rates on a basin-bounding normal fault will result in an increase in catchment area and thus in an increase of the sediment yield (Leeder & Jackson 1993; Friedman & Burbank 1995). The influx of footwall-derived erosional products into the basin will largely take place after a phase of tectonic activity because of the time required for erosion and sediment transport (e.g. Blair & Bilodeau 1988). Thus, in an extensional basin deposited under constant climatic conditions, an increase in sediment supply from the footwall will tentatively coincide roughly with a period of low subsidence rates. This will lead to a drastic reduction in A/S ratio and to enhanced progradation of footwall-derived material into the basin (e.g. Gordon & Heller 1993; Whipple & Trayler 1996). In the Kvamshesten basin, the transition from retrogradational to progradational systems tracts appears to record a switch from footwall- (i.e. E)-directed to hanging-wall- (i.e. W)-directed sediment dispersal. This conforms well with the above model. In more detail, the sedimentary response to tectonics may be modulated by a number of factors that produce rhythmic signals in the basin fill.

In the Kvamshesten basin, landslides occur at several stratigraphic levels (Osmundsen *et al.* 1998) and mass-wasting processes must have been active in the catchments during sedimentation. In mountainous catchments, landsliding has been suggested as an effective mechanism for sediment production (Burbank *et al.* 1996; Allen & Hovius 1998). In uplifting catchments, river incision leads to repeated slope instability and failure, and in humid climates, landsliding appears to keep interfluve slopes at a constant angle (Burbank *et al.* 1996). This has been interpreted as a long-term steady-state process in humid climates (Burbank *et al.* 1996), although periods of unsteady-state conditions must exist between periods of landslide activity. In semi-arid climates, deviations from steady-state conditions may be enhanced by the relative scarceness of precipitation and by the ephemeral nature of stream activity. Landsliding is often triggered by earthquakes (Allen & Hovius 1998), and earthquake recurrence intervals may thus produce variations in sediment production. Dependent on the amount of sediment storage in the catchment, the above mechanisms may give rise to rhythmic signals in the basin stratigraphy.

Scenarios for tectonic and climatic controls on basin architecture. The present dataset does not enable us to properly evaluate the effects of climate; thus, we envisage three alternative scenarios for the Kvamshesten basin as follows.

(1) Under a constant semi-arid climate, sediment production would fluctuate with (tectonically controlled) catchment size and tentatively with landslide production rate; the latter would be controlled by stream incision rate and by the length of earthquake recurrence intervals. In the basin, fan progradation distance would be controlled mainly by subsidence rate; high rates of subsidence would lead to storage of footwall-derived material in the proximal areas, whereas low subsidence rates would favour fan progradation (i.e. Whipple & Trayler 1996). Tentatively, footwall-derived erosional products reached the basin after a main phase of subsidence. The resulting stratigraphy would comprise series of CUFU units where at some distance away from the basin margins, the lower part of each unit mainly represents hanging-wall-derived material deposited during a period of high subsidence rates. The higher, coarser parts would correspondingly represent erosional products from the footwall deposited during a period of low subsidence rates. The uppermost, fining-upward parts of the CUFU unit would represent a situation where the influx of footwall-derived material no longer outpaced the accommodation creation rate, temporarily during renewed subsidence and footwall-directed rotation. A component of strike-slip probably accompanied each phase of subsidence along the northern, and at a later stage, also along the southeastern basin margin (Osmundsen *et al.* 1998). This may have modulated grain-size trends on a smaller scale through lateral stacking of fan segments, much as envisaged for the Hornelen basin (Steel & Gloppen 1980; Steel 1988).

(2) With a constant subsidence rate, fan progradation distance would depend on climatically driven fluctuations in sediment production and transport capacity, both of which would increase in more humid periods. If large volumes of sediment had been trapped in the catchment areas during drier periods, these would now be transported to the basin together with more recent erosional products. In the basin, fan progradation would result when sediment supply exceeded a threshold value limited by the subsidence rate. In this scenario, the large-scale coarsening-upward successions encountered in the Kvamshesten basin would represent periods dominated by relatively humid conditions. Independent evidence for such climatic fluctuations has, however, not been recorded in the basin.

(3) Tectonism may trigger climatic effects. In semi-arid climates, uplifting mountain ranges may receive higher amounts of precipitation than adjacent basins (e.g. De Boer *et al.* 1991). Thus, the uplifting footwall of a large-magnitude normal fault may experience an increase in precipitation following a period of high slip rates. The Devonian basins in western Norway were probably bordered by NE–SW-trending normal faults with uplifting footwall topography to the SE of the basins (Nilsen 1968; Steel *et al.* 1977; Steel & Gloppen 1980; Osmundsen *et al.* 1998; Steel 1988). Depending on their altitude, the uplifting footwall ranges may have served as effective moisture traps. A periodicity in the uplift rate would thus induce a tectonically driven cyclicity with respect to precipitation rate.

Differentiating between the three above scenarios is not obvious. All models do, however, require significant amounts of hanging-wall subsidence and footwall uplift. Periodic variations in these factors are likely to have occurred. The SE- and E-directed palaeocurrents inferred from parts of the fluvial succession indicate that large-scale fan progradation followed important phases of hanging-wall tilt. Thus, although climatic factors may have influenced sediment production and transport rates, tectonics must have constituted a fundamental driving mechanism for the observed variations in sediment input.

Concepts of sequence stratigraphy

In normal sequence stratigraphy, eustacy provides the driving mechanism for changes in the ratio between accommodation and sediment supply and thus for the formation of sequence-bounding unconformities (e.g. Vail 1975; Van Wagoner *et al.* 1988). In a continental basin, the mechanisms responsible for the production of unconformities are different and more complex; also, unconformities may be more difficult to recognize in the field and on seismic sections (Galloway 1989; Martinsen 1993; Folkestad & Steel in press). In basins where unconformities characterized by valley incision are lacking or are hard to identify, the distinction between a sequence-bounding unconformity and the base of a channel sandstone sheet may be impossible to define (e.g. Martinsen 1993).

The systems tract is a fundamental unit in sequence stratigraphy, and was originally defined

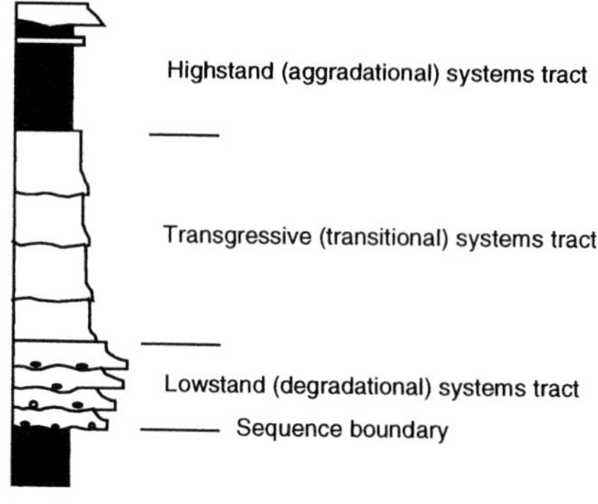

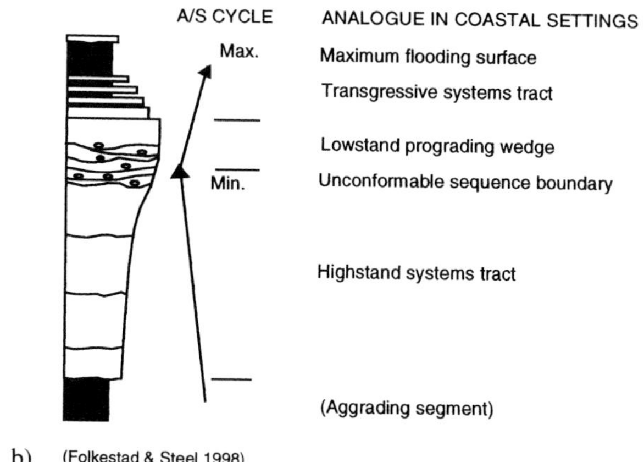

Fig. 12. (**a**) Fluvial sequence as proposed by a number of earlier workers (see text). (**b**) A/S cycle is based on multiparameter analysis of medium-scale CUPF units in the Hornelen basin (Folkestad 1995; Folkestad & Steel in press).

as a unit that links contemporaneous depositional systems (Brown & Fischer 1977). The 'fluvial sequence' as it has been defined in a number of recent publications is commonly displayed as a stratigraphic coloumn or strip log (Fig. 12a), where systems tracts are defined from specific associations of lithofacies that can be correlated with marine systems tracts. In this way, the fluvial response to sea-level changes can be evaluated (Posamentier & Vail 1988). This approach is meaningful only if a direct correlation with an established sequence framework can be made. In continental strata that cannot be directly correlated with marine deposits, the definition of systems tracts and sequences is generally based on: (1) unconformities; (2) stratigraphic markers suggested as continental analogues to the maximum flooding surface; (3) vertical changes in fluvial style as inferred from changes in the amount of (preserved) fine-grained sediment and by changes in the mobility of channel systems; (4) inferences of vertical changes in the ratio between accommodation creation rate and sediment supply rate (A/S ratio) based on the above observations; (5) grain-size and stacking patterns (e.g. Etheridge 1985; Posamentier & Vail 1988; Legarreta *et al.* 1993; Shanley & McCabe 1994; Olsen *et al.* 1995; Currie 1997; Borquin *et al.* 1998; Folkestad & Steel in press).

In a number of published models, the definition of systems tracts is closely tied to a particular succession of lithofacies (Fig. 12a). For example, floodbasin strata with intercalated anastomosing or meandering channel deposits are commonly interpreted to represent high sediment

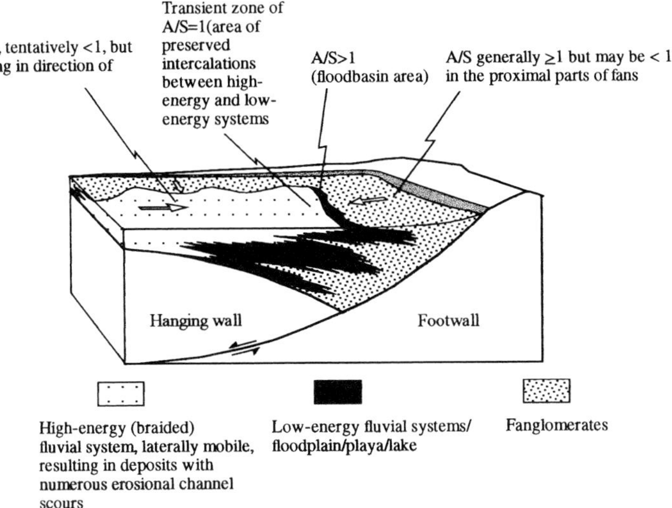

Fig. 13. Facies configuration in a hypothetical alluvial–fluvial basin deposited in the hanging wall of a normal fault. The decay of displacement away from the fault plane controls differential subsidence and therefore a lateral decrease in the accommodation creation rate away from the fault. The response of the fluvial system is expressed as an increase in the storage of overbank deposits in the area close to the basin-bounding fault and as a transition from laterally mobile (braided) channel belts in the distal parts of the handing wall to isolate channels in the area close to the fault breakaway. Tectonic rotation of alluvial fan deposits close to the breakaway will lead to progressive annihilation of basinwards sloping fan topography, and thus to retrogradational stacking of fan segments and to the intercalation of coarse (conglomeratic) deposits with overbank fine deposits in the breakaway area. The decay of subsidence away from the (normal) fault is not necessarily coupled with a reduction in sediment supply; hanging-wall-sourced fluvial sandstones as well as coarse debris shed from transfer faults (if present) will ensure abundant sediment supply also to areas where subsidence and thus the accommodation creation rate are low. Thus, the A/S ratio may be reversed away from the principal basin-bounding fault, leading to a change in stacking patterns from retrogradational to progradational. In the central basin area, this will lead to a change in the overall grain-size motif from fining to coarsening upward.

preservation potential; they are thus assigned to a highstand or aggradational systems tract. The lowstand or degradational systems tract is portrayed as a succession of amalgamated braided river deposits characterized by a relatively low sediment preservation potential. Although this type of model may work well locally, it may be in conflict with the definition of a systems tract; the linkage of contemporaneous depositional systems is not evident from the fluvial sequence as displayed in Fig. 12a. Thus, the changes in A/S ratio inferred from logs such as that in Fig. 12 may have local significance only. In particular, this must be the case in tectonically controlled basins (e.g. Prosser 1993).

Let us consider a simple, extensional half-graben basin (Fig. 13). In a direction parallel to the extension direction, the displacement on the basin-controlling normal fault decays away from the fault plane. Thus, a slip event will induce a lateral gradient in the accommodation creation rate in the hanging-wall block. Thus it follows that a lateral gradient in the accommodation creation rate will be reflected by lateral variations in fluvial style and eventually in the lithofacies distribution.

Theoretically, the A/S ratio may even be reversed across the basin, so that stacking patterns in the distal part of the basin may be progradational although fan segments close to the extensional fault breakaway are stacked in a retrogradational pattern. In particular, this situation may apply to extensional basins bounded by transfer faults (e.g. Eliet & Gawthorpe 1995). Progradational stacking patterns in the breakaway area must, however, be associated with progradational stacking in more distal parts of the basin as the rate of accommodation creation is always lower away from the extensional basin-bounding fault.

In summary, stacking patterns and grain-size motifs may vary considerably across a tectonically controlled basin, even in a stratigraphic interval produced through one period of subsidence at a constant slip rate. A systems tract constructed through a tectonically controlled basin must therefore contain: (1) lithofacies associations representing more than one depositional system, and (2) a variety of inferred A/S ratios dependent on position in the basin; furthermore, (3) inversion of the A/S ratio across the basin may give rise to variable stacking patterns and resultant grain-size motifs. In the following, we use the Kvamshesten Group to illustrate some of these relationships.

A sequence stratigraphic model for the Kvamshesten basin

The rhythmic organization and the pronounced variations in depositional style displayed by the rocks of the Kvamshesten Group do, together with the variations in stacking patterns displayed by the marginal fanglomerates, suggest a subdivision into stratigraphic sequences. The use of intrabasinal unconformities as sequence boundaries would, in our view, not be the ideal choice in the Kvamshesten basin, as at least two types of mechanisms were responsible for the production of unconformities (see below). This would complicate the geological interpretation of the sequence boundary. Moreover, channel scours that occur within multi-storey sandstone units are often characterized by relief that is similar to that of the basal scour beneath the unit. In addition, unconformities (and their potential correlative conformities) are difficult to trace for long distances in the basin.

In coastal deposits, an alternative to the use of unconformities as sequence boundaries is to place the sequence boundary at the maximum flooding surface (Galloway 1989). In the Kvamshesten basin, intervals dominated by siltstones and very fine sandstones are widespread and easy to trace. These do not always represent the maximum migration of the central fluvial system towards the principal (extensional) basin margin, but they are usually traceable for considerable distances toward the northern margin, where they intercalate with marginal conglomeratic facies. On the large as well as the medium scale, thick floodplain units separate FUP from CUP grain-size trends (Figs 9 and 10). Thus, where present, they appear to represent maximum turnaround intervals for the A/S ratio. Folkestad (1995) and Folkestad & Steel (in press) suggested that a continental analogue to the maximum flooding surface may be defined in the finest-grained deposits that cap individual CUFU units. We choose the tops of the finest-grained units as sequence boundaries because of their lateral continuation through significant parts of the basin fill.

We divide the Kvamshesten Group into three sequences (S1, S2 and S3, Fig. 14) based on the above criteria. The subdivision into retrogradational and progradational systems tracts is based on the stacking patterns of marginal fanglomerates as seen in sections subparallel to the general extension direction. The various systems tracts display grain-size motifs that are partly in and partly out of phase with the marginal stacking patterns, and we discuss possible explanations below.

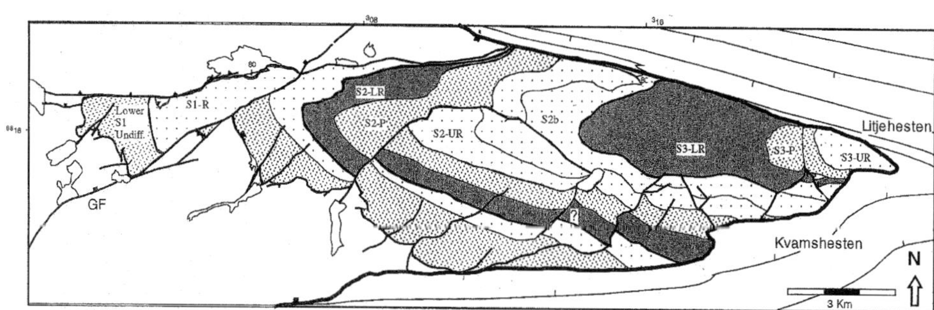

Fig. 14. Map of the Kvamshesten basin with subdivision into large-scale sequence and systems tracts (compare with Fig. 2 for lithofacies configuration). Thinning of S1 towards the northern margin, accommodated in particular by the onlap of lower parts of S1 as well as large parts of S1-R onto depositional basement, should be noted. A similar relationship id displayed by S2-LR. A pronounced shift in the basin configuration occurs with the introduction of S2-P and S2-UR, which are thick also along the present northern basin margin. From these levels, the basin configuration is apparently more symmetrical than at lower stratigraphic levels. The relationship with the basal unconformity cannot, however, be worked out east of the hanging-wall cutoff of basement. The relationship between S3-LR and S3-P is somewhat obscured by faulting in the eastern basin areas, but it is assumed that the conglomerates of S3-P covered a fairly large area at a stratigraphic level above S3-LR. In the massive parts of the SMFC, boundaries between sequences and systems tracts are tentative.

S1. The basal sequence in the Kvamshesten basin crops out in the western basin area (Fig. 14) and is highly asymmetrical because of the apparent NE-ward onlap of the lower parts of the SMFC. It can be divided into one broadly progradational (S1-P) and one retrogradational (S1-R) systems tract. As a result of the diachroneity between the SMFC and the NMFC, the large-scale stacking pattern in the NMFC is mainly retrogradational within S1. In the central fluvial sandstones, the retrogradational stacking patterns observed in S1-R are associated with a general fining-upward. In the upper parts of S1-R, the proportion of RGF increases rapidly relative to intercalated channel deposits and culminates in a thick unit dominated almost entirely by RGF. The upper boundary of S1 is placed at the top of this unit. Along the northern and southern basin margins, S1 generally onlaps basement in a generally eastward direction. Exact correlation of S1 across the Instelva fault is problematic, but the lowermost parts of the SMFC east of the Instelva Fault are included in S1 and constitute the lowest part of an eastward-thickening wedge (Figs 2 and 14). Thus, a NE-trending basement high probably existed in the central basin area during deposition of S1.

S2. S2 is characterized by eastward-thickening conglomeratic wedges on both the northern and southern basin margins. In the central basin area, S2 displays a large-scale coarsening upward into a thick conglomerate succession at intermediate stratigraphic levels, then fining upward into sandstone (Figs 2 and 14).

The S1–S2 boundary is overlain by the lower retrogradational systems tract (S2-LR), characterized in the central basin area by floodbasin deposits intercalated with single- and multistorey sandstone sheets. Towards the top of S2-LR, the amount and thickness of sheet sandstones increases rapidly, and S2-LR is characterized by general upward coarsening. The top of S2-LR is set within the MPSS-dominated succession below the conglomerates in Eitrenipa (Fig. 10) at a level where stacking patterns along the basin margins apparently change from retrogradational–aggradational to progradational. Correlation with marginal deposits is complicated by faulting and cover (lake) at this level. It appears clear, however, that within S2-LR, retrogradational marginal stacking patterns are associated with general upward coarsening in the central basin areas.

S2-LR is overlain by the progradational systems tract (S2-P). In the central basin area, S2-P is sandstone dominated in its lower parts but coarsens rapidly upward into a succession dominated by conglomerate (Fig. 10a). The conglomerates are cut by a reverse fault south of Kringlefjellet (Fig. 2), but a general correlation of the conglomerates in Eitrenipa with those in the Kringlefjellet can be made based on mapping of siltstone units stratigraphically below the conglomerates (Osmundsen *et al.* (1998) and Fig. 2). In the Kringlefjellet area, the conglomerates are intercalated with red, fine-grained units (Fig. 2), whereas this is not observed in the Eitrenipa area.

S2-P is overlain by a unit characterized by strongly retrogradational stacking patterns (S2-UR). In the Kringlefjellet area, back-stepping fan segments are intercalated with red siltstones and overlain by a thick RGF unit with intercalated, mainly single-storey channel deposits in its lower parts (Fig. 9b). In the Eitrenipa area, S2-UR is constituted by ICS and MPSS units. Thus, both S2-P and S2-UR are highly asymmetrical with respect to lithofacies configuration. It follows from our earlier definition of the sequence boundary that it is necessary to introduce a sub-sequence (S2b) for the stratigraphic interval between the top of the thick RGF units in the Kringlefjellet area and the S2–S3 boundary. Both S2-UR and S2b are characterized by strongly retrogradatonal stacking patterns and by fining-upward grain-size motifs. S2-UR is apparently associated with low-angle, eastward onlap onto the top of S2-P.

The uppermost parts of S2b in the south-eastern basin area is characterized by a thick unit dominated by RGF with single-storey channel deposits (Figs 8 and 10c). The S2–S3 boundary is set at the top of this unit (Fig. 14).

S3. S3 constitutes the uppermost large-scale sequence encountered in the Kvamshesten basin and is exposed in the basin's eastern parts (Fig. 14). In Fig. 9c, the architecture of S3 largely resembles that of S2, with S3-LR consisting of a coarsening-upward unit of intercalated floodplain, sheeted channel and channel mouth splay sandstones that can be correlated with retrogradational stacking patterns along the (northern) basin margin. In Fig. 9d, however, the CUP motif of S3-LR is apparently lacking. This is explained by the gradual eastward interfingering of MPSS and PCS with RGF at high stratigraphic levels (Fig. 2). Immediately west of Litjehesten, the progradational tendency becomes more pronounced in the marginal fanglomerates, and S3-LR passes into a progradational systems tract (S3-P). S3-P is characterized by a coarsening-upward succession of conglomerates and red fine sediments, capped by a thick conglomerate bed (Fig. 10d). The top of this bed marks the boundary between S3-P and S3-UR. In the latter, stacking patterns are retrogradational, and the rocks generally consist of conglomerates and red fine-grained sediments.

Generalized characteristics of systems tracts. The characteristics outlined above for sequences and systems tracts allow for the following generalizations.

(1) Each systems tract constitutes a well-defined stratigraphic interval that displays pronounced lateral and vertical variations in its configuration of lithofacies. Lateral variations in lithofacies are interpreted in terms of linking of various depositional systems. Thus, in the broad sense, our systems tracts are in accordance with the original definition (Brown & Fischer 1977). Vertical variations can be attributed to variations in the relative distribution of depositonal systems with time.

(2) Several systems tracts appear to contain lithofacies characterized by opposite general palaeocurrents; this reflects the linkage of depositional systems sourced in the hanging wall and the footwall, respectively.

(3) The retrogradational systems tracts record kilometre-scale eastward shifts of the boundary between the marginal fanglomerates and the central fluvial sandstones. Thus, they record major eastward shifts in the location of the basin's main depocentre(s). The retrogradational systems tract as defined here resembles the 'rift climax' systems tract of Prosser (1993).

(4) Grain-size motifs within the retrogradational systems tracts are partly in and partly out of phase with the marginal stacking patterns. This appears to be dependent on position in the basin. Thus the retrogradational systems tracts seem to be sensitive to lateral variations in A/S ratio. In the progradational systems tracts, the large-scale (CUP) grain-size motif is mainly in phase with the marginal stacking pattern.

Deformation above a ramp-flat detachment; implications for sedimentary architecture

Ramp-flat detachments. A fundamental principle in rock mechanics is that the dip of a normal fault depends on the strength of the rock within which it develops (e.g. Anderson 1951). A fault that cuts a rheologically layered (tectono-) stratigraphy may therefore develop with dips that vary downsection. In ancient orogenic belts, the crust contains numerous structures inherited from mountain building; these commonly include low-angle thrusts as well as steeper thrust ramps (e.g. Boyer & Elliott 1982). Reactivation of thrusts and related structures as ramp-flat detachments during subsequent extension has been inferred from field studies as well as from the interpretation of deep seismic reflection profiles (e.g. Cheadle *et al.* 1987; Axen 1993). At the onset of Devonian late- or post-orogenic extension, the overthickened Caledonian crust was both rheologically and structurally heterogeneous. Thus, it is highly probable that large-magnitude, late- to post-orogenic extensional faults would develop ramp-flat geometries. The present North Sea basin is largely underlain by extended Caledonian crust and in the North Sea area, deep-seated, ramp-flat extensional detachments have been interpreted from seismic sections, by, for example, Cheadle *et al.* (1987) and Gibbs (1987). The extensional detachments have been suggested by those workers to be inherited from Devonian large-magnitude extension.

It is generally believed that the hanging wall deforms internally to accommodate the geometry of the underlying, basin-controlling fault (e.g. Gibbs 1984, 1987). Thus, the geometry of the fault that controlled basin sedimentation may be inferred from onlap relationships, synsedimentary structures and thickness variations displayed by sedimentary units. At low stratigraphic levels in the Kvamshesten basin, the crest of a syndepositional basement high was oriented at a high angle to the principal extension direction and was associated with a corresponding depression to the (N)W. These observations are compatible with the presence of a rollover anticline–syncline pair in the basin during sedimentation. Rollover anticlines are commonly associated with listric normal faults, and tend to be monoclinal in shape unless the fault has a ramp-flat geometry at depth (e.g. Gibbs 1984). In the latter case, a rollover anticline–syncline pair develops in the hanging wall to accommodate the space problems associated with fault displacement (e.g. Gibbs 1984, 1987; McClay & Scott 1991; Schlische 1995). On the basis of onlap relationships towards basement and on the thickening and thinning of sedimentary units in sections parallel to the inferred extension direction, Osmundsen *et al.* (1998) interpreted the Kvamshesten basin as having formed in the hanging wall of a ramp-flat normal fault.

Sedimentation in the hanging wall of a ramp-flat extensional fault. If a rollover syncline–anticline pair develops during basin sedimentation, there are two principal areas where subsidence rates, and thus accommodation creation rates, reach maximum values; one is located proximal to the fault breakaway, the other in the hanging-wall syncline above the ramp (Fig. 15a). These areas are separated by the crest of the rollover anticline, where the rates of accommodation creation are lower than in the sinks on either side.

As accommodation creation rate is thought to be a main control on sediment preservation potential and thus fluvial style, fluvial architecture in the hanging wall of a ramp-flat detachment may be predicted as a function of differential subsidence. Our first approach is thus to regard the long-term sediment supply as a constant. Let us consider a closed basin being filled by sediment under a stable, semi-arid climate while moving down a ramp-flat detachment at a constant rate (Fig. 15b). The breakaway area and the hanging-wall syncline will be occupied largely by fluvial systems that accommodate high rates of vertical aggradation, i.e. floodplain or floodbasin deposits with intercalated isolated single-storey channel deposits or sheet sands. In areas close to the breakaway, downwearing of footwall block topography will induce input of coarse-grained, footwall-derived sediment. Tectonic rotation will cause footwall-directed migration of the floodbasin (Alexander *et al.* 1994) as the basinward slope of alluvial fans is counterbalanced by subsidence and tectonic rotation. The alluvial fans will thus be stacked in a retrogradational pattern that reflects progressive normal displacement and fault-block rotation.

With increasing distance from the fault, fluvial systems will be characterized by increasing lateral channel activity and decreasing sediment preservation potential. Thus, a transition is expected from thick floodbasin deposits through an area with intercalated floodbasin siltstones and sheet channel sandstones into amalgamated, multistorey sandstone units (Fig. 15b). The last style of deposition will characterize the crest of the rollover anticline, as this area experiences relatively low rates of accommodation creation. On the rollover anticline, low overall rates of subsidence and low sediment supply may cause erosion and formation of unconformities. Higher overall rates of subsidence and high rates of sediment supply may, however, prevent significant erosion of the anticlinal area. In the hanging-wall syncline, high rates of accommodation creation will favour deposition and preservation of floodbasin strata and low-energy channel deposits. During periods of decreasing A/S ratio, however, fluvial systems characterized by lateral migration may spread across the hanging-wall syncline so that a stratigraphy characterized by intercalated floodplain deposits and sheet sandstones is developed. Towards the distal parts of the hanging-wall, the sediment preservation potential will decrease and the hanging wall slope away from the syncline will be characterized by high-energy fluvial deposits that become coarser towards the source area.

In an extensional basin that deforms above a ramp-flat extensional fault, the rollover anticline as well as associated crestal collapse grabens will migrate in the direction of the footwall (McClay & Scott 1991). Thus in a section parallel to the extension direction, the basin stratigraphy will be characterized by successions that thin across the rollover anticline and are thicker in the breakaway and hanging-wall syncline areas. The resulting stratigraphy will display sequences stacked in an asymmetrical pattern, where stratigraphically upwards, the facies belts resulting from accommodation-related variations in fluvial style are skewed towards the footwall. We suggested earlier that in a half-graben basin, A/S ratios may be reversed within a systems tract as a result of the effects of differential subsidence. In the present setting, three reversals of the A/S ratio may take place along a particular stratigraphic interval (Fig. 15b and c). Surfaces characterized by $A = S$ would be located along the margins of the breakaway and hanging-wall syncline areas, separating areas characterized by $A > S$ from areas characterized by $A < S$. The $A = S$ surfaces would be dipping in the direction of the hanging-wall, reflecting the footwall-directed migration of the rollover. It is clear that any stratigraphic interval or systems tract would be cut by the $A = S$ surfaces, making the definition of a systems tract from one particular A/S relationship meaningless.

In summary, under a given climate, the architecture of individual systems tracts and sequences will be controlled largely by the geometry of the ramp-flat extensional fault. The dip of the breakaway will control the amount of subsidence per unit extension, and thus the accommodation creation rate in the breakaway area. The parts of systems tracts deposited in the hanging-wall syncline will depend largely on the dip of the ramp. A steep ramp will produce high subsidence rates in the synclinal area, and thus high rates of accommodation creation. Correspondingly, a low-angle ramp may produce subsidence rates that are only slightly higher than those experienced by the anticlinal area. The dynamics of the system described above predicts the migration of depositional sinks through the basin with time. In this respect, the model highlights the principle of sediment volume partitioning in sedimentary basins (Cross *et al.* 1993).

Applications to the Kvamshesten basin

A main difference between the stratigraphy of our model basin (Fig. 15b) and that of the Kvamshesten basin is the pronounced, large-scale rhythmicity displayed by the latter. In the sequence framework suggested by us for the Kvamshesten basin, rhythmicity is an important factor in the definition of systems tracts. We thus introduce an element of rhythmicity in our model basin, which may be imposed by tectonics, climate or a combination of the two (Fig. 15). The resulting stratigraphy can be divided into sequences and systems tracts along the lines discussed earlier (Fig. 15d). We now compare our model with the stratigraphic framework of the Kvamshesten basin.

Retrogradational systems tracts. An upsection decrease in plunge of the basinal syncline has been interpreted to reflect a fanning wedge in the Devonian strata eastwards from the central parts of the basin area (Osmundsen *et al.* 1998). In the western parts of the basin, sedimentary units become thicker towards the (N)W and onlap the basal unconformity eastward. This indicates that high stratigraphic levels were deposited in a position between the fault breakaway and the crest of the rollover anticline, whereas stratigraphic levels exposed west of the Instelva Fault were deposited in the proximal parts of the hanging-wall syncline (Osmundsen *et al.* 1998).

In our model basin (Fig. 15b and c), fine-grained floodbasin sediments deposited close to the breakaway are progressively overlain by higher-energy fluvial sandstones as the rollover anticline migrates through the hanging wall. The effect of the migrating anticline is that accommodation creation rates are reduced in areas that were previously part of the breakaway area (Fig. 15b and c). In the Kvamshesten basin, our model is supported by the westwards increase

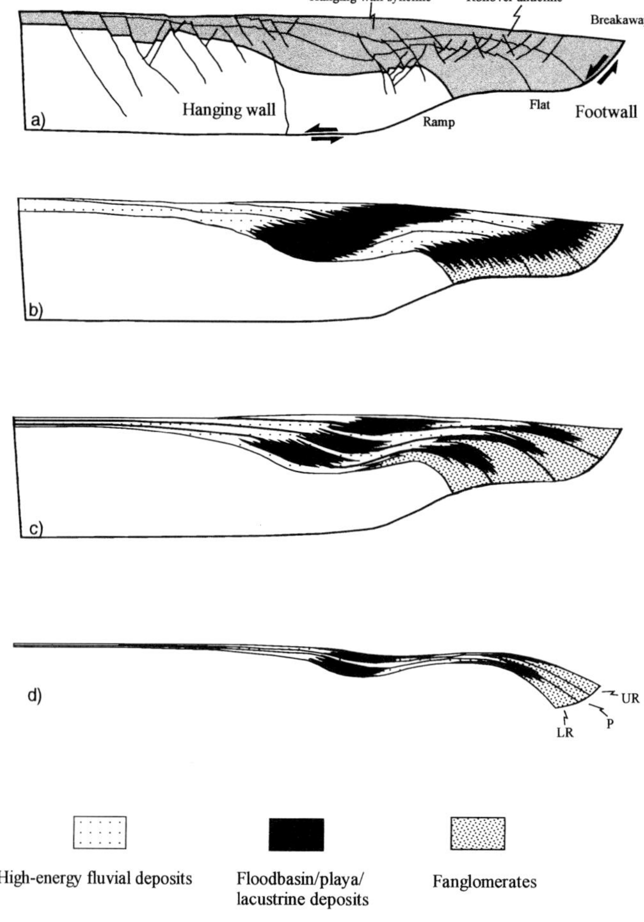

in the proportion of MPSS observed in the retrogradational systems tracts at high stratigraphic levels (S2-UR/S2b and S3-LR, Fig. 14). S3-LR is associated with retrogradational stacking patterns, but with grain-size motifs that change laterally. In Fig. 10c, S3-LR is characterized by a large-scale homogeneous to general coarsening-upward motif, whereas parts of the systems tract exposed further east appear to be fining upward (Fig. 9d). We interpret these observations to reflect the lateral decrease in A/S ratio away from the breakaway of the basin-controlling fault (see Fig. 15c). Our logs through S2-UR/S2b and S3-LR in the eastern parts of the basin (Figs. 9c and 10c) show large-scale upward fining followed by large-scale upward coarsening. The latter is related to the eastward migration of units dominated by multi-storey sandstone sheets (Figs 2 and 3). We interpret the large-scale FUP to reflect increasing A/S ratios in the breakaway area, whereas the large-scale CUP motif reflects a decreasing A/S ratio related to eastward migration of the rollover anticline (see Fig. 15c).

A similar situation will be encountered on the distal side of the rollover anticline. Here, enhanced deposition of fine-grained strata will take place in the hanging-wall syncline because of a relatively high A/S ratio. As the rollover syncline migrates towards the footwall, accommodation creation rates are progressively reduced above the previous site of the hanging-wall syncline. In the Kvamshesten basin, S1-UR and S2-LR displays large-scale grain-size variations similar to those described above for higher stratigraphic levels. The main differences between the two settings are that in the breakaway area, the succession thickens towards the basin-controlling fault, whereas on the other side of the rollover anticline, sedimentary units become thicker towards the axial plane trace of the hanging-wall syncline (Fig. 15b and c).

In our model basin (Fig. 15b and c), the trace of the rollover anticline is represented in the stratigraphy by a thick, amalgamated sandstone succession that is skewed in the direction of the footwall. We interpret the thick sandstone (mainly MPSS) succession exposed eastward

Fig. 15. (**a**) Stratal geometries and intrabasinal structure of basin deposited in the hanging wall of a ramp-flat extensional fault (simplified from sandbox experiment by McClay & Scott (1991)). (Note thinning of sedimentary units across the crest of the rollover anticline and thickening adjacent to the master fault breakaway and in the rollover syncline area.) The breakaway and rollover syncline areas are principal loci of subsidence characterized by higher rates of accommodation creation than the crest of the rollover anticline and the distal parts of the hanging wall. (**b**) Tentative facies distribution of continental, alluvial–fluvial basin in setting as outlined in (**a**) (intrabasinal faults removed for simplicity). The basin moves down the extensional fault at a constant rate under stable (semi-arid) climatic conditions. The distribution of sedimentary units is controlled by differential subsidence as the hanging wall deforms into a rollover anticline–syncline pair to accommodate the shape of the underlying basin-bounding fault. The asymmetrical stacking pattern displayed by the sedimentary units reflects the footwall-directed migration of the rollover as progressively younger parts of the basin pass the ramp. (See text for discussion.) (**c**) If basin formation is associated with periodic variations in slip rate or precipitation, the resulting stratigraphy will contain a rhythmic signal expressed by stacking patterns and grain-size variations. Low rates of accommodation creation will favour progradation of coarse-grained material sourced in the footwall of the basin-bounding fault. In more distal parts of the basin, low rates of accommodation creation will favour lateral expansion of high-energy fluvial systems. Increase in the rate of accommodation creation will lead to storage of fine-grained (overbank) or lacustrine deposits in the breakaway area and in the hanging-wall syncline, and to retrogradational stacking of marginal fan deposits. The rhythmicity described above may serve as a basis for a subdivision into stratigraphic sequences. In the present example, the sequence boundaries are placed at the top of the finest-grained rocks in the succession, corresponding to the maximum value for the ratio between accommodation creation and sediment supply (see text). (**d**) Each sequence can be divided into systems tracts based mainly on stacking patterns observed adjacent to the basin-bounding normal fault. Within each systems tract, the configuration of lithofacies as well as overall grain-size motifs will change laterally as a result of the non-uniformity imposed by differential subsidence. In the chosen framework, the sequence boundary is overlain by the lower retrogradational (LR) systems tract, followed by the progradational (P) and upper retrogradational (UR) systems tracts. The retrogradational systems tracts are characterized by retrogradational stacking patterns and thus represent a situation where $A/S > 1$ in proximal parts of the basin. However, as the A/S ratio depends on position in the basin, large-scale grain-size motifs may change from fining to coarsening upward away from the basin-bounding fault. The LR systems tract is probably the most sensitive in this respect, as it represents the lower part of an A/S decrease trend. The progradational systems tract (P) is characterized everywhere by progradational stacking patterns and a coarsening-upward grain-size motif. The geometry and thickness variations displayed by the systems tracts and sequences record the geometry of the rollover anticline–syncline pair and its migration towards the footwall during basin formation. Thus, the entire basin fill architecture reflects the geometry of the underlying basin-controlling fault.

from the Eitrenipa area (Figs 2 and 3) to reflect this situation.

The effect of intrabasinal faulting, and in particular of migrating crestal collapse grabens (Fig. 15a), will be to affect the migration of fluvial systems. Intrabasinal normal faults may temporarily trap hanging-wall-derived fluvial systems, postponing their migration towards the basin-controlling fault. When activity stops on a set of intrabasinal faults, footwall-directed migration may be relatively rapid. Osmundsen *et al.* (1998) explained local facies configuration in the area of the Selsvatn Fault system (Figs 2 and 3) in this way.

The apparent paradox represented by coarsening-upward, retrogradational systems tracts is readily explained in a continental extensional basin. Whereas in coastal settings, sediment transport is mainly from dry land into the sea (variably modulated by longshore drift), continental basins are characterized by a variety of sediment transport directions. A large part of the sediment supplied to continental half-graben basins is derived from the hanging-wall slope (e.g. Leeder & Gawthorpe 1987). Progressive rotation of the hanging wall will result in the migration of hanging-wall-sourced, high-energy fluvial systems across axial and proximal facies (Leeder & Gawthorpe 1987; Alexander *et al.* 1994). Axial facies often comprise lacustrine, playa or floodplain deposits depending on climate and on the maturity of the rift system with respect to axial drainage (Leeder & Gawthorpe 1987). Thus, in a simple half-graben basin (e.g. Fig. 13), large-scale fining upward will be succeeded by large-scale coarsening upward even with a constant supply of footwall-derived sediment. Large-scale coarsening upward may therefore not necessarily be related to a slowdown in subsidence rate (see Prosser 1993).

Progradational system tracts. The progradational systems tracts are characterized by progradational stacking patterns and general upward coarsening. S2-P and S3-P overlies retrogradational systems tracts where we have explained upwards coarsening by rollover anticline migration. Whereas this was related to effects of differential subsidence, the progradational systems tracts appear to represent a situation where $A/S < 1$ in large parts of the basin. Probably, this was a response to basin-wide decrease in accommodation creation rates, increase in the rate of footwall-derived sediment supply or both.

In the area close to the fault breakaway, the progradational systems tracts are characterized either by continuous conglomerate successions or by intercalation of conglomerates with floodbasin fine sediments (e.g. S3-P, Fig. 10d). Away from the breakaway, the conglomerates of S2-P intercalate with MPSS units consistent with positions closer to the crest of the rollover anticline (Figs 2 and 10a).

Variations normal to the general extension direction. Some of the lateral variations displayed by systems tracts in the Kvamshesten basin cannot be explained by 2D models parallel to the extension direction. In particular, S1, S2-P and S2-UR show architectural variations in the north–south direction. The eastward onlap of most of S1 onto basement can be explained in terms of onlap onto the rollover anticline from the hanging-wall side (Osmundsen *et al.* 1998). The NE-ward pinchout of the southern margin fan complex does, however, require another explanation. One possibility is that the earliest phase of basin formation was characterized by the growth of a NE-trending normal fault, which produced a radial onlap onto basement away from the area of maximum subsidence (e.g. Schlische 1991). In this setting, the lower parts of S1 may represent the northeastern parts of the initial basin, in accordance with the observation that the SMFC is thickest along the SW basin margin (Osmundsen *et al.* 1998). A consequence of this interpretation is that the originally deepest parts of the early basin was located to the SW of the preserved basin. The Kvamshesten basin was, however, shortened in a direction approximately normal to the general direction of extension. Part of the shortening has been inferred to have taken place contemporaneously with sedimentation (Bryhni & Skjerlie 1975; Chauvet & Seranne 1994; Osmundsen & Andersen in press). Thus, there is a possibility that the NE-ward pinchout of the SMFC was related to NE–SW-directed shortening, which may have produced a scoop-shaped flexure in the basement rocks. The thickest parts of the SMFC are, however, located along the NE-dipping flank of the basin syncline (Osmundsen *et al.* 1998). Thus, if folding was syndepositional at the lowest stratigraphic level, the geometry of the present basin syncline does not conform to that of the original fold. In S3-P, a fanning wedge relationship is observed onto the flank of an anticline that crops out in the western parts of Litjehesten (Fig. 11). Relationships similar to that in Fig. 11 have been documented from growth folds in foreland basins (Burbank *et al.* 1996). Additional support for synsedimentary folding may be given by the observation that at several stratigraphic levels, palaeocurrents are parallel to the fold axes. Floodbasin strata are locally confined to synclinal areas. This can be demonstrated at high levels in S1-UR, and in particular, in S3-UR in the Litjehesten area (Fig. 2). Here, floodbasin rocks crop out in a belt that follows the axial plane trace of the basin syncline. The above observations can be interpreted as due to enhanced rates of accommodation creation in the synclinal areas.

Some of the variations in north–south direction are more difficult to explain by effects of syndepositional shortening. The northward increase in the amount of RGF units observed in S2-P and in particular in S2-UR occurs across the Kringlefjellet reverse fault, and thick RGF rocks cap the crest of an anticline associated with the reverse fault. Apparently, S2-P and S2-UR record an increase in the accommodation creation rate along the northern basin margin relative to the area in the south. The increase in accommodation rate along the northern basin margin has tentatively been interpreted to reflect a change in slip direction on the basin-bounding faults (Osmundsen *et al.* 1998; see below). The shortening of the Kvamshesten and other Devonian basins in a direction roughly normal to the principal extension direction has been interpreted as due to a component of sinistral strike-slip during basin formation (Roberts 1983; Chauvet & Seranne 1994; Osmundsen *et al.* 1998; Krabbendam & Dewey 1998).

Conclusions

(1) The observations and inferences made in this work emphasize that considerations about the A/S ratio made from vertical sections are relevant to the localities involved but cannot be used to define systems tracts as such. The A/S ratio varies laterally within the systems tract as a function of differential subsidence, controlling the relative position of depositional systems through time.

(2) The Kvamshesten Group can be subdivided into a number of large-scale sequences and systems tracts based on (i) stacking patterns observed subparallel to the extension direction and (ii) large-scale grain-size variations. Grain-size variations are, however, partly out of phase with marginal stacking patterns, emphasizing the importance of differential subsidence, depocentre migration and multidirectional sediment transport in a tectonically controlled continental basin.

(3) The Kvamshesten basin was deposited in the hanging wall of a ramp-flat extensional fault. Thus, a rollover anticline–syncline pair accompanied basin formation. This had a pronounced effect on basin sedimentation expressed as the footwall-directed migration of two main depocentres; one positioned close to the fault breakaway, one positioned in the hanging-wall syncline above the ramp. The main depocentres were characterized by relatively high rates of accommodation creation and thus by storage of relatively large amounts of floodplain or floodbasin deposits. In the depocentre located closest to the breakaway of the basin-bounding fault, alluvial fan deposits were stacked in retrogradational and progradational sets, responding to the main periods of subsidence along the basin-bounding fault and subsequent sediment efflux from footwall catchments. The migrating crest of the rollover anticline was characterized mainly by high-energy fluvial deposition in laterally mobile (braided) channel systems. The resulting basin stratigraphy was characterized by facies belts that were progressively skewed towards the footwall upward in the stratigraphy. The trace of the migrating anticline is thus represented by multi-storey channel units stacked in several hundred metres thick successions. Syndepositional shortening in a north–south direction may to some degree have aided the location of, and the dominant flow directions in, the terminal river system. Locally, low-angle onlap relationships and facies variations observed in north–south section can be ascribed to syndepositional shortening.

We thank NORSK AGIP and Philips Petroleum Company for financial support.

References

ALEXANDER, J. A. & LEEDER, M. R. 1987. Active tectonic control on fluvial architecture. *In*: ETHERIDGE, F. G. & FLORES, R. M. (eds), *Fluvial Sedimentology*. Society of Economic Paleontologists and Mineralogists, Special Publications, **39**, 243–252.

——, BRIDGE, J. S., LEEDER, M. R., COLLIER, R. E. LL. & GAWTHORPE, R. L. 1994. Holocene meander-belt evolution in an active extensional basin, SW Montana. *Journal of Sedimentary Research*, **B64**, 542–559.

ALLEN, P. A. & HOVIUS, N. 1998. Sediment supply from landslide-dominated catchments: implications for basin-margin fans. *Basin Research*, **10**, 19–35.

ANDERSEN, T. B. 1998. Extensional tectonics in the Caledonides of southern Norway, an overview. *Tectonophysics*, **285**, 333–351.

—— & JAMTVEIT, B. 1990. Uplift of deep crust during orogenic extensional collapse: a model based on field studies in the Sogn–Sunnfjord region of Western Norway. *Tectonics*, **9**, 1097–1111.

ANDERSON, E. M. 1951. *The Dynamics of Faulting and Dyke Formation with Applications to Britain*. Olivier and Boyd, Edinburgh.

ASPHAUG, E. J. 1975. *Sedimentologi og stratigrafi i vestlige deler av Kvamshesten Devonbasseng.* Cand. Real thesis, University of Bergen.

AXEN, G. 1993. Ramp-flat detachment faulting and low-angle normal reactivation of the Tule springs thrust, southern Nevada. *Geological Society of America Bulletin*, **105**, 1076–1090.

BERRY, H. N., LUX, D. R., ANDERSON, A. & ANDERSON, T. B. 1995. Progressive exhumation during orogenic collapse as indicated by ^{40}Ar/^{39}Ar cooling ages from different structural levels, southwest Norway (Abstract). *Geonytt*, **22**, 20–21.

BJØRLYKKE, K. 1983. Subsidence and tectonics in late Precambrian and Paleozoic sedimentary basins of southern Norway. *Norges Geologiske Undersøkelse Bulletin*, **380**, 159–172.

BLAIR, T. C. & BILODEAU, W. L. 1988. Development of tectonic cyclothems in rift, pull-apart and foreland basins: sedimentary response to episodic tectonism. *Geology*, **16**, 517–520.

—— & MCPHERSON, J. G. 1994*a*. Alluvial fan processes and forms. *In*: ABRAHAMS, A. D. & PARSONS, A. J. (eds) *Geomorphology of Desert Environments*. Chapman & Hall, London, 354–402.

—— & —— 1994*b*. Alluvial fans and their natural distinctionfrom rivers based on morphology, hydraulic processes, sedimentary processes and facies assemblages. *Journal of Sedimentary Research*, **A64**, 450–489.

BORQUIN, S., RIGOLLET, C. & BOURGES, P. 1998. High-resolution sequence stratigraphy of an alluvial fan–fan delta environment: stratigraphic and geodynamic implications—an example from the Keuper Chaunoy Sandstones, Paris Basin. *Sedimentary Geology*, **121**, 207–237.

BOYER, S. M. & ELLIOT, D. W. 1982. Thrust systems. *American Association of Petroleum Geologists Bulletin*, **66**, 1196–1230.

BROWN, L. F. & FISCHER, W. L. 1977. Seismic-stratigraphic interpretation of depositional systems: examples from Brazilian rift and pull-apart basins. *In*: PAYTON, C. E. (ed.) *Seismic Stratigraphy—Applications to Hydrocarbon Exploration*. American Association of Petroleum Geologists, Memoirs, **26**, 213–248.

BRAATHEN, A. 1999. Kinematics of post-Caledonian polyphasal brittle faulting in the Sunnfjord Region, western Norway. *Tectonophysics*, **302**, 99–121.

BRYHNI, I. 1964*a*. Migrating basins on the Old Red Continent. *Nature*, **202**, 284–285.

—— 1964*b*. Relasjonen mellom senkaledonsk tektonikk og sedimentasjon ved Hornelens og Håsteinens Devon. *Norges Geologiske Undersøkelse, Bulletin*, **223**, 10–25.

—— & SKJERLIE, F. 1975. Syn-depositional tectonism in the Kvamshesten district (Old Red Sandstone), western Norway. *Geological Magazine*, **112**, 593–600.

BURBANK, D., LELAND, J., FIELDING, E., ANDERSON, R. S., BROZOVIC, N., REID, M. R. & DUNCAN, C. 1996. Bedrock incision, rock uplift and threshold hillslopes in the northwestern Himalayas. *Nature*, **379**, 505–510.

——, MEIGS, A. & BROZOVIC, N. 1996. Interactions of growing folds and coeval depositional systems. *Basin Research*, **8**, 199–223.

BURNS, B. A., HELLER, P. L., MARZO, M. & PAOLA, C. 1997. Fluvial response in a sequence stratigraphic framework: example from the Monserrat fan delta, Spain. *Journal of Sedimentary Research*, **67**, 311–321.

CHAUVET, A. & DALLMEYER, R. D. 1992. ^{40}Ar/^{39}Ar mineral dates related to Devonian extension in the southwest Scandinavian Caledonides. *Tectonophysics*, **210**, 2155–177.

—— & SERANNE, M. 1994. Extension-parallel folding in the Scandinavian Caledonides: implications for late-orogenic processes. *Tectonophysics*, **238**, 31–54.

CHEADLE, M. J., MCGEARY, S., WARNER, M. R. & MATTHEWS, D. H. 1987. Extensional structures on the western UK continental shelf: a review of evidence from deep seismic profiling. *In*: COWARD, M. P., DEWEY, J. F. & HANCOCK, P. L. (eds) *Continental Extensional Tectonics*. Geological Society, London, Special Publications, **28**, 19–33.

CROSS, T. A., BAKER, M. R., CHAPIN, M. A. *et al.* 1993. Applications of high-resolution sequence stratigraphy to reservoir analysis. *In*: ESCHARD, R. & DOLIGEZ, B. (eds) *Subsurface Reservoir Characterization from Outcrop Observations*. Technip, Paris, 11–33.

CROWELL, J. C. 1974. Origin of late Cenozoic basins in southern California. *In*: DICKINSON, W. R. (ed.), Tectonics and sedimentation. Society of Economic Paleontologists and Mineralogists, Special Publications, **22**, 190–204.

CURRIE, B. 1997. Sequence stratigraphy of nonmarine Jurassic–Cretaceous rocks, central Cordilleran foreland–basin system. *Geological Society of America Bulletin*, **109**, 1206–1222.

CUTHBERT, S. J. 1991. Evolution of the Devonian Hornelen Basin, Western Norway: new constraints from petrological studies of metamorphic clasts. *In*: MORTON, A. C., TODD, S. P. & HAUGHTON, P. D. W. (eds) *Developments in Sedimentary Provenance Studies*. Geological Society, London, Special Publications, **57**, 343–360.

DE BOER, P. L., PRAGT, J. S. J. & OOST, A. P. 1991. Vertically persistent sedimentary facies boundaries along growth anticlines and climatic control in the thrust-sheet-top, south Pyrenean Tremp–Graus foreland basin. *Basin Research*, **3**, 63–78.

EIDE, E. A., TORSVIK, T. H. & ANDERSEN, T. B. 1997. Ar–Ar geochronologic and paleomagnetic dating of fault breccias; characterization of late Paleozoic and early Cretaceous fault reactivation in Western Norway. *Terra Nova*, **9**, 135–139.

ELIET, P. P. & GAWTHORPE, R. L. 1995. Drainage development and sediment supply within rifts, examples from the Sperchios basin, central Greece. *Journal of the Geological Society, London*, **152**, 883–893.

ETHERIDGE, F. G. 1985. Modern alluvial fans and fan deltas. *In*: FLORES, R. M., ETHRIDGE, F. G.,

MIALL, A. D., GALLOWAY, W. E. & FOUCH, T. D. (eds) *Recognition of Fluvial Systems and their Resource Potential*. Society of Economic Paleontologists and Mineralogists, Short Course, **19**, 101–126.

FOLKESTAD, A. 1995. *Cyclicity in the Hornelen Basin (Devonian), Norway*. Cand. Scient. Thesis, University of Bergen, Bergen, Norway.

—— & STEEL, R. J. In press. *The alluvial cyclicity in the Hornelen Basin (Devonian, western Norway) revisited: a multi parameter sedimentary analysis and stratigraphic implications*. Norwegian Petroleum Society.

FOSSEN, H. 1992. The role of extensional tectonics in the Caledonides of South Norway. *Journal of Structural Geology*, **14**, 1033–1046.

FRIEDMANN, S. J. & BURBANK, D. W. 1995. Rift basins and supra detachment basins: intra-continental extenuation end-members. *Basin Research*, **7**, 109–127.

FRIEND, P. F. 1978. Distinctive features of some ancient river systems. *In*: MIALL, A. D. (ed.) *Fluvial Sedimentology*. Canadian Society of Petroleum Geologists, Memoirs, **5**, 531–542.

GALLOWAY, W. E. 1989. Genetic stratigraphic sequences in basin analysis I: architecture and genesis of flooding-bounded depositional units. *AAPG Bulletin*, **73**, 125–142.

GIBBS, A. D. 1984. Structural evolution of extensional basin margins. *Journal of the Geological Society, London*, **141**, 609–620.

—— 1987. Development of extension and mixed-mode sedimentary basins. *In*: COWARD, M. P., DEWEY, J. F. & HANCOCK, P. L. (eds) *Continental Extensional Tectonics*. Geological Society, London, Special Publications, **28**, 19–33.

GLOPPEN, T.-G. & STEEL, R. J. 1981. The deposits, internal structure and geometry in six alluvial fan–fan delta bodies (Devonian, Norway)—a study in the significance of bedding sequences in conglomerates. *In*: ETHERIDGE, F. & FLORES, R. M. (eds) *Recent and Ancient Non-marine Depositional Environments: Models for Exploration*. Society of Economic Paleontologists and Mineralogists, Special Publications, **31**, 49–69.

GORDON, I. and HELLER, P. L. 1993. Evaluating controls on basinal stratigraphy, Pine Valley, Nevada: implications for syntectonic deposition. *Geological Society of America Bulletin*, **105**, 47–55.

GRIFFIN, W. L., AUSTRHEIM, H., BRASTAD, K. *et al.* 1985. High-pressure metamorphism in the Scandinavian Caledonides. *In*: GEE, D. G. & STURT, B. A. (eds) *The Caledonide Orogen: Scandinavia and Related Areas*. Wiley, New York, 783–801.

HOSSACK, J. R. 1984. The geometry of listric normal faults in the Devonian basins of Sunnfjord, W. Norway. *Journal of the Geological Society, London*, **141**, 629–637.

JARVIK, E. 1949. On the Middle Devonian Crossopterygians from the Hornelen field in western Norway. *Universitetet i Bergen Aarbok, Naturvitenskapelig Rekke*, **8**.

KELLY, S. B. 1992. Milankovitch cyclicity recorded from Devonian non-marine sediments. *Terra Nova*, **4**, 578–584.

—— & OLSEN, H. 1993. Terminal fans—a review with reference to Devonian examples. *Sedimentary Geology*, **85**, 339–374.

KILDAL, E. 1970. *Berggrunnskart 1 : 250 000, Måløy*. Norges Geologiske Undersøkelse, Trondheim.

KOLDERUP, C. F. 1921. Kvamshestens devonfelt. *Bergens Museums Aarbok 1920–1921, Naturvidenskabelig Række*, **4**.

KRABBENDAM, M. & DEWEY, J. F. 1998. Exhumation of UHP rocks by transtension in the Western Gneiss Region, Scandinavian Caledonides. *In*: HOLDSWORTH, R. E., STRACHAN, R. A. & DEWEY, J. F. (eds) *Continental transpression and Transtension Tectonics*. Geological Society, London, Special Publications. **135**, 159–182.

KULLERUD, L., TØRUDBAKKEN, B. O. & ILEBEKK, S. 1986. A compilation of radiometric age determinations from the Western Gneiss Region, South Norway. *Norges Geologiske Undersøkelse, Bulletin*, **406**, 17–42.

LEEDER, M. R. & GAWTHORPE, R. L. 1987. Sedimentary models for extensional tilt-block/half-graben basins. *In*: COWARD, M. P., DEWEY, J. F. & HANCOCK, P. L. (eds) *Continental Extensional Tectonics*. Geological Society, London, Special Publications, **28**, 139–152.

—— & JACKSON, J. A. 1993. The interaction between normal faulting and drainage in active extensional basins, with examples from the western United States and central Greece. *Basin Research*, **5**, 79–102.

——, HARRIS T. & KIRKBY, M. J. 1998. Sediment supply and climate change: implications for basin stratigraphy. *Basin Research*, **10**, 7–18.

——, MACK, G. H. & SALYARD, S. L. 1996. Axial–transverse fluvial interactions in half-graben: Plio-Pleistocene Palomas Basin, southern Rio Grande Rift, New Mexico, USA. *Basin Research*, **12**, 225–241.

LEGARRETA, L., ULIANA, M. A., LAROTONDA, C. A. & MECONI, G. R. 1993. Approaches to nonmarine sequence stratigraphy—theoretical models and examples from Argentine basins. *In*: ESCHARD, R. & DOLIGEZ, B. (eds) *Subsurface Reservoir Characterization from Outcrop Observations*. Technip, Paris, 125–143.

MARTINSEN, O. J. 1993. Namurian (late Carboniferous) depositional systems of the Craven–Askrigg area, northern England: implications for sequence-stratigraphic models. *In*: POSAMENTIER, H. W., SUMMERHAYES, C. P., HAQ, B. U. & ALLEN, G. P. (eds) *Sequence Stratigraphy and Facies Associations*. International Association of Sedimentologists, Special Publications, **18**, 247–281.

MCCLAY, K. R. & SCOTT, A. D. 1991. Experimental models of hanging wall deformation in ramp-flat listric extensional detachment systems. *Tectonophysics*, **188**, 85–96.

MIALL, A. D. 1996. *The Geology of Fluvial Deposits: Sedimentary Facies, Basin Analysis and Petroleum Geology*. Springer, Berlin.

NILSEN, T. H. 1968. The relationship of sedimentation to tectonics in the Solund Devonian District of southwestern Norway. *Norges Geologiske Undersøkelse, Bulletin*, **259**, 1–108.

NORTON, M. G. 1986. Late Caledonian extension in western Norway: a response to extreme crustal thickening. *Tectonics*, **5**, 195–204.

—— 1987. The Nordfjord–Sogn detachment, W. Norway. *Norsk Geologisk Tidsskrift*, **67**, 93–106.

OLSEN, H. 1990. Astronomical forcing of meandering river behaviour: Milankovitch cycles in Devonian of East Greenland. *Palaeogeography, Palaeoclimatology, Palaeoecology*, **79**, 99–115.

OLSEN, T., STEEL, R., HØGSETH, K., SKAR, T. & RØE, S.-L. 1995. Sequential architecture in a fluvial succession: sequence stratigraphy in the upper Cretaceous Mesaverde Group, Price Canyon, Utah. *Journal of Sedimentary Research*, **B65**, 265–280.

OSMUNDSEN, P. T. (1996) *Late-orogenic structural geology and Devonian basin formation in Western Norway: a study from the hanging wall of the Nordfjord–Sogn Detachment in the Sunnfjord region*. Dr. Scient thesis, University of Oslo.

OSMUNDSEN, P. T. & ANDERSEN, T. B. 1994. Caledonian compressional and late-orogenic extensional deformation in the Staveneset area, Sunnfjord, Western Norway. *Journal of Structural Geology*, **16**, 1385–1401.

—— & ANDERSEN, T. B. 2000. The middle Devonian basins of western Norway: sedimentary response to large-scale transtensional tectonics? *Tectonophysics*, in press.

——, ——, MARKUSSEN, S. & SVENDBY, A. K. 1998. Tectonics and sedimentation in the hanging wall of a major extensional detachment: the Devonian Kvamshesten basin, western Norway. *Basin Research*, **10**, 213–234.

POSAMENTIER, H. W. & VAIL, P. R. 1988. Eustatic controls on clastic deposition II—sequence and systems tract models. *In*: WILGUS, C. K., HASTINGS, B. S., KENDALL, C. G. St C., POSAMENTIER, H. W., ROSS, C. A. & VAN WAGONER, J. C. (eds) *Sea-level Changes: an Integrated Approach*. Society of Economic Paleontologists and Mineralogists, Special Publications, **42**, 109–124.

PROSSER, S. 1993. Rift-related linked depositional systems and their seismic expression. *In*: WILLIAMS, G. D. & DOBB, A. (eds) *Tectonics and Seismic Sequence Stratigraphy*. Geological Society, London, Special Publications, **71**, 35–66.

ROBERTS, D. 1983. Devonian tectonic deformation in the Norwegian Caledonides and its regional perspectives. *Norges Geologiske Undersøkelse, Bulletin*, **380**, 85–96.

SADLER, S. P. & KELLY, S. B. 1993. Fluvial processes and cyclicity in terminal fan deposits: an example from the Late Devonian of southwest Ireland. *Sedimentary Geology*, **85**, 375–386.

SCHLISCHE, R. W. 1991. Half-graben filling models; new constraints on continental extensional basin development. *Basin Research*, **3**, 123–141.

—— 1995. Geometry and origin of fault-related folds in extensional settings. *AAPG Bulletin*, **79**, 1661–1678.

SCHWARZACHER, W. 1993. Cyclostratigraphy and the Milankovitch theory. *In*: *Developments in Sedimentology*, **52**, Elsevier, Amsterdam.

SÉRANNE, M. & SEGURET, M. 1987. The Devonian basins of western Norway: tectonics and kinematics of an extending crust. *In*: COWARD, M. P., DEWEY, J. F. & HANCOCK, P. L. (eds) *Continental Extensional Tectonics*. The Geological Society, London, Special Publications, **28**, 537–548.

——, CHAUVET, A., SEGURET, M. & BRUNEL, M. 1989. Tectonics of the Devonian collapse-basins of western Norway. *Bulletin de Société Géologique de France*, **8**, 489–499.

SHANLEY, K. W. & MCCABE, P. J. 1994. Perspectives on the sequence stratigraphy of continental strata. *AAPG Bulletin*, **78**, 544–568.

SKJERLIE, F. J. 1971. Sedimentasjon og tektonisk utvikling i Kvamshesten devonfelt, Vestlandet. *Norges Geologiske Undersøkelse, Bulletin*, **270**, 77–108.

STEEL, R. J. 1976. Devonian basins of western Norway—sedimentary response to tectonism and varying tectonic context. *Tectonophysics*, **36**, 207–224.

—— 1988. Coarsening-upwards and skewed fan bodies: symptoms of strike-slip and transfer fault movement in sedimentary basins. *In*: NEMEC, W. & STEEL, R. J. (eds) *Fan Deltas: Sedimentology and Tectonic Settings*. Blackie, Glasgow, 75–83.

—— & AASHEIM, S. 1978. Alluvial sand deposition in a rapidly subsiding basin (Devonian, Norway). *In*: MIALL, A. D. (ed.) *Fluvial Sedimentology*. Canadian Society of Petroleum Geologists, Memoirs, **5**, 385–413.

—— & GLOPPEN, T. G. 1980. Late Caledonian (Devonian) basin formation, western Norway: signs of strike-slip tectonics during infilling. *In*: READING, H. G. & BALLANCE, P. F. (eds) *Sedimentation in Oblique-slip Mobile Zones*. International Association of Sedimentologists, Special Publication, **4**, 79–103.

——, MÆHLE, S., NILSEN, H., RØE, S. L. & SPINNANGR, Å. 1977. Coarsening-upwards cycles in the alluvium of Hornelen Basin (Devonian), Norway: sedimentary response to tectonic events. *Geological Society of America Bulletin*, **88**, 1124–1134.

——, SIEDLECKA, A. & ROBERTS, D. 1985. The Old Red Sandstone basins of Norway and their deformation: a review. *In*: GEE, D. G. & STURT, B. A. (eds) *The Caledonian Orogen—Scandinavia and Related Areas*. Wiley, Chichester, 293–315.

TORSVIK, T. H., STURT, B. A., RAMSAY, D. M., BERING, D. & FLUGE, P. R. 1988. Paleomagnetism, magnetic fabrics and the structural style of the Hornelen Old Red Sandstone, western Norway. *Journal of the Geological Society, London*, **145**, 413–430.

——, ——, ——, KISCH, H. J. & BERING, D. 1986. The tectonic implications of Solundian (Upper

Devonian) magnetization of the Devonian rocks of Kvamshesten, western Norway. *Earth and Planetary Science Letters*, **80**, 337–347.

——, ——, SWENSSON, E., ANDERSEN, T. B. & DEWEY, J. F. 1992. Palaeomagnetic dating of fault rocks: evidence for Permian and Mesozoic movements and brittle deformation along the extensional Dalsfjord fault, western Norway. *Geophysical Journal International*, **109**, 565–580.

VAIL, P. R. 1975. Eustatic cycles from seismic data for global stratigraphic analysis. *AAPG, Bulletin*, **59**, 2198–2199.

VAN WAGONER, J. C., POSAMENTIER, H. W., MITCHUM, R. M., VAIL, P. R., SARG, J. F., LOUTIT, T. S. & HARDENBOL, J. 1988. An overview of fundamentals in sequence stratigraphy and key definitions. *In*: WILGUS, C. K, HASTINGS, B. S., KENDALL, C. G. St C., POSAMENTIER, H. W., ROSS, C. A. & VAN WAGONER, J. C. (eds) *Sea-level Changes: an Integrated Approach*. Society of Economic Paleontologists and Mineralogists, Special Publications, **42**, 39–45.

WHIPPLE, K. X. & TRAYLER, C. R. 1996. Tectonic control on fan size: the importance of spatially-variable subsidence rates. *Basin Research*, **8**, 351–366.

WILKS, W. H. & CUTHBERT, S. J. 1994. The evolution of the Hornelen Basin detachment system, western Norway: implications for the style of late-orogenic extension in the southern Scandinavian Caledonides. *Tectonophysics*, **238**, 1–30.

Early syndepositional tectonics of East Greenland's Old Red Sandstone basin

EBBE HARTZ

Department of Geology, Box 1047 Blindern, Oslo University, 0316 Oslo, Norway (e-mail: Ebbe.hartz@geologi.uio.no)

Abstract: East Greenland's Old Red Sandstone basin formed in the interior of a megacontinent created by the Caledonian collision of Baltica and Laurentia. The basin has been regarded as a typical late- to post-orogenic extensional basin, formed in a collapsing orogen. Cross-cutting faults that have extended the basin substrata have been explained as Riedel and anti-Riedel shears that formed during basin initiation, but no detailed structural analysis has been presented. Analysis of geometrical relationships between faults and folds, and their relative timing with respect to the syn-tectonic deposits suggest a new model for basin evolution. In eastern Greenland, orogenic collapse was initiated at least as early as *c.* 425 Ma. Thus, the preserved Devonian basins formed after *c.* 70 Ma of large-magnitude crustal extension. Thick successions of Mid-Devonian Old Red Sandstone were deposited in troughs controlled by orogen-parallel extension and east–west folding. Later north–south-trending extensional structures cut the basal deposits, as episodes of folding continued until Permian time. Folding continued for *c.* 175 Ma after the Caledonian continent–continent collision and it is thus unrealistic to suggest that internal Caledonian forces caused these 'late Caledonian spasms'. However, the collapsed orogen may have been thermally weakened, and the region was thus probably vulnerable to external forces resulting from the continuing orogenies along the rim of the megacontinent.

Following the definition of the largely Devonian Old Red Sandstones (ORS) in the British Isles (Conybeare & Phillips 1822; Miller 1841), similar rocks have been found on all continents. One of the largest exposed ORS areas occurs in the central East Greenland Caledonides, where it was first recognized by Nathorst (1901). Since then, this group of terrestrial basins has been the subject of continuing debates on basin evolution (e.g. Bütler 1959; Friend *et al.* 1983; Larsen & Bengaard 1991; Olsen & Larsen 1993*a*; Hartz *et al.* 1996). These interpretations have largely been based on the geometry of the basin deposits and to a lesser degree on detailed structural analysis. The purpose of this study is to evaluate tectonic models for basin initiation, by analysis of cross-cutting relationships between folds, faults and basal deposits. Perhaps the most interesting tectonic features of the basin are the complex relationships between folds and faults. The basin clearly post-dated the major Caledonian structures and is commonly referred to as a collapse feature (McClay *et al.* 1986; Larsen & Bengaard 1991; Olsen 1993; Hartz & Andresen 1995). This basic observation, however, leaves several aspects of the basin evolution open for discussion. A common feature with all the Old Red Sandstone basins in the North Atlantic region is the enigmatic interference between structures related to shortening (folds) and extension (normal faults). Throughout the region, this has been related to an orogen-parallel component of left-lateral shearing in the orogen (Vogt 1936; Harland 1965). East Greenland's Old Red Sandstone deposits provide unique information about the decay of the Caledonian orogen. The basin is large, well exposed at different stratigraphic levels, and the stratigraphy of the pre-Caledonian substrata is easily differentiated. This allows precise evaluation of fault displacements. Furthermore, slickensides are common on fault surfaces; this is essential for kinematic analysis. Perhaps most important, the ORS is overlain by well-exposed Upper Palaeozoic deposits. This allows subdivision into pre-, syn- and post-depositional tectonic phases. The structural analysis of folds and faults in this study is therefore preceded by a discussion of the diachronous nature of the basal deposits related to these structures.

From: Friend, P. F. & Williams, B. P. J. (eds). *New Perspectives on the Old Red Sandstone.* Geological Society, London, Special Publications, **180**, 537–555. 0305-8719/00/$15.00

Regional geology and previous studies

The East Greenland Caledonides can be divided into three major tectonostratigraphic units (Henriksen 1985) (Fig. 1):

(1) Crystalline basement, metasediments and migmatites; these were highly deformed and metamorphosed during the Caledonian Orogeny (Haller 1971) and are generally referred to as gneiss in this paper. They are separated from overlying rocks by a major late Caledonian detachment (Hartz & Andresen 1995).

(2) A folded and faulted *c*. 18 km thick sequence of Upper Proterozoic (the Eleonore Bay Supergroup; EBS) to Middle Ordovician, mostly shallow-marine sediments (Fränkl 1956; Sønderholm & Tirsgaard 1993), generally referred to as pre-Caledonian sediments in this paper. Caledonian granites and mafic dykes intrude these sediments (Haller 1971).

(3) Continental clastic sediments of Devonian to Carboniferous age (Nathorst 1901; Bütler 1957). In the coastal areas of East Greenland, these rocks are covered by a thick sequence of Upper Palaeozoic to Tertiary sedimentary and volcanic rocks (Koch & Haller 1971).

Early stratigraphic subdivisions were either lithostratigraphic, often based on the colour of the sediments (e.g. Kulling 1930), or biostratigraphic (e.g. Säve-Söderbergh 1934; Jarvik 1961). In contrast, Bütler's extensive study of the basin (e.g. 1959), led to a stratigraphic subdivision based on unconformities and their correlative conformities. Bütler's subdivision into five stratigraphic series was in general agreement with the biostratigraphic data, and was defined from unconformities related to folding events. Friend and co-workers further subdivided and partly renamed the basin, in the first modern sedimentological study of depositional facies of the basin (e.g. Friend *et al.* 1976, 1983; Nicholson & Friend 1976). They did, however, largely maintain Bütler's (1959) overall stratigraphic framework. Olsen & Larsen (1993*a*) conducted a basin-wide sedimentological investigation of lithofacies and depositional environment with emphasis on drainage and wind patterns and climate changes. Their outcrop study was combined with regional air-photo interpretation of the basin and its immediate basement (Bengaard 1991). These workers subdivided the basin into four groups and one formation (Fig. 2a and b) (Vilddal, Kap Kolthoff, Kap Graah, Celsius Bjerg and Harder Bjerg (the last is not present in the studied area) (Olsen & Larsen 1993*a*), but most importantly some of the new group boundaries were moved from unconformities and correlative conformities to lithological transitions (Fig. 2b). The new basin division was linked to five basin stages, subdivided into ten chronosomes or complexes, and generally related to tectonic phases along two basin-bounding left-lateral faults, the Western Fault Zone (WFZ) and the Eastern Fault Zone (Fig. 1) (Olsen 1993; Olsen & Larsen 1993*a*, *b*). The latter is today mainly cut by younger faults (Fig. 1). This model attributes the cross-cutting fault pattern in the basin substrata to transtensional NNW–SSE-trending left-lateral Riedel and east–west-trending right-lateral anti-Riedel shears along the north–south-trending WFZ and the folding of the basin deposits to transpression (Larsen & Bengaard 1991; Olsen 1993). This left-lateral fault system was suggested to extend from Northeastern Greenland, through the British Isles, to Labrador.

Researchers who favour a classic hiatus-based stratigraphic subdivision (Hartz *et al.* 1996, 1997, 1998; Gjeldvik *et al.* 1997; Slettemeås & Aubert 1998) have challenged this recent lithostratigraphic subdivision, and this view is elaborated below. Hartz *et al.* (1997, 1998) furthermore argued that the upper basin deposits are younger (*c*. 335 Ma) than previously suggested, based on $^{40}Ar/^{39}Ar$ plagioclase ages from basalts. Stemmerik & Bendix-Almgreen (1998) argued against this young age, on the basis of lithostratigraphic correlation between the basalts and sediments dated by pollen and spores. Marshall *et al.* (1999) used new detailed palynology data to illustrate the diachronous nature of the current lithostratigraphic correlation, and concluded that East Greenland's Old Red Sandstone deposits are Devonian in age.

Stratigraphy of the syntectonic basal deposits

Early models for basin evolution suggested that faults and folds generally become younger eastward, as do the preserved basin deposits (Bütler 1959; Friend *et al.* 1983). Olsen (1993) elaborated this view, dividing the basin evolution into four stages, of which the two youngest are discussed in this paper: (1) the Vilddal stage, characterized by activity along the WFZ and deposition primarily by eastward-draining alluvial fans; (2) the Kap Kolthoff–Kap Graah stage, when the EFZ was active, and north–south-trending folds formed as sandstones, bimodal volcanic rocks and local conglomerates were deposited on a south-dipping palaeoslope. Generally, syn-sedimentary north–south-trending folds continued to be active until the formation of the pronounced unconformity

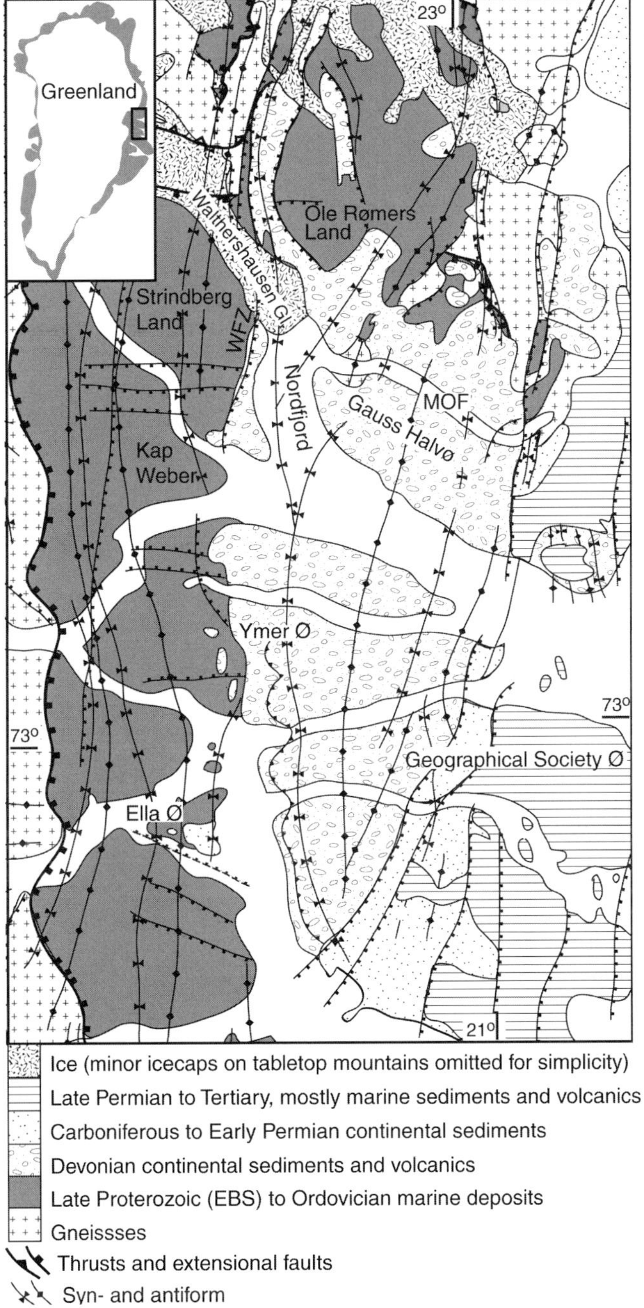

Fig. 1. (**a**) Geological map of central East Greenland, simplified from Haller (1970), Koch & Haller (1971), Bengaard (1991) and own studies. The insert shows Greenland, with the map marks as a frame. WFZ, Western Fault Zone; MOF, Moskusoksefjord.

marking the Late Permian marine transgression (Bütler 1959).

This lithostratigraphic subdivision is demonstrated here using a series of logs through the Vilddal and basal Kap Kolthoff Groups (*sensu* Olsen & Larsen 1993*a*). The stratigraphic examples are from the area around the Moskusoksefjord (MOF) Inlier and its northern

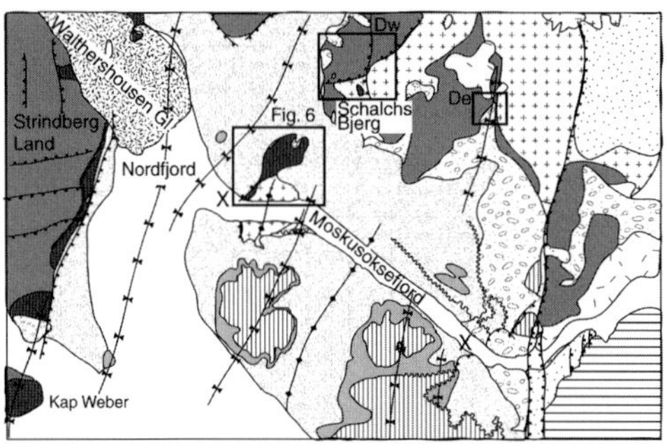

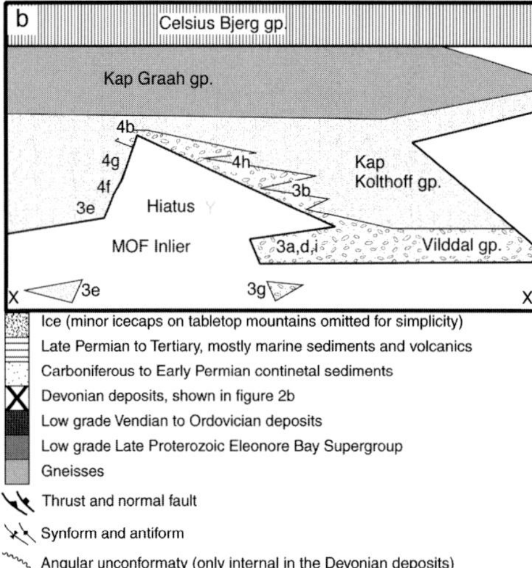

Fig. 2. (**a**) Simplified map of the Moskusoksefjord–Nordfjord area (See Fig. 1 for location). De and Dw, Dybendal East and West. (**b**) Schematic illustration of the current stratigraphic subdivision of the ORS in the MOF area, between X and X′ in (**a**). (Note that the sedimentary groups are highly diachronous.) The positions of the logs in Figs 3 and 4 are indicated.

continuation (Fig. 2a and b), because this region has been studied most extensively. The MOF Inlier is a ridge of high palaeotopography, consisting of gneisses, granites and pre-Caledonian sediments, and is in a sense a remnant of the 'true' Caledonian mountain range.

The MOF Inlier is surrounded by sandstones and conglomerates of the Vilddal and lower Kap Kolthof Groups (Olsen & Larsen 1993*a*), and the first (Vilddal) and the second (Kap Koltholf–Kap Graah) basin stages are defined in this area, for this study. This lithostratigraphic subdivision contrasts with earlier studies (Bütler 1959; Friend *et al.* 1983) in that the pronounced regional unconformities are no longer used to subdivide groups or even formations, and the groups thereby become highly diachronous. The following section presents examples of the basal deposits surrounding the MOF Inlier, illustrating the lithological heterogeneity and diachronous nature of these deposits.

Basal breccias

The lowermost unit within the group is a characteristic massive carbonate-boulder breccia. This is illustrated in the logs, where carbonate breccias occur locally above the unconformity in most localities (Figs. 3a, e, h, i and 4a, b, c, h). The carbonate breccias were previously defined as the 'Basis Group' and locally (SE MOF Inlier) named 'Konglomerat 1' (Bütler 1959) or 'Unit 1' (Friend *et al.* 1983).

Folded and deeply eroded basal deposits

The carbonate breccias are unconformably overlain by the Kap Bull Formation (the 'Konglomerat 2' of Bütler (1959) or 'Unit 2' of Friend *et al.* (1983)). The Kap Bull Formation occurs as a thick sequence of cyclic upward-coarsening sandstones (Fig. 3a and g) to boulder conglomerates (Fig. 3d) that onlap the crystalline rocks of the MOF Inlier, towards the NNW (Bütler 1959; Alexander-Marrack & Friend 1976; Olsen & Larsen 1993*a*; Hartz *et al.* 1996; Gjeldvik *et al.* 1997). The 2D log in Fig. 3j illustrates the cyclic vertical and lateral variations in deposits along the basal unconformity. Cycles containing trough cross-bedded sandstones and boulder conglomerates of probable alluvial fan origin occur from metres scale (as in Fig. 3j) to 100-m scale (Bütler 1959). Clasts in the boulder conglomerates consist primarily of locally derived crystalline rocks (white two-mica granite and red K-feldspar–biotite gneiss), whereas pebbles in the sandstone and fine conglomerates have a most distant origin (Fig. 3k). Palaeocurrent vectors in the alluvial fan deposits are directed away (south) from the unconformity, whereas the sandstones show more variable palaeocurrent directions (Alexander-Marrack & Friend 1976; Hartz *et al.* 1996; Gjeldvik *et al.* 1997). Deposits in the Kap Bull Formation are rotated by folding to dips ranging between 70° eastward to 25° westward, but typically dip *c.* 45° towards the east (Fig. 3g) (Bütler 1959; Alexander-Marrack & Friend 1976; Gjeldvik *et al.* 1997)

Deposits onlapping the Moskusoksefjord Inlier

The pronounced angular unconformity between folded and deeply eroded interbedded conglomerates and sandstones and the overlying subhorizontal debris flows mark the end of the most pronounced folding in the basin (Hudson Land phase 2 (Bütler 1959; Alexander-Marrack & Friend 1976). This unconformity previously marked the base of the Kap Kolthof Group; now, however, the lowermost overlying deposits are named the Genvejsdalen Member and incorporated in the deeply eroded Kap Bull Formation (Vilddal Group of Olsen & Larsen (1993*a*)). The unconformity can be examined in outcrop at one locality, where it is onlapped at a low angle by parallel and cross-bedded, relatively fine-grained conglomerates (Fig. 3b and f). Directly above these follows a massive unit of nearly monomict conglomerate composed entirely of red K-feldspar–biotite gneiss clasts, which is the rock type of the MOF Inlier directly below (Fig. 3k). The highly angular clasts at the base of deposits are often imbricated along their long axis, suggesting deposition by debris flow with a generally eastward transport direction (Fig. 3b). Upward and laterally away from the unconformity (eastward), the clasts become smaller and rounded, as the massive conglomerates interfinger with finer-grained bedded conglomerates and sandstones. In the measured vertical section, the last conglomerates occur 80 m above the unconformity (Fig. 3b).

The coarse fanglomerates exposed at the southeastern corner of the MOF Inlier are the most spectacular in the Vilddal Group. However, proximal fans of Vilddal conglomerate are common near the MOF Inlier and similar inliers near Dybendal (Bengaard 1991; Olsen & Larsen 1993*a*). There is no fundamental difference between the deposits described from the southern MOF and the proximal conglomerates further north. Basal carbonate breccias (Fig. 4a and b) typically compare to those south of the fjord (Fig. 3e and i). In the lowermost deposits, these breccias were folded and eroded before deposition of overlying sandstones and conglomerates (Fig. 4a and b), which onlap the MOF Inlier directly at higher stratigraphic levels. Where the conglomerates trace the eastern rim of the MOF Inlier they have a stratigraphic thickness of *c*. 1500 m and fine laterally into the sandstones of the Kap Kolthoff Group (Fig. 4b, c, g and h). A similar situation occurs near the Devonian topography in Dybendal (Gjeldvik *et al.* 1997) (Fig. 4e, f and i).

At the western side of the MOF Inlier carbonate breccias (Figs 3e and 4f), debris flow deposits (Fig. 3e and h) and bedded conglomerates occur at the base of the Kap Kolthoff Group proximal to the Inlier. These conglomerates were not included in the Vilddal Group by Olsen & Larsen (1993*a*), as their 'mirror-image' rocks at the eastern side of the MOF Inlier (Fig. 3b, d and i).

Lithostratigraphy as event and time-marker

Collectively, the above observations suggest that the Vilddal Group is highly diachronous, includes deposits affected by different structural events, and that it partly fines into the Kap Kolthoff Group. In addition, the Vilddal conglomerates, in occurring 800–1600 m above sea level at Northern Moskusoksefjord Inlier and in Dybendal (Fig. 4c), sit approximately 1–2 km stratigraphically above Kap Kolthoff deposits farther south at sea level along Moskusoksefjord, as the beds generally dip gently northwards (Fig. 4f). Thus, the current lithostratigraphic subdivision records facies variations, but cannot be used to separate tectonic events.

The Kap Bull Formation, which is the first depositional complex (chronosome) of Olsen (1993), serves well to illustrate this. The formation includes basal breccias together with alluvial fan deposits, and braided river and debris flow deposits that were eroded from different source areas (Figs 3 and 4), and that are separated by two distinct unconformities related to phases of east–west shortening (Bütler 1959; Alexander-Marrack & Friend 1976; Friend *et al.* 1983). Furthermore, palaeomagnetic poles vary significantly ($>10°$) across the second unconformity, suggesting that this hiatus represents a relatively long period of erosion and non-deposition (Hartz *et al.* 1997).

Structural evolution of the basal ORS

In earlier models, stratigraphic subdivisions were tied to tectonic events. The earliest tectonic models emphasized folding of the ORS basins (Bütler 1959) generally referred to as 'Late Caledonian spasms' (Haller 1971) (Fig. 5a). Friend *et al.* (1983) proposed a new model based on left-lateral wrench faulting as the underlying control of the East Greenland ORS basins. They suggested that faults along the western, and later along the eastern side of the basin, were active as left-lateral transtensional fault then as transpressional faults, thereby explaining alternating phases of extension and shortening. In contrast to the early shortening (fold)-oriented papers, more recent models for basin evolution have emphasized the extensional collapse of the orogen (McClay *et al.* 1986; Larsen & Bengaard 1991; Olsen 1993; Hartz & Andresen 1995). Larsen & Bengaard (1991) adapted the wrench model of Friend *et al.* (1983) and renamed the basin-bounding faults the Western and Eastern Fault Zones. They further suggested that the cross-cutting faults in the basin substrata could be explained as Riedel and anti-Riedel shears (Fig. 5d). This model involves a number of predictions: (1) faults are oriented NNW–SSE (left-lateral Riedel shears) and east–west (right-lateral anti-Riedel shears) (Fig. 5d); (2) the WFZ is a major left-lateral fault, active during the initial phases of basin evolution; (3) folds should be oriented NNE–SSW.

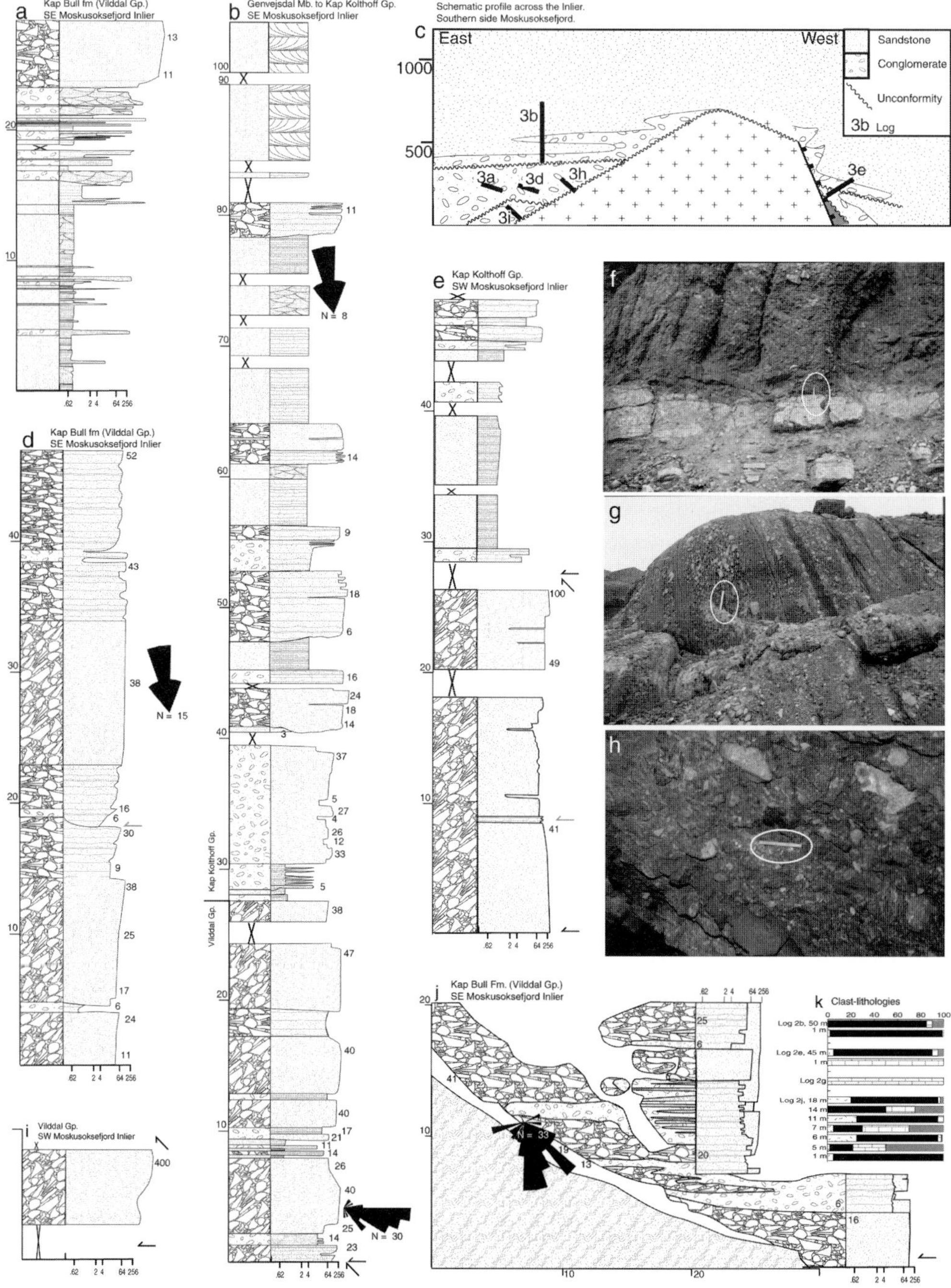

In the following sections of this paper, structural data from folds and faults are presented from the MOF Inlier, Dybendal, Strindberg Land and Ella Ø (Fig. 1), to test proposed tectonic models. Detailed mapping shows that although cross-cutting relationships between folds and faults are complex, they are still systematic. For the sake of clarity, the faults are grouped into three populations. F1 faults are syndepositional normal faults with very large displacements. These were typically active as orogen transverse faults (trending *c*. ESE), but are often rotated along folds to other apparent directions. F2 faults are typically orthogonal, showing a 'dog-leg' geometry, with an overall NNE trend. F2 faults both pre- and post-date folding. F3 faults are minor late faults, which are unconstrained upwards in age. These faults typically reactivate earlier structures, but generally consists of a conjugate pair of ESE-trending left-lateral faults and WSW-trending right-lateral faults.

Western Fault Zone at Strindberg Land

The WFZ marks the western side of the basin and plays a key role in all earlier models of basin evolution. In the northern part of the basin it has been variably interpreted as: (1) an east-dipping late Devonian normal fault cutting folded pre-Devonian sediments at a low angle (Bütler 1959); (2) a late Devonian left-lateral oblique normal fault (Friend *et al.* 1983); (3) a right-lateral strike-slip fault (Bengaard 1989); (4) an early to middle Devonian left-lateral strike-slip fault (Larsen & Bengaard 1991; Olsen 1993).

The WFZ can be traced from Ole Rømers Land and southwards across Strindberg Land, where it cuts into the Devonian deposits, to Ymer Ø (Fig. 1). From Ymer Ø, the WFZ continues below Kong Oscars Fjord. Further south, it is cut by younger faults.

Detailed mapping of several profiles across the WFZ confirm observations from classic studies (Bütler 1959; Friend *et al.* 1983). Basal breccias typical for the Vilddal Group directly onlap Ordovician carbonate rocks (Fig. 5b). Both the Devonian rocks and their substrata are folded. Similar onlap relationships occur at a small inlier in the Nordfjord Graben and along the western margin of the MOF Inlier (Fig. 6f), suggesting that the Nordfjord 'Graben' in fact is a pre- to syndepositional synform cut by younger normal faults. The hiatus between the Ordovician carbonate rocks and the lower Devonian deposits is the shortest hiatus observed in the central East Greenland Caledonides excluding major pre-depositional footwall uplift along the WFZ.

There is thus no evidence for a major pre- to syn-sedimentary fault zone juxtaposing Devonian deposits with their substrata at Strindberg Land. However, the Devonian deposits are cut by multiple sets of oblique extensional faults that trend in almost all directions. Three major fault zones (oriented *c*. 030/70) that occur *c*. 1 km into the basin (Fig. 5b), are instead suggested to be the WFZ, which can be traced north and southwards from Strindberg Land in a dog-leg pattern (strike 350°–030°) (Fig. 1). There are no obvious piercing points along this fault zone, and kinematic analysis therefore has to rely on other criteria.

The large core-faults in the WFZ show extensional offsets by drag folding (Fig. 5b), which alone accounts for *c*. 200 m of offset. Axes of the drag folds are calculated to plunge gently (*c*. 10°) NNE, suggesting (but not proving) a predominant dip-slip component. Slickensides on the fault surfaces show considerable dispersion (150° difference in rake on some single fault

Fig. 3. Sedimentary logs through the Vilddal Group and Kap Kolthoff Group (Olsen & Larsen 1993*a*) along the southern side of MOF. (Note the heterogeneity of the sediments in each group, and that each group is highly diachronous.) (**a**) Coarsening-upward sequence of the Kap Bull Fm. (**b**) The basal debris flows (Genvejsdalen Mb (Vilddal Gp)) above pronounced unconformity. Genvejsdalen Mb interfingering eastwards with the Kap Kolthoff Gp, which also occurs above the basal deposits. (**c**) Schematic profile across Inlier at the southern side of the fjord, showing approximate location of logs. (**d**) Massive conglomerate wedges in the Kap Bull Fm. (**e**) The lowermost deposits (Kap Kolthoff Gp) along the southwestern side of the Inlier. The carbonate breccia of the lowermost 26 m of the column occurs only south of south-dipping normal fault. North of the fault the conglomeratic sandstones (at 28 m) onlap the substrata directly, thereby illustrating the syn-extensional nature of the sediments. (**f**) The angular unconformity between the bedded pale conglomerates (granites and sedimentary clasts) and sandstones of the Kap Bull Fm overlapped by dark (gneiss clasts) mass flow conglomerates of the Genvejsdalen Mb (20 cm scale in ring). (**g**) Conglomerates and sandstones of the Kap Bull Fm rotated to a 70° dip towards the east (20 cm scale in ring). (**h**) Gneiss conglomerate at the base of the Kap Kolthoff Group at the SW side of the MOF Inlier (base log 2e). (**i**) Carbonate breccia of the 'Basis Gp' (20 cm scale in ring). (**j**) Lateral and vertical log showing the onlap between the Kap Bull Fm and basement. (Note how conglomerates interfinger with finer sediments on a scale of metres, similar to the megascale interfingering of the entire formation. The two coarseness scales represent the sediments along the unconformity, and laterally (SE) from it.) (**k**) Clast lithologies in logs (50 clasts counted at each locality).

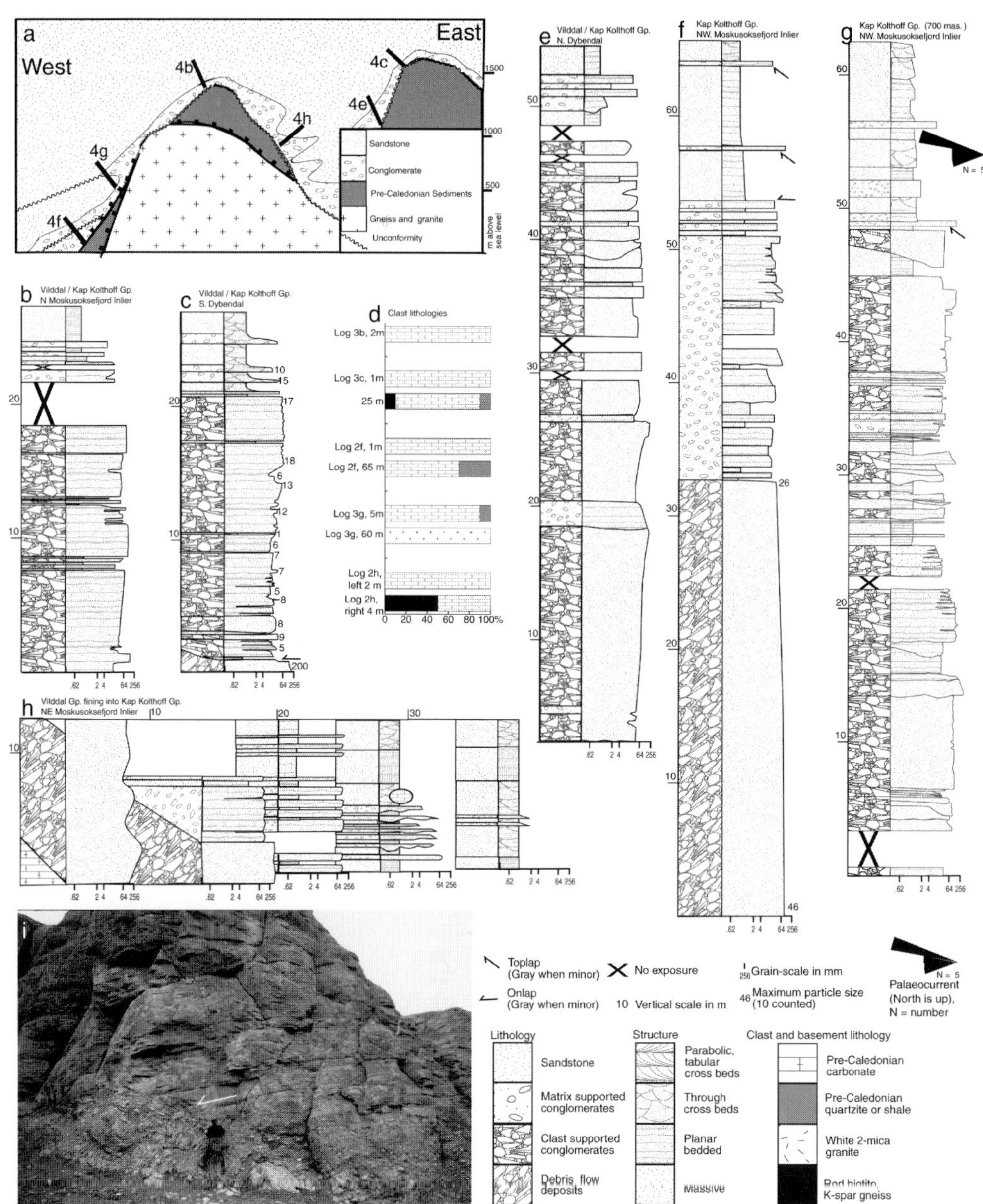

Fig. 4. Logs showing the conglomerates surrounding the northern side of the MOF Inlier fining into sandstones. In some localities (**b**, **c**, **e**, **h** and **i**) the conglomerates are mapped as Vilddal Gp, and in others (**f**, **g**) not (Olsen & Larsen 1993*a*). (**a**) Schematic cross-section of the northern side of MOF, showing the Inlier and Dybendal (see Figs 5 and 6 for localities). (**b**) Basal conglomerates at the northwestern side of the Inlier. (**c**) Syntectonic conglomerates shed from a west-dipping F2 fault scarp south of Dybendal. (**d**) Clast lithologies in logs (50 clasts counted at each locality). (**e**) Conglomerates (Vilddal Gp) along the eastern side of the Dybendal graben, interfingering with the Kap Kolthoff Gp. (**f**) Syntectonic deposits at the northwestern side of the Inlier (see Fig. 6e for setting). (**g**) Syntectonic deposits 700 m above sea level at the northwestern side of the Inlier. (**h**) Lateral logs along a partly reactivated fault scarp, at the NE side of the Inlier. (Note how the coarse carbonate breccias of the Vilddal Gp laterally fine into polymict conglomerates (with some outsized cobbles), to become sandstone (Kap Kolthoff Gp).) (**i**) Base of log (**c**) view towards SE. (Note the mass-flow deposits at the base, onlapped by bedded conglomerates that fine laterally, and fine upward into sandstone.)

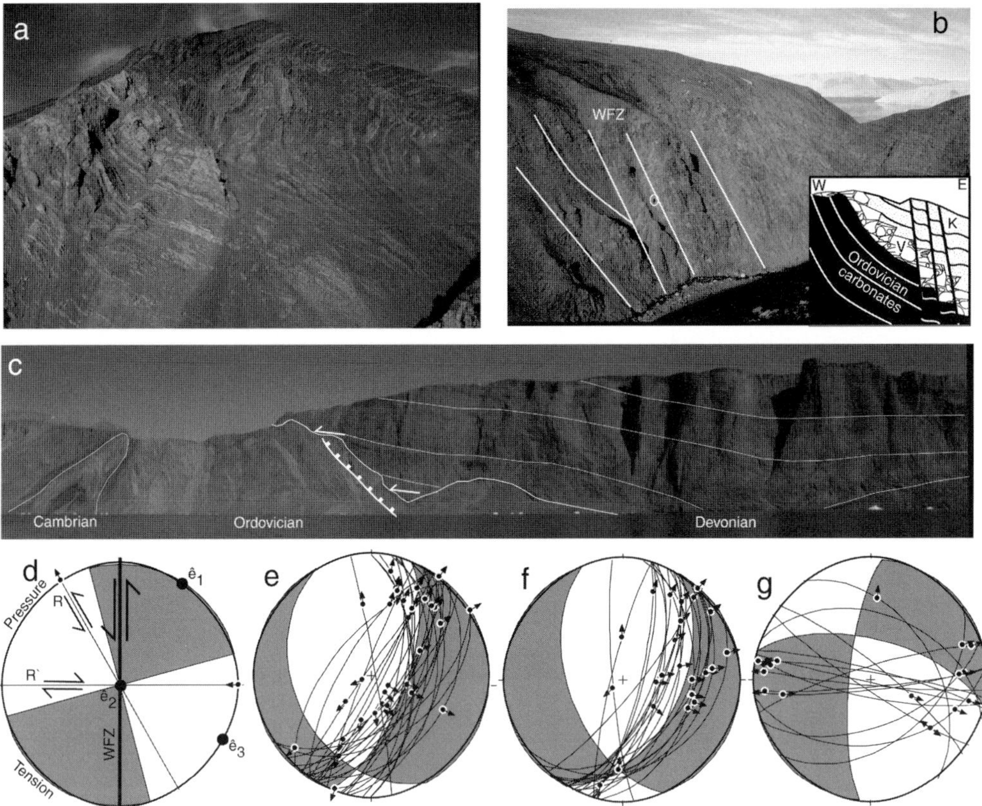

Fig. 5. (**a**) Folded Kap Kolthoff deposits at Geographical Society Ø, viewed towards north. Vergence is towards east. (**b**) Western Fault Zone at Strindberg Land is composed of three parallel fault zones, each oriented *c*. 020/70. One of these fault zones is shown in a view towards the NE. The fault zones are clearly post-depositional, and drag beds into parallelism with the fault planes. Person for scale (white ring) in the core of the fault zone. The small inserted figure illustrates the synsedimentary folding and post-depositional faulting. (**c**) Nathorst's (1901) classic locality at southern Ella Ø showing Devonian onlap against the substrata. A bedding-parallel fault is mapped directly in the field. Fault motion pre-dates the sedimentation above the upper pointer. (Note that the folds in the Devonian deposits are syn-depositional and die out upwards.) Stereonets (equal area, lower hemisphere) show faults with slickensides (small arrows), and fields of extension (dark) and shortening (white) (also illustrated by the convention that extension (ê) defines axes in the strain ellipsoid: ê1 > ê2 > ê3 in (**d**). Only faults with a clear interference relationship with other major faults or with Devonian deposits are presented. (**d**) Plots of hypothetical left-lateral Riedel shears (R), and right-lateral anti-Riedel shears (R′) (Larsen & Bengaard 1991), developed as a response to the north-trending WFZ. (**e**) Faults (F2) in the core of the WFZ. (**f**) Minor fault zones outside WFZ. (**g**) Young faults (F3) cross-cutting the WFZ. (Note that the fault patterns, even directly at the WFZ, do not describe an R–R′ pattern.)

planes). However, generally there is a normal component, and typically the slickensides vary from steeply oblique right lateral, through dip-slip to direct left-lateral (Fig. 5e). Weighting of slickenside data in a quantitative analysis of fault slip is complex (Marrett & Almendinger 1990); however, when each measurement is weighed equally, the WFZ is an oblique normal fault, with *c*. 20° left-lateral component. With an estimate of <1 km of dip-slip, based on the stratigraphic argument above, the left-lateral offset on the WFZ is 300–400 m (dip-slip × tan(90° − average rake)).

Data from minor faults in the Devonian deposits generally display the same properties as those from the main fault (Fig. 5f). Faults with an approximate east–west orientation cut the fault planes of the main WFZ (Fig. 5g). These faults typically show left-lateral displacements when striking north of east, and right-lateral displacement when striking south of east. This suggests that they are part of an orthorhombic strike-slip

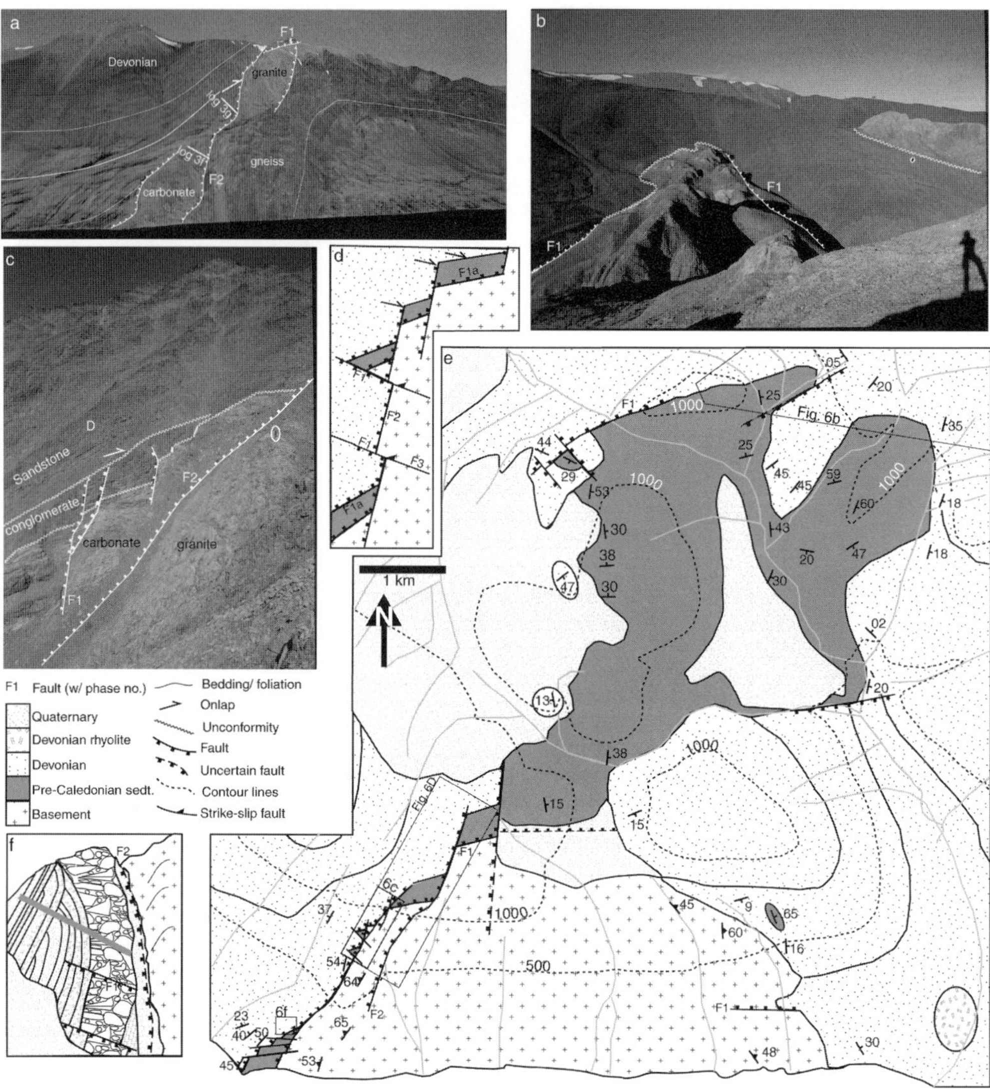

Fig. 6. Fault patterns near Moskusoksefjord (MOF) Inlier. (**a**) The southwestern side of the MOF Inlier. (Note folding of both gneisses and Devonian deposits.) (**b**) View towards ESE of the northernmost segment of the Inlier. (Note the 'palaeomountains' of carbonate deposits, onlapped along fault scarps by beds, often with a thin rim of Vilddal conglomerate fining into sandstones of the Kap Kolthoff Gp.) F1 faults cut by F2 faults, along the NW Inlier (see (**e**) for location). Person in white ring for scale. (**d**) Enlarged schematic map showing fault relationships (see (**e**) for location). (**e**) Map of the northern side of MOF Inlier. (**f**) View towards north syn-depositional F1 faults, and associated toplap unconformities, cut by the F2 fault. (Note the drag folding of the gneisses.) Position on log in Fig. 3(**f**) marked by grey line.

fault system. These faults are, however, relatively small scale because their offsets are generally less than the magnitude of the outcrop (< 10 m).

The WFZ cuts faults in the footwall with an approximate east–west trend (Fig. 1). Some of these faults show major offset and juxtapose Ordovician sediments with basement gneisses (Koch & Haller 1971). If a basement block has been uplifted along such faults before displacement along the WFZ, the apparent offset along the WFZ is impressive, yet the actual fault displacement, as estimated from the minimum stratigraphic offset, can be seen to be minor in Strindberg Land.

Syndepositional folding and faulting on Ella Ø

Ella Ø (Fig. 1a) was the first area in which the basal Devonian unconformity was described in East Greenland (Nathorst 1901), and the locality has since then been described in several stratigraphic studies (Bütler 1935; Yeats & Friend 1978; Larsen & Bengaard 1991; Marshall & Stephenson 1997). The important point to notice, for the present structural analysis, is that both folds and faults are syndepositional. The island displays an impressive sequence of upright, gently south-plunging synforms and antiforms. The easternmost synform is onlapped by Devonian conglomerate beds that are themselves gently folded, so that some beds thickens *c.* 25% in the core of the synform (Fig. 5c). This suggests that folding was syndepositional. The fjord south of Ella Ø is inferred to include a normal fault (Koch & Haller 1971). On the basis of stratigraphic correlation, the fault must have >5 km of vertical offset. The orientation of the fjord constrains the trend of the fault to approximately WNW, which is consistent with minor faults exposed on nearby land. Clast lithologies in the conglomerates (Yeats & Friend 1978), show that rock types present in the footwall of this fault become increasingly common upwards, and thereby represent a reverse stratigraphy of the footwall. This suggests that the NNE-dipping fault south of Ella Ø also was syndepositional, collectively illustrating synchronous NNE–SSW extension and WNW–ESE folding.

Cross-cutting faults at Moskusoksefjord Inlier

The MOF Inlier is a dome of basement rocks and pre-Caledonian (Vendian to Ordovician) sediments occurring within the ORS basin in west-central MOF (Figs 2a and 6a). The migmatitic gneisses have been uplifted from more than 18 km depth after anatexis at *c.* 425 Ma and before exposure and deposition in the Middle Devonian time (Hartz *et al.* 2000). In contrast, the crustal thickness lost by erosion of the Ordovician rocks before the Devonian deposition is relatively small. The northern part of the Inlier has been mapped in detail, with emphasis on fault relationships. With cross-cutting bedding and fault planes, piercing points can be accurately located, and three phases of offset can be documented.

F1 faults. The earliest faults mapped in the area (F1) typically cut the pre-Caledonian sediments the basal Devonian carbonate breccia (Figs 6b, c, d, e, f and 7a) and parts of the lower section of polymict Devonian conglomerates; here, F1 faults can clearly be shown to be syndepositional (Fig. 6c and f), but some of the F1 faults have been reactivated by F3 faults, thereby appearing to cut F2 faults (Fig. 6d). Folding along with the overlying sediments (Fig. 7b and d) has rotated the F1 faults. When rotated back to their syndepositional orientations, along with the overlying sediments, these faults appear to have been active as NNE- or SSW-dipping normal faults (Fig. 7c). Faults of this orientation are not the most obvious along the subparallel (east–west-trending) fjord; however, it is the east–west-trending faults or fault scarps that control the geometry of the structure. Other prominent examples of east–west-trending palaeo-fault scarps include the eroded ENE-trending horst along the north side of the Inlier (Fig. 6b), and similar structures along the south-dipping normal fault in the central part of the Inlier (Fig. 6e).

F2 faults. An F2 fault forms the western limit of the Inlier, where rocks in both the footwall and hanging wall have been drag folded (Fig. 6a). Above 1000 m elevation the sediments onlap the fault scarp. The syndepositional nature of the F2 fault is particularly well demonstrated south of the fjord, as the fault does not cut the higher Kap Kolthoff and Kap Graah Group deposits southwards across Gauss Halvø (Koch & Haller 1971) (Fig. 2a). The F2 faults are subvertical today (Fig. 7f), as a result of rotation by folding (Fig. 7e). Rotation back with the beds suggests that they were active with a shallower dip (Fig. 7g).

F3 faults. The F1 and F2 faults are cut by a set of conjugate strike-slip faults (F3), formed by steep SSW-dipping left-lateral faults and SSE-dipping right-lateral faults (Fig. 7i). Some of these reactivate pre-existing faults (Figs 6d and 7h). Piercing points (F2 faults and bedding), illustrate that the total offset typically was less than 20 m. That the displacement represents two separate events is also demonstrated by the observation that syn F1 conglomerates have no granite clasts and thus must have been rotated and juxtaposed to the granites by the F2 fault before deposition of the overlying sediments, which do contain granitic material.

Syn-extensional folding at Eastern Dybendal

The deep valley of Dybendal runs parallel to the MOF, *c.* 15 km farther north (Fig. 1b). Folded

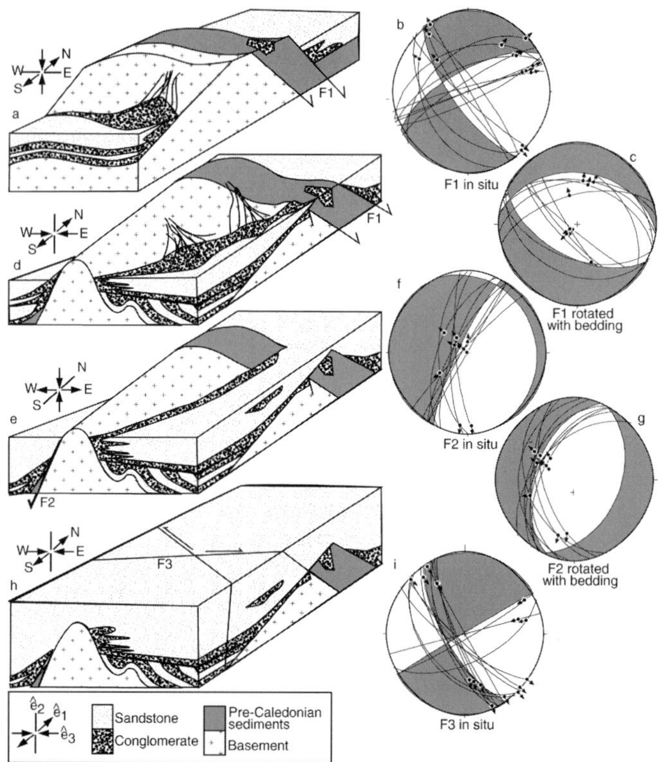

Fig. 7. Schematic block diagram for Moskusoksefjord Inlier illustrating the syn-extensional folding and faulting of the ORS basins in East Greenland. The diagram also illustrates the conglomates rimming the Inlier (typically the Vilddal Gp), and interfingering with sandstones (Kap Kolthof Gp). Stereonets show faults and slickensides, and associated areas of extension and compression (see Fig. 5 for explanation). The major early faults (F1) trend east–west (**a**), and were formed by orogen-parallel extension, during continued orogen-normal shortening. These faults are today rotated (**b**), but were active as ENE–WSW-trending normal faults (**c**). Faults and bedding are rotated in two stages along with unconformities (bedding in the lower section is 200/80, overlain by 225/75 and 223/34 in the upper section). (**d**) Basal conglomerates are folded and eroded, forming major top-lap unconformities in the basin, during continued east–west shortening and north–south extension. (**e**) Later the deposits are cut by F2 faults (**f** and **g**), formed by orogen-normal extension. The F2 faults are rotated back to active orientation along with the least rotated overlying sediments (223/34). (**h**) Conjugate F3 strike-slip faults cut all younger structures.

Devonian deposits occur in synclines of pre-Devonian carbonate rocks in the high walls of the valley (Fig. 8a), and the timing of structures can be interpreted by using the geometrical relationships between the sedimentary bedding, folds and faults.

F1 faults. The earliest faults (F1) here also trend *c*. ESE–WNW. Along the eastern side of the syncline, the F1 faults juxtapose carbonate rocks of the uppermost Eleonore Bay Supergroup with migmatitic gneisses that originally were separated by more than 10 km of sediments (Fig. 8a). F1 faults also occur along the western side of the syncline, where they cut the onlapped flank of the fold. In this locality, the F1 faults can be shown to be syndepositional because the tops of some of the uplifted carbonate footwall fault blocks have been eroded and onlapped as fault scarps (Fig. 8c and g), producing the curved boundary zone apparent in map view (Fig. 8b). The carbonate rocks below the Devonian deposits are folded more tightly than the onlapping Devonian deposits (Fig. 8b, c, g and h), and internal onlaps in the trough (Bütler 1957) suggest collectively that the structure formed as a growth-syncline. Large (up to 20 m) carbonate blocks occur in the basin deposits along the flanks of these folds, suggesting considerable palaeotopography (Fig. 8c). Because the sediments are synchronous with both the folding and the F1 faults, the east–west shortening and north–south extension must also

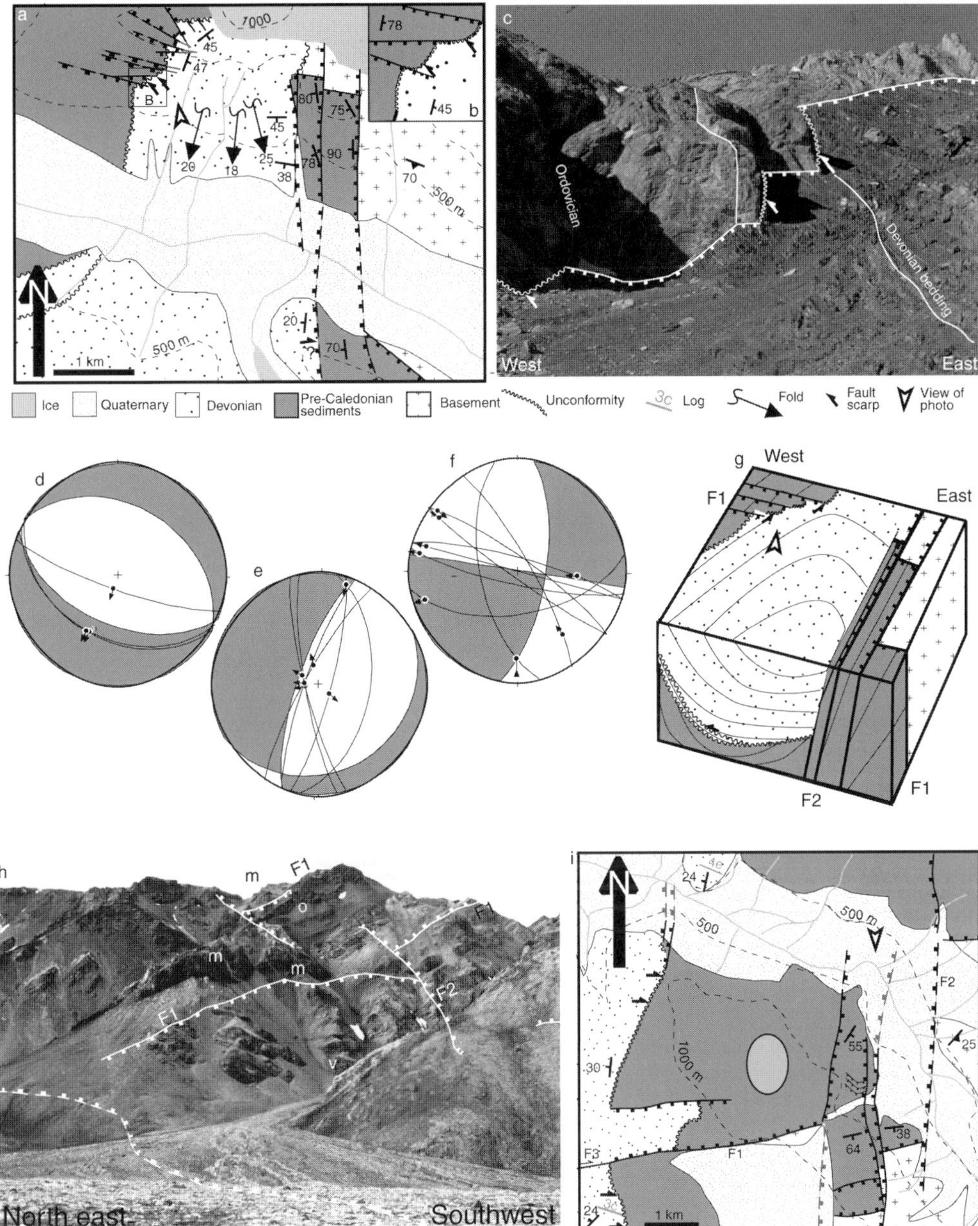

Fig. 8. Fault patterns in the Devonian substrata in east–central Dybendal (**a**) (including (**b**) enlarged map of F1 faults reactivated by F3 faults). (**c**) Northward view of the northernmost F1 faults at the NW corner of the basin. (Note the large blocks of carbonate deposits sitting in the basinal red beds and their 45° dip.) Stereonets show faults with slickensides and fields of strain. (**d**) F1 faults rotated back with bedding ($S_0 = 017/45$). Only the three fault scarps that clearly are synsedimentary are included. (**e**) F2 faults in their current position along the eastern side of the basin. The faults were probably less steep when they were active (overlying beds are folded); however, rotation cannot be quantified. (**f**) F3 faults cut the folded unconformity. Some of these faults reactivate the F1 faults. (**g**) Schematic block diagram illustrating the structural relationships between faults, folds and sediments. (**h**) F2 faults cutting F1 at the southern side of central Dybendal. (See text for detailed description.) The top of the mountain is about 700 m above the valley. m, migmatitic gneiss; O, Ordovician carbonate deposits; V, Vendian tillites. (**i**) Map of the same faults.

be broadly synchronous. When the bedding is rotated to horizontal, F1 faults rotate to trend ESE–WNW (Fig. 8d).

F2 faults. Along the eastern side of the syncline, the F1 faults are cut by west-dipping F2 faults (Fig. 8a). Devonian deposits at the northernmost extension of the faults directly overlie the F2 faults, and the conglomerates in this upper section contain clasts of crystalline rocks. The beds onlapping the F1 fault scarps at the western side of the syncline can be traced to the eastern side of the syncline, where they are cut by F2 faults, confirming the age relationship between the F1 and F2 faults. The F2 faults are subvertical, and even in some localities overturned to a steep reverse fault orientation, probably during the folding of the basin (Fig. 8e and g). Although the north–south-trending F2 faults dominate the map pattern, their stratigraphic offsets are far less than those of the F1 faults.

F3 faults. Conjugate sets of F3 faults cut the western side of the synform with left-lateral offset along ESE-trending faults, and right-lateral offset along SSE-trending faults (Fig. 8a and f). Some of these reactivate F1 faults. In map view the apparent lateral offset is large (*c*. 100 m) on some of these faults (Fig. 8b). However, normal (F1) offset of the east-dipping flank of the synform created this map pattern before F3 reactivation. Folding of F1 horsts and grabens would enhance this map effect. The lateral displacement on the F3 faults cannot directly be quantified; however, the faults cannot be traced across to the eastern side of the basin, suggesting that F3 displacements were relatively small.

Cross-cutting faults at west–central Dybendal

In west–central Dybendal (Fig. 8i and h) a cross-cutting pattern of faults again reveals a three-stage thinning of the Devonian substrata.

F1 faults. Migmatitic gneisses are juxtaposed against Ordovician carbonate deposits by NNW-dipping F1 faults (Fig. 8h and i). The pre-extensional stratigraphic separation between gneisses and Ordovician carbonate deposits (*c*. 18 km, (Fänkl 1956; Sønderholm & Tirsgaard 1993)) illustrates the considerable displacement along these faults. The F1 faults are syndepositional, as can be illustrated southwestwards from Dybendal to Schalks Bjerg (Fig. 2a), a 'palaeomountain' of carbonate deposits cut by several smaller east–west-trending grabens filled with basal breccias (Fig. 8i). This 'palaeomountain' (Schalks Bjerg), is terminated southward by a major south-dipping F1 fault, displacing Devonian and Ordovician rocks down to the south (Fig. 2a). Small-scale F1 faults are generally oriented ESE–WNW, but with some dog-leg kinks (trending between SE–NW and ENE–WSW).

F2 faults. East or west-dipping F2 faults cut the F1 faults. This relationship is visible on both sides of the valley, but is particularly obvious along the southern side of the valley (Fig. 8h–j). Further west, a steep, west-dipping F2 fault juxtaposes Devonian conglomerates against Cambro-Ordovician sediments (Gjeldvik *et al*. 1997). The fault cuts the deposits in the lower section of the wall in Dybendal, but is onlapped by coarse conglomerates as a fault scarp higher in the section (Figs 4c, i and 8i). The conglomerates interfinger with crossbedded sandstone to the west (Gjeldvik *et al*. 1997).

F3 faults. The F1 faults are, at least in one case, reactivated as steeply north-dipping right-lateral F3 faults (Fig. 8i). However, the lateral displacement (F3) is relatively minor (tens of metres) (Gjeldvik *et al*. 1997).

Tectonic history of East Greenland's ORS basin

From the first discovery of Devonian deposits in East Greenland it has been noted that the sediments occur in synclines in the substrata (Nathorst 1901), and that they are themselves folded (Bütler 1935, 1957, 1959; Friend *et al*. 1976, 1983; Olsen 1993; Olsen & Larsen 1993*b*) (Fig. 4c). However, such short-wavelength folds cannot, by themselves, have exhumed deep crustal rocks. Gneisses previously buried under almost 18 km of pre-Caledonian strata (Fränkl 1956; Sønderholm & Tirsgaard 1997) are overlain by Middle Devonian strata, and provide evidence of major extensional thinning. Extensional thinning began at least by *c*. 425 Ma (Hartz *et al*. 2000), so the exhumation rates must have been high (*c*. 0.5 cm a^{-1}), through Early Devonian time.

F1 faults. Major crustal extension may appear contradictory in a folded basin, but the structural data are in fact consistent. Synchronous SSW-trending folds and ESE–WNW-trending F1 normal faults occur at Ella Ø (Fig. 5c), in Dybendal (Fig. 8g) and in the MOF (Fig. 6a).

Furthermore, the Kap Bull Formation at the southwestern side of the MOF Inlier was probably deposited southwards from an F1 fault scarp or its rotated hanging wall (Hartz *et al.* 1996), and then folded and eroded before deposition of the Genvejsdalen Member (Bütler's (1959) Hudson Land phase 2). This again illustrates that the F1 faults and early folds are synchronous and syndepositional. It is likely that the F1 faults may have cut pre-existing extensional faults or detachments that had thinned the upper crust, but such a relationship has so far not been directly mapped in the field. The F1 faults thus formed by major NNE–SSW extension, during ESE–WNW shortening recorded by the folds (Fig. 7b).

F2 faults. The F2 faults trend approximately north–south, but with a regional dog-leg (orthorhombic) pattern (Figs 1b and 7e). Orthorhombic fault systems suggest an oblate strain ellipsoid, with vertical flattening, and extension oriented both north–south and east–west (maximum elongation) (Reches 1983). This suggest a decrease in east–west oriented stress after the F1 faults formed. The overall extension along the F2 faults is much less than that along the F1 faults. A particularly good example of this relationship is found along the WFZ (Fig. 5b). The structure predominantly represents an onlap contact against the flank of a synform, which (post-Kap Kolthoff Group) was activated later by the east-dipping WFZ with minor displacement (Bütler 1959; Friend *et al.* 1983). This is in contrast to a recent basin model, which suggested that the WFZ is a prominent strike-slip fault controlling the early phases of basin development (Larsen & Bengaard 1991; Olsen 1993). The fault lacks piercing points so precise lateral offset cannot be determined. Slickensides suggest that the WFZ does show some lateral movement, yet locally, both right- and left-lateral motion can be demonstrated (Fig. 5e). To what extent the slickensides represent a quantitatively accurate determination of strain is a matter of debate (e.g. Cashman & Ellis 1994), and some of the subhorizontal slickensides on the faults may result from F3 reactivation of the zone as argued below. However, taking the drag folding and slickensides at face value, the fault is an oblique normal fault with a minor left-lateral component of slip. East–west-trending faults west of the WFZ have been interpreted to represent faults that were dragged into parallelism (north–south) with the WFZ over a distance of 15 km on the basis of air-photo interpretation (Bengaard 1991). The curved faults are used as evidence for major left-lateral displacement along the WFZ (Larsen & Bengaard 1991). However, field examination of the area shows that much of the bedrock is under cover, and the curvature may be an artefact of cross-cutting faults, as previously mapped (Katz 1952). A flower structure at Kap Weber (Fig. 1a) has also been postulated as evidence for left-lateral faulting along the WFZ (Larsen & Bengaard 1991). However, judged from the fjord, this structure is an east-verging overturned fold cut by extensional faults, as mapped previously (Fränkl 1956).

F3 faults. Cross-cutting ESE-trending right lateral F3 faults are fairly common in the WFZ. Some ENE-striking left-lateral faults also cut the WFZ (Fig. 5g). However, the NNE-trending left-lateral F3 faults are oriented subparallel to the WFZ, and thus probably reactivated this zone, causing some (all?) of the sub-horizontal slickensides on the F2 planes. The conjugate F3 fault sets occur throughout the mapped area, but seldom with significant offset. These faults cut folds associated with the early Hudson Land phases, named by Bütler (1959). Interference with the younger folds has not been studied. Generally the conjugate F3 faults show NNE–SSW extension (Figs 5g and 7h, i), which compares with the orogen-parallel extension recorded by F1 faults.

Wrench faulting or strain partitioning

The north–south to NE–SW-trending folds and potentially both F1 and F3 faults could have been caused by overall left-lateral transtension and transpression along the orogen (Friend *et al.* 1983; Larsen & Bengaard 1991; Olsen 1993; Hartz *et al.* 1997). However, neither the timing nor the geometry of the faults fits the model of Riedel and anti-Riedel shears (Larsen & Bengaard 1991).

There are two end-member models for folding related to strike-slip faulting. One model involves wrench faulting, where folds first form with the axis 45° to the strike-slip faults, perpendicular to maximum shortening in the strain ellipsoid, and later rotate towards parallelism. The second model predicts fault-parallel folds and reverse faults caused by strain partitioning between strike-slip faulting (simple shear) and a fault-perpendicular component of shortening (pure shear) (Miller 1998). Minor rotation around vertical axes is recorded by palaeomagnetic data, and may relate to wrenching (Hartz *et al.* 1997). Further palaeomagnetic analysis is needed to constrain this effect;

however, it seems unrealistic that the north–south-trending fold structures, which in some cases continue for >200 km (Koch & Haller 1971), have rotated 45° around the vertical axis. Strain partitioning rather than wrenching thus seems to be the most realistic model for fold formation during left-lateral shearing along the orogen.

Regional comparisons

Collapse of the upper crust of the central East Greenland Caledonides led to the formation of a major east-dipping detachment fault that was active from *c.* 425 Ma (Hartz *et al.* 2000). After *c.* 45 Ma of this extensional collapse, with little preserved evidence of lasting deposition, the Devonian ORS basins formed in the hanging wall (east) of the detachment with orogen-parallel extension along east–west-trending F1 faults that formed during continued east–west shortening. Extensional faults with an orthogonal geometry suggest an oblate strain ellipsoid (Resches 1983). The orogen-parallel F2 faults thus record a decrease (or possible eastward migration) of east–west shortening, which is in accordance with earlier observations based on fold geometries (Bütler 1957, 1959; Friend *et al.* 1983; Olsen 1993).

These late phases of basin folding may relate to left-lateral faulting (Friend *et al.* 1983; Larsen & Bengaard 1991; Olsen 1993; Hartz *et al.* 1997), yet the specific structural model, or regional tectonic cause for these events is not immediately apparent. All the Devonian basins in the North Atlantic region show a left-lateral component of faulting (Vogt 1936) yet the geometries of the folding are very different. The Devonian basins in western Norway are folded with axes perpendicular to the orogen, probably resulting from constrictional strain (Fletcher & Bartley 1994) partitioned with strike-slip faulting (Chauvet & Séranne 1994; Hartz & Andresen 1997; Krabbendam & Dewey 1998; Osmundsen *et al.* 1998, this volume). In contrast, the ORS basins in the Laurentian Caledonides (Northern Britain (e.g. Bluck 1974), Svalbard (Harland 1965; Friend & Moody-Stuart 1972; McCann this volume) and Greenland (Bütler 1959)), are folded with axes parallel to the orogen, thereby favouring a model of direct strain partitioning between simple shear (strike-slip) and pure shear (folding).

The Caledonian Orogeny is related to closing of the Iapetus Ocean and collision of Baltica and Laurentia. The Caledonian continental collisional event peaked in Late Silurian time and the term 'late orogenic spasms' (Haller 1971) has therefore been used to describe Middle Devonian to Lower Permian structures (Bütler 1935). However, more recent work relates the Caledonian Orogeny to the closure of the Iapetus Ocean, and final underthrusting of Baltica below Laurentia, in a process that is suggested to have ended in Early Devonian time by delamination of the subducted orogenic root (Andersen & Jamtveit 1990). If there was no convergence between Laurentia and Baltica driven by Caledonian destruction of Iapetus after the main orogenic event, then models for the 'spasms' need an alternative external mechanism. By that time the East Greenland Caledonides had undergone a period of extreme crustal thinning, involving a thermally weakened lower crust that remained hot through Early Devonian time (Hartz *et al.* 2000). This thinned crust and high heat flow is also recorded in the ORS basins, by the widespread bimodal volcanic activity (Bütler 1935). Collectively, this would suggest that the Caledonian orogen was a weak zone, surrounded by colder and stiffer plates (Laurentia and Baltica). Such a zone would be vulnerable to long-distance transmission of deformation. In terms of time, the deformation recorded in East Greenland overlaps with the Devonian closure of the Rheic Ocean (Van der Voo 1988) and with the Late Palaeozoic Variscan–Alleghanian Orogeny related to the continuing amalgamation of Pangaea (Rast 1989). 'Internal' Caledonian causes of these folds seem unlikely compared with external causes.

Conclusion

In spite of nearly 100 years of research on East Greenland's ORS basins, many problems are unsolved, and stratigraphic and tectonic models have had relatively short 'shelf-lives'. The basins are, like all the ORS basins in the North Atlantic region, considered late Caledonian extensional basins, but are folded.

Coarse clastic sediments poured into these north–south-trending synforms as they formed, and, in some cases onlap major north- or south-dipping faults, suggesting that these early basins were initiated by orogen-parallel extension, during orogen-normal shortening (Fig. 7a and b). These basal deposits are today referred to as the Vilddal Group (Olsen & Larsen 1993*a*), a highly diachronous unit typically deposited proximal to steep palaeotopography (fault scarps of fold flanks). The north–south-trending faults that today dominate the map pattern crosscut the east–west-trending faults (Fig. 7b), and probably mark a decrease of, or eastward migration of the orogen-normal shortening.

The minor orthorhombic strike-slip faults that cut these two early fault sets may be related to the latest of these east–west shortening events.

'Late Caledonian spasms' (Haller 1971) fold sediments older than Late Permian in age. If these folds relate to continuing Caledonian destruction of Iapetus and convergence between Baltica and Laurentia, they would record a very long, 175 Ma, orogeny. However, sections of the Caledonian orogen may have been left thermally weakened following its collapse, and may therefore have been susceptible to the effects of the distant Acadian and Variscan–Alleghanian continental collisions.

Discussions, comments and co-operation in the field with A. Andresen, P. T. Osmundsen, G. Gjeldvik and A. McCann were vital for this study. Furthermore, reviewers A. G. Smith, P. F. Friend and A. Maloof are thanked for constructive criticism. I thank the Danish Polar Centre, and Sirius (the Danish military dog-sled patrol) for logistical assistance. Fieldwork was sponsored by the Norwegian Science Foundation, BAT and VISTA. P. Hoffman is thanked for inviting me to spend my sabbatical at Harvard University, where this paper was written.

References

ALEXANDER-MARRACK, P. D. & FRIEND, P. F. 1976. Devonian sediments of East Greenland, III, The eastern sequence, Vilddal Supergroup and part of the Kap Kolthoff Supergroup. *Meddelelser om Grønland*, **206**(3), 1–122.

ANDERSEN, T. B. & JAMTVEIT, B. 1990. Uplift of deep crust during orogenic extensional collapse: a model based on field studies in the Sogn–Sunnfjord region of Western Norway. *Tectonics*, **9**, 1097–1111.

BENGAARD, H.-J. 1989. *Geometrical and geological analysis of photogrammetrically measured deformed sediments of the fjord zone, central East Greenland.* Raport, Grønlands Geologiske Undersøgelser, **6**, 1–101.

—— 1991. Upper Proterozoic (Eleonore Bay Supergroup) to Devonian, central fjord zone, East Greenland (1:250 000). Grønlands Geologiske Undersøgelser, Copenhagen.

BLUCK, B. J. 1974. Sedimentation in the late orogenic basins: the Old Red Sandstone of the Midland Valley of Scotland. *In*: BOWES, D. R. & LEAKE, B. E. (eds) *Crustal Evolution of the Northwestern Britain and Adjoining Regions. Journal of the Geological Society, London, Special Issue*, **10**, 249–278.

BÜTLER, H. 1935. Some new investigations of the Devonian stratigraphy and the tectonics of East Greenland. *Meddelelser om Grønland*, **103**(2), 1–35.

—— 1957. Beobachtungen an der hauptbruchzone der Küste von zentral–ost Grönland. *Meddelelser om Grønland*, **160**(1), 1–79.

—— 1959. Das Old Red-Gebiet am Moskusoksefjord. *Meddelelser om Grønland*, **160**(5), 1–182.

CASHMAN, P. H. & ELLIS, M. A. 1994. Fault interaction may generate multiple slip vectors on a single fault surface. *Geology*, **22**, 1123–1126.

CHAUVET, A. & SÉRANNE, M. 1994. Extension-parallel folding in the Scandinavian Caledonides: implications for late-orogenic processes. *Tectonophysics*, **238**, 31–54.

CONYBEARE, W. D. & PHILLIPS, W. 1822. *Outlines of the Geology of England and Wales*, W. Phillips, London.

FLETCHER, J. M. & BARTLEY, J. M. 1994. Constrictional strain in a non-coaxial shear zone: implications for fold and rock fabric development, central Mojave metamorphic core complex, California. *Journal of Structural Geology*, **16**(4), 555–570.

FRÄNKL, E. 1956. Some general remarks on the Caledonian chain of East Greenland. *Meddelelser om Grønland*, **103**(11), 1–43.

FRIEND, P. F. & MOODY-STUART, M. 1972. Sedimentation of the Wood Bay Formation (Devonian) of Spitsbergen: regional analysis of the late orogenic basin. *Norsk Polarinstitutts Skrifter*, **157**, 1–77.

——, ALEXANDER-MARRACK, P. D., ALLEN, K. C., NICHOLSON, J. & YEATS, A. K. 1983. Devonian sediments of East Greenland. VI. Review of results. *Meddelelser om Grønland*, **206**(6), 1–96.

——, ——, NICHOLSON, J. & YEATS, A. K. 1976. Devonian sediments of East Greenland. I. Introduction, classification of sequences, petrographic notes. *Meddelelser om Grønland*, **206**(1), 1–56.

GJELDVIK, G., HARTZ, E. H., MCCANN, A., ANDRESEN, A. & OSMUNDSEN, P. T. 1997. The lower Middle Devonian deposits on Hudson Land–Gauss Halvø, East Greenland: a record of prolonged tectonic activity. *Geonytt*, **24**(1), 42.

HALLER, J. 1970. Tectonic map of East Greenland (1:500 000). An account of tectonism, plutonism, and volcanism in East Greenland, *Meddelelser om Grønland*, **171**(5), 1–286.

—— 1971. *Geology of the East Greenland Caledonides.* Interscience, New York.

HARLAND, W. B. 1965. The tectonic evolution of the Arctic–North Atlantic region. *Philosophical Transactions of the Royal Society of London, Series A*, **258**, 59–75.

HARTZ, E. H. & ANDRESEN, A. 1995. Caledonian sole thrust of central east Greenland: a crustal-scale Devonian extensional detachment? *Geology*, **23**(7), 637–640.

—— & —— 1997. From collision to collapse: complex strain permutations in the hinterland of the Scandinavian Caledonides. *Journal of Geophysical Research*, **102**(B11), 24697–24711.

——, ——, HODGES, K. V. & MARTIN, M. W. 2000. The Fjord Region Detachment Zone, a long-lived extensional fault in the East Greenland Caledonides. *Journal of the Geological Society, London*, **157**, 795–809.

——, OSMUNDSEN, P. T. & ANDRESEN, A. 1996. Structural control of the Devonian basin in the

East Greenland Caledonides. *Geologiska Föreningens i Stockholm Förhandlingar*, **118**, 37–38.

——, Torsvik, T. H. & Andresen, A. 1997. Carboniferous age for the East Greenland 'Devonian' basin: paleomagnetic and isotopic constraints on age, stratigraphy, and plate reconstructions. *Geology*, **25**(8), 675–678.

——, —— & —— 1998. Reply to comment: Carboniferous age for the East Greenland 'Devonian' basin: paleomagnetic and isotopic constraints on age, stratigraphy, and plate reconstructions. *Geology*, **26**, 285–286.

Henriksen, N. 1985. The Caledonides of central East Greenland 70–76 N. *In*: Gee, D. G. & Sturt, B. A. (eds) *The Caledonide Orogen, Scandinavia and Related Areas*, Wiley, Chichester, 1095–1013.

Jarvik, E. 1961. Devonian vertebrates. *In*: Raasch, G. O. (ed.) *Geology of the Arctic, Vol. 1*, Toronto University Press, Toronto, Ont., 197–204.

Katz, H. R. 1952. Geologie von Strindbergs Land (NE-Grønland). *Meddelelser om Grønland*, **111**, 1–150.

Koch, L. & Haller, J. 1971. Geological map of East Greenland 72°–76° N. *Meddelelser om Grønland*, **183**.

Krabbendam, M. & Dewey, J. F. 1998. Exhumation of UHP rocks by transtension in the Western Gneiss Region, Scandinavian Caledonides. *In*: Holdsworth, R. E., Strachan, R. A. & Dewey, J. F. (eds) *Continental Transpressional and Transtensional Tectonics*. Geological Society, London, Special Publications, **135**, 159–181.

Kulling, O. 1930. Stratigraphic studies of the geology of Northeast Greenland (preliminary report). *Meddelelser om Grønland*, **74**(13), 317–346.

Larsen, P. H. & Bengaard, H. J. 1991. The Devonian basin initiation in East Greenland: a result of sinistral wrench faulting and Caledonian extensional collapse. *Journal of the Geological Society London*, **148**, 355–368.

Marrett, R. & Allmendinger, R. W. 1990. Kinematic analysis of fault-slip data. *Journal of Structural Geology*, **12**(8), 973–986.

Marshall, J. E. A. & Stephenson, B. J. 1997. Sedimentological responses to basin initiation in the Devonian of East Greenland. *Sedimentology*, **44**, 407–419.

——, Astin, T. R. & Clack, J. A. 1999. East Greenland tetrapods are Devonian in age, *Geology*, **27**(7), 637–640.

McCann, A. 2000. Deformation of the Old Red Sandstone of NW Spitsbergen; links to the Ellesmerian and Caledonian orogenies. This volume.

McClay, K. R., Norton, M. G., Coney, P. & Davis, G. H. 1986. Collapse of the Caledonide orogen and the Old Red Sandstones. *Nature*, **323**, 147–149.

Miller, D. D. 1998. Distributed shear, rotation, and partitioned strain along the San Andreas fault, central California. *Geology*, **26**(10), 867–870.

Miller, H. (ed.) 1841. *The Old Red Sandstone; or New Walks in an Old Field*. Constable, Edinburgh.

Nathorst, A. G. 1901. Bidrag til nordøst Grønlands geologi. *Geologiska Föreningens i Stockholm Förhandlingar*, **23**(207), 275–306.

Nicholson, J. & Friend, P. F. 1976. Devonian sediments of East Greenland. V. The central sequence, Kap Graah Group and the Mount Celsius Supergroup. *Meddelelser om Grønland*, **206**(5), 117.

Olsen, H. 1993. Sedimentary basin analysis of the continental Devonian basin in North-East Greenland. *Bulletin Grønlands Geologiske Undersøgelser*, **168**, 80.

—— & Larsen, P.-H. 1993*a*. Lithostratigraphy of the continental Devonian sediments in North-East Greenland. *Bulletin Grønlands Geologiske Undersøgelse*, **165**, 108.

—— & —— 1993*b*. Structural and climatic controls on fluvial depositional systems: Devonian, North-East Greenland. *In*: Marzo, M. & Puigdefabregas, C. (eds) *Alluvial Sedimentation*. International Association of Sedimentologists, Special Publications, **17**, 401–423.

Osmundsen, P. T., Andersen, T. B., Markussen, S. & Svendby, A. K. 1998. Tectonics and sedimentation in the hangingwall of a major extensional detachment: the Devonian basin, western Norway. *Basin Research*, **10**, 213–234.

——, Bakke, B., Svendby, A. K. & Andersen, T. B. 2000. Architecture of the Middle Devonian Kvamshesten Group, western Norway: sedimentary response to deformation above a ramp-flat extensional fault. This volume.

Rast, N. 1989. The evolution of the Appalachian chain. *In*: Bally, A. W. & Palmer, A. R. (eds) *The Geology of North America—An Overview, A*. Geological Society of America, Boulder, CO, 323–348.

Reches, Z. 1983. Faulting in rocks in three-dimensional strain fields. *Tectonophysics*, **95**, 133–156.

Säve-Söderbergh, G. 1934. Further contributions to the Devonian stratigraphy of East Greenland. *Meddelelser om Grønland*, **96**(2), 1–72.

Slettemeås, T. & Aubert, H. G. 1998. Development of unconformities in the 'Old Red Sandstone' deposits of Eastern Moskusoksefjord, East Greenland. *In*: Frederiksen, K. S. & Thrane, K. (eds) *Symposium on Caledonian Geology in East Greenland*, GEUS, Copenhagen, 29.

Sønderholm, M. & Tirsgaard, H. 1993. Lithostratigraphic framework of the upper Proterozoic Eleonore Bay Supergroup of East and North Greenland. *Bulletin Grønlands Geologiske Undersøgelse*, **167**, 1–38.

Stemmerik, L. & Bendix-Almgreen, S. E. 1998. Comment on: Carboniferous age for the East Greenland 'Devonian' basin: paleomagnetic and isotopic constraints on age, stratigraphy, and plate reconstructions. *Geology*, **26**, 284

Van der Voo, R. 1988. Paleozoic paleogeography of North America, Gondwana and intervening displaced terranes—comparisons of paleomagnetism with paleoclimatology and biostratigraphic patterns. *Geological Society of America Bulletin*, **100**, 311–324.

VOGT, T. 1936. *Orogenesis in the Region of Paleozoic Folding of Scandinavia and Spitsbergen*. Report of the 16th International Geologic Congress, Washington, 1933. Washington, DC, 953–955.

YEATS, A. K. & FRIEND, P. F. 1978. Devonian sediments of East Greenland, IV, the Western sequence. Kap Kolthoff Supergroup of the western areas. *Meddelelser om Grønland*, **206**(4), 1–112.

Fossils from the Celsius Bjerg Group, Late Devonian sequence, East Greenland; significance and sedimentological distribution

J. A. CLACK & S. L. NEININGER

University Museum of Zoology, Downing Street, Cambridge CB2 3EJ, UK

(e-mail: j.a.clack@zoo.cam.ac.uk)

Abstract: Recent collections of fossils from the Upper Devonian sequence of East Greenland have increased our understanding of the origin and evolution of limbed vertebrates (tetrapods). An expedition in 1998 collected additional material to study the faunal and sedimentological context of the vertebrates from the Celsius Bjerg Group. Specimens collected include new, articulated material of ichthyostegids, a skull of the lungfish *Jarvikia* and the first record of acanthodian spines and scales. Invertebrates were found in these formations for the first time. Vertebrates occur most often in flood-scours, transported from elsewhere and usually disarticulated. Occasionally, more complete specimens occur, which may have been either desiccated before transport or come from more local environments.

Vertebrate fossils from the Upper Devonian sequence of East Greenland were first discovered by Nathorst and reported by Woodward (1901), but the fauna has been held in special regard by palaeontologists since it was found to include the tetrapod genus *Ichthyostega*. For many years this was the earliest known tetrapod (Säve-Söderbergh 1932*a*) and it has found its way into popular literature as well as textbooks and scientific journals as the 'four-legged fish' (Jarvik 1952, 1980, 1996). In 1952 it was joined by a second genus, *Acanthostega* (Jarvik 1952). The history of the expeditions to collect these vertebrates was documented in lively detail by Jarvik (1996) in his monograph on *Ichthyostega*.

Following a serendipitous find of more *Acanthostega* remains in 1970 by Nicholson (Nicholson & Friend 1976), an expedition in 1987 collected many more specimens of that animal from Nicholson's site (Clack 1988; Bendix-Almgreen *et al.* 1990). From these, almost the complete skeletal anatomy of *Acanthostega* has been described in recent years (Clack 1994*a*, 1998; Coates 1996; Ahlberg & Clack 1998), and it has provided some radical new views on the origin of tetrapods (Coates & Clack 1990, 1991, 1995; Clack & Coates 1995; Clack 1997*a*). A good popular account of the discovery, description and significance of this material has been given by Zimmer (1998). Phylogenetic analysis has almost always placed *Acanthostega* as more primitive (plesiomorphic) than *Ichthyostega*, and it has provided a much more informative model of an early tetrapod than has *Ichthyostega* (Coates 1996). Among the most significant discoveries were the multidigited limbs in the Greenland Devonian forms and in the Russian Devonian tetrapod *Tulerpeton* (Lebedev & Coates 1995) providing insight into the origin of limbs and digits (Coates 1994, 1995); the possession of internal gills by *Acanthostega* (Coates & Clack 1991), and the form of its stapes and braincase, which have thrown light on the evolution of the middle ear in tetrapods (Clack 1989, 1994*b*, *b*).

By contrast, despite its widely known status, *Ichthyostega* has proved problematic in many aspects of its anatomy, and it has so far yielded little insight into early tetrapod evolution. In many respects it is a highly specialized form with many apparently unique features, to which the epithets 'bizarre' or 'enigmatic' have been applied. Several key anatomical features remain unknown, such as the hand, the cervical vertebrae and the sacrum, and the braincase as interpreted by Jarvik (1996) appears not to resemble that of any other contemporary vertebrate. The braincase is currently being restudied by one of the authors (J.A.C.) and P. E. Ahlberg. Even the proportions of the limbs to each other and the rest of the body, and the stance and possible gait that these imply, are disputable, and have been reconstructed in different ways by different researchers (Jarvik 1952, 1980, 1996; Bjerring 1985; Coates & Clack 1995) (Fig. 1). Resolution of these questions is critical for studies of the acquisition of early tetrapod locomotion.

From: FRIEND, P. F. & WILLIAMS, B. P. J. (eds). *New Perspectives on the Old Red Sandstone.* Geological Society, London, Special Publications, **180**, 557–566. 0305-8719/00/$15.00

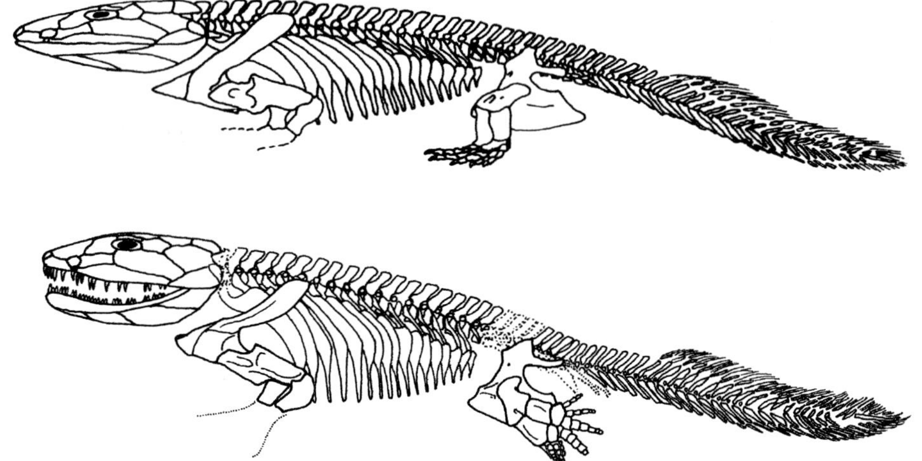

Fig. 1. Alternative reconstructions of *Ichthyostega*: (**a**) from Jarvik (1980), showing limbs restored as pentadactyl, and a stout hind limb supporting the body; (**b**) from Coates & Clack (1995), showing the reduced and paddle-like hindlimb and indicating unknown regions of the postcranium. Skull from Jarvik (1996).

Säve-Söderbergh (1932*a*) named four species of *Ichthyostega* as well as recognizing two unnamed ones ('spp. a and b') and the genus *Ichthyostegopsis*. All but one derive from the Eastern Plateau on the northern slope of Celsius Bjerg and from about the same altitude (Säve-Söderbergh 1932*b*). The exception comes from a site on the central part of the northern slope, not very far away. Given their close proximity and presumably similar horizon of origin, the differences Säve-Söderbergh used to separate them are most likely to be preservational. Jarvik (1996) suggested that they should all be regarded as *Ichthyostega stensioei* or cf. *stensioei*. Nevertheless, there are proportional and dentitional differences detectable in the existing material (pers. obs.), which have to be explained. Possibilities could be age or ontogenetic variation in a single population, stratigraphic separation of two or more species, or the existence of two morphs representing sexual or some other form of dimorphism. Only detailed study will resolve this issue.

Study of *Acanthostega* in particular has stimulated discovery of other Devonian tetrapods elsewhere in the world (Ahlberg 1991, 1995, 1998; Ahlberg *et al.* 1994; Daeschler *et al.* 1994), but most of these new genera are known only from fragments. *Acanthostega* and *Ichthyostega* remain anatomically the most primitive tetrapod taxa known from substantial numbers of fossils.

Other elements of this important vertebrate fauna have received less attention than the tetrapods, and were last reviewed by Bendix-Almgreen (1976). Numerous genera are represented in the Upper Devonian succession generally, but this paper is mainly concerned with those of the Aina Dal, Wimans Bjerg and Britta Dal Formations ('*Remigolepis* Group', Nicholson & Friend 1976) from the upper part of the Upper Devonian sequence. These formations are part of the Celsius Bjerg Group as defined by Olsen (1993) and Olsen & Larsen (1993) (Fig. 2). Here, the genera represented include the tetrapodomorph fish *Eusthenodon* (Jarvik 1985), the porolepiform *Holoptychius*, the long-snouted lungfishes *Soederberghia* and *Oervigia*, the short-snouted lungfish *Jarvikia*, and the placoderms *Remigolepis* and *Bothriolepis*. The placoderm *Groenlandaspis* occurs in the overlying '*Grönlandaspis* series' (Bendix-Almgreen 1976), equivalent to the Stensiö Bjerg Formation (Olsen 1993; Olsen & Larsen 1993). Acanthodian scales and the short-snouted lungfish *Nielsenia* have been found in the '*Phyllolepis* series' (Bendix-Almgreen 1976) but it is not clear whether they are from a horizon that equates to the more recently distinguished Agda or Elsa Dal Formations (Olsen & Larsen 1993) or one from the underlying Kap Graah sandstones. Actinopterygians have been found in the Stensiö Bjerg Formation ('*Grönlandaspis* series', Bendix-Almgreen 1976) above the Britta Dal Formation but not, so far, within it. A spine mentioned by Clack (1994*a*) as possibly belonging to a ctenacanth shark may have been misidentified and requires more detailed analysis.

Since 1976, studies of Devonian fishes, especially sarcopterygians, in other parts of the world have made significant progress (Schultze &

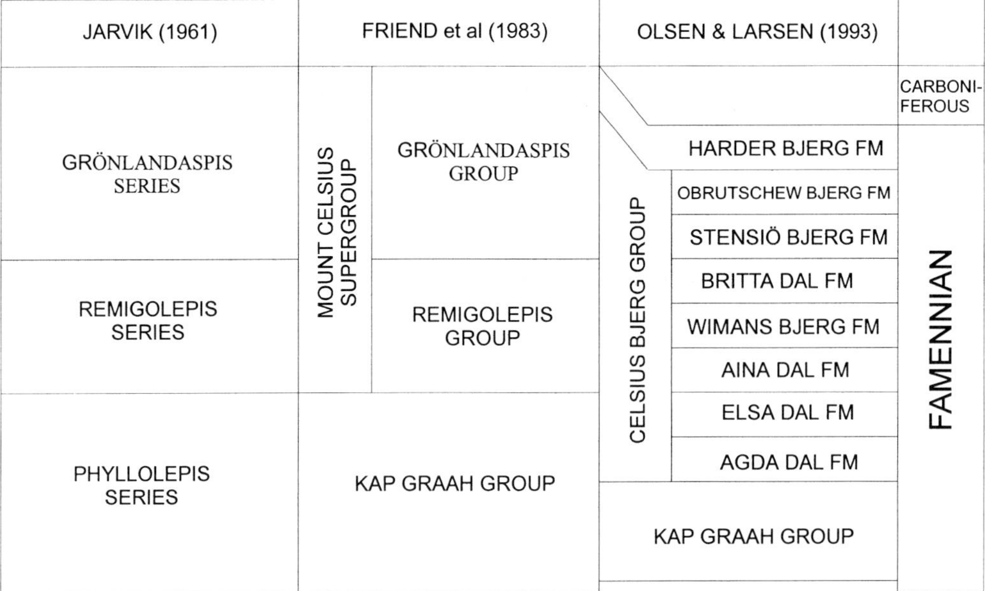

Fig. 2. Stratigraphy of the Fammenian deposits of central East Greenland, giving alternative nomenclatures and correlations.

Arsenault 1985; Vorobyeva & Schultze 1991; Lebedev 1995; Ahlberg & Johanson 1997; Johanson & Ahlberg 1997, 1998; Schultze & Chorn 1998) and ideas of their interrelationships have changed (Cloutier & Ahlberg 1997; Ahlberg & Johanson 1998). Reassessment of the fish faunas from the Upper Devonian of East Greenland would make a timely study, especially in comparison with those of the USA, the Baltic region, Russia and Australia.

No invertebrates have been reported from the Celsius Bjerg Group. Jarvik (1961) mentioned '*Estheria*' (*Asmuzzia*) from the Middle Devonian succession of the Vilddal Series (Givetian). Arthropod tracks and resting traces were described by Nicholson (Nicholson & Friend 1976) from the older Kap Graah Formation.

Many questions regarding the vertebrate faunas and the tetrapods in particular remained to be answered. For example, although *Acanthostega* has been described from an *in situ* horizon (Bendix-Almgreen *et al* 1990), interpretation of the sedimentary environment of the deposit may be challenged (Astin & Marshall, pers. comm.). Furthermore, although *Ichthyostega* remains have been recovered from an *in situ* horizon on Sederholm Bjerg (Jarvik (Johansson) 1935), the sedimentological context has not been described in detail. The sedimentological and stratigraphical relationships of the tetrapods and the other vertebrates to one another also remain unclear; reliable conclusions have been hampered by the fact that most specimens are derived from talus slopes and are of uncertain provenance. Additionally, although the sediments in which the vertebrates have been found have been much studied, there are still problems in relating the stratigraphical units of the main tetrapod localities to one another. Palaeoecological studies, stratigraphical distribution and taxonomic analysis of the vertebrates all depend to some degree on these problems being resolved. Most recently, the dating of the sediments has been disputed (Hartz, this volume).

During the field season of 1998 an expedition organized from the University of Cambridge (UK) under the auspices of GEUS (the Denmark and Greenland Geological Survey), revisited two of the vertebrate localities, to try to address some of these problems

Gauss Halvø

On Gauss Halvø, two formations preserve tetrapod remains. The lower Aina Dal Formation and the upper Britta Dal Formation, which are readily distinguishable (Fig. 2), are separated by the largely unfossiliferous Wimans Bjerg Formation. This triple division can be traced along the SW-facing slopes of Smith Woodward Bjerg, Stensiö Bjerg and Wiman Bjerg, fading out along the flanks of Obrutschew Bjerg

Fig. 3. Photograph of the Aina Dal Formation cliffs at the base of Stensiö Bjerg.

(an inappropriate transliteration of the name of the Russian palaeoichthyologist V. Obruchev). Along Stensiö and Wiman Bjergs, the Aina Dal Formation is exposed in a cliff varying from 10 to 25 m high, beneath which lie the talus deposits formed from it (Fig. 3). The top of the formation forms a plateau varying from about 70 to 100 m high, although on Smith Woodward Bjerg, it rises suddenly by faulting to 300 m. These three formations have been referred to in the past as the (lowermost) Red Division, Middle Grey and Upper Reddish Division of the *Remigolepis* Series (eg. Jarvik (Johansson 1935), although Säve-Söderbergh (1934) noted that the division was local to Gauss Halvø. Further details have been given by Olsen & Larsen (1993) (Fig. 2).

In 1987, several articulated specimens of *Ichthyostega* were found in the talus of the Aina Dal Formation, including the hind limb and tail specimen described by Coates & Clack (1990) as showing a seven-digited foot. Two laterally compressed skulls were also found. The specimens are preserved in a dark reddish fine-grained silty sandstone, and are among the best preserved and articulated of any *Ichthyostega* specimens. This dark horizon is sporadically distributed in the talus along Gauss Halvø. It occurs *in situ* near the top of the Aina Dal Formation, but appears unique to that level, in contrast to other sediment types, which are cyclically repeated. It can be seen as a dark band running about 5 m below the top of the formation according to Olsen & Larsen (1993), at the point where the resistant upper part of the formation is in contact with the more recessive lower part. On Stensiö Bjerg, we noted *Holoptychius* scales *in situ* in this horizon, and found restricted talus deposits that yielded placoderm armour. However, no more significant tetrapod remains were forthcoming.

Several other lithologies in the talus contain fragmentary fossil material. Vertebrate fossils are found in restricted bands at a few places along the base of the Aina Dal cliff, suggesting that they are weathering out of very laterally limited outcrops. Fish material is locally abundant, including semi-articulated remains of *Eusthenodon*, *Holoptychius* and lungfishes, occurring at one point in a bed of 100–150 mm thick well-jointed sandstone blocks of considerable surface area. Isolated *Holoptychius* scales were found at another point *in situ*, in massive, red, medium-grained sandstone at the base of the exposed Aina Dal cliff. Scales occur at three levels, but may all have been part of one flood event. Fossils apparently occur throughout the formation, but are only occasionally present in significant numbers, or in an articulated state.

The uppermost part of the Aina Dal Formation, where the tetrapod material appears to originate, consists of fine silty sandstones, exhibiting abundant cross lamination, climbing ripples and occasionally parallel lamination. Brecciated bedding, rootlets and mud-cracked surfaces are also present. Olsen & Larsen (1993) and Olsen (1993) have published logs of the sequence, and Marshall & Astin (pers. comm.) have more recently revisited the area and have collected further information to be published in due course.

In 1987 part of a large block preserving disarticulated tetrapod and lungfish fragments had been collected from Wiman Bjerg, but the bulk of the block had been left behind. Although the block was resting on a talus slope, its altitude suggested that it derived from the Britta Dal Formation. The 1998 expedition rediscovered this block and collected the remainder of the fossil-bearing layer from it. The tetrapod material represents more than one individual, and at least two sizes of animal are present. All of these are, however, among the smallest tetrapod specimens from the Upper Devonian succession of East Greenland. Lower jaws, jugal bones, palates, scapulocoracoids and two plates

Fig. 4. Photograph of the central part of the south side of Celsius Bjerg, showing the tetrapod-bearing sites marked by arrows. The log was taken almost directly above the more westerly site.

of articulated gastralia have been identified. They probably belong to *Acanthostega*, and if so, the assemblage should provide information on growth and ontogeny in this genus.

From the talus at the 1987 *Acanthostega* site, the 1998 expedition collected one specimen that appears to represent either a 'regurgita' or a ball of stomach contents, an assortment of disarticulated vertebrate remains in a confined area. They are much eroded but it is not clear how much of this occurred before burial. They include a large number of spines and scales identified as acanthodid acanthodians (Young, pers. comm.), eroded and broken fish scales (*Holoptychius* or lungfish) and an incomplete dermal bone (possibly an *Acanthostega* interclavicle). This is the first time that acanthodians have been recorded from the Celsius Bjerg Group of East Greenland, and confirms their presence. Some of the other spines may be from members of other acanthodian families or from chondrichthyans, but only histological studies could confirm which. The specimen raises the question of why these groups are not usually preserved in the known localities, and whether they, or indeed all the animals that have been found, were in fact living elsewhere, and were brought to these deposits only after transport for some distance.

Ymer Ø

The southern face of Celsius Bjerg is divided by two valleys into three parts, which Jarvik (in personal field notes) referred to as the western, central and eastern parts. We visited sites along the western and central parts, the eastern part consisting mainly of older and younger deposits. The Aina Dal and Wimans Bjerg Formations are not clearly evident in this locality, but Olsen & Larsen (1993) estimated their levels as lying below an altitude of 300 m. All the sites we explored were above 300–400 m, which according to Olsen and Larsen makes them equivalent to the Britta Dal and Stensiö Bjerg Formations.

Two sites yielded significant tetrapod fossils. The first was located at the eastern side of the valley dividing the western and central parts of the mountain, at an altitude of about 400 m (Fig. 4). The specimens were found in a restricted area of talus and clearly fell from a small outcrop somewhere directly above. Two of the specimens were isolated skulls, one of which is from an apparently small tetrapod, and one from an ichthyostegid of size and proportions characteristic of many previously collected (Fig. 5a). This latter specimen (MGUH (Geological Museum, University of Copenhagen) fn. 180) has potentially the best external surface of any *Ichthyostega* skull: most of the previously collected specimens are natural moulds accessible in external view only as latex peels. MGUH fn. 180 was lying in sediments filling a small flood-scour, with its ventral surface uppermost.

Other specimens from this site include two that include partial natural moulds of skulls with shoulder girdle, rib and limb elements in articulation. They are both rather poorly preserved as far as can be seen, having been much eroded, but the ribs seem to be of the broad overlapping type found in ichthyostegids. They may yield

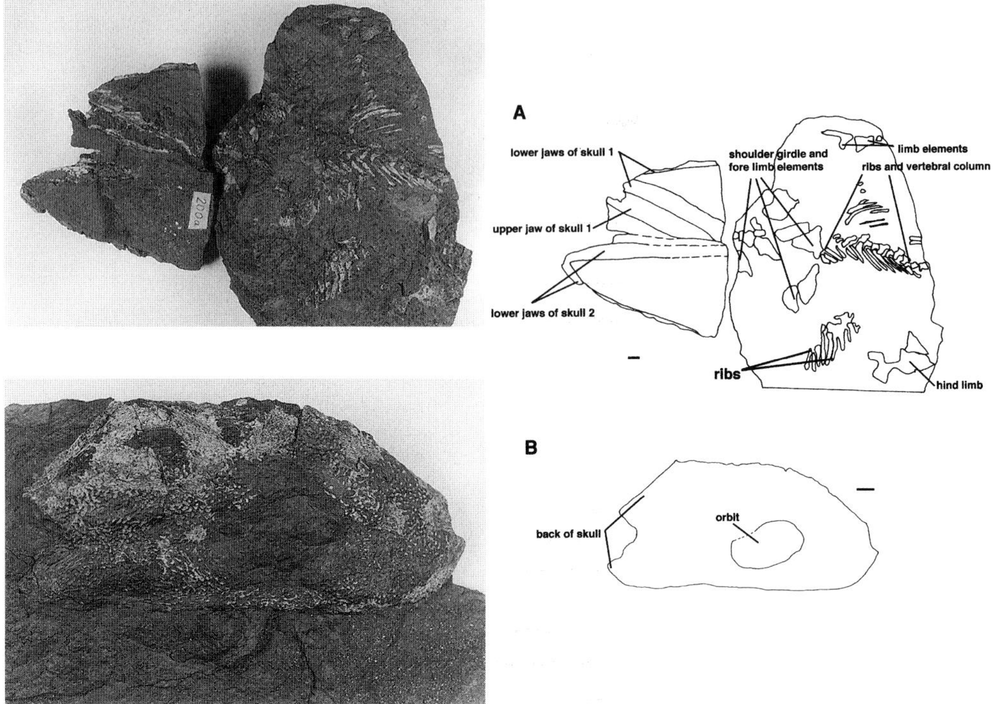

Fig. 5. Photographs with interpretive diagrams of two of the ichthyostegid tetrapod specimens from the south side of Celsius Bjerg. (**a**) Two individuals with skulls and associated postcrania; partially exposed and eroded elements can be seen on the surface, but much remains inside to be prepared. (**b**) An isolated skull of *Icthyostega* sp. in dorsal view. Scale bars represent 10 mm.

sufficient of the dentition to compare with other ichthyostegid specimens. They are both somewhat compressed laterally and recall specimen MGUH VP 6115, collected in 1947 from Smith Woodward Bjerg. That specimen appears to have been mummified before burial, and possibly before transport, and is the only one in which a head, body, forelimb, hindlimb and tail from the same individual are associated. However, there is room for doubt about whether the hindlimb really belongs. It is hoped that the new specimens might reveal parts of the forelimb, in particular the hand, so far unknown in ichthyostegids. Numerous other fragmentary tetrapod specimens were also found at this site, but require much preparation before they can be assessed.

The second tetrapod-bearing site lay on the front, south-facing slope of the central part of Celsius Bjerg, at about the mid-point, and at an altitude of 360 m. Several small blocks containing skeletal elements were found, but of most importance are two individuals that lie adjacent to one another. Two skulls are preserved in one block and at least two sets of postcranial elements in another (MGUH fn. 200, 301). Although the blocks cannot be matched exactly, the bones that they contain appear contiguous. Like the skull MGUH fn. 180 above, the two skeletons are lying ventral surface uppermost in a small flood-scour, although their disposition in articulation and close together suggests that they might not have been transported very far.

The postcranial elements contain shoulder, forelimb, hindlimb, rib and vertebral elements in articulation, so that the whole assembly represents two almost complete individuals (Fig, 5b). They are rather small compared with the majority of ichthyostegid specimens, but one of the skulls clearly shows a single postparietal bone, a unique feature (apomorphy) of *Ichthyostega*. These specimens should therefore yield unequivocal information about the body form and proportions of *Ichthyostega*, and potentially about inter- or intra-specific variation, and thus the evolution of its limbs.

A specimen collected in 1947 (MGUH VP 6088) from the south side of Celsius Bjerg was provisionally identified as a possible third taxon of tetrapod from the Upper Devonian sequence of East Greenland (Clack 1988). One of the few

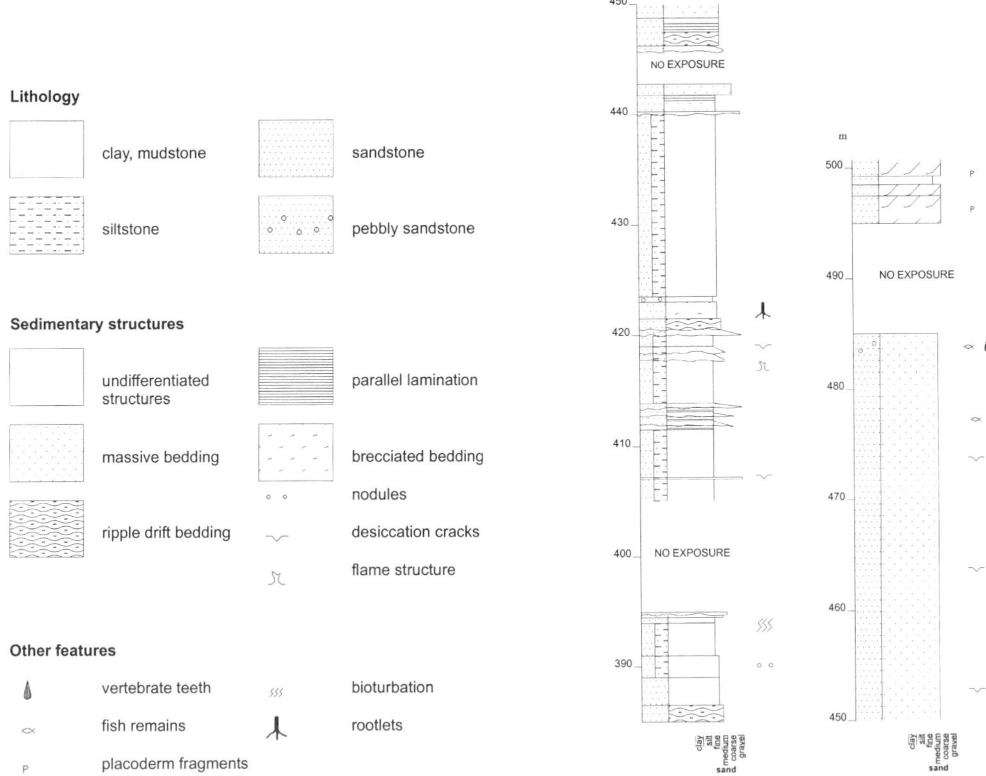

Fig. 6. Log of the sediments above the tetrapod site on the south side of Celsius Bjerg (see text for details of location).

points of anatomical overlap between the new specimens and MGUH VP 6088 might be in the dentition, though this must await preparation. If the dentitions emerge as similar, specimen MGUH VP 6088 would be a small *Ichthyostega* rather than a new genus. However, the existence of a small *Ichthyostega* might help resolve questions about variability within this genus. If the dentitions emerge as different, specimen MGUH VP 6088 may yet represent a new taxon.

One partial, isolated skull roof portion belongs to an animal having paired postparietals rather than the single one of *Ichthyostega*, but has proportions and ornament unlike that of *Acanthostega*. This may hint at the existence of another taxon, but is insufficiently complete to be more informative.

Several good specimens of fish were also recovered from each of these horizons, or in close proximity. A well-preserved lungfish skull along with some associated postcranial material of the little-known genus *Jarvikia* was found close by the pair of ichthyostegid skeletons.

A section up the eastern side of the valley dividing the western and central parts of the mountain was logged from about 400 to 500 m at GPS reading 73° 06′ N, 23° 21′ W (Fig. 6a and b). The section lies almost directly above the more westerly of the two tetrapod-bearing talus sites. Immediately below the base of the logged section are fine red sandstones, towards the top showing the type of vertical fracturing ('pencil cleavage') observed in the upper part of the Aina Dal Formation on Stensiö Bjerg. Above this lie *c.* 200 m of red and grey fine to coarse sandstones, alternating with muddier heavily fractured layers. The sandstones contain many scours and shallow gravel-lined channel features, commonly 2–3 m wide, interpreted as flood deposits. Desiccation (mud-cracked) surfaces are also present. Fish material was observed throughout the sequence, mostly in the form of isolated scales, but at *c.* 475 m, a partially articulated specimen of *Eusthenodon* was collected from the base of a scour structure, and a large jaw was seen *in situ* at 800 m. Although no tetrapod remains were found *in situ* here, at

480 m, large gravel-lined scours were observed containing fish scales and teeth; this site is directly above the first tetrapod locality. The sediments closely resemble those from the talus slope and is a possible origin of this material. At 500 m, placoderm fragments were identified *in situ*. Above this altitude, the section is covered by scree consisting of pale green and pink, coarser, sandstones. A few tens of metres to the west of the section, red sandstones spotted with green reducing patches crop out. At 800 m is a prominent outcrop of red sandstone. Asymmetrically rippled surfaces were observed indicating a palaeocurrent direction to the NE. Surface cracks, infilled with mud 10 cm deep and up to 3 cm in width were measured (Fig. 6).

The pale-coloured sandstones towards the top of the exposure are similar to those observed above the *Acanthostega* site on Stensiö Bjerg, which belong to the Stensiö Bjerg Formation of Olsen & Larsen (1993). This would suggest that the section beneath equates at least to the Britta Dal Formation, and may include Wimans Bjerg and Aina Dal equivalents. There is some disagreement about the exact boundaries of these formations, partly because the sequence consists of very similar facies types, repeating vertically but varying horizontally, without conspicuous marker beds (Bütler 1961). If the formational interpretation is correct, the ichthyostegid specimens collected here extend the known temporal range of these tetrapods from the Aina Dal formation into the Britta Dal Formation, making them contemporaneous with *Acanthostega*.

Most of the vertebrates from the Celsius Bjerg Group are disarticulated. Many of them appear to derive from flood-scours, deposited chaotically, but within distinct 'lenses'. Often the 'lens' will contain mainly a single taxon, as with the 1987 *Acanthostega* horizon. The occurrence of most vertebrate fossils, particularly the tetrapods, within flood-scours, suggests, as with the evidence from the acanthodian-containing 'regurgita', that the animals lived in a different environment from that in which the deposits of Gauss Halvø and Celsius Bjerg formed. Further evidence for this is the large size of some of the animals represented compared with the size of the scours in which they are found. Fish of 3 or 4 m in length are indicated by scales and jaws. Analysis of their palaeoecology is therefore rather difficult.

Olsen (1993) showed palaeocurrent direction plots for the Aina Dal and Britta Dal–Stensiö Bjerg Formations. The Aina Dal Formation shows a strong north to north-eastward current direction, whereas those in the overlying Britta Dal–Stensiö Bjerg Formations are much less strongly unidirectional. This could have implications for the source of the vertebrates, which may have had different geographical and thus environmental origins in these two time intervals. Olsen also suggested increasing aridification of the climate towards the end of Aina Dal times, which may have meant the periodic or seasonal drying of the streams where the tetrapods were living. The bodies were then washed into the flood basin during episodes of catastrophic flooding.

Invertebrates

Bivalves were found for the first time in the Celsius Bjerg Group, at about 400 m on the south side of Celsius Bjerg. They cannot be further identified at present. Trails, resting traces and worm casts identified as *Cruziana*, *Rusophycus* and *Gordia* (Jensen, pers. comm.) were found on a large (about 2 m^2) block on Smith Woodward Bjerg at about 300 m, in a lithology consistent with the Wimans Bjerg Formation. An arthropod trail occurs in a similar lithology found in 1987. Although similar traces have been recorded in earlier, Kap Graah deposits by Nicholson, this is the first time they have been found in the later Celsius Bjerg Group. They hint at a rich infauna and varied benthic invertebrates in the Wimans Bjerg Formation environment, which has previously been regarded as barren.

Questions for further study

Many questions relating to the fossils of the Celsius Bjerg Group remain unanswered. Of particular interest to vertebrate palaeontologists are those concerning the tetrapods. One of these is the current debate over dating of the sediments, and thus their age relative to other early tetrapod fossils. The arguments have been set out by Hartz (1997, this volume) and Marshall *et al.* (1999). They concern the anomalously young date obtained by radiometric methods (equivalent to mid-Viséan time, Hartz 1997) for basalts thought to lie just below the Celsius Bjerg Group, in contrast to the more conventional Famennian 2b date suggested by palynological correlations (Marshall *et al.* 1999).

Other questions, outlined above, are those of the postcranial anatomy, proportions, stance and gait of *Ichthyostega*, especially the still unknown digital count for the forelimb. Answers to these questions will have an impact on studies of the early evolution of tetrapod locomotion and terrestrialization, and the specimens found in 1998 may go some way towards answering them.

Many specimens of *Ichthyostega* skulls exist but no adequate systematic study of their alpha-taxonomy has been made. Careful examination of previous collections for sedimentary, locality and altitude data as well as anatomical characters should allow us to establish whether more than one species or body morph exists, especially in the light of the specimens collected in 1998. Indeed, such a study could have a reciprocal effect on clarifying the relative stratigraphy of the Celsius Bjerg Group. Correlation between the southwest Gauss Halvø sites and those of Celsius Bjerg itself produces one set of problems, but another is encountered in correlating those two with those of Sederholm Bjerg. Here, the majority of tetrapod fossils were collected at altitudes of 800–1000 m. Presumably this represents the Britta Dal Formation, but it is not certain.

Puzzles remain over the localities and environments of the vertebrates, the limited composition of the fauna, and the lack of invertebrate fossils. The fossils seem unrepresentative of the animals which must surely have been living not too far away, in much more diverse communities than we have good evidence for.

One of the continuing debates in the field of palaeoecology is whether Upper Devonian vertebrate sites represent purely freshwater deposits, as has been generally assumed, or whether there was some marine influence in any of them. Several studies now show that localities such as Escuminac Bay in Canada (Cloutier *et al.* 1996) varied from marine to marginal marine. Most of the vertebrate families represented in the Celsius Bjerg Group are known to occur in both freshwater and marine deposits (Schultze & Cloutier 1996). Although no evidence for marine influence has ever been suggested for the Celsius Bjerg Group formations, perhaps this idea should be borne in mind in future analyses.

Funding for the expedition was provided by the National Geographic Society grant number 6356–98, Newnham College Gibbs Fellowship and the Hans Gadow Fund, Department of Zoology, University of Cambridge. We thank P. Ahlberg for kindly translating Jarvik's field notes, and S. E. Bendix-Almgreen for co-operation and help in facilitating this expedition.

References

AHLBERG, P. E. 1991. Tetrapod or near tetrapod fossils from the Upper Devonian of Scotland. *Nature*, **354**, 298–301.

—— 1995. *Elginerpeton pancheni* and the earliest tetrapod clade. *Nature*, **373**, 420–425.

—— 1998. Postcranial stem tetrapod remains from the Devonian of Scat Craig, Morayshire, Scotland. *Zoological Journal of the Linnean Society*, **122**, 99–141.

—— & CLACK, J. A. 1998. Lower jaws, lower tetrapods—a review based on the Devonian genus *Acanthostega. Transactions of the Royal Society of Edinburgh: Earth Sciences* **89**, 11–46.

—— & JOHANSON, Z. 1997. Second tristichopterid (Sarcopterygii, Osteolepiformes) from the Upper Devonian of Canowindra, New South Wales, Australia, and phylogeny of the Tristichopteridae. *Journal of Vertebrate Palaeontology*, **17**, 653–673.

—— & JOHANSON, Z. 1998. Osteolepiforms and the ancestry of tetrapods. *Nature*, **395**, 792–794.

——, LUKŠEVIČS, E. & LEBEDEV, O. A. 1994. The first tetrapod finds from the Devonian (Upper Famennian) of Latvia. *Philosophical Transactions of the Royal Society of London, Series B*, **343**, 303–328.

BENDIX-ALMGREEN, S. E. 1976. Palaeovertebrate faunas of Greenland. *In*: ESCHER, A. & WATT, W. S. (eds) *Geology of Greenland.* Grønlands Geologiske Undersøgelse, Copenhagen, 536–573.

——, CLACK, J. A. & OLSEN, H. 1990. Upper Devonian tetrapod palaeoecology in the light of new discoveries in East Greenland. *Terra Nova*, **2**, 131–137.

BJERRING, H. C. 1985. Facts and thoughts on piscine phylogeny. *In*: FOREMAN, R. E., GORBMAN, A., DODD, J. M. & OLSSON, R. (eds) *Evolutionary Biology of Primitive Fishes.* Plenum, New York, 31–57.

BÜTLER, H. 1961. Devonian deposits of central East Greenland. *In*: RAASCH, G. O. (ed.) *Geology of the Arctic, Vol. 1.* University of Toronto Press, Toronto, Ont., 188–196.

CLACK, J. A. 1988. New material of the early tetrapod *Acanthostega* from the Upper Devonian of East Greenland. *Palaeontology*, **31**, 699–724.

—— 1989. Discovery of the earliest-known tetrapod stapes. *Nature*, **342**, 425–427.

—— 1994*a*. *Acanthostega gunnari*, a Devonian tetrapod from Greenland; the snout, palate and ventral parts of the braincase, with a discussion of their significance. *Meddelelser om Grønland: Geoscience*, **31**, 1–24.

—— 1994*b*. Earliest known tetrapod braincase and the evolution of the stapes and fenestra ovalis. *Nature*, **369**, 392–394.

—— 1997*a*. Devonian tetrapod trackways and trackmakers; a review of the fossils and footprints. *Palaeogeography, Palaeoclimatology, Palaeoecology*, **130**, 227–250.

—— 1997*b*. The evolution of tetrapod ears and the fossil record. *Brain, Behavior and Evolution*, **50**, 198–212.

—— 1998. The neurocranium of *Acanthostega gunnari* and the evolution of the otic region in tetrapods. *Zoological Journal of the Linnean Society*, **122**, 61–97.

—— & COATES, M. I. 1995. *Acanthostega*—a primitive aquatic tetrapod? *Bulletin du Muséum National d'Histoire Naturelle*, **17,** 359–372

CLOUTIER, R. & AHLBERG, P. E. 1997. Morphology, characters and the interrelationships of basal sarcopterygians. *In*: STIASSNY, M. L. J. &

PARENTI, L. (eds) *Interrelationships of fishes II*. Academic Press, London, 445–479.

——, LOBOZIAK, S., CANDILIER, A.-M. & BLIEK, A. 1996. Biostratigraphy of the Upper Devonian Escuminac Formation, easter Québec, Canada: a comparative study based on miospores and fishes. *Review of Palaeobotany and Palynology*. **93**, 191–215.

COATES, M. I. 1994. The origin of vertebrate limbs. *Development* Supplement **1994**, 169–180

—— 1995 Fish fins or tetrapod limbs—a simple twist of fate? *Current Biology*, **5**, 844–848.

—— 1996. The Devonian tetrapod *Acanthostega gunnari* Jarvik: postcranial anatomy, basal tetrapod relationships and patterns of skeletal evolution. *Transactions of the Royal Society of Edinburgh: Earth Sciences*, **87**, 363–421.

—— & CLACK, J. A. 1990. Polydactyly in the earliest known tetrapod limbs. *Nature*, **347**, 66–69.

—— 1991. Fish-like gills and breathing in the earliest known tetrapod. *Nature*, **352**, 234–236.

—— 1995. Romer's Gap—tetrapod origins and terrestriality. *Bulletin du Muséum National d'Histoire Naturelle*, **17**, 373–388

DAESCHLER, E. B., SHUBIN, N. H, THOMSON, K. S. & AMARAL, W. W. 1994. A Devonian tetrapod from North America. *Science*, **265**, 639–642.

HARTZ, E. 1997. Carboniferous age for the East Greenland 'Devonian' basin: paleomagnetic and isotopic constraints on age, stratigraphy and plate reconstructions. *Geology* **25**, 675–678.

—— 2000. Early syndepositional tectonics of East Greenland's Old Red Sandstone basin. This volume.

JARVIK, E. (JOHANSSON, A. E. V.) 1935. Upper Devonian fossiliferous localities in Parallel valley on Gauss Penninsula, East Greenland investigated in the summer of 1934. *Meddelelser om Grønland*, **96**(3), 1–37.

—— 1952. On the fish-like tail in the ichthyostegid stegocephalians. *Meddelelser om Grønland*, **114**, 1–90.

—— 1961. Devonian vertebrates. *In*: RAASCH, G. O. (ed.) *Geology of the Arctic, Vol. 1*. University of Toronto Press, Toronto, Ont., 197–204.

—— 1980. *Basic Structure and Evolution of Vertebrates, Vols 1 and 2*. Academic Press, New York.

—— 1985. Devonian osteolepiform fishes from East Greenland. *Meddelelser om Grønland: Geoscience*, **13**, 1–52.

—— 1996. The Devonian tetrapod *Ichthyostega*. *Fossils and Strata*, **40**, 1–206.

JOHANSON, Z. & AHLBERG, P. E. 1997. A new tristichopterid (Osteolepiformes: Sarcopterygii) from the Mandagery Sandstone (Late Devonian, Famennian) near Canowindra, NSW, Australia. *Transactions of the Royal Society of Edinburgh: Earth Sciences*, **88**, 39–68.

—— 1998. A complete primitive rhizodont from Australia. *Nature*, **394**, 569–573.

LEBEDEV, O. A. 1995. Morphology of a new osteolepidid from Russia. *Bulletin du Muséum National d'Histoire Naturelle*, **17**, 287–342.

—— & COATES, M. I. 1995. The postcranial skeleton of the Devonian tetrapod *Tulerpeton curtum* Lebedev. *Zoological Journal of the Linnean Society*, **113**, 307–348.

MARSHALL, J. E. A., ASTIN, T. R. & CLACK, J. A. 1999. The East Greenland tetrapods are Devonian in age. *Geology*, **27**(7), 637–640

NICHOLSON, J. & FRIEND, P. F. 1976. Devonian sediments of East Greenland. V. The central sequence, Kap Graah Group and Mount Celsius Supergroup. *Meddelelser om Grønland*, **206**, 1–117.

OLSEN, H. 1993. Sedimentary basin analysis of the continental Devonian basin in North-East Greenland. *Bulletin of the Grønlands Geologiske Undersøgelse*, **168**, 1–80.

—— & LARSEN, P.-H. 1993. Lithostratigraphy of the continental Devonian sediments in North-East Greenland. *Bulletin of the Grønlands Geologiske Undersøgelse*, **165**, 1–108.

SÄVE-SÖDERBERGH, G. 1932*a*. Preliminary note on Devonian stegocephalians from East Greenland. *Meddelelser om Grønland*, **98**, 1–211.

—— 1932*b*. Notes of the Devonian stratigraphy of East Greenland. *Meddelelser om Grønland*, **94**(4), 1–40.

—— 1934. Further contributions to the Devonian stratigraphy of East Greenland. *Meddelelser om Grønland*, **96**(2), 1–74.

SCHULTZE, H.-P. & ARSENAULT, M. 1985. The panderichthyid fish *Elpistostege*: a close relative of tetrapods? *Palaeontology*, **28**, 293–309.

—— & CHORN, J. 1998. Sarcopterygian and other fishes from the marine Upper Devonian of Colorado, USA. *Mitteilungen aus dem Museum für Naturkunde in Berlin, Geowissenschaftliche Reihe*, **1**, 53–72.

—— & CLOUTIER, R. 1996. Comparison of the Escuminac Formation ichthyofaunas with other late Givetian/early Frasnian ichthyofaunas. *In*: SCHULTZE, H.-P. & CLOUTIER, R. (eds) *Devonian Fishes and Plants of Miguasha, Québec, Canada*. Pfeil, Munich, 348–368.

VOROBYEVA, E. & SCHULTZE, H.-P. 1991. Description and systematics of panderichthyid fishes with comments on their relationship to tetrapods. *In*: SCHUTZE, H.-P. & TRUEB, L. (eds) *Origins of the Higher Groups of Tetrapods: Controversy and Consensus*. Cornell Publishing Associates, Ithaca, NY, 68–109

WOODWARD, A. S. 1901. Notes on some Upper Devonian fish-remains discovered by Professor A. G. Nathorst in East Greenland. *Bihang till Kungliga Svenska Vetenskapsakademiens Handlingar*, **1900**, **26**, 1–10.

ZIMMER, C. 1998. *At the water's edge*. Free Press, Simon and Schuster, New York.

Deformation of the Old Red Sandstone of NW Spitsbergen; links to the Ellesmerian and Caledonian orogenies

ANDREW J. McCANN

Department of Earth Sciences, University of Cambridge, Downing Street, Cambridge CB2 3EQ, UK

Present address: Statoil's Research Centre, Postuttak, N-7005 Trondheim, Norway (e-mail: ajmc@statoil.com)

Abstract: The Late Silurian?–Devonian fluvial deposits of northern Spitsbergen were deposited on basement with Caledonian and earlier metamorphic ages in which two distinct terranes are recognized (Biskayerhalvøya and Krossfjorden). These form part of the central of three major terranes in Svalbard, assembled during the Caledonian Orogeny. The subsequent geological history of the Svalbard area has been strongly influenced by the north-trending structures which were active as transcurrent fault zones at this time. The unconformable base of the Siktefjellet Group, a Late Silurian?–earliest Devonian sequence of coarse conglomerates and breccias, overlain by fluvial sandstones, is preserved only on the Biskayerhalvøya terrane, and the final juxtaposition of the two terranes (during the Haakonian sinistral strike-slip phase) is interpreted to post-date the deposition of these sediments. The Lochkovian Red Bay Group, a sequence of conglomerates, fluvial sandstones and siltstones, was deposited on both terranes. This has been mapped and correlated throughout the basin exposure, allowing the reconstruction of the tectonic history. Sedimentation was influenced by active faulting during deposition of the oldest Wulffberget Formation, but subsequent deposits show little evidence of this. Deposition was interrupted in latest Lochkovian time by renewed sinistral strike-slip faulting, which broke up an area of the basin into rotating fault blocks, across which about 30 km of extension occurred. This was followed by east–west shortening, which uplifted the Red Bay basin and underlying basement, developing large folds, locally with related thrusting. The Monacobreen phase is defined to involve this deformation. The Andrée Land Group reflects a subsequent renewal of subsidence, and re-establishment of an extensive fluvial basin, occupying an area east of the inverted Red Bay basin. Conglomeratic units that overlie the Red Bay Group are interpreted as the products of the reworking of the uplifted Red Bay basin and its basement. The Latest Devonian–Earliest Carboniferous Svalbardian phase again involved east–west shortening, with limited strike-slip faulting, but it is difficult to discriminate these effects from the Monacobreen phase in the Siktefjellet and Red Bay groups. A review of North Atlantic and Arctic Devonian basins shows that during deposition of the Red Bay and Andrée Land groups, the tectonics of Svalbard was more similar to that of the developing Ellesmerian orogen, than to that of the collapsing Caledonian orogen. A model is proposed that links the repeated extension and shortening seen from Early Devonian time in north Spitsbergen to anticlockwise rotation of the Chukotka–Alaska plate, about an axis near the position of Svalbard during Ellesmerian collision, coupled with minor Caledonian-related strike-slip movement along reactivated fault zones.

The Old Red Sandstone (ORS) deposits of Spitsbergen, the main island of the Svalbard archipelago, attracted the attention of scientific expeditions a century ago (e.g. Holtedahl 1914), particularly for their rich vertebrate faunas, and the basic stratigraphic outline was established by the 1920s (e.g. review by Føyn & Heintz (1943)). As the northernmost exposures of the North Atlantic ORS, these deposits are an important link between the Caledonian-related basins in the UK, East Greenland and Norway, and the Devonian succession of northern Canada. In Early Devonian time, Svalbard was situated off the northeast margin of Laurentia in a position along-strike from the Scandian collision zone of the Caledonides. It was also east along-strike from the Franklinian Trough, which would close in the Ellesmerian Orogeny culminating in earliest Carboniferous time. This position implies that the Devonian basin development

From: FRIEND, P. F. & WILLIAMS, B. P. J. (eds). *New Perspectives on the Old Red Sandstone.* Geological Society, London, Special Publications, **180**, 567–584. 0305-8719/00/$15.00

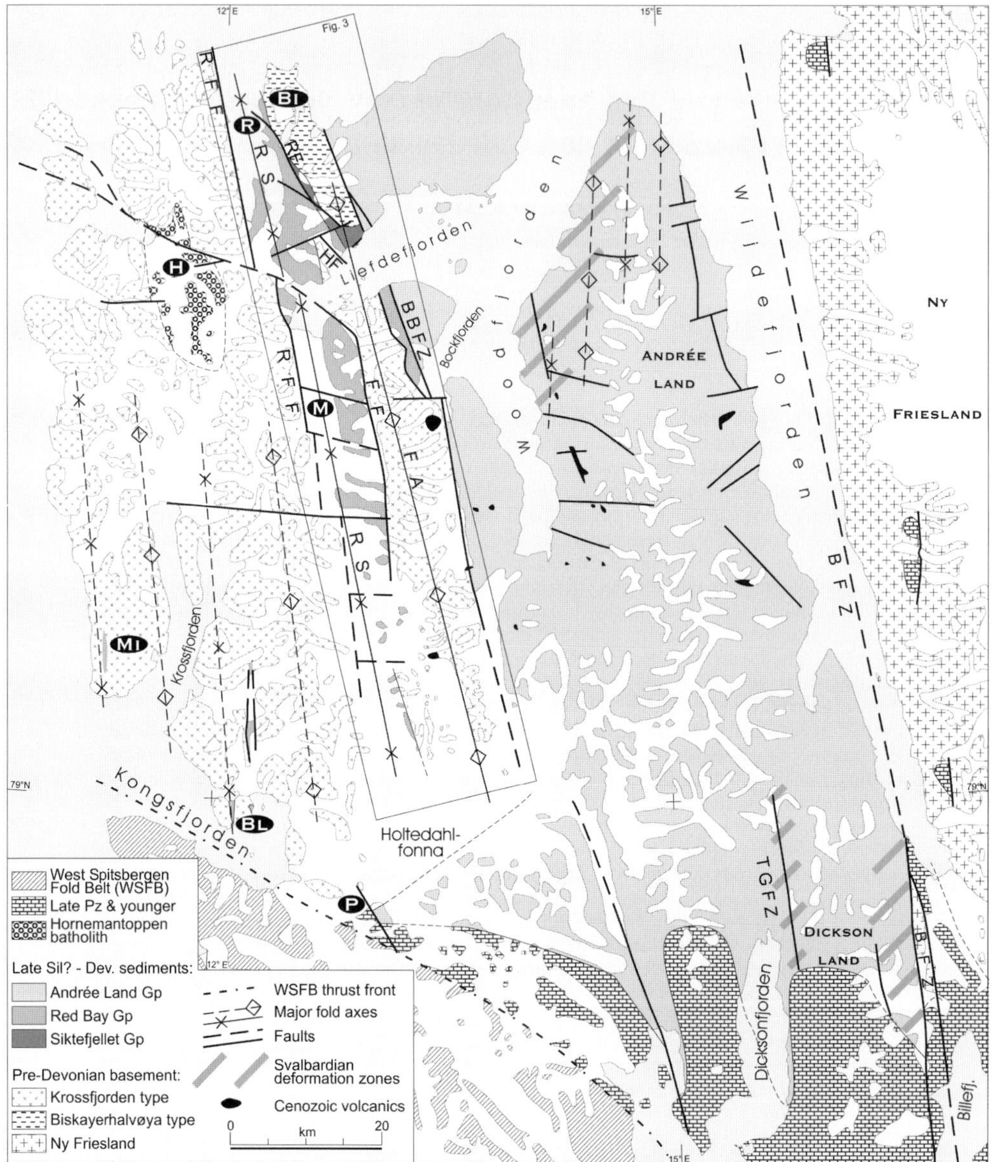

Fig. 1 Geological and structural outline of NW Spitsbergen. Ice cover is shown in white. Location of Fig. 2 is shown by the rectangle. Compiled from Hjelle & Lauritzen (1982), Piepjohn (1994) and this study. Major structures: FA, Friedrichbreen Anticline; BFZ, Billefjorden Fault Zone; BBFZ, Breibogen–Bockfjorden Fault Zone; FF, Friedrichbreen Fault; HF, Hannabreen Fault; RFF, Raudfjorden Fault; RF, Rabotdalen Fault; RS, Raudfjorden Syncline; TGFZ, Triungen-Grønhorgdalen Fault Zone. Locations referred to in the text: BI, Biskayerhalvøya; H, Hornemantoppen; M, Monacobreen; MI, Mitrahalvøya; BL, Blomstrandhalvøya; P, Pretender; R, Raudfjorden.

and deformation history was related to the tectonics of both these orogens.

Three major stratigraphic sequences have been recognized since Gee & Moody-Stuart (1966) identified the lowermost unit, the Siktefjellet Group. Deposits of that group and the overlying Red Bay Group are exposed along the shores of Raudfjorden and Liefdefjorden, and south along the Monacobreen glacier (Fig. 1). The youngest Andrée Land Group is exposed to the east across central northern Spitsbergen. The predominantly fluvial deposits of the Red Bay and Andrée Land groups were the subject of detailed sedimentological and stratigraphic work (Friend 1961), and the comprehensive interpretation of the basin development during deposition of the Wood Bay Formation (Andrée Land Group), by Friend & Moody-Stuart (1972), laid a foundation for much of the subsequent sedimentological work on the Old Red Sandstone elsewhere.

Gee (1972) defined the Haakonian tectonic phase as a late-Caledonian sinistral strike-slip episode along major NNW–SSE-trending structures including the Raudfjorden Fault and the Breibogen–Bockfjorden Fault Zone. Folding related to this faulting was interpreted as having caused the unconformity between the Siktefjellet and Red Bay groups. Deposition of the Red Bay Group and its subsequent deformation were ascribed to the late Haakonian phase. The Svalbardian phase, defined by Vogt (1928), involved folding and faulting that affected the whole of the Devonian basin before deposition of unconformable Early Carboniferous sediments (Piepjohn, this volume).

The area of Red Bay Group exposure along the Monacobreen glacier presents severe access problems and the only previous work there was of a reconnaissance nature (Gjelsvik 1979 and unpub. report 1996). All other published work on the stratigraphy, sedimentology and tectonic history of the Red Bay Group has been based on data from the Raudfjorden–Liefdefjorden area, which includes only 40% of the total exposure. The fieldwork for this study, carried out with helicopter support in the summers of 1994 and 1995, was part of the Norsk Polarinstitutt geological mapping programme. New data from this southern area require a new look at previous stratigraphic and tectonic interpretations, and provide good evidence for an episode of deformation between the deposition of the Red Bay and Andrée Land groups.

Regional setting

Svalbard lies at the northwestern corner of the Barents Shelf, on the edge of the Eurasian Plate, approximately halfway between the north coast of Norway and the North Pole. Before the Tertiary opening of the North Atlantic, Svalbard lay off the northeast margin of Greenland (its position essentially unchanged from latest Devonian time), and its Late Palaeozoic and Mesozoic history is closely related to the Wandel Sea basin of that area (Håkansson & Stemmerik 1984). The pre-Carboniferous history of Svalbard, however, involves a number of different crustal terranes, which had distinct geological histories before the Caledonian Orogeny (Harland & Wright 1979).

The present-day structure of Svalbard is dominated by north–south trends in the distribution of units, controlled by north–south structures, coupled with a general plunge towards the south, where the youngest rocks are exposed (Fig. 1). The mainly continental Devonian basin-fill is exposed in NW Spitsbergen, bounded for much of its extent by the Breibogen–Bockfjorden and Billefjorden Fault Zones. The central Andrée Land block includes predominantly fluvial sediments of Pragian to Famennian age (the Andrée Land Group), which were deformed during the Svalbardian phase, before Late Tournaisian time (Piepjohn *et al.* this volume). West of the Breibogen–Bockfjorden Fault Zone, two older groups of deposits are preserved, the Siktefjellet Group (of Late Silurian? to Early Devonian age) and the Red Bay Group (of Lochkovian age). These are bounded against pre-Devonian basement along the Raudfjorden Fault, but small patches of Devonian deposits are also found further west. The Devonian basin extends south beneath younger cover, as seen from seismic reflection profiles (e.g. Nøttvedt 1994). An isolated area of ORS exposure is also found around Hornsund, in south Spitsbergen, within the West Spitsbergen Fold Belt.

Pre-ORS basement

The basement below the ORS is exposed from the west coast to the Breibogen–Bockfjorden Fault Zone. Further east it is covered by the fill of the Andrée Land basin, but it is exposed again east of the Billefjorden Fault Zone in Ny Friesland (Fig. 1). Problems in correlating different basement areas on Svalbard, and comparisons with other areas along the Laurentian margin, have led to a number of terrane interpretations, involving large-scale lateral movements along sinistral fault zones within and west of Svalbard (Harland & Wright 1979; Gee 1986; Ohta *et al.* 1989). The Western Province of Harland (1985) has affinities to the Franklinian belt of North Greenland and also the allochthonous Pearya terrane. The Eastern Province (Ny Friesland and Nordaustlandet) is comparable with the Eleonore Bay Group and associated rocks of central East Greenland, and the Central Province has a provenance along the intervening section of the margin. Such interpretations require major sinistral strike-slip movement along the Laurentian margin, interpreted either as the result of oblique collision between Laurentia and Baltica (Harland 1985), or as a consequence of lateral extrusion during orthogonal collision (Gee 1986; Gee & Page 1994; Ohta 1994). Harland (1985, 1997) considered that major movement continued along the Billefjorden Fault Zone during Devonian time, but Gee & Page (1994) concluded that terrane assembly was completed by earliest Devonian time.

Although these models all consider the basement underlying the ORS basins in NW Spitsbergen to be part of one major crustal block, there are different views on its internal subdivision. Harland & Wright (1979) and Ohta *et al.* (1989) both emphasized the Raudfjorden Fault as a boundary between contrasting basement types, whereas Gee (1986) separated the eastern part of the area between Raudfjorden and Liefdefjorden as having a distinct tectonothermal history (Fig. 1). This Biskayerhalvøya terrane includes the Biskayerfonna Group of garnet schists and amphibolites and the Richarddalen Complex with retrogressed eclogites (Gee 1966*a*, *b*). Cooling of the Biskayerfonna Group after eclogite facies metamorphism occurred from 437 to 412 Ma (Dallmeyer *et al.* 1990). U–Pb zircon ages from granites and gabbros in the Richarddalen Complex show a crystallization age of 965–955 Ma (Peucat *et al.* 1989), and subsequent peak eclogite metamorphism is thought to have occurred in early Ordovician time (*c.* 470 Ma) (Gromet & Gee 1998). The Breibogen–Bockfjorden Fault limits exposure of the Biskayerhalvøya terrane to the east, whereas the western boundary against the Siktefjellet and Red Bay groups is faulted or unconformable. Similar rocks are exposed on islands in Liefdefjorden, south of where the Siktefjellet Group is exposed, so the western limit probably follows the line of the Rabotdalen and Hannabreen faults, joining the Breibogen–Bockfjorden Fault under Liefdefjorden.

The basement exposed directly west of the Hannabreen Fault (Fig. 1), and across to the west coast of Spitsbergen comprises metamorphosed supracrustal rocks (the Krossfjorden Group) underlain by migmatite gneisses and granitoid intrusions (Gee & Hjelle 1966; Hjelle 1979). This is here referred to as the Krossfjorden terrane. Metamorphic grade varies from chlorite to garnet zone northwards through the supracrustal rocks. There is considerable retrogressive overprint (Hjelle 1979) and U–Pb zircon ages suggest that the gneisses may be Grenvillian in age (Balašov *et al.* 1995). The granitoids, which tend to be elongated in a NNW direction, partly intrude the gneisses and zircon ages of 423 ± 22 Ma have been reported for some (Balašov *et al.* 1996). The youngest intrusion is the Hornemantoppen batholith (Fig. 1) with a whole-rock Rb–Sr age of 414 ± 10 Ma (Hjelle 1979; Balašov *et al.* 1996). Three deformation phases have been distinguished in the metasediments, the latest involving large-scale folding, evident by the structure of the marble-dominated uppermost unit (Gee & Hjelle 1966; Hjelle 1979).

ORS sediments completely cover the basement east of the Breibogen–Bockfjorden Fault Zone, but evidence from boulder clasts in the Siktefjellet Group suggests that this basement may have a distinct tectono-thermal history (Hellman *et al.* 1998), implying that this fault zone is one of the most important terrane boundaries in Svalbard.

ORS stratigraphy

Three unconformity-bounded stratigraphic groups make up the ORS (the Liefde Bay Supergroup of Friend *et al.* (1997)) in northern Spitsbergen. The stratigraphy used here (Fig. 2) is modified from earlier schemes (Friend 1961; Murašov & Mokin 1976; Friend *et al.* 1997) because of new data acquired during mapping of the area south of Liefdefjorden (McCann 1997). The majority of deposits are terrestrial, chiefly fluvial, and each group represents a generally fining-upward sequence.

The Siktefjellet Group (of Late Silurian?–Lochkovian age) is exposed along a 12 km strip between Liefdefjorden and Raudfjorden. The basal unconformity has a relief of many tens of metres with deep weathering into the basement, and the unit is found only above rocks of the Biskayerhalvøya terrane. The Lilljeborgfjellet Formation comprises up to 400 m of very coarse, chaotic breccias and better stratified, locally sandy, conglomerates. Clast types include quartzites, schists, augen gneiss, eclogite, marble and quartz–feldspar porphyry. All but the last of these are typical of the local Biskayerhalvøya type basement; the porphyry has an age (*c.* 1740 Ma) and geochemistry that exclude a source from any known part of northwest Svalbard (Gjelsvik 1991; Hellman *et al.* 1998). These clasts may have been derived from the basement buried below Andrée Land, east of the Breibogen–Bockfjorden Fault Zone (Hellman *et al.* 1998). There is a general fining upward to the overlying trough cross-bedded sandstones of the Albertbreen Formation, which are at least 1100 m thick on Siktefjellet. This unit contains plant fragments and miospores suggesting an Early Devonian age (Murašov & Mokin 1976).

The overlying Red Bay Group (of Lochkovian age) also comprises a generally fining-upward sequence, but there is also a fining northward

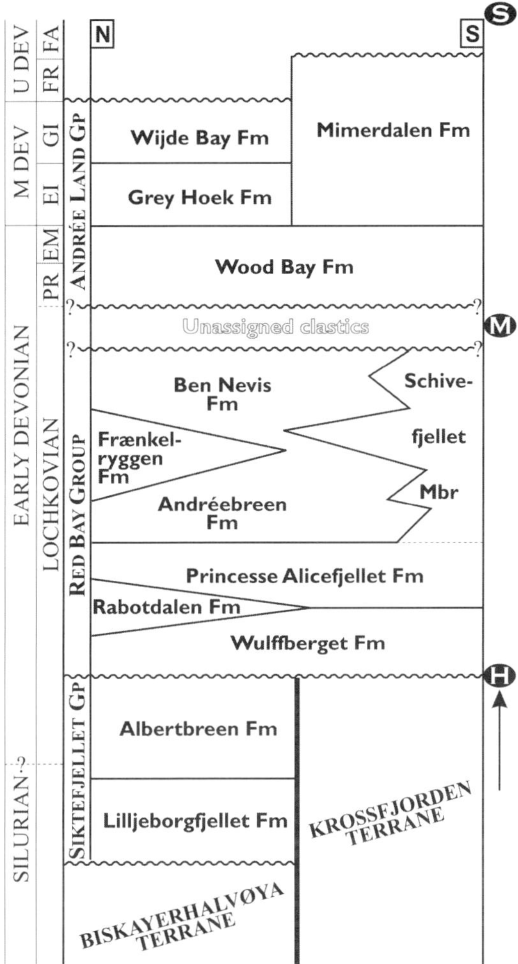

Fig. 2. Composite stratigraphic scheme and nomenclature of the Old Red Sandstone of northern Spitsbergen, after Friend (1961) and Murašov & Mokin (1976) with a new subdivision of the Red Bay Group in this study. The Schivefjellet Mbr interfingers with the Andréebreen and Ben Nevis formations in the north of the basin, but is interpreted as continuous with the higher parts of the Princesse Alicefjellet Fm to the south. The timing of the Haakonian (H), Monacobreen (M) and Svalbardian (S) deformation phases is marked in the right-hand column.

across the basin. It is preserved between the Raudfjorden and Breibogen–Bockfjorden faults, but similar deposits are also locally found capping the basement west from here. Where it does not overlie the Siktefjellet Group or the Biskayerhalvøya terrane, the lowest Wulffberget Formation (alluvial fan and braided river conglomerates) almost invariably rests on deeply weathered basement marbles and marble forms the commonest clast type. Lacustrine and sandy fluvial deposits (probably inter-fan) are found in the Raudfjorden area (Rabotdalen Formation). The overlying Princesse Alicefjellet Formation is a quartz-dominated braided river conglomerate unit that is found throughout the basin. Palaeocurrent evidence suggests a source from the south and the unit is up to 2000 m thick in its southernmost exposures. In the north, it becomes much thinner but is also found interfingering with higher stratigraphic units.

The Andréebreen (sandy braided rivers) and Frænkelryggen (incised sandy channels, with extensive overbank deposits) formations represent continued rising of base level, accompanied by persistent higher-energy deposition in the south of the basin. This fining trend was reversed during deposition of the Ben Nevis Formation by the return to widespread braid plain deposition, with only restricted floodplain development. This is reflected by the northward progradation of the

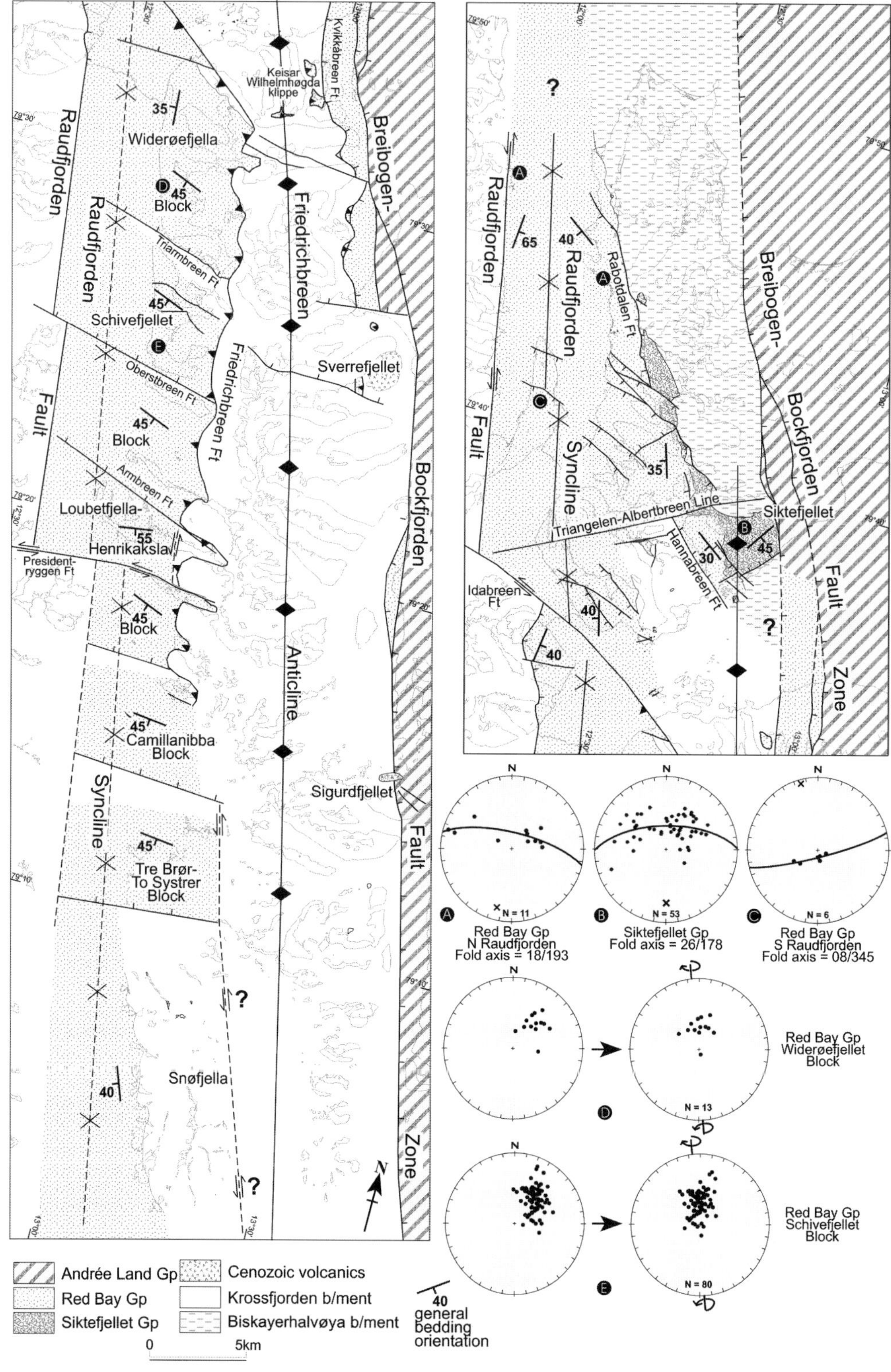
Raudfjorden Fault
Raudfjorden Syncline
Friedrichbreen Anticline
Breibogen-Bockfjorden Fault Zone
Keisar Wilhelmhøgda klippe
Kvikkåbreen Ft
Widerøefjella Block
Triarmbreen Ft
Schivefjellet
Oberstbreen Ft
Friedrichbreen Ft
Sverrefjellet
Armbreen Ft
Loubetfjella-Henrikakslaa
President-ryggen Ft
Block
Camillanibba Block
Sigurdfjellet
Tre Brør-To Systrer Block
Snøfjella
Rabotdalen Ft
Triangelen-Albertbreen Line
Siktefjellet
Hannabreen Ft
Idabreen Ft
Red Bay Gp N Raudfjorden Fold axis = 18/193
Siktefjellet Gp Fold axis = 26/178
Red Bay Gp S Raudfjorden Fold axis = 08/345
Red Bay Gp Widerøefjellet Block
Red Bay Gp Schivefjellet Block
N = 11
N = 53
N = 6
N = 13
N = 80
Andrée Land Gp
Red Bay Gp
Siktefjellet Gp
Cenozoic volcanics
Krossfjorden b/ment
Biskayerhalvøya b/ment
40 general bedding orientation
0 5km

Princesse Alicefjellet Formation braided system. There is little evidence for syn-sedimentary faulting during deposition of the Red Bay Group, except in the Wulffberget Formation, which is interpreted to have filled in a faulted basin floor topography.

The Andrée Land Group (of Pragian to Famennian age) is exposed between the Breibogen–Bockfjorden and Billefjorden fault zones across central northern Spitsbergen, and some higher units are also found further west, south of Holtedahlfonna (Fig. 1). Previous interpretations have assumed continuity of deposition from the Red Bay Group, but evidence for an unconformity between these units is discussed below. Friend & Moody-Stuart (1972) showed that the Wood Bay Formation was deposited by low- and high-sinuosity rivers within a floodplain environment, with marine incursions in the north. A dominant northwards axial drainage pattern was also fed from the west, and Gee & Moody-Stuart (1966) showed that the position of the Breibogen–Bockfjorden Fault Zone was at least a significant topographical break, if not an active basin margin fault during deposition of the Wood Bay Formation. The Grey Hoek and Wijde Bay formations both comprise sandstones with shales and siltstones, deposited in coastal, possibly intertidal settings (Worsley 1972). The youngest Mimerdalen Formation is found only in the southeast of the Andrée Land basin, and includes local units of fluvial sandstones and shales, with an upper conglomeratic unit, indicating reworking of older ORS deposits. The Billefjorden Fault Zone did not form a basin margin during deposition (the basin probably covered an extensive area east of here) but became a faulted margin before deposition of the Tournaisian Billefjorden Group, which cuts across deformed Andrée Land Group sediments in Dickson Land.

Structural description

Siktefjellet Group and the Biskayerhalvøya terrane

The Siktefjellet Group forms a plunging anticline in Siktefjellet, north of Liefdefjorden and an elongated, WSW-dipping, exposure continues northwards, east of Raudfjorden (Fig. 3). The basal unconformity is invariably on basement of the Biskayerhalvøya terrane, and at only one exposure are these deposits found west of the proposed terrane boundary along the Rabotdalen and Hannabreen faults. The final juxtaposition of the Biskayerhalvøya and Krossfjorden terranes to their present position is therefore interpreted to post-date the deposition of the Siktefjellet Group. The outcrop on the west side of Hannabreen (Konglomeratryggen) is fault bounded and no primary relationship with either basement terrane can be observed. It lies immediately west of the interpreted line of the Hannabreen Fault and may be a faulted sliver from movement along this fault, or possibly reworked Siktefjellet Group material.

From the Siktefjellet anticline, dip readings from Albertbreen Formation exposures around the western and southern slopes give some scatter, but a definable fold axis of 26/178 (Fig. 3), rather than the SW plunge described by Gee & Moody-Stuart (1966). They suggested that the SW-plunging anticline was the result of pre-Red Bay Group NW–SE shortening (during the Haakonian phase), later tilted west with the Red Bay Group. This southerly plunging fold, however, is compatible with the large-scale folding also seen in the Red Bay Group, and lies along-strike from the Friedrichbreen Anticline, south of Liefdefjorden (Fig. 3). The generally low angle of the unconformity between the Siktefjellet and Red Bay groups and the small amount of reworking of the older sediments into the younger units are further evidence against significant folding and uplift between their deposition. This suggests that the folding of the Siktefjellet Group was the result of the same shortening episode that affected the Red Bay Group.

Large-scale folding

Throughout the basement exposed in NW Spitsbergen, a series of *c.* N10W trending folds can be traced, with a wavelength in the order of 15–20 km (Fig. 1) (Gee & Hjelle 1966; Hjelle &

Fig. 3. Outline geological map and structural elements of the Raudfjorden–Monacobreen area (from this study and Dallmann *et al.* (1998)). The top of the left map overlaps with the bottom of the right map; location shown in Fig. 1. Thin lines mark the boundaries between exposed ridges and ice or water. Shading has been applied to areas where the stratigraphic units are interpreted to continue under the ice or water. Stereonets (**A**)–(**C**) show bedding measurements from the units and areas indicated. The folding of the Siktefjellet Gp is parallel to that of the Red Bay Gp in the Raudfjorden Syncline. The greater southerly tilt is due to pre-folding flexuring, but the folding is interpreted to be coeval with the folding of the Red Bay Gp (see text). Stereonets (**D**) and (**E**) show present-day bedding measurements (left) within the tilted fault blocks shown. By back-rotating around the dip (28/265) of the west limb of the Friedrichbreen Anticline (right) it is seen that when developed, the fault blocks dipped south.

Lauritzen 1982). Despite the complex structure of the basement, marbles of the upper unit (Generalfjella Formation) are commonly found in the synclinal cores of these folds, and on Mitrahalvøya and Blomstrandhalvøya, remnants of sediments similar to the Wulffberget Formation are also found, unconformably on the marbles (Gjelsvik 1974; Thiedig & Manby 1992; Piepjohn pers. comm.).

The Red Bay Group is preserved in a syncline (the Raudfjorden syncline) running south from Raudfjorden along Monacobreen to Holtedahlfonna, partially faulted along both limbs and offset sinistrally along the Idabreen Fault (Fig. 3). Generally only the eastern limb is exposed, which has led to interpretations of the structure, particularly south of Liefdefjorden, as being a west-tilted half-graben rather than a faulted syncline (e.g. Hjelle & Lauritzen 1982; Piepjohn 1994).

To the east, south of Liefdefjorden, basement is exposed in the adjacent Friedrichbreen Anticline, with thrust fault-bounded Red Bay Group sediments on the eastern limb. This anticline is continuous with the structure of the Siktefjellet Group north of Liefdefjorden, except that the Siktefjellet anticline plunges south. Further north on Biskayerhalvøya, it is not easy to distinguish this late long-wavelength folding within the complex basement structure, but the exposed east limb of the Raudfjorden syncline and lack of Devonian cover suggest that the anticlinal axis may continue through there.

These fold structures affecting the ORS sediments are coaxial with the folds affecting the basement to the west and appear to form a single system, implying that this basement folding also post-dates the ORS deposits.

Rotated fault blocks

On the east limb of the Raudfjorden syncline, south of Liefdefjorden, the Red Bay Group appears to be broken up by a series of *c*. W10N trending faults downthrowing to the NNE (Fig. 3). These faults separate rotated fault blocks within which the dip is rather constant and there is little minor faulting. Apart from the fault along Triarmbreen, these structures have been inferred from the major stratigraphic offsets (many hundreds of metres) between ridges separated by glaciers. The orientation of these faults and the amount of displacement across them are reasonably well constrained by the narrow shape of the ice-covered areas and correlation of distinctive units within the stratigraphy.

The northern margin of the Widerøefjella block is flexed and exposes the basal Red Bay Group unconformity (Fig. 3). The dip rotates from 30° NW here to a constant 35–45° SSW in the southern three-quarters of the block. A stratigraphic thickness of over 3500 m is exposed through this block, including the greatest mapped thickness (2500 m) of the Ben Nevis Formation. Cross-cutting NW–SE and WSW–ESE faults are found in the northern part of the block, but there is little internal deformation in the south.

The Schivefjellet Block again exposes a thick Red Bay Group stratigraphy with the underlying basement, and the stratigraphic separation across the intervening Triarmbreen Fault therefore exceeds 3500 m. The fault line is exposed only as a change of scree type, but when mapped it appears to be steeply NNE dipping. Bedding dip within this fault block is 35–45° SSW and it is cut by WNW–ESE faults in its northern part. Correlation of the Princesse Alicefjellet Formation (Schivefjellet Member) conglomerates across Oberstbreen implies a fault along here, downthrowing 500–600 m to the north. There may be other hidden faults cutting the southern part of this block, but the stratigraphic separation across the Armbreen Fault to the Loubetfjella Block must still be 3000–4000 m.

The same pattern of deformation continues through two further blocks as far south as Tre Brør. South of here, the Red Bay Group exposures of Snøfjella dip 45–60° W, on the eastern limb of the Raudfjorden Syncline.

Despite having large displacements, the block-bounding faults mapped along Monacobreen cannot be directly traced west of the Raudfjorden Fault or east into the basement of the Friedrichbreen Anticline. They must therefore be bounded by other structures.

Friedrichbreen Fault

The contact between the Red Bay Group and basement in the west limb of the Friedrichbreen Anticline has previously been mapped in places as an unconformity, an extensional fault (Piepjohn 1994), both (Manby & Lyberis 1992) or a reverse fault (Gjelsvik 1979). Tracing this structure south, it can be shown to be a reverse fault, here named the Friedrichbreen Fault (Fig. 3). In some places it is exposed only as a flat basement surface with remnants of sediments, but mapping shows that it cuts up through the Red Bay Group stratigraphy within the rotated fault blocks. A spectacular exposure on Henrikaksla shows intense reverse faulting and folding in the hanging-wall sediments, with fold axes trending approximately N15E. Construction of structure contours between exposures shows it to be a single fault dipping *c*. 30° W over a distance of *c*. 30 km.

West of Liefdefjorden, the line of the Raudfjorden Fault and the margins of the Hornemantoppen batholith are offset by 3.5–4 km sinistrally on a fault running along the Idabreen glacier (Figs 1 and 3). A similar offset is seen in the map trace of the Raudfjorden Syncline axis, but the Friedrichbreen Anticline does not appear to be offset across Liefdefjorden. The map trace of the Friedrichbreen Fault rotates to the NW down slope on the southern shore of the fjord and its continuation projects towards the Idabreen Fault. The sense and magnitude of slip on these two structures are compatible with the interpretation that the Idabreen Fault forms a lateral ramp to the Friedrichbreen thrust.

Thrust klippen are found on the crest and eastern limb of the Friedrichbreen Anticline, south of Liefdefjorden (Gee & Moody-Stuart 1966; Piepjohn 1994), involving rocks interpreted to be parts of the upper Red Bay Group (Gjelsvik & Ilyes 1991), in which bedding measurements describe north–south to NNE–SSW fold axes. Similar folding is found east of the Kvikkåbreen Fault, which is an extensional fault cutting through an earlier thrust, parts of which are preserved in the footwall. Piepjohn (1994) also described intense west-vergent folding and thrusting in Red Bay Group sediments from coastal sections just north of here. The relationship between these west-vergent thrust structures and the Friedrichbreen Fault is unclear as they do not interact, but they appear to be similar, out-of-the-syncline, shortening structures.

Andrée Land Group

Exposure of the Andrée Land Group is bounded to the west by the Breibogen–Bockfjorden Fault Zone from Breibogen to Holtedahlfonna, but no continuation of this fault has been found south of Holtedahlfonna, where flat-lying Andrée Land Group sediments are exposed (Fig. 1). Similarly, the Raudfjorden Syncline and Friedrichbreen Anticline cannot be traced through these exposures. Only on the mountain Pretender are Andrée Land Group sediments found directly on basement (Andresen, pers. comm.), and it appears that the Andrée Land Group was deposited unconformably across the basement and Red Bay Group structure.

The general structure of Andrée Land is a major north–south-trending anticline with limbs dipping gently towards Woodfjorden and Wijdefjorden. Within this are zones of intense shortening, with tight asymmetrical chevron and box folds and associated thrusts, along the west coast of Woodfjorden (Piepjohn 1994), north of Dicksonfjorden, and west of the Billefjorden Fault Zone (Piepjohn 1994; McCann & Dallmann 1996) (Fig. 1).

Basin development and deformation

Manby & Lyberis (1992) interpreted the development of the Devonian basin of northern Spitsbergen as a single lithospheric stretching event, with the Siktefjellet and Red Bay groups deposited in the syn-rift phase and the Andrée Land Group deposited during post-rift thermal subsidence. Many other workers, however, have considered that major or minor tectonic events may have separated the three stratigraphic groups (e.g. the Siktefjellet–Red Bay Group Haakonian phase of Gee (1972)), and have emphasized the importance of sinistral strike-slip faulting (e.g. Friend *et al.* 1997). The basin as a whole suffered inversion in Late Devonian–Early Carboniferous time, during the Svalbardian event (Vogt 1928). This has been interpreted as the culmination of a major strike-slip episode, related to Devonian sinistral shearing elsewhere in the Caledonides (Harland *et al.* 1974; Harland 1985), but this model has been challenged, in favour of orthogonal east–west shortening (Lamar *et al.* 1986; Manby *et al.* 1994).

Haakonian deformation phase

The contrast in clast composition between the basal units of the Siktefjellet and Red Bay groups, and the fact that the former is only found deposited on the Biskayerhalvøya terrane, suggests that this was not juxtaposed to its present position against the Krossfjorden terrane until after the deposition of the Siktefjellet Group. The Rabotdalen and Hannabreen faults, although they are not linked as structures affecting the Red Bay Group, effectively define the boundary between the two basement blocks and are therefore interpreted to mark the suture between them (Piepjohn 1994). This must continue NNW within Raudfjorden and SSE across Liefdefjorden, east of the exposed basement on the south coast.

Folding of the Siktefjellet Group appears to be the same as that which affected the Red Bay Group, but the moderate plunge may be the result of earlier southerly tilting of the sediments in this area related to the terrane juxtaposition (see stereonets Fig. 3). The proposed model involves sinistral strike-slip along the Rabotdalen–Hannabreen Fault, to bring the two terranes together after deposition of the Siktefjellet

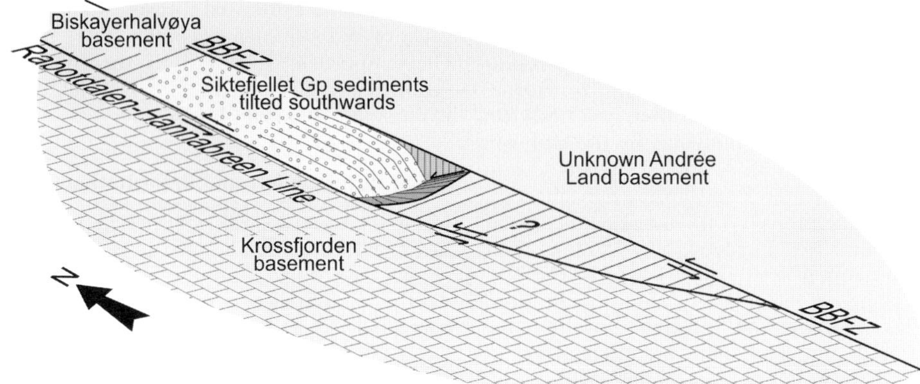

Fig. 4. Model for the juxtaposition of the two basement terranes and local preservation of the Siktefjellet Gp cover on the Biskayerhalvøya terrane during the Haakonian phase. The main boundary is a sinistral strike-slip fault along the Rabotdalen–Hannabreen fault line. Overstep to another fault, possibly along the present Breibogen–Bockfjorden Fault Zone, may explain the resultant south-dipping structure of the sediments against an extensional cross fault. (Partly after Friend *et al.* (1997)).

Group (Fig. 4). As the Breibogen–Bockfjorden Fault Zone may have been an important terrane boundary (Hellman *et al.* 1998), the Rabotdalen–Hannabreen Fault is interpreted as a splay from this structure. The tilting of the sediments could have been the result of a change in strike of the main fault, with extension occurring across the fault now exposed at the foot of Siktefjellet, in a releasing bend.

Monacobreen phase

Harland (1961) first noted that doming of the Friedrichbreen Anticline also affected the Red Bay Group, and suggested that some of the folding of the NW Spitsbergen basement occurred in Late Devonian time (Svalbardian phase). Gee (1972) included this folding plus some faulting of the Red Bay Group as the later stages of the Haakonian phase, thus pre-dating the Pragian base of the Andrée Land Group. Despite this, many researchers have assumed continuous development of the two stratigraphic groups, particularly from palaeontological evidence. The effects of the inter Red Bay–Andrée Land Group deformation, here named the Monacobreen phase, can now be interpreted in greater detail.

Sinistral strike-slip faulting. Correlation of the stratigraphy and the consistent structure within the tilted fault blocks exposed along the eastern side of Monacobreen show that this deformation was post-sedimentary here. However, as this may have been a localized effect, it does not exclude the possibility that deposition was continuing elsewhere in the basin.

At present, the five identified rotated blocks sit in the hanging wall of the Friedrichbreen Fault in the west limb of the Friedrichbreen Anticline. Movement across the fault and the folding would have caused westward tilting in the hanging wall and is responsible for the present-day bedding orientation within the blocks, which mostly dips WSW or SW (see below). Figure 3 shows bedding dip measurements from the Widerøefjella and Schivefjellet Blocks, back-rotated about the dip of the Friedrichbreen Fault (30° W). This shows that, before thrust faulting and folding, the structure was a series of south-dipping fault blocks.

The structure that developed at this time can be considered in a rotating domino model, where extensional faults bound rigid crustal blocks, and extension is accommodated by linked rotation of adjacent blocks (Fig. 5). Assuming rigid block rotations, the extension factor can be calculated from the initial dip of the block-bounding faults and the final angle of dip of originally horizontal strata (Jackson & McKenzie 1983). On average, bedding in the blocks in question is consistently around 40–45°. If it is assumed that the bounding faults initiated at 60° from horizontal, they must also have rotated by 40–45° giving a final orientation of 15–20° from horizontal. These figures give an extension factor of 2.5–3.3. The length of the extended area is at present about 45 km, from the coast of Liefdefjorden to a position south of Tre Brør (Fig. 3). The extension factors imply that this distance was

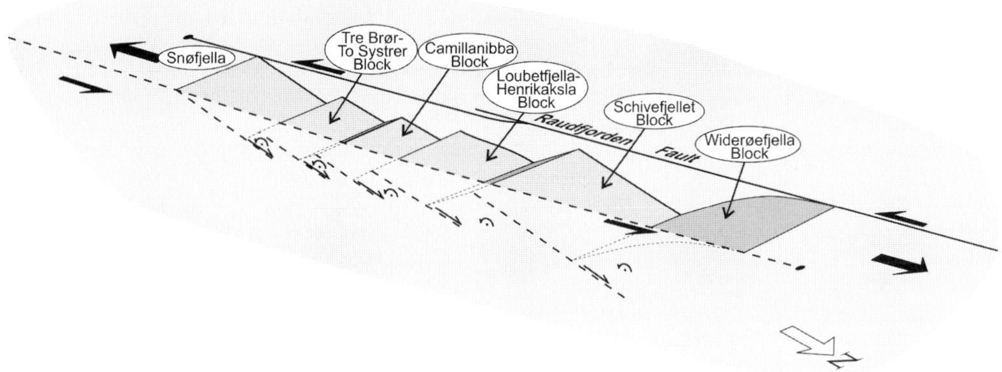

Fig. 5. Schematic model for the pull-apart origin of the tilted fault blocks along Monacobreen, during the Monacobreen phase. Transfer of sinistral strike-slip between the Raudfjorden Fault and a now hidden fault to the east resulted in extension of the overlap area. From estimates of the rotation angles, the north–south extension was about 30 km.

initially 14–18 km and the total extension was 27–31 km.

The bounding faults of the blocks in this area are not well exposed. Between the Widerøefjella and Schivefjellet blocks, a fault has been mapped along Triarmbreen, dipping up to 80° NNE. This is too steep to have been the bounding fault between these blocks responsible for their full rotation. Similarly, the steep fault mapped in the basement of NE Loubetfjella cannot have been active throughout the extension. These faults probably developed when the initial block-bounding faults had rotated to too low an angle to be easily reactivated.

The exposed parts of these blocks are bounded along-strike by the Raudfjorden and Friedrichbreen faults. None of the exposed or proposed bounding faults to the domino blocks can be directly traced into the basement to the west or east. This is clear from mapping of the basement structure west of the Raudfjorden Fault, which has been affected by other through-going structures (e.g. the Oberstbreen and Presidentryggen Faults), but not by any of the major ones bounding the domino blocks (Fig. 3).

The Raudfjorden Fault has a very straight exposure over at least 50 km, but its continuation south of the Presidentryggen Fault (Fig. 3) is uncertain. Many maps show a continuation hidden by the ice west of Snøfjella, but this has been suggested on the assumption that the west-dipping sediments there form a half-graben against the fault (e.g. Hjelle & Lauritzen 1982). If these rather represent the western limb of a syncline, as appears to be the case, there is no need to infer a fault through here, and therefore it could die out in this region. It is reasonable to assume that the Raudfjorden Syncline continues through here, and that the erosion level has cut below the Devonian sediments that formed the western limb.

A model to explain the narrow, elongate area affected by the domino block extension is that it was bounded by strike-slip faults, forming an overstep zone between them (Fig. 5). Where slip is transferred between two parallel strike-slip faults, extension or shortening of the overlap zone may result, depending on the sense of slip on the faults and the direction of overstep. Where extension results, it is commonly accommodated by conjugate sets of normal faults dipping towards the centre of the extended zone, but examples with a single fault polarity are also known (e.g. the Dead Sea Basin, Manspeizer 1985). In the present case, it is proposed that extension was achieved by domino-style rotation of fault-bounded blocks about horizontal axes, between the Raudfjorden Fault and a fault to the east, now hidden. There is no evidence of continued sedimentation during this extension, but the preservation of the Red Bay Group here is evidence of subsidence accompanying this extension.

It is suggested here that displacement was transferred from the Raudfjorden Fault to another, parallel strike-slip fault on the east side of the rotated block zone. In the northern section, this may be hidden in the footwall of the Friedrichbreen Fault, and not exposed in the hanging wall. Gjelsvik (1979) described and illustrated a steep NNW–SSE-trending fault cutting a col northeast of Snøfjella. Although no further interpretation was made and the area was not visited in the present study, it is suggested that this could be part of the eastern strike-slip fault.

It is possible that a steep north–south fault separating tilted Devonian rocks from basement on eastern Loubetfjella is also part of this strike-slip structure (Fig. 3). This would be the only place where it is preserved in the hanging wall of the Friedrichbreen Fault. It does not continue south to Presidentryggen because of the sinistral offset on the east–west fault here. At the northern limit of the extended overlap zone (Widerøefjella Block, Fig. 3), the present-day west- and NW-dipping structure suggests that this area was not tilted southwards before being folded and therefore may mark a flexure at one end of the extended region. The eastern strike-slip fault would therefore not be expected to be found north of here.

The NW–SE-trending extensional faults that cut the Red Bay Group between Raudfjorden and Liefdefjorden were probably also initiated at this time (Fig. 3). They post-date deposition of all the Red Bay Group units, and their concentration in the area north of the strike-slip overstep zone, described above, suggests a relationship to it. If these are dip-slip faults, the inferred extension direction (*c*. NE–SW) fits well with sinistral strike-slip faulting along the north–south-trending Raudfjorden Fault to the west.

At the end of this strike-slip phase, a series of tilted blocks lay in the area of Monacobreen, downthrown against the less deformed basement-cover sequence to the east and west. Extensional faulting had also affected the area immediately north of here.

East–west shortening. The development of the Friedrichbreen Anticline and Raudfjorden Syncline post-dated the strike-slip related extension, and probably also involved deformation across the whole area to the west. The train of folds has wavelengths of 15–20 km and amplitudes in the order of 3 km. A simple estimate of the shortening this represents gives about 5%. Over the width of the exposed region involved (*c*. 55 km), this would give a shortening of about 3 km.

The timing of this large-scale folding is constrained by the overstep of Andrée Land Group sediments seen south of Holtedahlfonna (Fig. 1). Although not exposed, the Friedrichbreen Anticline and Raudfjorden Syncline would be expected to continue south under the Wood Bay Formation nunataks, in which there is no evidence of this folding. The base of these sediments rest on pre-Devonian basement on the west limb of the Raudfjorden Syncline at Pretender (Fig. 1).

The Idabreen strike-slip fault, which is interpreted as a lateral fault to the Friedrichbreen reverse fault, offsets the axis of the Raudfjorden Syncline and so post-dates the main folding. The position of the Friedrichbreen Fault suggests that it is related to the folding as an out-of-the-syncline thrust, transferring the area west of here eastwards up the western limb of the Friedrichbreen Anticline.

The thrust sheet in the east side of the Friedrichbreen Anticline, in which the klippen of Keisar Wilhelmhøgda lie (Fig. 3), may also have been emplaced during the same shortening, but in this case were transported towards the west. The relationship of Devonian rocks thrust over basement requires that the former were first dropped down to the east, probably along the line of the Breibogen–Bockfjorden Fault (Piepjohn 1994). This extension may have accompanied the development of the strike-slip pull-apart zone along the Raudfjorden Fault and therefore would probably have been transtensional also.

The location and strike of both the Friedrichbreen and Keisar Wilhelmhøgda thrusts were probably controlled by the buttressing effect of north–south faults separating the basement of the Friedrichbreen Anticline from sediments downthrown on either side. These discontinuities may have interfered with further amplification of the earlier folds, and led to the breakthrough of low-angle faults.

Piepjohn (1994) suggested that the thrust on Keisar Wilhelmhøgda was emplaced during the Latest Devonian to Earliest Carboniferous Svalbardian phase, rather than late Lochkovian time, as suggested here. Piepjohn's interpretation, however, was made without the evidence of earlier shortening further west. There is a marked difference in intensity of deformation between these Red Bay Group rocks and the adjacent parts of the Andrée Land Group, east of the Breibogen–Bockfjorden Fault, which could be explained in two ways. Either the thrusting predated the Andrée Land Group or it represents a Svalbardian-age detachment at a deeper level than the exposed Andrée Land Group rocks, which were carried passively above it (Piepjohn 1994).

It has been thought that a continuous succession from the upper Red Bay Group to the Andrée Land Group was exposed at Sigurdfjellet (Føyn & Heintz 1943) (Fig. 3). A polymict conglomeratic unit, including clasts of typical basement and Red Bay Group lithologies, is overlain here by lower Andrée Land Group rocks. Similar conglomerates are found directly west of the Breibogen–Bockfjorden Fault in places north of here, and it is likely that these deposits represent reworking of the deformed Red Bay Group.

Palaeontological evidence suggests that the Red Bay Group was deposited during Lochkovian time, and that the lowest exposed levels of the Andrée Land Group were already being deposited by earliest Pragian time (Blieck *et al.* 1987), so the time scale for the development and subsequent inversion of the Red Bay basin must have been very short, as the Lochkovian epoch has been considered to span only about 5 Ma (Tucker & McKerrow 1995). Whether the emplacement of the Friedrichbreen and Keisar Wilhelmhøgda thrusts occurred at this stage or during the Svalbardian phase remains uncertain.

The pre-Andrée Land Group inversion episode, named here as the Monacobreen phase, has been alluded to by previous workers, but it involved greater shortening and had a much greater influence on the basin development than realized before. It has generally been assumed that much of the deformation of the Red Bay Group occurred in the Latest Devonian to Earliest Carboniferous Svalbardian phase, which also affected the younger Devonian deposits.

Svalbardian phase

Sedimentation in the Andrée Land basin continued until at least early Frasnian time, probably into Famennian time (Piepjohn *et al.* this volume), but there was a major unconformity and change in the configuration of depositional basins by earliest Carboniferous time. The unconformity to the Tournaisian Billefjorden Group forms the southern limit of Devonian exposure, striking across southern Dickson Land to the area south of Holtedahlfonna (Fig. 1). The recognition of intensely deformed Devonian rocks below this unconformity, and the juxtaposition with basement rocks east of the Billefjorden Fault Zone, has long been seen as the result of an important folding episode, named the Svalbardian phase by Vogt (1928).

The nature and causes of this basin inversion episode have been discussed by a number of workers, particularly focusing on the Billefjorden Fault Zone (BFZ) where it is exposed between Wijdefjorden and Billefjorden (Fig. 1) (Harland *et al.* 1974; Lamar *et al.* 1986; Manby *et al.* 1994; McCann & Dallmann 1996). This eastern margin to the Devonian basin was reactivated at this time, along a Caledonian sinistral shear zone (Harland *et al.* 1974), which may have also been active during Red Bay Group times (any record of sedimentation at this time is buried). The fault zone itself exposes a strip of pre-Devonian basement, up to 6 km wide, with intense brittle sinistral shearing and a number of later sub-parallel extensional and reverse faults. The bounding reverse fault to the Andrée Land basin is steeply (*c.* 70°) east-dipping.

Immediately to the west, Devonian sediments are affected by tight, upright to steeply east-dipping chevron folds, cut by a number of west-vergent thrusts. This intense deformation zone is about 4 km wide, and open folds separate it from another narrow deformed zone with steep contractional structures, about 10 km further west, the Triungen-Grønhorgdalen Fault Zone (TGFZ) (Fig. 1) (Friend 1962; Harland *et al.* 1974; McCann & Dallmann 1996). These narrow zones of intense deformation typify the Svalbardian structure and Piepjohn (1994) has described another major zone along western Andrée Land (Svalbardian deformation zones in Fig. 1), but open folding also affects the whole basin area.

The location of these deformation zones is probably controlled by reactivation of underlying basement structures, as is seen along the Billefjorden Fault Zone; Piepjohn (1994) suggested that these may be bounding faults to Red Bay Group grabens. The orientation of these structures is important in the assessment of the overall tectonic control. Harland *et al.* (1974) interpreted the Svalbardian phase as a sinistral transpressional episode, as they had mapped Devonian fold axes at an oblique angle clockwise to the Billefjorden Fault Zone. This was used to explain the juxtaposition of the Devonian basin with the Ny Friesland basement terrane. Manby *et al.* (1994), however, reported only structures indicative of east–west shortening, and suggested that the brittle strike-slip shearing in the basement was of Late Silurian to Early Devonian age, and that the Late Devonian shortening was perpendicular to the BFZ.

The Billefjorden Fault Zone, where exposed, does trend north–south but on a larger scale this is atypical as it generally trends about N10°W (McCann & Dallmann 1996). This is parallel to the major faults to the west of Andrée Land (Raudfjorden and Breibogen–Bockfjorden faults) and east–west shortening would therefore imply a sinistral component of slip across these faults. McCann & Dallmann (1996) showed that the footwall structures of the BFZ actually trend slightly east of north, and Piepjohn (1994) showed that the Svalbardian structures of Andrée Land are also indicative of WNW–ESE shortening, although further west there is a change in strike to parallel the Breibogen–Bockfjorden Fault Zone. The shortening direction during the Svalbardian inversion was therefore oblique to the bounding faults and probably involved a component of sinistral strike-slip across them. Strain was predominantly dip-slip, but oblique shortening was responsible for part of the 4 km of apparent dip-slip offset (Harland *et al.* 1974) across the steep Balliolbreen Fault, along the western margin of the BFZ. Along the western basin margin, the buttressing effect of basement in the footwall rotated the shortening direction to be more perpendicular to the Breibogen–Bockfjorden Fault Zone.

Discussion and links to the Ellesmerian and Caledonian orogens

Interpretation of the regional significance of the Devonian tectonic history outlined here requires comparison with areas to the south, along the Caledonides and to the west, through the Ellesmerian orogen. Closure of Iapetus and the Scandian collision between Baltica and Laurentia was at least partly oblique, involving large-scale sinistral movement between their margins (e.g. Torsvik *et al.* 1996), as reflected in the transpression of northeast Greenland (Holdsworth & Strachan 1991). At about the same time as the culmination of collision in the Late Silurian, the Pearya terrane was emplaced against the northern margin of the North American craton. Affinities with the North Atlantic Caledonides suggest that it may have been derived by major sinistral strike-slip movement around the northeastern corner of Laurentia (Trettin 1987). This would be the first of a series of terranes transported around this margin, including the terranes of Svalbard, also derived from eastern Laurentia (Fig. 6A) (e.g. Harland 1985). The Eastern terrane of Svalbard records intense Scandian shortening and metamorphism (Ny Friesland Orogen, Harland *et al.* 1992), but the Central and Western terranes were less affected and may have been derived from a more foreland position. Gee (1996) interpreted the line of the Scandian suture to pass through the Barents Sea, southeast of Svalbard, so these terranes can be seen as having been assembled along fault zones splaying off the main Caledonian strike-slip zone. At this time, shortening of the Franklinian Trough had not commenced, so there were no hindrances to terrane transport into this area. The Siktefjellet Group probably represents deposition into an extensional basin during continued terrane transport. Extension across these crustal blocks is compatible with the sinistral movement between Laurentia and Baltica. Subsequent juxtaposition of the Biskayerhalvøya and Krossfjorden terranes, recorded as the Haakonian phase deformation, may have been linked to movement along the Breibogen–Bockfjorden Fault Zone and marks the end of large-scale strike-slip movements in Svalbard. This was also probably the time of cessation of movement along the Billefjorden Fault Zone between the Eastern and Central province. This does not, however, exclude continued movement between Svalbard and Greenland.

The Red Bay Group was deposited in an extensive deep basin, but local evidence does not constrain the tectonic controls on subsidence. Extension across northern Svalbard is compatible with continued sinistral strike-slip along the Caledonian suture southeast of here. Inversion of the Red Bay basin in the Monacobreen phase started with renewed sinistral strike-slip faulting, but with a displacement only of the order of 30 km. This reactivation of major faults in Svalbard is probably the last remnant of a tectonic link with the Caledonides.

The North Greenland Fold Belt (NGFB) is a deformed Late Proterozoic–Early Palaeozoic passive margin sequence along-strike from the Franklinian Mobile Belt, which was deformed during accretion of Pearya in Late Silurian time. A thick sequence of Silurian turbidites was derived from the rising Caledonides to the east, and this deposition may have continued into Devonian time (Hurst & Surlyk 1982). These were deformed in the Ellesmerian Orogeny sometime during the Devonian period, and the oldest overlying sediments are of Moscovian age (Stemmerik & Håkansson 1991). The Ellesmerian Orogeny is recorded by the deformation of the Franklinian Mobile Belt and the NGFB, and also by the development of a Mid- and Late Devonian foreland basin in the Canadian Arctic Archipelago (Embry 1988). The orogeny was probably the result of collision between the Greenland and Canadian cratons and the Chutkotka–Alaska plate (Fig. 6B–D), now lying on the opposite side of the Mesozoic Canada Basin (e.g. Ziegler 1988). The Devonian foreland basin was deformed diachronously, migrating southwest (Embry 1988), and Ellesmerian deformation in North Greenland also increases westward (Soper & Higgins 1990). This implies that the collision was progressive from east to west and may have involved anticlockwise rotation of the Chukotka–Alaska plate around an axis not far from the position of Svalbard.

The shortening seen during the Monacobreen phase in Svalbard, with folding and thrusting, probably occurred around the time of initiation

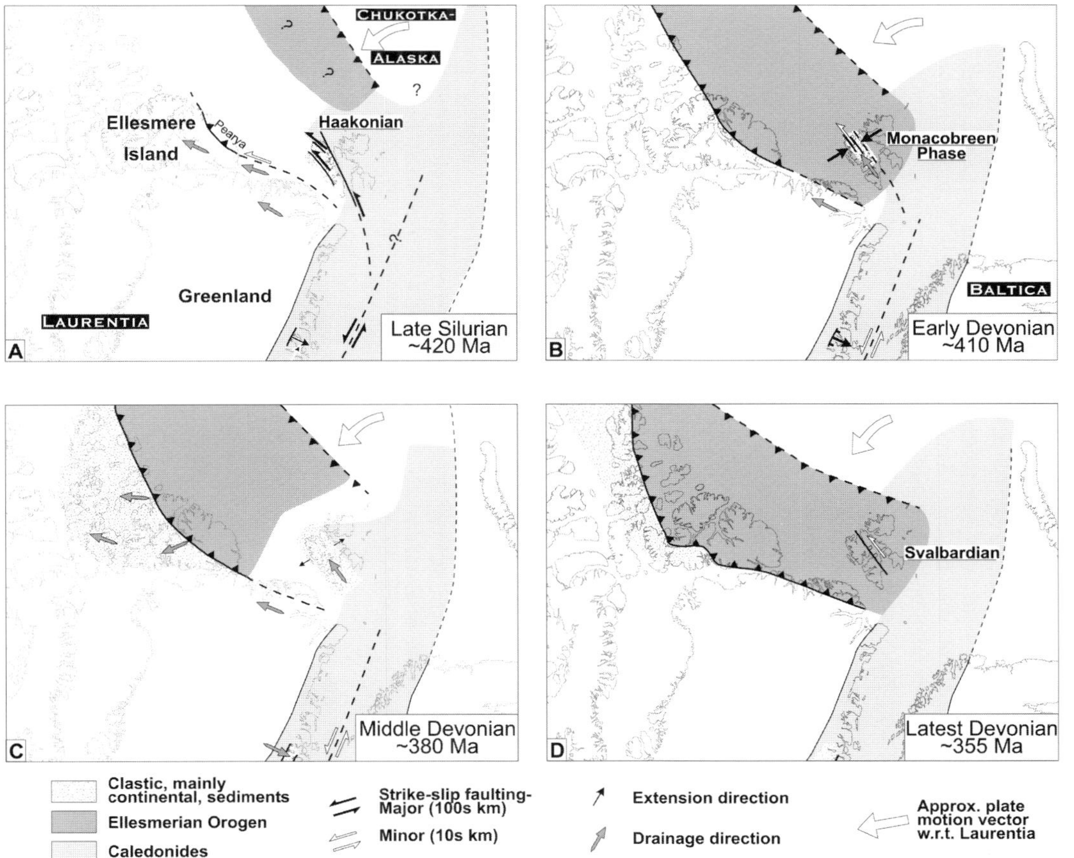

Fig. 6. Plate reconstructions showing the relationship between the Devonian tectonic events of Svalbard and the tectonics of the Caledonian and Ellesmerian orogens. Anticlockwise rotation of the Chukotka–Alaska plate caused progressive deformation across northern Laurentia, increasing in intensity and occurring later further west. Svalbard occupied a position that may have been near the pole of rotation between this plate and Laurentia. Drift of this pole may have been responsible for the repeated extension and shortening history seen in the ORS of Svalbard, combined with decreasing strike-slip faulting related to the Caledonian collision. (Partly after Rowley & Lottes (1988), Soper *et al.* (1992) and Torsvik *et al.* (1996)).

of Ellesmerian collision (Fig. 6B). The small degree of shortening compared with further west was a consequence of the rotational nature of the collision and Svalbard's position near the rotation axis. The subsequent extension and development of the Andrée Land basin on Svalbard, throughout Mid- and Late Devonian time, occurred during continued shortening in the Ellesmerian orogen not far west of here. This can also be explained by the rotation axis being near Svalbard, at this time having moved slightly to the north so that Svalbard lay within the arc of extension (Fig. 6C).

Old Red Sandstone deposition was finally halted by the Svalbardian deformation, which was sub-parallel to and synchronous with the culmination of the Ellesmerian Orogeny. Within the rotating plate model suggested here, the final closure of the Franklinian Trough may have resulted when the two plate margins became parallel, and convergence peaked and then ceased (Fig. 6D). The timing of the deformation in the North Greenland Fold Belt is only constrained to be of post-Silurian and pre-Moscovian age. It is possible, therefore, that this deformation was restricted to Late Devonian time, as on Svalbard, rather than having been progressive throughout the Devonian period as further west.

Conclusions

This paper has briefly reviewed the deformation history of the Old Red Sandstone of northern Spitsbergen, which can be summarized as follows:

(1) The Siktefjellet Group was deposited during latest Silurian or earliest Devonian time on the Biskayerhalvøya Terrane before final juxtaposition to its present position.

(2) Early Devonian terrane juxtaposition (the Haakonian phase) against the Krossfjorden Terrane was achieved by sinistral strike-slip faulting along the line of the Rabotdalen and Hannabreen faults. The Siktefjellet Group was preserved and flexed down towards the south against a cross fault. This marked the end of major Caledonian-related shear movements through this part of Svalbard.

(3) The Red Bay Group was deposited in an extensive fluvial basin, with sediments chiefly sourced from the south. A succession of over 3.5 km was deposited before renewed strike-slip faulting cut through the basin.

(4) At the end of Lochkovian time, sinistral strike-slip faulting along the Raudfjorden Fault and a parallel fault to the east broke up the Red Bay basin in an overstep zone between them. About 30 km of north–south extension occurred across a set of rotating fault blocks, which were tilted by up to 45°. This first part of the Monacobreen deformation phase, may reflect renewed tectonic links with the Caledonian Orogeny, but was followed by shortening related to the initiation of collision against the northern Laurentian margin, at the beginning of the Ellesmerian Orogeny.

(5) As Ellesmerian collision continued, Svalbard was extended again, and the Andrée Land basin developed. This may have been due to the pole of rotation between Chukotka–Alaska and Laurentian having migrated north of Svalbard.

(6) The culmination of Ellesmerian collision at the end of the Devonian period is reflected in Svalbard by the intense shortening and inversion of the whole ORS basin in the Svalbardian deformation.

This work formed part of a PhD study at the University of Cambridge, for which the support of the Cambridge Arctic Shelf Programme (CASP) and Norsk Polarinstitutt (NP) is gratefully acknowledged. Thanks are due to W. K. Dallmann, T. Gjelsvik, Y. Ohta (NP), M. Miloslavskij, A. Teben'kov (PMGRE, Lomonosov), K. Piepjohn (Münster) and D. Gee (Uppsala) for collaboration in the field, and to P. F. Friend and W. B. Harland for fruitful discussions in Cambridge. Thorough reviews by D. Gee and K. Piepjohn improved the original manuscript.

References

BALAŠOV, JU. A., PEUCAT, J. J., TEBEN'KOV, A. M., OHTA, Y., LARIONOV, A. N. & SIROTKIN, A. N. 1996. Additional Rb–Sr and single-grain zircon datings of Caledonian granitoid rocks from Albert I Land, northwest Spitsbergen. *Polar Research*, **15**, 153–165.

——, TEBEN'KOV, A. M., OHTA, Y., LARIONOV, A. N., SIROTKIN, A. N., GANNIBAL, L. F. & RYÜNGENEN, G. I. 1995. Grenvillian U–Pb zircon ages of quartz porphyry and rhyolite clasts in a metaconglomerate at Vimsodden, southwestern Spitsbergen. *Polar Research*, **14**, 291–302.

BLIECK, A., GOUJET, D. & JANVIER, P. 1987. The vertebrate stratigraphy of the Lower Devonian (Red Bay Group and Wood Bay Formation) of Spitsbergen. *Modern Geology*, **11**, 197–217.

DALLMANN, W. K., PIEPJOHN, K., MCCANN, A. J., SIROTKIN, A. N., MILOSLAVSKIJ, M. JU., OHTA, Y. & GJELSVIK, T. 1998. *Woodfjorden*. Geological map of Svalbard 1:100 000, sheet B5G, preliminary edition. Norsk Polarinstitutt, Oslo.

DALLMEYER, R. D., PEUCAT, J. J. & OHTA, Y. 1990. Tectonothermal evolution of contrasting metamorphic complexes in northwest Spitsbergen (Biskayerhalvøya): evidence from $^{40}Ar/^{39}Ar$ and

Rb–Sr mineral ages. *Geological Society of America Bulletin*, **102**, 653–663.

EMBRY, A. F. 1988. Middle–Upper Devonian sedimentation in the Canadian Arctic Islands and the Ellesmerian Orogeny. *In*: MCMILLAN, N. J., EMBRY, A. F. & GLASS, D. J. (eds) *Devonian of the World, Vol. 2*. Canadian Society of Petroleum Geologists, Memoirs, **14**, 15–28.

FØYN, S. & HEINTZ, A. 1943. The Downtonian and Devonian vertebrates of Spitsbergen. *In*: *The English–Norwegian–Swedish Expedition 1939, VIII. Geological Results*. Skrifter om Svalbard og Ishavet, **85**, 1–51.

FRIEND, P. F. 1961. The Devonian stratigraphy of North and Central Vestspitsbergen. *Proceedings of the Yorkshire Geological Society*, **33**, 77–118.

—— 1962. *Devonian rocks of Northern Spitsbergen*. PhD thesis, University of Cambridge.

——, HARLAND, W. B., ROGERS, D., SNAPE, I. & THORNLEY, S. 1997. Late Silurian and Early Devonian stratigraphy and probable strike-slip tectonics in Northwestern Spitsbergen. *Geological Magazine*. 134, 459–479.

—— & MOODY-STUART, M. 1972. Sedimentation of the Wood Bay Formation (Devonian) of Spitsbergen: Regional Analysis of a Late Orogenic Basin. *Norsk Polarinstitutt Skrifter*, **157**, 1–77.

GEE, D. G. 1966*a*. *The structural geology of the Biskayerhuken Peninsula, North Spitsbergen*. PhD thesis, University of Cambridge.

—— 1966*b*. A note on the occurrence of eclogites in Spitsbergen. *Norsk Polarinstitutt Årbok*, **1964**, 240–241.

—— 1972. Late Caledonian (Haakonian) movements in northern Spitsbergen. *Norsk Polarinstitutt Årbok*, **1970**, 92–101.

—— 1986. Svalbard's Caledonian terranes reviewed. *Geologiska Föreningens i Stockholm Förhandlingar*, **108**, 284–286.

—— 1996. Barentia and the Caledonides of the High Arctic. *Geologiska Föreningens i Stockholm Förhandlingar*, **118** (Jubilee Issue), A32–A33.

—— & HJELLE, A. 1966. On the crystalline rock of northwest Spitsbergen. *Norsk Polarinstitutt Årbok*, **1964**, 31–45.

—— & MOODY-STUART, M. 1966. The base of the Old Red Sandstone in central north Haakon VII Land, Vestspitsbergen. *Norsk Polarinstitutt Årbok*, **1964**, 57–68.

—— & PAGE, L. M. 1994. Caledonian terrane assembly on Svalbard: new evidence from $^{40}Ar/^{39}Ar$ dating in Ny Friesland. *American Journal of Science*, **294**, 1166–1186.

GJELSVIK, T. 1974. A new occurrence of Devonian rocks in Spitsbergen. *Norsk Polarinstitutt Årbok*, **1972**, 23–28.

—— 1979. The Hecla Hoek ridge of the Devonian Graben between Liefdefjorden and Holtedahlfonna, Spitsbergen. *In*: WINSNES, T. S. (ed.) *The Geological Development of Svalbard during the Precambrian, Lower Palaeozoic, and Devonian*. Norsk Polarinstitutt Skrifter, **167**, 63–71.

—— 1991. Composition and provenance of the Lilljeborgfjellet Conglomerate, Haakon VII Land, Spitsbergen. *Polar Research*, **9**, 141–154.

—— & ILYES, R. 1991. Distribution of Late Silurian(?) and Early Devonian grey–green sandstones in the Liefdefjorden–Bockfjorden area, Spitsbergen. *Polar Research*, **9**(1), 77–87.

GROMET, L. P. & GEE, D. G. 1998. An evaluation of the age of high-grade metamorphism in the Caledonides of Biskayerhalvøya, NW Svalbard. *Geologiska föreningens i Stockholm förhandlingar*, **120**, 199–208.

HÅKANSSON, E. & STEMMERIK, L. 1984. Wandel Sea Basin—The North Greenland equivalent to Svalbard and the Barents Shelf. *In*: SPENCER, A. M., JOHNSEN, S. O., MØRK, A. *et al.* (eds) *Petroleum Geology of the North European Margin*. Norwegian Petroleum Society, Oslo; Graham & Trotman, London, 97–107.

HARLAND, W. B. 1961. An outline structural history of Spitsbergen. *In*: RAASCH, G. O. (ed.) *Geology of the Arctic, Vol. 1*. University of Toronto Press, Toronto, Ont., 68–132.

—— 1985. Caledonide Svalbard. *In*: GEE, D. G. & STURT, B. A. (eds) *The Caledonide Orogen—Scandinavia and Related Areas*. Wiley, Chichester, 999–1016.

—— 1997. *The Geology of Svalbard*. Geological Society, London, Memoirs, **17**.

—— & WRIGHT, N. J. R. 1979. Alternative hypothesis for the pre-Carboniferous evolution of Svalbard. *In*: WINSNES, T. S. (ed.) *The Geological Development of Svalbard during the Precambrian, Lower Palaeozoic, and Devonian*. Norsk Polarinstitutt Skrifter, **167**, 89–117.

——, CUTBILL, J. L., FRIEND, P. F. *et al.* (eds) 1974. The Billefjorden Fault Zone, Spitsbergen: the Long History of a Major Tectonic Lineament. *Norsk Polarinstitutt Skrifter*, **161**, 1–72.

——, SCOTT, R. A., AUCKLAND, K. A. & SNAPE, I. 1992. The Ny Friesland Orogen, Spitsbergen. *Geological Magazine*, **129**, 679–708.

HELLMAN, F. J., GEE, D. G., GJELSVIK, T. & TEBENKOV, A. M. 1998. Provenance and tectonic implications of Palaeoproterozoic (*c.* 1740 Ma) quartz porphyry clasts in the basal Old Red Sandstone (Lilljeborgfjellet Conglomerate Formation) of northwestern Svalbard's Caledonides. *Geological Magazine*, **135**, 755–768.

HJELLE, A. 1979. Aspects of the geology of northwest Spitsbergen. *In*: WINSNES, T. S. (ed.) *The Geological Development of Svalbard during the Precambrian, Lower Palaeozoic, and Devonian*. Norsk Polarinstitutt Skrifter, **167**, 37–62.

—— & LAURITZEN, Ø. 1982. *Geological map of Svalbard 1 : 500 000, sheet 3G, Spitsbergen northern part*. Norsk Polarinstitutt Skrifter, **154C**.

HOLDSWORTH, R. E. & STRACHAN, R. A. 1991. Interlinked system of ductile strike slip and thrusting formed by Caledonian sinistral transpression in northeastern Greenland. *Geology*, **19**, 510–513.

HOLTEDAHL, O. 1914. New features in the geology of northwestern Spitzbergen. *American Journal of Science*, **37**, 415–424.

HURST, J. M. & SURLYK, F. (eds) 1982. *Stratigraphy of the Silurian turbidite sequence of North Greenland.* Grønlands Geologiske Undersøgelse Bulletin, **145**.

JACKSON, J. A. & MCKENZIE, D. P. 1983. The geometrical evolution of normal fault systems. *Journal of Structural Geology*, **5**, 471–482.

LAMAR, D. L., REED, W. E. & DOUGLASS, D. N. 1986. The Billefjorden Fault Zone, Spitsbergen: is it part of a major Late Devonian transform? *Geological Society of America Bulletin*, **97**, 1083–1088.

MANBY, G. M. & LYBERIS, N. 1992. Tectonic evolution of the Devonian Basin of northern Svalbard. *In*: DALLMANN, W. K., ANDRESEN, A. & KRILL, A. (eds) *Post-Caledonian Tectonic Evolution of Svalbard. Norsk Geologisk Tidsskrift*, **72**, 7–19.

——, ——, CHOROWICZ, J. & THIEDIG, F. 1994. Post-Caledonian tectonics along the Billefjorden fault zone, Svalbard, and implications for the Arctic region. *Geological Society of America Bulletin*, **106**, 201–216.

MANSPEIZER, W. 1985. The Dead Sea Rift: impact of climate and tectonism on Pleistocene and Holocene sedimentation. *In*: BIDDLE, K. T. & CHRISTIE-BLICK, N. (eds) *Strike-slip Deformation, Basin Formation, and Sedimentation.* Society of Economic Paleontologists and Mineralogists, Special Publications, **37**, 143–158.

MCCANN, A. J. 1997. *Structure and sedimentology of the Late Silurian-Early Devonian basins of NW Spitsbergen.* PhD thesis, University of Cambridge.

—— & DALLMANN, W. K. 1996. Reactivation history of the long-lived Billefjorden Fault Zone in north central Spitsbergen, Svalbard. *Geological Magazine*, **133**, 63–84.

MURAŠOV, L. G. & MOKIN, JU. I. 1976. Stratigraficeskoe rasclelenie devonskih otlozenij o. Spicbergen (Stratigraphic subdivision of Devonian deposits of the island of Spitsbergen). *In*: SOKOLOV, V. N. (ed.) *Geologija Sval'barda.* NIIGA, Leningrad, 78–91.

NØTTVEDT, A. 1994. Post-Caledonian sediments on Spitsbergen. *In*: EIKEN, O. (ed.) *Seismic Atlas of Western Svalbard.* Norsk Polarinstitutt Meddelelser, **130**, 40–48.

OHTA, Y. 1994. Caledonian and Precambrian history in Svalbard: a review, and an implication of escape tectonics. *Tectonophysics*, **231**, 183–194.

——, DALLMEYER, R. D. & PEUCAT, J. J. 1989. Caledonian terranes in Svalbard. *In*: DALLMEYER, R. D. (ed.) *Circum-Atlantic Paleozoic Orogens*/ Geological Society of America Special Paper, **230**, 1–15.

PEUCAT, J. J., OHTA, Y., GEE, D. G. & BERNARD-GRIFFITHS, J. 1989. U–Pb, Sr and Nd evidence for Grenvillian and latest Proterozoic tectonothermal activity in the Spitsbergen Caledonides, Arctic Ocean. *Lithos*, **22**, 275–285.

PIEPJOHN, K. 1994. *Tektonische Evolution der Devongräben (Old Red) in NW-Svalbard* (The tectonic evolution of the Devonian (Old Red) graben of NW Svalbard). Doktor Grades thesis, Westphalian Wilhelms-University of Münster.

—— 2000. The Svalbardian–Ellesmerian deformation of the Old Red Sandstone and the pre-Devonian basement in NW Spitsbergen (Svalbard). This volume.

——, BRINKMANN, L., GREWING, A. & KERP, H. 2000. New data on the age of the uppermost ORS and the lowermost post-ORS strata in Dickson Land (Spitsbergen) and implications for the age of the Svalbardian deformation. This volume.

ROWLEY, D. B. & LOTTES, A. L. 1988. Plate-kinematic reconstructions of the North Atlantic and Arctic: Late Jurassic to Present. *Tectonophysics*, **155**, 73–120.

SOPER, N. J. & HIGGINS, A. K. 1990. Models for the Ellesmerian mountain front in North Greenland: a basin margin inverted by basement uplift. *Journal of Structural Geology*, **12**, 83–97.

——, STRACHAN, R. A., HOLDSWORTH, R. E., GAYER, R. A. & GREILING, R. O. 1992. Sinistral transpression and the Silurian closure of Iapetus. *Journal of the Geological Society, London*, **149**, 871–880.

STEMMERIK, L. & HÅKANSSON, E. 1991. Carboniferous and Permian history of the Wandel Sea Basin, North Greenland. *In*: PEEL, J. S. & SØNDERHOLM, M. (eds) *Sedimentary Basins of North Greenland.* Grønlands Geologiske Undersøgelse Bulletin, **160**, 141–151.

THIEDIG, F. & MANBY, G. M 1992. Origins and deformation of post-Caledonian sediments on Blomstrandhalvøya and Lovénøyane, northwest Spitsbergen. *In*: DALLMANN, W. K., ANDRESEN, A. & KRILL, A. (eds) *Post-Caledonian Tectonic Evolution of Svalbard.* Norsk Geologisk Tidsskrift, **72**, 27–33.

TORSVIK, T. H., SMETHURST, M. A., MEERT, J. G. *et al.* 1996. Continental break-up and collision in the Neoproterozoic and Palaeozoic—a tale of Baltica and Laurentia. *Earth-Science Reviews*, **40**, 229–258.

TRETTIN, H. P. 1987. Pearya: a Composite Terrane with Caledonian Affinities in Northern Ellesmere Island. *Canadian Journal of Earth Sciences*, **24**, 224–245.

TUCKER, R. D. & MCKERROW, W. S. 1995. Early Paleozoic chronology—a review in light of new U–Pb zircon ages from Newfoundland and Britain. *Canadian Journal of Earth Sciences*, **32**, 368–379.

VOGT, T. 1928. Den norske fjellkjedes revolusjonshistorie. *Norsk Geologisk Tidsskrift*, **10**, 97–115.

WORSLEY, D. 1972. Sedimentological observations on the Grey Hoek Formation of northern Andrée Land, Spitsbergen. *Norsk Polarinstitutt Årbok*, **1970**, 102–111.

ZIEGLER, P. A. 1988. Laurussia—The Old Red Continent. *In*: MCMILLAN, N. J., EMBRY, A. F. & GLASS, D. J. (eds) *Devonian of the World.* Canadian Society of Petroleum Geologists, Memoirs, **14**, 15–48.

The Svalbardian–Ellesmerian deformation of the Old Red Sandstone and the pre-Devonian basement in NW Spitsbergen (Svalbard)

KARSTEN PIEPJOHN

Geologisch-Paläontologisches Institut der Westfälischen Wilhelms-Universität Münster, Corrensstraße 24, D-48149 Münster, Germany (e-mail: piepjoh@uni-muenster.de)

Abstract: In NW Spitsbergen, the Late Silurian to Late Devonian infill of an Old Red Sandstone (ORS) basin was affected by west-vergent folding and west-directed thrusting during the Early Carboniferous (Tournaisian) Svalbardian deformation. The brittle, predominantly compressional structures of the Svalbardian Fold-and-Thrust Belt are concentrated along at least five narrow, more or less north–south-trending zones. Three zones are exposed in the Devonian infill of the ORS basin. The involvement of the post-Caledonian ?Late Silurian to Earliest Devonian Viggobreen weathering zone and deposits Early Devonian in two thrust zones within the basement of the western basin margin indicates that the Svalbardian deformation also affected the basement areas along the west coast of NW Spitsbergen. Structures of the Svalbardian Fold-and-Thrust Belt are exposed within an area at least 100 km wide between the Billefjorden Fault Zone in the east and the west coast of NW Spitsbergen. Therefore, the Svalbardian deformation represents a much more important fold belt than previously recognized. On the basis of the timing, the large extent and the orientation of the fold-and-thrust zones, the Svalbardian Fold-and-Thrust Belt appears to represent the eastern continuation of the Ellesmerian Fold Belt in North Greenland and the Canadian Arctic Archipelago.

In NW Spitsbergen, a 70 km wide and 160 km long, NNW–SSE-trending Old Red Sandstone (ORS) basin is exposed (Fig. 1), which contains a great thickness (8000 m is the sum of the greatest thicknesses of the various formations; Friend & Moody-Stuart 1972) of clastic sediments of Late Silurian to Late Devonian age (e.g. Holtedahl 1914; Vogt 1938; Føyn & Heintz 1943; Friend 1961; Gee & Moody-Stuart 1966; Murašov & Mokin 1979; Piepjohn *et al.* this volume). To the west and the east, it is bounded by Caledonian and older basement rocks along the Raudfjorden Fault (Frebold 1935) and the Billefjorden Fault Zone (Harland *et al.* 1974) (Fig. 1). In the basement areas west of the ORS basin, north–south-trending occurrences of Lower Devonian sediments are exposed northeast of Kongsfjorden on Blomstrandhalvøya (Gjelsvik 1974; Thiedig & Manby 1992), at Løvlandfjellet (Hjelle 1979) and west of Krossfjorden on Mitrahalvøya (Piepjohn *et al.* 1997*b*; Thielemann & Thiedig 1997).

The ORS in Spitsbergen can be divided into two successions. The lower ORS is represented by coarse-clastic deposits of the ?Upper Silurian Siktefjellet Group (Gee & Moody-Stuart 1966) and the Lower Devonian Red Bay Group (Holtedahl 1914). These units consist of thick sandstones and conglomerates and are exposed in the western areas of the ORS basin.

After deposition of the lower ORS, the Red Bay deposits were downfaulted within narrow north–south-trending zones during the late Haakonian (Gee 1972) or Monacobreen phase (McCann this volume). This tectonic phase is interpreted as having been controlled by extensional block-faulting (Piepjohn 1994, 1997) or by sinistral strike-slip movements (McCann this volume).

The upper succession of the ORS is dominated by fine-clastic sediments of the Andrée Land Group (Harland *et al.* 1974) and forms the main part of the basin between the Breibogen Fault in the west and the Billefjorden Fault Zone in the east (Fig. 1). It is divided into the Wood Bay, Grey Hoek and Wijde Bay formations (Holtedahl 1914) and the Mimerdalen Formation (Vogt 1938) (Fig. 2).

The red and partly green silt- and sandstones of the *c.* 3000 m thick Wood Bay Formation (Friend & Moody-Stuart 1972; Gee 1972) are exposed in the entire basin between Reinsdyrflya and Dickson Land. In northern Andrée Land, they are conformably overlain by dark and light grey shales, silt- and sandstones of the Middle

From: Friend, P. F. & Williams, B. P. J. (eds). *New Perspectives on the Old Red Sandstone.* Geological Society, London, Special Publications, **180**, 585–601. 0305-8719/00/$15.00

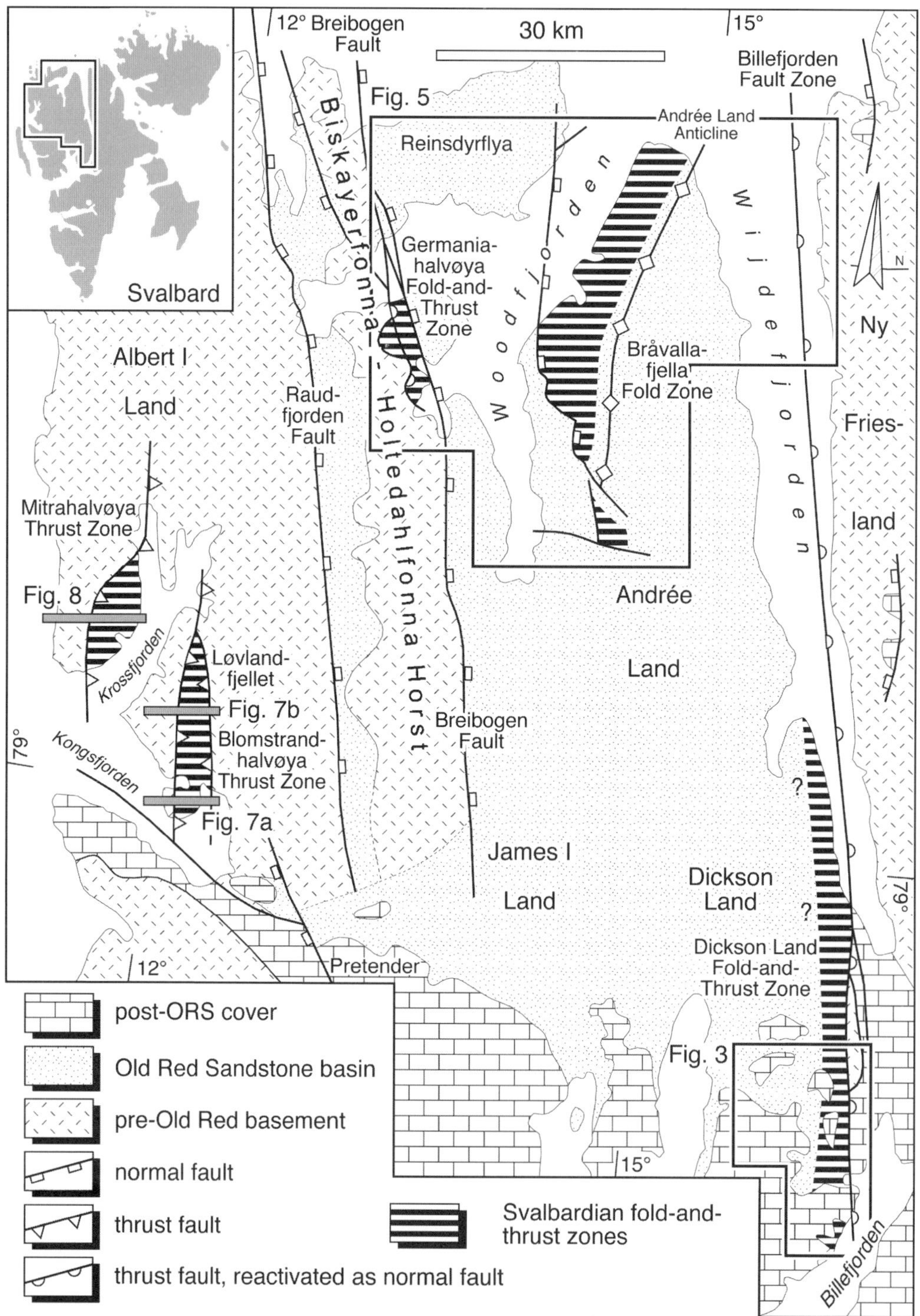

Fig. 1. Tectonic map of NW Spitsbergen showing the fold-and-thrust zones of the Svalbardian deformation in the ORS basin and the western basin margin in NW Spitsbergen (redrawn from Piepjohn & Thiedig (1997)).

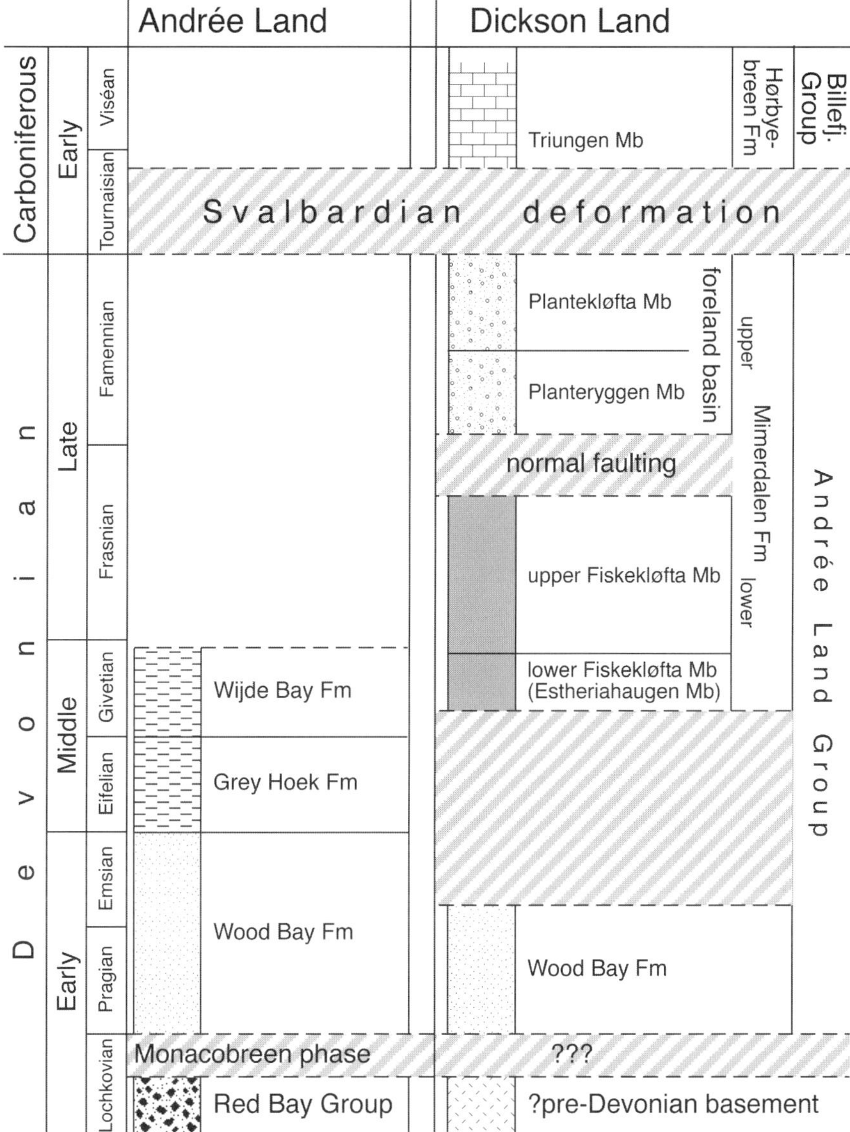

Fig. 2. Stratigraphic division of the upper Old Red Sandstone (Andrée Land Group) and of tectonic events in northern Spitsbergen, after Holtedahl (1914), Vogt (1941), Føyn & Heintz (1943), Friend (1961), Friend & Moody-Stuart (1972), Murašov & Mokin (1979), Pčelina *et al.* (1986), Piepjohn *et al.* (1997*a*), McCann (this volume) and Piepjohn *et al.* (this volume).

Devonian Grey Hoek and Wijde Bay formations (Heintz 1937; Nilsson 1941; Føyn & Heintz 1943) (Fig. 2).

In Dickson Land, the Wood Bay Formation is unconformably overlain by silt- and sandstones of the Fiskekløfta Member (Fiskekløfta Formation, Friend 1961) of the lower Mimerdalen Formation (Figs 2 and 3). The former Estheriahaugen Formation (Friend 1961) is interpreted to represent the lower part of the Fiskekløfta Member (Piepjohn *et al.* 1997*a*). In this paper, it is suggested to be equivalent to the Wijde Bay Formation in northern Andrée Land (Fig. 2).

After deposition of the Fiskekløfta Member, normal faulting along north–south-trending faults affected Wood Bay and Fiskekløfta deposits at the southern slope of Hugindalen (Fig. 3) (Piepjohn *et al.* 1997*a*). It was followed

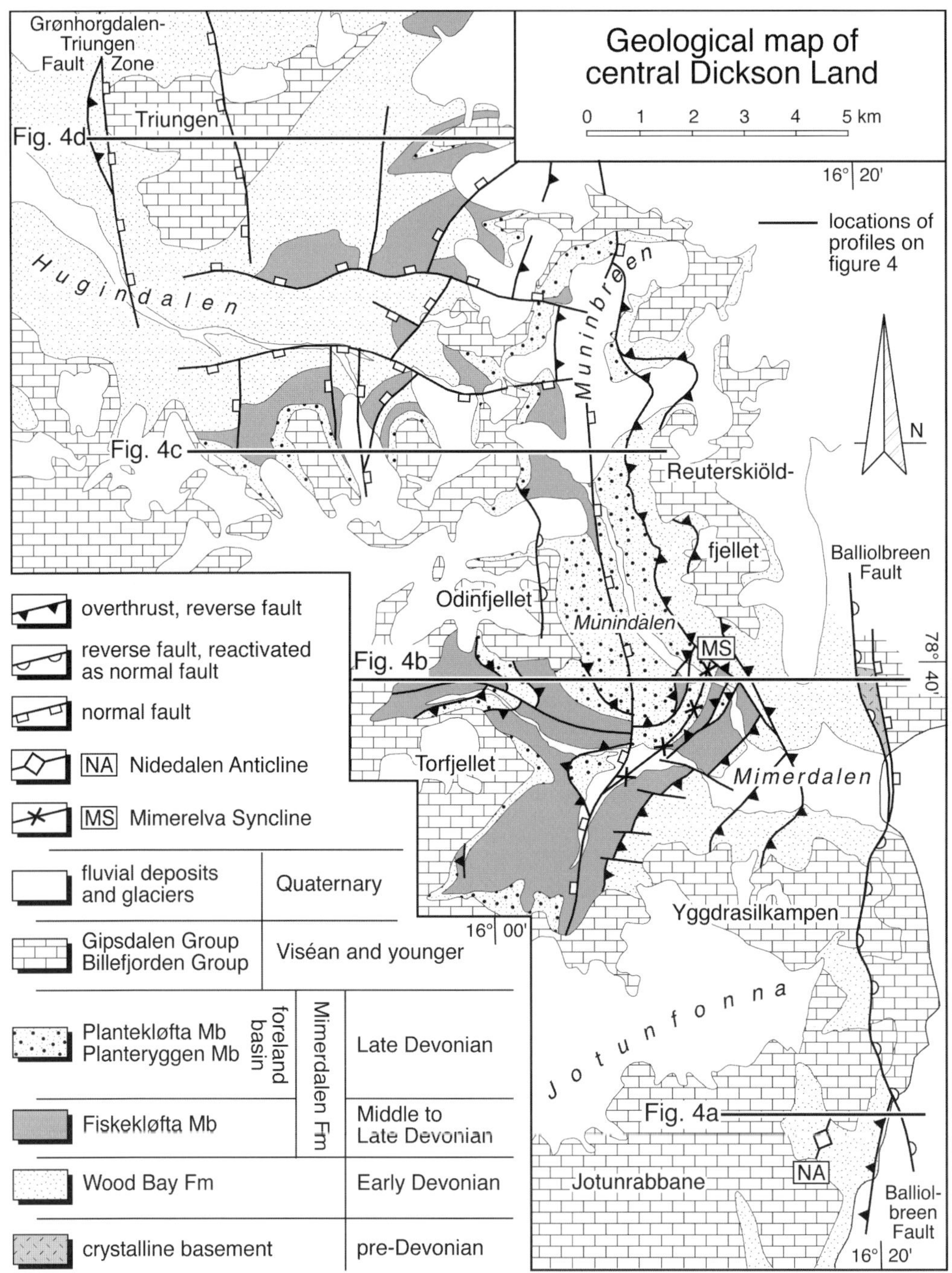

Fig. 3. Geological map of the Dickson Land Fold-and-Thrust Zone to the west of the Billefjorden Fault Zone in central Dickson Land, redrawn from Piepjohn *et al.* (1997*a*) (for location see Fig. 1).

by the deposition of siltstones, sandstones and conglomerates of the Planteryggen and Plantekløfta members (Friend 1961) of the upper part of the Mimerdalen Formation which unconformably overlie the faulted Fiskekløfta Member (Figs 3 and 4) (Piepjohn *et al.* 1997*a*).

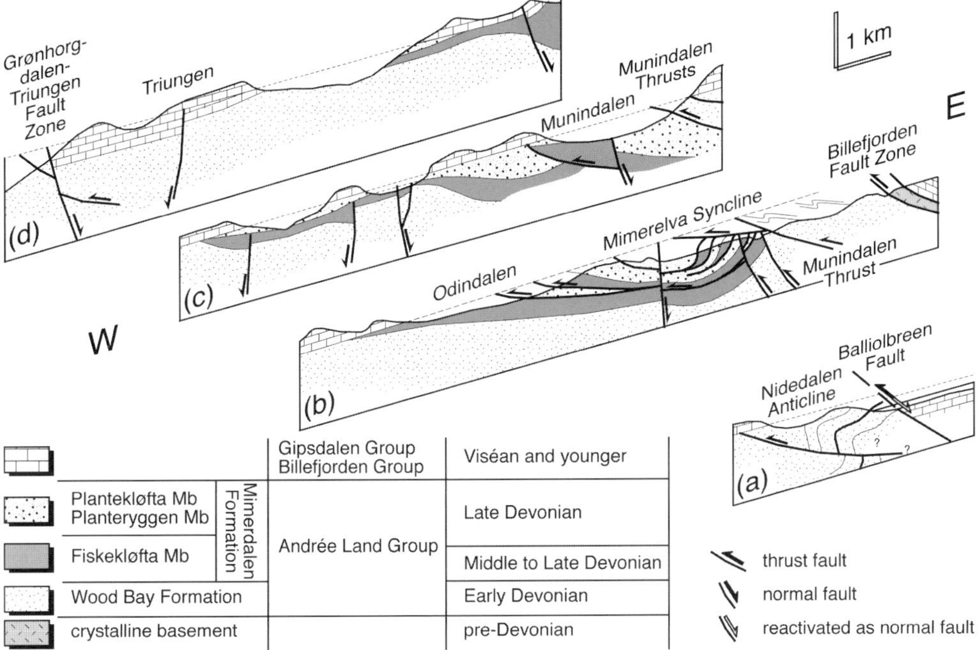

Fig. 4. Cross-sections through the Dickson Land Fold-and-Thrust Zone, redrawn from Piepjohn *et al.* (1997*a*) (for location of profiles see Fig. 3).

The infill of the ORS basin in NW Spitsbergen is affected by folding and thrusting, which was first mentioned by Holtedahl (1914), who described fold structures in the northern part of the ORS basin, west of Wijdefjorden. Stensiø (1918) observed that folded and thrust-faulted Devonian sediments to the west of Billefjorden are unconformably overlain by Middle Carboniferous limestones. In this area, Vogt (1928) defined the Late Devonian Svalbardian folding, because deformed Old Red sediments are overlain by Early Carboniferous clastic deposits in places parallel to the Billefjorden Fault Zone.

The Svalbardian deformation did not affect the entire infill of the Devonian basin in NW Spitsbergen (Friend & Moody-Stuart 1972; Piepjohn 1994; McCann & Dallmann 1996; Piepjohn & Thiedig 1997). Within large areas in James I Land, in western Dickson Land and to the west of Woodfjorden (Fig. 1) the ORS is not folded. In the northern part of the ORS basin, a broad, SSW-plunging anticline–syncline pair on Reinsdyrflya (Piepjohn 1994) and the large-scale wide-spanned Andrée Land Anticline (Burov & Semevskij 1979) between Woodfjorden and Wijdefjorden are exposed (Fig. 1).

Besides these undeformed to weakly folded areas, the Svalbardian deformation is concentrated in distinct approximately north–south-trending zones (Friend & Moody-Stuart 1972; Piepjohn 1994). In this paper, three zones of intense crustal shortening are described within the ORS basin (Dickson Land Fold-and-Thrust Zone, Bråvallafjella Fold Zone and Germaniahalvøya Fold-and-Thrust Zone) (Fig. 1). Outside the ORS basin, the Svalbardian deformation can be recognized in the Blomstrandhalvøya and Mitrahalvøya thrust zones in the Kongsfjorden and Krossfjorden regions on the western basement margin of the ORS basin (Fig. 1) (Piepjohn & Thiedig 1997).

The Svalbardian deformation was, until now, thought to be of Late Devonian age (e.g. Vogt 1928; Harland *et al.* 1974; Lauritzen *et al.* 1989) or was interpreted to have taken place during a Late Frasnian to Early Famennian hiatus (Harland 1997). Recent palynological investigations of the uppermost ORS unit and the lowermost post-ORS deposits show that the tectonic stages of the Svalbardian deformation that affected the infill of the ORS basin took place during Tournaisian times (Piepjohn *et al.* this volume).

The Svalbardian fold-and-thrust zones within the ORS basin

Dickson Land Fold-and-Thrust Zone

The eastern margin of the ORS basin in NW Spitsbergen is formed by the Billefjorden Fault Zone (BFZ) (Fig. 1) (Harland *et al.* 1974). It represents a major structural lineament in Spitsbergen, which has been reactivated several times since at least Caledonian times (McWhae 1953; Harland *et al.* 1974; Lamar *et al.* 1986; Manby 1990; Manby & Lyberis 1992; Manby *et al.* 1994; McCann & Dallmann 1996). The major fault of the BFZ (Balliolbreen Fault, Harland *et al.* 1974) separates the ORS in the west and the pre-Devonian basement of Ny Friesland in the east (Holtedahl 1914, 1925; Orvin 1940; Harland *et al.* 1974) (Fig. 3).

Directly to the west of the Balliolbreen Fault, deformed Devonian sediments have been reported (e.g. Harland *et al.* 1974; Lamar *et al.* 1986; Manby *et al.* 1994; Lamar & Douglass 1995; McCann & Dallmann 1996; Michaelsen *et al.* 1997; Piepjohn *et al.* 1997*a*). Parallel to the fault, intense folding and thrusting of the Svalbardian deformation is concentrated in the Dickson Land Fold-and-Thrust Zone. To the south, Middle Carboniferous platform carbonates overlie unconformably the folded and thrust-faulted Devonian sediments.

The Dickson Land Fold-and-Thrust Zone represents the only zone within the Svalbardian Fold-and-Thrust Belt in which a succession of at least three tectonic stages can be observed (Michaelsen *et al.* 1997; Piepjohn *et al.* 1997*a*), as follows

Stage 1. In central Mimerdalen, the Mimerdalen Formation is affected by décollements subparallel to the bedding, which repeat the lower members of the Mimerdalen Formation three times (Fig. 4, profile b). This field observation was supported by Pčelina *et al.* (1986), who detected tectonic repetitions of the lower members in several boreholes of the Russian mining company Trust Arctikugol. The décollements of this stage 1 thrusting are characterized by flat-ramp geometries, and the vergences of small-scale, fault-related folds indicate transport directions to the WNW.

Stage 2. Following stage 1, the entire Devonian succession was affected by the formation of kilometre-scale, more or less WNW-vergent fold structures which also involved the stage 1 décollements and thrust-stacked Mimerdalen sediments (Fig. 4, profile b).

In central Mimerdalen, deposits of the Early Devonian Wood Bay Formation and the Middle to Late Devonian Mimerdalen Formation are folded into the Mimerelva Syncline (Figs 3 and 4) (Michaelsen *et al.* 1997). The vertical to overturned eastern short limb of the syncline is at least 2 km thick. The gently east-dipping western long limb passes into the flat-lying sediments of the undeformed Devonian sequence to the west of the Dickson Land Fold-and-Thrust Zone (Fig. 4, profile b). In the south, the Mimerelva Syncline is unconformably overlain by Middle Carboniferous limestones. In the north, it is overthrust by the Early Devonian Wood Bay Formation.

The second large-scale fold structure is represented by the Nidedalen Anticline (Piepjohn *et al.* 1997*a*) below Middle Carboniferous limestones in Nidedalen southeast of Jotunfonna (Figs 3 and 4, profile a). It consists of red beds of the Wood Bay Formation with a gently WNW-dipping eastern long limb and an at least 1 km thick, partly overturned western short limb.

The formation of the large-scale fold structures was followed by thrusting along steeply ESE-dipping reverse faults cutting through the Mimerelva Syncline (Fig. 4, profile b). These faults probably represent accommodation thrusts during the continuous shortening of stage 2.

Stage 3. The last stage of the Svalbardian deformation is characterized by the formation of thrusts, which have already been described by Stensiø (1918) north of Mimerdalen. The 30–40° east-dipping Munindalen thrusts (McCann & Dallmann 1996; Piepjohn *et al.* 1997*a*) cut through the entire eastern short limb of the Mimerelva Syncline (Michaelsen *et al.* 1997) (Figs 3 and 4, profiles b and c) separating the Mimerelva Syncline in the footwall and the Nidedalen Anticline in the hanging wall. The lower Munindalen Thrust carries Early Devonian sediments over the Plantekløfta Member of the upper Mimerdalen Formation (Vogt 1941; Friend 1961), which is of a Late Famennian age (Piepjohn *et al.* this volume). The intensity of the deformation rapidly decreases to the west. The westernmost exposed stage 3 structure in this area, the Triungen–Grønhorgdalen Fault Zone (McCann & Dallmann 1996), is located some 12 km west of the Balliolbreen Fault (Figs 3 and 4, profile d). It represents a minor east-dipping reverse fault, which was reactivated as a normal fault after the Svalbardian deformation.

The crustal shortening within the Dickson Land Fold-and-Thrust Zone can be roughly estimated to be in the range of 50% (Piepjohn *et al.* 1997*a*): the restoration of the Mimerelva

Syncline, the reverse faults and Munindalen thrusts suggests that the original section of 13 km was reduced to 7 km after the shortening.

Bråvallafjella Fold Zone (northern Andrée Land)

Schenk (1937) described intensely folded sediments of the Grey Hoek and Wijde Bay formations in northern Andrée Land. Here, a 5–10 km wide and 35 km long fold zone is exposed between Woodfjorden in the west and the Andrée Land Anticline in the east (Figs 1 and 5) (Piepjohn 1994; Piepjohn & Thiedig 1997; Dallmann *et al.* 1998*a*, *b*).

The NNE–SSW-trending Bråvallafjella Fold Zone is characterized by a succession of kink folds of hundreds-of-metres and kilometre scale (Piepjohn 1994). Along the west coast of northern Andrée Land, the kilometre-scale Bråvallafjella Syncline, with a vertical fold axial plane and steeply WNW- and ESE-dipping limbs (Figs 5 and 6a) forms the western part of the exposed fold zone. On the coastal plains, the bedding of the fold limbs can be traced about 1000 m perpendicular to the strike, indicating the large extent of this structure. The NNE–SSE-trending Bråvallafjella Syncline can be followed about 25 km along-strike. In the southern parts, the kink fold passes into a broad syncline, which is truncated by an east–west fault of the West Andrée Land Fault Zone (Fig. 5).

The eastern limb of the syncline consists of second-order kink folds of hundreds-of-metres scale, and the enveloping surface of the folds climbs towards the Andrée Land Anticline (Fig. 6a). In this area, the fold axial planes steeply dip to the ESE. Further to the east, the kink fold structures pass into broad open anticlines and synclines, and finally into the Andrée Land Anticline (Figs 5 and 6a).

The western limb of the Bråvallafjella Syncline in the southern area is also characterized by kink fold structures of hundreds-of-metres scale. In the west, it is truncated by a north–south fault of the West Andrée Land Fault Zone (Fig. 5).

The fold axes generally strike NNE–SSW and plunge moderately to the SSW (Piepjohn 1994, 1997) (Fig. 5). The sandstones are affected by a steeply ESE- and WNW-dipping cleavage whereas the silt- and mudstones are characterized by intense pencil cleavage (Piepjohn 1994).

Recent mapping (Dallmann *et al.* 1998*a*, *b*) showed that the Bråvallafjella Fold Zone is also affected by a small number of predominantly west-directed thrust faults. These flat-lying or east-dipping faults cut through the large-scale kink fold structures (Fig. 6b) suggesting that the thrust faulting took place after the folding. Compared with the Dickson Land Fold-and-Thrust Zone, they probably represent stage 3 structures, whereas the folding can be related to stage 2 of the Svalbardian deformation.

Germaniahalvøya Fold-and-Thrust Zone

Directly to the east of the crystalline basement of the Biskayerfonna–Holtedahlfonna Horst (Harland 1997) (Fig. 1) a narrow, NNW–SSE trending occurrence of deformed Lower Devonian sandstones is exposed between the Breibogen Fault (Gee & Moody-Stuart 1966) in the west and the Fotkollen Fault (Piepjohn 1994) in the east (Fig. 5). This Germaniahalvøya Fold-and-Thrust Zone is characterized by WSW-vergent folds of hundreds-of-metres scale with gently ENE-dipping long limbs and steeply WSW-dipping short limbs (Fig. 6c) (Piepjohn 1994, 1997). The short limbs and anticlines are partly truncated by gently ENE-dipping thrust faults with transport directions to the WSW.

To the west of the Breibogen Fault, three klippen of Early Devonian sandstones are carried over the crystalline rocks of the Biskayerfonna–Holtedahlfonna Horst (Gee & Moody-Stuart 1966; Piepjohn 1994) (Fig. 6c). The sandstones in these klippen are strongly tectonized. Small-scale duplex structures and shear zones indicate transport directions of the klippen to the west (Piepjohn 1994) similar to the fold-and-thrust zone east of the Breibogen Fault. Mica schists of the basement on the islands north of Germaniahalvøya are affected by east-directed, brittle thrust faults and duplex structures on a tens-of-metres scale. Additionally, a west-dipping contact between the basement and conglomerates of the Early Devonian Red Bay Group is thrust faulted at one locality on the north coast of Liefdefjorden. These observations suggest that the basement of the Biskayerfonna–Holtedahlfonna Horst is partly involved in the Svalbardian deformation.

To the east, the Germaniahalvøya Fold-and-Thrust Zone is bounded along the Fotkollen Fault (Piepjohn 1994) by the red beds of the Wood Bay Formation (Figs 5 and 6c). The sediments gently dip to the east and contain very few compressional Svalbardian structures except for some small-scale thrust faults and flexures. However, the eastern continuation of the Germaniahalvøya Fold-and-Thrust Zone is suggested to be downfaulted along the Fotkollen Fault and to be situated below the sediments of the Wood Bay Formation (Piepjohn 1994) (Fig. 6c).

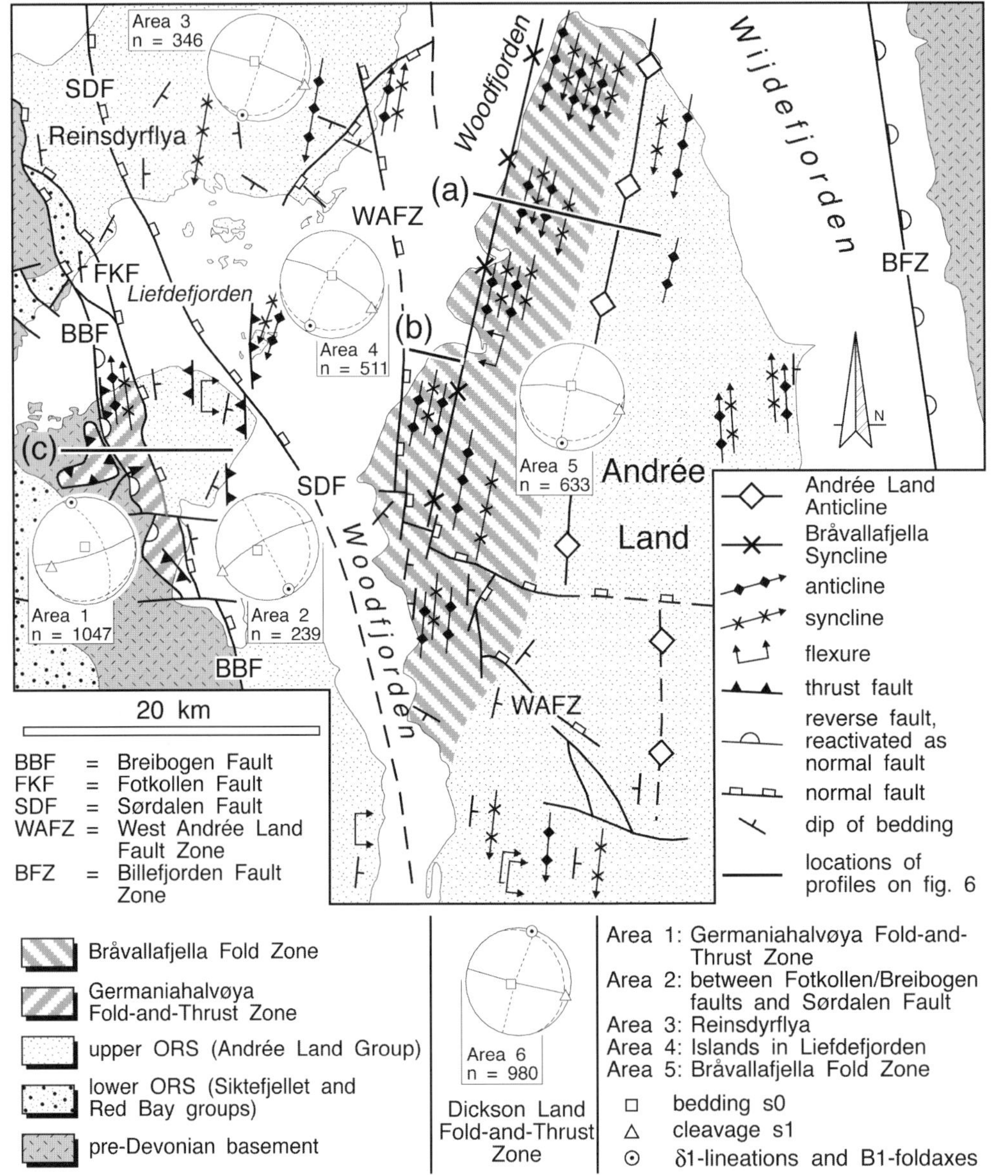

Fig. 5. Tectonic map of the northern part of the ORS basin of NW Spitsbergen, redrawn from Piepjohn (1994) (for location see Fig. 1).

The shortening in the northern part of the ORS basin between the Breibogen Fault and the Andrée Land Anticline can be carefully estimated to be in the range of 10% (Piepjohn 1994, 1997): the restoration of the 80 km long profile suggests that the original section was 8–10 km longer before the shortening, which is concentrated within the Bråvallafjella and Germaniahalvøya fold-and-thrust zones.

Svalbardian thrust zones on the western basement margin of the ORS basin

Blomstrandhalvøya Thrust Zone

On the western basement margin of the ORS basin of NW Spitsbergen, two north–south-trending thrust zones are exposed in phyllites and marbles of the pre-Devonian basement (Fig. 1).

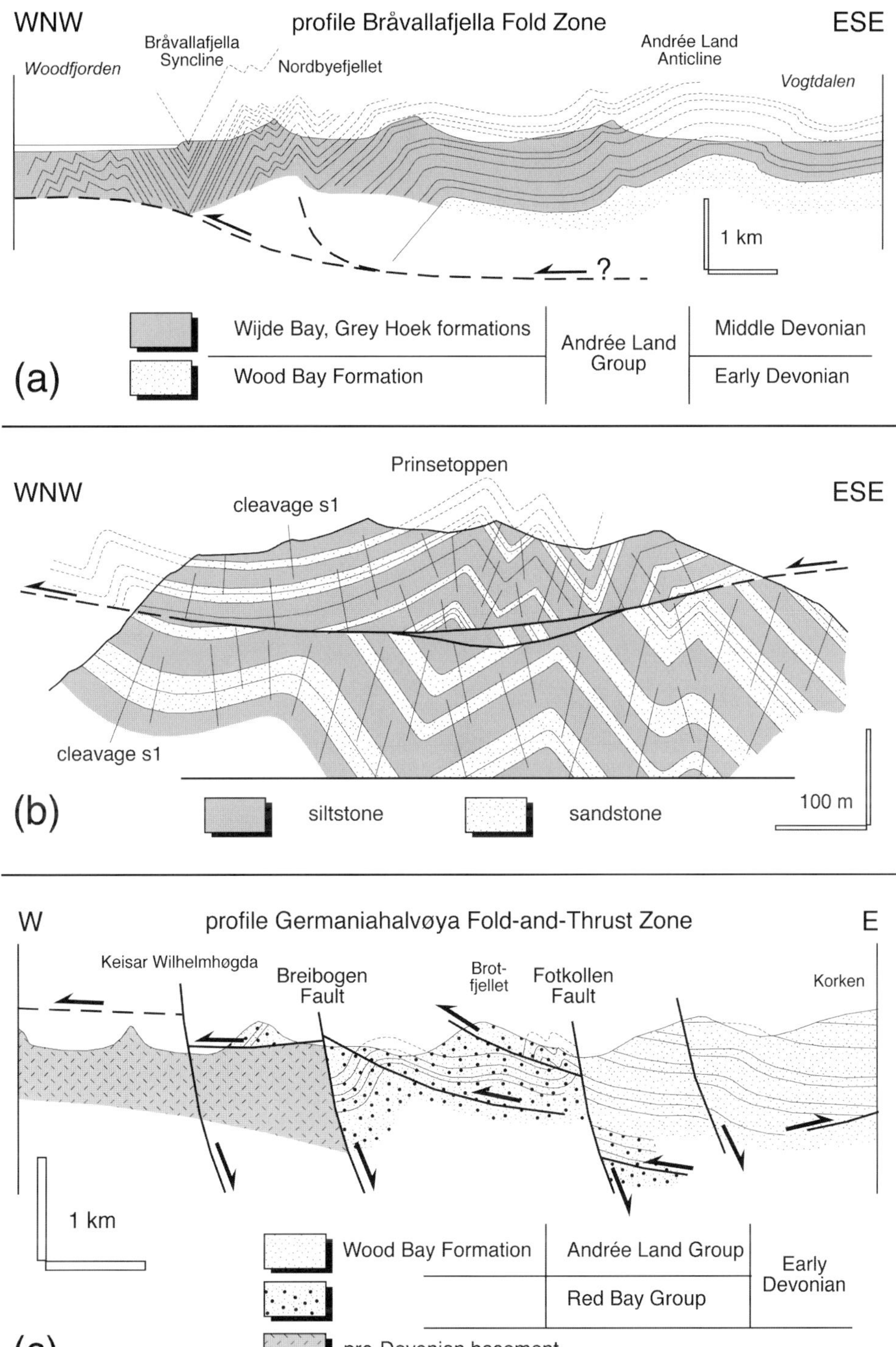

Fig. 6. Cross-sections within the northern part of the ORS basin of NW Spitsbergen (for location see Fig. 5). (**a**) cross-section through the Bråvallafjella Fold Zone. (**b**) cross-section through the western limb of the Bråvallafjella Syncline. (**c**) cross-section through the Germaniahalvøya Fold-and-Thrust Zone.

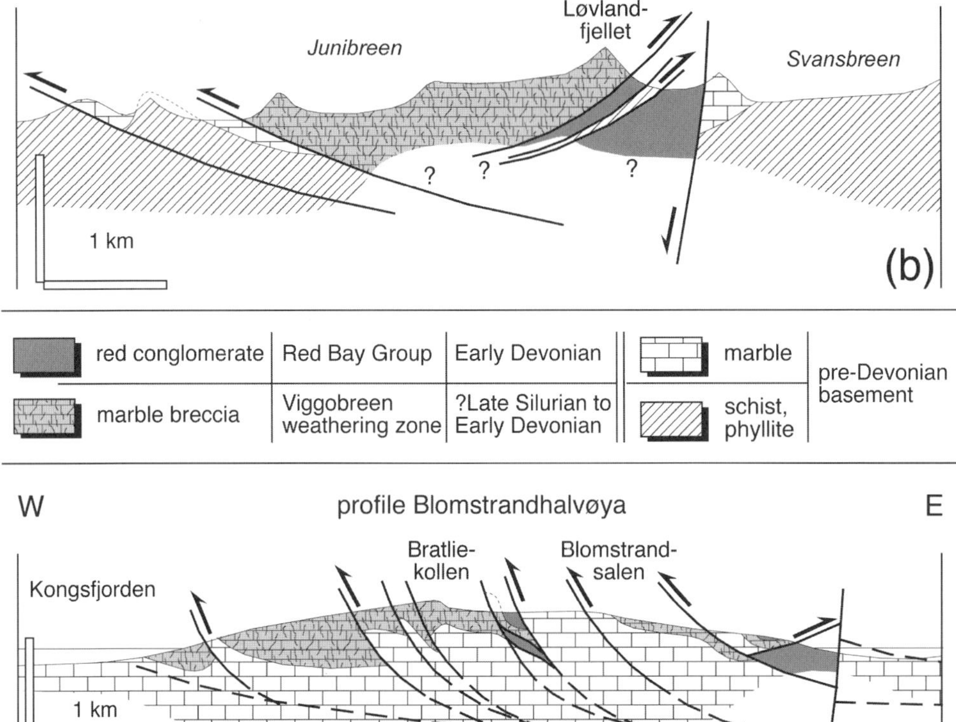

Fig. 7. Cross-sections through the Blomstrandhalvøya Thrust Zone (for location see Fig. 1). (**a**) Profile at Blomstrandhalvøya. (**b**) Profile at Løvlandfjellet.

On Blomstrandhalvøya, occurrences of Early Devonian sandstones and conglomerates of the Red Bay Group (Gjelsvik 1974) are present as imbricate slices within basement marbles (Thiedig & Manby 1992; Kempe *et al.* 1997).

The pre-Devonian marbles and Devonian red beds on Blomstrandhalvøya are affected by west-directed, brittle and east-dipping reverse faults forming large imbricate fans (Kempe *et al.* 1997) or the preserved lower part of a large-scale duplex structure (Fig. 7*a*). At Bratliekollen, the Devonian sediments and the underlying marbles are folded into a medium-scale west-vergent fold (Thiedig & Manby 1992) (Fig. 7*a*). In the eastern part of Blomstrandhalvøya, marbles are thrust over Devonian conglomerates to the east along a west-dipping back thrust (Hjelle & Lauritzen 1982; Kempe *et al.* 1997).

The involvement of sandstones and conglomerates indicates a post-Early Devonian age of the thrust faulting (Thiedig 1988; Thiedig & Manby 1992; Kempe *et al.* 1997). Buggisch *et al.* (1994) and Kempe *et al.* (1997) described karst fillings within marbles on Blomstrandhalvøya that are not affected by thrust faulting. The karst sediments contain a Bashkirian to Early Moscovian conodont fauna (Buggisch *et al.* 1994) suggesting that the age of the thrust faulting along the Blomstrandhalvøya Thrust Zone occurred between Early Devonian and the Mid-Carboniferous time and therefore can be assigned to the Svalbardian deformation but not to the formation of the Tertiary West Spitsbergen Fold-and-Thrust Belt (Buggisch *et al.* 1994).

The basement marbles are often characterized by intense brecciation and reddening probably representing a fossil weathering surface (Gjelsvik 1974, 1979; Manby & Lyberis 1992; Thiedig & Manby 1992; Piepjohn 1994). This Viggobreen weathering zone (Piepjohn & Thiedig 1997) post-dates the Caledonian Orogeny and is overlain by the Lower Devonian Red Bay Group (Gjelsvik 1974, 1979; Thiedig & Manby 1992), indicating a ?Late Silurian to Earliest Devonian age of the weathering (Piepjohn & Thiedig 1997). On Blomstrandhalvøya, unweathered marbles in the eastern part are thrust over weathered marbles in the western part (Kempe *et al.* 1997) (Fig. 7*a*). This observation shows that most of the thrust faults in the basement marbles post-date the Viggobreen weathering zone (Fig. 7*a*) and thus can also be related to the Svalbardian deformation.

The northern continuation of the Blomstrandhalvøya Thrust Zone on the mainland is exposed in the Løvlandfjellet area (Fig. 1). Here, a narrow exposure of thick, red conglomerates is bounded by a north–south-trending normal fault in the east and by a west-dipping thrust in the west (Hjelle 1979) (Fig. 7b). Along this thrust, Devonian red beds are overthrust to the east by thick marbles, which are affected by red weathering of the ?Late Silurian to Earliest Devonian Viggobreen weathering zone. This thrust fault is interpreted to represent the northern continuation of the back thrust on Blomstrandhalvøya (Hjelle & Lauritzen 1982) (Fig. 7*a*, *b*). In the western part of the outcrop, a small slice of basement rocks is present in the Devonian red beds between two west-dipping thrusts (Fig. 7b). To the west of Løvlandfjellet, a gently east-dipping thrust fault is exposed (Lange & Hellebrandt 1997). Thus, the area between this thrust and the back thrust to the east represents a large-scale pop-up structure (Fig. 7b), suggesting that the Svalbardian deformation in this area took place at an upper-crustal level.

Mitrahalvøya Thrust Zone

On Mitrahalvøya to the west of Krossfjorden (Fig. 1), flat-lying detachments are developed in dolomites, marbles and phyllites of the crystalline basement (Fig. 8). The Mitrahalvøya Thrust Zone is characterized by four thrust sheets, which are bounded by flat-lying or gently east-dipping thrust faults (Piepjohn *et al.* 1997*b*). The entire structure of the thrust zone is interpreted as a large-scale, west-directed duplex structure between the lowest thrust and the Willeberget Thrust Sheet, which overlies both the Tromsdalen and Talusfjellet thrust sheets (Peletz *et al.* 1997) (Fig. 8a).

Until now, it was not known whether these structures were formed in the Caledonian Orogeny or in a post-Caledonian event. Piepjohn *et al.* (1997*b*) and Thielemann & Thiedig (1997) described intense weathering phenomena in the marbles and dolomites on Mitrahalvøya. Partly reddish breccias cut through the Caledonian penetrative foliation and can be correlated with the post-Caledonian Viggobreen weathering zone. The weathered dolomites are overlain by unweathered phyllites at Scoresbyfjellet and Willeberget (Fig. 8b), indicating that the emplacement of the phyllites took place after the weathering in ?Late Silurian to Earliest Devonian times (Piepjohn *et al.* 1997*b*).

This is supported by the discovery of two small occurrences of post-Caledonian conglomerates on the western and southern slopes of Scoresbyfjellet. The structural position of the conglomerates below the basement phyllites (Fig. 8b) suggests that the phyllites have been carried over the conglomerates after the deposition of the latter in the Early Devonian time (Peletz *et al.* 1997; Piepjohn *et al.* 1997*b*; Thielemann & Thiedig 1997). Therefore, the Mitrahalvøya Thrust Zone can also be assigned to the Svalbardian deformation representing the westernmost exposed areas of the Svalbardian Fold-and-Thrust Belt.

Similar to the Blomstrandhalvøya Thrust Zone, a pop-up structure north of Scoresbyfjellet and the transport of thrust sheets over weathered basement rocks indicate that the formation of the Mitrahalvøya Thrust Zone took place in the uppermost crust.

Discussion

The overstep of the Wood Bay Formation onto the western basement margin of the ORS basin SE of Kongsfjorden (Pretender mountain, Orvin 1940; Friend & Moody-Stuart 1972; Hjelle 1993) (Fig. 1) suggests that the formation of the Biskayerfonna–Holtedahlfonna Horst took place before the deposition of the Wood Bay Formation (Piepjohn 1994, 1997; McCann this volume). That would mean that the Breibogen Fault as the eastern boundary fault of the basement horst was already active during the Monacobreen phase. Therefore, the location of the Germaniahalvøya Fold-and-Thrust Zone could have been determined by the pre-existing basement block as a ramp during the more or less west-directed Svalbardian deformation (Piepjohn 1994, 1997). This is supported by the Blomstrandhalvøya and Mitrahalvøya thrust zones, in which slices of lower ORS deposits are involved. The narrow, north–south-trending shape of the Devonian slices suggests that the Red Bay Group was downfaulted in graben-like structures. During the Svalbardian deformation, the boundary faults could have been reactivated as thrust faults. This could also explain that in parts (younger) Devonian sediments are thrust over (older) pre-Devonian basement rocks.

Although the tectonic stages of the Svalbardian deformation affected the deposits of

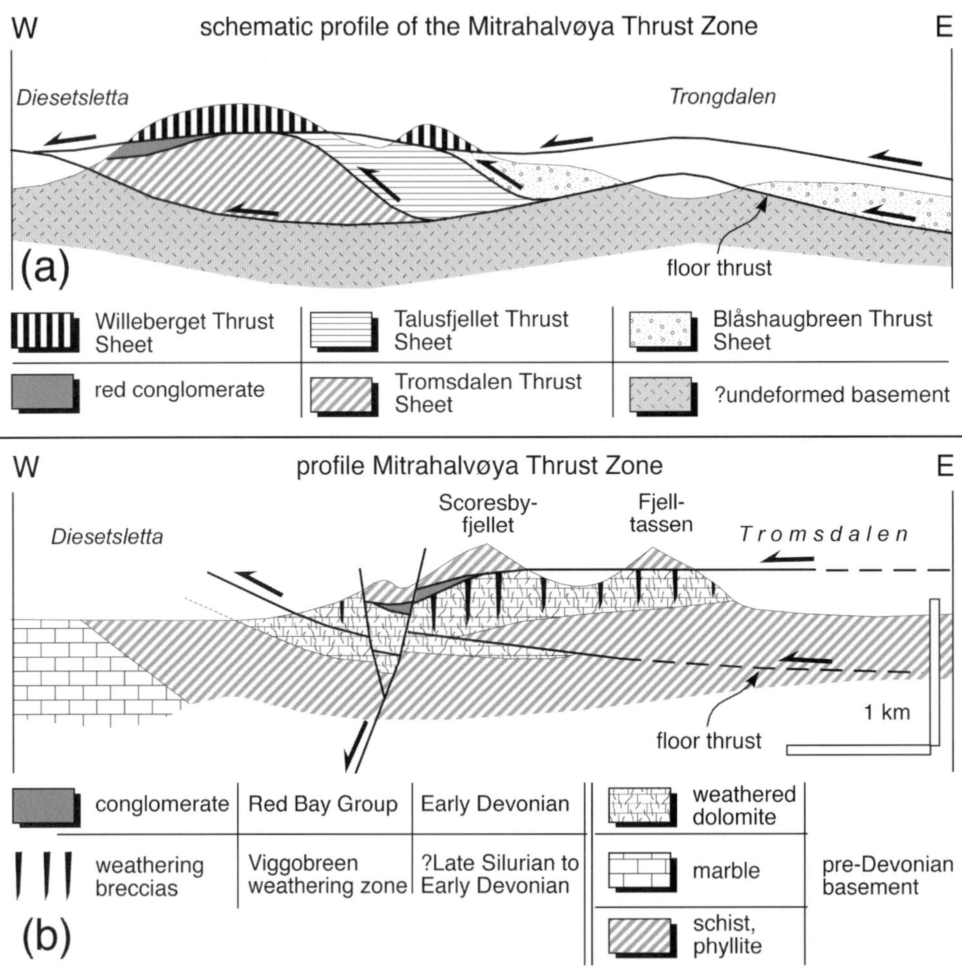

Fig. 8. Cross-sections through the Mitrahalvøya Thrust Zone (for location see Fig. 1). (**a**) Schematic profile of the interpreted duplex structure of the Mitrahalvøya Thrust Zone. (**b**) Profile at Scoresbyfjellet.

the Mimerdalen Formation in Earliest Carboniferous time (Piepjohn *et al.* this volume), the onset of Svalbardian movements appears to have taken place simultaneously with the deposition of the Famennian Planteryggen and Plantekløfta members, which may have formed in a foreland basin. Both units were deposited after a stage of normal faulting along north–south-trending faults and contain the first conglomerates in the ORS basin since deposition of the coarse-clastic Lower Devonian Red Bay Group.

The clasts within the conglomerates are dominated by green sandstones of the lower Wood Bay Formation and basement quartzites. The latter and the absence of clasts of red siltstone and sandstone of the upper Wood Bay Formation argue against a source area in the west. The shales of the Plantekløfta Member contain numerous groups of *in situ* trunks of *Archeosigillaria* of up to 40 cm length, which are mostly tilted to the west. This indicates transport directions from the east of the Billefjorden Fault Zone and suggests that the Famennian forests have been affected and disturbed by sudden events of mud flow-like sedimentation. These observations were supported by Vogt (1938) and Friend (1961), who interpreted the coarse-grained sediments, particularly in the Plantekløfta Member, as being the result of an early uplift to the east of the Billefjorden Fault Zone leading to erosion of the Devonian cover and the exhumation of the basement in this area during initial stages of the Svalbardian deformation.

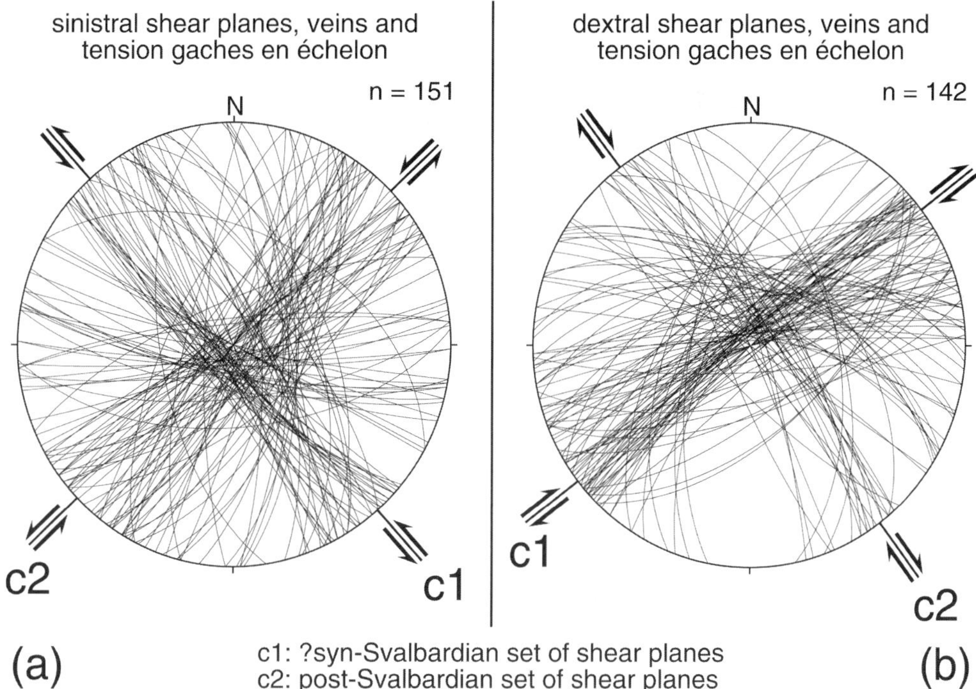

Fig. 9. Schmidt net projection (great circles) of (**a**) sinistral and (**b**) dextral faults, shear planes, veins, kink bands and tension gashes en échelon in the northern part of the ORS basin.

It is still a matter of debate whether the Svalbardian deformation was controlled by major sinistral strike-slip movements (Harland *et al.* 1974) or by east–west compression (McWhae 1953; Lamar *et al.* 1986; Manby & Lyberis 1992; Manby *et al.* 1994; Piepjohn 1994; Lamar & Douglass 1995). The NNE–SSW trending fold axes, δ-lineations and cleavage in central Dickson Land, northern Andrée Land and on Reinsdyrflya are oriented at a low angle to the major fault and seem to confirm sinistral transpression along the Billefjorden Fault Zone (Figs 3 and 5). In particular, the WNW–ESE-trending dextral faults cutting through the overturned short limb of the Mimerelva Syncline (Fig. 3) could be an indication of strike-slip movements along the Balliolbreen Fault.

On the other hand, the NNW–SSE-striking fold axes, δ-lineations and cleavage planes to the west of the Sørdalen Fault (Fig. 5) cannot be related to a sinistral displacement and transpression along the more or less north–south-trending Billefjorden Fault Zone. In addition, the lack of en échelon structures as pointed out by Lamar *et al.* (1986), Manby & Lyberis (1992) and Manby *et al.* (1994), as well as the absence of strike-slip faults parallel to the Billefjorden Fault Zone, argue against major strike-slip movements.

Measurements of sinistral (Fig. 9a) and dextral (Fig. 9b) faults, shear planes, veins, tension gashes en échelon, and kink bands between the Germaniahalvøya and Bråvallafjella zones demonstrate the overlap of two sets of strike-slip fabric elements in the northern part of the ORS basin. The origin of these two sets is unclear. Field observations show that one set post-dates the Svalbardian deformation: some of the NW–SE-trending dextral veins and kink-bands c2 (Fig. 9b) cut through and displace the Svalbardian cleavage. The conjugate component of c2 may be represented by the NE–SW-trending sinistral planes (Fig. 9a). This means that only the set c1 of NW–SE sinistral (Fig. 9a) and NE–SW dextral (Fig. 9b) planes could be related to a possible syn-Svalbardian transpression. However, the 45° orientation of these elements to the principal displacement zone along the Billefjorden Fault Zone does not show a pattern typical for strike-slip regimes.

There is no evidence for strike-slip in the Germaniahalvøya, Blomstrandhalvøya and Mitrahalvøya zones. Here, the structures of the Svalbardian deformation are characterized by

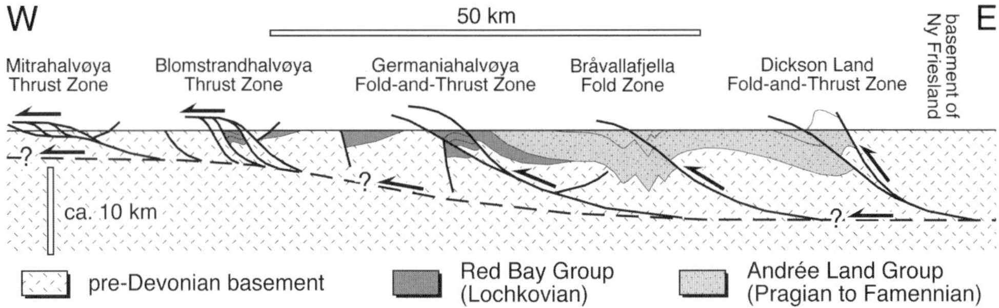

Fig. 10. Schematic east–west cross-section through the five fold-and-thrust zones of the Svalbardian Fold-and-Thrust Belt between the Billefjorden Fault Zone in the east and the coast of NW Spitsbergen in the west.

west-directed thrust-faulting and by compression orthogonal to a pre-existing basement block (Germaniahalvøya Fold-and-Thrust Zone). In the northern part of the ORS basin and in the Dickson Land Fold-and-Thrust Zone, the thrusts mostly trend parallel to the Billefjorden Fault Zone. It is possible that the eastern part of the study area was affected by oblique compression as proposed by McCann & Dallmann (1996) during the stage 2 folding of the Svalbardian deformation. The orientation and the large extent of the Svalbardian stage 3 thrusting between the Billefjorden Fault Zone and the west coast of Spitsbergen, however, suggest that the Svalbardian deformation was dominated by orthogonal compression.

Conclusions and regional implications

The investigations in NW Spitsbergen clearly show that the Svalbardian deformation is more than a local folding event. The structures of the Svalbardian Fold-and-Thrust Belt can be found within a zone, more than 100 km wide, between the Billefjorden Fault Zone in the east and the west coast of NW Spitsbergen (Fig. 1) (Piepjohn & Thiedig 1997). The five fold-and-thrust zones of the Svalbardian deformation are characterized by brittle faulting that occurred in the upper crust. The two thrust zones in the western basement margin of the basin demonstrate that the crystalline basement was involved in the Svalbardian deformation. In addition, the thrusting of the basement over the eastern margin of the Devonian basin along the Billefjorden Fault Zone suggests that the basement successions to the east of the ORS basin were also involved into the Svalbardian deformation.

The shape of the narrow, more-or-less north–south-trending fold-and-thrust zones indicates that their development was probably controlled by older structures in the basement as proposed by Friend & Moody-Stuart (1972). Piepjohn (1994, 1997) suggested that these controlling structures are represented by fault-bounded horsts and grabens, which have been formed during the Monacobreen phase (McCann this volume) in the time interval between the sedimentation of the lower ORS (Red Bay Group) and the upper ORS (Andrée Land Group).

The involvement of the basement on the western margin and along the Billefjorden Fault Zone suggests that the Svalbardian fold-and-thrust zones are floored by a basal detachment zone within the basement below the ORS basin (Fig. 10). The formation of the fold-and-thrust zones within the ORS basin and on the western basement margin could be controlled by large splay thrusts ramping from the detachment in front of basement faults or horst structures and cutting up-section (Piepjohn 1994, 1997) (Fig. 10). This indicates that the basement of Ny Friesland to the east of the Dickson Land Fold-and-Thrust Zone could also be involved in the Svalbardian deformation.

The similarity of the structural trends and the corresponding time intervals of the Late Devonian to Early Carboniferous Ellesmerian Orogeny in the Canadian Arctic Islands and in North Greenland (Thorsteinsson & Tozer 1957, 1960) and the Tournaisian Svalbardian Fold-and-Thrust Belt (Piepjohn *et al.* this volume) suggest that both fold belts can be correlated (e.g. Christie 1979; Manby & Lyberis 1992). The Ellesmerian deformation took place after Mid-Devonian time on southern Ellesmere Island (Mayr *et al.* 1994) and between Latest Devonian and Earliest Carboniferous time on NW Devon Island (Mayr *et al.* 1998). In particular, the Tournaisian deformation in large areas of the Canadian Arctic Islands, which affects the Late Devonian Okse Bay Formation during the Ellesmerian Orogeny *sensu stricto* (Trettin 1991), corresponds very closely to the timing of the Svalbardian deformation.

The field observations show that the structures of the Svalbardian Fold-and-Thrust Belt are not restricted to the ORS basin of NW Spitsbergen (Fig. 10). The Svalbardian deformation also affected the western basement areas along the west coast of NW Spitsbergen in the vicinity of the recent continental margin of the Barents Shelf. Thus, the Svalbardian Fold-and-Thrust Belt could be interpreted as the eastern continuation of the Ellesmerian Fold Belt and its formation as the result of the approach and the docking of Svalbard north of Greenland during Earliest Carboniferous time.

Financial support from the German Research Foundation (DFG) is gratefully acknowledged (Projects Th 126/17-3, Th 126/22, Th 126/27-1, Pi 330/1-1). The author would also like to express his thanks to the Alfred-Wegener-Institute (AWI) for material support and especially to the Norsk Polarinstitutt (Oslo) for logistic support and transportation in Spitsbergen. Finally, I would like to thank A. J. McCann, P. T. Osmundsen and K. Saalmann for critical reading of the manuscript and correction of the English text.

References

BUGGISCH, W., PIEPJOHN, K., THIEDIG, F. & VON GOSEN, W. 1994. Middle Carboniferous conodont fauna from Blomstrandhalvøya (NW-Spitsbergen): implications on the age of post-Devonian karstification and the Svalbardian deformation. *Polarforschung*, **62**(2–3), 83–90.

BUROV, Y. P. & SEMEVSKIJ, D. V. 1979. The tectonic structure of the Devonian Graben (Spitsbergen). The Geological Development of Svalbard during the Precambrian, Lower Palaeozoic, and Devonian. *Norsk Polarinstitutt Skrifter*, **167**, 239–248.

CHRISTIE, R. L. 1979. The Franklinian Geosyncline in the Canadian Arctic and its relationship to Svalbard. The Geological Development of Svalbard during the Precambrian, Lower Palaeozoic, and Devonian. *Norsk Polarinstitutt Skrifter*, **167**, 263–314.

DALLMANN, W. K., PIEPJOHN, K., MCCANN, A. J., SIROTKIN, A. N., MILOSLAVSKIJ, M. J., OHTA, Y. & GJELSVIK, T. 1998*a*. Geological map of Svalbard 1:100 000, sheet B5G Woodfjorden. Norsk Polarinstitutt, preliminary edition.

——, —— & OHTA, Y. 1998*b*. Geological map of Svalbard 1:100 000, sheet B4G Reinsdyrflya. Norsk Polarinstitutt, preliminary edition.

FØYN, S. & HEINTZ, A. 1943. The Downtonian and Devonian vertebrates of Spitsbergen. *In*: The English–Norwegian–Swedish Expedition 1939, VIII. Geological Results. Skrifter om Svalbard og Ishavet, **85**.

FREBOLD, H. 1935. Geologie von Spitzbergen, der Bäreninsel, des König Karl- und Franz-Joseph-Landes. *In*: KRENKEL, E. (ed.) *Geologie der Erde*. Borntraeger, Berlin, 1–195.

FRIEND, P. F. 1961. The Devonian stratigraphy of north and central Vestspitsbergen. *Proceedings of the Yorkshire Geological Society*, **33**, 77–118.

—— & MOODY-STUART, M. (eds) 1972. Sedimentation of the Wood Bay Formation (Devonian) of Spitsbergen: Regional Analysis of a Late Orogenic Basin. *Norsk Polarinstitutt Skrifter*, **157**.

GEE, D. G. 1972. Late Caledonian (Haakonian) movements in northern Spitsbergen. *Norsk Polarinstitutt Årbok*, **1970**, 92–101.

—— & MOODY-STUART, M. 1966. The base of the Old Red Sandstone in central north Haakon VII Land, Vestspitsbergen. *Norsk Polarinstitutt Årbok*, **1964**, 57–68.

GJELSVIK, T. 1974. A new occurrence of Devonian rocks in Spitsbergen. *Norsk Polarinstitutt Årbok*, **1962**, 50–54.

—— 1979. The Hecla Hoek ridge of the Devonian Graben between Liefdefjorden and Holtedahlfonna, Spitsbergen.The Geological Development of Svalbard during the Precambrian, Lower Palaeozoic, and Devonian. *Norsk Polarinstitutt Skrifter*, **167**, 63–71.

HARLAND, W. B. 1997. *The Geology of Svalbard*. Geological Society, London, Memoirs, **17**.

——, CUTBILL, J. L., FRIEND, P. F. *et al.* 1974. The Billefjorden Fault Zone, Spitsbergen: the Long History of a Major Tectonic Lineament. *Norsk Polarinstitutt Skrifter*, **161**.

HEINTZ, A. 1937. *Die Downtonischen und Devonischen Vertebraten von Spitzbergen. VI.* Lunaspis-*Arten aus dem Devon Spitzbergens*. Skrifter om Svalbard og Ishavet, **72**.

HJELLE, A. 1979. Aspects on the geology of north-west Spitsbergen. The Geological Development of Svalbard during the Precambrian, Lower Palaeozoic, and Devonian. *Norsk Polarinstitutt Skrifter*, **167**, 37–62.

—— 1993. *Geology of Svalbard*. Norsk Polarinstitutt Polarhåndbok, **7**.

—— & LAURITZEN, Ø. 1982. *Geological map of Svalbard 1:500 000, sheet 3G, Spitsbergen northern part*. Norsk Polarinstitutt Skrifter, **154c**.

HOLTEDAHL, O. 1914. On the Old Red Sandstone Series of northwestern Spitsbergen. *In*: Congress Report. XIIth Session. International Geological Congress, Ottawa, Canada, 707–712.

—— 1925. Some points of structural resemblance between Spitsbergen and Great Britain, and between Europe and North America. *Avhandlingar utgitt av det Norske Videnskapakademi i Oslo*, **4**, 1–20.

KEMPE, M., NIEHOFF, U., PIEPJOHN, K. & THIEDIG, F. 1997. Kaledonische und svalbardische Entwicklung im Grundgebirge auf der Blomstrandhalvøya, NW-Spitzbergen. *Münstersche Forschungen zur Geologie und Paläontologie*, **82**, 121–128.

LAMAR, D. L. & DOUGLASS, D. N. 1995. *Geology of an area astride the Billefjorden Fault Zone, northern Dickson Land, Svalbard*. Norsk Polarinstitutt Skrifter, **197**.

——, REED, W. E. & DOUGLASS, D. N. 1986. The Billefjorden Fault Zone, Spitsbergen: is it part of a major Late Devonian transform? *Geological Society of America Bulletin*, **97**, 1083–1088.

LANGE, M. & HELLEBRANDT, B. 1997. Geologie, Petrographie und Tektonik des südwestlichen Haakon VII Landes, Nordwest-Spitzbergen. *Münstersche Forschungen zur Geologie und Paläontologie*, **82**, 99–119.

LAURITZEN, Ø., ANDRESEN, A, SALVIGSEN, O. & WINSNES, T. 1989. *Geological map of Svalbard 1:100 000, Sheet C8G Billefjorden*. Norsk Polarinstitutt Temakart, **5**.

MANBY, G. M. 1990. The petrology of the Harkerbreen Group, Ny Friesland, Svalbard: protoliths and tectonic significance. *Geological Magazine*, **127**, 155–172.

—— & LYBERIS, N. 1992. Tectonic evolution of the Devonian Basin of northern Svalbard. *In*: DALLMANN, W. K., ANDRESEN, A. & KRILL, A. (eds) *Post-Caledonian Tectonic Evolution of Svalbard. Norsk Geologisk Tidsskrift*, **72**, 7–19.

——, ——, CHOROWICZ, J. & THIEDIG, F. 1994. Post-Caledonian tectonics along the Billefjorden Fault Zone, Svalbard, and its implications for the Arctic region. *Geological Society of America Bulletin*, **106**, 201–216.

MAYR, U., DE FREITAS, T., BEAUCHAMP, B. & EISBACHER, G. 1998. *The Geology of Devon Island north of 76°, Canadian Arctic Archipelago*. Geological Survey of Canada Bulletin, **526**.

——, PACKARD, J. J., GOODBODY, Q. H., OKULITCH, A. V., RICE, R. J., GOODARZI, F. & STEWART, K. R. (1994). *The Phanerozoic Geology of Southern Ellesmere and North Kent Islands, Canadian Arctic Archipelago*. Geological Survey of Canada Bulletin, **470**.

MCCANN, A. 2000. Deformation of the Old Red Sandstone of NW Spitsbergen; links to the Ellesmerian and Caledonian orogenies, this volume.

—— & DALLMANN, W. K. 1996. Reactivation history of the long-lived Billefjorden Fault Zone in north central Spitsbergen, Svalbard. *Geological Magazine*, **133**(1), 63–84.

MCWHAE, J. R. H. 1953. The major fault zone of Central Vestspitsbergen. *Quarterly Journal of the Geological Society, London*, **108**, 209–232.

MICHAELSEN, B., BRINKMANN, L. & PIEPJOHN, K. 1997. Struktur und Entwicklung der svalbardischen Mimerelva Synkline im zentralen Dickson Land, Spitzbergen. *Münstersche Forschungen zur Geologie und Paläontologie*, **82**, 203–214.

MURAŠOV, L. G. & MOKIN, J. L. 1979. Stratigraphic subdivision of the Devonian deposits of Spitsbergen. The Geological Development of Svalbard during the Precambrian, Lower Palaeozoic, and Devonian. *Norsk Polarinstitutt Skrifter*, **167**, 249–261.

NILSSON, T. 1941. *The Downtonian and Devonian vertebrates of Spitsbergen. Order Antiarchi*. Skrifter om Svalbard og Ishavet, **82**.

ORVIN, A. K. 1940. *Outline of the Geological History of Spitsbergen*. Skrifter om Svalbard og Ishavet, **78**.

PČELINA, T. M., BOGAČ, S. I. & GAVRILOV, B. P. 1986. Novye dannye po litostratigrafii devonskich otloženij rajona Mimerdalen archipelaga Špicbergen (New data on the lithostratigraphy of the Devonian deposits of the region of Mimerdalen of the archipelago of Svalbard). *In*: *Geologija osadočnogo čechla archipelaga Špicbergen (Geology of the Sedimentary Blanket of the Archipelago of Spitsbergen)*. Sevmorgeologija, Leningrad, 7–19.

PELETZ, G., GREVING, S. & THIEDIG, F. 1997. Der tektonische Bau des Überschiebungsgürtels auf der Mitrahalvøya, Albert I Land, NW-Spitzbergen. *Münstersche Forschungen zur Geologie und Paläontologie*, **82**, 79–86.

PIEPJOHN, K. 1994. *Tektonische Evolution der Devongräben (Old Red) in NW-Svalbard*. PhD thesis, University of Münster.

—— 1997. Erläuterungen zur Geologischen Karte 1:150.000 des Woodfjorden-Gebietes (Haakon VII Kand, Andrée Land), NW-Spitzbergen, Svalbard. *Münstersche Forschungen zur Geologie und Paläontologie*, **82**, 15–37.

—— & THIEDIG, F. 1997. Geologisch-tektonische Evolution NW-Spitzbergens im Paläozoikum. *Münstersche Forschungen zur Geologie und Paläontologie*, **82**, 215–233.

——, BRINKMANN, L., DIßMANN, B., GREWING, A., MICHAELSEN, B. & KERP, H. 1997*a*. Geologische und tektonische Entwicklung des Devons im zentralen Dickson Land, Spitzbergen. *Münstersche Forschungen zur Geologie und Paläontologie*, **82**, 175–202.

——, ——, GREWING, A. & KERP, H. 2000. New data on the age of the uppermost ORS and the lowermost post-ORS strata in Dickson Land (Spitsbergen) and implications for the age of the Svalbardian deformation. This volume.

——, GREVING, S., PELETZ, G., THIELEMANN, T., WERNER, S. & THIEDIG, F. 1997*b*. Kaledonische und svalbardische Entwicklung im kristallinen Basement auf der Mitrahalvøya, Albert I Land, NW-Spitzbergen. *Münstersche Forschungen zur Geologie und Paläontologie*, **82**, 53–72.

SCHENK, E. 1937. Kristallin und Devon im nördlichen Spitzbergen. *Geologische Rundschau*, **28**(1–2), 112–124.

STENSIØ, E. 1918. Zur Kenntnis des Devons und des Kulms an der Klaas Billenbay, Spitsbergen. *Bulletin of the Geological Institution of the University of Uppsala*, **16**, 65–80.

THIEDIG, F. 1988. Post-Caledonian thrust structures on Blomstrandhalvøya, Kongsfjorden, NW Spitzbergen. *In*: DALLMANN, W. K., OHTA, Y. & ANDRESEN, A. (eds) *Tertiary Tectonics of Svalbard, Extended Abstracts from Symposium held in Oslo 26 and 27 April 1988*. Norsk Polarinstitutt Rapportserie, **46**, 15–16.

—— & MANBY, G. M. 1992. Origins and deformation of post-Caledonian sediments on Blomstrandhalvøya and Lovénøyane, NW-Spitzbergen. *In*: DALLMANN, W. K., ANDRESEN, A. & KRILL, A. (eds) *Post-Caledonian Tectonic Evolution of Svalbard. Norsk Geologisk Tidsskrift*, **72**, 27–35.

THIELEMANN, T. & THIEDIG, F. 1997. Paläozoisch-postkaledonische Sedimente auf Mitrahalvøya,

NW-Spitzbergen. *Münstersche Forschungen zur Geologie und Paläontologie*, **82**, 87–98.

THORSTEINSSON, R. & TOZER, E. T. 1957. Geological investigations in Ellesmere and Axel Heiberg Islands, 1956. *Arctic*, **10**, 2–31.

—— & —— 1960. *Summary Account of Structural History of the Canadian Arctic Archipelago since Precambrian Time. Geological Survey of Canada, Papers*, **60–7**.

TRETTIN, H. P. 1991. Summary (Silurian–Early Carboniferous deformational phases and associated metamorphism and plutonism, Arctic Islands). *In*: TRETTIN, H. P. (ed.) *Geology of the Innuitian Orogen and Arctic Platform of Canada and Greenland*. Geological Survey of Canada, Geology of Canada, **3**, 337–338.

VOGT, T. 1928. Den norske fjellkjedes revolusjonshistorie. *Norsk Geologisk Tidsskrift*, **10**, 97–115.

—— 1938. The stratigraphy and tectonics of the Old Red formations of Spitsbergen. *Abstracts of the Proceedings of the Geological Society, London*, **1343**, 88.

—— 1941. Geology of a Middle Devonian Cannel coal from Spitsbergen. *Norsk Geologisk Tidsskrift*, **21**, 1–12.

New data on the age of the uppermost ORS and the lowermost post-ORS strata in Dickson Land (Spitsbergen) and implications for the age of the Svalbardian deformation

KARSTEN PIEPJOHN[1], LARS BRINKMANN[2], ANKE GREWING[2] & HANS KERP[2]

[1]*Geologisch-Paläontologisches Institut der Westfälischen Wilhelms-Universität Münster, Corrensstraße 24, D-48149 Münster, Germany (e-mail: piepjoh@uni-muenster.de)*

[2]*Forschungsstelle für Paläobotanik der Westfälischen Wilhelms-Universität Münster, Hindenburgplatz 57–59, D-49143 Münster, Germany*

Abstract: In NW Spitsbergen, the infill of a large Old Red Sandstone (ORS) basin was affected by the Svalbardian deformation shortly after the sedimentation of the uppermost ORS units. In the Billefjorden area, along the eastern margin of the basin, folded and thrust-faulted Devonian deposits are unconformably overlain by undeformed Carboniferous clastic sediments and platform carbonate deposits. To re-examine the age of the Svalbardian deformation, samples for palynological investigations were taken from the youngest deformed ORS strata and the oldest post-Svalbardian sediments. The results of palynological investigations show that the folded and thrust-faulted uppermost ORS unit, the Plantekløfta member of the Mimerdalen Formation (Andrée Land Group), is Late Famennian in age. The lowermost undeformed and unconformably overlying post-ORS unit, the Triungen member of the Hørbyebreen Formation (Billefjorden Group), is ?Late Tournaisian to Viséan but not Famennian in age. Thus, the compressional west-directed folding and thrusting of the Svalbardian deformation took place after Late Famennian and before Late Tournaisian time.

In NW Spitsbergen, a 70 km wide and 160 km long Old Red Sandstone (ORS) basin ('fossa magna', De Geer 1909) is bounded by pre-Devonian crystalline basement rocks along the Raudfjorden Fault (Gee & Moody-Stuart 1966) to the west and the Billefjorden Fault Zone (BFZ) (Harland *et al.* 1974) to the east (Fig. 1). To the south, the basin plunges below the mostly undeformed post-Devonian succession of central Spitsbergen.

The main ORS basin to the east of the Breibogen Fault (Gee & Moody-Stuart 1966) (Fig. 1) mainly consists of the fine-clastic upper ORS of the Andrée Land Group (Harland *et al.* 1974; Friend 1981). Between Austfjorden and Billefjorden, the eastern margin of the ORS basin of NW Spitsbergen is exposed in central Dickson Land (Fig. 1). Here, the Balliolbreen Fault (Harland *et al.* (1974); main fault of the Billefjorden Fault Zone) separates Devonian sediments with an unknown base in the west and pre-Devonian basement rocks in the east (Stensiø 1918; Orvin 1940; Harland *et al.* 1974).

In the Munindalen and Hugindalen areas (Fig. 2), the Andrée Land Group consists of Lower Devonian green and red siltstones and sandstones of the Wood Bay Formation (Holtedahl 1914; Føyn & Heintz 1943; Friend 1961), which is unconformably overlain by grey–green and red siltstones, sandstones and conglomerates of the Mimerdalen Formation (Vogt 1938; Friend 1961). After the deposition of the Plantekløfta member (Vogt 1941; Friend 1961) (uppermost Mimerdalen Formation), which represents the uppermost ORS unit in Spitsbergen (Friend 1961), the basin infill was affected by west-directed folding and thrusting, which was defined as the Svalbardian folding (of Late Devonian age) by Vogt (1928). In the study area (Fig. 2), the folded and thrust-faulted Devonian deposits (both the Wood Bay and Mimerdalen formations) are unconformably overlain by horizontal, undeformed Carboniferous sediments (Stensiø 1918; Vogt 1928; Orvin 1940) of the clastic Triungen member of the Hørbyebreen Formation (Cutbill & Challinor 1965; Cutbill *et al.* 1976) and platform carbonate deposits of the Wordiekammen Formation (Gee *et al.* 1953).

The age of the Svalbardian deformation has been unclear up to now, because contradictory

From: FRIEND, P. F. & WILLIAMS, B. P. J. (eds). *New Perspectives on the Old Red Sandstone.* Geological Society, London, Special Publications, **180**, 603–609. 0305-8719/00/$15.00

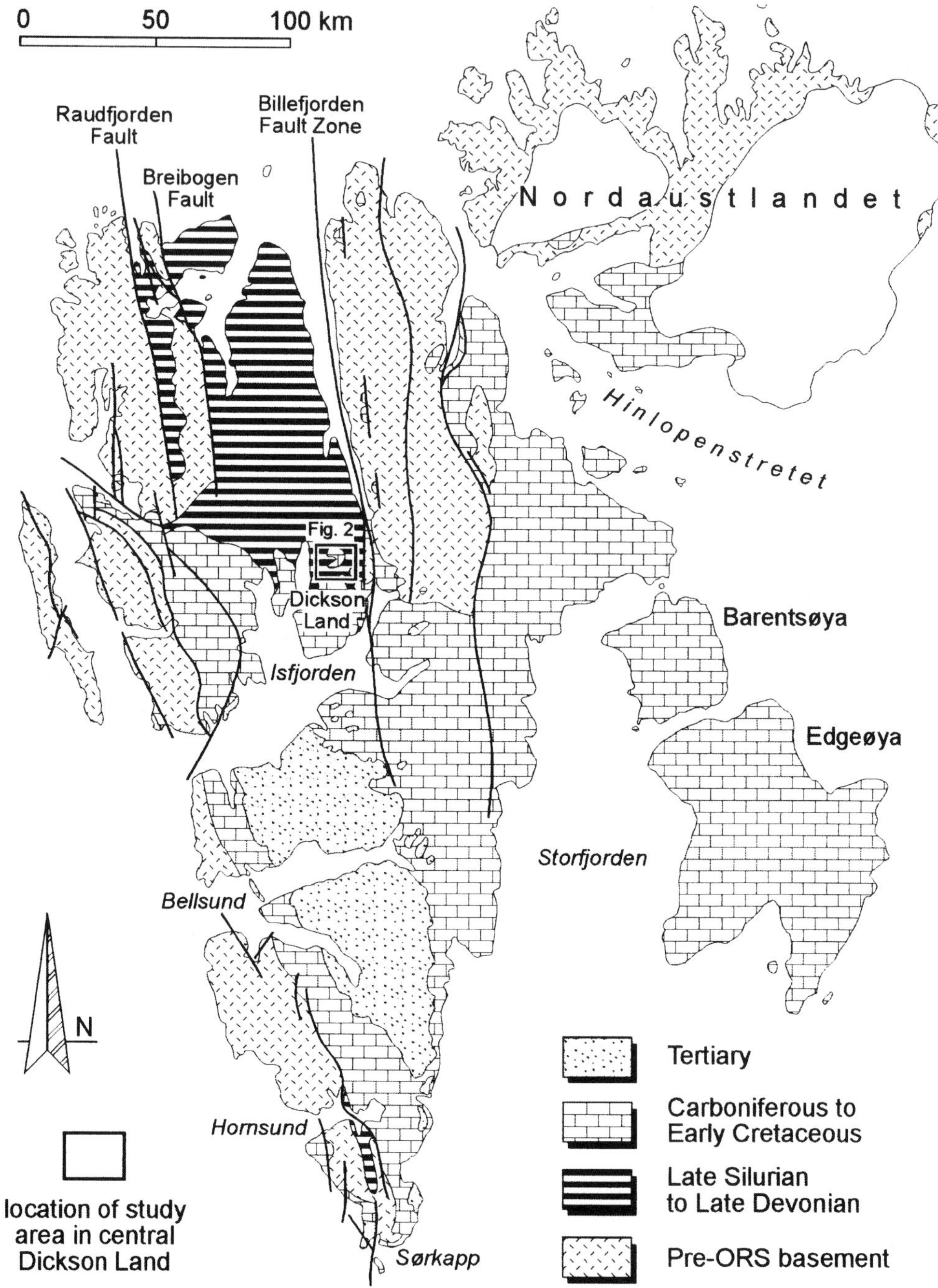

Fig. 1. Geological map of Svalbard, after Winsnes (1988) and Hjelle (1993).

interpretations of palynological data by different workers resulted in an apparent overlap in the time of deposition of the underlying unit (Plantekløfta member) and the overlying unit (Triungen member). The Plantekløfta member of the uppermost Mimerdalen Formation was suggested to be Famennian in age by Høeg (1942), Vigran (1964), Allen (1965, 1967, 1973) and Pčelina *et al.* (1986), and to be Early Carboniferous in age by Murašov & Mokin

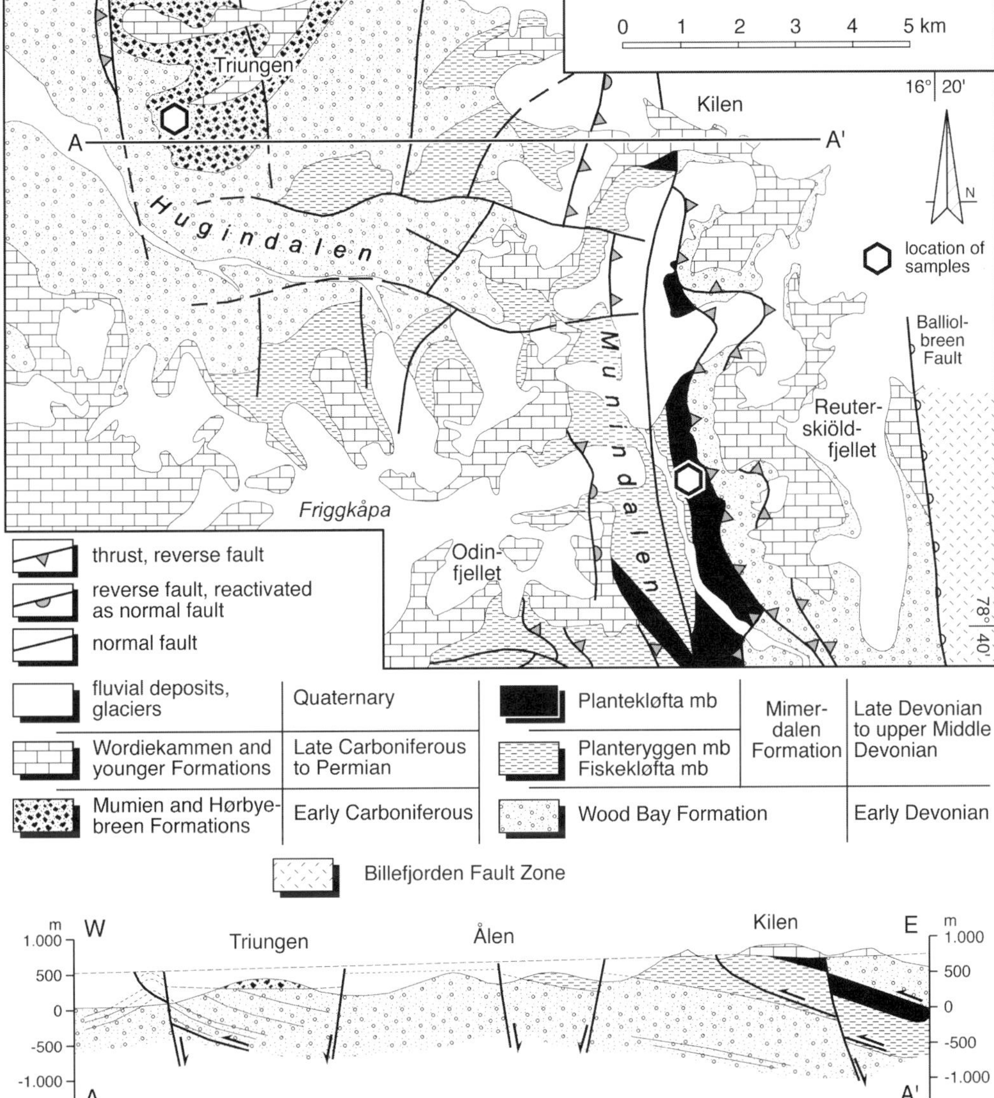

Fig. 2. Geological map and profile of the Mimerdalen and Hugindalen areas in central Dickson Land (Spitsbergen), after Piepjohn *et al.* (1997).

(1979). The unconformably overlying Hørbyebreen Formation was dated as being Famennian to Viséan in age (Vigran 1994; Dallmann 1999) with deposition of the basal Triungen member during Famennian time (Vigran 1994; Dallmann 1999), which is the same time interval as supposed for the Plantekløfta member. This overlap does not leave time for the Svalbardian folding and thrusting event. Therefore, the main task of the mapping programme of the seventh Münsteraner Spitzbergen Expedition in 1996 was to collect palynological material from the uppermost Mimerdalen and the lowermost Hørbyebreen formations to investigate this contradiction and to establish a more precise age for the youngest ORS sediments, the oldest post-ORS sediments, and the Svalbardian deformation.

Plantekløfta member (uppermost Mimerdalen Formation)

The siltstones, sandstones and conglomerates and the rich flora of the Mimerdalen Formation (uppermost Andrée Land Group) have been already described and partly named by previous workers (Nathorst 1910; Stensiø 1918; Vogt 1938, 1941; Nilsson 1941; Høeg 1942). The formation can be divided into the Estheriahaugen, Fiskekløfta, Planteryggen and Plantekløfta members (Friend 1961; Pčelina *et al.* 1986). The Plantekløfta member contains a rich macroflora (Høeg 1942) and consists of an intercalation of dark siltstones, shales and coarse-grained thick conglomerates (Friend 1961; Murašov & Mokin 1979). It is exposed only on the two sides of Munindalen (Fig. 2).

Stensiø (1918) described an east-dipping thrust fault in Munindalen, which carried Lower Devonian rocks over those of younger Devonian age. In this area (eastern slope of Munindalen west of Reuterskiöldfjellet, Fig. 2), our palynological samples were taken from the uppermost exposed shales of the Plantekløfta member, which have been overthrust from the east by sandstones of the Lower Devonian Wood Bay Formation along the lower Munindalen thrust (Fig. 2) (McCann & Dallmann 1996; Michaelsen *et al.* 1997; Piepjohn *et al.* 1997).

The samples from the uppermost parts of the Plantekløfta member contain both-well preserved and poorly preserved, reworked spores. The entire spore assemblage represents a wide temporal range between Early and Late Devonian time (Fig. 3). The occurrence of *Retispora lepidophyta* is important because its short stratigraphic range between Late Famennian time and the boundary to the Carboniferous (Streel & Loboziak 1996) indicates a Latest Famennian age of the uppermost Plantekløfta member. *R. lepidophyta* has been found in very small numbers, as was confirmed by Streel & Loboziak (1996), who described spore assemblages containing less than 1% *R. lepidophyta*. In addition, Kedo (1962) described the occurrence of *R. lepidophyta* together with *Archaeozonotriletes variabilis* and *Lophozonotriletes* as characteristic of Famennian time. This corresponds to the situation in the uppermost Plantekløfta member.

The occurrence of clasts within the thick conglomerates that are derived from the green sandstones of the Early Devonian Wood Bay Formation (Vogt 1941; Friend 1961) shows that the deposits of the Plantekløfta member partly consist of reworked older Devonian sandstones. This reassortment of pre-Plantekløfta sediments explains the mixing of spores from different Devonian units and the poor preservation of older spores. The youngest observed spores (*R. lepidophyta*) indicate a maximum age of the Plantekløfta member of Late Famennian time. If the specimens of *Retispora lepidophyta* have been reworked, the Plantekløfta member could be slightly younger than Late Famennian time (i.e. Early Carboniferous), but it could not be older than the Late Famennian time.

Triungen member (lowermost Hørbyebreen Formation)

North of Hugindalen, the basal Triungen member of the Hørbyebreen Formation (lowermost Billefjorden Group) (Cutbill &

	Lochkovian	Pragian	Emsian	Eifelian	Givetian	Frasnian	Famennian
Samarisporites	—	—	—	—	—		
Dictyotriletes	—	—	—	—	—	—	—
Chelinospora	—	—	—	—	—		
Calamospora	—	—	—	—	—	—	—
Retusotriletes	—	—	—	—	—	—	—
Ancyrospora			—	—	—	—	
Geminospora			—	—	—	—	
Archaeozonotriletes			—	—	—	—	—
Convolutispora			—	—	—	—	—
Grandispora			—	—	—	—	—
Perotrilites			—	—	—	—	—
Verrucosisporites			—	—	—	—	—
Acinosporites				—	—		
Hystricosporites				—	—	—	—
Apiculiretusispora					—		
Cymbosporites					—		
Lophozonotriletes					—	—	—
Brochotriletes						—	—
Retispora lepidophyta							—

Fig. 3. Table of the spore genera found in the uppermost exposed Plantekløfta member at the eastern slope of Munindalen and their age relationship after Chaloner (1967).

Challinor 1965) unconformably overlies gently folded Lower Devonian red siltstones and sandstones of the Wood Bay Formation (Fig. 2). The Triungen member represents the oldest post-Svalbardian deposits and consists of mudstones, siltstones, sandstones and conglomerates with intercalated coal seams. At Triungen, this unit is *c.* 100 m thick. Our palynological samples were taken from the lowermost coal seam 30–40 m above the unconformity on the southwestern slope of Triungen (Fig. 2).

Investigation of spores within Lower Carboniferous sediments (Culm) in the Billefjorden area have previously been carried out by Hughes & Playford (1960), Bharadway & Venkatachala (1962) and Playford (1962). Playford (1962) described two characteristic spore assemblages in the Culm succession of Spitsbergen: he distinguished a Rarituberculatus assemblage in the Tournaisian and an Aurita assemblage in the Viséan to Early Namurian time interval.

In the coal seam at Triungen, several species can be assigned to the Aurita assemblage (Fig. 4). However, some typical species such as *Murospora aurita*, *Tumulispora appendica*, *Cirratriradites elegans* and *Foveosporites insculptus* are not as frequent as expected for species that are characteristic for an assemblage. Compared with this, the high number of *Densosporites* genera is conspicuous, representing up to 60% of the spores. This observation is supported by Playford (1962), who described a coal seam at Svenbreen (just north, outside the study area) that contains a high proportion of *Densosporites* and *Reticulatisporites cancellatus* relative to the number of the genus *Murospora*. He suggested that this phenomenon is caused by an independent (coal-) flora that was isolated from the other Early Carboniferous floras.

Very small numbers (0.25–0.75%) of *Tumulispora rarituberculata* and *Tumulispora variverrucata* in the sample are characteristic of the Rarituberculatus assemblage (Fig. 4) (Playford 1962). They are described as being typical Tournaisian spores (Van der Zwan 1979). The occurrence of these forms in the Triungen samples along with *Murospora aurita* is problematic because they indicate different ages. This contradiction can be explained if the species of the Rarituberculatus assemblage are interpreted as reworked spores that have been resedimented in the lowermost coal seam of the Triungen member. The Aurita assemblage material suggests a Viséan age for the lowermost coal seam of the Triungen member and not a Famennian age as proposed by Vigran (1994) and adopted by Dallmann (1999). Although there are numerous species that also occur during

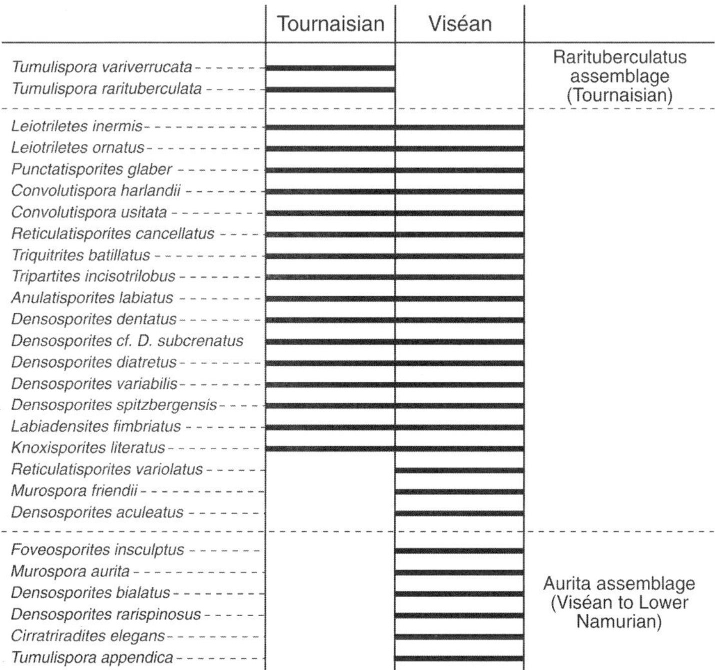

Fig. 4. Table of the spore species found in the lowermost coal seam of the Triungen member southwest of Triungen and their stratigraphic position in the Lower Carboniferous of Dickson Land after Playford (1962).

the Tournaisian time but continue through Viséan (Fig. 4), a third of the species appear at the base of the Viséan sequence and also suggest that the deposition of the coal seam took place during the Viséan time.

Conclusions

The deposition of the uppermost Plantekløfta member took place before the Svalbardian deformation. The occurrence of *Retispora lepidophyta* clearly indicates a Late Famennian age of the uppermost Devonian deposits of the ORS in Spitsbergen. The observation that the Plantekløfta member was overthrust by the Lower Devonian Wood Bay Formation along the lower Munindalen thrust and that it is situated within the core of a kilometre-scale west-vergent fold-structure south of Munindalen (Michaelsen *et al.* 1997; Piepjohn *et al.* 1997) indicates that the main tectonic stage of the compressional Svalbardian deformation was of a post-Late Famennian age.

The occurrence of green sandstone clasts of the Wood Bay Formation within the conglomerates of the Plantekløfta member suggests that the onset of the Svalbardian deformation had already started during the deposition of the conglomerates in Famennian time as a result of uplift in the area to the east of the Billefjorden Fault Zone (Vogt 1938; Friend 1961). Thus, the narrow stripe of the upper Mimerdalen Formation parallel to the BFZ represents a foreland basin that is related to the first movements of the Svalbardian deformation east of the Billefjorden Fault Zone (Piepjohn *et al.* 1997).

The upper time limit on the Svalbardian deformation is represented by the deposition of the unconformably overlying Triungen member. The spore assemblage indicates a Viséan age for the lowermost coal seam of this member. Taking into account that the basal conglomerates below the coal seam are slightly older, the upper time limit on the Svalbardian deformation may be upper Tournaisian time.

These palynological investigations suggest that the main Svalbardian deformation was restricted to Tournaisian time. Earlier stages of the Svalbardian event were related to the formation of the foreland basin of the upper Mimerdalen Formation just west of the BFZ already during Famennian time.

The authors would like to thank the German Research Foundation (DFG) for financial support (Project Pi 330/1-1). We would also like to express our thanks to the Norsk Polarinstitutt (Oslo) for material and logistic support, especially for the transportation by boat and helicopter to Dickson Land and back to Longyearbyen. We also gratefully acknowledge B. Dißmann and B. Michaelsen for their co-operation and many discussions in the field. We are grateful to P. F. Friend, A. J. McCann and P. T. Osmundsen for remarks and suggestions which improved the manuscript.

References

ALLEN, K. C. 1965. Lower and Middle Devonian spores of North and Central Vestspitsbergen. *Palaeontology*, **8**, 678–748.

—— 1967. Spore assemblages and their stratigraphical application in the Lower and Middle Devonian of North and Central Vestspitsbergen. *Palaeontology*, **10**, 280–297.

—— 1973. Further information on the Lower and Middle Devonian spores from Dickson Land, Spitsbergen. *Norsk Polarinstitutt Årbok*, **1971**, 7–15.

BHARADWAY, D. C. & VENKATACHALA, B. S. 1962. Spore assemblage out of a Lower Carboniferous shale from Spitsbergen. *Palaeobotanist*, **10**(1961), 18–47.

CHALONER, W. C. 1967. Spores and land plant evolution. *Review of Palaeobotany and Palynology*, **1**, 83–93.

CUTBILL, J. L. & CHALLINOR, A. 1965. Revision of the stratigraphical scheme for the Carboniferous and Permian rocks of Spitsbergen and Bjørnøya. *Geological Magazine*, **102**, 418–439.

——, HENDERSON, W. G. & WRIGHT, N. J. R. 1976. The Billefjorden Group (Early Carboniferous) of Central Spitsbergen. *In*: HARLAND, W. B., PICKTON, C. A. G., WRIGHT, N. J. R., CROXTON, C. R., SMITH, D. G., CUTBILL, J. L. & HENDERSON, W. G. (eds) *Some Coal-Bearing Strata in Svalbard.* Norsk Polarinstitutt Skrifter, **164**, 57–89.

DALLMANN, W. K. (ed.) 1999. *Lithostratigraphic Lexicon of Svalbard: review and recommendations for nomenclature use. Upper Palaeozoic Quaternary Bedrock.* Stratigrafisk Komitê for Svalbard (SKS), Committee on the Stratigraphy of Svalbard, Norwegian Polar Institute, Norway.

DE GEER, G. 1909. Some leading lines of dislocation in Spitsbergen. *Geologiska Föreningens i Stockholm Förhandlingar*, **31**, 199–208.

FØYN, S. & HEINTZ, A. 1943. The Downtonian and Devonian vertebrates of Spitsbergen. *In*: The English–Norwegian–Swedish Spitsbergen Expedition 1939, VIII. Geological Results. Skrifter om Svalbard og Ishavet, **85**, 1–52.

FRIEND, P. F. 1961. The Devonian stratigraphy of north and central Vestspitsbergen. *Proceedings of the Yorkshire Geological Society*, **33**, 77–118.

—— 1981. Devonian sedimentary basins and deep faults of the northernmost Atlantic borderlands. *In*: KERR, J. W., FERGUSSON, A. J. & MACHAN, L. C. (eds) *Geology of the North Atlantic Borderlands*. Canadian Society of Petroleum Geologists, Memoirs, **7**, 149–166.

GEE, D. G. & MOODY-STUART, M. 1966. The base of the Old Red Sandstone in central north Haakon VII

Land, Vestspitsbergen. *Norsk Polarinstitutt Årbok*, **1964**, 57–68.

——, HARLAND, W. B. & MCWHAE, J. R. H. 1953. Geology of central Vestspitsbergen. Part I. Review of the geology of Spitsbergen with special reference to central Vestspitsbergen. Part II. Carboniferous to Lower Permian of Billefjorden. *Transactions of the Royal Society of Edinburgh*, **63**, 299–356.

HARLAND, W. B., CUTBILL, J. L., FRIEND, P. F. *et al.* 1974. The Billefjorden Fault Zone, Spitsbergen: the Long History of a Major Tectonic Lineament. *Norsk Polarinstitutt Skrifter*, **161**.

HJELLE, A. 1993. *Geology of Svalbard*. Norsk Polarinstitutt Polarhåndbok, **7**.

HØEG, O. A. 1942. *The Downtonian and Devonian Flora of Spitsbergen*. Skrifter om Svalbard og Ishavet, **83**.

HOLTEDAHL, O. 1914. On the Old Red Sandstone Series of northwestern Spitsbergen. *In*: *Congress Report. XIIth Session*. International Geological Congress, Ottawa, Canada, 707–712.

HUGHES, N. F. & PLAYFORD, G. 1960. Palynological reconnaissance of the Lower Carboniferous of Spitsbergen. *Micropalaeontology*, **7**, 27–44.

KEDO, G. I. 1962. Spore assemblages of Upper Famennian and Tournaisian deposits at the boundaries of the Devonian and Carboniferous in the Pripyat Depression. *In*: FLADKOV, A. H. (ed.) *First International Conference on Palynology. Extended Reports of Soviet Palynologists*. Izdatel'stvo Akademii Nauk SSSR, Moscow, 73–76.

MCCANN, A. J. & DALLMANN, W. K. 1996. Reactivation history of the long-lived Billefjorden Fault Zone in north central Spitsbergen, Svalbard. *Geological Magazine*, **133**(1), 63–84.

MICHAELSEN, B., BRINKMANN, L. & PIEPJOHN, K. 1997. Struktur und Entwicklung der svalbardischen Mimerelva Synkline im zentralen Dickson Land, Spitzbergen. *Münstersche Forschungen zur Geologie und Paläontologie*, **82**, 203–214.

MURAŠOV, L. G. & MOKIN, J. L. 1979. Stratigraphic subdivision of the Devonian deposits of Spitsbergen. *Norsk Polarinstitutt Skrifter*, **167**, 249–261.

NATHORST, A. G. 1910. Beiträge zur Geologie der Bären-Insel, Spitzbergens und des König-Karl-Landes. *Bulletin of the Geological Institution of the University of Uppsala*, **10**, 261–416.

NILSSON, T. 1941. *The Downtonian and Devonian Vertebrates of Spitsbergen. VII. Order Antiarchi*. Skrifter om Svalbard og Ishavet, **82**.

ORVIN, A. K. 1940. *Outline of the Geological History of Spitsbergen*. Skrifter om Svalbard og Ishavet, **78**.

PČELINA, T. M., BOGAČ, S. I. & GAVRILOV, B. P. 1986. Novye dannye po litostratigrafii devonskich otloženij rajona Mimerdalen archipelaga Špicbergen (New data on the lithostratigraphy of the Devonian deposits of the region of Mimerdalen of the archipelago of Svalbard). *In*: *Geologija osadočnogo čechla archipelaga Špicbergen (Geology of the Sedimentary Blanket of the Archipelago of Spitsbergen)*. Sevmorgeologija, Leningrad, 7–19.

PIEPJOHN, K., BRINKMANN, L., DIßMANN, B., GREWING, A., MICHAELSEN, B. & KERP, H. 1997. Geologische und tektonische Entwicklung des Devons im zentralen Dickson Land, Spitzbergen. *Münstersche Forschungen zur Geologie und Paläontologie*, **82**, 175–202.

PLAYFORD, G. 1962. Lower Carboniferous microfloras of Spitsbergen, Part I and II. *Palaeontology*, **5**(1963), 550–678.

STENSIØ, E. 1918. Zur Kenntnis des Devons und des Kulms an der Klaas Billenbay, Spitsbergen. *Bulletin of the Geological Institution of the University of Uppsala*, **16**, 65–80.

STREEL, M. & LOBOZIAK, S. 1996. Middle and Upper Devonian miospores. *In*: JANSONIUS, J. & MCGREGOR, D. C. (eds) *Palynology: Principles and Applications*. American Association of Stratigraphic Palynologists Foundation, **2**. Applications, 575–587.

VAN DER ZWAN, C. 1979. *Aspects of Late Devonian and Early Carboniferous palynology of southern Ireland*. PhD thesis, University of Utrecht.

VIGRAN, J. O. 1964. *Spores from Devonian Deposits, Mimerdalen, Spitsbergen*. Norsk Polarinstitutt Skrifter, **132**.

—— 1994. *Palynology of Upper Devonian to basal Permian rocks of the Arctic*. IHU report **23.1438.00/18/94**.

VOGT, T. 1928. Den norske fjellkjedes revolusjonshistorie. *Norsk Geologisk Tidsskrift*, **10**, 97–115.

—— 1938. The stratigraphy and tectonics of the Old Red formations of Spitsbergen. *Abstracts of the Proceedings of the Geological Society, London*, **1343**, 88.

—— 1941. Geology of a Middle Devonian cannel coal from Spitsbergen. *Norsk Geologisk Tidsskrift*, **21**, 1–12.

WINSNES, T. 1988. *Bedrock map of Svalbard and Jan Mayen 1:1 000 000*. Norsk Polarinstitutt Temakart, **3**.

Index

Page numbers in *italic* refer to figures or tables.

Acadian (mid-Devonian) deformation 37, 44, 49–50, 192, 391
Acadian Orogeny 33, 50, 52, 63, 185, 312
Acanthostega tetrapod remains 557–9, 561, 564
Achanarras Sandwick Fish Beds 474, 479, *480*
aeolian deposition
  Ballydavid Formation 166
  Beenmore Sandstone Formation 172–3, 175
  Farran Formation 161, 166
  Kilmurry Formation 178, 179, 190, 207, 209
  Sauce Creek Formation 168
Agda Formation (East Greenland) 558
Aglaophyton major 454
Aglaophyton sp. 449, 455, 456
Aiculiretusispora sp. 361
Aina Dal Formation (East Greenland) 558, 559–60, *560*, 563, 564
Albertbreen Formation 570, 573
Albion Sands Formation 352–3
Ambitisporites sp. 366
Ancyrospora ancyrea 477, 478, *478*, 479, 480
Ancyrospora ancyrea cf. var. *brevispinosa 478*
Ancyrospora simplex 289, 325
Ancyrospora sp. 476
Andrée Land Anticline 589, 591–2
Andrée Land Group *35*, 42, 46, 568–9, 573, 575–6, 578–9, 585, 603, 605–6
Andréebreen Formation 571
Aneurospora sp. 361, 364, 366
Anglo-Welsh Basin 37, 44, 51–2, 371, 379
Annascaul Formation *150*, 187, 202
AOM (Amorphous Organic Matter) 326–7, 475, 479
Archaeopteris hibernica 333, 340
Archaeozonotriletes variabilis 606
Archanodon sp. 89, 93
Archaoperrisaccus ovalis 291
Archeosigillaris sp. 596
Ardane Formation 249
Ardaturrish Member 311
Arenicolites sp. *76*, 78, 99–101
Argyll Group 111
Armorica 32, 34, 51
Arran, Lower Old Red Sandstone conglomerates 423
Asteroxylon mackiei 454
Asteroxylon sp. *447*, 449, 456
astronomical rhythmicity (precession and obliquity cycles) 10, 520–1
Auchtitench Formation 428
Auroraspora cf. *hyalina* 291
Auroraspora macromanifestus 478
Auroraspora micromanifestus 478

Ballinskelligs Sandstone Formation *188*, 249, 275, *276*, 292
Ballydavid Formationn *150*, 154, 156, 158, 165, 166–7, 168
  cyclic deposition 167
  fluvial-aeolian architecture *157*, 165, 167, 169
  stratigraphy 166
Ballyferriter Formation 132, *150*, *188*
Ballymastocker Basin 40, 109–21
Ballymore Formation *150*, 153, 187, *188*, 206, 218
Ballynane Formation *150*, 202
Ballyoughteragh Member, Farran Formation 154, 158, 165
Ballyroe Group 155, *188*, 189
Ballytrasna Formation 251
Baltica 32–3, 47, 417, 505
Barr Group 463
Battery Point Formation *35*, 36, *62*, 63–82, *63*
Beaconites antarticus 376
Bealtra Volcanic Breccia 277, *279*, 282–3
Beara Peninsula 231–4, 303, 305, 306, 307–11, *308*, 312–13
  basin faulting 311, 312
  Black Ball Head Pipe 309–10, 311, 312, 313
  Caha Mountain Formation 275
  crustal stretching 313
  depocentre 224, 233
  pipe-like intrusions (lamprophyre affinities) 309–10, 312
  sills and dykes 307–8
Beenaman Conglomerate Formation *150*, 154, 170–8, *173*, 179
  clast lithotypes 176
Beenmore Sandstone Formation 154, *155*, 170–3, 175, 176–8
  fluvial-aeolian sedimentation 173–7
  Lower Member 173, 175
  Upper Member 173, 175
Beenreagh basalts 250, 277
Bellewstown Terrane 216
Ben Nevis Formation 571, 574
Benan Conglomerate 465
Bennaunmore Volcanic Centre 272–4, 277, 294, 306
Billefjorden Group 579
bioturbation
  Devonian rivers and coasts 88–9, 93, 99–101
  Moor Cliffs Formation 376–7, 381–2
  Pantymaes Quarry 393, 397
  Skolithos-type vertical burrows 393
Biskayerfonna Group 570
Biskayerfonna-Holtedahlfonna Horst 591, 595
Biskayerhalvoya terrane 570–1, 575, *576*, 580, 582
Blasket Islands 124, *124*, 125, *127*, *128*, 130, *134*
Boat Cove Member, Bull's Head Formation 132, 133, 135, 141
Bothriolepis sp. 270, 293, 298, 330, 558

Brandon Point Anticline 154, 170, 173
Bray Group 202
Bristol Channel Fault Zone (BCFZ) 403, *405*, 406
Bristol Channel Landmass 403
Britta Dal Formation (East Greenland) 558, 559, 560, 561, 564–5
Brownstones Group *35*, 37
Bull's Head Formation 38, 132–3, *133*, *134*, 135, 140, *150*, *188*
  lithofacies *134*, 167
Bythotrephis sp. 334, 336

Caha Mountain Formation 275, *276*, 293, 312, 321
Caherbla Group *35*, 39–40, 218, 243, 246, 262, 298
  age determination 155, 180, 207
  geology *210*
  palaeogeography *172*, 178–80, *211*
  stratigraphy 154–5, 170, 186–90, 190, 207
Calamospora atava 478, *478*
Calcrete morphology and karst development, Milton Ness 485–99
  boulder calcrete 493–5, *494*, 499
  calcrete accumulation rates 496
  calcrete profile 487
    vertical zonation of pedogenic carbonates 487, *488*, 499
  carbonate dissolution rates 496
  distribution of Old Red Sandstone in Midland Valley *486*
  geological setting 486–7, 495
  hardpan calcrete 487–91, *488*, *492*
    clotted micrite 490, *491*
    displacive calcite growth 487–8, *490*, 493, 494, 499
    floating grain texture 487, *489*, 490, *491*, 493
    microspar fringes on detrital grains 490, *491*
    pisolites 489, *490*, 493
    silica laminations 489, *490*
  karst solution cavities 485, 487, 491–3, *492*, 494, 495
  karst/calcrete association 495–6
  palaeocurrents 487
  rhizoliths 485, 487, 492–3, 496–9, *498*, 499
  river drainage systems 487
calcrete nodules and horizons (pedogenesis) 393–4, 396–9
  Lower Old Red Sandstone 406, 426
Caledonides 32–4, *33*, 47, 49–50, 537–8, 552, 580–2, *581*
Calyculiphyton sp. 456
Cap-aux-Os Member, Battery Point Formation 36, *62*, 63–82, *63*
Carrigduff Group *35*, 155, 170, 173, 177, *188*, 189, 244, 257, 264
Carrigduff Volcanic Member, Coumshingaun Conglomerate Formation 251
Castlehaven Formation 247, 248, *250*, 251, 263, 319
Catskill foreland basin *see* Devonian rivers and coasts, Eastern USA
Celsius Bjerg Group (East Greenland) 538, 558–9, 561, 564–5
Central Terrane 216–17
Chapel Point Calcrete 373
Chelinospora concinna 289, *290*, 325, *329*, 479
Cheviot Hills lavas 420, 428
Chloritic Sandstone Formation, Munster Basin *150*, *188*, 250, 271, 273–5, 277, 306, 328
  fluvial environment 294
  miospore assemblage 289–91, *290*
Chondrites sp. 130, 321
chronostratigraphy, Old Red Sandstone (Devonian) 23–6, *24 see also* Devonian time scales
Chutkotka–Alaske Plate 580, *581*
'Cirratriradites' avius 478
Cirratriradites elegans 607
'Cirratriradites' monogrammos 477, *478*, 479
Clear Island 320–1, *321*, 330
Clogher Head Formation 127–9, 130, 131, 132, 138–40, *150*
Clousta Volcanic series 473–4
Cockburnspath Formation 486
Comeragh Conglomerate-Sandstone Group 246–7
Contagisporites optivus 477, *478*, 479
Converrucosisporites liratus 325
Cooksonia 379
Cooksonia pertoni 356, 366
Coolnahorna Volcanic Member, Coumshingaun Conglomerate
Formation 251
Cooscrawn Tuff Bed, Ballymore Formation 153, 180
Coosglass Member, Farran Formation 154, 165
Coosgorrib Conglomerate Formation *150*, 155, 156, 158, 165, 168, *175*, 180
Coralliferous Group 344, 352–3
Cornuodus cf. *C. longibasis 463*
Corsewall Conglomerate 466
Cosheston Group 37, 407, 412
Coumeenoole Formation 38–9, 135, 138, *150*, *188*, 206
Coumshingaun Conglomerate Formation 251
Couverrucosisporites sp. 291
Cowie Formation 459
Craighead Limestone 465
Crana Quartzite Formation 112, 115, *118*
Cranford Limestone Formation 112
Crawton Basin 40, 49
Crawton Group 459, 461
Crawton Sub-basin 459
Crenaticaulis verruculosus 74
Crinan Subgroup 112
Cristatisporites mediconus 478
Cristatisporites triangulatus 479
Croaghmarhin Formation 131, *131*, 140–1, *150*
Crossgate-Burnside sediment bars 426–8, *427*
Curlew Basin 40
Cushendall Basin 40, 41, 49, 52
Cyclostigma kiltorcense 333, 334, 335, 338, 340
Cyclotites lenticulata 198
Cymatiosphaera sp. 327, *329*

Dalradian basement 111–12, 420, 424, *425*, 426, 461
  palaeomagnetism 116–17
  sediment source for Lower Old Red Sandstone basins 112, 114, 119, 420, 424–5, 430, 434
Dartmouth Group *35*, 36, 51, 242

Dawsonites sp. 356
Decoroproetus? sp. 463
Deheubarthia sp. 355
Densosporites 607
Densosporites concinnus *477*, *478*, 479
Densosporites devonicus *478*, 479
Derryreag miospore assemblage 291
Devil's Punch Bowl rhyolite 274
Devonian (Old Red Sandstone) chronostratigraphy 23–6, *24 see also* Devonian time scales
Devonian rivers and coasts, Eastern USA 85–104
bioturbation 88–9, 93, 99–101
Catskill foreland basin 42, 85–104
coastal strata 95–102, *97*, *98*
cyclicity of sedimentation 85–7
environment of deposition 86–7, *86*, 89–90, 93–4, 99–102
large-scale sequences 102–4
lithostratigraphic correlation 87
non-marine strata 87–95
main channel deposits 87–92, *92*
mathematical modelling 92
overbank deposits 92–4
stratal geometry *88*
variations 94–5, *94*, *95*
palaeocurrents 88, 91–2, 94, 97, 98–101
palaeoshoreline 93, 94–5, 99
palaeosols 93–4, 100
palynostratigraphy 87
plant remains 88–9, 93, 99, 100–1
sandstone-mudstone sequences 101–2
sea-level changes 42, 102–4
sedimentology of fluvial deposits *92*
stratigraphy *86*
Devonian sediments at Ballymastucker 109–21
Dalradian basement stratigraphy 111–12, *111*, *118*
depositional framework 119–20
depth of burial 118–19, 120–1
faults and fractures *110*, 114–16, *117*, 119, *119*
fluid inclusion data 117–19, 120, *120*, 121
geological setting *110*, 111–12, *113*
Leannan Fault system 109, 110–11, 112, 114–15, 118–21
Riedel shears 112, 115, 119
lithofacies and depositional environment 112–14, *114*, *115*, *116*, 120–1
palaeomagnetic data 116–17, 121
structural chronology 119
Upper Palaeozoic (Lower ORS) sediments 112–14
Devonian time scales 1–17
analytical methods
ID-MS (isotope dilution mass spectrometry) 1, 15, 17
SHRIMP (sensitive high-mass resolution ion microprobe) 1, 7–8, 11–12, 14–15, 17
chronostratigraphy 2
evolution of Devonian time scale 10–15, *13*, *14*
geological time scales database 3–8
fission-track ages 5–7
geological context of data points 3–5
glauconite ages 4–5
non U-Pb isotopic systems 5–7, 17
stratigrahically discordant intrusion ages *6*
stratigraphically concordant isotopic dates 3, *5*
U-Pb zircon dating 7–8, 15–17, *16*
miospore zonation *2*
recent data compilation 15–17
time-scale construction methodology 8–10
best-fit time line method 9, *9*
faunal evolution and biozone durations 10
graphic correlation technique 10
orbital precession and obliquity cycles 10
scanning age procedure 9, *9*
Diaphanospore reticulata 325
Dictyotriletes craticulatus 289, *290*, 325
Dictyotriletes perlotus 289, 325
Diducites sp. 337
Dingle Basin 38–40, 49, 51, 216–18, 224–5, 228, 230, 232–3, 235, 243, 246, 249, 264
palaeogeography *195*
sedimentological evidence of hinterland granite 233
Dingle Basin conglomerates, basin setting and provenance 185–218
Glashabeg Formation *193*, 202–6, *205*
clast composition 203–5
depositional environment 203
palaeohydraulic scale *197*, 203
provenance 205–6
sedimentary facies 203
Inch Conglomerate Formation 207–16
clast composition 211–14
depositional environment 179, 209
geochemistry (Sm-Nd and Rb-Sr isotopes) 214, *215*
palaeohydraulic scale 209–11
provenance 214–16
sedimentary facies 207–9
jasper 205–6
regional context 216–18
basin-forming mechanism 216
early Devonian hinterland geology 216
Siluro-Devonian tectonic history 216–18
sedimentary facies *193*
Slea Head Formation *193*, 206–7, *209*
clast composition 207
depositional environment 206
palaeohydraulic scale *197*, 206–7
provenance 207
sedimentary facies 206
stratigraphy of Dingle and Caherbla Groups 186–90
structure 190–2, *191*
Acadian transpression 192
Trabeg Conglomerate Formation 192–202, *193*, *199*, *200*
clast composition 196–8, 212
depositional environment 192–5
geochemistry (Sm-Nd isotopes) 198–201, *201*
limestone clast fauna 198
palaeohydraulic scale 195–6, *197*
provenance 201–2
sedimentary facies 192

Dingle Basin, early development of 123–43
basin evolution 138–43, *139*
active subduction phase 138
post-subduction thermal subsidence 140
strike-slip fault-controlled subsidence 140–1
subduction termination phase 138–43
Dingle Bay Lineament 135
Dingle Group 132–8, *139*
overview 135–8
stratigraphy and depositional facies 124, *126*, 132–5, 140–1
Dunquin Group 125–32, *128*, *139*
hydrothermal metamorphic event 131–2, 140
igneous petrology and magmatic evolution 131
stratigraphy and depositional facies 124, 125–31, *126*, *129*, 138–40
intrabasinal faulting 140–1
lavas and pyroclastics 125, 127–31, 138, 140
plate reconstruction *125*
tectonic setting *126*, *129*, 141–3
Dingle Bay-Galtee Fault Zone 44, 223–4, 228–30, 232–5, 243, 244, 246, 257, 263, 272, 319
Dingle Group *35*, 123, 124, 132–8, 140–1, 149–53, *150*, 155, 180, 225
age 153, 187–8
lithostratigraphy *189*
stratigraphy 158, 186–90
structure 190–2 *see also* Northwest Dingle Domain
Dingle Peninsula 123–4, 186–218
basin evolution *175*
chronostratigraphy *188*
geological setting *124*, *127*, *131*, *189*
structural history 190–2, *191*
volcano-tectonism *130* *see also* Northwest Dingle Domain (ORS)
Diplocraterion sp. *76*, 78, 100–1
Ditton Group *35*, 37
Doon Lava Member, Ballytrasna Formation 251
Doulus Conglomerate 243, 249
Dounans Limestone 468
Downton Group *35*, 37, 52
Drepanodus arcuatus 465
Drepanodus robustus 463, *465*
Drepanophycus sp. 67, 74
Drepanophycus spinaeformis sp. 74
Drom Point Formation 130–1, *131*, 140, *150*
Duneaton Formation 428
Dunquin Group *35*, 38, 123, 125–32, *128*, *129*, *131*, 149, 186, *188*, 205, 224
hydrothermal event 131–2
igneous petrology and magmatic evolution *130*, 131
stratigraphy 155, 187, 190
stratigraphy and depositional facies 124, 125–32, *129*, 138–40
structure 192
volcanic suite, trace elements characteristics 142–3

Easdale Subgroup 111–12
Eask Formation *133*, 135–8, 149, *150*, 187, *188*
lithofacies *136*
East Greenland Basin, syndepositional tectonics 537–53
conglomerates 540–1, *544*, 547
diachronous basal deposits 537–8, 540–1, 552
Eastern Fault Zone 538, 542
extensional thinning 550–2
fault patterns *546*, *549*
fault populations 543, 547, 548–51, *548*, *549*
Moskusoksefjord (MOF) Inlier 539–41, *539*, *540*, 543, *543*, *544*, *546*, *548*, 550–1
previous studies 538
regional comparisons 552
regional geology 538, *539*
sedimentary logs *542*, *545*
stratigraphic frameworks 538
stratigraphy of syntectonic basal deposits 538–41, *540*
structural evolution of basal ORS 541–50
cross-cutting faults 547, *549*, 550
syn-extensional folding, Eastern Dybendal 547–50, *548*
syndepositional folding and faulting, Ella O 547
Western Fault Zone 543–6
tectonic history 550–2
wrench faulting or strain partitioning 551
Western Fault Zone 538, *539*, 542–6, *545*, 551
drag folds 543
slickensides 537, 543–5, 551
East Greenland Upper Devonian fossils 557–65
invertebrates 564
lithofacies 563–4, *563*
radiometric dating 564
stratigraphy *559*
tetrapod localities
Gauss Halvo 559–61
Ymer O 561–4
trace fossils 559, 564
East Mendips inlier volcanic suite 142–3
Eastern Avalonia 32–4, 47, 49–52, 123, 138, 141–3, 245, 343–4, 352
Eday Flagstone Formation 479, *480*
Eday Marl Formation *480*
Eleonore Bay Supergroup 538, 548, 569
Ellesmerian Orogeny 32, 580, 582
Elsa Dal Formation (East Greenland) 558
Emphanisporites annulatus 357
Emphanisporites decoratus 361
Emphanisporites micrornatus 359–63, *360*
Emphanisporites protophanus 356
Emphanisporites rotatus 357, *477*, *478*
Emphanisporites splendens 357
Enagh Tuff Bed *see* Keel-Enagh Tuff Bed
Eoplacognathus lindstroemi 463, *463*
ephemeral fluvial systems 158–61, 167, 168, 169, 180
erg migration
Ballydavid Formation 167
Beenmore Sandstone Formation 175, 177
Farran Formation 161, 165, 166
Sauce Creek Formation 168
Escuminac Formation, Canada *35*, 36, 330
Eskine-Knocknagullion tuff 250
Eusthenodon 558, 560

extensional subsidence in Munster Basin 239–64
basin fill stratigraphy *244*, *245*, 246–51, *249*
crustal stretching factors 263, 264
crustal thinning 229, 232, 234–5
decompaction and back-stripping 251–3, *255*, *256*
methodology 251–2, *251*
results 252–3, *252*
Dingle Bay-Galtee Fault Zone 243, 244, 246, 257, 263
erosion rates 263
extensional faulting 229, 230, 232–4
flexural cantilever model 253
fluvial dispersal systems 44, 246–51, *247*
footwall to hanging-wall correlations 248–9, *249*
forward models 241–3, 253–9, *257*
northern Dingle-Clear Island 255–7, *258*, *259*
Slieve Phelim-Ardmore 257–9, *260*, *261*
geological setting *241*
Leinster Terrane 245–6
ORS accumulation *262*
palaeogeography *248*, *250*
palinspastic restoration 242
pre-Variscan structure 239
pre-Variscan template 241–6
previous models 240–1, 242
sub-Munster crustal basement 244–6
syn-ORS magmatism 249–51
syn-rift alluvial systems *247*, 262
tectonic setting *240*, 241–6
thermal syn-rift subsidence 259–62, 264
Variscan deformation 239, 242–6

Fahan Syncline 132–3, 135, 141
Fanad Boulder Bed 111
Farran Formation *150*, 153, 156–66, 167
aeolian deposition 156, *156*, 158, 161, 165–6
Ballyoughteragh Member 154, 158, 165
Coosglass Member 154, 165
ephemeral fluvial sedimentation 158–61
lateral facies variations *159*, 168
stratigraphy 158
tectonic influence on cyclic deposition 165–6
faults and lineaments
Balliolbreen Fault 590, 597, 603
Bantry Bay Fault 311
Benton Fault 37, 344, 373, 403, 404, 406, 407, 412
Billefjorden Fault Zone 569, 570, 573, 579–80, 585, 589, 590, 596–8, 603, 608
Breibogen Fault 569, 585, 591–2, 603
Breibogen-Bockfjorden Fault Zone 569, 570–1, 573, 576, *576*, 578–80
Caherconree Fault 178, 243, 255, 257, 264
Cahermore-Castletownbere Fault 310, 311–12
Carreg Cennen Disturbance 391, 395, 398
Castle Archdale-Omagh Fault 41
Church Stretton Lineament 391
Church Stretton-Carreg Cennen-Llandefaelog Fault Zone 37
Coomnacronia Fault 243, 246, 249, 257, 262–3, 274, 277–5, 292–3, 311, 312
Cork-Kenmare Line 40, 224, 228–9, 232, 234–5
Dalsfjord Fault *504*, 506
Dingle Bay Fault (Lineament) 38, 147, 154, 169, *175*, 177, 178, 179, 186, *187*, 190, 216, 255, 257
Dingle Bay-Galtee Fault *see* Dingle Bay-Galtee Fault Zone
Dun an Oir Fault 135, 154, 188
Dunmanus-Castletown Fault 223, 244, 246–8, 252, 253, 257, 263
Eastern Fault Zone, East Greenland Basin 538
Fohernamanagh Fault 154–5, 156, 158, 168, 169–70, 177
Fotkollen Fault 591
Fradingmullach Fault 468
Friedrichbreen thrust 578–9
Glen of Aherlow Fault 243, 249, 258
Glen Fumart Fault 468
Great Glen Fault 41, 48, 49
Gruinart Fault system 119–20
Hannabreen Fault 570, 573, 575
Heol Senni fault 391
Highland Boundary Fault *see* Highland Boundary Fault
Idabreen Fault 574, 575, 578
Instelva Fault *504*, 506
Johnson Thrust 344
Keisar Wilhelmhogda thrust 578–9
Kenmare River Fault 311, 312, 328
Kenmare Thrust 312
Killarney-Mallow Fault *see* Killarney-Mallow Fault
Kringlefjellet Fault *517*
Kvikkabreen Fault 575
Leannan Fault 40, 109, 110–12, 114–15, 118–20
Llandyfaelog Fault 373
Melby Fault 473, 479, 481
More-Hitra Fault 42
Munindalen thrust 590–1, 606, 608
Musselwick Fault 344, 373
North Kerry lineament 38, 147, 175, 176, 190, 230, 235, 255
Oberstbreen Fault 577
Presidentryggen Fault 577
Rabotdalen Fault 570, 573, 575
Rabotdalen-Hannabreen Fault 575–6, *576*, 582
Raudfjorden Fault 569, 570, 571, 574–5, 577–8, *577*, 582, 603
Ritec Fault 37, 44, 344, *372*, 373, 375, 401, 403, 404
St Magnus Bay Fault 473, 481
Slievenamuck Fault 243, 258
Sordalen Fault 597
South Galtees Fault 244, 257
Southern Uplands Fault 40, 49, 429, 463, 465, 486
Sulma Water Fault 473
Swansea Valley Disturbance 391
Triarmbreen Fault 574
Triungen-Gronhorgdalen Fault Zone 579, 590
Vale of Neath Disturbance 391
Walls Boundary Fault 473–4, 481
Welsh Borderland Fault System 37, 391
Wenall Fault 344, 350
West Andrée Land Fault Zone 591
Western Fault Zone, East Greenland Basin 538

Faveosporites insculptus 607
Favosites alveolaris 198
Feohanagh Anticline 154, 156, 166, 167, 169
Ferriter's Cove Formation 123, 127, 128, 138, *150*
Fintona Basin 40, 41, 49
Fintona Group *35*, 41
fish beds, west Iveragh succession 275, *276*, 293
fish remains
Eusthenodon 560, 563
Holoptychius sp. 403, *403*, 560
Fiskeklofta Member, Mimerdalen Formation 587, 588
Flerant Formation *35*, 36
fluvial dispersal systems
Chloritic Sandstone-Gortanimill System 271, 294–5, *294*
Galtee System 44, 246
Gun Point System 247, 248, 271, *294*
Iveragh System 44, 246
Sherkin System 44, 247
fluvial-aeolian sequences
Ballydavid Formation 165, 167
Beenaman Conglomerate Formation 174, 176
Beenmore Sandstone Formation 175, 176, 177
Farran Formation 158, 160
Kilmurry 178
Sauce Creek Formation 167
Smerwick Group 169
Foilcoagh Bay Beds, Sherkin Formation 251, 319, 320
anoxic conditions of deposition 327–8
Conodont Zone 325
lithofacies 321–3, *322*
marine influence 327–30
palaeoenvironments 321, 323–5
palynology 321, *322*, 323, 324, 325–7, *329*
pyritized spores 327–8
Foilnamahagh Formation 123, 125–6, 138, *150*
Fraenkelryggen Formation 571
Franklinian Mobile Belt 580
Freshwater East Formation 373
Freshwater West Formation 380
Friedrichbreen Anticline 573, 574–6, 578
Friedrichbreen Fault 574–5, 576–8

Galtees-Ballyhoura Mountains anticline 242
Garvock Group *422*, 426
Gaspé Bay, Quebec *see* land plants
Gauss Complex 46
Geminospora boleta 291
Geminospora lemurata 289, *290*, 291, 325, *329*, *477*, *478*, 479
Geminospora plicata 325
Geminospora sp. *477*, *478*
Generalfjella Formation 574
Genvejsdalen Member, Vilddal Group *543*, 551
geochronology (U-Pb) in Munster Basin 269–98, *270*
basin duration 291, 298
biostratigraphy 289–91
Derryreag microflora 291
Moll's Gap Quarry microflora 289–91
Reenagaveen microflora 289, *290*
Sauripterus sp. 293
tetrapod ichnofauna (Valentia Island) 270, *279*, 293, 298
chronostratigraphy 270, *296*, 297
cyclicities 297
depocentre sedimentation rate 291
Dingle Basin 270, 297–8
Dingle Bay-Galtee Fault Zone 269
Dunmanus-Castletown Fault 269
geochemistry 277–83, *280–4*, 298
discrimination diagrams *277*, *281–4*, 282
geochronology 284–9
analytical methods 285
U-Pb results 285–9, *287*, *288*, 298
zircon morphology 285, *286*
geohistory of Iveragh region of basin *295*, 296–7
geological setting 271–2, *271*
Moll's Gap Quarry miospore assemblage 270
stratigraphic-structural context of dated levels 272–7, *278*, *279*
Enagh Tuff Bed 277, *279*
Horses Glen Lower Tuffs 274
Keel Tuff Bed 274–5, *276*
Killeen Tuffs 273–4
Lough Guitane Volcanic Complex 272–4
Moll's Gap Quarry 274
Reenagaveen microflora 275–7, *279*
subsidence rate 291, 297, 298
geochronometric dating 319
Glandahalin Formation *150*, *250*
Glashabeg Formation *150*, 154, 187, *188 see also* Dingle Basin conglomerates
Glen Turret outlier (Lower Old Red Sandstone) 420, 424, 429
Gondwana 32, 34, 47
Gorgonisphaeridium sp. 327, *329*
Gosslingia breconensis 455, 456
Gosslingia sp. 356
Grampian Orogeny 32, 49
Gramscatho Basin 242
Grandispora cf. *megaformis* 289, *290*
Grandispora inculta 289, 325
Grandispora saetosa 325
Grandispora tomentosa 325
Grangegeeth Terrane 216
Gray Sandstone Group 37, *150*, 343–53
facies associations *350*
palaeogeography *344*
trace fossil assemblage 348, *349*
Great Conglomerate 428
Grey Hoek Formation 573, 585–7, 591
Greywacke Conglomerate 428, 463, 467
Groenlandaspis 558
Gupton Formation 403, 406

Hangman Sandstone Group 44
Hasteinen Basin 45
Haysites cf. *thomasi* 198
Heterolithic Member, Bull's Head Formation 132–3, *134*, 135, 141

Highland Border Complex 418, 422, 423, 424, *425*, 459, 461, 468
Highland Boundary Fault 40, 48, 49, 52, 418, 420, *422*, 423, 424, *425*, 430, 486
Hitra Basin 47
Hitra Formation 42
Holcospirifer bigugosus 138
Holoptychius sp. 402, *403*, *409*, 558, 560
Horbyebreen Formation 605, 606–8
Hornelen Basin 45, 503, 505, *507*, 514, 519
Horneophyton lignieri 361, 453–4
Horneophyton sp. 449
Horses Glen Lower Tuffs 274, 277, 285, 286–8, 293–4, 298
Horses Glen Upper Tuffs 274
Horses Glen Volcanic Centre *271*, 272–4, 293–4, *294*, 306
Huvenia sp. 67, 74
Hystricosporites delectabilis 325

Iapetus Ocean 32–3, 49, 123, 465, 580
Iapetus Suture 32–3, 38, 40, 52, 123, *124*, 141–2, 186, *186*
Iapetus Suture Zone 147, 186, 190, 216, 218
  chronostratigraphy 216–18, *217*
  geology *187*
Ichthyostega tetrapod remains 557–60, *558*, 561–5, *562*
Ichthyostegopsis tetrapod remains 558
Igneous Conglomerate 467
igneous rocks in Ireland (Upper Palaeozoic) *304*, *307*
Inch Conglomerate Formation *150*, 178, 179, *188*, 190, 243, 246, 249 *see also* Dingle Basin conglomerates
Indotriradites explanatus 337
Inishnabro Formation 141, *150*
Inishvickillane Formation 128, 130, *150*
Inishvickillane Island 130, 138
Insculptospora confossa *477*, 478, *478*
Islandeady Group 41
Islay Subgroup 111
Iveragh Peninsula 190, 313
Iveragh-Derrynasaggarts region 271–2

Jarvikia 558, 563

Kap Bull Formation 540–1, *543*, 550–1
Kap Graah Group 538, 558–9
Kap Kolthoff Group 538–41, *543*, *544*, *545*, *548*
Keadew Formation 40–1
Keel Tuff Bed *see* Keel-Enagh Tuff Bed
Keel-Enagh Tuff Bed 218, 243, *271*, 274–5, 277, 282, 282–3, 285, 288–9, 306, 310
  age 170, 190, 293, 298
  correlation of Keel and Enagh Tuffs 291–4, *292*, 298
  stratigraphical context *276*
Kenmare Syncline 297, 312
Kilcrohane Anticline 273, 274
Kilcullen Group 202
Killarney-Mallow Fault 223, 225, 228–30, 233–5, 242, 246, 250, 272, 273
Killeen Tuffs 273, 274, 277, 285–6, 293, 298
Killeen Volcanic Centre *271*, 272–3, 306, 310
Kilmurry Sandstone Formation *150*, 178, *188*, 190, 207, 209
  aeolian deposits 178, 179, 190, 207, 209
Kiltorcan Formation 249, 333–40
Kiltorcan Formation, palaeoenvironment of plant-bearing horizons 333–40
  miospores 334, 337–9
  New Quarry 333, *334*, 335–6, *336*
  Old Plant Quarry 333, *334*, 335
  palynofacies assemblages *338*, *339*
  pedogenic slickensides 335–6
  Roadstone Quarry 333, *334*, 336–7, *336*
  sedimentological facies analysis 334–7, *335*, *336*
Kinnesswood Formation 485
Kinsale Formation 253, 311
Kirkcolm Formation 468
Krossfjorden Group 570, 573
Krossfjorden Terrane 570, 575, 580, 582
Kvamshesten Basin 45, 503–6, *507*, *508*, *509*, *524* *see also* Kvamshesten Group architecture
Kvamshesten Group architecture (mid-Devonian) 503–31
  climatic fluctuations 520
  geological setting 505–6, *507*
    Devonian basins of western Norway *504*, 505–6
    Kvamshesten basin 506, *507*, *508*, *509*
  Milankovitch time-band 520
  sedimentary architecture of the Kvamshesten Group 506–16
    applications to Kvanshesten basin 528–31
    central fluvial sandstones 509–13
    concepts of sequence stratigraphy 521–4, *523*, *524*
    deformation above ramp-flat detachment 526–8, *529*, 531
    landsliding 520–1
    lithofacies associations and architecture 506, *510*, *511*, *512*, 516–19, *516*, *517*
    marginal fan complexes 506–9, *511*
    progradation and marginal facies retreat 520–1
    rhythmicity and rhythmites 513–15, 520–1, 528
    sequence stratigraphic model, Kvamshesten basin 524–6, *524*
    systems tracts (fluvial sequences) 522–6, *522*, *524*, 528–30, 531
    tectonic contributions to rhythmicity 520–1
    unconformities 515–16, 524

Lack Formation *150*, *188*, 190
Lanark Basin 40
Lanark Group 462
land plants (Early Devonian), Gaspé Bay, Québec 61–82, *62*
  facies associations 63, *73*, 80–2
  marine influence 69, 70, 74, 77–80
  methodology 62
  palaeocurrents 66–7, 68–9, 71–2, 73–5, *74*, 77
  palaeoecology 70–1, 77–8, 80–2, *80*, *81*
  palaeoenvironments 63, 68–70, 74, 76–7, 78, 80–2, *80*, *81*
  palaeosols 74, 79

plant remains 67–8, 70–1, 74, *75*, 80–2
roots 67, 74, 76, 82
quantitative interpretation of thick sandstones 71–2
sedimentology 63–78, *64*, *73*
thick sandstone bodies 64–72, *65*, *66*
thin sandstones and mudstones *67–72*, 72–8
stratigraphic setting 63
trace fossils 76, *76*, 78
vertical variations 78–80, *79*
Landing Place Formation 130, *150*
Laurentia-Baltica collision 33–4, 47, 51, *186*, 552, 569, 580
Laurentia-Eastern Avalonia 32–3, 52, 123, 141, 185
Laurentian plate 32, 47, 417
Leinster Batholith 202
Leinster Massif 201, 207, 216, 412
Leinster Terrane 38, 202, 216–17, 245–6
Leiosphaeridia sp. 327, *329*
Lepidocaris sp. 444, 447, 457
Lepidodendropsis aff. *L. hirmeri* 334, 336
Liefde Bay Supergroup 570
Lilljeborgfjellet Formation 570
limestone clasts in Lower Old Red Sandstone conglomerates 459–69
biostratigraphy *462*, 463, *465*
chronostratigraphy *461*
conodont colour alteration index 465–7
conodont faunas 461, 463, *463*, 465, *465*, *466*, 467–8
depth of burial 465
faunal similarity analysis 463–5, *467*
fossiliferous limestone clasts 459–63
Lanark Basin 459, 462–3, 465
Pentlands Sub-basin 463
Strathmore Basin 459–62, 465
geological setting *460*
lithostratigraphy *461*
source of southerly-derived conglomerates 467–8
terrane analysis 468
Lindsway Bay Formation 352–3
Lingula sp. *77*
Lintrathen outlier (Lower Old Red Sandstone) 420, *422*, 424, 429
Lintrathen porphyry (ignimbrite) 420, *422*, 430
Llanddeusant Formation 391
Llanishen Conglomerate 37
Lobster Pot Tuff 349–50
Lophonozontriletes media 325, *329*
Lophozonotriletes 606
Lorne Basin 40
Lough Guitaine volcanism, Munster Basin 249–50, 271, 294, 306, 310, 311
Lough Guitane Volcanic Complex 272–4, 293–4, 298
Volcanic Complex stratigraphy *273*
Lough Guitane-west Iveragh correlation 293–4, *294*
Lough Slat Conglomerate Formation *150*, *188*, 248, 249
Lower Dingle Group, stratigraphy and depositional facies 132–8
Lower Eday Sandstone Formation 479, *480*
Lower Farran Member, Farran Formation 154, 165
Lower Limestone Shales *150*, *188*, 401, 403
Lower Old Red Sandstone, Pantymaes Quarry *see* Pantymaes Quarry
Lower Old Red Sandstone, Midland Valley 417, 418–19, 422–30, 431
calcrete 406, 426, 431
conglomerates 420, 423–4, 426, 428, 431, 434
Dalradian basement provenance 420
interbedded lavas 420, 422, 423, 428, 431
Lanark basin 420, 428–9
lavas interbedded with conglomerates *422*, 423, 428
palaeoflow directions 420, *421*, *422*, 428
recycled sediments (clasts) 420, 423–4, 431, 434
rejuvenation of source by faulting 423, 431, 432
sediment bars, Garvock Group 426–8, *427*
sediment sources 420, 424–5
sedimentary basins 420, 422, 423, 431
strike-slip fault control 423, 428, 430, 432
Strathmore basin 420, *422*, 423, 424, 426, 429
Lower Old Red Sandstone sedimentology and floodplain architecture *see* Moor Cliffs Formation, Lower Old Red Sandstone
Ludlow Bone Bed 37
Lycopodium cernuum 456

Magdalen Basin (Eastern Canada)
fault adjustments to magma intrusion 311
magmatic evolution 311, 312
magmatism in Munster Basin 303–14, *304*, *305*, *307*
basin inversion 306, 313
basin sedimentation and tectonic control 306
basin subsidence 313
buried granite 228, 232–4
crustal stretching and lithospheric thinning 312–13, 314
geochemical discriminant diagrams 308–9, *309*
magmatic evolution 308–9, 311, 312–13, 314
mantle melts 232
metamorphic grade 303, 306
radiometric dating (K-Ar) 305
thermal effects of intrusions 306, 308
tuff layers 310–11, *310*, 313
Variscan deformation 303, 305, 313
see also Beara Peninsula
Malbaie Formation *35*, 36, 179
Mangerton Anticline *271*, 272–3, 274, 297
Margie Limestone Formation 468
marine sedimentation transitional to Old Red Sandstone, Pembrokeshire 343–53
environment of deposition 344
regional setting 343–4, *344*
sedimentology of valley fills 345–8
soft-sediment deformation 347–8, *349*
spatial distribution of incised valleys 348–50
tidal channels 350
valley fill 5 and Lower Old Red Sandstone 350–2
Marloes Peninsula 343, 344, 348, 350–3
Meguma Terrane 50
Melby Formation 474, *475*, 481
Meunsteria sp. 100
Micrhystridium sp. 327
Micrornatus newportensis zone 391

Midland Valley *see* Old Red Sandstone of Midland Valley
Midland Valley Basin 49, 51–2
Midland Valley Terrane 459–69
Milford Haven Group 37, 405
Mill Cove Formation 128, 129–30, 131, 138, 140, *150*
Milton Ness *see* calcrete morphology
Mimerdalen Formation 573, 585–8, 590, 596, 603, 605–6, 608
Mimerelva Syncline 590, 597
miospores
  Carrigduff Group 155, 189
  Dingle Group 153, 187–8
  Foilcoagh Bay Beds 325–6
  Kiltorcan Formation assemblages 334, 337–9
  Moll's Gap Quarry 289–91, *290*, 294, 298
  Munster Basin assemblages 310–11
  Red Marl Group 391
  Skrinkle Sandstone 409–10
  Slea Head Formation 124, 188
miospores (Givetian), Walls Group 473–81
  environment of deposition 479, *480*
  geological setting *474*, *475*
  material and methods 474–6, *475*, *476*
  miospore assemblage 476–9
  palynofacies and miospore palaeoecology 479–80
  previous palaeontological studies 474
  significance of Givetian age 480–1
Moll's Gap Quarry 270
  miospore assemblage 289–91, *290*, 294, 298
Moor Cliffs Formation, Lower Old Red Sandstone 371
  carbonate nodules (pedogenic) 373, 377–8, *379*, 381–3
  climate 379–81, 383, 385
  floodplain facies association 380, 381–3, *385*
    aggredation 383
  fluvial facies association 380–1
    channel systems 380, 384, 385
  lithofacies *375*
    conglomerates 375–6
    fine-grained sediments 376
    sandstones 376
    volcanic ash 376
  palaeosols (Vertisols) 373, 377–9, 381–5
    variations in development 383–5, *384*
  vertical cyclicity 384, *394*
  sedimentology 371–3, 375–9
  stratigraphy 371–3, *374*
  structural and tectonic setting 373–5
  trace fossils and bioturbation 376–7, 381–2
  vegetation, influence on fluvial processes 379–80, 383, 385
Morte Slates 330
Moygara Formation 40
Munindalen Thrust 590–1, 606, 608
Munster Basin *see* extensional subsidence in Munster Basin; geochronology (U-Pb) in Munster Basin; magmatism in Munster Basin
Munster Basin development 50, 52, *175*, 223–35
  Ambassador Meelin No. 1 well 228–9
  basin fill décollement 227, 229, 235
  depocentres 224, 233–4
  granite-cored horst 232, 233, 235
  mantle melts 232
  regional tectonic synthesis 232–5
  rhythmites 520
  tectonic evolutionary model *231*
  underlying granite 232–3, 235
  wide-angle seismic study *see* VARNET–96
Munster Basin marine incursion 224, 319–31, 412
  fluvial drainage pattern 328–30, 412
  mid-Frasnian sea-level maxima 328–31
  palaeogeography 328–31
  stratigraphy 320–1, *320*
Murospora aurita 607
Murospora sp. 607

Nidedalenn Anticline 590
Nier Sandstone Group 251
Nordfjord-Sogn detachment 33, 45, 48, 505
  geology and associated basins *504*
North Devon Basin 44, 50
North Greenland Fold Belt 580
Northern Appalachians 50 *see also* Devonian rivers and coasts; land plants (Early Devonian)
Northwest Dingle Domain (ORS), Dingle Peninsula 147–81
  Ballydavid Formation 166–7
  chronostratigraphy *152*
  Dingle Group 149–53, 155
    age determination 153
  Farran Formation *see* Farran Formation
  geology *148*, *149*
  lithostratigraphy *148*, *149*, *151*, *153*
  Pointagare Group *see* Pointagare Group
  previous work 148–9
  Sauce Creek Formation *150*, 154, *155*, 158, 166, 167–8
  Smerwick Group 149, *151*, *155*, 158
    age determination 154
    lateral facies variations *160*
    palaeogeography *162–4*
    stratigraphy 153–5, 166, 167
    structural relations with adjacent sequences *151*
Nothia sp. 449

Ochil-Sidlaw anticline 422
Old Head Sandstone Formation 253, 311, 328, 330
Old Red Sandstone basins, kinematics and dynamics 29–52
  basin-forming mechanisms 29–31, *30*
  developments in ORS understanding 31–2
  North Atlantic ORS basins 34–46, *35*
    Anglo-Welsh 37–8
    Baltic States 45–6
    Catskill foreland basin 42
    Dingle 38–40
    East Greenland 46
    Kvamshesten 45
    Maritime Canada 35–6
    Midland Valley of Scotland 40–1
    Munster 42–4

North Devon-South Wales 44
Norway 41–2, 45
Orcadian Basin 41, 44–5
South Devon-Trevone Basin 36
Spitsbergen 42, 46
orogenic framework and timing 32–4
Caledonides 32–4, *33*
Ellesmerides *33*, 34
Variscides *33*, 34
plate boundary and orogenic context 46–52
Late Caledonian flexural basins 49–50
Orcadian basin, NE Scotland 48–9
orogen-scale sediment dispersal 51–2
Scottish Midland Valley 49
syn- to post-Scandian basins 47–8
Syn-Variscan extensional basins 50–1
Old Red Sandstone deformation, NW Spitsbergen 567–82
Andrée Land Group 575
basin development and deformation 575–80
Haakonian deformation phase 569, 573, 575–6, 580, 582
Monacobreen phase 576–9, 580, 582
Svalbardian phase 579–82
Biskayerhalvoya terrane 570–1, 573, 575, *576*
Friedrichbreen Fault 574–5, 576–8
links to Ellesmerian and Caledonian orogens 580–2, *581*
metamorphism 570
radiometric dating 570
regional setting *568*, 569–73
environment of deposition 571–3
ORS stratigraphy 570–3, *571*
pre-ORS basement 569–70
Siktefjellet Group 569, 570–1, 573, *573*, 574, 585
structure *568*, 573–5, *573*
Friedrichbreen Anticline 573, 574–6
large-scale folding 573–4
rotated fault blocks 574, 576–8, *576*
terrane movements 580
Old Red Sandstone of Midland Valley 49, 417–34
Ayr-Ochil-Sidlaw volcanic axis 420, 429, 434
Dalradian block convergent on Midland Valley 424, 426, 430, 434
Dalradian (Southern Highland) 420
distribution *418*, *419*, *486*
flanking blocks 420
large-scale drainage 426, 427–8, 430, 431, 434
outliers (Glen Turret and Lintrathen) 420, 424, 429
Southern Uplands 420
Strathmore syncline 418, 419, 422–6, *422*, 434
upward-maturing sequence 432, 434 *see also* Lower Old Red Sandstone; Upper Old Red Sandstone
Orcadian Basin 44–5, 48–9, 481
marine incursion 330
microfloras 479, 480
Mid-Devonian stratigraphy *480*

palaeocurrents
Ballymastucker 113–14, 120
Beenmore Sandstone Formation 177
Cap-aux-Os Member, Gaspé Bay 66–7, 68–9, 71–2, 73–5, *74*, 77
Catskill foreland basin 42, 88, 91–2, 94, 97, 98–101
Dingle Basin 127, 135, 138, 205, 206, 233
East Greenland Basin 540, 564
Gauss Complex 46
Gray Sandstone Group, Marloes Peninsula 349, 350, 352
Kvamshesten Group 505, 509, 518, 519, 526
Lanark Basin 462
Milton Ness 487
Moor Cliffs Formation 373, 376, 380
Pointagare Group 175
Sherkin Formation 328
Skrinkle Sandstone 405, 406, 407, *407*, 409, 411, *515*
palaeoenvironments
Foilcoagh Bay Beds 323–5
Kiltorcan Formation 334, 336–40
Pantymaes Quarry 392–9
Red Marls, central South Wales 391
Sherkin Formation 321
Skrinkle Sandstone 407, 409
Palaeofavosites rugosus 198
palaeomagnetism 32, 47, 48, *48*, 116–17, 551
Palaeonitella sp. 444, 454
palaeosols, Cap-aux-Os Member 74
palaeosols, Catskill facies, Eastern USA 93–4
palaeosols, Moor Cliffs Formation
characteristic features 377–9, *379*
pedogenic slickensides 335–6, 373, 377–9, 382, 383
variations in development 381, 383–5
Vertisols 373, 377–9, 381–5, *384*
palaeowinds 166, 169, 175, 176, 177, 178, 179
palynology, Sherkin Formation 325
palynomorphs 326–7
Foilcoagh Bay Beds 325–7
Red Marl Group 391
Panderodus sulcatus 465
Pantymaes Quarry 389–99, *389*, *390*
architecture of Mudstone Facies Association 396–9
architecture of Sandstone Facies Association 393–6, 397–9
channel complexes 393–7
fluvial development 397–9, *398*
Mudstone Facies Association 389, *392*, *396*
palaeoenvironment 392–9
plant material, Pantymaes 391, 393, 394, 397
Sandstone Facies Association 389, *392*, *396*
sedimentary facies 392–3, 397–9
tectonic control of fluvial development 397–9, *398*
Paraconglomerate Member, Bull's Head Formation 132, 133, 135
Parka decipiens 391, 393, 394
Pearya terrane 34, 569, 580
Pelecypodichnus sp. 68
Penstrowed Grits Formation 350
Pentamerus oblongus 198
Periodon aculeatus 461, 463, *463*
Phycodes sp. 101
phytoclasts 326–7
Pickwell Down Sandstone *35*, 44

Pigeon Rock Formation 244
Pinus contorta 456
Planolites sp. 78, 100–1, 348, 351
Planteklofta Member, Mimerdalen Formation 588, 590, 596, 603–5, 605–8
Planteryggen Member, Mimerdalen Formation 588, 596, 606
Pointagare Group 154–5, *155*, 168–9, 170–80, 188–9, *188*
  age determination 170, 180
  Beenaman Conglomerate Formation 170–3, *173*, 179
  Beenmore Sandstone Formation 170–3, *172*
  depositional synthesis 176–8
  fluvial-aeolian succession 172–3, 174–5, 176–8
  lateral facies variations *171*
  palaeocurrents 175
  palaeogeography *172*, 178–80
  stratigraphy 149, 167, 170
Polonodus sp. 461, *463*
Port Askaig Tillite Formation 111
Portmagee Anticline 275, 277
prasinophyte phycoma 327
prasinophytes 327
Princesse Alicefjellet Formation 571–2, 574
Protopanderodus graeai 463, *465*
Protopanderodus varicostatus 465
Prototaxites sp. 68, 71, 81
Pseudosphaerexochus? sp. 463
Psilophyton spp. 67, 74
Pterygotus gaspesiensis 77
Punctatisporites minutus 291
Punctatisporites planus 291
Punctatisporites sp. 289
Pygodus anserinus 463, *463*
Pygodus serra 463, *463*

Quartzite Conglomerate 467

Rabotdalen Formation 571
radiometric dating
  East Greenland Basin 538
  Enagh Tuff Bed 190
  Fintona Basin 41
  Munster Basin 305, 310
  ORS basement, NW Spitsbergen 570
Raudfjorden Syncline 574–5, 577–8
Rawns Conglomerate Formation *35*, 44
Reacaslagh Conglomerate 249
Red Bay Group *35*, 42, 568–70, 573–82, *573*, 585, 591, 594, 595–6
Red Cliff Formation 351, 352–3
Red Marl Group (Dittonian) 391
Reenadrolaun Tuff 277, *279*, 282–3, 293
Reenagaveen miospore assemblage 289, *290*, 293, 298
Remigolepis 558
Remigolepis Series 558, 560
Renalia hueberi 67
Reticulatisporites cancellatus 607
Retispora lepidophyta 337, 338, 606, 608
Retispora macroreticulata 337
Retusotriletes distinctus 478
Retusotriletes pychovii 289, *290*, 325
Retusotriletes rotundus 478
Retusotriletes rugulatus 477, *478*
Retusotriletes simplex 291
Retusotriletes sp. 361, 364
Rhabdosporites langii 478, 479
Rhabdosporites parvulus 289, 325
Rhacophyton sp. 339
Rhacopteris sp. 336
Rheic Ocean 34, 51, 217, 328, 552
Rhipidium cf. *hibernicum* 205
Rhipidium hibernicum 131, 138
Rhynia gwynne-vaughanii 446, 453, 454, 455
Rhynia sp. *447*, 449
Rhynie cherts, palaeoecology and plant succession 439–57, *440*
  borehole materials 440–2
  depositional environments 444
  favoured conditions of Rhynie plants 453–5
  hydrothermal systems, environments 446–9
  lithologies *441*, 442–4
    chert 442, *442*, *445*
  palaeontology *443*, *450*
  plant occurrences and successions 449, *452*, *453*
  previous palaeoecological interpretations 442
  radiometric age 439
  silicification and plant accumulation 444–6
    preservation states 444
  spore assemblages 448, 449–53, 455
  typicality of Rhynie assemblage 455–7
Ribband Group 202, 207
Richarddalen Complex 570
Ridgeway Conglomerate 37, 380, 401, 403, 405
Riedel shears 112, 115, 119, 538, 542, *545*, 551
Ringerike Group 41, 47
Ronas Hill Granite 481
Rooks Cave Tuff 375, 376, 381, 384
roots and rhizomes
  Catskill facies, Eastern USA 88–9, 93, 99–101
  Gaspé Bay 67, 74, 76, 82
  Milton Ness 496–9, *497*
Rosscarbery Anticline 320
Rosslare Complex 214
Rosslare Terrane 216
Rugospora bricei 289, *290*, 291, 325, *329*

St Finan's Sandstone Formation 271, 275, 277, 292
  fish beds 275
  fluvial environment 294
  lithofacies *188*, *279*
St Maughans Formation 391
Salopella sp. 361, 364, 366
Samarisporites triangulatus 289, *290*, 325, *329*
Sandness Formation 473–8, *475*, 480–1, *480*
Sandsting Granite Complex 473, 476, 481
Sandy Haven Formation 373
Sauce Creek Formation *150*, 154, *155*, 167–8
  fluvial-aeolian deposits 167, 168, 169
  stratigraphy 158, 166, 167
Sawdonia ornata 67, 74, *75*
Scandian Orogeny 33, 45, 47, 52

Schivefjellet rotated block 574, 576, 577
Sciadophyton spp. 67, 74, *75*, *80*, 456
Scout Hill Flags *35*, 49
Senni Beds 37
Lower Old Red Sandstone (Breconian) *35*, 389, 391, 394, 399
Sennicaulis sp. 356
Shannon Trough 232
Sherkin Formation 242, 251, 263, 319, 320, 328
Chondrites sp. 321
palaeocurrents 328
palaeoenvironment 321
palynology 325
Shinnel Formation 465, 467–8
Siktefjellet anticline 573–4
Siktefjellet Group *35*, 42, 569, 570–1, 573, *573*, 574–5, *576*, 580, 582, 585
Simpson coefficient of similarity 463, *466*, *467*, 468
Skolithos sp. 68, 74, 76, *76*, 78, 99–101, 348, *349*, 351
Skomer Basin 344, 353
Skomer Volcanic Group 142, 344, 352
Skrinkle Sandstone fault-bounded basin fill *35*, 44, 401–15
axial basin fill 405–6, 412
Broad Haven fault block 403, *404*
Carboniferous transgression 407–11, *409*
channel deposits 405, 407
conglomerate clasts 406–7, *407*
Conglomerate Member, West Angle Formation *410*, *411*
depositional environments *412*, *414*, *415*
geological setting 401–5, *402*
Gupton Formation 403, 405, *405*, 406, *409*, 411, 412
lithostratigraphy *404*
Lower Limestone Shales 401, 408–9
Lower Sandstone Member, Gupton Formation 405–6, *405*, *406*
palaeocurrents 405, 406, 407, *407*, 409, 411, *415*
palaeoenvironment 407, 409
palaeogeography *402*
Red-Grey Member, West Angle Formation 407–8, 410–11, *412*, *413*, *414*
Ridgeway Conglomerate 401, *405*, 406
Ritec Fault 401, 403, 404, 406, 407, 411–12
spore data 409–10
Stackpole Sandstone Member, Gupton Formation *405*, 406, *407*, *408*
stratigraphy 403, *403*, 405
Tenby-Angle fault block 403, *404*, 405, 407, 411
Townsend Tuff Bed 403, *404*
transverse basin fill 406–7, 412
West Angle Formation 403, 406, 407–8, *407*, *411*, 412
Winsle fault block 403, *404*, 411, 412
Slea Head Formation 124, 149, *150*, 153, 187, *188* *see also* Dingle Basin conglomerates
Slieve Mish Group *35*, 44, 155, *188*, 189–90, 248, 249
Slieve Tooey Quartzite Formation 109, 111, 114–15, *118*
Slievenamuck Formation 246
Smerwick Group 149, *151*, *155*, 158, 188, *188*, 202
age determination 154, 180
depositional synthesis 168–70
fluvial-aeolian processes 156, 169–70
lateral facies variations *160*
palaeogeography *162–4*
stratigraphy 153–5, 166, 167, 170, 173
Smerwick Group Terrane 154, 169, *175*
Soederberghia 558
Solund Basin 45, 505
South Munster Basin 42, 224, 328
Cork-Kenmare Line 224
depocentre 234
rifting phase 253, 262
Variscan deformation 225, 227–9
Southern Uplands 467–8
erosional history 420, 428
Spelaeotriletes spp. 338
Spenopteris sp. 333
Spermolithus devonicus 333–4
Sphaerexochus sp. 463
Sphenopteris hookeri 333
Spinodus spinatus 463, *463*
Spirophyton sp. 101
Spongiophyton sp. 68, 71, 81
Stenopora fribosa 198
Stensiö Bjerg Formation (East Greenland) 558, 561, 564
Stinchar Limestone 463, 464, 465, 467
Stonehaven Group 459
Strachanognathus parvus 463, *463*
Strathmore Basin 40, 49, 420, *422*, 423, 424, 426, 429, 459–62, 465
Strathmore Group 423, 426
Strathmore syncline 418, 419, 422–6, *422*, *425*
Streelispora newportensis 366–8
Stromness Flagstone Formation 479
Strophomena depressa 198
Svalbardian deformation of Old Red Sandstone, NW Spitsbergen 585–99
age of deformation 589
Biskayerfonna-Holtedahlfonna Horst 591, 595
Blomstrandhalvoya Thrust Zone 589, 594–5, *598*
Bravallafjella Fold Zone 589, 591, *593*, *598*
Bravallafjella Syncline 591, *593*
Dickson Land Fold-and-Thrust Zone 589, *589*, 590–1, 598, *598*
geological setting 585–8, *586*, *588*
Germaniahalvoya Fold-and-Thrust Zone 589, 591–2, *593*, 595, 598, *598*
Mitrahalvoya Thrust Zone 589, 595, *596*, *598*
stratigraphy 585–8, *587*
Svalbardian Fold-and-Thrust Belt 590–2, 598
tectonic maps *586*, *592*
Viggobreen weathering zone 594–5
Svalbardian deformation, palynological dating 603–8
geological setting 603–5, *604*, *605*
Planteklofta Member 605–6, *606*
Triungun Member 606–8, *607*
Swanshaw Formation 428

Tarella sp. 355
Temeside Formation 385
Templetown Conglomerate Formation 250
Tenby-Angle basin 401

Termon Pelite Formation 112, 114
tetrapod remains, ichthyostegid 564
Thecia sp. 198
Theic Ocean 34
Thursophyton sp. 480
Toe Head Formation 253, 319
Tornquist Line 33
Tortilicaulis offaeus 360, 365
Tortilicaulis sp. 366
Tortworth inlier volcanic suite 142
Townsend Tuff Bed 373, 403, *404*
Trabane Member, Bull's Head Formation 132, 133, *134*, 135, 141
Trabeg Conglomerate Formation 39, 135, 138, 141, *150*, 169, 187, *188 see also* Dingle Basin conglomerates
trace fossils
  East Greenland 559, 564
  Gaspé Bay 76, *76*, 78
  Gray Sandstone Group 348, *349*
  Moor Cliffs Formation 376–7, 381–2
  Pantymaes Quarry 391, 393, 397, 398
Trentishoe Formation *35*, 44
Trevone Basin 36, 51–2, 242
Trichopherophyton sp. 449
Trichopherophyton teuchansii 454–5
Triglochin maritima 455, 456
Trileites langii 478
Trimerophyton 67, 74
Trimerophyton robustius 75
Triungen Member, Horbyebreen Formation 603–5, 606–8
Tulerpeton tetrapod remains (Russia) 557
Tumulispora appendica 607
Tumulispora rarituberculata 607
Tumulispora variverrucata 607
Tweedale Member, Wrae Limestone 468

Upper Eday Sandstone Formation *480*
Upper Farran Member, Farran Formation 154, 165
Upper Gaspé Limestone Group *35*, 36
Upper Old Red Sandstone, Midland Valley 417, 418, 430–1, 431–2
  caliche 430
  palaeoflow directions 430, 431
  sediment bar (Seamill) *432*
  sediment bars 430–1
  upward-fining sequence 430–1, *433*, 434
  Usk Basin 344
  Uskiella spargens 355

Valentia Harbour magmatism, Munster Basin 249, 306–7, 310, 311–13
Valentia Slate Formation *188*, 249, 271, 275, *276*, 319
  fish beds 275, *276*, 293
  Keel-Enagh Tuff Bed 274–5, *276*, 277, 292–3
  lithofacies *279*
  miospore assemblage 289, *290*
  Puffin Sound Tuff Bed 249, 275
  Valentia Harbour-Beginish Island volcanics 277
Variscan deformation
  dated (K-Ar) 272
  Munster Basin 225, 227–9, 234–5, 239, 242–6, 272, 303, 305, 313
  Pembroke Peninsula 403–5
Variscan Orogeny 32, 34, 47, 552
VARNET–96 seismic survey, Munster Basin 221–35
  data acquisition 225
  data limitations 226
  interpretation of velocity models *226*, 227–32
    Line A 228–30
    Line B 230–2, *231*
  methodology and modelling 225–6 *see also* Munster Basin development
vegetation
  Milton Ness 492–3, 496–9
  Moor Cliffs Formation 379–80, 383
vegetation on Old Red Sandstone Continent 355–68
  genus *Emphanisporites* 356–8, *367*
  mesofossils and spores 358–65
    cf. *Horneophyton* sp. 358–9, *359*, *363*, 364
    Emphanisporites cf. *micrornatus* 359–63
    Emphanisporites sp. A *362*, 363–5, *364*
  methodologies 356
  new material 355–6
  palaeoecology and taphonomy 365–8
  preparation procedures 358
Verruccosisporites nitidus 338
Verruciretusispora dubia 478
Verruciretusispora sp. 289, *290*
Verrucosisorites bulliferus 289, *290*, 291, 325
Verrucosisporites premnus 477, *478*
Verrucosisporites scurrus 478
Veryhacium sp. 327
Videospora glabrimarginata 325
Vilddal Supergroup (East Greenland) *35*, 46, 538, 540–1, 543, *543*, *544*, *548*, 552, 559

Walliserodus costatus 465
Walliserodus nakhholmensis 465
Walls Formation 473–4, *475*, 476–8, 480–1
Walls Group 473–4, *474*, 476
Welsh Basin 343, 350, 352–3
  structural controls on bathymetry *344*
Wenall Fault 403
West Angle Formation 403, 406
Western Avalonia 32–3, 50
Wideroefjella rotated block 574, 576–8
Wijde Bay Formation 573, 585–7, 591
Wimans Bjerg Formation (East Greenland) 558, 559, 564
Winsle Inlier 352
Wood Bay Formation *35*, 51, 569, 573, 578, 585–7, 590, 591, 595–6, 603, 606–8
Woolhope Basin 344
Wordiekammen Formation 603
Wrae Limestone 461, 464, 465, 467–8
Wulffberget Formation 571, 573, 574

York River Formation 36

Zosterophyllum llanoveranum 355